Artificial Intelligence (AI) Applied in Civil Engineering

Artificial Intelligence (AI) Applied in Civil Engineering

Editors

Nikos D. Lagaros
Vagelis Plevris

MDPI • Basel • Beijing • Wuhan • Barcelona • Belgrade • Manchester • Tokyo • Cluj • Tianjin

Editors
Nikos D. Lagaros
National Technical University
of Athens
Greece

Vagelis Plevris
Qatar University
Qatar

Editorial Office
MDPI
St. Alban-Anlage 66
4052 Basel, Switzerland

This is a reprint of articles from the Topic published online in the open access journals *Applied Sciences* (ISSN 2076-3417), *CivilEng* (ISSN 2673-4109), *AI* (ISSN 2673-2688), *Symmetry* (ISSN 2073-8994), and *Mathematics* (ISSN 2227-7390) (available at: https://www.mdpi.com/topics/Artificial_intelligence_Civil_Engineering).

For citation purposes, cite each article independently as indicated on the article page online and as indicated below:

LastName, A.A.; LastName, B.B.; LastName, C.C. Article Title. *Journal Name* **Year**, *Volume Number*, Page Range.

ISBN 978-3-0365-5083-1 (Hbk)
ISBN 978-3-0365-5084-8 (PDF)

Contents

About the Editors

Nikos D. Lagaros

Dr. Nikos D. Lagaros is the Dean of the School of Civil Engineering, National Technical University of Athens (NTUA), Greece, and Professor of Structural Optimization for the Institute of Structural Analysis and Antiseismic Research. He is the Director of the Digital Technology Laboratory for Technical Projects. Prior moving to NTUA, he was Assistant Professor at the University of Thessaly, Greece. He also served as Visiting Professor in the Department of Biological Engineering at MIT, and in the Department of Mechanical Engineering at McGill University. Dr. Lagaros is an active member of the computational mechanics research community, and his publication track record includes more than 140 peer-reviewed journal papers, 10 books and 25 book chapters, and he has presented his work at numerous international venues. His h-index is 43 according to Google Scholar, with his work receiving more than 6000 citations.

Vagelis Plevris

Dr. Vagelis Plevris is an Associate Professor of Structural Engineering in the Department of Civil and Architectural Engineering, Qatar University. He holds a bachelor's degree in Civil Engineering from the National Technical University of Athens (NTUA), an MSc from NTUA on "Structural Design and Analysis of Structures", an MBA from Athens University of Economics and Business and a PhD in Computational Mechanics from NTUA. His research areas cover the fields of finite element analysis, earthquake engineering, optimum design of structures, reliability and probabilistic analysis of structures, and neural networks and their applications in engineering.

applied sciences

Editorial

Artificial Intelligence (AI) Applied in Civil Engineering

Nikos D. Lagaros [1] and Vagelis Plevris [2,*]

1 Institute of Structural Analysis and Antiseismic Research, School of Civil Engineering, National Technical University of Athens, Heroon Polytechneiou 9, 157 80 Zographou, Greece; nlagaros@central.ntua.gr
2 Department of Civil and Architectural Engineering, Qatar University, Doha P.O. Box 2713, Qatar
* Correspondence: vplevris@qu.edu.qa

Citation: Lagaros, N.D.; Plevris, V. Artificial Intelligence (AI) Applied in Civil Engineering. *Appl. Sci.* **2022**, *12*, 7595. https://doi.org/10.3390/app12157595

Received: 5 June 2022
Accepted: 27 July 2022
Published: 28 July 2022

1. Introduction

In recent years, artificial intelligence (AI) has drawn significant attention with respect to its applications in several scientific fields, varying from big data handling to medical diagnosis. The use of AI is already present in our daily lives with several uses, such as personalized ads, virtual assistants, autonomous driving, etc. Not surprisingly, AI methodologies have found a wide range of uses and applications in engineering fields, including civil and structural engineering [1,2], with impressive results [3–5]. Figure 1 shows the research articles related to AI published in the field of civil engineering. In particular, these are results from the Scopus database (www.scopus.com, obtained on 2 June 2022), using the query *"TITLE-ABS-KEY (("artificial intelligence" or "AI") and ("civil" or "structural" or "transportation" or "geotechnical" or "hydraulic" or "environmental" or "construction" or "shm" or "structural health")) and PUBYEAR > 1999 and (LIMIT-TO (SUBJAREA, "ENGI"))"*, which returned 14,059 document results in total (for years from 2000 to 2022). The increase in AI studies with great acceleration shows that the use of AI in civil engineering is gaining momentum and will keep increasing in the coming years, bringing new innovations and applications.

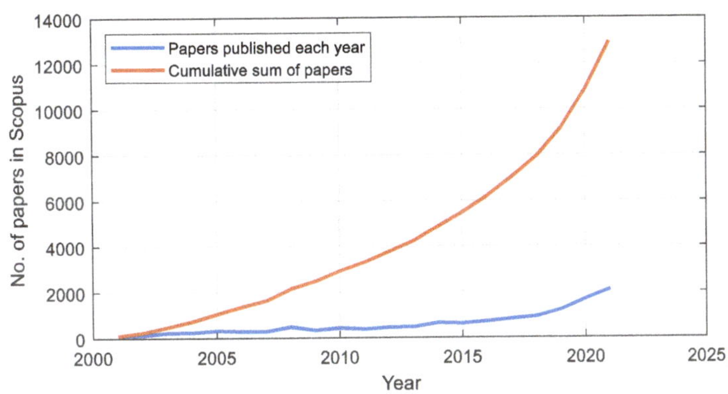

Figure 1. Published articles (in Scopus) using AI in civil engineering-related fields (2000–2021).

This research topic contains applications and recent advances of AI in civil engineering problems, promoting cross-fertilization between these scientific fields. In particular, the focus is on hybrid studies and applications related to structural engineering, transportation engineering, geotechnical engineering, hydraulic engineering, environmental engineering, coastal and ocean engineering, structural health monitoring, as well as construction management.

2. Contributions

The research topic has been quite successful, gathering 35 contributions in total, from 19 different countries around the world, covering a broad range of topics related to the applications of AI in civil engineering. Three MDPI journals participated by cross-listing the research topic. Most of the articles (29) were published in the "Applied Sciences" journal, while 3 of them were published in "Mathematics" and another 3 in "Symmetry".

The articles are divided into 6 groups, as follows: (i) Optimization methods and applications (7 articles), (ii) Combined machine learning and optimization methodologies (2 articles), (iii) Machine learning in identification problems (3 articles), (iv) Applications of convolutional neural networks (8 articles), (v) Combined and multiple AI-based methodologies (6 articles), and (vi) Other AI-based methods, formulations, and applications (9 articles). A brief description of each article, for every category, is presented in the following sections.

2.1. *Optimization Methods and Applications*

Rosso et al. [6] propose an enhanced multi-strategy Particle Swarm Optimization (PSO) variant to solve constrained problems with a different approach to the classical penalty function technique. The authors propose several improvements to the original algorithm, including a new local search operator based on the Evolutionary Strategy (ES). Li et al. [7], propose an optimization approach with a parallel updated particle swarm optimization (PUPSO) algorithm aiming at minimizing the objective function of the levelized cost of energy of the prestressed concrete–steel hybrid wind turbine towers. This is conducted in a life cycle perspective which represents the direct investments, labor costs, machinery costs, and the maintenance costs.

Cucuzza et al. [8] study the size and shape optimization of a guyed radio mast for radiocommunications, using the genetic algorithm (GA) and carrying out both static and dynamic analyses considering the action of wind, ice, and seismic loads. Guo et al. [9], propose the use of GA, correlation analysis, and two parametric design methods (floor plan generation method and component selection method) for optimizing the building performance of prefabricated buildings.

Uray et al. [10] use the Taguchi method integrated hybrid harmony search algorithm, carry out a statistical investigation of the optimum values for the control parameters of the harmony search algorithm and examine their effects on the best solution. The new hybrid method has been successfully applied to different real-world engineering optimization problems. Sarjamei et al. [11] use the Gold Rush Optimization (GRO) algorithm for the optimal design of real-scale symmetric structures under frequency constraints. The efficacy of the concept of cyclic symmetry to minimize the needed time is assessed with three examples, including Disk, Silo, and Cooling Tower.

Bao et al. [12], investigate the decision-making problem of pavement maintenance prioritization considering both quality and cost. They consider a linear optimization model that maximizes maintenance quality with limited maintenance costs and a multi-objective optimization model that maximizes maintenance quality while minimizing maintenance costs. These models are employed in making decisions for actual pavement maintenance using sequential quadratic programming and GA.

2.2. *Combined Machine Learning and Optimization Methodologies*

3D printing is already established in the production processes of several industries while more are continuously being added. Lately, parametric design has become popular in the architectural design literature, while topology optimization has become part of the design procedure of various industries. Kallioras and Lagaros [13] propose MLGen, a novel generative design framework which integrates machine learning (ML) into the generative design practice. Several benchmark topology optimization problems are examined to show the ability of MLGen to efficiently handle different engineering problems.

In order to deal with dynamic traffic flow, adaptive traffic signal controls using reinforcement learning are being studied. Gu et al. [14] propose a reinforcement learning-based signal optimization model with constraints. The model maintains the sequence of typical signal phases and considers the minimum green time. It is trained using Simulation of Urban MObility (SUMO), a microscopic traffic simulator and it is evaluated in a virtual environment similar to a real road with multiple intersections.

2.3. Machine Learning in Identification Problems

In recent years, deep learning-based detection methods have been successfully applied to pavement crack detection problems. In this field, Li et al. [15] propose a method to improve the accuracy of crack identification by combining a semantic segmentation and edge detection model. Their work is inspired by the U-Net semantic segmentation network and holistically nested edge detection network. A side-output part is added to the U-Net decoder that performs edge extraction and deep supervision. A network model is proposed, combining two tasks that can output the semantic segmentation results of the crack image and the edge detection results of different scales. The model can also be used for other tasks that need both semantic segmentation and edge detection. On the topic of concrete structures, Liu and Li [16] propose an improved self-organizing mapping (SOM) neural network (NN) model to solve the problem of intelligent detection of damage to modern concrete structures under complex constraints. The method is based on a small number of samples and the use of a self-developed 3D laser scanning system. The improved SOM model method fully combines the network topology and its unique image features and can accurately identify structural damage, contributing to the realization of high-precision intelligent health monitoring of damage to modern concrete structures.

In railway engineering, the performance of the passing train and the structural state of the track bed are common concerns regarding the safe operation of the subway. Monitoring the vibration response of the track bed structure and identifying abnormal signals within it can help address these concerns. In this direction, Li et al. [17] propose an unsupervised learning-based methodology for identifying the abnormal signals of the track beds detected by the ultra-weak fiber optic Bragg grating sensing array. The experimental results demonstrate that the established unsupervised learning network and the selected metric for quantifying error sequences can serve the threshold selection well, based on the receiver operating characteristic curve.

2.4. Applications of Convolutional Neural Networks

In earthquake engineering, the analysis of site seismic amplification characteristics is one of the most important tasks of seismic safety evaluation. Yang et al. [18] propose a new prediction method for the amplification characteristics of local sites, using a CNN combined with real-time seismic signals. The CNN is used to establish the relationship between the amplification factors of local sites and eight parameters, while the training and testing samples are generated through observed and geological data. The results show that the CNN method can provide a powerful tool for predicting the amplification factors of local sites both for recorded and unrecorded positions. Yan et al. [19] propose a measurement method of bridge vibration by unmanned aerial vehicles (UAVs) combined with convolutional neural networks (CNNs) and the Kanade–Lucas–Tomasi (KLT) optical-flow method. The KLT optical-flow method is used to track the target points on the structure and the background reference points in the video to obtain the coordinates of these points on each frame, while the characteristic relationship between the reference points and the target points is learned by a CNN according to the coordinates of the reference points and the target points. The objective is to correct the displacement time–history curves of target points containing the false displacement caused by the UAV's egomotion.

Based on the features of cracks, Wang et al. [20] propose the concept of a crack key point as a method for crack characterization and establish a model of image crack detection based on the reference anchor points method, named KP-CraNet. The accuracy of the model recognition is controllable and can meet both the pixel-level requirements and the efficiency needs of engineering. The results show that the method can improve crack detection quality and has a strong generalization ability. Dos Santos Junior et al. [21] propose an architecture for segmenting cracks in facades with Deep Learning (DL) that includes an image pre-processing step. The authors also propose the Ceramic Crack Database, a set of images to segment defects in ceramic tiles. The proposed model can adequately identify the crack even when it is close to or within the grout.

Blockage of culverts by transported debris materials is the salient contributor to originating urban flash floods, with conventional hydraulic modeling having no success in addressing the problem. Iqbal et al. [22] explore a new dimension to investigate the issue by proposing the use of intelligent video analytics (IVA) algorithms for extracting blockage-related information. Their research aims to automate the process of manual visual blockage classification of culverts from a maintenance perspective by remotely applying DL models. On the other hand, Calton and Wei [23] use transfer learning on three advanced NNs, ResNet, MobileNet, and EfficientNet, and apply techniques for damage classification and damaged object detection to a post-hurricane image dataset comprised of damaged buildings from the coastal region of the southeastern USA. The dataset includes 1000 images for the classification model with a binary classification structure containing classes of floods and non-floods and 800 images for the object detection model with four damaged object classes, i.e., damaged roof, damaged wall, flood damage, and structural damage.

Lin et al. [24] aim at the long-term (24–72 h ahead) prediction of wind power with a mean absolute percentage error of less than 10% by using the Temporal Convolutional Network (TCN) algorithm of DL networks. In their experiment, they perform TCN model pretraining using historical weather data and the power generation outputs of a wind turbine from a Scada wind power plant in Turkey.

Chen et al. [25] propose a text-mining-based accident causal classification method based on a relational graph convolutional network (R-GCN) and pre-trained bidirectional encoder representation from transformers (BERT). The proposed method avoids preprocessing such as stop word removal and word segmentation, but also avoids tedious operations, while the dependence of BERT retraining on computing resources can also be avoided.

2.5. Combined and Multiple AI-Based Methodologies

Some of the research works use multiple AI-based methodologies, either for comparison purposes or in a combined way to achieve better results. In particular, Benbouras et al. [26] elaborate on a new alternative model for predicting the bearing capacity of piles based on eleven new advanced ML methods, in order to overcome the problems of the time-consuming and costly traditional methods. The modeling phase uses a database of 100 samples collected from different countries. Additionally, eight relevant factors are selected in the input layer based on recommendations from the literature. Su et al. [27] propose a data processing framework that uses a long short-term memory (LSTM) model coupled with an attention mechanism to predict the deformation response of a dam structure. The results of the case study show that, of all tested methods, the proposed coupled method performs best. In addition, it was found that temperature and water level both have significant impacts on dam deformation and can serve as reliable metrics for dam management.

Zhao et al. [28] use a sparrow search algorithm to improve a backpropagation NN, and an Elman NN and support vector regression models to predict the thickness of an excavation damaged zone. The proposed model can provide a reliable reference for the thickness prediction of an excavation-damaged zone and is helpful in the risk management of roadway stability. Ma et al. [29] investigate the performance of the extreme gradient boosting (XGboost) method in predicting multiclass of clay sensitivity, and the ability of the synthetic minority over-sampling technique (SMOTE) in addressing imbalanced categories of clay sensitivity. The results reveal that XGBoost shows the best performance in the multiclassification prediction of clay sensitivity.

In transportation engineering, Xiang et al. [30] propose a two-phase approach in an effort to predict highway passenger volume. The datasets subsume highway passenger volume and impact factors of urban attributes. The findings provide useful information for guiding highway planning and optimizing the allocation of transportation resources. Cheng et al. [31] use smartcard data from the bus system to identify important variables that affect passenger flow. These data are combined with other influential variables to establish an integrated-weight time-series forecast model. The results show that the model can improve passenger flow forecasting based on three bus routes with three different series of time data.

2.6. Other AI-Based Methods, Formulations, and Applications

Kruachottikul et al. [32] aim to improve collaboration on bridge inspections that typically require the involvement of many people, personal judgement, and extensive travel to survey bridges across the country of Thailand. One major challenge is to standardize human judgement. To address this, the authors develop a user-centric bridge visual defect quality control mobile application to improve collaboration and assist field technicians to conduct visual defect inspections. Based on nonlinear finite element numerical simulation and synergistic theory, the cooperative control problems of the bridge–subgrade transition section are studied in the work by Zhang et al. [33]. Huang et al. [34] propose a data-driven reinforcement-learning (RL)-based approach to achieve automatic bucket-filling. An automatic bucket-filling algorithm based on Q-learning is developed to enhance the adaptability of the autonomous scooping system. A nonlinear, non-parametric statistical model is also built to approximate the real working environment using the actual data obtained from tests.

Chen et al. [35] summarize the main factors affecting the large deformation of soft rock tunnels, including the lithology combination, weathering effect, and underground water status, by reviewing the typical cases of largely-deformed soft rock tunnels. The method can be used to invert the geological parameters of the surrounding rock mass for a certain point, which can provide important mechanical parameters for the design and construction of tunnels. Lin et al. [36] introduce a modern space remote sensing technology, InSAR, as a direct observable for the slope dynamics. The InSAR-derived displacement fields and other in situ geological and topographical factors are integrated, and their correlations with landslide susceptibility are analyzed. Moreover, multiple ML approaches are applied with the goal to construct an optimal model between these complicated factors and landslide susceptibility. Zenkour et al. [37] introduce the thermoelastic coupled response of an unbounded solid with a cylindrical hole under a traveling heat source and harmonically altering heat. A refined dual-phase-lag thermoelasticity theory is used for this purpose. A generalized thermoelastic coupled solution is developed by using Laplace's transforms technique.

Heo et al. [38] highlight that many human resources are needed on the research and development (R&D) process of AI and discuss factors to consider in the current method of development. Labor division of a few managers and numerous ordinary workers as a form of the light industry appears to be a plausible method of enhancing the efficiency of AI R&D projects. Inspired by the powerful ability of NNs in the field of representation learning, Xie et al. [39] design a hierarchical generative embedding model (HGE) to map nodes into latent space automatically. Then, with the learned latent representation of each node, they propose an HGE-GA algorithm to predict influence strength and compute the top-K influential nodes. Extensive experiments on real-world attributed networks demonstrate the outstanding superiority of the proposed HGE model and HGE-GA algorithm compared with the state-of-the-art methods, verifying the effectiveness of the proposed model and algorithm. Xie et al. [40] incorporate a co-embedding model for KG embedding, which learns low-dimensional representations of both entities and relations in the same semantic space. To address the issue of neglecting uncertainty for KG components, they propose a variational auto-encoder that represents KG components as Gaussian distributions.

Author Contributions: All authors contributed to every part of the research described in this paper. All authors have read and agreed to the published version of the manuscript.

Funding: This research has been financed by the ADDOPTML project: "ADDitively Manufactured OPTimized Structures by means of Machine Learning" (No: 101007595).

Acknowledgments: This research has been supported by the ADDOPTML project: "ADDitively Manufactured OPTimized Structures by means of Machine Learning" (No: 101007595) belonging to the Marie Skłodowska-Curie Actions (MSCA) Research and Innovation Staff Exchange (RISE) H2020-MSCA-RISE-2020. Their support is highly acknowledged.

Conflicts of Interest: The authors declare no conflict of interest.

References

1. Plevris, V.; Tsiatas, G. Computational Structural Engineering: Past Achievements and Future Challenges. *Front. Built Environ.* **2018**, *4*, 21. [CrossRef]
2. Tsiatas, G.C.; Plevris, V. Editorial: Innovative Approaches in Computational Structural Engineering. *Front. Built Environ.* **2020**, *6*, 39. [CrossRef]
3. Papadrakakis, M.; Lagaros, N.D. Soft computing methodologies for structural optimization. *Appl. Soft Comput. J.* **2003**, *3*, 283–300. [CrossRef]
4. Lu, X.; Plevris, V.; Tsiatas, G.; De Domenico, D. Editorial: Artificial Intelligence-Powered Methodologies and Applications in Earthquake and Structural Engineering. *Front. Built Environ.* **2022**, *8*, 43. [CrossRef]
5. Lagaros, N.D.; Tsompanakis, Y. (Eds.) *Intelligent Computational Paradigms in Earthquake Engineering*; Idea Group Publishing: Hershey, PA, USA, 2006.
6. Rosso, M.M.; Cucuzza, R.; Aloisio, A.; Marano, G.C. Enhanced Multi-Strategy Particle Swarm Optimization for Constrained Problems with an Evolutionary-Strategies-Based Unfeasible Local Search Operator. *Appl. Sci.* **2022**, *12*, 2285. [CrossRef]
7. Li, Z.; Chen, H.; Xu, B.; Ge, H. Hybrid Wind Turbine Towers Optimization with a Parallel Updated Particle Swarm Algorithm. *Appl. Sci.* **2021**, *11*, 8683. [CrossRef]
8. Cucuzza, R.; Rosso, M.M.; Aloisio, A.; Melchiorre, J.; Giudice, M.L.; Marano, G.C. Size and Shape Optimization of a Guyed Mast Structure under Wind, Ice and Seismic Loading. *Appl. Sci.* **2022**, *12*, 4875. [CrossRef]
9. Guo, J.; Li, M.; Jiang, Z.; Wang, Z.; Zhou, Y. Optimized Design of Floor Plan and Components of Prefabricated Building with Energy-Cost Effect. *Appl. Sci.* **2022**, *12*, 3740. [CrossRef]
10. Uray, E.; Carbas, S.; Geem, Z.W.; Kim, S. Parameters Optimization of Taguchi Method Integrated Hybrid Harmony Search Algorithm for Engineering Design Problems. *Mathematics* **2022**, *10*, 327. [CrossRef]
11. Sarjamei, S.; Massoudi, M.S.; Sarafraz, M.E. Frequency-Constrained Optimization of a Real-Scale Symmetric Structural Using Gold Rush Algorithm. *Symmetry* **2022**, *14*, 725. [CrossRef]
12. Bao, S.; Han, K.; Zhang, L.; Luo, X.; Chen, S. Pavement Maintenance Decision Making Based on Optimization Models. *Appl. Sci.* **2021**, *11*, 9706. [CrossRef]
13. Kallioras, N.A.; Lagaros, N.D. MLGen: Generative Design Framework Based on Machine Learning and Topology Optimization. *Appl. Sci.* **2021**, *11*, 12044. [CrossRef]
14. Gu, J.; Lee, M.; Jun, C.; Han, Y.; Kim, Y.; Kim, J. Traffic Signal Optimization for Multiple Intersections Based on Reinforcement Learning. *Appl. Sci.* **2021**, *11*, 10688. [CrossRef]
15. Li, P.; Xia, H.; Zhou, B.; Yan, F.; Guo, R. A Method to Improve the Accuracy of Pavement Crack Identification by Combining a Semantic Segmentation and Edge Detection Model. *Appl. Sci.* **2022**, *12*, 4714. [CrossRef]
16. Liu, J.; Li, K. Research on an Improved SOM Model for Damage Identification of Concrete Structures. *Appl. Sci.* **2022**, *12*, 4152. [CrossRef]
17. Li, S.; Qiu, Y.; Jiang, J.; Wang, H.; Nan, Q.; Sun, L. Identification of Abnormal Vibration Signal of Subway Track Bed Based on Ultra-Weak FBG Sensing Array Combined with Unsupervised Learning Network. *Symmetry* **2022**, *14*, 1100. [CrossRef]
18. Yang, X.; Chen, Y.; Teng, S.; Chen, G. A Novel Method for Predicting Local Site Amplification Factors Using 1-D Convolutional Neural Networks. *Appl. Sci.* **2021**, *11*, 11650. [CrossRef]
19. Yan, Z.; Jin, Z.; Teng, S.; Chen, G.; Bassir, D. Measurement of Bridge Vibration by UAVs Combined with CNN and KLT Optical-Flow Method. *Appl. Sci.* **2022**, *12*, 5181. [CrossRef]
20. Wang, D.; Cheng, J.; Cai, H. Detection Based on Crack Key Point and Deep Convolutional Neural Network. *Appl. Sci.* **2021**, *11*, 11321. [CrossRef]
21. Junior, G.; Ferreira, J.; Millán-Arias, C.; Daniel, R.; Junior, A.; Fernandes, B. Ceramic Cracks Segmentation with Deep Learning. *Appl. Sci.* **2021**, *11*, 6017. [CrossRef]
22. Iqbal, U.; Barthelemy, J.; Li, W.; Perez, P. Automating Visual Blockage Classification of Culverts with Deep Learning. *Appl. Sci.* **2021**, *11*, 7561. [CrossRef]
23. Calton, L.; Wei, Z. Using Artificial Neural Network Models to Assess Hurricane Damage through Transfer Learning. *Appl. Sci.* **2022**, *12*, 1466. [CrossRef]

24. Lin, W.-H.; Wang, P.; Chao, K.-M.; Lin, H.-C.; Yang, Z.-Y.; Lai, Y.-H. Wind Power Forecasting with Deep Learning Networks: Time-Series Forecasting. *Appl. Sci.* **2021**, *11*, 10335. [CrossRef]
25. Chen, Z.; Huang, K.; Wu, L.; Zhong, Z.; Jiao, Z. Relational Graph Convolutional Network for Text-Mining-Based Accident Causal Classification. *Appl. Sci.* **2022**, *12*, 2482. [CrossRef]
26. Benbouras, M.A.; Petrişor, A.-I.; Zedira, H.; Ghelani, L.; Lefilef, L. Forecasting the Bearing Capacity of the Driven Piles Using Advanced Machine-Learning Techniques. *Appl. Sci.* **2021**, *11*, 10908. [CrossRef]
27. Su, Y.; Weng, K.; Lin, C.; Chen, Z. Dam Deformation Interpretation and Prediction Based on a Long Short-Term Memory Model Coupled with an Attention Mechanism. *Appl. Sci.* **2021**, *11*, 6625. [CrossRef]
28. Zhao, G.; Wang, M.; Liang, W. A Comparative Study of SSA-BPNN, SSA-ENN, and SSA-SVR Models for Predicting the Thickness of an Excavation Damaged Zone around the Roadway in Rock. *Mathematics* **2022**, *10*, 1351. [CrossRef]
29. Ma, T.; Wu, L.; Zhu, S.; Zhu, H. Multiclassification Prediction of Clay Sensitivity Using Extreme Gradient Boosting Based on Imbalanced Dataset. *Appl. Sci.* **2022**, *12*, 1143. [CrossRef]
30. Xiang, Y.; Chen, J.; Yu, W.; Wu, R.; Liu, B.; Wang, B.; Li, Z. A Two-Phase Approach for Predicting Highway Passenger Volume. *Appl. Sci.* **2021**, *11*, 6248. [CrossRef]
31. Cheng, C.-H.; Tsai, M.-C.; Cheng, Y.-C. An Intelligent Time-Series Model for Forecasting Bus Passengers Based on Smartcard Data. *Appl. Sci.* **2022**, *12*, 4763. [CrossRef]
32. Kruachottikul, P.; Cooharojananone, N.; Phanomchoeng, G.; Kovitanggoon, K. Development of a User-Centric Bridge Visual Defect Quality Control Assisted Mobile Application: A Case of Thailand's Department of Highways. *Appl. Sci.* **2021**, *11*, 9555. [CrossRef]
33. Zhang, Y.; Li, R.; Chen, J. The Cooperative Control of Subgrade Stiffness on Symmetrical Bridge—Subgrade Transition Section. *Symmetry* **2022**, *14*, 950. [CrossRef]
34. Huang, J.; Kong, D.; Gao, G.; Cheng, X.; Chen, J. Data-Driven Reinforcement-Learning-Based Automatic Bucket-Filling for Wheel Loaders. *Appl. Sci.* **2021**, *11*, 9191. [CrossRef]
35. Chen, C.; Li, T.; Ma, C.; Zhang, H.; Tang, J.; Zhang, Y. Hoek-Brown Failure Criterion-Based Creep Constitutive Model and BP Neural Network Parameter Inversion for Soft Surrounding Rock Mass of Tunnels. *Appl. Sci.* **2021**, *11*, 10033. [CrossRef]
36. Lin, Y.-T.; Chen, Y.-K.; Yang, K.-H.; Chen, C.-S.; Han, J.-Y. Integrating InSAR Observables and Multiple Geological Factors for Landslide Susceptibility Assessment. *Appl. Sci.* **2021**, *11*, 7289. [CrossRef]
37. Zenkour, A.M.; Mashat, D.S.; Allehaibi, A.M. Thermoelastic Coupling Response of an Unbounded Solid with a Cylindrical Cavity Due to a Moving Heat Source. *Mathematics* **2022**, *10*, 9. [CrossRef]
38. Heo, S.; Han, S.; Shin, Y.; Na, S. Challenges of Data Refining Process during the Artificial Intelligence Development Projects in the Architecture, Engineering and Construction Industry. *Appl. Sci.* **2021**, *11*, 10919. [CrossRef]
39. Xie, L.; Huang, H.; Du, Q. A Hierarchical Generative Embedding Model for Influence Maximization in Attributed Social Networks. *Appl. Sci.* **2022**, *12*, 1321. [CrossRef]
40. Xie, L.; Huang, H.; Du, Q. A Co-Embedding Model with Variational Auto-Encoder for Knowledge Graphs. *Appl. Sci.* **2022**, *12*, 715. [CrossRef]

applied sciences

MDPI

Article

Enhanced Multi-Strategy Particle Swarm Optimization for Constrained Problems with an Evolutionary-Strategies-Based Unfeasible Local Search Operator

Marco Martino Rosso [1], Raffaele Cucuzza [1,*], Angelo Aloisio [2] and Giuseppe Carlo Marano [1]

1 DISEG, Department of Structural, Geotechnical and Building Engineering, Politecnico di Torino, Corso Duca degli Abruzzi, 24, 10128 Turin, Italy; marco.rosso@polito.it (M.M.R.); giuseppe.marano@polito.it (G.C.M.)

2 Civil Environmental and Architectural Engineering Department, Università degli Studi dell'Aquila, Via Giovanni Gronchi n.18, 67100 L'Aquila, Italy; angelo.aloisio1@univaq.it

* Correspondence: raffaele.cucuzza@polito.it

Abstract: Nowadays, optimization problems are solved through meta-heuristic algorithms based on stochastic search approaches borrowed from mimicking natural phenomena. Notwithstanding their successful capability to handle complex problems, the No-Free Lunch Theorem by Wolpert and Macready (1997) states that there is no ideal algorithm to deal with any kind of problem. This issue arises because of the nature of these algorithms that are not properly mathematics-based, and the convergence is not ensured. In the present study, a variant of the well-known swarm-based algorithm, the Particle Swarm Optimization (PSO), is developed to solve constrained problems with a different approach to the classical penalty function technique. State-of-art improvements and suggestions are also adopted in the current implementation (inertia weight, neighbourhood). Furthermore, a new local search operator has been implemented to help localize the feasible region in challenging optimization problems. This operator is based on hybridization with another milestone meta-heuristic algorithm, the Evolutionary Strategy (ES). The self-adaptive variant has been adopted because of its advantage of not requiring any other arbitrary parameter to be tuned. This approach automatically determines the parameters' values that govern the Evolutionary Strategy simultaneously during the optimization process. This enhanced multi-strategy PSO is eventually tested on some benchmark constrained numerical problems from the literature. The obtained results are compared in terms of the optimal solutions with two other PSO implementations, which rely on a classic penalty function approach as a constraint-handling method.

Keywords: particle swarm optimization (PSO); multi-strategy PSO; self-adaptive evolutionary strategies (ES); local search operator; constraints handling

Citation: Rosso, M.M.; Cucuzza, R.; Aloisio, A.; Marano, G.C. Enhanced Multi-Strategy Particle Swarm Optimization for Constrained Problems with an Evolutionary-Strategies-Based Unfeasible Local Search Operator. *Appl. Sci.* **2022**, *12*, 2285. https://doi.org/10.3390/app12052285

Academic Editors: Nikos D. Lagaros, Vagelis Plevris and Jong Wan Hu

Received: 13 January 2022
Accepted: 15 February 2022
Published: 22 February 2022

1. Introduction

In optimization problems, the aim is optimizing certain mathematical functions, called Objective Functions (OF) $f(x)$. These problems can be divided into single-objective or multi-objective problems, depending on the number of OFs, and a further subdivision for single-objective problems is based on the presence of constraints. Unconstrained problems are defined as:

$$\min_{x \in \Omega} \{f(x)\} \tag{1}$$

meanwhile, constrained problems are defined as:

$$\begin{aligned} &\min_{x \in \Omega} \{f(x)\} \\ \text{s.t.} \quad & g_q(x) \leq 0 \quad \forall q = 1, \ldots, n_q \\ & h_r(x) = 0 \quad \forall r = 1, \ldots, n_r \end{aligned} \tag{2}$$

where $x = \{x_1, \ldots, x_j, \ldots, x_n\}^T$ is the design vector whose terms are the parameters to be optimized. The search domain is a multidimensional space Ω based on the admissible intervals of values for each j-th variable, which are defined by its lower and upper bounds $[x_j^l, x_j^u]$. This detects a box-type hyper-rectangular search space Ω, which is typically defined as the Cartesian product (denoted by the $\times$ symbol) among the admissible intervals:

$$\Omega = [x_1^l, x_1^u] \times \ldots \times [x_j^l, x_j^u] \times \ldots \times [x_n^l, x_n^u] \tag{3}$$

The constraints in (2) can belong to two different categories: inequality $g_q(x)$ and/or equality $h_r(x)$ constraints. Each equality constraint can be easily converted into a couple of inequality constraints; therefore, without any loss of generality, it is possible to consider only inequality constraints in (2), i.e., $g_p(x) \leq 0$, where $p = 1, \ldots, n_q, n_{q+1}, \ldots, n_p$, being $n_p = n_q + 2n_r$.

The adoption of evolutionary algorithms (EAs) has received much more attention in recent years because of their successful capability to handle complex optimization problems. This is addressed mainly to the fact that they do not require any first-order (gradient) or second-order (Hessian) information coming from the problem to be solved, which is conversely a prerogative of the traditional gradient-based mathematical search approaches. Furthermore, the quite simple implementation of EAs has determined their rapid spread, and they have immediately become an attractive tool among practitioners. Among the many alternatives available nowadays, the genetic algorithm (GA) proposed by J. Holland in the 1970s [1] still represents one of the most popular population-based tools, which tries to simulate the biological evolutionary process of a set of candidates solutions mimicking the biological Darwinian Theory. This is realized by adopting specific pseudo-random-based operators such as crossover, mutation, and selection in order to reproduce the long-term process of evolution in a population with the survival of the fittest individuals [2]. In the last two decades, the adoption of metaheuristic algorithms in many engineering applications highlighted their successful capabilities to deal with real-world constrained problems [3–8], e.g., dealing with structural design [9–12] and structural optimization tasks [13–16].

In the framework of EAs, a more recent but already well-known approach is the particle swarm optimization (PSO) algorithm. It was mentioned by Kennedy and Eberhart [17] in 1995 for the first time, and then it rapidly became widespread during the following years. Contributions from the Scientific Community have not ended yet, and still nowadays there is active research about this topic to improve the search operators and the performances. The PSO is also a population-based algorithm which takes inspiration from the study of the behavioural models of birds flocking or fish schooling, whose individuals explore the natural environment in order to find and reach some source of food. Similarly, the algorithm tries to evolve a particle swarm of candidate solutions in the domain search space in order to find the optimum. The PSO was originally developed to face unconstrained problems, but it was later adapted to also solve constrained problems exploiting specific strategies.

The following section presents a brief review of the PSO mechanisms, and the main adopted strategies to solve constrained problems are mentioned. After that, the description of the proposed enhanced multi-strategy PSO method is illustrated. Finally, the authors try to merge several state-of-the-art concepts to obtain an improved PSO algorithm to successfully handle constrained problems with a non-penalty based approach. The novel contributions of this article can be summarized as follows:

- PSO implementation with the main state-of-art improvements, adopting a multi-strategy approach. In this way, the algorithm attempts to avoid wasting many iterations when the algorithm stalls or is trapped in local minima, etc.;
- A non-penalty approach for constraint handling which instead exploits information of swarm positions in terms of the objective function and the actual degree of constraint violation to guide the swarm evolution;

- A novel unfeasible local search operator is presented to help the PSO when it stalls in an unfeasible region quite close to the actual feasible one. This local search operator relies on the meta-heuristic, self-adaptive Evolutionary Strategy (ES) approach, which does not require any other further arbitrary parameter.

In a different recent contribution of the authors [18], some further novel approaches to deal with constraints have been presented, considering a hybridization of the PSO with a machine learning support vector machine. However, the current paper presents a completely different approach based on handling constraints directly based on information which can be retrieved from the swarm positions in terms of objective function and constraints violations. Finally, the enhanced multi-strategy PSO is successfully tested on some benchmark constrained mathematical problems from the literature compared with other PSO implementations that adopt more standard penalty-based constraint handling techniques. In conclusion, the proposed multi-strategy PSO has been validated on real-world case studies, considering some literature on three-dimensional truss design structural optimization problems.

2. Review of PSO and Constraint Handling Approaches

The PSO algorithm was directly inspired by biological behavioral models of birds flockings, school fishing or swarming of insects. In nature, these animals adopt a collective behaviour to ensure their survival, even though each individual acts as an intelligent independent entity making its own decisions. Mimicking this trend, Kennedy and Eberhart in 1995 proposed a first model of the PSO algorithm [17]. The PSO algorithm encodes a population of candidate solutions in the search space, which is composed of a certain number N of intelligent agents. Although the latter can independently move inside the domain, in order to ensure an emerging intelligent collective behaviour toward the optimum, the dynamic movement of each agent is affected by some information obtained from the swarm. One of the first proposed methods is related to a Newtonian dynamics perspective, in which each i-th particle (with $i = 1, \dots, N$, where N is the population size) is completely defined by its position ${}^{k}\boldsymbol{x}_i$ and its velocity ${}^{k}\mathbf{v}_i$ at the k-th generation. The velocity is thus updated taking into account two main kinds of information: First, the self-cognitive memory of each particle, which is related to the so far best visited position ${}^{k}\boldsymbol{x}_i^{Pb}$ (cognitive term) and, second, the attraction toward the other particles' best visited positions ${}^{k}\boldsymbol{x}^{Gb}$ (social term). Therefore, the position and the velocity of the i-th particle in the next $k+1$ iteration can be written as:

$$ {}^{(k+1)}\mathbf{v}_i = {}^{k}\mathbf{v}_i + c_1\, {}^{(k+1)}\boldsymbol{r}_{1i} * \left[{}^{k}\boldsymbol{x}_i^{Pb} - {}^{k}\boldsymbol{x}_i\right] + c_2\, {}^{(k+1)}\boldsymbol{r}_{2i} * \left[{}^{k}\boldsymbol{x}_i^{Gb} - {}^{k}\boldsymbol{x}_i\right], \tag{4} $$

$$ {}^{(k+1)}\boldsymbol{x}_i = {}^{k}\boldsymbol{x}_i + \tau\ {}^{(k+1)}\mathbf{v}_i \quad (\tau = 1), \tag{5} $$

where the symbol $*$ denotes the term-by-term vector multiplication (Hadamard product, [19]), and the positive scalar acceleration factors c_1 and c_2 are denoted as *cognitive* and *social* parameters, respectively. The terms ${}^{(k+1)}\boldsymbol{r}_{1i}, {}^{(k+1)}\boldsymbol{r}_{2i} = rand[0,1]$ are two random weights of the social and cognitive terms, respectively. These terms are fundamental for the purpose to introduce some randomized behaviour inside this quite deterministic model with the aim of enhancing the exploration capabilities of the model. The cognitive term is also denoted as *pbest*, whereas the social term is denoted as *gbest* when it is referred to the best global visited position so far among all the particles of the swarm. This explains why this latter strategy is also known as the *gbest PSO* model [20]. Later studies revealed that a good practise is to protect the cohesion of the swarm by restricting the velocity component to a maximum value, typically assumed as $\mathrm{v}^{\max} = \gamma(\boldsymbol{x}^u - \boldsymbol{x}^l)/\tau$, where $\tau = 1$ is a time-related parameter, whereas $\gamma \in [0.1, 1]$ (generally set to 0.5) defines how far a particle can move starting from its current position [21]. The typical stopping criterion of the PSO is generally set as a maximum number of iterations k_{max}. However, a predetermined maximum number of iterations for each problem is not usually known in advance, therefore, one can refer to the suggestions of [22] or conduct experimental trial and error tuning of the minimum k_{max}, which allows one to achieve the optimum, reducing the overall computational cost. Later on, for the sake of improving the exploration capacity of the swarm, [23] introduced an inertia weight term ${}^k w$ multiplied to the current kth velocity in the update rule (4). This parameter can be a constant or a variable with respect to the iterations flow, e.g., from an initial value ${}^0 w$ to a final one ${}^L w$ with a linearly decreasing law, but there are also many other variants in [20]. The performance of the algorithm is strongly affected by the choice of the parameters such as the swarm size N, usually set in a range of $[20, 100]$ with $n \leq 30$, or such as the acceleration factors, which are usually assumed statically fixed to $c_1 = c_2 = 2$ [21]. In this study, it is assumed which of all of them are constant values equal to $c_1 = c_2 = 2$, ${}^0 w = 0.90$ and ${}^L w = 0.40$ [24].

One of the most important aspects to enhance the PSO performances is to improve the way in which the information are exchanged among the particles. With efficient information sharing, the swarm can exhibit a better collective convergent behaviour. The information exchange is related to the structure of the neighbourhood of each particle, which is denoted as *neighbourhood topology*. This kind of implementation is also called a local PSO model or simply *lbest model* to differentiate it from the classical so-called global PSO model or simply *gbest model* [1,20,21]. The classical gbest model approach can also be regarded as a neighbourhood strategy in which the neighbourhood is composed of the entire population. In this sense, the swarm is denoted as fully informed or fully connected. A schematic graphical representation of the swarm with the information flows is depicted in Figure 1a. The main negative aspect of this latter strategy is the greater inclination to premature convergence. If the global attractor gbest is entrapped in a local minimum, the entire swarm may probably fall down in the same local minimum without a sufficient exploration capability. The enhancement of the PSO was performed by a counter-intuitive approach which relies on slowing down the rapid convergence attitude of the PSO through channelling and limiting the information exchange, the neighbourhood concept indeed [1,20,21]. In the lbest models, it is necessary to define, firstly, the structure of the neighbourhood which controls the way in which the particles are interconnected and, secondly, the size of the neighbourhood which affects the influence of the swarm on each particle [1]. Considering the most popular time-invariant neighbourhood topologies, the ring topology is one of the easiest to be implemented, and it has also been adopted in the present study. As illustrated in Figure 1b,c, each particle in this topology forms a neighbourhood considering the nearest particles (nearest indices in a vector of positions), resulting in an ideal circular interconnection. The total number of the particles which belongs to the neighbourhood is denoted as radius R, as depicted in Figure 1b, in which $R = 2$, and (c), in which $R = 4$. These methods can be implemented considering that each particle in the numerical vector has a unique index, therefore, each particle can unequivocally be selected to enter in a neighbourhood through its index [25], as schematically depicted in Figure 2. A very great number of different neighbourhood topologies were developed in the last decades as showed in [25,26]. Some other implementations also involve a dynamic update of the neighbourhood size, which identifies new types of lbest models which are denoted as multi-populations or multi-swarm PSO, such as in [27].

State of the Art of Constraint Handling

In order to adapt EAs to deal with constrained problems, several strategies were developed by the scientific community. As a matter of fact, constraint handling is a big challenge because it is related to find the optimal point respecting all the constraints, and therefore, the algorithms may be able to deal with unfeasible solutions in an efficient way. Despite several studies (e.g., [28]) demonstrating that PSO has a good convergence rate, it was originally proposed to solve unconstrained optimization problems, such as many other Soft Computing techniques. The implementation of some effective constraint-handling mechanisms is a crucial issue for all biologically inspired optimizers [29–32]. The several strategies developed have been classified by different authors into basically five main categories (see, for instance, the state-of-the-art review by [30,33,34]):

- Penalty-functions-based methods;
- Methods based on special operators and representations;
- Methods based on repair algorithms;
- Methods based on the separation between OFs and constraints;
- Hybrid methods.

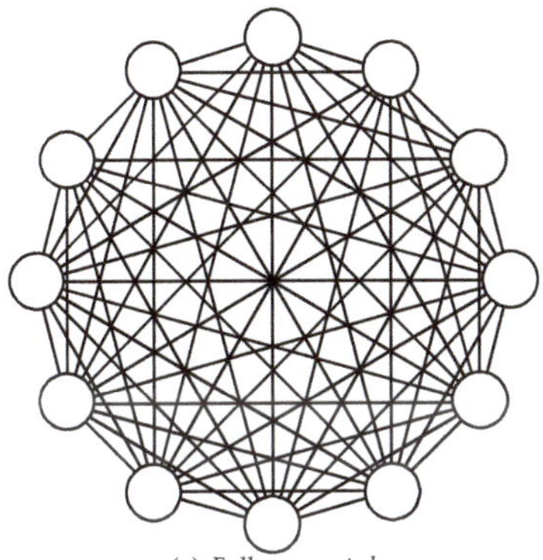

(**a**) *Fully-connected*

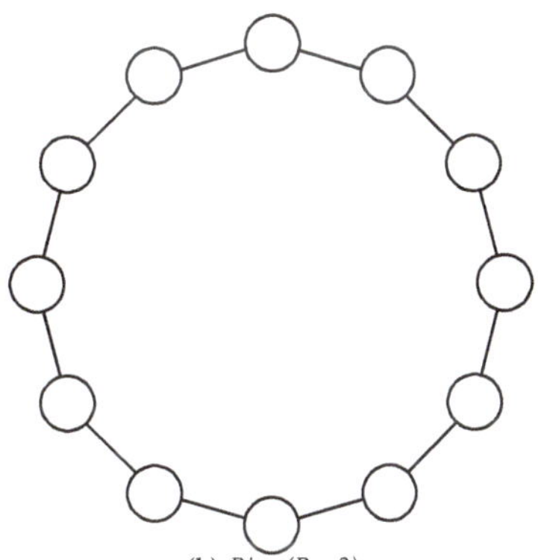

(**b**) *Ring* (R = 2)

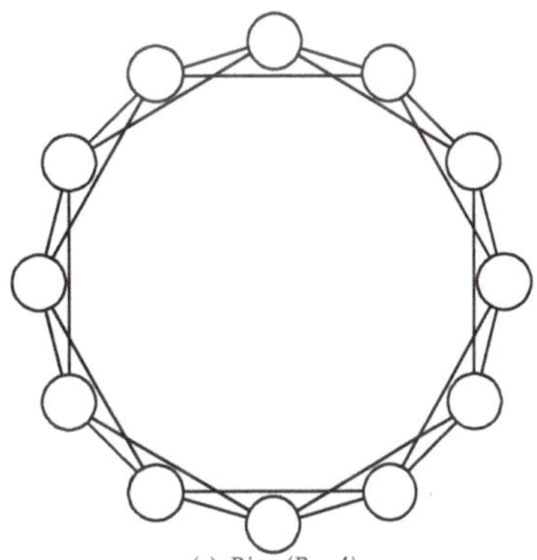

(**c**) *Ring* (R = 4)

Figure 1. Some examples of PSO Neighborhood Topologies.

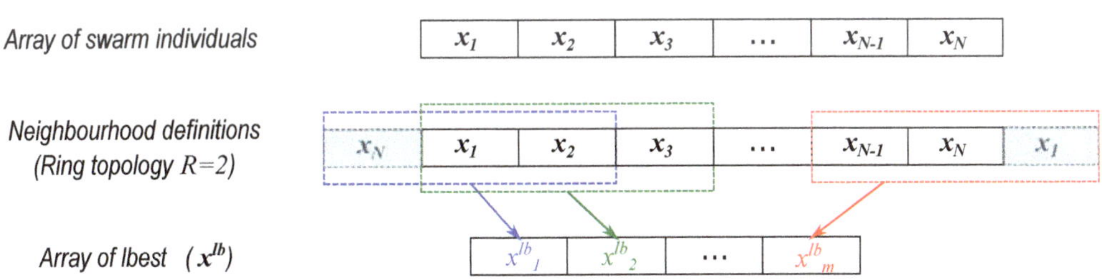

$$x^{lb}_{j} = \min\{OF(\boldsymbol{x_i}) \mid \forall\, \boldsymbol{x_i} \text{ in the neighbourhood}\}$$

Figure 2. Graphical schematization of the Ring topology implementation (R = 2).

The most adopted method due to its simplicity is the exterior penalty approach which allows to convert the problem in an unconstrained version [35,36]. Many different approaches such as the death, static, dynamic, or adaptive penalty functions have been proposed in time, e.g., one can refer to [35]. A proper choice of the constraint-handling mechanism affects the performance of the algorithm, and one of the critical issues to take into account is the preservation of the diversity of the population. The brutal elimination of the unfeasible particles, such as in the death penalty rule, can jeopardize the exploration performances due to a loss of information [30,37]. In general, the penalty approach rely on the evaluation of a factor that applies a certain penalty to the OF, depending on the degree of violation and the number of violated constraints. Therefore, the constrained OF $f(\boldsymbol{x})$ is transformed into an analogous unconstrained OF $\phi(\boldsymbol{x})$:

$$\min_{\boldsymbol{x}\in\Omega}\{\phi(\boldsymbol{x}))\} = \min_{\boldsymbol{x}\in\Omega}\{f(\boldsymbol{x}) + H(\boldsymbol{x})\} \tag{6}$$

where $H(\boldsymbol{x})$ is the penalty function, whose specific definition depends on the strategy adopted. If the penalty is constant during the iterations, it is a *static penalty function*, while if it is changing at each iteration, it is addressed as a *dynamic penalty function*. These two techniques are the most popular tools in structural optimization, see, for instance, the papers by Hasançebi et al. [38] and Dimopoulos [39].

In the case of static-penalty-based techniques, the equivalent unconstrained problem is formulated with a static penalty factor $H_s(\boldsymbol{x})$ that is generally expressed as follows (see [40,41]):

$$H_s(\boldsymbol{x}) = w_1 H_{NVC}(\boldsymbol{x}) + w_2 H_{SVC}(\boldsymbol{x}) \tag{7}$$

where H_{NVC} is the number of constraints that are violated by the particle $\boldsymbol{x}$, H_{SVC} is the sum of all violated constraints, and $w1$ and $w2$ are static control parameters of the penalty scheme:

$$H_{SVC}(\boldsymbol{x}) = \sum_{p=1}^{n_p} \max\{0, g_p(\boldsymbol{x})\} \tag{8}$$

The numerical values adopted by Parsopoulos and Vrahatis [40] are $w_1 = w_2 = 100$. In the present research, some standard penalty PSO approaches are adopted for making comparisons with the enhanced PSO version, which is presented in the following section. For these PSOs with penalty approaches, $w_1 = 0$ and $1000 < w_2 < 10{,}000$ have been assumed, depending on the analysed problem. Depending on the values of w_1 and w_2, it is possible to set the level of severity of the constraint violations: In case of extremely high control parameters, the penalty is called the *death penalty*, and it tries to completely avoid any kind of research inside the unfeasible region, even if the number of violated constraints is rather limited.

The popularity of the penalty function technique is due to its simple implementation, and it strongly enhances the performance of an algorithm that is trying to solve constrained optimization problems. To improve the effectiveness of the penalty factor, a penalty function which changes the weight of the penalty during the iterations is also adopted in the current study. Indeed, it is possible to better control the search space of the particles with this latter dynamic approach, allowing a more relaxed constraint handling at the beginning and an increasing penalty value approaching the end of the available iterations. Firstly proposed by Parsopoulos and Vrahatis [42], it has recently been adopted by Barakat and Altoubat [43] for the optimum design of RC water tanks. To this end, the (7) is readily modified as follows:

$$\min_{\boldsymbol{x}\in\Omega}\{f(\boldsymbol{x}) + {}^{k}h H_d(\boldsymbol{x})\} \tag{9}$$

in which ${}^{k}h$ is a dynamic penalty whose numerical value was evaluated as ([42,43]):

$$^{k}h = \sqrt{k} \tag{10}$$

and $H_d(\boldsymbol{x})$ is the dynamic penalty factor:

$$H_d(\boldsymbol{x}) = \sum_{p=1}^{n_p} \theta_p(\boldsymbol{x})[\max\{0, g_p(\boldsymbol{x})\}]^{\gamma_p(\boldsymbol{x})} \tag{11}$$

Typical assignments for the penalty parameters are (see, for instance, [42,43]):

$$\theta_p(\boldsymbol{x}) = \begin{cases} 10 & \text{if } \max\{0, g_p(\boldsymbol{x})\} \leq 0.001 \\ 20 & \text{if } 0.001 < \max\{0, g_p(\boldsymbol{x})\} \leq 0.100 \\ 100 & \text{if } 0.100 < \max\{0, g_p(\boldsymbol{x})\} \leq 1.000 \\ 300 & \text{otherwise.} \end{cases} \tag{12}$$

$$\gamma_p(\boldsymbol{x}) = \begin{cases} 1 & \text{if } \max\{0, g_p(\boldsymbol{x})\} \leq 1 \\ 2 & \text{otherwise.} \end{cases} \tag{13}$$

It is evident that dynamic penalty methods require a larger number of control parameters in comparison to the static one. Considering ^{k}h as defined in (10), in the present paper, the dynamic penalty factor is assumed to have:

$$10 < H_d(\boldsymbol{x}) < 1000 \tag{14}$$

The evaluation of a proper penalty is a fundamental passage to achieve a good solution of an optimization problem: Ideally, it should be set as low as possible to avoid high computational efforts and problems arising when the global optimum is close to the constraint. Indeed, if the optimum is at the boundary and the penalty is too high, the element which is attracted by that area is immediately pushed back when the boarder is crossed. This mechanism is avoided by adopting a low penalty that is not too severe in case of small violations and also allows a good investigation in such critical areas. However, if the penalty is too low and it does not contrast the constraint violation properly, a lot of effort will be spent in the unfeasible region, providing no useful information for the minimization purpose.

3. Enhanced PSO with a Multi-Strategy Implementation and Hybridisation with an ES-Based Operator

In the present work, starting from the standard Newtonian-dynamics-based PSO approach proposed by Kennedy and Eberhart (1995) in [17], an enhanced PSO is implemented adopting some of the most well-known available strategies in literature and adding a special operator in order to increase the search performance of the standard version. The various strategies are merged together, and the flowchart of the implemented algorithm is illustrated in Figure 3.

At first, the initial population is generated randomly in the hyper-rectangle search space, adopting the Latin Hypercube Sampling (LHS) to generate an initial population with minimum correlation between samples [44]. Thereafter, for each particle, the OF and the constraints are evaluated defining the level of violation of each constraint. Each particle is addressed to a specific aim according to their violation value. If none of the constraints are violated, this particle is labelled as feasible, and it will be addressed to minimize the objective function. Otherwise, if it violates at least one constraint, it is labelled as unfeasible, and it will try to find the right path to minimize the constraint violation. If more than one constraint is violated, only the maximum violation is considered at that point. Therefore, it is possible to assume that each particle is able to see only the envelope of the maximum violations for all points in the solution space. For this reason, the current approach has been named as a "multi-strategy" PSO. In this way, it is not necessary to define some arbitrary violation penalty factor because the code directly relies on the envelope of the violation of the constraints in a particle position at a certain iteration number. After the first population is randomly sampled and evaluated, the role and the aim of each particle have been defined, and the swarm evolution cycle can start, as illustrated in Figure 3. The evolutionary phase of the PSO involves the Velocity update according to the before mentioned formulation (4) and the Position update according to Equation (5). After that, the cognitive memory (pbest) of each particle is updated if a better feasible position is reached with respect to the previous iterations, and the local best attractor (lbest) and the best position for the current generation (gbest) are also updated. The termination criterion is encountered when a predefined maximum number (k_{max}) of iterations is reached.

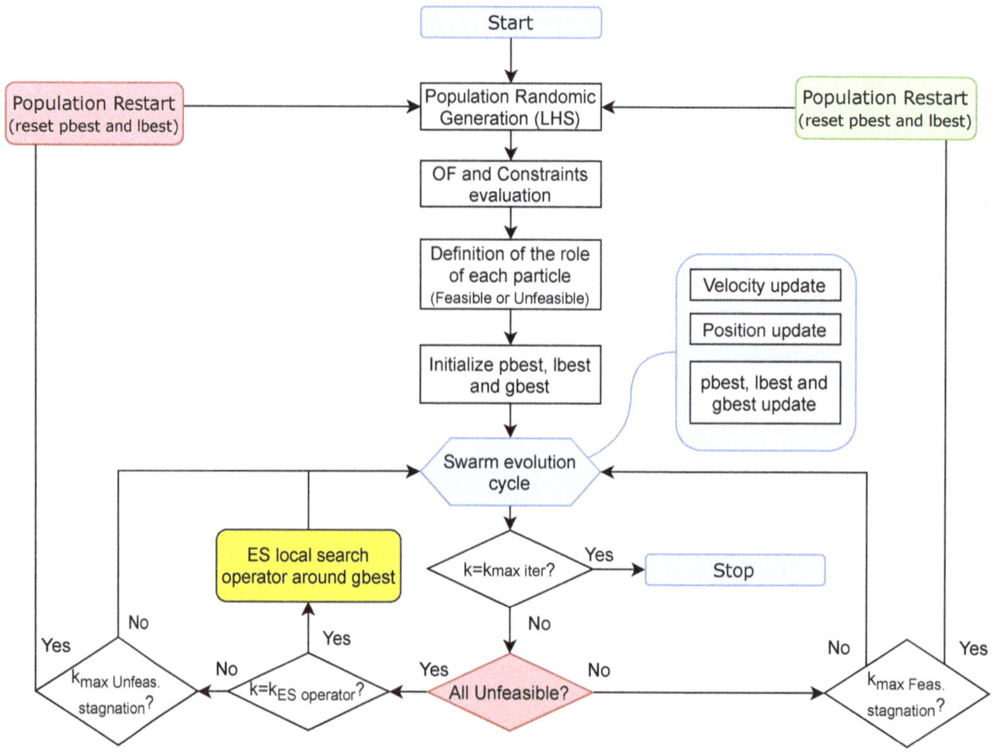

Figure 3. Enhanced PSO multi-strategy flowchart.

It may happen that the feasible region is quite little and narrow with respect to the entire search space; therefore, after some iterations, the swarm also may not have found the feasible region yet. Since the swarm has so far minimized the constraint violation, the swam has probably converged to an unfeasible point with the minimum value of constraint violation, and the feasible region may be located relatively close to that point. This fact suggests that by enhancing the local exploration around the so far unfeasible gbest founded point, the algorithm could be able to identify the feasible search space. Therefore, if the swarm has stalled to an unfeasible point for a number $k = k_{ES\ operator}$ of iterations, a local search operator based on the Evolutionary Strategy approach is thus performed. The Evolutionary Strategy (ES) algorithm is another famous paradigm of the classical EAs based on Darwinian Selection and it was developed by Ingo Rechenberg and Hans-Paul Schwefel at the Technical University of Berlin around the 1960s [1,45]. Without entering deeper into the details of this algorithm, it is necessary to recall that this is a population-based method which relies on the survival of the fittest members. Starting from a parent population, the best individuals have a greater chance to be selected and evolve, forming a certain number of offspring which are generated throughout a slight mutation in the genome of the selected parents. The degree of mutation is governed by a mutation step, which is usually drawn by a Gaussian normal distribution $N(0,\sigma)$, in which σ is also known as the *mutation step* size [45,46]. In formulae, it is possible to express that each gene of a selected parent x_i undergoes a mutation procedure which produces a new offspring's gene equal to $x_i + N(0,\sigma)$. Then, the parents and the offspring will compete for survival, and only the best individuals will survive to the next generation. The main advantage of ES is that it is based on a single parameter to be tuned, the mutation step σ. Many variants of ES were developed in recent decades as mentioned in [45], but the self-adaptation strategy (also denoted as σSA-ES or simply SA-ES [47–49]) is taken into account in the current study. To perform an SA-ES, it is necessary to consider a new representation for the individuals. From a practical point of view, when the parent genome is slightly mutated, if the generated offspring is better in terms of OF evaluation, this offspring will probably survive to the next generation, and it will probably spread its improved genome in the next iterations. Based on this observation, the mutation step can also be added to the original genome of the parent chromosome, giving a new individual representation such as $(x_1,\ldots,x_n,\sigma)$. In this way, not only the genes but also the mutation step undergoes the mutation operator. Thus, if a better offspring is obtained, it will survive and spread its chromosome information, which now implicitly takes into account a new adaptive mutation step. Therefore, in an indirect manner, good individuals will also generate good mutation steps which will be adaptively tuned during the next generations. The above-mentioned approach is known in the literature as SA-ES with uncorrelated mutation with one step size [46,48]. When a number of different mutation steps are considered, one for each gene in the chromosome, such as $(x_1,\ldots,x_n,\sigma_1,\ldots,\sigma_n)$, the adaptive ES strategy is called SA-ES with uncorrelated mutation with n step size [46,48]. It is now clear that the main advantage to introduce the ES local search operator to the current enhanced PSO implementation is due to the fact that it can be implemented without manually tuning other parameters because they are self-tuned by the algorithm itself. For example, in [50], a hybridization of the PSO with ES was performed to enhance the classical velocity update with an adaptive update of the inertia weight and the acceleration factors. For the sake of completeness, there are more sophisticated self-adaptive approaches which take into account also the correlations among the various step sizes associated with the various genes, which are named as SA-ES with correlated mutation [46,48] or covariance matrix adaptation CMA-ES [47,48,51]. In the current study, the SA-ES with uncorrelated mutation with n step size operator is integrated with the PSO inside a local search operator in order to try to locate the feasible region if the swarm stalls to an unfeasible point for $k_{\text{ES operator}} = 10$ iterations. From the unfeasible gbest starting point $\boldsymbol{x}^{Gb,\text{unfea}}$, a population of $N_p = 50$ parent points is sampled from a multivariate Gaussian mixture model in which each component has mean equal to the gbest's i-th component, $x_i^{Gb,\text{unfea}}$, and covariance equal to a first attempt mutation step σ_i. Each i-th mutation step is defined by:

$$\sigma_i = |\tau \cdot N(0,1)| \tag{15}$$

i.e., the absolute value of the product of a random number sampled from a normal standard distribution $N(0,1)$ multiplied to a learning rate parameter τ, which is suggested in [47] to be assumed as $1/\sqrt{N_p}$. Then, a first population of $N_o = 100$ mutated offspring is generated by randomly selected parents adopting a mutation scheme in which the i-th new mutation step size component is updated as:

$$\sigma_{i,\text{off}} = \max(0, |\sigma_i + N(0,1)|). \tag{16}$$

Thereafter, a new offspring point is obtained by adding to the parent position the mutated vector sampled by the multivariate Gaussian mixture model with a mean equal to a zero array and covariance equal to the mutation step size vector updated as above. Subsequently, the mutated offspring are added to the parent population, and the best N_p individuals are selected to survive to the next iteration in terms of constraints violations (or in case of feasible points in term of OF). In the ES jargon, this approach is called the $\mu + \lambda-$ES strategy because the μ (N_p) parents will compete with both each other and also new λ (N_o) offspring, but finally, only μ individuals will survive, whereas the others will be discarded [47]. This mechanism resembles the steady-state approach of other EAs likewise in the genetic algorithm GA [1]. The ES operator could theoretically perform a maximum number of local iterations equal to $k_{\text{max,Local}} = 50$, but in the case that a feasible point is found, the ES evolutionary cycle is interrupted. This new feasible point is thus set up as the gbest of the previous PSO swarm, which remained in a sort of standby state while the local ES operator was in action. In summary, the PSO cycle, which has entered in the ES operator due to the fact that it stalled for $k_{\text{ES operator}} = 10$ iterations on an unfeasible gbest point, can now restart again as usual with an improved knowledge provided by a new feasible posed gbest point found by the local search ES operator. The numerical example Problem g06, whose statement is in the Appendix A (Sickle Problem [52]), has been depicted in Figures 4 and 5 to graphically show the enhanced multi-strategy PSO procedure. Each swarm particle is able to see only the sub-figures (a), (c), and (e) of Figure 4 when its position is inside the feasible region (with the role to minimize the OF); otherwise, it is able to see only the landscape produced by the constraint envelope, subfigures (b), (d), and (f) of Figure 4. After 10 stagnations on the unfeasible gbest point (black cross in sub-figures (a), (b), (c), and (d) of Figure 4), the ES operator was performed. It generated a local population of points near the unfeasible gbest point, which are colored as purple if they are unfeasible or green if they are feasible. Then, this population evolves with the before explained SA-ES approach until at least one point falls inside the feasible region (which is the space between the two blue parabolas) or the maximum number of local iterations is reached. In that specific case, at the first local iteration, some feasible points were already found. Therefore, the best individuals in term of OFs was selected among the green points of Figure 4c,d, and then the PSO could continue its evolutionary cycles until the maximum number of iterations were reached ($k_{max} = 500$). The history of the optimal solution found during the PSO iterations is depicted in Figure 5.

For some very hard problems, it may also happen that after the action of the ES local search operator, the feasible region is not found. In that case, the PSO starts the evolution cycle again with the same unfeasible gbest point for some other iterations until the feasible region is found. Otherwise, when the iterations reach a total number of unfeasible stagnations $k_{\text{max Unfeas Stagn}} = 15$, the complete reset of the population is performed. In practise, the algorithm completely restarts again from the first point of the flowchart, as shown in Figure 3. Therefore, the hope is that a completely new random sampling of the initial swarm will generate a new initial configuration which may find this time the right path to the optimal solution of the optimization problem.

On the contrary, when the PSO normally finds the feasible region and it optimizes the solution until it reaches a gbest which stagnates for a certain number of iterations $k_{\text{max Feas Stagn}} = 50$, the population is restarted as well. This is due to the fact that the so far found optimal solution could be a local minimum. If there is a certain number of iterations left before reaching the maximum PSO available iterations, $k < k_{\text{max}}$, the swarm is thus restarted again from the first step of the PSO flowchart. In that case, all the memories of the population are reset (pbests and lbests), but the so far found optimal solution (gbest) remains unchanged, unless a better solution in terms of OF is found from the new restarted-swarm exploration phase.

In the following section, the enhanced multi-strategy PSO has been tested on some constrained numerical benchmark literature problems, and the results are compared with two PSO implementations, which adopt a typical penalty approach.

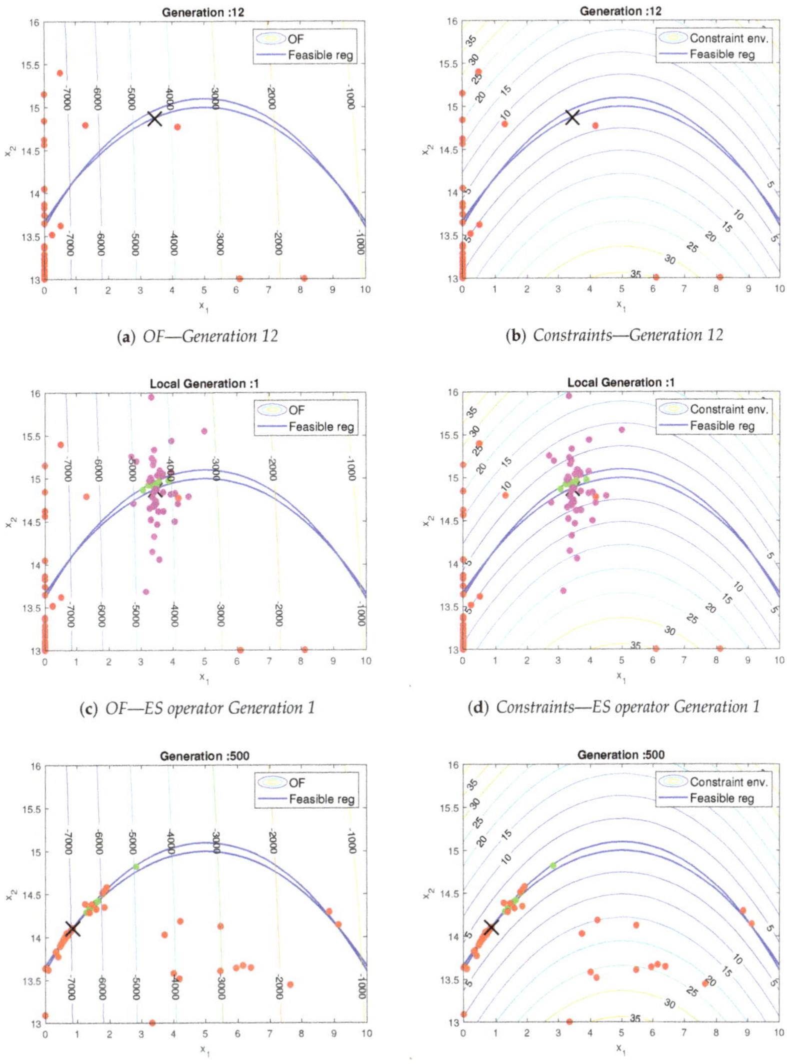

(**a**) *OF—Generation 12* (**b**) *Constraints—Generation 12*

(**c**) *OF—ES operator Generation 1* (**d**) *Constraints—ES operator Generation 1*

(**e**) *Generation 500* (**f**) *Generation 500*

Figure 4. Example Problem g06, see the Appendix A (Sickle Problem [52]); (**a**,**b**) the OF and constraints envelope contour representations, respectively at generation 12. The black cross marker is the unfeasible gbest, the red dots are the swarm points. (**c**,**d**) After 10 unfeasible stagnations, the ES local search operator generate a local search population (purple dots) to find the feasible region (green dots). (**e**,**f**) the OF and constraints envelope contour representations, respectively, at the final generation 500. The black cross marker is the feasible gbest point, the red ones are the particles in a unfeasible region, and the green ones are the particle inside the feasible region.

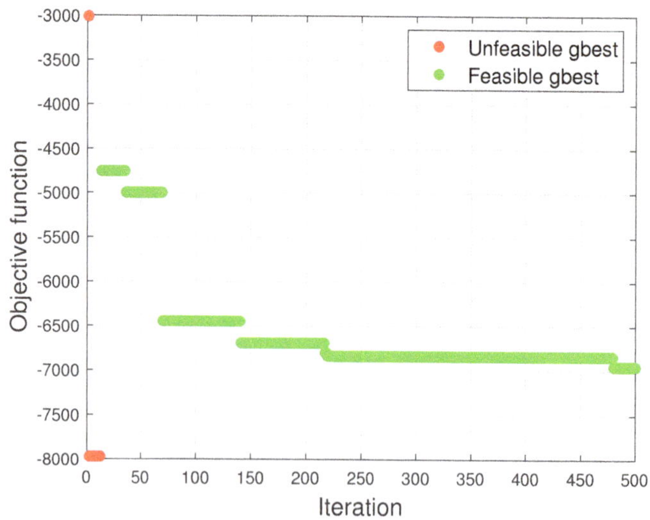

Figure 5. Example Problem g06, see Appendix A (Sickle Problem [52]); Objective function history of the gbest (optimal solution).

4. Numerical Test and Comparisons

The new enhanced multi-strategy PSO illustrated in the previous section was implemented in a Matlab environment and some numerical constrained benchmark tests from the literature were analysed. In particular, the statements of the mathematical constrained problems were taken from [53], in which a total of 13 constrained problems are illustrated. In the current study, only some problems were considered, in particular, the problems with inequalities constraints only were analysed. As stated before, the PSO does not perform very well with equality constraints despite some strategies being proposed in literature to convert each equality constraint into a couple of equivalent inequality constraints. For the sake of completeness, the selected problem statements are also reported in the Appendix A of the present paper. In order to make some comparisons with the other more classical constraint handling approaches, the current enhanced multi-strategy PSO is compared with a more classic penalty approach. For this purpose, the PSO code proposed by [54] was adopted and modified in order to take into consideration both a static penalty approach as previously mentioned in (6) and also with a dynamic penalty as in (9). The penalty factors were properly tuned problem by problem in order to obtain the optimal results. The swarm size was set to $N = 100$, and the maximum allowable iterations were fixed to $k_{\max} = 500$ for all the PSOs considered. The comparisons shown in Table 1 are developed from the results obtained by 50 independent runs and making comparisons among best and worst results and the mean and standard deviation of the OF from the dataset of the 50 final results for the 3 different PSOs. The results in Table 1 produced by the enhanced multi-strategy PSO are satisfactory for the selected numerical problems, and they are generally consistent if compared with the theoretical results and with the other penalty-based PSO implementations. This proves the effectiveness of the current enhanced PSO implementation to deal with constrained optimization problems without the tedious calibration of too many arbitrary parameters. Because of these initial promising results, future works should therefore include some other numerical applications and some engineering practical optimization problems.

Table 1. Selected numerical benchmark examples taken from [53] and comparisons of the final results for 50 runs among the enhanced multi-strategy PSO (*PSO_MS*), the PSO with static penalty (*PSO_ST*), and the PSO with dynamic penalty (*PSO_DYN*).

Problem g01	*PSO_MS*	*PSO_ST*	*PSO_DYN*
optimum		−15.000	
best OF	−15.000	−15.000	−15.0
worst OF	−12.002	−12.000	−12.000
mean	−14.443	−13.938	−13.920
std	0.89478	1.4333	1.4546
Problem g02	*PSO_MS*	*PSO_ST*	*PSO_DYN*
optimum		0.803619	
best OF	0.80357	0.80146	0.79358
worst OF	0.60963	0.52013	0.38285
mean	0.75896	0.70105	0.66597
std	0.063604	0.07356	0.087006
Problem g04	*PSO_MS*	*PSO_ST*	*PSO_DYN*
optimum		−30,665.539	
best OF	−30,666.0	−30,666.0	−31,207.0
worst OF	−30,666.0	−30,665.0	−30,137.0
mean	−30,666.0	−30,665.0	−31,138.2
std	2.20e-05	0.86587	252.2036
Problem g06	*PSO_MS*	*PSO_ST*	*PSO_DYN*
optimum		−6961.81388	
best OF	−6961.8	−6973.0	−6963.0
worst OF	−6958.4	−6973.0	−6963.0
mean	−6960.7	−6973.0	−6963.0
std	0.97521	0.0000	0.0000
Problem g07	*PSO_MS*	*PSO_ST*	*PSO_DYN*
optimum		24.3062091	
best OF	24.426	25.034	24.477
worst OF	27.636	30.203	30.112
mean	25.4129	28.508	27.043
std	1.1209	1.4351	1.8821
Problem g08	*PSO_MS*	*PSO_ST*	*PSO_DYN*
optimum		0.095825	
best OF	0.095825	0.095825	0.095825
worst OF	0.095825	0.095825	0.095825
mean	0.095825	0.095825	0.095825
std	6.96e-17	6.77e-17	7.10e-17
Problem g09	*PSO_MS*	*PSO_ST*	*PSO_DYN*
optimum		680.6300573	
best OF	680.64	680.63	680.63
worst OF	680.98	680.72	680.73
mean	680.73	680.66	680.66
std	0.079365	0.017526	0.018915

Table 1. *Cont.*

Problem g12	*PSO_MS*	*PSO_ST*	*PSO_DYN*
optimum		1.0	
best OF	1.0	1.0	1.0
worst OF	1.0	1.0	1.0
mean	1.0	1.0	1.0
std	0.0000	2.12e-15	0.0000

5. Structural Optimization on Literature Benchmarks

In this final part, some well-acknowledged structural engineering optimization problems from the literature have been adopted for evaluating the performances of the proposed multi-strategy PSO algorithm with the unfeasible local search operator. In the analysed benchmarks, the multi-strategy PSO has been compared with other optimization strategies, i.e., the PSO with static and dynamic penalty inspired by the code of [54] and with the GA from Matlab's built-in code functions. Structural optimization problems can be mainly grouped into three main categories [55]: the *size optimization*, where the aim is to find the optimal size of the structural elements; the *shape optimization*, in which the design variables govern the structural shape; the *topology optimization*, which is the more complex because it involves the modification of the structural typology and morphology. These problems could be tackled separately or even combined. Mainly focusing on the contribution of [56], in the current study, three different truss design constrained size optimization problems have been analysed. The main goal of truss design problems is to minimize the total weight w of the structure, which is indirectly connected to the material consumption volume amount and thus to the cost of the structure [55]. Indeed, adopting a certain material with unit weight ρ_i, the main goal results in seeking for the optimal cross-sectional areas A_i to be devoted to every structural element in the design domain. A first constraint is represented by the box-constraint related to the admissible range of cross section area values to be adopted $A_i \in [A_i^{\mathrm{LB}}, A_i^{\mathrm{UB}}]$. Thereafter, at least two other inequality constraints have to be considered. The first one is related to the respectfulness of the maximum allowable stress σ_{adm} in each truss member (resistance-side constraint) and the second one is referred to the respectfulness of a maximum displacement threshold δ_{adm} (deformation-side constraint). The general formulation of the truss design problem can be stated as follows:

$$\begin{aligned} \min_{x\in\Omega} \quad & f(x) = \sum_{i=1}^{N_el} \rho_i L_i A_i \\ \text{s.t.} \quad & A_i^{\mathrm{LB}} \le A_i \le A_i^{\mathrm{UB}} \\ & \sigma_i \le \sigma_{\mathrm{adm}} \\ & \delta \le \delta_{\mathrm{adm}} \end{aligned} \tag{17}$$

where N_el is the total number of elements in the truss design domain and L_i is the actual length of each member. The material adopted in the current study is structural steel with unit weight of $\rho_i = \rho = 0.1\ \mathrm{lb/in^3}$ ($1\ \mathrm{lb/in^3}$ is equal to $0.0276799\ \mathrm{kg/cm^3}$) and Young's modulus of 10^7 psi (1 psi is equal to 0.00689476 MPa).

5.1. Ten-Bar Truss Design Optimization

The first problem analysed is referred to as a 10 bar truss cantilever structure, as depicted in Figure 6. In the cantilever structure, each member has been labelled with a number from 1 to 10. The cantilever span is in total 720 inches (1 inch is equal to 25.4 mm), and the depth is 360 in. The truss structure is loaded by 2 downward forces of 100 kips each (1 kips is equal to 4.4482 kN). The design vector considers cross-section areas as continuous variables belonging to the a close interval $[0.1, 35]\ \mathrm{in}^2$. The maximum allowable deflection both in horizontal and vertical direction for every node has been set to

$\delta_{adm} = \pm 2$ in, whereas the maximum allowable stress is equal to $\sigma_{adm} = \pm 25$ ksi. In total, 100 independent executions have been performed, and the mean and standard deviation of the OFs have been calculated. A population size of 50 particles and a maximum iterations number of 500 have been set both for the multi-strategy PSO and the GA. For the PSO with penalty approaches, 500 particles have been set as the swarm size because of their very poor results when only 50 particles have been considered. The optimization results obtained are reported in Table 2, which compares the multi-strategy PSO with the PSO with static penalty (PSO-Static), with dynamic penalty (PSO-Dynamic), and with GA. It is worth noting that the penalty approaches fail dreadfully, in this case, to deal with real-life structural design problems, whereas the proposed multi-strategy PSO algorithm produces good results which are comparable with the GA and quite close to the actual unknown optimum solution.

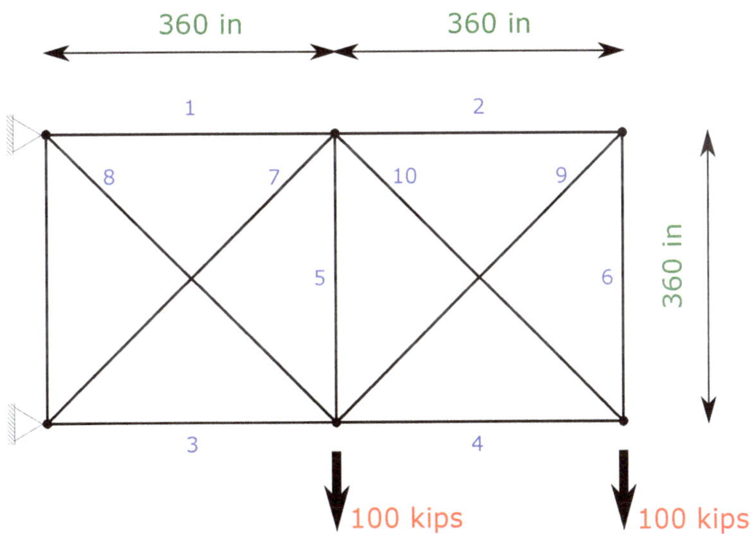

Figure 6. Graphical representation of the 10 bar truss design optimization problem.

5.2. *Twenty-Five-Bar Truss Design Optimization with Multi-Load Cases Conditions*

The second structural optimization problem analysed is referred to as the 25 bar three-dimensional truss tower structure, as depicted in Figure 7. In plan view, the tower footprint is a square of side 200 in, which tapers to 75 in at an elevation of 100 in, and finally reaches the maximum elevation at 200 in from the ground. The structural nodes have been labelled with a number from 1 to 10. The design vector considers the cross section areas of each member as continuous variables belonging to the close interval $[0.01, 3.40]$ in^2. The cross-sectional areas have been gathered into eight groups, as depicted in Figure 8, in order to reduce the dimensionality of the design vector. The maximum allowable displacement has been set to $\delta_{adm} = \pm 0.35$ in in every direction, whereas the maximum allowable stress of each member has been to $\sigma_{adm} = \pm 40$ ksi. Furthermore, the current structural problem takes into account two different load cases during the optimization process, as shown in Figure 7. In total, 100 independent executions have been performed, and the mean and standard deviation of the OFs have been calculated. A population size of 50 particles and a maximum iterations number of 500 have been set both for the multi-strategy PSO and the GA. For the PSO with penalty approaches, 500 particles have been set as the swarm size because of their very poor results when only 50 particles have been considered. The optimization results obtained are reported in Table 3, which compares the multi-strategy PSO with the PSO with the static penalty (PSO-Static), with the dynamic penalty (PSO-Dynamic), and with the GA. It is worth noting that, even in this case, the penalty approaches dreadfully fail to deal with real-life structural design problems, whereas the proposed multi-strategy PSO algorithm produces good results which are comparable with the GA and quite close to the actual optimum solution.

Table 2. Ten-bar truss design example: results comparisons for 100 runs among the enhanced multi-strategy PSO (*PSO-MS*), the PSO with static penalty (*PSO-Static*), and the PSO with dynamic penalty (*PSO-Dynamic*) and GA.

	Cross-Section [in^2]				
Element	**Ref. Sol. from [56]**	**PSO-Static**	**PSO-Dynamic**	**GA**	**PSO-MS**
1	28.920	29.6888	30.3092	30.145	30.372
2	0.100	18.3211	14.7464	0.100	0.110
3	24.070	19.9891	16.5717	22.466	23.644
4	13.960	18.2381	25.1945	15.112	15.391
5	0.100	2.3404	4.5489	0.101	0.101
6	0.560	20.8674	26.1207	0.543	0.496
7	21.950	21.1805	32.2698	21.667	20.984
8	7.690	16.0851	0.2168	7.577	7.410
9	0.100	6.0845	7.5871	0.100	0.103
10	22.090	25.5632	23.524	21.695	21.378
best OF [lb]	5076.310	6141.986	6333.035068	5063.250	5063.328
worse OF [lb]	-	8415.134	8675.749551	5144.148	5229.108
mean [lb]	-	7294.455	7501.394582	5079.744	5076.473
std. dev. [lb]	-	516.7823	475.3885728	14.1194	24.8666

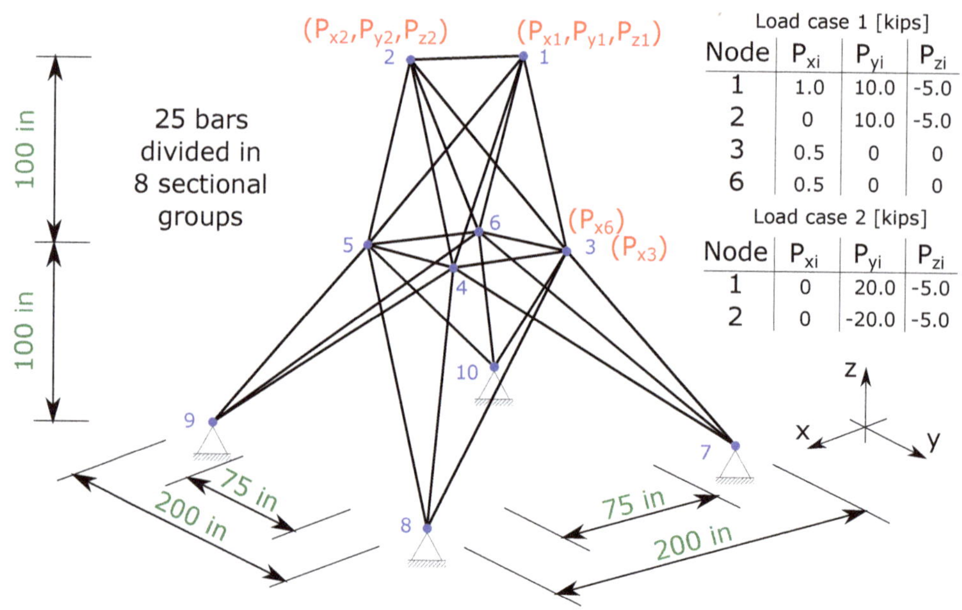

Figure 7. Graphical representation of the 25 bar truss design optimization problem.

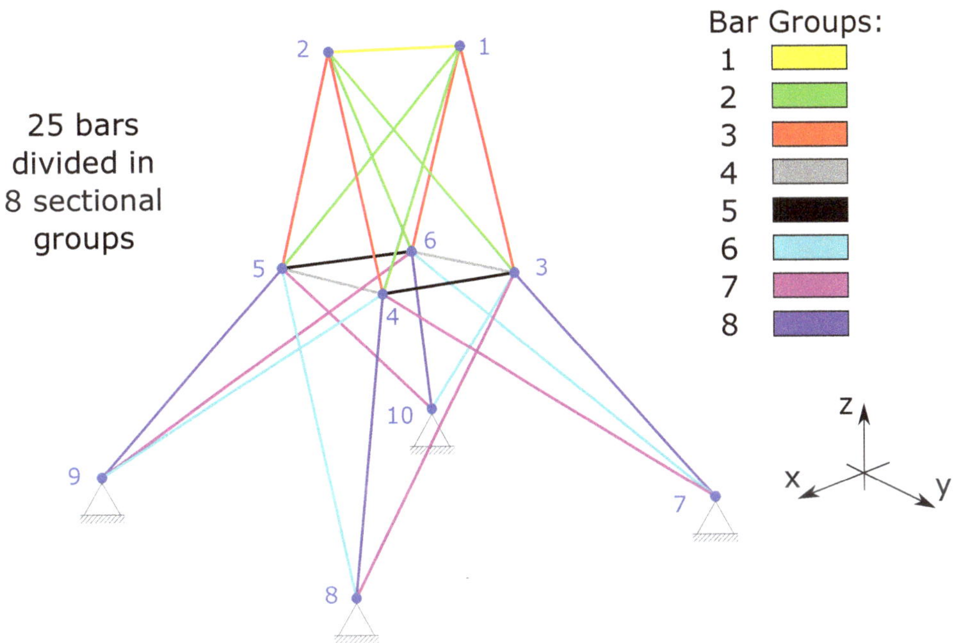

Figure 8. Graphical representation of the 8 bar groups in which are collected all the members of the 25 bar truss design optimization problem.

Table 3. Twenty-five bar truss design example: results comparisons for 100 runs among the enhanced multi-strategy PSO (*PSO-MS*), the PSO with static penalty (*PSO-Static*), and the PSO with dynamic penalty (*PSO-Dynamic*) and GA.

	Cross-Section [in^2]				
Bar Group	**Ref. Sol. from [56]**	**PSO-Static**	**PSO-Dynamic**	**GA**	**PSO-MS**
1	0.100	2.054	1.116	0.010	0.011
2	1.800	2.675	2.670	2.023	1.976
3	2.300	1.402	1.942	2.941	2.989
4	0.200	3.388	0.166	0.010	0.010
5	0.100	0.204	0.342	0.010	0.011
6	0.800	0.453	1.985	0.671	0.690
7	1.800	1.274	1.976	1.673	1.689
8	3.000	0.048	2.345	2.694	2.654
best OF [lb]	546.010	568.186	596.058	545.236	545.249
worse OF [lb]	-	100,583.118	22,954.297	557.755	552.378
mean [lb]	-	1673.393	1122.518	547.828	546.003
std. dev. [lb]	-	9991.0201	3129.3192	2.0743	0.7879

5.3. Seventy-Two-Bar Truss Design Optimization with Multi-Load Cases Conditions

The last structural optimization problem analysed in the current study is referred to as a 72 bar three-dimensional truss tower structure, as depicted in Figure 9. In plan view, the tower footprint is a square of side 120 in, with 4 modular floors, each of them with a height of 60 in. The structural nodes have been labelled with a number from 1 to 20. The design vector considers the cross-sectional areas of each member as continuous variables belonging to the close interval $[0.1, 3.0]$ in^2. There are 18 bars inside each modular floor which can be grouped in 4 groups, as depicted in Figure 10. Therefore, since there are 4 floors, the cross-sectional areas have been parametrized into 16 groups in total in order to reduce the dimensionality of the design vector. The maximum allowable displacement has been set to $\delta_{\text{adm}} = \pm 0.25$ in in every direction, whereas the maximum allowable stress of each member has been set to $\sigma_{\text{adm}} = \pm 25$ ksi. Furthermore, the current structural problem takes into account two different load cases during the optimization process, as shown in Figure 9. In total, 100 independent executions have been performed, and the mean and standard deviation of the OFs have been calculated. A population size of 50 particles and a maximum iterations number of 500 have been set both for the multi-strategy PSO and the GA. For the PSO with penalty approaches, 500 particles have been set as the swarm size because of their very poor results when only 50 particles have been considered. The optimization results obtained are reported in Table 4, which compares the multi-strategy PSO with the PSO with the static penalty (PSO-Static), with the dynamic penalty (PSO-Dynamic), and with the GA. Similarly to the previous cases, it is worth noting that the penalty approaches dreadfully fail to deal with real-life truss design structural optimization problems, whereas the proposed multi-strategy PSO algorithm produces good results which are comparable with the GA and quite close to the actual optimum solution. It is worth noting that the mean value and the best one are very close to the reference optimal solution from [56]. The final solution is even characterized by a low standard deviation among the 100 algorithm runs, demonstrating that the multi-strategy PSO is able to reach the optimal results in a more reliable way, reducing the uncertainties and scattering of the final results.

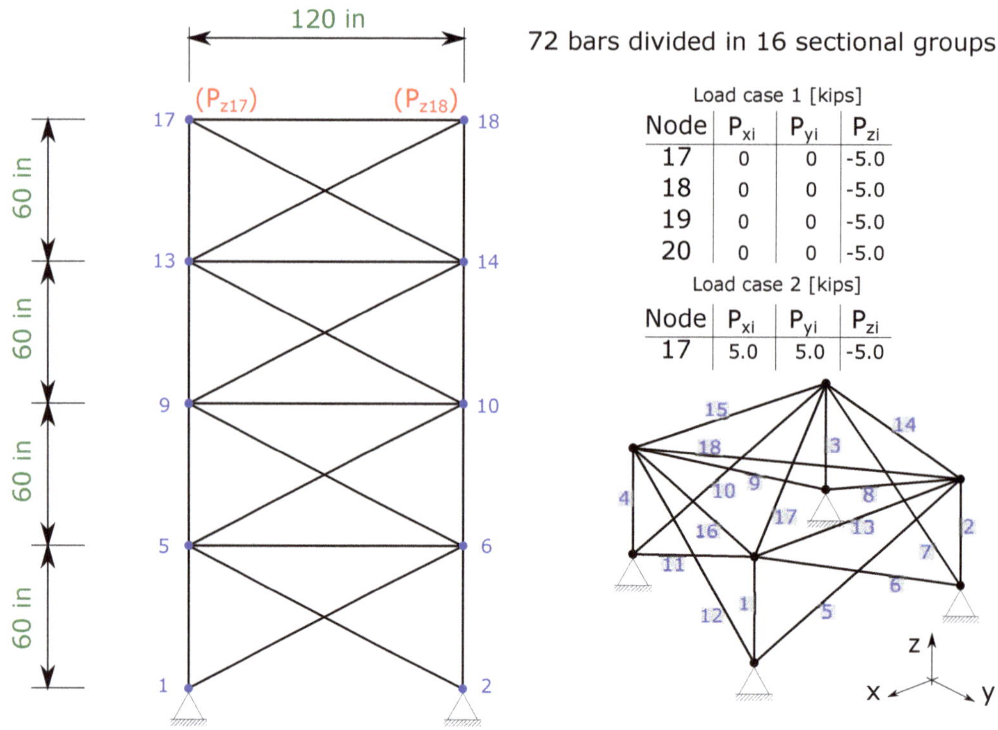

Figure 9. Graphical representation of the seventy-two bars truss design optimization problem.

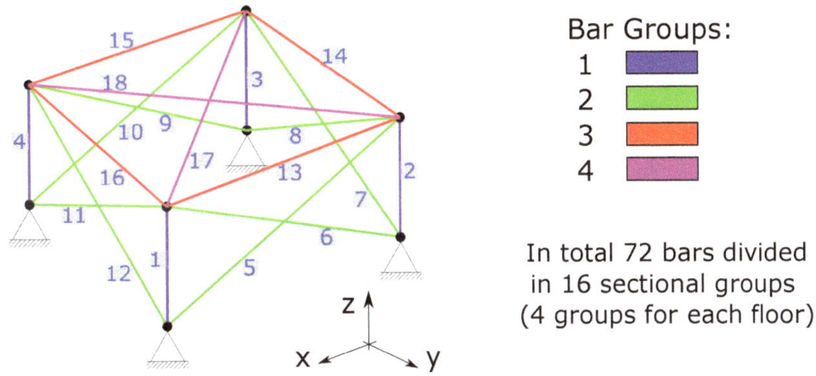

Figure 10. Graphical representation of the four bar groups in which are collected the members inside one module of the seventy-two bars truss design optimization problem.

Table 4. Seventy-two bars truss design example: results comparisons for 100 runs among the enhanced multi-strategy PSO (*PSO-MS*), the PSO with static penalty (*PSO-Static*) and the PSO with dynamic penalty (*PSO-Dynamic*) and GA.

	Cross-Section [in^2]				
Bar Group	**Ref. Sol. from [56]**	**PSO-Static**	**PSO-Dynamic**	**GA**	**PSO-MS**
1	2.026	2.176	0.746	1.801	1.856
2	0.533	0.661	0.539	0.545	0.523
3	0.100	2.686	0.523	0.100	0.100
4	0.100	1.771	2.660	0.100	0.100
5	1.157	1.662	2.316	1.311	1.301
6	0.569	0.276	1.051	0.511	0.519
7	0.100	0.158	0.642	0.100	0.100
8	0.100	0.986	2.370	0.100	0.100
9	0.514	0.271	0.757	0.531	0.539
10	0.479	1.240	0.793	0.520	0.507
11	0.100	0.517	0.453	0.100	0.100
12	0.100	0.378	1.754	0.107	0.101
13	0.158	0.119	2.236	0.157	0.157
14	0.550	0.794	1.677	0.534	0.540
15	0.345	1.363	0.824	0.386	0.403
16	0.498	1.190	0.830	0.561	0.564
best OF [lb]	379.310	629.108	662.148	380.150	379.753
worse OF [lb]	-	1054.764	1110.795	400.147	381.541
mean [lb]	-	874.024	854.233	383.377	380.150
std. dev. [lb]	-	88.8254	82.1187	3.7299	0.2766

6. Discussion

In the previous sections, it has been demonstrated that the proposed multi-strategy PSO algorithm provided quite interesting results. Foremost, focusing on numerical benchmark problems, the multi-strategy PSO technique has been compared with two other traditional PSO implementations which adopt the penalty function approaches to deal with constraints. The three algorithms have been executed 50 independent times for each numerical problem stated in the Appendix A, and the final results have been collected in Table 1. The optimization results have been presented in terms of the best solution, the worst solution, the mean of the OF values, and the standard deviation of the final results.

These parameters evidence the scattering in the found solutions by the various algorithms. Specifically, the standard deviation parameter gives a direct insight into the degree of failure of the meta-heuristic algorithm to find the known benchmark solutions among the independent executions. In particular, the multi-strategy PSO presents in general lower values of a standard deviation compared with PSO-penalty methods, or at least the same order of magnitude. Furthermore, the multi-strategy PSO appears to be a more reliable algorithm because, focusing, e.g., on the problem g06, despite the standard deviation of the PSO-penalty being zero, they fail to reach the optimum solution. This fact highlights that, notwithstanding that the penalty functions method is very simple and easy to implemented, in general, it does not always represent the best approach to successfully deal with every kind of problem. Indeed, e.g., in problem g06, the nil value of the standard deviation actually points out how the penalty method provides a quite deterministic PSO algorithm which is trivially entrapped in the same local optimum among the independent runs, jeopardizing the potentialities of the stochastic search.

On the other hand, focusing on real-world engineering structural optimization problems, the multi-strategy PSO algorithm has revealed its powerful capabilities to deal with complex, combinatorially demanding, and highly non-linear optimization problems. For the sake of completeness, in these problems, a further comparison has been provided by the GA algorithm from the Matlab environment. This latter comparison is extremely relevant because it allows for performing a more objective evaluation which relies on a completely different implementation with respect to the PSO framework only. The optimization results of the 10 bar truss, 25 bar truss, and 72 bar truss problems have been reported in Tables 2–4, respectively. In all the analysed cases, the multi-strategy PSO provided very interesting results, which are really close or even better to the reference solution obtained from [56]. The penalty method revealed their weakness when dealing with these kinds of highly non-linear problems because they provided mean solutions quite far from the reference one and even more scattered when considering the standard deviation values. In conclusion, the proposed multi-strategy PSO algorithm provides an enhanced and more reliable implementation because it results in lower standard deviation values than the GA ones, at least in the last two problems hereby analysed, which are the most complex and computationally demanding.

7. Conclusions

The research and developments in the EAs field to solve optimization problems are continuously increasing because of their lack of mathematical proofs and also because the perfect algorithm to solve any kind of problem does not exist. Therefore, in the present study, a new variant of the PSO has been implemented for the purpose of studying a different way to deal with constrained optimization problems. In fact, the standard version of the PSO [17] lacked a proper mechanism to deal with constrained problems, and in literature [30,33,34], there are at least five main kinds of constraint-handling approaches. The so far most extensively used method in many different practical applications is the penalty function method. The main disadvantage of this technique is that it requires the user to tediously tune some arbitrary penalty factors, which is not always an easy task. In the current study, for the purpose of enhancing the performance of the standard version of the algorithm, the most important state-of-the-art improvements are also implemented, such as the inertia weight [23] and the neighbourhood topology [25]. Furthermore, in order to avoid a penalty-based approach, the violation degree of the constraints is directly exploited to define the aim of a particle which has to minimize this violation if it lies in the unfeasible region. Otherwise, if a particle lies in the feasible region, this particle is dedicated to minimize the OF. Another improvement is given by a local search self-adaptive ES operator, which takes action if the feasible region is not found by the PSO for a certain number of iterations. This allows the algorithm to spread the exploration around the so far unfeasible best solution found, which may be very close to the feasible region, if it is located in near this point. If the ES operator successfully finds the feasible region, this allows it to boost the PSO, giving it an important hint on where the feasible region is located, as demonstrated in Figure 4. If the local operator fails to identify the feasible region, the swarm has probably been entrapped in a local unfeasible minimum quite far from the feasible region. Consequently, only a new randomly resampled swarm may probably find the right path to the feasible region and thus to the real optimum. This new enhanced PSO appears to be noticeably effective compared to other PSO algorithms which adopt a more traditional penalty-function-based method, as shown in Table 1. Outstanding results have been pointed out in the structural optimization benchmark analysed in the current study, which involves three truss design problems from the literature. The proposed PSO effectively dealt with real-life optimization problems, much better than traditional penalty approaches, and reached results comparable and competitive with other state-of-the-art implementations such as the GA.

Although the PSO algorithm already possesses two kinds of memories (cognitive and social), most of the information about the swarm visited positions is discarded, and a better exploitation of the past particles positions remains to be fully determined. In another recent work [18], a first promising step in that direction has been already made. In [18], the PSO has been hybridized with a machine learning algorithm, the support vector machine (SVM). The SVM has been trained on the dataset composed by all the visited swarm positions in order to build a predictive model which is able to learn where the feasible and the unfeasible regions are located in the search domain. The improvement in the managing information provided by the swarm positions during all the iterations allowed the algorithm to reduce the search space extension and considerably improve the PSO's performance. In future studies, another promising direction can be a hybridization with the estimation distribution algorithm (EDA) [57], which relies on building and updating a complex probability distribution model of the search space domain, and therefore, it is potentially able to give considerably much more information about the fitness landscape with respect to a simple blind sampling inside the search space.

Author Contributions: Conceptualization, M.M.R. and G.C.M.; methodology, M.M.R. and G.C.M.; software, M.M.R. and R.C.; validation, R.C. and A.A.; formal analysis, A.A.; investigation, M.M.R.; resources, M.M.R. and G.C.M.; data curation, R.C. and A.A.; writing—original draft preparation, M.M.R.; writing—review and editing, R.C. and A.A.; visualization, R.C. and A.A.; supervision, G.C.M.; All authors have read and agreed to the published version of the manuscript.

Funding: This research was supported by project MSCA-RISE-2020 Marie Skłodowska-Curie Research and Innovation Staff Exchange (RISE)—ADDOPTML (ntua.gr).

Institutional Review Board Statement: Not applicable.

Informed Consent Statement: Not applicable.

Data Availability Statement: The data used to support the findings of this study are available from the corresponding author upon reasonable request.

Acknowledgments: The authors would like to thank anonymous reviewers for their valuable comments and suggestions in revising the paper. The authors would like to thank G.C. Marano and the project ADDOPTML for funding supporting this research.

Conflicts of Interest: The authors declare no conflict of interest.

Appendix A. Test Functions Constrained Problems

In the following, the statements of the selected benchmark numerical problems, taken by [53], which were tested in the present work are exposed.

1. **Problem g01**
   Minimize:

$$f(x) = 5\sum_{i=1}^{4} x_i - 5\sum_{i=1}^{4} x_i^2 - \sum_{i=5}^{13} x_i$$

Subject to:

$$g_1(\boldsymbol{x}) = 2x_1 + 2x_2 + x_{10} + x_{11} - 10 \le 0$$
$$g_2(\boldsymbol{x}) = 2x_1 + 2x_3 + x_{10} + x_{12} - 10 \le 0$$
$$g_3(\boldsymbol{x}) = 2x_2 + 2x_3 + x_{11} + x_{12} - 10 \le 0$$
$$g_4(\boldsymbol{x}) = -8x_1 + x_{10} \le 0$$
$$g_5(\boldsymbol{x}) = -8x_2 + x_{11} \le 0$$
$$g_6(\boldsymbol{x}) = -8x_3 + x_{12} \le 0$$
$$g_7(\boldsymbol{x}) = -2x_4 - x_5 + x_{10} \le 0$$
$$g_8(\boldsymbol{x}) = -2x_6 - x_7 + x_{11} \le 0$$
$$g_9(\boldsymbol{x}) = -2x_8 - x_9 + x_{12} \le 0$$

where the search space is defined as $0 \le x_i \le 1$ $(i = 1,\ldots,9)$, $0 \le x_i \le 100$ $(i = 10, 11, 12)$, $0 \le x_{13} \le 1$. The optimum is located at $\boldsymbol{x}^* = [1; 1; 1; 1; 1; 1; 1; 1; 1; 3; 3; 3; 1]$, where $f(\boldsymbol{x}) = -15$.

2. **Problem g02**
Maximize:

$$f(\boldsymbol{x}) = \left| \frac{\sum_{i=4}^{n} \cos^4(x_i) - 2\prod_{i=1}^{n} \cos^2(x_i)}{\sqrt{\sum_{i=1}^{n} i x_i^2}} \right|$$

Subject to:

$$g_1(\boldsymbol{x}) = 0.75 - \prod_{i=1}^{n} x_i \le 0$$

$$g_2(\boldsymbol{x}) = \sum_{i=1}^{n} x_i - 7.5n \le 0$$

where $n = 20$ and the search space is defined as $0 \le x_i \le 10$ $(i = 1,\ldots,n)$. The optimum OF is $f(\boldsymbol{x}) = 0.803619$.

3. **Problem g04**
Minimize:

$$f(\boldsymbol{x}) = 5.3578547x_3^2 + 0.8356891x_1x_5 + 37.293239x_1 - 40792.141$$

Subject to:

$$g_1(\boldsymbol{x}) = 85.334407 + 0.0056858x_2x_5 + 0.0006262x_1x_4 + 0.0022053x_3x_6 \le 92,$$
$$g_2(\boldsymbol{x}) = -85.334407 - 0.0056858x_2x_5 - 0.0006262x_1x_4 + 0.0022053x_3x_6 \le 0,$$
$$g_3(\boldsymbol{x}) = 80.51249 + 0.0071317x_2x_5 + 0.0029955x_1x_2 + 0.0021813x_3^2 - 110 \le 0,$$
$$g_4(\boldsymbol{x}) = -80.51249 - 0.0071317x_2x_5 - 0.0029955x_1x_2 - 0.0021813x_3^2 + 90 \le 0,$$
$$g_5(\boldsymbol{x}) = 9.300961 + 0.0047026x_3x_5 + 0.0012547x_1x_3 + 0.0019085x_3x_4 - 25 \le 0,$$
$$g_6(\boldsymbol{x}) = -9.300961 - 0.0047026x_3x_5 - 0.0012547x_1x_3 - 0.0019085x_3x_4 + 20 \le 0,$$

where the search space is defined as $78 \le x_1 \le 102$ and $33 \le x_2 \le 45$ and $27 \le x_3, x_4, x_5 \le 45$. The optimum is located at $\boldsymbol{x}^* = [78, 33, 29.995256025682, 45, 36.775812905788]$, where $f(\boldsymbol{x}) = -30,665.539$.

4. **Problem g06**
Minimize:

$$f(\boldsymbol{x}) = (x_1 - 10)^3 + (x_2 - 20)^3$$

Subject to:

$$g_1(\boldsymbol{x}) = -(x_1-5)^2-(x_2-5)^2+100 \leq 0$$
$$g_2(\boldsymbol{x}) = (x_1-6)^2-(x_2-5)^2-82.81 \leq 0$$

where the search space is defined as $13 \leq x_1 \leq 100$ and $0 \leq x_2 \leq 100$. The optimum is located at $\boldsymbol{x}^* = [14.095; 0.84296]$, where $f(\boldsymbol{x}^*) = -6961.81388$.

5. **Problem g07**
Minimize:

$$f(\boldsymbol{x}) = x_1^2+x_2^2+x_1x_2-14x_1-16x_2+(x_3-10)^2+4(x_4-5)^2+(x_5-3)^2 \\ +2(x_6-1)^2+5x_7^2+7(x_8-11)^2+2(x_9-10)^2+(x_{10}-7)^2+45$$

Subject to:

$$g_1(\boldsymbol{x}) = -105+4x_1+5x_2-3x_7+9x_8 \leq 0$$
$$g_2(\boldsymbol{x}) = 10x_1-8x_2-17x_7+2x_8 \leq 0$$
$$g_3(\boldsymbol{x}) = -8x_1+2x_2+5x_9-2x_{10}-12 \leq 0$$
$$g_4(\boldsymbol{x}) = 3(x_1-2)^2+4(x_2-3)^2+2x_3^2-7x_4-120 \leq 0$$
$$g_5(\boldsymbol{x}) = 5x_1^2+8x_2+(x_3-6)^2-2x_4-40 \leq 0$$
$$g_6(\boldsymbol{x}) = x_1^2+2(x_2-2)^2-2x_1x_2+14x_5-6x_6 \leq 0$$
$$g_7(\boldsymbol{x}) = 0.5(x_1-8)^2+2(x_2-4)^2+3x_5^2-x_6-30 \leq 0$$
$$g_8(\boldsymbol{x}) = -3x_1+6x_2+12(x_9-8)^2-7x_{10} \leq 0$$

where the search space is defined as $-10 \leq x_i \leq 10$ $(i = 1, \ldots, 10)$. The optimum OF is $f(\boldsymbol{x}^*) = 24.3062091$.

6. **Problem g08**
Maximize:

$$f(\boldsymbol{x}) = \frac{\sin^3(2\pi x_1)\sin 2\pi x_2}{x_1^3(x_1+x_2)}$$

Subject to:

$$g_1(\boldsymbol{x}) = x_1^2-x_2+1 \leq 0$$
$$g_2(\boldsymbol{x}) = 1-x_1+(x_2-4)^2 \leq 0$$

where the search space is defined as $0 \leq x_1, x_2 \leq 10$. The optimum is located at $\boldsymbol{x}^* = [1.2279713; 4.2453733]$, where $f(\boldsymbol{x}^*) = -0.0958250414$.

7. **Problem g09**
Minimize:

$$f(\boldsymbol{x}) = (x_1-10)^2+5(x_2-12)^2+x_3^4+3(x_4-11)^2 \\ +10x_5^6+7x_6^2+x_7^4-4x_6x_7-10x_6-8x_7$$

Subject to:

$$g_1(\boldsymbol{x}) = -127+2x_1^2+3x_2^4++x_3+4x_4^2+5x_5 \leq 0$$
$$g_2(\boldsymbol{x}) = -282+7x_1+3x_2+10x_3^2+x_4-x_5 \leq 0$$
$$g_3(\boldsymbol{x}) = -196+23x_1+x_2^2+6x_6^2-8x_7 \leq 0$$
$$g_4(\boldsymbol{x}) = 4x_1^2+x_2^2-3x_1x_2+2x_3^2+5x_6-11x_7 \leq 0$$

where the search space is defined as $-10 \leq x_i \leq 10$ $(i = 1, \ldots, 7)$. The optimum OF is $f(x^*) = 680.6300573$.

8. **Problem g12**
Maximize:

$$f(x) = \frac{100 - (x_1 - 5)^2 - (x_2 - 5)^2 - (x_3 - 5)^2}{100}$$

Subject to:

$$g(x) = (x_1 - p)^2 + (x_2 - q)^2 + (x_3 - r)^2 - 0.0625 \leq 0$$

where the search space is defined as $0 \leq x_i \leq 10$ $(i = 1, 2, 3)$ and $p, q, r = 1, 2, \ldots, 7$. The optimum OF is $f(x^*) = -1$.

References

1. Martí, R.; Pardalos, P.M.; Resende, M.G.C. *Handbook of Heuristics*, 1st ed.; Springer Publishing Company, Incorporated: Berlin/Heidelberg, Germany, 2018. [CrossRef]
2. Lagaros, N.D.; Papadrakakis, M.; Kokossalakis, G. Structural optimization using evolutionary algorithms. *Comput. Struct.* **2002**, *80*, 571–589. [CrossRef]
3. Marano, G.; Trentadue, F.; Greco, R. Stochastic optimum design criterion of added viscous dampers for buildings seismic protection. *Struct. Eng. Mech.* **2007**, *25*, 21–37. [CrossRef]
4. Pelliciari, M.; Marano, G.; Cuoghi, T.; Briseghella, B.; Lavorato, D.; Tarantino, A. Parameter identification of degrading and pinched hysteretic systems using a modified Bouc–Wen model. *Struct. Infrastruct. Eng.* **2018**, *14*, 1573–1585. [CrossRef]
5. Xue, J.; Lavorato, D.; Bergami, A.; Nuti, C.; Briseghella, B.; Marano, G.; Ji, T.; Vanzi, I.; Tarantino, A.; Santini, S. Severely damaged reinforced concrete circular columns repaired by turned steel rebar and high-performance concrete jacketing with steel or polymer fibers. *Appl. Sci.* **2018**, *8*, 1671. [CrossRef]
6. Greco, R.; Marano, G. Identification of parameters of Maxwell and Kelvin-Voigt generalized models for fluid viscous dampers. *JVC/J. Vib. Control* **2015**, *21*, 260–274. [CrossRef]
7. Di Trapani, F.; Tomaselli, G.; Sberna, A.P.; Rosso, M.M.; Marano, G.C.; Cavaleri, L.; Bertagnoli, G. Dynamic Response of Infilled Frames Subject to Accidental Column Losses. In Proceedings of the 1st Conference of the European Association on Quality Control of Bridges and Structures, Padua, Italy, 29 August–1 September 2021; Pellegrino, C., Faleschini, F., Zanini, M.A., Matos, J.C., Casas, J.R., Strauss, A., Eds.; Springer International Publishing: Cham, Switzerland, 2022; pp. 1100–1107.
8. Asso, R.; Cucuzza, R.; Rosso, M.M.; Masera, D.; Marano, G.C. Bridges Monitoring: An Application of AI with Gaussian Processes. In Proceedings of the 14th International Conference on Evolutionary and Deterministic Methods for Design, Optimization and Control; Institute of Structural Analysis and Antiseismic Research National Technical University of Athens, Athens, Greece, 28–30 June 2021. [CrossRef]
9. De Domenico, D.; Qiao, H.; Wang, Q.; Zhu, Z.; Marano, G. Optimal design and seismic performance of Multi-Tuned Mass Damper Inerter (MTMDI) applied to adjacent high-rise buildings. *Struct. Des. Tall Spec. Build.* **2020**, *29*, e1781. [CrossRef]
10. De Tommasi, D.; Marano, G.; Puglisi, G.; Trentadue, F. Morphological optimization of tensegrity-type metamaterials. *Compos. Part B Eng.* **2017**, *115*, 182–187. [CrossRef]
11. Sardone, L.; Rosso, M.M.; Cucuzza, R.; Greco, R.; Marano, G.C. Computational Design of Comparative Models and Geometrically Constrained Optimization of A Multi Domain Variable Section Beam Based on Timoshenko Model. In Proceedings of the 14th International Conference on Evolutionary and Deterministic Methods for Design, Optimization and Control, Athens, Greece, 28–30 June 2021; Institute of Structural Analysis and Antiseismic Research National Technical University of Athens: Athens, Greece, 2021. [CrossRef]
12. Cucuzza, R.; Rosso, M.M.; Marano, G. Optimal preliminary design of variable section beams criterion. *SN Appl. Sci.* **2021**, *3*, 745. [CrossRef]
13. Cucuzza, R.; Costi, C.; Rosso, M.M.; Domaneschi, M.; Marano, G.C.; Masera, D. Optimal strengthening by steel truss arches in prestressed girder bridges. In *Proceedings of the Institution of Civil Engineers—Bridge Engineering*; Thomas Telford Ltd.: London, UK, 2021; pp. 1–21. [CrossRef]
14. Fiore, A.; Marano, G.; Greco, R.; Mastromarino, E. Structural optimization of hollow-section steel trusses by differential evolution algorithm. *Int. J. Steel Struct.* **2016**, *16*, 411–423. [CrossRef]
15. Aloisio, A.; Pasca, D.P.; Battista, L.; Rosso, M.M.; Cucuzza, R.; Marano, G.; Alaggio, R. Indirect assessment of concrete resistance from FE model updating and Young's modulus estimation of a multi-span PSC viaduct: Experimental tests and validation. *Elsevier Struct.* **2022**, *37*, 686–697. [CrossRef]
16. Marano, G.; Trentadue, F.; Petrone, F. Optimal arch shape solution under static vertical loads. *Acta Mech.* **2014**, *225*, 679–686. [CrossRef]

17. Kennedy, J.; Eberhart, R. Particle swarm optimization. In Proceedings of the ICNN'95—International Conference on Neural Networks, Perth, Australia, 27 November–1 December 1995; Volume 4, pp. 1942–1948. [CrossRef]
18. Rosso, M.M.; Cucuzza, R.; Di Trapani, F.; Marano, G.C. Nonpenalty Machine Learning Constraint Handling Using PSO-SVM for Structural Optimization. *Adv. Civ. Eng.* **2021**, *2021*, 6617750. [CrossRef]
19. Plevris, V. Innovative Computational Techniques for the Optimum Structural Design Considering Uncertainties. Ph.D. Thesis, Institute of Structural Analysis and Seismic Research, School of Civil Engineering, National Technical University of Athens (NTUA), Athens, Greece, 2009. [CrossRef]
20. Sengupta, S.; Basak, S.; Peters, R.A. Particle Swarm Optimization: A Survey of Historical and Recent Developments with Hybridization Perspectives. *Mach. Learn. Knowl. Extr.* **2019**, *1*, 157–191. [CrossRef]
21. Quaranta, G.; Lacarbonara, W.; Masri, S. A review on computational intelligence for identification of nonlinear dynamical systems. *Nonlinear Dyn.* **2020**, *99*, 1709–1761. [CrossRef]
22. Li, B.; Xiao, R. The Particle Swarm Optimization Algorithm: How to Select the Number of Iteration. In Proceedings of the Third International Conference on Intelligent Information Hiding and Multimedia Signal Processing (IIH-MSP 2007), Kaohsiung, Taiwan, 26–28 November 2007; pp. 191–196. [CrossRef]
23. Shi, Y.; Obaiahnahatti, B. A Modified Particle Swarm Optimizer. In Proceedings of the 1998 IEEE International Conference on Evolutionary Computation, IEEE World Congress on Computational Intelligence (Cat. No. 98TH8360), Anchorage, AK, USA, 4–9 May 1998; Volume 6, pp. 69–73. [CrossRef]
24. Perez, R.; Behdinan, K. Particle swarm approach for structural design optimization. *Comput. Struct.* **2007**, *85*, 1579–1588. [CrossRef]
25. Medina, A.; Toscano Pulido, G.; Ramírez-Torres, J. *A Comparative Study of Neighborhood Topologies for Particle Swarm Optimizers*; IJCCI: Pasadena, CA, USA, 2009; pp. 152–159.
26. Schmitt, B.I. Convergence Analysis for Particle Swarm Optimization. Ph.D. Thesis, FAU University Press, Erlangen, Nürnberg, Germany, 2015.
27. Liang, J.; Suganthan, P. Dynamic Multi-Swarm Particle Swarm Optimizer with a Novel Constraint-Handling Mechanism. In Proceedings of the 2006 IEEE International Conference on Evolutionary Computation, Vancouver, BC, Canada, 16–21 July 2006; pp. 9–16. [CrossRef]
28. Kennedy, J.; Eberhart, R.C. *Swarm Intelligence*; Morgan Kaufmann Publishers Inc.: San Francisco, CA, USA, 2001.
29. Deb, K. An efficient constraint handling method for genetic algorithms. *Comput. Method Appl. Mech. Eng.* **2000**, *186*, 311–338. [CrossRef]
30. Coello Coello, C.A. Theoretical and numerical constraint-handling techniques used with evolutionary algorithms: A survey of the state of the art. *Comput. Methods Appl. Mech. Eng.* **2002**, *191*, 1245–1287. [CrossRef]
31. Wang, Y.; Cai, Z.; Zhou, Y.; Fan, Z. Constrained optimization based on hybrid evolutionary algorithm and adaptive constraint-handling technique. *Struct. Multidiscip. Optim.* **2008**, *37*, 395–413. [CrossRef]
32. Mezura-Montes, E. *Constraint-Handling in Evolutionary Optimization*; Springer: Berlin/Heidelberg, Germany, 2009; Volume 198. [CrossRef]
33. Koziel, S.; Michalewicz, Z. Evolutionary Algorithms, Homomorphous Mappings, and Constrained Parameter Optimization. *Evol. Comput.* **1999**, *7*, 19–44. [CrossRef]
34. Michalewicz, Z.; Fogel, D. *How to Solve It: Modern Heuristics*; Springer: Berlin/Heidelberg, Germany, 2008.[CrossRef]
35. Rezaee Jordehi, A. A review on constraint handling strategies in particle swarm optimisation. *Neural Comput. Appl.* **2015**, *26*, 1265–1275. [CrossRef]
36. Kohler, M.; Vellasco, M.M.; Tanscheit, R. PSO+: A new particle swarm optimization algorithm for constrained problems. *Appl. Soft Comput.* **2019**, *85*, 105865. [CrossRef]
37. Mezura-Montes, E.; Coello, C. A simple multimembered evolution strategy to solve constrained optimization problems. *IEEE Trans. Evol. Comput.* **2005**, *9*, 1–17. [CrossRef]
38. Hasançebi, O.; Çarbaş, S.; Doğan, E.; Erdal, F.; Saka, M. Performance evaluation of metaheuristic search techniques in the optimum design of real size pin jointed structures. *Comput. Struct.* **2009**, *87*, 284–302. [CrossRef]
39. Dimopoulos, G.G. Mixed-variable engineering optimization based on evolutionary and social metaphors. *Comput. Methods Appl. Mech. Eng.* **2007**, *196*, 803–817. [CrossRef]
40. Parsopoulos, K.; Vrahatis, M. Unified Particle Swarm Optimization for Solving Constrained Engineering Optimization Problems. In *International Conference on Natural Computation*; Springer: Berlin/Heidelberg, Germany, 2005; Volume 3612, pp. 582–591.[CrossRef]
41. Coello, C. Self-adaptive penalties for GA-based optimization. In Proceedings of the 1999 Congress on Evolutionary Computation-CEC99 (Cat. No. 99TH8406), Washington, DC, USA, 6–9 July 1999; Volume 1, pp. 573–580. [CrossRef]
42. Parsopoulos, K.; Vrahatis, M. Particle Swarm Optimization Method for Constrained Optimization Problem. *Intell. Technol.-Theory Appl. New Trends Intell. Technol.* **2002**, *76*, 214–220.
43. Barakat, S.A.; Altoubat, S. Application of evolutionary global optimization techniques in the design of RC water tanks. *Eng. Struct.* **2009**, *31*, 332–344. [CrossRef]
44. Monti, G.; Quaranta, G.; Marano, G. Genetic-Algorithm-Based Strategies for Dynamic Identification of Nonlinear Systems with Noise-Corrupted Response. *J. Comput. Civ. Eng.* **2010**, *24*, 173–187. [CrossRef]

45. Beyer, H.G.; Schwefel, H.P. Evolution strategies—A comprehensive introduction. *Nat. Comput.* **2002**, *1*, 3–52. [CrossRef]
46. Eiben, A.; Smith, J. *Introduction To Evolutionary Computing*; Springer: Berlin/Heidelberg, Germany, 2003; Volume 45. [CrossRef]
47. Beyer, H.G. Toward a Theory of Evolution Strategies: Self-Adaptation. *Evol. Comput.* **1995**, *3*, 311–347. [CrossRef]
48. Fister, I., Jr.; Fister, I. *On the Mutation Operators in Evolution Strategies*; Springer: Cham, Switzerland, 2015; Volume 18, pp. 69–89. [CrossRef]
49. Hansen, N. An analysis of mutative σ-self-adaptation on linear fitness functions. *Evol. Comput.* **2006**, *14*, 255–275. [CrossRef]
50. Miranda, V.; Fonseca, N. EPSO-evolutionary particle swarm optimization, a new algorithm with applications in power systems. In Proceedings of the IEEE/PES Transmission and Distribution Conference and Exhibition, Yokohama, Japan, 6–10 October 2002; Volume 2, pp. 745–750.
51. Kramer, O. Evolutionary self-adaptation: A survey of operators and strategy parameters. *Evol. Intell.* **2010**, *3*, 51–65. [CrossRef]
52. Simionescu, P.A.; Beale, D.; Dozier, G.V. Constrained optimization problem solving using estimation of distribution algorithms. In Proceedings of the 2004 Congress on Evolutionary Computation (IEEE Cat. No. 04TH8753), Portland, OR, USA, 19–23 June 2004; Volume 1, pp. 296–302.
53. Long, W.; Liang, X.; Huang, Y.; Chen, Y. A hybrid differential evolution augmented Lagrangian method for constrained numerical and engineering optimization. *Comput.-Aided Des.* **2013**, *45*, 1562–1574. [CrossRef]
54. Alam, M. *Codes in MATLAB for Particle Swarm Optimization*; Research Gate Indian, Institute of Technology Roorkee: Roorkee, India, 2016. [CrossRef]
55. Christensen, P.; Klarbring, A. *An Introduction to Structural Optimization*; Springer: Berlin/Heidelberg, Germany, 2008; Volume 153. [CrossRef]
56. Camp, C.; Farshchin, M. Design of space trusses using modified teaching–learning based optimization. *Eng. Struct.* **2014**, *62–63*, 87–97. [CrossRef]
57. Pelikan, M.; Hauschild, M.W.; Lobo, F.G. *Estimation of Distribution Algorithms*; Springer: Berlin/Heidelberg, Germany, 2015; pp. 899–928. [CrossRef]

applied sciences

MDPI

Article

Hybrid Wind Turbine Towers Optimization with a Parallel Updated Particle Swarm Algorithm

Zeyu Li [1,*], Hongbing Chen [2], Bin Xu [3,4] and Hanbin Ge [3,5]

1 College of Civil Engineering, Hunan University, Changsha 410082, China
2 School of Civil and Resource Engineering, University of Science and Technology Beijing, Beijing 100083, China; hongbingchen@ustb.edu.cn
3 College of Civil Engineering, Huaqiao University, Xiamen 361021, China; binxu@hqu.edu.cn (B.X.); gehanbin@meijo-u.ac.jp (H.G.)
4 Key Laboratory for Intelligent Infrastructure and Monitoring of Fujian Province, Huaqiao University, Xiamen 361021, China
5 Department of Civil Engineering, Meijo University, Nagoya 468-8502, Japan
* Correspondence: lizeyu@hnu.edu.cn

Abstract: The prestressed concrete–steel hybrid (PCSH) wind turbine tower, characterized by replacing the lower part of the traditional full-height steel tube wind turbine tower with a prestressed concrete (PC) segment, provides a potential alterative solution to transport difficulties and risks associated with traditional steel towers in mountainous areas. This paper proposes an optimization approach with a parallel updated particle swarm optimization (PUPSO) algorithm which aims at minimizing the objective function of the levelized cost of energy (LCOE) of the PCSH wind turbine towers in a life cycle perspective which represents the direct investments, labor costs, machinery costs, and the maintenance costs. Based on the constraints required by relevant specifications and industry standards, the geometry of a PCSH wind turbine tower for a 2 MW wind turbine is optimized using the proposed approach. The dimensions of the PCSH wind turbine tower are treated as optimization variables in the PUPSO algorithm. Results show that the optimized PCSH wind turbine tower can be an economic alternative for wind farms with lower LCOE requirements. In addition, compared with the traditional particle swarm optimization (PSO) algorithm and UPSO algorithm, the proposed PUPSO algorithm can enhance the optimization computation efficiency by about 60–110%.

Keywords: prestressed concrete–steel hybrid (PCSH) wind turbine tower; optimal design; parallel updated particle swarm optimization (PUPSO) algorithm; wind; earthquake; levelized cost of energy (LCOE)

Citation: Li, Z.; Chen, H.; Xu, B.; Ge, H. Hybrid Wind Turbine Towers Optimization with a Parallel Updated Particle Swarm Algorithm. *Appl. Sci.* **2021**, *11*, 8683. https://doi.org/10.3390/app11188683

Academic Editor: Nikos D. Lagaros

Received: 14 August 2021
Accepted: 12 September 2021
Published: 17 September 2021

1. Introduction

The wind turbine tower, as the structure supporting the wind turbine, represents a highly significant component of wind turbine systems and accounts for approximately 30% of the overall investment in onshore installations [1]. With the increase in unit power capacity of wind turbines, the heights of wind turbine towers have increased for the purpose of capturing wind energy efficiently, as wind profiles are strong and steady at higher elevations [2–4]. In recent years, wind turbine towers with a height of over 100 m have been widely employed in practice alongside increasing investment [5]. Many wind farms have been developed or are under construction in mountainous areas in the mainland of China after decades of wind farm development in plain areas. The transportation of segmental steel tubes and long blades to the top of mountains is a challenging task with risks. Moreover, the construction of temporary transportation roads with large turning radii in mountains leads to additional investment and environmental destruction [6]. The traditional steel-tubular wind turbine tower systems are typical soft supporting systems, and it is hard to meet the stiffness requirements of large capacity wind turbines due to the limitation of steel-tube diameter transportation.

In recent years, the prestressed concrete–steel hybrid (PCSH) wind turbine tower has been proposed to overcome the difficulty of transportation and the limitation of the structural mechanical behavior of traditional steel tubular towers. Compared with the full-height steel tubular tower, the PCSH wind turbine tower results in a lower center of gravity and higher flexural stiffness. The use of concrete leads to a lower sensitivity to fluctuations in steel prices [2]. Moreover, by replacing parts of the steel tubular segments with prestressed concrete (PC), the total cost of the PCSH tower system can be decreased while the design–servicing life of PC is much longer than steel. This leads to reasonable life-cycle cost savings and decreases in the levelized cost of energy (LCOE) in a life-cycle perspective. The development of the PCSH wind turbine tower has received great attention in recent years. Singh [7] investigated concrete construction for wind energy towers and highlighted the advantages of concrete as the major construction material for wind turbine towers. Seidel [8] compared a steel and concrete hybrid tower with a steel tower and concluded that hybrid towers are an effective alternative to traditional steel towers, can be built at nearly every site, and help overcome transportation issues caused by mountains or other terrains.

The optimization algorithm plays key roles in realizing the economical results that withstand the most demanding functional requirements arising during their service life [9]. Hani et al. [10] proposed and tested five different optimization strategies for a 100 kW wind turbine system considering the natural frequencies as the most representative objective function. Uys et al. [11] used optimization to calculate the least cost of a steel wind turbine tower that meets the structural demands and emphasized the influence of ring stiffeners. Nicholson et al. [12,13] redesigned wind turbine towers with a generalized reduced gradient (GRG) method and analyzed how individual design variables affected the objective function of a hybrid wind turbine tower. Employing the genetic algorithm (GA), Ma et al. [14] optimized a 100 m PC tower system for a 5 MW wind turbine and discussed the advantages of a PC wind turbine tower. Oest et al. [15] explored three different state-of-the-art analytical gradient-based optimization approaches to minimize the mass of a jacket structure for wind turbines considering fatigue and frequency constraints and provided insight into critical structural and modeling parameters. Adopting GA, Chen et al. [16] suggested that the optimal height of the concrete segment should be 80.5 m for one 120 m PCSH wind turbine tower. The safety factors of the tower are improved and the total construction cost can be reduced by about 20% after optimization. However, the optimization result is easily stuck at locally optimal values and the material utilization ratio of the optimization result is comparatively low. Different from the direct investment or construction cost for a wind turbine system investigated in the above studies, it is more important to minimize its LCOE in a life-cycle perspective, which is treated as the objective function for geometry optimization for PCSH wind turbine towers in this study.

Particle swarm optimization (PSO) has proven to be a powerful method for optimization problems [17]. Adopting PSO, Poitras et al. [18] investigated the optimum floor configuration by minimizing the total mass or cost while satisfying all design criteria. Ye et al. [19] conducted a comprehensive investigation on cold-formed steel beam designs using PSO techniques. Luo et al. [20] proposed a computational approach based on PSO to obtain the lower bound of the buckling load of shell structures with geometric imperfections. Based on a PSO algorithm, Xu et al. [21] optimized the active control strategy for machinery-equipment-induced structural vibrations. Tsiptsis et al. [22] carried out structural optimization employing isogeometric tools in PSO for a two-dimensional truss or a frame tower. Farias et al. [23] introduced a new hybrid algorithm based on PSO and GA to find optimal fiber orientation of stiffened laminated composite panels to reach their maximum buckling load. Kaveh and Eslamlou optimized a series of usual-size skeletal structures by transplanting a harmony search-based mechanism to particle swarm optimization with an aging leader and challengers (HALC-PSO) and multistage particle swarm optimization (MSPSO) and obtained satisfactory results [24].

In this paper, in order to enhance the computation efficiency for the geometry optimization of the PCSH wind turbine towers, a parallel updated PSO (PUPSO) method is proposed to optimally design a PCSH wind turbine tower subjected to both wind and seismic excitations, considering the constraints of load-carrying capacity, fatigue, stability, natural frequency, and maximum top displacement. It employs the LCOE as the objective function. The proposed approach is used to optimally design a 2 MW PCSH wind turbine tower with a design height of 77.5 m as an alternative to the traditional steel tubular tower. The optimal result is evaluated by utilization ratio of the tower. Results show that the PUPSO algorithm efficiently optimizes the PCSH wind turbine towers when compared with the traditional particle swarm optimization (PSO) algorithm and the LCOE of the optimized PCSH wind turbine significantly decreases when compared with the benchmark wind turbine tower. The height of the steel segment of the optimized PUPSO tower is recommended to be 30% of total height of the PCSH wind turbine tower. Compared with the original PCSH tower, the increased utilization rates of both PC and steel segments illustrate the effectiveness of the PUPSO algorithm. Moreover, the fundamental natural frequency of the optimized PCSH wind turbine tower increases significantly when compared with that of the original wind turbine tower.

2. Effects of Wind and Earthquake Excitations on Wind Turbine Tower

Under normal operation, wind power generation systems are subjected to wind loads and are also affected by earthquakes in seismically active areas over their service life. These effects are of importance to the performance, durability, and safety of wind turbine towers. In this study, the PCSH wind turbine tower is geometrically optimized with the consideration of both wind and earthquake loads. The effects of both wind and earthquakes on the PCSH wind tower system are discussed in the following sections.

2.1. Wind Load Applied to the PCSH Wind Turbine Towers

2.1.1. Aerodynamic Load Determination

Due to differences in wind pressure, the aerodynamic wind load applied on the top of a wind turbine tower is usually calculated under four different working conditions, including the annual average wind speed, nominal wind speed, cut-out wind speed, and extreme wind speed [25]. The aerodynamic load can be determined by the following equations [26,27]:

$$F_1 = C_p \rho {V_a}^2 \pi R^2 \tag{1}$$

$$F_2 = C_p \rho {V_n}^2 \pi R^2 \tag{2}$$

$$F_3 = 0.4 C_p \rho {V_c}^2 \pi R^2 \tag{3}$$

$$F_4 = 0.5 C_t \rho {V_e}^2 A \tag{4}$$

$$\rho = 0.00125 e^{-0.0001 z^3} \tag{5}$$

where F_1 is the wind load under the annual average wind speed, C_p is the wind energy utilization coefficient which can take the value of 4/9 for an ideal wind turbine but 0.4 is chosen for the PCSH wind turbine tower in this study, ρ is the density of air, V_a is the annual average wind speed, R is the impeller radius of the wind turbine, F_2 is the wind load applied to the turbine under the nominal wind speed, V_n is the nominal wind speed, F_3 is the wind load under the cut-out wind speed, V_c is the cut-out wind speed, F_4 is the wind load under the strongest wind speed in 50 years, the drag coefficient $C_t = 1.6$, V_e is the extreme wind speed, A is the projection of the blades in the plane perpendicular to the direction of the wind, and z is the height.

2.1.2. Pitching Moment

The pitching moment, M_P, caused by inhomogeneity in the wind speed can be calculated by the following Equation (6) [13]:

$$M_P = \frac{4}{27}\frac{\rho}{B}\pi R^3\left(V_1^2 - V_2^2\right) \tag{6}$$

where B is the number of blades and V_1 and V_2 are the wind speeds at locations 1 and 2, respectively, as illustrated in Figure 1.

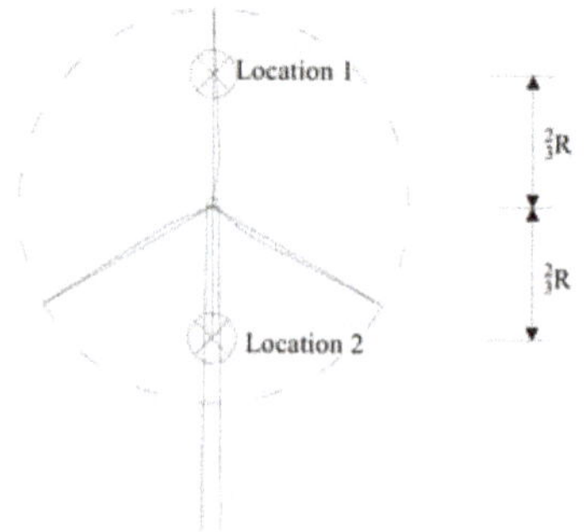

Figure 1. Computational locations for V_1 and V_2.

2.1.3. Deflecting Torque

The deflecting torque on the wind turbine tower is mainly caused by the generator impeller. The equation for deflecting torque T can be simplified as [27]:

$$T = 0.23\rho V_c{}^2\pi R^2 e_h \tag{7}$$

where e_h is the horizontal distance between the center of the hub and the center of the tower.

2.2. Wind Load Acting on the Tower

According to the load code for the design of building structure GB50009-2012 [28], the characteristic value of the wind load can be calculated with the following equations:

$$\omega_k = \beta_z\mu_s\mu_z\omega_0 \tag{8}$$

$$\beta_z = 1 + 2g_f I_{10} B_z\sqrt{1+R_f^2} \tag{9}$$

$$R_f = \sqrt{\frac{\pi}{6\zeta_1}\frac{x_1^2}{\left(1+x_1^2\right)^{4/3}}} \tag{10}$$

$$x_1 = \frac{30f_1}{\sqrt{k_w\omega_0}} \tag{11}$$

$$B_z = k_{fr}H^{a_1}\rho_x\rho_z\frac{\phi_1(z)}{\mu_z} \tag{12}$$

$$\rho_z = \frac{10\sqrt{H+60e^{-H/60}-60}}{H} \tag{13}$$

where ω_0 is the basic wind speed at a height of 10 m, β_z is the wind-induced vibration factor, μ_s is the wind load shape coefficient and μ_z is the wind pressure height coefficient, g_f equals to 2.5, I_{10} is the nominal turbulence intensity, B_z is background component of fluctuating wind load, R_f is the resonance component of the fluctuating wind load, ζ_1 is the damping ratio and is equal to 0.03 in this paper, f_1 is the first-order natural frequency,

k_w is the surface roughness correction coefficient and is equal to 1.0 in this paper, H is the height of the tower, ρ_x is the horizontal correlation coefficient and equals 1.0 due to the small width of the windward side of the tower, ρ_z is the vertical correlation coefficient, and the $k_{fr}, a_1, \phi_1(z)$ can be determined according to GB 50009-2012 [28].

The lateral static force is applied along the height of the tower as a distributed load. The force of the tower section at height i owing to the wind can be calculated as [28]:

$$F_i = \omega_k A_i \tag{14}$$

where F_i is the wind force of the tower section at height i and A_i is the wind pressure area of the section.

2.3. Additional Bending Moment

The additional bending moment, M_e, at the top of the wind turbine tower can be calculated according to Equation (10):

$$M_e = m_e g e_e \tag{15}$$

where g is the acceleration of gravity, m_e is the weight of the equipment at the top of the tower, including the blades, nacelle, hub, etc., and e_e is the distance between the center of the equipment and the center of the tower.

2.4. Earthquake Effect

In order to consider the effect of earthquakes on the wind turbine tower, it is reasonable to model the tower structure as a mass-lumped structure. The natural frequencies are determined for calculating the earthquake effect on the wind turbine tower [29]. For the PCSH wind turbine structure, the tower is simplified as a five degrees-of-freedom (DOF) model with five lumped masses as shown in Figure 2 [30]. The lumped mass on the top of the wind turbine tower is the largest because of the existence of blades, nacelle, hub, etc. The other lumped mass is determined by the distributed mass along the tower. The bending stiffness of the model changes with the height of the tower. Based on the seismic influence coefficient curve, the earthquake effect can be estimated by the mode–superposition response spectrum method [31].

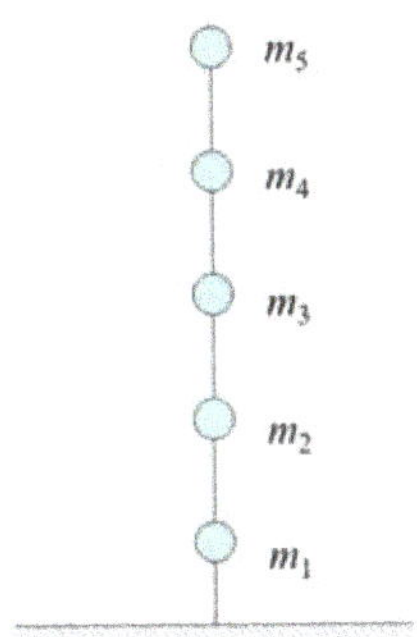

Figure 2. Simplified tower model.

2.5. Load Combination

Referring to the relevant literatures [28,32,33], the load combinations are given in Table 1. In this table, WL is the wind load on the tower, DL is the dead load, and EQ is the effect of an earthquake.

Table 1. Load combinations.

Load Combinations	Load Factors
Ultimate 1	$1.4 \times DL + 1.4 \times 0.2 \times WL + 1.3EQ$
Ultimate 2	$1.0 \times DL + 1.4 \times WL$
Service 1	$1.0 \times DL + 0.2 \times WL + 1.0 \times EQ$
Service 2	$1.0 \times DL + 1.0 \times WL$

3. Design Constraints for Optimization of the PCSH Tower

The PCSH wind turbine tower can be modelled as a typical cantilever beam with variable cross sections. The following assumptions are made during the geometry optimization of the PCSH wind turbine tower in this paper. The bottom of the tower is fixed on the ground, and a concentrated mass representing the blades, nacelle, hub, and top part of the tower is attached at the top of the tower while the distributed mass along the tower is simulated by four lumped masses [30]. The nonlinearity of both PC and steel materials is not considered. The stress concentration around the door opening and the connection between the concrete and steel are not considered in the geometry optimization model due to the fact that a local strengthening measure is adopted around the door opening [16]. The optimization analysis is only performed in the fore-and-aft direction.

3.1. Constraints on the Steel Tubular Segment

3.1.1. Local Buckling

According to the code for the design of chimneys GB50051-2013 [34], the following condition should be satisfied in order to avoid local buckling of the steel tubular tower:

$$\frac{N_i}{A_{ni}} + \frac{M_i}{W_{ni}} \leq \sigma_{crt} \tag{16}$$

where M_i is the design maximum bending moment of a cross section i, N_i is the design axial tension or pressure associated with M_i, A_{ni} is the net cross-sectional area of a cross section i, and W_{ni} is the net cross-sectional resistance moment of the cross section. The local buckling critical stress of the steel segment $\sigma_{crt} = 0.4\frac{E}{k}\frac{t}{D}$, E is the elastic modulus of steel, k is the regulation factor of local bearing strength, t is the thickness of the segment, and D is the outer diameter of the segment.

3.1.2. Overall Stability

According to the code for the design of steel structures GB50017-2003 [35], the monolithic stability should fulfill the following requirement:

$$\frac{N_i}{\varphi A_{bi}} + \frac{M_i}{W_{bi}(1 - 0.8N_i/N_{Ex})} \leq f_t \tag{17}$$

where A_{bi} is the gross cross-sectional area of cross section i, W_{bi} is the gross cross-sectional resistance moment of cross section i, φ is the coefficient of stability of the axial compression members of level cross section i, f_t is the yield strength value of the steel segment, and N_{Ex} is the Euler critical load.

3.1.3. Load-Carrying Capacity

According to Agbayani [36], the following constraints need to be considered:

$$f_{cu} \leq \phi_c F_{cn} \tag{18}$$

$$f_{vu} \leq \phi_v F_{vn} \tag{19}$$

$$f_{Tu} \leq \phi_T F_{Tn} \tag{20}$$

$$f_{vu}/(\phi_v F_{vn}) + f_{Tu}/(\phi_T F_{Tn}) \leq 1 \tag{21}$$

and

For $f_{Tu}/(\phi_T F_{Tn}) \leq 0.2$:

$$f_{cu}/(\phi_c F_{cn}) \leq 1 \tag{22}$$

For $f_{Tu}/(\phi_T F_{Tn}) > 0.2$:

$$\left[f_{cu}/(\phi_c F_{cn})\right]^2 + \left[f_{vu}/(\phi_v F_{vn}) + f_{Tu}/(\phi_T F_{Tn})\right]^2 \leq 1 \tag{23}$$

where f_{cu} is the compression stress of the steel segment, $\phi_c = 0.9$, F_{cn} is the nominal compressive strength, f_{vu} is the transverse shear of the steel segment, $\phi_v = 0.9$, F_{vn} is the nominal shear strength of the steel segment, but F_{vn} should not exceed $F_y/\sqrt{3}$, f_{Tu} is the torsion of the steel segment, $\phi_T = 0.9$, T_u is the design torsional moment, and F_{Tn} is the nominal torsional strength of the steel segment.

3.1.4. Fatigue

The supporting structures for wind turbines are usually subjected to variable amplitude stress cycles caused by wind over their service life. As a result, the investigation of fatigue strength is of considerable significance for the design of wind-turbine-supporting structures.

According to the code for the design of steel structures GB50017-2017 [35], the allowable stress range of fatigue can be calculated by the equation:

$$[\Delta\sigma] = \left(\frac{C}{n}\right)^{\frac{1}{\beta}} \tag{24}$$

where n is the number of stress cycles and C and β can be determined by the code for the design of steel structures.

The Weibull Distribution function is commonly used to represent the wind speed frequency distribution. Based on the wind data for a given site, a method for estimating the wind speed frequency distribution is used [37]. The wind speed over 5.29×10^8 cycles for a 20-year fatigue design life of a wind farm can be synthesized [32]. The stress amplitude of the steel tubular tower segment can be determined based on the wind turbine tower model, the probability distribution, and the rain-flow counting method [38]. Fatigue assessment can be performed according to the amplitude, Miner rule, and code for the design of steel structures [35]. The equivalent stress range of the variable amplitude fatigue $\Delta\sigma_e$ can be identified with the following equation:

$$\Delta\sigma_e = \left[\frac{\sum n_l(\Delta\sigma_l)^{\beta}}{\sum n_l}\right]^{\frac{1}{\beta}} \tag{25}$$

where $\sum n_l$ is the life expectancy of the structure expressed in the number of stress cycles and n_l is the number of stress cycles of stress range $\Delta\sigma_l$ during the expected lifespan of the structure.

3.2. Constraints on the PC Segments

3.2.1. Load-Carrying Capacity

The minimum concrete compressive stress is set to be larger than zero. According to the code for design of high-rising structures GB50135-2006 [39] and code for the design of concrete structures GB50010-2010 [40], the following conditions must be fulfilled for compressive load-carrying capacity:

$$0 < \sigma_c < f_c \tag{26}$$

$$\frac{V_u}{1.2tD} + \frac{T_u}{W_t} < 0.7 f_t + 0.05 \frac{N_{p0}}{1.2tD} \tag{27}$$

where σ_c is the concrete stress, f_c are the concrete axial compressive load-carrying capacity, V_u is the design shear force, T_u is the design torsional moment, W_t is the torsional section modulus,f_t is the concrete axial tensile load-carrying capacity, N_{p0} is the concrete normal prestressing force of the cross section, t is the thickness of the segment, and D is the outer diameter.

3.2.2. Fatigue

According to GB50010-2010 [40], the following constraints for concrete fatigue stress must be fulfilled:

$$\sigma^f_{cc,max} \leq f^f_c \tag{28}$$

$$\Delta\sigma^f_p \leq \Delta f^f_{py} \tag{29}$$

where $\sigma^f_{cc,max}$ is the maximum concrete compressive stress of a cross section, f^f_c is the axial compressive fatigue strength, $\Delta\sigma^f_p$ is the prestressed reinforcement stress amplitude, and Δf^f_{py} is the fatigue stress amplitude limit of prestressed reinforcement.

3.2.3. Geometry Constraint

According to GB50135-2006 [39], the thinnest thickness of the wall t_{min} (mm) should fulfill the following Equation (30) and be thicker than 180 mm:

$$t_{min} = 100 + 0.01D \tag{30}$$

3.3. *Other Constraints*

3.3.1. Natural Frequency

To avoid resonance of the PCSH tower caused by the rotation of wind turbine blades, there should be a 10% safety margin between the natural frequencies of the whole system and the excitation frequencies of the rotating turbine blades. The value of natural frequencies of the tower system should be away from the blade passing frequency and the blade rotor frequency [41].

3.3.2. Maximum Top Displacement

To avoid excessive vibration and displacement, the maximum deflection at the top of the PCSH tower is restricted [10]:

$$\frac{W_{max}}{W_{al}} < 1 \tag{31}$$

$$\frac{\theta_{max}}{\theta_{al}} < 1 \tag{32}$$

where W_{max} is the maximum top deflection, W_{al} is the allowable deflection, θ_{max} is the maximum rotation angle of the top section, and θ_{al} is the allowable rotation angle of the top section. According to GB50135-2006 [39], $W_{al} = H/100$ and $\theta_{al} = 5°$, where H is the height of the wind turbine tower.

4. PUPSO Approach with the Objective Function of LCOE

4.1. *Updated Partial Swarm Optimization (UPSO) Approach*

With the development of intelligent optimization algorithms, solving engineering computing problems by simulating biological behavior is becoming increasingly popular in a series of practical applications [42]. In this paper, the geometry optimization problem of PCSH wind turbine towers can be expressed as the following equations:

$$\begin{cases} Z_{target} = \min f(\boldsymbol{x}) = \min f((x_1, x_2, \cdots, x_n)^{\mathrm{T}}) \\ c(\boldsymbol{x}) = [h_1(\boldsymbol{x}), h_2(\boldsymbol{x}), \cdots, h_n(\boldsymbol{x})]^{\mathrm{T}} \leq 0 \end{cases} \tag{33}$$

where x is an N-dimensional vector to represent the particle, Z_{target} is the optimal target, $f(x)$ is a function to calculate the total cost of the tower and has been described above, and $c(x)$ is the vector of constraints functions. Both $f(x)$ and $c(x)$ are nonlinear functions.

The particles are operated by the following equations:

$$\begin{cases} v_q^{k+1} = w \times v_q^k + c_1 \times \xi \times \left(x_{q(best)} - x_q^k\right) \\ \quad + c_2 \times \xi \times \left(x_{g(best)} - x_q^k\right) \\ x_q^{k+1} = x_q^k + v_q^{k+1} \end{cases} \tag{34}$$

in which v_q^k and x_q^k are the speed and position, respectively, of the qth particle in the kth loop, w is the inertia weight, c_1 and c_2 are the learning factors of the algorithm, $x_{q(pbest)}$ is the position of the optimal point of the qth particle in the cycles from 1st to kth, $x_{g(best)}$ is the position of the optimal point of all particles in the periods from 1st to kth, and ξ is an uniformly distributed random number within (0, 1).

A penalty term in the fitness valuation process is added to coordinate the movement of particles within the feasible region and ensure that the wind turbine tower design fulfills the design constraints.

Because the basic PSO algorithm usually encounters premature convergence issues, it is first updated in this paper. The updated PSO (UPSO) is carried out as follows.

1. Weight function's learning factor

The algorithm with a weight function's learning factor [43] is adopted in this paper to speed up the computation. Unlike the traditional PSO algorithm, the learning factor and inertia weight can be calculated as:

$$\begin{cases} w = w_{min} + (w_{max} - w_{min}) exp(-20(m/M)^6) \\ c_1 = 0.5w^2 + w + 0.5 \\ c_2 = 2.5 - c_1 \end{cases} \tag{35}$$

where m is the number of iterations, M is the maximum number of iterations, w is the inertia weight, w_{min} is the minimum inertia weight, w_{max} is the maximum inertia weight, and c_1 and c_2 are the learning factors.

2. Random perturbation

To avoid premature convergence, a random operator is introduced to the optimization process in this paper [23]. The fitness variance of particles is defined as the following equation:

$$\sigma^2 = \sum_{i=1}^{n} \left(\frac{f_q - f_{avg}}{f_N}\right)^2 \tag{36}$$

where f_q is the fitness of the qth particle; f_{avg} is the average value of fitness of particles; and f_N is the normalized scaling factor, which can be calculated by the following equation:

$$f_N = \begin{cases} max(|f_q - f_{avg}|), max(|f_q - f_{avg}|) > 1 \\ 1, others \end{cases} \tag{37}$$

The mutation probability p_m can be calculated by the following equation:

$$p_m = \begin{cases} \zeta, \sigma^2 < \sigma_d^2 \\ 0, others \end{cases} \tag{38}$$

where ζ takes the values within [0.1, 0.3] and σ_d^2 is set to be 0.15.

For the purpose of mutating the operator $x_{g(best)}$ in the kth loop, random perturbation is adopted according to the following equation:

$$x_{gm(best)} = x_{g(best)} \times (1 + 0.5\eta) \tag{39}$$

where η follows a Gaussian distribution, $x_{gm(best)}$ is the position of the optimal point of all particles in the periods from 1st to kth after mutation, and $x_{g(best)}$ is the position of the optimal point of all particles in the periods from 1st to kth.

4.2. Objective Function

With the start of bidding in wind power markets, LCOE, as the world's most commonly used index to evaluate the cost of electricity, has been favored by participants involved in wind power projects. Bruck et al. suggests that LCOE can be used as a basis for setting appropriate power purchasing agreement terms [44]. Based on the LCOE method, Myhr et al. studied the influence of deployable operating depth and other factors on offshore wind power platforms [45]. Khojasteh et al. proposed and optimized a distributed generation by adding a shroud to the wind turbine and assessed it by LCOE [46]. In this paper, the LCOE is chosen as the objective function and can be calculated by Equation (40):

$$LCOE = \frac{\sum_{t=1}^{n} \frac{I_t + M_t + F_t}{(1+r)^t}}{\sum_{t=1}^{n} \frac{E_t}{(1+r)^t}} \tag{40}$$

where I_t are the investment expenditures in year t (including financing); M_t is the operations and maintenance expenditures in year t; F_t is the fuel expenditures in year t; E_t is the electricity generation in year t; r is the discount rate; and n is the life of the system. There are four types of wind energy resource areas in China and the LCOE of each area needs to be calculated separately.

Table 2 shows the costs and fees of the wind farm, which are estimated based on official files and engineering experiences [16]. Management expenses and measure expenses are 33.3% and 15.3% of the total labor and machinery costs, respectively. However, due to the difficulties of considering extra costs during the project, such as wind curtailment, transportation, and road construction, these factors are not considered in the optimization.

The geometry of a PCSH wind turbine tower, including the heights of the PC and steel segments, is optimized in this paper. Figure 3 shows the cost evaluation flow chart for the PCSH tower. As illustrated in Figure 3, the function of LCOE can be determined accordingly when the design is completed.

Table 2. The comprehensive cost of LCOE.

Title	Item	Unit Price
Direct cost	Concrete	600 yuan/m^3
	Reinforcement	5500 yuan/ton
	Prestressing steel strand	14,390 yuan/ton
	Sheeting	50 yuan/m^2
	Timber support	15 yuan/m^2
	Metallic pipe	679 yuan/100 m
	Flange	40,000 yuan/pcs
	Q345	1000 yuan/ton
Labor cost and mechanical cost	Reinforcement	1500 yuan/ton
	Prestressing steel strand	1000 yuan/ton
	Sheeting	300 yuan/100 m^2
	Timber support	15 yuan/m^2
	concrete	60 yuan/m^3

Table 2. *Cont.*

Title	Item	Unit Price
Project condition	Installed capacity	50 MW
	Equipment fee	5200 yuan/kW
	Other cost	1200 yuan/kW
	Annual cost during operation	80 yuan/kW (Year 1–5) 120 yuan/kW (Year 6–20)
	construction period	1 a
	Loan-to-value ratio	80%
	Depreciation life	20 a
	Ratio of remaining value	5%
	Length of maturity	15 a
	Interest rate	4.9%

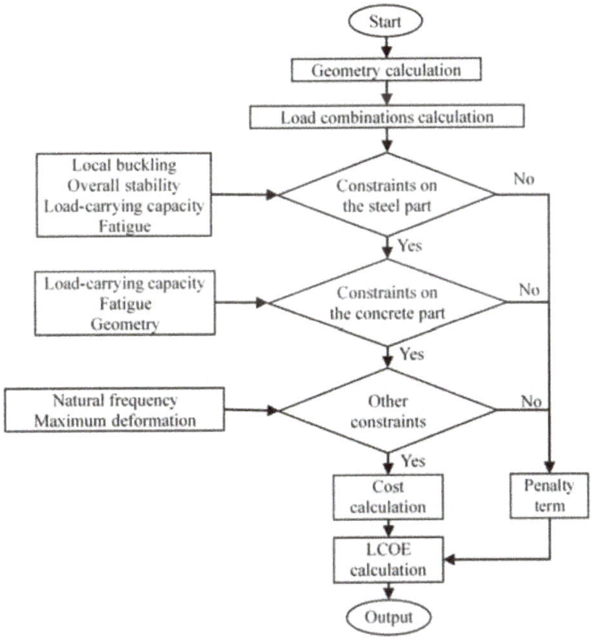

Figure 3. Flow chart of cost evaluation for the PCSH tower.

When the particle violates the constraint, measures need to be taken to orchestrate the motion of the particle. As shown in Figure 3, the penalty term is set to 0.5, which coordinates the movement of particles within the feasible region and cannot be treated as the ultimate goal of cost calculation. When a constraint violation occurs, the checking procedure is interrupted directly and the penalty term is employed to evaluate the fitness value for enhancing the computing efficiency of the optimal algorithm.

4.3. Optimization Variables

In this paper, the influence of geometric dimensions is considered to achieve a more economical design in the form of LCOE. The independent variables are shown in Figure 4 and their ranges are listed in Table 3. The ranges of the variables are set according to engineering and design experience. The thickness of each steel segment is assumed to be constant along the height direction. The range of the thickness of the steel section can be narrowed down with increasing design experience and determined in the cost estimate function to reduce the optimization variables and to accelerate the optimization computation. Based on the assumption that the length of each steel section is basically the same, the number of flanges is determined by the length and stress condition of the steel tower. For the PC segments, the thickness and diameter of the bottom cross section should not be smaller than the upper cross section of the segment.

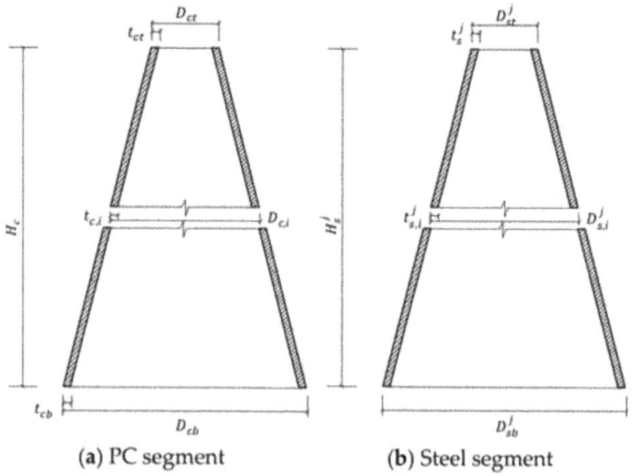

Figure 4. Design parameters of the tower.

Table 3. Variables and their ranges.

Variable	Range
Length of the j^{st} steel section H_s^j (mm)	500–70,000
Thickness of the j^{st} steel section t_s^j (mm)	10–25
Outer diameter of the top end of the j^{st} steel section D_{st}^j	2686 (j = 1) D_{sb}^{j-1} ($j > 1$)
Outer diameter of the bottom end of the j^{st} steel section D_{sb}^j	$> D_{st}^j$
Steel segments	1–3
Length of the concrete section H_c	7500–77,000
Thickness of the top end of the concrete part t_{ct} (mm)	180–500
Thickness of the bottom end of the concrete part t_{cb} (mm)	t_{ct}-500
Outer diameter of the top end of the concrete part D_{ct} (mm)	-
Outer diameter of the bottom end of the concrete part D_{cb} (mm)	$> D_{ct}$
Area of prestressed reinforcement (mm^2)	31,150–62,300

Apart from the aforementioned variables, there are some known dependent variables. The total length of the tower is 77.5 m. Therefore, the length of the steel segment determines the length of the concrete segments. The length of each steel section is the longitudinal dimension of the whole steel segment divided by the number of steel segments. The diameter of the steel tubular segment at the top of the tower is determined by the design of the nacelle and hub. Hence, their values are constant during the optimization. To simplify the problem and construction process, it is assumed that the generatrix of every section of the tower is a straight-line segment rather than a curved segment to easily determine the dimension of the tower at any height.

4.4. Flow Chart of PUPSO Algorithm

The above mentioned UPSO algorithm is carried out in a sequential form and the optimization process is usually time-consuming when the speed and position of a large number of particles are updated. In this study, further efforts are made to improve the computational efficiency by proposing a parallel UPSO (PUPSO) approach, where the computing body is divided into several concurrent tasks on the basis of different particles when evaluating the objective function. The flow chart of the PUPSO algorithm is shown in Figure 5.

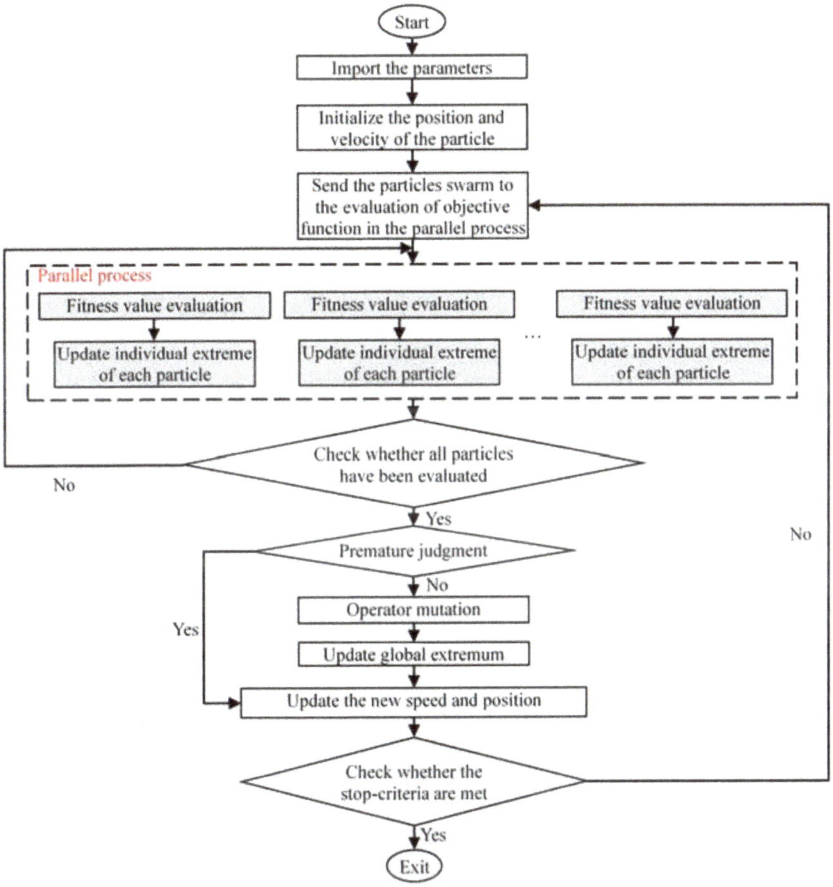

Figure 5. Flow chart of the proposed PUPSO.

The PUPSO algorithm starts by reading the initial conditions of the PCSH tower model, including the number of particles, the optimization parameters, and the termination condition. The termination condition for this implementation is the maximum number of iterations. Then, the particle swarm is generated randomly in the range mentioned in Section 4.3 and sent into the fitness value function. Based on the fitness value function, the fitness value of every particle can be determined and fed back to the PUPSO algorithm. Based on the returned values, the parameters are modified as described in Section 4.1. Then, new particles are generated according to the modified parameters and sent to the next loop. The optimal solution is then obtained after a number of cycles.

5. Optimization for PCSH Wind Turbine Tower

5.1. Design Parameters

It is assumed that the wind farm is built in a mountainous area, and site information and the parameters of the wind turbine studied in this paper are listed in Table 4 [25,47]. In the PCSH wind turbine tower, the upper steel part of the tower is made of Q345 steel and the lower PC part is made of C50 concrete. The height of the wind turbine tower is 77.5 m. The material properties are determined by GB 50010-2010.

Table 4. Parameters of the wind turbine.

Wind Turbine Parameters	Value
Generator model	XE93-2000
Rated power	2 MW
Rotor diameter	93.4 m
Nacelle and hub weight	80 t
Distance from gravitational center of the nacelle and hub to the center of tower	3000 mm
Weight of blades	48.5 t
Distance from gravitational center of the blades to the center of tower	4864 mm
IEC wind zone	IECIIIA
Annual average wind speed	7.5 m/s
Cut-in wind speed	3 m/s
Nominal wind speed	11 m/s
Cut-out wind speed	25 m/s
Extreme wind speed	52.5 m/s
Rotational speed	23 rpm
Maximum turbulence intensity	0.18

The parameters used in the PUPSO approach are listed in Table 5.

Table 5. Parameters of the PUPSO approach.

Parameter	Value
w_{max}	0.9
w_{min}	0.4
M	50
N	30
Penalty term	0.5
ζ	0.3

5.2. Optimization Results for the PCSH Wind Turbine Tower

5.2.1. LCOE Optimization

The relationship between the LCOE under the category IV wind energy resource area and the number of iterations is illustrated in Figure 6. According to Figure 6, by the use of the proposed PUPSO optimization approach, the LCOE of the PCSH tower defined above decreases clearly with the iteration of the approach and the minimization of the objective function is realized when the number of the iteration reaches 31. The LCOE decreases sharply in the first iterations because the algorithm in this paper strengthens the searching space diffusion and heightens the weight of particle optimization in the early stage and the weight of global optimum in the later iterations. The LCOE also drops fast in the early stage of the PSO optimization. However, the optimal result of PSO is inferior to that of the PUPSO algorithm proposed in this paper after the process is iterated four times.

The LCOE of four types of wind energy resource areas are presented in Table 6. Compared with the LCOE of the original wind turbine tower, the LCOE of the optimized wind turbine tower reduced by about 4% due to the reduction in construction costs. Theoretically, if the LCOE is higher than the electricity price, the project is not economically feasible. Therefore, the optimized PCSH wind turbine tower can increase profits and make it economically possible to build wind farms in areas with lower electricity prices.

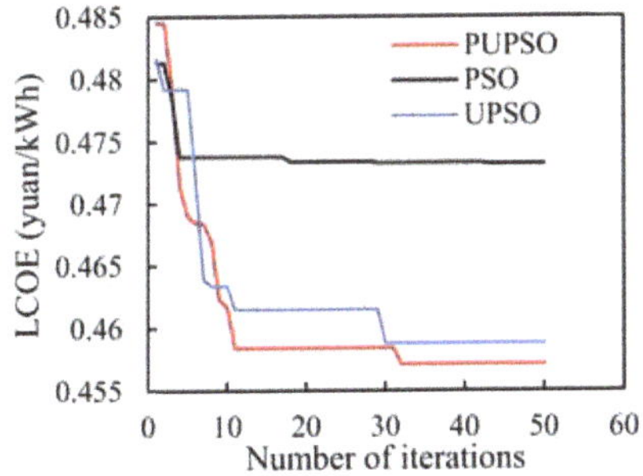

Figure 6. Optimization of costs of PCSH wind tower as a function of iteration number.

Table 6. LCOE measurement.

Category	Equivalent Available Duration (h)	Electricity Price in 2019 (Yuan/kWh)	LCOE for the Benchmark PCSH Tower (Yuan/kWh)	LCOE for the Optimized PCSH Tower (Yuan/kWh)
I	2850	0.34	0.3613	0.3474
II	2600	0.39	0.3874	0.3722
III	2500	0.43	0.3993	0.3835
IV	2000	0.52	0.4769	0.4571

The optimization rates, that is, the ratio of the difference of variables before and after optimization to the value before optimization, using the PUPSO are shown in Figure 7. It can be seen that the variables illustrated in the figure are less than zero, which means the variables are smaller than they were before the optimization. As the number of iterations increases, the variable tends to decrease.

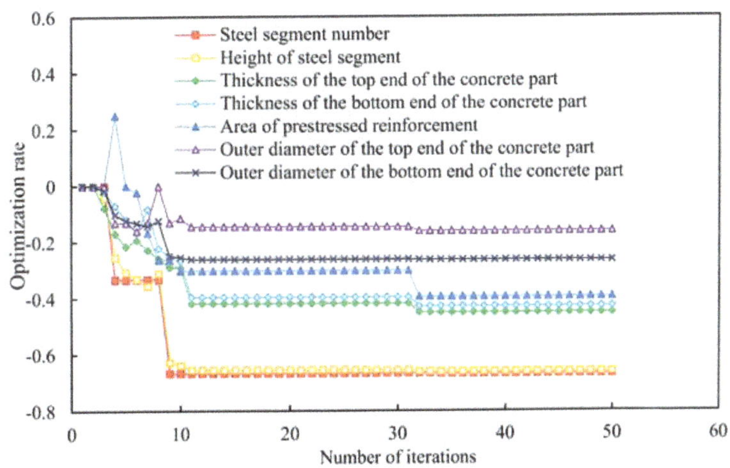

Figure 7. The optimization ratio of variables.

The comparison of corresponding dimensions for the PCSH wind turbine tower is listed in Table 7. Compared with the original design, the height of the upper steel segment is greatly reduced to 22 m and approximately 30% of the total height of the PCSH tower while the steel segment number is 1. The thickness of the tower, including the steel segment and concrete segment, is also reduced, which decreases the material consumption of the PCSH tower.

Table 7. PCSH tower dimension before and after optimization.

Tower	Variable	Before Optimization	After Optimization
Steel tube segment	Segment	3	1
	t_s^1 (mm)	14	10
	D_{st}^1 (mm)	2686	2686
	D_{sb}^1 (mm)	3485	3296
	H_s^1 (mm)	21,500	22,000
	t_s^2 (mm)	18	-
	D_{st}^2 (mm)	3485	-
	D_{sb}^2 (mm)	4046	-
	H_s^2 (mm)	20,000	-
	t_s^3 (mm)	20	-
	D_{st}^3 (mm)	4046	-
	D_{sb}^3 (mm)	4400	-
	H_s^3 (mm)	20,000	-
PC segment	t_{ct} (mm)	500	270
	t_{cb} (mm)	500	285
	H_c (mm)	16,000	55,500
	D_{ct} (mm)	4878	3549
	D_{cb} (mm)	6900	5800
	Prestressed duct number	36	36
	Prestressed reinforcement	$8\Phi^S1 \times 7$ (d = 15.2 mm)	$7\Phi^S1 \times 7$ (d = 12.7 mm)
	Prestressed reinforcement area (mm^2)	40,320	24,872

5.2.2. Utilization Ratio Comparison

In order to evaluate the utilization of both concrete and steel material of the optimized PCSH wind turbine tower and the effectiveness of the approach, the material utilization ratio as the ratio of the actual to maximum allowable performance values is determined. Figure 8 shows the maximum constraint activity of all cases for the optimized design. According to Figure 8, it can be seen that all utilization rates are less than one but greater than zero and local buckling for the steel section and fatigue damage for the concrete section are prominent. Therefore, no constraint was violated and the safety of the structure is ensured. The utilization rate at the upper part of the steel segment is less than that at the lower part of the steel segment due to the fact that D_{st}^1 is not optimized and is determined by the wind turbine and that the thickness of the steel section is constant along the height direction. The maximum utilization rate of the concrete segment is close to one along with the height, which means that the optimal result is close to the global optimal solution.

The maximum utilization ratio of the PCSH tower before and after optimization are listed in Table 8. According to Table 8, the utilization ratio of steel and concrete has been enhanced significantly. That both maximum utilization ratios of prestressing bars are close to one means the prestressing bars are fully used. The change of prestressed reinforcement is mainly due to the change of structural internal forces caused by the change of structural dimensions. The maximum utilization ratio for the load-carrying capacity of the PC segment at windward side is below zero, which means that the windward side of the tower is compressed rather than tensioned under the impact of prestress. The maximum utilization rate of the optimized PCSH wind turbine tower is close to one, which illustrated the effectiveness of the PUPSO algorithm.

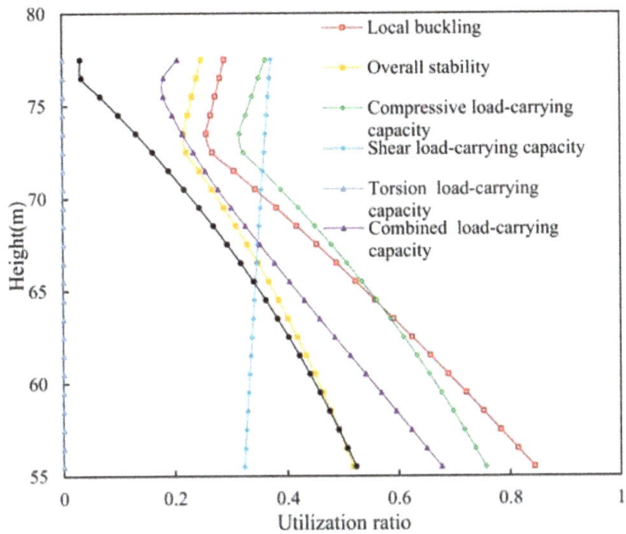

(a) Steel section

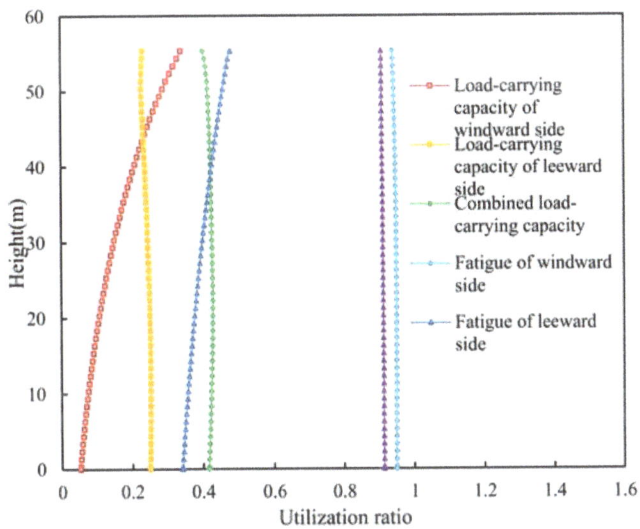

(b) Concrete section

Figure 8. The utilization ratio for constraints along the tower.

5.2.3. Fundamental Natural Frequency Comparison

The rotating speed of the rotor in the rated power is 23 rpm. Therefore, the corresponding rotational frequency is 0.38 Hz and the blade passing frequency is 1.15 Hz. The natural frequency of the different tower is listed in Table 9. The fundamental natural frequency of the original wind turbine tower is 0.45 Hz and the natural frequency of the optimized PCSH wind turbine tower is 0.56 Hz. The natural frequency of the PCSH tower is higher than that of the original tower. Moreover, the natural frequency of the proposed PCSH wind turbine tower has a safety margin of 0.18 Hz away from the rotational frequency and

0.59 Hz away from the blade passing frequency, which means a better dynamic behavior compared with the original design.

Table 8. The maximum utilization ratio of PCSH tower before and after optimization.

Tower	Maximum Utilization Ratio	Before Optimization	After Optimization
Steel segment	Local buckling	0.45	0.84
	Overall stability	0.40	0.52
	Compressive load-carrying capacity	0.53	0.76
	Shear load-carrying capacity	0.15	0.37
	Torsion load-carrying capacity	0.00075	0.0013
	Combined load-carrying capacity	0.29	0.67
	Fatigue	0.41	0.52
PC segment	Load-carrying capacity of windward side	0.064	0.34
	Load-carrying capacity of leeward side	0.48	0.25
	Combined load-carrying capacity	0.091	0.45
	Fatigue of windward side	0.56	0.94
	Fatigue of leeward side	0.21	0.48
	Fatigue of prestressing bar	0.93	0.91

Table 9. Natural frequency comparison of different tower.

Tower	Frequency (Hz)
Before optimization	0.45
After optimization	0.56

5.2.4. Weight Comparison

The weight of the wind turbine tower before and after optimization are listed in Table 10. Due to the reduction in the proportion of steel sections, the consumption of steel is reduced by about 82% and the weight of PC segment is increased by about 56%. The optimized design greatly reduces the steel consumption of the tower. The weight of the structure is increased by about 27%, which strengthens the anti-overturning capacity of the structure.

Table 10. Weight comparison of different tower.

Weight	Before Optimization	After Optimization
Steel segment (t)	90	16
PC segment (t)	338	528
Total (t)	428	544

5.2.5. Computation Efficiency Comparison

To verify the effectiveness of the proposed PUPSO algorithm for the optimization of the PCSH wind turbine tower, the comparison of computational times at different number of cycles between the PUPSO approach and the UPSO computation is shown in Table 11. When the cycle was 5, the computation time was saved by 49% and 38%, respectively. When the cycle was 10, the computation time was saved by 47% and 51%, respectively. It can be seen that by the use of the proposed PUPSO approach, the optimization computation efficiency was clearly enhanced.

Table 11. Comparison of computational time with two computing method.

Cycle Number	PSO Computation (s)	UPSO Computation (s)	PUPSO Computation (s)
5	55,074	44,912	27,845
10	99,352	107,553	52,427

The comparison of computational time for PSO, UPSO, and PUPSO is shown in Table 12. The proposed PUPSO algorithm has better optimization abilities compared with the PSO and UPSO. As shown in Table 12, the computation time of the proposed PUPSO can be saved by 51% and 53% when compared with the PSO and UPSO, respectively. In conclusion, compared with the PSO and UPSO algorithm, the PUPSO algorithm can improve the optimization efficiency by 60–110%. The reason for the large difference in calculation efficiency is that computing time will be saved when a constraint violation occurs as the checking procedure is interrupted directly and employs the penalty term, as shown in Figure 3. Not only does the approach speed up the calculation efficiency but it also avoids premature convergence as much as possible.

Table 12. Comparison of computational time with three computing methods.

PSO Computation (s)	UPSO Computation (s)	PUPSO Computation (s)
432,759	451,480	212,801

6. Conclusions

Based on the PUPSO algorithm, a geometry optimization approach for PCSH wind turbine towers has been proposed in this paper. During the optimization procedure, several working conditions, including wind and earthquakes as well as combinations of these factors, are considered. The LCOE of the PCSH tower is treated as the objective function and the geometry variables for the optimization of the PCSH tower include the dimensions of the PC and steel segments of the PCSH wind turbine tower. Based on this analysis, a geometrically optimal result was obtained, and the following findings can be made:

1. The proposed PUPSO algorithm performs better when compared with the traditional PSO algorithm and the UPSO. The computation time is greatly reduced by using parallel algorithms. Fulfilling the design constraints of relevant specifications and industry standards, the PUPSO algorithm provides an optimal design for the PCSH wind turbine towers with considerably improved computational efficiency.
2. The levelized cost of energy (LCOE) of the PCSH wind turbine tower in a life cycle perspective is considered as the objective function as an alternative to the direct investment. The LCOE of the optimized PCSH wind turbine clearly decreases when compared with the benchmark tower and increases the material utilization rate of the tower. The optimized PCSH wind turbine tower can be an economic alternative for wind farms with lower LCOE requirements. The height of the steel segment of the optimized PUPSO tower is recommended to be 30% of the total height of the PCSH wind turbine tower.
3. The optimized tower can provide better dynamic behavior to avoid the resonance caused by wind turbine excitation.
4. The optimization results for PCSH wind turbine towers provide valuable references in practice for PCSH wind turbine tower design in mountainous areas. This paper, based on a linear hypothesis and limited deformation, has been conducted as the preliminary optimization. Because of the nonlinearity present in prestressed concrete towers, nonlinear calculations should be investigated in the future.

Author Contributions: Conceptualization, B.X. and H.G.; methodology, B.X.; software, H.C.; validation, Z.L.; investigation, Z.L.; data curation, H.C.; writing—original draft preparation, Z.L.; writing—review and editing, B.X. and H.G.; visualization, Z.L.; supervision, H.G.; funding acquisition, B.X. All authors have read and agreed to the published version of the manuscript.

Funding: This research was funded by National Natural Science Foundation of China (NSFC) under Grants No. 51478175, the Scientific Research Funds of Huaqiao University (Grant number 605-50Y18016), and the International Science and Technology Cooperation and Exchange Fund Projects of Ministry of Science and Technology of China (2014DFE70230).

Institutional Review Board Statement: Not applicable.

Informed Consent Statement: Not applicable.

Data Availability Statement: Data are contained within in the article.

Acknowledgments: The authors gratefully acknowledge the support provided to the corresponding author through the National Natural Science Foundation of China (NSFC) under Grants No. 51478175, the Scientific Research Funds of Huaqiao University (Grant number 605-50Y18016), and the International Science and Technology Cooperation and Exchange Fund Projects of Ministry of Science and Technology of China (2014DFE70230).

Conflicts of Interest: The authors declare no conflict of interest.

References

1. Quilligan, A.; O'Connor, A.; Pakrashi, V. Fragility analysis of steel and concrete wind turbine towers. *Eng. Struct.* **2012**, *36*, 270–282. [CrossRef]
2. Lantz, E.; Roberts, O.; Nunemaker, J.; Edgar, D.; Dykes, K.; Scott, G. *Increasing Wind Turbine Tower Heights: Opportunities and Challenges*; National Renewable Energy Laboratory (NREL): Denver, CO, USA, 2019.
3. Yadav, K.K.; Gerasimidis, S. Instability of thin steel cylindrical shells under bending. *Thin Walled Struct.* **2019**, *137*, 151–166. [CrossRef]
4. Yadav, K.K.; Gerasimidis, S. Imperfection insensitive thin cylindrical shells for next generation wind turbine towers. *J. Constr. Steel Res.* **2020**, *172*, 106228. [CrossRef]
5. de Lana, J.A.; Júnior, P.A.A.M.; Magalhães, C.A.; Magalhães, A.L.M.A.; de Andrade Junior, A.C.; de Barros Ribeiro, M.S. Behavior study of prestressed concrete wind-turbine tower in circular cross-section. *Eng. Struct.* **2021**, *227*, 111403. [CrossRef]
6. Jay, A.; Myers, A.T.; Torabian, S.; Mahmoud, A.; Smith, E.; Agbayani, N.; Schafer, B.W. Spirally welded steel wind towers: Buckling experiments, analyses, and research needs. *J. Constr. Steel Res.* **2016**, *125*, 218–226. [CrossRef]
7. Singh, A. Concrete construction for wind energy towers. *Indian Concr. J.* **2007**, *81*, 43–49.
8. Seidel, M. Experiences with two of the world's largest wind turbine towers. In Proceedings of the 2003 EWEC European Wind Energy Conference and Exhibition, Madrid, Spain, 6–19 June 2003; pp. 16–19.
9. Kaveh, A.; Ghazaan, M.I. *Meta-Heuristic Algorithms for Optimal Design of Real-Size Structures*; Springer: Cham, Switzerland, 2018; pp. 5–6.
10. Negm, H.M.; Maalawi, K.Y. Structural design optimization of wind turbine towers. *Comput. Struct.* **2000**, *74*, 649–666. [CrossRef]
11. Uys, P.; Farkas, J.; Jarmai, K.; Van Tonder, F. Optimisation of a steel tower for a wind turbine structure. *Eng. Struct.* **2007**, *29*, 1337–1342. [CrossRef]
12. Nicholson, J.C. *Design of Wind Turbine Tower and Foundation Systems: Optimization Approach*; The University of Iowa: Iowa City, IA, USA, 2011.
13. Nicholson, J.C.; Arora, J.S.; Goyal, D.; Tinjum, J.M. Multi-Objective Structural Optimization of Wind Turbine Tower and Foundation Systems using Isight: A Process Automation and Design Exploration Software. In Proceedings of the 10th World Congress on Structural and Multidisciplinary Optimization, Orlando, FL, USA, 19–24 May 2013; pp. 1–10.
14. Ma, H.; Meng, R. Optimization design of prestressed concrete wind-turbine tower. *Sci. China Technol. Sci.* **2014**, *57*, 414–422. [CrossRef]
15. Oest, J.; Sandal, K.; Schafhirt, S.; Stieng, L.E.S.; Muskulus, M. On gradient-based optimization of jacket structures for offshore wind turbines. *Wind Energy* **2018**, *21*, 953–967. [CrossRef]
16. Chen, J.; Li, J.; He, X. Design optimization of steel–concrete hybrid wind turbine tower based on improved genetic algorithm. *Struct. Des. Tall Spec. Build.* **2020**, *29*, e1741. [CrossRef]
17. Sahib, N.M.; Hussein, A. Particle Swarm Optimization in Managing Construction Problems. *Procedia Comput. Sci.* **2019**, *154*, 260–266. [CrossRef]
18. Poitras, G.; Lefrançois, G.; Cormier, G. Optimization of steel floor systems using particle swarm optimization. *J. Constr. Steel Res.* **2011**, *67*, 1225–1231. [CrossRef]
19. Ye, J.; Hajirasouliha, I.; Becque, J.; Eslami, A. Optimum design of cold-formed steel beams using Particle Swarm Optimisation method. *J. Constr. Steel Res.* **2016**, *122*, 80–93. [CrossRef]
20. Luo, L.; He, W.; Zhang, X. PSO-Based Approach for Buckling Analysis of Shell Structures with Geometric Imperfections. *Math. Probl. Eng.* **2019**, *2019*, 1–8. [CrossRef]
21. Xu, J.; Zhang, T.-Y.; Huang, W.; Hu, M.-Y.; Qin, J.-W.; Zu, X.-C. Active control for machinery equipment induced structural vibration using $H\infty$ criterion and PSO technique. *Eng. Rev.* **2019**, *39*, 21–28. [CrossRef]
22. Tsiptsis, I.N.; Liimatainen, L.; Kotnik, T.; Niiranen, J. Structural optimization employing isogeometric tools in Particle Swarm Optimizer. *J. Build. Eng.* **2019**, *24*, 100761. [CrossRef]
23. Moradi, S.; Vosoughi, A.R.; Anjabin, N. Maximum buckling load of stiffened laminated composite panel by an improved hybrid PSO-GA optimization technique. *Thin-Walled Struct.* **2021**, *160*, 107382. [CrossRef]
24. Kaveh, A.; Eslamlou, A.D. *Metaheuristic Optimization Algorithms in Civil Engineering: New Applications*; Springer: Cham, Switzerland, 2020; pp. 24–34.

25. Niu, J. Research on the Structural Performance of Prestressed Concrete-Steel Hybrid Wind Turbine Tower. Master's Thesis, Hunan University, Changsha, China, 2014.
26. Chen, X. Static loads and dynamic loads of wind turbine tower. *Wind Power Gener.* **1989**, *3*, 44–49.
27. Zhao, W.; Cao, P.; Chen, J. The Research of Load Calculation Method and Loads Combination about Wind Turbine Tower. *Spec. Struct.* **2020**, *4*, 80–83.
28. SAC. *GB50009-2012: Load Code for the Design of Building Structures*; Standardization Administration of China: Beijing, China, 2012.
29. Dai, K.; Wang, Y.; Huang, Y.; Zhu, W.D.; Xu, Y.F. A Modified Stochastic Subspace Identification Method for Estimating Natural Frequencies. In Proceedings of the ASME 2016 International Mechanical Engineering Congress and Exposition, Phoenix, Arizona, USA, 11–17 November 2016.
30. Ma, H.; Zhang, D. Seismic Response of a Prestressed Concrete Wind Turbine Tower. *Int. J. Civ. Eng.* **2016**, *14*, 1–11. [CrossRef]
31. SAC. *GB 50011-2010: Code for Seismic Design of Buildings*; Standardization Administration of China: Beijing, China, 2010.
32. LaNier, M.W. *LWST Phase I Project Conceptual Design Study: Evaluation of Design and Construction Approaches for Economical Hybrid Steel/Concrete Wind Turbine Towers; 28 June 2002–31 July 2004*; National Renewable Energy Lab.: Golden, CO, USA, 2005. [CrossRef]
33. He, Y.-S.; Wang, J.; Zhao, H.-Y.; Li, Z.-L.; Xing, K.; Yao, S.; Zeng-Jun, H. Research of load calculation and load combination for wind turbine generator system of beach-shallow sea. *Build. Sci.* **2012**, 120–127.
34. SAC. *GB 50051-2013: Code for Design of Chimneys*; Standardization Administration of China: Beijing, China, 2013.
35. SAC. *GB 50017-2017: Code for Design of Steel Structure*; Standardization Administration of China: Beijing, China, 2017.
36. Agbayani, N.A. A Technical Overview of ASCE/AWEA RP2011: Recommended Practice for Compliance of Large Land-Based Wind Turbine Support Structures. In Proceedings of the Structures Congress 2014, Boston, MA, USA, 3–5 April 2014; pp. 1759–1770.
37. Justus, C.; Hargraves, W.; Mikhail, A.; Graber, D. Methods for estimating wind speed frequency distributions. *J. Appl. Meteorol.* **1978**, *17*, 350–353. [CrossRef]
38. Huo, T.; Tong, L. An approach to wind-induced fatigue analysis of wind turbine tubular towers. *J. Constr. Steel Res.* **2020**, *166*, 105917. [CrossRef]
39. SAC. *GB 50135-2006: Code for Design of High-Rising Structures*; Standardization Administration of China: Beijing, China, 2007.
40. SAC. *GB 50010-2010: Code for Design of Concrete Structures*; Standardization Administration of China: Beijing, China, 2010.
41. Grünberg, J.; Göhlmann, J. *Concrete Structures for Wind Turbines*; John Wiley & Sons: Berlin, Germany, 2013.
42. Liu, X. *Introduction to Artificial Intelligence: Approaches and Systems*; National Defense Industry Press: Beijing, China, 2008.
43. Zhao, Y.; Fang, Z. Particle swarm optimization algorithm with weight function's learning factor. *J. Comput. Appl.* **2013**, *33*, 2265–2268. [CrossRef]
44. Bruck, M.; Sandborn, P.; Goudarzi, N. A Levelized Cost of Energy (LCOE) Model for Wind Farms that Includes Power Purchase Agreement (PPA) Energy Delivery Limits. *Renew. Energy* **2016**, *122*, 131–139. [CrossRef]
45. Myhr, A.; Bjerkseter, C.; Ågotnes, A.; Nygaard, T.A. Levelised cost of energy for offshore floating wind turbines in a life cycle perspective. *Renew. Energy* **2014**, *66*, 714–728. [CrossRef]
46. Khojasteh, H.; Noorollahi, Y.; Tahani, M.; Masdari, M. Optimization of Power and Levelized Cost for Shrouded Small Wind Turbine. *Inventions* **2020**, *5*, 59. [CrossRef]
47. Product Introduction of XE93-2000. Available online: http://wp.china-nengyuan.com/member_product/21420.html (accessed on 20 April 2021).

applied sciences

MDPI

Article

Size and Shape Optimization of a Guyed Mast Structure under Wind, Ice and Seismic Loading

Raffaele Cucuzza [1], Marco Martino Rosso [1], Angelo Aloisio [2,*], Jonathan Melchiorre [1], Mario Lo Giudice [1] and Giuseppe Carlo Marano [1]

1 Department of Structural, Geotechnical and Building Engineering, Politecnico di Torino, Corso Duca Degli Abruzzi, 24, 10128 Turin, Italy; raffaele.cucuzza@polito.it (R.C.); marco.rosso@polito.it (M.M.R.); jonathan.melchiorre@polito.it (J.M.); logiudice.mario@polito.it (M.L.G.); giuseppe.marano@polito.it (G.C.M.)

2 Civil Environmental and Architectural Engineering Department, Università Degli Studi dell'Aquila, via Giovanni Gronchi n.18, 67100 L'Aquila, Italy

* Correspondence: angelo.aloisio1@univaq.it

Abstract: This paper discusses the size and shape optimization of a guyed radio mast for radiocommunications. The considered structure represents a widely industrial solution due to the recent spread of 5G and 6G mobile networks. The guyed radio mast was modeled with the finite element software SAP2000 and optimized through a genetic optimization algorithm (GA). The optimization exploits the open application programming interfaces (OAPI) SAP2000-Matlab. Static and dynamic analyses were carried out to provide realistic design scenarios of the mast structure. The authors considered the action of wind, ice, and seismic loads as variable loads. A parametric study on the most critical design variables includes several optimization scenarios to minimize the structure's total self-weight by varying the most relevant parameters selected by a preliminary sensitivity analysis. In conclusion, final design considerations are discussed by highlighting the best optimization scenario in terms of the objective function and the number of parameters involved in the analysis.

Keywords: guyed mast; structural optimization; genetic algorithm; structural design

Citation: Cucuzza, R.; Rosso, M.M.; Aloisio, A.; Melchiorre, J.; Lo Giudice, M.; Marano, G.C. Size and Shape Optimization of a Guyed Mast Structure under Wind, Ice and Seismic Loading. *Appl. Sci.* **2022**, *12*, 4875. https://doi.org/10.3390/app12104875

Academic Editors: Nikos D. Lagaros and Vagelis Plevris

Received: 17 April 2022
Accepted: 9 May 2022
Published: 11 May 2022

1. Introduction

Guyed masts are extensively used in the telecommunications industry, and the size, shape, and topology optimization can significantly benefit their transportation and installation. The main loads acting on guyed mast structures arise from wind [1,2], earthquakes [3–6], sudden rupture of guys [7], galloping of guys [8], and sudden ice shedding from ice-laden guy wires [9].

Their optimization must fulfil several requirements under ultimate and service limit states [10]. Specifically, service limit states are crucial for guyed mast structures due to high-amplitude oscillations caused by their high deformability. In some cases, these vibrations have led to a signal loss caused by excessive displacement and rotation of the antennas and, in other cases, have resulted in permanent deformation or failure. Therefore, size optimization of the guyed mast structure represents a challenging task since the increment of the performance ratio of the materials should be counterbalanced by an adequate lateral stiffness to reduce high-vibration drawbacks [11].

Saxena [12] reported several happenings where heavy icing combined with moderate wind resulted in severe misalignment of towers and complete failure. Novak et al. [13] showed that ice accumulation on some parts of the guy wires and moderate winds could lead to the guy galloping, resulting in unacceptable stress levels throughout the structure. The main topics investigated in the field of guyed structures can be summarized as follows:

- Structural design. Several researchers investigated the dynamic response of guyed mast structures through experimental tests and numerical modeling to derive design approaches and recommendations [14–16]. In particular, there are studies dealing with the dynamic identification and accurate estimate of the wind loads [17–21].

- Nonlinear dynamics. The proneness to global and local instabilities challenged several scholars to estimate and predict the nonlinear behaviour of guyed masts [22–26].
- Structural optimization. The need for guyed structures that are easy to install and transport challenged several scholars to optimize their shape in order to reduce the structural mass without reducing the lateral stiffness and prevent instability phenomena [27].
- Structural control. There are some attempts of control methods to reduce vibrations in mast-like structures [28–30]. Among others, Blachowski [31] proposed the use of a hydraulic actuator to control cable forces in guyed masts using Kalman filtering.

This paper tackles the size and shape optimization of guyed mast structures. A video of the considered structure is available in Supplementary Material. Since the first attempts by Bell and Brown [32], many engineers attempted to optimize guyed masts under wind loads using deterministic global optimization algorithms. However, as pointed out by [27], this approach leads to local optimum points, since each design variable was considered separately. Thornton et al. [33] and Uys et al. [34] proposed general procedures for optimizing steel towers under dynamic loads. To the author's knowledge, Venanzi and Materazzi [35] were the first to implement a multi-objective optimization method for guyed mast structures under wind loads using the stochastic simulated annealing algorithm for size optimization. The objective function implemented by [35] included the sum of the squares of the nodal displacements and the in-plan width of the structure. Zhang and Li [36] attempted to achieve both shape and size optimization in a two-step procedure using the ant colony algorithm (ACA). Cucuzza et al. [37] proposed an alternative approach in which the multi-objective optimization problem has been reduced to a single-objective problem through suitable parameters. Luh and Lin [38] were challenged in achieving the topology, size, and shape optimization of guyed masts using a modification of the binary particle swarm optimization (PSO) and the attractive and repulsive particle swarm optimization.

This paper discusses the size optimization of guyed masts using a genetic algorithm (GA) by considering different design scenarios (e.g., Cucuzza et al. [37] and Manuello et al. [39]). Kaveh and Talatahari [40] noticed that the particle swarm optimization (PSO) is more effective than ACA and the harmony search scheme for optimizing truss structures. However, Deng et al. [41] and Guo and Li [42] were successful in optimizing tapered masts and transmission towers using modifications of genetic algorithms (GA). Moreover, Belevivcius et al. [27] attempted the topology-sizing optimization problem of the guyed mast as a single-level single-objective global optimization problem using GAs.

Therefore, given the numerous successful solutions of guyed masts using GAs, the authors chose to investigate the size optimization of a guyed mast structure using GAs. Following [35], this paper focuses on the size optimization by considering eight possible design scenarios. The purpose of the present paper is two-fold. Firstly, this work aims at achieving a size optimization on a real application case adopting structural verification according to Eurocode 3. During the load evaluation phase, detailed analyses have been conducted, including wind, ice, and seismic actions and the verifications against instabilities. Secondly, the computational intelligence procedure adopted by the authors allowed the investigation of several scenarios simultaneously. As a result, the parameters that mainly affected the design process have been detected to provide preliminary indications to engineers in the practical design of similar structural typologies. Furthermore, the considered case study may represent a benchmark case for validating the reliability and accuracy of alternative numerical approaches. Therefore, the paper is organized as follows. After the case study description and the FE model, the authors introduce the first numerical results and the outcomes of the size and shape optimization.

2. Case Study

The considered structure is a guyed radio mast. It is a thin, slender, vertical structure sustained by tension cables fixed to the ground and typically arranged at 120° between each other.

The main body is a single central column made of tube profiles or truss systems when a high elevation must be reached, see Figure 1. More than one set of cables is placed at different elevations to prevent instability phenomena. Guyed towers are usually built for meteorological purposes or to support radio antennas, such as the one considered in this research. In particular, this structure can be used for a limited time during an event or maintenance of primary transmission towers. Therefore, it is also called a temporary base transceiver station (BTS), typically adopted to supply the immediate service. Sporting events, concerts, motor racing, military camps, and emergency events are typical examples of temporary BTS applications. The BTS is usually mounted on a moveable platform called the shelter.

The considered structure is located in Bassano Del Grappa, in the north of Italy, at a 129 m elevation from the sea level. The surrounding area is low-urbanized, with no relevant obstacles to the wind loads. The total height of the mast is 30.00 m. It is sustained by a central pole where 21 cables are fixed, see Figure 2. Other structural elements with rectangular cross-sections are used to create truss systems connecting cables and the central pole.

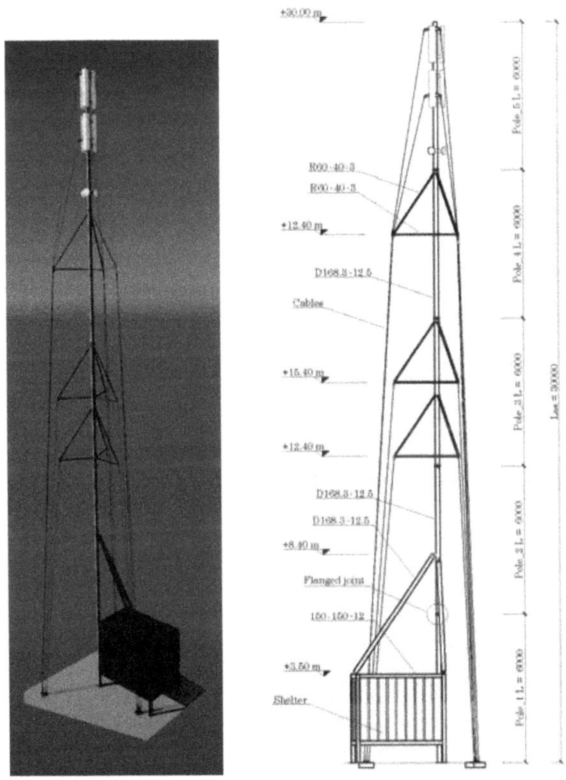

Figure 1. (**a**) Render model realized using Tekla Structures. (**b**) Technical drawing of the structure investigated with dimensions in mm.

The central pole consists of five circular hollow steel profiles with flanged joints and 6 m in length. All connections are bolted, as well as those connecting the cables to the pole. The shelter is a steel box devoted to partially sustaining the structure and hosting electronic equipment. It is usually mounted on a moveable platform.

Figure 2. Pictures and details of the considered structure.

3. Load Analysis

This section details the loads acting on the structures, from the dead to the variable loads. According to the Italian Standard Regulation NTC2018, the load combinations of the actions have been evaluated at the ultimate limit state (ULS) and, for seismic conditions, at the life safety (LS) limit state. In Appendix A, Table A4 illustrates the most critical combinations for both static and dynamic configurations. Partial safety factors γ and combination coefficients ψ were adopted in order to consider maximization (positive sign) or minimization (negative sign) of effects both for vertical and horizontal actions.

3.1. Dead Loads

The structure is made of steel S355, whose mechanical stress-strain behaviour is depicted in Figure 3, and the characteristics are listed here: f_{us} = 510 MPa, f_{ys} = 355 MPa, E_s = 210,000 MPa, which are the ultimate and yielding stresses and Young's modulus, respectively.

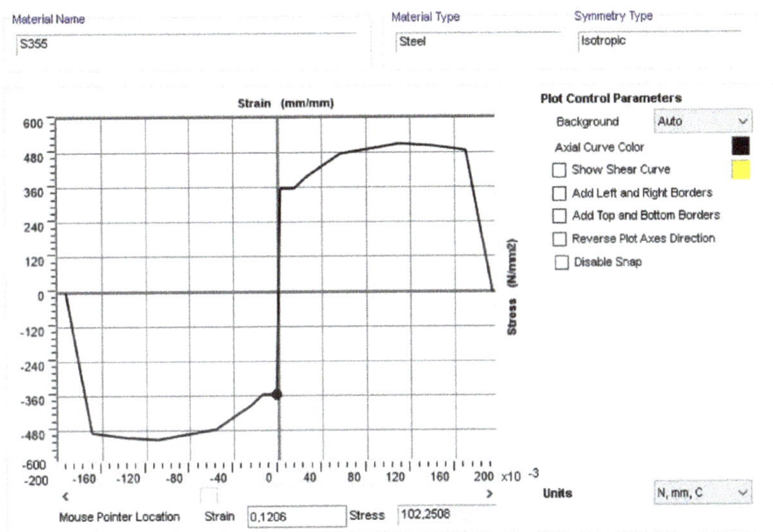

Figure 3. Mechanical stress–strain behaviour of steel S355 implemented in SAP2000.

The cables are made of galvanized steel consisting of 6 strands (216 wires) with an independent metal core (49 wires). The main characteristics are illustrated in Table 1.

Table 1. Technical specifications of the steel ropes.

Steel Ropes (Cables)				
Model			6 × 36WS + IWRC/265 wires	
Construction pattern			6 × (14 + (7 + 7) + 7 + 1) + (7 × 7)	
Winding direction Material			right cross galvanized steel	
Resistance			1170 N/mm^2–180 kg/mm^2	
Cable diameter	Weight	Area	Wire diameter	Load to failure
[mm]	[kg/m]	[mm^2]	[mm]	[kN]
16	1.36	173.25	0.91	161
18	1.67	212.74	1.03	204
20	2.02	257.32	1.14	252
22	2.41	307.01	1.26	305

The structure investigated consists of a few types of elements, as indicated in Table 2. Dead loads are calculated from the weight per unit length of each member.

Table 2. Computation of the dead loads.

Computation of Dead Loads					
Profile [mm]		w [kg/m]	Length [m]	n°	Wtot [kg]
Circular	D168.3 × 12.5	48	6	5	1440
	D168.3 × 12.5	48	5.65	2	543
Rectangular	60 × 40 × 3	4.35	3.16	9	124
	60 × 40 × 3	4.35	1.8	9	71
	100 × 40 × 3	6.13	0.45	6	17
Rope	D16	1.3667	12.45	3	51
	D16	1.3667	15.44	3	63
	D16	1.3667	24.43	9	300
	D16	1.3667	5.76	3	24
	D16	1.3667	8.46	3	35
				2651 Kg	

The non-structural dead loads originate from the wiring weight and the steel ladder for inspection and maintenance. This load results in 0.3 kN/m. Antennas and parabolas represent the weight of the equipment. Two groups of three antennas are located at 26.00 and 29.25 m in height, with a 120° in mutual spacing. The first one is the model AOC4518R7v06 produced by Huawei®. The second one is the model 6888670N manufactured by Amphenol®. Finally, there are three parabolas located at 23.15 m height, spaced 120° apart from each other, 30 cm in diameter. Tables 3 and 4 detail the weight of the equipment and the non-structural dead loads.

Table 3. Weight of equipment, H, W, and D stand for height, width, and depth.

Typology	Model	No	Elevation [m]	H×W×D [mm]	Self-Weight [kg]	Clamps [kg]	Total [kg]
Antenna	AOC4518R7v06	3	29.25	1509 × 469 × 206	39.3	2 × 5.8	153
Antenna	6888670N	3	26	1997 × 305 × 163	32	2 × 3.9	119
Parabola	n.d	3	23.15	Diameter = 300	15	2.2	51.6

Table 4. Non-structural dead loads.

Item	qk [kN/m]	Qk [kN]
Steel ladder, other	0.3	-
Antenna	-	1.53
Antenna	-	1.19
Parabolas	-	0.52

3.2. Variable Loads

In this section, the detailed load modeling phase, for each variable load considered, is described. With specific reference to the wind action evaluation, the drag and lift forces are calculated according to the CNR-DT 207 R1/2018 [43]. The relationship between inertia and viscous forces, i.e., how wind load impacts to the surface, is taken into account with the Reynold's number R_e with the following expression:

$$R_e(z) = \frac{l \cdot v_m(z)}{\nu} \quad (1)$$

where z is the elevation, l is the characteristic length, v_m is the averaged wind speed, while ν is the kinematic viscosity of air ($\nu = 15 \times 10^{-6}$ m^2/s).

3.2.1. Maintenance and Repairing Loads

Following the Italian national recommendations [44], it is supposed that a typical situation of inspection or maintenance is performed by an operator working on the steel ladder. A concentrated load of 120 kg is applied at the top of the tower. Despite that, it is reasonable to believe that the operator could work by using a basket elevator, without loading the structure.

3.2.2. Wind Loads

The wind action was evaluated according to the Italian recommendations in [43]. Firstly, the peak kinetic pressure (q_p) was evaluated as follows:

$$q_p = \frac{1}{2} \cdot \rho \cdot v_r^2 \cdot c_e(z) \tag{2}$$

where p is the kinetic pressure, while:

- ρ is the air density;
- v_r^2 is the reference wind velocity;
- c_e is the exposure coefficient, varying with the elevation z of the structure.

For this purpose, the equivalent longitudinal or drag forces, f_D, and transverse or lift force, f_L, are evaluated as follows:

$$f_{drag} = q_p(z) \cdot l \cdot c_{drag}; \qquad f_{lift} = q_p(z) \cdot b \cdot c_{lift} \tag{3}$$

where

- $q_p(z)$ is the *peak kinetic pressure* evaluated at height z;
- l is the characteristic element size;
- b is the reference transverse dimension of the section;
- c_{drag} and c_{lift} are the longitudinal and transverse dynamic coefficients.

Drag D and Lift L forces are reported in Tables A2 and A3.

3.2.3. Ice Load

Ice and snow attached to the structural surface can significantly increase the variable loads in flexible and light structures. In particular, the radio mast is very sensitive to changes in the wind-exposed surface. In addition, the ice covering can increase the volume and the surface of the structural elements more than twice due to the low thickness of the central pole. The recommendations in [43] provide several scenarios for ice coverings. In the absence of more detailed evaluations, it is customary to consider an ice sleeve formation that is 12.5 mm thick. After the estimate of the wind loads, the influence of the ice sleeve formation on the structure is considered by assuming an additional exposed surface equal to 15% of the original one.

3.2.4. Seismic Action

Seismic action is evaluated according to the Italian seismic hazard map [44]. A linear dynamic analysis with seismic elastic response spectrum corresponding to the service limit state was carried out. Specifically, seismic actions are considered as acting independently in the X and Y plane directions.

The elastic response spectrum considered in the analysis was calculated by considering the topographic category of the site and geometry of the building (Figure 4). The first 33 vibration modes of the structure are included in the analysis, to reach 85% of the total participating mass according to the national regulations in [44]. The mass participating ratios are listed in Appendix A.

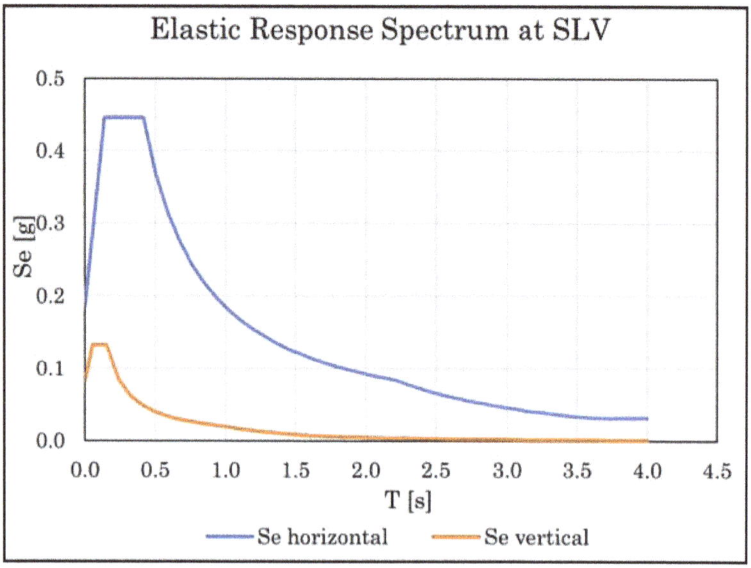

Figure 4. Elastic response spectrum corresponding to the service limit state (SLV), where *Sa* is the spectral acceleration.

4. Finite Element Modeling

The structural model was developed using two different element types: beams and cables. Beam elements model the main pole and all structural elements except for the cables. They possess the geometric and material properties of structural elements. The beam elements are used to model the main pole and secondary elements. Moreover, except for the main pole, rotation releases are applied at the ends in order to consider no flexural rigidity, as occurring for trussed structures.

Cable elements are used to simulate the steel ropes. Cable elements undergo large displacements that give rise to geometric nonlinearities. Therefore, the equilibrium of the cables is considered in the deformed configuration using SAP2000. As a result, the structural behaviour of guyed towers can be highly nonlinear, especially for low pre-tension cables, which are prone to large displacements. On the contrary, the nonlinear behaviour becomes less pronounced by increasing the pre-tension, resulting in high compression levels and minor instability effects. This paper considers the envelope of the maximum and minimum responses associated with each load condition.

Figure 5 plots the three modes with a higher mass participation ratio. These are the 10th, 11th, and 12th modes obtained from the dynamic analysis of the mast structure with the dead loads. On the contrary, the first modes arising from the dynamic analysis have lower mass participation factors and are characterized by local deformation of the structural elements. The 10th, 11th, and 12th modes are the first modes exhibiting the global deformation of the mast structure.

X and Y identify the in-plane orthogonal directions. The 10th and 11th modes have an approximate 26% mass participation factor in the Y and X directions, respectively. The natural period is very low and at approximately 0.4 s. The 13th mode is mainly torsional with nearly a 7 and 4% mass participation factor in the X and Y directions.

Figure 6 shows the positive (in dark green and purple) and negative (in red and light green) maximum and minimum envelopes of the axial, shear forces, and bending moments acting on the structural elements. Figure 7 plots the performance ratios of all structural elements except for the cables. The performance ratio is the ratio between the maximum stress in the structural element and the yielding stress. The performance ratios are defined by the colour map next to Figure 7. The plots highlight the presence of a structural element in the first half of the central pole with a high-performance ratio, depicted in yellow. The first section of the central pole has a low performance ratio, lower than 0.25. After the section with a performance ratio in the range 0.4–0.65, the following sections fall in the range 0.25–0.4 and are coloured in green. The top sections of the central pole are not significantly stressed, with a performance ratio of 0–0.25. The bracings have low stress, plotted in cyan, with performance ratios of 0–0.25.

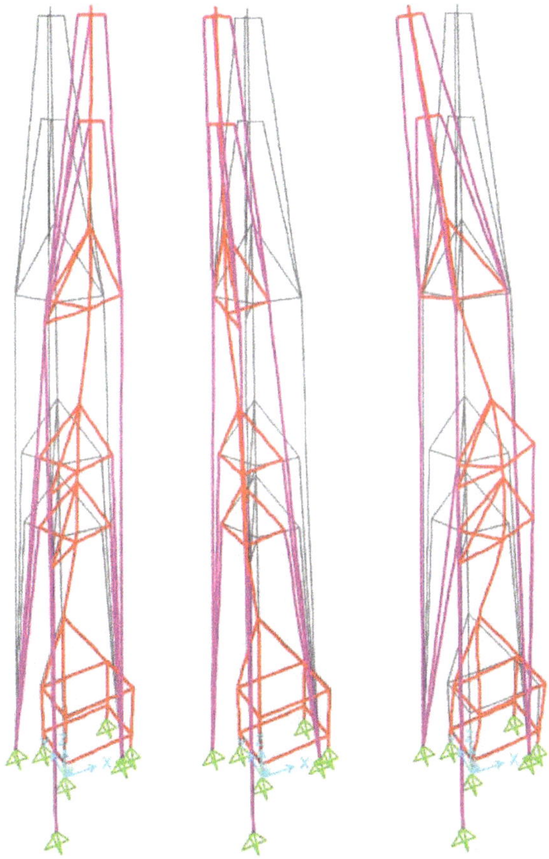

(a) (b) (c)

Figure 5. (**a**) Mode 10th—Ts = 0.437 s—mass participation ratio X = 9.6%, Y = 26.2%; (**b**) Mode 11th—Ts = 0.434 s—mass participation ratio X = 26.4% Y = 9.2%; (**c**) Mode 12—Ts = 0.206 s—mass participation ratio X = 7.2% Y = 4.4%.

Figure 8 shows the maximum displacements in the X (u_1), Y (u_2) directions and their combination (u_t) at the service limit state. The maximum displacement is located at the top of the tower, in particular at joint 6 (z = 30.00 m), with a maximum displacement equal to u_t = 18.7 mm.

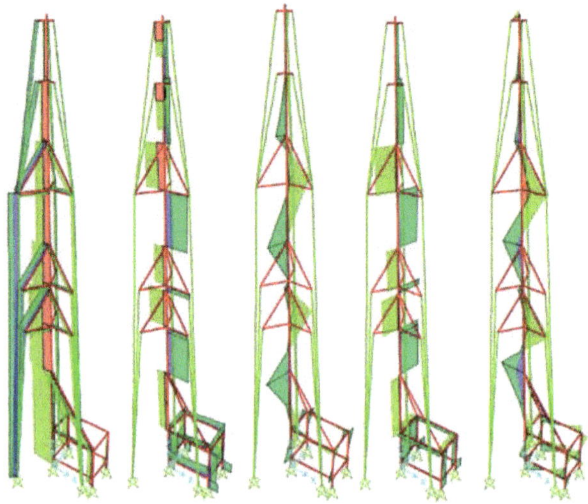

(a) (b) (c) (d) (e)

Figure 6. (**a**) Axial force, (**b**) shear force (V_2), (**c**) bending moment (M_2), (**d**) shear force (V_3), (**e**) bending moment (M_3).

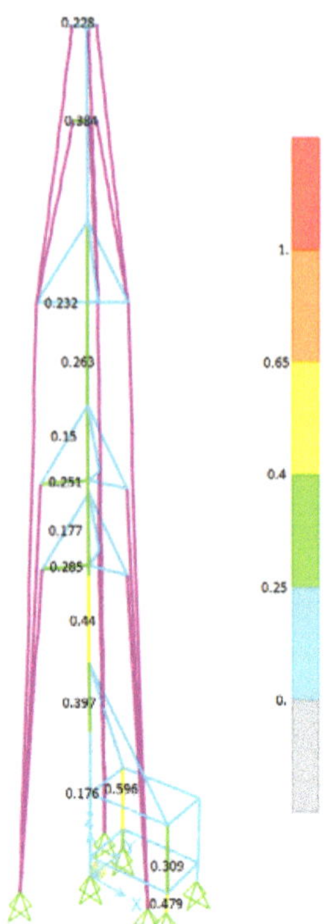

Figure 7. Performance ratios of the pole before optimization. Cables are depicted with magenta colour because their performance ratios are not included in the current representation.

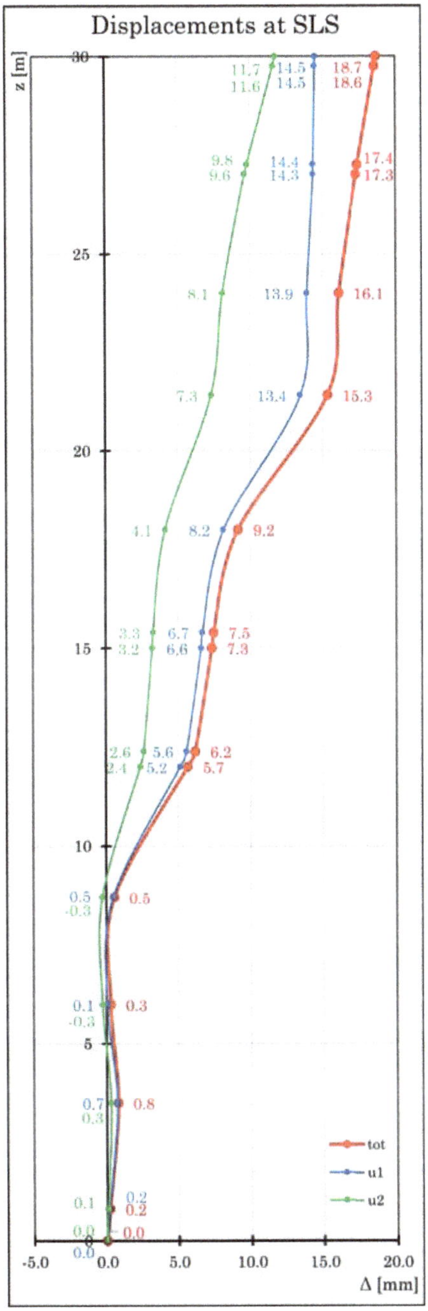

Figure 8. Displacements vs. elevation at the service limit state in two in-plane orthogonal directions (u_1, u_2) and their combination (tot).

5. Structural Optimization

In optimization problems, the main goal is to find the best conditions in terms of the optimal set of design parameters collected in the design vector x, which minimizes an objective function (OF) $f(x)$ [45–47]. These problems can be categorized into single-objective or multi-objective based on the number of OFs involved, and a further classification is based on the presence (or not) of constraints [48–50]. In the structural optimization field, it is common to deal with constrained optimization, whose general statement is [51]:

$$\begin{aligned} &\min_{x\in\Omega}\{f(x)\} \\ \text{s.t.}\quad & g_q(x)\leq 0 \quad \forall q=1,\ldots,n_q \\ & h_r(x)=0 \quad \forall r=1,\ldots,n_r \end{aligned} \tag{4}$$

where $x = \{x_1,\ldots,x_j,\ldots,x_n\}^T$ is the design vector to be optimized, whose terms are limited into a hyper-rectangular multidimensional box-type search space domain of interest denoted as Ω, given by the Cartesian product of the range of interest of each j-th of each design variable bounded in $[x_j^l, x_j^u]$, $\Omega = [x_1^l, x_1^u] \times \ldots \times [x_j^l, x_j^u] \times \ldots \times [x_n^l, x_n^u]$. The term $g_q(x)$ in (4) denotes inequality constraints whereas $h_r(x)$ are equality ones, which further reduce the feasible search space inside Ω. In structural optimization, it is typical to deal with inequality constraints, and a common goal is to minimize the global cost of the structure. Since this involves many terms, the main attempt is minimizing the self-weight of the structure, indirectly connected to material cost, i.e., material usage and natural resources consumption [51]. Several strategies have been developed over the years to handle constraints [52–54]. In the present work, the penalty function-based approach was implemented due to its simplicity, allowing converting the problem with OF $f(x)$ into a new unconstrained version $\phi(x)$:

$$\min_{x\in\Omega}\{\phi(x))\} = \min_{x\in\Omega}\{f(x)+H(x)\} \tag{5}$$

where $H(x)$ is the penalty function. Adopting a static-penalty strategy, $H(x)$, assume this form [55,56]

$$H_s(x) = w_1 H_{NVC}(x) + w_2 H_{SVC}(x) \tag{6}$$

where H_{NVC} is the number of violated constraints and H_{SVC} is the sum of all violations:

$$H_{SVC}(x) = \sum_{p=1}^{n_p} \max\{0, g_p(x)\} \tag{7}$$

w_1 and w_2 are the violation control parameters, whose numerical values are assumed equal to $w_1 = w_2 = 100$ following [55].

In the current study, the authors carried out a parametric study on the design variables of the guyed mast. This fact has led to eight different scenarios, summarized in Table 5. In addition, the starting initial values of the design parameter are listed in Table 6, while the general optimization workflow is illustrated in Figure 9. To compare the results, the focus is related only to the performance ratios PR of the central pole of the guyed radio mast, being the pole the most stressed element. It consists of five segments 6.00 m long with the same cross-section. Thus, starting from the ground level:

1. $Pole_1$ (0.00 to 6.00 m);
2. $Pole_2$ (6.00 to 12.00 m);
3. $Pole_3$ (12.00 to 18.00 m);
4. $Pole_4$ (18.00 to 24.00 m);
5. $Pole_5$ (24.00 to 30.00 m).

Starting with a constant diameter of the cross-section for the pole, at the end of the optimization, it is advisable to find a tapered solution following a linear relationship with the height, as represented in Figure 10f. Accordingly, it is possible to shape the pole cross-section with two design variables described by the bottom Φ_i and top Φ_f diameters. In the following, the different scenarios obtained from the parametric study based on the design variables involved in the optimization problem are described:

- Scenario A: this scenario involves the diameter Φ, as a sole variable, in the attempt to reduce the material consumption with a constant pole cross-section diameter with the height, as illustrated in Figure 10a.
- Scenario B: this scenario attempts to refine the previous case by adopting a tapered solution for the pole, by using the bottom Φ_i and the top Φ_f diameters, as represented in Figure 10b.
- Scenario C: further improvements are considered concerning scenario B by adding the cable pre-stressing force F as a variable of the optimization, as represented in Figure 10c.
- Scenario D: further improvements are considered to scenario B by using a unique value for the pole thickness t of the tapered elements of the pole, as represented in Figure 10c.
- Scenario E: further improvements are considered with respect to scenario B by optimizing both cable pre-stressing force F with a unique value of thickness t for the tapered elements of the pole, as represented in Figure 10e.
- Scenario F: from the structural analysis of scenario E, it is possible to point out how the linear law for the tapering forces to use a larger section where it is not necessary. Elements 2 and 3 are the most stressed ones. Therefore it is possible to further refine scenario E by considering a thickness value for the intermediate pole elements t_{inter} and a different thickness for the other extremal pole elements t_ends.
- Scenario G: in this scenario, the five different thickness values only have been governed for every pole element $\{t_1, t_2, t_3, t_4, t_5\}$ for a constant diameter solution with height, as depicted in Figure 10f.
- Scenario H: in this last scenario, a complete approach involves both the tapered solution by governing the initial bottom Φ_i and the final top Φ_f diameters, the five values of thickness for every pole element $\{t_1, t_2, t_3, t_4, t_5\}$, and even the cable pre-stressing force.

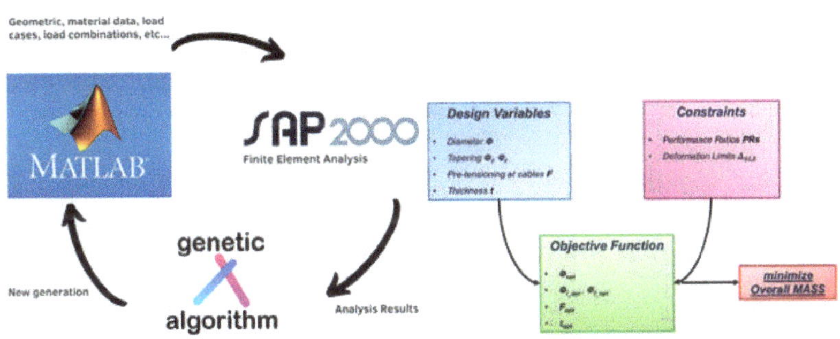

Figure 9. Workflow of the optimization problem.

Table 5. Parametric study on the design variables involved and summary of the different scenarios.

Scenario	No. Parameters
$A(\Phi)$	1
$B(\Phi_i, \Phi_f)$	2
$C(\Phi_i, \Phi_f, F)$	3
$D(\Phi_i, \Phi_f, t)$	3
$E(\Phi_i, \Phi_f, t, F)$	4
$F(\Phi_i, \Phi_f, t_{ends}, t_{inter}, F)$	5
$G(t_1, t_2, t_3, t_4, t_5)$	5
$H(\Phi_i, \Phi_f, t_1, t_2, t_3, t_4, t_5)$	8

Table 6. Total mass of the main pole.

Parameter	Measure	Value
Φ_0	[mm]	168.3
t_0	[mm]	12.5
L	[mm]	6000
Mass	[kg]	288
no elements	[-]	5
Total Mass	[kg]	1440

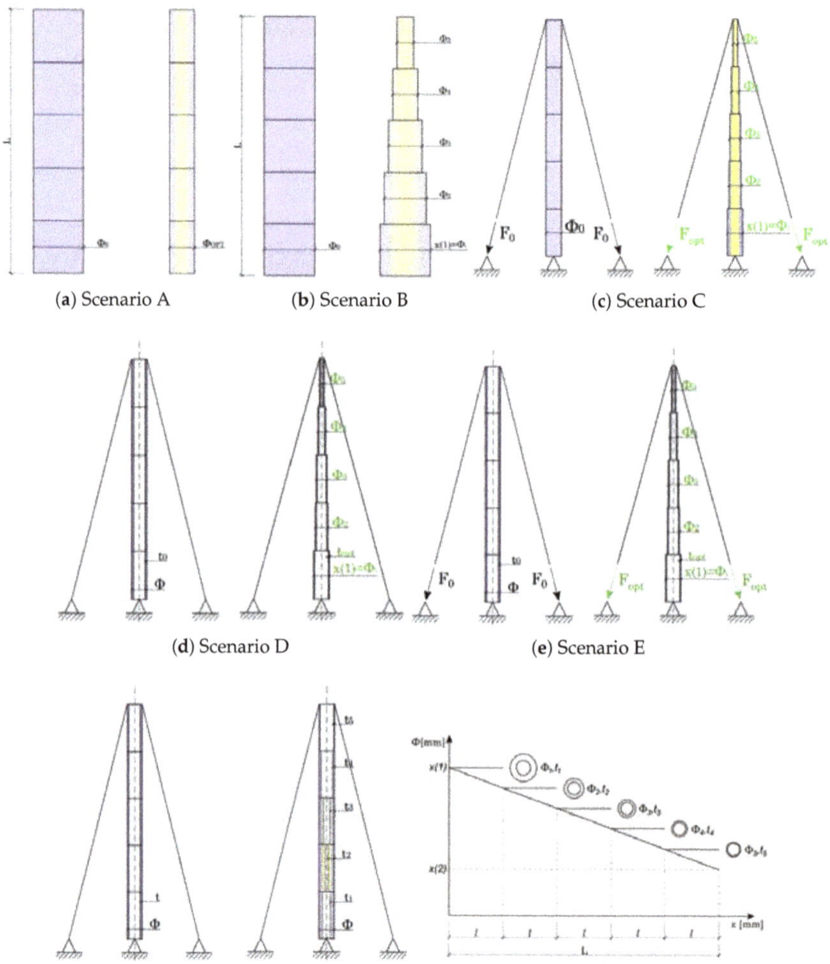

(**a**) Scenario A (**b**) Scenario B (**c**) Scenario C

(**d**) Scenario D (**e**) Scenario E

(**f**) Scenario F (**g**) Scenario G

Figure 10. Parametric study on the design variables involved and representation of the different scenarios described in Table 5.

Constraints Involved in the Structural Optimization Problem

The optimization problem statement is reported in (4) and the constraints were treated with the penalty-based approach illustrated in (5), by converting the constrained problem into an equivalent unconstrained one. The resolution of the optimization task considers the structural design assessment required by national and international codes to ensure the safety of constructions. In particular, the structural verifications derive from Eurocode 3

(EN 1993-1-1: 2005) and are referred to the ultimate limit state (ULS). The design verifications include tensile, compression, and buckling verification, and a combined assessment, such as the interaction capacity according to Annex B of the Eurocode 3:

$$\frac{D}{C} = \frac{N_{Ed}}{\frac{\chi_y A f_{yk}}{\gamma_{M1}}} + k_{yy} \frac{M_{y,Ed}}{\frac{\chi_{LT} W_{pl,y} f_{yk}}{\gamma_{M1}}} + k_{yz} \frac{M_{z,Ed}}{\frac{W_{pl,z} f_{yk}}{\gamma_{M1}}} \leq 1 \tag{8}$$

$$\frac{D}{C} = \frac{N_{Ed}}{\frac{\chi_z A f_{yk}}{\gamma_{M1}}} + k_{zy} \frac{M_{y,Ed}}{\frac{\chi_{LT} W_{pl,y} f_{yk}}{\gamma_{M1}}} + k_{zz} \frac{M_{z,Ed}}{\frac{W_{pl,z} f_{yk}}{\gamma_{M1}}} \leq 1 \tag{9}$$

where D stands for the demand and C stands for the capacity of the structure. Specifically, N_{Ed} is the acting axial force, whereas $M_{y,Ed}$ and $M_{z,Ed}$ represent the acting bending moments in the two principal directions of a planar local reference system centered on the cross section center of gravity. A is the cross section area of the pole, $W_{pl,y}$ and $W_{pl,z}$ are the plastic section modulus in the two principal directions, f_{yk} is the yielding strength of the steel, whereas γ_{M1} is the partial safety factor for instability conditions, equal to 1.05 from the Italian National Annex. χ_{LT} is the reduction factor for lateral–torsional buckling, whereas kyy, kyz, kzy, and kzz are interaction factors whose values are derived according to two alternative approaches based on Annex A (method 1)and Annex B (method 2). The global structural deformation referred to the service limit state (SLS) has also been considered by verifying the top displacement of the mast. Specific recommendations for guyed mast structures are missing in national and international codes. Therefore, the authors adopted the suggestions defined in the Italian Technical Code NTC2018 (D.M.17/01/2018) reported in Chapter 4.2.4.2.2 Table 4.2.XIII related to limitations of lateral displacements of steel multi-storey frame structures. These limitations express a threshold condition in terms of the total height of the structure H:

$$\delta_{\text{SLS,top}} \leq \delta_{\text{SLS,top,lim}} = \frac{H}{500} = \frac{30000\ \text{mm}}{500} = 60\ \text{mm} \tag{10}$$

Since this condition is specific for steel multi-storey frame structures, the authors will assume this value as a reasonable choice to ensure service life assessment and preservation of working conditions of the telecommunication guyed mast tower. In the next section, a discussion on the results is carried out.

6. Results and Discussion

The paper compares the outcomes of the size and shape optimization in eight different scenarios, distinguished by different design variables. Scenario A is associated with the worst improvement of the structural performance since a single diameter is used for the central pole. Additionally, industrial steel profiles do not cover all possible ranges of the diameter. Improvements in the structural performance and weight reduction are achieved in the following scenarios when the search space becomes larger by increasing the number of design variables.

Scenario B introduces the tapering of the central pole with a linear variation from the bottom to the top. In this case, the optimal solution is affected by intermediate sections, which are more stressed. Consequently, the end cross-sections are over-estimated. In response to that, Scenario F introduces the linear tapering of the tube thickness t_{ends},t_{inter} to enhance the performance of the optimal solution. Parallelly, in Scenario G, five different thicknesses are adopted (t_1, t_2, t_3, t_4, t_5), and the results are analogue to case F. Therefore, the thickness of the steel members is a suitable optimization parameter. At the same time, the diameter alone is not capable of returning attractive solutions because a linear interpolation trend is used. In addition, lower and upper limits were imposed for d and t. In particular, for this kind of structure, a minimum diameter $d_{min} \geq 100$ mm and a minimum thickness $t_{min} \geq 3$ mm was imposed.

The cross-section area depends on the square of the thickness. Therefore, small changes in t significantly affect the resulting area. Conversely, if the diameter is the sole search space, despite being tapered linearly with height, even significant modifications may not produce notable improvements. Still, the increment of design variables involved in the structural optimization typically increases the computational efforts. However, the scenario with the highest number of variables was characterized by an average time iteration close to 18s, using a computer with average performance. The computational effort cost of the optimization procedure strongly depends on the machine performance, no convergence issues occur. Table 7 lists the average values of performance ratio obtained from the eight optimization scenarios. All scenarios were collected in terms of number of parameters involved during the analysis. Table 7 proves that the increment in the number of design variables is associated with higher performance ratios. The target of the optimization achieves the best weight reduction, fully exploiting the structural material, without exceeding the ultimate and service limit states. Table 7 lists three sets of performance ratios: the initial one before optimization, the optimized, and the one obtained using commercial steel profiles, called the design performance ratio. The averaged performance ratio is equal to 28% before optimization. It significantly increases from scenario A, nearly 45%, to scenario G with 68%.

Table 7. Averaged performance ratios obtained in each optimization scenario.

No Parameters	PR Initial	PR Optimized	PR Design
	[%]	[%]	[%]
1		45.7	40.5
2		39.5	43.1
3		50.5	50.6
4	28.0	54.4	58
5		65.8	60.2
8		68	66

Essentially related to PR, mass reduction gives an idea about how much lighter (or heavier) the structure becomes due to the optimization process. It directly provides an estimate of cost savings.

Therefore, the results in Table 8 are consistent with the ones in terms of performance ratios, shown in Table 7.

Table 8. Mass values before/after optimization and after proper approximation (design) using commercial steel profiles.

No Parameters	Initial Mass [kg]	Optimized Mass [kg]	Design Mass [kg]
1		1003	1176
2		1051	1111
3		803	818
4	1440	574	588
5		403	453
8		385	408

Figure 11 shows the optimization results for the Scenario G, in term of the performance ratio obtained by averaging the performance ratios for each structural element. The results for all scenarios are reported in Appendix A. Scenario G, depicted below, exhibits higher values of the performance ratios. This fact becomes become more evident for poles 2, 3, and 4. In these cases, the performance ratios, associated with the design solutions, achieved values equal or greater than the optimized one due to the approximation of the design section adopted. In the post-processing phase, in fact, the optimized section chosen by the list of the FE software was manually edited since the structural constraint violation or the maximum performance ratio was not reached during to the optimization process.

Moreover, in Table 9, the optimized design section for different independent iterations and the proposed industrial solutions according to product list, provided by the software, are listed. As expected, the mass reduction achieved during the optimization process results higher than the design solution due to the approximation issue. For the proposed scenario, the iteration (N_{trial}) that guarantees the best objective function is the second. In Appendix A, the graphical (through histogram charts) and numerical representation (through tables) of the optimization result for each scenario are provided. In order to provide an overview of the objective function trend, the performance ratios and mass reduction for each scenario were collected into Figures 12 and 13. The mentioned values were obtained for each scenario, making an average of the results, before and after optimization, independently, for each steel profile composing the central pole.

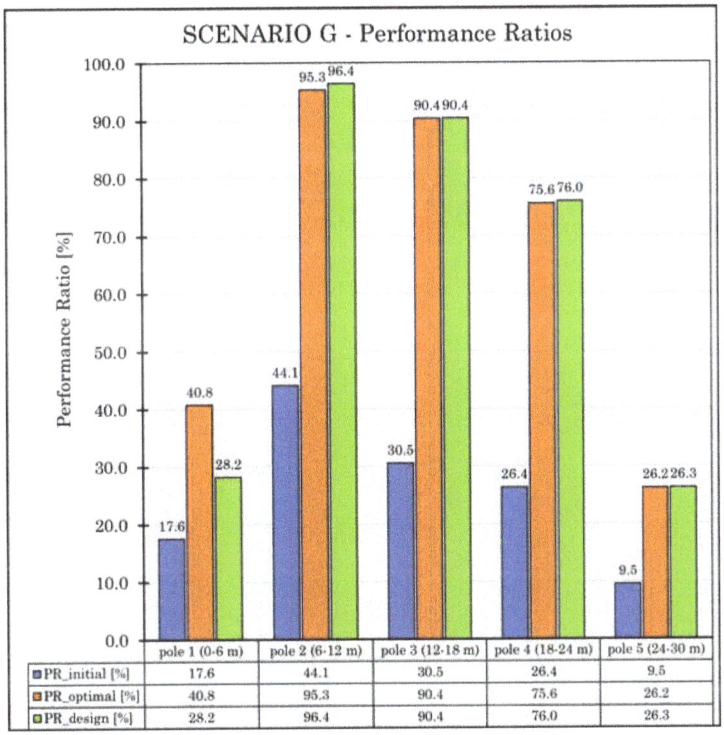

Figure 11. Scenario G.—PRs trend. In blue—the performance ratios of each pole before optimization are illustrated, otherwise orange for the optimized solution. In green—PRs at a design configuration according to the product list.

Figure 12 highlights an almost monotone increment of the performance ratios to the number of design variables. Interestingly, for a number of variables $n \geq 5$, no significant improvements are achieved. Figure 13 emphasizes an important reduction of structural mass as the design variables increase. Once again, n = 5 represents trade-off. If the number of variables exceed 5, no significant improvements are observed.

Figures 12 and 13 show a comparison between each scenario in terms of the average performance ratio and mass reduction, respectively. Figure 12 highlights the difference with the initial state, which has an average performance ratio PR_0 = 25.6%. An evident improvement is achieved for scenarios that include the thickness t as the design variable.

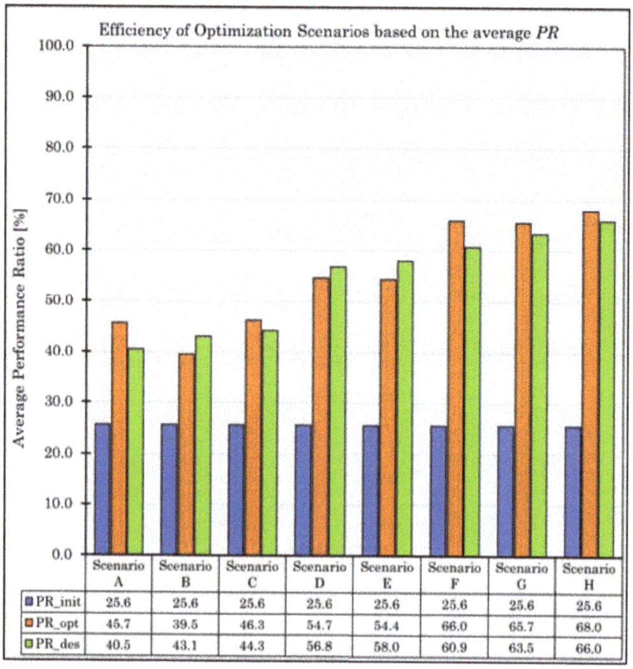

Figure 12. In blue, orange, and green, the average PRs, respectively, at the initial condition, after optimization, and design solution.

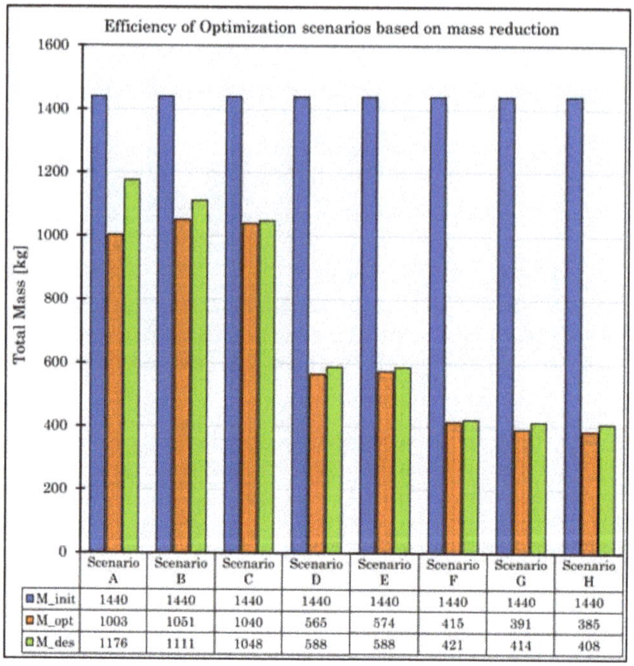

Figure 13. Increasing the number of design variables, the final mass becomes gradually smaller, until 385 kg (scenario H).

Table 9. Scenario G results: optimized solutions for the different independent executions (Ntrial) and proposed industrial one, according to the product list.

SCENARIO G—Optimized Solution					
Element	d [mm]	t [mm]	L [mm]	Mass [Kg]	
Pole 1 (0–6 m)	168.3	3	6000	73	
Pole 2 (6–12 m)	168.3	4	6000	97	
Pole 3 (12–18 m)	168.3	3	6000	73	
Pole 4 (18–24 m)	168.3	3	6000	73	
Pole 5 (24–30 m)	168.3	3	6000	73	
	Total Mass [kg]		Σ	391	
Mass variation [kg]	−1050	Mass variation [%]		−72.88	
SCENARIO G—Design proposed according to the product list					
Element	d [mm]	t [mm]	L [mm]	Mass [Kg]	
Pole 1 (0–6 m)	168.3	4	6000	97	
Pole 2 (6–12 m)	168.3	4	6000	97	
Pole 3 (12–18 m)	168.3	3	6000	73	
Pole 4 (18–24 m)	168.3	3	6000	73	
Pole 5 (24–30 m)	168.3	3	6000	73	
	Total Mass [kg]		Σ	414	
Mass variation [kg]	−1026	Mass variation [%]		−71.22	
Ntrial = 3					
t_1 [mm]	t_2 [mm]	t_3 [mm]	t_4 [mm]	t_5 [mm]	OF [kN]
3	4	4	3	3	34.985
3	4	3	3	3	34.751
3	4	3	4	3	34.985

In particular, from Scenarios D, E, F, G, H, the average performance ratios exceed 50%, resulting in a more than 40% difference compared to the initial state. Figure 12 shows that the commercial profiles are sufficient to accommodate the optimized solution. An exception is noticeable in Scenario A because the optimization is performed using just one diameter Φ, which is optimal for a few parts of the structure, while others are "over-fitted", resulting in a decrease of the performance ratios −28.4% and an increase of structural mass (+173 kg), as shown in Figure 13.

Similarly, a monotonic increment of the structural mass at the end of the optimization process is evident from Figure 13. In this case, the tonnage decreases with the increasing of the parameter's number. There is an overall mass reduction of about −67.5% (−972 kg) from scenario D to H. In scenarios A, B, and C, the thickness t of structural members is not considered. Therefore, the mass loss is not satisfactory, at about −28.4% (−409 kg). The choice of the best scenario should depend on one of the five situations described above (from D to H) related to the better PRs gain and mass loss.

7. Conclusions

In this paper, a guyed radio mast's size and shape optimization process was carried out to identify the equilibrium solution that guarantees the lighter optimized model, verifying strength, instability, and deformation requirements. The paper considers a detailed evaluation of the variable loads according to the Eurocodes recommendations. Furthermore, the OAPI was used to perform a structural analysis with the finite element software SAP2000 by considering the non-linearity of the cables. The optimization was carried out using a genetic optimization algorithm. Eight scenarios (labeled from A to H) were investigated, considering different arrangements of the geometric characteristics of the central pole and cables. The input parameters were increased from Scenario A to H to achieve the best fitness value of the self-weight. From Scenario A to H, the mass reduction index generally increased with the computational effort except in scenarios B and E, in which the input parameter did not represent the best vector design for the structural optimization. At this stage, the best design solution was evaluated from the database of cross-sections inside the finite element software. Though Scenario A provides the worst structural solution in terms of objective function, it represents the most convenient optimization strategy due to its low computational effort; on the contrary, Scenario H exhibits the best fitness value with the lowest self-weight, but it represents the most time-consuming solution. The best solution is achieved when the thickness values of each member, which, composed of the central pole, are included in the optimization process. An improvement of the structural behaviour against instability is observed with increasing thickness. This verification is critical for this structure, mainly subjected to normal stresses resulting from self-weight and pre-stressing cable force. The entire optimization process seems to not be sensible to the pole diameter, chosen as the input parameter of the design vector. Although the final results of the FEM analyses are based on the Italian standards, other codes (e.g., Eurocodes, American code, etc.) can be selected from the SAP2000 settings. However, since no detailed analysis was carried out and many standards are based on the semi-probabilistic approach, the final results should be similar, even with different code formulations. Nevertheless, the partial safety factors involved in load combinations remain quite the same from the numerical point of view, regardless of the followed code.

In future developments, the authors will attempt to replace circular hollow sections with built-up steel solutions to achieve the best structural performance and assemblage procedures. Especially for higher structures, guyed radio masts generally consist of a truss skeleton. Another possible development could be a structural optimization for a cable-stayed radio antenna adopting other optimization strategies, such as particle swarm optimization, PSO, and the evolution differential algorithm (EDA), which could be less time-consuming. Finally, it could perform a typological optimization by managing the position of the cable connection, trying to find the best attachment points.

Supplementary Materials: The following are available online at https://www.mdpi.com/article/10.3390/app12104875/s1.

Author Contributions: Conceptualization, R.C., M.M.R. and J.M.; methodology, R.C., M.M.R., A.A., J.M., M.L.G. and G.C.M.; software, M.L.G., M.M.R. and R.C.; validation, R.C., M.L.G., J.M. and A.A.; formal analysis, R.C., J.M. and A.A.; investigation, A.A. and R.C.; resources, R.C., M.M.R. and G.C.M.; data curation, R.C. and A.A.; writing—original draft preparation, A.A., R.C. and M.M.R.; writing—review and editing, J.M., R.C., M.L.G. and A.A.; visualization, M.M.R., R.C. and A.A.; supervision, G.C.M. All authors have read and agreed to the published version of the manuscript.

Funding: This research was supported by project MSCA-RISE-2020 Marie Skłodowska-Curie Research and Innovation Staff Exchange (RISE)—ADDOPTML (ntua.gr).

Institutional Review Board Statement: Not applicable.

Informed Consent Statement: Not applicable.

Data Availability Statement: The data used to support the findings of this study are available from the corresponding author upon reasonable request.

Acknowledgments: The authors would like to thank the anonymous reviewers for their valuable comments and suggestions in revising the paper. The authors would like to thank G.C. Marano and the project ADDOPTML for funding/supporting this research.

Conflicts of Interest: The authors declare no conflict of interest.

Appendix A

Table A1. Drag and lift forces according to [36] at ULS in [Kg/m].

Wind Action (Drag D, Lift L) at ULS						
z (m)	**Drag_1**	**Lift_1**	**Drag_2**	**Lift_2**	**Drag_2**	**Lift_3**
1	3.92	6.5	6.01	5.2	6.85	1.95
2	3.92	6.5	6.01	5.2	6.85	1.95
3	3.92	6.5	6.01	5.2	6.85	1.95
4	3.92	6.5	6.01	5.2	6.85	1.95
5	3.92	6.5	6.01	5.2	6.85	1.95
6	4.2	6.97	6.44	5.58	7.34	2.09
7	4.44	7.37	6.81	5.9	7.76	2.21
8	4.66	7.73	7.15	6.19	8.14	2.32
9	4.85	8.05	7.44	6.44	8.48	2.42
10	5.03	8.35	7.71	6.68	8.79	2.5
11	5.19	8.61	7.96	6.89	9.07	2.58
12	5.34	8.86	8.19	7.09	9.33	2.66
13	5.48	9.09	8.4	7.27	9.57	2.73
14	5.61	9.31	8.6	7.45	9.8	2.79
15	5.73	9.51	8.79	7.61	10.01	2.85
16	5.85	9.7	8.97	7.76	10.22	2.91
17	5.96	9.88	9.13	7.91	10.41	2.97
18	6.06	10.06	9.29	8.04	10.59	3.02
19	6.16	10.22	9.45	8.18	10.76	3.07
20	6.25	10.38	9.59	8.3	10.92	3.11
21	6.34	10.53	9.73	8.42	11.08	3.16
22	6.43	10.67	9.86	8.54	11.23	3.2
23	6.51	10.81	9.99	8.65	11.38	3.24
24	6.59	10.94	10.11	8.75	11.52	3.28
25	6.67	11.07	10.23	8.86	11.66	3.32
26	6.75	11.19	10.35	8.96	11.79	3.36
27	6.82	11.31	10.46	9.05	11.91	3.39
28	6.89	11.43	10.56	9.14	12.03	3.43
29	6.96	11.54	10.67	9.23	12.15	3.46
30	7.02	11.65	10.77	9.32	12.27	3.5

Table A2. Drag and lift forces according to [36] at SLS in [Kg/m].

Wind Action (Drag D, Lift L) at SLS						
z (m)	**Drag_1**	**Lift_1**	**Drag_2**	**Lift_2**	**Drag_2**	**Lift_3**
1	2.29	3.81	3.52	3.05	4.01	1.14
2	2.29	3.81	3.52	3.05	4.01	1.14
3	2.29	3.81	3.52	3.05	4.01	1.14
4	2.29	3.81	3.52	3.05	4.01	1.14
5	2.29	3.81	3.52	3.05	4.01	1.14
6	2.31	3.84	3.54	3.07	4.04	1.15
7	2.32	3.86	3.57	3.09	4.06	1.16
8	2.34	3.88	3.58	3.1	4.08	1.16
9	2.34	3.89	3.6	3.11	4.1	1.17

Table A2. *Cont.*

Wind Action (Drag D, Lift L) at SLS						
z (m)	Drag_1	Lift_1	Drag_2	Lift_2	Drag_2	Lift_3
10	2.35	3.9	3.61	3.12	4.11	1.17
11	2.36	3.92	3.62	3.13	4.12	1.17
12	2.37	3.93	3.63	3.14	4.13	1.18
13	2.37	3.94	3.64	3.15	4.14	1.18
14	2.38	3.94	3.64	3.15	4.15	1.18
15	2.38	3.95	3.65	3.16	4.16	1.19
16	2.39	3.96	3.66	3.17	4.17	1.19
17	2.39	3.96	3.66	3.17	4.17	1.19
18	2.39	3.97	3.67	3.18	4.18	1.19
19	2.4	3.98	3.67	3.18	4.19	1.19
20	2.4	3.98	3.68	3.19	4.19	1.19
21	2.4	3.99	3.68	3.19	4.2	1.2
22	2.4	3.99	3.69	3.19	4.2	1.2
23	2.41	4	3.69	3.2	4.21	1.2
24	2.41	4	3.7	3.2	4.21	1.2
25	2.41	4	3.7	3.2	4.21	1.2
26	2.41	4.01	3.7	3.21	4.22	1.2
27	2.42	4.01	3.71	3.21	4.22	1.2
28	2.42	4.01	3.71	3.21	4.23	1.2
29	2.42	4.02	3.71	3.21	4.23	1.21
30	2.42	4.02	3.72	3.22	4.23	1.21

Table A3. Modal participating mass ratios.

Modal Participating Mass Ratios				
n. Modes	Period (s)	Frequence (Hz)	Part. Mass X (%)	Part. Mass Y [%]
1	3.99	0.251	0.0	0.28
2	3.99	0.251	0.83	0.0
3	3.99	0.251	0.0	0.55
4	3.473	0.288	0.0	0.25
5	3.473	0.288	0.75	0.0
6	3.472	0.288	0.0	0.5
7	2.929	0.341	0.0	2.52
8	2.925	0.342	0.0	0.05
9	2.916	0.343	2.55	0.0
10	0.437	2.290	9.58	26.24
11	0.434	2.304	26.41	9.16
12	0.206	4.853	7.16	4.44
13	0.203	4.934	5.64	4.78
14	0.155	6.437	27.38	0.78
15	0.144	6.935	3.67	26.10
16	0.116	8.584	0.14	0.00
17	0.116	8.595	0.00	0.32
18	0.106	9.436	2.35	9.82
19	0.057	17.410	0.03	0.08
20	0.057	17.442	0.01	0.01
21	0.054	18.573	0.48	0.45
22	0.050	20.163	0.03	0.00
23	0.047	21.496	0.00	0.00

Table A3. *Cont.*

Modal Participating Mass Ratios				
n. Modes	**Period (s)**	**Frequence (Hz)**	**Part. Mass X (%)**	**Part. Mass Y (%)**
24	0.046	21.516	0.00	0.00
25	0.036	27.906	0.27	0.11
26	0.035	28.313	0.04	0.13
27	0.032	31.224	0.01	0.48
28	0.032	31.722	0.20	0.00
29	0.031	32.590	0.01	0.06
30	0.024	40.831	0.00	0.00
31	0.024	40.836	0.00	0.00
32	0.023	42.600	12.32	0.00
33	0.022	44.518	0.01	10.59

Table A4. Load combination.

Load Combination	
ULS Max_1	$1.3 \cdot G_1 + 1.5 \cdot G_2 + 1.5 \cdot Wind_1 + 1.5 \cdot 0.5 \cdot Ice_1 + 1.5 \cdot 0 \cdot Q_M$
ULS $Max1_2$	$1.3 \cdot G_1 + 1.5 \cdot G_2 + 1.5 \cdot Q_M + 1.5 \cdot 0.6 Wind_1 + 1.5 \cdot 0.2 Ice_1$
ULS Min_1	$1 \cdot G_1 + 0.8 \cdot G_2 + 1.5 \cdot Wind_1 + 1.5 \cdot 0.5 Ice_1 + 1.5 \cdot 0 Q_M$
$Quake_1$	$E + G_1 + G_2 + 0 \cdot Wind_1 + 0 Ice_1 + 0 Q_M$
ULS Max_2	$1.3 \cdot G_1 + 1.5 \cdot G_2 + 1.5 \cdot Wind_2 + 1.5 \cdot 0.5 \cdot Ice_2 + 1.5 \cdot 0 \cdot Q_M$
ULS $Max2_1$	$1.3 \cdot G_1 + 1.5 \cdot G_2 + 1.5 \cdot Q_M + 1.5 \cdot 0.6 Wind_2 + 1.5 \cdot 0.2 Ice_2$
ULS Min_1	$1 \cdot G_1 + 0.8 \cdot G_2 + 1.5 \cdot Wind_2 + 1.5 \cdot 0.5 Ice_2 + 1.5 \cdot 0 Q_M$
$Quake_2$	$E + G_1 + G_2 + 0.8 \cdot Wind_2 + 1.5 Ice_2 + 1.5 Q_M$

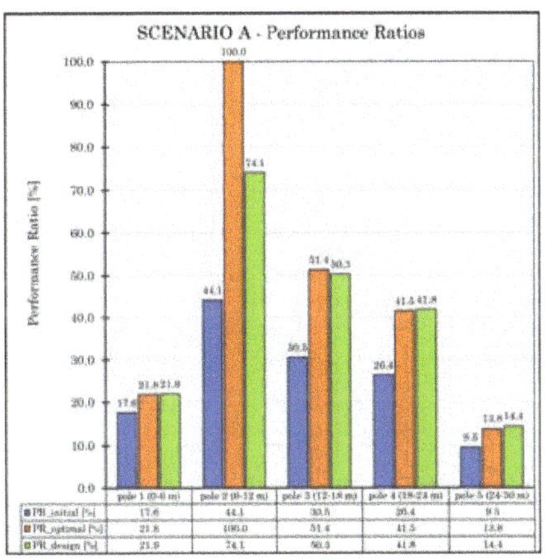

Figure A1. Cont.

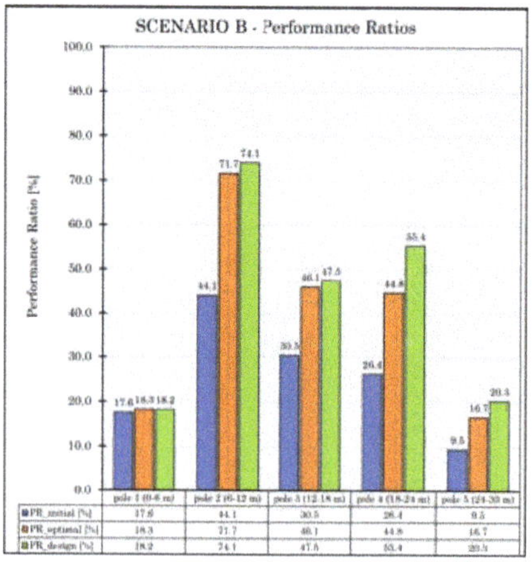

Figure A1. Scenarios A, B. In blue, orange, and green, the average PRs, respectively, at the initial condition, after optimization, and the design solution.

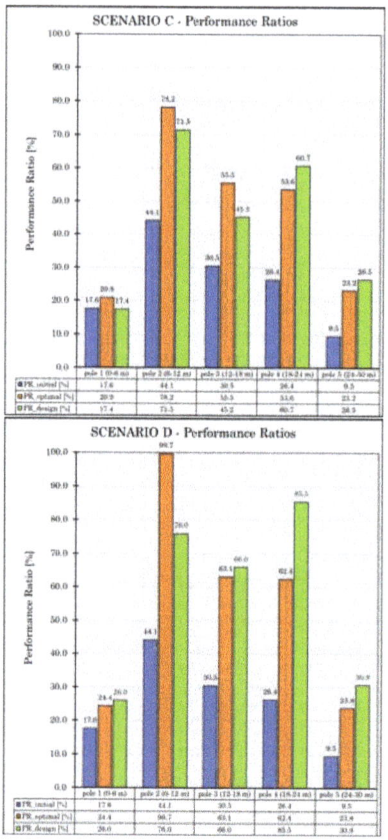

Figure A2. Scenarios C, D. In blue, orange, and green, the average PRs, respectively, at the initial condition, after optimization, and the design solution.

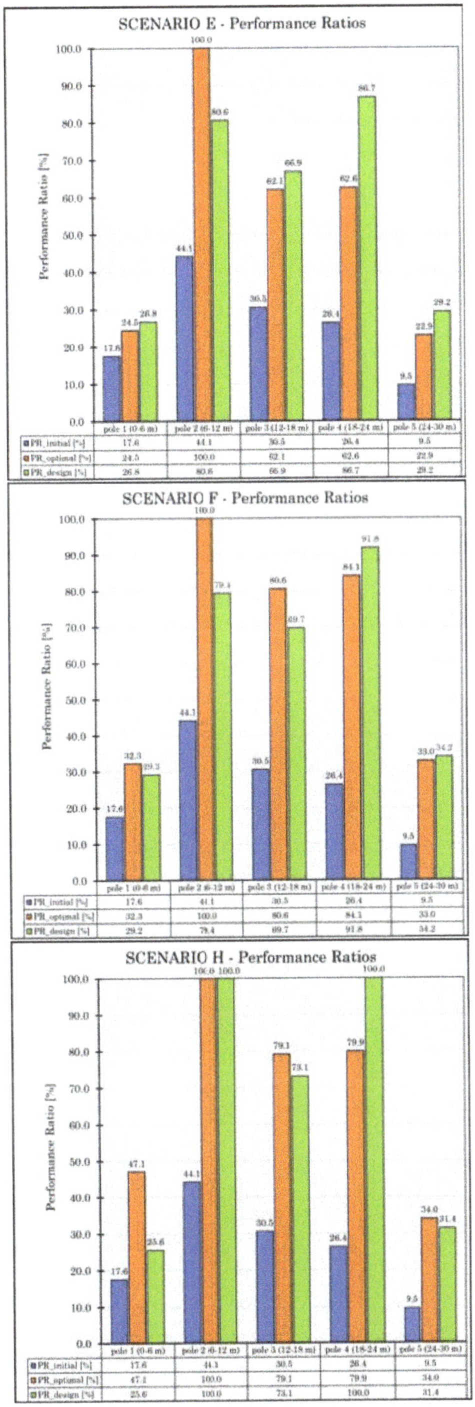

Figure A3. Scenarios E, F, H. In blue, orange and green, the average PRs, respectively, at the initial condition, after optimization, and the design solution.

Table A5. Scenario A results: optimized solutions for the different independent executions (Ntrial) and the proposed industrial one, according to the product list.

SCENARIO A—Optimized Solution				
Element	d [mm]	t [mm]	L [mm]	Mass [Kg]
Pole 1 (0–6 m)	121	12.5	6000	201
Pole 2 (6–12 m)	121	12.5	6000	201
Pole 3 (12–18 m)	121	12.5	6000	201
Pole 4 (18–24 m)	121	12.5	6000	201
Pole 5 (24–30 m)	121	12.5	6000	201
	Total Mass [kg]		Σ	1003
Mass variation [kg]	−437	Mass variation [%]		−30.36
SCENARIO A—Design proposed according to the product list				
Element	d [mm]	t [mm]	L [mm]	Mass [Kg]
Pole 1 (0–6 m)	139.7	12.5	6000	235
Pole 2 (6–12 m)	139.7	12.5	6000	235
Pole 3 (12–18 m)	139.7	12.5	6000	235
Pole 4 (18–24 m)	139.7	12.5	6000	235
Pole 5 (24–30 m)	139.7	12.5	6000	235
	Total Mass [kg]		Σ	1176
Mass variation [kg]	−264	Mass variation [%]		−18.36

Ntrial = 5	
Φ_{opt} [mm]	OF [kN]
121	40.758
121	40.758
121	40.758
121	**40.758**
122	40.849

Table A6. Scenario B results: optimized solutions for the different independent executions (Ntrial) and the proposed industrial one according to the product list.

SCENARIO B—Optimized Solution				
Element	d [mm]	t [mm]	L [mm]	Mass [Kg]
Pole 1 (0–6 m)	149	12.5	6000	252
Pole 2 (6–12 m)	138	12.5	6000	231
Pole 3 (12–18 m)	126	12.5	6000	210
Pole 4 (18–24 m)	115	12.5	6000	189
Pole 5 (24–30 m)	103	12.5	6000	168
	Total Mass [kg]		Σ	1051
Mass variation [kg]	−389	Mass variation [%]		−27.02
SCENARIO B—Design proposed according to product list				
Element	d [mm]	t [mm]	L [mm]	Mass [Kg]
Pole 1 (0–6 m)	168.3	12.5	6000	288
Pole 2 (6–12 m)	139.7	12.5	6000	235
Pole 3 (12–18 m)	139.7	12.5	6000	235
Pole 4 (18–24 m)	114.3	12.5	6000	188

Table A6. *Cont.*

Pole 5 (24–30 m)	101.6	12.5	6000	165
	Total Mass [kg]		Σ	1111
Mass variation [kg]	−329	Mass variation [%]		−22.84
	Ntrial = 5; best solutions			
	Φ_i [mm]		Φ_f [mm]	OF [kN]
	148		94	41.248
	146		103	41.466
	148		94	41.248
	146		103	41.466
	149		**92**	**41.230**

Table A7. Scenario C results: optimized solutions for the different independent executions (Ntrial) and the proposed industrial one according to the product list.

	SCENARIO C—Optimized solution			
Element	d [mm]	t [mm]	L [mm]	Mass [Kg]
Pole 1 (0–6 m)	147	12.5	6000	249
Pole 2 (6–12 m)	136	12.5	6000	228
Pole 3 (12–18 m)	125	12.5	6000	208
Pole 4 (18–24 m)	114	12.5	6000	188
Pole 5 (24–30 m)	103	12.5	6000	167
	Total Mass [kg]		Σ	1040
Mass variation [kg]	−400	Mass variation [%]		−27.79
	SCENARIO C—Design proposed according to the product list			
Element	d [mm]	t [mm]	L [mm]	Mass [Kg]
Pole 1 (0–6 m)	168.3	12.5	6000	288
Pole 2 (6–12 m)	139.7	12.5	6000	235
Pole 3 (12–18 m)	139.7	12.5	6000	235
Pole 4 (18–24 m)	114.3	10	6000	154
Pole 5 (24–30 m)	101.6	10	6000	135
	Total Mass [kg]		Σ	1048
Mass variation [kg]	−392	Mass variation [%]		−27.22
	Ntrial = 5			
Φ_i [mm]		Φ_f [mm]	F [kN]	OF [kN]
152		92	1.8	41.393
151		92	1.4	41.339
149		92	1	41.230
156		92	2.4	41.610
147		**92**	**0.8**	**41.121**

Table A8. Scenario D results: optimized solutions for the different independent executions (Ntrial) and the proposed industrial one according to the product list.

SCENARIO D—Optimized solution				
Element	d [mm]	t [mm]	L [mm]	Mass [Kg]
Pole 1 (0–6 m)	161	6	6000	138
Pole 2 (6–12 m)	147	6	6000	125
Pole 3 (12–18 m)	133	6	6000	113
Pole 4 (18–24 m)	120	6	6000	101
Pole 5 (24–30 m)	106	6	6000	89
		Total Mass [kg]	Σ	565
Mass variation [kg]	−875	Mass variation [%]		−60.75
SCENARIO D—Design proposed according to the product list				
Element	d [mm]	t [mm]	L [mm]	Mass [Kg]
Pole 1 (0–6 m)	168.3	6	6000	144
Pole 2 (6–12 m)	168.3	6	6000	144
Pole 3 (12–18 m)	139.7	6	6000	119
Pole 4 (18–24 m)	114.3	6	6000	96
Pole 5 (24–30 m)	101.6	6	6000	85
		Total Mass [kg]	Σ	588
Mass variation [kg]	−853	Mass variation [%]		−59.20
Ntrial = 5				
Φ_i [mm]		Φ_f [mm]	t [mm]	OF [kN]
161		92	6	**36.465**
146		117	7	37.389
162		92	6	36.491
162		92	6	36.491
163		92	6	36.517

Table A9. Scenario E results: optimized solutions for the different independent executions (Ntrial) and the proposed industrial one according to the product list.

SCENARIO E—Optimized Solution				
Element	d [mm]	t [mm]	L [mm]	Mass [Kg]
Pole 1 (0–6 m)	165	6	6000	141
Pole 2 (6–12 m)	150	6	6000	128
Pole 3 (12–18 m)	135	6	6000	115
Pole 4 (18–24 m)	121	6	6000	102
Pole 5 (24–30 m)	106	6	6000	89
		Total Mass [kg]	Σ	574
Mass variation [kg]	−866	Mass variation [%]		−60.13
SCENARIO E—Design proposed according to the product list				
Element	d [mm]	t [mm]	L [mm]	Mass [Kg]
Pole 1 (0–6 m)	168.3	6	6000	144
Pole 2 (6–12 m)	168.3	6	6000	144
Pole 3 (12–18 m)	139.7	6	6000	119

Table A9. *Cont.*

Pole 4 (18–24 m)	114.3	6	6000	96
Pole 5 (24–30 m)	101.6	6	6000	85
	Total Mass [kg]		Σ	588
Mass variation [kg]	−853	Mass variation [%]		−59.20
	Ntrial = 5; best solutions			
Φ_i [mm]	Φ_f [mm]	t [mm]	F [kN]	OF [kN]
150	97	7.8	1.3	37.766
153	112	6.4	1.6	36.964
165	91	6	2.3	36.552
160	91	7	1.3	37.287
139	104	8.8	1.3	38.337

Table A10. Scenario F results: optimized solutions for the different independent executions (Ntrial) and the proposed industrial one according to the product list.

SCENARIO F—Optimized Solution					
Element	d [mm]	t [mm]	L [mm]	Mass [Kg]	
Pole 1 (0–6 m)	157	4	6000	91	
Pole 2 (6–12 m)	144	6	6000	122	
Pole 3 (12–18 m)	131	4	6000	75	
Pole 4 (18–24 m)	118	4	6000	67	
Pole 5 (24–30 m)	105	4	6000	60	
	Total Mass [kg]		Σ	415	
Mass variation [kg]	−1025	Mass variation [%]		−71.16	
SCENARIO F—Design proposed according to the product list					
Element	d [mm]	t [mm]	L [mm]	Mass [Kg]	
Pole 1 (0–6 m)	168.3	4	6000	97	
Pole 2 (6–12 m)	168.3	5	6000	121	
Pole 3 (12–18 m)	139.7	4	6000	80	
Pole 4 (18–24 m)	114.3	4	6000	65	
Pole 5 (24–30 m)	101.6	4	6000	58	
	Total Mass [kg]		Σ	421	
Mass variation [kg]	−1019	Mass variation [%]		−70.75	
Ntrial = 3; best solutions					
Φ_i [mm]	Φ_f [mm]	t_{ends} [mm]	t_{inter} [mm]	F [kN]	OF [kN]
155	92	4	7	3.2	35.141
157	92	4	6	0.9	34.993
151	92	4	7	1.3	35.058

Table A11. Scenario H results: optimized solutions for the different independent executions (Ntrial) and the proposed industrial one according to the product list.

SCENARIO H—Optimized Solution								
Element					d [mm]	t [mm]	L [mm]	Mass [Kg]
Pole 1 (0–6 m)					158	3	6000	69
Pole 2 (6–12 m)					146	6	6000	124
Pole 3 (12–18 m)					133	4	6000	76
Pole 4 (18–24 m)					121	4	6000	69
Pole 5 (24–30 m)					108	3	6000	47
					Total Mass [kg]		Σ	385
Mass variation [kg]					−1055	Mass variation [%]		−73.27
SCENARIO H—Design proposed according to the product list								
Element					d [mm]	t [mm]	L [mm]	Mass [Kg]
Pole 1 (0–6 m)					168.3	5	6000	121
Pole 2 (6–12 m)					168.3	4	6000	97
Pole 3 (12–18 m)					139.7	4	6000	80
Pole 4 (18–24 m)					139.7	3	6000	61
Pole 5 (24–30 m)					114.3	3	6000	49
					Total Mass [kg]		Σ	408
Mass variation [kg]					−1032	Mass variation [%]		−71.65
Ntrial = 3								
Φ_i [mm]	Φ_i [mm]	t_1 [mm]	t_2 [mm]	t_3 [mm]	t_4 [mm]	t_5 [mm]	F [kN]	OF [kN]
164	109	4	5	4	3	3	0.9	34.789
167	111	3	6	4	3	3	2	34.839
158	96	3	6	4	4	3	2	34.695

References

1. Law, S.; Bu, J.; Zhu, X.; Chan, S.L. Wind characteristics of Typhoon Dujuan as measured at a 50 m guyed mast. *Wind Struct.* **2006**, *9*, 387–396. [CrossRef]
2. Sparling, B.; Wegner, L. Estimating peak wind load effects in guyed masts. *Wind Struct.* **2007**, *10*, 347–366. [CrossRef]
3. Hensley, G.M.; Plaut, R.H. Three-dimensional analysis of the seismic response of guyed masts. *Eng. Struct.* **2007**, *29*, 2254–2261. [CrossRef]
4. Amiri, G.G. Seismic sensitivity indicators for tall guyed telecommunication towers. *Comput. Struct.* **2002**, *80*, 349–364. [CrossRef]
5. Liu, C.; Fang, D.; Zhao, L. Reflection on earthquake damage of buildings in 2015 Nepal earthquake and seismic measures for post-earthquake reconstruction. *Structures* **2021**, *30*, 647–658. [CrossRef]
6. Sun, X.; Tao, X.; Duan, S.; Liu, C. Kappa (k) derived from accelerograms recorded in the 2008 Wenchuan mainshock, Sichuan, China. *J. Asian Earth Sci.* **2013**, *73*, 306–316. [CrossRef]
7. Buchholdt, H.; Moossavinejad, S.; Iannuzzi, A. Non-Linear Dynamic Analysis of Guyed Masts Subjected to Wind and Guy R Uptures. *Proc. Inst. Civ. Eng.* **1986**, *81*, 353–395. [CrossRef]
8. Ballaben, J.S.; Guzmán, A.M.; Rosales, M.B. Nonlinear dynamics of guyed masts under wind load: Sensitivity to structural parameters and load models. *J. Wind Eng. Ind. Aerodyn.* **2017**, *169*, 128–138. [CrossRef]
9. Gerstoft, P.; Davenport, A. A simplified method for dynamic analysis of a guyed mast. *J. Wind Eng. Ind. Aerodyn.* **1986**, *23*, 487–499. [CrossRef]
10. de Souza, R.R.; Miguel, L.F.F.; Lopez, R.H.; Miguel, L.F.F.; Torii, A.J. A procedure for the size, shape and topology optimization of transmission line tower structures. *Eng. Struct.* **2016**, *111*, 162–184. [CrossRef]
11. Saudi, G. Structural assessment of a guyed mast through measurement of natural frequencies. *Eng. Struct.* **2014**, *59*, 104–112. [CrossRef]
12. Saxena, R.; Popplewell, N.; Trainor, P.; Shah, A. Vibrations of complex guyed towers. In Proceedings of the 12th Biennial Conference on Mechanical Vibration and Noise Control, Montreal, QC, Canada, 17–21 September 1989.
13. Novak, M.; Tanaka, H.; Davenport, A.G. Vibration of towers due to galloping of iced cables. *J. Eng. Mech. Div.* **1978**, *104*, 457–473. [CrossRef]

14. Wahba, Y.M.; Madugula, M.K.; Monforton, G.R. Shake Table for Dynamic Testing of Guyed Towers. In *Building to Last*; ASCE: Reston, VA, USA, 1997; pp. 353–357.
15. Madugula, M.K.; Wahba, Y.M.; Monforton, G.R. Dynamic response of guyed masts. *Eng. Struct.* **1998**, *20*, 1097–1101. [CrossRef]
16. Luzardo, A.C.; Parnás, V.E.; Rodríguez, P.M. Guy tension influence on the structural behavior of a guyed mast. *J. Int. Assoc. Shell Spat. Struct.* **2012**, *53*, 111–116.
17. Davenport, A.; Sparling, B. Dynamic gust response factors for guyed towers. *J. Wind Eng. Ind. Aerodyn.* **1992**, *43*, 2237–2248. [CrossRef]
18. Harikrishna, P.; Annadurai, A.; Gomathinayagam, S.; Lakshmanan, N. Full scale measurements of the structural response of a 50 m guyed mast under wind loading. *Eng. Struct.* **2003**, *25*, 859–867. [CrossRef]
19. Gioffrè, M.; Gusella, V.; Materazzi, A.; Venanzi, I. Removable guyed mast for mobile phone networks: Wind load modeling and structural response. *J. Wind Eng. Ind. Aerodyn.* **2004**, *92*, 463–475. [CrossRef]
20. Clobes, M.; Peil, U. Unsteady buffeting wind loads in the time domain and their effect on the life-cycle prediction of guyed masts. *Struct. Infrastruct. Eng.* **2011**, *7*, 187–196. [CrossRef]
21. Pezo, M.L.; Bakić, V.V. Numerical determination of drag coefficient for guyed mast exposed to wind action. *Eng. Struct.* **2014**, *62*, 98–104. [CrossRef]
22. Sparling, B.F. The Dynamic Behavior of Guys and Guyed Masts in Turbulent Winds. Ph.D. Thesis, Western University, London, ON, Canada, 1995.
23. Wahba, Y.; Madugula, M.; Monforton, G. Evaluation of non-linear analysis of guyed antenna towers. *Comput. Struct.* **1998**, *68*, 207–212. [CrossRef]
24. Madugula, M.K. *Dynamic Response of Lattice Towers and Guyed Masts*; ASCE Publications: Reston, VA, USA, 2001.
25. Orlando, D.; Gonçalves, P.B.; Rega, G.; Lenci, S. Nonlinear dynamics and instability as important design concerns for a guyed mast. In *IUTAM Symposium on Nonlinear Dynamics for Advanced Technologies and Engineering Design*; Springer: Berlin/Heidelberg, Germany, 2013; pp. 223–234.
26. Ballaben, J.S.; Rosales, M.B. Nonlinear dynamic analysis of a 3D guyed mast. *Nonlinear Dyn.* **2018**, *93*, 1395–1405. [CrossRef]
27. Belevičius, R.; Jatulis, D.; Šešok, D. Optimization of tall guyed masts using genetic algorithms. *Eng. Struct.* **2013**, *56*, 239–245. [CrossRef]
28. Gawronski, W.; Bienkiewicz, B.; Hill, R. Wind-induced dynamics of a deep space network antenna. *J. Sound Vib.* **1994**, *178*, 67–77. [CrossRef]
29. Fujino, Y.; Warnitchai, P.; Pacheco, B. Active Stiffness Control of Cable Vibration. *J. Appl. Mech.* **1993**, *60*, 948–953. [CrossRef]
30. Lacarbonara, W.; Ballerini, S. Vibration mitigation of guyed masts via tuned pendulum dampers. *Struct. Eng. Mech. Int. J.* **2009**, *32*, 517–529. [CrossRef]
31. Błachowski, B. Model based predictive control of guyed mast vibration. *J. Theor. Appl. Mech.* **2007**, *45*, 405–423.
32. Bell, L.C.; Brown, D.M. Guyed tower optimization. *Comput. Struct.* **1976**, *6*, 447–450. [CrossRef]
33. Thornton, C.H.; Joseph, L.; Scarangello, T. Optimization of tall structures for wind loading. *J. Wind Eng. Ind. Aerodyn.* **1990**, *36*, 235–244. [CrossRef]
34. Uys, P.; Farkas, J.; Jarmai, K.; Van Tonder, F. Optimisation of a steel tower for a wind turbine structure. *Eng. Struct.* **2007**, *29*, 1337–1342. [CrossRef]
35. Venanzi, I.; Materazzi, A. Multi-objective optimization of wind-excited structures. *Eng. Struct.* **2007**, *29*, 983–990. [CrossRef]
36. Zhang, Z.Q.; Li, H.N. Two-level optimization method of transmission tower structure based on ant colony algorithm. In *Advanced Materials Research*; Trans Tech Publications Ltd.: Freienbach, Switzerland, 2011; Volume 243, pp. 5849–5853.
37. Cucuzza, R.; Costi, C.; Rosso, M.M.; Domaneschi, M.; Marano, G.C.; Masera, D. Optimal strengthening by steel truss arches in prestressed girder bridges. In *Proceedings of the Institution of Civil Engineers—Bridge Engineering*; Thomas Telford Ltd.: London, UK, 2021; pp. 1–21. [CrossRef]
38. Luh, G.C.; Lin, C.Y. Optimal design of truss-structures using particle swarm optimization. *Comput. Struct.* **2011**, *89*, 2221–2232. [CrossRef]
39. Manuello Bertetto, A.; Marano, G. Numerical and dimensionless analytical solutions for circular arch optimization. *Eng. Struct.* **2022**, *253*, 113360. [CrossRef]
40. Kaveh, A.; Talatahari, S. Particle swarm optimizer, ant colony strategy and harmony search scheme hybridized for optimization of truss structures. *Comput. Struct.* **2009**, *87*, 267–283. [CrossRef]
41. Deng, Z.Q.; Zhang, Y.; Huang, H.L.; Li, B. Parametric optimization for a tapered deployable mast in an integrated design environment. In *Advanced Materials Research*; Trans Tech Publications Ltd.: Freienbach, Switzerland, 2012; Volume 346, pp. 426–432.
42. Guo, H.; Li, Z. Structural topology optimization of high-voltage transmission tower with discrete variables. *Struct. Multidiscip. Optim.* **2011**, *43*, 851–861. [CrossRef]
43. delle Ricerche, C.N. Istruzioni per la valutazione delle azioni e degli effetti del vento sulle costruzioni. *CNR-DT* **2009**, *207*, 2008.
44. Mordà, N.; Mancini, A. *Norme Tecniche per le Costruzioni (NTC 2018) D. Min. Infrastrutture e Trasporti 17 Gennaio 2018*; Ministero delle Infrastrutture e dei Trasporti: Roma, Italy, 2018.
45. Melchiorre, J.; Bertetto, A.M.; Marano, G.C. Application of a Machine Learning Algorithm for the Structural Optimization of Circular Arches with Different Cross-Sections. *J. Appl. Math. Phys.* **2021**, *9*, 1159–1170. [CrossRef]

46. Rosso, M.M.; Cucuzza, R.; Aloisio, A.; Marano, G.C. Enhanced Multi-Strategy Particle Swarm Optimization for Constrained Problems with an Evolutionary-Strategies-Based Unfeasible Local Search Operator. *Appl. Sci.* **2022**, *12*, 2285. [CrossRef]
47. Rosso, M.M.; Cucuzza, R.; Di Trapani, F.; Marano, G.C. Nonpenalty machine learning constraint handling using PSO-svm for structural optimization. *Adv. Civ. Eng.* **2021**, *2021*, 6617750. [CrossRef]
48. Rao, S.S. *Engineering Optimization: Theory and Practice*; John Wiley & Sons: Hoboken, NJ, USA, 2019.
49. Aloisio, A.; Pasca, D.P.; Battista, L.; Rosso, M.M.; Cucuzza, R.; Marano, G.; Alaggio, R. Indirect assessment of concrete resistance from FE model updating and Young's modulus estimation of a multi-span PSC viaduct: Experimental tests and validation. *Elsevier Struct.* **2022**, *37*, 686–697. [CrossRef]
50. Sardone, L.; Rosso, M.M.; Cucuzza, R.; Greco, R.; Marano, G.C. Computational Design of Comparative models and geometrically constrained optimization of a multi-domain variable section beam based on Timoshenko model. In Proceedings of the EUROGEN2021, 14TH ECCOMAS Thematic Conference on Evolutionary and Deterministic Methods for Design, Optimization and Control, Athens, Greece, 28–30 June 2021. [CrossRef]
51. Christensen, P.W.; Klarbring, A. *An Introduction to Structural Optimization*; Springer Science & Business Media: Berlin/Heidelberg, Germany, 2008; Volume 153.
52. Coello Coello, C.A. Theoretical and numerical constraint-handling techniques used with evolutionary algorithms: A survey of the state of the art. *Comput. Methods Appl. Mech. Eng.* **2002**, *191*, 1245–1287. [CrossRef]
53. Koziel, S.; Michalewicz, Z. Evolutionary Algorithms, Homomorphous Mappings, and Constrained Parameter Optimization. *Evol. Comput.* **1999**, *7*, 19–44. [CrossRef] [PubMed]
54. Michalewicz, Z.; Fogel, D. *How to Solve It: Modern Heuristics*; Springer Science & Business Media: Berlin/Heidelberg, Germany, 2008.
55. Parsopoulos, K.; Vrahatis, M. Unified Particle Swarm Optimization for Solving Constrained Engineering Optimization Problems. In *International Conference on Natural Computation*; Springer: Berlin/Heidelberg, Germany, 2005; Volume 3612, pp. 582–591. [CrossRef]
56. Coello, C. Self-adaptive penalties for GA-based optimization. In Proceedings of the 1999 Congress on Evolutionary Computation-CEC99 (Cat. No. 99TH8406), Washington, DC, USA, 6–9 July 1999; IEEE: Hoboken, NJ, USA, 1999; Volume 1, pp. 573–580. [CrossRef]

applied sciences

MDPI

Article

Optimized Design of Floor Plan and Components of Prefabricated Building with Energy-Cost Effect

Juanli Guo [1], Mingchen Li [2,*], Zixin Jiang [2], Zhoupeng Wang [2] and Yangkong Zhou [2]

[1] School of Architecture, Tianjin University, Tianjin 300072, China; guojuanli@tju.edu.cn
[2] Tianjin International Engineering Institute, Tianjin University, Tianjin 300072, China; jiangzx@tju.edu.cn (Z.J.); wangzhoupeng@tju.edu.cn (Z.W.); zyk36@tju.edu.cn (Y.Z.)
* Correspondence: limingchen@tju.edu.cn; Tel.: +86-13059077720

Abstract: Optimizing building performance and economic benefits through feedback in building design is a hot topic in current academic research. However, few studies on prefabricated buildings have been undertaken in this field. Meanwhile, the methodology used for achieving optimized solutions is still poor. In this paper, genetic algorithms and correlation analysis are employed and two parametric design methods—i.e., the floor plan generation method and the component selection method—are proposed for the modularity of the prefabricated buildings. Taking a typical high-rise building in Tianjin as an example, correlation analyses are performed on the basis of the two proposed methods to enhance the depth of the optimized finding approach. The outcome of this research demonstrates the feasibility of the proposed numerical approach, which can produce the optimized floor plan and construction set under the local conditions. This also reveals that the shape coefficient and window-to-wall ratio are strongly correlated with the energy performance of a building, which can help architects to pursue optimized design solutions in the schematic design process.

Keywords: building energy saving; prefabricated building; genetic algorithm; parametric design; multi-objective optimization; correlation analysis

Citation: Guo, J.; Li, M.; Jiang, Z.; Wang, Z.; Zhou, Y. Optimized Design of Floor Plan and Components of Prefabricated Building with Energy-Cost Effect. *Appl. Sci.* **2022**, *12*, 3740. https://doi.org/10.3390/app12083740

Academic Editors: Edyta Plebankiewicz and Jürgen Reichardt

Received: 14 February 2022
Accepted: 5 April 2022
Published: 8 April 2022

1. Introduction

The issue of global energy consumption is in the spotlight today. The U.S. Information Administration presents a future scenario in which global energy consumption will increase by nearly 50% over the next 30 years or so [1]. Building energy consumption accounts for 21.7% of the national energy consumption in China [2], and there is still work to be done to reduce overall energy consumption in the building industry.

Cost limits, of course, are crucial factors in restricting building energy efficiency [3]. The optimization of energy efficiency without the consideration of cost may result in high incremental costs and the inability to promote applications. As a result, energy usage and cost should be considered throughout the optimization process.

Many earlier researchers have investigated this topic using the genetic algorithm technique. I2n 2002, Caldas et al. used evolutionary algorithms on the DOE2.1E platform to improve the arrangement and size of windows in public buildings to lower the yearly energy consumption [4]. Ferrara et al. optimized near-zero-energy buildings to achieve a low economic cost [5]. Thalfeldt et al. identified the design priorities for cold-climate building facades [6]. However, related research has revealed that diverse optimization targets have conflicting characteristics [7], suggesting that focusing on a single goal is unsuitable in particular optimization schemes. As a result, an increasing number of researchers are attempting to adapt the multi-objective optimum design technique to the purpose of the architecture. With illumination and energy consumption as the optimization targets, Khoroshiltseva et al. employed modeFRONTIER and Daysim to optimize the spacing and angle of the sunshade for the south window of an office [8]. To complete the façade design while guaranteeing building performance, Mohammadjavad et al. exploited the twin aims of lighting and heating to optimize the curtain wall design parameters (surface angle) of an office building [9]. By using the Grasshopper platform, Cheng Sun et al. achieved the performance optimization of a large public building focusing on energy, cost, and daylight [10]. Shaoqing Gou et al. used the Energyplus and JEPlus platforms to create an architectural design plan for a residential project in Shanghai to improve indoor thermal comfort and lower energy consumption [11].

However, the basic models commonly employed in related studies are primarily used for non-assembled buildings—i.e., the optimized solutions often find it difficult to meet the standardization, modularity, and modulization needs of prefabricated buildings.

The trend of building industrialization has been evident in recent years. With the deepening of the concept of green and sustainable development, prefabricated buildings are receiving more and more attention from the domestic and international construction community because of their standardization, energy efficiency, and economy [12–14]. Prefabricated building envelopes can be selected to have an appropriate envelope structure based on the building orientation, climate conditions, and economic costs, among other factors, in order to achieve low energy costs, thanks to their modular design, factory manufacturing, and assembly construction [15]. The application of this approach to the design process, as well as the successful combination of genetic algorithm and assembly construction, is the focus of this study. In summary, the previous studies also had the following shortcomings:

- Few scholars have applied the synergy of energy consumption and cost to prefabricated buildings.
- The models of the former studies can mainly be divided into two categories: one is a generic model with similar characteristics to that obtained from our research (it is usually a city building and is used to propose some common optimization conclusions [4,8–10]); the other is generally a specific model, usually for a public building, and the findings primarily relate to the renovation and refurbishment of the building [5,6,11]. However, there are numerous phases in the architectural design process, including conceptual design, preliminary design, and detailed design [16]. In this article, we think that applying the two models to conceptual and preliminary design is most beneficial.
- Many of the articles in this area end up focusing on the optimization results, while in practice designers tend to make changes based on these. These articles tend to lack any discussion of which parts need to be changed to have less impact on the optimization results.

Therefore, this paper proposes two design methods based on the genetic algorithm to take building energy consumption into consideration: a floor plan generation method for the conceptual design process and a complement selection method for the preliminary design. The designer can then use the results of the optimization and parameter correlation analysis as a theoretical basis to make further modifications to the computer-aided optimization design. The total workflow of the two design methods is shown in Figure 1.

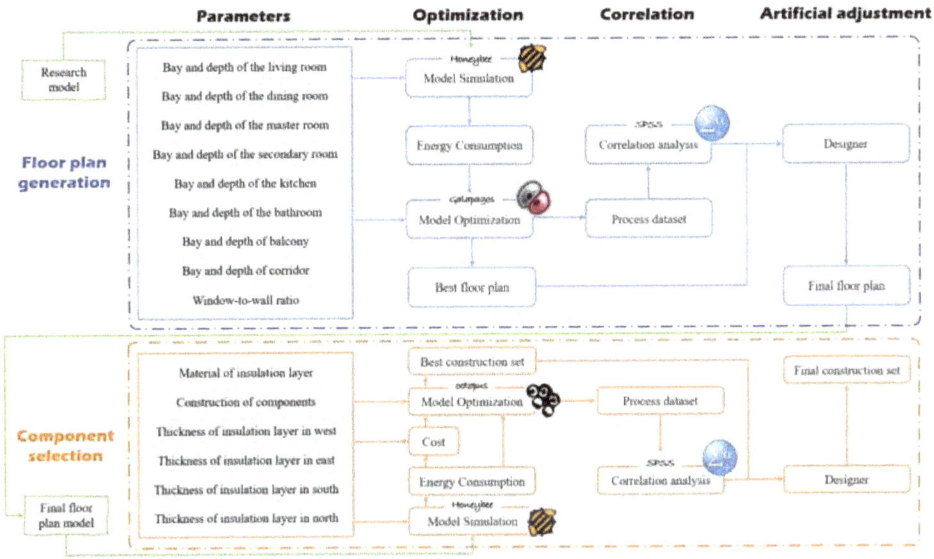

Figure 1. Optimized design workflow based on floor plan generation and component selection.

2. Methods

2.1. Optimization Method

The genetic algorithm (GA) was utilized as the optimization approach in this work; it has been frequently used in similar papers in recent years [17]. Genetic algorithms, which were researched and proposed by Professor Holland of Michigan University in 1975, are based on the theory of biological evolution and incorporate the evolutionary concepts of reproduction, hybridization, mutation, competition, and selection into the optimization process to achieve global optimization [18]. Figure 2 depicts the optimization concept [19]. This study uses the Galapagos and Octopus plug-ins integrated with the Grasshopper platform. Galapagos is a GA component that comes with the new version of Grasshopper and can perform optimization solutions for a single objective with a simple operation, fast computation, and easy convergence. However, it has the drawback of being able to optimize solutions for only one goal. Developed by the University of Applied Arts Vienna, Austria, and Bollinger + Grohmann Engineering, Germany, the Octopus plug-in is a Grasshopper component that combines Pareto frontier solution sets and GA for the optimization of multiple objectives. Pareto frontier solution sets can provide a basis for analyzing the trade-offs made between design objectives [20].

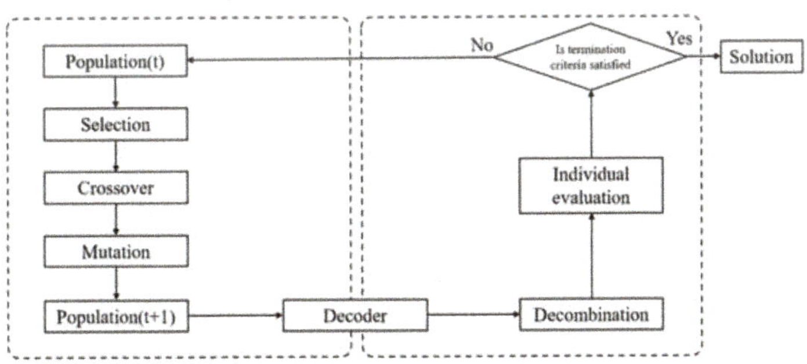

Figure 2. Genetic algorithm calculation process.

2.2. Correlation Analysis Method

2.2.1. Standard Regression Coefficient

The size of the absolute value of the standard regression coefficient, which is the regression coefficient obtained after eliminating the impact between the objective and the units of the parameters, directly represents the degree of effect of the parameters on the objective [21]. Its regression model can be expressed as Function (1):

$$Y = \sum_{i=1}^{N} \beta_i \frac{X_i - \overline{X}}{\sigma_X} + \varepsilon \tag{1}$$

where β_i is the ith parameter's standard regression coefficient, $\overline{X}$ is the ith parameter's mean, σ_X is the ith parameter's standard deviation, ε is a constant, and N is the number of parameters.

We need to test the problem of multicollinearity suggested by Frisch in 1934 over the course of the investigation [22]. To detect multicollinearity, a variety of approaches are used, including partial correlation coefficient, tolerance, variance inflation factor (VIF), and conditional index [23]. The variance inflation factor measured by the SPSS software is used to assess the aforementioned problem in this study. A result of greater than one and less than ten generally suggests that the problem does not exist [24].

2.2.2. Pearson Correlation Coefficient

The Pearson correlation coefficient, which is used to estimate the correlation between X and Y variables, can be calculated using Function (2). The coefficient takes values in the range of $[-1, 1]$: the closer it is to 1, the more likely the two variables are positively correlated; the closer it is to -1, the more likely it is that the two variables are negatively correlated; a value of 0 indicates that the two variables are uncorrelated. The article was followed up with calculations conducted using SPSS to obtain the Pearson correlation coefficient.

$$\rho_{X,Y} = \frac{cov(X,Y)}{\sigma_X \sigma_Y} \tag{2}$$

where $cov(X,Y)$ is the covariance between X and Y and σ_X and σ_Y are their respective standard deviations.

3. Model, Parameters, and Objectives

3.1. Model

In this study, the thresholds of room bays, depths (Table 1), and the laws of arrangement of each functional space (Figure 3) were summarized from several sets of house types, which were based on the prototype of prefabricated high-rise (one-staircase, two-family) commercial houses with a PC frame shear wall structure in Tianjin. After this, random values in the parameter range were used to form the base floor plan.

Table 1. Building geometric parameter threshold.

Geometric Parameters	Parameter Thresholds
Bay depth of the living room (m × m)	(2.8–3.3) × (4.0–6.0)
Bay depth of the dining room (m × m)	(2.8–3.3) × (4.0–6.0)
Bay depth of the master bedroom (m × m)	(2.4–3.0) × (3.2–3.6)
Bay depth of the secondary bedroom (m × m)	(2.1–2.4) × (2.4–2,9)
Bay depth of the kitchen (m × m)	(1.5–2.1) × (1.95–3.1)
Bay depth of the bathroom (m × m)	(1.5–1.8) × (1.8–2.65)
Bay depth of the balcony (m × m)	(1.8–3.3) × (1.2–1.5)
Bay depth of the corridor (m × m)	—
Window-to-wall ratio of each room	0.1–0.9
Interleaved length between each functional space (m)	−1.8–1.8

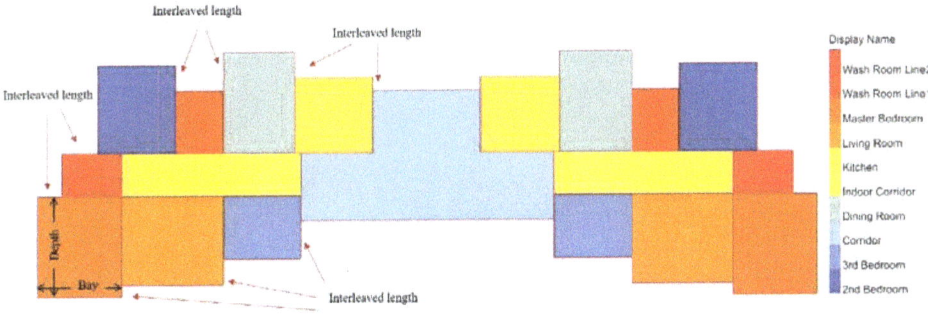

Figure 3. The laws of arrangement of each functional space.

Following that, specific common attributes are assumed to finish the model's development, and their values are provided in Table 2. After the attributes are specified, a complete 3D building model (Figure 4) can be created, which is the default model used in this study.

Table 2. The value of the attributes used in the optimization process.

The Name of the Attributes	Value
Height between floors	3 m
Number of floors	30
Envelope walls	Default values of constructs and materials
Operation schedules, equipment load	Related standards [25,26]

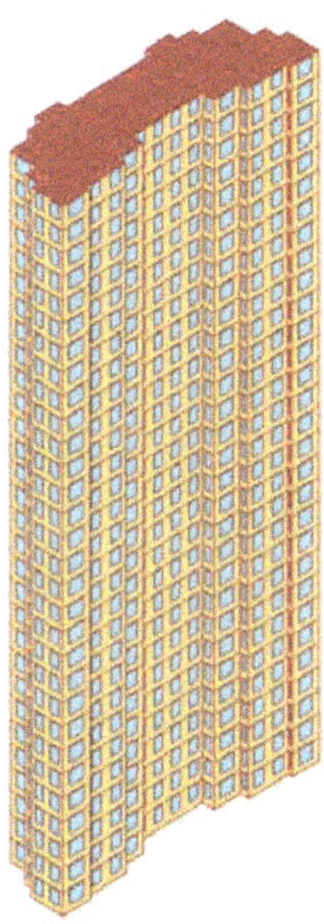

Figure 4. Three-dimensional axonometric drawing of the building model.

3.2. Parameters

3.2.1. Parameters of the Floor Plan Generation Method

The control variables required for the floor plan generation method are the window-to-wall ratio of each orientation, the bays and depths of each functional space, and the interleaved length between them. The research object of this paper is the prefabricated building. One of the core tasks to promote the degree of industrialization of the building is to adopt a modular coordinated system to achieve the universality and interchangeability of the components [27]. Therefore, a novel qualification method that considers the modulus of prefabricated buildings is proposed when defining the control variables of space types. This method can reduce the range of control variables based on the modulus to enhance the optimization efficiency and ensure the reasonable size of the prefabricated buildings. The objective function can be seen in Function (3).

$$\begin{aligned} x &= [x_1, x_2, x_3, \cdots, x_n]^T \\ k_i &= [1, 2, \cdots, \tfrac{b_i - a_i}{jM}]^T \\ \min \quad y_{ec} &= f(x) = f(x_1, x_2, \cdots, x_n) \\ s.t. \quad &\begin{cases} a_m \le x_m \le b_m \quad m \in \{1,2,3,4\} \\ x_i \in \{x | x = a_i + jM \cdot k_i\} \end{cases} \end{aligned} \tag{3}$$

where x is the vector of control variables; k_i is the step vector of x_i control variables; x_m is the window-to-wall ratio for each orientation; a_m and b_m are the minimum and maximum values of the corresponding window-to-wall ratio; x_i is the interleaved length of the bay, depth, and spaces, $i \in \{5, 6 \cdots n\}$, m; j is a constant in dimensional transformation which is used to expand the step length of the building modulus in a single transformation, $j \in \mathbb{N}^+$; and M is the basic modulus of the building at the dimensional transformation of the prefabricated building (M = 0.1 m), $\frac{b_i - a_i}{jM} \in \mathbb{Z}$.

3.2.2. Parameters of the Component Selection Method

The parameters used in this optimization are the material of the insulation layer, the thickness of the insulation layer, and the construction of the wall components, which can be divided into qualitative and quantitative indexes. The range of values shown Table 3 are determined by summarizing after researching manufacturers.

Table 3. Ranges of the component selection method's parameters.

Parameters of Components	Ranges
Material of insulation layer	{Extruded polystyrene, expanded polystyrene, foamed polyurethane, rock wool}
Construction of components	{External insulation, internal insulation, sandwich insulation}
Thickness of insulation layer in each direction (m)	0–0.5

In the optimization process of the component selection method, the physical properties and cost of each material are considered as attributes, assuming that they are constant throughout the construction phase. The specific attributes are shown in Table 4. In the optimization process, material and construction are qualitative indicators, and there are few desirable types. The optimal solution can be selected using the exhaustive method. On the basis of this solution, the thickness of the insulation material for each orientation is used as the optimization variable, while the total building energy consumption and total cost of insulation material are used as the optimization objectives for the next optimization step.

Table 4. The value of the attribute of the optimized process.

Material	Density kg/m^3	Specific Heat Capacity J/kg·°C	Heat Transfer Coefficient W/(m^2·K)	Material Price CNY/m^3
Extruded polystyrene	35	1380	0.033	450
Expanded polystyrene	25	1380	0.042	280
Foamed polyurethane	30	1380	0.027	650
Rock wool	150	1220	0.0045	260

3.3. Objectives

3.3.1. Objective of the Floor Plan Generation Method

The goal energy consumption in the floor plan generation design technique is confined to the use phase, since most buildings consume around 70% of their total energy over their whole life cycle during their use phase [28]. Lighting energy consumption and equipment energy consumption are not affected by control variables [29]. The annual energy consumption (y_{ec}) per unit area of the building optimized in this study can be calculated using Function (4):

$$y_{ec} = \frac{E_h + E_c}{A} \tag{4}$$

where E_h is the annual heating energy consumption of the buildings, kWh; E_c is the annual cooling energy consumption of the buildings, kWh; and A is the gross floor area of the buildings, m^2.

3.3.2. Objective of Component Selection Method

g_{ec} is the same as the expression of y_{ec} in the floor plan generation method. Instead, only the cost of materials used during the construction phase is considered. The changes in the dimensions of the structural material will affect the energy use of buildings. We assume that the structural material cannot be changed during the selection process, which means that the cost of the structural material is constant for the same floor plan. In order to simplify the calculation, the material cost during construction is considered only as the insulation construction cost (g_{ic}), which can be expressed through Function (5).

$$g_{ic} = \sum_{i=1}^{4} C_i d_i S_i \tag{5}$$

where i from 1 to 4 are the four orientations of the buildings; C_i is the price of the insulation board used for each facade orientation, CNY/m^3; d_i is the thickness of the insulation board used in each direction of the external wall, m; and S_i is the total area of each facade orientation, m^2.

The optimization objective function can be expressed as Function (6).

$$\begin{aligned} x &= [x_1, x_2, x_3, \cdots, x_n]^T \\ \min \quad g &= g(x) = \{g_{ec}(x), g_{ic}(x)\} \\ s.t. \quad x &\in B = \{x | h_s(x) \leq 0, s = 1, 2, \cdots, p\} \end{aligned} \tag{6}$$

where x is the vector of control variables; g is the vector of objectives; and $h_s(x)$ is the sth constraint of the vector x, from which the feasible domain B is formed.

After completing the above settings, we can obtain the details of the changes in parameters and properties for each phase of the complete workflow, as shown in Figure 5.

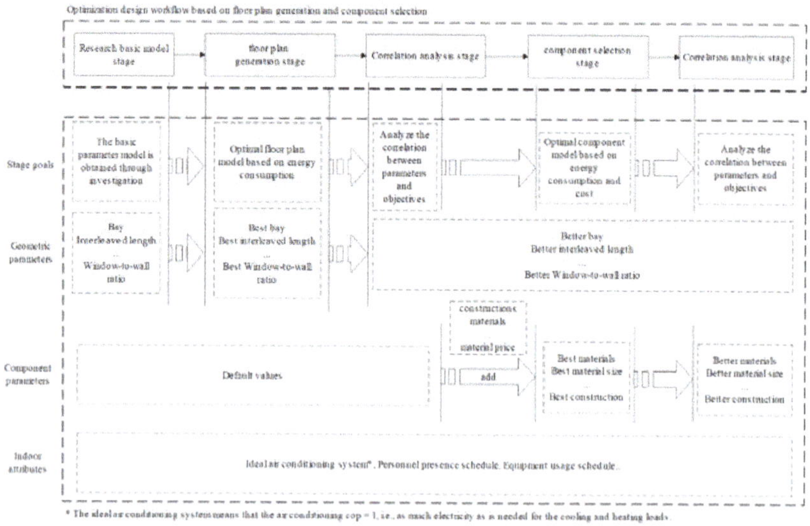

Figure 5. Goals, parameters, and attributes of the different phases of the floor plan generation and component selection methods.

4. Results and Discussion

Two design methods were applied on the above case and then the correlation between some parameters or attributes with the energy consumption was analyzed.

4.1. Optimization Results

In the floor plan generation method, the specific range of values for window-to-wall ratio, bays, depths, and interleaved length parameters could be determined using Function (3). In this case, $j = 3$. The threshold was adjusted appropriately according to the limitation of $\frac{b_i - a_i}{jM} \in \mathbb{Z}$. Finally, the bays, depths, and interleaved lengths were limited to a small range of values. The Galapagos parameters were set as stated in Table A1 during the optimization process, and convergence was mostly obtained at around 80 iterations, with the optimization ending after around 120 generations. Table 5 shows a comparison of the model before and after the plan optimization with improved energy consumption as the aim. Table A1 in the Appendix A shows the detail of optimization technique used. It can be seen that after the optimization, the building energy consumption and interleaved length between each functional space are reduced.

Table 5. Model comparison before and after optimization.

Number of Iterations	0	120
Plan shape		
Energy consumption (kWh/m^2)	86.18	18.11

In the optimization of the component selection method, the model was inherited from the previous optimization step without any modification and then the enumeration method was used to optimize the material and structure. The Octopus parameters were set as stated in Table A3 during the optimization process. The step size of the thickness in the optimization process was 0.05 m and the distribution was between 0 and 0.15 m. The curves of the energy consumption and cost for different insulation materials with different constructions is shown in Figure 6.

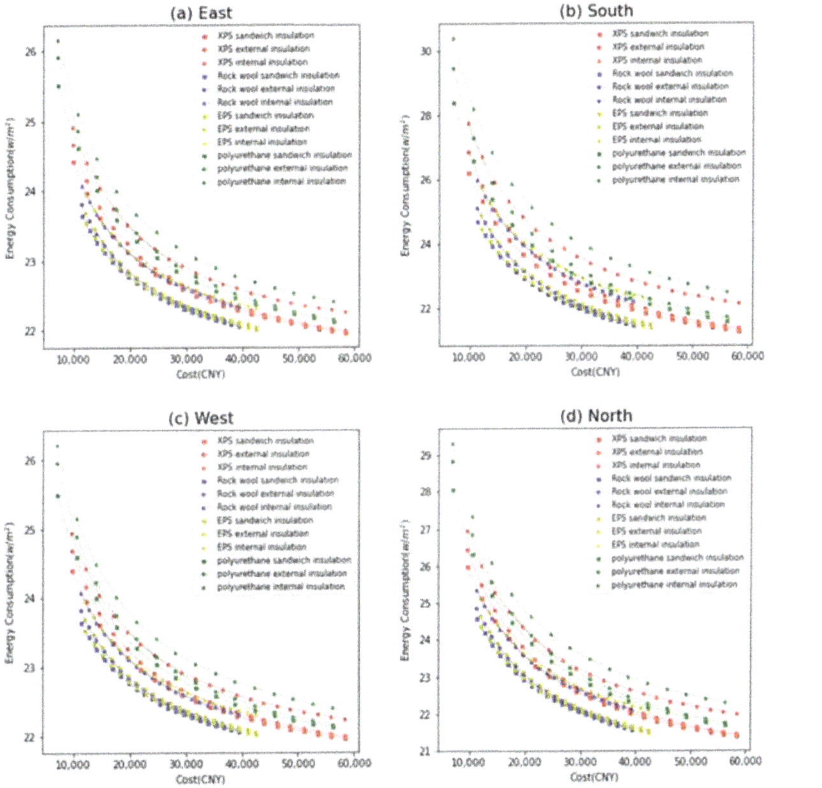

Figure 6. Energy consumption and cost charts of various orientations and different insulation materials.

The energy-cost charts for the four orientations show the same pattern. Rock wool board material and polystyrene board material were closer to the 0-coordinate point of the coordinate system—i.e., they were better than the other insulation materials in terms of their reducing energy consumption and cost. Considering the cost of fire protection, rock wool board was selected as the optimal material in the next dual-objective optimization. From Figure 6, it can be seen that the two innermost curves are for sandwich insulation, which indicates that the effect of sandwich insulation is better than that of external insulation and internal insulation. Thus, sandwich insulation was chosen as the construction method for the optimization determination.

The thickness was selected as the optimization object, and basic convergence was achieved after 10 iterations. The Pareto frontier solution set (Figure 7) was derived after reaching the maximum number of iterations, and correlation plots between insulation thickness, total energy consumption, and total cost for each orientation were derived (Figure 8).

From the Pareto frontier solution set, it can be seen that the cost kept increasing and the total energy consumption kept decreasing as the total thickness increased within a certain thickness range, in accordance with the objective law.

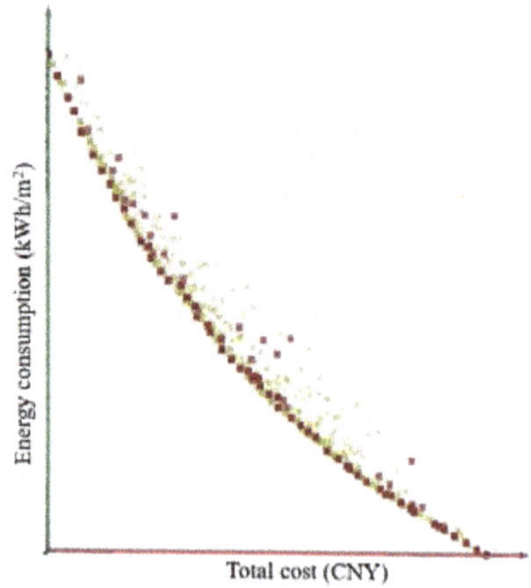

Figure 7. Pareto frontier solution set.

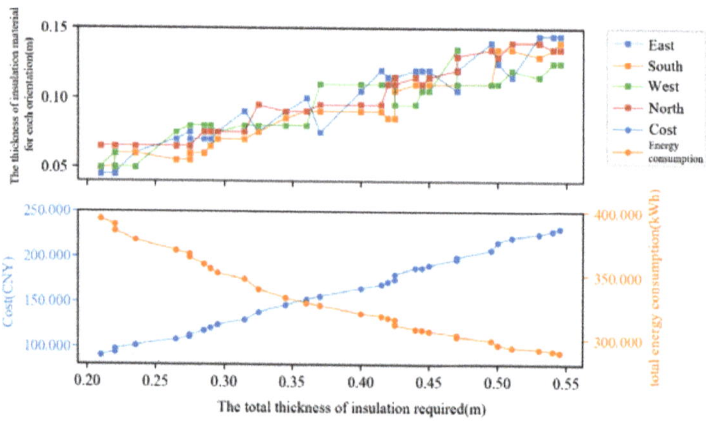

Figure 8. Chart of thickness–total energy consumption–cost.

4.2. Correlation Analysis Results

4.2.1. Correlation Analysis of Floor Plan Generation Method

Eleven influencing factors were selected as variables for regression analysis from the parameters that may affect energy consumption—namely, interleaved length, east window wall ratio, south window wall ratio, west window wall ratio, north window wall ratio, east exterior wall area, south exterior wall area, west exterior wall area, north exterior wall area, shape coefficient, and aspect ratio. After exporting 3680 sets of data from the optimization process and eliminating 3 sets of invalid data, we obtained 3677 sets of valid data. The validity of the selected variables was first determined by testing the multicollinearity; if the variance inflation factor (VIF) of each variable was tested to be less than 10 (Table 6), this meant that each variable had a certain degree of independence. A significance test was then performed, yielding an overall p-value of 0.0001 less than 0.05—i.e., the proposed model was valid at a 95% confidence interval. Additionally, the p-value for each variable (Table 7) was less than 0.05—i.e., each variable was significant at a 95% confidence interval. The larger the standard regression coefficient of a variable is, the more important the variable is under the same condition. The variables are ranked in Table 7, and it can be seen that parameters such as the shape coefficient and the window-to-wall ratio are more important than the area of the exterior walls of each orientation—i.e., when making adjustments, the radiant area of the walls and windows of each orientation can be appropriately altered while controlling the shape coefficient and the window-to-wall ratio.

Table 6. VIF value of each parameter.

	Aspect Ratio	East Window-to-Wall Ratio	West Window-to-Wall Ratio	Interleaved Length	South Window-to-Wall Ratio	South Exterior Wall Area	Shape Coefficient [1]	North Exterior Wall Area	North Window-to-Wall Ratio	West Exterior Wall Area	East Exterior Wall Area
VIF	7.79	5.71	5.61	5.33	4.62	4.37	4.21	3.94	3.67	2.22	2.2

[1] Shape coefficient: the ratio of the exterior area of a building in contact with the outdoor atmosphere to the volume it encloses.

Table 7. Standard regression coefficient of each parameter after standard regression.

	Shape Coefficient	North Window-to-Wall Ratio	South Window-to-Wall Ratio	West Window-to-Wall Ratio	East Window-to-Wall Ratio	Interleaved Length	South Exterior Wall Area	Length-Width Ratio	North Exterior Wall Area	East Exterior Wall Area	West Exterior Wall Area
p-value	0	0	0	0	0	0	0	0	0	0	0.012
Coefficient	0.492	0.221	0.209	0.168	0.142	0.071	0.0465	0.029	0.020	0.015	0.004

4.2.2. Correlation Analysis of Component Selection Method

In the phase of correlation analysis, only the relationship between the percentage of the thickness of the insulation in each orientation to the total insulation thickness and the total energy consumption is required, given that the types of material and construction have already been determined (Table 8).

Table 8. Comparison of the model before and after the generation of building shape.

Thickness of Insulation Layer in Each Direction/Total Thickness	Pearson Correlation
East ratio	−0.427
West ratio	0.361
South ratio	−0.252
North ratio	0.318

A negative correlation could be observed between the insulation thickness in the east/south directions and total energy consumption—i.e., increasing the proportion of the insulation in the east and south directions will decrease the total energy consumption; conversely, increasing the proportion of the insulation in the west and north directions will increase the total energy consumption. From a correlation point of view, in order to reduce the total building energy consumption, the insulation of east- and south-oriented buildings should be appropriately increased and the insulation of west- and north-oriented buildings should be reduced under a certain cost limit.

5. Conclusions

1. Simulation-based single-objective or multi-objective optimization can be performed for prefabricated buildings. Unlike traditional buildings, the building modulus and component selection need to be considered in the optimization process. This not only meets the demand for the standardization of prefabricated buildings, but also increases the speed of optimization computation through reducing the number of values taken from parameters.
2. A novel, modular parametric modeling approach was proposed and applied in the floor plan generation method. After this, the optimal generation of prefabricated high-rise buildings in Tianjin was completed based on this method. The correlation between each parameter and energy consumption was also studied, and it was concluded that the shape coefficient and window-to-wall ratio are the main factors affecting the energy consumption of the buildings in Tianjin.
3. A preliminary component selection method based on computer simulation was proposed—i.e., the component selection for the prefabricated building was mainly

carried out to determine the construction of exterior walls, the selection of insulation materials, and the thickness of the insulation layer. By optimizing the generated models, it was finally concluded that sandwich insulation constructions and rock wool board insulation materials should be selected for buildings in Tianjin. According to the correlation analysis, the thickness of the insulation material in the east and south directions should be increased under a certain cost limit in order to reduce the total energy consumption of buildings.

Author Contributions: Conceptualization, J.G.; methodology, M.L.; software, M.L.; validation, J.G.; formal analysis, Z.J.; investigation, M.L.; resources, Z.J.; data curation, Z.J.; writing—original draft preparation, M.L., Z.W. and Y.Z.; writing—review and editing, M.L., Z.W. and Y.Z.; visualization, M.L., Z.J. and Z.W.; supervision, J.G.; project administration, J.G.; funding acquisition, J.G. All authors have read and agreed to the published version of the manuscript.

Funding: This work was supported by the project funded by Tianjin Natural Science Foundation (Grant No. 18JCQNJC08200).

Conflicts of Interest: The authors declare no conflict of interest.

Appendix A

Table A1. Parameter setting of Galapagos.

Project	The Maximum Number of Iterations	Population Size	Multiplier of Initial Boost
Value	120	30	2
Project	Proportion of retained elites	Crossover ratio	
value	5%	75%	

Table A2. Tianjin area floor plan generation process results.

Number of Iterations	0	4	8	12
Plan shape				
Energy consumption (kWh/m^2)	86.18	70.54	63.13	51.97
Number of iterations	15	19	23	27
Plan shape				
Energy consumption (kWh/m^2)	42.86	34.52	28.09	27.55
Number of iterations	31	35	39	43

Table A2. *Cont.*

Number of Iterations	**0**	**4**	**8**	**12**
Plan shape				
Energy consumption (kWh/m^2)	27.14	26.78	26.21	25.95
Number of iterations	47	51	55	59
Plan shape				
Energy consumption (kWh/m^2)	24.51	24.19	24.01	23.88
Number of iterations	62	66	70	74
Plan shape				
Energy consumption (kWh/m^2)	23.31	22.89	22.45	22.13
Number of iterations	78	82	85	89
Plan shape				
Energy consumption (kWh/m^2)	21.95	21.15	20.92	20.3
Number of iterations	93	97	101	105

Table A2. *Cont.*

Number of Iterations	0	4	8	12
Plan shape				
Energy consumption (kWh/m^2)	19.9	19.6	19.22	18.87
Number of iterations	108	112	116	120
Plan shape				
Energy consumption (kWh/m^2)	18.73	18.66	18.55	18.11

Table A3. Parameter setting of Octopus.

Project	The Maximum Number of Iterations	Population Size	Multiplier of Initial Boost
Value	50	60	50%
Project	Ratio of variation	Crossover ratio	
Value	50%	80%	

References

1. Park, C.; Kim, C.; Lee, S.; Lim, G.; Lee, S.; Choi, Y. Effect of Control Strategy on Performance and Emissions of Natural Gas Engine for Cogeneration System. *Energy* **2015**, *82*, 353–360. [CrossRef]
2. China Construction Energy Conservation Association 2020 China Building Energy Consumption Research Report. Available online: https://www.cabee.org/site/content/24021.html (accessed on 14 February 2022).
3. Long, H. Inquiry of restrictive factors for green building development. *Jiangxi Build. Mater.* **2021**, *2*, 189–191.
4. Caldas, L.; Norford, L. A Design Optimization Tool Based on a Genetic Algorithm. *Autom. Constr.* **2002**, *11*, 173–184. [CrossRef]
5. Ferrara, M.; Fabrizio, E.; Virgone, J.; Filippi, M. A Simulation-Based Optimization Method for Cost-Optimal Analysis of Nearly Zero Energy Buildings. *Energy Build.* **2014**, *84*, 442–457. [CrossRef]
6. Thalfeldt, M.; Pikas, E.; Kurnitski, J.; Voll, H. Facade Design Principles for Nearly Zero Energy Buildings in a Cold Climate. *Energy Build.* **2013**, *67*, 309–321. [CrossRef]
7. Znouda, E.; Ghrab-Morcos, N.; Hadj-Alouane, A. Optimization of Mediterranean Building Design Using Genetic Algorithms. *Energy Build.* **2007**, *39*, 148–153. [CrossRef]
8. Khoroshiltseva, M.; Slanzi, D.; Poli, I. A Pareto-Based Multi-Objective Optimization Algorithm to Design Energy-Efficient Shading Devices. *Appl. Energy* **2016**, *184*, 1400–1410. [CrossRef]
9. Mandavinejad, M.; Nazar, N.S. Daylightophil High-Performance Architecture: Multi-Objective Optimization of Energy Efficiency and Daylight Availability in BSk Climate. In Proceedings of the International Conference—Alternative and Renewable Energy Quest (areq 2017), Barcelona, Spain, 1–3 February 2017; Amer, M., Sotoca, A., Nasselli, F., Mohareb, N., Alalouch, C., Eds.; Elsevier Science Bv: Amsterdam, The Netherlands, 2017; Volume 115, pp. 92–101.
10. Sun, C.; Liu, Q.; Han, Y. Many-Objective Optimization Design of a Public Building for Energy, Daylighting and Cost Performance Improvement. *Appl. Sci.* **2020**, *10*, 2435. [CrossRef]

11. Gou, S.; Nik, V.M.; Scartezzini, J.-L.; Zhao, Q.; Li, Z. Passive Design Optimization of Newly-Built Residential Buildings in Shanghai for Improving Indoor Thermal Comfort While Reducing Building Energy Demand. *Energy Build.* **2018**, *169*, 484–506. [CrossRef]
12. Bingbing, Z. Application analysis of domestic and foreign application. *Tiles Bricks* **2017**, *9*, 47–48. [CrossRef]
13. Yan, J. prefabricated building standardization helps to promote the overall development of residential industrialization. *Archit. Des. Manag.* **2015**, *32*, 44–45.
14. Goldfinch, D.A. Health Centre Design; Prefabricated Construction. *Hosp. Health Manag.* **1949**, *12*, 11–14. [PubMed]
15. Yu, J.; Yang, C.; Tian, L.; Liao, D. A Study on Optimum Insulation Thicknesses of External Walls in Hot Summer and Cold Winter Zone of China. *Appl. Energy* **2009**, *86*, 2520–2529. [CrossRef]
16. Jin, R.; Zhong, B.; Ma, L.; Hashemi, A.; Ding, L. Integrating BIM with Building Performance Analysis in Project Life-Cycle. *Autom. Constr.* **2019**, *106*, 102861. [CrossRef]
17. Kheiri, F. A Review on Optimization Methods Applied in Energy-Efficient Building Geometry and Envelope Design. *Renew. Sust. Energ. Rev.* **2018**, *92*, 897–920. [CrossRef]
18. Ming, Z.; Shudong, S. *Genetic Algorithm Principle and Application*; National Defense Industry Press: Beijing, China, 1999; p. 77.
19. Jie, S. Application Research on Green Building Techniques Based on Grasshopper. Master's Thesis, South China University of Technology, Guangzhou, China, 2012.
20. Mattson, C.A.; Messac, A. Pareto Frontier Based Concept Selection under Uncertainty, with Visualization. *Optim. Eng.* **2005**, *6*, 85–115. [CrossRef]
21. Haiyan, W.; Fangting, Y.; Lu, L. Comparison and Application of Standardization Coefficient and Delay Comparison Coefficient. *Quant. Econ. Tech. Econ. Res.* **2006**, *9*, 150–155.
22. Farrar, D.E.; Glauber, R.R. Multicollinearity in Regression Analysis: The Problem Revisited. *Rev. Econ. Stat.* **1967**, *49*, 92–107. [CrossRef]
23. Xia, X.; Xingguo, W. Multi-contiguous linear geometric explanation in linear regression. *Stat. Decis.* **2021**, *37*, 46–51. [CrossRef]
24. Multicollinearity Test Example Using SPSS. Available online: http://www.spsstests.com/2015/03/multicollinearity-test-example-using.html (accessed on 16 March 2022).
25. *JGJ/T449-2018*; Standard for Green Performance Calculation of Civil Buildings. Ministry of Housing and Urban-Rural Development of the People's Republic of China, China Architecture & Building Press: Beijing, China, 2018.
26. *GB50176-2016*; Code for Thermal Design of Civil Building. Ministry of Housing and Urban-Rural Development of the People's Republic of China, China Architecture & Building Press: Beijing, China, 2016.
27. Yan, K. China Residential Standardization History and Prospect. *China Constr.* **2007**, *6*, 22–24.
28. Noori, M.; Ghattas, R.; Gregory, J.; Miller, T.; Olivetti, E.; Greene, S. Life Cycle Assessment for Residential Buildings: A Literature Review and Gap Analysis Rev. 1. 2016. Available online: https://www.researchgate.net/publication/309034390_Life_Cycle_Assessment_for_Residential_Buildings_A_Literature_Review_and_Gap_Analysis_Rev_1/citations (accessed on 16 March 2022).
29. Zhenyu, Y.; Fei, L.; Yu, Z.; Wei, X.; Dening, S. Application of Multi-Objective Optimization Method Based on Multi-Objective Optimization Design of Near Zero Energy Consumption Building. *Archit. Sci.* **2019**, *35*, 8–15. [CrossRef]

mathematics

Article

Parameters Optimization of Taguchi Method Integrated Hybrid Harmony Search Algorithm for Engineering Design Problems

Esra Uray [1,*], Serdar Carbas [1,2], Zong Woo Geem [3,*] and Sanghun Kim [4]

1 Department of Civil Engineering, KTO Karatay University, Konya 42020, Turkey; scarbas@kmu.edu.tr
2 Department of Civil Engineering, Karamanoglu Mehmetbey University, Karaman 70200, Turkey
3 College of IT Convergence, Gachon University, Seongnam 13120, Korea
4 Department of Civil and Environmental Engineering, Temple University, Philadelphia, PA 19122, USA; sanghun.kim@temple.edu
* Correspondence: esra.uray@karatay.edu.tr (E.U.); geem@gachon.ac.kr (Z.W.G.)

Abstract: Performance of convergence to the optimum value is not completely a known process due to characteristics of the considered design problem and floating values of optimization algorithm control parameters. However, increasing robustness and effectiveness of an optimization algorithm may be possible statistically by estimating proper algorithm parameters values. Not only the algorithm which utilizes these estimated-proper algorithm parameter values may enable to find the best fitness in a shorter time, but also it may supply the optimum searching process with a pragmatical manner. This study focuses on the statistical investigation of the optimum values for the control parameters of the harmony search algorithm and their effects on the best solution. For this purpose, the Taguchi method integrated hybrid harmony search algorithm has been presented as an alternative method for optimization analyses instead of sensitivity analyses which are generally used for the investigation of the proper algorithm parameters. The harmony memory size, the harmony memory considering rate, the pitch adjustment rate, the maximum iteration number, and the independent run number of entire iterations have been debated as the algorithm control parameters of the harmony search algorithm. To observe the effects of design problem characteristics on control parameters, the new hybrid method has been applied to different engineering optimization problems including several engineering-optimization examples and a real-size engineering optimization design. End of extensive optimization and statistical analyses to achieve optimum values of control parameters providing rapid convergence to optimum fitness value and handling constraints have been estimated with reasonable relative errors. Employing the Taguchi method integrated hybrid harmony search algorithm in parameter optimization has been demonstrated as it is a reliable and efficient manner to obtain the optimum results with fewer numbers of run and iteration.

Keywords: hybrid harmony search algorithm; Taguchi method; algorithm control parameter optimization; engineering design problems; reinforced cantilever retaining wall design

Citation: Uray, E.; Carbas, S.; Geem, Z.W.; Kim, S. Parameters Optimization of Taguchi Method Integrated Hybrid Harmony Search Algorithm for Engineering Design Problems. *Mathematics* **2022**, *10*, 327. https://doi.org/10.3390/math10030327

Academic Editor: Alfredo Milani

Received: 4 December 2021
Accepted: 14 January 2022
Published: 21 January 2022

1. Introduction

The well-functioning optimum designs, which aim to reach stable and economic or productive mechanisms in engineering regulation, are based on mathematical theorems and approaches. While optimization methods were applied by Newton, Lagrange, Cauchyeski, and so on for smaller-sized problems in ancient times, today produce the solutions with improved or hybrid versions of the optimization algorithms for large-size complex engineering designs. In this conjuncture, metaheuristic optimization algorithms that enable them to achieve reasonable solutions in a shorter time have been commonly employed in complicated engineering designs, since environmental and global phenomena due to developing technology and increasing population have been raised in the last two decades [1]. Although each of them adopts a different process and texture within itself, many effective and robust metaheuristic optimization algorithms hitherto have been developed dealing with better optimization processes than previous ones.

Metaheuristic optimization methods are the algorithms that generate solutions to large-scale design optimization problems which are inspired by natural events such as swarms (bird, fish, etc.), physics, evolution, or uniqueness [2]. Major metaheuristic optimization algorithms improved by mimicking the characteristics and feeling of swarms that try to survive and meet some needs such as nutrition, defense, and migration in nature are the ant colony optimization (ACO) [3], the particle swarm optimization (PSO) [4], the artificial bee colony algorithm (ABC) [5], and the whale optimization algorithm (WOA) [6]. While the gravitational search algorithm (GSA) [7] and big bang-big crunch algorithm (BB-BC) [8] are evaluated as based on physics, optimization methods such as the cuckoo search algorithm (CSA) [9], the firefly algorithm (FA) [10], and the bat algorithm (BA) [11] are inspired by animals' nature. The differential evolution (DE) [12] and the biogeography-based optimization (BBO) [13] are based on evolution concepts such as the genetic algorithm (GA) [14] and the simulated annealing algorithm (SA) [15].

As different from the other algorithms the harmony search algorithm (HSA) presented by Geem et al. [16] is based on the music and mimics the process of finding the best harmony of the notes performed by musicians' intuition. The HSA, which is a powerful and effective optimization method because of its simple algorithm scheme, gives fast results, has an easy-to-apply algorithm, has been exceedingly employed by the researchers for design optimization analysis. Thanks to the implementation of the algorithm to design optimization problems effectively and convergence achievement of optimum solutions, hybrid and improved versions of HSA have been employed in the several fields of civil, electrical, industrial, software, mechanical engineering, scheduling, clustering, networking, image processing, and so on [2].

To boost the convergence performance of metaheuristic algorithms and their capacity to produce solutions with fewer iterations, improved versions [17,18] and hybrid versions of algorithms [19–21] combined each other have been proposed by researchers. In Figure 1a, the number and the percentage of conducted studies considering hybrid optimization algorithms in literature [22] are demonstrated as a comparison graph by years. It is seen that the usage of hybrid metaheuristic optimization algorithms has increased considerably in the last two decades. The distribution of hybrid optimization studies for the different fields has been examined by utilizing the Web of Science database and obtained results are given in Figure 1b [22]. Although other studies except for the fields given in the figure correspond to 78% out of whole fields, it is obvious that the hybrid optimization algorithm studies have a considerable extent of usage in, especially multidisciplinary engineering studies. Hybrid HSA studies included setting algorithm parameters and hybridization of HSA with other metaheuristic algorithms as well as collocation of the artificial intelligence algorithms, which are depicted in Figure 1c as the result of a comprehensive literature survey [23]. According to the graph, harmony search hybrid optimization studies carried out in the last five years being 72% out of all studies published between 2008 and 2021 shows that hybrid studies of HSA substantially have been preferred by researchers. Results given in Figure 1 belong to all types of studies such as research articles, proceeding papers, early access, book chapters, review articles, and so on.

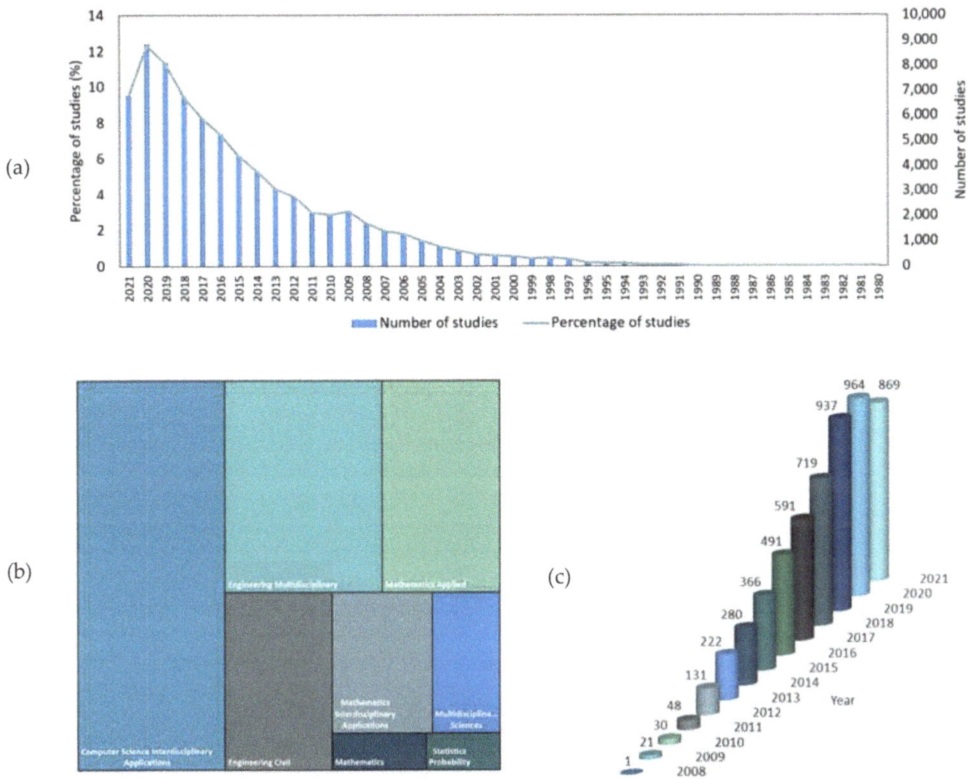

Figure 1. Web of Science citation report studies in literature following: (**a**) Change between published years of hybrid optimization studies and numbers with percentages; (**b**) Distribution of hybrid optimization studies according to fields; (**c**) Change between published years of hybrid and based on harmony search optimization studies and numbers.

In the literature, the number of optimization studies carried out utilizing metaheuristic algorithms and their improved or hybrid versions so far is mainly due to the researcher's effort to reach better convergence to the optimum solution. The literature survey has been demonstrated the popularity of these algorithms in applying engineering design problems even for real-size complex ones. While the possibility of finding new solutions has increased by adding some algorithm parameters to the optimum search process, formed mathematical expressions combining two or more optimization algorithms effectively enable to reach optimum results. Even though new or hybrid versions of metaheuristic algorithms have been suggested in this manner, investigating the reasonable values of the current algorithm parameters is an important issue for convergence to the optimum with fewer computational attempts. To investigate reasonable factors, design comprehensive sensitivity analyses have been performed considering different values of algorithm control parameters [24]. Although sensitivity analyses are employed as a path for the researcher to converge to the optimum result, it takes time because it follows a trial-and-error method. In addition, it can't guarantee the appropriate value of a parameter when it is closest to the optimum solution. In most of the studies using metaheuristic algorithms, the values of the algorithm parameters are chosen by referring to the studies in the literature. As it may vary depending on discrete-continuous design variables, constraint-unconstrained cases, and the size of the current design problem with the numbers of design variables and constraints and so on in the search for the appropriate value of the metaheuristic algorithm parameter, it would be a better manner to find algorithm parameter values considering the current handled optimization problem. According to a study presented by Uray et al. [25] which investigated optimum values of the scatter search algorithm parameters by the Taguchi method, it has been seen that it is possible to estimate the statistically appropriate values of the algorithm parameters according to a selected objective.

One of the statistical experimental design methods commonly utilized to investigate the parameter effect on the quality in the manufacturing or design process of goods is the Taguchi method [26,27]. Thanks to this successful and robust design method, it can estimate the optimum value of considering effective parameters according to specific response values depending on desired aim. Studies for Taguchi method hybridization of the metaheuristic optimization algorithms such as the simulated annealing (SA) algorithm [28], the genetic algorithm (GA) [29,30], and particle swarm optimization (PSO) [31] are instances to overcome problems encountered in their field and obtained better results. In the study, which is conducted shape optimization design by employing Taguchi method hybrid version with the HSA, more optimum design variable values have been acquired regardless of the investigating for optimum values of the HSA parameters [32]. In the study which used statistical mathematical models improved by considering the Taguchi method employed as objective function and design constraints, the optimum design of the cantilever retaining wall has been investigated via HSA [33]. According to the extensive research results in the literature, no study has been found in which the optimum values for the number of runs and the number of maximum iterations with optimum HSA control parameters have been investigated based on the Taguchi method with different engineering problems.

Thus, in this study, some of the considered complex benchmark engineering design optimization problems and a real-size engineering design optimization problem have been employed to examine HSA parametric effect and to investigate the optimum values of algorithmic parameters. In this scope, statistical and optimization analyses to be presented in this paper have been conducted as follows:

- The effect of variable run values on finding the optimum solution by employing different complex benchmark engineering design optimization problems and a real-size engineering design problem, frequently considered in optimization analyzes in the literature has been investigated;
- Taguchi method integrated hybrid harmony search algorithm (TIHHSA) has been generated based on the HSA and Taguchi method, namely the proposed hybridization can be defined as initial optimization for optimum algorithm parameter values of HSA;
- The effect of HSA parameters on the objective function and the optimum number of runs and maximum iterations with optimum HSA control parameters have been examined for different engineering optimization design problems utilizing TIHHSA;
- Whether the variation of the optimum values of the HSA parameters depending on the nature of the engineering design optimization problem has been evaluated;
- According to accomplished optimum results for engineering design optimization problems, the robustness, and the benefits of TIHHSA presented a new method have been interpreted and evaluated with previously reported studies in the literature.

2. Materials and Methods

2.1. Complex Benchmark Engineering Design Optimization Problems

In this section, the welded beam design (WBD), the pressure vessel design (PVD), the gear train design (GTD), and the speed reducer design (SRD) engineering design optimization problems demonstrated in Figure 2 have been presented with their design variables, constraints, and objective functions.

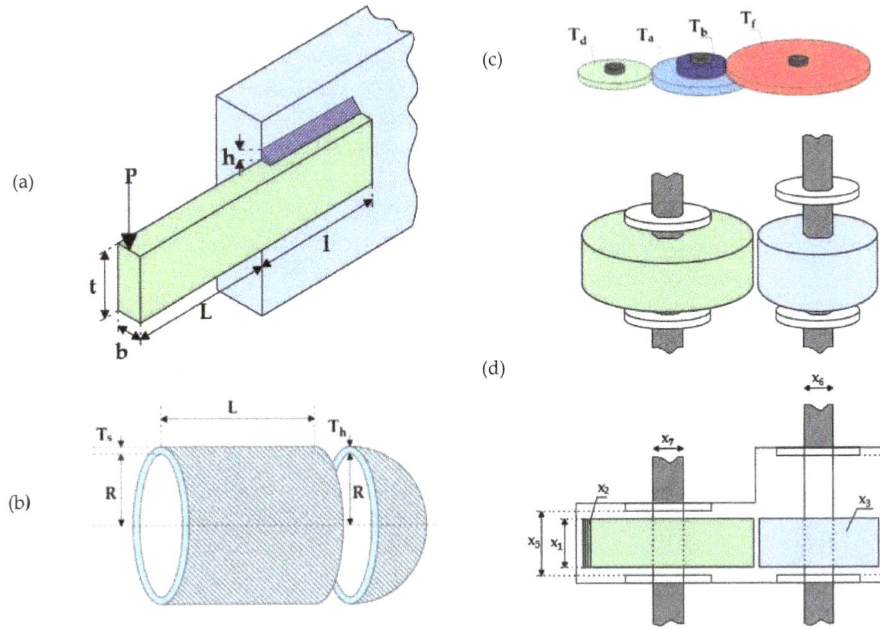

Figure 2. Engineering design optimization problems: (**a**) Welded beam design problem (WBD); (**b**) Pressure vessel design problem (PVD); (**c**) Gear train design problem (GTD); (**d**) Speed reducer design problem (SRD).

2.1.1. Welded Beam Design Problem

The first considered benchmark engineering design example is the design of the weld joint of thickness h and length l between the bar and beam with cross-section b x t and the total length L + l (Figure 2a). The welded beam benchmark engineering design problem (WBD) [34], which considers the shear stress in the weld (τ), bending stress in the beam (σ), buckling load on the bar (Pc), end deflection of the beam (δ), and side constraints for the minimum cost, are employed in the optimization analyses (Equation (A1), Appendix A section). Design variables are the thickness of the weld (h) as x_1 within the range [0.1 in., 2 in.], the length of the welded joints (l) as x_2 within the range [0.1 in., 10 in.], the width of the beam (t) as x_3 within the range [0.1 in., 10 in.], and the thickness of the beam (b) as x_4 within the range [0.1 in., 2 in.].

2.1.2. Pressure Vessel Design Problem

The pressure vessel design (PVD) problem [35], in which a cylindrical pressure vessel is capped with hemispherical heads at both ends of a vertical cylindrical shell by using the welded joint, is demonstrated in Figure 2b. For PVD, the thickness of the shell (T_s), the thickness of the head (T_h), inner radius (R), and the length of the cylindrical section of the vessel (without including the head) are treated as the design variables. It is one of the well-known complex engineering design optimization problems that gives its minimum cost, including material, forming, and welding, under the influence of constraints such as the thickness of heads and shell, certain values of working pressure, volume, and shell length.

While the design variables of R (x_3) and L (x_4) are taken as continuous, which are between ranges in [10 in., 200 in.], the discrete integer design variables as multiples of 0.0625 between ranges in [(0.0625 $\times$ 99) in., (0.0625 $\times$ 99) in.] are considered for T_s (x_1) and T_h (x_2), due to the available thicknesses of rolled steel plates (Equation (A2), Appendix A section).

2.1.3. Gear Train Design

Sandgren [35] introduced the gear train design (GTD) with discrete and integer design variables, then it has been treated as an engineering design optimization problem to research the numbers of teeth on each gear with the desired gear ratio. The output shaft's angular velocity ratio to the input shaft's angular velocity should be close to 1/6.931 for the desired gear ratio. In the GTD problem, each design variable corresponds to T_a (x_1), T_b (x_2), T_d (x_3), and T_f (x_4), which takes a value between 12 and 60 as an integer due to considering the number of them (Figure 2c).

The objective function without constraints, which aims to minimize the difference between desired gear ratio and the current gear ratio, is given by Equation (A3) (Appendix A section).

2.1.4. Speed Reducer Design

Speed reducer design (SRD), one of the complex benchmark engineering design optimization problems, was first studied by Golinski [36]. The SRD problem satisfies eleven constraints at the minimum gear box's weight and is accepted as a benchmark for the new metaheuristic optimization methods. The design consists of gears between the engine and propeller working at its most efficient speed of rotating with seven design variables. In the design problem demonstrated in Figure 2d, face width, b (x_1), teeth module, m (x_2), number of pinion teeth (x_3), shaft length 1 (x_4), shaft length 2 (x_5), shaft diameter 1 (x_6), and shaft diameter 2 (x_7) are considered as design variables.

Design variables of the design problem are determined following ranges, [2.6 cm, 3.6 cm] is for x_1, [0.7 cm, 0.8 cm] is for x_2, [17 pieces, 28 pieces] is for x_3, [7.3 cm, 8.3 cm] is for x_4 and x_5, [2.9 cm, 3.9 cm] is for x_6, and [5.0 cm, 5.5 cm] is for x_7. Mathematical formulations for the objective function and the constraints include the limits on the bending stress of the gear teeth, surface stress, transverse deflections of shafts 1 and 2 due to transmitted force, and stresses in shafts 1 and 2 (Equation (A4), Appendix A section).

2.2. Real-Size Engineering Design Optimization Problem

In today's world, where obtaining the most economical designs in a short time gains importance, metaheuristic optimization algorithms have become an alternative method. In this context, Afzal et al. [37] have reported that hundreds of retaining wall design optimization studies for solving such real-life designs were conducted in the literature. In geotechnical engineering, the design of a cantilever retaining wall is a complex engineering problem used to provide stability against lateral soil loads that happen between two soil levels. Furthermore, the trial-error method utilized in the traditional wall design is challenging, and finding the safe design is time-consuming considering many iterations due to the existence of various soil and slope properties.

The reinforced concrete cantilever retaining wall design (RCRW) (Figure 3) has been selected as a real-size engineering design optimization problem because of the abovementioned cases. In investigating optimum RCRW designs, Building Code Requirements for Structural Concrete (ACI 318-05) and commentary (ACI 318R-05) [38] have been considered as design provisions for stable and safe design. Arranged mathematical expressions by investigating some of the optimum RCRW studies in the literature [39–43] have been presented in this section. In the design problem demonstrated in Figure 3a, base width (x_1), toe extension, (x_2), stem bottom width (x_3), stem top width (x_4), base thickness (x_5), key distance from toe (x_6), key width (x_7), key thickness (x_8), vertical steel area in the stem per unit length of the wall (x_9), horizontal steel area of the toe slab (x_{10}), horizontal steel area of the heel slab (x_{11}), and vertical steel area of the shear key per unit length of the wall (x_{12}) are considered as design variables in the design optimization of an RCRW. The RCRW design stability conditions taken as design constraints in the optimization process are checked according to acting loads on the wall demonstrated in Figure 3b for geotechnical external and internal reinforced concrete stability conditions.

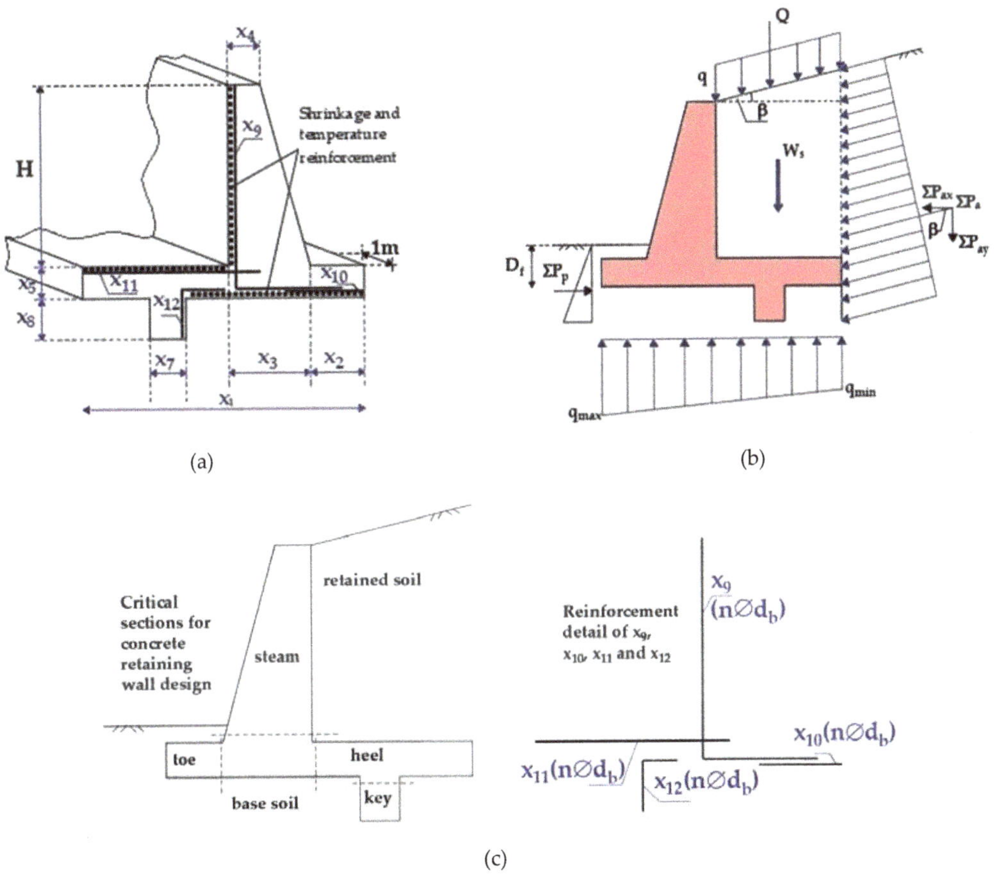

Figure 3. Reinforced concrete cantilever retaining wall design (RCRW): (**a**) Design variables; (**b**) Acting loads on retaining wall; (**c**) Design details.

The steel areas (As) of x_9, x_{10}, x_{11}, and x_{12} design variables have been determined with the number (n) and diameter (d_b) of the rebar. The steel areas for x_9, x_{10}, x_{11}, and x_{12} design variables have been determined by considering together the number and diameter of reinforcement for the stem, toe, heel, and key of the wall, respectively.

By employing the limit bounds of the design variables tabulated in Table 1 the design space has been formed. Input parameters utilized for geotechnical and design as RCRW design problem are demonstrated in Table 2.

Mathematical formulations of sliding, overturning, and bearing capacity safety factors detailed given in Equation (A5) (Appendix A section) have been utilized to satisfy of geotechnical external stability of the wall [44].

In terms of providing internal reinforced concrete stability, the flexural strengths ($Mn_{s,t,h,k}$) resistance to design moments ($Md_{s,t,h,k}$) have been examined for four critical cross-sections; (i) the section linked stem to base slab, (ii) the initial section of the toe extension from the stem, (iii) initial section of heel extension from the stem, (iv) the section linked the key to base slab (Figure 3c). In the same way, the design shear forces ($Vd_{s,t,h,k}$) should be safely fulfilled by nominal shear strength ($Vn_{s,t,h,k}$) at critical cross-sections of the wall. The nominal shear and flexural strengths for the critical cross-sections of the wall have been computed via Equation (A6) (Appendix A section) [38]. In Equation (A6), b is the width of the section (1000 mm), d is the height of the section, and a is the depth of the equivalent rectangular stress block. Design shear forces ($Vd_{s,t,h,k}$) and moments ($Md_{s,t,h,k}$) due to lateral soil and surcharge loads given Figure 3b at critical cross-sections, which are stem, toe, heel, and key have been determined by utilizing Equation (A7) (Appendix A section), respectively [39,41,45,46].

Table 1. The design variables and limit bounds for wall dimensions and reinforcement.

Design Variables		Lower Bound	Upper Bound
x_1 (m)		1.96	5.50
x_2 (m)		0.65	1.16
x_3 (m)		0.25	0.50
x_4 (m)		0.25	0.50
x_5 (m)		0.40	0.50
x_6 (m)		1.96	5.50
x_7 (m)		0.20	0.50
x_8 (m)		0.20	0.50
	n (piece)	3	30
x_9, x_{10}, x_{11}, x_{12}	d_b (mm)	10	30
	As (cm^2)	2.356	212.0575

Table 2. Input parameters for optimization analyses of RCRW.

Input Parameters	Symbol	Value	Unit
Stem height	H	4.5	m
Surcharge load	q	30	kPa
Backfill slope	β	0	°
Internal friction angle of the retained soil and the base soil	$Ø_r$ and $Ø_b$	36 and 34	°
Unit weight of retained soil, base soil, and concrete	γ_r, γ_b, and γ_c	17.5, 18.5, 23.5	kPa
Cohesion of base and retained soils	c_b and c_r	0	kPa
Depth of soil in front of the wall	D_f	0.75	m
Terzaghi bearing capacity factors for $Ø_b$ = 34° [45]	N_c, N_q, N_γ	52.64, 36.50, 38.04	–
The factor of safety for sliding and overturning stability	SF_{ss} and SF_{so}	1.50	–
The factor of safety for bearing capacity	SF_{sb}	3.00	–
Reinforcing steel yield strength	f_y	400	MPa
Concrete compressive strength	f_c	21	MPa
Concrete cover	c_c	0.07	m
Shrinkage and temperature reinforcement percentage	ρ_{st}	0.002	–
Nominal strength coefficient for the flexural moment	ϕ_m	0.90	–
Nominal strength coefficient for shear force	ϕ	0.75	–
Reinforcement location factor (1.0 for concrete below < 30.48 cm)	ψ_t	1.00	–
Coating factor (for uncoated bars)	ψ_e	1.00	–
Lightweight aggregate concrete factor (1.0 for normal-weight conc.)	λ	1.00	–
Cost of steel and concrete	C_s and C_c	0.4 and 40	\$/kg and \$/m^3

Randomly selected steel areas ($As_{s,t,h,k}$) from the design pool for calculating flexural strengths ($Mn_{s,t,h,k}$) should be greater than the minimum steel area ($As_{mins,t,h,k}$) and smaller than the maximum steel area ($As_{maxs,t,h,k}$). Furthermore, obtained reinforcement bar lengths in the optimum design should be satisfied bond strength as minimum development length ($Ld_{s,t,h,k}$) or minimum hook development length ($Ldh_{s,t,h,k}$) for all members according to design code. The abovementioned design criteria with required details have been demonstrated by Equation (A8) (Appendix A section) [38].

The mathematical expression of an RCRW design categorized as one of the most challenging real-size engineering design optimization problems is given in Equation (A9) (Appendix A section). The equation is presented each of stem (s), toe (t), heel (h), and key (k) critical cross-sections of RCRW. Due to the necessity of satisfying external and internal stability conditions RCRW optimum design is a complex engineering problem with 12 design variables and 26 design and side constraints.

2.3. Harmony Search Algorithm

The harmony search algorithm which is applied to many complex and real-size engineering design problems successfully is based on the principle of finding the best harmony with the notes played by the musicians in an orchestra [16]. In this process of reaching the best harmony, each musician may play notes or choose any notes randomly from whole possible combinations in their mind, which correspond to design space filled with different values of design variables. The harmony memory (HM) matrix, which stores the design variables values of the problem in the algorithm, is created by mimicking the situation which plays the notes from the musicians' minds in the music-making process. The HM matrix is depicted by Equation (1).

$$HM = \begin{bmatrix} x_{(1,1)} & x_{(2,1)} & \cdots & \cdots & x_{(N_{var}-1,1)} & x_{(N_{var},1)} \\ x_{(1,2)} & x_{(2,2)} & \cdots & \cdots & x_{(N_{var}-1,2)} & x_{(N_{var},2)} \\ \cdots & \cdots & \cdots & \cdots & \cdots & \cdots \\ \cdots & \cdots & \cdots & \cdots & \cdots & \cdots \\ x_{(1,HMS-1)} & x_{(2,HMS-1)} & \cdots & \cdots & x_{(N_{var}-1,HMS-1)} & x_{(N_{var},HMS-1)} \\ x_{(1,HMS)} & x_{(2,HMS)} & \cdots & \cdots & x_{(N_{var}-1,HMS)} & x_{(N_{var},HMS)} \end{bmatrix}_{HMS \times N_{var}} \quad (1)$$

where HMS and N_{var} correspond to harmony memory matrix size and the number of design variables, respectively.

A new harmony (solution vector) improves by considering three cases which are memory consideration, random selection, and pitch adjustment mechanisms. In memory consideration and a random selection, whether selecting the note in the mind of the musician or not is decided according to the value of harmony memory consideration rate (HMCR), which is an algorithm control parameter. If a random number (rnd (0,1)) assigned in the algorithm is smaller than HMCR, a harmony is selected from HM. Otherwise, a random harmony is selected from the design space with the possibility of (1−HMCR). The probabilistic process of updating for each design variable value depending on HMCR, where x_i' is the new solution vector and X_i is a random selection from the defined range of design variables in design space, is given by Equation (2).

$$x_i' = \begin{cases} x_i' \in \{x_i^1, x_i^2, x_i^3, \ldots, x_i^{HMS}\} & \text{w.p.(HMCR)} \\ x_i' \in X_i = \{x_i(1), x_i(2), \ldots, x_i(K)\} & \text{w.p.}(1 - \text{HMCR}) \end{cases} \quad (2)$$

Similar to the process of achieving the best harmony by tuning each musical instrument appropriately, the pitch adjustment mechanism in the algorithm is applied considering pitch adjustment rate (PAR) if HMCR possibility is valid in updating the value of the current design variable. If assigned new rnd (0,1) value for current design variable is smaller than PAR, design variable is updated according to possibility HMCRxPAR which is given in Equation (3). Otherwise, the updating process is not applied (1 − PAR) [16,47].

$$x_i' = \begin{cases} x_i(k+m) & \text{w.p. HMCRxPARx} \times \text{PAR} \times 0.5 \\ x_i(k-m) & \text{w.p. HMCR} \times \text{PAR} \times 0.5 \\ x_i(k) & \text{w.p. HMCR} \times (1 - \text{PAR}) \end{cases} \quad (3)$$

Here, $x_i(k)$ and m correspond to the kth element in X_i and neighboring value (usually is taken a value of 1), respectively.

In addition, the steps of classical HSA are itemized as follows:

- Step 1: HSA is initialized by determining the constant algorithm parameters (HMS, HMCR, PAR, and maximum iteration number) and generating design space with design variable values according to range limitation;
- Step 2: HM matrix is formed randomly by selecting from design space;
- Step 3: Improvisation of a new HM matrix conceiving memory consideration, random selection, and pitch adjustment mechanisms is carried out;

- Step 4: HM matrix is updated depending on whether a better solution is obtained, and then the worst solution is drawn from HM by replacing the better one;
- Step 5: Until the current iteration is reached the predefined maximum iteration number, Step 3 and Step 4 are repeated. If it is conducted HSA is ended.

2.4. Taguchi Method Background

By determining the proper orthogonal array for the current problem is possible to limit the number of analyses required for pre-research in the Taguchi method [26,27], which is based on statistical and a robust design manner. Thus, the orthogonal array which has a specific array configuration with an extraordinary set of Latin Squares reduces research costs and allows parametric analysis with fewer trials [48]. The general representation of orthogonal array is $L_d(a)^k$ or Ld mean that d is the total number of trials; a is the number of levels; k is the number of parameters; L is the type of the orthogonal array. In the Taguchi design, initially, the parameters that are assumed to be effective on the response value are determined, and the appropriate orthogonal arrays tabulated in Table 3 are selected according to the definite number of parameters and the number of levels.

Table 3. Orthogonal array.

Ld $L_d(a)^k$	L4	L4	L8	L8	L9	L9	L9	L18	L16	L16	L16	L16	L25	L25	L25	L25
d	4	4	8	8	9	9	9	18	16	16	16	16	25	25	25	25
k	2	2	4	5	2	3	4	5	2	3	4	5	2	3	4	5
a	2	3	2	2	3	3	3	3	4	4	4	4	5	5	5	5

Finding the best combination of parameters from the cluster which is formed with different levels of the parameter is possible by using Taguchi Method with less trial, contrary to performing all analyses as in the full factorial design. For instance, if it is desired to investigate the parameter effect and the optimum values of the parameters in a design problem that has five parameters with four levels, 1024 (4^5) trials are required in a full factorial design demonstrated schematically in Figure 4. The data set of 1024 combinations is repeated with each other with a specific rule and in order. The harmony search algorithm must be run to find the best combination from between 1024 algorithm parameter combinations, including all values of design parameters with their levels. Here, the best combination means the minimum value of the objective function for the best-acquired value of parameters. In Figure 4, P_{mn} (m = 1, ... ,k; n = 1, ... ,a) corresponds to design parameters with their levels, which have an impact on the response. The number of design parameters (k) and their levels (m) have been taken as five and four, respectively.

However, only 16 trials are performed, which is sufficient to predict the desired results with an acceptable error employing the L16 orthogonal array according to the Taguchi method. Calculating the Signal/Noise (S/N) ratio is another of two important steps in the Taguchi design method. The Signal/Noise ratio (S/N) is described by Taguchi to decrease variance and is used as performance criteria in experiment design. The S/N ratio is divided into three depending on the purpose of application; smaller is better, nominal is best, larger is better as given in, respectively, Equations (4)–(6).

$$S/N = -10\log\left[\sum\left(Y^2\right)/n\right] \quad (4)$$

$$S/N = -10\log\left[\sum(\overline{Y})/\sigma^2\right] \quad (5)$$

$$S/N = -10\log\left[\sum\left(1/Y^2\right)/n\right] \quad (6)$$

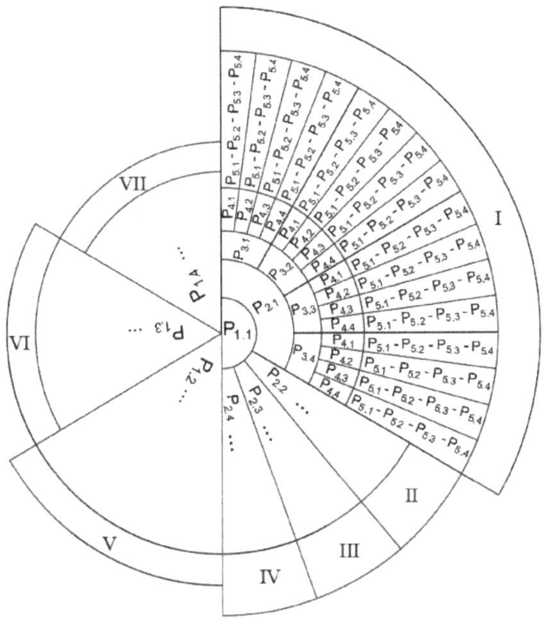

Figure 4. Full factorial design combinations.

Here, Y is the response value; n is the number of repetitions; $\overline{Y}$ and σ are the arithmetic mean and the standard deviation of the Y values. The S/N ratios are determined for the obtained Y values by the Taguchi designs. By employing the ascertained S/N ratios, the arithmetic mean of all S/N ratios (η) is calculated and then η_{ij} is determined via Equation (7). While η_{ij} informs about the change depending on the response value in all levels of each parameter, $(S/N)_{ij}$ is the sum of the S/N ratios whose levels are equal to j for the ith design parameter. The effect of the parameters on the selected response value is determined by the variance value which shows the distance of the numbers in the series to the mean of all the numbers in the series. The variance (v_i), which is defined as the sum of the squares of the deviations of the data from the arithmetic mean, is calculated according to the η_i values, and variance analyses (ANOVA) are performed (Equation (8)). In the Taguchi approach, the prediction of the response value ($\eta_{prediction}$) by considering the value that has the most influential parameter level on the design for each parameter is ascertained via Equation (9), which η_{pi} is the average S/N ratio value in the estimated optimum parameter level for the current parameter. The relative error (ε) is calculated by employing the predicted response value ($\eta_{prediction}$) and the real value of response (η_{real}) which is acquired by substituting the predicted optimum parameter level of parameters by Equation (10).

$$\eta_{ij} = \frac{(S/N)_{ij}}{a} \ (i = 1, \ldots, k;\ j = 1, \ldots, a) \quad (7)$$

$$v_i^2 = \left[\sum_{i=1}^{j} \left(\eta_{ij} - \eta \right)^2 \right] \frac{1}{a-1} \quad (8)$$

$$\eta_{prediction} = \eta + \sum_{i=1}^{k} \left(\eta_{pi} - \eta \right) \quad (9)$$

$$\varepsilon = \frac{\eta_{prediction} - \eta_{real}}{\eta_{real}} \times 100 \quad (10)$$

2.5. A New Hybrid Method Based on Taguchi for Optimum Values of Algorithm Parameters

In this study, a novel Taguchi method integrated hybrid harmony search algorithm (TIHHSA) has been presented that enables optimum algorithm parameter values by statistically predicting the best fitness value. The TIHHSA flowchart, which explains of forming Taguchi design matrix, initializing HSA, and performing Taguchi analyses, is depicted as three sections in Figure 5.

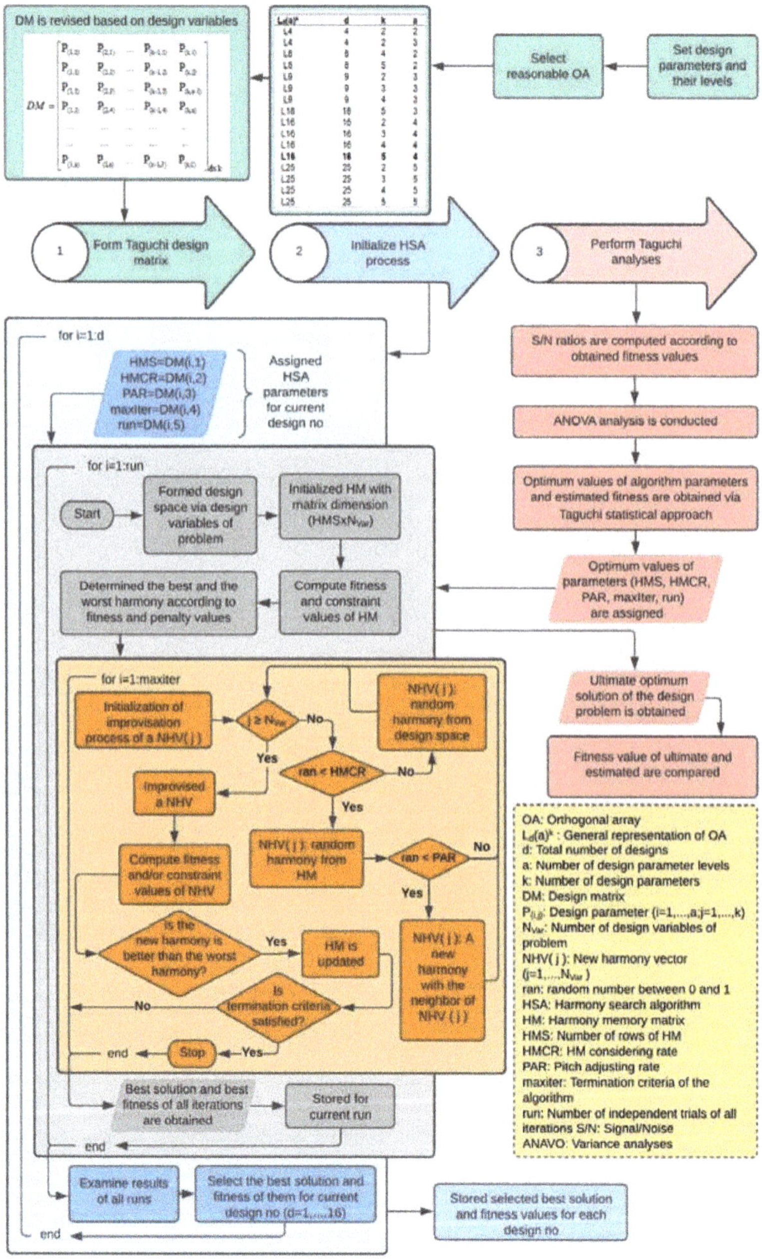

Figure 5. Flowchart of Taguchi Method Integrated Hybrid Harmony Search Algorithm (TIHHSA).

2.5.1. Forming Taguchi Design Matrix

Finding the best combination is possible by using Taguchi Method with the least trial, contrary to performing 1024 analyses such as full factorial design which is involved formed by design parameters with their parameter levels. Initially, the design parameters and their defined ranges which affect the response value of the optimization problem to investigate the best combinations that give the minimum fitness (objective) value are specified according to the process of forming the Taguchi design matrix given in Figure 5.

As the Taguchi design matrix (DM) is not created randomly or repetitively according to a certain rule, the parameter and its parameter level given differently for each design in the orthogonal array table are considered for generating DM. In this study, the harmony memory size (HMS), the harmony memory consideration rate (HMCR), the pitch adjustment rate (PAR), the maximum iteration number (MAXITER), and the independent run number of whole iterations (RUN) have been accepted as design parameters that directly affect the convergence rate to the desired fitness value. Parameter levels have been set as 20, 30, 40, 50 for HMS, 0.80, 0.85, 0.90, 0.95 for HMCR, 0.10, 0.20, 0.30, 0.40 for PAR, 2000, 4000, 6000, 8000 for MAXITER and 30, 100, 500, 1000 for RUN. To generate the Taguchi design matrix (DM), the appropriate orthogonal array L16 ($L_{16}(4)^5$) from Table 3 has been chosen by five different design parameters, each of which has four levels. If an example is given for the creation of the 8 of design no in the DM, the parameter levels take as 2 for HMS, 4 for HMCR, 3 for PAR, 2 for MAXITER, and 2 for the RUN as seen in Table 4. Accordingly, design no 8 is generated by selecting 30 for HMS, 0.95 for HMCR, 0.30 for PAR, 2000 for MAXITER, and 100 for the RUN. Similarly, the other designs are formed according to parameter level for the current parameter.

Table 4. L16 orthogonal array and Taguchi design matrix.

Design No	Design Parameters with Levels					DM				
	P1	P2	P3	P4	P5	HMS	HMCR	PAR	MAXITER	RUN
1	1	1	1	1	1	20	0.80	0.10	2000	30
2	1	2	2	2	2	20	0.85	0.20	4000	100
3	1	3	3	3	3	20	0.90	0.30	6000	500
4	1	4	4	4	4	20	0.95	0.40	8000	1000
5	2	1	2	3	4	30	0.80	0.20	6000	1000
6	2	2	1	4	3	30	0.85	0.10	8000	500
7	2	3	4	1	2	30	0.90	0.40	2000	100
8	2	4	3	2	1	30	0.95	0.30	4000	30
9	3	1	3	4	2	40	0.80	0.30	8000	100
10	3	2	4	3	1	40	0.85	0.40	6000	30
11	3	3	1	2	4	40	0.90	0.10	4000	1000
12	3	4	2	1	3	40	0.95	0.20	2000	500
13	4	1	4	2	3	50	0.80	0.40	4000	500
14	4	2	3	1	4	50	0.85	0.30	2000	1000
15	4	3	2	4	1	50	0.90	0.20	8000	30
16	4	4	1	3	2	50	0.95	0.10	6000	100

2.5.2. Initializing HSA Process

The process of HSA is performed by employing pre-defined design variables, design constraints, and objective function of complex benchmark engineering design optimization problems which are the welded beam (WBD), the pressure vessel (PVD, the gear train (GTD), and the speed reducer (SRD) engineering design problems and the real-size reinforced concrete cantilever retaining wall (RCRW) design.

In the optimization process of 16 designs, the algorithm parameters are assigned by considering DM tabulated in Table 4. End of the optimization process the best fitness values are acquired as response values for each row of DM and engineering design optimization problems.

2.5.3. Performing Taguchi Analyses

In this section, the S/N ratios of response value have been calculated via Equation (4), which is given for smaller is the better purpose, for WBD, PVD, GTD, SRD, and RCRW engineering design optimization problems, separately. The graphs that show the variation between the determined η_{ij} values via Equation (7) and the design parameter with their levels were drawn. The variance (v) values of design parameters by employing η and η_i values according to Equation (8) and the parameter effect (PE) on response value based on the sum of squares for the design parameters are specified.

And then the verification analyses are performed vis a vis estimated optimum values of design parameters which are suggested for $\eta_{prediction}$ value depicted in Equation (9). By assigning the HMS, HMCR, PAR, MAXITER, and RUN optimum values that come from the Taguchi approach results to HSA, optimization analyses are conducted again for each design optimization problem and η_{real} is determined. The relative error (ε) which is a reliability criterion of the Taguchi design is calculated by Equation (10).

3. Design Experiments and Results

In this section, the optimization analysis results reached by the HSA and the proposed TIHHSA method have been given with the aim of investigating different engineering optimization design problem's characteristic effect on the fitness value. In the optimization analyses performed with HSA, the variation of the fitness values achieved for different numbers of run values according to the characteristics of different engineering optimization problems has been examined. The robustness of the proposed TIHHSA method has been evaluated for engineering optimization design problems and the optimum values of the algorithm parameters have been estimated with the HMS, HMCR, PAR, MAXITER, and RUN effects obtained from the variance analyses (ANOVA).

3.1. Optimization Analyses of Engineering Design Problems and Real-Size Engineering Design Optimization Problem

In this section, the optimization analyses result of the welded beam (WBD), the pressure vessel (PVD), the gear train (GTD), and the speed reducer (SRD) benchmark engineering design problems and the real-size reinforced concrete cantilever retaining wall (RCRW) design have been presented. In the optimization analysis through harmony search algorithm (HSA) [16], the algorithm parameters are selected as HMS = 20, HMCR = 0.90, and PAR = 0.35 [47]. Deb's rules [49] are implemented as a constraint-handling strategy. The best solution is determined according to penalty values of all constraints with the fitness values. The best solution is evaluated according to the current fitness value if solutions have the same penalty value or no penalty.

Different run cases (R30, R100, R500, R1000) have been chosen to investigate the effect of variable run values on the minimum objective function, in the optimization analyses. HS algorithm is performed until maximum iteration numbers reach 30,000. This process has been repeated for different independent runs as 30, 100, 500, and 1000. In the evaluation of the results, while the best iteration number (BIN) corresponds to the iteration in which no more minimum objective function value is yielded with ongoing analysis, the best run number (BRN) is the best-obtained fitness value among all runs for each case. The best, the worst, the mean, the standard deviation (StD), and the median values of the minimum objective function (fitness value) have been determined for BRN and BIN, separately.

Achieved statistical evaluations of optimization analyses satisfied the constraints are tabulated in Table 5 for WBD, PVD, GTD, and SRD.

Table 5. Statistical results for f(x) optimum values of engineering design optimization problems.

Case		Run							Iteration				
		BRN	Best	Mean	Worst	StD	Median	BIN	Best	Mean	Worst	StD	Median
WBD ($)	R30	14(47%)	1.85149	2.62202	3.67302	0.463237	2.57699	28,383(95%)	1.85149	2.10940	7.12472	0.295862	1.97920
	R100	33(33%)	1.80231	2.71590	4.23073	0.467682	2.51094	22,548(75%)	1.80231	1.95674	7.35794	0.309656	1.82127
	R500	34(7%)	1.75598	2.70736	4.48978	0.500434	2.61591	21,170(71%)	1.75598	2.16143	4.28967	0.312273	2.32239
	R1000	276(28%)	**1.74026**	2.69101	4.89958	0.492996	2.61779	28,817(96%)	1.74026	1.76394	2.36984	0.104493	1.74136
PVD($)	R30	21(70%)	6089.66	6815.66	7428.54	404.478	6856.34	16,355(55%)	6089.66	6342.58	43,582.4	2212.31	6094.59
	R100	13(13%)	6195.10	6970.5	7502.81	367.191	7038.74	24,997(83%)	6195.10	6436.25	44,557.6	1508.82	6228.33
	R500	216(43%)	6000.09	6865.0	7497.80	415.43	6898.75	23,513(78%)	6000.09	6265.77	55,383.2	1979.37	6051.16
	R1000	795(80%)	**5959.86**	6887.8	7557.39	400.214	6919.51	24,449(82%)	5959.86	6417.04	25,980.4	1918.94	6065.6
GTD (unitless)	R30	7 (23%)	1.54505×10^{-10}	4.7459×10^{-8}	5.3303×10^{-7}	1.0461×10^{-7}	1.6106×10^{-8}	512 (2%)	1.54505×10^{-10}	9.0999×10^{-8}	4.7841×10^{-5}	2.0475×10^{-6}	1.5451×10^{-10}
	R100	77 (77%)	$\mathbf{2.70086 \times 10^{-12}}$	5.8049×10^{-8}	7.7986×10^{-7}	1.2036×10^{-7}	1.3531×10^{-8}	12,552 (42%)	2.70086×10^{-12}	8.2076×10^{-7}	6.1311×10^{-3}	7.0792×10^{-5}	2.7009×10^{-12}
	R500	34 (7%)	$\mathbf{2.70086 \times 10^{-12}}$	4.9250×10^{-8}	1.3811×10^{-6}	1.0699×10^{-7}	1.8274×10^{-8}	210 (1%)	2.70086×10^{-12}	2.4972×10^{-5}	1.0236×10^{-2}	4.8766×10^{-4}	2.7009×10^{-12}
	R1000	198 (20%)	$\mathbf{2.70086 \times 10^{-12}}$	4.5731×10^{-8}	1.0883×10^{-6}	9.0341×10^{-8}	1.8274×10^{-8}	10 (0%)	2.70086×10^{-12}	2.7009×10^{-12}	2.7009×10^{-2}	7.5126×10^{-26}	2.7009×10^{-12}
SRD (kg)	R30	5(17%)	2994.93	2999.97	3090.64	17.1989	2996.53	22,048(73%)	2994.93	3008.39	4933.98	66.3806	2996.80
	R100	59(59%)	**2994.79**	2996.69	3001.84	1.21818	2996.37	23,877(80%)	2994.79	3002.48	3260.32	28.4670	2997.22
	R500	244(49%)	2994.90	2997.30	3098.08	6.27332	2996.60	25,612(85%)	2994.90	3010.85	5137.55	91.5543	2995.26
	R1000	112(11%)	2994.84	2997.23	3092.00	4.29835	2996.55	20,879(70%)	2994.84	3006.64	5430.07	88.7313	3000.99

NOTE: The bolded values are the best fitness.

According to Table 5, the minimum fitness value of the WBD problem has been found for the R1000 case as \$1.74026 in 28,817 iterations, which corresponds to 96% of the optimum search process performed with 30,000 iterations. After this value was accomplished for 276 runs out of 1000 ($276/1000 \times 100 = 28\%$), no more minimum fitness was not reached during continuing runs. For the other run cases (R30, R100, R500), while the optimum values have been achieved on the average of 80% of 30,000 iterations, it is seen that the average of 38% of all runs is enough for this search process.

It is clearly shown that the minimum fitness value of PVD has been gained for the R1000 case as \$5959.86 in 82% out of 30,000 iterations and 795 runs out of 1000 (80%). While the minimum fitness values have been found in an average of 70% of the optimum search process, an average of 42% of independently operated runs has been sufficient to acquire the minimum objective function value for R30, R100, and R500 cases.

Though increasing the run number has been caused that the algorithm obligates to investigate more optimum value, the only best fitness value of GTD problem has revealed as 2.70086×10^{-12} in the case of R100. While the algorithm has been reached the best fitness value in 210 and 10 trials out of 30,000 iterations, it is seen that 34 and 198 runs are enough to find the optimum solution for R500 and R1000 cases, respectively.

When the values tabulated for the SRD problem are examined, it is observed that the best fitness value (2994.79 kg) has been just attained for the R100 case alike for the GTD problem, even though the optimization process for more runs which is ongoing. The searching process for the minimum fitness value of the SRD problem has been performed with 30,000 iterations and the best value has been found in 23,877 iterations out of entire iterations that means 80% of the process. The more minimum fitness value has not been reached for increasing runs.

Since the algorithm cannot reach a better solution after reaching the best solution, the process is conducted again to possibly find a more minimum result with different runs. It is observed that more fitness values have been generally yielded with increasing runs for different design problems. The fitness values of WBD and PVD design problems have been achieved for the R1000 case when given statistical result tables are examined. The fitness value of GTD and SRD design problems has been obtained for the R100 case in optimization analyses which are seen that the fitness value is not changing with continuing analysis any longer.

It has been seen from the statistical results given in Table 6 for the optimum cost (RCRW1) and the optimum weight (RCRW2) of the real-size engineering design optimization problem (RCRW) that the optimum results are not achieved at the equal runs for different objective functions of the same optimization problem. The optimum results have been reached in the R500 case for RCRW1 as \$179.449/m and the R100 case for RCRW2 as 5883.61 kg/m. For RCRW1, the optimization process has been completed in 98% iterations out of whole iterations and 359(72%) run out of 500 runs. In contrast with RCRW1, it has been seen that this process for RCRW2 is conducted at the time when is operated in 25,444 iterations out of 30,000 and 38 runs out of 500 runs.

Table 6. Statistical results for f(x) optimum values of real-size RCRW engineering design optimization problem.

	Case	Run						Iteration					
		BRN	Best	Mean	Worst	StD	Median	BIN	Best	Mean	Worst	StD	Median
RCRW1 (\$/m)	R30	11(37%)	180.082	185.85	194.525	4.52073	185.329	29,447(98%)	180.082	186.376	627.563	21.5973	181.305
	R100	76(76%)	179.842	186.153	198.7	4.39706	186.333	29,975(99%)	179.842	186.275	468.5	21.6613	181.156
	R500	359(72%)	**179.449**	186.049	200.756	4.72074	185.306	29,405(98%)	179.449	184.779	480.697	21.1682	180.267
	R1000	379(38%)	179.693	186.064	200.572	4.6882	185.019	23,496(78%)	179.693	184.553	462.683	19.8148	179.699
RCRW2 (kg/m)	R30	4(13%)	5886.67	5964.16	6411.61	128.141	5898.14	26,196(87%)	5886.67	5987.62	9578.21	348.742	5894.12
	R100	38(38%)	**5883.61**	5962.77	6302.54	108.939	5910.18	25,444(85%)	5883.61	5964.5	11,125.5	358.102	5884.06
	R500	319(64%)	5883.64	5958.45	6764.28	126.388	5903.05	22,311(74%)	5883.64	6007	9570.89	401.738	5892.82
	R1000	735(74%)	5884.09	5955.51	6966.85	115.767	5901.38	28,442(96%)	5884.09	6005.54	9984.8	449.616	5884.82

NOTE: the bolded values are the best fitness values.

The minimum objective function value comparison between the previously reported studies in the literature which were conducted by utilizing different metaheuristic algorithms and the obtained results of design optimization problems are given for WBD, PVD, GTD, and SRD in Figure 6a [35,50–70]. In the same figure, general (b) and zoomed (c) view iteration history graphs of the best solution (fmin) for the best run among all runs are demonstrated. Detailed comparisons of results for design problems of WBD, PVD, GTD, and SRD with the studies in the literature are given for optimum solutions in Tables A1–A4 and for constraints in Table A5 of the Appendix A section.

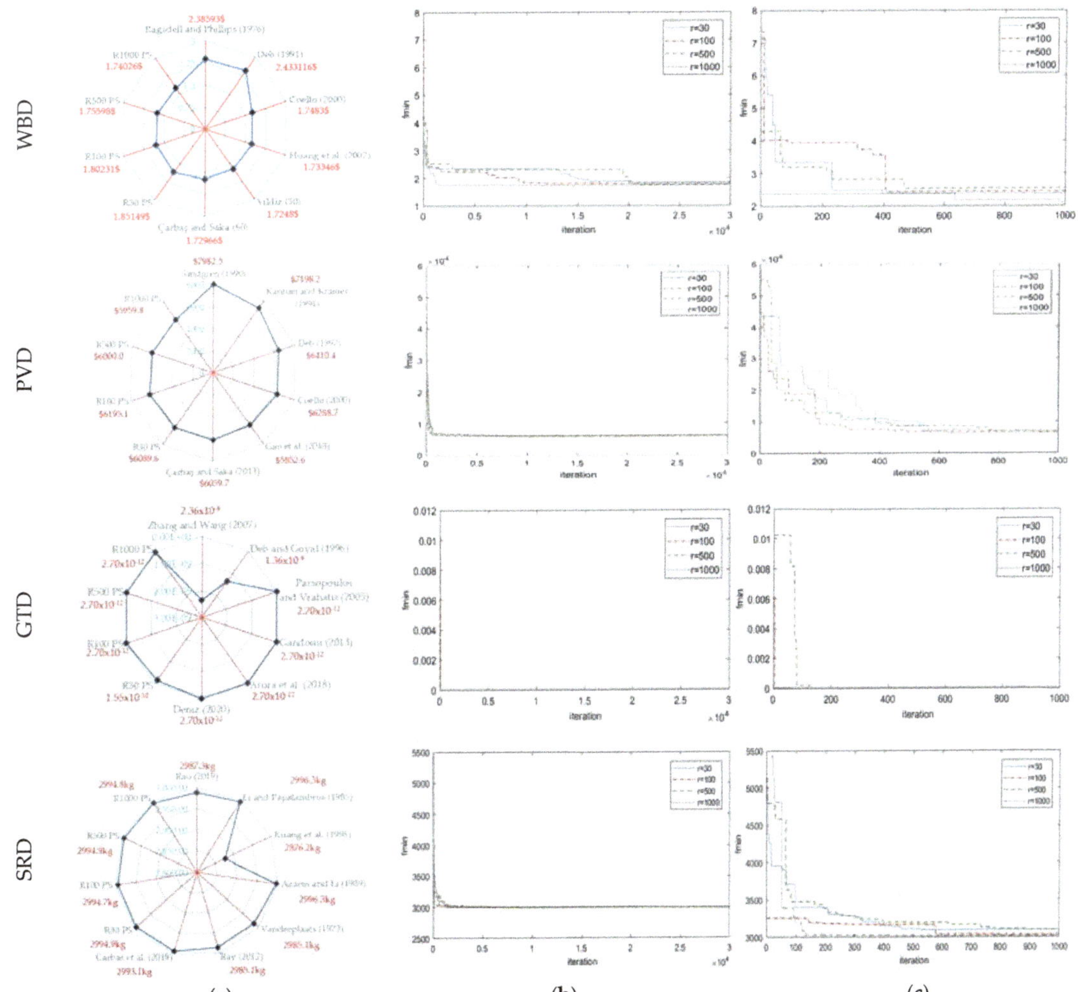

Figure 6. Optimization analyses result for WBD, PVD, GTD, and SRD, respectively: (**a**) Comparisons graphs between the best fitness values and the literature studies; (**b**) General view iteration history graphs; (**c**) Zoomed view iteration history graphs.

It is seen that the optimal fmin value ($1.74026) acquired with the HSA algorithm when is R1000 case in this study is approximately 1% greater than the value ($1.7248) yielded with the hybrid Taguchi harmony search algorithm [54] according to comparison given in Figure 6a for WBD. The fmin value acquired as $5959.86 in the R1000 case for PVD is above the best fitness value with 2% ($5852.6394) presented according to the study by Gao et al. [58], which utilized the HSA with the bandwidth improvisation to the pitch adjustment rate. The reached optimum value of GTD in the R30 case which is 2.70086×10^{-12} is the same as the best values presented in the literature [61–64]. The acquired optimum value (2994.79 kg) for SRD is higher than the best fitness value (2876.22 kg) presented in the literature study [66] with 4% obtained according to the Taguchi-aided optimization search method.

The comparison between obtained minimum objective function values of RCRW designs of the optimum cost (RCRW1) and the optimum weight (RCRW2) and the optimization study of retaining wall design in the literature are shown in Figure 7a [40]. The optimum search process at the run value which is the best solution (fmin) for R30, R100, R500, and R1000 is given as the general view and zoomed view of iteration history graphs in Figure 7b,c, respectively. Detailed results with the comparison of the optimum values in the literature for RCRW1 and RCRW2 designs are listed in Tables A6 and A7 of the Appendix A section, respectively.

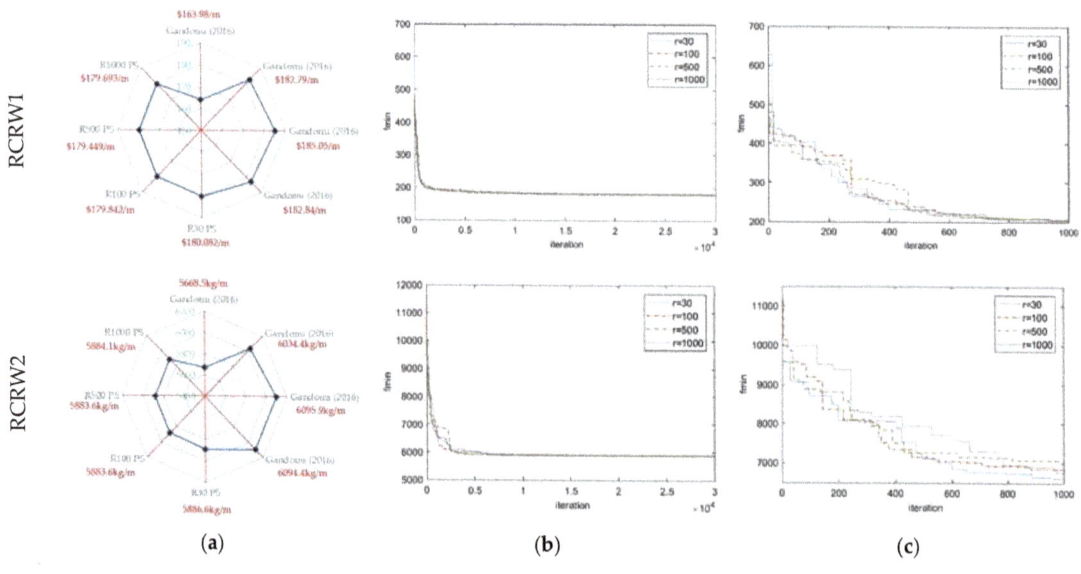

Figure 7. Optimization analyses result for the optimum cost and the optimum weight of RCRW designs, respectively: (**a**) Comparisons graphs between the best fitness values and the literature study; (**b**) General view iteration history graphs; (**c**) Zoomed view iteration history graphs.

According to the comparison of optimum designs with the literature studies given in Figure 7a, it is obvious that RCRW1 and RCRW2 optimum objective function values ($179.4495/m and 5883.61 kg/m) are greater than the best values ($163.98/m and 5668.5 kg/m) presented by Gandomi et al. [40] as 9% and 4%, respectively. In this study, it has been determined that the optimum solutions yielded by the HSA for two different objective functions of a real size engineering design optimization problem are close to the literature values which were reached by using different algorithms in general for comparison. Although it is presented as the best solution acquired with the biogeography-based optimization (BBO) algorithm considering the optimization problem, it is not specified whether the optimum solution of the 26 constraints utilized in the study is provided. Since it is important to obtain the best solution that provides all constraints of the optimization problem, the values of the constraints accomplished for this study, in which the mathematical model of the optimization problem is compared, are given in Table A8 of the Appendix A section.

Conducted analyses have been shown that the appropriate numbers of iteration and the independent run of entire iterations, which formed the extent of the acquiring optimum process, are significant to reaching the best fitness value instead of many or fewer numbers of them. The reaching process of the maximum iteration number defined as the run is accepted 30 times in the literature of optimization studies and the most minimum fitness value satisfying the design constraints is presented as the optimum result. The minimum fitness values have been yielded by operating different design problems with different runs when the number of the runs is greater than 30 according to the results given in the tables. It observed that while the more minimum fitness values generally are obtained with the increasing runs in some engineering design problems, the minimum result may not be found with larger runs too in some of them. This brings to the fore the necessity that the number of executions of the optimization algorithm may have an optimum value.

3.2. *Taguchi Analyses*

The best combination, which provided the best fitness value for the welded beam design (WBD), the pressure vessel design (PVD), the gear tear design (GTD), the speed reducer design (SRD), the reinforced concrete cantilever retaining wall design (RCRW1) optimization problems have been investigated in terms of design parameters effective on the searching optimum solutions via the Taguchi method integrated hybrid harmony search algorithm (TIHHSA) as visualized in Figure 5.

3.2.1. Part I: Investigation of Five Optimum Design Parameter Values with Effect on the Fitness Value

For the abovementioned aim, by considering different values of design parameters, the harmony memory size (HMS), the harmony memory consideration rate (HMCR), the pitch adjustment rate (PAR), maximum iteration number (MAXITER), and the independent run number of the whole iterations (RUN), 16 designs given in Table 4 formed according to L16 orthogonal array have been performed. By utilizing obtained the minimum objective function values, f(x), for each design optimization problem, the Signal/Noise ratios (S/N), defined in Equation (4) with the aim of the case of smaller is better, have been calculated and listed with the response values of different engineering design optimization problems in Table 7.

Table 7. Response values (f(x)) and S/N ratios of response values.

Design No	f(x)					S/N				
	WBD ($)	PVD ($)	GTD (Unitless)	SRD (kg)	RCRW1 ($/kg)	WBD	PVD	GTD	SRD	RCRW1
1	2.1880	6595.36	9.94×10^{-11}	3002.24	196.841	−6.8010	−76.38	200.052	−69.5489	−45.882
2	1.8766	6315.66	2.70×10^{-12}	2996.59	183.264	−5.4672	−76.01	231.37	−69.5325	−45.262
3	1.8656	6040.75	2.70×10^{-12}	2996.09	181.325	−5.4165	−75.62	231.37	−69.5311	−45.169
4	1.8073	5985.52	2.70×10^{-12}	2996.03	180.652	−5.1405	−75.54	231.37	−69.5309	−45.137
5	1.8383	6039.20	2.70×10^{-12}	2995.6	183.881	−5.2881	−75.62	231.37	−69.5297	−45.291
6	1.8528	6014.33	2.70×10^{-12}	2995.81	181.395	−5.3564	−75.58	231.37	−69.5303	−45.173
7	2.1053	6116.35	2.31×10^{-11}	3002.82	189.297	−6.4661	−75.73	212.736	−69.5506	−45.543
8	2.5046	6389.99	2.70×10^{-12}	3001.70	187.417	−7.9747	−76.11	231.37	−69.5474	−45.456
9	2.1041	6257.68	2.70×10^{-12}	2996.93	185.566	−6.4615	−75.93	231.37	−69.5335	−45.370
10	2.0103	6385.19	9.94×10^{-11}	2998.29	186.013	−6.0654	−76.10	200.052	−69.5375	−45.391
11	1.7921	6105.73	2.70×10^{-12}	2997.21	183.491	−5.0673	−75.71	231.37	−69.5343	−45.272
12	2.0951	6286.05	2.70×10^{-12}	3000.60	188.117	−6.4240	−75.97	231.37	−69.5442	−45.489
13	1.9126	6136.63	2.70×10^{-12}	2998.17	190.651	−5.6324	−75.76	231.37	−69.5371	−45.605
14	2.0021	6169.29	2.70×10^{-12}	3002.63	195.777	−6.0298	−75.80	231.37	−69.550	−45.835
15	1.9833	6186.53	2.70×10^{-12}	2998.66	183.548	−5.9477	−75.83	231.37	−69.5386	−45.275
16	2.1366	6147.29	2.70×10^{-12}	2997.24	183.776	−6.5943	−75.77	231.37	−69.5344	−45.286
η	2.0047	6198.22	1.61×10^{-11}	2998.50	186.313	−6.0083	−75.84	226.29	−69.5382	−45.402

The rank (R) which indicates the order of design parameters effect from largest to smallest have been accomplished by using η_{ij} values for each design optimization problem. The sum of squares (SS), variance (v), and rank (R) values acquired from ANOVA analyses are demonstrated in Table 8. The Taguchi method which is a fractional factorial design is a saturated model [71,72]. It means that all degrees of freedom are used in the estimation. For this reason, p values are not given in Table 8 as no residual error occurs in the Taguchi design with $L_{16}(4)^5$. While the RUN is the most effective parameter being that the biggest variance value having for WBD, PVD, and GTD designs, the MAXITER is the most important parameter for SRD and RCRW1 in reaching minimum objective function. Although the possibility of obtaining the more minimum or the most minimum fitness value is triggered by extending the optimization process with more iteration such as SRD and RCRW1 designs, it occurs the outcome that unavailability of no more optimum values with continuing analyses and needed for a new independent run process such as WBD, PVD, and GTD problems.

Table 8. Results of variance analyses of WBD, PVD, GTD, SRD, and RCRW1 optimization problems.

Optimization Problem	Evaluation Criteria	Design Parameter				
		HMS	HMCR	PAR	MAXITER	RUN
WBD	SS	0.6492	1.7416	1.2039	1.1445	4.0607
	v	0.216309	0.580439	0.401491	0.381525	1.35359
	R	5	2	3	4	1
PVD	SS	0.0754	0.0865	0.0188	0.1542	0.4473
	v	0.02515	0.028833	0.006269	0.051393	0.149109
	R	4	3	5	2	1
GTD	SS	164.432	164.432	456.416	456.416	654.917
	v	54.8032	54.8032	152.065	152.065	218.268
	R	4	5	2	3	1
SRD	SS	4.44×10^{-5}	9.58×10^{-6}	4.53×10^{-5}	6.15×10^{-4}	1.36×10^{-4}
	v	1.48×10^{-5}	3.21×10^{-6}	1.51×10^{-5}	2.05×10^{-4}	4.55×10^{-5}
	R	4	5	3	1	2
RCRW1	SS	0.052	0.1184	0.0349	0.4879	0.0535
	v	0.0173	0.0395	0.0116	0.1626	0.0179
	R	4	2	5	1	3

It is observed from the variance results that the PAR and the HMS have an average or minimal effect with rank values of 3 or 5 and 3, 4 or 5. In improvising a new solution of the HSA, if the assigned random number is smaller than HMCR, the PAR is compared to a new random number. In satisfying this condition, the solution is improved, and its new fitness value is determined. As including the PAR in this process depends on the possibility of an assigned random number, it is commented that the PAR may not be a much effective parameter to find the minimum except for the GTD problem. The PAR is the second effective design parameter on the best fitness value for GTD which is an unconstrained design problem. The HMS design parameter may not be the most critical one since the solutions of HM become the same each other with increasing iteration for each different run to reach the fmin.

In addition, the prediction of the response value ($\eta_{prediction}$), and the optimum parameter combination have been determined separately for each design problem (Table 9). The real response values (η_{real}) which are specified by considering the optimum parameter combination have been obtained with verification analyses (Table 10).

Table 9. Verification analyses data and results with optimum values of design parameters.

Optimization Problem	Optimum Parameter Combination	fmin ($\eta_{prediction}$)
WBD	HMS1-HMCR3-PAR2-MAXITER4-RUN4	$1.63817
PVD	HMS2-HMCR3-PAR4-MAXITER4-RUN4	$5813.73
GTD	HMS4-HMCR4-PAR2-MAXITER2-RUN3	2.60398×10^{-12}
SRD	HMS1-HMCR1-PAR2-MAXITER3-RUN3	2994.16 kg
RCRW1	HMS1-HMCR3-PAR2-MAXITER4-RUN3	$177.724/m

Table 10. Statistical results from verification analyses of engineering design optimization problems.

Case	Run						Iteration					
	BRN	Best (η_{real})	Mean	Worst	StD	Median	BIN	Best (η_{real})	Mean	Worst	StD	Median
WBD	202/1000(20%)	1.7455	2.8743	5.10816	0.54932	2.8190	2598/8000 (32%)	1.7455	1.82	9.44763	0.418665	1.7455
PVD	897/1000(89%)	6054.14	7000.4	8272.08	427.637	7032.49	7828/8000 (98%)	6054.14	7153.31	42,950.5	3947.25	6180.14
GTD	2/500 (0.4%)	2.70086×10^{-12}	5.4247×10^{-8}	1.38114×10^{-6}	1.34484×10^{-7}	1.31252×10^{-8}	312/4000 (8%)	2.70086×10^{-12}	2.99998×10^{-5}	5.5068×10^{-3}	3.7964×10^{-3}	2.70086×10^{-12}
SRD	399/500 (80%)	2995.97	3004.43	3022.88	4.62559	3003.84	5746/6000 (96%)	2995.97	3062.32	5322.68	296.145	2996.6
RCRW1	425/500 (85%)	181.035	191.431	206.659	5.36727	191.363	7740/8000 (97%)	181.035	199.648	386.834	38.1441	189.368

Since the optimum values of MAXITER and RUN have been accomplished as their maximum values (MAXITER = 6000 and 8000 and RUN = 500 and 1000) for different optimization design problems except for GTD, it is concluded that finding more fitness values are needed more research process for constrained optimization design problems. Generally, it is detected from the yielded results that the optimum values of algorithm parameters of HSA (HMS, HMCR, PAR) have been altered according to the characteristics of the design problem. In cases with smaller HMS values (20, 30) for WBD, PVD, and RCRW1, the large HMCR value (0.90) has increased the probability that the new solutions improvised in the algorithm is selected from the HM, while the new solution has been randomly selected from the design space with the possibility of small HMCR value (0.80) for SRD. The GTD unconstrainted design problem whose optimum values are obtained for HMS = 50 and HMCR = 0.95 shows that the optimum search has been sufficient with fewer iterations and runs due to the different characteristics of the design optimization problem and its small size.

3.2.2. Part II: Investigation of Four Optimum Design Parameter Values with Effect on the Fitness Value

It is apparent from analyses that the most effective algorithm parameter for reaching the best fitness value is mostly the RUN with S/N ratios and a change percentage of parameter effect for the 5P case. Furthermore, the minimum objective function value is estimated via the TIHHSA, when the optimum RUN value equals 1000 for WBD and PVD and 500 for GTD, SRD, and RCRW1 problems. For this reason, the S/N ratios, variance, and optimum Taguchi parameter values have been repeated by analyses which are taken as fix values whose optimum RUN value for four parameters (the 4P case) to reasonably observe the parameter effect of the other design parameters. According to DM given in the first section of Table 11, optimization analyzes have been performed and then response values given in the second section of Table 11 have been obtained.

Table 11. Response values (f(x)) and S/N ratios of response values.

DM							f(x)					S/N				
					RUN											
Design No	HMS	HMCR	PAR	MAXITER	WBD PVD	GTD SRDR CRW1	WBD ($)	PVD ($)	GTD (Unitless)	SRD (kg)	RCRW1 ($/kg)	WBD	PVD	GTD	SRD	RCRW1
1	20	0.80	0.10	2000	1000	500	1.8162	6105.21	2.70×10^{-12}	3000.31	190.785	−5.1831	−75.71	231.37	−69.54	−45.611
2	20	0.85	0.20	4000	1000	500	1.7882	6029.48	2.70×10^{-12}	2996.04	183.739	−5.0485	−75.61	231.37	−69.53	−45.284
3	20	0.90	0.30	6000	1000	500	1.8059	5978.28	2.70×10^{-12}	2996.04	182.141	−5.1340	−75.53	231.37	−69.53	−45.208
4	20	0.95	0.40	8000	1000	500	1.8331	6033.37	2.70×10^{-12}	2996.22	180.155	−5.2637	−75.61	231.37	−69.53	−45.113
5	30	0.80	0.20	6000	1000	500	1.7931	6055.08	2.70×10^{-12}	2996.34	182.660	−5.0721	−75.64	231.37	−69.53	−45.233
6	30	0.85	0.10	8000	1000	500	1.7869	5999.42	2.70×10^{-12}	2996.01	181.975	−5.0420	−75.56	231.37	−69.53	−45.200
7	30	0.90	0.40	2000	1000	500	1.8419	6127.22	2.70×10^{-12}	3000.97	189.144	−5.3055	−75.75	231.37	−69.55	−45.536
8	30	0.95	0.30	4000	1000	500	1.7892	6052.60	2.70×10^{-12}	2997.24	183.061	−5.0532	−75.64	231.37	−69.53	−45.252
9	40	0.80	0.30	8000	1000	500	1.9165	6003.01	2.70×10^{-12}	2996.23	184.049	−5.6501	−75.57	231.37	−69.53	−45.299
10	40	0.85	0.40	6000	1000	500	1.8280	6044.31	2.70×10^{-12}	2996.86	182.456	−5.2397	−75.63	231.37	−69.53	−45.223
11	40	0.90	0.10	4000	1000	500	1.8945	6098.92	2.70×10^{-12}	2997.09	182.774	−5.5498	−75.71	231.37	−69.53	−45.238
12	40	0.95	0.20	2000	1000	500	2.1363	6034.29	2.70×10^{-12}	3002.13	189.270	−6.5932	−75.61	231.37	−69.55	−45.542
13	50	0.80	0.40	4000	1000	500	1.9349	6087.48	2.70×10^{-12}	2997.00	190.329	−5.7333	−75.69	231.37	−69.53	−45.590
14	50	0.85	0.30	2000	1000	500	1.9848	6213.94	2.70×10^{-12}	3001.98	199.838	−5.9544	−75.87	231.37	−69.55	−46.014
15	50	0.90	0.20	8000	1000	500	1.8826	6073.32	2.70×10^{-12}	2995.93	180.585	−5.4950	−75.67	231.37	−69.53	−45.134
16	50	0.95	0.10	6000	1000	500	2.0482	6091.36	2.70×10^{-12}	2997.44	182.249	−6.2276	−75.69	231.37	−69.54	−45.213
				η			1.8800	6064.21	2.70×10^{-12}	2997.74	185.326	−5.4716	−75.66	231.37	−69.54	−45.356

In contrast to the Taguchi design with $L_{16}(4)^5$, obtained p values with SS and R values which are due to the reduction of the number of parameters for the Taguchi design with $L_{16}(4)^4$ are given in Table 12. The variance values (Table 12) have been specified by utilizing the S/N ratios (the third section of Table 11) have been calculated with the aim of smaller is the best. MAXITER is the most effective algorithm parameter according to variance and rank values for PVD, SRD, and RCRW1 design optimization problems. While the MAXITER has the first rank value for SRD and RCRW1 problems whose sizes are higher than the others due to the number of design constraints and design variables, it hasn't been a critical factor for the WBD problem. The HMS and HMCR design parameters are the first and second effective factors, respectively. It is noticed that the HMCR and PAR design parameters, which are included in the process of reaching the best solution according to the random number assigned in the algorithm, generally have lower variance. Conducting the statistical analysis with four design parameters instead of five has not shown reasonable and changing results for the GTD problem which is unconstrained and has a relatively small problem size.

Table 12. Results of variance analyses of WBD, PVD, GTD, SRD, and RCRW1 optimization problems.

Optimization Problem	Evaluation Criteria	Design Parameter			
		HMS	HMCR	PAR	MAXITER
WBD	SS	1.803926	0.53773	0.061213	0.45212
	ν	0.60131	0.17924	0.020407	0.150711
	p	0.100039	0.353229	0.901733	0.405384
	R	1	2	4	3
PVD	SS	0.031716	0.001684	0.003575	0.040544
	ν	0.0105718	0.000560958	0.00119166	0.0135146
	p	0.405366	0.971217	0.921773	0.332366
	R	2	4	3	1
SRD	SS	1.98×10^{-5}	1.33×10^{-6}	1.20×10^{-6}	5.93×10^{-6}
	ν	6.57677×10^{-6}	4.40671×10^{-6}	4.02287×10^{-7}	1.97717×10^{-4}
	p	0.046127	0.077733	0.658534	0.000333
	R	3	2	4	1
RCRW1	SS	0.094191	0.092708	0.050628	0.598937
	ν	0.0314006	0.0309005	0.0168764	0.199651
	p	0.141871	0.144468	0.272054	0.012293
	R	2	3	4	1

In Table 13, the prediction of the response value ($\eta_{prediction}$), and the optimum parameter combination have been demonstrated for each design problem. In verification analyses, the minimum objective function values (η_{real}) with the statistical evaluations have been obtained for estimated optimum values of design parameters (Table 14). While the optimum value of MAXITER has been found as its maximum level as 4 (8000) for PVD, SRD, and RCRW1, its optimum level is 2 (4000) and 1 (2000) for WBD and GTD, respectively. It is observed that the optimum level of HMS is equal to 1 (20) for all optimization design problems except for WBD to reach the minimum objective function. Consequently, it is concluded that the optimum values of HSA parameters act upon properties of the considered design optimization problem although the convergence of the minimum fitness value has increased with many iterations.

Table 13. Verification analyses data and results with optimum values of design parameters.

Optimization Problem	Optimum Parameter Combination	fmin ($\eta_{prediction}$)
WBD	HMS2-HMCR2-PAR4-MAXITER2	\$1.7291
PVD	HMS1-HMCR4-PAR2-MAXITER4	\$5973.13
GTD	HMS1-HMCR1-PAR1-MAXITER1	2.70086×10^{-12}
SRD	HMS1-HMCR1-PAR2-MAXITER4	2995.11 kg
RCRW1	HMS1-HMCR3-PAR2-MAXITER4	\$177.842/m

Table 14. Statistical results from verification analyses of engineering design optimization problems.

Case	Run						Iteration					
	BRN (%)	Best (η_{real})	Mean	Worst	StD	Median	BIN (%)	Best (η_{real})	Mean	Worst	StD	Median
WBD	373/1000 (37%)	1.78312	1.88853	6.22216	0.320969	1.82613	3561/4000 (89%)	1.78312	2.82897	4.61973	0.492371	2.74472
PVD	881/1000 (88%)	6005.19	7125.64	8728.03	503.107	7169.81	7870/8000 (%98)	6005.19	7442.32	67288.7	3342.35	6395.62
GTD	89/500 (%18)	2.70086×10^{-12}	5.4247×10^{-8}	1.38114×10^{-6}	1.34484×10^{-7}	1.31252×10^{-8}	1250/2000 (63%)	2.70086×10^{-12}	2.99998×10^{-5}	5.5068×10^{-3}	3.7964×10^{-3}	2.70086×10^{-12}
SRD	195/500 (40%)	2995.63	3002.75	3275.14	13.0823	3001.31	7187/8000 (90%)	2995.63	3037.28	5194.7	234.639	2998.37
RCRW1	460/500 (92%)	180.301	190.495	206.994	5.28013	190.4	7841/8000 (98%)	180.301	204.729	385.939	48.688	188.459

4. Discussion

In this section, yielded results from optimization and statistical analyses by utilizing proposed TIHHSA with different design parameters have been evaluated in terms of different design optimization problems. In addition, the examination of the change of design parameters for different design problems in terms of 5P and 4P cases has been made according to the comparison graphics given in Figures 8 and 9. By utilizing S/N ratios, the η_{ij} values have been determined and the variation between response value and design parameter levels are demonstrated for each design optimization problem according to five design parameters (the 5Pcase) and four design parameters (the 4P case) as shown in Figure 8a–e.

The S/N ratios which are control parameters in the Taguchi design supply information about the variation of the design parameters in different levels. In variation evaluations of η_{ij} values based on S/N ratios, three features have been observed. The first feature is belonging to the WBD problem with four design variables and seven design constraints and the PVD problem with four design variables and three design constraints and it is detected that there is an important change between S/N ratios of 5P and 4P. It is assumed that the contributions of other design parameters (HMS, HMCR, PAR, and MAXITER) are perceived more clearly, since the RUN parameter, which has the most variation in the 5P case, is taken as constant for its optimum value in the 4P case. The second feature, which has not emerged any change from 5P to 4P, is observed for the GTD problem with four design variables and no design constraints. It has been interpreted as no change because there are no design constraints, and it is a relatively small-size design optimization problem. The third feature has been determined for large-size design optimization problems which are SRD with seven design variables and eleven design constraints and RCRW1 with twelve design variables and twenty-six design constraints. As it has been apparent in the figures, since the RUN design parameter is not significantly effective in the 5P case, there is no significant change in other parameters in the 4P case.

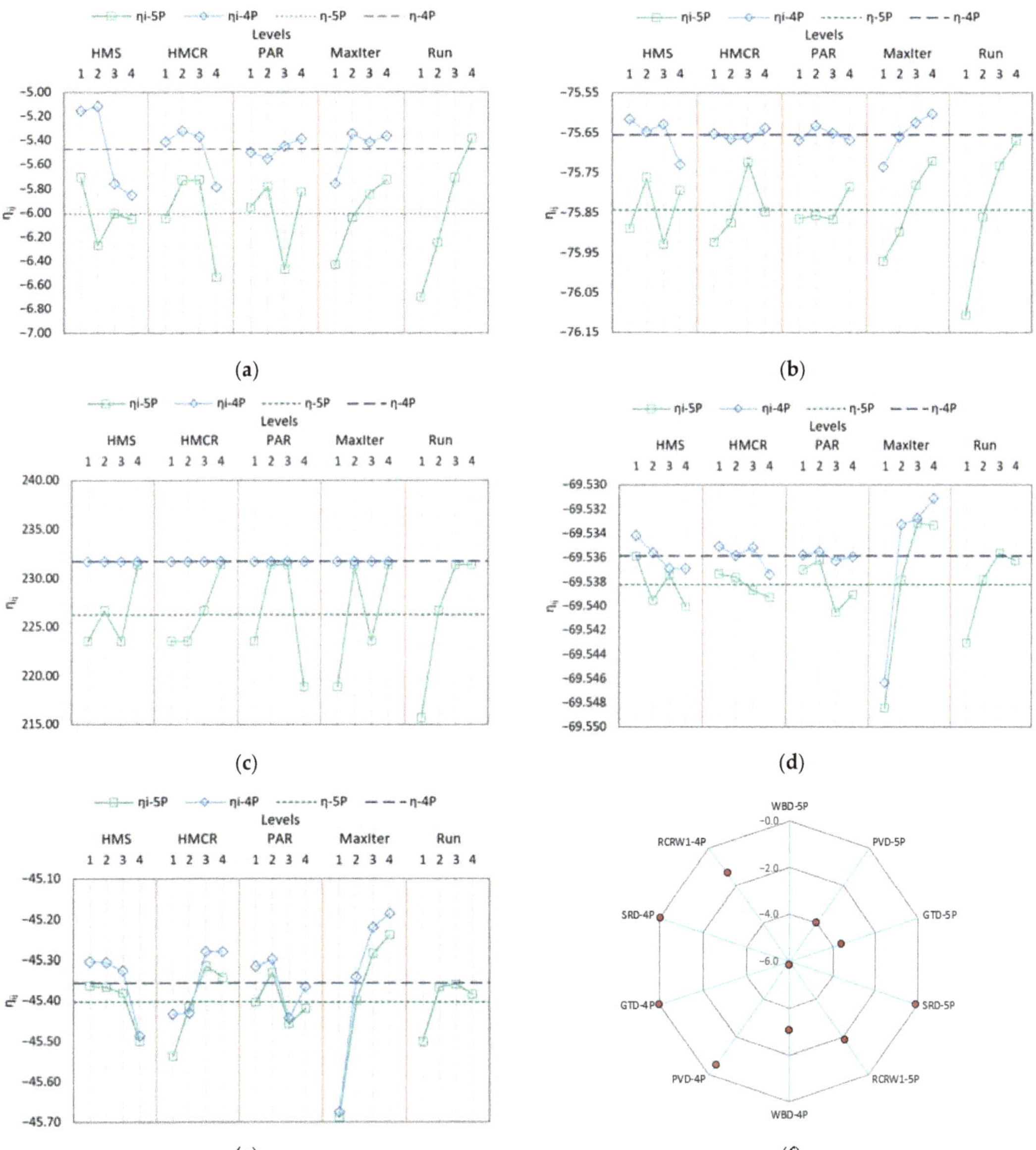

Figure 8. The change between the η_{ij} values and the design parameter with their levels for: (**a**) WBD; (**b**) PVD; (**c**) GTD; (**d**) SRD; (**e**) RCRW1 and (**f**) Relative error (ε, %) of all optimization design problems.

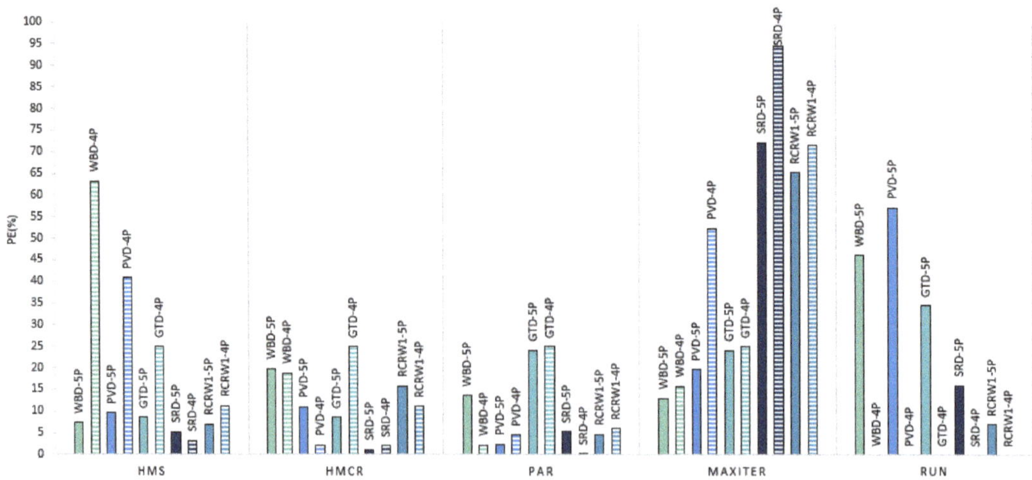

Figure 9. PE values for 4P and 5P cases of all optimization design problems in terms of design parameters.

The parameter percentage (PE, %) values which are based on the sum of squares values acquired from variance analyses are demonstrated for 4P and 5P cases of all design optimizations problems in Figure 9. In the 4P case where the RUN design parameter is taken as constant for depending on optimum values of the current design optimization problem, an increase has been monitored in the PE values yielded for the HMS, HMCR, PAR, and MAXITER design parameters compared to the 5P case. The PE values of the RUN design parameter in the 5P case have been obtained 46%, 57%, 35%, 16%, and 7% for WBD, PVD, GTD, SRD, and RCRW1 problems, respectively. When the PE values 4P and 5P cases have been compared in terms of changes in MAXITER values, it has been seen that the PE values have increased from 13% to 16% for WBD, from 20% to 52% for PVD, from 24% to 25% for GTD, from 72% to 95% for SRD and from 65% to 72% for RCRW1 problem. The HMS, HMCR, PAR, MAXITER and RUN design parameters have been the most effective factor for cases of WBD-4P (63%), GTD-4P (25%), GTD-4P (25%), SRD-4P (95%), and PVD-5P (57%), respectively.

When the optimum design parameter combinations estimated for the 4P and 5P cases of different design optimization problems (Figure 10) are compared, it is detected that a generalization cannot be made because it changes depending on the nature of the design optimization problem. This result has recognized that taking into account the different number of design variables, the process of reaching the best solution providing many design constraints has occurred differently for especially HSA design parameters. While the η_{real} values for 5P case have been accomplished as \$1.7455, \$6054.14, 2.70×10^{-12}, 2995.97 kg and 181.035 (\$/m), they have been found for 4P case as \$1.78312, \$6005.14, 2.70×10^{-12}, 2995.63 kg, and 180.301(\$/m) for WBD, PVD, GTD, SRD, and RCRW1 problems, respectively. Except for the WBD problem with 2% of change, it has not been found more minimum objective function values. When the relative error (ε, %) values, which are calculated by using η_{real} and $\eta_{prediciton}$, for all design optimization problems and the cases have been examined, the maximum ε value has been marked as 6% (Figure 8f).

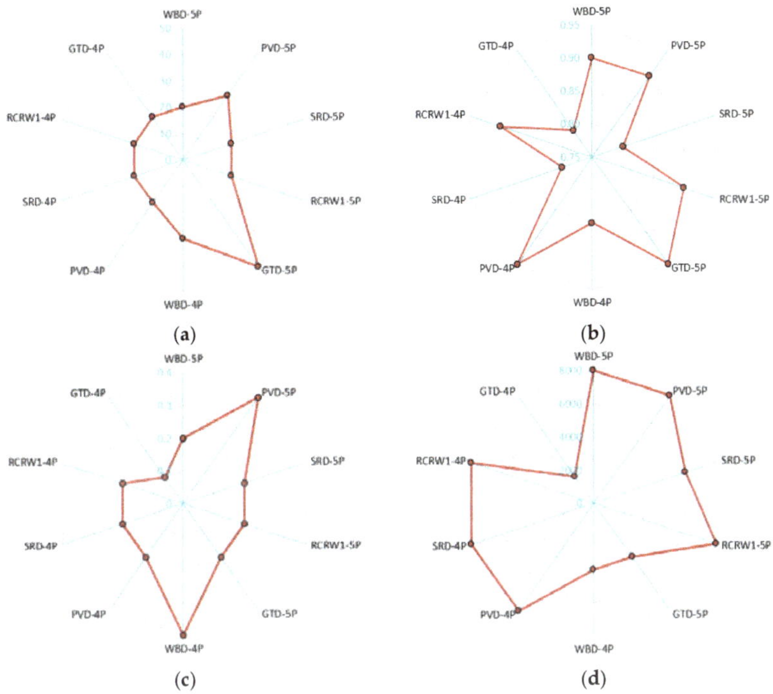

Figure 10. Parametric investigation of design parameters for all cases and optimization problems in terms of optimum values for; (**a**) HMS; (**b**) HMCR; (**c**) PAR; (**d**) MAXITER.

A comparison of the best fitness values presented in this study with those reported in the literature [73], which was shared the best fitness values of HSA and their variant, has been conducted for the WBD design optimization problem. When the mentioned study has been examined in terms of the best fitness value and the optimum values of the HS algorithm, it is seen that fmin value ($1.72489123) is obtained for HMS = 8, HMCR = 0.80, PAR = 0.30, and MaxIter = 200,000. In this study, the fmin value has been estimated as $1.7291 and has been found as $1.7455 with verification analyses at HMS = 20, HMCR = 0.95, PAR = 0.20, MAXITER = 8000, and RUN = 1000. These results show that it is possible to convergence to the best fitness value with fewer iterations in the optimization process. Besides, it is concluded that the fmin value ($1.7455) is reasonable according to compared with the optimum values acquired for other heuristic optimization methods given in Table A1 of the Appendix A section. Being almost the close fmin values eachother show that the Taguchi method is an alternative and effective in estimating the optimum algorithm parameter values of HSA.

The optimum values of MAXITER and RUN have been found as their maximum values (MAXITER = 6000 and 8000 and RUN = 500 and 1000) for different optimization design problems, except for GTD. While the f(x) minimum values of WBD and PVD for HMS = 20, HMCR = 0.90, PAR = 0.35, MAXITER= 30,000 and RUN = 1000 have been obtained as $1.74026 and $5959.86, the same values of GTD and SRD for HMS = 20, HMCR = 0.90, PAR = 0.35, MAXITER = 30,000 and RUN = 100 have been found as 2.70086×10^{-12} and 2994.79 kg (Table 5). While fmin (η_{real}) of WBD $1.7455 which is quite close to $1.74026 has been found for HMS = 20, HMCR = 0.90, PAR = 0.20, MAXITER = 8000 and RUN = 1000, fmin (η_{real}) of PVD $6054.14 which is quite close to $5959.86 too has been found for HMS = 30, HMCR = 0.90, PAR = 0.40, MAXITER = 8000 and RUN = 1000.

5. Conclusions

In this study, optimum values of harmony search algorithm (HSA) design parameters, which are the harmony memory size (HMS), the harmony memory consideration rate (HMCR), the pitch adjustment rate (PAR), maximum iteration number (MAXITER), and the independent run number of whole iterations (RUN), have been investigated for complex benchmark engineering design problems (the welded beam (WBD), the pressure vessel (PVD), the gear train (GTD), and the speed reducer (SRD)) and complicated real-size reinforcement cantilever retaining wall (RCRW) design problem. To examine the optimum values of algorithm design parameters, the Taguchi method integrated hybrid harmony search algorithm (TIHHSA) has been presented as a new hybrid method based on the Taguchi Method which is a statistical-based experiment procedure utilized in boosting algorithmic quality. In addition, the effect of algorithm design parameters on the best fitness value and characteristics of the optimization problem has been studied. The results yielded according to the optimum algorithm design parameters and the best fitness values, whose values do not change with repetitive statistical, and optimization analyzes for different engineering design optimization problems, are as follows;

Accomplished results from the Taguchi analyses show that converging to the best fitness value is possible with fewer iteration numbers in a shorter time;

- The obtained estimations have a reasonable relative error in determining optimum values of algorithm design parameters without performing many trials;
- It has been seen that the optimum values of the algorithm design parameters vary depending on the nature of the design optimization problem, which includes the number of design variables, the number of design constraints, exposure to the constraints.
- Instead of taking into account the value of the algorithm parameter proposed for characteristically different optimization problems in the literature, it has been concluded that using the optimum values yielded statistically according to the nature of the problem is an effective and prosperous manner in converging to the optimum.
- Instead of the trial-error method, which is time-consuming and exhaustive, it has been concluded that the newly proposed TIHHSA is a robust and reliable method for estimating the optimum algorithm parameter values of the harmony search metaheuristic optimization technique in a shorter time without conducting sensitivity analyses which are utilized to increase convergence rate in the solution of the design optimization problem.

Author Contributions: Conceptualization, E.U.; methodology, E.U. and S.C.; software, E.U.; validation, S.C.; writing—original draft preparation, E.U.; writing—review and editing, E.U, S.C., Z.W.G. and S.K.; supervision, Z.W.G. and S.K.; funding acquisition, Z.W.G. All authors have read and agreed to the published version of the manuscript.

Funding: This work was supported by a National Research Foundation of Korea (NRF) grant funded by the Korean government (MSIT) (2020R1A2C1A01011131). This research was also supported by the Energy Cloud R&D Program through the National Research Foundation of Korea (NRF) funded by the Ministry of Science, ICT (2019M3F2A1073164).

Institutional Review Board Statement: Not applicable.

Informed Consent Statement: Not applicable.

Data Availability Statement: Not applicable.

Conflicts of Interest: The authors declare no conflict of interest.

Appendix A

$$
\begin{aligned}
&\text{Design variables}: \vec{x} = [h, l, t, b]^T \\
&\text{Minimize}:\ f\left(\vec{x}\right) = 1.10471x_1^2x_2 + 0.04811x_3x_4(14.0 + x_2) \\
&\text{Subject to}:\ g_1\left(\vec{x}\right) = \tau\left(\vec{x}\right) - \tau_{max} \le 0 \\
&g_2\left(\vec{x}\right) = \sigma\left(\vec{x}\right) - \sigma_{max} \le 0 \\
&g_3\left(\vec{x}\right) = x_1 - x_4 \le 0 \\
&g_4\left(\vec{x}\right) = 0.10471x_1^2 + 0.04811x_3x_4(14.0 + x_2) - 5.0 \le 0 \\
&g_5\left(\vec{x}\right) = 0.125 - x_1 \le 0 \\
&g_6\left(\vec{x}\right) = \delta\left(\vec{x}\right) - \delta_{max} \le 0 \\
&g_7\left(\vec{x}\right) = P - P_c\left(\vec{x}\right) \le 0 \\
&\text{where } \tau\left(\vec{x}\right) = \sqrt{(\tau')^2 + 2(\tau')(\tau'')\tfrac{x_2}{2R} + (\tau'')^2}\ \tau' = \tfrac{P}{\sqrt{2}x_1x_2}\ \tau'' = \tfrac{MR}{J} \\
&M = P\left(L + \tfrac{x_2}{2}\right)\ R = \sqrt{\tfrac{x_2^2}{4} + \left(\tfrac{x_1+x_3}{2}\right)^2}\ \sigma\left(\vec{x}\right) = \tfrac{6PL}{x_4x_3^2}\ \delta\left(\vec{x}\right) = \tfrac{4PL^3}{Ex_3^3x_4} \\
&J = 2\left\{\sqrt{2}x_1x_2\left[\tfrac{x_2^2}{12} + \left(\tfrac{x_1+x_3}{2}\right)^2\right]\right\} \\
&P_c\left(\vec{x}\right) = \tfrac{4.013E\sqrt{x_3^2x_4^6}}{6L^2}\left(1 - \tfrac{x_3}{2L}\sqrt{\tfrac{E}{4G}}\right) \\
&P = 6000\text{lb}\ L = 14\text{in.}\ E = 30 \times 10^6\text{psi}\ G = 12 \times 10^6\text{psi} \\
&\tau_{max} = 13,600\text{psi}\ \sigma_{max} = 30,000\text{psi}\ \delta_{max} = 0.25\text{in.}
\end{aligned}
\tag{A1}
$$

$$
\begin{aligned}
&\text{Design variables}: \vec{x} = [T_s, T_h, R, L]^T \\
&\text{Minimize}:\ f\left(\vec{x}\right) = 0.6224x_1x_3x_4 + 1.7781x_2x_3^2 + 3.1661x_1^2x_4 + 19.8621x_1^2x_3 \\
&\text{Subject to}:\ g_1\left(\vec{x}\right) = -x_1 + 0.0193x_3 \le 0 \\
&g_2\left(\vec{x}\right) = -x_2 + 0.00954x_3 \le 0 \\
&g_3\left(\vec{x}\right) = -\pi x_3^2x_4 - \tfrac{4}{3}\pi x_3^3 + 1,296,000 \le 0 \\
&g_4\left(\vec{x}\right) = x_4 - 240 \le 0
\end{aligned}
\tag{A2}
$$

$$
\begin{aligned}
&\text{Design variables}: \vec{x} = [T_a, T_b, T_d, T_f]^T \\
&\text{Minimize}:\ f\left(\vec{x}\right) = \left(\tfrac{1}{6.931} - \tfrac{x_2x_3}{x_1x_4}\right)^2
\end{aligned}
\tag{A3}
$$

$$
\begin{aligned}
&\text{Design variables}: \vec{x} = [x_1, x_2, x_3, x_4, x_5, x_6, x_7]^T \\
&\text{Minimize}:\ f\left(\vec{x}\right) = 0.7854x_1x_2^2(3.3333x_3^3 + 14.9334x_3 - 43.0934) - 1.5079x_1\left(x_6^2 + x_7^2\right) \\
&+7.4777\left(x_6^3 + x_7^3\right) + 0.7854\left(x_4x_6^2 + x_5x_7^2\right) \\
&\text{Subject to}:\ g_1\left(\vec{x}\right) = 27x_1^{-1}x_2^{-2}x_3^{-1} \le 1.0 \\
&g_2\left(\vec{x}\right) = 397.5x_1^{-1}x_2^{-2}x_3^{-2} \le 1.0 \\
&g_3\left(\vec{x}\right) = 1.93x_4^3x_2^{-1}x_3^{-1}x_6^{-4} \le 1.0 \\
&g_4\left(\vec{x}\right) = 1.93x_5^3x_2^{-1}x_3^{-1}x_7^{-4} \le 1.0 \\
&g_5\left(\vec{x}\right) = \left(745^2x_4^2x_2^{-2}x_3^{-2} + 16.9 \times 10^6\right)/110^2x_6^6 \le 1.0 \\
&g_6\left(\vec{x}\right) = \left(745^2x_5^2x_2^{-2}x_3^{-2} + 157.5 \times 10^6\right)/85^2x_7^6 \le 1.0 \\
&g_7\left(\vec{x}\right) = x_2x_3/40 \le 1.0\ g_8\left(\vec{x}\right) = 5x_2/x_1 \le 1.0\ g_9\left(\vec{x}\right) = x_1/12x_2 \le 1.0 \\
&g_{10}\left(\vec{x}\right) = (1.5x_6 + 1.9)x_4^{-1} \le 1.0\ g_{11}\left(\vec{x}\right) = (1.1x_7 + 1.9)x_5^{-1} \le 1.0
\end{aligned}
\tag{A4}
$$

$$\begin{aligned}
&F_{ss} = \frac{\left(\sum W_c+\sum W_s+Q+\sum P_{Ay}\right)\tan\left(\frac{2}{3}O_b\right)+\frac{2}{3}x_1c_b+\sum P_P}{\sum P_{Ax}}\\
&F_{so} = \frac{\sum W_ix_i+\sum P_{Ay}x_{PAy}}{\sum P_{Ax}x_{PAx}}\ F_{sb} = \frac{q_{ult}}{q_{max}}\\
&\sum W_c = [x_1x_5 + Hx_4 + 0.5(x_3 - x_4)H + x_7x_8]\gamma_c\\
&\sum W_s = [H + 0.5\tan(\beta)(x_1 - x_2 - x_3)](x_1 - x_2 - x_3)\gamma_r\\
&\sum P_{Ax} = [q + 0.5\gamma_rK_a[\tan(\beta)(x_1 - x_2 - x_3) + H + x_5]][\tan(\beta)(x_1 - x_2 - x_3) + H + x_5]K_a\cos\beta\\
&\sum P_{Ay} = [q + 0.5\gamma_rK_a[\tan(\beta)(x_1 - x_2 - x_3) + H + x_5]][\tan(\beta)(x_1 - x_2 - x_3) + H + x_5]K_a\sin\beta\\
&\sum P_P = 0.5\gamma_b(D_f + x_8)^2K_p + 2c(D_f + x_8)^2\sqrt{K_p}\\
&K_a = \cos\beta\frac{\cos\beta-\sqrt{(\cos\beta)^2-(\cos O_r)^2}}{\cos\beta+\sqrt{(\cos\beta)^2-(\cos O_r)^2}}\ K_p = \tan^2\left(45 + \frac{O_b}{2}\right)\\
&q_{ult} = c_bN_c + \gamma_b(D_f + x_8)N_q + 0.5\gamma_bx_1N_\gamma\\
&q_{\substack{min\\max}} = \frac{\left(\sum W_c+\sum W_s+Q+\sum P_{Ay}\right)\tan\left(\frac{2}{3}O_b\right)+\frac{2}{3}x_1c_b+\sum P_P}{x_1}\left(1 \pm \frac{6e}{x_1}\right)\\
&e = \frac{\sum W_ix_i+\sum P_{Ay}x_{PAy}-\sum P_{Ax}x_{PAx}}{\left(\sum W_c+\sum W_s+Q+\sum P_{Ay}\right)\tan\left(\frac{2}{3}O_b\right)+\frac{2}{3}x_1c_b+\sum P_P}
\end{aligned} \tag{A5}$$

$$\begin{aligned}
&Vn_{s,t,h,k} = \phi 0.17\sqrt{f_c}bd\\
&Mn_{s,t,h,k} = \phi_m A_sf_y\left(d - \frac{a}{2}\right)
\end{aligned} \tag{A6}$$

$$\begin{aligned}
&Md_s = 1.7\left[qK_a\cos\beta\frac{((x_1-x_2-x_3)\tan\beta+H)^2}{2} + K_a\gamma_r\cos\beta\frac{((x_1-x_2-x_3)\tan\beta+H)^3}{6}\right]\\
&Vd_s = 1.7\left[qK_a\cos\beta(H_s + H - d_s) + K_a\gamma_r\cos\beta\frac{((x_1-x_2-x_3)\tan\beta+H-(x_5-c_c))^2}{2}\right]\\
&Md_t = \left[1.7\left((q_{max} - q_{min})\frac{x_1-x_2+q_{min}x_1}{6x_1} + \frac{q_{max}}{3}\right) - 0.9(\gamma_cx_5 + \gamma_rD_f)\right]x_2^2\\
&Vd_t = \left[1.7\left((q_{max} - q_{min})\frac{x_1-x_2+q_{min}x_1}{2x_1} + \frac{q_{max}}{2}\right) - 0.9(\gamma_cx_5 + \gamma_rD_f)\right](l_t - (x_5 - c_c))\\
&Md_h = \left[\left(\frac{1.7q+1.4\gamma_cx_5+1.4\gamma_rH}{2}\right) + \left(\frac{1.4W_5}{3}\right) - \left(\frac{q_1+2q_{min}}{6}\right)\right](x_1 - x_2 - x_3)^2\\
&Vd_h = \left[\begin{array}{l} 1.7q + 1.4\left(\gamma_cx_5 + \gamma_rH + \frac{W_5}{2}\left(1 + \frac{(x_5-c_c)+q_{min}(x_1-x_2-x_3)}{x_1-x_2-x_3}\right)\right)\\ -0.9\left(\frac{q_{min}}{2} + \frac{(q_{max}-q_{min})(x_1-x_2-x_3+(x_5-c_c)+q_{min}x_1)}{2x_1}\right)\end{array}\right](l_h - (x_5 - c_c))\\
&Md_k = \left[\frac{K_p\gamma_bx_8^3}{3}\right]\ Vd_k = \left[\frac{K_p\gamma_b(x_8-(x_7-c_c))+K_p\gamma_bx_8}{2}\right](x_8 - (x_7 - c_c))
\end{aligned} \tag{A7}$$

$$\begin{aligned}
&As_{min} = 0.25\frac{\sqrt{f_c}}{f_y}bd \geq 1.4\frac{bd}{f_y}\\
&As_{max} = 0.75 \times 0.85\frac{f_c}{f_y}\beta_1\left(\frac{600}{600+f_y}\right)bd\\
&L_d = \begin{cases} \frac{12f_y\psi_t\psi_e\lambda}{25\sqrt{f_c}} \geq 300\text{mm for } d_b < 19\text{mm}\\ \frac{12f_y\psi_t\psi_e\lambda}{20\sqrt{f_c}} \geq 300\text{mm for } d_b \geq 19\text{mm}\end{cases}\\
&L_{dh} = \left(\frac{0.24f_y}{\sqrt{f_c}}\right)d_b \geq 150\text{mm or } 8d_b
\end{aligned} \tag{A8}$$

$$\begin{aligned}
&\text{Design variables}: \vec{x} = [x_1, x_2, x_3, x_4, x_5, x_6, x_7, x_8, x_9, x_{10}, x_{11}, x_{12}]^T\\
&\text{Minimize}: f_{cost}(\vec{x}) = C_sW_{st} + C_cV_c\\
&f_{weight}(\vec{x}) = W_{st} + 100V_c\gamma_c\\
&\text{Subject to}: g_1(\vec{x}) = 1 - \frac{F_{ss}}{SF_{ss}}\ g_2(\vec{x}) = 1 - \frac{F_{so}}{SF_{so}} \leq 0\ g_3(\vec{x}) = 1 - \frac{F_{sb}}{SF_{sb}} \leq 0\ g_4(\vec{x}) = -q_{min} \leq 0\\
&g_{[5-8]}(\vec{x}) = \frac{Md_{case}}{Mn_{case}} - 1 \leq 0\ g_{[9-12]}(\vec{x}) = \frac{Vd_{case}}{Vn_{case}} - 1 \leq 0\ g_{[13-16]}(\vec{x}) = \frac{As_{mincase}}{As_{case}} - 1 \leq 0\\
&g_{[17-20]}(\vec{x}) = \frac{As_{case}}{As_{maxcase}} - 1 \leq 0\ g_{21}(\vec{x}) = \frac{x_2+x_3}{x_1} - 1 \leq 0\ g_{22}(\vec{x}) = \frac{x_6+x_7}{x_1} - 1 \leq 0\\
&g_{23}(\vec{x}) = \left(\frac{Ld_s}{x_5-c_c} - 1\right)\text{or}\left(\frac{Ldh_s}{x_5-c_c} - 1\right) \leq 0\ g_{24}(\vec{x}) = \left(\frac{Ld_t}{x_1-x_2-c_c} - 1\right)\text{or}\left(\frac{12db_t}{x_5-c_c} - 1\right) \leq 0\\
&g_{25}(\vec{x}) = \left(\frac{Ld_h}{x_2+x_3-c_c} - 1\right)\text{or}\left(\frac{12db_h}{x_5-c_c} - 1\right) \leq 0\ g_{26}(\vec{x}) = \left(\frac{Ld_k}{x_5-c_c} - 1\right)\text{or}\left(\frac{Ldh_k}{x_5-c_c} - 1\right) \leq 0
\end{aligned} \tag{A9}$$

Table A1. Optimum values and comparison of the best solutions in literature for WBD.

	Optimum Solutions	x_1(h)(in.)	x_2(l) (in.)	x_3(t) (in.)	x_4(b) (in.)	f(x) ($)
Literature	Ragsdell and Phillips [50]	0.24550	6.1960	8.2730	0.2455	2.38593
	Deb [51]	0.2489	6.173	8.1789	0.2533	2.433116
	Coello [52]	0.2088	3.4205	8.9975	0.21	1.7483
	Huang et al. [53]	0.203137	3.542998	9.033498	0.206179	1.73346
	Yildiz [54]	0.20573	3.47042	9.03649	0.205735	1.7248
	Çarbaş and Saka [55]	0.203907	3.499898	9.063898	0.205594	1.72966
Case [PS]	R30	0.206741	3.65285	8.54856	0.231265	1.85149
	R100	0.171535	4.42418	8.98313	0.208289	1.80231
	R500	0.198864	3.66442	8.94678	0.209895	1.75598
	R1000	0.19823	3.64539	9.02857	0.206407	1.74026
	WBD-5P	0.195872	3.70387	9.07235	0.205574	1.7455
	WBD-4P	0.188171	3.95948	8.91723	0.21133	1.78312

[PS] Present study.

Table A2. Optimum values and comparison of the best solutions in literature for PVD.

	Optimum Solutions	$x_1(T_s)$ (in.)	$x_2(T_h)$ (in.)	x_3(R) (in.)	x_4(L) (in.)	f(x) ($)
Literature	Sandgren [35]	1.125	0.625	48.97	106.72	7982.5
	Kannan and Kramer [56]	1.25	0.625	50	120	7198.20
	Deb [57]	0.9375	0.50	48.329	112.679	6410.381
	Coello [52]	0.8125	0.4375	40.3239	200.0	6288.7445
	Gao et al. [58]	0.75	0.375	38.8441	221.612	5852.639
	Çarbaş and Saka [55]	0.8125	0.4375	42.09845	176.6366	6059.7143
Case [PS]	R30	0.876366	0.434563	45.3293	140.344	6089.66
	R100	0.915835	0.454142	47.2497	121.872	6195.1
	R500	0.833985	0.413952	43.1577	163.94	6000.09
	R1000	0.814181	0.403799	42.1533	176.032	5959.86
	PVD-5P	0.84119	0.430252	43.5749	159.245	6054.14
	PVD-4P	0.822121	0.409189	42.367	173.44	6005.19

[PS] Present study.

Table A3. Optimum values and comparison of the best solutions in literature for GTD.

	Optimum Solutions	x_1(Ta) (piece)	x_2(Tb) (piece)	x_3(Td) (piece)	x_4 (Tf) (piece)	Gear ratio	f(x) (unitless)
Literature	Zhang and Wang [59]	43	16	19	49	0.1442	2.36×10^{-9}
	Deb and Goyal [60]	33	14	17	50	0.1442	1.362×10^{-9}
	Parsopoulos and Vrahatis [61]	43	16	19	49	0.1442	2.701×10^{-12}
	Gandomi [62]	43	16	19	49	0.1442	2.701×10^{-12}
	Arora et al. [63]	43	16	19	49	0.1442	2.701×10^{-12}
	Deniz [64]	43	16	19	49	0.1442	2.701×10^{-12}
Case [PS]	R30	44	13	21	43	0.144292	1.54505×10^{-10}
	R100	43	16	19	49	0.144281	2.70086×10^{-12}
	R500	49	16	19	43	0.144281	2.70086×10^{-12}
	R1000	49	16	19	43	0.144281	2.70086×10^{-12}
	GTD-5P	49	16	19	43	0.144281	2.70086×10^{-12}
	GTD-4P	49	16	19	43	0.144281	2.70086×10^{-12}

[PS] Present study.

Table A4. Optimum values and comparison of the best solutions in literature for SRD.

	Optimum Solutions	x_1 (cm)	x_2 (cm)	x_3 (piece)	x_4 (cm)	x_5 (cm)	x_6 (cm)	x_7 (cm)	f(x) (kg)
Literature	Li and Papalambros [65]	3.50	0.70	17.00	7.30	7.71	3.3500000	5.2900000	2996.30977
	Kuang et al. [66]	3.60	0.70	17.00	7.30	7.80	3.4000000	5.0000000	2876.22
	Azarm and Li [67]	3.50	0.70	17.00	7.30	7.71	3.3500000	5.2900000	2996.30978
	Vanderplaats [68]	3.50	0.70	17.00	7.30	7.30	3.3502145	5.2865176	2985.15188
	Ray [69]	3.50	0.70	17.00	7.30	7.30	3.3502145	5.2865176	2985.15188
	Carbas et al. [70]	3.50	0.70	17.00	7.17984	7.70889	3.35009	5.28668	2993.13917
Case [PS]	R30	3.5001	0.700016	17.0017	7.30052	7.71562	3.35025	5.28667	2994.93
	R100	3.50014	0.700021	17.0002	7.30117	7.71637	3.35053	5.28667	2994.79
	R500	3.50029	0.700019	17.0001	7.3009	7.71572	3.35053	5.28681	2994.90
	R1000	3.50025	0.700016	17.0004	7.30034	7.71652	3.35036	5.28673	2994.84
	SRD-5P	3.50184	0.700073	17.0008	7.3036	7.72133	3.35036	5.28679	2995.97
	SRD-4P	3.50006	0.700006	17.0034	7.30108	7.71868	3.3516	5.28677	2995.63

[PS] Present study.

Table A5. Constraint values of WBD, PVD, and SRD optimization problems.

		R30	R100	R500	R1000	5P	4P
WBD	$g_1(x)$	−8.821	−0.086	−7.948	−7.525	−65.745	−32.706
	$g_2(x)$	−178.142	−14.670	−1.819	−45.079	−213.275	−7.722
	$g_3(x)$	−0.025	−0.037	−0.011	−0.008	−0.010	−0.023
	$g_4(x)$	−3.317	−3.338	−3.400	−3.414	−3.407	−3.368
	$g_5(x)$	−0.082	−0.047	−0.074	−0.073	−0.071	−0.063
	$g_6(x)$	−0.235	−0.235	−0.235	−0.236	−0.236	−0.235
	$g_7(x)$	−2211.802	−202.416	−330.005	−55.908	−1.921	−446.574
PVD	$g_1(x)$	−0.002	−0.002	−0.004	−0.001	−0.001	0.000
	$g_2(x)$	−0.002	−0.002	−0.003	−0.002	−0.002	−0.015
	$g_3(x)$	−89.466	−89.466	−636.323	−9.038	−412.835	−498.618
	$g_4(x)$	−99.656	−99.656	−118.128	−76.060	−63.968	−80.755

		R30	R100	R500	R1000	5P	4P
SRD	$g_1(x)$	0.92592	0.92598	0.92595	0.92595	0.92536	0.92587
	$g_2(x)$	0.80178	0.80190	0.80188	0.80187	0.80134	0.80165
	$g_3(x)$	0.50085	0.50086	0.50081	0.50079	0.50141	0.50012
	$g_4(x)$	0.09535	0.09539	0.09536	0.09539	0.09555	0.09545
	$g_5(x)$	0.99994	0.99944	0.99944	0.99974	0.99975	0.99752
	$g_6(x)$	0.99998	0.99998	0.99982	0.99991	0.99985	0.99987
	$g_7(x)$	0.29754	0.29751	0.29751	0.29751	0.29755	0.29756
	$g_8(x)$	0.99999	0.99999	0.99994	0.99995	0.99958	0.99999
	$g_9(x)$	0.41667	0.41667	0.41669	0.41669	0.41684	0.41667
	$g_{10}(x)$	0.94861	0.94859	0.94862	0.94866	0.94824	0.94882
	$g_{11}(x)$	0.99996	0.99987	0.99997	0.99986	0.99924	0.99958

Table A6. Optimum values of RCRW design for the optimum cost (RCRW1).

	Optimum Solutions	x_1 (m)	x_2 (m)	x_3 (m)	x_4 (m)	x_5 (m)	x_6 (m)	x_7 (m)	x_8 (cm^2)	x_9 (cm^2)	x_{10} (cm^2)	x_{11} (cm^2)	x_{12} (cm^2)	f(x) ($/kg)
Literature	Gandomi [40]	2.709	1	0.412	0.25	0.4	2.455	0.2	0.2	21.9911	11.7809	11.7809	4.7124	163.98
	Gandomi [40]	2.727	1.035	0.36	0.28	0.401	2.274	0.293	0.296	32.1699	13.3517	13.8544	8.6394	182.79
	Gandomi [40]	2.816	0.988	0.447	0.294	0.422	2.223	0.367	0.203	21.9911	15.2681	15.2681	12.7234	185.05
	Gandomi [40]	2.694	0.836	0.403	0.27	0.405	2.346	0.227	0.445	23.7504	12.7234	22.6195	26.1380	182.84
Case [PS]	R30	3.3439	1.1598	0.3526	0.2504	0.4	2.5435	0.2002	0.2	21.2999	14.3212	18.8016	7.1532	180.082
	R100	3.3431	1.1596	0.3917	0.25	0.4001	2.4623	0.2001	0.2002	18.8963	14.2586	18.9194	9.5172	179.842
	R500	3.3351	1.1593	0.4418	0.25	0.4001	3.0523	0.2004	0.2002	16.6776	14.3189	16.557	7.2319	179.449
	R1000	3.3394	1.1593	0.3916	0.2501	0.4	2.7275	0.2004	0.2003	18.691	14.392	18.7638	7.2583	179.693
	RCRW1−5P	3.36761	1.15955	0.391877	0.250366	0.400094	2.84595	0.200248	0.201545	18.7689	14.4128	18.7273	10.0995	181.035
	RCRW1−4P	3.34757	1.15992	0.394711	0.250081	0.400157	2.37294	0.202173	0.201494	18.7718	14.1765	18.7523	9.34978	180.301

[PS] Present study.

Table A7. Optimum values of RCRW design for the optimum weight (RCRW2).

	Optimum Solutions	x_1 (m)	x_2 (m)	x_3 (m)	x_4 (m)	x_5 (m)	x_6 (m)	x_7 (m)	x_8 (cm^2)	x_9 (cm^2)	x_{10} (cm^2)	x_{11} (cm^2)	x_{12} (cm^2)	f(x) ($/kg)
Literature	Gandomi [40]	2.709	1	0.412	0.25	0.4	2.455	0.2	0.2	21.9911	11.7809	11.7809	4.7124	5668.5
	Gandomi [40]	2.727	1.035	0.36	0.28	0.401	2.274	0.293	0.296	32.1699	13.3517	13.8544	8.6394	6034.4
	Gandomi [40]	2.816	0.988	0.447	0.294	0.422	2.223	0.367	0.203	21.9911	15.2681	15.2681	12.7234	6095.9
	Gandomi [40]	2.694	0.836	0.403	0.27	0.405	2.346	0.227	0.445	23.7504	12.7234	22.6195	26.1380	6094.4
Case[PS]	R30	3.342	1.1599	0.2505	0.2504	0.4	3.1292	0.2001	0.2001	35.1479	14.1715	21.5672	7.0299	5886.67
	R100	3.3426	1.16	0.2501	0.2501	0.4	3.0709	0.2001	0.2002	35.1127	15.0607	21.2321	7.0677	5883.61
	R500	3.343	1.1581	0.25	0.25	0.4001	3.0355	0.2001	0.2002	35.247	14.4184	21.3267	7.3429	5883.64
	R1000	3.3434	1.1594	0.2501	0.25	0.4	3.1304	0.2001	0.2001	35.2741	14.1095	21.2558	9.3741	5884.09
	RCRW1−5P	2.709	1	0.412	0.25	0.4	2.455	0.2	0.2	21.9911	11.7809	11.7809	4.7124	5668.5
	RCRW1−4P	2.727	1.035	0.36	0.28	0.401	2.274	0.293	0.296	32.1699	13.3517	13.8544	8.6394	6034.4

[PS] Present study.

Table A8. Constraint values of RCRW designs for optimum cost and optimum weight.

	Optimum Cost (RCRW1)						Optimum Weight (RCRW2)			
	R30	R100	R500	R1000	5P	4P	R30	R100	R500	R1000
$g_1(x)$	−0.0462	−0.043	−0.037	−0.0421	−0.0497	−0.0441	−0.0536	−0.0537	−0.0543	−0.0541
$g_2(x)$	−0.638	−0.6334	−0.6224	−0.6305	−0.6613	−0.6376	−0.6468	−0.6472	−0.648	−0.6484
$g_3(x)$	−2.5539	−2.569	−2.5824	−2.5638	−2.6425	−2.5813	−2.5106	−2.5112	−2.5092	−2.5137
$g_4(x)$	−0.1129	−0.1141	−0.0221	−0.0199	−1.8384	−0.3677	−0.0072	−0.0142	−0.0149	−0.0925
$g_5(x)$	−0.0001	−0.0004	0.0000	−0.0002	−0.0009	−0.0085	−0.0633	−0.0615	−0.061	−0.0615
$g_6(x)$	−0.4023	−0.4058	−0.4103	−0.4057	−0.4115	−0.4066	−0.3937	−0.4242	−0.3951	−0.3944
$g_7(x)$	−0.0326	−0.0701	−0.0052	−0.0728	−0.0531	−0.07	−0.062	−0.0482	−0.0464	−0.0472
$g_8(x)$	−0.9964	−0.9973	−0.9964	−0.9964	−0.9974	−0.9973	−0.9964	−0.9964	−0.9964	−0.9973
$g_9(x)$	−0.3955	−0.4637	−0.5333	−0.4635	−0.464	−0.4684	−0.1169	−0.1154	−0.115	−0.1154
$g_{10}(x)$	−0.4032	−0.4065	−0.4107	−0.4064	−0.4096	−0.407	−0.3956	−0.3955	−0.3964	−0.3959
$g_{11}(x)$	−0.065	−0.0845	−0.1132	−0.0858	−0.0767	−0.0847	−0.0179	−0.0175	−0.0167	−0.017
$g_{12}(x)$	−0.9936	−0.9935	−0.9936	−0.9935	−0.9934	−0.9936	−0.9935	−0.9935	−0.9935	−0.9935
$g_{13}(x)$	−0.4189	−0.2727	−0.0625	−0.2729	−0.2724	−0.2671	−0.7519	−0.7523	−0.7524	−0.7523
$g_{14}(x)$	−0.0097	−0.0095	−0.0095	−0.0097	−0.0095	−0.0093	−0.0097	−0.0618	−0.0095	−0.0097
$g_{15}(x)$	−0.2573	−0.2571	−0.151	−0.2573	−0.2571	−0.257	−0.3504	−0.3408	−0.3406	−0.3408
$g_{16}(x)$	−0.0087	−0.2569	−0.0077	−0.0077	−0.3028	−0.2492	−0.0092	−0.0092	−0.0092	−0.2569
$g_{17}(x)$	−0.6471	−0.7181	−0.7813	−0.718	−0.7182	−0.7202	−0.1734	−0.1721	−0.1718	−0.1721
$g_{18}(x)$	−0.7929	−0.793	−0.793	−0.7929	−0.793	−0.793	−0.7929	−0.7814	−0.793	−0.7929
$g_{19}(x)$	−0.7239	−0.724	−0.7585	−0.7239	−0.724	−0.724	−0.6844	−0.689	−0.689	−0.689
$g_{20}(x)$	−0.7931	−0.7241	−0.7934	−0.7934	−0.7059	−0.7269	−0.793	−0.793	−0.793	−0.7241
$g_{21}(x)$	−0.5477	−0.536	−0.5199	−0.5356	−0.5393	−0.5356	−0.578	−0.5781	−0.5788	−0.5784
$g_{22}(x)$	−0.1795	−0.2036	−0.0247	−0.1232	−0.0954	−0.2308	−0.0038	−0.0214	−0.0321	−0.0039
$g_{23}(x)$	−0.2382	−0.3654	−0.3654	−0.3652	−0.3654	−0.3655	−0.3652	−0.3652	−0.3654	−0.3652
$g_{24}(x)$	−0.6364	−0.6365	−0.6365	−0.6364	−0.6365	−0.6365	−0.6364	−0.4909	−0.6365	−0.6364
$g_{25}(x)$	−0.6364	−0.6365	−0.6365	−0.6364	−0.6365	−0.6365	−0.4182	−0.5636	−0.5638	−0.5636
$g_{26}(x)$	−0.1113	−0.3654	−0.1115	−0.1113	−0.1115	−0.3655	−0.1113	−0.1113	−0.1115	−0.3652

References

1. Houssein, E.H.; Mahdy, M.A.; Shebl, D.; Mohamed, W.M. A Survey of Metaheuristic Algorithms for Solving Optimization Problems. *Stud. Comput. Intell.* **2021**, *967*, 515–543. [CrossRef]
2. Dubey, M.; Kumar, V.; Kaur, M.; Dao, T.P. A Systematic Review on Harmony Search Algorithm: Theory, Literature, and Applications. *Math. Probl. Eng.* **2021**, *2021*, 5594267. [CrossRef]
3. Dorigo, M.; Gambardella, L.M. Ant Colonies for the Travelling Salesman Problem. *BioSystems* **1997**, *43*, 73–81. [CrossRef]
4. Kennedy, J.; Eberhart, R. Particle Swarm Optimization. In Proceedings of the ICNN'95—International Conference on Neural Networks, Perth, WA, Australia, 27 November–1 December 1995; IEEE: Piscataway, NJ, USA, 1995; Volume 4, pp. 1942–1948.
5. Karaboga, D. *An Idea Based on Honey Bee Swarm for Numerical Optimization*; Technical Report TR06; Erciyes University: Kayseri, Turkey, 2005.

6. Mirjalili, S.; Lewis, A. The Whale Optimization Algorithm. *Adv. Eng. Softw.* **2016**, *95*, 51–67. [CrossRef]
7. Rashedi, E.; Nezamabadi-pour, H.; Saryazdi, S. GSA: A Gravitational Search Algorithm. *Inf. Sci.* **2009**, *179*, 2232–2248. [CrossRef]
8. Erol, O.K.; Eksin, I. A New Optimization Method: Big Bang–Big Crunch. *Adv. Eng. Softw.* **2006**, *37*, 106–111. [CrossRef]
9. Rajabioun, R. Cuckoo Optimization Algorithm. *Appl. Soft Comput.* **2011**, *11*, 5508–5518. [CrossRef]
10. Yang, X.S. Firefly Algorithms for Multimodal Optimization. In *International Symposium on Stochastic Algorithms. SAGA 2009: Stochastic Algorithms: Foundations and Applications*; Watanabe, O., Zeugmann, T., Eds.; Lecture Notes in Computer Science; Springer: Berlin/Heidelberg, Germany, 2009; Volume 5792, pp. 169–178. ISBN 978-3-642-04943-9. [CrossRef]
11. Yang, X.-S. A New Metaheuristic Bat-Inspired Algorithm. In *Nature Inspired Cooperative Strategies for Optimization*; González, J.R., Pelta, D.A., Cruz, C., Terrazas, G., Natalio, K., Eds.; Springer: Berlin/Heidelberg, Germany, 2010; Volume 284, pp. 65–74.
12. Storn, R.; Price, K. Differential Evolution—A Simple and Efficient Heuristic for Global Optimization over Continuous Spaces. *J. Glob. Optim.* **1997**, *11*, 341–359. [CrossRef]
13. Simon, D. Biogeography-Based Optimization. *IEEE Trans. Evol. Comput.* **2008**, *12*, 702–713. [CrossRef]
14. Goldberg, D.E.; Kuo, C.H. Genetic Algorithms in Pipeline Optimization. *J. Comput. Civ. Eng.* **1987**, *1*, 128–141. [CrossRef]
15. Kirkpatrick, S.; Gelatt, C.D.; Vecchi, M.P. Optimization by Simulated Annealing. *Science* **1983**, *220*, 671–680. [CrossRef]
16. Geem, Z.W.; Kim, J.H.; Loganathan, G.V. A New Heuristic Optimization Algorithm: Harmony Search. *Simulation* **2001**, *76*, 60–68. [CrossRef]
17. Hakli, H.; Ortacay, Z. An Improved Scatter Search Algorithm for the Uncapacitated Facility Location Problem. *Comput. Ind. Eng.* **2019**, *135*, 855–867. [CrossRef]
18. Mahdavi, M.; Fesanghary, M.; Damangir, E. An Improved Harmony Search Algorithm for Solving Optimization Problems. *Appl. Math. Comput.* **2007**, *188*, 1567–1579. [CrossRef]
19. Liu, Y. Hybrid Particle Swarm Optimizer for Constrained Optimization Problems. *Qinghua Daxue Xuebao/J. Tsinghua Univ.* **2013**, *53*, 242–246.
20. Fesanghary, M.; Mahdavi, M.; Minary-Jolandan, M.; Alizadeh, Y. Hybridizing Harmony Search Algorithm with Sequential Quadratic Programming for Engineering Optimization Problems. *Comput. Methods Appl. Mech. Eng.* **2008**, *197*, 3080–3091. [CrossRef]
21. Sheikholeslami, R.; Khalili, B.G.; Sadollah, A.; Kim, J.H. Optimization of Reinforced Concrete Retaining Walls via Hybrid Firefly Algorithm with Upper Bound Strategy. *KSCE J. Civ. Eng.* **2016**, *20*, 2428–2438. [CrossRef]
22. Search for Hybrid Optimization in Web of Science. Available online: https://www.webofscience.com/wos/woscc/summary/19ecf7e7-1092-459a-bc98-e94ff827d659-0de1a535/relevance/1 (accessed on 22 October 2021).
23. Search for Hybrid Harmony Search in Web of Science. Available online: https://www.webofscience.com/wos/woscc/summary/f92b392b-51ef-41e5-9784-2f785de062f3-0f0bc6d0/relevance/1 (accessed on 22 October 2021).
24. Akay, B.; Karaboga, D. A Modified Artificial Bee Colony Algorithm for Real-Parameter Optimization. *Inf. Sci.* **2012**, *192*, 120–142. [CrossRef]
25. Uray, E.; Hakli, H.; Carbas, S. Statistical Investigation of the Robustness for the Optimization Algorithms. In *Nature-Inspired Metaheuristic Algorithms for Engineering Optimization Applications*; Carbas, S., Toktas, A., Ustun, D., Eds.; Springer: Singapore, 2021; pp. 201–224.
26. Taguchi, G. *Introduction to Quality Engineering: Designing Quality into Products and Processes*; The Organization: Tokyo, Japan, 1986; ISBN 9283310837.
27. Taguchi, G.; Chowdhury, S.; Wu, Y. *Taguchi's Quality Engineering Handbook*; John Wiley & Sons, Inc.: Hoboken, NJ, USA, 2005; ISBN 9780471413349.
28. Chao, S.-M.; Whang, A.J.-W.; Chou, C.-H.; Su, W.-S.; Hsieh, T.-H. Optimization of a Total Internal Reflection Lens by Using a Hybrid Taguchi-Simulated Annealing Algorithm. *Opt. Rev.* **2014**, *21*, 153–161. [CrossRef]
29. Tsai, J.; Liu, T.; Chou, J. Hybrid Taguchi-Genetic Algorithm for Global Numerical Optimization. *IEEE Trans. Evol. Comput.* **2004**, *8*, 365–377. [CrossRef]
30. Cheng, B.W.; Chang, C.L. A Study on Flowshop Scheduling Problem Combining Taguchi Experimental Design and Genetic Algorithm. *Expert Syst. Appl.* **2007**, *32*, 415–421. [CrossRef]
31. Jia, X.; Lu, G. A Hybrid Taguchi Binary Particle Swarm Optimization for Antenna Designs. *IEEE Antennas Wirel. Propag. Lett.* **2019**, *18*, 1581–1585. [CrossRef]
32. Yildiz, A.R.; Öztürk, F. Hybrid Taguchi-Harmony Search Approach for Shape Optimization. *Stud. Comput. Intell.* **2010**, *270*, 89–98. [CrossRef]
33. Uray, E.; Tan, Ö.; Çarbaş, S.; Erkan, H. Metaheuristics-Based Pre-Design Guide for Cantilever Retaining Walls. *Tek. Dergi* **2021**, *32*, 10967–10993. [CrossRef]
34. Rao, S. *Engineering Optimization: Theory and Practice*; John Wiley & Sons: Hoboken, NJ, USA, 1996.
35. Sandgren, E. Nonlinear Integer and Discrete Programming in Mechanical Design Optimization. *J. Mech. Des.* **1990**, *112*, 223–229. [CrossRef]
36. Jan Golinski Optimal Synthesis Problems Solved by Means of Nonlinear Programming and Random Methods. *J. Mech.* **1970**, *5*, 287–309. [CrossRef]
37. Afzal, M.; Liu, Y.; Cheng, J.C.P.; Gan, V.J.L. Reinforced Concrete Structural Design Optimization: A Critical Review. *J. Clean. Prod.* **2020**, *260*, 120623. [CrossRef]

38. ACI Committee. *Building Code Requirements for Structural Concrete (ACI 318-05) and Commentary (ACI 318R-05)*; American Concrete Institute: Farmington Hills, MI, USA, 2005.
39. Akin, A.; Saka, M.P. Optimum Design of Concrete Cantilever Retaining Walls Using the Harmony Search Algorithm. *Civ. -Comp Proc.* **2010**, *93*, 1–21. [CrossRef]
40. Gandomi, A.H.; Kashani, A.R.; Roke, D.A.; Mousavi, M. Optimization of Retaining Wall Design Using Evolutionary Algorithms. *Struct. Multidiscip. Optim.* **2016**, *55*, 809–825. [CrossRef]
41. Kalemci, E.N.; İkizler, S.B.; Dede, T.; Angın, Z. Design of Reinforced Concrete Cantilever Retaining Wall Using Grey Wolf Optimization Algorithm. *Structures* **2020**, *23*, 245–253. [CrossRef]
42. Uray, E.; Çarbaş, S.; Erkan, İ.H.; Tan, Ö. Parametric Investigation for Discrete Optimal Design of a Cantilever Retaining Wall. *Chall. J. Struct. Mech.* **2019**, *5*, 108. [CrossRef]
43. Uray, E.; Çarbaş, S.; Erkan, İ.H.; Olgun, M. Investigation of Optimal Designs for Concrete Cantilever Retaining Walls in Different Soils. *Chall. J. Concr. Res. Lett.* **2020**, *11*, 39. [CrossRef]
44. Camp, C.; Akin, A. Design of Retaining Walls Using Big Bang–Big Crunch Optimization. *J. Struct. Eng.* **2012**, *138*, 438–448. [CrossRef]
45. Das, B.M.; Sivakugan, N. *Principles of Foundation Engineering*, 9th ed.; Cengage Learning: Boston, MA, USA, 2017; ISBN 9781305970953.
46. Gandomi, A.H.; Kashani, A.R.; Roke, D.A.; Mousavi, M. Optimization of Retaining Wall Design Using Recent Swarm Intelligence Techniques. *Eng. Struct.* **2015**, *103*, 72–84. [CrossRef]
47. Lee, K.S.; Geem, Z.W. A New Meta-Heuristic Algorithm for Continuous Engineering Optimization: Harmony Search Theory and Practice. *Comput. Methods Appl. Mech. Eng.* **2005**, *194*, 3902–3933. [CrossRef]
48. Rankovic, N.; Rankovic, D.; Ivanovic, M.; Lazic, L. A New Approach to Software Effort Estimation Using Different Artificial Neural Network Architectures and Taguchi Orthogonal Arrays. *IEEE Access* **2021**, *9*, 26926–26936. [CrossRef]
49. Deb, K. An Efficient Constraint Handling Method for Genetic Algorithms. *Comput. Methods Appl. Mech. Eng.* **2000**, *186*, 311–338. [CrossRef]
50. Ragsdell, K.M.; Phillips, D.T. Optimal Design of a Class of Welded Structures Using Geometric Programming. *J. Manuf. Sci. Eng. Trans. ASME* **1976**, *98*, 1021–1025. [CrossRef]
51. Deb, K. Optimal Design of a Welded Beam via Genetic Algorithms. *AIAA J.* **1991**, *29*, 2013–2015. [CrossRef]
52. Coello Coello, C.A. Use of a Self-Adaptive Penalty Approach for Engineering Optimization Problems. *Comput. Ind.* **2000**, *41*, 113–127. [CrossRef]
53. Huang, F.Z.; Wang, L.; He, Q. An Effective Co-Evolutionary Differential Evolution for Constrained Optimization. *Appl. Math. Comput.* **2007**, *186*, 340–356. [CrossRef]
54. Yildiz, A.R. Hybrid Taguchi-Harmony Search Algorithm for Solving Engineering Optimization Problems. *Int. J. Ind. Eng. Theory Appl. Pract.* **2008**, *15*, 286–293.
55. Carbas, S.; Saka, M.P. Efficiency of Improved Harmony Search Algorithm for Solving Engineering Optimization Problems. *Iran Univ. Sci. Technol.* **2013**, *3*, 99–114.
56. Kannan, B.K.; Kramer, S.N. An Augmented Lagrange Multiplier Based Method for Mixed Integer Discrete Continuous Optimization and Its Applications to Mechanical Design. *J. Mech. Des.* **1994**, *116*, 405–411. [CrossRef]
57. Deb, K. GeneAS: A Robust Optimal Design Technique for Mechanical Component Design. *Evol. Algorithms Eng. Appl.* **1997**, 497–514. [CrossRef]
58. Gao, L.; Zou, D.; Ge, Y.; Jin, W. Solving Pressure Vessel Design Problems by an Effective Global Harmony Search Algorithm. In Proceedings of the 2010 Chinese Control and Decision Conference, CCDC 2010, Xuzhou, China, 26–28 May 2010; pp. 4031–4035. [CrossRef]
59. Zhang, C.; Awang, H.-P.(Ben). Mixed-Discrete Nonlinear Optimization with Simulated Annealing. *Eng. Optim.* **2007**, *21*, 277–291. [CrossRef]
60. Deb, K.; Deb, K.; Goyal, M. A Combined Genetic Adaptive Search (GeneAS) for Engineering Design. *Comput. Sci. Inform.* **1996**, *26*, 30–45.
61. Parsopoulos, K.E.; Vrahatis, M.N. Unified Particle Swarm Optimization for Solving Constrained Engineering Optimization Problems. *Lect. Notes Comput. Sci.* **2005**, *3612*, 582–591. [CrossRef]
62. Gandomi, A.H.; Yang, X.-S.; Alavi, A.H. Cuckoo Search Algorithm: A Metaheuristic Approach to Solve Structural Optimization Problems. *Eng. Comput.* **2013**, *29*, 17–35. [CrossRef]
63. Arora, S.; Singh, S.; Yetilmezsoy, K. A Modified Butterfly Optimization Algorithm for Mechanical Design Optimization Problems. *J. Braz. Soc. Mech. Sci. Eng.* **2018**, *40*, 21. [CrossRef]
64. Ustun, D. An Enhanced Adaptive Butterfly Optimization Algorithm Rigorously Verified on Engineering Problems and Implemented to ISAR Image Motion Compensation. *Eng. Comput.* **2020**, *37*, 3543–3566. [CrossRef]
65. Li, H.L.; Papalambros, P. A Production System for Use of Global Optimization Knowledge. *J. Mech. Transm. Autom. Des.* **1985**, *107*, 277–284. [CrossRef]
66. Kuang, J.K.; Rao, S.S.; Chen, L. Taguchi-Aided Search Method for Design Optimization of Engineering Systems. *Eng. Optim.* **1998**, *30*, 1–23. [CrossRef]
67. Azarm, S.; Li, W.-C. Multi-Level Design Optimization Using Global Monotonicity. *J. Mech. Transm. Autom.* **1989**, *111*, 259–263. [CrossRef]

68. Vanderplaats, G.N. *Conmin, a Fortran Program for Constrained Function Minimization: User's Manual*; Technical Memorandum TM X-62282; NASA Ames Research Center: Moffett Field, CA, USA, 1973.
69. Ray, T. Golinski's Speed Reducer Problem Revisited. *AIAA J.* **2012**, *41*, 556–558. [CrossRef]
70. Çarbas, S.; Tunca, O.; Yıldızel, S. Contemporary Optimization Assessment of Complicated Engineering Problems. In Proceedings of the International Conference on Mathematical Studies and Applications, Karaman, Turkey, 4 October 2018; pp. 61–66.
71. Bissell, A.F. Interpreting Mean Squares in Saturated Fractional Designs. *J. Appl. Stat.* **2006**, *16*, 7–18. [CrossRef]
72. Chen, Y.; Chan, C.K.; Leung, B.P.K. An Analysis of Three-Level Orthogonal Saturated Designs. *Comput. Stat. Data Anal.* **2010**, *54*, 1952–1961. [CrossRef]
73. Li, H.; Shih, P.C.; Zhou, X.; Ye, C.; Huang, L. An Improved Novel Global Harmony Search Algorithm Based on Selective Acceptance. *Appl. Sci.* **2020**, *10*, 1910. [CrossRef]

symmetry

MDPI

Article

Frequency-Constrained Optimization of a Real-Scale Symmetric Structural Using Gold Rush Algorithm

Sepehr Sarjamei, Mohammad Sajjad Massoudi * and Mehdi Esfandi Sarafraz

Department of Civil Engineering, West Tehran Branch, Islamic Azad University, Tehran 1468763785, Iran; sarjamei.sepehr@wtiau.ac.ir (S.S.); sarafraz.m@wtiau.ac.ir (M.E.S.)
* Correspondence: massoudi.ms@wtiau.ac.ir

Abstract: The optimal design of real-scale structures under frequency constraints is a crucial problem for engineers. In this paper, linear analysis, as well as optimization by considering natural frequency constraints, have been used for real-scale symmetric structures. These structures require a lot of time to minimize weight and displacement. The cyclically symmetric properties have been used for decreasing time. The structure has been decomposed into smaller repeated portions termed substructures. Only the substructure elements are needed when analyzing and designing with the concept of cyclic symmetries. The frequency constrained design of real-scale structures is a complex optimization problem that has many local optimal answers. In this research, the Gold Rush Optimization (GRO) algorithm has been used to optimize weight and displacement performances due to its effectiveness and robustness against uncertainties. The efficacy of the concept of cyclic symmetry to minimize the time calculated is assessed by three examples, including Disk, Silo, and Cooling Tower. Numerical results indicate that the proposed method can effectively reduce time consumption, and that the GRO algorithm results in a 14–20% weight reduction of the problems.

Keywords: structural optimization; frequency constraints; cyclic symmetry; Gold Rush Optimization algorithm

Citation: Sarjamei, S.; Massoudi, M.S.; Esfandi Sarafraz, M. Frequency-Constrained Optimization of a Real-Scale Symmetric Structural Using Gold Rush Algorithm. *Symmetry* **2022**, *14*, 725. https://doi.org/10.3390/sym14040725

Academic Editors: Jan Awrejcewicz and Sergei D. Odintsov

Received: 21 February 2022
Accepted: 31 March 2022
Published: 2 April 2022

1. Introduction

In vibrational analysis, the optimal design of real-scale symmetric structures under frequency constraints is a crucial problem. Since the modal properties of a structure determine its dynamic behavior, the frequency constraints and the capacity to adjust the values of natural frequencies are sensitive items in the analysis and design. Concerning the frequency constraints, including non-convex search spaces, sophisticated methods are needed [1]. Since the frequency constrained design of large-scale structures is a complex optimization problem with many local optima, an appropriate optimization technique is usually required. Among the research conducted to optimize the design of structures under frequency constraints, the following studies can be briefly reviewed.

Using laws of momentum and energy between collisions bodies, Kaveh and Mahdavi [2] introduced a new Colliding Bodies Optimization algorithm (CBO). Kaveh and Mahdavi [3] looked into the effectiveness of CBO for the problem and conducted parametric research on its internal characteristics. Enhanced Colliding Bodies Optimization (ECBO) introduced by Kaveh and Ilchi Ghazaan [4] improved the function of the CBO algorithm. ECBO uses memory to save some optimal solutions. Enhanced Colliding Bodies Optimization (ECBO) was used by Kaveh and Ilchi Ghazaan [5] to demonstrate the algorithm's efficiency in frequency-constrained structural optimization. Song and Zhang [6] assessed the wind deflection of a railway catenary in a crosswind under frequency constraints, based on wind tunnel tests and a nonlinear finite element model. Ho-Huu et al. [7] proposed a new version of the Differential Evolution (DE) method called Roulette Wheel Selection-Elitist-Differential Evolution (ReDE), which employs elitism in the selection phase using the

Roulette Wheel Selection technique. Lieu et al. [8] proposed the Adaptive hybrid Evolutionary Firefly Method by combining the differential evolution (DE) algorithm and the Firefly Algorithm (FA) (AHEFA). Tejani and Mirjalili [9] used Symbiotic Organisms Search (SOS) to optimize the size of space trusses. SOS is based on the biological interactions between organisms in an ecosystem. Kaveh and Dadras [10] have introduced a chaotic version of a newly-established metaheuristic algorithm called the Water Strider Algorithm (WSA) to tackle this problem. Kaveh and Ilchi Ghazaan [1] used the ECBO method to optimize large-scale dome trusses with frequency limitations, incorporating multi-stage cascading techniques. The possibilities of the Vibrating Particles System (VPS), an algorithm inspired by the damped oscillation of a single degree of freedom system, to cope with large-scale dome trusses were examined by Kaveh and Ilchi Ghazaan [11]. To handle a large number of variables, Kaveh and Ilchi Ghazaan [12] combined the VPS technique with multi-design variable configuration (Multi-DVC) cascade optimization, as well as employing an upper bound strategy (UBS) to reduce computing time. Weight optimization of truss structures with different frequency constraints was investigated by Carvalho et al. [13]. Rao [14] created the Teaching-Learning-Based Optimization (TLBO) algorithm based on a traditional school learning. Kar et al. [15] proposed a Craziness-based Particle Swarm Optimization (CRPSO), which they used. They employed cardinality constraints and frequency constraints to limit the maximum number of distinct cross-sectional areas, lowering the cost of selecting a different cross-section of elements and weights of structures. The Charged System Search (CSS) algorithm was introduced using principles from physics and mechanics [16]. Furthermore, they utilized a combination of governing Coulomb law from electrostatics and the Newtonian laws of mechanics. The (CSS) algorithm and its enhanced version (ECSS) are being used to optimize various truss structures [17]. To improve the CSS algorithm's convergence time, Jalili and Talatahari [18] devised a hybrid Charged System Search (CSS) method with a Migration-based Local Search (MBLS) mechanism. The effectiveness of the proposed hybrid approach was proved in their research by proving the optimum design of many benchmark truss instances with frequency constraints. For the best design of large-scale cyclically symmetric dome trusses with frequency constraints, Kaveh and Zolghadr [19] employed the Cyclical Parthenogenesis Algorithm (CPA). They used the block diagonalization technique to divide the domes' repeated patterns into smaller parts, lowering the computer time necessary for the analysis. Liu et al. [20] added the vision search radius for each fruit fly as well as an enhanced Deb (IDeb) rule to handle the limitations to the Fruit Fly Optimization Algorithm (FOA) utilizing a memory-based search strategy. They used this technique to optimize truss structures with frequency constraints, demonstrating that the new algorithm finds better answers. To modify the attractiveness and light absorption coefficients of FA, Kaveh, and Javadi [21] used two chaotic maps, namely Logistic and Gaussian maps. These chaotic algorithms were used to optimize large-scale domes that have various frequency constraints.

Real-scale structures need a lot of effort to find the modal parameters, whereas symmetric structures can be solved rapidly. Wang [22] optimized the real-scale bridge cables under frequency constraints. For tackling rotationally periodic structures, Williams [23] presented an accurate eigen solution technique. The component mode technique was used by Tran [24] for vibration analysis of cyclic symmetry systems. He used a scaled finite element approach for cyclically symmetric domain heat transport and structural mechanics difficulties [25,26]. For optimal structural analysis, graph theory [27,28] has been suggested as a helpful solution. Kaveh and Koohestani [29] created graph models for ordinary finite element meshes. In the free vibration analysis of cyclically repeated structures, Kaveh and Rahami [30] used block circulant matrices. Kaveh and Rahami [31] proposed a method for efficiently computing graph product-generated repeating structures. Using the force method, Koohestani [32] proposed an orthogonal self-stress matrix for quickly evaluating cyclically symmetric space truss designs. Koohestani [33] applied the properties of symmetry in graph theory to finite and boundary elements. For the free vibration analysis of cyclic symmetry, Koohestani [34] implemented the decomposition of extended Eigen problems.

This research aimed to optimize the design of real-scale symmetric structures under frequency constraints using the GRO metaheuristic algorithm. It has been practically impossible to optimize real-scale symmetric structures in the previous research using meta-heuristic algorithms due to the large volume and time of calculations. This study hypothesized that the abilities of the cyclic symmetric concept could reduce the time and volume of calculations. Using the metaheuristic algorithm and the concept of cyclic symmetry, real-scale structures such as a Disk, Silo, and Cooling Tower have been investigated. Furthermore, the results are discussed and compared with CSS and TLBO algorithms.

2. Materials and Methods

In this section, the methodology of the frequency constraint, cyclically symmetric formulation, and optimization algorithm are introduced.

2.1. Methodology of the Frequency Constraint Optimization Problem

This problem aims to find the optimal design for structures with real-scale cyclic symmetry and multiple frequency constraints, where the cross-sectional area of the structural members is constantly changing in the search space. In such problems, size optimization minimizes weight while satisfying the constraints. The mathematical formula of the problem can be expressed as Equation (1).

$$\begin{aligned} & Find\ X = [x_1, x_2, \ldots, x_{nDV}],\ x_i \in R_i \\ & \text{to minimize } P(X) = f_{penalty}(X) \times W(X); \\ & \text{subject to}: \begin{cases} \omega_j \le \omega_j^*\ for\ some\ natural\ frequencies\ j \\ \omega_K \le \omega_k^*\ for\ some\ natural\ frequencies\ k \end{cases} \quad x_i^L \le x_i \le x_i^U \\ & R_i = \{x_i \,|x_i \in [x_i^L, x_i^U]\} \end{aligned} \tag{1}$$

where the vector X contains the design variables (sections), the ith design variable is x_i. (nDV) is the number of design variables according to the grouping of elements. $W(X)$ shows the weight of the structure. The penalty approach is used to consider the constraints, in which the cost function is defined as that which must be minimized. $f_{penalty}(X)$ is a penalty function. When certain constraints are violated in a particular solution, the penalty function $f_{penalty}(X)$ artificially increases the weight of the structure by taking values more significant than one. Also, ω_j is the jth natural frequency of the structure, ω_j^* is the upper limit, ω_k is the kth natural frequency of the structure, and ω_k^* is the lower limit. x_i^L and x_i^U are the lower and upper bounds of the design variable x_i. R_i is the allowable range of the ith design variable. The design variable x_i can be constantly changed in R_i. The weight of the structure is calculated from Equation (2).

$$W(X) = \sum_{i=1}^{nDV} x_i \sum_{j=1}^{nm\ (i)} \rho_j L_j \tag{2}$$

In which $nm(i)$ is the number of members allocated to the ith element. ρ_j and L_j are the material density and the length of the jth member, respectively. The penalty function is defined in Equation (3).

$$f_{penalty}(X) = (1 + \varepsilon_1.v)^{\varepsilon 2} \tag{3}$$

For a particular solution, v shows the sum of violations, defined in Equation (4).

$$v = \sum_{i=1}^{s} v_i \tag{4}$$

In which S is the number of frequency constraints. Values of v_i can be considered as shown in Equation (5).

$$v_i = \begin{cases} 0 & \text{if the } i\text{th constraint is satisfied} \\ \left|1 - \frac{\omega_i}{\omega_i^*}\right| & else \end{cases} \tag{5}$$

In this study, ε_1 and ε_2 are calculated from Equation (6) to create a suitable balance between the algorithm's exploration and exploitation.

$$\begin{aligned} \varepsilon_1 &= 1.5 + 0.5 \times \frac{NSA}{MaxNSA} \\ \varepsilon_2 &= 1.5 + 1.5 \times \frac{NSA}{MaxNSA} \end{aligned} \tag{6}$$

where NSA is the current analysis number and $MaxNSA$ is the total number of structural analyses for the optimization process and is defined as the criterion for the optimization termination. As the value of NSA increases, the values of ε_1 and ε_2 grow. It can be inferred that the algorithm explores the search space in the early stages but in the final stages tends to choose solutions without violations. Equation (6) helps the algorithm search near a low-cost solution as a final design and converge to reduce errors.

2.2. Cyclically Symmetric Formulation

One of the general tasks in engineering is to determine the natural frequency of structures on a real scale. The system displays a specific pattern in structures with cyclic symmetry. The rotations of several repeating identical units (called substructures) along the central axis make up a cyclically symmetric structure [29]. Structural analysis is performed on only one of the substructures using the cyclic symmetry concept. Except at support nodes, points and elements are numbered in each substructure from top to bottom. The pattern of stiffness and mass matrices is obtained by using the right way of numbering nodes, which is the same as the Canonical Form F matrix as shown:

$$K = \begin{bmatrix} K_{11} & K_{12} & & & K_{21} \\ K_{21} & K_{11} & K_{12} & & \\ & . & . & . & \\ & & K_{21} & K_{11} & K_{12} \\ K_{12} & & & K_{21} & K_{11} \end{bmatrix}_{n\times n} \qquad M = \begin{bmatrix} m_{11} & m_{12} & & & m_{21} \\ m_{21} & m_{11} & m_{12} & & \\ & . & . & . & \\ & & m_{21} & m_{11} & m_{12} \\ m_{12} & & & m_{21} & m_{11} \end{bmatrix}_{n\times n}$$

In which n is the number of repetitive substructures, $K_{12} = K_{21}^t$, and $m_{12} = m_{21}^t$. Given this, determining the natural frequency of the cyclically symmetric structure is a matter of total eigenvalue, as shown in Equation (7):

$$K\phi = M\phi D \tag{7}$$

In Equation (7), the matrix D is diagonal. The values of matrix D are λ_i general eigenvalues. The T_i periods of the substructure are obtained from Equation (8).

$$T_i = 2\pi/\sqrt{\lambda_i} \qquad i = 1, \ldots, n \tag{8}$$

φ_i the eigenvector corresponds to the ith eigenvalues, and Equation (7) is rewritten as follows for each substructure Equation (9):

$$K\varphi_i = \lambda_i \, M \, \varphi_i \qquad i = 1, \ldots, n \tag{9}$$

The mass and stiffness matrix for each substructure is as follows in Equation (10) to reduce computing time using the canonical concept:

$$\begin{aligned} K &= I \otimes k_{11} + H \otimes k_{12} + H^t \otimes k_{21} \\ M &= I \otimes m_{11} + H \otimes m_{12} + H^t \otimes m_{21} \end{aligned} \tag{10}$$

In Equation (10), I and H are an n × n identity matrix and a special and important matrix, respectively, with the desired pattern presented as follows:

$$I = \begin{bmatrix} 1 & & & & \\ & 1 & & & \\ & & \cdot & & \\ & & & \cdot & \\ & & & 1 & \\ & & & & 1 \end{bmatrix}_{n\times n} \qquad H = \begin{bmatrix} 0 & 1 & & & & \\ & 0 & 1 & & & \\ & & \cdot & \cdot & & \\ & & & \cdot & 1 & \\ & & & & 0 & 1 \\ 1 & & & & & 0 \end{bmatrix}_{n\times n}$$

A full description of this section can be found in Kaveh [29].

2.3. Optimization Algorithm

In this study, the GRO, CSS, and TLBO algorithms are used for Equation (1). These algorithms have been adopted due to their ability to optimize truss structures in previous studies. The algorithms are stated below:

2.3.1. Gold Rush Optimization (GRO) Algorithm

Massoudi and Sarjamei created a GRO algorithm [35] based on the power of human thinking and decision making, and which will be called a Gold Rush Optimization. The GRO algorithm is a population-based evolutionary algorithm with a higher convergence speed than other optimization algorithms. The aim is to find the place of gold. Firstly, a group of people called operators stand in a random spot of search space. Every operator uses a device (metal detector) to find gold. In every stage, the operators move altogether and listen to the sound until they hear an increase in the sound and then stop at that point. Every operator would also listen to the sounds produced by other devices and constantly monitor if any other devices create a louder sound. At each stage, the group moves to the place of the loudest sound. In the end, the exact location of the gold is determined. Three parameters α, β, and γ indicate the probability of moving towards the loudest sound or moving away from it. The parameters α, β, and γ in the interval [0–1] are selected.

Level 1: Initialization

Each operator stands randomly in one spot inside the search space as represented in Equation (11). lb_i and ub_i are the lower and upper bounds of a domain (search space). *rand* in the interval [0–1] is a random number, and N is the number of operators.

$$location_i^{(0)} = lb_i + (ub_i - lb_i) * rand, i = 1, 2, \ldots, N \tag{11}$$

Level 2: Monitoring-Choosing the best locations

SOP is an operator who is successful in finding the optimal location. In this step, SOP should be generated. At the end of every iteration, the top ten percent of operators should be chosen and kept in the SOP.

Level 3: Fitness-distance

The analysis of the loudness of every sound (rate), operator with the most probability to extract gold, is calculated from Equation (12):

$$\text{rate}(i) = \frac{D_i}{\rho} * \frac{sound(highest\ volume) - sound(i)}{(sound(highest\ volume) - sound(lowest\ volume) + \varepsilon)} \tag{12}$$

The epsilon (ε) is a small positive number to avoid singularities. To prevent errors from environmental, the coefficients, ρ, and D_i represented in Equation (13) are used. The indices i and j indicate the current position of the two operators.

$$\rho = 2 - \frac{iter}{max_{iter}} \quad , \quad D_i = \sqrt{(x_i - x_j)^2 + (y_i - y_j)^2 + \ldots} \tag{13}$$

Level 4: Think-Decisions-move

In this step, every operator will create completely different selections based on a mix of sounds represented in Equation (14).

$$new\ location(i) = location(i) + md \times [(rate(j) - rate(i)) * (\ location(j) - location(i)) * rand] \tag{14}$$

The coefficients md means move direction determined from Equation (15):

$$md = \begin{cases} +1 \Rightarrow towards\ a\ loudest\ sound? & \alpha > rand \\ -1 \Rightarrow away\ from\ a\ loudest\ sound? & \alpha < rand \end{cases} \tag{15}$$

Level 5: Correct location

If the location obtained in Equation (14) does not meet the problem's constraints, Equation (16) is utilized to generate new locations. β and γ coefficients are selected as $0 < \beta < \gamma < 1$.

$$new\ location(i) = \begin{cases} choose\ a\ neighboring\ location & rand < \beta \\ select\ a\ new\ location\ randomly & \beta < \ rand < \gamma \\ do\ not\ move & \gamma < rand \end{cases} \tag{16}$$

Level 6: Termination

Steps 4 to 6 are eventually repeated in a loop until one of the following terminating conditions is met:

1. The maximum number of tries.
2. There has been no noticeable change in the optimal location.
3. The gap between the SOP function's values and the obtained most optimal answer is smaller than a pre-determined expected threshold. The parameters in the interval [0–1] are selected.
4. If the difference between the best and worst location's objective values is smaller than a given accuracy.

In this study, the GRO algorithm is used for Equation (1). The optimum amount of weight of an element is obtained. The algorithm is performed using MATLAB.

The flowchart of the GRO algorithm is illustrated in Figure 1.

2.3.2. Charged System Search (CSS) Algorithm

The Charged System Search (CSS) algorithm was created by Kaveh and Talatahari [16,36] as an efficient population-based metaheuristic using some physics and mechanics concepts, and it has been effectively applied to a variety of structural optimization problems [37–40]. CSS is based on the electrical Coulomb laws and the Newtonian rules of mechanics. Each agent in this algorithm is a charged particle (CP) with a fixed radius. The charge of a particle's magnitude q_i is calculated based on its quality from Equation (17):

$$q_i = \frac{fit(i) - fitworst}{fitbest - fitworst} \qquad i = 1, 2, \ \ldots, \ N \tag{17}$$

The best and the worst fitness of all the particles are $fitbest$ and $fitworst$, respectively; $fit(i)$ represents the fitness of the agent i, and N is the total number of CPs.

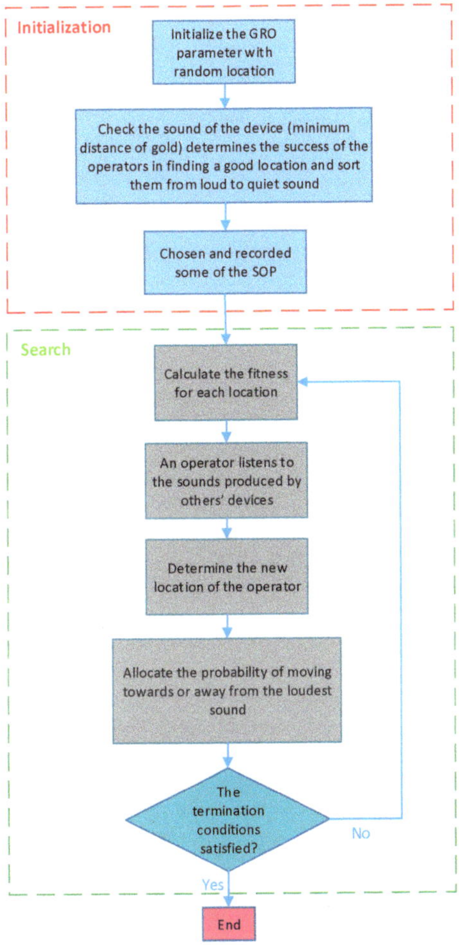

Figure 1. The Gold Rush Optimization algorithm flowchart.

Between two charged particles, the separation distance r_{ij} is defined from Equation (18):

$$r_{ij} = \frac{\|X_i - X_j\|}{\|(X_i - X_j)/2 - X_{best}\| + \varepsilon} \tag{18}$$

Here X_i and X_j are the positions of ith and the jth of charged particles. The position of the best charge particles is X_{best}. The epsilon (ε) is a small positive number to avoid singularities.

Each good particle generates an electric field that attracts other electrically charged things from Equation (19):

$$p_{ij} = \begin{cases} 1 & \frac{fit(i)-fitbest}{fit(j)-fit(i)} > rand \vee fit(j) > fit(i) \\ 0 & else \end{cases} \tag{19}$$

As a result, charged particles can interact with another one depending on their fitness values and separation distance from Equation (20):

$$F_j = q_j \sum_{i,i\neq j} \left(\frac{q_i}{a^3} r_{ij}.i_1 + \frac{q_i}{r_{ij}^2}.i_2 \right) p_{ij} (X_i - X_j) \quad \left\langle \begin{array}{l} j = 1,2, \ldots, N \\ i_1 = 1, i_2 = 0 \Leftrightarrow r_{ij} < a \\ i_1 = 0, i_2 = 1 \Leftrightarrow r_{ij} \geq a \end{array} \right. \tag{20}$$

The resultant force acting on the jth charged particle is F_j. (a) the radius of the charged sphere is set to unity.

The new position and velocity of each CP are computed by Equation (21):

$$\begin{array}{ll} X_{j,new} &= rand_{j1}.k_a.\frac{F_j}{m_j}.\Delta t^2 + rand_{j2}.k_v.V_{j,old}.\Delta t + X_{j,old} \\ V_{j,new} &= \frac{X_{j,new} - X_{j,old}}{\Delta t} \end{array} \tag{21}$$

K_a and K_v are the acceleration coefficient and the velocity coefficient, respectively; $rand_{j1}$ and $rand_{j2}$ in the interval [0–1] are random numbers. m_j is the mass of the charged particles, and Δt is the time step set to one. Electrostatics laws are used to calculate the magnitude of the resultant force, whereas Newtonian mechanics laws are used to define the quality of the movement.

2.3.3. Teaching-Learning-Based Optimization (TLBO) Algorithm

Rao [14] created the Teaching-Learning-Based Optimization (TLBO) algorithm based on traditional school learning. The influence of a teacher on students and the effect of students on each other are the two stages of this algorithm. The population of random solutions in TLBO was dubbed students or learners at the start and initialized the population size (P_n). In TLBO, the regular distribution of marks received by pupils is considered as the performance of the class in learning or the teacher's performance in instructing. In each iteration, the best learner or most intelligent student with the best goal function is designated as the instructor. Students are updated iteratively to find the best solution in two phases: the first is based on the knowledge that transferred from a teacher (teacher phase), and the second is based on interaction with other students (interaction phase) (learner phase).

In the first phase (teacher phase), the mean of each design variable is calculated. The best solution that will act as a teacher is given by Equation (22).

$$X_{teacher} = X_{f(x)=\min} \tag{22}$$

$f\,(X)$ is the objective function, and X is a design variable.

The critical difference between the two normal distributions is the mean value (M), which means that a better instructor will teach pupils with higher average scores. In the teacher phase, TLBO enhances other pupils by utilizing the difference between the instructor's knowledge and the intermediate knowledge of all students. Modified solution based on the best solution is given by Equation (23):

$$X_{new,i} = X_{old,i} + r_i(M_{new} - T_F M_i) \tag{23}$$

r_i in the interval [0–1] is a random number. Determining the change in mean value is a teaching factor (T_F) and is explained in Equation (24).

$$T_F = round[1 + rand(0,1)\{2-1\}] \tag{24}$$

rand in the interval [0–1] is a random number.

In the second phase (learner phase), each student's knowledge is derived from their position in the search space. Students can also improve themselves by conversing with another student after the teacher has finished teaching. The mathematical expression is explained as follows (25).

$$\begin{cases} X_{new,i} = X_{old,i} + r_i(X_i - X_j) & f(X_i) < f(X_j) \\ X_{new,i} = X_{old,i} + r_i(X_j - X_i) & else \end{cases} \quad (25)$$

TLBO increases each student's knowledge after contact with another randomly selected student throughout the learner phase

In this study, three numerical examples, including a Disk, Silo, and Cooling Tower, have been studied to evaluate the efficiency of the proposed method. In numerical examples, the results of optimal design by the GRO algorithm are evaluated and compared with two other famous algorithms.

3. Numerical Examples

In this section, the effectiveness of the concept of cyclic symmetry for minimizing the time required is assessed by three examples with continuous environments, including Disk, Silo, and Cooling tower. It is assumed that the number of frequencies in Equation (1) is equal to the number of degrees of freedom of the substructure. Structural properties which should be considered in all problems are listed in the following: the number of elements of the substructure is 30, the number of nodes of the substructure is 31, the number of repetitions of the substructure is 60, the number of elements of the structure is 1800, and the number of nodes of the structure is 1860. Material properties of these structures are: modulus of elasticity (E) = 2.4 × 10^7 (kN/m^2), mass per unit volume (ρ) = 2.4 (kNs2m-4), Poisson's ratio (ν) = 0.2. Examples have used a flat thin-shell element. This element is obtained by combining two elements, Q4 and DKQ. The isoperimetric four-node quadrilateral Q4 [41] element and the Discrete Kirchhoff Quadrilateral DKQ [42] element. The quadrilateral flat shell element has 24 degrees of freedom (6 for each node). The structure's weight is obtained from Equation (1). Furthermore, CSS and TLBO, two well-known algorithms in structural engineering problems, are used for optimization. The whole structure is then modeled in MATLAB using the approach mentioned above. In all examples, due to the random nature of meta-heuristic algorithms, each algorithm was run 20 times independently. The best result from 20 independent performances was reported as the best answer. Calculations were performed on the first-generation Intel Corei3 CPU. Frequency and displacement constraints were considered. Frequency constraints are considered to control the structure's dynamic behavior.

3.1. Disk

The first numerical example is a Disk, as shown in Figure 2. It is simply supported where R = 0.1 and R = 10.

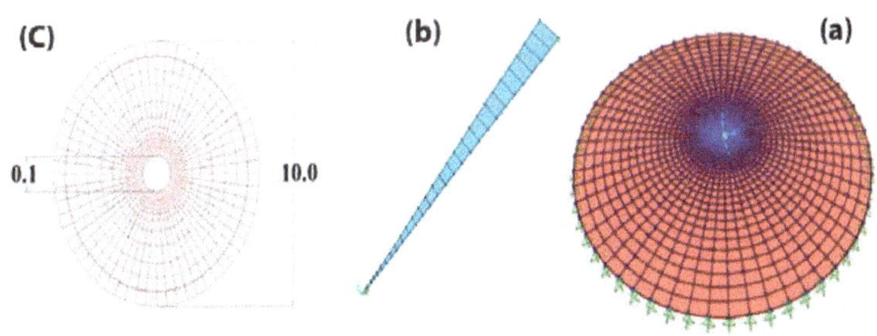

Figure 2. (**a**) A Disk discretized by quadrilateral shell finite elements, (**b**) substructure of Disk, (**c**) Disk dimensions.

The coordinate of the nodes of the Disk's substructure is presented in Table 1. This substructure is rotated 60 times around the center to create a Disk. The whole structure uses a four-node quadrilateral flat shell element. In each element, the active mass is considered at the nodes.

Table 1. Coordinates of the nodes of the Disk.

Node Number	Coordinates (x, y, z)	Node Number	Coordinates (x, y, z)	Node Number	Coordinates (x, y, z)
1	(0.1, 0, 0)	12	(1.5052, 0, 0)	23	(5.4865, 0, 0)
2	(0.1213, 0, 0)	13	(1.7607, 0, 0)	24	(5.9762, 0, 0)
3	(0.1639, 0, 0)	14	(2.0374, 0, 0)	25	(6.4871, 0, 0)
4	(0.2277, 0, 0)	15	(2.3355, 0, 0)	26	(7.0194, 0, 0)
5	(0.3129, 0, 0)	16	(2.6548, 0, 0)	27	(7.5729, 0, 0)
6	(0.4194, 0, 0)	17	(2.9955, 0, 0)	28	(8.1478, 0, 0)
7	(0.5471, 0, 0)	18	(3.3574, 0, 0)	29	(8.7439, 0, 0)
8	(0.6961, 0, 0)	19	(3.7407, 0, 0)	30	(9.3613, 0, 0)
9	(0.8665, 0, 0)	20	(4.1452, 0, 0)	31	(10.0, 0, 0)
10	(1.0581, 0, 0)	21	(4.5710, 0, 0)		
11	(1.2710, 0, 0)	22	(5.0181, 0, 0)		

In this example, the optimization is performed once by considering the frequency constraint and next by considering both frequency and displacement constraints to evaluate the efficiency of the proposed method:

First constraint: first and third frequencies assumed to be less than 0.155 and 0.149, respectively ($\omega_1 < 0.155, \omega_3 < 0.149$).

Second constraint: first and third frequencies assumed to be less than <0.155 and <0.149, respectively ($\omega_1 < 0.155, \omega_3 < 0.149$), and the displacement of node number two under a load of 10,000 kN in the Y direction is less than <0.0103.

The thickness member of the substructure elements in this example is the main variable obtained according to the constraints considered by meta-heuristic algorithms. The specified range for the thickness of structural elements is (0.25–0.35 m) in both cases of ten and thirty variables.

In this example, to analyze a cyclically symmetric structure, only the matrix with dimensions 186, equal to the substructures, is calculated. Only once the calculation for the frequency of the structure without using the concept of cyclic symmetry was stopped after about 4 h without any result. The calculation time using the proposed method is significantly reduced.

For Disk evaluation, the coefficient of the GRO algorithm was $\alpha = 0.7, \beta = 0.5, \gamma = 0.8$. For 100 operators working in the Disk, a maximum of 500 repetitions was determined as a termination condition in the example. For this purpose, the structure was optimized with ten elements or variables (three elements in one group) and once with thirty variables, as shown in Figure 3.

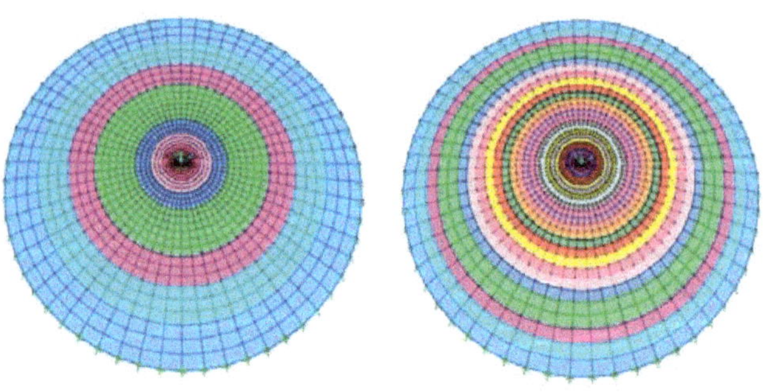

Figure 3. Disk discretized by ten and thirty quadrilateral shell finite elements.

The structure's weight with an initial thickness of 0.35 m is 263.3873 kN. The results are presented in Tables 2 and 3. The results in Table 2 show that, under the first and second constraint, the structure's weight is 210.6800 kN (20.0113% weight reduction) and 225.1152 kN (14.5307% weight reduction) by using the GRO algorithm, respectively, indicating the GRO algorithm's very intimate performance and better efficacy compared with the other two algorithms.

Table 2. Optimal design results for Disk with ten variables.

Group of Element	Constraint One			Constraint Two		
	GRO	**CSS**	**TLBO**	**GRO**	**CSS**	**TLBO**
1	0.3223	0.25432	0.32055	0.32432	0.31181	0.30816
2	0.31358	0.26921	0.28618	0.29656	0.32971	0.32527
3	0.35	0.30995	0.28575	0.31167	0.32477	0.33627
4	0.27218	0.28793	0.27218	0.30292	0.3174	0.30797
5	0.29664	0.27156	0.31867	0.27249	0.26037	0.30335
6	0.28693	0.30716	0.27789	0.25351	0.27004	0.27181
7	0.27102	0.3094	0.27254	0.32775	0.25913	0.27129
8	0.30973	0.30414	0.34035	0.30485	0.29696	0.32472
9	0.27882	0.29335	0.2535	0.32963	0.29731	0.33487
10	0.26383	0.25606	0.28665	0.27671	0.33245	0.28385
Weight (kN)	210.68	214.0883	215.5785	225.1152	227.3797	227.8022
Weight reduction (percentage)	20.0113	18.7173	18.1515	14.5307	13.671	13.5106

Table 3. Optimal design results for Disk with thirty variables.

Group of Element	Constraint One			Constraint Two		
	GRO	**CSS**	**TLBO**	**GRO**	**CSS**	**TLBO**
1	0.32895	0.34863	0.3236	0.27153	0.28859	0.31611
2	0.3392	0.27333	0.27333	0.29466	0.26096	0.28818
3	0.2831	0.2831	0.31302	0.27043	0.31731	0.27222
4	0.30319	0.2946	0.34495	0.29997	0.33555	0.29043
5	0.25	0.28761	0.28761	0.29314	0.30404	0.25802
6	0.34779	0.32129	0.34832	0.32587	0.33136	0.34974
7	0.30548	0.30548	0.29013	0.33109	0.3306	0.28823
8	0.26898	0.26898	0.33433	0.3434	0.2793	0.27152
9	0.29968	0.33482	0.28475	0.32096	0.29732	0.25
10	0.2951	0.30222	0.29329	0.3207	0.27777	0.34179
11	0.34994	0.29251	0.28194	0.33587	0.32934	0.32385
12	0.28303	0.30634	0.25284	0.32524	0.34933	0.28928
13	0.34992	0.31479	0.25787	0.33255	0.3445	0.30287
14	0.33866	0.28642	0.26022	0.33715	0.34862	0.31572
15	0.31011	0.27181	0.25	0.25431	0.30349	0.28225
16	0.2935	0.2935	0.29923	0.25576	0.27672	0.33805
17	0.30241	0.34981	0.31378	0.25793	0.29611	0.30806
18	0.32113	0.33979	0.28981	0.29638	0.32797	0.30032
19	0.28916	0.32681	0.35	0.31963	0.25141	0.3022
20	0.258	0.258	0.25008	0.31484	0.2888	0.26151
21	0.33324	0.30111	0.35	0.26566	0.3161	0.35
22	0.28162	0.30381	0.31155	0.30963	0.29594	0.26372
23	0.30371	0.30371	0.33676	0.35	0.29573	0.25857
24	0.29568	0.29568	0.29316	0.26058	0.31643	0.34493
25	0.26007	0.26007	0.26432	0.3004	0.34709	0.28111

Table 3. *Cont.*

Group of Element	Constraint One			Constraint Two		
	GRO	CSS	TLBO	GRO	CSS	TLBO
26	0.27469	0.29141	0.27469	0.33481	0.27642	0.3143
27	0.25478	0.27388	0.29988	0.28506	0.3321	0.33521
28	0.34821	0.34837	0.32301	0.34921	0.32673	0.28613
29	0.26503	0.26503	0.26624	0.30049	0.30046	0.32646
30	0.30911	0.32668	0.33041	0.25829	0.25107	0.31654
Weight (kN)	220.815	225.397	225.4451	226.4353	227.5151	231.0375
Weight reduction (percentage)	16.1634	14.4237	14.4055	14.0295	13.6196	12.2822

The results in Table 3 show that, under the first constraint, the structure's weight is 220.8150 kN (16.1634% weight reduction) by using the GRO algorithm, indicating its very intimate performance and better efficacy compared with the other two algorithms. However, under the second constraint, the structure's weight is 226.4353 kN (14.0295% weight reduction) by using the GRO and 227.5151 kN (13.6196% weight reduction) by using the CSS, indicating the GRO and CSS algorithms' very intimate performances, and a better efficacy compared with the TLBO algorithms. Tables 2 and 3 show the calculated thicknesses obtained by all algorithms. As can be seen, the thicknesses are in a suitable range, indicating the correct operation of the algorithms in finding the optimal answers in the search space.

3.2. *Silo*

The second numerical example is a Silo, as shown in Figure 4. It is simply supported at Z = 0.

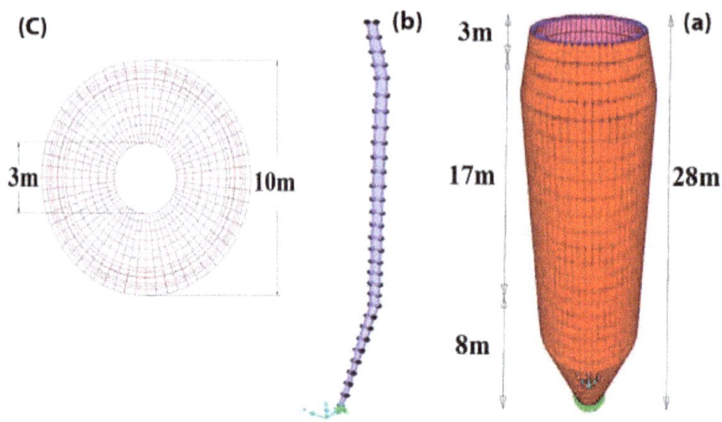

Figure 4. (**a**) A Silo discretized by quadrilateral shell finite elements, (**b**) substructure of Silo (**c**), Silo dimensions.

The coordinate of the nodes of Silo's substructure is presented in Table 4. This substructure is rotated 60 times around the center to create a Silo. The whole structure uses a four-node quadrilateral flat shell element. In each element, the active mass is considered at the nodes.

In this example, the optimization is performed once by considering the frequency constraint and next by considering both frequency and displacement constraints to evaluate the efficiency of the proposed method:

First constraint: first and third frequencies assumed to be less than 0.49 and 0.3, respectively ($\omega_1 < 0.49, \omega_3 < 0.3$).

Second constraint: first and third frequencies assumed to be less than <0.49 and <0.3, respectively ($\omega_1 < 0.49, \omega_3 < 0.3$), and the displacement of node number one under a load of 10,000 kN in the Y direction is less than <0.327.

Table 4. Coordinates of the nodes of the Silo.

Node Number	Coordinates (x, y, z)	Node Number	Coordinates (x, y, z)	Node Number	Coordinates (x, y, z)
1	(4, 0, 28)	12	(5, 0, 18)	23	(4.61111, 0, 7.111)
2	(4.25, 0, 27.25)	13	(5, 0, 17)	24	(4.22222, 0, 6.22222)
3	(4.5, 0, 26.5)	14	(5, 0, 16)	25	(3.83333, 0, 5.33333)
4	(4.75, 0, 25.75)	15	(5, 0, 15)	26	(3.4444, 0, 4.44444)
5	(5, 0, 25)	16	(5, 0, 14)	27	(3.05556, 0, 3.55556)
6	(5, 0, 24)	17	(5, 0, 13)	28	(2.66667, 0, 2.66667)
7	(5, 0, 23)	18	(5, 0, 12)	29	(2.27778, 0, 1.77778)
8	(5, 0, 22)	19	(5, 0, 11)	30	(1.88889, 0, 0.88889)
9	(5, 0, 21)	20	(5, 0, 10)	31	(1.5, 0, 0)
10	(5, 0, 20)	21	(5, 0, 9)		
11	(5, 0, 19)	22	(5, 0, 8)		

The thickness member of the substructure elements in this example is the main variable obtained according to the constraints considered by meta-heuristic algorithms. The specified range for the thickness of structural elements is (0.25–0.35 m) in both cases of ten and thirty variables.

In this example, to analyze a cyclically symmetric structure, only the matrix with dimensions 186, equal to the dimensions of the substructure, is calculated. Only once was the calculation for the frequency of the structure without using the concept of cyclic symmetry was stopped after about 4 h without any result. The calculation time using the proposed method is significantly reduced.

For Silo evaluation, the coefficient of the GRO algorithm was $\alpha = 0.7, \beta = 0.5, \gamma = 0.8$. For 120 operators working in the Silo, a maximum of 600 repetitions was determined as a termination condition in the example. For this purpose, the structure was optimized with ten elements or variables (three elements in one group) and once with thirty variables, as shown in Figure 5.

Figure 5. Silo discretized by ten and thirty quadrilateral shell finite elements.

The structure's weight with an initial thickness of 0.35 m is 673.2002 kN. The results are presented in Tables 5 and 6. Table 5 shows that the optimal weights obtained by GRO and CSS algorithms were almost similar. Under the first constraint, the structure's weight is 555.6478 kN (17.7588% weight reduction) by using the GRO algorithm and 553.9053 kN (17.7206% weight reduction) by using the CSS algorithm, indicating their very similar performance and a better efficacy compared with the TLBO algorithms. Under the second constraint, the GRO algorithm obtained a weight of 584.8135 kN (13.1293% weight reduction), indicating its very intimate performance and also a better efficacy compared with the CSS and TLBO algorithms. Moreover, the calculated thicknesses obtained by all algorithms are in a suitable range indicating the correct operation of the algorithms in finding the optimal answers in the search space.

Table 5. Optimal design results for Silo with ten variables.

Group of Element	Constraint One			Constraint Two		
	GRO	CSS	TLBO	GRO	CSS	TLBO
1	0.32025	0.32399	0.33612	0.29212	0.34293	0.27209
2	0.27501	0.27337	0.28823	0.33717	0.32757	0.33174
3	0.26468	0.27171	0.28095	0.3301	0.29867	0.33572
4	0.27503	0.27634	0.27058	0.2539	0.29358	0.27366
5	0.29596	0.29512	0.30204	0.33178	0.29467	0.35
6	0.30478	0.30104	0.30587	0.29637	0.28063	0.29641
7	0.26461	0.26172	0.2746	0.2635	0.30085	0.26175
8	0.32331	0.32247	0.32201	0.34564	0.30107	0.32959
9	0.25977	0.25875	0.27422	0.27857	0.33176	0.31459
10	0.31475	0.3161	0.32023	0.31055	0.32948	0.30645
Weight (kN)	553.6478	553.9053	567.7535	584.8135	589.7962	591.6707
Weight reduction (percentage)	17.7588	17.7206	15.6635	13.1293	12.3892	12.1107

Table 6. Optimal design results for Silo with thirty variables.

Group of Element	Constraint One			Constraint Two		
	GRO	CSS	TLBO	GRO	CSS	TLBO
1	0.25297	0.34543	0.32236	0.32116	0.30579	0.28185
2	0.33252	0.32762	0.3451	0.26957	0.32116	0.3034
3	0.30635	0.33772	0.29503	0.31626	0.35	0.25899
4	0.34707	0.25129	0.30513	0.32934	0.35	0.26117
5	0.28937	0.32143	0.29793	0.27499	0.34743	0.26362
6	0.28224	0.30234	0.34672	0.27977	0.27941	0.31786
7	0.26436	0.29812	0.25715	0.34766	0.31038	0.29951
8	0.25862	0.28594	0.32884	0.33849	0.25	0.26897
9	0.25809	0.3413	0.34648	0.33029	0.30293	0.2995
10	0.34392	0.2599	0.34856	0.28625	0.25	0.26476
11	0.32459	0.33119	0.29741	0.26383	0.32394	0.25549
12	0.32583	0.30706	0.32901	0.31237	0.33197	0.33507
13	0.29277	0.33953	0.27285	0.33736	0.25	0.30605
14	0.29757	0.26708	0.26982	0.26647	0.25	0.34296
15	0.2946	0.33958	0.30887	0.3001	0.35	0.31966
16	0.29474	0.25971	0.27594	0.31944	0.34774	0.30827
17	0.33942	0.32193	0.26614	0.30722	0.32049	0.33153
18	0.28089	0.31204	0.30539	0.26862	0.25	0.3379
19	0.3167	0.30271	0.25261	0.27952	0.28465	0.34889
20	0.31116	0.25531	0.26994	0.26276	0.25	0.25005
21	0.3238	0.29538	0.29521	0.34741	0.31815	0.33654

Table 6. *Cont.*

Group of Element	Constraint One			Constraint Two		
	GRO	CSS	TLBO	GRO	CSS	TLBO
22	0.25836	0.31413	0.33671	0.28007	0.32062	0.31125
23	0.27941	0.26399	0.30183	0.31973	0.35	0.34899
24	0.26723	0.27624	0.28075	0.32454	0.26019	0.30276
25	0.34343	0.28986	0.32634	0.33977	0.33982	0.29795
26	0.35	0.26216	0.25405	0.25345	0.25	0.33013
27	0.33301	0.25092	0.33797	0.27689	0.31054	0.27278
28	0.25373	0.33171	0.27102	0.28472	0.35	0.2998
29	0.26444	0.30078	0.27342	0.29952	0.35	0.34008
30	0.31457	0.32356	0.34425	0.31369	0.30959	0.30746
Weight (kN)	576.8934	577.6467	579.7768	580.8192	583.9202	584.0356
Weight reduction (percentage)	14.3058	14.1939	13.8775	13.7227	13.262	13.2449

Table 6 shows that under the first and second constraint, the structure's weight is 576.8934 kN (14.3058% weight reduction) and 580.8192 kN (13.7227% weight reduction), respectively, by using the GRO algorithms, indicating the GRO algorithms' very intimate performance and better efficacy compared with the other two algorithms. The CSS and TLBO algorithms had a lower ability to optimize the structure's weight. Moreover, the calculated thicknesses obtained by all algorithms are in a suitable range indicating the correct operation of the algorithms in finding the optimal answers in the search space.

3.3. Cooling Tower

The third numerical example is a Cooling Tower, as shown in Figure 6. It is simply supported at Z = 0.

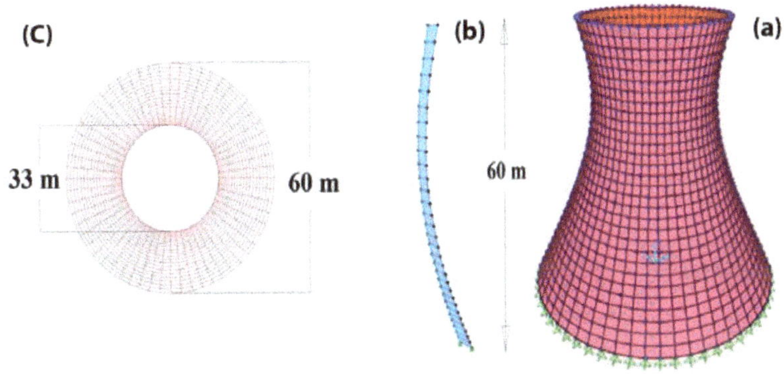

Figure 6. (**a**) A Cooling Tower discretized by quadrilateral shell finite elements, (**b**) substructure of Cooling Tower, (**c**) Cooling Tower dimensions.

The coordinates of the nodes of the Cooling Tower's substructure are presented in Table 7. This substructure is rotated 60 times around the center to create the Cooling Tower. The whole structure uses a four-node quadrilateral flat shell element. In each element, the active mass is considered at the nodes.

In this example, the optimization is performed once by considering the frequency constraint and next by considering both frequency and displacement constraints to evaluate the efficiency of the proposed method:

First constraint: first and third frequencies assumed to be less than 0.3 and 0.28, respectively ($\omega_1 < 0.3, \omega_3 < 0.28$).

Second constraint: first and third frequencies assumed to be less than <0.3 and <0.28, respectively ($\omega_1 < 0.3$, $\omega_3 < 0.28$), and the displacement of node number one under a load of 10,000 kN in the Y direction is less than <0.024.

Table 7. Coordinates of the nodes of the Cooling Tower.

Node Number	Coordinates (x, y, z)	Node Number	Coordinates (x, y, z)	Node Number	Coordinates (x, y, z)
1	(30, 0, 0)	12	(19.4492, 0, 22)	23	(13.9331, 0, 44)
2	(28.9589, 0, 2)	13	(18.6435, 0, 24)	24	(13.9331, 0, 46)
3	(27.929, 0, 4)	14	(17.8796, 0, 26)	25	(14.0328, 0, 48)
4	(26.9115, 0, 6)	15	(17.163, 0, 28)	26	(14.2302, 0, 50)
5	(25.9079, 0, 8)	16	(16.5, 0, 30)	27	(14.5214, 0, 52)
6	(24.9199, 0, 10)	17	(15.8972, 0, 32)	28	(14.9007, 0, 54)
7	(23.9493, 0, 12)	18	(15.3616, 0, 34)	29	(15.3616, 0, 56)
8	(22.9985, 0, 14)	19	(14.9007, 0, 36)	30	(15.8972, 0, 58)
9	(22.0699, 0, 16)	20	(14.5214, 0, 38)	31	(16.5, 0, 60)
10	(21.1665, 0, 18)	21	(14.2302, 0, 40)		
11	(20.2916, 0, 20)	22	(14.0328, 0, 42)		

The thickness member of the substructure elements in this example is the main variable obtained according to the constraints considered by meta-heuristic algorithms. The specified range for the thickness of structural elements is (0.25–0.35 m) in both cases of ten and thirty variables.

In this example, to analyze a cyclic symmetric structure, only the matrix with dimensions 186, equal to the dimensions of the substructure, is calculated. Only once was the calculation of the frequency of the structure without using the concept of cyclic symmetry stopped after about 4 h without any result. The calculation time using the proposed method is significantly reduced.

For Cooling Tower evaluation, the coefficient of the GRO algorithm was $\alpha = 0.7$, $\beta = 0.5$, $\gamma = 0.8$. For 150 operators working in the Cooling Tower, a maximum of 750 repetitions was determined as a termination condition in the example.

For this purpose, the structure was optimized with ten elements or variables (three elements in one group) and once with thirty variables, as shown in Figure 7.

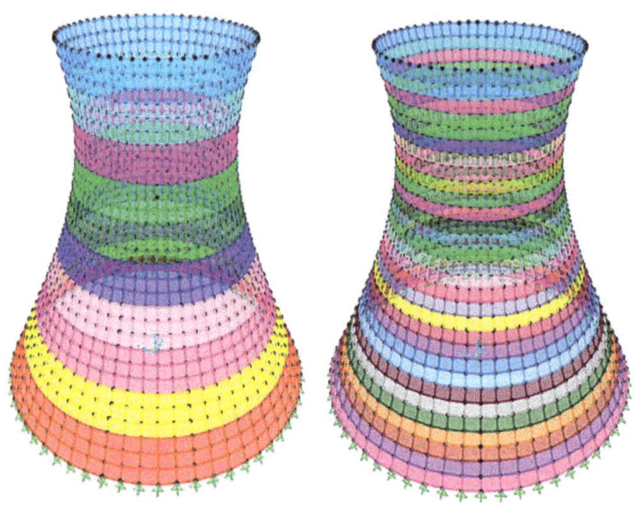

Figure 7. Cooling Tower discretized by ten and thirty quadrilateral shell finite elements.

The structure's weight with an initial thickness of 0.35 m is 6.3602 × 10^3 kN. The results are presented in Tables 8 and 9. Table 8 shows that the optimal weights obtained by GRO and CSS algorithms were almost similar. Under the first constraint, the structure's weight is 5.1639 × 10^3 kN (18.8092% weight reduction) using the GRO algorithm and 5.1791 × 10^3 kN (18.5702% weight reduction) by using the CSS algorithm, indicating their very close performances and better efficacy compared with the TLBO algorithms. Under the second constraint, the structure's weight is 5.2833 × 10^3 kN (16.9319% weight reduction) by using the GRO algorithm, indicating the GRO algorithm's very intimate performance and a better efficacy compared with the other two algorithms. Moreover, the calculated thicknesses obtained by all algorithms are in a suitable range indicating the correct operation of the algorithms in finding the optimal answers in the search space.

Table 8. Optimal design results for Cooling Tower with ten variables.

Group of Element	Constraint One			Constraint Two		
	GRO	CSS	TLBO	GRO	CSS	TLBO
1	0.33742	0.33688	0.33612	0.34202	0.32676	0.33282
2	0.2702	0.26562	0.28823	0.32097	0.31465	0.34823
3	0.29306	0.28943	0.28095	0.31782	0.27892	0.28108
4	0.27632	0.27001	0.27058	0.30344	0.26085	0.27296
5	0.32973	0.33358	0.30204	0.3077	0.2742	0.29914
6	0.32264	0.32705	0.30587	0.31144	0.34497	0.32055
7	0.25114	0.25375	0.2746	0.26305	0.26624	0.29247
8	0.30594	0.31358	0.32201	0.25248	0.31324	0.26423
9	0.25114	0.25	0.27422	0.26438	0.3196	0.30321
10	0.25238	0.25322	0.32023	0.28449	0.2583	0.34176
Weight (kN)	5.16 × 10^3	5.18 × 10^3	5.43 × 10^3	5.28 × 10^3	5.36 × 10^3	5.57 × 10^3
Weight reduction (percentage)	18.8092	18.5702	14.6017	16.9319	15.674	12.4084

Table 9. Optimal design results for Cooling Tower with thirty variables.

Group of Element	Constraint One			Constraint Two		
	GRO	CSS	TLBO	GRO	CSS	TLBO
1	0.3459	0.34619	0.34586	0.30839	0.31638	0.32904
2	0.33143	0.33243	0.3321	0.29527	0.33477	0.34493
3	0.33312	0.33098	0.33066	0.2964	0.27562	0.28275
4	0.34452	0.34429	0.34489	0.32947	0.2517	0.31712
5	0.27742	0.27807	0.2784	0.348	0.27984	0.29386
6	0.28809	0.2895	0.29123	0.29577	0.3173	0.33335
7	0.25415	0.25776	0.25804	0.33183	0.25603	0.32688
8	0.26287	0.26121	0.26145	0.34692	0.29207	0.26672
9	0.28618	0.28212	0.28212	0.34079	0.30397	0.33619
10	0.27722	0.27711	0.27741	0.33078	0.32918	0.34898
11	0.27621	0.27786	0.27854	0.25715	0.34789	0.30144
12	0.34381	0.34673	0.34718	0.25984	0.25573	0.33842
13	0.32408	0.32499	0.32521	0.28145	0.25559	0.3088
14	0.30331	0.30355	0.304	0.3264	0.3405	0.26547
15	0.29603	0.29459	0.2941	0.26143	0.27699	0.26998
16	0.3446	0.34786	0.3489	0.33107	0.28078	0.29069
17	0.343	0.34412	0.344	0.2536	0.27745	0.32487
18	0.32094	0.32502	0.32554	0.33921	0.30914	0.33255
19	0.27975	0.27827	0.27781	0.32665	0.25043	0.32899
20	0.26122	0.25572	0.25603	0.25729	0.25043	0.28185

Table 9. *Cont.*

Group of Element	Constraint One			Constraint Two		
	GRO	CSS	TLBO	GRO	CSS	TLBO
21	0.26689	0.26755	0.26834	0.26037	0.33234	0.3034
22	0.2644	0.26331	0.26222	0.25246	0.25907	0.25899
23	0.2775	0.27809	0.27915	0.34578	0.34298	0.26117
24	0.25	0.25061	0.25003	0.25425	0.26278	0.26362
25	0.25827	0.25381	0.25396	0.26124	0.343	0.31786
26	0.25265	0.25273	0.25312	0.26699	0.28866	0.29951
27	0.25904	0.2615	0.26037	0.25202	0.33224	0.26897
28	0.27787	0.28072	0.28161	0.27124	0.34494	0.2995
29	0.25462	0.2533	0.25552	0.25193	0.29461	0.26476
30	0.27682	0.28003	0.28003	0.27306	0.33866	0.25549
Weight (kN)	5.20×10^3	5.206×10^3	5.210×10^3	5.23×10^3	5.47×10^3	5.38×10^3
Weight reduction (percentage)	18.2321	18.1472	18.0718	17.729	14.0341	15.4067

Table 9 shows that the optimal weights obtained by GRO and CSS algorithms were almost similar. Under the first constraint, the structure's weight is 5.2006×10^3 kN (18.2321% weight reduction) using the GRO algorithm and 5.2060×10^3 kN (18.1472% weight reduction) by using the CSS algorithm, indicating their very close performances and a better efficacy compared to the TLBO algorithms. Under the second constraint, the structure's weight is 5.2326×10^3 kN (17.7290% weight reduction) by using the GRO algorithm, indicating the GRO algorithm's very intimate performance, and a better efficacy compared with the other two algorithms. Moreover, the calculated thicknesses obtained by all algorithms are in a suitable range indicating the correct operation of the algorithms in finding the optimal answers in the search space.

4. Checking the Frequencies and Mode Shapes

To ensure that the frequency constraints are satisfied, limited frequencies with optimal results were considered and are presented in Table 10. In all examples, the constraints were adequately satisfied, and the frequencies were approximately close to the limit values with an average difference of 0.02%.

Table 10. Constrained natural frequencies of structures (Hz).

Structure	Frequency	Limited Frequencies	Ten Variable	Thirty Variable
			GRO	GRO
Disk	ω_1	0.29	0.2900	0.2900
	ω_2	0.27	0.2700	0.2700
Silo	ω_1	0.49	0.4900	0.4900
	ω_3	0.3	0.3000	0.3000
Cooling Tower	ω_1	0.3	0.3000	0.3000
	ω_3	0.28	0.2800	0.2800003

Mode shapes of the frequencies and the optimal schemes obtained by the GRO algorithm in the first and third modes are shown in Figures 8–10. In structures, the first mode showed a sway-type shape, and the third mode had a vertical displacement.

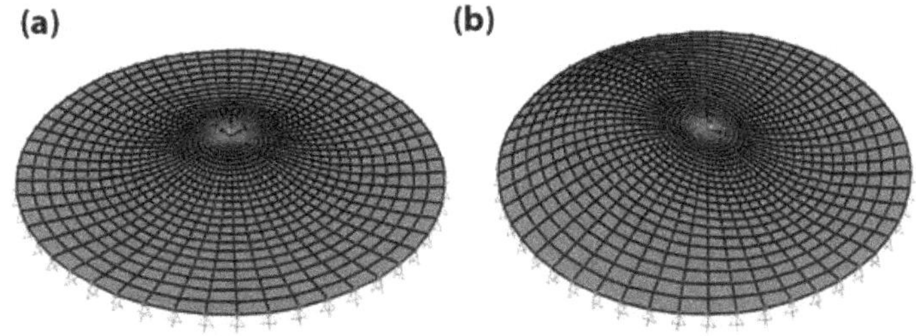

Figure 8. The mode shapes of the Disk. (**a**) First mode shape. (**b**) Third mode shape.

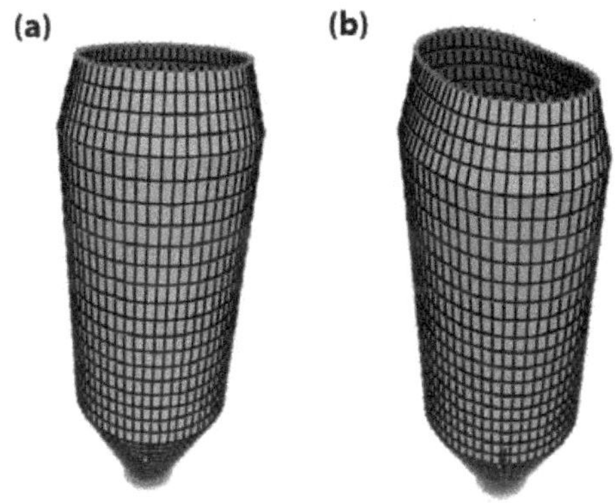

Figure 9. The mode shapes of the Silo. (**a**) First mode shape. (**b**) Third mode shape.

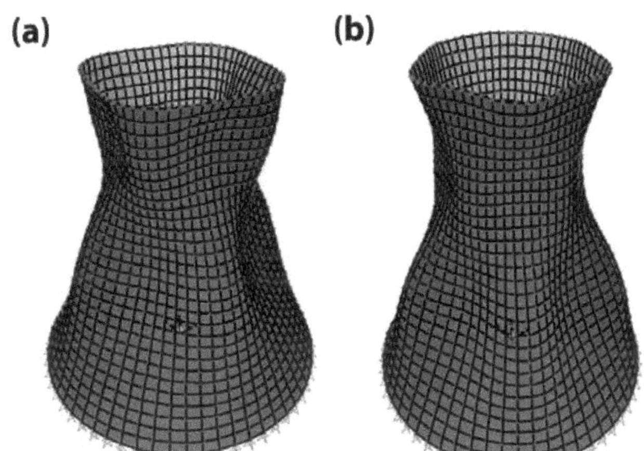

Figure 10. The mode shapes of the Cooling Tower. (**a**) First mode shape. (**b**) Third mode shape.

5. Conclusions

Problems of optimizing structures with cyclic symmetry, considering the frequency constraints and multiple displacements, including non-convex search spaces, are among the most challenging issues in civil engineering. In this study, a design optimization was performed using GRO meta-heuristic algorithms to deal with this problem. It is practically impossible to use algorithms to optimize real-scale structures due to the large volume and the calculations time. The concept of cyclic symmetry was used to reduce the volume and time of analyses. By decomposing the main structure into repetitive patterns called substructures, the number of calculations was significantly reduced. To evaluate the performance of the proposed method, Disk, Silo, and Cooling Tower were modeled at real scale to confirm the ability of the concept of cyclic symmetry and the GRO algorithm. To better evaluate the performance of this algorithm, two well-known and widely used algorithms, CSS and TLBO, were used for comparison. According to the obtained results, the GRO algorithm was stable and well performed in finding optimal answers. Results show GRO reduces the structure's weight by 14–20% with good accuracy in finding global optimal designs. The algorithm correctly satisfies the constraints and shows that using the concept of cyclic symmetry is an efficient and useful solution in reducing computation time for analyzing symmetric structures. However, it does not apply to nonlinear analyses. As a new application of the concept of cyclic symmetry, it can be used to ellipsoid symmetry problems. The authors intend to implement the proposed scheme for optimizing ellipsoid structures, considering the frequency constraints and multiple displacements.

Author Contributions: S.S.: Conceptualization, methodology, software, validation, analysis, investigation, resources, data curation, writing—original draft preparation, writing—review. M.S.M.: Conceptualization, methodology, resources, data curation, review—editing, supervision, project administration. M.E.S.: resources, review—editing. All authors have read and agreed to the published version of the manuscript.

Funding: This research received no external funding.

Institutional Review Board Statement: Not applicable.

Informed Consent Statement: Not applicable.

Data Availability Statement: Not applicable.

Conflicts of Interest: The authors declare no conflict of interest.

References

1. Kaveh, A.; Ilchi Ghazaan, M. Optimal design of dome truss structures with dynamic frequency constraints. *Struct. Multidiscip. Optim.* **2016**, *53*, 605–621. [CrossRef]
2. Kaveh, A.; Mahdavi, V.R. Colliding Bodies Optimization method for optimum discrete design of truss structures. *Comput. Struct.* **2014**, *70*, 1–12. [CrossRef]
3. Kaveh, A.; Mahdavi, V.R. Colliding-Bodies Optimization for Truss Optimization with Multiple Frequency Constraints. *J. Comput. Civ. Eng.* **2015**, *29*, 4014078. [CrossRef]
4. Kaveh, A.; Ilchi Ghazaan, M. Enhanced colliding bodies optimization for design problems with continuous and discrete variables. *Adv. Eng. Softw.* **2014**, *77*, 66–75. [CrossRef]
5. Kaveh, A.; Mahdavi, V.R. A hybrid CBO–PSO algorithm for optimal design of truss structures with dynamic constraints. *Appl. Soft Comput.* **2015**, *34*, 260–273. [CrossRef]
6. Song, Y.; Zhang, M.; Øiseth, O.; Rønnquist, A. Wind deflection analysis of railway catenary under crosswind based on nonlinear finite element model and wind tunnel test. *Mech. Mach. Theory* **2022**, *168*, 104608. [CrossRef]
7. Ho-Huu, V.; Nguyen-Thoi, T.; Truong-Khac, T.; Le-Anh, L.; Vo-Duy, T. An improved differential evolution based on roulette wheel selection for shape and size optimization of truss structures with frequency constraints. *Neural Comput. Appl.* **2018**, *29*, 167–185. [CrossRef]
8. Lieu, Q.X.; Do, D.T.T.; Lee, J. An adaptive hybrid evolutionary firefly algorithm for shape and size optimization of truss structures with frequency constraints. *Comput. Struct.* **2018**, *195*, 99–112. [CrossRef]
9. Tejani, G.; Savsani, V.J.; Mirjalili, S.; Patel, V.K. Truss optimization with natural frequency bounds using improved symbiotic organisms search. *Knowl.-Based Syst.* **2018**, *143*, 162–178. [CrossRef]
10. Kaveh, A.; Dadras Eslamlou, A. Water strider algorithm: A new metaheuristic and applications. *Structures* **2020**, *25*, 520–541. [CrossRef]
11. Kaveh, A.; Ilchi Ghazaan, M. Vibrating particles system algorithm for truss optimization with multiple natural frequency constraints. *Acta Mech.* **2017**, *228*, 307–322. [CrossRef]
12. Kaveh, A.; Ilchi Ghazaan, M. A new hybrid meta-heuristic algorithm for optimal design of large-scale dome structures. *Eng. Optim.* **2018**, *49*, 235–252. [CrossRef]
13. Carvalho, J.P.G.; Lemonge, A.C.C.; Carvalho, É.C.R.; Hallak, P.H.; Bernardino, H.S. Truss optimization with multiple frequency constraints and automatic member grouping. *Struct. Multidiscip. Optim.* **2018**, *56*, 547–577. [CrossRef]

14. Rao, R.V. *Teaching Learning Based Optimization Algorithm and Its Engineering Applications*; Springer: London, UK, 2016; ISBN 978-3-319-22731-3.
15. Kar, R.; Mandal, D.; Mondal, S.; Ghoshal, S.P. Craziness based Particle Swarm Optimization algorithm for FIR band stop filter design. *Swarm Evol. Comput.* **2012**, *7*, 58–64. [CrossRef]
16. Kaveh, A.; Talatahari, S. A novel heuristic optimization method: Charged system search. *Acta Mech.* **2010**, *3*. [CrossRef]
17. Kaveh, A.; Zolghadr, A. Shape and Size Optimization of Truss Structures With Frequency Constraints Using Enhanced Charged System Search Algorithm. *Asian J. Civ. Eng.* **2011**, *12*, 487–509.
18. Jalili, S.; Talatahari, S. Optimum Design of Truss Structures Under Frequency Constraints using Hybrid CSS-MBLS Algorithm. *KSCE J. Civ. Eng.* **2018**, *22*, 1840–1853. [CrossRef]
19. Kaveh, A.; Zolghadr, A. Optimal design of cyclically symmetric trusses with frequency constraints using cyclical parthenogenesis algorithm. *Adv. Struct. Eng.* **2018**, *21*, 739–755. [CrossRef]
20. Liu, S.; Zhu, H.; Chen, Z.; Cao, H. Frequency-constrained truss optimization using the fruit fly optimization algorithm with an adaptive vision search strategy. *Eng. Optim.* **2020**, *52*, 777–797. [CrossRef]
21. Kaveh, A.; Javadi, S.M. Chaos-based firefly algorithms for optimization of cyclically large-size braced steel domes with multiple frequency constraints. *Comput. Struct.* **2019**, *214*, 28–39. [CrossRef]
22. Wang, D.; Sun, M.; Ma, R.; Shen, X. Numerical Modeling of Ice Accumulation on Three-Dimensional Bridge Cables under Freezing Rain and Natural Wind Conditions. *Symmetry* **2022**, *14*, 396. [CrossRef]
23. Williams, F.W. An algorithm for exact eigenvalue calculations for rotationally periodic structures. *Int. J. Numer. Methods Eng.* **1986**, *23*, 609–622. [CrossRef]
24. Tran, D.M. Component mode synthesis methods using partial interface modes: Application to tuned and mistuned structures with cyclic symmetry. *Comput. Struct.* **2009**, *87*, 1141–1153. [CrossRef]
25. He, Y.; Yang, H.; Deeks, A.J. On the use of cyclic symmetry in SBFEM for heat transfer problems. *Int. J. Heat Mass Transf.* **2014**, *71*, 98–105. [CrossRef]
26. He, Y.; Yang, H.; Xu, M.; Deeks, A.J. A scaled boundary finite element method for cyclically symmetric two-dimensional elastic analysis. *Comput. Struct.* **2013**, *120*, 1–8. [CrossRef]
27. Kaveh, A. *Optimal Analysis of Structures by Concepts of Symmetry and Regularity*; Springer: Vienna, Austria, 2013; ISBN 9783709115657.
28. Kaveh, A. *Computational Structural Analysis and Finite Element Methods*; Springer: Vienna, Austria, 2014; ISBN 978-3-319-02963-4.
29. Kaveh, A.; Koohestani, K. Formation of graph models for regular finite element meshes. *Comput. Assist. Mech. Eng. Sci.* **2009**, *16*, 101–115.
30. Kaveh, A.; Rahami, H. Block circulant matrices and applications in free vibration analysis of cyclically repetitive structures. *Acta Mech.* **2011**, *217*, 51–62. [CrossRef]
31. Kaveh, A.; Rahami, H. An efficient analysis of repetitive structures generated by graph products. *Int. J. Numer. Methods Eng.* **2010**, *84*, 108–126. [CrossRef]
32. Koohestani, K. An orthogonal self-stress matrix for efficient analysis of cyclically symmetric space truss structures via force method. *Int. J. Solids Struct.* **2011**, *48*, 227–233. [CrossRef]
33. Koohestani, K. Exploitation of symmetry in graphs with applications to finite and boundary elements analysis. *Int. J. Numer. Methods Eng.* **2012**, *90*, 152–176. [CrossRef]
34. Koohestani, K. On the decomposition of generalized eigenproblems for the free vibration analysis of cyclically symmetric finite elementmodels. *Int. J. Numer. Methods Eng.* **2010**, *82*, 359–378. [CrossRef]
35. Sarjamei, S.; Massoudi, M.S.; Esfandi Sarafraz, M. Gold Rush Optimization Algorithm. *Iran Univ. Sci. Technol.* **2021**, *11*, 291–327.
36. Kaveh, A.; Talatahari, S. Optimal design of skeletal structures via the charged system search algorithm. *Struct. Multidiscip. Optim.* **2010**, *41*, 893–911. [CrossRef]
37. Kaveh, A.; Talatahari, S. Charged system search for optimum grillage system design using the LRFD-AISC code. *J. Constr. Steel Res.* **2010**, *66*, 767–771. [CrossRef]
38. Kaveh, A.; Talatahari, S. Geometry and topology optimization of geodesic domes using charged system search. *Struct. Multidiscip. Optim.* **2011**, *43*, 215–229. [CrossRef]
39. Talatahari, S.; Kaveh, A.; Mohajer Rahbari, N. Parameter identification of Bouc-Wen model for MR fluid dampers using adaptive charged system search optimization. *J. Mech. Sci. Technol.* **2012**, *26*, 2523–2534. [CrossRef]
40. Kaveh, A.; Talatahari, S. Charged system search for optimal design of frame structures. *Appl. Soft Comput.* **2012**, *12*, 382–393. [CrossRef]
41. Cook, R.D.; Malkus, D.S.; Plesha, M.E.; Witt, R.J. *Concepts and Applications of Finite Element Analysis*, 4th ed.; John Wiley & Sons: New York, NY, USA, 2001; ISBN 978-0-471-35605-9.
42. Batoz, J.; Tahar, M.B. Evaluation of a new quadrilateral thin plate bending element. *Int. J. Numer. Methods Eng.* **1982**, *18*, 1655–1677. [CrossRef]

applied sciences

MDPI

Article

Pavement Maintenance Decision Making Based on Optimization Models

Shitai Bao [1], Keying Han [1], Lan Zhang [2], Xudong Luo [3] and Shunqing Chen [3,*]

[1] College of Natural Resources and Environment, South China Agricultural University, 483 Wushan Road, Guangzhou 510642, China; bst100@scau.edu.cn (S.B.); hankeying@stu.scau.edu.cn (K.H.)
[2] School of Foreign Studies, South China Agricultural University, 483 Wushan Road, Guangzhou 510642, China; zhanglan9416@stu.scau.edu.cn
[3] Center of R&D, Augur Intelligent Technology Co., Ltd., 1029 Gaopu Road, Tianhe Dist., Guangzhou 510663, China; luoxd@augurit.com
* Correspondence: chensq@augurit.com; Tel.: +86-020-37020096; Fax: +86-020-37020432

Abstract: Pavement maintenance prioritization considering both quality and cost is an important decision-making problem. In this paper, the actual pavement condition index of city roads was calculated using municipal patrol data. A linear optimization model that maximized maintenance quality with limited maintenance costs and a multi-objective optimization model that maximized maintenance quality while minimizing maintenance costs were developed based on the pavement condition index. These models were subsequently employed in making decisions for actual pavement maintenance using sequential quadratic programming and a genetic algorithm. The results showed that the proposed decision-making models could effectively address actual pavement maintenance issues. Additionally, the results of the single-objective linear optimization model verified that the multiobjective optimization model was accurate. Thus, they could provide optimal pavement maintenance schemes for roads according to actual pavement conditions. The reliability of the models was investigated by analyzing their assumptions and validating their optimization results. Furthermore, their applicability in pavement operation-related decision making and preventive maintenance for roads of different grades was confirmed.

Keywords: pavement maintenance and rehabilitation; decision optimization; pavement condition index; linear optimization; genetic algorithm

Citation: Bao, S.; Han, K.; Zhang, L.; Luo, X.; Chen, S. Pavement Maintenance Decision Making Based on Optimization Models. *Appl. Sci.* **2021**, *11*, 9706. https://doi.org/10.3390/app11209706

Received: 9 September 2021
Accepted: 15 October 2021
Published: 18 October 2021

1. Introduction

Continuous urbanization in China has significantly increased the cumulative mileage of roads, which are considered the arteries of economic and social activities at the national and local levels. As an essential component of road infrastructure, a pavement is a type of hard surface made from durable surface material, which is able to withstand traffic and harmful environments. Because of increasing traffic volumes with heavy loads and the impacts of adverse environments, regular pavement maintenance is necessary to repair damage and mitigate degradation. A sufficient budget should be allocated to maintain the pavement at an appropriate condition, but insufficient budget is the primary obstacle of pavement maintenance [1]. When conducting large-scale pavement maintenance, it is important to consider the maintenance costs while ensuring that maintenance quality meets the requisite standards. The selection of a proper pavement maintenance scheme is a multiobjective optimization task. Therefore, determining the optimal maintenance scheme is crucial for decision making in pavement maintenance [2,3]. Decision making in pavement maintenance is a complicated, nondeterministic polynomial time (NP)-hard problem [4].

Maintenance personnel generally make judgments based on their personal experience when facing pavement maintenance tasks. Typical considerations include determination of which roads need maintenance, measures to be adopted for pavement maintenance, estimation of the maintenance cost, and the deadline for completing the maintenance. However, decision making based on human experience cannot control pavement maintenance costs or evaluate whether the maintenance quality meets the relevant standards. This can subsequently lead to deterioration of road structures or pose challenges in the assessment of maintenance outcomes. Therefore, assessment and prediction of road deterioration [5] as well as establishment of the relationship between maintenance schemes and corresponding influencing factors [6] are crucial in decision making regarding pavement maintenance and fund allocation.

Recently, various decision-making methods for pavement maintenance have been proposed. These methods generally aim at improving quality and reducing costs under diverse scenarios. The most common methods include the analytic hierarchy process (AHP), decision making based on a clustering algorithm, and decision making based on a genetic algorithm (GA). The AHP method was first used for decision making in pavement maintenance by Saaty [7]. Farhan and Fwa proposed a pavement maintenance prioritization method based on AHP and developed a three-level hierarchy model to determine the priority ranking of different pavement maintenance schemes [8]. Li et al. developed an AHP-based pavement maintenance priority model by assigning weights to the factors influencing pavement conditions [9]. However, the selection of the indices and the determination of the index weights in the AHP method remained subjective. Thus, the method of selecting an optimal solution from all available plans determines the effectiveness of the final decision.

In recent years, with the development of machine learning and corresponding algorithms, some researchers began studying pavement maintenance decisions based on machine learning. Han et al. proposed a decision-making framework for pavement maintenance that combined a clustering algorithm with the PageRank algorithm [10]. In this framework, the pavements are first grouped into clusters by the clustering algorithm. Then, the maintenance ranking of the clustered road samples is determined according to the road conditions. Hafez et al. proposed a decision-making algorithm for pavement maintenance based on pattern recognition [11]. This method helps determine the optimal maintenance and repair plans for low-volume paved roads. GAs and traditional mathematical programming are often used to solve road M&R planning problems based on multiobjective optimization (MOO), but they also have their limitations. Elhadidy et al. [12] proposed a two-objective optimization model balancing minimum action costs and maximum conditions for used road networks and developed a GA-based procedure for solving the MOO problem. Hadiwardoyo et al. [13] described the development of a genetic algorithm based on multiobjective programming for pavement and investigated the optimal maintenance strategy options applied as function of road surface distress conditions with a lack of information related to monitoring data and evaluation. Sindi and Agbelie [14] explored the expected accuracy rates of network treatment options through a multiobjective optimization methodology that utilized a GA and mixed-integer programming; their method was capable of effectively assigning pavement maintenance tasks under certain conditions. Hafez et al. [15] performed large-scale optimization to compare the existing maintenance policy with an alternative strategy. Specifically, they incorporated a GA into the optimization model to address the issues associated with specific optimization constraints and the limitations related to low-volume roads during the optimization process. Alqaili et al. offered a new multiobjective stochastic algorithm called the integer search algorithm (ISA) [1]. The ISA and GA were applied to improve the performance condition rating (PCR) of pavement in developing countries and achieved this by maximizing the condition of the pavement with minimum costs at specified constraints. Santos et al. [3] proposed a novel adaptive hybrid GA that incorporated local search techniques to improve the overall efficiency and effectiveness of the search. Hosseininasab et al. [16] proposed two multiobjective evolutionary approaches to solve problems of road construction within a reasonable time. These approaches were combinations of different techniques, such as

GA, NSGA-II, the Frank–Wolfe algorithm, and the ordered logit model. The NSGA-II and its improved algorithms are popular for solving MOO problems [17,18]. To address the high computation complexity associated with pavement maintenance at the network level, Hankach et al. [19] developed a model to reduce the search space and formulated the original problem as a generalized assignment problem, which was a well-known problem in mathematical optimization. Ahmed et al. [20] proposed a chaotic particle swarm optimization algorithm to find the optimal solution for pavement maintenance. This algorithm could effectively resolve the maintenance and rehabilitation issues associated with flexible pavements. Ameri and Jarrahi [21] used condition indicators in the form of normalized values and developed technical constraints in a linear integer programming model to improve network-level pavement maintenance and rehabilitation planning. Chen et al. [22] proposed a network-level pavement maintenance and rehabilitation optimization model considering the costs of user travel time and vehicle fuel consumption. The model optimized the asphalt pavement performance evaluation method, including 11 different combinations, which could be easily extended to the study of more complex road networks considering other factors. Mataei et al. [23] proposed a model based on the cloud decision tree (CDT) theory, which included a general decision-making model and various decision trees for every province of the country.

As evident from the foregoing discussion, considerable research has been conducted to improve GAs and to utilize them for decision making in pavement maintenance because of their effectiveness in resolving multiobjective optimization problems. In this study, the pavement maintenance problem was introduced and formulated as a multiobjective decision-making problem. Subsequently, two optimization models were developed based on actual road conditions and applied to a real-world case. Sequential quadratic programming and the NSGA-II algorithm were used to solve the two models. Finally, the characteristics and the optimization space of the proposed models were analyzed based on the results obtained. Future plans for extending the proposed model were discussed in depth.

2. Problem Formulation

Pavement maintenance usually involves decision making at the macro level for selecting roads to be maintained under objective constraints and decision makers' priorities. In China, pavement maintenance is usually subject to territorial management within each administrative region. For example, each district and county has its own department responsible for decision making during pavement maintenance at the macro level. Typical decision-making tasks involved in pavement maintenance include the formulation of an annual pavement maintenance scheme, budget application, and quality control. When making pavement maintenance decisions in real scenarios, the degree of deterioration of the road network in a certain region is first evaluated based on daily patrol data. Pavement deterioration can be quantified as the pavement condition index (PCI), which is a numerical rating of pavement condition based on the type and severity of distresses observed on the pavement surface. PCI is represented by a numerical value between 0–100, where 0 is the worst condition and 100 is the best.

Subsequently, under the premise of considering the total maintenance cost and time, the annual maintenance scheme is formulated to improve the condition of roads with low PCIs. Thus, the maintenance of regional road networks must take multiple objectives into account, such as minimum maintenance cost and optimum maintenance quality. Therefore, determining a suitable approach to formulating an efficient maintenance scheme is a decision-making challenge.

The multiobjective maintenance problem of a road network in a specific region can be described mathematically as follows. Given the number of roads (N), the area of each road (A) that requires maintenance, the PCI of each road, and the total budget available for pavement maintenance (C) in a certain region, m roads ($m \leq N$) are prioritized for maintenance among the total N roads, such that (1) the total maintenance cost is less than or equal to C, (2) the total maintenance cost is minimized, and (3) the maintenance quality is maximized. Selecting the appropriate roads based on human experience results in 2^N combination schemes, because the decision variable is a Boolean representing whether a road is included in the maintenance schedule. Therefore, N cannot be excessively large, for when N is large, it is infeasible to select the final roads manually. This problem is NP-hard. To resolve it, heuristic algorithms were researched and developed aiming at an approximate optimal solution that could be a particularly sensible choice. In this study, a decision-making optimization model for pavement maintenance was solved by using NSGA-II. Optimum maintenance schemes were obtained that could ensure the maximal maintenance quality with minimal maintenance cost.

Figure 1 shows the overall procedure of decision-making optimization for pavement maintenance. The foundation is to evaluate road condition by pavement condition index, and the key is to establish optimization models of road maintenance. The models can be applied to provide optimal schemes for multiobjective decision making in regional road network maintenance. Many optimization models have been widely used, as found from engineering management and research literature, but these models of optimizing pavement maintenance have not been fully examined. In this paper, we examined how these models could be extended for our research goals and demonstrated how they could be used in pavement maintenance for an application area.

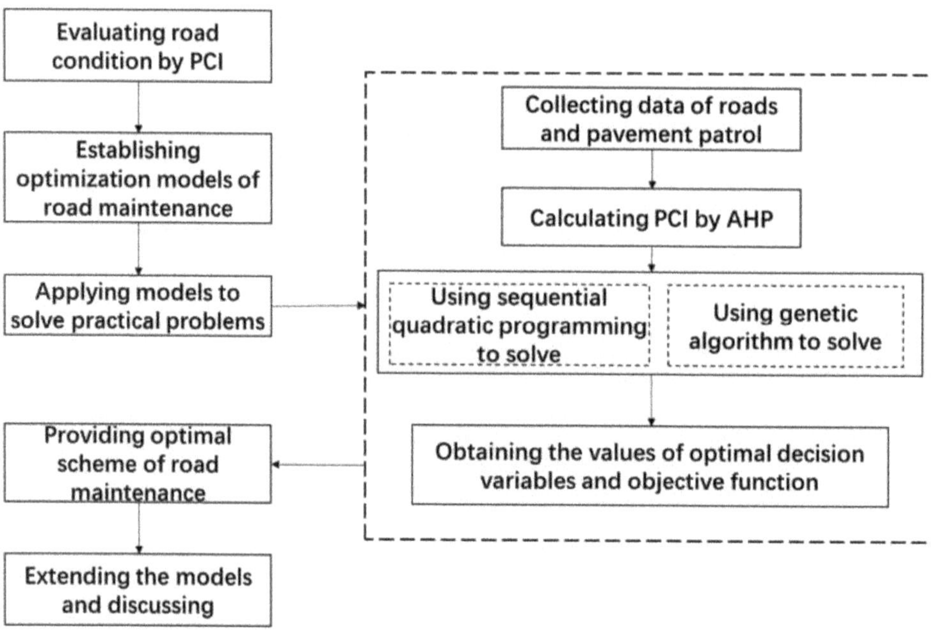

Figure 1. Overall flowchart of decision-making optimization for pavement maintenance.

3. Model Construction

The construction of the optimization models involved multiple steps, such as selecting the variables for decision making, setting up the model parameters, determining the objective function, and establishing the constraint conditions. In this section, two optimization models are established to resolve the road network maintenance problem.

3.1. Maximum Maintenance Quality–Limited Budget Model

To achieve optimal maintenance quality under a limited maintenance budget, a maximum maintenance quality–limited budget (MMQLB) model is built. The MMQLB model is a single-objective linear optimization model and is expressed as follows.

$$maxZ = \sum_{i=1}^{n} (PCI_i' - PCI_i)x_i \tag{1}$$

$$s.t.\begin{cases} \sum_{i=1}^{n} x_i Area_i \cdot P \leq Budget \\ x_i = 1 \quad\quad \forall PCI_i < D \end{cases} \tag{2}$$

In these equations, x_i denotes the decision-making variable of the model, which represents whether the ith road requires maintenance. For each road, there exist two states in terms of whether maintenance is required. Setting x_i to 1 or 0 indicates that the road does or does not require maintenance at present, respectively.

Equation (1) represents the objective function of this model, where n is the total number of roads and $PCI_i' - PCI_i$ represents the level of improvement in PCI (i.e., the maintenance quality) for a specific road. PCI_i' indicates the PCI of the road after maintenance and is set to 100 in this model; in other words, the road is assumed to be in perfect condition after maintenance. In contrast, PCI_i represents the original PCI of the ith road before maintenance. Equation (2) represents the constraint condition of the model, where $Area_i$ represents the area (m^2) of the ith road that requires maintenance. P is a constant that represents the maintenance cost per square meter of the road and is set as 200 CNY/m^2. *Budget* represents the total budget available for pavement maintenance, and D is a constant that represents the maintenance threshold of the road. When the PCI of a specific road drops below this threshold value, the road is judged to be of poor quality and require maintenance. Here, the default value of D is set to 70.

PCI is selected as the parameter in the objective function because it is an important index for evaluating whether the road is in a satisfactory condition according to the Chinese "technical code of maintenance for urban road" [24]. In addition, PCI plays an important role in finalizing the maintenance scheme [25]. In this study, asphalt pavement is selected as the maintenance object. The equation for calculating the PCI of asphalt pavement is as follows:

$$PCI = 100 - \sum_{i=1}^{n} \sum_{j=1}^{m} DP_{ij}\, \omega_{ij} \tag{3}$$

In this equation, PCI ranges from 0 to 100; n is the total number of main deterioration types (here set as 4 for asphalt pavement exhibiting cracks, deformation, loosening, and other forms of deterioration); m is the total number of deterioration subtypes included in each main deterioration type; and DP_{ij} is the deduction caused by the jth subtype of deterioration in the ith main type. ω_{ij} is the weight of the jth subtype of deterioration in the ith main type, which can be calculated by Equations (4) and (5) as follows:

$$\omega_{ij} = 3.0u_{ij}^3 - 5.5u_{ij}^2 + 3.5u_{ij} \tag{4}$$

$$u_{ij} = \frac{DP_{ij}}{\sum_{ij=1}^{m} DP_{ij}} \tag{5}$$

When computing the DP_{ij} value, first, the actual deterioration density of a pavement, the ratio of the total area of the subtype j deterioration to that of the pavement, is calculated. The actual deterioration density is compared with the deterioration density of each deterioration subtype in Table 1, and DP_{ij} is found to be proportional to two numbers in two adjacent columns in Table 1. The u_{ij} is calculated by Equation (5) and w_{ij} by Equation (4) based on the DP_{ij} in Table 2. For example, the PCI of AnQing Rd. in the research area is calculated as shown in Table 2. The paved area of AnQing road is about 26,204.1 m^2, with 1746.94 m of length and 4 lanes. The deterioration density is the ratio of the deteriorated area of each subtype and the paved area of AnQing road. The final PCI of AnQing road is 68.63, subtracting the sum of multiple of DP_{ij} and ω_{ij} from 100.

Table 1. Deductions from pavement condition index (PCI) associated with different subtypes of deterioration on asphalt pavement.

Deterioration Type		Deterioration Density (%)					
		0.01	0.1	1	10	50	100
Crack	Slippage cracks	3	5	8	16	38	48
	Alligator cracks	5	8	10	20	45	70
	Spalling	8	10	15	30	55	80
Deformation	Subsidence	3	5	12	25	47	63
	Rut	2	7	12	25	45	55
	Upheaval	3	10	15	30	52	65
Loosening	Pit	10	15	25	40	65	72
	Edge failure	2	4	8	15	30	40
	Stripping	2	5	8	15	35	45
Others	Poor frame	3	8	12	12	12	12
	Damage of repaired section	2	5	8	15	25	33

Table 2. An example of calculating PCI using AnQing Rd.

Deterioration Type		Deterioration Area (m^2)	Deterioration Density (%)	DP_{ij}	u_{ij}	ω_{ij}	$\sum_{i=1}^{n}\sum_{j=1}^{m} DP_{ij}\omega_{ij}$
Crack	Slippage cracks	56	0.21	5.38	0.080	0.247	1.33
	Alligator cracks	0	0	0.00	0.000	0.000	0.00
	Spalling	300	1.14	15.24	0.228	0.547	8.34
Deformation	Subsidence	100	0.38	11.11	0.166	0.443	4.92
	Rut	0	0	0.00	0.000	0.000	0.00
	Upheaval	0	0	0.00	0.000	0.000	0.00
Loosening	Pit	60	0.23	16.43	0.245	0.572	9.40
	Edge failure	0	0	0.00	0.000	0.000	0.00
	Stripping	980	3.74	10.13	0.151	0.414	4.19
Others	Poor frame	0	0	0.00	0.000	0.000	0.00
	Damage of repaired section	488	1.86	8.67	0.129	0.367	3.19
Sum		1984		66.96			31.37

As mentioned previously, in this model, the road is assumed to be in perfect condition after maintenance. Therefore, the maintenance quality can be represented by $100 - PCI_i$. In other words, the larger the value of $100 - PCI_i$, the higher the maintenance quality.

3.2. Minimum Budget–Maximum Maintenance Quality Model

Optimal maintenance of multiple roads should seek to minimize the maintenance cost while maximizing the maintenance quality. To this end, a multiobjective optimization model named the minimum budget–maximum maintenance quality (MBMMQ) model is constructed and expressed by the following equations:

$$maxZ_1 = \sum_{i=1}^{n} \left(PCI_i' - PCI_i\right) x_i \tag{6}$$

$$minZ_2 = \sum_{i=1}^{n} x_i Area_i \cdot P \tag{7}$$

$$s.t.\ x_i = 1\ \forall PCI_i < D \tag{8}$$

In Equations (6) and (7), the function Z_1 represents the maintenance quality, which equals the summation of improved PCI for all involved roads, whereas Z_2 represents the total maintenance cost. In this model, PCI_i' is set as 100 to indicate that roads are assumed to be in perfect condition after maintenance. PCI_i denotes the original PCI of the ith road, and x_i is the state variable, which indicates whether the ith road requires maintenance. The value of x_i can be set only as 1 or 0, which signifies that the road does or does not require maintenance, respectively. Finally, $Area_i$ and P have the same definitions as those in Equation (2). In Equation (8), D is the maintenance threshold; when the PCI of a specific road drops below threshold value, it is compulsory to maintain that road. Here, D is set to 70.

4. Model Application

This section elaborates on the application of the proposed models in decision making for a real-world pavement maintenance problem. Using actual data, the global optimal results were obtained from the model.

4.1. Dataset

In total, 149 roads under the jurisdiction of the Shushan District, a county-level district in Hefei, Anhui (China), were taken as the research objects. The deterioration data of these roads in 2019 were collected, and the maintenance area and premaintenance PCI were calculated by Equation (3). The roads were divided into three grades based on the National Standard of Road Classification: 9 expressways, 39 main roads, and 101 branch roads. For brevity, only a few roads in the three grades and their premaintenance PCI are listed in Table 3. Among the 149 roads, the average, minimum, and maximum PCI were 88.43, 67.89, and 100, respectively. There were 5 roads with PCI less than 70 and 26 roads with PCI equaling 100. As can be observed from the data, some roads were in poor condition, while others were in satisfactory condition. This reflects the level of diversity of the road dataset chosen for the study.

Table 3. PCI and other attributes of roads in Shushan District.

FID	Name	Grade	Maintenance Area (m^2)	PCIi
1	AnQing Rd.	Main road	1984	68.63
2	BaiYanWan Rd.	Branch road	87	98.75
3	BanDao Rd.	Branch road	364	80.36
4	YanHe N. Rd.	Branch road	298	86.08
5	North 1st Ring Exp.	Expressway	1002	75.63
……	……	……	……	……
70	YanHe S. Rd.	Branch road	236	88.98
71	South 1st Ring Exp.	Expressway	320	78.69
72	NingXi Rd.	Branch road	0	100.00
73	NongXin Rd.	Branch road	125	92.52
74	PiHe Rd.	Main road	695	71.31
75	QianShan Rd.	Expressway	789	78.95
……	……	……	……	……
145	Changjiang Middle Rd.	Main road	302	86.36
146	ZhenXin Rd.	Branch road	0	100.00
147	ZhiWuYuan E. Rd.	Branch road	123	90.20
148	ZhiWuYuan S. Rd.	Branch road	231	89.54
149	HaiTang Rd.	Main road	33	99.15

4.2. Solving the Models

Prior to solving the models, the premaintenance PCI, maintenance area, maintenance cost per unit area, and maintenance threshold for each road were first fed into the model as input parameters. Subsequently, the optimization models were implemented, and the values of the decision-making variables and the objective functions were calculated by using sequential quadratic programming and the NSGA-II algorithm. Finally, the optimal pavement maintenance scheme was determined based on the results obtained from the model. The first model was solved using the sequential quadratic programming method.

Because the multiobjective model could be solved by a GA, the second model was solved by using the NSGA-II algorithm in our research. The calculation steps involved in the NSGA-II algorithm are as follows. First, multiple sets of optimization solutions for pavement maintenance are randomly generated and are referred to as the solution population. Subsequently, these solution sets are ranked based on the maximum pavement maintenance quality and the minimum pavement maintenance cost calculated for each set of solutions. This process is also known as rapid nondominated sorting of the solution population. Next, a new generation of solutions is obtained through selection, crossover, and mutation of the solution population. The new solutions are the offspring of the first-generation solutions. Starting from the second generation, the parent and offspring populations are combined to perform rapid nondominated sorting. Simultaneously, the crowding degrees are calculated for each individual in the nondominated layer. Suitable individuals are then selected according to their crowdedness and the nondominated relationship to form a new parent population. Finally, a new offspring population is generated through selection, crossover, and mutation of the new parent population.

The final solution set of the objective function can be obtained by repeating the aforementioned process continuously until the maximum number of iterations is reached. NSGA-II can select multiple nondominated individuals that form an optimal set of trade-off solutions called the Pareto set. The pavement maintenance scheme can then be formulated according to the solution results. Some researchers have proposed simpler coding methods to represent complex evolutionary phenomena. These methods can realize heuristic searches in the complex search space and can determine the global optimal solution of the objective function with high probability through a simplified genetic process.

4.3. Results of the Models

Table 4 shows the optimal solutions obtained by the MMQLB model with different budgets. The best pavement maintenance quality provided by the model was 1724.29, which equaled the sum of the improved PCI for all roads; meanwhile, the maintenance cost was 10.199 million CNY. Because of actual budget limitations, some roads had to be eliminated from the full maintenance scheme. The maintenance quality was 154.51 with the cost = 1,567,200 CNY if only the worst roads under 70 PCI were to be maintained. The optimal maintenance quality was 324.861 with a cos t = 1,999,600 CNY, and the maximum maintenance was 1025.2 under a 5 million CNY budget. The maximum improvement to PCI was 1479.52 under an 8 million CNY budget. The optimal maintenance was 1702.64, and the cost was 9,991,800 CNY under a budget of 10 million CNY. Table 5 shows the PCI improvements to 149 roads in the optimal scheme obtained by MMQLB model under the 5 million CNY budget limitation. According to the optimal scheme, the PCI improvements of each road equaled the difference between 100 and the original PCI if a road was selected to be maintained; otherwise, its PCI improvement was zero.

Table 4. Optimal solutions obtained by MMQLB model with different budgets.

Optimal Solutions	Maintenance Quality (Sum of Improved PCI)	Maintenance Cost (CNY)
Necessary maintenance (PCI < 70)	154.51	1,567,200
Optimal maintenance with 2 million CNY budget	324.861	1,999,600
Optimal maintenance with 5 million CNY budget	1025.2	4,967,400
Optimal maintenance with 8 million CNY budget	1479.52	7,998,600
Optimal maintenance with 10 million CNY budget	1702.64	9,991,800
Full maintenance without budget constraint	1724.29	10,199,000

Table 5. PCI improvements of 149 roads according to the optimal scheme obtained by the MMQLB model with a 5 million CNY budget limitation.

Road Number (from Left to Right)	PCI Improvements of Each Road (Difference between the Output PCI from the Model and Original PCI of Each Road)									
No. 1–10	31.37	0.00	19.64	13.92	0.00	0.00	17.86	27.14	0.00	0.00
No. 11–20	2.31	18.65	0.00	0.00	30.45	0.00	0.00	2.13	0.00	18.68
No. 21–30	23.11	16.79	0.00	2.86	13.32	0.00	0.00	2.11	0.00	0.00
No. 31–40	30.47	0.00	26.32	0.00	0.00	2.63	18.19	0.00	0.00	0.00
No. 41–50	0.00	1.76	0.00	32.11	17.74	0.00	19.94	26.46	13.35	25.43
No. 51–60	0.00	0.00	0.00	0.00	10.35	7.78	0.00	0.00	0.00	25.46
No. 61–70	0.00	28.88	0.00	0.00	0.00	0.00	0.00	0.00	0.00	0.00
No. 71–80	11.02	21.31	0.00	7.48	28.69	0.00	0.00	0.00	0.00	24.70
No. 81–90	25.46	0.00	0.00	30.11	0.00	0.00	19.91	21.84	19.40	0.00
No. 91–100	0.00	9.78	0.00	0.00	0.00	0.00	7.37	0.00	12.11	13.10
No. 101–110	21.37	0.00	0.00	0.00	10.13	0.00	9.61	10.32	7.68	14.32
No. 111–120	11.22	6.99	9.24	0.00	15.46	0.00	0.00	8.41	0.00	0.00
No. 121–130	0.00	23.90	0.00	10.61	0.00	0.00	0.00	0.00	0.00	0.00
No. 131–140	3.31	11.76	9.42	0.00	0.00	0.00	0.00	0.00	0.00	0.00
No. 141–149	0.73	0.00	15.76	0.00	11.39	13.64	0.00	9.80	10.46	—

The optimal maintenance scheme of 1702.64 PCI improvement under a 10 million CNY budget is shown in Figure 2 using a thematic map of 149 roads in the Shushan District. All roads with a PCI smaller than 70 required maintenance according to the constraint set in the MMQLB model. Each road was labelled by its FID and name in the map. For the roads in red, the corresponding decision-making variables were calculated to be 1 under the optimization model, implying that these roads had to be maintained. For the roads in blue, the decision-making variable was calculated to be 0 under the optimization model, signifying that these roads did not require maintenance.

Next, Figure 3 shows the resulting Pareto front obtained by the MBMMQ model. The objective function was solved by the NSGA-II algorithm and converged to stable values at approximately the 3000th generation. The first objective function was maximizing maintenance quality in the x-axis direction, and the second was minimizing maintenance cost in the y-axis direction. The solution set included many optimal solutions, represented by star points, which made up the Pareto front. Three solutions are highlighted along with their objective function values in Figure 3.

The two optimal models were applied to the integrated management system for municipal facilities in Shushan Dist., Hefei, Anhui (China). Since 2018, the Municipal Engineering Management Office in Shushan Dist. used the system for pavement decision making. The total improved PCI of road maintenance increased by 15% in 2019, while the budget for maintenance remained almost the same.

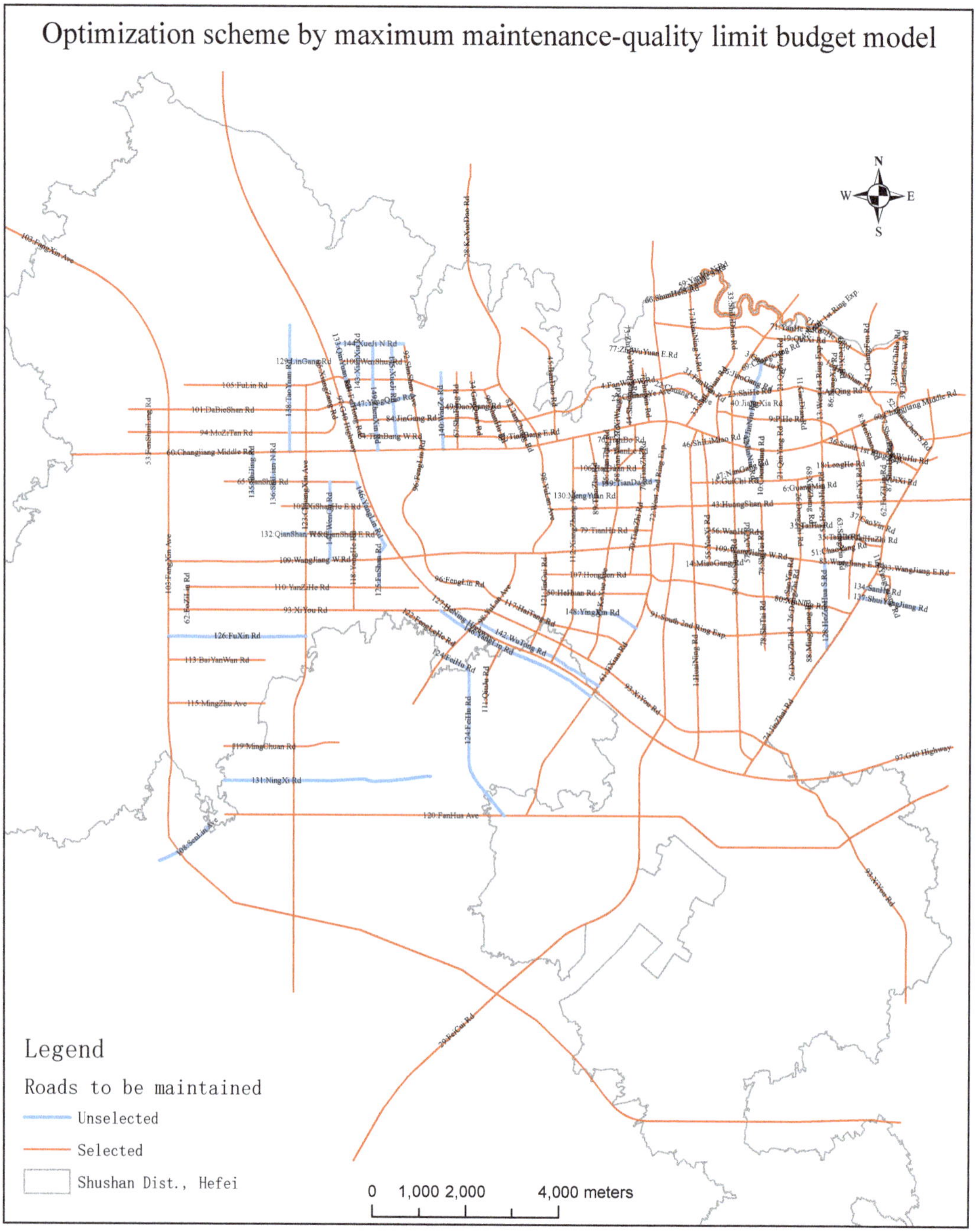

Figure 2. Pavement maintenance scheme obtained using the MMQLB model.

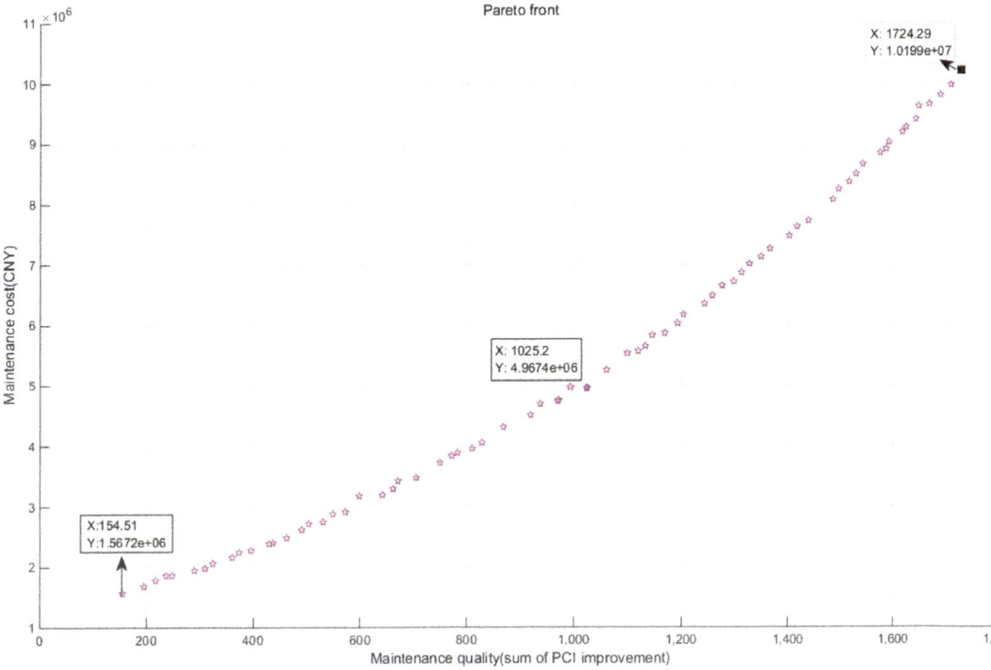

Figure 3. Pareto front of the MBMMQ model solved by genetic algorithms.

5. Discussion

In this section, we discuss the suitability of the developed models, the validation of the obtained results, and extensions of the multiobjective model.

5.1. Analysis of Model Suitability

The theoretical suitability of the proposed models was analyzed as follows. A linear optimization model must satisfy the following implicit assumptions.

(1) Assumption of linearization: a function is considered linear [26] when the following equation is satisfied:

$$f(x_i) = \sum_{i=1}^{n} c_i x_i + K \quad (9)$$

In the equation, K and c_i are constants. The objective functions of the two proposed models can be converted to the form shown in Equation (9). Therefore, these objective functions can be considered as linear functions.

(2) Assumptions of proportionality: A change in the decision-making variable should cause the objective function to change proportionally. Because the relationship between the decision-making variable and objective function can be expressed in the form given in Equation (13) for both models, the assumption of proportionality is satisfied for the decision-making variable and the objective function.

(3) Assumption of additivity: The value of the objective function should equal the sum of the individual contributions from each decision-making variable to the objective function. This assumption is satisfied for all models, as is evident from their equations.

(4) Assumption of the fixed parameter: All the parameters are fixed constants, and no random factor is introduced into the analysis. The PCI used in all models is also a fixed parameter. Therefore, this assumption is also satisfied.

5.2. Validation of Results Obtained from the Models

The proposed MMQLB model is a single-objective linear optimization model that can provide an exact result. Its result is a maintenance scheme that maximizes the maintenance quality under a limited maintenance budget, such as some key schemes shown in Table 4. Meanwhile, the MBMMQ model, which is a biobjective optimization model, was solved by NSGA-II, a heuristic algorithm. In Figure 3, each point represents an optimal maintenance scheme solved by the biobjective optimization model for 149 roads. Among the optimal maintenance schemes suggested by the MBMMQ model, three schemes were highlighted along with their objective function values, and the costs exactly matched the results obtained by the MMQLB model, satisfying the corresponding cost constraints. The best PCI improvement was 1724.29, with a cost of 10,199,000 CNY; the maximum PCI improvement was 1025.2, with a cost of 4,967,400 CNY; and the necessary PCI improvement was 154.51, with a cost of 1,567,200 CNY. Some solutions of MMQLB were included in the solution set of MBMMQ. This suggests that the results of the MMQLB model verified the part results of the MBMMQ model in this case. For this reason, we believe that the solutions found by the NSGA-II were indeed optimal and that they formed the Pareto front for our biobjective maintenance decision-making problem. The overall shape of the Pareto front indicated the same trend of increasing sum of minimal maintenance cost with the total improved quality. The Pareto front for our problem was discrete, but more importantly, it had a nonconvex shape. This was clearly shown when sum of PCI improvement reached around 1000, where the front took an obvious turn. This suggests that it is often difficult to use a heuristic algorithm to obtain an exact convex Pareto front for a multiobjective decision-making problem.

5.3. Extension of the Multiobjective Model

This section focuses on extension of the multiobjective model. Considering different road grades, expressways have a higher maintenance priority than main roads. Moreover, main roads usually have a higher maintenance priority than branch roads. The MBMMQ model was extended accordingly, as follows:

$$maxZ_1 = \sum_{i=1}^{n} (PCI_i' - PCI_i) w_i x_i \quad (10)$$

$$minZ_2 = \sum_{i=1}^{n} w_i x_i Area_i \cdot P \quad (11)$$

Here, w_i denotes the grade of road i, which is set to 1.5, 1, or 0.5 when road i is an expressway, a main road, or a branch road, respectively. Based on this revised model, the effects of the different road grades and maintenance schemes were further determined, as shown in Figure 4. Compared with the original solution, the maintenance costs and quality are slightly different, which was primarily due to different road PCI values and maintenance areas between expressways and branch roads. These results indicated that the revised model is suitable for multiple road grades.

Because of constraints on pavement maintenance, a road may be only partially maintained, and its PCI may be lower 100 after maintenance. PCI_i' could replace x_i as a decision-making variable, which represents the PCI of the road after maintenance. Based on the MBMMQ model, the third MOO model is proposed as follows.

$$maxZ_1 = \sum_{i=1}^{n} (PCI_i' - PCI_i) \quad (12)$$

$$minZ_2 = \sum_{i=1}^{n} \frac{Area_i \cdot P \cdot PCI_i'}{PCI_i} \quad (13)$$

$$s.t.\ PCI_i \le PCI_i' \le 100 \quad (14)$$

In this model, PCI_i' is selected as the variable, which can vary from PCI_i to 100. The constants PCI_i, $Area_i$, and P are the same as those in Equation (2). Therefore, when calculating the maintenance cost for a specific road, PCI_i'/PCI_i should be incorporated into Equation (13) to consider the effect of PCI improvement on the maintenance area. This model is suitable to maintain quantitatively each selected road and can be solved by NSGA-II.

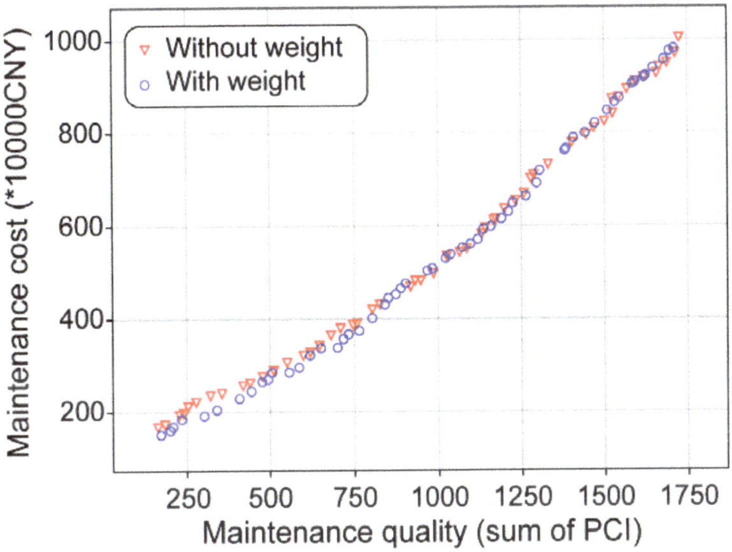

Figure 4. Comparison between the pavement condition index (PCI) values of the roads before and after maintenance calculated by the MBMMQ model.

6. Conclusions

In multiobjective decision making of pavement M&R, some main research fields have emerged, such as sustainable pavement maintenance management, optimization of decision models with high-dimensional objectives, nonlinear planning, and high-dimensional variables, which have received increasing attention [27]. Here, actual decision-making problems of pavement maintenance were investigated, and an overall procedure for decision-making optimization was proposed. First, PCI was calculated based on the deterioration type and area to evaluate road conditions using the AHP method. Then, two models, named MMQLB and MBMMQ, were developed. The first model was a single-objective linear optimization model, whereas the other was a MOO model. These models were employed for a real-world case involving the maintenance of 149 roads in Shushan District, Hefei, China. First, the pavement condition data of these roads were recorded and imported into the models. The models were then solved using sequential quadratic programming and a GA. Finally, appropriate pavement maintenance schemes were established based on the solutions of the optimization models. Among the optimal maintenance schemes of the MBMMQ model, some optimal values in the pair of maintenance quality and cost exactly matched the results obtained by the MMQLB model with the corresponding cost constraints. This indicated that the MMQLB model could achieve similar decision making to the MBMMQ model by setting different maintenance constraints. According to one nonlinear constraint of MMQLB, it could be improved to substitute the nonlinear formulation and reduce the difficulty of solving the problem in future research. In practical scenarios, the pavement maintenance cost is also dependent on the time taken and priority levels of the road grades. Therefore, the parameters used in the proposed models will be further extended in future studies to reflect real-world scenarios better and to broaden the applicability of the models.

Author Contributions: Conceptualization and funding, S.C.; methodology and formal analysis, S.B.; software and validation, K.H. and X.L.; writing—original draft preparation, K.H. and X.L.; writing—review and editing, S.B. and L.Z.; visualization, K.H. All authors have read and agreed to the published version of the manuscript.

Funding: This work was supported by Guangzhou Sci-Tech under Grant 201803030024 and 202103050001.

Institutional Review Board Statement: Not applicable.

Informed Consent Statement: Informed consent was obtained from all subjects involved in the study.

Data Availability Statement: The raw data of roads and their maintenance cannot be shared at this time, as it belongs to the Hefei Municipal Engineering Management Office. The processed data for solving the multiobjective models in this study are available upon request by contacting the corresponding author.

Acknowledgments: We would like to extend our sincere gratitude and appreciation to the Municipal Engineering Management Office in Shushan Dist., Hefei for providing invaluable support and project cooperation.

Conflicts of Interest: The author declares no conflict of interest.

Disclosure Statement: No potential competing interest is reported by the authors.

References

1. Alqaili, A.; Qais, M.; Al-Mansour, A. Integer Search Algorithm: A New Discrete Multi-Objective Algorithm for Pavement Maintenance Management Optimization. *Appl. Sci.* **2021**, *11*, 7170. [CrossRef]
2. Wang, F.; Zhang, Z.; Machemehl, R.B. Decision-Making Problem for Managing Pavement Maintenance and Rehabilitation Projects. *Transp. Res. Rec. J. Transp. Res. Board* **2003**, *1853*, 21–28. [CrossRef]
3. Santos, J.; Ferreira, A.; Flintsch, G.W. An Adaptive Hybrid Genetic Algorithm for Pavement Management. *Int. J. Pavement Eng.* **2019**, *20*, 266–286. [CrossRef]
4. Syan, C.S.; Ramsoobag, G. Maintenance Applications of Multi-Criteria Optimization: A Review. *Reliab. Eng. Syst. Saf.* **2019**, *190*, 106520. [CrossRef]
5. Chopra, T.; Parida, M.; Kwatra, N.; Chopra, P. Development of Pavement Distress Deterioration Prediction Models for Urban Road Network Using Genetic Programming. *Adv. Civ. Eng.* **2018**, *2018*, 1–15. [CrossRef]
6. Fani, A.; Golroo, A.; Mirhassani, S.A.; Gandomi, A.H. Pavement Maintenance and Rehabilitation Planning Optimization under Budget and Pavement Deterioration Uncertainty. *Int. J. Pavement Eng.* **2020**, *2020*, 1–11.
7. Saaty, T.L. *Fundamentals of Decision Making and Priority Theory with the Analytic Hierarchy Process*; RWS Publications: Pittsburgh, PA, USA, 2000.
8. Farhan, J.; Fwa, T.F. Pavement Maintenance Prioritization Using Analytic Hierarchy Process. *Transp. Res. Rec. J. Transp. Res. Board* **2009**, *2093*, 12–24. [CrossRef]
9. Li, H.; Ni, F.; Dong, Q.; Zhu, Y. Application of Analytic Hierarchy Process in Network Level Pavement Maintenance Decision-Making. *Int. J. Pavement Res. Technol.* **2018**, *11*, 345–354. [CrossRef]
10. Han, C.; Fang, M.; Ma, T.; Cao, H.; Peng, H. An Intelligent Decision-Making Framework for Asphalt Pavement Maintenance Using the Clustering-PageRank Algorithm. *Eng. Optim.* **2020**, *52*, 1829–1847. [CrossRef]
11. Hafez, M.; Ksaibati, K.; Atadero, R.A. Optimizing Expert-Based Decision-Making of Pavement Maintenance Using Artificial Neural Networks with Pattern-Recognition Algorithms. *Transp. Res. Rec. J. Transp. Res. Board* **2019**, *2673*, 90–100. [CrossRef]
12. Elhadidy, A.A.; Elbeltagi, E.E.; Ammar, M.A. Optimum Analysis of Pavement Maintenance Using Multi-Objective Genetic Algorithms. *HBRC J.* **2015**, *11*, 107–113. [CrossRef]
13. Sigit, H.; Hadiwardoyo, P.; Correia, A.G.; Pereira, P. Pavement Maintenance Optimization Strategies for National Road Network in Indonesia Applying Genetic Algorithm. *Procedia Eng.* **2017**, *210*, 253–260. [CrossRef]
14. Sindi, W.; Agbelie, B. Assignments of Pavement Treatment Options: Genetic Algorithms versus Mixed-Integer Programming. *J. Transp. Eng. Part B Pavements* **2020**, *146*, 04020008. [CrossRef]
15. Hafez, M.; Ksaibati, K.; Atadero, R.A. Applying Large-Scale Optimization to Evaluate Pavement Maintenance Alternatives for Low-Volume Roads Using Genetic Algorithms. *Transp. Res. Rec. J. Transp. Res. Board* **2018**, *2672*, 205–215. [CrossRef]
16. Hosseininasab, S.M.; Shetab-Boushehri, S.N.; Hejazi, S.R.; Karimi, H. A Multi-Objective Integrated Model for Selecting, Scheduling, and Budgeting Road Construction Projects. *Eur. J. Oper. Res.* **2018**, *271*, 262–277. [CrossRef]
17. Deb, K.; Pratap, A.; Agarwal, S.; Meyarivan, T. A Fast and Elitist Multi Objective Genetic Algorithm: NSGA-II. *IEEE Trans. Evol. Comput.* **2002**, *6*, 182–197. [CrossRef]
18. Deb, K.; Jain, H. An Evolutionary Many-Objective Optimization Algorithm Using Reference-Point-Based Non Dominated Sorting Approach, Part I: Solving Problems with Box Constraints. *IEEE Trans. Evol. Comput.* **2014**, *18*, 577–601. [CrossRef]

19. Hankach, P.; Lorino, T.; Gastineau, P. A Constraint-Based, Efficiency Optimisation Approach to Network-Level Pavement Maintenance Management. *Struct. Infrastruct. Eng.* **2019**, *15*, 1450–1467. [CrossRef]
20. Ahmed, K.; Al-Khateeb, B.; Mahmood, M. Application of Chaos Discrete Particle Swarm Optimization Algorithm on Pavement Maintenance Scheduling Problem. *Clust. Comput.* **2019**, *22*, 4647–4657. [CrossRef]
21. Ameri, M.; Jarrahi, A. An Executive Model for Network-Level Pavement Maintenance and Rehabilitation Planning Based on Linear Integer Programming. *J. Ind. Manag. Optim.* **2020**, *16*, 795–811. [CrossRef]
22. Leilei, C.; Zepeng, F.; Pengfei, L.; Zhendong, Q.; Adolfo, P. Optimization Model of Network-Level Pavement Maintenance Decision considering User Travel Time and Vehicle Fuel Consumption Costs. *Adv. Civ. Eng.* **2021**, *2021*, 4699838. [CrossRef]
23. Behrouz, M.; Moghadas, N.F.; Hamzeh, Z. Pavement maintenance and rehabilitation optimi-zation based on cloud decision tree. *Int. J. Pavement Res. Technol.* **2021**, *14*, 740–750. [CrossRef]
24. Ministry of Housing and Urban-Rural Development of the People's Republic of China. *Industry Standard CJJ 36–2016*; Technical Code of Maintenance for Urban Road; Ministry of Housing and Urban-Rural Development of the People's Republic of China: Beijing, China, 2016.
25. Shah, Y.U.; Jain, S.S.; Tiwari, D.; Jain, M.K. Development of Overall Pavement Condition Index for Urban Road Network. *Procedia Soc. Behav. Sci.* **2013**, *104*, 332–341. [CrossRef]
26. Hyers, D.H. On the Stability of the Linear Functional Equation. *Proc. Natl. Acad. Sci. USA* **1941**, *27*, 222–224. [CrossRef]
27. Wang, C.; Mulian, Z. Multi-objective optimization for pavement maintenance and rehabilitation decision-making: A critical review and future directions. *Autom. Constr.* **2021**, *130*, 103840. [CrossRef]

applied sciences

MDPI

Article

MLGen: Generative Design Framework Based on Machine Learning and Topology Optimization

Nikos Ath. Kallioras † and Nikos D. Lagaros *,†

Institute of Structural Analysis and Antiseismic Research, School of Civil Engineering, National Technical University of Athens, 9, Heroon Polytechniou Str., Zografou Campus, GR-15780 Athens, Greece; nkallio@mail.ntua.gr
* Correspondence: nlagaros@central.ntua.gr; Tel.: +30-210-772-2625
† These authors contributed equally to this work.

Abstract: Design and manufacturing processes are entering into a new era as novel methods and techniques are constantly introduced. Currently, 3D printing is already established in the production processes of several industries while more are continuously being added. At the same time, topology optimization has become part of the design procedure of various industries, such as automotive and aeronautical. Parametric design has been gaining ground in the architectural design literature in the past years. Generative design is introduced as the contemporary design process that relies on the utilization of algorithms for creating several forms that respect structural and architectural constraints imposed, among others, by the design codes and/or as defined by the designer. In this study, a novel generative design framework labeled as MLGen is presented. MLGen integrates machine learning into the generative design practice. MLGen is able to generate multiple optimized solutions which vary in shape but are equivalent in terms of performance criteria. The output of the proposed framework is exported in a format that can be handled by 3D printers. The ability of MLGen to efficiently handle different problems is validated via testing on several benchmark topology optimization problems frequently employed in the literature.

Keywords: generative design; machine learning; topology optimization; long short-term networks; ant colony optimization

Citation: Kallioras, N.A.; Lagaros, N.D. MLGen: Generative Design Framework Based on Machine Learning and Topology Optimization. *Appl. Sci.* **2021**, *11*, 12044. https://doi.org/10.3390/app112412044

Academic Editor: Chiara Bedon

Received: 16 November 2021
Accepted: 14 December 2021
Published: 17 December 2021

1. Introduction

Generally speaking, structural optimization can be distinguished into three categories: topology, shape and sizing optimization [1,2]. Topology optimization refers to a mathematical procedure that aims to identify the optimal shape, in terms of structural performance, of a structural system when subjected to specific load and support conditions. This is achieved by optimizing the topological placement of a specific quantity of material into the design domain. Apart from structural performance, topology optimization can be used for optimization with respect to sizing and shape criteria as well. The application of such approaches helps, among others, to create structural systems that are very close to their optimal shape, and thus can be considered a supporting procedure in the conceptual design phase [3,4].

With the term generative design, a design exploration process is defined. The basic idea of this process is that engineers/designers in general introduce their design goals and constraints imposed by design codes, etc., into the generative design framework, along with other parameters, such as performance demands, material properties, etc. The generative design framework should be able to examine all or most of the possible solutions of the specific problem, and produce design alternatives of equivalent performance and criteria values.

During the last decade, research on modern soft computing methods drew significant attention [5], mainly due to the excessive amount of data generated. This led to the development of new methods, able to handle large amounts of data and extremely complex problem definitions. The scope of this study is to examine and combine modern machine learning methods with topology optimization procedures aiming to generate multiple forms in the concept of generative design. Recently, there is growing interest for applying deep learning techniques in topology optimization, mainly for accelerating its procedure [6]. MLGen, a generative design procedure based on machine learning integrated with topology optimization, is presented herein. More specifically, MLGen is formulated by combining the topology optimization solution approach of solid isotropic material with penalization [7–9], long short-term memory networks [10], image filtering techniques and derivative-free search algorithms [11].

The layout of this study begins with a short introduction on structural topology optimization together with some notes on the solid isotropic material with penalization (SIMP) approach in Section 2, followed by a description of long short-term memory (LSTM) networks in Section 3. Subsequently, in Section 4 the proposed MLGen (machine learning-based generative design) framework is described in detail. Section 5 presents the results of the numerical tests performed in order to assess its efficiency to generate multiple optimized solutions.

2. Topology Optimization

In this part of the study, a short introduction of structural topology optimization (STO) is presented: the general mathematical concept used to formulate the STO problem, together with theory and implementation issues of the solution process. STO is a mathematical formulation of an optimization problem that aims to optimally distribute material within a specific design domain, subjected to given series of loading, boundary conditions and constraints. The goal of STO is to maximize the structural performance of the system. This is achieved by optimizing the allocation of material within the finite element mesh of the design domain.

2.1. Problem Formulation

The definition of the structural optimization problem requires introducing: a function (known as objective function) that refers to the criterion to be optimized, design unknowns (known as design variables), and state quantities. The criterion F that is to be optimized represents an objective metric that could either be maximized or minimized. Such an objective could be the material volume or the stiffness of the structural system. Moreover, the structural design domain and the state quantities associated to the criterion need to be defined. The design variable x represents unknown parameters that are required in order to describe the design of the structure; it could represent, for example, the geometry. The state quantity y represents the structural response quantities, which may represent strain, stress or displacement.

$$\begin{aligned} \min_{x \in S} F(x, y(x)) \\ subject \quad to: \\ state \quad constraints \quad on \quad y(x) \end{aligned} \tag{1}$$

where the objective function $F(x)$ (e.g., stiffness of the structure, and material volume) is to be optimized, and S is the design space of the unknowns' vector x. The state quantities depend on the design variable $y(x)$. The objective function is subjected to the design variables and state quantities constraints to lead the optimization process to the desired solution. A so-called state function $g(y)$ that corresponds to a certain state quantity can be devised, for example, as a deformation quantity along a certain degree of freedom (DoF). This state function $g(y)$ can be integrated as a constraint function to the structural optimization problem, where it is commonly expressed such that $g(y) \leq 0$, for example, the case where $g(y)$ is expressed by the displacement quantities vector $g(U(x))$ in a discrete finite element problem. In STO problems, the design variable x represents the presence or absence of material in a specific part of the design domain. It might involve features such as the size and/or the number of holes in the design domain. The general formulation of a STO problem is summarized below:

$$\begin{aligned} &\min_{x\in[0,1]} F(x) \\ &subject \quad to: \\ &K(x)\cdot U(x) = P \\ &g(x) \leq 0 \end{aligned} \tag{2}$$

where x refers the vector of unknowns, i.e., the density values of the finite elements, K denotes the global stiffness matrix of the structural system, vectors P and U contain the loads and displacements, respectively, and $g(x)$ refers to the vector of constraint functions (volume fraction, etc.).

2.2. The Solid Isotropic Material with Penalization (SIMP) Approach

In the past, several approaches have been proposed for solving the STO problem. The main ones are as follows [12]: (i) level-set method, (ii) density method, (iii) phase field method, (iv) topological derivative method, and (v) evolutionary method. SIMP, proposed in 90 s [7–9], is the most well-known representative of the density method and is commonly used for dealing with the STO problem. The structural system's compliance C is the most widely adopted performance indicator used in STO problem formulations. If n finite elements are used to discretize the design domain Ω, the distribution of material over Ω is denoted by x_i that are the density values of each finite element, $i \in [1,\ldots,n]$ and $x_i \in (0,1]$. When $x_i \approx 0$, it denotes no material to the ith finite element, while if $x_i = 1$, then the ith finite element is fully filled. Thus, Equation (2) becomes as follows:

$$\begin{aligned} &\min_{x\in(0,1]} C(x) = U(x)^T \cdot P \\ &subject \quad to: \\ &K(x)\cdot U(x) = P \\ &\frac{V(x)}{V_0} = V_t \end{aligned} \tag{3}$$

where $C(x)$ denotes the compliance of the structural systems for specific material distribution x, while $V(x)$, V_0 and V_t express the volume corresponding to density vector x, the initial volume for $x = x_0$ and the targeted volume of the optimized domain, respectively. According to SIMP, Young's modulus E is correlated based on a power law expression to the density value x_i of each element of the FE discretization as follows:

$$E_x(x_i) = x_i^p \times E_0 \iff K_x(x_i) = x_i^p \times K_{i,0} \tag{4}$$

where the penalization coefficient p is often set to $p = 3$, while $K_x(x_i)$ and $K_{i,0}$ denote the local stiffness matrix of the ith element when its density is equal to x_i and its original stiffness matrix, respectively. In the literature, various search algorithms were combined with SIMP for solving the problem of Equation (3); the method of moving asymptotes (MMA) and the optimality criteria (OC) algorithm are the most commonly used ones.

3. Long Short-Term Memory Networks

Long short-term memory (LSTM) networks were created as a variant of recurrent neural networks (RNNs) with a differentiated architecture. RNNs were firstly introduced in 1990 [13,14]. They were inspired by typical feedforward networks, but were tweaked in order to efficiently handle data that were in the form of time steps. In order to achieve that, RNNs are equipped with recurrent (feedback) connections as well. RNNs are able to operate not only on single data patterns (such as images), but also on sequences of data, such as video or speech. Due to this new architecture, RNNs presented the ability to efficiently handle time-series problems. A representation of an RNN is depicted in Figure 1 [15]. These networks were found to be difficult to train, as they were prone to numerical instability problems such as the exploding and/or vanishing gradient. This led to the development of the LSTM networks and the so-called bi-directional LSTMs [16] that are used in this study. LSTMs include an extra memory value that is trained on a specific number of time steps. This temporary memory acts as a hidden state that is cycled over the time steps.

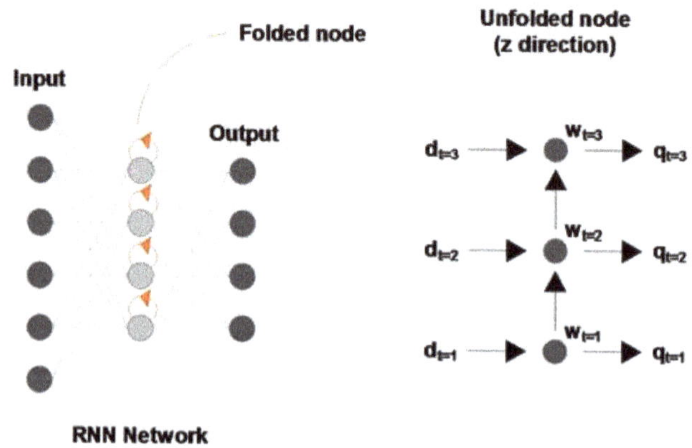

Figure 1. Recurrent neural network.

A typical LSTM architecture contains the input layer, a series of hidden layers (according to the depth of the network) and finally, the output layer. The size of the input layer depends on the problem and, specifically, the size of the input vector. The hidden layers present the special architecture of LSTM networks, the memory cells. These cells are characterized by three nodes, known as gates. The first gate, named input, is responsible for introducing a new value to the cell. The second one, called the forget gate, decides which previous state will be dropped by the network. The last one, known as the output gate, decides on the cell's output value. In a time series problem, whenever a new step is introduced to the network, all gates are fed with the new input along with the output value of the previous step that was stored in the networks' memory. Finally, the size of the output layer also depends on the problem. For example, in a classification problem with two classes, the size of the output layer would be equal to two [17]. As LSTMs were gaining more research attention, a new architecture of LSTMs was introduced, named bi-directional LSTMs. The novelty in this approach was that they include two LSTMs functioning at the same time but in reverse directions (forward and backward) and are connected by a merging function. Due to this form, the hidden state repeatedly extracts knowledge from both past and future information fed to the network [18]. Bi-directional LSTMs are very efficient when dealing with problems of time series, as they can take advantage of past and future input values simultaneously.

4. The MLGen Framework Description

This section presents the proposed methodology MLGen, together with its components. The goal of MLGen is to support the designing engineer with a tool able to automatically create a large number of equivalent designs that can act as an inspiration trigger. The proposed shapes are created via algorithms and it is ensured that all respect the goals and restrictions set by the user/designer. As a result, the combination of a number of methods is necessary. The proposed framework incorporates topology optimization but it is worth pointing out that the goal of MLGen is not to identify the optimal shape but to be able to discover a large number of differentiated designs that present near-to-optimal behavior. The methods used and the workflow of MLGen are presented in the following sections.

4.1. Framework Architecture

MLGen is a generative design framework that incorporates several methods in order to deliver automatically generated shapes. The methods used are a topology optimization algorithm (SIMP), neural networks (LSTM), image filtering, metaheuristics (ACO) and 3D printing support. The first three are used for generating a plethora of shapes, metaheuristics are used for visualizing the final result (not as optimizers) and 3D printing is for experimentally producing the outputs of MLGen. The flowchart of MLGen is presented in Figure 2.

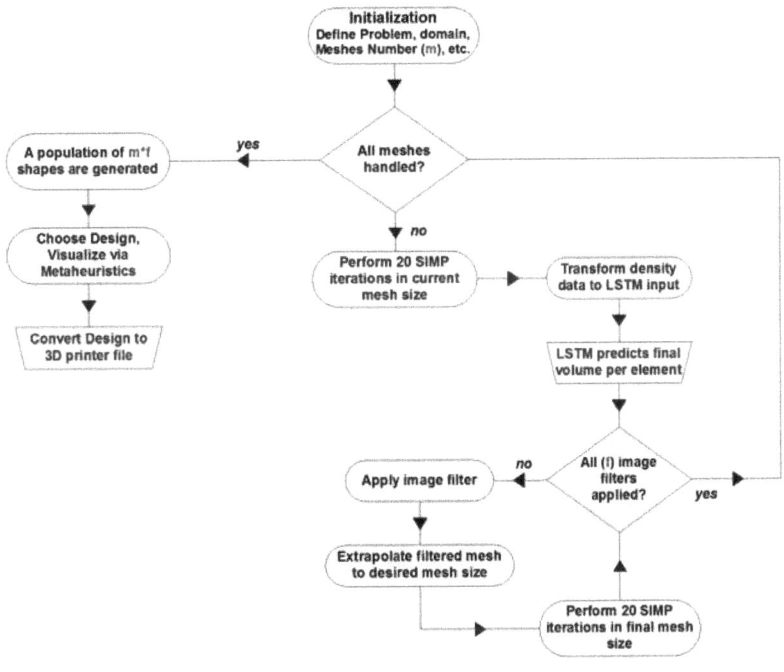

Figure 2. MLGen flowchart.

The first step of executing MLGen is the initialization step. The user is called to define the problem by choosing the original mesh dimensions, FE population, loading and support conditions. Additionally the number $m - 1$ and the FE population of less dense domains need to be decided as well. Once this step is completed, SIMP performs 20 iterations on all defined domains. Via this step, density data are exported for all FEs in the m domains and these data are then transformed according to Equation (7). The trained LSTM is then used for predicting the outcome of each FE of all domains used in the application. The output of the LSTM is an estimation of the final density distribution in all the m domains. Each of the m outputs is then filtered with the use of f number of image filters, such as Gaussian, edge detection, etc. Following that, each filtered domain is then extrapolated according to the FE center weight coordinates' proximity to the FE population of the desired output. Finally, the extrapolated domains are then passed into SIMP, which performs a maximum of 20 iterations for fine-tuning of the output. Once this procedure is finished, a population of $m \times f$ shapes is generated. Once all shapes are created and the designer chooses the one that is most preferable, a metaheuristic algorithm is used for visualizing the result design. Lastly, an algorithm is used for transforming the SIMP output into a type of file that can be used as an input for 3D printing. This formulation can be applied in 2D and 3D problems as well. As the volume classification is performed per finite element, it can be performed on both cases. What differentiates in the application in 2D and 3D models is the image filtering process. While for 2D cases, it is only performed once, in 3D cases it is executed z times, where z is the dimension of the mesh on the 3rd dimension. The z outputs of the filtering part are then joined to formulate the 3D shapes that are fine-tuned by SIMP.

4.2. Design and Training of the LSTM Network

In this part of the work, a detailed description of the network architecture, the training procedure and the database creation is presented. As previously stated, in the literature, LSTM networks are used successfully in several time-series classification problems. In this study, the scope for using the LSTM networks is to derive the density class of every element of the FE discretization. This class is defined according to the final one that the SIMP approach would have calculated, after multiple iterations, at convergence. It is worth mentioning that the prediction does not use information regarding the support conditions, the location of the element in the mesh of the design domain, the loading conditions or the desired volume fraction. The only information used is the density value of each element in the first 20 iterations of the SIMP approach in the original problem. This strategy is chosen in order to assure that the same trained network can offer high quality results without needing to be retrained regardless of the problem definition.

LSTMs present the ability to handle inputs that, apart from a single value per time-step, also include additional values (features) per time step. Assuming that SIMP is applied on a test example with a population of finite elements equal to $nx \times ny = m$ finite elements and a number of k iterations are performed until convergence is achieved, then a matrix D containing all densities $d_{i,j}$ per FE is formulated as follows [19]:

$$D = \begin{bmatrix} d_{1,1} & d_{1,2} & \cdot & \cdot & d_{1,k} \\ d_{2,1} & d_{2,2} & \cdot & \cdot & d_{2,k} \\ \cdot & \cdot & \cdot & \cdot & \cdot \\ \cdot & \cdot & \cdot & \cdot & \cdot \\ d_{m,1} & d_{m,2} & \cdot & \cdot & d_{m,k} \end{bmatrix} \tag{5}$$

It is noted that in the remaining part of this work, the density of an FE is defined as d rather than x, which is used in Section 2. In an attempt to improve the performance of the network, a decision is made to increase the features of the input instead of just using the density of one FE. The additional features that are used as inputs are the density values of neighboring finite elements of the element to be classified. In detail, apart from the density of the FE whose final density is to be classified, the densities of extra 24 FEs that are the closest to it in terms of center weights coordinates are also used as features of the networks' inputs. For example, in case of an element $e(i,j)$, the density recording input at the tth iteration is presented in Equation (6).

$$D^t_{i,j} = \begin{bmatrix} d_{i-2,j-2} \\ d_{i-2,j-1} \\ \cdot \\ \cdot \\ d_{i,j} \\ \cdot \\ \cdot \\ d_{i+2,j+1} \\ d_{i+2,j+2} \end{bmatrix} \tag{6}$$

The formulation of the database for m finite elements containing their density values over the optimization process until convergence, where the first t iterations of the SIMP approach out of the T are needed for convergence, is presented in Equation (7).

$$\left[\begin{array}{cccc|ccc|c} D_1^1 & D_1^2 & \dots & D_1^t & D_1^{t+1} & \dots & D_1^{T-1} & d_1^T \\ D_2^1 & D_2^2 & \dots & D_2^t & D_2^{t+1} & \dots & D_2^{T-1} & d_2^T \\ \vdots & \vdots & \ddots & \vdots & \vdots & \ddots & \vdots & \vdots \\ D_{m-1}^1 & D_{m-1}^2 & \dots & D_{m-1}^t & D_{m-1}^{t+1} & \dots & D_{m-1}^{T-1} & d_{m-1}^T \\ D_m^1 & D_m^2 & \dots & D_m^t & D_m^{t+1} & \dots & D_m^{T-1} & d_{m,T} \end{array}\right] \tag{7}$$
$$m \quad InputVectors \qquad\qquad Output$$

A schematic representation of the features used can also be seen in Figure 3. It is worth pointing out that in the case of finite elements located close to the boundaries of the design domain in which there are no neighboring elements in one or more directions, padding of the necessary rows and/or columns is performed. The volume fractions of the generated FEs are set equal to the volume of their existing neighboring element.

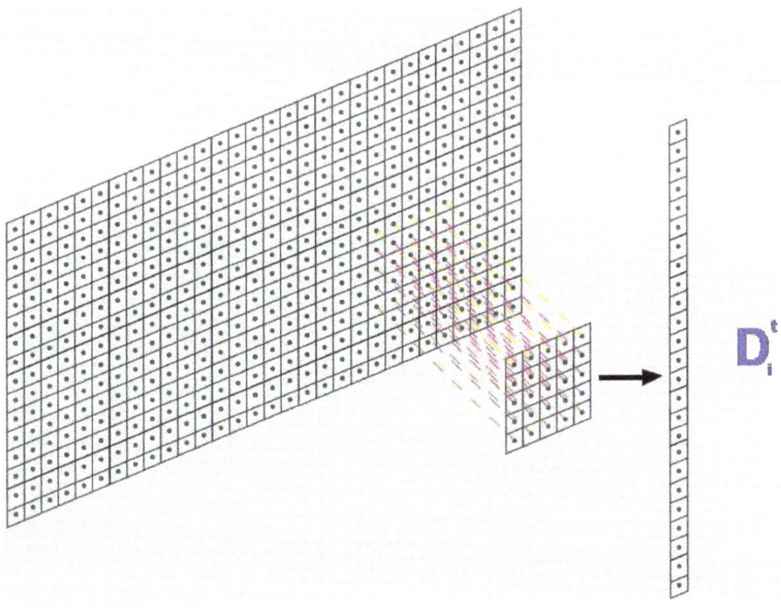

Figure 3. Creating density input vector.

As previously stated, the LSTM network is used for classifying the time series of the volume fluctuation of each finite element in the design domain. In order for the LSTM to be to able to perform this classification, the network has to be trained on a number of samples of finite elements. The training database is formulated by performing topology optimization via the SIMP approach on a specific problem with varying the FE mesh discretization in terms of its denseness. In detail, the test example chosen is the cantilever beam shown in Figure 4.

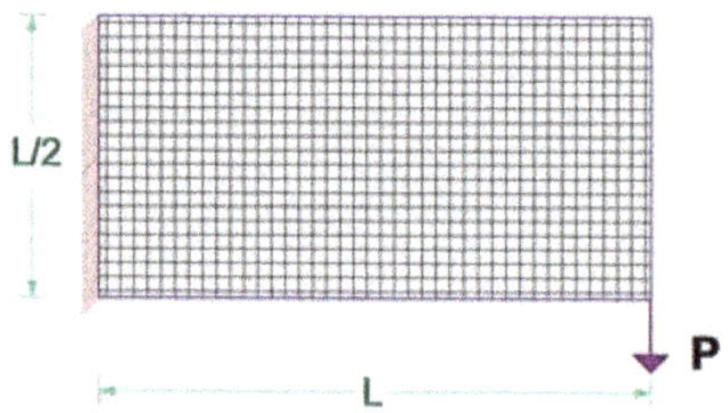

Figure 4. Creating density input vector.

The support condition can be described as fully fixed boundary conditions at the left part of the design domain along the vertical y axis, covering the height of the domain, while the loading conditions are a single concentrated load at the bottom right corner of the domain, facing towards the negative part of the y axis. The desired volume fraction is set equal to $V_t = 40\%$.

A population of 20 different FE mesh discretizations varying drastically on the number of the finite elements is chosen equal to [100, 500, 1000, 3000, 6000, 10,000, 20,000, 40,000, 60,000, 100,000] elements, and each mesh is generated with $ne_y/ne_x = 1/2$ and $ne_y/ne_x = 1/3$ height to length ratios. For each FE mesh discretization, a separate topology optimization problem is defined. After all problems are optimized by the SIMP approach, the density time series of all finite elements for all problems are saved and are divided into classes according to their final density value. The population of finite elements time series created is equal to 510,000 samples. The separation bounds of the classes are presented in the following Equation (8).

$$d_{i,k} \in \begin{cases} [0, 0.1] \Rightarrow d_{i,k} = 0 \\ (0.1, 0.9] \Rightarrow d_{i,k} = 0.50 \\ (0.9, 1] \Rightarrow d_{i,k} = 1 \end{cases} \tag{8}$$

The populations of the time-series samples in each class are equal to [240,000, 132,000, 138,000], respectively. The database is divided into a training and testing sample with an analogy of 80–20%. The first 20 steps of SIMP are used as the length of the network input. The network used in this study is a bi-directional LSTM network. The complete architecture is defined by five different layers: the input layer, the bi-directional LSTM, a fully connected layer, a soft-max layer and the output class layer. The network is built in Matlab and training is implemented both in CPU and GPU environments. In the training process, 10 epochs are run with 2000 iterations per epoch. The CPU and the GPU that are used for performing the training part are an Intel i9-9920x and an NVidia Titan RTX; in both cases, RAM is equal to 64 GB. In the case of CPU usage, the training needs 5500 s to complete, while in the case of GPU usage, the training time is reduced by a factor of more than 10, as it only needs 460 s. The accuracy achieved by both training procedures is equal to 97.3%.

4.2.1. Metaheuristics

In the past, metaheuristic algorithms were used as optimizers in several problems presented in the literature or in real life. One group of these algorithms are mainly inspired by nature mimicking procedures, such as ant colony optimization (ACO) [20], particle swarm optimization (PSO) [21], river formation dynamics (RFD) [22], pity beetle algorithm (PBA) [23], etc. In this work, the goal of using a metaheuristic algorithm is not to perform the actual optimization part, as this is handled by the SIMP approach supported by either OC [24] or MMA [25] algorithms. The goal of the metaheuristic algorithm is to assist in the visualization of the shape proposed by MLGen. In detail, information of the density of each finite element in the domain is used by the metaheuristic algorithm for deciding on which finite element to go to, moving from the current FE. For example, in the ACO algorithm, a number of ants are randomly placed inside the search space. On each step of the algorithm, each ant decides to move from a current node to another one according to the amount of pheromones on the trail connecting the two nodes. The amount of pheromones on the trail depends on the distance between the two nodes and the likelihood of this trail to be chosen by other ants as well [20].

In the MLGen implementation, particles (ants) are not randomly placed inside the search space (domain). They are placed on FEs that should definitely have the presence of particles. Such FEs are the ones where the loads and supports are placed. Instead of using pheromones, the attraction of a node depends on the density it presents according to SIMP and the likelihood of a connection to be chosen depends on the density value of the end node, divided by the normalized distance between the start and the end node. As a result, ACO is transformed into a design visualization method.

4.2.2. 3D Printing of MLGen Output

In order to produce the output of MLGen methodology, connecting the result exported by MLGen with a 3D printer is needed. For this reason, a bibliographic survey of methods for translating topology optimization results into files used by 3D printers is performed. The most common file types proposed are $*.stl$ and $*.obj$ files. Such methods were introduced for use with SIMP, BESO and level-set [3,26–28]. In the current work, the algorithm proposed by Liu [28] for exporting $*.stl$ files out of the results of the SIMP approach, named Top3DSTL, is selected, as it is found to be very efficient, while the methods used are more similar to the ones implemented in the MLGEn framework. Top3DSTL uses cubic representation and the export is in binary file format.

5. Numerical Tests

In order to validate the performance of the MLGen framework, five, known from the literature, benchmark test examples of topology optimization problems are selected. The selected test examples employ different loading and support conditions, while the volume fraction is also differentiated in each of them. In all of the test examples, seven models are used with the number of the FEs being equal to [1000, 5000, 7000, 10,000, 20,000, 25,000, 75,000] respectively. The iterations of SIMP in the original models are set equal to 20, while 23 different image filters are used. As a result, a population of 161 different shapes are generated for each test example with the population of the FEs being equal to 75,000. The final (fine-tuning) iterations of SIMP implemented for every design proposed out of the previous image filtering step is set equal to 20.

The definitions of dimensions along with the loading and support conditions of the five selected test examples are presented below. In addition, out of the different designs obtained 24 variants of the optimized designs are provided for each test example. For all five test examples, a sensitivity filter with radius equal to 3 is used for the SIMP implementation.

5.1. Test Example A

The design domain for the case of test example A can be seen in Figure 5 that is of rectangular shape with dimensions $L \times L/2$; it is supported by two fixed joints with the first one located in the bottom left corner on the x axis and the second one also placed on the bottom edge along the x axis at a distance of $0.6 \times L$ from the other support. The loading condition refers to a load distributed along the top edge of the x axis applied toward the vertical axis y. For this test example, the desired volume fraction is set equal to $V_t = 40\%$, while the structured FE discretization ratio is equal to $ne_y/ne_x = 1/2$. Out of the 161 optimized designs that are generated by means of the MLGen methodology for this test example, a randomly chosen sample of 24 optimized designs can be seen in Figure 6. As it can be observed, the MLGen methodology generates multiple different, but equivalent in terms of performance, optimized designs.

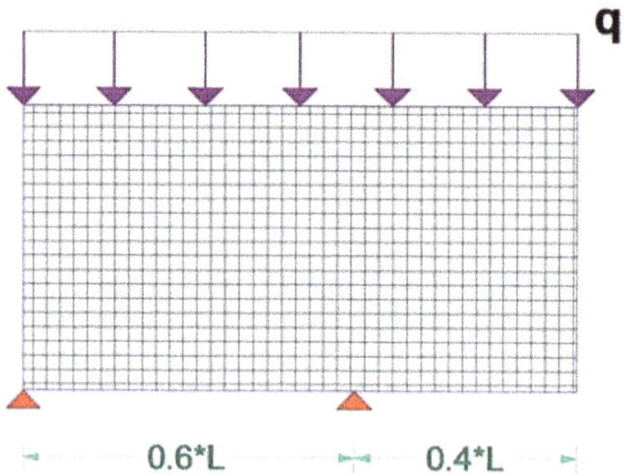

Figure 5. Definition of test example A.

Figure 6. Results for test example A.

5.2. Test Example B

In test example B—the design domain that can be seen in Figure 7 that is of *lamda* shape with dimensions $L \times L/2$—fully fixed boundary conditions at the top left edge along the x axis are implemented, while a concentrated point load P at the right edge of the design domain is applied along the vertical y axis, defining the loading conditions. With respect to the structured FE discretization ratio, $ne_y/ne_x = 1/2$ while the target volume percentage is set equal to $V_t = 35\%$. A set of 24 randomly chosen images of the optimized results of this example as produced by MLGen can be seen in Figure 8. Similar to the previous test example, these are of equivalent value with respect to the design criteria chosen during the optimization problem formulation.

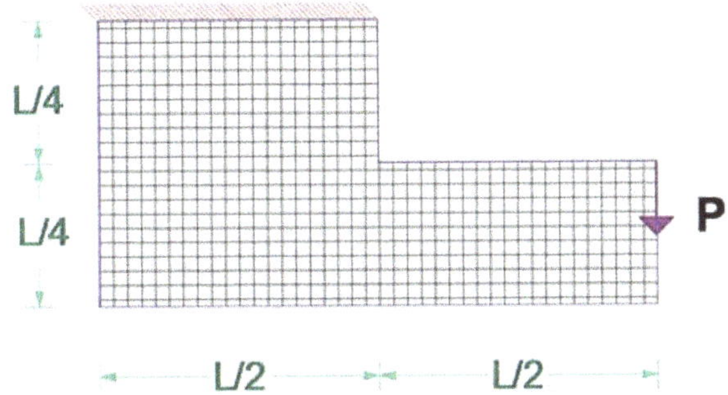

Figure 7. Definition of test example B.

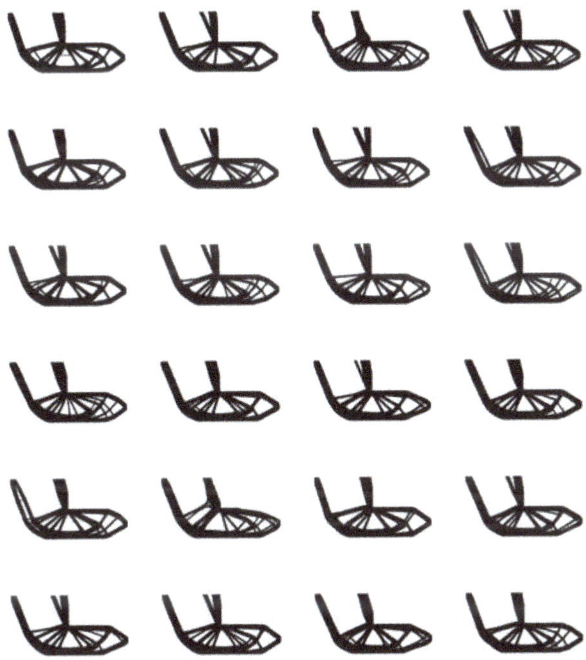

Figure 8. Results for test example B.

5.3. Test Example C

In Figure 9, the design domain that is also of rectangular shape with dimensions $L \times L/2$ and the corresponding loading and support conditions for the third test example C are presented. In detail, five fixed joints are located at the bottom edge of the design domain and they are equally distributed along the x axis. A load distributed along the horizontal x axis at the top edge of the design domain is applied toward the vertical y axis.The target volume percentage is set equal to $V_t = 25\%$, and the structured FE discretization ratio is equal to $ne_y/ne_x = 1/2$. The randomly chosen set of 24 optimized solution for this test example out of the 161 ones generated through the implementation of MLGen methodology is presented in Figure 10.

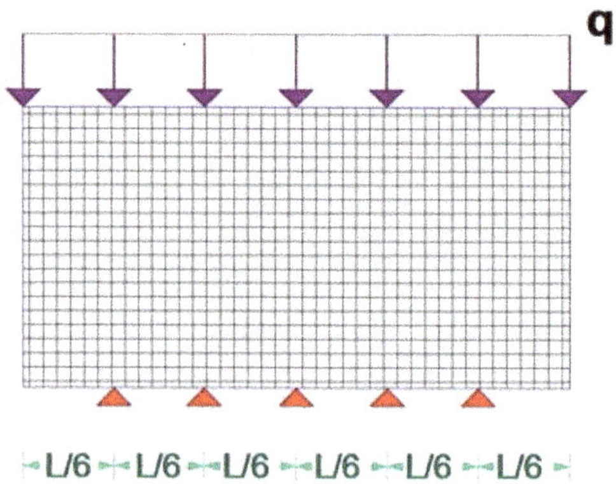

Figure 9. Definition of test example C.

Figure 10. Results for test example C.

5.4. Test Example D

The next example is labeled test example D, and the corresponding design domain is presented in Figure 11 that is also of rectangular shape with dimensions $L \times L/2$. The support conditions correspond to two fixed points at the bottom edges located at the two opposite corners of the design domain, while the loading conditions refer to a concentrated load at the middle of the bottom edge applied along the vertical y axis. The desired volume fraction is equal to $V_t = 45\%$, and the structured FE discretization ratio is equal to $ne_y/ne_x = 1/2$. The corresponding set of randomly chosen optimized shapes by means of the MLGen methodology is presented in Figure 12.

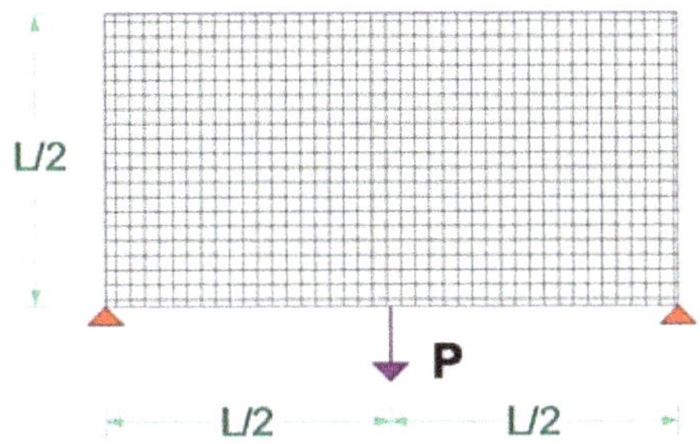

Figure 11. Definition of test example D.

Figure 12. Results for test example D.

5.5. Test Example E

The final test example presented in this work can be seen in Figure 13, where the design domain is also of rectangular shape with dimensions $L \times L/2$. The loading conditions refer to two concentrated loads at the top and bottom right corners of the left edge of the design domain facing upwards and downwards, respectively. The support conditions are defined as fully fixed boundary conditions along the left vertical side of the domain. With respect to the structured FE discretization ratio, $ne_y/ne_x = 1/3$ while the final volume fraction is set equal to $V_t = 50\%$. A randomly chosen sample of the optimized results produced by means of the MLGen methodology is presented in Figure 14.

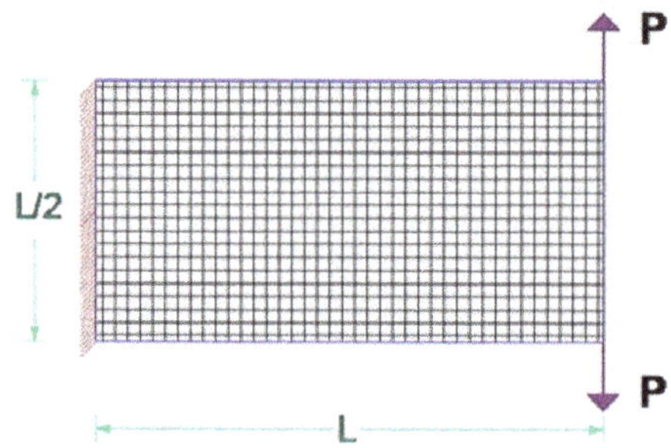

Figure 13. Definition of test example E.

Figure 14. Results for test example E.

The parameters of all test examples previously described can be seen in Table 1.

Table 1. Test examples parameters.

Example	Load Type	Load Num.	Support Type	Support Num.	Volume
A	Distributed	1	Fixed joints	2	40%
B	Concentrated	1	Fully fixed	1	35%
C	Distributed	1	Fixed joints	5	25%
D	Concentrated	1	Fixed joints	2	45%
E	Concentrated	2	Fully fixed	1	50%

6. Discussion

In this work, the application of MLGen in several topology optimization problems with varying parameters is presented in order to validate its performance against differently defined problems in terms of loading and support condition, volume percentage, etc. As presented, MLGen is capable of creating a large number of different shapes regardless of problem parameters. MLGen is successful in generating differentiated shapes, regardless of the type and population of supports and loads used. It is also able to efficiently handle problems, even in the case of a very low volume percentage equal to 25% as in test example C or of a high volume percentage (50%) as presented in test example E. It is also worth pointing out that the above results are acquired in less than 3 min per generated shape. It is logical though that the time needed is analogous to the total number of finite elements in the mesh discretization of the final sample. It must also be noted that no symmetry is imposed in any of the test examples.

7. Conclusions

MLGen is proposed as a framework for performing automated shape generation under specific design rules and constraints, acting as a support tool for engineers and designers. To achieve that, the incorporation of several methods is needed. Topology optimization supported by its solution approach (SIMP) is used for generating a shape that satisfies all design goals and constraints set by the designer. Machine learning (LSTM) is used for accelerating the solution procedure along with imposing a shape differentiation. Image filtering is used for further increasing the number of shapes generated, while metaheuristics (ACO) are used for shape visualization. As the LSTM is trained on classifying density time-series of independent finite elements, the trained network can be applied to any example, regardless of the type and population of the loading conditions, mesh sizes, support conditions, SIMP filtering type and value, whether it is a 2D or 3D problem, the density of the FE mesh discretization, etc. There are several points for future research. The incorporation of different machine learning techniques could further increase the number of proposed shapes for each test example. It is also very important to work on reducing the necessary iterations of the SIMP approach both for the initial ones as well as the fine-tuning ones, as those iterations represent the biggest work load of the procedure. Additionally, the incorporation of other topology optimization approaches, such as level-set [29] and BESO [30], is also worth examining. Finally, a web application where MLGen can be freely used will be developed by the authors in the future.

Author Contributions: Conceptualization, N.A.K. and N.D.L.; methodology, N.A.K. and N.D.L.; software, N.A.K.; validation, N.A.K.; formal analysis, N.A.K.; investigation, N.A.K.; writing—original draft preparation, N.A.K. and N.D.L.; writing—review and editing, N.A.K. and N.D.L.; visualization, N.A.K.; supervision, N.D.L.; project administration, N.D.L.; funding acquisition, N.A.K. All authors have read and agreed to the published version of the manuscript.

Funding: This research was co-financed by Greece and the European Union MIS-5033021.

Institutional Review Board Statement: Not applicable.

Informed Consent Statement: Not applicable.

Acknowledgments: This research is co-financed by Greece and the European Union (European Social Fund—ESF) through the Operational Programme «Human Resources Development, Education and Lifelong Learning» in the context of the project "Reinforcement of Postdoctoral Researchers—2nd Cycle" (MIS-5033021), implemented by the State Scholarships Foundation (IKY).

Ευρωπαϊκή Ένωση
European Social Fund

Operational Programme
Human Resources Development,
Education and Lifelong Learning
Co-financed by Greece and the European Union

ΕΣΠΑ
2014-2020

Conflicts of Interest: The authors declare no conflict of interest.

Abbreviations

The following abbreviations are used in this manuscript:

RNN	Recurrent Neural Networks
LSTM	Long Short-Term Memory
FE	Finite Element
SIMP	Solid Isotropic Material with Penalization
STO	Structural Topology Optimization
TO	Topology Optimization
TOP	Topology Optimization Problem

References

1. Lagaros, N.D.; Papadrakakis, M.; Kokossalakis, G. Structural optimization using evolutionary algorithms. *Comput. Struct.* **2002**, *80*, 571–589. [CrossRef]
2. Lagaros, N.; Plevris, V.; Papadrakakis, M. Multi-objective design optimization using cascade evolutionary computations. *Comput. Methods Appl. Mech. Eng.* **2005**, *194*, 3496–3515. [CrossRef]
3. Kazakis, G.; Kanellopoulos, I.; Sotiropoulos, S.; Lagaros, N.D. Topology optimization aided structural design: Interpretation, computational aspects and 3D printing. *Heliyon* **2017**, *3*, e00431. [CrossRef]
4. Lagaros, N.D.; Vasileiou, N.; Kazakis, G. A C# code for solving 3D topology optimization problems using SAP2000. *Optim. Eng.* **2018**. [CrossRef]
5. Papadrakakis, M.; Lagaros, N.D. Soft computing methodologies for structural optimization. *Appl. Soft Comput.* **2003**, *3*, 283–300. [CrossRef]
6. Zhang, Z.; Li, Y.; Zhou, W.; Chen, X.; Yao, W.; Zhao, Y. TONR: An exploration for a novel way combining neural network with topology optimization. *Comput. Methods Appl. Mech. Eng.* **2021**, *386*, 114083. [CrossRef]
7. Bendsøe, M.P. Optimal shape design as a material distribution problem. *Struct. Optim.* **1989**, *1*, 193–202. [CrossRef]
8. Zhou, M.; Rozvany, G. The COC algorithm, Part II: Topological, geometrical and generalized shape optimization. *Comput. Methods Appl. Mech. Eng.* **1991**, *89*, 309–336. [CrossRef]
9. Mlejnek, H. Some aspects of the genesis of structures. *Struct. Optim.* **1992**, *5*, 64–69. [CrossRef]
10. Rumelhart, D.E.; Hinton, G.E.; Williams, R.J. *Learning Internal Representations by Error Propagation*; Technical Report; California Univ San Diego La Jolla Inst for Cognitive Science: La Jolla, CA, USA, 1985.
11. Fister, I., Jr.; Yang, X.S.; Fister, I.; Brest, J.; Fister, D. A brief review of nature-inspired algorithms for optimization. *arXiv* **2013**, arXiv:1307.4186.
12. Sigmund, O.; Maute, K. Topology optimization approaches. *Struct. Multidiscip. Optim.* **2013**, *48*, 1031–1055. [CrossRef]
13. Rumelhart, D.E.; Hinton, G.E.; Williams, R.J. Learning representations by back-propagating errors. *Nature* **1986**, *323*, 533. [CrossRef]
14. Schmidhuber, J. Netzwerkarchitekturen, Zielfunktionen und Kettenregel. Ph.D. Thesis, Technische Universität München, München, Germany, 1993.
15. Kallioras, N. Reduced Order Models and Machine Learning in Analysis and Optimum Design of Structures. Ph.D. Thesis, National Technical University of Athens, Athens, Greece, 2019.
16. Hochreiter, S.; Schmidhuber, J. Long short-term memory. *Neural Comput.* **1997**, *9*, 1735–1780. [CrossRef]
17. Fischer, T.; Krauss, C. Deep learning with long short-term memory networks for financial market predictions. *Eur. J. Oper. Res.* **2018**, *270*, 654–669. [CrossRef]
18. Tai, K.S.; Socher, R.; Manning, C.D. Improved Semantic Representations From Tree-Structured Long Short-Term Memory Networks. In *Proceedings of the 53rd Annual Meeting of the Association for Computational Linguistics and the 7th International Joint Conference on Natural Language Processing (Volume 1: Long Papers)*; Association for Computational Linguistics: Beijing, China, 2015; pp. 1556–1566. [CrossRef]
19. Kallioras, N.A.; Kazakis, G.; Lagaros, N.D. Accelerated topology optimization by means of deep learning. *Struct. Multidiscip. Optim.* **2020**, *20*, 21–36. [CrossRef]
20. Dorigo, M.; Stutzle, T. *Ant Colony Optimization*; The MIT Press: Cambridge, MA, USA, 2004.
21. Kennedy, J.; Eberhart, R.C. Particle swarm optimization. In Proceedings of the IEEE International Conference on Neural Networks, Perth, WA, Australia, 27 November–1 December 1995; pp. 1942–1948.
22. Rabanal, P.; Rodríguez, I.; Rubio, F. Using river formation dynamics to design heuristic algorithms. In *International Conference on Unconventional Computation*; Springer: Berlin/Heidelberg, Germany, 2007, pp. 163–177.
23. Kallioras, N.A.; Lagaros, N.D.; Avtzis, D.N. Pity beetle algorithm—A new metaheuristic inspired by the behavior of bark beetles. *Adv. Eng. Softw.* **2018**, *121*, 147–166. [CrossRef]
24. Christensen, P.W.; Klarbring, A. *An Introduction to Structural Optimization*; Springer: Berlin/Heidelberg, Germany, 2008; Volume 153.
25. Svanberg, K. The method of moving asymptotes—A new method for structural optimization. *Optim. Syst. Theory* **1987**, *24*, 359–373. [CrossRef]

26. Vogiatzis, P.; Chen, S.; Zhou, C. An Open Source Framework for Integrated Additive Manufacturing and Level-Set-Based Topology Optimization. *J. Comput. Inf. Sci. Eng.* **2017**, *17*, 041012. [CrossRef]
27. Brackett, D.; Ashcroft, I.; Hague, R. Topology optimization for additive manufacturing. In *2011 International Solid Freeform Fabrication Symposium*; University of Texas at Austin: Austin, TX, USA, 2011.
28. Liu, K.; Tovar, A. An efficient 3D topology optimization code written in Matlab. *Struct. Multidiscip. Optim.* **2014**, *50*, 1175–1196. [CrossRef]
29. Allaire, G.; Jouve, F.; Toader, A.M. A level-set method for shape optimization. *Comptes Rendus Math.* **2002**, *334*, 1125–1130. [CrossRef]
30. Querin, O.; Steven, G.; Xie, Y. Evolutionary structural optimisation (ESO) using a bidirectional algorithm. *Eng. Comput.* **1998**, *15*, 1031–1048. [CrossRef]

applied sciences

MDPI

Article

Traffic Signal Optimization for Multiple Intersections Based on Reinforcement Learning

Jaun Gu [1], Minhyuck Lee [1], Chulmin Jun [1,*], Yohee Han [2], Youngchan Kim [2] and Junwon Kim [2]

[1] Department of Geoinformatics, University of Seoul, Seoul 02504, Korea; umseakind2@uos.ac.kr (J.G.); lmhll123@uos.ac.kr (M.L.)
[2] Department of Transportation Engineering, University of Seoul, Seoul 02504, Korea; yeohee@gmail.com (Y.H.); yckimm@uos.ac.kr (Y.K.); mirageno21@uos.ac.kr (J.K.)
* Correspondence: cmjun@uos.ac.kr

Abstract: In order to deal with dynamic traffic flow, adaptive traffic signal controls using reinforcement learning are being studied. However, most of the related studies are difficult to apply to the real field considering only mathematical optimization. In this study, we propose a reinforcement learning-based signal optimization model with constraints. The proposed model maintains the sequence of typical signal phases and considers the minimum green time. The model was trained using Simulation of Urban MObility (SUMO), a microscopic traffic simulator. The model was evaluated in the virtual environment similar to a real road with multiple intersections connected. The performance of the proposed model was analyzed by comparing the delay and number of stops with a reinforcement learning model that did not consider constraints and a fixed-time model. In a peak hour, the proposed model reduced the delay from 3 min 15 s to 2 min 15 s and the number of stops from 11 to 4.7 compared to the fixed-time model.

Keywords: traffic signal optimization; reinforcement learning; adaptive traffic signal control; multiple intersections; Deep Q-network

Citation: Gu, J.; Lee, M.; Jun, C.; Han, Y.; Kim, Y.; Kim, J. Traffic Signal Optimization for Multiple Intersections Based on Reinforcement Learning. *Appl. Sci.* **2021**, *11*, 10688. https://doi.org/10.3390/app112210688

Academic Editors: Nikos D. Lagaros, Vagelis Plevris and Paola Pellegrini

Received: 20 August 2021
Accepted: 10 November 2021
Published: 12 November 2021

1. Introduction

Traffic signal control plays an essential role in city management because traffic congestion brings economic, environmental, and social disadvantages. Traffic signal control aims to minimize congestion by determining the optimal values of parameters such as the cycle length and phases duration [1,2]. In many areas, the traffic signal control systems based on a fixed-time model are still in use [3–5]. While these systems are easy to implement, they cannot respond flexibly to dynamic traffic flows [6,7].

To quickly respond to variety in the traffic environment, signal control systems should be able to choose their own actions without waiting for instructions from a central computer [8]. Therefore, reinforcement learning models are being studied that allow the traffic signal controller to receive realtime data around the intersection, such as traffic volume and vehicle speed, and change signal appropriately for the given traffic situation [9,10]. If the above sentence is expressed in reinforcement learning terms, the controller is the agent, the data input to the controller is the state, the controller's decision is the action, and the benefit provided to the agent according to the action is called reward. The goal of reinforcement learning is to maximize the future reward that an agent can obtain [11].

2. Literature Review

Reinforcement learning is the most recently used algorithm in the field of signal control research. However, most studies have not considered the constraints applied to a real-world intersection or tested in a local area such as a single intersection. Touhbi et al. (2017) analyzed the possibility of using the Q-Learning algorithm for adaptive traffic signal control [12]. The Q-Learning algorithm was helpful in resolving traffic congestion compared to the fixed-time model, but different results were obtained depending on the definition of reward and various traffic volumes. Liang et al. (2019) proposed a DQN-based signal control model [13]. The state was defined as a grid-type location and the speed of vehicles around the intersection. The reward was the difference in accumulated waiting time between learning cycles, and the action was to select one of phases. Wang et al. (2019) proposed a model based on the assumption that data are collected by a loop detector [14]. Since the state used as input to the model is not a data format that can be obtained, a method for converting the data acquired through the detector into data useful for the model was presented. Gong et al. (2019) proposed a cooperative learning method in which signal controllers at adjacent intersections share a state they can observe with each other [15]. Using this, a traffic signal optimization model for multiple intersections was proposed.

Chu et al. (2019) pointed out the limitations of the centralized reinforcement learning model and suggested a way to optimize a large-scale road network by placing the model at each intersection [16]. As the algorithm of the model, A2C (Advantage Actor Critic) was proposed, and the state of different scales was delivered to the model with city-unit traffic volume data and actual observable traffic flow data. Egea et al. (2020) pointed out the limitations of the realtime response of the existing adaptive signal control method and suggested a reinforcement learning-based signal control model as an alternative [17]. Efficiency was evaluated through various indices for the compensation that is judged to have the greatest impact on the model's performance. Rasheed et al. (2020) introduced a multiagent-based reinforcement learning algorithm [10]. The model was designed to solve high-level problems such as dynamic traffic volume through cooperation between agents. As a result of the simulation, it was shown that the travel time was reduced through the proposed model.

In general, traffic signal control has constraints such as the sequence of phases is fixed, and the minimum green time is given. However, in related studies, the constraints are ignored in consideration of only mathematical optimization such as delay minimization [18–21]. If the phase sequence is random or does not give a minimum green time, this can cause over-waiting for vehicles and confuse drivers [16,22]. In this study, we propose a model that can compensate for the problems that arise when the reinforcement learning-based signal control models proposed in related studies are applied to actual road networks. The proposed model maintains the same signal order as the fixed-time model and gives the minimum green time. Therefore, it can be applied to the actual signal controller. The performance of the proposed model was evaluated in a simulation environment depicting real roads connected with multiple intersections. In order to evaluate the effect of constraints, a comparative simulation between the proposed model and the reinforcement learning model that did not consider the constraints was performed.

3. Methods

3.1. Learning Process

Figure 1 shows the learning process of the reinforcement learning-based traffic signal optimization model. The microscopic simulation environment was implemented by Simulation of Urban MObility (SUMO). The model received the realtime traffic flow of the intersection as a state. Based on this, the action determined by the model was implemented at the intersection by SUMO's signal controller. Traffic flow changed due to signal control was delivered to the model as a reward. By repeating this process, the model learned an optimized signal pattern that minimized traffic congestion, such as the vehicles' delay and the number of vehicles stopped. The reinforcement learning model was designed based on Deep Q-network (DQN). Since the given problem was a classification that selected an action appropriate for the situation, the SoftMax Function was applied to the activation function of the output layer.

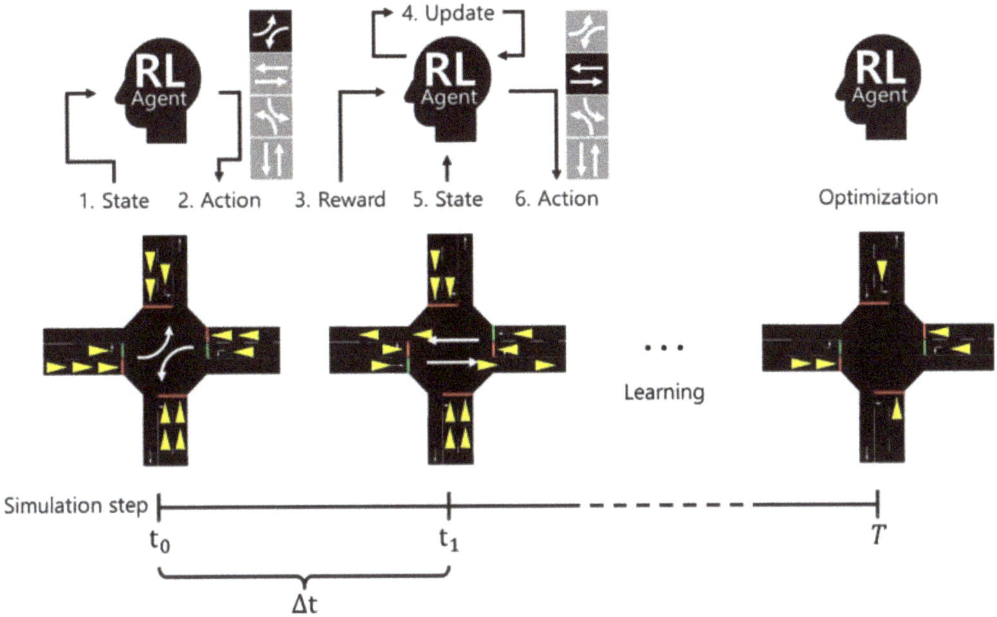

Figure 1. Learning process of the reinforcement learning-based traffic signal control model.

3.2. State

The state is defined as Equation (1). Q_t represents the queue of each lane at time t. At an intersection with k incoming lanes, q_t^i represents the number of vehicles stopped in the incoming lane i at time t. P_t indicates which phase is currently on. When there are n phases, if the jth phase is active, $p_j = 1$, otherwise $p_j = 0$. d_t means the elapsed time of the currently turned-on phase.

$$\begin{aligned} S_t &= [Q_t, P_t, d_t] \\ Q_t &= \left[q_t^1, q_t^2, \ldots, q_t^k\right] \\ P_t &= [p_1, p_2, \ldots, p_n] \end{aligned} \quad (1)$$

Figure 2 shows an example of traffic control at a single intersection. There were 10 incoming lanes into the intersection, and phase 4 was in progress for 10 s. Stopped vehicles are marked in red. Therefore, in the order of the lanes, $Q_t = [0, 1, 0, 0, 3, 1, 2, 0, 0, 0]$ and with phase 4 on, $P_t = [0, 0, 0, 1]$ and accordingly gave a green signal to the 3rd, 4th, 8th, and 9th lanes. Finally, the phase lasted for 10 s, $d_t = [10]$.

3.3. Action

The action was to select whether to keep the current phase ($A_t = 0$) or change to the next phase ($A_t = 1$). Since the proposed model has a constraint that the sequence of the phases is maintained, actions such as returning to the previous phase or skipping the next phase were impossible. In addition, the proposed model included a minimum green time, therefore, every phase must be active once per signal cycle. In Figure 2, the agent decided whether to keep phase 4 or change to phase 1. If the minimum green time for phase 4 was 15 s, the agent could only select to keep phase 4 because d_t is still 10 s.

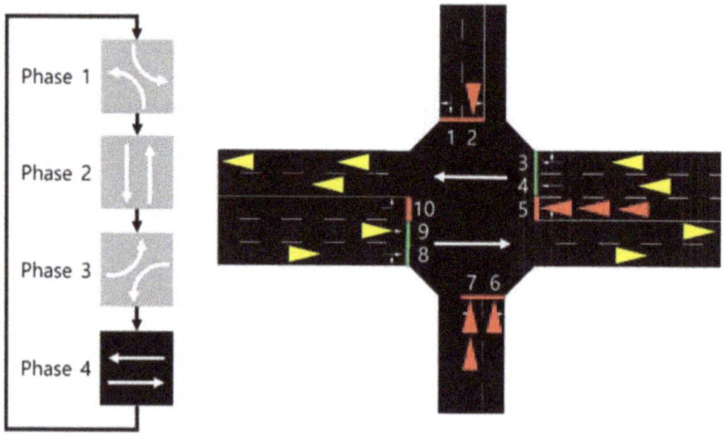

Figure 2. Example of traffic signal control at a single intersection.

3.4. Reward

The traffic flow changed during the time interval Δt by the action A_t was given as a reward to the model. Reward $R_{t+\Delta t}$ is defined as Equation (2). $q^i_{t+\Delta t}$ is defined the number of vehicles stopped in the incoming lane i at time step $t + \Delta t$ and $f^o_{t,t+\Delta t}$ is defined as the number of vehicles located in the outgoing lane among the vehicles passing through the intersection between time step t and time step $t + \Delta t$. Accordingly, the reward $R_{t+\Delta t}$ was defined as the number of passing vehicles compared to stopped vehicles. As the number of stopped vehicles decreased and the number of passing vehicles increased, the reward was increased.

$$R_{t+\Delta t} = \sum_{l=1} (f^o_{t,t+\Delta t}) / \sum_{i=1} \left(q^i_{t+\Delta t}\right) \tag{2}$$

4. Simulation

The performance of the proposed model, the reinforcement learning-based comparison model excluding constraints, and the fixed-time model, PASSER II, were analyzed for two scenarios. The training of the proposed model and the comparison model was carried out by generating random traffic under the same conditions. When the accumulated waiting time or reward no longer decreased and converged, it was judged that learning was complete. Table 1 shows the number of repeated episodes until learning was complete. The simulation time of scenario 2 was based on the period of the acquired traffic data. For the proposed model to respond flexibly to realtime traffic flow, the time interval Δt was set to 3 s. Agents meaning signal controllers are equal to the number of intersections. In this simulation, it was assumed that the signal controller could obtain the traffic situation in realtime using the vehicle detectors.

Table 1. Parameters used in the model.

Parameters	Scenario 1	Scenario 2
Number of episodes	100	160
Simulation time T of one episode (second)	3600	18,000
Time interval Δt (second)	3	3
Learning rate	0.0001	0.0001
Number of intersections	2	6

4.1. Scenario 1

In Scenario 1, performance evaluation of the proposed model and the fixed-time model was performed before comparison with the model without constraints. Figure 3 is a simple road structure used in the evaluation. Two intersections were connected, and the traffic volume is indicated on each lane. The same phase sequence was applied to both intersections. In the fixed-time model, the duration of each phase was calculated sequentially as 14, 56, 13, and 27 s. Therefore, the cycle length was 110 s.

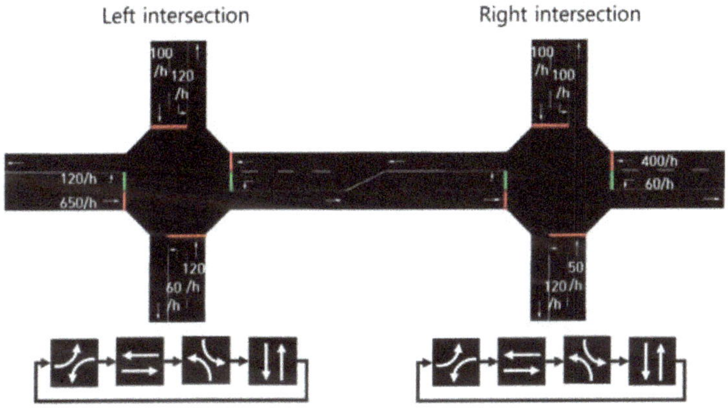

Figure 3. Simple network with two intersections.

Figure 4 shows the duration ratio for each phase of the proposed model and fixed-time model. Since the proposed model dynamically responded to realtime traffic flow, the duration of the phase changed with each signal cycle. The average phase duration ratio appeared similar to the optimization result of the fixed-time model. Compared to the fixed-time model, the proposed model reduced the average delay per vehicle from 40 s to 30 s and the average number of stops per vehicle from 2.5 to 2 times.

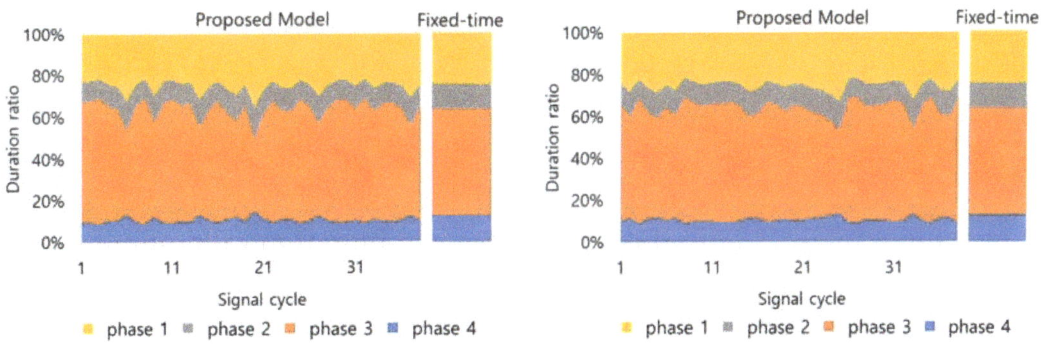

(a) (b)

Figure 4. Phase duration ratio of scenario 1: (**a**) Left intersection; (**b**) Right intersection.

4.2. Scenario 2

In Scenario 2, a comparison model without constraints was added to analyze the performance of the proposed model. Figure 5 show a road network with six continuous intersections in the real world. The total length of the main road with 6 intersections is 2.5 km, and the distance between intersections is 650 m at the maximum and at least 70 m, with an average of 400 m. The model was evaluated by applying the collected traffic data. From 4 to 6 pm, 8200 vehicles were created, from 6 to 7 pm, 6200 vehicles were created, and from 7 to 9 pm, 6400 vehicles were created. The least traffic volume was at intersection 3, with 2000 vehicles per hour. On the other hand, the highest traffic volume was intersection 6, with 3800 vehicles per hour. The average speed was 70 km/h. As measure of effectiveness, the cumulative delay, and the cumulative number of stops at each intersection, and the average delay and the average number of stops per vehicle were considered. In addition, the proposed model set the cycle length equal to the fixed model to maintain the set offset at each intersection.

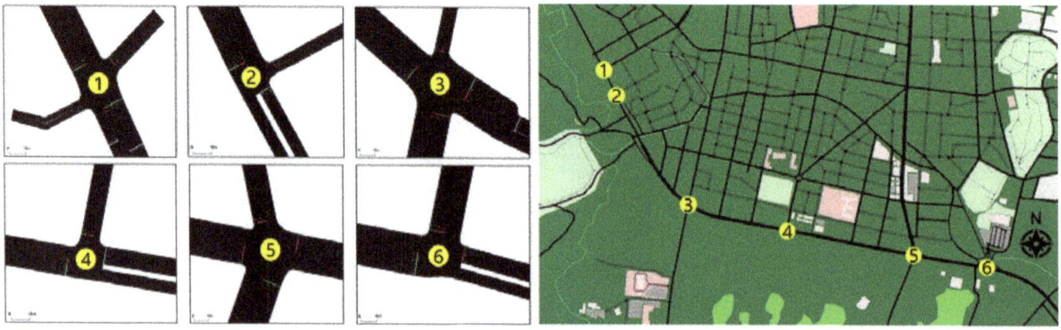

Figure 5. Complex network with six intersections.

Figure 6 is the same as Figure 4, it shows the phase duration ratio of the proposed model and fixed-time model. The comparison model could not calculate the duration ratio, because the phase sequence was not constant. The cycle length was set to 160 s. From the 41st cycle to the 59th cycle, it was set to 180 s because there was heavy traffic. Unlike Scenario 1, the average duration ratio of the proposed model was different from that of fixed-time model. The proposed model showed a tendency to give longer green time to the main road with a lot of traffic.

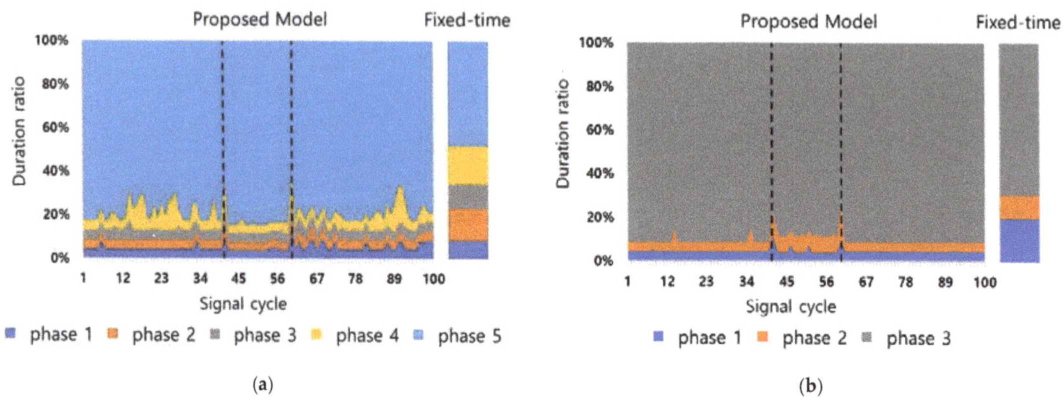

Figure 6. *Cont.*

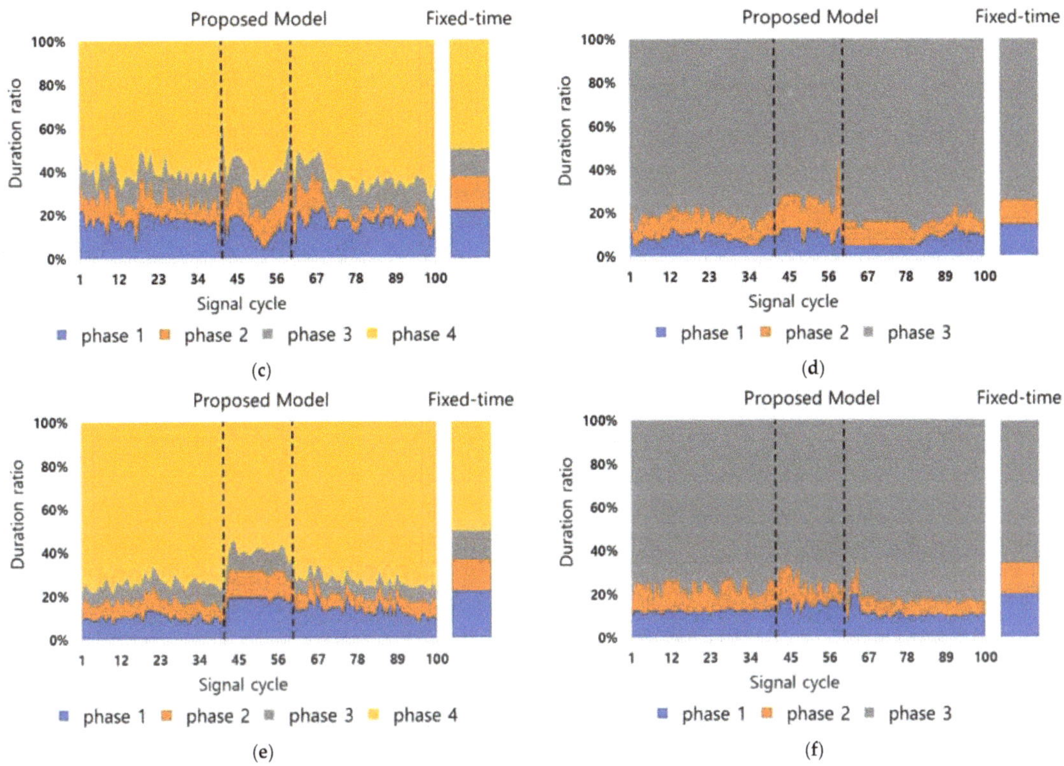

Figure 6. Phase duration ratio of scenario 2: (**a**) Intersection 1; (**b**) Intersection 2; (**c**) Intersection 3; (**d**) Intersection 4; (**e**) Intersection 5; and (**f**) Intersection 6. The time point at which the cycle length changes is indicated by a dotted line. The first cycle length is 160 s and changes to 180 s after the first dotted line, and then changes back to 160 s after the second dotted line.

Figure 7 shows the cumulative delay and number of stops for each intersection during the simulation. At all intersections except for intersection 5, the performance of the comparison model without constraints was the best. However, the number of stops of the proposed model and the comparison model was similar. The proposed model reduced delay by up to 88% to at least 31%, and the number of stops by up to 95% to at least 46% compared to the fixed-time model. The fixed-time model had the worst congestion at intersection 3, and the comparative model had the worst congestion at intersection 5. The proposed model had the longest delay at intersection 3, and the highest number of stops at intersection 5.

Figure 8 compares the average delay and the average number of stops per vehicle of each model by time period. Compared with the fixed-time model, the reinforcement learning-based models showed excellent performance. The delay decreased by 48% for the proposed model and 55% for the comparison model compared to the fixed model. The number of stops decreased by 67% for the proposed model and 73% for the comparison model compared to the fixed model.

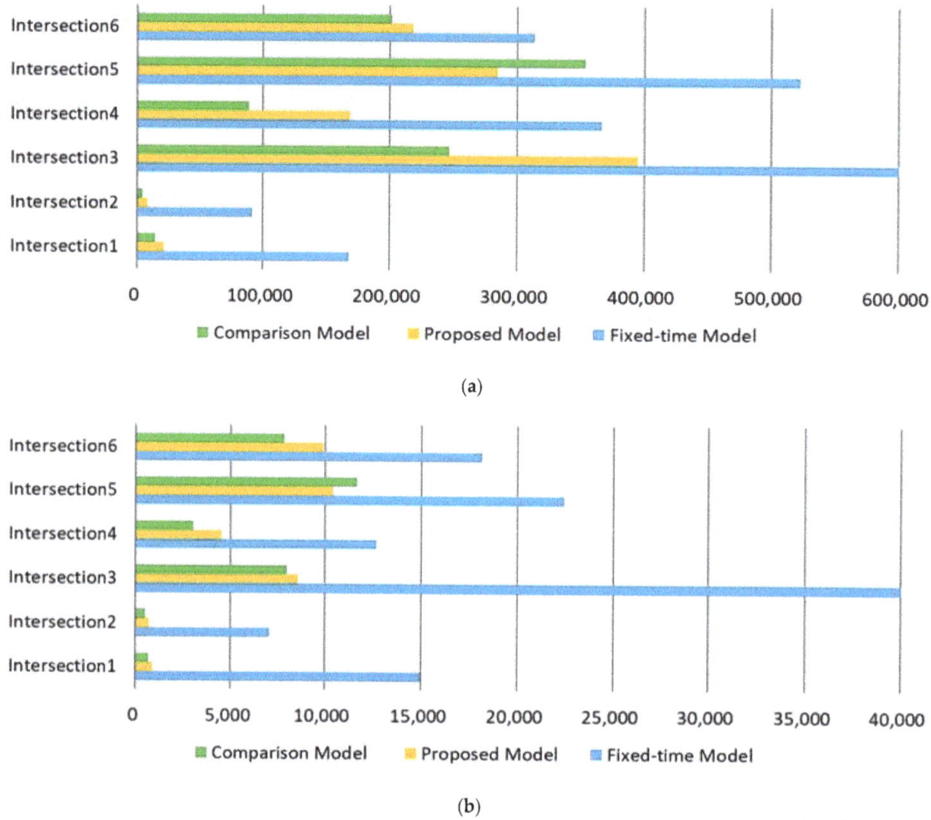

Figure 7. Measure of effectiveness by intersections: (**a**) Cumulative delay; (**b**) Cumulative number of stops.

From 6–7 pm, the traffic volume increased by 33% compared to other time periods. Accordingly, the delay and the number of stops also increased. In the peak hour, the fixed model waited for an average of 3 min and 15 s and the number of stops was 11 times, whereas in the proposed model, the delay was reduced to 2 min and 15 s, and the number of stops was reduced by more than half to 4.7. In addition, the number of stops of the proposed model was 0.7 times from 4 to 6 pm when the traffic volume was low. It was an ideal result that the number of stops was less than 1 when passing 6 intersections on a 2.5 km road.

Figure 9 shows the signal pattern calculated for each model as a space-time diagram. It shows the signal patterns of each intersection and the trajectories of vehicles accordingly. In the case of the fixed-time model, the average travel time per vehicle was 250 s, while the comparison model decreased by about 60 s to 190 s. The proposed model decreased by about 30 s to 220 s. However, for the comparison model, the phase duration was irregular and short.

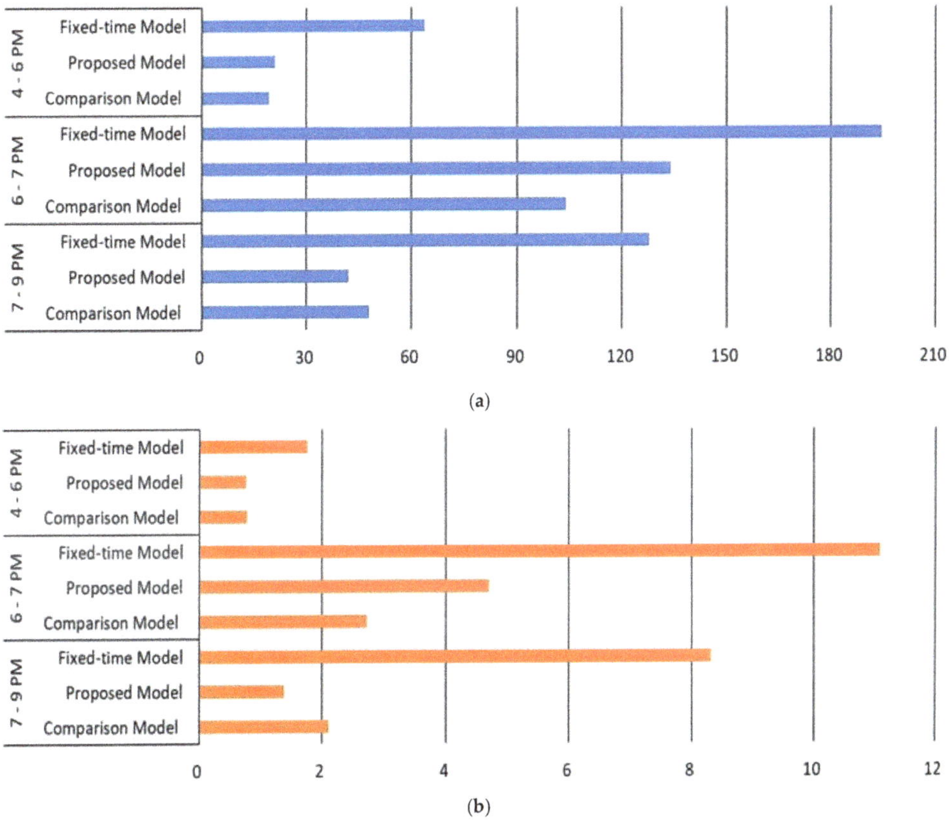

Figure 8. Measure of effectiveness: (**a**) Average delay per vehicle; (**b**) Average number of stops per vehicle.

Figure 10 shows the traffic situation when the proposed model and the comparison model were applied to each intersection. The color of the road indicates the average speed of the vehicles. In the comparison model, traffic congestion occurred at intersection 5. On the other hand, the proposed model can be seen as a solution to the congestion at intersection 5. However, some congested sections occurred between intersection 3 and intersection 4.

During the entire simulation, the comparison model showed 13% shorter delay and 17% fewer stops than the proposed model. Although the comparison model had the best performance, the proposed model also showed sufficiently ideal results. In addition, the comparative model calculated an irregular signal pattern, while the proposed model calculates a realistic signal pattern. Therefore, the proposed model would be the best in terms of applying it to real-world intersections.

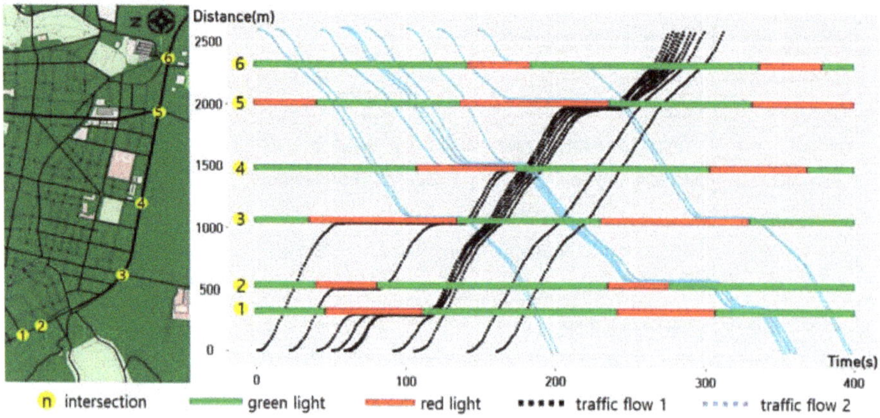

(a)

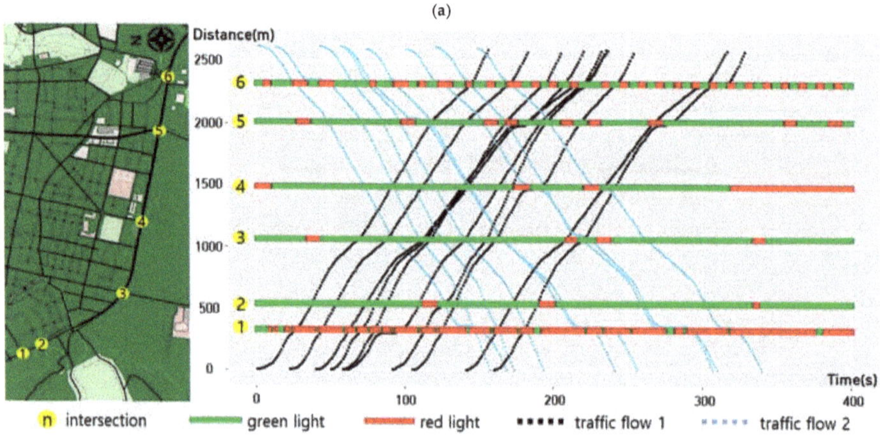

(b)

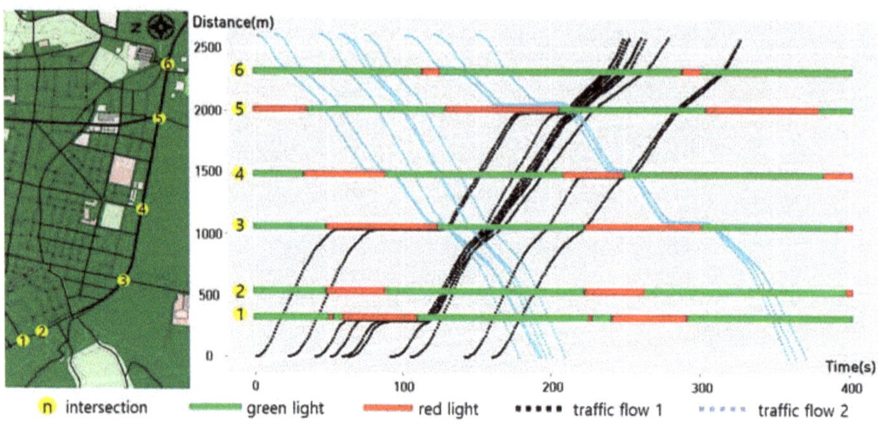

(c)

Figure 9. Time-space diagram: (**a**) Fixed-time model; (**b**) Comparison model; and (**c**) Proposed model.

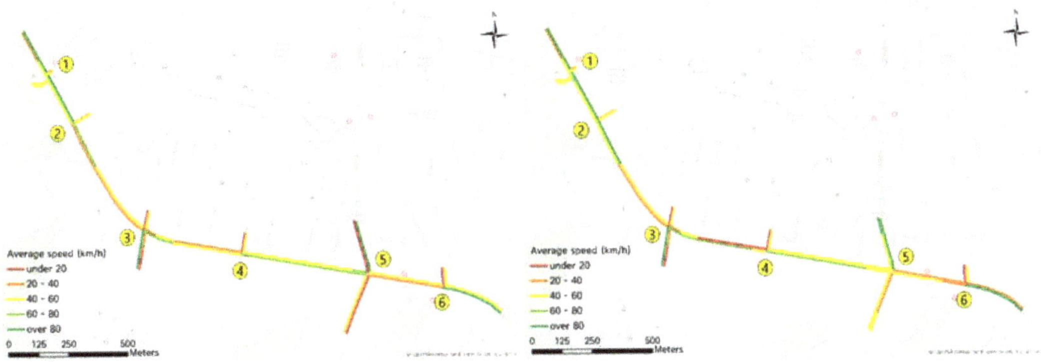

(a) (b)

Figure 10. Traffic condition: (**a**) Comparison model; (**b**) Proposed model.

5. Conclusions

In this study, traffic signal control based on a reinforcement learning algorithm was proposed to minimize traffic congestion. Early reinforcement learning-based signal control research focused on mathematical optimization, and when the model was applied to the road, excessively waiting vehicles and confused drivers could have occurred. Therefore, this study proposed a reinforcement learning-based traffic signal control model by applying the constraint that fixed the sequence of the pre-planned phases and provided a minimum green time.

Simulations of the proposed model and the comparison model without constraints and fixed-time model were performed in two scenarios. The scenarios included multiple intersections, and the delay and the number of stops were compared. Compared with the fixed-time model, the reinforcement learning-based models showed excellent performance. Although the comparison model showed the best performance, the proposed model also showed ideal results. Unlike the comparison model, the proposed model will show the best performance when applied to real world intersections, because it calculates realistic signal patterns.

Even if the simulation environment is based on reality, implementation will not be exactly the same. Therefore, it is necessary to test the model on real-world roads in future research. To this end, more constraints and data for safe road driving and simulations in various types of road networks will be required.

Author Contributions: Writing—original draft preparation, J.G.; visualization, M.L.; writing—review and editing, C.J.; methodology, Y.H. and J.K.; project administration, Y.K. All authors have read and agreed to the published version of the manuscript.

Funding: This research was supported by the Road Traffic Authority grant funded by the Korea government [grant number 1325163906].

Institutional Review Board Statement: Not applicable.

Informed Consent Statement: Not applicable.

Data Availability Statement: Not applicable.

Conflicts of Interest: The authors declare no conflict of interest.

References

1. Han, Y.; Kim, Y. Spatiotemporal congestion recognition index to evaluate performance under oversaturated conditions. *KSCE J. Civil Eng.* **2019**, *23*, 3714–3723. [CrossRef]
2. Yang, S.; Yang, B.; Wong, H.-S.; Kang, Z. Cooperative traffic signal control using multi-step return and off-policy asynchronous advantage actor-critic graph algorithm. *Knowl. Based Syst.* **2019**, *183*, 104855. [CrossRef]
3. Aslani, M.; Mesgari, M.S.; Seipel, S.; Wiering, M. Developing adaptive traffic signal control by actor–critic and direct exploration methods. *Proc. Inst. Civ. Eng. Transp.* **2019**, *172*, 289–298. [CrossRef]
4. Aslani, M.; Stefan, S.; Marco, W. Continuous residual reinforcement learning for traffic signal control optimization. *Can. J. Civ. Eng.* **2018**, *45*, 690–702. [CrossRef]
5. Mannion, P.; Duggan, J.; Howley, E. An experimental review of reinforcement learning algorithms for adaptive traffic signal control. *Auton. Road Transp. Support Syst.* **2016**, *4*, 47–66. [CrossRef]
6. Li, L.; Lv, Y.; Wang, F.-Y. Traffic signal timing via deep reinforcement learning. *IEEE/CAA J. Autom. Sin.* **2016**, *3*, 247–254.
7. Ge, H.; Song, Y.; Wu, C.; Ren, J.; Tan, G. Cooperative deep q-learning with q-value transfer for multi-intersection signal control. *IEEE Access* **2019**, *7*, 40797–40809. [CrossRef]
8. Al Islam, S.B.; Hajbabaie, A. Distributed coordinated signal timing optimization in connected transportation networks. *Transp. Res. Part C Emerg. Technol.* **2017**, *100*, 272–285. [CrossRef]
9. Mousavi, S.S.; Schukat, M.; Howley, E. Traffic light control using deep policy-gradient and value-function-based reinforcement learning. *IET Intell. Transp. Syst.* **2017**, *11*, 417–423. [CrossRef]
10. Rasheed, F.; Yau, K.L.A.; Low, Y.C. Deep reinforcement learning for traffic signal control under disturbances: A case study on Sunway city, Malaysia. *Future Gener. Comput. Syst.* **2020**, *109*, 431–445. [CrossRef]
11. Sutton, R.S.; Barto, A.G. *Reinforcement Learning: An Introduction*; MIT Press: Cambridge, MA, USA, 2018.
12. Touhbi, S.; Babram, M.A.; Nguyen-Huu, T.; Marilleau, N.; Hbid, M.L.; Cambier, C.; Stinckwich, S. Adaptive traffic signal control: Exploring reward definition for reinforcement learning. *Procedia Comput. Sci.* **2017**, *109*, 513–520. [CrossRef]
13. Liang, X.; Du, X.; Wang, G.; Han, Z. Deep reinforcement learning for traffic light control in vehicular networks. *Mach. Learn.* **2018**, *68*, 1–11. [CrossRef]
14. Wang, S.; Xie, X.; Huang, K.; Zeng, J.; Cai, Z. Deep reinforcement learning-based traffic signal control using high-resolution event-based data. *Entropy* **2019**, *21*, 744. [CrossRef]
15. Gong, Y.; Abdel-Aty, M.; Cai, Q.; Rahman, M.S. Decentralized network level adaptive signal control by multi-agent deep reinforcement learning. *Transp. Res. Interdiscip. Perspect.* **2019**, *9*, 10306–10316. [CrossRef]
16. Chu, T.; Wang, J.; Codecà, L.; Li, Z. Multi-agent deep reinforcement learning for large-scale traffic signal control. *IEEE Trans. Intell. Transp. Syst.* **2019**, *21*, 1086–1095. [CrossRef]
17. Egea, A.C.; Howell, S.; Knutins, M.; Connaughton, C. Assessment of Reward Functions for Reinforcement Learning Traffic Signal Control under Real-World Limitations. In Proceedings of the 2020 IEEE International Conference on Systems, Man, and Cybernetics (SMC), Totonto, ON, Canada, 11–14 October 2020; pp. 965–972.
18. Ozan, C.; Baskan, O.; Haldenbilen, S.; Ceylan, H. A modified reinforcement learning algorithm for solving coordinated signalized networks. *Transp. Res. Part C Emerg. Technol.* **2015**, *54*, 40–55. [CrossRef]
19. Kim, D.; Jeong, O. Cooperative traffic signal control with traffic flow prediction in multi-intersection. *Sensors* **2020**, *20*, 137. [CrossRef] [PubMed]
20. Yuan, J.; Abdel-Aty, M.; Gong, Y.; Cai, Q. Real-time crash risk prediction using long short-term memory recurrent neural network. *Transp. Res. Rec.* **2019**, *2673*, 314–326. [CrossRef]
21. Zhao, Y.; Liang, Y.; Hu, J.; Zhang, Z. Traffic Signal Control for Isolated Intersection Based on Coordination Game and Pareto Efficiency. In Proceedings of the 2019 IEEE Intelligent Transportation Systems Conference, Auckland, New Zealand, 27–30 October 2019; pp. 3508–3513. [CrossRef]
22. Kühnel, N.; Theresa, T.; Kai, N. Implementing an adaptive traffic signal control algorithm in an agent-based transport simulation. *Procedia Comput. Sci.* **2018**, *130*, 894–899. [CrossRef]

applied sciences

MDPI

Article

A Method to Improve the Accuracy of Pavement Crack Identification by Combining a Semantic Segmentation and Edge Detection Model

Peigen Li [1], Haiting Xia [1,2,*], Bin Zhou [3], Feng Yan [1] and Rongxin Guo [1]

1 Yunnan Key Laboratory of Disaster Reduction in Civil Engineering, Faculty of Civil Engineering and Mechanics, Kunming University of Science and Technology, Kunming 650500, China; pgli@stu.kust.edu.cn (P.L.); yanfengkmust@163.com (F.Y.); guorx@kust.edu.cn (R.G.)
2 Faculty of Civil Aviation and Aeronautics, Kunming University of Science and Technology, Kunming 650500, China
3 Yunnan Jiantou Boxin Engineering Construction Center Test Co., Ltd., Kunming 650217, China; fang@stu.kust.edu.cn
* Correspondence: haiting.xia@kust.edu.cn

Abstract: In recent years, deep learning-based detection methods have been applied to pavement crack detection. In practical applications, surface cracks are divided into inner and edge regions for pavements with rough surfaces and complex environments. This creates difficulties in the image detection task. This paper is inspired by the U-Net semantic segmentation network and holistically nested edge detection network. A side-output part is added to the U-Net decoder that performs edge extraction and deep supervision. A network model combining two tasks that can output the semantic segmentation results of the crack image and the edge detection results of different scales is proposed. The model can be used for other tasks that need both semantic segmentation and edge detection. Finally, the segmentation and edge images are fused using different methods to improve the crack detection accuracy. The experimental results show that mean intersection over union reaches 69.32 on our dataset and 61.05 on another pavement dataset group that did not participate in training. Our model is better than other detection methods based on deep learning. The proposed method can increase the MIoU value by up to 5.55 and increase the MPA value by up to 10.41 when compared to previous semantic segmentation models.

Keywords: convolutional neural network; crack detection; semantic segmentation; edge detection

Citation: Li, P.; Xia, H.; Zhou, B.; Yan, F.; Guo, R. A Method to Improve the Accuracy of Pavement Crack Identification by Combining a Semantic Segmentation and Edge Detection Model. *Appl. Sci.* **2022**, *12*, 4714. https://doi.org/10.3390/app12094714

Academic Editor: Luís Picado Santos

Received: 24 March 2022
Accepted: 4 May 2022
Published: 7 May 2022

1. Introduction

Highway pavements are affected by many factors such as the natural environment, load conditions, structural combinations, materials, construction techniques, and technical levels, which can produce various types of distress. With the construction of highways, pavement maintenance has begun increasing sharply. Accurate pavement distress detection results can provide reliable and effective technical support for pavement maintenance management decision making, improve highway pavement service performance, and reduce traffic accidents. However, traditional manual detection methods are often affected by subjective judgment in detecting highway pavement distress. There were considerable errors and low detection efficiencies. Therefore, automatic distress recognition and feature measurement of collected pavement images are the mainstream means of pavement detection.

The adoption of information management technology is an inevitable way to improve the level of highway maintenance management and realize efficient and orderly organization and management. For example, for common cracks on the highway, the development of an effective pavement crack identification algorithm can evaluate the pavement condition in advance and provide the basic data for maintenance decision making for the highway maintenance management department. The commonly used equipment for collecting pavement crack information include digital cameras, depth cameras, and lasers. Many researchers have studied pavement performance by using the images taken by digital cameras. At this stage, it has been applied to pavement crack detection [1], asphalt mixture crack detection [2], and concrete elements deformation tests [3]. Many researchers have recently begun to apply depth imaging technology to pavement detection engineering [4–6]. Unlike the traditional 2D camera, the depth camera can obtain depth information and provide the color image details in the 2D camera [7]. In addition, laser scanning is often used to detect pavement damage [8,9]. Laser scanning technology was used to extract cracks in concrete [10]. Although the depth camera and laser scanner can extract the three-dimensional information of the pavement and more accurately identify the distresses. The use of these two devices is limited due to the high purchase cost of the equipment, the complex post-processing process of 3D data, and inconvenient daily maintenance.

Pavement cracks usually appear as curved configurations with different widths in an image. They can be characterized by the edge detection and image segmentation methods in computer vision. In the ideal case, for such deep cracks with good continuity and no other noise interference, the traditional method can efficiently segment the crack from the image. Lu et al. [11]. proposed a new double-threshold algorithm to obtain detailed information on the crack number and width. Peng et al. [12]. proposed a triple-threshold pavement crack detection method using a random structured forest. However, in an actual detection task, different types of pavement types, shadows, and foreign objects will lead to a decline in the detection accuracy of the traditional methods. In addition to the automatic threshold segmentation method, there are crack detection methods based on spatial filtering and wavelet analysis; however, they have some disadvantages such as high requirements for equipment, complex operation, and environmental impact [13–16].

In recent years, convolutional neural networks (CNNs) have been proposed and applied to computer vision tasks such as image classification [17–19], target detection [20–22], and semantic segmentation [23–25]. Simultaneously, a CNN-based method has also been applied to pavement distress detection. Hoanga et al. [26] demonstrated the performance of the traditional and intelligent methods based on CNN in the pavement crack detection task. The experimental results show that the CNN-based crack detection methods are promising alternatives to regular methods. Majidifard et al. [27] developed a hybrid model by integrating the Yolo and U-Net models to classify pavement distresses and simultaneously quantify their severity. Jia et al. [28]. proposed a method based on Deeplabv3+ and a pixel-level quantization algorithm for crack detection. Park et al. [29] The CNN composed segmentation and classification modules to extract pavement cracks and remove the elements interfering in the image. Flah et al. [30] proposed a nearly automated detection model based on image processing and deep learning to detect defects in areas where concrete structures usually cannot enter. In summary, traditional methods based on digital image processing have been widely used in pavement damage detection and have laid a theoretical foundation for methods based on deep learning. Methods based on deep learning have strong potential, are more accurate and convenient than traditional methods, and will be the mainstream methods for detecting pavement distress in the future.

In the pavement crack detection task, the semantic segmentation model can be used to calculate the area occupied by cracks. It predicts the cracks pixel-by-pixel and segments the cracks from the image. The existing neural network models perform very well in the defect detection task, similar to the pavement crack detection task. For example, the U-Net semantic segmentation network was applied to the defect detection task in the industry [31–34]. Inspired by the above methods and the U-Net network structure, we herein improve the U-Net convolutional neural network and apply it to crack identification in complex pavement conditions. When measuring the characteristics of cracks, calculating the width is necessary. The width calculation is related to the edge line, and the edge detection algorithm is used to extract it. Classical edge detection algorithms in computer vision include the Roberts operator, Sobel operator, and Canny operator [35]. These classical algorithms have also been applied to crack detection tasks. Wang et al. [36] designed a local adaptive algorithm for Otsu threshold segmentation and proposed an improved Sobel operator to extract crack edge lines. Qiang et al. [37] proposed an adaptive Canny edge detection algorithm that achieved good results in crack detection. In addition to these modified traditional algorithms, some edge detection algorithms based on deep learning are also present.

Holistically nested edge detection (HED) [38] and side-output residual networks (SRN) [39] are two relatively new edge detection networks, and both adopt the method of deep supervision to improve the training effect. Liu et al. [40] continued the idea of deep supervision and proposed a Deepcrack for crack detection in multiple scenes. Similar to Heider et al. [41], by combining the two networks of HED and U-Net, we proposed an end-to-end method for coast and coastline detection. Traditional edge detection algorithms are easily disturbed by environmental factors. Especially in the pavement surface images, factors such as rough surfaces, vehicle shadows, water stains, and uneven lighting brightness affect the edge detection accuracy. In addition, the edge detection algorithm cannot recognize the meaning of objects inside and outside the edge line; however, the combination of semantic segmentation and edge detection results can solve this problem.

Therefore, a fusion model is proposed to segment cracks and simultaneously identify crack edge lines. The model uses a U-Net structure for image segmentation. It continues the idea of deep supervision in the HED and SRN networks. As the model uses the side-output method for edge line detection, it is called a side-output U-Net (SoUNet).

The remainder of this paper is organized as follows. The second section presents the proposed network model and the evaluation indicators in detail. The third section describes the collection and production of data and introduces the process and details of training. Section four provides the numerical results and intuitive prediction results. Our model was also compared with existing methods. The final section provides concluding statements.

2. Proposed Method

2.1. Model Architecture

Based on the U-Net semantic segmentation network model, we herein improve it and add a side-output module. We call the network model SoUNet. The traditional U-Net has a residual connected encoder–decoder architecture. The encoder part can obtain the low-resolution feature map after downsampling the high-resolution input image many times. This part is mainly used to extract the image features, and each layer is called the feature extraction layer. The decoder part includes several operations of feature concatenation, convolution, and deconvolution. It enlarges the low-resolution image outputted by the encoder through deconvolution, concatenates the same resolution image outputted by each feature extraction layer, and finally outputs the binary image through activation.

The structure of SoUNet is divided into two parts: the basic U-Net structure and the side-output structure. The structure of the network is shown in Figure 1. The first part is a semantic segmentation task. We removed the last 3 layers of VGG16 and used the first 13 layers as the encoder, which contained 13 convolution layers and 4 max-pooling layers. The max-pooling layer can downsample high-resolution images into low-resolution images, and there are five resolutions from high to low. The max-pooling layer enables the network model to learn semantic features at different resolutions and improve the learning efficiency of the model. The decoder includes nine convolution layers and four deconvolution layers. The deconvolution layer can restore the low-resolution feature map to a high-resolution one, and the feature map of the same resolution requires feature fusion in the decoder. In the entire U-Net structure, the kernel size of each convolution layer and deconvolution layer was set to 3×3. The rectified linear unit (Relu) was used as the activation function after the convolution layer. Only the last convolution layer uses a 1×1 kernel size, followed by the sigmoid activation function layer. The sigmoid function activates the input image after passing through the encoder–decoder structure. The final output image is output 1, and its size is the same as that of the input image. Output 1 is the result of the semantic segmentation task and is the probability map. The value of each pixel is between 0 and 1, indicating the probability that the pixel belongs to a category. The area with a high pixel value is a crack, and the area with a low pixel value is the background. We used 0.5 as the global threshold to transform the obtained probability map into a binary image.

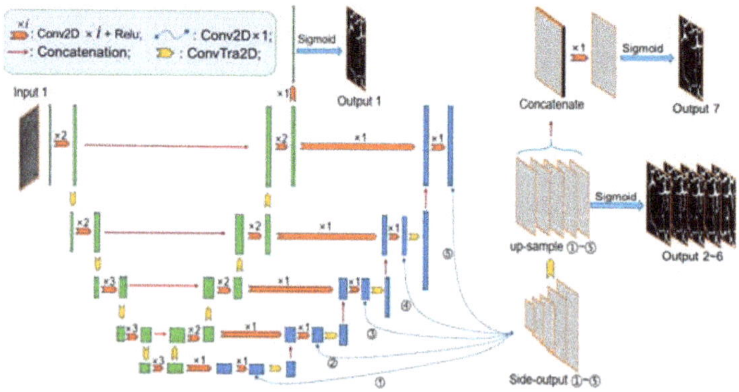

Figure 1. Illustration of our proposed Side-output U-Net architecture.

The second part of SoUNet is the side-output module, which performs the edge detection task, as shown in Figure 2. We extract the feature maps of different resolutions in the decoder and make them pass through two convolution layers of 3 × 3 kernel size. After enlarging the size, the lower-resolution feature map was deconvoluted and fused with the higher-resolution feature map. The feature maps of five resolutions were obtained by convolution, and then they were processed by convolution with a 1 × 1 kernel size. The feature map of each resolution was restored to the original image size after the deconvolution operation. Therefore, the side-output module was divided into five stages, corresponding to five feature maps of different scales. Five types of feature maps with the original size are sent to the sigmoid function for activation, and five images are denoted as Outputs 2–6. In addition, the feature maps of the five resolutions are fused into one size. It is sent to the sigmoid activation function after it passes through the convolution layer with a 1 × 1 kernel size. Finally, output the image called Output 7. The maximum ODS value was taken as the segmentation threshold to generate a binary image.

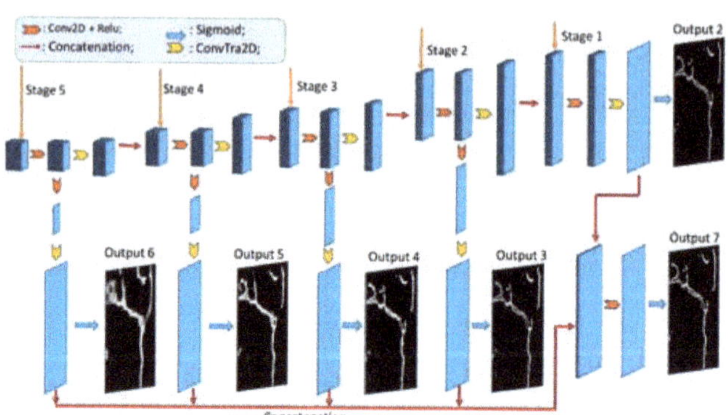

Figure 2. Illustration of a side-output module.

In this paper, we introduce a batch normalization (BN) layer [42] into the network architecture. When the depth of the network model gradually deepens, the model is more sensitive to changes in hyperparameters, and the model becomes more challenging to train.

However, the BN operation enables the model to be trained with a large learning rate. It reduces the requirements of parameter initialization, decays the oscillation of the loss function, and accelerates the training process. The ReLU function is used as the activation function in each convolution block of the middle architecture. The network was constructed in the order of convolution layer, BN layer, and ReLU layer.

By observing the pixel value distribution of the original crack image and the labeled image, it was observed that most cracks were composed of internal and edge areas. The inspectors captured photographs of pavement cracks with a monocular camera, which was mounted at the rear of the detection vehicle and had a fixed shooting distance and angle. Therefore, image quality is easily affected by the pavement environment. Identifying the crack width and length for road sections with limited daylighting conditions and rough surfaces is difficult. The crack gradually transits from the edge area to the internal area of the image. This means that the crack is composed of the inside and the edge. As shown in Figure 3, the cracks affected by environmental factors are divided into two distinct areas.

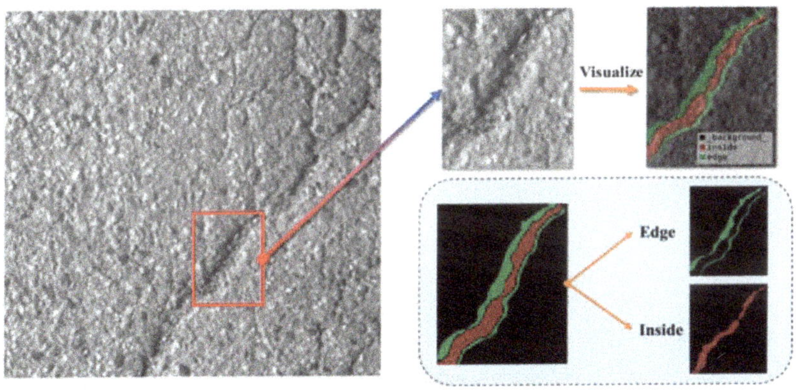

Figure 3. Cracks of rough pavement are divided into two parts: the edge and the interior.

SoUNet can output both segmentation results and edge line results for the input image. The linear fusion of the two results can effectively improve image segmentation accuracy. Output 1 is the crack segmentation image, and Outputs 2–7 are the crack edge line images. The detection accuracy can be improved by linear fusion of Output 1 and Outputs 2–7, respectively. In addition, the refinement method of guided filtering can improve the identification accuracy of the network [41,43]. Output 1 and Output 3 are the input image and guide image, respectively, and it sets the parameters of the guide filter as the kernel radius $r = 5$ and the penalty $\varepsilon = 1 \times 10^{-6}$. Therefore, the following three methods must be considered. These are (1) adding a BN layer to the network, (2) linear fusion of output results, and (3) processing the output results via guided filtering. We compared the segmentation accuracy of these methods in Section 4.2.

2.2. Loss Function

The purpose of image segmentation is to segment the cracks from the background. In the labeled image, the pixel value of the crack is 1, and the pixel value of the background is 0. It outputs the probability that each pixel is a crack after the input image passes through the encoder and decoder. The network model is more likely to extract the background in the training process because the area of the crack accounts for a small proportion of the entire image, which is less than 10% in most images. The imbalance of categories leads to a decline in the segmentation effect. We apply the loss function in HED [38] that can self-adaptively balance positive and negative samples. This cross-entropy loss function with category balance is defined by Equation (1):

$$L(\hat{y}) = -\beta \sum_{j \in Y_-} \log \hat{y}_j - (1-\beta) \sum_{j \in Y_+} \log(1 - \hat{y}_j) \quad (1)$$

The predicted pixel is $\hat{y}$ for a single-input image. There are $\beta = |Y_+|/|Y|$ and $1-\beta = |Y_-|/|Y|$ on the corresponding labeled image. $|Y_+|$ and $|Y_-|$ represent the pixels of the crack and background areas, respectively, and $|Y|$ represents the total number of pixels. This loss function can be used for segmentation and edge detection tasks, which are unbalanced categories.

2.3. Metrics

In the field of computer vision, MIoU and MPA have extensively used evaluation indicators for semantic segmentation tasks. Many conventional image segmentation algorithms use the mean intersection over union (MIoU) and the mean pixel accuracy (MPA) as evaluation indicators [23–25,34,44–46].

Accuracy indicators adopted in the training process: The MIoU) can be used as the evaluation metrics for the image segmentation task of unbalanced category samples. It is also an accuracy indicator for monitoring the training process, as shown in Equation (2):

$$MIoU = \frac{1}{k+1} \sum_{i=0}^{k} \frac{N_{ii}}{\sum_{j=0}^{k} N_{ij} + \sum_{j=0}^{k} N_{ji} - N_{ii}} \quad (2)$$

The intersection union (IoU) is the ratio of the overlapping part to the merged part of the two regions. This is a general measurement method for semantic segmentation tasks. $k+1$ is defined as the number of categories to be classified, where $k+1$ is 2 (the types include the fracture area and background area). N_{ii} is the number of pixels that are predicted correctly, N_{ij} is the number of pixels that class i is predicted as class j, and N is the total number of pixels. We use pixel error to monitor the training process for the edge detection task, as shown in Equation (3):

$$Pixel\ Error = \sum_{i=0}^{k} \sum_{j=0}^{k} \frac{N_{ij}}{N} \ (i \neq j) \quad (3)$$

Other accuracy indicators: After the model was trained, the prediction accuracy was evaluated on the test set. In addition to using the MIoU evaluation for crack segmentation results, the MPA can also be used. It calculates the average value of the percentage of correctly predicted pixels for each category, as shown in Equation (4):

$$MPA = \frac{1}{k+1} \sum_{i=0}^{k} \frac{N_{ii}}{\sum_{j=0}^{k} N_{ij}} \quad (4)$$

We use OIS-F and ODS-F to evaluate the boundary detection results. The training process and training results will be evaluated and presented in Sections 3.4 and 4.2, respectively.

3. Experiment

3.1. Image Collection

The image data of pavement distresses used in this experimental study were provided by the Yunnan Highway Science and Technology Research Institute. There are mainly net-shaped cracks, longitudinal cracks, and transverse cracks in the image data. Fatigue failure is the most common source of net-shaped cracks. The asphalt pavement structure eventually loses its bearing capacity due to repeated vehicle loads, and fatigue failure occurs. Uneven subgrade settlement and fatigue failure are the principal causes of longitudinal cracks. They will eventually develop into net-shaped cracks if not maintained. The most typical causes of transverse cracks are temperature changes and reflection cracks. Transverse cracks grow from top to bottom due to low-temperature shrinking. Reflection cracks develop from the bottom up, penetrating the road structure. The information offer basis for subsequent maintenance work.

We used the Teledyne Dalsa S3-24-02k40, which is a high-response, high-speed linear array industrial digital camera with a 2048 × 2048 picture resolution. A Camera Link is included within the camera. It can sustain a fast transmission speed while dealing with enormous amounts of picture data and high bandwidth needs. At the same time, the camera's improved user interface makes data collecting personnel's following image processing job easier. After the images are gathered on-site, the cracks are manually identified as mesh cracks, longitudinal cracks, and transverse cracks, and then images including single cracks, multiple cracks, and mesh cracks are picked.

It contains 3000 pictures collected by the road detection vehicle, with a pixel resolution of 2048 × 2048, and the format is a single-channel gray image. The images were collected at the K1209 + 080 − K1210 + 096 Xiuhe section of the No. 326 State Road and K1904 + 350 − K1902 + 300 Lanma section of the No. 248 State Road. Figure 4 shows the information about the roads. The selected road section included both cement and asphalt pavements. Owing to the influence of the driving load and natural environment, there are different types of cracks on the pavement. These complex data contents cause some difficulties in the crack identification task. We attempted to classify the degree of distress in the original road image using a convolution neural network. However, owing to shadows, water stains, and other foreign objects in the image, the identification accuracy can only reach around 75%, which does not accomplish the desired impact. We plan to improve the distress categorization method, as well as the accuracy and automation of pavement detection, in the future study. Water stains are caused by a portion of the road surface becoming wet. Many provinces are connected by the No. 326 State Road, and the No. 248 State Road, and trucks are frequently seen on the route. The sprinkler must constantly cool the heat brake pads and tires to guarantee driving safety. Wet strip tire imprints are frequently observed on the road. Some trucks will also be transporting wet goods, resulting in some partial wetness on the road surface. These create certain challenges for the task of detecting pavement cracks using digital images.

Figure 4. Selected detection part of the No. 326 State Road and the No. 248 State Road.

The convolutional neural network model we constructed can only train images with a pixel resolution of 256 × 256 due to the computing capability of the computer. The open-source computer vision software OpenCV is used to resize the image to match it with the network model's input. The original collected images were pretreated. The image with a pixel resolution of 2048 × 2048 was cropped to the image with a pixel resolution of 512 × 512, which is one-sixteenth of the original image. The image needs to be resized to 256 × 256 pixels to match the input port of the network model. If we immediately compress the image of 2048 × 2048 pixels to 256 × 256 pixels, the original image's crack information is significantly lost, resulting in a decrease in identification accuracy. The original image and the processing procedure are depicted in Figure 5.

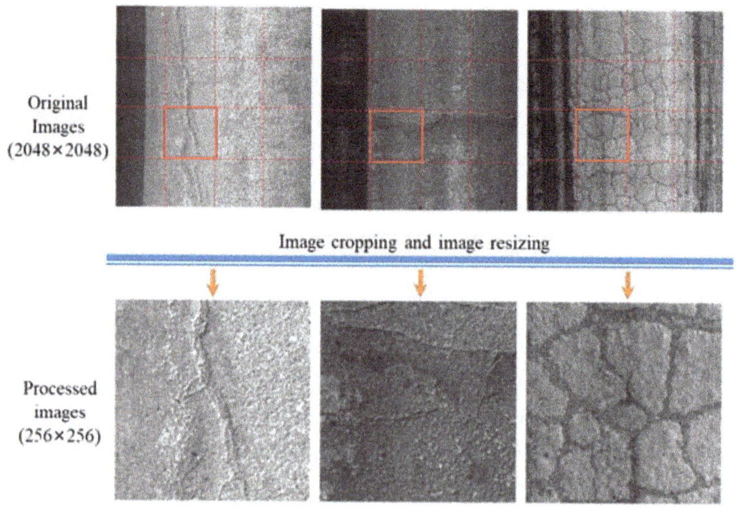

Figure 5. The original images and the processed images.

3.2. Image Dataset

The open-source tool LabelMe [47] was downloaded for semantic annotation, obtained from GitHub [48]. The annotated information is saved as a JSON file containing the marked image name, labeled type, coordinate points, and others. Extracting the information in a file can generate a binary image for training. The original image, manually labeled crack, and crack edge images are shown in Figure 6. Six hundred images with cracks were selected randomly from the dataset for the pixel-level annotation. The dataset included 420 images as the training set, 120 as the validation set, and 60 as the test set. The ratio of the training set, validation set, and test set was 7:2:1. Table 1 lists the percentages of the crack and non-crack pixels in the dataset. The table shows that the crack images only account for a small number, and the task is image segmentation with an unbalanced category. The labeled dataset includes asphalt pavement and cement pavement, and some images contain interference factors of water stains and shadow changes. Figure 7 shows labeled images in different environments. Table 2 shows the proportion of asphalt pavement and cement pavement images in the dataset and the proportion of images in different environments.

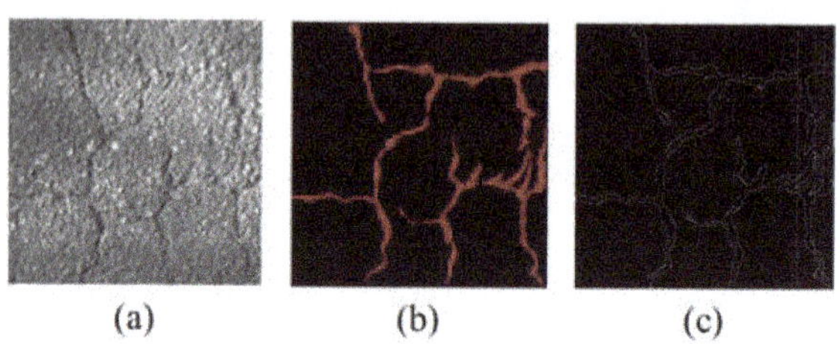

Figure 6. Annotation at pixel level using LabelMe tool: (**a**) original image; (**b**) labeled crack binary image; (**c**) labeled crack edge binary image.

Table 1. Proportion of crack and non-crack annotations in the dataset.

	Quantity	Crack Pixels (%)	Non-Crack Pixels (%)
Training data	420	6.79	93.21
Validation data	120	4.14	95.86
Test data	60	6.52	93.48

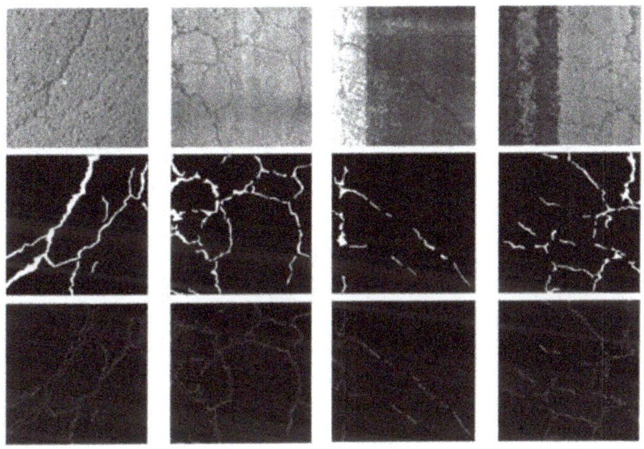

Figure 7. Labeled images under different conditions: (**a**) asphalt pavement; (**b**) cement pavement; (**c**) shadow interference; (**d**) water stain interference.

Table 2. Proportion of images in different pavement types and environmental conditions.

	Pavement		Environment				
Types	Concrete	Asphalt	Normal Brightness	Low Brightness	High Brightness	Shadow	Water Stain
Percentage (%)	21.6	78.4	81.8	13.0	5.2	9.2	20.7

3.3. Training Details

The training platform was performed on a workstation with an Intel(R) Core i9-10900k CPU and an NVIDIA 3090, 24G GPU. This study uses TensorFlow, which is Google's open source deep learning framework, to build and train the network. The software configuration was as follows: Windows 10, CUDA 11.1, cuDNN-v8.0.4, TensorFlow-GPU-2.4, and Python 3.8.

A total of 420 labeled images were taken as the training set, and the data of eight images in each batch were input into the SoUNet network after shuffling the training set. In the training process, the cross-entropy loss function with category balance in Equation (1) is used as the loss function. The adaptive moment estimation (Adam) optimizer [49] was selected for optimization. The optimizer adjusts the learning rate in the training process and changes the weight parameters and bias values in the network. The initial learning rates were set to 1×10^{-3}, 1×10^{-4}, and 1×10^{-5}, respectively, and the training epochs were set to 300. The accuracy indicators monitored during training are the MIoU value and pixel error, respectively.

3.4. Training Process

The model was trained after setting the parameters, and the entire training process was monitored. Figure 8 shows the training process of the model for different learning rates. It includes the variation curves of four variables measured on the training set, which are seven loss values, overall loss values, MIoU, and pixel error. Figure 8a–c show that the seven loss values continue to decline under different learning rates. We chose to stop training at 300 rounds to prevent overfitting. As shown in Figure 8d–f. When the initial learning rate was set to 1×10^{-4}, the overall loss value of the network decreased the fastest in the training process and reached the lowest value at the end of the training. Simultaneously, the MIoU value and pixel error measured in the training set can reach the optimal value. Therefore, the most effective model was selected in the training process when the learning rate was set to 1×10^{-4}.

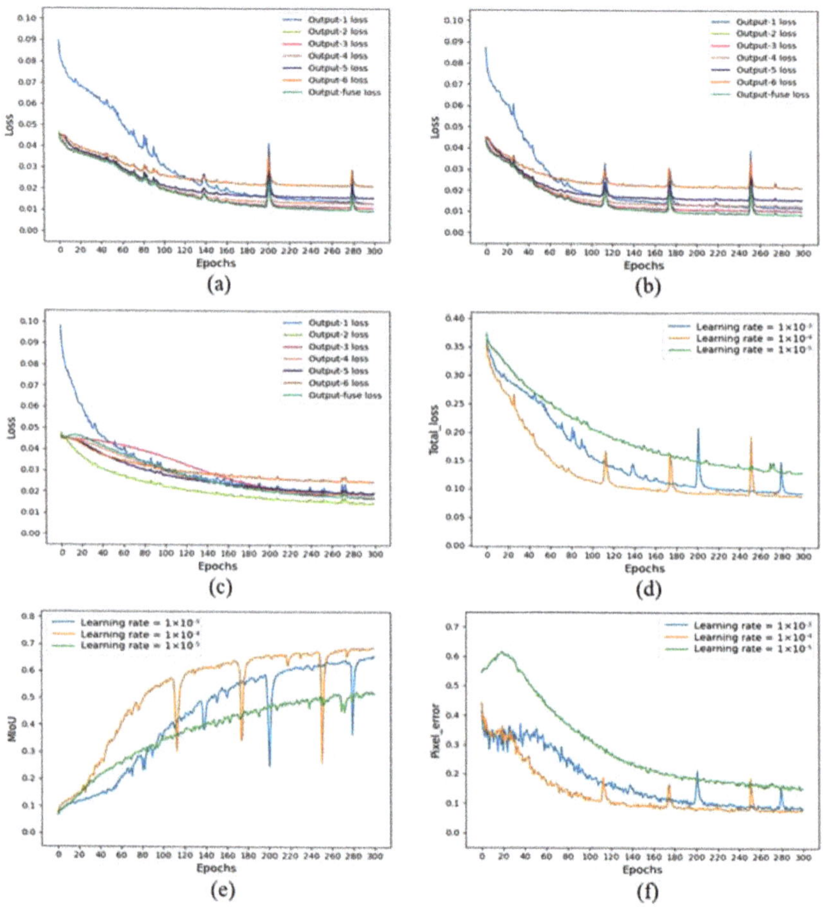

Figure 8. Training process of network model: (**a**–**c**) are the variation curves of 7 loss values with the epochs when the learning rate is equal to 1×10^{-3}, 1×10^{-4}, and 1×10^{-5}, respectively; (**d**) Variation curve of overall loss value with the epochs under different learning rates; (**e**) Variation curve of MIoU value measured on the training set; (**f**) Variation curve of pixel error measured on the training set.

4. Training Result and Comparison

4.1. Training Result

TensorFlow 2 has the function of saving the optimal model. The best network model for the validation set was extracted. This model was used to predict the test set. Figure 9 shows the crack segmentation results and edge extraction results of SoUNet for different types of pavement images. The model has a good segmentation effect on a single crack of both asphalt pavement and cement pavement. The effect of edge detection is normal, but the two edge lines tend to overlap for areas with narrow widths.

The model performs well for multiple cracks and net-shaped cracks, but there are many problems such as noise points, incomplete segmentation, and blurred areas. The recognition effect of the model on the cement pavement image is good, and there is more noise and missed detections in the recognition results of asphalt pavement. Figure 10 shows the prediction results under the interference of water stains and shadows. There are many cases of noise and missed detection in areas with water stains, and other missed detections occur at the borders of the shadows. Water stains have a greater impact on the prediction results. In general, the proposed model was effective. It shows a certain potential in detecting images with interference, and the MIoU is greater than 50%. According to the data statistics in Table 2, only 20.7% of the images contained water stains, and only 9.2% of the images contained shadow interference in our dataset. In the future, we can increase the number of such data and use more images with different interferences to participate in the training process to enhance the accuracy and robustness of recognition.

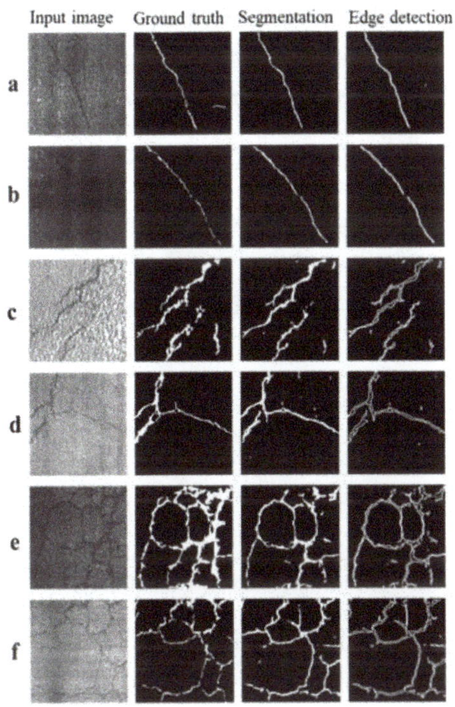

Figure 9. Identification results of different pavement types: (**a**) single crack in asphalt pavement; (**b**) single crack in cement pavement; (**c**) multiple cracks in asphalt pavement; (**d**) multiple cracks in cement pavement; (**e**) net-shaped crack in asphalt pavement; (**f**) net-shaped crack in cement pavement.

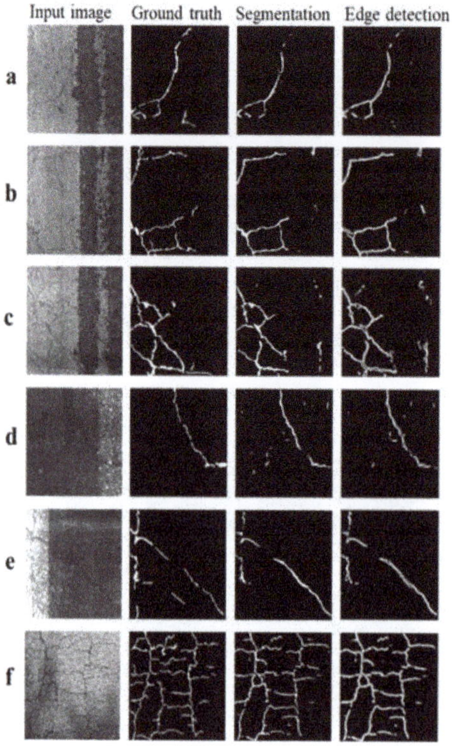

Figure 10. Identification results under the interference of different environmental conditions: (**a–c**) are single crack, multiple cracks, and net-shaped crack in water interference, respectively; (**d–f**) are single crack, multiple cracks, and net-shaped crack in shadow interference, respectively.

4.2. Evaluated Model

SoUNet can output both the crack segmentation image and the crack edge line image. The optimal model was extracted to predict the test set, and the segmentation and edge detection images were output. An input image corresponds to one segmentation image and six edge line images. One segmented image was linearly fused with the other six edge line images to optimize the segmentation results. The fusion images and fusion results are presented in Figure 11. Table 3 lists the MIoU, mean pixel accuracy, ODS-F, and OIS-F measured using different methods. SoUNet-Output-1 is the output of the semantic segmentation network in SoUNet, which is the image of Output 1. SoUNet-Fusion-ij is the linear fusion of outputs i and j. The numerical value shows that the linear fusion of the semantic segmentation results and edge line detection results can effectively improve the crack segmentation accuracy. The MIoU value increased by 2.47%, and the MPA value increased by 9.58%. SoUNet-Fusion-13 has high MIoU and MPA values and is the most stable under various accuracies from the result of the comprehensive comparison. The results are compared with those of other semantic segmentation models in Section 4.3.

We selected 30 crack pictures from the test set and measured the width of the initial position, middle position, and end position of the crack. The measurement direction is perpendicular to the crack trend, as shown in Figure 12a. The same method is used to measure the crack width in the label image, SoUNeT result image, and U-NET result image, respectively. Taking the crack width measured in the label image as the actual width, the width error statistical charts of SoUNet and U-NET are obtained. As shown in Figure 12b, S1–3 in the figure shows the SoUNet initial position, middle position, and end position of the crack, respectively. U1–3 indicates the U-Net initial position, middle position, and end position of the crack, respectively. It can be seen from Figure 12b that the width error measured by SoUNet is about 2 pixels, and the measurement error is less than that of U-Net.

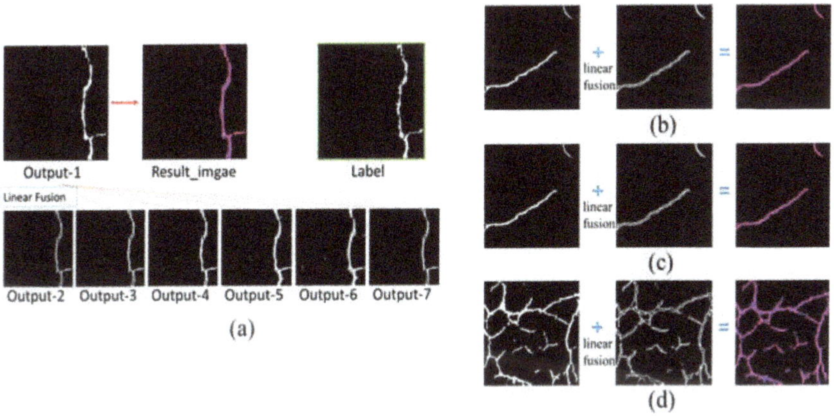

Figure 11. Fusion process and result of segmented image and edge image: (**a**) Linear fusion process of output images for (**b**) single crack, (**c**) multiple cracks, (**d**) net-shaped crack.

Table 3. Evaluation results of the model.

Methods	Metrics			
	MIoU	MPA	ODS-F	OIS-F
SoUNet-Output-1	67.17	72.31	—	—
SoUNet-Fusion-12	**69.64**	78.25	31.52	32.99
SoUNet-Fusion-13	69.32	80.33	**33.14**	**34.11**
SoUNet-Fusion-14	68.29	81.54	32.08	33.15
SoUNet-Fusion-15	65.92	**81.89**	29.46	30.63
SoUNet-Fusion-16	60.39	81.51	25.66	26.73
SoUNet-Fusion-17	69.42	80.14	33.08	34.08

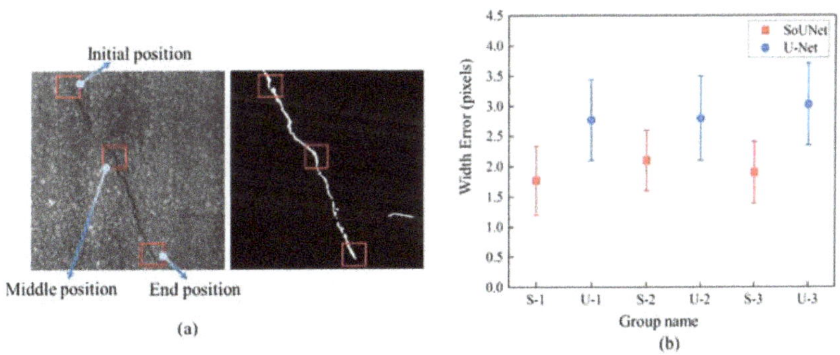

Figure 12. Statistics of the crack's width at different positions: (**a**) the width of the initial position, middle position, and end position of the crack; (**b**) the width error statistical charts of SoUNet and U-NET.

4.3. Comparative Study

To test the performance of SoUNet, we selected four methods based on deep learning for comparative study: (1) SegNet [24] is a fully convolutional network, which was used for semantic segmentation. It has also been proposed for crack identification of concrete pavement, asphalt pavement, and bridge deck [44]; (2) HED [38], which is an edge detection model with high performance that can also be used for crack detection; (3) VGG16-U-Net [45]: U-Net is a high-performance semantic segmentation network [25]. Its improved structure, VGG16 U-Net, has been used to detect surface defects in concrete and asphalt [46]. Comparing the recognition performance of the following strategies for the proposed SoUNet model is necessary: (1) SoUNet-Basic: The basic side-output U-Net structure, the side-output part plays the role of deep supervision and improves the model learning efficiency; (2) SoUNet-BN: Adding a batch normalization layer based on SoUNet-Basic. The BN layer can accelerate the training process; (3) SoUNet-GF: A and B are taken as the original image and guide image, respectively, from the outputs of SoUNet-BN and then perform the guided filter operation; (4) SoUNet-Fusion: This is the same as SoUNet-Fusion-13 in Table 3 of Section 4.2.

Deep-learning-based methods can be applied to image recognition tasks, but these methods are only suitable for specific scenes and tasks in most cases. Poor generalization performance is one of the main drawbacks of these methods. To further test the generalization performance of SoUNet in the crack detection task, the FISSURES dataset [50] was downloaded. This dataset is similar to our dataset. The preprocessing method in Section 3.1 is used to process the dataset and make those sizes suitable for the network model. Finally, they were sent to the trained model to view the results. Table 4 shows the evaluation results of the seven methods on the two datasets. Our test set is divided in Section 3.2, accounting for one-tenth of the original dataset. None of the images for the prediction evaluation in this section participated in the training process. The linear fusion method performs better than the other methods on its own test set and FISSURES dataset.

Figure 13 shows the prediction results of the seven methods on our dataset. In the case of no interference, the segmentation integrity of SoUNet-fusion is better than that of other methods, and the noise produced is less than other results. In addition, our method performs well on rough asphalt pavement that is difficult to identify, and the segmentation results are relatively complete, but there are some false positive areas and a small number of noise points. Shadows and water stains are not misjudged as cracks, but the segmentation accuracy decreases, and the results are incomplete. Figure 14 shows the test results for the FISSURES dataset. The asphalt pavement in the FISSURES dataset was relatively flat, but the crack depth was shallow, and the width was narrow, so the noise of the segmentation result was relatively small. Segmentation integrity is investigated in this section. SoUNet-fusion has good segmentation integrity in the images of single cracks, multiple cracks, and net-shaped cracks. There were relatively few misjudged areas. When interference occurs, the crack area can still be completely segmented.

Table 4. Evaluation and comparison results of different methods on two datasets.

Datasets	**Our Test Datasets**		**FISSURES Datasets**	
Metrics	**MIoU**	**MPA**	**MioU**	**MPA**
SegNet	63.77	69.92	56.34	60.65
HED	64.56	70.70	58.30	65.86
VGG16-U-Net	66.99	74.57	59.12	67.66
SoUNet-Basic	67.46	75.59	59.15	68.45
SoUNet-BN	68.46	74.65	60.07	65.56
SoUNet-GF	68.41	77.28	61.04	67.81
SoUNet-Fusion	**69.32**	**80.33**	**61.05**	**68.60**

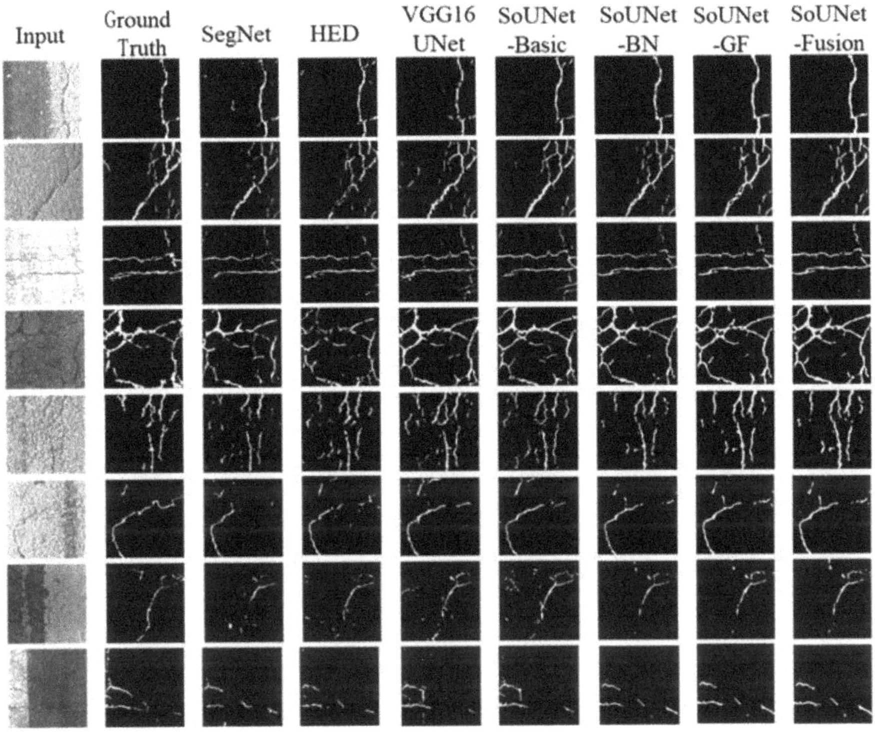

Figure 13. Comparison of prediction results on our dataset.

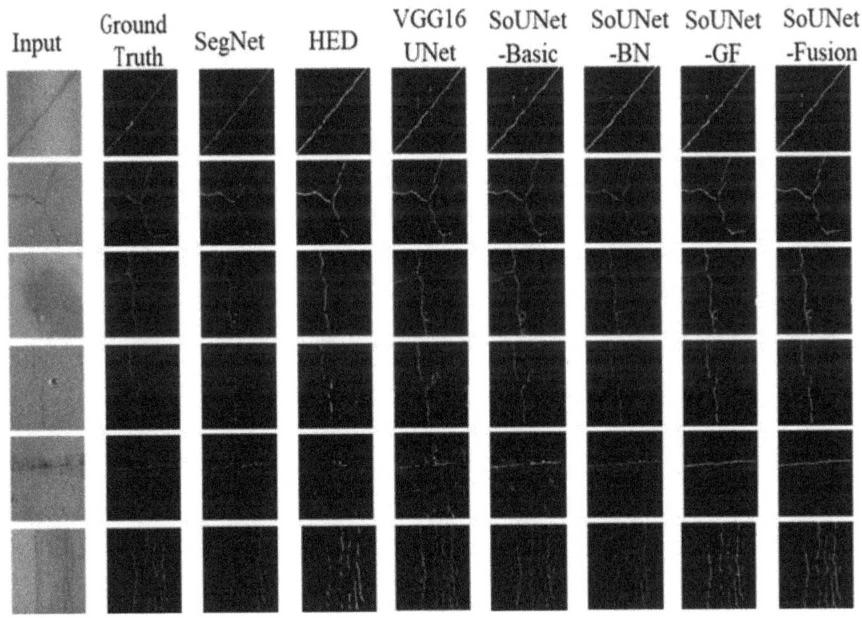

Figure 14. Comparison of prediction results on the FISSURES dataset.

5. Conclusions

In this paper, we introduce a model that can simultaneously perform semantic segmentation and edge detection. The proposed convolutional neural network SoUNet was used to output the crack segmentation images and crack edge images. Finally, the two output results were linearly fused to improve detection accuracy. When compared to previous semantic segmentation models, our method can increase the MioU value by up to 5.55 and increase the MPA value by up to 10.41.

The semantic segmentation part of SoUNet is based on the U-Net structure of the vgg16. The convolution feature map on each scale was fused in pairs, and the low-resolution fusion feature map was further fused to a higher resolution after passing through the convolution layer. The edge detection part extracts the feature map of each scale based on the U-Net. The low-resolution feature maps were trained and fused to the high-resolution features, and the crack edge image was outputted. The edge detection part is also the side-output part of the entire network. In addition, the crack dataset contains the pavement surface of cement and asphalt, and it also contains images of water stains and shadows. Therefore, the dataset is closer to the actual situation. The experimental results demonstrate that the edge detection part of the proposed method achieves ODS-F 33.14, OIS-F 34.11 on our dataset. Its MioU, the semantic segmentation evaluation indicator, reaches a value of 69.32. Both the intuitive and numerical results are better than those of other segmentation methods based on deep learning. The experimental results also show that SoUNet performs well in rough asphalt pavement images, is less affected by water stains and shadows, and has the potential to deal with multi-interference pavement conditions.

In the future, we plan to develop a new pavement detection network that is more accurate for identifying types of pavement cracks. We will enrich the pavement dataset and add crack images of various scenes to make the dataset closer to the actual situation. In addition, we will also use the model for other tasks that need both semantic segmentation and edge detection, such as pit contour detection in pavement distress detecting task, road edge line detection in an automatic driving task, and diseased organ contour recognition in picture medicine.

Author Contributions: Conceptualization, H.X.; methodology, P.L.; software, P.L.; validation, B.Z.; investigation, B.Z. and F.Y.; resources, F.Y.; writing—original draft preparation, P.L.; writing—review & editing, H.X.; visualization, P.L.; supervision, R.G.; project administration, H.X.; funding acquisition, R.G. All authors have read and agreed to the published version of the manuscript.

Funding: This work was supported by the National Natural Science Foundation of China (11862008).

Institutional Review Board Statement: Not applicable.

Informed Consent Statement: Not applicable.

Data Availability Statement: Data sharing not applicable.

Conflicts of Interest: The authors declare no conflict of interest.

Abbreviations

MIoU	Mean intersection over union
MPA	Mean pixel accuracy
ODS-F	Optimal dataset scale F-score
OIS-F	Optimal image scale F-score
CNN	Convolutional neural network
HED	Holistically nested edge detection network
SegNet	A deep convolutional encoder-decoder architecture for image segmentation
U-Net	U-shaped Convolutional networks for image segmentation
BN	Batch normalization
GF	Guided filter operation

References

1. Ayenu-Prah, A.; Attoh-Okine, N. Evaluating Pavement Cracks with Bidimensional Empirical Mode Decomposition. *EURASIP J. Adv. Signal Process.* **2008**, *2008*, 861701. [CrossRef]
2. Doll, B.; Ozer, H.; Rivera-Perez, J.J.; Al-Qadi, I.L.; Lambros, J. Investigation of viscoelastic fracture fields in asphalt mixtures using digital image correlation. *Int. J. Fract.* **2017**, *205*, 37–56. [CrossRef]
3. Tan, Y.Q.; Zhang, L.; Guo, M.; Shan, L. Investigation of the deformation properties of asphalt mixtures with DIC technique. *Constr. Build. Mater.* **2012**, *37*, 581–590.
4. Grabowski, D.; Szczodrak, M.; Czyzewski, A. Economical methods for measuring road surface roughness. *Metrol. Measurem. Syst.* **2018**, *25*, 533–549.
5. Jahanshahi, M.R.; Jazizadeh, F.; Masri, S.F.; Becerik-Gerber, B. Unsupervised Approach for Autonomous Pavement-Defect Detection and Quantification Using an Inexpensive Depth Sensor. *J. Comput. Civ. Eng.* **2013**, *27*, 743–754. [CrossRef]
6. Cui, X.; Zhou, X.; Lou, J.; Zhang, J.; Ran, M. Measurement method of asphalt pavement mean texture depth based on multi-line laser and binocular vision. *Int. J. Pavement Eng.* **2017**, *18*, 459–471. [CrossRef]
7. Ni, Z.; Shen, Z.; Guo, C.; Xiong, G.; Nyberg, T.; Shang, X.; Li, S.; Wang, Y. The Application of the Depth Camera in the Social Manufacturing: A review. In Proceedings of the 2016 IEEE International Conference on Service Operations and Logistics, and Informatics, Beijing, China, 10–12 July 2016; IEEE: New York, NY, USA, 2016; pp. 66–70.
8. Rahkonen, J.; Jokela, H. Infrared Radiometry for Measuring Plant Leaf Temperature during Thermal Weed Control Treatment. *Biosyst. Eng.* **2003**, *86*, 257–266. [CrossRef]
9. Tsai, Y.-C.J.; Li, F. Critical Assessment of Detecting Asphalt Pavement Cracks under Different Lighting and Low Intensity Contrast Conditions Using Emerging 3D Laser Technology. *J. Transp. Eng.* **2012**, *138*, 649–656. [CrossRef]
10. Janowski, A.; Nagrodzka-Godycka, K.; Szulwic, J.; Ziółkowski, P. Modes of Failure Analysis in Reinforced Concrete Beam Using Laser Scanning and Synchro-Photogrammetry How to apply optical technologies in the diagnosis of reinforced concrete elements? In Proceedings of the International Conference on Advances in Civil, Structural and Environmental Engineering—ACSEE-2014, Zurich, Switzerland, 21–22 September 2014.
11. Lu, C.; Yu, J.; Leung, C.K.Y. An improved image processing method for assessing multiple cracking development in Strain Hardening Cementitious Composites (SHCC). *Cem. Concr. Compos.* **2016**, *74*, 191–200. [CrossRef]
12. Peng, C.; Yang, M.; Zheng, Q.; Zhang, J.; Wang, D.; Yan, R.; Wang, J.; Li, B. A triple-thresholds pavement crack detection method leveraging random structured forest. *Constr. Build. Mater.* **2020**, *263*, 120080. [CrossRef]
13. Mardasi, A.G.; Wu, N.; Wu, C. Experimental study on the crack detection with optimized spatial wavelet analysis and windowing. *Mech. Syst. Signal Process.* **2018**, *104*, 619–630. [CrossRef]
14. Lakshmi, K. Detection and quantification of damage in bridges using a hybrid algorithm with spatial filters under environmental and operational variability. *Structures* **2021**, *32*, 617–631. [CrossRef]
15. Liebold, F.; Maas, H.-G. Advanced spatio-temporal filtering techniques for photogrammetric image sequence analysis in civil engineering material testing. *ISPRS J. Photogramm. Remote Sens.* **2016**, *111*, 13–21. [CrossRef]
16. Li, Q.; Zou, Q.; Zhang, D.; Mao, Q. FoSA: F* Seed-growing Approach for crack-line detection from pavement images. *Image Vis. Comput.* **2011**, *29*, 861–872. [CrossRef]
17. Simonyan, K.; Zisserman, A. Very deep convolutional networks for large-scale image recognition. *arXiv* **2014**, arXiv:1409.1556.
18. He, K.; Zhang, X.; Ren, S.; Sun, J. Deep Residual Learning for Image Recognition. In Proceedings of the 2016 IEEE Conference on Computer Vision and Pattern Recognition (CVPR), Las Vegas, NV, USA, 27–30 June 2016; pp. 770–778. [CrossRef]
19. Huang, G.; Liu, Z.; Van Der Maaten, L.; Weinberger, K.Q. Densely Connected Convolutional Networks. In Proceedings of the 2017 IEEE Conference on Computer Vision and Pattern Recognition (CVPR), Honolulu, HI, USA, 21–26 July 2017; pp. 2261–2269. [CrossRef]
20. Ren, S.; He, K.; Girshick, R.; Sun, J. Faster R-CNN: Towards Real-Time Object Detection with Region Proposal Networks. *IEEE Trans. Pattern Anal. Mach. Intell.* **2017**, *39*, 1137–1149. [CrossRef] [PubMed]
21. Redmon, J.; Divvala, S.; Girshick, R.; Farhadi, A. You Only Look Once: Unified, Real-Time Object Detection. In Proceedings of the 2016 IEEE Conference on Computer Vision and Pattern Recognition (CVPR), Las Vegas, NV, USA, 27–30 June 2016; pp. 779–788. [CrossRef]
22. Liu, W.; Anguelov, D.; Erhan, D.; Szegedy, C.; Reed, S.; Fu, C.; Berg, A.C. SSD: Single Shot MultiBox Detector. In *Computer Vision—ECCV 2016*; Lecture Notes in Computer Science; Springer: Cham, Switzerland, 2016; Volume 9905, pp. 21–37. [CrossRef]
23. Long, J.; Shelhamer, E.; Darrell, T. Fully Convolutional Networks for Semantic Segmentation. In Proceedings of the 2015 IEEE Conference on Computer Vision and Pattern Recognition (CVPR), Boston, MA, USA, 7–12 June 2015; pp. 640–651. [CrossRef]
24. Badrinarayanan, V.; Kendall, A.; Cipolla, R. SegNet: A Deep Convolutional Encoder-Decoder Architecture for Image Segmentation. *IEEE Trans. Pattern Anal. Mach. Intell.* **2017**, *39*, 2481–2495. [CrossRef] [PubMed]
25. Ronneberger, O.; Fischer, P.; Brox, T. *U-Net: Convolutional Networks for Biomedical Image Segmentation*; Springer International Publishing: Berlin/Heidelberg, Germany, 2015. [CrossRef]
26. Nhat-Duc, H.; Nguyen, Q.; Tran, V. Automatic recognition of asphalt pavement cracks using metaheuristic optimized edge detection algorithms and convolution neural network. *Autom. Constr.* **2018**, *94*, 203–213. [CrossRef]
27. Hamed, M.; Peng, J.; Yaw, A.; William, B. Pavement Image Datasets: A New Benchmark Dataset to Classify and Densify Pavement Distresses. *Transp. Res. Rec. J. Transp. Res. Board* **2020**, *2674*, 328–339. [CrossRef]

28. Ji, A.; Xue, X.; Wang, Y.; Luo, X.; Xue, W. An integrated approach to automatic pixel-level crack detection and quantification of asphalt pavement. *Autom. Constr.* **2020**, *114*, 103176. [CrossRef]
29. Park, S.; Bang, S.; Kim, H.; Kim, H. Patch-Based Crack Detection in Black Box Images Using Convolutional Neural Networks. *J. Comput. Civ. Eng.* **2019**, *33*, 04019017. [CrossRef]
30. Flah, M.; Suleiman, A.R.; Nehdi, M.L. Classification and quantification of cracks in concrete structures using deep learning image-based techniques. *Cem. Concr. Compos.* **2020**, *114*, 103781. [CrossRef]
31. Pratt, L.; Govender, D.; Klein, R. Defect detection and quantification in electroluminescence images of solar PV modules using U-net semantic segmentation. *Renew. Energy* **2021**, *178*, 1211–1222. [CrossRef]
32. Lin, D.; Li, Y.; Prasad, S.; Nwe, T.L.; Dong, S.; Oo, Z.M. CAM-guided Multi-Path Decoding U-Net with Triplet Feature Regularization for Defect Detection and Segmentation. *Knowl. Based Syst.* **2021**, *228*, 107272. [CrossRef]
33. Rong-qiang, L.; Ming-hui, L.; Jia-chen, S.; Yi-bin, L. Fabric Defect Detection Method Based on Improved U-Net. *J. Phys. Conf. Ser.* **2021**, *1948*, 012160. [CrossRef]
34. Zhong, Q.; Zhang, J.; Xu, Y.; Li, M.; Shen, B.; Tao, W.; Li, Q. Filamentous target segmentation of weft micro-CT image based on U-Net. *Micron* **2021**, *146*, 102923. [CrossRef]
35. Deriche, R. Using Canny's criteria to derive a recursively implemented optimal edge detector. *Int. J. Comput. Vis.* **1987**, *1*, 167–187. [CrossRef]
36. Wang, Y.; Zhang, J.Y.; Liu, J.X.; Zhang, Y.; Chen, Z.P.; Li, C.G.; He, K.; Yan, R.B. Research on Crack Detection Algorithm of the Concrete Bridge Based on Image Processing. *Procedia Comput. Sci.* **2019**, *154*, 610–616. [CrossRef]
37. Qiang, S.; Guoying, L.; Jingqi, M.; Hongmei, Z. An Edge-Detection Method Based on Adaptive Canny Algorithm and Iterative Segmentation Threshold. In Proceedings of the 2016 2nd International Conference on Control Science and Systems Engineering, Singapore, 27–29 July 2016; pp. 64–67. [CrossRef]
38. Xie, S.; Tu, Z. Holistically-Nested Edge Detection. *Int. J. Comput. Vis.* **2017**, *125*, 3–18. [CrossRef]
39. Wei, K.; Jie, C.; Jianbin, J.; Guoying, Z.; Qixiang, Y. SRN: Side-Output Residual Network for Object Reflection Symmetry Detection and Beyond. *IEEE Trans. Neural Netw. Learn. Syst.* **2021**, *32*, 1881–1895.
40. Liu, Y.; Yao, J.; Lu, X.; Xie, R.; Li, L. DeepCrack: A deep hierarchical feature learning architecture for crack segmentation. *Neurocomputing* **2019**, *338*, 139–153. [CrossRef]
41. Konrad, H.; Lichao, M.; Celia, B.; Andreas, D.; Xiang, Z. HED-UNet: Combined Segmentation and Edge Detection for Monitoring the Antarctic Coastline. *IEEE Trans. Geosci. Remote Sens.* **2021**, *60*, 4300514. [CrossRef]
42. Sergey, L.; Christian, S. Batch normalization: Accelerating deep network training by reducing internal covariate shift. In Proceedings of the 32nd International Conference on International Conference on Machine Learning, Lille, France, 6–11 July 2015; pp. 448–456.
43. He, K.; Sun, J.; Tang, X. Guided Image Filtering. *IEEE Trans. Pattern Anal. Mach. Intell.* **2013**, *35*, 1397–1409. [CrossRef] [PubMed]
44. Chen, T.; Cai, Z.; Zhao, X.; Chen, C.; Liang, X.; Zou, T.; Wang, P. Pavement crack detection and recognition using the architecture of segNet. *J. Ind. Inf. Integr.* **2020**, *18*, 100144. [CrossRef]
45. Wen, Z.; Wang, H.; Yuan, H.; Liu, M.; Guo, X. A method of pulmonary embolism segmentation from CTPA images based on U-net. In Proceedings of the 2019 IEEE 2nd International Conference on Computer and Communication Engineering Technology (CCET), Beijing, China, 16–18 August 2019; pp. 31–35. [CrossRef]
46. Li, D.; Duan, Z.; Hu, X.; Zhang, D. Pixel-Level Recognition of Pavement Distresses Based on U-Net. *Adv. Mater. Sci. Eng.* **2021**, *2021*, 5586615. [CrossRef]
47. Russell, B.C.; Torralba, A.; Murphy, K.P.; Freeman, W.T. LabelMe: A Database and Web-Based Tool for Image Annotation. *Int. J. Comput. Vis.* **2008**, *77*, 157–173. [CrossRef]
48. Wada, K. LabelMe. Github. 2019. Available online: https://github.com/wkentaro (accessed on 15 March 2021).
49. Kingma, D.P.; Ba, J. Adam: A Method for Stochastic Optimization. *arXiv* **2014**, arXiv:1412.6980.
50. Chambon, S.; Moliard, J. Automatic Road Pavement Assessment with Image Processing: Review and Comparison. *Int. J. Geophys.* **2011**, *2011*, 989354. [CrossRef]

applied sciences

MDPI

Article

Research on an Improved SOM Model for Damage Identification of Concrete Structures

Jinxin Liu [1] and Kexin Li [2,*]

[1] School of Medical Technology, BeiHua University, Jilin City 132000, China; liujinxin1990@163.com
[2] School of Civil Engineering and Transportation, BeiHua University, Jilin City 132000, China
* Correspondence: likenefu@126.com

Abstract: In order to solve the problem of intelligent detection of damage of modern concrete structures under complex constraints, an improved self-organizing mapping (SOM) neural network model algorithm was proposed to construct an accurate identification model of concrete structure damage. Based on the structure and algorithm of the SOM network model, the whole process of the core construction of the concrete structure damage identification network model is summarized. Combined with the damage texture characteristics of concrete structures, through the self-developed 3D laser scanning system, an improved method based on a small number of samples to effectively improve the effectiveness of network input samples is proposed. Based on the principle of network topology map analysis and its image characteristics, a SOM model improvement method that can effectively improve the accuracy of the network identification model is studied. In addition, based on the reactive powder concrete bending fatigue loading test, the feasibility and accuracy of the improved method are verified. The results show that the improved SOM concrete structure damage identification model can effectively identify unknown neuron categories in a limited sample space, and the identification accuracy of the SOM network model is improved by 4.69%. The proposed improved SOM model method fully combines the network topology and its unique image features and can accurately identify structural damage. This research contributes to the realization of high-precision intelligent health monitoring of damage to modern concrete structures. In addition, it is of great significance for the timely detection, identification and localization of early damage to structures.

Keywords: damage identification; neural network; concrete structure; improved SOM

Citation: Liu, J.; Li, K. Research on an Improved SOM Model for Damage Identification of Concrete Structures. *Appl. Sci.* **2022**, *12*, 4152. https://doi.org/10.3390/app12094152

Academic Editor: Dario De Domenico

Received: 26 March 2022
Accepted: 19 April 2022
Published: 20 April 2022

1. Introduction

Structural damage detection research is one of the most critical research contents in Structural Health Monitoring (SHM) [1–4]. As the relevant technology for structural damage detection, pattern recognition processes various forms of structural damage information to carry out structural damage analysis and is an important part of information science and artificial intelligence. Selecting an intelligent detection method suitable for practical engineering, combining damage indicators with feature-level and decision-level data, thereby simplifying calculation and inference time, and realizing efficient and automated intelligent evaluation are key issues that need further research [5–8].

The neural network has the learning ability to deal with nonlinear problems, strong fault tolerance and robustness [9,10]. Damage identification based on the neural network is based on the physical parameters or dynamic parameters of the structure in different states of health. The parameters sensitive to structural damage are selected as the input of the neural network [11]. The neural network is trained with a large number of damage cases in numerical simulations. Finally, the mature network is trained to realize automatic damage recognition based on the real structural response [12,13]. Scholars choose various pattern recognition techniques for in-depth research on structural damage recognition, such as

Fuzzy Algorithms, Support Vector Machines, etc. [14–16]. Bowen et al. (2021) proposed an integrated framework for data augmentation in a structural health monitoring system using machine learning algorithms [17]. Antonio M et al. (2021) summarize the main methods for detecting, localizing and characterizing damage through algorithms and metrics in structural health monitoring, using electromechanical impedance spectroscopy [18]. Alberto et al. (2021) propose a global methodology for damage detection based on a recently developed version of the Negative Selection Algorithm [19].

With the continuous development of intelligent technology, various network models have achieved good research results and are alternately used in research in various fields [20–22]. Islam M et al. (2022), Roberto et al. (2021) and Hong et al. (2021) used a convolutional neural network to build a network model for structural damage recognition [23–25]. Peng et al. (2021) proposed to construct a back-propagation neural network model in vibration signal analysis [26]. Wang et al. (2021) studied a probabilistic neural network model for damage status assessment of steel truss bridge joints [27]. Barbosa et al. (2021) proposed to carry out research on structural damage identification through a support-vector machine neural network model [28]. Sadeghi et al. (2021) constructed general regression neural network model for damage identification of steel–concrete composite beams [29]. Jersson X et al. (2021) proposed the use of a supervised self-organizing map in structural health monitoring [30]. Fu et al. (2022) used SOM to develop damage pattern recognition and crack propagation prediction [31]. In addition, Sofi, A et al. (2022) comprehensively summarized the application of artificial neural network models in structural health monitoring [32]. At present, the research field of damage identification of concrete structures is facing difficulties. The reason is that, compared with other industries, steel structure and concrete engineering are the products of industrialization, and the degree of mechanization, automation, intelligence and informatization of infrastructure is still relatively backward [33]. The complex constraints faced by engineering structures make the number of damage samples that can be extracted extremely limited, and the identification accuracy cannot meet the needs of engineering [34,35].

A SOM neural network effectively preserves the network topology. It can obtain higher identification accuracy based on the limited damage sample space, effectively reducing the demand for the space sample size for the establishment of an intelligent damage monitoring system [36]. However, the traditional SOM network model algorithm has certain limitations. In the damage identification research, based on the traditional SOM algorithm model, the selected data is normalized by manual measurement and then directly input to the network model as an input sample [37]. However, the extraction process of input samples is often affected by subjective factors, which seriously affects the efficiency of network recognition. At the same time, the core of damage identification research, based on the SOM algorithm model, is to analyze the spatial sample layout form after the network model is learned [38]. This layout form is used as the topology structure of the network model. By continuously reorganizing the arrangement of the topology structure, each weight vector is located in the cluster center of the input vector [39]. When the SOM neural network training is completed, cluster analysis will be performed according to the spatial layout of the training data. Based on this spatial layout, both the topological position of the winning neuron and the spatial distribution of each neuron can be obtained. Therefore, the analysis of the topological spatial layout of neurons directly determines the SOM neural network clustering results. The core purpose of the SOM neural network for damage identification of concrete structures is to realize the classification of damage types, that is, to perform cluster analysis on the spatial structure constructed by damage indicators. However, in the process of damage clustering based on the traditional SOM algorithm model, there are often problems such as unclear damage categories, low damage identification efficiency and inaccurate identification caused by the difficulty in comparing the depth of the topology map. Therefore, in this paper, the research on the improved SOM model algorithm for damage identification of concrete structures is carried out. The self-developed 3D laser scanning system is used to obtain structural damage images and an improved method for constructing input samples of SOM network model based on the gray level co-occurrence matrix and digital feature screening is proposed. Based on the principle of network topology map analysis and the characteristics of grayscale images of topology maps, an improved SOM topology map analysis algorithm is proposed. The improved SOM algorithm model was applied to the bending fatigue test of reactive powder concrete. Based on the improvement of the recognition accuracy and the test effect, the validity of the proposed improved algorithm model is verified.

2. Self-Organizing Map

In order to improve the SOM model, the network structure and network algorithm are analyzed.

2.1. Network Structure

The self-organizing map is an unsupervised feed-forward neural network model, in which neurons compete and cooperate with each other to identify pattern sets. The structure of the SOM model is shown in Figure 1.

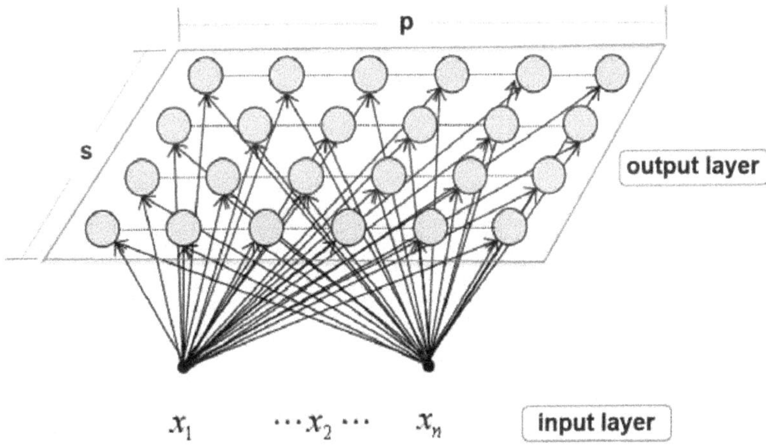

Figure 1. Model of SOM Network Structure.

The input layer consists of n self-organizing neurons (x_1, x_2, x_3, ... , x_{n-1}, x_n). The competition layer consists of a 2D planar (s × p) array of n input vector maps. The network model identifies pattern categories for a given data set by continuously adjusting the connection weights of low-dimensional to high-dimensional network nodes.

2.2. Network Algorithm

As the core of model construction, the network algorithm is the key to the self-organization and the mapping characteristics of the model. The SOM neural network algorithm includes network initialization, input vector setting, etc.

I. **Initialize.** Generally, the weight vector will be given any value in the interval [0, 1], represented by W_i. The learning rate is η.

II. **Set input vector input.** The input vector is the network model training sample:

$$X_n = [x_1, x_2, x_3, \cdots x_{(n-1)}, x_{(n)}]^T \quad (1)$$

III. **Derive Euclidean Distance.** W_{ij} represents the weight between the input layer neuron i, and the mapping layer neuron j. Derive the Euclidean distance between the input vector and the weight vector to get the specific position of the neuron. The Euclidean distance is calculated as:

$$d_i(t) = \|X - W_j\| = \sqrt{\sum_{i=1}^{n} [x_i(t) - w_{ij}(t)]^2} \quad (2)$$

IV. **Label the winning neuron.** The winning neuron position is the position of the neuron with the minimum Euclidean distance between the input vector and the weight vector. The input vector is denoted by X , the winning neuron is denoted by c, Then its calculation formula is:

$$\|X - W_c\| = \min_i \|X - W_c\|, i = 1,2,3,\cdots n-1,n \quad (3)$$

V. **Adjust weights.** Correct the input neuron and the neuron connection weights in the neighborhood according to Equation (3):

$$\Delta w_{ij} = w_{ij}(t+1) - w_{ij}(t) = \eta(t)[x_i(t) - w_{ij}(t)] \quad (4)$$

Among, $\eta(t)$ is the learning rate at t, $\eta(t) \in [0, 1]$, $\eta(t)$ gradually decreases with time, Inversely proportional to t, its expression is:

$$\eta(t) = 0.2 \times (1 - \frac{t}{1000}) \quad (5)$$

VI. Calculate the output value O_k:

$$O_k = f(\min\|X - W_c\|) \quad (6)$$

Among, f represents the function that takes the smallest Euclidean distance.

Determine whether the output results meet the requirements. If the result meets the classification requirements, output the category; if the result does not meet the category requirements, return to step (2) to continue learning until the judgment result is met. Output and end learning.

3. Improved SOM Damage Identification Method

Based on the core steps in the construction of the damage identification network model, the research on the improvement method of the SOM network model is carried out.

3.1. Construction of Damage Identification Model

In order to establish the damage identification model of concrete structure, the structure and algorithm characteristics of the SOM neural network are analyzed according to the performance requirements of the model. Its core steps include the selection of input samples, the setting of network parameters, the judgment of winning neurons and the analysis of topological graphs. Figure 2 is the overall process diagram of the construction method of the damage identification algorithm model for concrete structures.

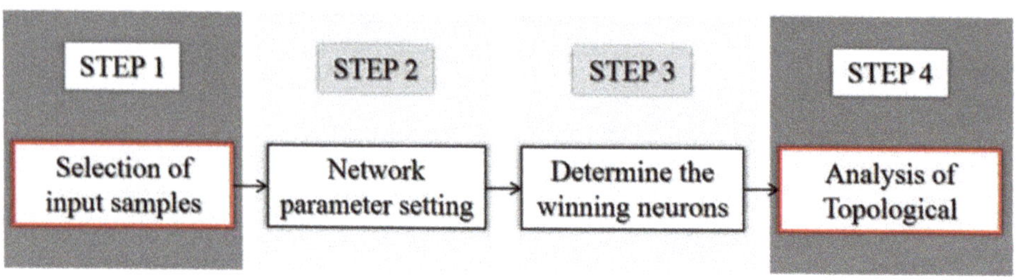

Figure 2. The whole process of model building.

Figure 2 shows the whole process of structural damage identification based on the SOM neural network, which is the core framework system of the research. Analysis of Figure 2 shows that the selection of input samples is the first step in the construction of the network model, which directly determines the network structure and is an important factor affecting the efficiency of network operation. As the last key step of the network model, the analysis of the topology structure directly determines the specific category of each neuron and is a key factor affecting the accuracy of network recognition. Therefore, in order to improve the recognition performance of the SOM network for concrete structure damage, the research will mainly focus on these two parts to improve the SOM neural network model.

3.2. SOM Improvement Method

a. Selection of input samples

In order to reduce the interference of complex factors such as environment and humans, a method based on machine vision is proposed to obtain input samples. The damage signal is collected based on the vision sensor, and the initial sample is extracted by the feature extraction algorithm. In order to reduce the requirement for the number of input samples, the input samples that can effectively characterize the damage characteristics are automatically screened based on statistical theory. The improved SOM model and its input sample selection process is shown in Figure 3.

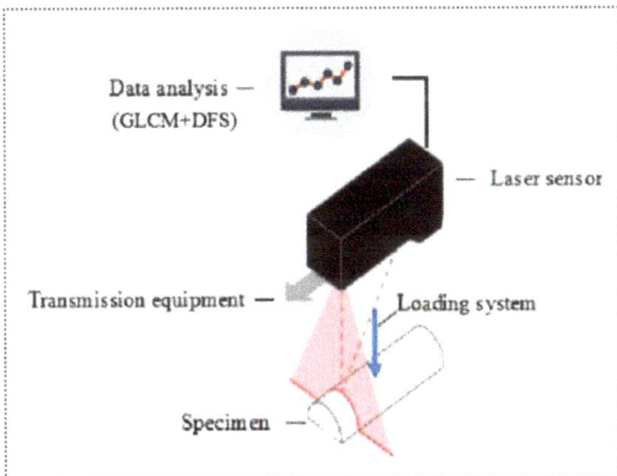

Figure 3. Improved method for selecting input samples.

An input sample for constructing a damage identification network model based on 3D laser scanning technology is proposed, as shown in Figure 3. First, a 3D image of the specimen under the loading system is acquired by adding 1D transmission equipment to a 2D laser sensor. Then, the initial samples are extracted by constructing the gray level co-occurrence matrix (GLCM) of structural damage. Finally, in order to further improve the effectiveness of the damaged samples, based on the digital feature screening (DFS), the feature parameters are selected as the input samples of the network model.

b. Analysis of topology map

In order to accurately identify the damage category information contained in the topology map image, according to the characteristics of the topology map image, a network model optimization method, the topology grayscale (TOP-G) algorithm, is proposed. Figure 4 shows the flow of the TOP-G algorithm.

- The first step is to determine the grayscale of the topology map:

First, determine the number L of connection polygons between neurons. The gray level of the topology map is determined according to the number of L, and the gray value range of the pixels in the topology map should be $[0, L]$. Thus, it is judged that the gray level of the image is $g = L = 2^n$, and it is deduced that $n = log_2 L$, $g = 2^{[n]}$, where $[n]$ represents the value of n is the smallest integer that exceeds the value of n. Then the obtained $g = 2^{[n]}$ is the gray level of the topology map.

- The second step is to grayscale the topological distance map:

According to the gray level of the topological map, the topological distance color image is converted into a grayscale image, which is called a topological grayscale map.

- The third step is to create a sliding window:

Suppose the number of neurons is m, create a sliding window, label L_1–L_m and assign grayscale values g_1–g_m to each neighborhood polygon in the topological grayscale map.

- The fourth step is to discriminate the category of neurons:

The gray values gi of all neighboring neurons of the unknown neuron i are extracted, compared and sorted. Determine whether g_i is the largest gray value in the neighborhood. If so, the neuron connecting the neighborhood polygon is a class, and the output i belongs to this class; If not, rejudge until the attribution category of all unknown neurons is determined, and the result is output.

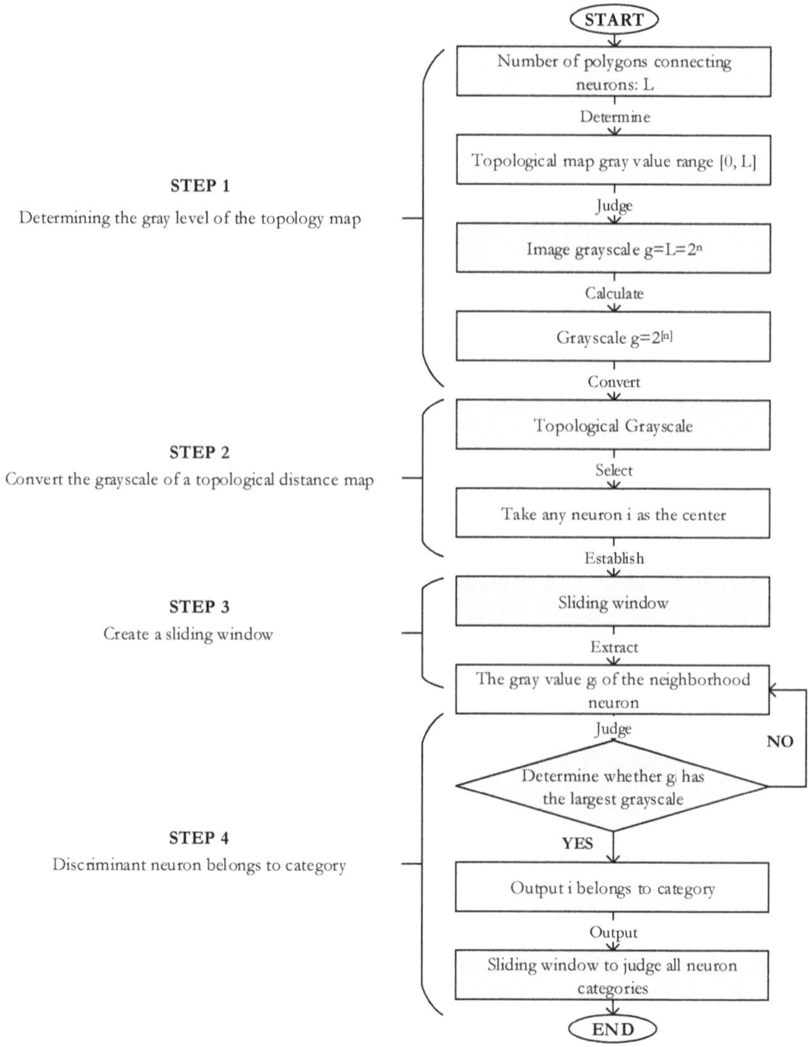

Figure 4. The flow of the TOP-G algorithm.

4. Experiments and Results Analysis

In order to verify that the improved SOM neural network model can effectively improve the recognition accuracy of the network model, based on the self-developed 3D laser scanning system, a network model for the recognition of bending fatigue damage of reactive powder concrete was established.

4.1. Selection of Input Samples of RPC Bending Fatigue Damage Identification Model

There is no obvious change in the appearance of the specimen before loading in the RPC bending fatigue test. When the loading force reaches 70% to 80% of the ultimate bending strength, initial cracks appear in the mid-span accompanied by the sound of steel fibers being pulled out, and damage images are obtained during this process. It was observed that the flexural strength value did not decrease with the occurrence of mid-span cracks in the specimen until the steel fibers in the crack section were completely pulled out, and the specimen lost its bearing capacity and declared failure. With the development of the experimental phenomenon, a three-dimensional model of the concrete specimen was obtained. Figure 5 shows the entire process of acquiring images during 3D damaged specimen loading, and Figure 6 is the obtained three-dimensional model of microcrack damage.

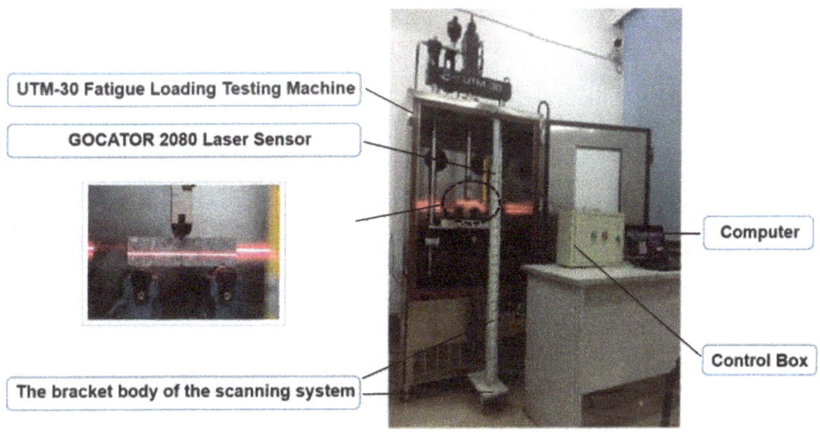

Figure 5. Damage 3D image acquisition system.

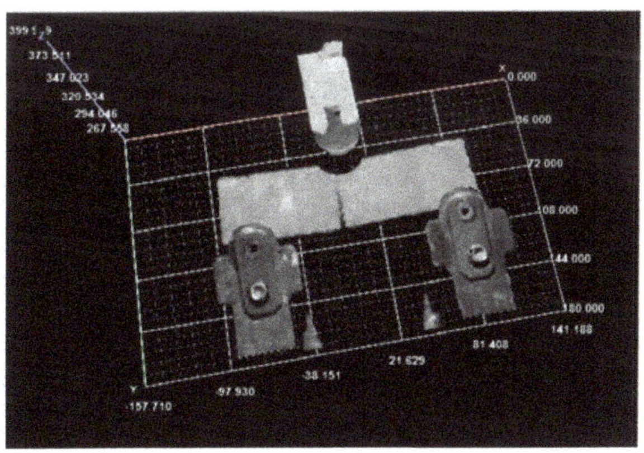

Figure 6. 3D model of the model of micro-crack damage.

In order to reduce the amount of data and improve the recognition efficiency, redundant information is removed based on a 3D point cloud projection algorithm and median filtering. Based on the GLCM, the damage model input samples are extracted. The image gray level g = 128 is constructed, the generation step size is d = 1, and the generation direction θ takes the gray level co-occurrence matrix of 0°, 45°, 90° and 135°. The 14 feature parameters such as angular second moment and correlation are extracted. In order to improve the quality of the input samples, the $\boldsymbol{P_1}$ (angular Second Moment), $\boldsymbol{P_2}$ (entropy), $\boldsymbol{P_3}$ (inertia moment), $\boldsymbol{P_4}$ (correlation), $\boldsymbol{P_5}$ (inverse difference moment), and $\boldsymbol{P_6}$ (variance) are screened out as standard samples based on the DFS method. Table 1 shows the damage texture properties represented by the input sample.

Table 1. Damage texture properties represented by the input sample.

Input Sample	Sample Name	Characterized Properties
P_1	ASM	Uniformity
P_2	ENT	Complexity
P_3	INM	Stability
P_4	COR	Correlation
P_5	IDM	Volatility
P_6	VAR	Circularity

By analyzing Table 1, the damage texture features represented by each input sample can be clearly grasped

4.2. Parameter Setting of RPC Bending Fatigue Damage SOM Network Model

Figure 7 shows the structure of the network model for the flexural fatigue damage identification of RPC.

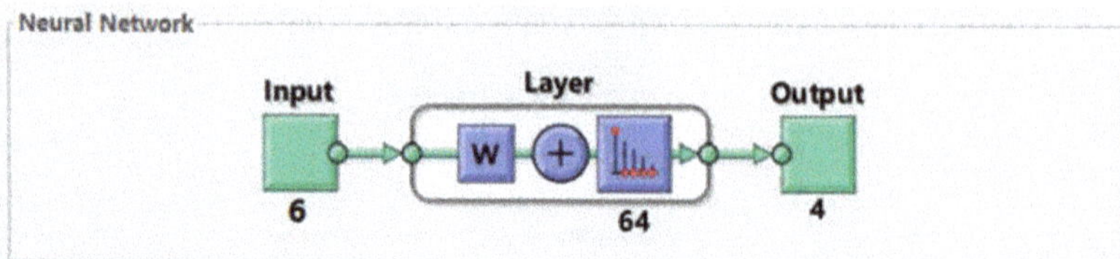

Figure 7. The structure of the network model.

As shown in Figure 7, the selected six feature parameters are used as the SOM network input vector $[P_1, P_2, P_3, P_4, P_5, P_6]$, the SOM network competition layer is set to 8 × 8 = 64 neurons and the network model output is four categories of damage.

In the parameter setting of the SOM network model for concrete structure damage, the number of training steps directly affects the network clustering performance. In order to improve the clustering efficiency, the optimal number of training steps is obtained. After determining the structure of the network model, select different steps for training and observe the performance changes of the network model. Using the step increment as a variable, analyze the clustering results of the network model. The statistics are shown in Table 2 for the clustering results under different training steps. Selecting the training steps with the fewest steps can not only satisfy the sample classification, but also ensure the clustering speed.

Table 2. Clustering results for different training steps.

Training Steps	Clustering Results			
	Honeycomb	Hole	Sag	Crack
10	55	37	37	55
50	43	37	37	55
100	43	1	37	37
200	**49**	**1**	**16**	**64**
500	49	1	16	64
1000	49	1	16	64

When the number of training steps is set to 10, 50, 100, 200, 500 and 1000, the classification effect of the network model is shown in Table 1. When the number of training steps is 10, the damage diagnosis model is initially established, and the damage is divided into two categories; as the number of training steps increases, when the number of training steps is 50 and 100, the recognition accuracy is further improved, and the injuries are divided into three categories; when the number of training steps reaches 200, the four injury types are completely distinguished; continue to increase the number of training steps to 500 and 1000, and the damage classification results are the same, which is not practical. Therefore, 200 training steps were chosen as the optimal value for the damage identification model.

4.3. Determining the Winning Neurons of RPC Bending Fatigue Damage Model

In order to further verify the accuracy of acquiring neurons when the number of training steps is 200, the topology map of the winning neuron positions of the damage type is output, as shown in Figure 8.

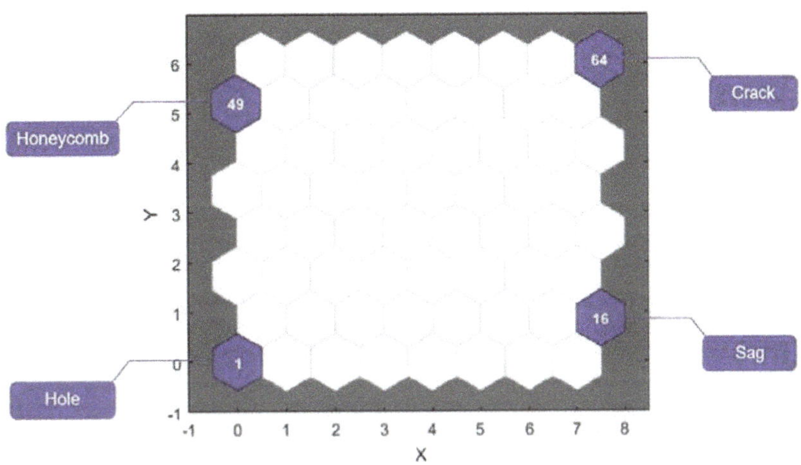

Figure 8. Winning neuron topology.

Figure 8 shows the topological structure of the winning neurons under the optimal number of steps. The gray–blue hexagons in the figure are the topological positions of the winning neurons, which shows that the types of damage are clearly distinguished. Combined with the topological location map information of the winning neurons in Table 1 and Figure 8, it is inferred that honeycombs, holes, sags and cracks correspond to winning neurons numbered 49, 1, 16 and 64, respectively. From the obtained topological positions and the number of winning neurons, the basis for the cluster analysis of the network model is basically obtained. However, further analysis of the network model is required to obtain the specific damage type for each neuron.

4.4. Neuron Topology Analysis for RPC Bending Fatigue Damage SOM Network Model

In order to obtain the damage category information corresponding to each neuron, the clustering results of the network model were analyzed. Obtain the topological structure distance map of the structure damage identification network model, as shown in Figure 9. The small gray squares in the figure represent neurons and the straight lines between them represent straight-line connections between neurons. The distance between neurons is obtained by the Euclidean distance formula. The hexagons connect the neurons, with the color depth representing the distance between neurons. The colors are from dark to light, indicating that the distance between neurons is from far to near. It can be inferred that the neurons with a light color have high similarity, and the difference between them is low; while the neurons with a dark color have low similarity, and the difference between them is large.

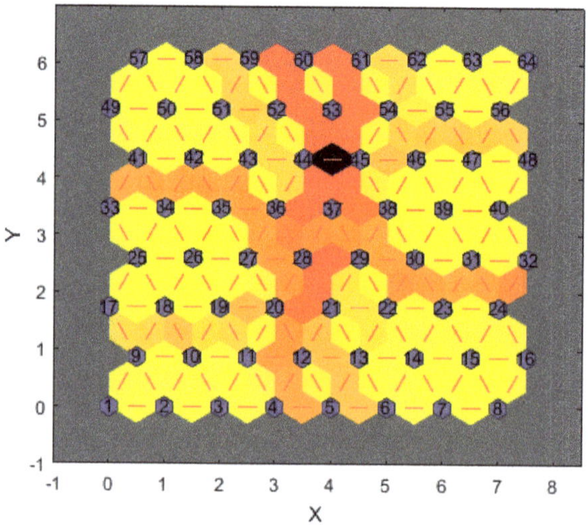

Figure 9. Topological distance graph of neurons.

Based on the analysis of the traditional comparison method, among the 64 input neurons of the damage identification network model, the damage types of 61 neurons correspond to the damage types of the standard samples. For example, for neurons 36, 41, 42, 43, 44, 50, 51, 52, 57, 58 and 59, their damage types may correspond to the winning neuron number 49, which corresponds to the honeycomb damage type. However, neuron 37 is between the hole and sag damage states, and is far away from neurons 53 and 60, corresponding to other unknown damage types. In order to clearly present the damage type corresponding to each neuron, a corresponding relationship table between the damage types and sample classification numbers is constructed as shown in Table 3.

Table 3. Correspondence between damage types and samples.

Damage Type	Sample Classification Number
Honeycomb	36, 41, 42, 43, 44, **49**, 50, 51, 52, 57, 58, 59
Hole	**1**, 2, 3, 4, 9, 10, 11, 17, 18, 19, 20, 25, 26, 27, 33, 34, 35
Sag	5, 6, 7, 8, 11, 12, 13, 14, 15, **16**, 21, 22, 23, 24, 29
Crack	30, 31, 32, 38, 39, 40, 45, 46, 47, 48, 54, 55, 56, 61, 62, 63, **64**
Unknown type	**37, 53, 60**

By analyzing Table 2, it can be seen that the model can obtain the damage classification of almost all neurons, and the recognition accuracy rate is as high as 95.31%. However, there are still cases where unknown neurons cannot be associated with their type of injury. In order to further improve the recognition accuracy of the network model, the TOP-G algorithm is used to determine the type of unknown neuron damage. The analysis process is shown in Figure 10.

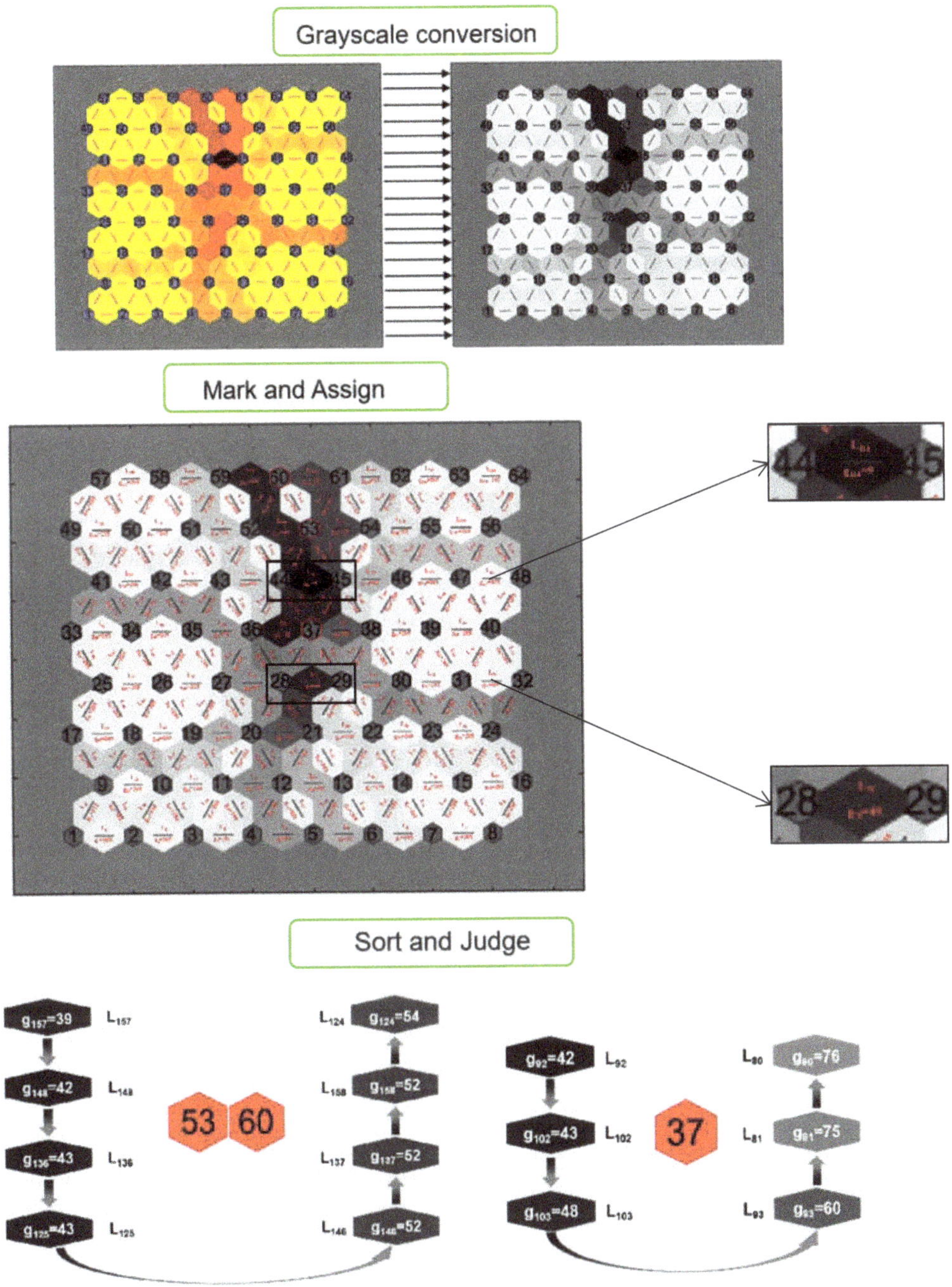

Figure 10. Determining unknown neuron category based on TOP-G algorithm.

First, determine the gray level of the topology map, $L = 161$, and the gray value range is [0, 161], then the gray level of the topology map image is $g = 2$ [8]; Then, at this grayscale, the topological distance color map is converted to a grayscale image; Create a sliding window, mark the polygon as L_1–L_{161}, assign the gray value g_1–g_{161}; Finally, the gray values are sorted from large to small, and the judgment is made according to the sorting result. Neurons 53 and 60 are connected by a neighborhood polygon number 147, with a grayscale value of 205, which can be seen as a class. The gray values of the eight neighborhood polygons are sorted by gray value, and the neighborhood polygon No.124 has the largest gray value, which connects the neuron number 45 and the neuron number 53. Therefore, it is determined that neurons 53 and 60 correspond to crack damage. Neuron 37 corresponds to six neighborhood polygons, and their gray levels are sorted. The gray value of the polygon in the neighborhood of No. 80 is the largest, which is connected to the neuron of No. 24, corresponding to sag damage. Therefore, it was judged that neuron No. 37 corresponds to sag damage. Therefore, all unknown neuron damage categories are determined based on the topology grayscale algorithm, which further improves the network identification accuracy. Compared with the traditional SOM model, the identification accuracy was improved by 4.69%.

4.5. Testing of Improved Algorithm Models

In order to further verify the detection effect of the improved SOM neural network model, the classification results of the detection samples were obtained. The classification labels and sample numbers of winning neurons corresponding to honeycombs, holes, sags and cracks are shown in Table 4. Figure 11 shows the classification results of the detected samples.

Table 4. Test sample.

Damage Type	Winning Neuron Classification Label	Sample Serial Number
Cracks	38	1, 2, 3, 4, 5, 6, 7, 8, 9, 10
Holes	16	11, 12, 13, 14, 15, 16, 17, 18, 19, 20
Honeycombs	23	21, 22, 23, 24, 25, 26, 27, 28, 29, 30
Sags	7	31, 32, 33, 34, 35, 36, 37, 38, 39, 40

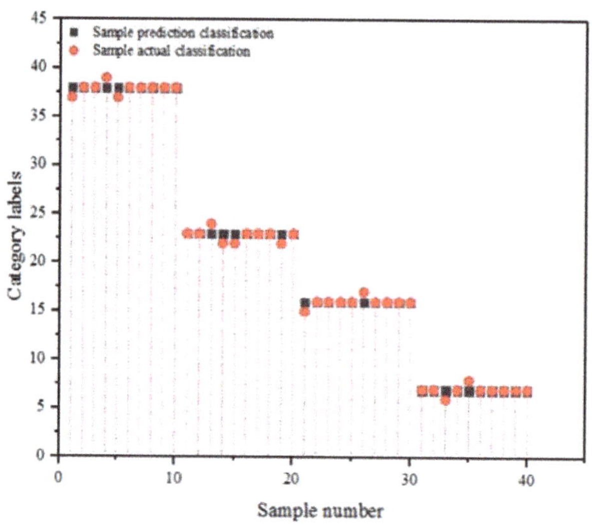

Figure 11. Classification results of the detected samples.

According to the analysis of Table 3 and Figure 11, the winning neuron corresponding to the test sample is consistent with the actual damage category, and the test sample corresponds to the actual sample type. The damage type corresponding to each sample can be detected based on the improved SOM neural network model.

5. Discussion

The core content of this paper is the improved of SOM algorithm model in structural damage identification. The ultimate goal of structural health monitoring research is to detect damage as early as possible in order to provide appropriate measures to avoid disaster. It is worth noting that the research object of this paper is micro-damage, and the size of the damage is usually less than 1 mm, which mainly depends on the accuracy of the image acquisition device (laser ranging sensor).

Therefore, the significance of this research is not limited to providing an improved SOM neural network model with a higher recognition accuracy based on a small number of samples. Research can help to effectively identify and even detect and locate damage information in the budding stage of damage, which is of great significance for the timely detection of early structural damage.

6. Conclusions

Taking the four core steps of constructing the SOM concrete structure damage identification network model as the main line, the network SOM algorithm improvement research is carried out and the following conclusions are obtained:

- Combined with the self-developed 3D laser scanning system and GLCM theory, the input sample selection method of the SOM network is improved;
- Based on the principle of the network topology map analysis and its image characteristics, the concept of the topology grayscale map and the TOP-G algorithm method, and process for the SOM topology map analysis are proposed for the first time;
- Based on the active powder concrete bending fatigue loading test, the damage (cracks, sags, honeycombs and holes) identification research of the improved SOM algorithm model was carried out.

Author Contributions: Conceptualization, J.L.; Data curation, J.L.; Formal analysis, J.L.; Funding acquisition, J.L.; Investigation, J.L.; Methodology, J.L.; Project administration, J.L.; Resources, J.L.; Software, J.L.; Supervision, K.L.; Validation, K.L.; Visualization, J.L.; Writing—original draft, J.L.; Writing—review & editing, K.L. All authors have read and agreed to the published version of the manuscript.

Funding: This research was funded by PhD research start-up project of Beihua University "Research on Fault Diagnosis of Building Structure Based on Improved SOM Neural Network", grant number 160321009.

Institutional Review Board Statement: Not applicable.

Informed Consent Statement: Not applicable.

Data Availability Statement: The data presented in this study are available from the corresponding author.

Conflicts of Interest: The authors declare no conflict of interest.

References

1. Irfan, M.S.; Khan, T.; Hussain, T.; Liao, K.; Umer, R. Carbon coated piezoresistive fiber sensors: From process monitoring to structural health monitoring of composites—A review. *Compos. Part A Appl. Sci. Manuf.* **2021**, *141*, 106236. [CrossRef]
2. Clementi, F.; Formisano, A.; Milani, G.; Ubertini, F. Structural Health Monitoring of Architectural Heritage: From the past to the Future Advances. *Int. J. Archit. Herit.* **2021**, *15*, 1–4. [CrossRef]
3. Mishra, M.; Lourenço, P.B.; Ramana, G.V. Structural health monitoring of civil engineering structures by using the internet of things: A review. *J. Build. Eng.* **2021**, *48*, 103954. [CrossRef]
4. Farrar, C.R.; Worden, K. An introduction to structural health monitoring. *Philos. Trans. R. Soc. A Math. Phys. Eng. Sci.* **2007**, *365*, 303–315. [CrossRef] [PubMed]

5. Gerhardt, L. Pattern recognition and machine learning. *IEEE Trans. Autom. Control* **2003**, *19*, 461–462. [CrossRef]
6. Kot, P.; Muradov, M.; Gkantou, M.; Kamaris, G.S.; Hashim, K.; Yeboah, D. Recent Advancements in Non-Destructive Testing Techniques for Structural Health Monitoring. *Appl. Sci.* **2021**, *11*, 2750. [CrossRef]
7. Sony, S.; Dunphy, K.; Sadhu, A.; Capretz, M. A systematic review of convolutional neural network-based structural condition assessment techniques. *Eng. Struct.* **2020**, *226*, 111347. [CrossRef]
8. Sakiyama, F.I.H.; Lehmann, F.A.; Garrecht, H. Structural Health Monitoring of Concrete Structures using Fibre Optic Based Sensors: A Review. *Mag. Concr. Res.* **2019**, *73*, 174–194. [CrossRef]
9. Wu, Z.; Pan, S.; Chen, F.; Long, G.; Zhang, C.; Philip, S.Y. A Comprehensive Survey on Graph Neural Networks. *IEEE Trans. Neural Netw. Learn. Syst.* **2019**, *32*, 4–24. [CrossRef]
10. Hou, R.; Xia, Y. Review on the new development of vibration-based damage identification for civil engineering structures: 2010–2019. *J. Sound Vib.* **2020**, *491*, 115741. [CrossRef]
11. Wang, S.; Li, J.; Luo, H.; Zhu, H. Damage identification in underground tunnel structures with wavelet based residual force vector. *Eng. Struct.* **2019**, *178*, 506–520. [CrossRef]
12. Alavi, A.H.; Hasni, H.; Lajnef, N.; Chatti, K.; Faridazar, F. Damage detection using self-powered wireless sensor data: An evolutionary approach. *Measurement* **2016**, *82*, 254–283. [CrossRef]
13. Martinez-Luengo, M.; Kolios, A.; Wang, L. Structural health monitoring of offshore wind turbines: A review through the Statistical Pattern Recognition Paradigm. *Renew. Sustain. Energy Rev.* **2016**, *64*, 91–105. [CrossRef]
14. Won, J.; Park, J.W.; Jang, S.; Jin, K.; Kim, Y. Automated Structural Damage Identification Using Data Normalization and 1-Dimensional Convolutional Neural Network. *Appl. Sci.* **2021**, *11*, 2610. [CrossRef]
15. Abdeljaber, O.; Avci, O.; Kiranyaz, M.S.; Boashash, B.; Sodano, H.; Inman, D.J. 1-D CNNs for Structural Damage Detection: Verification on a Structural Health Monitoring Benchmark Data. *Neurocomputing* **2018**, *275*, 1308–1317. [CrossRef]
16. Lin, C.J.; Wu, N.J. An ANN Model for Predicting the Compressive Strength of Concrete. *Appl. Sci.* **2021**, *11*, 3798. [CrossRef]
17. Tan, X.; Sun, X.; Chen, W.; Du, B.; Ye, J.; Sun, L. Investigation on the data augmentation using machine learning algorithms in structural health monitoring information. *Struct. Health Monit.* **2021**, *20*, 2054–2068. [CrossRef]
18. Tenreiro, A.F.G.; Lopes, A.M.; da Silva, L.F. A review of structural health monitoring of bonded structures using electromechanical impedance spectroscopy. *Struct. Health Monit.* **2022**, *21*, 228–249. [CrossRef]
19. Barontini, A.; Masciotta, M.G.; Amado-Mendes, P.; Ramos, L.F.; Lourenço, P.B. Negative selection algorithm based methodology for online structural health monitoring. *Eng. Struct.* **2021**, *229*, 111662. [CrossRef]
20. Pathirage, C.S.N.; Li, J.; Li, L.; Hao, H.; Liu, W.; Ni, P. Structural damage identification based on autoencoder neural networks and deep learning. *Eng. Struct.* **2018**, *172*, 13–28. [CrossRef]
21. Bao, Y.; Tang, Z.; Li, H.; Zhang, Y. Computer vision and deep learning–based data anomaly detection method for structural health monitoring. *Struct. Health Monit.* **2019**, *18*, 401–421. [CrossRef]
22. Rajadurai, R.S.; Kang, S.T. Automated Vision-Based Crack Detection on Concrete Surfaces Using Deep Learning. *Appl. Sci.* **2021**, *11*, 5229. [CrossRef]
23. Mantawy, I.M.; Mantawy, M.O. Convolutional neural network based structural health monitoring for rocking bridge system by encoding time-series into images. *Struct. Control Health Monit.* **2022**, *29*, e2897. [CrossRef]
24. Miorelli, R.; Fisher, C.; Kulakovskyi, A.; Chapuis, B.; Mesnil, O.; D'Almeida, O. Defect sizing in guided wave imaging structural health monitoring using convolutional neural network. *NDT E Int.* **2021**, *122*, 102480.
25. Zhang, B.; Hong, X.; Liu, Y. Deep Convolutional Neural Network Probability Imaging for Plate Structural Health Monitoring Using Guided Waves. *IEEE Trans. Instrum. Meas.* **2021**, *70*, 1–10. [CrossRef]
26. Gu, J.; Peng, Y.; Lu, H.; Chang, X.; Cao, S.; Chen, G.; Cao, B. An optimized variational mode decomposition method and its application in vibration signal analysis of bearings. *Struct. Health Monit.* **2021**, 1–22. [CrossRef]
27. Zhan, J.; Wang, C.; Fang, Z. Condition Assessment of Joints in Steel Truss Bridges Using a Probabilistic Neural Network and Finite Element Model Updating. *Sustainability* **2021**, *13*, 1474. [CrossRef]
28. Finotti, R.P.; Barbosa, F.d.S.; Cury, A.A.; Pimentel, R.L. Numerical and Experimental Evaluation of Structural Changes Using Sparse Auto-Encoders and SVM Applied to Dynamic Responses. *Appl. Sci.* **2021**, *11*, 11965. [CrossRef]
29. Sadeghi, F.; Yu, Y.; Zhu, X.; Li, J. Damage identification of steel-concrete composite beams based on modal strain energy changes through general regression neural network. *Eng. Struct.* **2021**, *244*, 112824. [CrossRef]
30. Angulo-Saucedo, G.A.; Leon-Medina, J.X.; Pineda-Muñoz, W.A.; Torres-Arredondo, M.A.; Tibaduiza, D.A. Damage Classification Using Supervised Self-Organizing Maps in Structural Health Monitoring. *Sensors* **2022**, *22*, 1484. [CrossRef]
31. Sun, J.; Chen, X.; Fu, Z.; Lacidogna, G. Damage Pattern Recognition and Crack Propagation Prediction for Crumb Rubber Concrete Based on Acoustic Emission Techniques. *Appl. Sci.* **2021**, *11*, 11476. [CrossRef]
32. Sofi, A.; Regita, J.J.; Rane, B.; Lau, H.H. Structural health monitoring using wireless smart sensor network—An overview. *Mech. Syst. Signal Process.* **2022**, *163*, 108113. [CrossRef]
33. Yapar, O.; Basu, P.K.; Volgyesi, P.; Ledeczi, A. Structural health monitoring of bridges with piezoelectric AE sensors. *Eng. Fail. Anal.* **2015**, *56*, 150–169. [CrossRef]
34. Kang, F.; Liu, J.; Li, J.; Li, S. Concrete dam deformation prediction model for health monitoring based on extreme learning machine. *Struct. Control Health Monit.* **2017**, *24*, e1997. [CrossRef]

35. Amezquita-Sanchez, J.P.; Adeli, H. Signal Processing Techniques for Vibration-Based Health Monitoring of Smart Structures. *Arch. Comput. Methods Eng.* **2016**, *23*, 1–15. [CrossRef]
36. Zhang, Y.; Simsek, M.; Kantarci, B. Empowering Self-Organized Feature Maps for AI-Enabled Modeling of Fake Task Submissions to Mobile Crowdsensing Platforms. *IEEE Internet Things J.* **2020**, *8*, 1334–1346. [CrossRef]
37. Li, K.; Wang, J.; Qi, D. Research on crack identification and monitoring method of concrete structure based on G-S-G. *J. Vib. Shock.* **2020**, *39*, 101–108.
38. Czarnecki, S.; Shariq, M.; Nikoo, M.; Sadowski, Ł. An intelligent model for the prediction of the compressive strength of cementitious composites with ground granulated blast furnace slag based on ultrasonic pulse velocity measurements. *Measurement* **2021**, *172*, 108951. [CrossRef]
39. Li, K.; Wang, J.; Qi, D. The development and application of an original 3D laser scanning: A precise and nondestructive structural measurements system. *Frat. Integrità Strutt.* **2020**, *14*, 386–397. [CrossRef]

symmetry

Article

Identification of Abnormal Vibration Signal of Subway Track Bed Based on Ultra-Weak FBG Sensing Array Combined with Unsupervised Learning Network

Sheng Li [1], Yang Qiu [2], Jinpeng Jiang [1], Honghai Wang [1], Qiuming Nan [1,*] and Lizhi Sun [3]

1 National Engineering Research Center of Fiber Optic Sensing Technology and Networks, Wuhan University of Technology, Wuhan 430070, China; lisheng@whut.edu.cn (S.L.); jiangjp2812@whut.edu.cn (J.J.); wanghh@whut.edu.cn (H.W.)
2 School of Information Engineering, Wuhan University of Technology, Wuhan 430070, China; qiuyang@whut.edu.cn
3 Department of Civil and Environmental Engineering, University of California, Irvine, CA 92697, USA; lsun@uci.edu
* Correspondence: nqm0723@whut.edu.cn

Abstract: The performance of the passing train and the structural state of the track bed are the concerns regarding the safe operation of subways. Monitoring the vibration response of the track bed structure and identifying abnormal signals within it will help address both of these concerns. Given that it is difficult to collect abnormal samples that are symmetric to those of the normal state of the structure in actual engineering, this paper proposes an unsupervised learning-based methodology for identifying the abnormal signals of the track beds detected by the ultra-weak fiber optic Bragg grating sensing array. For an actual subway tunnel monitoring system, an unsupervised learning network was trained by using a sufficient amount of vibration signals of the track bed collected when trains passed under normal conditions, which was used to quantify the deviations caused by anomalies. An experiment to validate the proposed procedures was designed and implemented according to the obtained normal and abnormal samples. The abnormal vibration samples of the track beds in the experiment came from two parts and were defined as three levels. One part of it stemmed from the vibration responses under the worn wheels of a train detected during system operation. The remaining abnormal samples were simulated by superimposing perturbations in the normal samples. The experimental results demonstrated that the established unsupervised learning network and the selected metric for quantifying error sequences can serve the threshold selection well based on the receiver operating characteristic curve. Moreover, the discussion results of the comparative tests also illustrated that the average results of accuracy and F1-score of the proposed network were at least 11% and 13% higher than those of the comparison networks, respectively.

Keywords: signal anomaly detection; subway track bed; distributed vibration; unsupervised learning network; attention mechanism; ultra-weak fiber optic Bragg grating

Citation: Li, S.; Qiu, Y.; Jiang, J.; Wang, H.; Nan, Q.; Sun, L. Identification of Abnormal Vibration Signal of Subway Track Bed Based on Ultra-Weak FBG Sensing Array Combined with Unsupervised Learning Network. *Symmetry* **2022**, *14*, 1100. https://doi.org/10.3390/sym14061100

Academic Editor: Igor V. Andrianov

Received: 10 May 2022
Accepted: 25 May 2022
Published: 27 May 2022

Publisher's Note: MDPI stays neutral with regard to jurisdictional claims in published maps and institutional affiliations.

1. Introduction

Generally speaking, in-service engineering structures are always in two symmetrical operating states, normal and abnormal. Although the probability of occurrence of the structural abnormal state is relatively low, tracking and monitoring the service status of subway track beds before catastrophic accidents is of great significance to ensure the safe operation of trains. The traditional inspection regime is usually labor-intensive and can be significantly expensive for rail operation management [1]. Although various types of rail inspection vehicles integrated with ultrasonic methods [2], eddy current [3], infrared thermography [4], laser scanning [5], and other non-destructive testing equipment have improved the efficiency of inspection, they still have difficulty meeting the frequent inspection needs of the entire subway line. The non-contact monitoring method based on machine vision combined with diverse deep learning models has improved the intelligence of the detection of track bed issues to a certain extent [6,7]. However, this approach needs to rely on sufficient samples to perform supervised learning-based training. The performance usually only meets the local detection, which is susceptible to the interior environment of the subway tunnel.

By deploying the ultra-weak fiber optic Bragg grating (FBG) array cable along the track bed in a subway tunnel, the structural vibration response caused by passing trains through the monitoring area can be gleaned. Given the relationship between the vibration response and the structural state, Li et al. [8] reported that the similarity of the vibration responses detected by the ultra-weak FBG array cable can be utilized to evaluate the structural state of the subway track bed. Gan et al. [9] presented that ground intrusion events above subway lines can be located and identified based on distributed vibration responses of ultra-weak FBG sensing arrays. Moreover, the research on train tracking and personnel intrusion identification based on distributed vibration response also illustrated that the ultra-weak FBG sensing array can effectively collect the vibration of the subway track bed [10]. Thus, despite the asymmetry existing in the data amount collected for the abnormal and normal states, it is possible to discover the security risks of the track bed or the passing train by analyzing the abnormal vibration signal of the track bed structure monitored by the ultra-weak FBG sensing array.

In the research area of abnormal signal detection, Tao et al. [11] completed the fault detection of the wind turbine by using the gray correlation algorithm combined with the support vector regression model. Qiu et al. [12] proposed that abnormal patterns of vibration response of a real long-span cable-stayed bridge can be distinguished based on statistical feature dimensionality reduction, forward difference data augmentation, and random forest. Li et al. [13] designed a deep convolutional neural network to classify and identify simulated damage signals for a scaled-down bridge model. Tang et al. [14] proposed a data anomaly detection method based on a convolutional neural network combined with a computer vision technique. Li et al. [15] discussed the degradation failure detection for the mechanical equipment in a data-driven manner through deep neural networks. Abid et al. [16] proposed a deep-SincNet-based learning network that was able to automatically learn fault features from the motor current. However, the above-mentioned studies primarily focus on the field of supervised learning, whose performance depends heavily on the support of adequate and balanced samples with labels. Given the complexity of accumulating a certain scale of anomalous signals with labels in practical engineering, current supervised learning-based methods are often difficult to transplant and apply directly, especially when the types of training samples are extremely asymmetric.

In the study of anomaly detection based on unsupervised learning, Hautamaki et al. [17] employed the idea of the K-neighbor algorithm and used the average distance of each sample to its K nearest sample points as a reference to discriminate abnormal samples. He et al. [18] performed cluster-based anomaly detection with predefined outliers. Li et al. [19] conducted anomaly detection for multivariate time series based on clustering through complex hyperparameter setting. Saari et al. [20] adopted the one-class support vector machine (SVM) to model the density distribution of samples to determine whether the data was abnormal. Zong et al. [21] completed the anomaly data detection based on a deep autoencoding Gaussian mixture model (DAGMM), which was verified by four public benchmark datasets. Purohit et al. [22] further extended the DAGMM method to anomaly detection in acoustic signals. Pei et al. [23] proposed a method for network traffic anomaly detection through self-coding of long short-term memory (LSTM) networks. Malhotra et al. [24] revealed that the LSTM-based encode–decode can detect the abnormal data of the sensor. Although these unsupervised learning-based methods have demonstrated their respective effectiveness in specific scenarios, there are few research cases involving high-dimensional time series. Hence, the existing methods reviewed may not be suitable for capturing the time dependence of high-dimensional vibration signals of subway track beds.

To address the previous research deficiencies, the motivation of this paper is to propose a method for identifying abnormal vibration signals of track beds that does not rely on the idea of supervised learning. During the implementation of the method, an unsupervised learning network used to learn the features of normal vibration signals of subway track bed structures in terms of the time-series correlation of vibration sequences was established. Based on the designed network, it is possible to quantify the results of the predicted error sequences due to abnormal input signals. Then, the identification of abnormal vibration responses of track bed structures can be performed based on the appropriate threshold derived from the quantification results. The experimental results of identifying abnormal signals based on the proposed procedure in an actual subway were reported. The procedure for identifying anomalous vibration sequences detected by the ultra-weak FBG array makes up the second part of this paper, followed by the experimental details of the design and arrangement used to validate the proposed method. Finally, the effectiveness of identifying the target signals is assessed based on the accuracy and F1-score. Furthermore, the performance superiority of the selected indicator used for quantifying prediction error sequences and the proposed unsupervised learning network are discussed through comparative tests.

2. Procedure for Abnormal Signal Identification

As shown in Figure 1, the proposed procedure for abnormal signal identification primarily includes three steps: unsupervised learning network establishment based on a training dataset of the normal state of the track bed, calculation quantification for prediction errors, and the selection of a threshold for identifying anomalies. The premise of the proposed procedure is that a large number of track bed vibration signals can be collected when passing trains under normal conditions. Under this premise, an unsupervised learning network with satisfactory accuracy can be established for subsequent error calculation and quantification. As well as that, it is easy to retain enough normal samples to participate in setting the threshold for identifying anomalies.

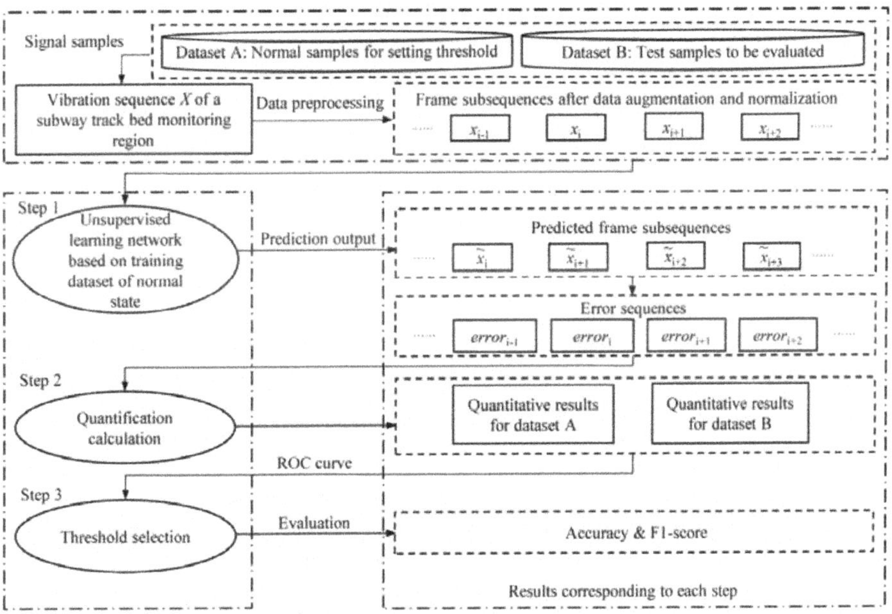

Figure 1. The proposed abnormal signal identification process.

Once the unsupervised learning network is established, the identification procedure first requires data preprocessing on the retained normal dataset A and dataset B to be evaluated. Each vibration sample X from dataset A or B will in turn undergo data augmentation and normalization before being put into the unsupervised learning network. Data augmentation is performed by directly dividing the original input signal into equal subsequences. To eliminate the amplitude differences in the subsequences, each input from data augmentation is normalized to a range of −1–1. In the first step, the predicted frame $\widetilde{x}_{i+1}$ of the frame x_i is obtained through the established unsupervised learning network, where x_i is a certain frame subsequence preprocessed by data augmentation and normalization. Then, the prediction $error_{i+1}$ for each sampling frame subsequence can be generated by Equation (1) from the result of each sampling point of the subsequence.

$$error_{i+1} = \sqrt{\sum_{j=1}^{n} \left(x_{i+1}^{j} - \widetilde{x}_{i+1}^{j}\right)^2} \quad (1)$$

where n represents the sampling length of the frame subsequence. Specifically, the processing details of the subway track bed vibration signal with time series characteristics are shown in Figure 2. For any input sequence, the main vibration characteristics caused by the action of the train axle in each sample are retained and the sampling length of the vibration response is set to match the multiple of the *frame_length*. Then the input sequence is divided into subsequences of length *frame_length*. The first *frame_length* is taken as the initial input sequence *input_seq1*. Next, a length of *frame_length* slides to construct a new input sequence *input_seq2*. Here, *input_seq2* is also regarded as the prediction target *out_seq1* of *input_seq1*. Based on Equation (1), the divergence between the prediction result *out_seq1* of *input_seq1* and *input_seq2* is the prediction error of the current frame. By repeating the above process, the error of each frame in any input sequence can be obtained.

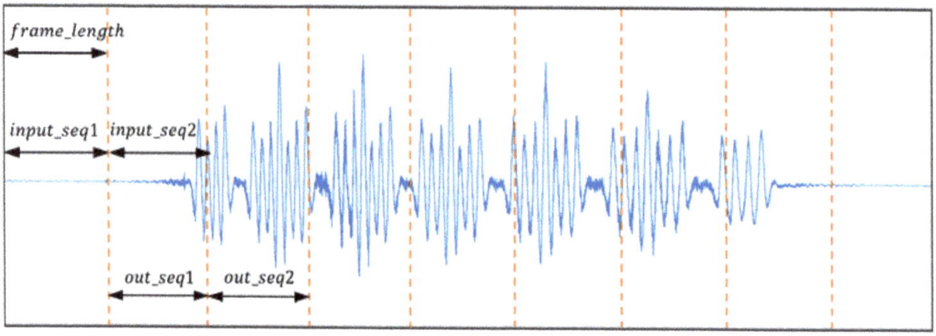

Figure 2. Processing details of vibration sequences used for generating prediction error.

To quantify the predicted error sequence is the idea of the second step. Here, moving average processing is first used to reduce the unstable interferences in the predicted error sequence. Then the maximum is taken in the moving average result to represent the error sequence. In the last step, the quantitative results of datasets A and B derived from the second step participate in setting the threshold and then the threshold is utilized to judge whether the sample in dataset B is abnormal. To ensure the credibility of the threshold-based judgment, the receiver operating characteristic (ROC) curve is adopted to search the candidate threshold. That is, the candidate threshold should satisfy the conditions that make false positive rate (FPR) zero and true positive rate (TPR) maximum. The rationale for determining a reasonable threshold from the ROC curve can be found in [25].

3. Design and Arrangement of the Experiment

3.1. Vibration Signal Acquisition of Subway Track Bed

During the subway operation, there are many opportunities to obtain the track bed vibration signal of the normal state. Instead, it is hard to capture anomaly vibration samples caused by structural deterioration of the track bed or the failure of the train wheel, if there are no suitable tracking ways. As shown in Figure 3, an ultra-weak FBG sensing cable was fixed along the track bed in an actual subway tunnel to collect the structural vibration response caused by passing trains. Three underground stations were covered along the monitoring area of the subway line, with a total length of nearly three kilometers. According to the spatial resolution of the probes in the sensing optic fiber, more than 500 consecutive regions monitored along the track bed can feedback the structural vibration response based on the interrogated address of the light interference [26]. When a train passed, the structural vibration response triggered in each monitoring area was collected at a 1 kHz sampling rate. The collected data was transmitted to the remote monitoring center for processing by the demodulator and server.

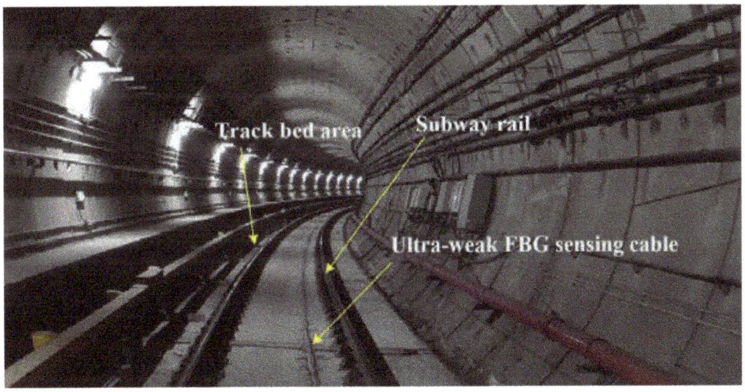

Figure 3. Field layout of ultra-weak FBG sensing array for acquiring vibration responses of track beds.

Based on the continuous running of the monitoring system described above, a train with operation risk due to wheel wear was identified. Compared to the normal state, a set of worn wheels in the first compartment resulted in changes in vibration signals of track beds when the train passed. As shown in Figure 4, three monitoring regions are randomly selected to depict the normal and abnormal vibration signals of the track beds. The left part of Figure 4 shows the vibration responses of the track beds in the three selected monitoring regions as the train without safety hazards passed through the subway line. The abnormal vibration signals corresponding to the same three monitoring regions when subjected to the action of the train with worn wheels are revealed in the right part of Figure 4. As can be seen from Figure 4, there are similarities between the normal vibration signals in different monitoring regions, which have been pointed out in [8]. As well as that, the abnormal signals at different locations exhibit some common features. The occurrence position of the abnormal signal in the overall response to the vibration of the track bed excited by the train is closely related to the compartment where the worn wheels are located. In addition, the duration of the abnormal disturbances is almost the same, accounting for about 20% of the total length of the track bed vibration response sequence caused by the passing train with worn wheels.

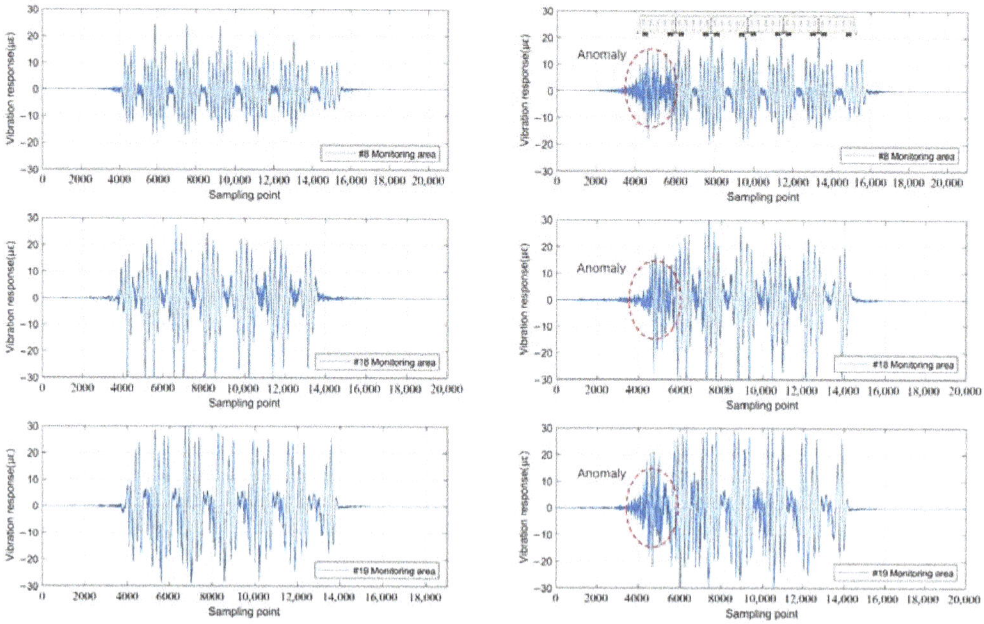

(a) **(b)**

Figure 4. Typical vibration responses of track beds caused by the (**a**) safety train and (**b**) train with worn wheels.

In addition to the above-mentioned abnormal vibration signals of the track bed structure caused by worn wheels, to obtain more abnormal signal samples, this paper considers adding perturbation to the actual normal vibration signals to simulate abnormal signal samples. The track bed vibration signals at the time of the ground intrusion event detected by the ultra-weak FBG sensing array were chosen as the source signal for adding perturbation. Specifically, the vibration response of the track bed structure under excavator ground intrusion was adopted to complement the anomaly dataset for the following experiment. The details of the vibration response of the selected real cases can be found in [9]. In the specific implementation, as shown in Figure 5, the sub-sequences of the adding perturbation are randomly selected from the source signal caused by the ground intrusion. The extracted sub-sequences are then randomly superimposed somewhere in the normal vibration samples in a certain proportion. Here, in addition to taking the proportion of anomalous perturbations observed in Figure 4 as a reference, the simulated anomaly proportion also takes into account fluctuations of 10% above and below 20%.

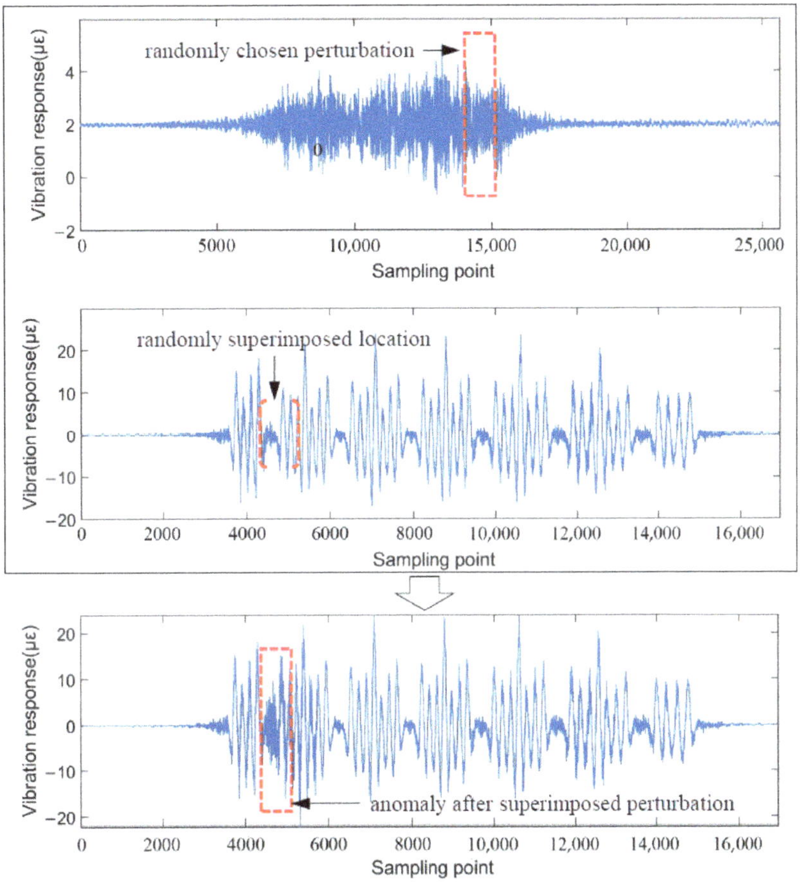

Figure 5. An example of the process of generating the simulated anomaly sample.

3.2. Composition and Division of the Experimental Dataset

Based on the above vibration signal acquisition strategy, the composition and division of the raw experimental dataset are shown in Figure 6. The dataset was derived from the operation monitoring system and the previous field test results, both measured by the ultra-weak FBG sensing array. The vibration signal samples of the track bed in normal and abnormal states are 260 and 147, respectively. For the abnormal signal samples, anomalies consist of three levels according to the proportion of superimposed disturbances. Here, taking 147 normal vibration signals as benchmarks that are different from the normal samples in Figure 6, simulated anomalies of different levels were constructed. Moreover, maintaining the data balance among the three-level abnormal samples to reduce the effect of the asymmetry of the sample size on the training effect was considered.

The raw normal vibration signals were divided into two parts to perform training and tests based on the commonly used ratio [27] of 8:2. Here, the training set was used to train the proposed unsupervised learning network, and the split test set of normal samples was used to participate in identifying whether the no-label signals are abnormal, that is, to assess the recognition effect of the proposed method on abnormal vibration signals. Considering the signal sampling rate and the current configuration of the experimental hardware environment that was composed of a graphics processing unit (GPU) core (GTX 1080 Ti) with 12 2.20 GHz processors (Intel Xeon E5-2650 v4), the *frame_length* was set to 500 to generate sub-sequences and to perform the model error prediction. Although the *frame_length* remains equal, there were duration differences in the raw vibration response of each track bed area due to the effects of train speed and weight. Therefore, the three-level anomaly dataset with the same original sample size in Figure 6 had different enhanced scales after being divided according to the consistent *frame_length*. After data augmentation [28] by dividing the original sample into equal sub-sequences, the detailed scales of the training set and test set are given in Table 1.

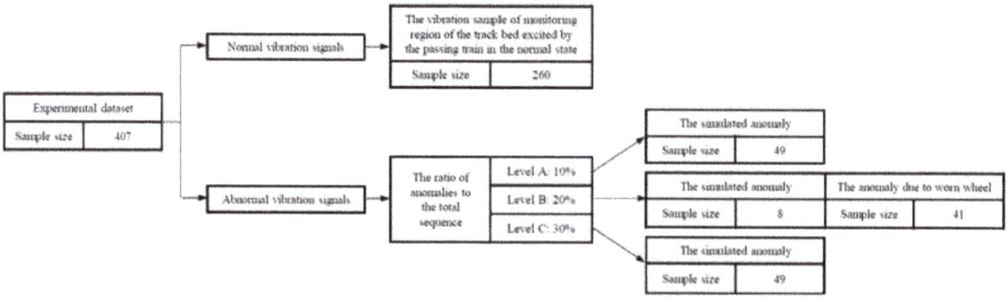

Figure 6. Raw dataset composition and division for the experiment.

Table 1. The division results of the experimental dataset after data augmentation.

Dataset	Size of Normal Sample	Size of Abnormal Sample		
		Level A	Level B	Level C
Training set	6741	-	-	-
Test set	2194	1971	1962	1968

3.3. *Establishment of the Unsupervised Learning Network*

Based on the study in [29] and the current experimental hardware environment, as shown in Figure 7, an unsupervised network based on CNN and LSTM and considering the attention mechanism (CNN-LSTM-AM) was constructed to obtain the error sequence result under step two in Figure 1. Here, the training set of 6741 normal samples as defined in Table 1 was used to train the CNN-LSTM-AM network. The dimension of the network input is determined by the length of the subsequence of *frame_length* 500 after data augmentation and the batch size was set as 64. The goal of network training was to ensure that the prediction error exhibited a rather weak fluctuation, that is, to ensure that the difference between the predicted sequence and the input normal sample was small. After repeated testing and parameter tuning based on grid searching [30], the established network consisted of two one-dimensional convolutional layers, two LSTM layers, and one attention mechanism layer, in which each convolutional layer contained 128 convolution kernels, and each LSTM layer contained 500 unit cells. The CNN and LSTM layers were used to obtain the local spatial features and time-series correlations of the vibration signals of the track bed under normal conditions, respectively. The kernel size of each CNN layer was 1 and a dilated convolution operation with a dilation factor of 10 was used to make the first convolutional layer have a stronger receptive field of the input sequence. The attention mechanism layer automatically learned the importance of each hidden layer. To meet the training objective, RMSprop [31] was selected as the optimization algorithm of the network.

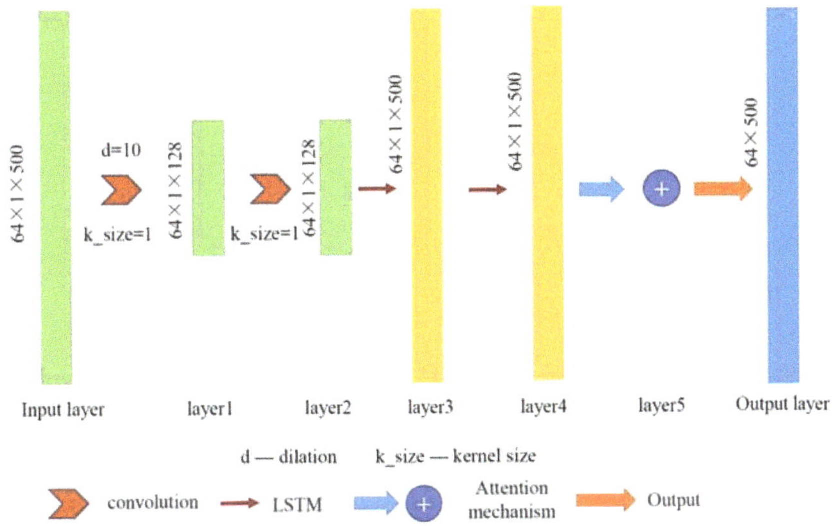

Figure 7. The proposed CNN-LSTM-AM network architecture with 5 hidden layers.

4. Results and Discussion

4.1. Result Analysis

The error sequence results of the three-level anomaly samples in Table 1, after being processed by the unsupervised network described in Figure 7, revealed that for about 80% of the test samples, the average of the main peak of the error sequence of the samples exceeded 6, while the rest was about 2. Figure 8 depicts two randomly selected samples with anomalies and their corresponding predicted error sequences, where the red dotted line represents the anomalous part. Although the main peak of the error sequence can indicate the anomaly to some extent, pseudo main peaks are also observed in the error sequence as shown in Figure 8b.

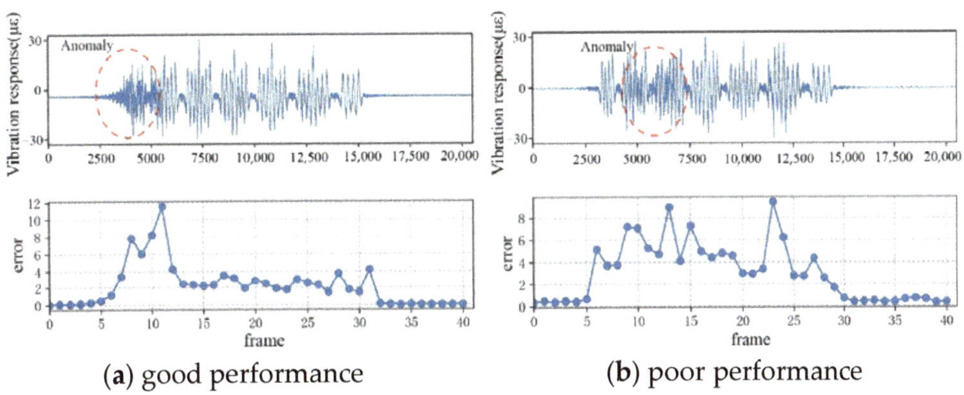

Figure 8. Randomly selected (**a**,**b**) anomaly signals and their corresponding prediction error sequences.

To reduce the interference of the pseudo peak, calculation quantification was performed for the predicted error sequence and the results are shown in Figure 9. The quantitative results in each subplot in Figure 9 are composed of 49 anomaly signal samples and 70 retained normal signal samples. Here, the aforementioned retained normal signal samples were used to participate in the subsequent threshold settings based on the ROC curve.

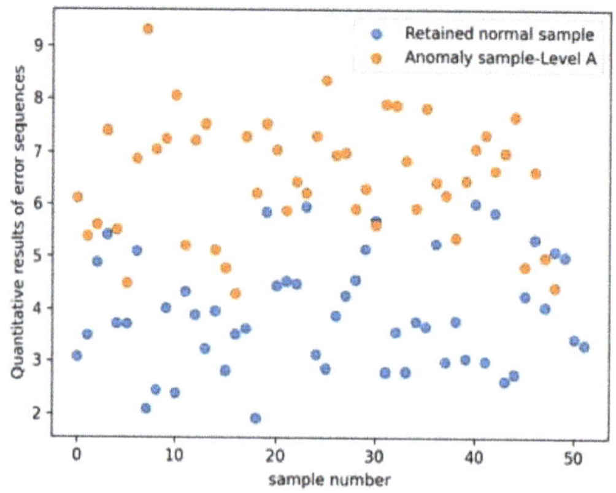

(**a**) Level A

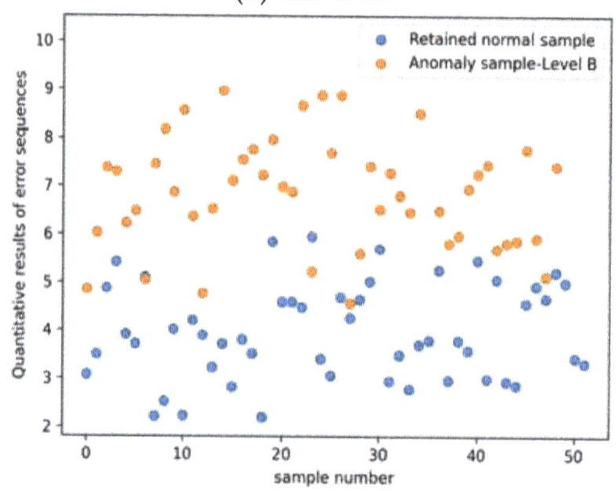

(**b**) Level B

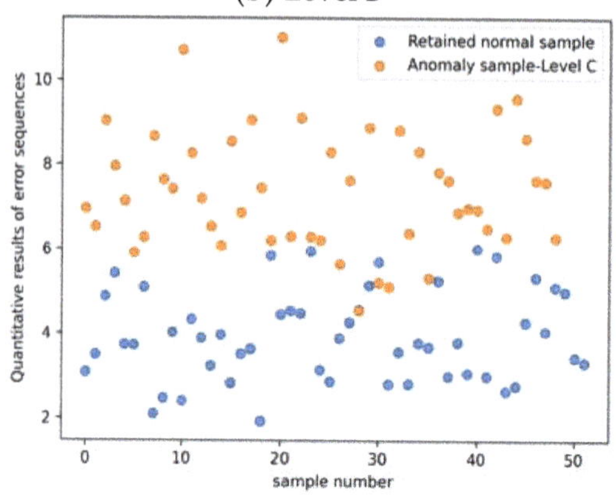

(**c**) Level C

Figure 9. Quantitative results of error sequences of (**a**–**c**) three-level anomaly samples and retained normal samples.

The quantitative results in Figure 9 were set to the tentative threshold in the order from small to large to calculate the corresponding FPR and TPR. The results of FPR and TPR were represented by the ROC curve shown in Figure 10. Then, the candidate threshold used to distinguish the anomaly in Figure 9 can be determined. Here, the purpose of using the quantified result in the maximum TPR state corresponding to the FPR of zero as the candidate threshold was primarily to reduce the likelihood of normal samples being identified as abnormal.

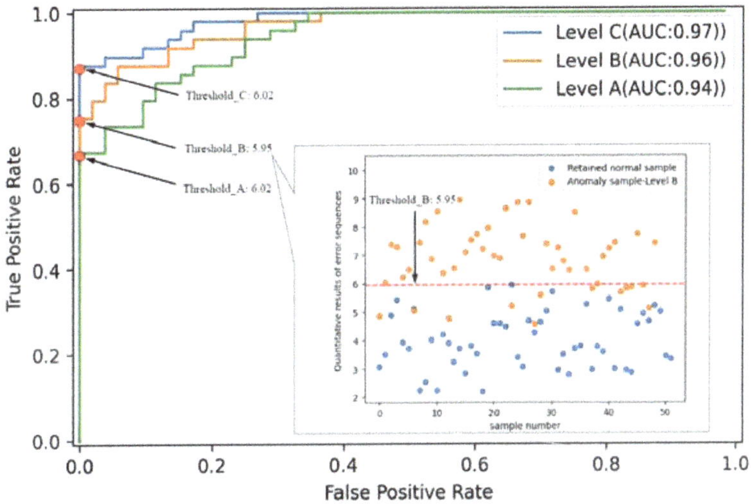

Figure 10. ROC curve based on quantitative results of error sequences.

The accuracy and F1-score in Table 2 quantified the experimental results based on the proposed method. The thresholds of the three levels set in Figure 10 can ensure that the recognition accuracy of the three types of abnormal samples in the designed experiment was no less than 0.84. With the increase in the proportion of anomalies, the accuracy of identifying anomalies and the F1-score gradually increased, reaching the highest of 0.94 and 0.93, respectively. The result not only conforms to the understanding that the more significant the anomaly is, the easier it is to identify, but also displays the capability for identifying abnormal signals based on the proposed procedure. In terms of the efficiency, under the hardware configuration and sample size described in this paper, the proposed anomaly recognition algorithm took about 10 min and 0.24 s in stages of the training and recognition, respectively.

Table 2. Quantitative evaluation of experimental results.

Anomaly Type	Accuracy	F1-Score
Level A	0.84	0.80
Level B	0.88	0.86
Level C	0.94	0.93

4.2. Discussion Based on Comparative Testing

To further evaluate the effectiveness of the proposed method, both the quantification metric of the error sequence and the established unsupervised learning network were in turn compared with other approaches. Here, the allocation of datasets for the comparison tests was consistent with that presented earlier in this paper. In addition to quantifying the error sequence by choosing the maximum processed after the moving average, the effects of identifying anomalous signals when representing the error sequence with the maximum, mean, and root mean square (RMS) were also discussed. Figure 11 shows the ROC curves derived from the four metrics of quantifying the error sequence. The results concerning the area under the ROC curve (AUC) based on the different metrics in Figure 11 all display good performance in terms of the aggregate measure represented by AUC. Although the method of threshold selection based on comparative metrics was reasonable and competitive, the proposed quantification indicator has the highest TPR when the FPR is zero. That is, the results in Figure 11 mean that the threshold based on the quantitative processing proposed in this paper can better identify abnormal samples, and the comparison results based on the accuracy and F1-score shown in Figure 12 further confirm this inference. For three levels of abnormal samples, the indicator adopted in this paper can guarantee higher accuracy and F1-score than that of the other three metrics.

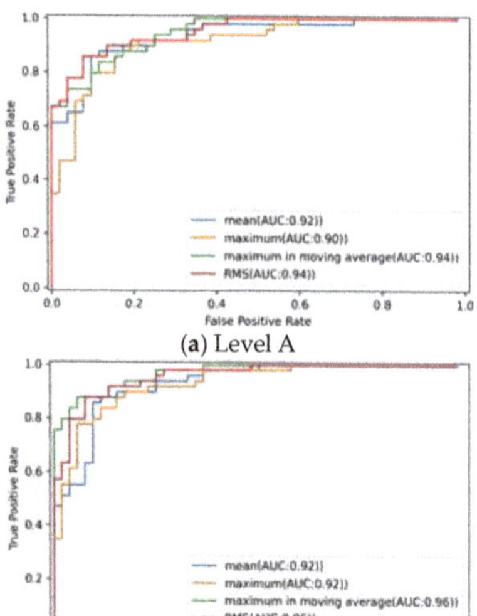

(**a**) Level A

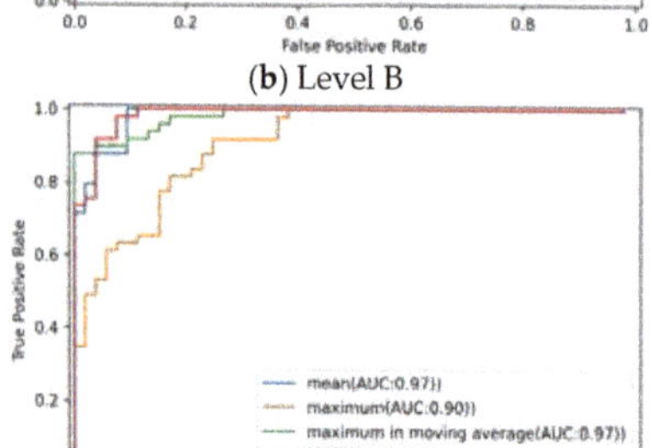

(**b**) Level B

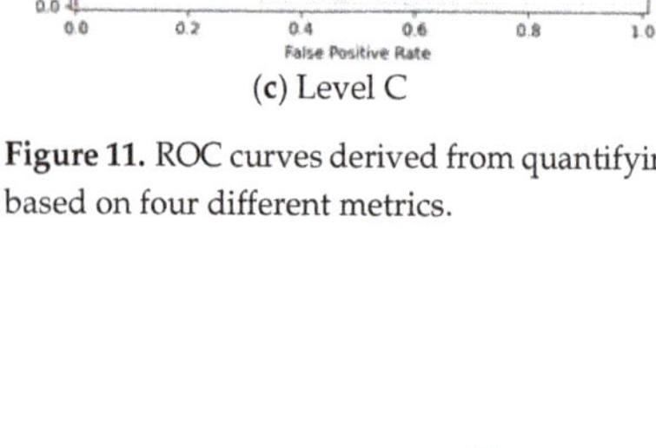

(**c**) Level C

Figure 11. ROC curves derived from quantifying error sequences of (**a**–**c**) three-level anomaly samples based on four different metrics.

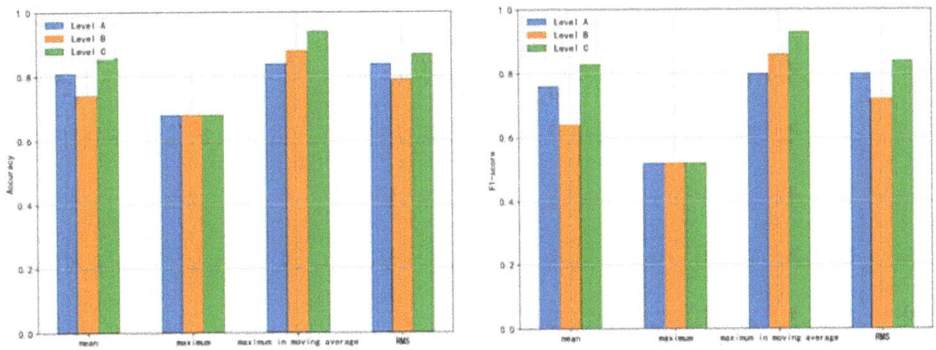

(a) (b)

Figure 12. Comparison results of (**a**) accuracy and (**b**) F1-score from four metrics for quantifying error sequences.

Three typical unsupervised networks (one-class SVM, DAGMM, and LSTM encoder-decoder (LSTM-ED)) which have been reported [20,21,24] to be suitable for anomaly recognition were selected for performance comparison with the CNN-LSTM-AM network proposed in this paper. To fully compete with the proposed network, the most appropriate key candidate hyperparameters sets [22,32] for one-class SVM, DAGMM, and LSTM-ED were derived through trial and error based on random searching [33]. The comparison results shown in Figure 13 revealed that one-class SVM was the least effective. This phenomenon may be related to the relatively lower algorithm complexity of one-class SVM compared with other methods. Although the detection performance of the three comparison methods improved in terms of accuracy and F1-score as the degree of anomaly increased, the proposed network still outperformed the effect of the best DAGMM among the comparison methods. Specifically, for three-level anomaly samples, the average results of accuracy and F1-score of the CNN-LSTM-AM network were 11% and 13% higher than that of DAGMM, respectively. Especially for abnormal samples at level A with the shortest anomaly duration, the network established in this paper led by 24% and 29% in the accuracy and F1-score, respectively.

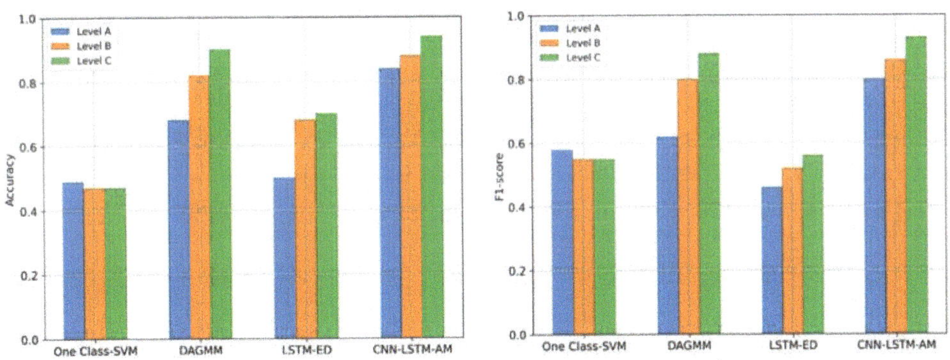

(a) (b)

Figure 13. Comparison results of (**a**) accuracy and (**b**) F1-score from four unsupervised learning networks.

5. Conclusions

The present work indicates that when sufficient normal vibration samples are guaranteed for training the network and auxiliary detection, anomalous vibration signals in the track bed response of subway tunnels detected by ultra-weak FBG sensing array can be identified based on the idea of unsupervised learning. Due to the difficulty of obtaining adequate abnormal samples, the pattern diversity of different levels of abnormal samples based on found and simulated cases in this paper is not rich enough. Thus, the recognition effect of this method on unknown samples with significant type differences collected at the same period needs to be further tested and improved. However, given that the source and location of the superimposed sub-sequence in each simulated abnormal sample are based on random selection, there is reason to believe that the proposed method has strong robustness in the actual engineering. Moreover, some other limitations are worth noting. Although the recognition effects were verified experimentally, the established unsupervised learning network architecture in the proposed method still needs to be further improved by more unknown types and degrees of abnormal events in the future. Moreover, the results of comparison experiments demonstrate that both the current quantification indicator for the error sequences and the established networks have better performance, but whether there is a more appropriate solution to quantify error sequences and the influence of the normal sample size on the identification effect require further investigation. For the above concerns, it is necessary to further investigate the performance of the method proposed in this paper in future work when more typical anomaly samples can be obtained. Additionally, the feasibility and effectiveness of the procedures reported in this paper need to be validated in other areas with similar needs.

Author Contributions: Conceptualization, S.L.; data curation, Y.Q.; funding acquisition, S.L. and J.J.; investigation, H.W.; methodology, Y.Q. and Q.N.; project administration, H.W.; resources, S.L.; supervision, L.S.; validation, S.L.; writing—original draft, Y.Q.; writing—review and editing, S.L. and Q.N. All authors have read and agreed to the published version of the manuscript.

Funding: This research was funded by the National Natural Science Foundation of China, grant numbers 61875155 and 52108472, and the Natural Science Foundation of Hubei Province, grant number 2020CFB144.

Institutional Review Board Statement: Not applicable.

Informed Consent Statement: Not applicable.

Data Availability Statement: The data presented in this study are available on request from the corresponding author.

Acknowledgments: The research work reported in this paper was supported by the National Engineering Research Center of Fiber Optic Sensing Technology and Networks, Wuhan University of Technology, and the Smart Nanocomposites Laboratory, University of California, Irvine.

Conflicts of Interest: The authors declare no conflict of interest.

References

1. Kaewunruen, S.; Osman, M.H.; Hao Cheng Eric, W. Risk-based maintenance planning for rail fastening systems. *ASCE-ASME J. Risk Uncertain. Eng. Syst. Part A Civ. Eng.* **2019**, *5*, 04019007. [CrossRef]
2. Poudel, A.; Lindeman, B.; Wilson, R. Current practices of rail inspection using ultrasonic methods: A review. *Mater. Eval.* **2019**, *77*, 871–883.
3. Alvarenga, T.A.; Carvalho, A.L.; Honorio, L.M.; Cerqueira, A.S.; Filho, L.M.A.; Nobrega, R.A. Detection and classification system for rail surface defects based on eddy current. *Sensors* **2021**, *21*, 7937. [CrossRef] [PubMed]
4. Kim, J. Non-destructive characterization of railway materials and components with infrared thermography technique. *Materials* **2019**, *12*, 4077. [CrossRef]
5. Lou, Y.; Zhang, T.; Tang, J.; Song, W.; Zhang, Y.; Chen, L. A fast algorithm for rail extraction using mobile laser scanning data. *Remote Sens.* **2018**, *10*, 1998. [CrossRef]
6. Gibert, X.; Patel, V.M.; Chellappa, R. Deep multitask learning for railway track inspection. *IEEE Trans. Intell. Transp. Syst.* **2017**, *18*, 153–164. [CrossRef]
7. James, A.; Jie, W.; Xulei, Y.; Chenghao, Y.; Ngan, N.B.; Yuxin, L.; Yi, S.; Chandrasekhar, V.; Zeng, Z. TrackNet—A deep learning based fault detection for railway track inspection. In Proceedings of the 2018 International Conference on Intelligent Rail Transportation, ICIRT 2018, Singapore, 12–14 December 2018.
8. Li, S.; Zuo, X.; Li, Z.; Wang, H.; Sun, L. Combining SDAE network with improved DTW algorithm for similarity measure of ultra-weak FBG vibration responses in underground structures. *Sensors* **2020**, *20*, 2179. [CrossRef]
9. Gan, W.; Li, S.; Li, Z.; Sun, L. Identification of ground intrusion in underground structures based on distributed structural vibration detected by ultra-weak FBG sensing technology. *Sensors* **2019**, *19*, 2160. [CrossRef]
10. Nan, Q.; Li, S.; Yao, Y.; Li, Z.; Wang, H.; Wang, L.; Sun, L. A novel monitoring approach for train tracking and incursion detection in underground structures based on ultra-weak FBG sensing array. *Sensors* **2019**, *19*, 2666. [CrossRef]
11. Tao, L.; Siqi, Q.; Zhang, Y.; Shi, H. Abnormal detection of wind turbine based on SCADA data mining. *Math. Probl. Eng.* **2019**, *2019*, 5976843. [CrossRef]
12. Qiu, Y.; Jing, L.; Li, S. Bridge anomaly data identification method based on statistical feature mixture and data augmentation through forwarding difference. In Proceedings of the 2021 3rd International Conference on Advances in Civil Engineering, Energy Resources and Environment Engineering, Qingdao, China, 28–30 May 2021.
13. Li, S.; Sun, L. Detectability of bridge-structural damage based on fiber-optic sensing through deep-convolutional neural networks. *J. Bridge Eng.* **2020**, *25*, 04020012. [CrossRef]
14. Tang, Z.; Chen, Z.; Bao, Y.; Li, H. Convolutional neural network-based data anomaly detection method using multiple information for structural health monitoring. *J. Struct. Control Health Monit.* **2019**, *26*, e2296. [CrossRef]
15. Li, Z.; Wang, Y.; Wang, K. A deep learning driven method for fault classification and degradation assessment in mechanical equipment. *Comput. Ind.* **2019**, *104*, 1–10. [CrossRef]
16. Abid, F.B.; Sallem, M.; Braham, A. Robust interpretable deep learning for intelligent fault diagnosis of induction motors. *IEEE Trans. Instrum. Meas.* **2019**, *69*, 3506–3515. [CrossRef]
17. Hautamäki, V.; Kärkkäinen, I.; Fränti, P. Outlier detection using k-nearest neighbour graph. In Proceedings of the 17th International Conference on Pattern Recognition, ICPR 2004, Cambridge, UK, 23–26 August 2004.
18. He, Z.; Xu, X.; Deng, S. Discovering cluster-based local outliers. *Pattern Recogn. Lett.* **2003**, *24*, 1641–1650. [CrossRef]
19. Li, J.; Izakian, H.; Pedrycz, W.; Jamal, I. Clustering-based anomaly detection in multivariate time series data. *Appl. Soft. Comput.* **2021**, *100*, 106919. [CrossRef]
20. Saari, J.; Strömbergsson, D.; Lundberg, J.; Thomson, A. Detection and identification of windmill bearing faults using a one-class support vector machine (SVM). *Meas. J. Int. Meas. Confed.* **2019**, *137*, 287–301. [CrossRef]
21. Zong, B.; Song, Q.; Min, M.R.; Cheng, W.; Lumezanu, C.; Cho, D.; Chen, H. Deep autoencoding Gaussian mixture model for unsupervised anomaly detection. In Proceedings of the 6th International Conference on Learning Representations, Vancouver, BC, Canada, 30 April–3 May 2018.
22. Purohit, H.; Tanabe, R.; Endo, T.; Suefusa, K.; Nikaido, Y.; Kawaguchi, Y. Deep Autoencoding Gmm-Based Unsupervised Anomaly Detection In Acoustic Signals And Its Hyperparameter Optimization. *arXiv* **2009**, arXiv:2009.12042. Available online: https://arxiv.org/abs/2009.12042 (accessed on 20 May 2022).
23. Pei, J.; Zhong, K.; Jan, M.A.; Li, J. Personalized federated learning framework for network traffic anomaly detection. *Comput. Netw.* **2022**, *209*, 108906. [CrossRef]
24. Malhotra, P.; Ramakrishnan, A.; Anand, G.; Vig, L.; Agarwal, P.; Shroff, G. LSTM-based encoder-decoder for multi-sensor anomaly detection. *arXiv* **2016**, arXiv:1607.00148. Available online: https://arxiv.org/abs/1607.00148 (accessed on 8 May 2022).
25. Taylor, A.; Leblanc, S.; Japkowicz, N. Anomaly detection in automobile control network data with long short-term memory networks. In Proceedings of the 3rd IEEE International Conference on Data Science and Advanced Analytics, DSAA 2016, Montreal, PQ, Canada, 17–19 October 2016.
26. Luo, Z.; Wen, H.; Guo, H.; Yang, M. A time-and wavelength-division multiplexing sensor network with ultra-weak fiber Bragg gratings. *Opt. Express* **2013**, *21*, 22799–22807. [CrossRef]
27. Fonseca, A.; Cabral, B. Designing a neural network from scratch for big data powered by multi-node GPUs. *Smart Innov. Syst. Technol.* **2019**, *136*, 1–19.

28. Abdeljaber, O.; Avci, O.; Kiranyaz, M.S.; Boashash, B.; Sodano, H.; Inman, D.J. 1-D CNNs for structural damage detection: Verification on a structural health monitoring benchmark data. *Neurocomputing* **2018**, *275*, 1308–1317. [CrossRef]
29. Liu, Y.; Liu, P.; Wang, X.; Zhang, X.; Qin, Z. A study on water quality prediction by a hybrid CNN-LSTM model with attention mechanism. In Proceedings of the International Conference on Smart Transportation and City Engineering 2021, Chongqing, China, 26–28 October 2021.
30. Lameski, P.; Zdravevski, E.; Mingov, R.; Kulakov, A. SVM parameter tuning with grid search and its impact on reduction of model over-fitting. *Lect. Notes Comput. Sci.* **2015**, *9437*, 464–474.
31. Huk, M. Stochastic optimization of contextual neural networks with RMSprop. In Proceedings of the Intelligent Information and Database Systems–12th Asian Conference, ACIIDS 2020, 3rd IEEE International Conference on Data Science and Advanced Analytics, Phuket, Thailand, 23–26 March 2020.
32. Buslim, N.; Rahmatullah, I.L.; Setyawan, B.A.; Alamsyah, A. Comparing bitcoin's prediction model using GRU, RNN, and LSTM by hyperparameter optimization grid search and random search. In Proceedings of the 2021 9th International Conference on Cyber and IT Service Management, CITSM 2021, Virtual, Bengkulu, Indonesia, 22–23 September 2021.
33. Bergstra, J.; Bengio, Y. Random search for hyper-parameter optimization. *J. Mach. Learn. Res.* **2012**, *13*, 281–305.

applied sciences

MDPI

Article

A Novel Method for Predicting Local Site Amplification Factors Using 1-D Convolutional Neural Networks

Xiaomei Yang, Yongshan Chen, Shuai Teng and Gongfa Chen *

School of Civil and Transportation Engineering, Guangdong University of Technology, Guangzhou 510006, China; xmyang@gdut.edu.cn (X.Y.); 2111909059@mail2.gdut.edu.cn (Y.C.); 1112009002@mail2.gdut.edu.cn (S.T.)
* Correspondence: gongfa.chen@gdut.edu.cn; Tel.: +86-136-6248-3527

Abstract: The analysis of site seismic amplification characteristics is one of the important tasks of seismic safety evaluation. Owing to the high computational cost and complex implementation of numerical simulations, significant differences exist in the prediction of seismic ground motion amplification in engineering problems. In this paper, a novel prediction method for the amplification characteristics of local sites was proposed, using a state-of-the-art convolutional neural network (CNN) combined with real-time seismic signals. The amplification factors were computed by the standard spectral ratio method according to the observed records of seven stations in the Lower Hutt Valley, New Zealand. Based on the geological exploration data from the seven stations and the geological hazard information of the Lower Hutt Valley, eight parameters related to the seismic information were presumed to influence the amplification characteristics of the local site. The CNN method was used to establish the relationship between the amplification factors of local sites and the eight parameters, and the training samples and testing samples were generated through the observed and geological data other than the estimated values. To analyze the CNN prediction ability for amplification factors on unrecorded domains, two CNN models were established for comparison. One CNN model used about 80% of the data from 44 seismic events of the seven stations for training and the remaining data for testing. The other CNN model used the data of six stations to train and the remaining station's data to test the CNN. The results showed that the CNN method based on the observation data can provide a powerful tool for predicting the amplification factors of local sites both for recorded positions and for unrecorded positions, while the traditional standard spectral ratio method only predicts the amplification factors for recorded positions. The comparison of the two CNN models showed that both can effectively predict the amplification factors of local ground motion without records, and the accuracy and stability of predictions can meet the requirements. With increasing seismic records, the CNN method becomes practical and effective for prediction purposes in earthquake engineering.

Keywords: amplification factor; ground motion; 1-D convolutional neural network; site amplification

Citation: Yang, X.; Chen, Y.; Teng, S.; Chen, G. A Novel Method for Predicting Local Site Amplification Factors Using 1-D Convolutional Neural Networks. *Appl. Sci.* **2021**, *11*, 11650. https://doi.org/10.3390/app112411650

Academic Editor: Nikos D. Lagaros

Received: 3 November 2021
Accepted: 3 December 2021
Published: 8 December 2021

1. Introduction

The seismic amplification effects in earthquake-prone areas need to be considered in building or structure designs. The relationship between the site condition and seismic ground motion has been researched for over one hundred years [1]. Pioneer researchers gathered a great deal of observational evidence to establish this relationship in the earlier studies [2]. Subsequently, many researchers [3–10] attempted to evaluate the amplification characteristics of strong ground motions at a given site according to the acceleration records. For unrecorded locations, it is common to rely on the regression relationship obtained from the recorded results. This approach is regarded as reliable because the earthquake records [11] include all the influences of the earthquake source, transmission path and site features. However, for many local site amplification zones with no ground motion records, a simple regression relationship based on a large-sized site and inadequate data seems unreasonable.

To overcome the lack of strong motion data to estimate the amplification factors at many local zones, an alternative to the seismic record method needs to be developed. The microtremor-based method, which is an empirical method, was first used by Kanai [12] and further developed by Nakamura [13] for the site amplification analysis of cases without seismic records. By measuring the ambient noise or microtremors, the method can obtain the experimental transfer functions and predict the amplification factors of the site. Although the microtremor-based method can improve the surface amplification prediction for the area without seismic records, the research shows that there are significant differences in the predictions of the amplification characteristics under strong earthquakes. Signal processing has also become a key to prediction. The existing methods of processing signal data include wavelet transform methods [14–16], Fourier transform methods [17] and so on. As an alternative to the seismic record method, some researchers [18,19] have used site response simulations based on the wave propagation theory and numerical methods to estimate the transfer functions and predict ground motions in particular regions. Compared with the empirical methods, numerical simulations [20] based on the wave propagation theory can establish an analytical model for the seismic response of a local site and adjust the parameters to predict the amplification factor of the site. However, simulation models still employ some major simplifications at present. Considering the heterogeneous structures of a local site [21,22], seismologists [23–25] usually establish more realistic 2-D or 3-D wave propagation models to predict the ground motion and site amplification factors. These 2-D or 3-D models can reflect the effect of more site geology information. However, it is difficult to model seismic ground motions and predict site amplification in simulations of engineering problems, due to the high computational cost and complex simulation technology [26–28]. To improve prediction of the surface amplification characteristics in local areas, including basin regions, more and more local site stations have been established around the world for obtaining seismic observation records. Since the number of earthquake observation records in local domains has been increasing over the years, many researchers across the world prefer to develop amplification prediction models in a specific local site.

Whether employing regression data from the observation records or computational simulations with predefined functions to predict the ground motion amplification factors, the aim is to develop reasonable prediction models or equations for the surface ground motion amplification factors in complex local sites. Due to the complex local geological conditions, unpredictable site conditions and other factors of seismic propagation it is difficult for traditional regression methods to achieve predictions consistent with the observations [25]. To establish a reasonable regression relationship for a local site, more effort is needed to seek more accurate methods for predicting ground motion characteristics. With the continuous rapid development of artificial intelligence methods, more and more scholars are committed to applying various machine learning techniques to ground motion prediction [29]. These methods are applied to the problems solved by traditional regression methods. Recently, new algorithms, such as the Bayesian method, clustering methods and neural network methods have been gradually developed for ground motion predictions [30,31]. For example, a new method of seismic site classification was proposed using HVSR [32] curves and a neural network [33]. For the aspect of attenuation prediction, Kuok and Yuen [29] proposed an effective generalized learning network for nonparametric spatial modeling to predict the ground motion attenuation law for Wenchuan earthquake in China. The M5 tree method [6], heterogeneous Bayesian learning and the back-propagation neural network (BPNN) were used by Mu and Yuen [34] and Kaveh and Kim [6,29], respectively, to predict the ground motion, improving the previous learning algorithm and providing a solid support for the automatic prediction of basin ground motion. As a new machine learning algorithm, the convolutional neural network (CNN) with convolution layers overcomes the disadvantages of traditional BPNNs. Compared with the typical machine learning models such as BPNNs, the CNN has dimension reduction [35] and strong feature extraction ability, fuses multiple inputs and converges fast. Furthermore, it can extract some important features more effectively. In the feature extraction stage, the CNN can learn directly from the original input data and optimize the features in the training stage, then grasp the feature information more accurately. As real seismic records in a local domain contain many kinds of potential seismic characteristics, the combination of multiple features when training a CNN is likely to result in a better method. It can be seen that using the CNN method to predict the amplification characteristics of ground motion offers a new direction in this research field.

In this paper, based on the growing number of ground motion data for the densely distributed local stations and the detailed geology data for Lower Hutt Valley, New Zealand, the CNN method was applied to establish novel prediction models to predict the surface seismic amplification factors in this local area. The seismic observation records of 44 earthquakes from seven stations were selected, and seven CNN models were created. The CNN models were used to identify the surface amplification factors for the basin with strong earthquake observation data.

2. Materials and Methods

2.1. Geological Condition

The Lower Hutt Valley is a middle-sized sedimentary basin located in the Wellington Metropolitan area, New Zealand. The valley is about 10 km wide and 35 km long and is surrounded by hills in the east and the north, while it is open to Wellington Bay in the west. Figure 1a displays a location map of the valley, and a detailed map of a small zone 10 km long and 5 km wide is shown in Figure 1b.

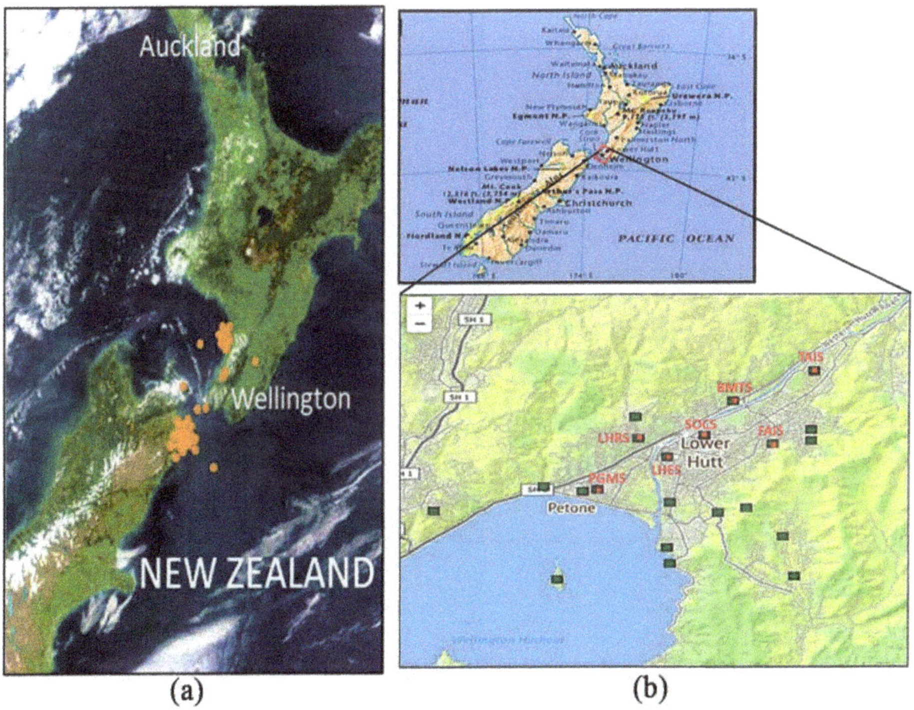

Figure 1. Location maps. (**a**) Locations of earthquake focuses; (**b**) station locations in Lower Hutt basin in New Zealand.

According to the geological exploration data provided by New Zealand's Institute of Geological and Nuclear Sciences (GNS), a local site in the Lower Hutt Valley with a length of 8400 m and a width of 5600 m was selected. There were four irregular layers of sediments over the bedrock. The S-wave velocity varied from 175 m/s to 300 m/s in the top layer, 300 m/s to 330 m/s in the second layer, 330 m/s to 500 m/s in the third layer and 500 m/s to 1500 m/s in the bottom layer. The S-wave velocity of the bedrock layer was 1500 m/s. As shown in Figure 1b, 7 accelerometers were located in the selected zone: the accelerometer stations BMTS, LHES, FAIS, TAIS, PGMS and SOCS were located in the soft soil zone and the station LHRS in the hard rock zone. The details of the soil and bedrock characteristics for the seven stations are listed in Table 1. The 1-D soil layer histogram of the seven stations was extracted from the 3-D geological structure of the lower Hutt gorge, as shown in Figure 2. As the soil properties are directly related to the amplification characteristics of the surface, the first-order frequencies and the equivalent shear wave velocities of the 1-D soil layer at each station were calculated. The results are shown in Table 1.

Table 1. Station information.

Station	Latitude (S)	Longitude (E)	Thickness of Soil Layer (m)	V_{30} (m/s)
LHRS	41°12′17″	174°53′35″	0.000	1500.000
BMTS	41°11′29″	174°55′34″	93.330	200.830
LHES	41°12′42″	174°54′12″	217.780	215.330
FAIS	41°12′27″	174°56′24″	62.220	206.000
TAIS	41°10′35″	174°58′12″	280.000	235.000
PGMS	41°13′28″	174°52′46″	130.670	236.000
SOCS	41°12′15″	174°54′57″	311.110	240.170

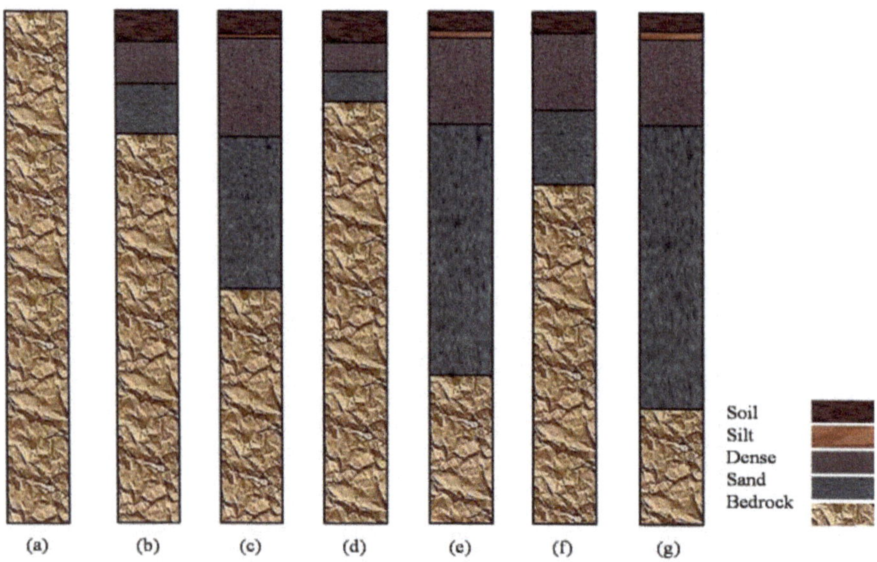

Figure 2. Borehole records. (**a**) LHRS; (**b**) BMTS; (**c**) LHES; (**d**) FAIS; (**e**) TAIS; (**f**) PGMS; (**g**) SOCS.

According to the geological hazard information network of New Zealand (https://www.geonet.org.nz, accessed on: 27 June 2021), the NS and EW components of the earthquake records from the bedrock to the soil layers of the 7 stations were used, as shown in Table 1. Since each station had two horizontal components in the seismic records, the records first needed to be processed. Figure 3 shows the two horizontal components (SV and SH) at each station.

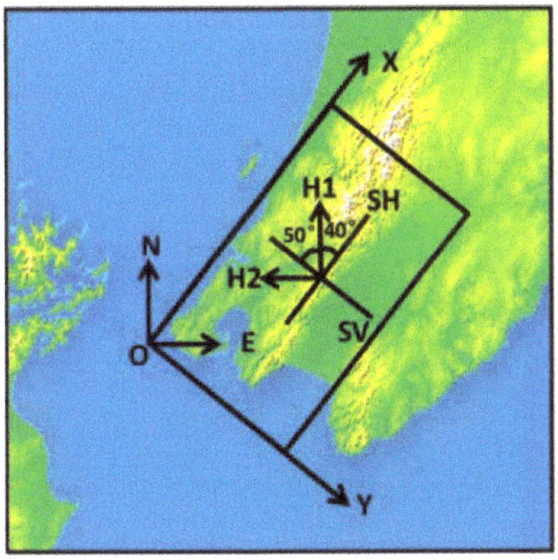

Figure 3. Seismic records from the Lower Hutt basin in New Zealand.

The original data of the horizontal observation records provided by the New Zealand geological disaster information network were recorded in two horizontal directions, where horizontal axis 1 was along the north direction (H1) and horizontal axis 2 along the west direction (H2). The angle between the X-direction of the Lower Hutt basin model and the geographical east-west direction was about 50°. The horizontal observation records H1 and H2 for the basin can be decomposed into SV and SH directions. The acceleration time history in the SH direction can be determined by Equation (1):

$$SH = H_1COS40^\circ - H_2COS50^\circ \quad (1)$$

2.2. Preparation of the Data

2.2.1. Amplification Factor

To estimate the surface amplification of the soil response with respect to the bedrock response, frequency-dependent seismic amplification factors were used to compute the soil effect of this valley surface. The standard spectral ratio [35] was used to calculate the amplification factors of the site location. This is defined as the spectral ratio of a sedimentary site with respect to a nearby bedrock reference site. A high-pass filtering program (JMTEST) method was used to smooth selected acceleration time histories, and then a fast Fourier transform (FFT) computer program was employed to obtain Fourier spectra of the acceleration time history at each soil station and bedrock station. For each earthquake, the spectral ratio of the Fourier spectra at each soil station to the spectrum at the bedrock station was used to estimate the frequency-dependent amplification factors, AFF (amplification Fourier factors) [5], from Equation (2):

$$AFF = \frac{F_{SH}{}^{surface}}{F_{SH}{}^{bedrock}} \quad (2)$$

where F_{SH} is the Fourier amplitude spectrum in the SH (S-wave horizon) direction at the soil surface or bedrock.

Then, for each group of acceleration time histories at soil stations, a mean frequency-dependent amplification factor was computed from the average AFF at a given frequency value [36], using Equation (3):

$$\overline{AFF} = \frac{\sum_{i=1}^{n} AFF_i}{n} \quad (3)$$

2.2.2. Surface Amplification Factors at Lower Hutt Valley

The seismic response records of 44 earthquakes in the Lower Hutt Valley were selected to calculate the amplification factors for all 6 soil stations, according to Equations (1) and (2).

The amplified spectra were obtained by calculating the ratio of the acceleration records of the six stations in the soft layer vs. the bedrock station (LHRS), and parts of the results are shown in Figure 4. The amplification factors at the six stations were also obtained by calculating the average of the amplified spectrum. Table 2 shows some of the average amplification factors calculated using Equation (2).

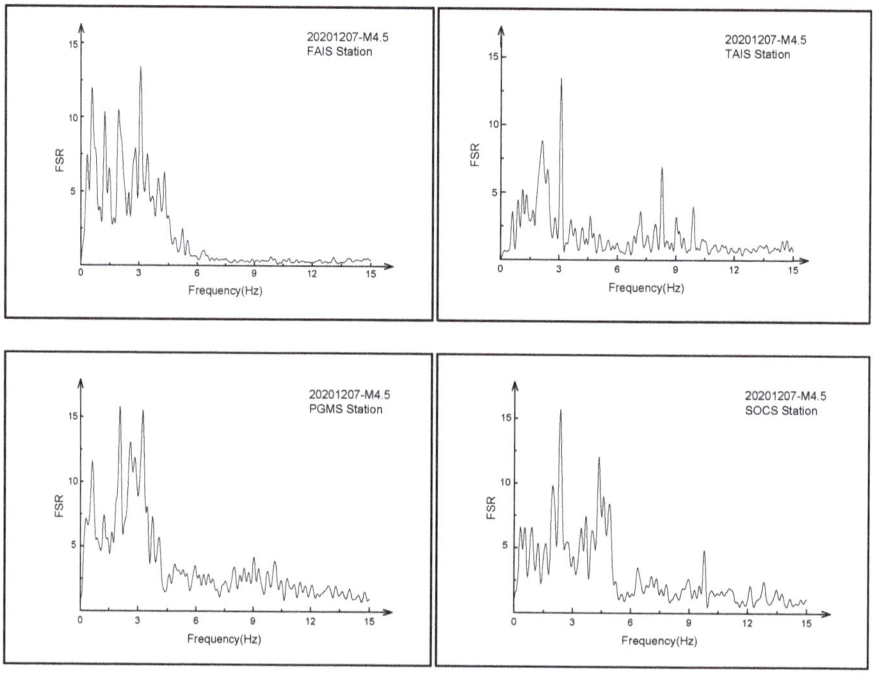

Figure 4. Sediment-to-rock Fourier spectral ratios (FSRs) of stations.

Table 2. Amplification factors for earthquakes at 7 stations.

Station Magnitude	LHRS	BMTS	LHES	FAIS	TAIS	PGMS	SOCS
M4.5	1.00	1.78	2.56	1.90	3.10	1.95	3.04
M5.8	1.00	2.01	2.81	1.91	3.65	2.06	3.24
M5.1	1.00	1.61	3.42	1.81	2.98	2.06	3.27
M4.0	1.00	1.26	2.09	1.42	2.72	1.50	2.38
M4.3	1.00	—	2.78	1.93	3.06	2.20	3.33
M4.8	1.00	2.96	2.45	—	2.70	2.21	3.85
M4.5	1.00	2.77	2.08	—	2.50	1.99	2.86
M4.0	1.00	2.49	2.36	—	2.45	2.48	2.93
M4.7	1.00	1.55	2.31	2.28	3.06	2.05	2.57
M4.6	1.00	3.10	2.39	2.13	2.71	2.27	3.16
M4.3	1.00	1.86	2.76	1.91	2.87	2.45	3.42
M4.8	1.00	1.71	2.38	1.89	—	1.98	—

2.3. 1-D Convolutional Neural Network

As shown in Figure 5, a standard CNN model usually includes the input layers, convolution layers, a fully connected layer and an output layer. The input data were transferred through a series of layers (two convolution layers, activation function, loss function). Finally, the mapping calculation obtained the amplification factors. Specifically, the input of the 1-D CNN was either a 1 × N or an N × 1 array. As shown in Figure 5, an N × 1 array was passed through a series of convolution layers and the fully connected layer. Then, the amplification factors for the local ground motion were obtained in the output layer.

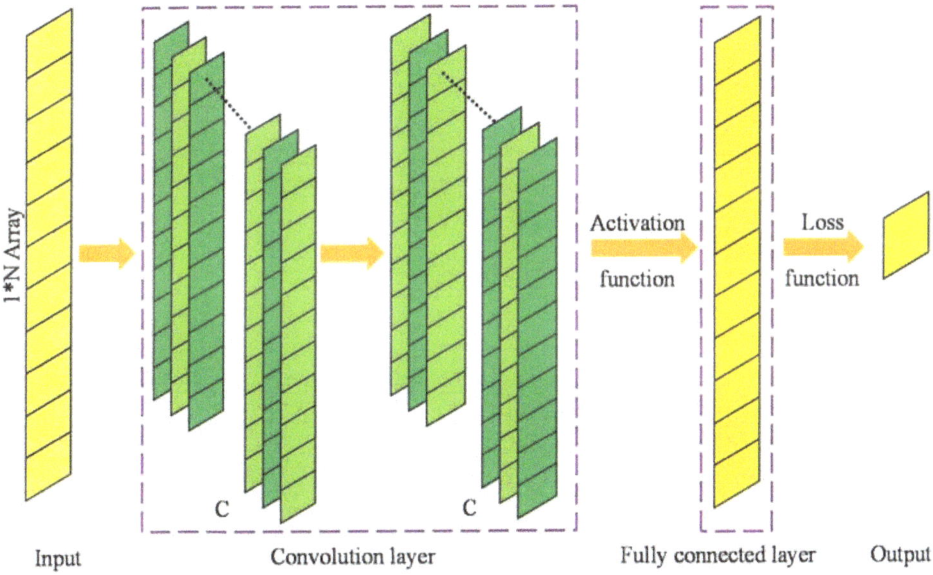

Figure 5. One-dimensional CNN model for predicting site amplification factors.

As shown in Figure 6, the convolution process was to multiply each element in the convolution kernel with the corresponding element in a sub-region (e.g., green box or blue dotted box) of the input data of the convolution layer and to sum the products to obtain an element in the feature map. Each time, the sub-region moved down by one step, and the process was repeated until all elements of the input data were involved. Finally, the convolution operation formed a new array (i.e., the feature map).

The activation function has a non-linear ability that gives the neural network better learning capabilities. Commonly used activation functions include the sigmoid, tanh, ReLU (rectified linear unit) and LeakyReLU functions, as shown in Figure 7. The LeakyReLU activation function was used in this study, as it is much faster than the sigmoid function and the tanh function.

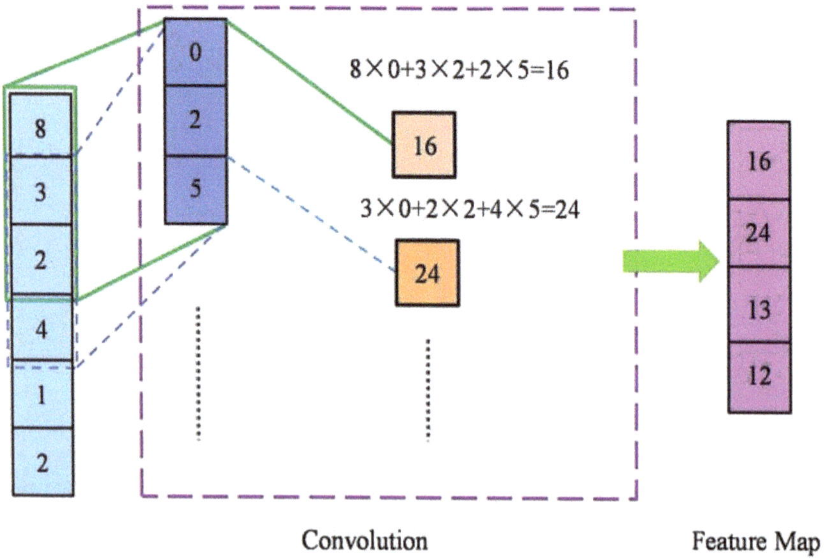

Figure 6. Based on comparisons of the three activation functions, the ReLU activation function was used in our model as it was much faster than the sigmoid and tanh functions.

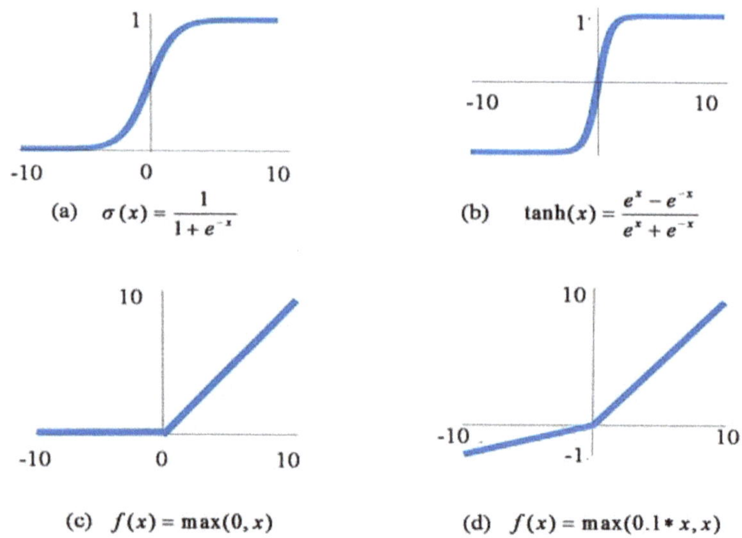

Figure 7. Activation function: (**a**) sigmoid; (**b**) tanh; (**c**) ReLU; (**d**) LeakyReLU.

Loss Function

The loss function was employed to evaluate the convergence of the trained CNN. It was embedded in the output layers. In this paper, the convergence of the network training process was evaluated by the validation samples. The loss function (RMSE, root mean squared error) was as given in Equation (4).

$$RMSE = \sqrt{\frac{1}{N}\sum_{t=1}^{N}(predicted_t - label)^2} \quad (4)$$

where N is the number of validation samples.

2.4. Design of the Models

In this section, a novel method based on the CNN model was established to predict the ground motion amplification. Compared with the traditional classical regression model that can usually use only two variables for the regression task, the CNN can use multivariate regression techniques to predict the results.

2.4.1. Parameters of the Models

In this section, the CNN method was established to predict the local site amplification based on earthquake observation records. The sampling data used in the proposed CNN models involved 44 seismic events. The input data of the samples were extracted from 40 seismic incidents recorded at 7 stations (Table 3). The geological exploration data of the 7 stations were provided by New Zealand's Institute of Geological and Nuclear Sciences (GNS) and the observation records were downloaded from the New Zealand's geohazard information network (GeoNet) (https://www.geonet.org.nz, accessed on: 27 June 2021). The output data of the samples were directly computed using Equation (2) from the downloaded observation records. A series of input variables of the proposed CNN model were specified: the station latitude, longitude, magnitude, focal depth, epicentral distance, soil layer thickness, V_{30} (equivalent shear wave velocity at a calculated depth of 30 m in the overlying soil layer) and the observation records of the bedrock PGAr (peak ground acceleration of rock). Based on a number of the input variables and output variables of the samples, an effective and reasonable CNN model was trained, and a CNN-based prediction method was obtained. With the trained CNN model, the ground motion amplification factors of the Lower Hutt Valley could be predicted, and the amplification characteristics of this engineering site could be investigated.

Table 3. Design of CNN-FSPA.

	CNN-FSPA
Total	44 earthquakes (274)
Training	40 earthquakes (246)
Validation	40 earthquakes (246)
Testing	4 earthquakes (28)

A 1-D CNN structure was established using the 'Deep Learning Toolbox' of MATLAB (MathWorks Inc., Natick, MA, USA), including convolution layers, activation layers (leaky ReLU activation function) and one fully connected layer (FC). The convolution kernel size and layer number were adjusted according to different situations. The CNN structure was designed according to the ground motion situation considered in this study. The weights of the CNN and the internal parameters were adjusted to achieve the best local site amplification prediction effect. The data were preprocessed and then input into the CNN for network training. Part of the dataset was used for network training, and the other part was used for testing. Furthermore, padding was employed to preserve the information on the edges.

2.4.2. Description of the Models

Figure 8 shows the location map of the 7 stations (blue points) and 5 selected unrecorded location points (yellow points). The CNN models were based on the sample datasets obtained from the nearest 44 earthquake records collected from the 7 stations. Two kinds of CNN prediction models were trained, with sample parameters as shown in Tables 3 and 4.

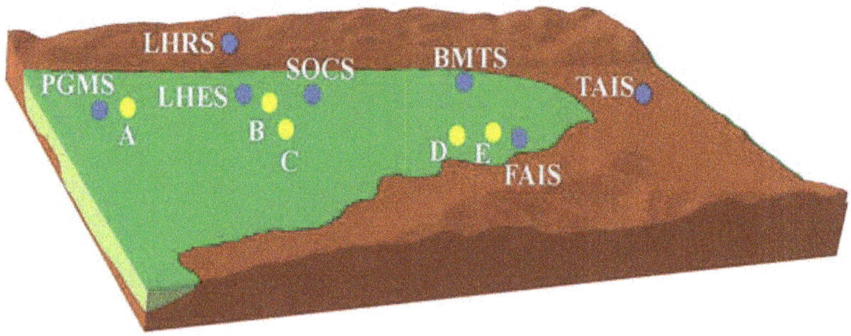

Figure 8. Location map.

Table 4. Design of CNN-PSPA.

	CNN-PSPA					
Total	**All Stations (274)**	**All Stations (274)**	**All Stations (274)**	**All Stations (274)**	**All Stations (274)**	**All Stations (274)**
Training	Non-SOCS (237)	Non-PGMS (233)	Non-TAIS (235)	Non-FAIS (242)	Non-LHES (235)	Non-BMTS (232)
Validation	(237)	(233)	(235)	(242)	(235)	(232)
Testing (PSPA)	SOCS (37)	PGMS (41)	TAIS (39)	FAIS (32)	LHES (39)	BMTS (42)

(1) CNN-FSPA (full-station predicted amplification) model: in this model, 40 of the 44 earthquake records were used for training and the remaining 4 for testing the model. The detailed information of CNN-FSPA is shown in Table 3.

(2) CNN-PSPA (part-station predicted amplification) models: in these models, data from 6 of the 7 stations were used for training and the data from the other station were used for testing. A total of 6 CNN sub-models were built, namely, non-SOCS, non-PGMS, non-TAIS, non-FAIS, non-LHES and non-BMTS. The detailed information of CNN-PSPA is shown in Table 4.

In the CNN-FSPA model, the comparisons between the observed value and the predicted value were used for testing the CNN's prediction abilities. In the CNN-PSPA models, the trained model was used to analyze the prediction error of the site amplification factors of unrecorded locations. The station data that were not included in the training samples, were used for testing the prediction abilities of the CNN-PSPA models for prediction of unrecorded locations.

3. Results and Discussion

In the following, the amplification factors of the seven stations obtained using Equation (3) were used to conduct an error analysis between the real values and predicted values of the amplification factors. The error of the amplification factor is defined in Equation (5).

$$error = \frac{|predicted - observed|}{observed} \times 100\% \tag{5}$$

3.1. Comparisons with BPNN Models

The CNN models were trained for predicting the site amplification factors of the Lower Hutt Valley. To demonstrate the advantages of the CNN models in predicting the site amplification factors, traditional BPNN models were also trained on similar data for comparison. These models were named BPNN-FSPA and BPNN-PSPA.

3.1.1. Effect of Parameters on Different Models

First, the differences between the CNN and BPNN models were considered for studying the effect of the parameters. The four groups of training parameters for the CNN and BPNN models are listed in Table 5. In the BPNN models, the tests referred to different hidden layers, as shown in Table 5. The errors of the prediction results are shown in Figure 9 for the CNN models and Figure 10 for the BPNN models.

Table 5. Four tests with different parameters.

Models	CNN		BPNN
	Kernel Size	**Kernel Number**	**Hidden Layer Size**
Test 1	[2 1] [3 1]	190 310	5
Test 2	[2 1] [3 1]	190 320	6
Test 3	[2 1] [3 1]	198 320	7
Test 4	[2 1] [3 1] [4 1]	190 320 322	8

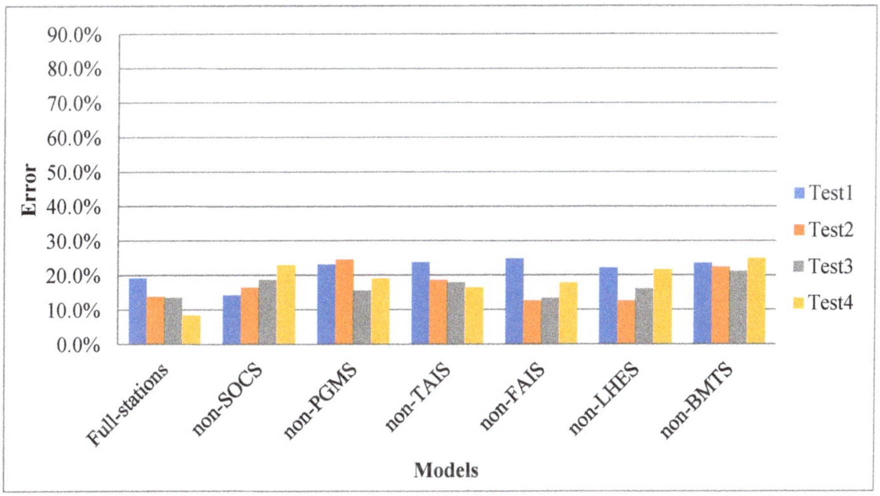

Figure 9. Errors of CNN predictions.

Figure 9 shows that with different model parameters, the prediction results for the CNN models did not differ much, and the training effect was relatively stable. In contrast, the BPNN training results appeared to be quite variable for all the tests and were extremely unstable with respect to the different model parameters, as shown in Figure 10. Figure 10 also shows that Test 1 had the worst testing results and the highest error rate. Furthermore, the variability of the red histogram was the largest, showing that Test 2 was the most unstable. Test 3 was relatively stable and its error rate was the lowest. Furthermore, in the non-FAIS model, the training effect was encouraging, except in Test 1. Among the four tests, the most stable BPNN model was the non-SOCS model.

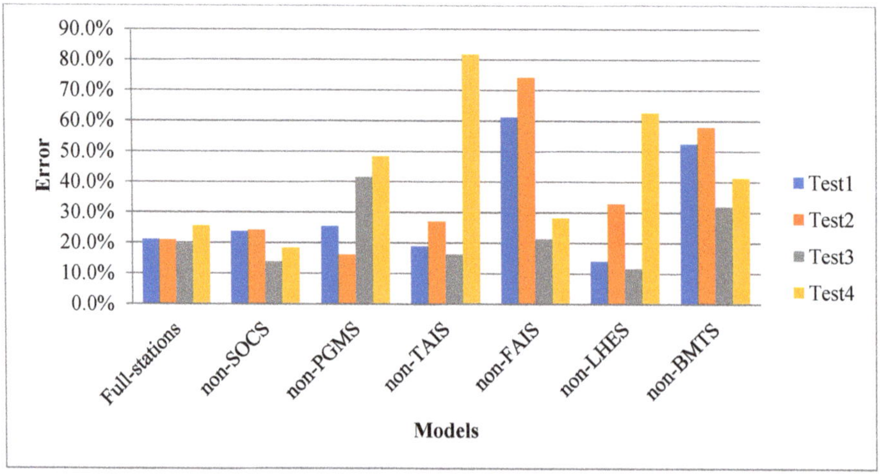

Figure 10. Errors of BPNN predictions.

As shown from the analyses of the CNN models and BPNN models, the parameters could be further optimized for the two kinds of models. Compared to the BPNN models, the CNN models had better accuracy and stability with different model parameters. To ensure the accuracy and stability of the prediction results, the BPNN should avoid the Test 1 parameters with a high error rate and the Test 2 parameters with poor stability.

3.1.2. Comparisons of Prediction Results

From Figures 9 and 10, the worst and the best CNN and BPNN models, respectively, were selected for comparison. The best CNN model, except for CNN-FSPA, was the non-FAIS model with Test 2 parameters and the best BPNN model, except for BPNN-FSPA, was the non-LHES model with Test 3 parameters. The worst CNN model was the non-BMTS model with Test 4 parameters and the worst BPNN model was the non-TAIS model with Test 4 parameters. The prediction results and errors for four earthquakes (M4.5, M5.6, M5.1 and M4.0) at four stations (FAIS, LHES, BMTS and TAIS) are shown in Table 6.

Table 6. Comparisons of CNNs and BPNNs.

		CNN-PSPA			BPNN-PSPA		
Model	**Magnitude**	**Observed (FAIS)**	**Predicted (FAIS)**	**Error**	**Observed (LHES)**	**Predicted (LHES)**	**Error**
Best testing	M4.5	2.86	2.48	13.2%	2.05	2.51	18.3%
	M5.6	1.89	1.98	4.7%	2.55	2.69	5.2%
	M5.1	1.90	1.93	1.7%	2.80	3.04	8.3%
	M4.0	1.81	2.12	17.3%	3.42	2.53	26.0%
Model	**Magnitude**	**Observed (BMTS)**	**Predicted (BMTS)**	**Error**	**Observed (TAIS)**	**Predicted (TAIS)**	**Error**
Worst testing	M4.5	2.79	1.99	28.6%	2.34	3.85	64.4%
	M5.6	2.45	2.30	6.1%	3.56	7.30	104.8%
	M5.1	1.78	1.77	0.7%	3.10	4.61	48.6%
	M4.0	2.01	1.81	9.9%	3.65	5.33	45.9%

As shown in Table 6, all the errors of the CNN models were lower than those of the corresponding BPNN models for the same earthquakes. In addition, the worst testing model (non-BMTS) of the CNNs had an error rate of 25.0% and the corresponding non-TAIS model of the BPNNs had an error rate of 81.6%.

It can also be seen from Table 6 that the error difference between the best and the worst predictions of the CNNs was much lower than for the BPNNs. Comparisons between the best and worst of the CNN-PSPA models showed that the errors of the results did not differ much. Furthermore, the earthquakes with magnitude 5.1 gave the best results, with little fluctuation in this CNN-PSPA. The CNNs showed better stability and less sensitivity to the parameters. In contrast, the difference between the best and worst tests of the BPNNs was significant. The worst of the BPNNs showed both a larger variance and the largest error of 104.8%, appearing in the M5.6 earthquakes. The BPNN-PSPA test for a magnitude of 5.6 showed the difference between the worst test with the largest error rate of 104.8% and the best test with the smallest error rate of 5.2%. It can be seen that the BPNN models were very unstable and were adversely affected by changing the parameters. The CNN models outperformed the corresponding BPNN models in terms of test stability.

Both the stability and the accuracy analyses showed that the CNNs were significantly better than the BPNNs.

3.2. Prediction Results of the CNN-FSPA Model

As the last section stated, the CNN models showed better prediction ability for the site amplification factors than the traditional BPNN models. In the following two sections, the CNN-FSPA and CNN-PSPA models are discussed in detail with regard to parameter optimization and prediction results. Based on the optimized parameters, the comparatively better prediction model was trained for predicting the site amplification factors of the Lower Hutt Valley.

3.2.1. Testing of Different Parameters

The CNN-FSPA model was trained using 40 earthquakes for training and 4 earthquakes for testing. The average test errors of the CNN-FSPA model for different CNN parameters are listed in Table 7.

Table 7. No padding and padding of the CNN models.

No Padding			Padding		
Kernel Size	Kernel Number	Error	Kernel Size	Kernel Number	Error
[2 1]	192	18.3%	[2 1]	192	19.7%
[2 1] [3 1]	190 320	13.0%	[2 1] [3 1]	190 320	12.6%
[2 1] [3 1] [4 1]	190 320 330	13.3%	[2 1] [3 1] [4 1]	190 320 330	12.8%

Table 7 shows that by varying the models with and without padding, it was found that the models with more than one convolutional layer showed lower error rates and better predictions than their counterparts without padding. The prediction results of the CNNs with different convolutional layers and convolutional kernels are also listed in Table 7. For the CNNs with one convolutional layer, the errors of the test results were 18.3–19.7%. For the CNNs with two convolutional layers, the error rate was 12.6–13.0%. For the CNNs with three convolutional layers, the error rate was slightly higher at 12.8–13.3%. Overall, the testing error could be reduced by about 1% using the padding function of the CNN models with two or three convolutional layers.

The test with the lowest error was selected to give the training parameters for determining the CNN structure. The specific structural parameters of the CNNs are shown in Table 8. There were five layers with two convolutional layers in this test. In the first convolutional layer, 190 convolutional kernels with a size of 2×1 were used to process the input data. In the second convolutional layer, the number of convolutional kernels was 320 with a size of 3 × 1.

Table 8. Structural parameters of the 1-D CNNs.

Layer	Type	Kernel No.	Kernel Size	Stride	Padding	Activation
1	Input	None	None	None	None	None
2	Convolution (C1)	190	2 × 1	1	1	Leaky ReLU
3	Convolution (C2)	320	3 × 1	1	1	Leaky ReLU
4	FC	None	None	None	None	None
5	Output	None	None	None	None	None

Selection of the appropriate parameters according to the actual data had a significant effect on the training of the CNN model.

3.2.2. Comparisons with Recorded Results

With the two convolutional layers shown in Table 8, the CNN-FSPA model gave better prediction results than the other models with different parameters. Therefore, the selected CNN-FSPA model was used to predict the amplification factors of the seven known station locations. The prediction results for earthquakes with magnitudes of 4.0, 4.5, 5.1 and 5.6 are listed in Table 9. The comparisons of the predicted and observed results are also shown in Table 9 and Figure 11.

Table 9. Prediction results for the CNN-FPSA.

Station	Earthquakes	Observed	Predicted	Error	Average Error
LHRS	M4.5	1.00	1.02	2.0%	8.5%
	M5.6	1.00	0.99	1.0%	
	M5.1	1.00	1.02	2.0%	
	M4.0	1.00	1.03	2.9%	
BMTS	M4.5	2.45	1.79	26.9%	
	M5.6	1.78	1.89	5.8%	
	M5.1	2.01	1.76	12.4%	
	M4.0	1.61	1.55	3.7%	
LHES	M4.5	2.05	2.23	8.1%	
	M5.6	2.56	2.69	5.1%	
	M5.1	2.81	2.93	4.1%	
	M4.0	3.42	2.66	22.2%	
FAIS	M4.5	2.86	2.30	19.6%	
	M5.6	1.90	1.95	2.6%	
	M5.1	1.91	1.94	1.5%	
	M4.0	1.81	1.88	3.7%	
TAIS	M4.5	3.56	3.08	13.5%	
	M5.6	3.10	3.15	1.6%	
	M5.1	3.65	3.23	11.5%	
	M4.0	2.98	2.37	20.5%	
PGMS	M4.5	1.83	2.10	12.9%	
	M5.6	1.95	2.04	4.4%	
	M5.1	2.06	2.02	1.9%	
	M4.0	2.06	1.72	16.5%	
SOCS	M4.5	2.50	2.61	4.2%	
	M5.6	3.04	3.00	1.3%	
	M5.1	3.24	3.16	2.5%	
	M4.0	3.27	2.57	21.4%	

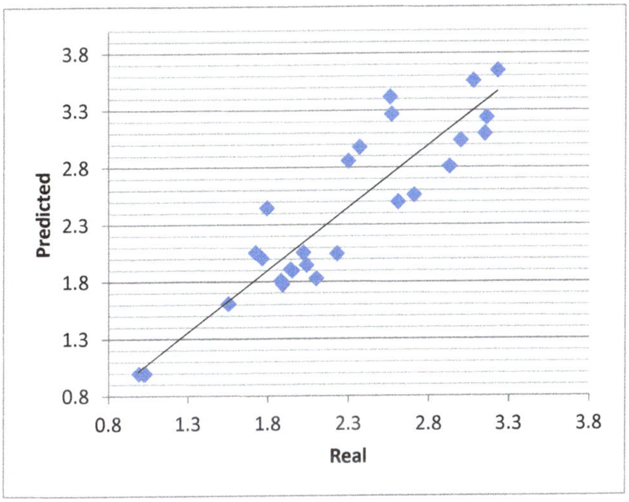

Figure 11. Comparison of predicted and observed results.

Table 9 shows that the total average prediction error was 8.5%, which is relatively low. In addition, it can clearly be seen from Table 9 that most of the prediction errors for small earthquakes such as M4.0 and M4.5 were rather high, probably due to the effect of anomalous amplification of seismic waves by the soil layer. In contrast, the predictions of the middle-sized earthquakes such as M5.1 and M5.6 were better. Furthermore, it can be observed that the stations FAIS, PGMS and SOCS (but not the bedrock station LHES) had low prediction errors. This indicates that the CNN-FSPA model had good testing results for those three stations.

As shown in the Figure 11, the scatter points were evenly and regularly distributed around the black oblique line. This shows that this model was able to provide a good prediction.

3.3. Prediction Results of CNN-PSPA Models

The CNN-PSPA models were also trained for predicting the site amplification factors of the Lower Hutt Valley. The CNN-PSPA models included six sub-models, described in Section 2.4.2. Unlike the trained CNN-FSPA model, which included the information for the testing samples, the trained CNN-PSPA models did not include the information for the testing samples. Since there was no information for test location at all in the training samples, the errors of the prediction results at test locations could be regarded as similar to those for the other unrecorded locations used for the reliability analysis of this type of model. In the following, the six CNN-PSPA models are used to discuss their prediction ability via error analysis. The errors of the amplification factors were also computed using Equation (4).

3.3.1. Optimal Parameters for CNN-PSPA Models

Since the accuracy and the stability in the training of the CNN models were mainly dependent on the parameters of the convolutional layers and convolutional kernels, error values computed with different parameters were used to optimize the CNN-PSPA models. In order to construct a more optimal CNN-based model, four groups with different convolutional kernel sizes and kernel numbers were selected for evaluating the prediction errors. The average errors of the six CNN-PSPA models with different parameters are shown in Table 10. The differences in the average error between the maximum and minimum for each CNN-PSPA model are also listed in Table 10.

Table 10. Results for CNN-PSPA with different parameters.

Models	Predicted Station	Kernel Size	Kernel Number	Ave. Error	Error Fluctuation
non-SOCS	SOCS	[2 1] [3 1] [2 1] [3 1] [2 1] [3 1] [2 1] [3 1]	190 310 190 315 190 318 190 320	16.5% 18.5% 14.2% 22.9%	8.7%
non-PGMS	PGMS	[2 1] [3 1] [2 1] [3 1] [2 1] [3 1] [2 1] [3 1]	190 320 196 310 196 320 198 320	23.1% 19.0% 24.5% 15.4%	9.1%
non-TAIS	TAIS	[2 1] [3 1] [4 1] [2 1] [3 1] [4 1] [2 1] [3 1] [4 1] [2 1] [3 1] [4 1]	190 320 322 190 320 325 190 320 330 192 320 330	16.5% 18.6% 17.9% 20.8%	4.3%
non-FAIS	FAIS	[2 1] [3 1] [2 1] [3 1] [2 1] [3 1] [2 1] [3 1]	190 315 190 320 196 320 190 326	22.4% 12.6% 13.2% 17.9%	9.8%
non-LHES	LHES	[2 1] [3 1] [2 1] [3 1] [2 1] [3 1] [2 1] [3 1]	188 320 190 320 192 320 196 320	22.1% 12.2% 16.0% 21.6%	9.9%
non-BMTS	BMTS	[2 1] [3 1] [2 1] [3 1] [2 1] [3 1] [2 1] [3 1]	190 320 192 320 192 330 196 320	21.1% 22.4% 25.0% 23.6%	3.9%

Table 10 shows that the average error values varied with the different parameters for all of the six CNN-PSPA models. The error differences of the six models ranged from 14.3%, 15.5%, 17.9%, 12.6%, 12.6%, and 21.2% to 23.0%, 24.5%, 23.9%, 22.5%, 22.1%, and 25.0%. Regarding the errors of the six models, the maximum change was 9.9% in the non-LHES model, and the minimum change was 3.9% in the non-BMTS model.

The non-BMTS model was the least sensitive to the CNN parameters, while the non-LHES model was the most influenced by the parameters. From Table 10, the optimal values of the six models were extracted from the four group parameters for subsequent predictions. Among the six CNN-PSPA models, the best prediction model was the non-LHES model, which had a relative error of 12.2%. The model with the biggest error of all the six models was the non-BMTS model, which had an average error of 21.1%. The accuracy and stability of the trained CNN-PSPA models remained at a reasonable level for different network parameters.

Table 10 shows that the sensitivity of the test results was inversely proportional to the error rate of the results. For example, the non-LHES model had the highest error fluctuation among the six CNN-PSPA models when trained with different parameters, but it had the lowest error value among the models. Compared to the five other models, it was more sensitive to the parameters within the control range. However, the models with higher errors did not fluctuate as much. This led to stable prediction results by the CNN-PSPA model, even with different training parameters.

3.3.2. Comparisons with Observed Results

The non-LHES model, with the best test results among all the six CNN-PSPA models, was selected for predicting all the site amplification factors of 39 earthquakes at the LHES station location, where five earthquake records were not included. The prediction results for the 39 earthquakes are listed in Table 11. The site amplification factors obtained from the observation records are listed in Table 11 for comparison. By comparing with the corresponding observation results, the errors were also calculated for all the incidents, as shown in Table 11.

Table 11. Prediction results for non-LHES.

Number	Earthquake	Epicentral Distance (km)	Depth (km)	CNN-PSPA			Average Error	
				Observed (LHES)	Predicted (LHES)	Error		
1	M4.0	104	40	3.42	2.51	26.5%		
2	M4.0	86	31	2.18	2.41	10.6%		
3	M4.1	113	11	3.03	2.38	21.5%		
4	M4.1	84	10	2.78	2.50	10.1%		
5	M4.1	57	12	3.04	2.45	19.4%		
6	M4.1	42	12	2.36	2.49	5.5%		
7	M4.2	113	6	3.83	2.70	29.5%		
8	M4.2	15	26	2.32	2.60	12.1%		
9	M4.3	72	11	2.73	2.45	10.3%		
10	M4.3	79	16	2.09	2.23	6.7%	12.7%	
11	M4.3	85	32	2.39	2.28	4.6%		
12	M4.4	98	10	2.60	2.51	3.5%		
13	M4.4	76	5	2.71	2.50	7.7%		
14	M4.4	87	28	2.47	2.28	7.7%		
15	M4.5	16	24	2.67	2.30	13.9%		
16	M4.5	106	36	2.05	2.39	16.5%		
17	M4.5	85	30	2.45	2.30	6.1%		
18	M4.5	79	7	3.20	2.55	20.3%		
19	M4.5	86	30	2.55	2.31	9.4%		
20	M4.6	116	32	2.31	2.42	4.8%		12.2%
21	M4.6	82	12	3.31	2.63	20.5%		
22	M4.7	81	33	2.36	2.21	6.4%		
23	M4.7	57	13	3.23	2.59	19.8%		
24	M4.8	72	11	2.59	2.39	7.7%		
25	M4.8	74	12	2.78	2.39	14.0%		
26	M4.8	84	54	2.76	2.59	6.2%		
27	M4.8	74	9	2.91	2.42	16.8%		
28	M5.0	86	36	2.63	2.34	11.0%		
29	M5.0	50	13	2.73	2.28	16.5%	11.7%	
30	M5.0	76	8	2.99	2.43	18.7%		
31	M5.1	78	5	2.81	2.52	10.4%		
32	M5.2	121	8	2.90	2.66	8.3%		
33	M5.4	82	34	2.14	2.24	4.7%		
34	M5.5	74	17	2.38	2.15	9.7%		
35	M5.5	91	15	3.65	2.97	18.6%		
36	M5.6	85	7	2.56	2.42	5.5%		
37	M5.6	82	13	2.58	2.38	7.8%		
38	M5.8	85	37	2.11	2.48	17.5%		
39	M6.2	104	34	2.57	2.81	9.3%		

Table 11 shows that the total average prediction error was 12.2%, which was relatively low. In addition, it can clearly be seen from Table 11 that most of the prediction errors for small earthquakes such as M4.0 and M4.5 were rather high, probably due to the effect of anomalous amplification of seismic waves by the soil layer. In contrast, the predictions for the medium-sized earthquakes such M5.1 and M5.6 were better. Furthermore, low prediction errors for stations FAIS, PGMS and SOCS (but not the bedrock station LHES) were observed. This indicates that the CNN-FSPA model showed good training for those three stations. Compared with the CNN-FSPA model in Table 9, the average error of the CNN-PSPA model was slightly higher by 3.7%.

In addition, the errors for small earthquakes with large epicenter and source depths, such as earthquakes with serial numbers 1, 3, 7 and 16, were significantly higher than those for other earthquakes. However, for medium earthquakes with large epicentral distances and focal depths, such as earthquakes with serial numbers 20, 32 and 39, errors were lower and predictions were satisfactory.

3.3.3. Comparison with CNN-FSPA Model

To estimate the ability of CNN-PSPA prediction models, all the six trained models (non-SOCS, non-PGMS, non-TAIS, non-FAIS, non-BMTS and non-LHES) were used to predict the site amplification factors at the corresponding stations. For example, the trained non-SOCS model predicted the site amplification factor at the SOCS station. The test-sample data were selected using the four earthquakes that were also used to test the CNN-FSPA model. All error values for the site amplification factors at the six soil stations were calculated from the six trained CNN-PSPA models. The corresponding errors for the same four earthquakes at the six station locations based on the trained CNN-FSPA models were also predicted for comparisons. All the errors of the predicted values for these seven CNN models are listed in Table 12.

Table 12. Errors for CNN-FSPA and CNN-PSPA.

Magnitude	Predicted	CNN-PSPA		CNN-FSPA
		Models	Error	Error
M4.5	SOCS	non-SOCS	21.7%	4.2%
	PGMS	non-PGMS	29.9%	12.9%
	TAIS	non-TAIS	37.9%	13.5%
	FAIS	non-FAIS	16.8%	19.6%
	BMTS	non-BMTS	12.1%	26.9%
	LHES	non-LHES	16.5%	8.1%
M5.6	SOCS	non-SOCS	16.6%	1.3%
	PGMS	non-PGMS	21.7%	4.4%
	TAIS	non-TAIS	1.8%	1.6%
	FAIS	non-FAIS	4.7%	2.6%
	BMTS	non-BMTS	21.0%	5.8%
	LHES	non-LHES	5.5%	5.1%
M5.1	SOCS	non-SOCS	10.2%	2.5%
	PGMS	non-PGMS	16.5%	1.9%
	TAIS	non-TAIS	13.5%	11.5%
	FAIS	non-FAIS	1.7%	1.5%
	BMTS	non-BMTS	20.8%	12.4%
	LHES	non-LHES	10.4%	4.1%
M4.0	SOCS	non-SOCS	10.0%	21.4%
	PGMS	non-PGMS	21.3%	16.5%
	TAIS	non-TAIS	12.7%	20.5%
	FAIS	non-FAIS	17.3%	3.7%
	BMTS	non-BMTS	12.2%	3.7%
	LHES	non-LHES	26.5%	22.2%

Table 12 shows that since the prediction errors of the CNN-FSPA model were lower than those of the CNN-PSPA model, the CNN-FSPA model was significantly better than the CNN-PSPA model. This phenomenon shows that the more station data that are included in the CNN training, the better the CNN model is trained. The included test information could influence the accuracy and credibility of the prediction results. Table 12 also shows that although the trained CNN-PSPA model had higher errors than the trained CNN-FSPA model in any prediction, it had relatively smaller differences, demonstrating that reliable amplification factors could be obtained even using the trained CNN-PSPA model.

Table 12 also shows that, as for the predictive regularities of the CNN-FSPA models, the CNN-PSPA predictions for the medium-sized earthquakes such as M5.1 and M5.6 were better than those for micro-earthquakes such as M4.0 and M4.5. The non-FAIS model showed outstanding predictive power for medium earthquakes, even though it exhibited a relatively high error for the micro-earthquakes such as M4.0 and M4.5. In contrast, for the CNN-FSPA model, it should be noted that the non-FAIS and non-BMTS models for M4.5 and the non-TAIS model for M4.0 unexpectedly exhibited higher error rates than the CNN-PSPA model.

3.4. Comparisons of Unrecorded Locations

To predict the amplification characteristics of unrecorded locations, five randomly selected site points, other than those shown in Figure 8, named A, B, C, D and E, were used to predict their amplification factors. With the earthquakes with magnitude M5.0 as an example, the seven trained CNN models with the best testing results, namely, CNN-FSPA, non-SOCS, non-PGMS, non-TAIS, non-FAIS, non-BMTS and non-LHES, were used to calculate the amplification factors of the five points. The prediction results of the amplification factors at these five unrecorded points by the seven CNN-models are shown in Figure 12.

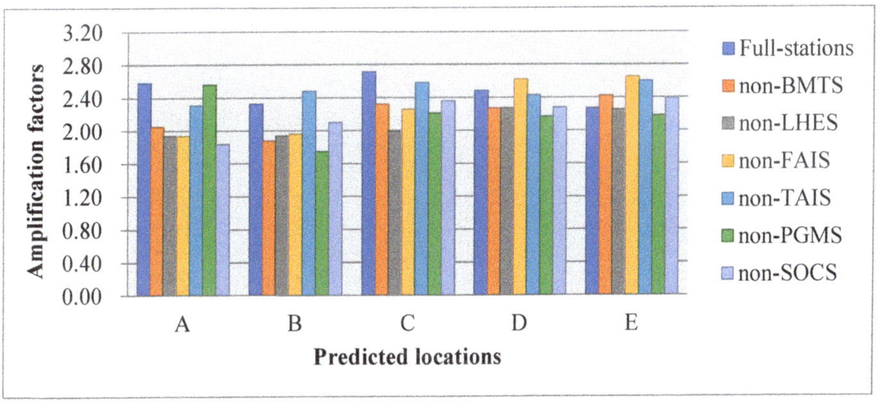

Figure 12. Prediction of the amplification factors of unrecorded points by seven models.

Figure 12 shows that the amplification factors varied in the order E > D > C > A > B. As shown in Figure 12, all the five points, A, B, C, D and E, exhibited a large amplification factor when the CNN-FSPA model was used for prediction. If only the CNN-PSPA model (which showed better training) was used for prediction, the prediction results of the non-FAIS and non-LHES models for the five points were similar.

However, the predicted amplification factor using the CNN-PSPA model was still rather small compared to that of the CNN-FSPA model. The reason for this was its insufficient training quantity and incomplete learning. Therefore, the task of predicting the amplification factor of unknown points should be performed in a series of steps. These inspiring results could provide a reference for seismic intensity setting in practical projects.

For discussing the differences in the prediction results, the relative error of two types of CNNs was used, as given by Equation (6):

$$Error_{PSPA} = \frac{|Value_{\text{CNN-PSPA}} - Value_{\text{CNN-FSPA}}|}{|Value_{\text{CNN-FSPA}}|} \times 100\% \tag{6}$$

where $Error_{PSPA}$ denotes the difference rate between the CNN-PSPA and the CNN-PSPA, and $Value_{\text{CNN-PSPA}}$ and $Value_{\text{CNN-FSPA}}$ are the predicted amplification factors of the CNN-PSPA model and CNN-FSPA model, respectively. The error values of the predicted amplification factors of CNN-PSPA for the five points (A, B, C, D and E) for an earthquake with M5.0 are shown in Table 13.

Table 13. Relative errors of CNN-PSPA.

Points \ Models	CNN-PSPA					
	Non-SOCS	**Non-PGMS**	**Non-TAIS**	**Non-FAIS**	**Non-LHES**	**Non-BMTS**
A	26.8%	8.6%	10.5%	24.9%	24.9%	20.6%
B	9.8%	24.8%	6.1%	15.8%	16.7%	19.3%
C	13.0%	18.5%	4.9%	16.7%	26.6%	14.5%
D	8.2%	12.6%	2.2%	5.2%	8.6%	8.6%
E	5.0%	3.9%	12.7%	14.4%	0.9%	6.2%

Compared with the CNN-FSPA model, the CNN-PSPA models clearly showed a high error rate at point A, indicating that more attention is required. In addition, for the predictions in Table 13, it can be seen that the amplification factors of points D and E showed good predictions and small fluctuations in each model. Therefore, the CNN-FSPA model could predict the amplification factor of an unrecorded location well.

4. Conclusions

In this paper, a 1-D CNN method was used to predict the ground motion amplification of the Lower Hutt Valley. Based on a relatively small sample dataset, two kinds of CNN models were established and applied, to predict the amplification factors of the local zone. The prediction results were inspiring.

Based on the above results, the following conclusions were drawn.

Compared with BNPP models, the trained CNN models were influenced little by different training parameters, leading to a more stable trained model.

The CNN-FSPA model could effectively predict the station amplification factors from the existing data; the accuracy was 91.5%, which was 15.8% higher than that of the corresponding BPNN-FSPA model.

Both CNN-FSPA and CNN-PSPA models could effectively predict the amplification factors of local ground motion.

The comparisons of the CNN-FSPA and CNN-PSPA models showed that, by increasing the number of earthquakes and the training samples, a prediction model of the ground surface amplification could be established based on strong earthquake observations.

5. Discussion

(1) For the data processing, we used the Fourier transform method, and also tried other data processing methods such as the wavelet transform method, to find more accurate data processing methods for subsequent research.

(2) Since it is difficult for numerical simulations to contain all the seismic information, in this study we used a combination of seismic records and CNN to predict the amplification factors with an increased quantity of seismic data. In the future, we will also use a combination of simulations and observed data to improve the prediction accuracy via migration learning.

(3) The current trend is to use deep learning to solve some prediction problems. With the construction of regional station networks and the increase in training samples, we will continue to collect seismic information from other sites, such as Ashighara Valley and Ohba Valley in Japan, the Volvi basin in Greece and so on. Therefore, using machine learning to predict earthquakes has good prospects for development.

Author Contributions: X.Y. contributed toward the conceptualization, methodology, investigation, formal analysis, original draft preparation, software, visualization, and data curation. Y.C. contributed toward the conceptualization, methodology, investigation, formal analysis, original draft preparation, and supervision. S.T. and G.C. contributed toward the investigation, formal analysis methodology, investigation, and review and editing. All authors have read and agreed to the published version of the manuscript.

Funding: The financial supports provided by the National Key R&D Program of China (project No. 2018YFC1504004) was greatly appreciated.

Institutional Review Board Statement: Not applicable.

Informed Consent Statement: Not applicable.

Data Availability Statement: Some or all data, models, or codes generated or used during the study are available from the corresponding author by request.

Acknowledgments: The geological data provided by New Zealand's Institute of Geological and Nuclear Sciences (GNS) was greatly appreciated.

Conflicts of Interest: The authors declare no conflict of interest.

References

1. Sánchez-Sesma, F.; Crouse, C.B. Effects of site geology on seismic ground motion: Early history. *Earthq. Eng. Struct. Dyn.* **2015**, *44*, 1099–1113. [CrossRef]
2. Trifunac, M.D. Site conditions and earthquake ground motion–A review. *Soil Dyn. Earthq. Eng.* **2016**, *90*, 88–100. [CrossRef]
3. Markham, C.S.; Bray, J.D.; Macedo, J.; Luque, R. Evaluating nonlinear effective stress site response analyses using records from the Canterbury earthquake sequence. *Soil Dyn. Earthq. Eng.* **2016**, *82*, 84–98. [CrossRef]
4. Harmsen, S.C. Determination of site amplification in the Los Angeles urban area from inversion of strong-motion records. *Bull. Seismol. Soc. Am.* **1997**, *87*, 866–887.
5. Hartzell, S. Variability in nonlinear sediment response during the 1994 Northridge, California, earthquake. *Bull. Seismol. Soc. Am.* **1998**, *88*, 1426–1437.
6. Kaveh, A.; Bakhshpoori, T.; Hamzeh-Ziabari, S.M. Derivation of New Equations for Prediction of Principal Ground-Motion Parameters using M5′ Algorithm. *J. Earthq. Eng.* **2016**, *20*, 910–930. [CrossRef]
7. Pezzo, E.D.; Martino, S.D.; Parrinello, M.T.; Sabbarese, C. Seismic site amplification factors in Campi Flegrei, Southern Italy. *Phys. Earth Planet. Inter.* **1993**, *78*, 105–117. [CrossRef]
8. Sandikkaya, M.A.; Akkar, S.; Bard, P.Y. A Nonlinear Site-Amplification Model for the Next Pan-European Ground-Motion Prediction Equations. *Bull. Seismol. Soc. Am.* **2013**, *103*, 19–32. [CrossRef]
9. Yu, J. The Choice of Reference Sites for Seismic Ground Amplification Analyses: Case Study at Parkway, New Zealand. *Bull. Seismol. Soc. Am.* **2003**, *93*, 713–723. [CrossRef]
10. Zaslavsky, Y.; Shapira, A. Experimental study of topographic amplification using the Israel Seismic Network. *J. Earthq. Eng.* **2000**, *4*, 43–65. [CrossRef]
11. Ferrari, R.D.; Ferretti, G.; Barani, S.; Pepe, G.; Cevasco, A. On the role of stiff soil deposits on seismic ground shaking in western Liguria, Italy: Evidences from past earthquakes and site response. *Eng. Geol.* **2017**, *226*, 172–183. [CrossRef]
12. Kanai, K.; Tanaka, T. On Microtremors. VIII. *Bull. Earthq. Res. Inst.* **1961**, *39*, 97–114.
13. Nakamura, Y.A. Method for Dynamic Characteristics Estimation of Subsurface Using Microtremor on Ground Surface. *Q. Rep. Rtri.* **1989**, *30*, 25–33.
14. Sadiq, M.T.; Yu, X.; Yuan, Z.; Fan, Z.; Rehman, A.U.; Li, G.; Xiao, G. Motor imagery EEG signals classification based on mode amplitude and frequency components using empirical wavelet transform. *IEEE Access* **2019**, *7*, 127678–127692. [CrossRef]
15. Sadiq, M.T.; Yu, X.; Yuan, Z.; Aziz, M.Z. Motor imagery BCI classification based on novel two-dimensional modelling in empirical wavelet transform. *Electron. Lett.* **2020**, *56*, 1367–1369. [CrossRef]
16. Sadiq, M.T.; Yu, X.; Yuan, Z.; Zeming, F.; Rehman, A.U.; Ullah, I.; Li, G.; Xiao, G. Motor imagery EEG signals decoding by multivariate empirical wavelet transform-based framework for robust brain–computer interfaces. *IEEE Access* **2019**, *7*, 171431–171451. [CrossRef]
17. Bernal, D. Amplification factors for inelastic dynamic p–Δ effects in earthquake analysis. *Earthq. Eng. Struct. Dyn.* **1987**, *15*, 635–651. [CrossRef]
18. Tropeano, G.; Soccodato, F.M.; Silvestri, F. Re-evaluation of code-specified stratigraphic amplification factors based on Italian experimental records and numerical seismic response analyses. *Soil Dyn. Earthq. Eng.* **2018**, *110*, 262–275. [CrossRef]
19. Glinsky, N.; Bertrand, E.; Régnier, J. Numerical simulation of topographical and geological site effects. Applications to canonical topographies and Rognes hill, South East France. *Soil Dyn. Earthq. Eng.* **2019**, *116*, 620–636. [CrossRef]
20. Kamai, R.; Abrahamson, N.A.; Silva, W.J. Nonlinear Horizontal Site Amplification for Constraining the NGA-West2 GMPEs. *Earthq. Spectra* **2014**, *30*, 1223–1240. [CrossRef]
21. Athanasopoulos, G.A.; Pelekis, P.C.; Leonidou, E.A. Effects of surface topography on seismic ground response in the Egion (Greece) 15 June 1995 earthquake. *Soil Dyn. Earthq. Eng.* **1999**, *18*, 135–149. [CrossRef]

22. Riga, E.; Makra, K.; Pitilakis, K. Aggravation factors for seismic response of sedimentary basins: A code-oriented parametric study. *Soil Dyn. Earthq. Eng.* **2016**, *91*, 116–132. [CrossRef]
23. Wald, D.J.; Graves, R.W. The seismic response of the Los Angeles basin, California. *Bull. Seism. Soc. Am.* **1998**, *88*, 337–356.
24. Khanbabazadeh, H.; Iyisan, R.; Ansal, A.; Hasal, M.E. 2D non-linear seismic response of the Dinar basin, TURKEY. *Soil Dyn. Earthq. Eng.* **2016**, *89*, 5–11. [CrossRef]
25. Stamati, O.; Klimis, N.; Lazaridis, T. Evidence of complex site effects and soil non-linearity numerically estimated by 2D vs 1D seismic response analyses in the city of Xanthi. *Soil Dyn. Earthq. Eng.* **2016**, *87*, 101–115. [CrossRef]
26. Semblat, J.F.; Duval, A.M.; Dangla, P. Numerical analysis of seismic wave amplification in Nice (France) and comparisons with experiments. *Soil Dyn. Earthq. Eng.* **2000**, *19*, 347–362. [CrossRef]
27. Meza-Fajardo, K.C.; Varone, C.; Lenti, L.; Martino, S.; Semblat, J.F. Surface wave quantification in a highly heterogeneous alluvial basin: Case study of the Fosso di Vallerano valley, Rome, Italy. *Soil Dyn. Earthq. Eng.* **2019**, *120*, 292–300. [CrossRef]
28. Abdullah, S.M. On linear site amplification behavior of crustal and subduction interface earthquakes in Japan: (1) regional effects, (2) best proxy selection. *Bull. Earthq. Eng.* **2018**, *17*, 119–139.
29. Kim, S.; Hwang, Y.; Seo, H.; Kim, B. Ground motion amplification models for Japan using machine learning techniques. *Soil Dyn. Earthq. Eng.* **2020**, *132*, 106095. [CrossRef]
30. Yuen, K.V.; Ortiz, G.A.; Huang, K. Novel nonparametric modeling of seismic attenuation and directivity relationship. *Comput. Methods Appl. Mech. Eng.* **2016**, *311*, 537–555. [CrossRef]
31. Güllü, H.; Erçelebi, E. A neural network approach for attenuation relationships: An application using strong ground motion data from Turkey. *Eng. Geol.* **2007**, *93*, 65–81. [CrossRef]
32. Yaghmaei-Sabegh, S.; Rupakhety, R. A new method of seismic site classification using HVSR curves: A case study of the 12 November 2017 Mw 7.3 Ezgeleh earthquake in Iran. *Eng. Geol.* **2020**, *270*, 105574. [CrossRef]
33. Yaghmaei-Sabegh, S.; Tsang, H.-H. A new site classification approach based on neural networks. *Soil Dyn. Earthq. Eng.* **2011**, *31*, 974–981. [CrossRef]
34. Mu, H.Q.; Yuen, K.V. Ground Motion Prediction Equation Development by Heterogeneous Bayesian Learning. *Comput. Civ. Infrastruct. Eng.* **2016**, *116*, 761–776. [CrossRef]
35. Sadiq, M.T.; Yu, X.; Yuan, Z. Exploiting dimensionality reduction and neural network techniques for the development of expert brain–computer interfaces. *Expert Syst. Appl.* **2021**, *164*, 114031. [CrossRef]
36. Hough, S.E.; Altidor, J.R.; Anglade, D.; Given, D.; Yong, A. Localized damage caused by topographic amplification during the 2010 M7.0 Haiti earthquake. *Nat. Geosci.* **2010**, *3*, 778–782. [CrossRef]

applied sciences

MDPI

Article

Measurement of Bridge Vibration by UAVs Combined with CNN and KLT Optical-Flow Method

Zhaocheng Yan [1], Zihan Jin [1], Shuai Teng [1], Gongfa Chen [1,*] and David Bassir [2,3]

1 School of Civil and Transportation Engineering, Guangdong University of Technology, Guangzhou 510006, China; 13612318991@163.com (Z.Y.); jin15155955794@163.com (Z.J.); 1112009002@mail2.gdut.edu.cn (S.T.)
2 Centre Borelli, ENS-University of Paris-Saclay, 91190 Gif-sur-Yvette, France; david.bassir@utbm.fr
3 UTBM, IRAMAT UMR 7065-CNRS, Rue de Leupe, CEDEX, 90010 Belfort, France
* Correspondence: gongfa.chen@gdut.edu.cn; Tel.: +86-136-6248-3527

Abstract: A measurement method of bridge vibration by unmanned aerial vehicles (UAVs) combined with convolutional neural networks (CNNs) and Kanade–Lucas–Tomasi (KLT) optical-flow method is proposed. In this method, the stationary reference points in the structural background are required, a UAV is used to shoot the structure video, and the KLT optical-flow method is used to track the target points on the structure and the background reference points in the video to obtain the coordinates of these points on each frame. Then, the characteristic relationship between the reference points and the target points can be learned by a CNN according to the coordinates of the reference points and the target points, so as to correct the displacement time–history curves of target points containing the false displacement caused by the UAV's egomotion. Finally, operational modal analysis (OMA) is used to extract the natural frequency of the structure from the displacement signal. In addition, the reliability of UAV measurement combined with CNN is proved by comparing the measurement results of the fixed camera and those of UAV combined with CNN, and the reliability of the KLT optical-flow method is proved by comparing the tracking results of the digital image correlation (DIC) and KLT optical-flow method in the experiment of this paper.

Keywords: bridge vibration; unmanned aerial vehicles; Kanade–Lucas–Tomasi optical-flow method; operational modal analysis; convolutional neural networks

Citation: Yan, Z.; Jin, Z.; Teng, S.; Chen, G.; Bassir, D. Measurement of Bridge Vibration by UAVs Combined with CNN and KLT Optical-Flow Method. *Appl. Sci.* **2022**, *12*, 5181. https://doi.org/10.3390/app12105181

Academic Editor: Giuseppe Lacidogna

Received: 26 April 2022
Accepted: 19 May 2022
Published: 20 May 2022

1. Introduction

The long-term use of bridges may lead to structural damages; hence, it is necessary to detect damages regularly. Vibration measurement is an important step in structural damage detections. In recent years, some noncontact-measurement methods have been proposed, such as Global Positioning System (GPS) [1] and laser Doppler vibrometer (LDV) [2], to replace the traditional contact-measurement methods (such as acceleration sensors [3] and strain gauges [4]). However, the GPS is of low accuracy [5] and LDV is costly and time-consuming [6]. With the development of computer-vision technology, digital image correlation (DIC) is more and more widely used in bridge vibration measurement [7,8]. Compared with GPS and LDV, DIC technology has the advantages of low cost, high precision, and high efficiency. DIC is also used for deformation and displacement measurement of other engineering structures [9]. However, the measurement accuracy of the DIC method is limited due to the errors caused by pixel interpolation [10]. Also as a computer-vision method, the optical-flow method is widely used in bridge vibration measurement, and its accuracy has been confirmed [11,12].

The Kanade–Lucas–Tomasi (KLT) optical-flow method [13] is proposed on the basis of the Lucas–Kanade optical-flow method [14]. The concept of optical flow was first proposed by Gibson [15], and represents the velocity of a moving object in a time-varying image. According to the idea of optical flow, the KLT optical-flow method matches and tracks the feature points of two adjacent frames to obtain the motion information of the feature points. The KLT optical-flow method is widely used in tracking the feature points of different scenes (buildings, grasslands, etc.) to validate its reliability [16] in measuring vibration of bridge models [12].

In order to adapt to various measurement environments (such as cross-river bridges, etc.), in recent years, UAVs have been used to replace fixed cameras for bridge measurement [17,18]. UAVs are also gradually applied to structural crack detection, displacement measurement, and damage inspection of bridges [19–21]. However, due to egomotion of UAVs during their flights, the measured displacement includes not only the displacement of the measured structure, but also the false displacement caused by UAV egomotion. The common method to eliminate the false displacement is homography transformation [22,23], but this requires that four or more static reference points are in the same plane of the target points, which is difficult to achieve in actual measurement. The random components in the signals collected by UAVs may be suppressed by a differential filtering method [24] to obtain the structural modal parameters [25]. Although the bridge modal parameters can be extracted by this method without using any reference point, it does not obtain the real displacement time–history curves of the target points. The correction method of three-dimensional reconstruction is proposed to obtain the intrinsic and extrinsic parameters of the UAV cameras by using Zhang's method [26], so as to recover the 3D world coordinates of the structure to obtain its true displacement [27]. However, this method requires that the plane of the reference points is parallel to that of the target points and the distance between the two planes is known, which is difficult to achieve in practical measurement. The correction of UAV images by neural networks has attracted much attention in recent years. The method of correcting UAV images with a radial basis function (RBF) neural network is proposed [28]. The selected control points are used by this method as the training samples of the network. The corrected images can be obtained after the UAV images are inputted into the network. The two-layer feedforward neural network (FNN) was used to learn the characteristic relationship between the reference background and target points from the video with no structural motion [29]. The target-point coordinates can be estimated through the reference background when the structure is vibrating, which can be used to determine the homography transformation matrix of each frame image, so as to obtain the real coordinates of the target points.

A measurement method of bridge vibration by UAVs based on the KLT optical-flow method and a CNN is proposed in this paper. In this method, the KLT optical-flow method is used to accurately track the target points, and the multilayer CNN is used to effectively learn the characteristic relationship between the reference points and the target points to eliminate the false displacement of the UAV, so as to obtain the real displacement time–history curves of the structure, and the natural frequency of the model is extracted by OMA [30,31]. Finally, the measurement results of DIC and fixed camera are used as references to validate the reliability of the method proposed by this paper.

2. Methods

Firstly, the video of a static structure taken by a UAV is converted into a set of continuous digital image sequences stored in the frame form. The coordinates of reference points and target points are obtained by DIC and KLT optical-flow method to train the CNN. Then, the structure is excited and the image sequences are captured by a fixed camera and UAV. The target points are tracked with DIC technology and KLT optical-flow method, respectively, to obtain the displacement time–history curves (the one obtained by the UAV needs to be corrected by a CNN). The natural frequency of the structure is then extracted by OMA. The technical flowchart is shown in Figure 1.

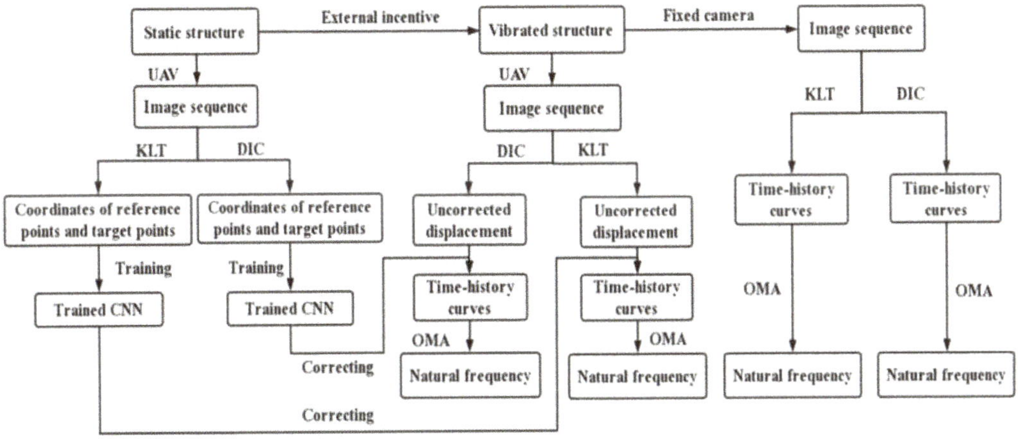

Figure 1. Technical flowchart.

2.1. KLT Optical-Flow Method

2.1.1. Assumptions of Optical-Flow Method

The optical-flow method is based on two assumptions: brightness constancy and small motion; that is, the pixel value of the same points between frames are unchanged and the motion of the points is small. Figure 2 shows three target points in two adjacent images. The position of the target points in the second image can be determined by finding the point whose pixel value is consistent with the target points in the first image.

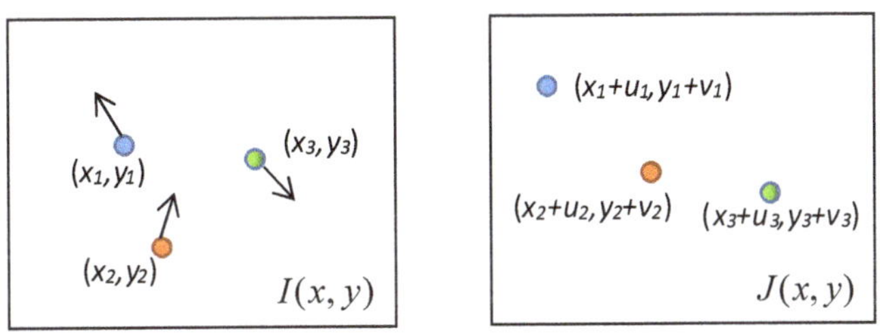

Figure 2. Motion of points.

2.1.2. KLT Optical-Flow Method

In addition to the two assumptions, the KLT optical-flow method [13] also assumes the spatial consistency; that is, the adjacent pixels in the previous frame are also adjacent in the next frame.

Suppose $I_1(x, y)$ and $I_2(x, y)$ are two adjacent images; a fixed size window, W, centered on the position of a target point is established in the first image. All pixels in W move (d_x, d_y) between the two images, and (x, y) are the coordinates of pixels. Let $\boldsymbol{d} = [d_x \ d_x]^T$, $\mathbf{X} = [x \ y]^T$. The $\boldsymbol{d}$ closest to the actual value can be obtained by minimizing the following expression:

$$\varepsilon = \iint_W [I_2(\mathbf{X}) - I_1(\mathbf{X} - \boldsymbol{d})]^2 \omega(\mathbf{X}) d\mathbf{X}. \tag{1}$$

where $\omega(\boldsymbol{X})$ is a weighting function. In the simplest case, $\omega(\boldsymbol{X}) = 1$. Alternatively, ω could be a Gaussian-like function to emphasize the central area of the window. Move the centers of I_1 and I_2 by $\boldsymbol{d}/2$ to obtain

$$\varepsilon = \iint_W \left[I_2\left(\boldsymbol{X}+\frac{\boldsymbol{d}}{2}\right) - I_1\left(\boldsymbol{X}-\frac{\boldsymbol{d}}{2}\right) \right]^2 \omega(\boldsymbol{X}) d\boldsymbol{X}. \quad (2)$$

Set the partial derivative of ε with respect to $\boldsymbol{d}$ as 0:

$$2\iint_W \left[I_2\left(\boldsymbol{X}+\frac{\boldsymbol{d}}{2}\right) - I_1\left(\boldsymbol{X}-\frac{\boldsymbol{d}}{2}\right) \right] \left[\frac{\partial I_2\left(\boldsymbol{X}+\frac{\boldsymbol{d}}{2}\right)}{\partial \boldsymbol{d}} - \frac{\partial I_1\left(\boldsymbol{X}-\frac{\boldsymbol{d}}{2}\right)}{\partial \boldsymbol{d}} \right] \omega(\boldsymbol{X}) d\boldsymbol{X} = 0. \quad (3)$$

The following formula can be obtained from Taylor's expansion:

$$I_2\left(\boldsymbol{X}+\frac{\boldsymbol{d}}{2}\right) \approx I_2(\boldsymbol{X}) + \frac{d_x}{2}\frac{\partial I_2}{\partial x}(\boldsymbol{X}) + \frac{d_y}{2}\frac{\partial I_2}{\partial y}(\boldsymbol{X}). \quad (4)$$

$$I_1\left(\boldsymbol{X}-\frac{\boldsymbol{d}}{2}\right) \approx I_1(\boldsymbol{X}) - \frac{d_x}{2}\frac{\partial I_1}{\partial x}(\boldsymbol{X}) - \frac{d_y}{2}\frac{\partial I_1}{\partial y}(\boldsymbol{X}). \quad (5)$$

Substitution of Equations (4) and (5) into Equation (3) leads to

$$\iint_W \left[I_2(\boldsymbol{X}) - I_1(\boldsymbol{X}) + \boldsymbol{p}^T \boldsymbol{d} \right] \boldsymbol{p}(\boldsymbol{X}) \omega(\boldsymbol{X}) d\boldsymbol{X} = 0. \quad (6)$$

where

$$\boldsymbol{p} = \left[\begin{array}{cc} \frac{\partial}{\partial x}\left(\frac{I_1+I_2}{2}\right) & \frac{\partial}{\partial y}\left(\frac{I_1+I_2}{2}\right) \end{array} \right]^T. \quad (7)$$

The following equation can be obtained from Equation (6):

$$Z\boldsymbol{d} = e. \quad (8)$$

where $Z = \iint_W \boldsymbol{p}(\boldsymbol{X})\boldsymbol{p}^T(\boldsymbol{X})\omega(\boldsymbol{X})d\boldsymbol{X}$, $e = \iint_W [I_1(\boldsymbol{X}) - I_2(\boldsymbol{X})]\boldsymbol{p}(\boldsymbol{X})\omega(\boldsymbol{X})d\boldsymbol{X}$.

Equation (8) can be solved by an iterative method to obtain the value of $\boldsymbol{d}$. When the value of e is less than the set threshold, the approximate solution of $\boldsymbol{d}$ can be obtained.

In this paper, the process of tracking the target point on the bridge model by the KLT optical-flow method is as follows:

1. Select the target point in the initial image;
2. Based on the local template of each target point, the vector of the point between adjacent frames can be found [13];
3. The tracking effect is judged in each image to optimize the result for each target point.

2.1.3. KLT Optical-Flow Method under Pyramid

It can be seen from Section 2.1.2 that if the assumption of small motion is not met, the Taylor expansion of Equations (4) and (5) cannot be carried out. The image pyramid [32] referred by the KLT optical-flow method is used in this paper, which can effectively solve the above problem. Suppose there is an 800 × 800 image, the displacement range of the target point in this image is 32 × 32. Now the image is reduced to 400 × 400, the displacement range of the target point is reduced to 16 × 16, and the assumption of small motion can be met again according to this principle.

The specific principle of KLT optical-flow method under the pyramid is as follows. For each image, the original image is taken as layer 0, and the image reduced by 2^L times in length and width is taken as layer L. The obtained image is superimposed from bottom to top to generate the Gaussian pyramid shown in Figure 3. The displacement value of the target point on the highest layer is calculated in the way of the previous section, which is taken as the initial value of the optical-flow calculation of the next layer to calculate the accurate displacement value of this layer. The calculated displacement value is transmitted to the next layer again, so as to calculate to the lowest layer (level 0) to obtain the real displacement value.

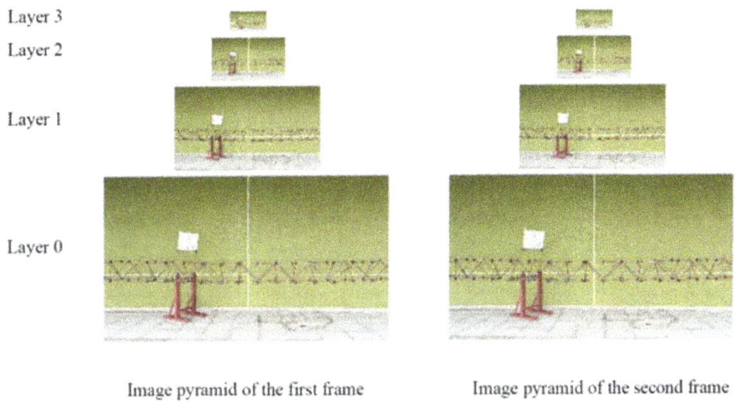

Figure 3. Image pyramid.

2.2. DIC Technology

The principle of DIC method is shown in Figure 4: $I_2(x, y)$ is the deformed image of the image $I_1(x, y)$; S_1 centered on (x, y) and S_2 centered on (x', y') are two windows with the same size, S, which are established at $I_1(x, y)$ and $I_2(x, y)$, respectively. Where $x' = x + \Delta x$, $y' = y + \Delta y$. The correlation between two windows can be expressed as [33]:

$$C(\Delta x, \Delta y) = \frac{\iint_S I_1(x, y) I_2(x + \Delta x, y + \Delta y) dxdy}{\sqrt{\iint_S I_1^2(x, y) dxdy} \cdot \sqrt{\iint_S I_2^2(x + \Delta x, y + \Delta y) dxdy}}. \quad (9)$$

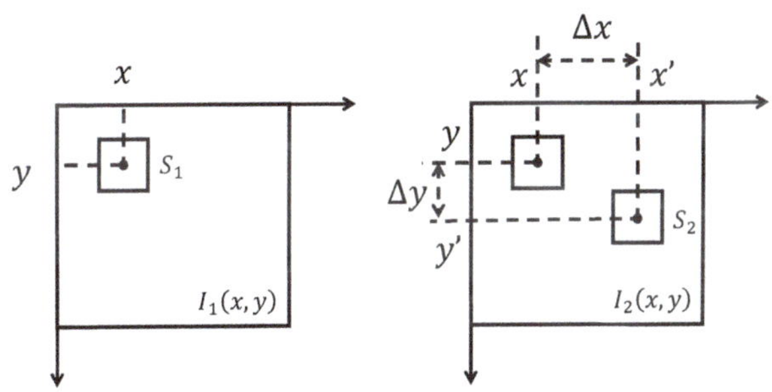

Figure 4. Principle of DIC method.

The real displacement of (x, y) between two images, $(\Delta x, \Delta y)$, can be obtained when the function value of $C(\Delta x, \Delta y)$ is the maximum.

A fixed-size window is used by both DIC and KLT optical-flow methods to search target points in the whole image range; however, the two methods are different in tracking target points. As shown in Equation (1), the target tracking of KLT optical-flow method is based on the residual error of a fixed-size window between two frames, while that of DIC is based on the correlation of a fixed-size window between two frames, as shown in Equation (9). Moreover, the implementation process of KLT optical-flow method is accompanied by the image-pyramid technology, which tracks the points at multiple levels of resolution of an image to optimize tracking effect to improve the tracking accuracy.

2.3. Convolutional Neural Networks

The false displacement caused by UAV measurement can be removed by a CNN. As a feedforward neural network, each neuron of the CNN only extracts the local information of the input data, and the information are collected at a higher level of the network to obtain the global information [34]. The complexity of the model is reduced by weight sharing [35], which accelerates the computing speed and improves the calculation accuracy. The following is an introduction to the function layer involved in this paper.

The convolution layer is the core of the whole CNN, and the convolution process is shown in Figure 5. Suppose that there is a 2 × 2 convolution kernel, the sub area consistent with the size of convolution kernel is found in the input data. An element of the new matrix can be obtained by multiplying and summing each corresponding element in the sub area and convolution kernel, and a new matrix can be generated by navigating all of the input data in steps according to the above method.

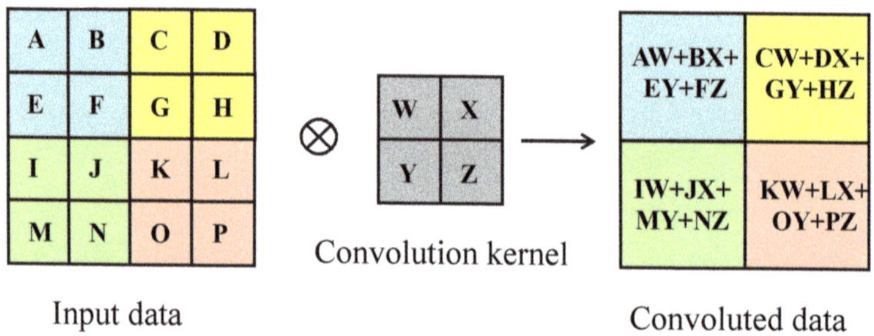

Figure 5. Convolution process.

The feature-extraction ability of the network and the approximation ability of complex functions can be enhanced by activation-function layer. The activation function used in this paper is Leaky Relu, and its expression is as follows:

$$f(x) = \begin{cases} x, & x \geq 0 \\ scale \times x, & x < 0 \end{cases}. \quad (10)$$

The regression layer generally deals with regression problems, and its loss function is expressed as follows:

$$J = \frac{1}{2}\sum_{i=1}^{R}\frac{(t_i - y_i)^2}{R}. \quad (11)$$

where R is the number of samples, t_i is the target output, and y_i is the prediction output. The value of J will decrease with the training of the network until it converges to the target value.

The network used in this paper is a deep-convolution neural network including input layer, convolution layer, activation-function layer, full-connection layer, normalization layer, and output layer. The architecture of the CNN is shown in Figure 6.

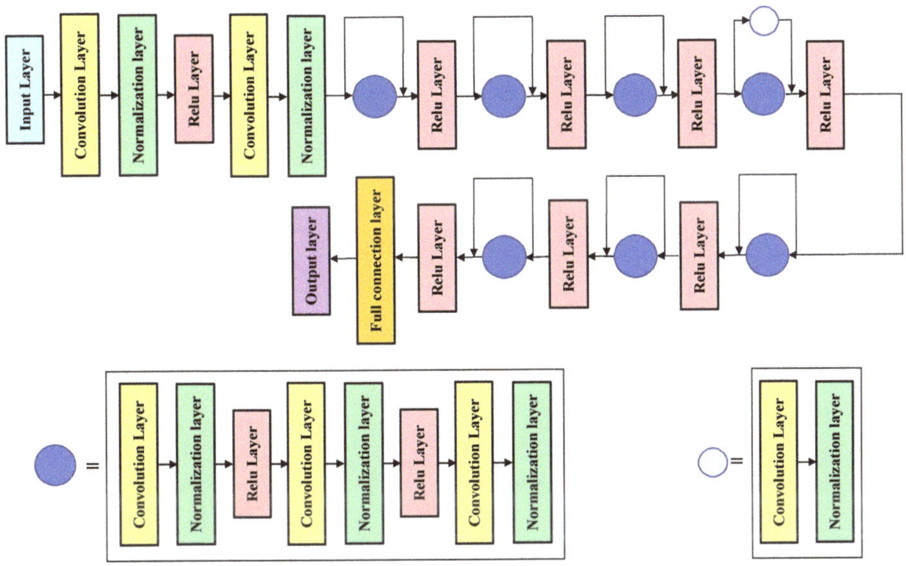

Figure 6. Architecture of the CNN.

2.4. Operational Modal Analysis

In a multiple-degree-of-freedom system, the expression of the response transmissibility $T_{io}(\omega)$ [36] is

$$T_{io}(\omega) = \frac{X_i(\omega)}{X_o(\omega)}. \tag{12}$$

where $X_i(\omega)$ and $X_o(\omega)$ are the Fourier transformations of $x_i(t)$ and $x_o(t)$ that are the time–history response signals at degrees of freedom i and o. The expression of the power spectral density (PSD) transmissibility [37] is

$$\hat{T_{io}}(\omega) = \frac{S_{i,o}(\omega)}{S_{o,o}(\omega)} = \frac{X_i(\omega)X_o^*(\omega)}{X_o(\omega)X_o^*(\omega)} = T_{io}(\omega). \tag{13}$$

where $S_{o,o}(\omega)$ is the self-power spectral density of $x_o(t)$, $S_{i,o}(\omega)$ is the cross-power spectral density of $x_i(t)$ and $x_o(t)$, and $X_o^*(\omega)$ is the conjugate complex number of $X_o(\omega)$. In this paper, the displacement time–history curves are obtained by KLT optical-flow method and DIC, which are used to obtain the PSD curves. The natural frequencies of the structure are obtained from the peaks of the self PSD curve, while the modal shape can be obtained from the ratios of the PSD transmissibility at corresponding degrees of freedom.

3. Experiment

3.1. Experimental Equipment

The bridge model used in the experiment is a spatial steel-frame structure with a total length of 9.8 m and 29 nodes (Figure 7a,b). Figure 7c,d show the hollow round bar (Q235 steel), bolts, and connecting balls constituting the model, respectively, in which the lengths of red and yellow hollow round members are 365 mm and 215 mm, respectively. The width and height of each span of the model are 0.35 m.

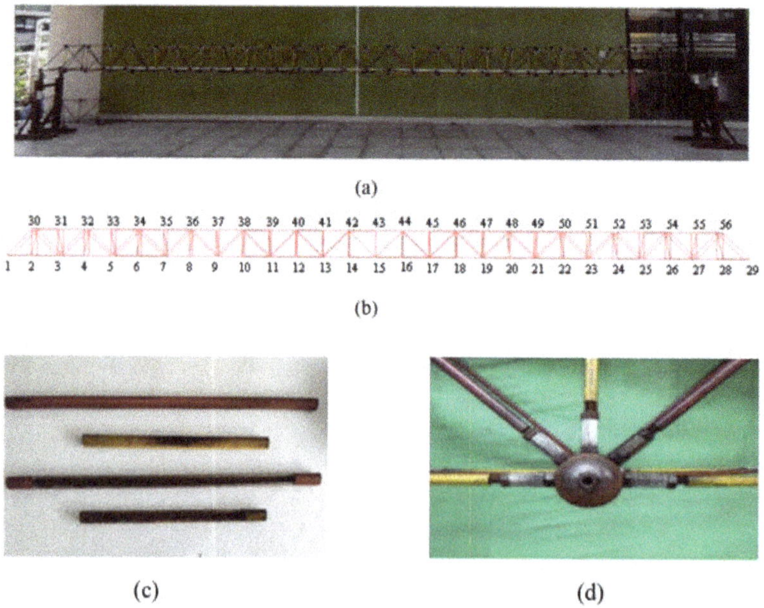

Figure 7. Experimental model: (**a**) bridge model; (**b**) model node; (**c**) hollow round bar; (**d**) bolt-connecting balls.

The images of the vibrational model are collected by the fixed camera (D5300, Nikon Corporation, Japan) in Figure 8a and the quad-rotor UAV (Mavic Air 2, Da-Jiang Innovations, Shenzhen, China) in Figure 8b, which are placed 2 m away from the model [8]. The acquisition frequency of the UAV is 30 frames/s and the image resolution is 3840 × 2160. In order to correct the measurement of the UAV, 5 reference points are selected on the cardboard at 45° to the plane of the model; the other 5 reference points are selected on the back wall. Hence, all reference points are not in the same plane. The experimental layout is shown in Figure 9b.

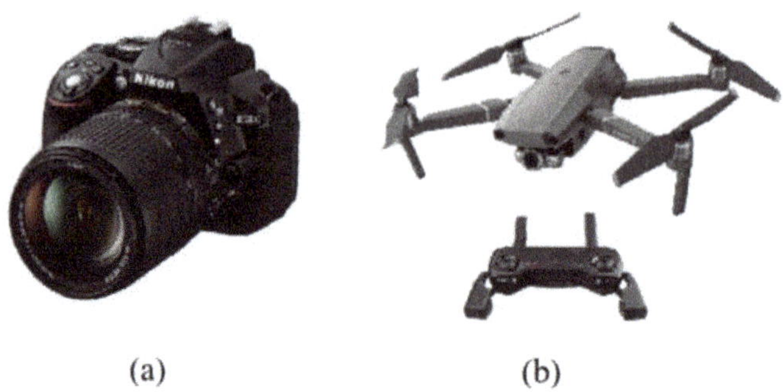

Figure 8. Image-acquisition equipment: (**a**) fixed camera; (**b**) UAV.

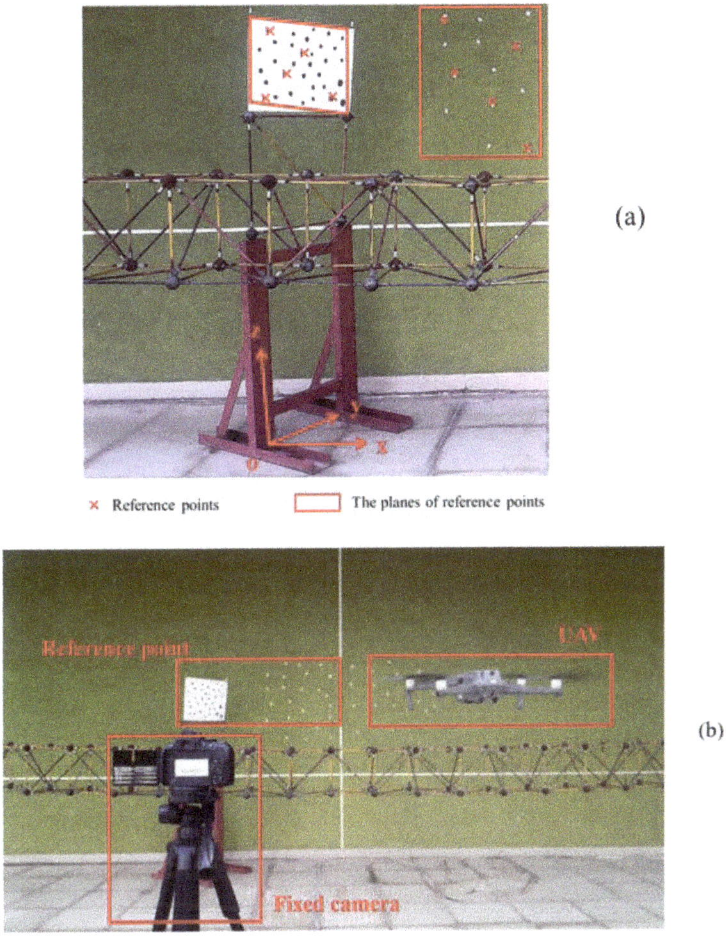

Figure 9. Reference points and experimental layout: (**a**) reference points; (**b**) experimental layout.

3.2. Experimental Scheme

To validate the reliability of vibration measurement of the bridge model by UAV combined with CNN and KLT optical-flow method, the measurement results of the fixed camera are taken as a reference. Moreover, the measurement results of DIC are taken as a reference to validate whether KLT optical-flow method is feasible to replace DIC in bridge vibration measurement.

Firstly, the fixed camera and UAV are placed about 2 m away from the bridge model, and the UAV is used to shoot the static model (the shooting time is 8 min). The reference-point coordinates tracked by DIC and KLT optical-flow method are used as the input of CNN, and the target-point coordinates are used as the network output to train the network. Secondly, the bridge model is excited to vibrate, the fixed camera and UAV are used to shoot (the shooting time is 80 s for 5 times of excitation), and the points are tracked by DIC and KLT optical-flow method again. The reference-point coordinates measured by the UAV are still taken as the input of CNN, the false displacement of the target points caused by the UAV will be outputted by the trained CNN according to the characteristic relationship between the reference points and the target points, and the real displacements of the target points measured by the UAV are obtained by subtracting the output value from the uncorrected displacement of the target point. The natural frequency of the model is obtained by OMA from the time–history curves. The specific process is shown in Figure 1, and the process of correcting displacement measured by a UAV through the CNN is shown in Figure 10. Finally, the tracking effects of DIC and KLT optical-flow method are compared under the fixed-camera and UAV measurements to validate the feasibility of KLT optical-flow method replacing DIC, and the results of fixed camera and UAV are compared under the KLT optical-flow method to validate the reliability of UAV data combined with CNN and KLT optical-flow method. The DIC, KLT optical-flow method, and CNN are realized using MATLAB (MathWorks Inc., Natick, MA, USA) software.

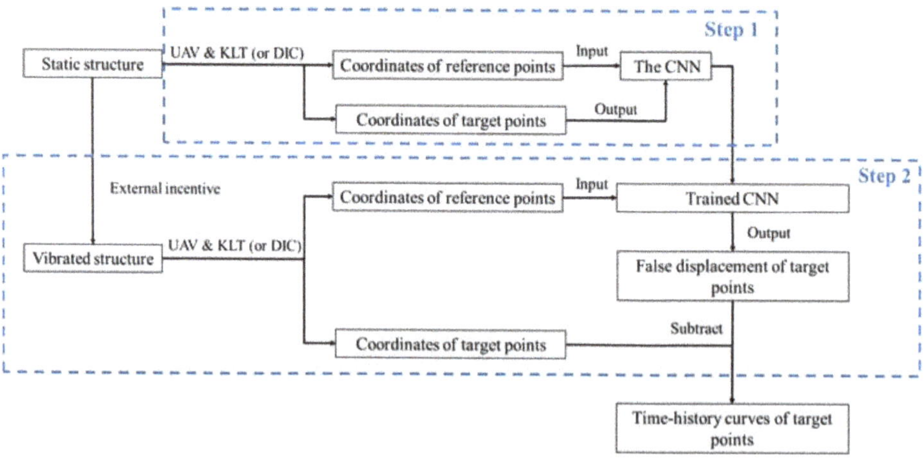

Figure 10. Process of displacement correction by CNN.

4. Experimental Results and Analysis

The bridge model used in this paper has the largest displacement at node 15 in the middle of the span. Therefore, in order to better display the results, node 15 is selected as the target point in this paper (See Appendix A for the measurement results of node 5 and node 10).

4.1. Comparisons for DIC and KLT Method

The displacement results of the bridge model measured by fixed camera are showed in Figure 11, which demonstrates that the displacement time–history curves obtained by the DIC and KLT optical-flow method have very similar characteristics of free-vibration attenuation, and the maximum amplitude is about 5.7 mm. The ideal displacement curve $x'(t)$ can be obtained by smoothing the displacement time–history curve $x(t)$ through locally weighted regression (LOWESS [38]). The difference between $x(t)$ and $x'(t)$ is the error curve, and the average of the error curve reflects the smoothness of the displacement time–history curve measured by the DIC or KLT optical-flow method, which is defined as SDTH in this paper. The SDTHs of the DIC and KLT optical-flow method are 0.048 and 0.031, respectively. In addition, the displacement time–history curve obtained by DIC has obvious burrs, which illustrates its low measurement accuracy. In contrast, the displacement diagram obtained by the KLT optical-flow method is smoother, indicating that DIC is not as accurate as the KLT optical-flow method.

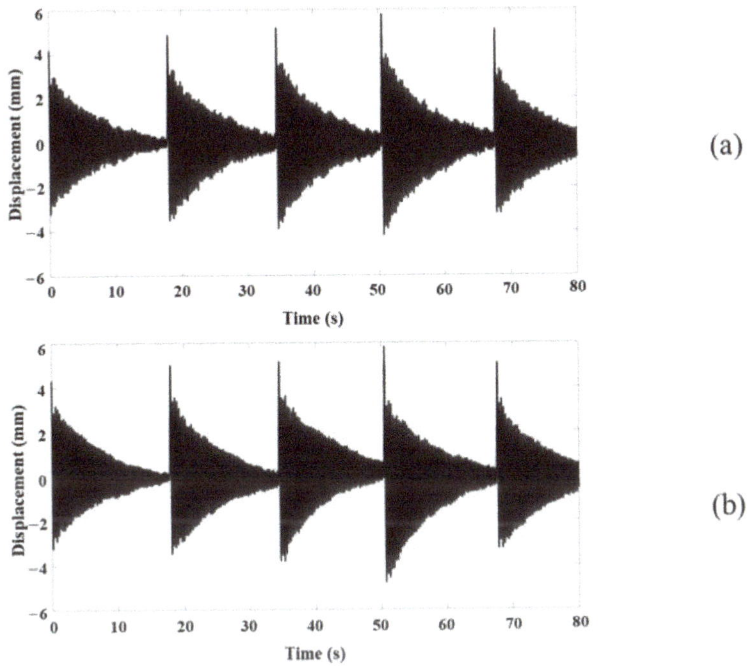

Figure 11. Measurement results of vibration model with fixed camera: (**a**) results of DIC; (**b**) results of KLT optical-flow method.

As shown in Figure 12, it is the displacements of the static point on the model measured by the fixed camera and obtained by the DIC and KLT optical-flow method. Theoretically, the curve should be a straight line with displacement of 0, so that the standard deviation (STD) of the curve can reflect the accuracy of the two methods. The STDs of the DIC and KLT optical-flow method are 0.075 and 0.061, respectively, which indicate that the KLT optical-flow method is more accurate than DIC.

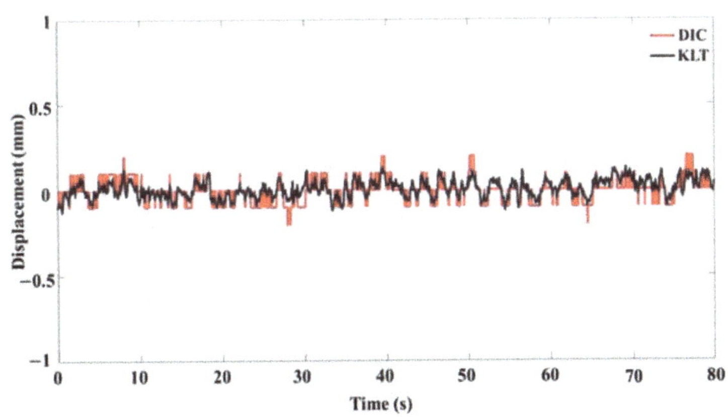

Figure 12. Displacements of the static point obtained by the DIC and KLT optical-flow method.

The displacement signals in Figure 11 are processed into PSD curves. The peak of the PSD curve corresponds to the natural first frequency of the model, as shown in Figure 13. It can be seen from the figure that the natural frequencies extracted by the two methods are 3.281 Hz and 3.282 Hz, respectively, and the relative error is 0.03%. It shows that the KLT optical-flow method is feasible to replace DIC to extract the natural frequency of the structure under fixed camera.

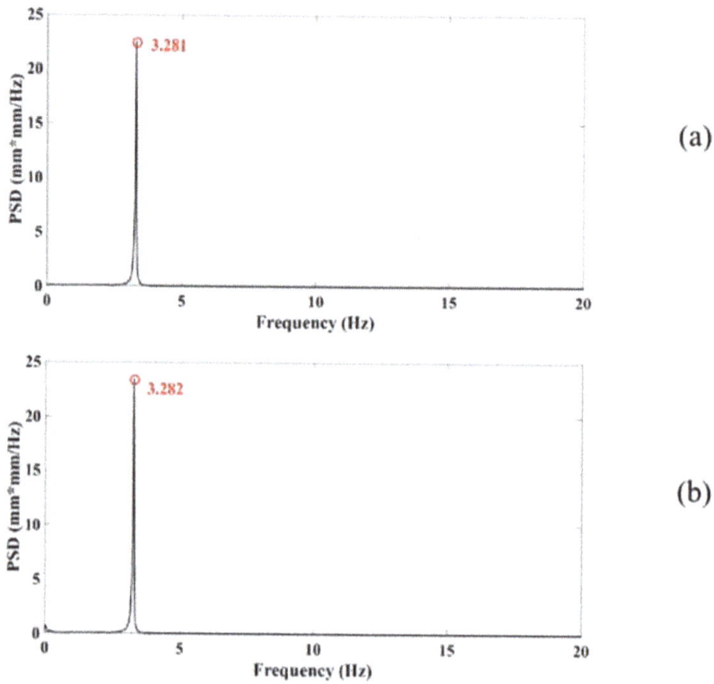

Figure 13. PSD obtained by DIC and KLT optical-flow method: (**a**) results of DIC; (**b**) results of KLT optical-flow method.

4.2. Comparisons of UAV Measurement

The uncorrected displacement measured by the UAV and the false displacement predicted by the CNN are shown in Figure 14. The figure shows that the uncorrected displacement curves measured by the UAV have serious drift.

The true displacements of the structure are shown in Figure 15, which are obtained by subtracting the false displacement of the target points predicted by the CNN from the uncorrected displacement of the target points. The corrected displacement curve obtained by DIC has more drift than that by the KLT optical-flow method. In addition, it can be seen from Figure 15 that each wave of displacement time–history curves fluctuates at $Displacement = 0$. The greater the drift of the curve, the farther the mean line of each wave is from $Displacement = 0$. The drift degree of the displacement time–history curves can be represented by the root mean square (RMS [39]). The RMSs of the DIC and KLT optical-flow methods are 0.063 and 0.025, respectively, which illustrates the low tracking accuracy of DIC, and it proves the stability of the KLT optical-flow method.

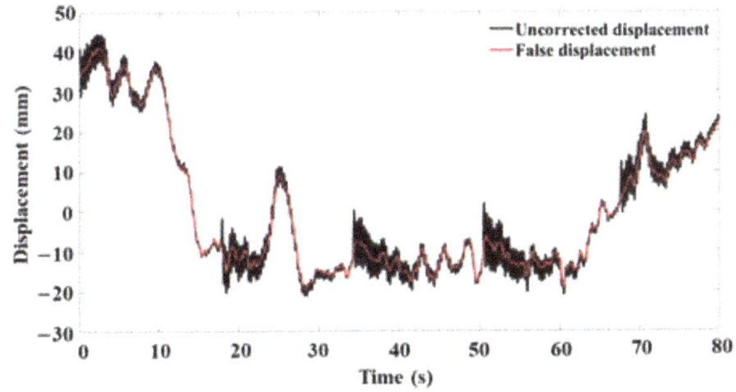

(a)

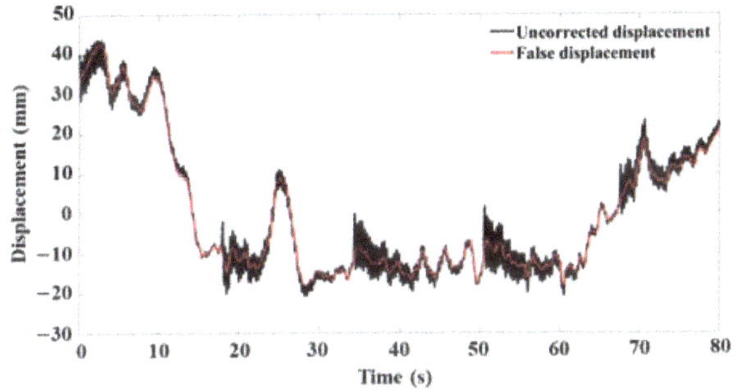

(b)

Figure 14. Uncorrected displacement measured by UAV and false displacement obtained by CNN: (**a**) results of DIC; (**b**) results of KLT optical-flow method.

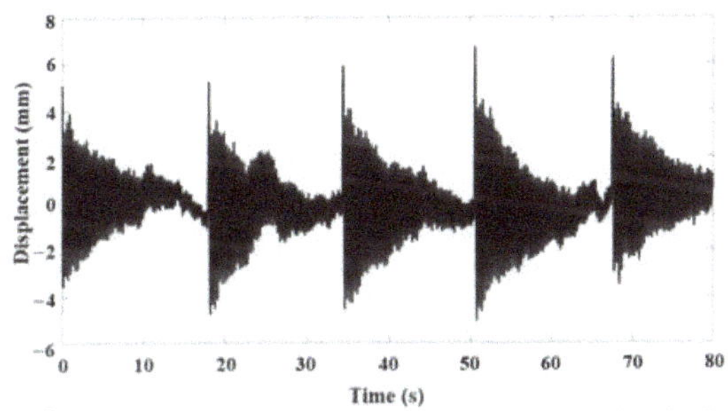

(a)

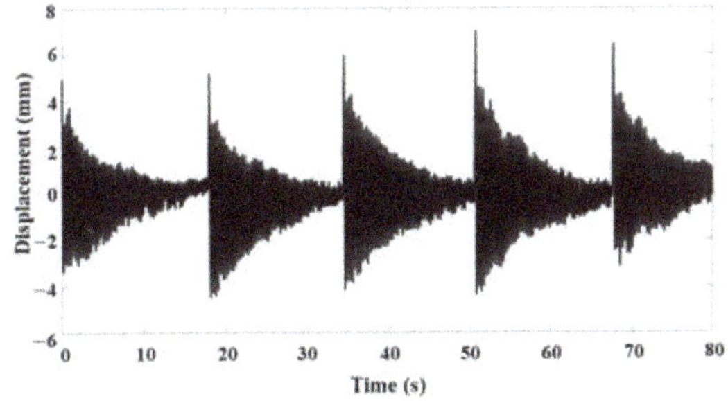

(b)

Figure 15. *Cont.*

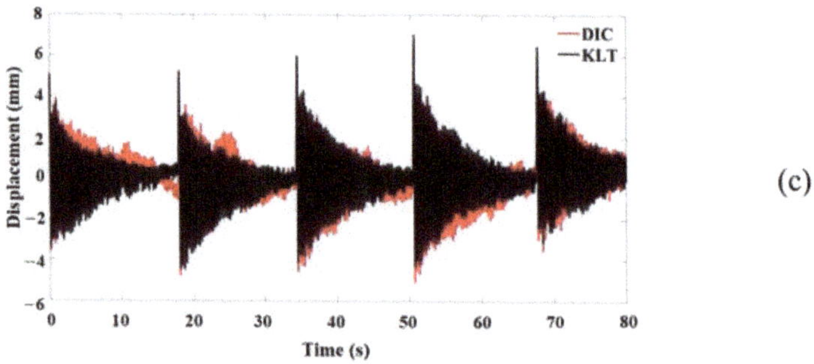

Figure 15. Comparisons of correction effect of CNN: (**a**) results of DIC; (**b**) results of KLT optical-flow method; (**c**) comparisons.

The above displacement signals in Figure 15 are processed into PSD curves, as shown in Figure 16. The natural frequencies extracted for the two displacement signals are 3.285 Hz and 3.286 Hz, respectively, the relative error is 0.04%. It shows that the KLT optical-flow method is feasible to replace DIC to extract the natural frequency of the structure under UAV.

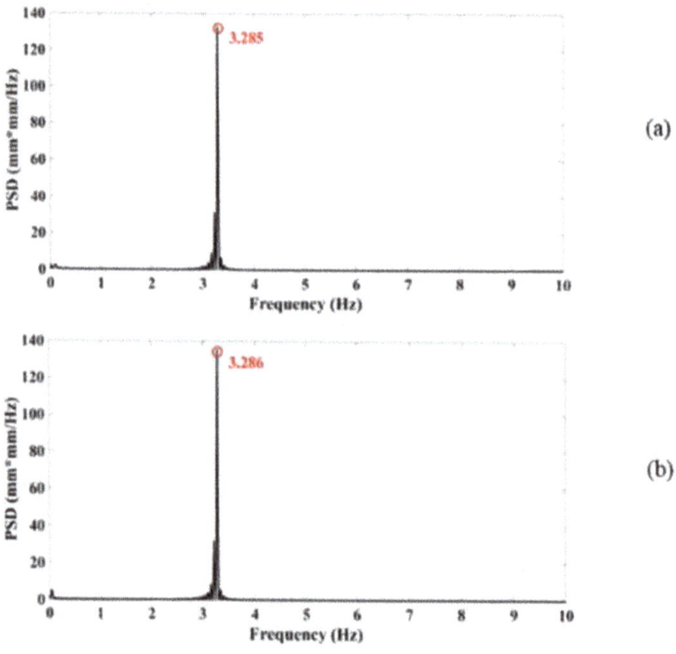

Figure 16. PSD measured by UAV: (**a**) results obtained by DIC; (**b**) results obtained by KLT.

4.3. Comparisons between Fixed Camera and UAV

The comparisons of the measurement results of the fixed camera and UAV processed by the KLT optical-flow method are shown in Figure 17, which demonstrates that the corrected displacement curve measured by UAV and the displacement curve measured by fixed camera have very similar free-vibration characteristics. By calculating the time response assurance criterion (TRAC) between the corrected displacement measured by the UAV and the displacement measured by the fixed camera, the correction effect of CNN can be judged [10]. The higher the TRAC, the higher the degree of consistency, and the better the correction effect. The TRAC between displacement measured by the UAV and that measured by the fixed camera is 0.996, which shows their high consistency. Combined with the results of Section 4.2, it can be proved that the displacement of bridge model measured by the UAV combined with the CNN and KLT optical-flow method is reliable.

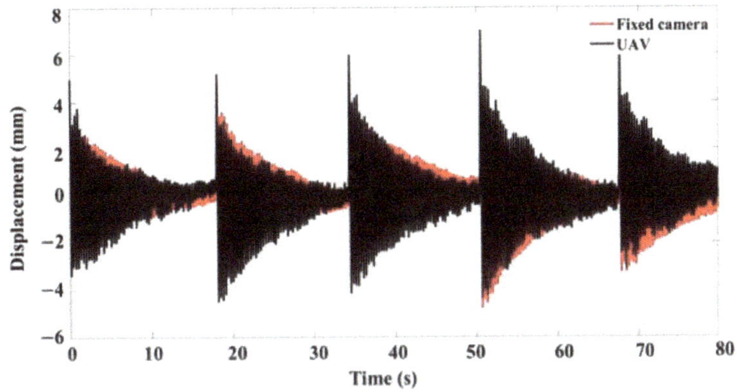

Figure 17. Comparisons of results of fixed camera and UAV under KLT optical-flow method.

5. Discussion and Conclusions

5.1. Discussion

5.1.1. Discussion of the Proposed Method

The experimental results show that the displacement of the bridge model measured by a UAV combined with the CNN and KLT optical-flow method is reliable. The displacement signals obtained by the KLT optical-flow method under fixed camera and UAV are more stable than those of DIC, and the extracted structural natural frequency of the KLT optical-flow method is basically consistent with that extracted by DIC, which shows the feasibility of replacing DIC with the KLT optical-flow method in bridge vibration measurement. The difference of the tracking effect between the two methods is mainly caused by the image pyramid of the KLT optical-flow method, which tracks target points at different levels of resolutions of an image to improve tracking accuracy. Under the KLT optical-flow method, the displacement curve measured by UAV combined with CNN is very close to that of the fixed camera, and the natural frequencies obtained by the two methods are basically the same, which shows the feasibility of the method of correcting displacements measured by UAV with CNN proposed in this paper. In the experiment, the 10 reference points are from two different planes (a cardboard and a wall), which shows that the reference points from the same plane are not required for the correction method of this paper, unlike other correction methods. That is one of the highlights of the method proposed in this paper. Moreover, in order to ensure the assumption of constant brightness of the KLT optical-flow method, the measurement can be carried out on cloudy days or when the light condition is stable. Since the measurement duration of the proposed method is about 10 min, the measurement time needs to be determined according to the actual situation.

5.1.2. Follow-Up Study

Although impressive results were obtained in this paper, there are still some problems that need follow-up research:

1. The reference point is an important part of the correction method proposed in this paper, so its influence on the correction effect will be studied in the follow-up.
2. There may be many influencing factors of accuracy of the proposed method in the actual bridge measurement, such as the wind condition and the actual measurement distance; therefore, the method in this paper will be verified in combination with the actual bridge-measurement results in the follow-up.

5.2. Conclusions

In this paper, a measurement method of bridge vibration by UAVs based on a CNN and KLT optical-flow method is proposed. The effectiveness of the UAV measurement results corrected by a CNN is validated by comparing them with the measurement results of a fixed camera, and the measurement results of DIC are taken as a reference to prove that the KLT optical-flow method is feasible to replace DIC in bridge vibration measurement.

As most of actual bridges cross rivers or valleys, which causes great inconvenience to the measurement, the method introduced in this paper has great prospects in measuring bridge vibration from the perspective of feasibility and accuracy.

Author Contributions: Conceptualization, G.C. and Z.J.; methodology, Z.Y., G.C. and S.T.; software, Z.Y.; validation, Z.Y.; Data curation, Z.Y.; investigation, Z.Y. and G.C.; resources, G.C. and Z.J.; writing—original draft preparation, Z.Y.; writing—review and editing, G.C., Z.J. and S.T.; supervision, G.C. and D.B. All authors have read and agreed to the published version of the manuscript.

Funding: This research received no external funding.

Institutional Review Board Statement: Not applicable.

Informed Consent Statement: Not applicable.

Data Availability Statement: The data used to support the findings of this study are available from the corresponding author upon request.

Conflicts of Interest: The authors declare no conflict of interest.

Appendix A

In order to verify that the conclusions of this paper are valid in the case of large, medium and small displacement, the measurement results of node 5 and node 10 are shown in the figures below.

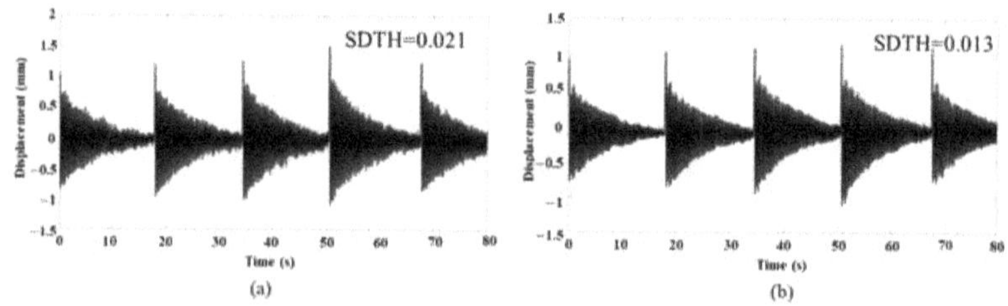

Figure A1. Measurement results of node 5 with fixed camera: (**a**) results of DIC; (**b**) results of KLT optical-flow method.

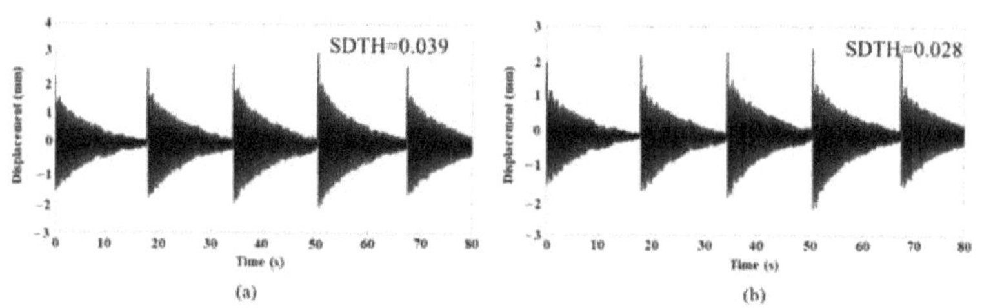

Figure A2. Measurement results of node 10 with fixed camera: (**a**) results of DIC; (**b**) results of KLT optical-flow method.

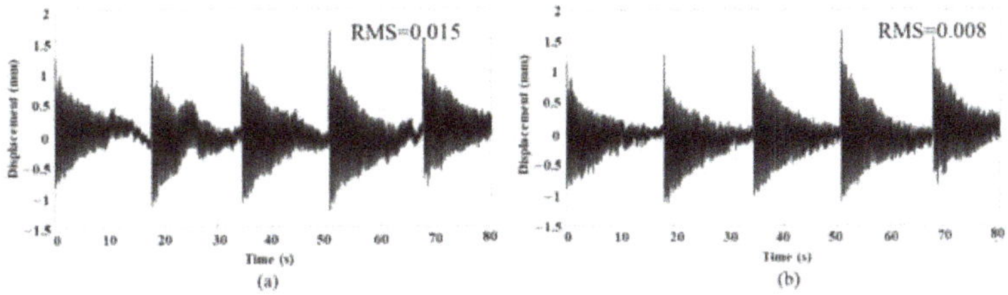

Figure A3. Measurement results of node 5 with UAV: (**a**) results of DIC; (**b**) results of KLT optical-flow method.

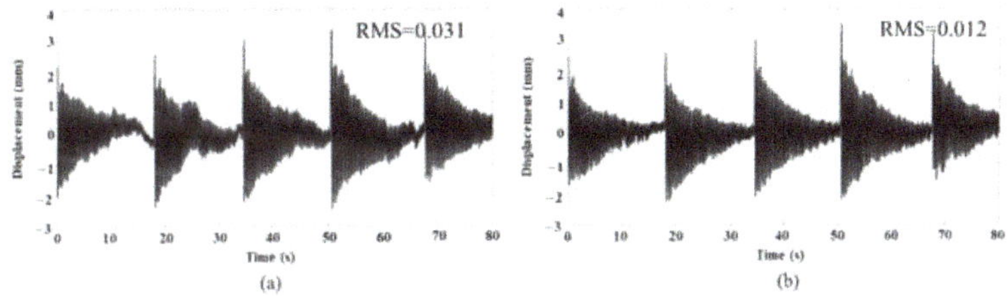

Figure A4. Measurement results of node 10 with UAV: (**a**) results of DIC; (**b**) results of KLT optical-flow method.

References

1. Xi, R.; Jiang, W.; Meng, X.; Chen, H.; Chen, Q. Bridge monitoring using BDS-RTK and GPS-RTK techniques. *Measurement* **2018**, *120*, 128–139. [CrossRef]
2. Siringoringo, D.M.; Fujino, Y. Noncontact Operational Modal Analysis of Structural Members by Laser Doppler Vibrometer. *Comput. Aided Civ. Infrastruct. Eng.* **2010**, *24*, 249–265. [CrossRef]
3. Park, K.T.; Kim, S.H.; Park, H.S.; Lee, K.W. The determination of bridge displacement using measured acceleration. *Eng. Struct.* **2005**, *27*, 371–378. [CrossRef]
4. Kovačič, B.; Kamnik, R.; Štrukelj, A.; Vatin, N. Processing of Signals Produced by Strain Gauges in Testing Measurements of the Bridges. *Procedia Eng.* **2015**, *117*, 795–801. [CrossRef]
5. Yi, T.H.; Li, H.N.; Gu, M. Experimental assessment of high-rate GPS receivers for deformation monitoring of bridge. *Meas. J. Int. Meas. Confed.* **2013**, *46*, 420–432. [CrossRef]
6. Reu, P.L.; Rohe, D.P.; Jacobs, L.D. Comparison of DIC and LDV for practical vibration and modal measurements. *Mech. Syst. Signal Process.* **2017**, *86*, 2–16. [CrossRef]
7. Pjsa, B.; Fb, B.; Plc, D.; Pjt, B.; Pmgpm, B. Experimental measurement of bridge deflection using Digital Image Correlation—ScienceDirect. *Procedia Struct. Integr.* **2019**, *17*, 806–811.
8. Chen, G.; Wu, Z.; Gong, C.; Zhang, J.; Sun, X. DIC-Based Operational Modal Analysis of Bridges. *Adv. Civ. Eng.* **2021**, *2021*, 6694790. [CrossRef]
9. Yoneyama, S. Basic principle of digital image correlation for in-plane displacement and strain measurement. *Adv. Compos. Mater.* **2015**, *25*, 1–19. [CrossRef]
10. Chen, G.; Liang, Q.; Zhong, W.; Gao, X.; Cui, F. Homography-based measurement of bridge vibration using UAV and DIC method. *Measurement* **2020**, *170*, 108683. [CrossRef]
11. Deng, G.; Zhou, Z.; Shao, S.; Chu, X.; Jian, C. A Novel Dense Full-Field Displacement Monitoring Method Based on Image Sequences and Optical Flow Algorithm. *Appl. Sci.* **2020**, *10*, 2118. [CrossRef]
12. Zhu, J.; Lu, Z.; Zhang, C. A marker-free method for structural dynamic displacement measurement based on optical flow. *Struct. Infrastruct. Eng.* **2020**, *18*, 84–96. [CrossRef]
13. Shi, J.; Tomasi, C. Good Features to Track. In Proceedings of the Proceedings/CVPR, IEEE Computer Society Conference on Computer Vision and Pattern Recognition, San Francisco, CA, USA, 6 August 2002; Volume 600, pp. 593–600.
14. Lucas, B.D. An Iterative Image Registration Technique with an Application to Stereo Vision (DARPA). In Proceedings of the 7th International Joint Conference on Artificial Intelligence, Vancouver, BC, Canada, 24–28 August 1981; Volume 81, pp. 674–679.
15. Review, B. The Perception of the Visual Worldby James J. Gibson. *J. Philos.* **1951**, *48*, 788–789.
16. Liu, F.; Yundong, W.U.; Cai, G.; Chen, S.; Science, S.O.; University, J. Research of Feature Point Tracking Algorithm for UAV Video Image Based on KLT. *J. Jimei Univ.* **2017**, *22*, 73–80.
17. Reagan, D.; Sabato, A.; Niezrecki, C. Feasibility of using digital image correlation for unmanned aerial vehicle structural health monitoring of bridges. *Struct. Health Monit.* **2018**, *17*, 1056–1072. [CrossRef]

18. Ellenberg, A.; Kontsos, A.; Moon, F.; Bartoli, I. Bridge related damage quantification using unmanned aerial vehicle imagery. *Struct. Control Health Monit.* **2016**, *23*, 1168–1179. [CrossRef]
19. Hyunjun, K.; Junhwa, L.; Eunjong, A.; Soojin, C.; Myoungsu, S.; Sung-Han, S. Concrete Crack Identification Using a UAV Incorporating Hybrid Image Processing. *Sensors* **2017**, *17*, 2052.
20. Yoon, H.; Shin, J.; Spencer, B.F. Structural Displacement Measurement Using an Unmanned Aerial System. *Comput. Aided Civ. Infrastruct. Eng.* **2018**, *33*, 183–192. [CrossRef]
21. Mandirola, M.; Casarotti, C.; Peloso, S.; Lanese, I.; Brunesi, E.; Senaldi, I. Use of UAS for damage inspection and assessment of bridge infrastructures. *Int. J. Disaster Risk Reduct.* **2022**, *72*, 102824. [CrossRef]
22. Dong, C.Z.; Bas, S.; Catbas, N. A completely non-contact recognition system for bridge unit influence line using portable cameras and computer vision. *Smart Struct. Syst.* **2019**, *24*, 617–630.
23. Yoneyama, S.; Ueda, H. Bridge Deflection Measurement Using Digital Image Correlation with Camera Movement Correction. *Mater. Trans.* **2012**, *53*, 285–290. [CrossRef]
24. Zhu, X.; Chen, Z.; Tang, C.; Mi, Q.; Yan, X. Application of two oriented partial differential equation filtering models on speckle fringes with poor quality and their numerically fast algorithms. *Appl. Opt.* **2013**, *52*, 1814–1823. [CrossRef] [PubMed]
25. Zhang, J.; Wu, Z.; Chen, G.; Liang, Q. Comparisons of Differential Filtering and Homography Transformation in Modal Parameter Identification from UAV Measurement. *Sensors* **2021**, *21*, 5664. [CrossRef] [PubMed]
26. Zhang, Z. A Flexible New Technique for Camera Calibration. *IEEE Trans. Pattern Anal. Mach. Intell.* **2000**, *22*, 1330–1334. [CrossRef]
27. Wu, Z.; Chen, G.; Ding, Q.; Yuan, B.; Yang, X. Three-Dimensional Reconstruction-Based Vibration Measurement of Bridge Model Using UAVs. *Appl. Sci.* **2021**, *11*, 5111. [CrossRef]
28. Ruian, L.; Nan, L.; Beibei, Z.; Tingting, C.; Ninghao, Y. Geometry correction Algorithm for UAV Remote Sensing Image Based on Improved Neural Network. *IOP Conf.* **2018**, *322*, 072002.
29. Weng, Y.; Shan, J.; Lu, Z.; Lu, X.; Spencer, B.F. Homography-based structural displacement measurement for large structures using unmanned aerial vehicles. *Comput. Aided Civ. Infrastruct. Eng.* **2021**, *36*, 1114–1128. [CrossRef]
30. Brownjohn, J.; Magalhaes, F.; Caetano, E.; Cunha, A. Ambient vibration re-testing and operational modal analysis of the Humber Bridge. *Eng. Struct.* **2010**, *32*, 2003–2018. [CrossRef]
31. Yan, W.J.; Ren, W.X. An Enhanced Power Spectral Density Transmissibility (EPSDT) approach for operational modal analysis: Theoretical and experimental investigation. *Eng. Struct.* **2015**, *102*, 108–119. [CrossRef]
32. Bouguet, J.Y. Pyramidal implementation of the Lucas Kanade feature tracker. *Intel Corp. Microprocess. Res. Labs* **2000**. Available online: http://citeseerx.ist.psu.edu/viewdoc/summary?doi=10.1.1.185.585 (accessed on 25 April 2022).
33. Sutton, M.A.; Cheng, M.; Peters, W.H.; Chao, Y.J.; Mcneill, S.R. Application of an optimized digital correlation method to planar deformation analysis. *Image Vis. Comput.* **1986**, *4*, 143–150. [CrossRef]
34. Hubel, D.H.; Wiesel, T.N. Receptive fields, binocular interaction and functional architecture in the cat's visual cortex. *J. Physiol.* **1962**, *160*, 106–154. [CrossRef] [PubMed]
35. Lecun, Y.; Bottou, L. Gradient-based learning applied to document recognition. *Proc. IEEE* **1998**, *86*, 2278–2324. [CrossRef]
36. De Vriendt, C.; Guillaume, P. The use of transmissibility measurements in output-only modal analysis. *Mech. Syst. Signal Process.* **2007**, *21*, 2689–2696. [CrossRef]
37. Yan, W.J.; Ren, W.X. Operational Modal Parameter Identification from Power Spectrum Density Transmissibility. *Comput.-Aided Civ. Infrastruct. Eng.* **2012**, *27*, 202–217. [CrossRef]
38. Cleveland, W.S. LOWESS: A Program for Smoothing Scatterplots by Robust Locally Weighted Regression. *Am. Stat.* **1981**, *35*, 54. [CrossRef]
39. Margaliot, M.; Hespanha, J.P. Root-mean-square gains of switched linear systems: A variational approach. *Automatica* **2008**, *44*, 2398–2402. [CrossRef]

applied sciences

MDPI

Article

Detection Based on Crack Key Point and Deep Convolutional Neural Network

Dejiang Wang *, Jianji Cheng and Honghao Cai

Department of Civil Engineering, Shanghai University, Shanghai 200444, China; jianjicheng@shu.edu.cn (J.C.); seekercai@shu.edu.cn (H.C.)
* Correspondence: djwang@shu.edu.cn

Abstract: Based on the features of cracks, this research proposes the concept of a crack key point as a method for crack characterization and establishes a model of image crack detection based on the reference anchor points method, named KP-CraNet. Based on ResNet, the last three feature layers are repurposed for the specific task of crack key point feature extraction, named a feature filtration network. The accuracy of the model recognition is controllable and can meet both the pixel-level requirements and the efficiency needs of engineering. In order to verify the rationality and applicability of the image crack detection model in this study, we propose a distribution map of distance. The results for factors of a classical evaluation such as accuracy, recall rate, F1 score, and the distribution map of distance show that the method established in this research can improve crack detection quality and has a strong generalization ability. Our model provides a new method of crack detection based on computer vision technology.

Keywords: crack detection; deep convolutional neural network; object detection; crack key point; fusion and filtration of features

Citation: Wang, D.; Cheng, J.; Cai, H. Detection Based on Crack Key Point and Deep Convolutional Neural Network. *Appl. Sci.* **2021**, *11*, 11321. https://doi.org/10.3390/app112311321

Academic Editors: Nikos D. Lagaros and Vagelis Plevris

Received: 20 October 2021
Accepted: 25 November 2021
Published: 29 November 2021

1. Introduction

Cracks are critical flaws that affect the behavior and durability of structures, which can have a negative effect on structural safety. Due to the inevitability and general of cracks on the surface of concrete structures, the search for efficient and low-cost crack detection of concrete has been important in structural damage identification. There are two main directions for the research on crack detection methods: the one is through sensors to test a static and dynamic response of the structure, based on which, the position and depth of a crack are identified [1–3]; the other is through image processing techniques to provide the position and other information about a crack [4,5].

Image-based methods are simple and effective, so they have gained extensive attention. Computer image processing and vision technology, as well as the upgrading of computing hardware and image-based crack detection methods, especially those based on deep convolutional neural networks, have undergone unprecedented development.

Classical image crack detection methods, such as segmentation by a threshold [6], the edge detection algorithm [7,8], and the morphological filtering method [9], not only identify cracks effectively but also assess parameters such as crack length and width. However, their main work is focused on image processing. Crack detection remains a manual process with low efficiency.

To improve the efficiency of detection, researchers have introduced machine learning to deal with crack features and have established a classifier to realize automatic crack detection [10–12]. Crack detection methods of traditional machine learning algorithms combined with image processing techniques have been applied in this area.

Machine learning has broadened the idea of applying computer vision methods for defect detection and condition assessment in civil engineering [13] and has brought about new research directions for all types of detection, including crack detection. Many researchers, using these latest machine learning algorithms, have continued to propose novel image crack recognition models [14–17].

Relying on manual extraction of the characteristics of cracks to realize the crack detection of an image cannot meet the needs of a project due to the complex information contained in the actual crack images. In image- and video-based ML approaches for structural health monitoring, differences in illumination, rotation, and the angle of the camera can significantly affect the final results [18,19]. To meet the requirements for crack detection in practical engineering, automatic learning algorithms based on crack features, especially deep convolutional neural network algorithms, have become a research hotspot. These methods eliminate the first image processing step of most traditional methods, and, based on original crack images, can directly extract crack features and detect cracks through automatic learning models.

Hinton [20] first proposed the concept of deep learning, which has gained extensive attention in the machine learning area. Models based on deep learning began to emerge [21–23]. Using the same dataset, Dorafshan [24] compared the concrete crack detection results of classic edge detection with a deep convolutional neural network (DCNN). The results showed that DCNN had advantages in terms of accuracy, detection speed, and resolution.

Based on the image segmentation algorithm, many detection methods have been proposed [25–28]. Those methods with high accuracy obtain a good detection effect, especially for crack width. However, the above methods are all based on image segmentation algorithms, which require a huge amount of work of pixel-level marking on pictures yet still do not reach the expected accuracy in some cases. A large number of diverse crack training samples are also usually required to achieve better detection results [29].

In recent years, object detection methods based on points have been emerging. Zhou [30] modeled an object as a single point. Hei [31] raised corner detection, while Duan [32] established a method through center pooling and cascade corner pooling, three of which have inspired the research on crack detection method based on key points. Although the method Lee [33] established is still based on pixels, the detection result is more targeted at predicting crack areas rather than pixels. This method has been an inspiration for further research on crack detection based on crack key points that is effective and suitable for engineering.

Crack detection methods based on deep learning depend on the extensity of the training set and the validity of their algorithm. In terms of obtaining crack image data, the current information era ensures easy access to a huge number of surface crack images, so the crack detection methods based on deep learning are reasonably trustworthy. There are two main directions for crack detection algorithms. One regards crack detection as an object segmentation task by image segmentation algorithms to classify and predict pixels and finally output binary images [34]. This method can precisely predict the location and width of cracks but requires detailed and accurate annotation of crack images. However, crack detection in practical engineering works does not require pixel-level positioning. Another strategy is to simplify the problem. This method partitions the crack image and detects crack individually on each patch before stitching them together to locate the cracks [22]. This method simplifies the annotation and can use object sorting algorithms with high convergence. Nonetheless, this method causes two problems: the lack of integrity of cracks when partitioning and the loss of the global feature of the crack image, reducing the generalizability and noise immunity of the model.

Compared with the usual visual computer tasks, crack detection poses three notable challenges. Firstly, cracks are typical linear objects. Secondly, the commonly existing environmental noise causes confusion, so the crack detection results are sensitive to global features. This crack detection problem is relatively severe, influencing the detection results and invalidating the traditional image processing algorithms. Third, the task is substantially a binary classification problem with only two options, having a crack or not. Therefore, only these specificities are considered when designing and modifying a conventional deep convolutional neural network model. The accuracy and suitability of the crack detection method are therefore called into question. Some studies have also shown that the crack feature information of different scales greatly influences the crack identification effect.

For this purpose, we established a crack detection model that considers the cracks' linear characteristics and maximizes the use of their global characteristics. Based on the cracks' linear characteristics, this method can identify cracks as long as the key points on the cracks are identified. Considering the network advantages of traditional computer vision tasks, this research work uses the separation and fusion of the global and local features of cracks to construct a KP-CraNet model for crack detection. Finally, evaluation criteria are set to evaluate the effectiveness and suitability of the model. The numerical experiment has proven that our crack detection model, KP-CraNet, showed a relatively strong detection ability with great potential for further improvement.

2. Crack Characterization

2.1. Crack Key Point

Cracks are gaps with a certain width presented as melanic pixel point sets in a crack image in physical space. We try to realize the second-level evaluation task proposed in [35]—that is, to find the expression of the geometric position of crack damage. Most of the current image-based crack detection methods predict cracks based on pixel points and use the binary image to present cracks, such as the surface crack [36] shown in Figure 1a, and its detection result binary image is Figure 1b. The crack binary image detection takes full advantage of the image pixel information and precisely expresses the location, length, width, and other information about a crack. However, most crack detection works do not require pixel-level information about a crack in reality. Additionally, for deep learning based on image pixels, the supervised training of the model requires a large number of highly accurate cracks that are manually marked on the image to create a pixel-based crack dataset. The amount of work required to mark every pixel of the image is enormous. The representation of cracks based on crack key points, as shown in Figure 1c, requires the ligature of the adjacent key points to represent the crack location information. Crack key points do not have to be in the crack but can be just near it with a distance required by the engineering detection accuracy. Therefore, there can be more than one crack key point set. For example, the crack is shown in Figure 2a can define two types of key points set in Figure 2b,c. The non-uniqueness of crack key points set may lead to some problems, which will be discussed later in the article.

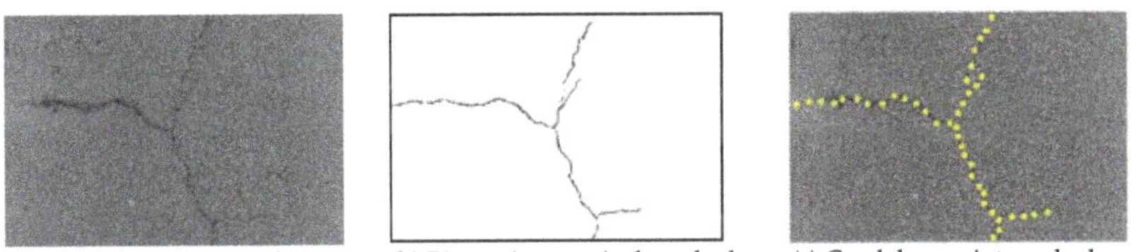

(**a**) Surface crack original image (**b**) Binary image pixel marked (**c**) Crack key point marked

Figure 1. Two characterizations of cracks.

For a given image, the crack key point set can be defined through Equation (1) based on the crack features:

$$P = \{p_1, p_2, p_3, \cdots, p_N\}, \tag{1}$$

where N is the number of key points; any element in the set is a subset as given by Equation (2):

$$p_i = \left\{s_i^{\text{front}}, s_i^{\text{back}}, (px_i, py_i)\right\} \tag{2}$$

The subscript i is the sequence number of a crack key point; s_i^{front} is the sequence number of its preceding key point, and s_i^{back} is the latter; and (px_i, py_i) is the pixel coordinate of the crack key point. The sequence numbers of the preceding and latter key points demonstrate the connection type of crack key points, the ligature of which provides the location information of the crack.

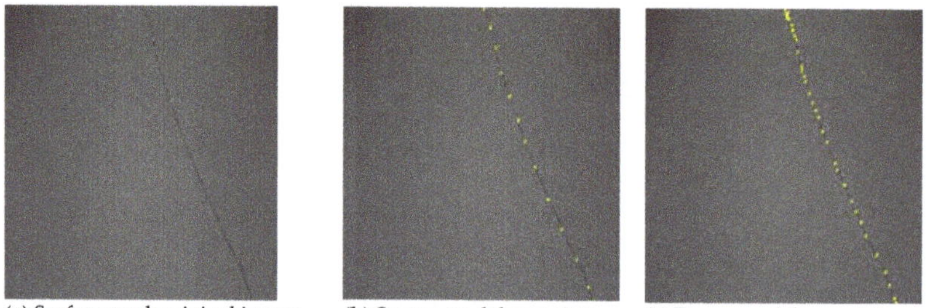

(**a**) Surface crack original image (**b**) Sparse crack key points (**c**) Dense crack key points

Figure 2. Different crack key points can represent the same crack.

2.2. *Reference Anchor Point Method*

The non-uniqueness of crack key points means three problems need to be solved. Firstly, the crack detection model clarifies the relevance of detected key points to predict cracks correctly, regardless of their non-uniqueness. Secondly, whether sparse or dense, the key points are distributed in the crack area; thirdly, we have to control the number of detected crack key points, because it prejudices the model training when there are too many or too few.

2.2.1. Set-Distance Scattering

When detecting cracks, the actual pixel distance of crack key points is hard to set to a universal standard. For cracks with high sinuosity, their key points should be dense, while for cracks with low sinuosity, such as linear types, several key points should be enough.

To avoid the crack detection results being influenced by the sparse key points, it is important to scatter the key points evenly to keep the distances between adjacent key points the same. If the distance between any two adjacent key points is smaller than a certain threshold, no operation is needed; if it is longer, new key points need to be found on their ligature at the same distance as the others to be part of the original crack key point set, forming a new one, to make sure the distance between any two adjacent key points is smaller than a threshold. This process is shown in Figure 3.

For adjacent key points, when their ligature is shorter than a pre-set threshold, no operation needs to be done; when it is longer than the pre-set threshold, then the ligature of the two points needs to be scattered to insert points at equal distance, as shown as Equation (3):

$$M = \left\lceil \frac{\mathrm{dis}(p_i, p_j)}{\varepsilon_s} \right\rceil, \tag{3}$$

where $\mathrm{dis}(p_i, p_j)$ is the pixel distance between adjacent crack key points p_i and p_j, while symbol $\lceil\rceil$ means rounding up to an integer.

We detected the $M - 1$ inserted points to determine whether they were crack key points, and traverse preceding and subsequent key point pairs to form a new crack key point set, as in Equation (4):

$$P^{\mathrm{label}} = \{(x_1, y_1), (x_2, y_2), \cdots, (x_{N^{\mathrm{label}}}, y_{N^{\mathrm{label}}})\}, \tag{4}$$

where N^{label} is the number of crack key points after set-number scattering in the current image, while P^{label} is the crack key point set in a single image of the model we proposed.

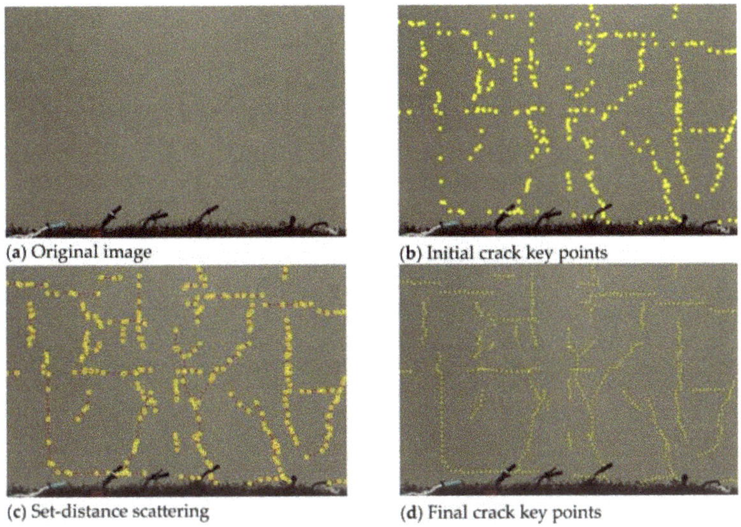

Figure 3. The process and result of set-distance scattering.

2.2.2. Set Reference Anchor Point

Although the sparsity of crack key points can be controlled through set-distance scattering, which reduces the error caused by the density of crack key points to an extent, the number of crack key points is still a problem in real-life detection. If the number of crack key points is too low, it may result in an inability to identify all cracks accurately; if the number is too large, it is very computationally intensive. Therefore, this research proposes a reference anchor point method for crack identification.

The reference anchor point method is derived from the anchor mechanism of the prediction box in the faster R-CNN model [37]; in this research, the anchors were considered the prediction points for crack detection. We laid out the anchor points on the image in advance and detected whether each anchor point was near a crack key point to keep only the nearest anchor points as the final detection results. This method can effectively solve the latter two problems mentioned above.

We set the reference anchor point as in Equation (5):

$$p^{\text{anchor}} = \{(x_1, y_1), (x_1, y_1), \cdots, (x_{N^{\text{anchor}}}, y_{N^{\text{anchor}}})\}, \tag{5}$$

where N^{anchor} is the number of the anchor point of a single image, and N^{anchor} is related to the anchor point layout. The distance between the adjacent two anchor points S is the only parameter for determining the anchor point location, i.e., the anchor point stride. The reference anchor point method is used to set anchor points in an image based on a certain pixel distance in advance, as the red dots show in Figure 4. Thus, crack detection has turned into a matter of calculating the probability value of reference anchor points as crack key points, and when the probability value is higher than a certain threshold, the anchor point will be regarded as a detected crack key point.

The number of anchor points varies according to the actual situation, enabling the reference anchor point method to accurately detect cracks based on anchor points in different densities. For example, the crack area is first determined using anchor points with large step lengths, and then the local crack area obtained is refined to make a more accurate prediction of the cracks using dense anchor points, thus establishing a multiscale crack identification method.

The interval between the anchor step and the crack key determines the crack identification accuracy. If the anchor step is in pixel units, this crack-critical point method is almost equivalent to the pixel-based image crack identification method (conventional image segmentation method). They differ only in terms of the labeling error in crack labeling and whether the pixel width of the crack is considered. On the other hand, if the entire cracked image has only one crack key, the method is equivalent to the image classification task, which divides the whole image into areas with or without cracks. Thus, the image segmentation and image classification methods are special cases of the crack key point method.

Figure 4. Reference anchor layout.

2.3. The Determination of Positive and Negative Sample Point

During the model training for crack detection, the crack key points need to be marked in the training set images as marking points. The training process involves the model learning to detect whether a reference anchor point is a crack key point (also called a marked point). Not all crack key points coincide with reference anchor points, so reference anchor points can be classified into three types: positive sample points, negative sample points, and general anchor points. A positive sample point is defined as the closest anchor point to the crack key points. For each crack key point $p_j^{\text{label}}(j = 1, 2, 3, \cdots)$, its positive sample point is shown in Equation (6).

$$p_{ji}^{\text{anchor}} = \min_i(\text{dis}\,(s_{ij})), \tag{6}$$

where $\text{dis}\,(s_{ij})$ is the distance between the anchor point p_i^{anchor} and the crack key point p_j^{label}.

However, to maintain the convergence of the model training, all anchor points whose distance to a crack key point is shorter than a certain threshold, ε_P, will be set as a positive sample point. So, the positive sample point is set as in Equation (7):

$$E^\text{T} = \left\{ p_i^{\text{anchor}} \middle| \text{dis}_{ij} < \varepsilon_\text{P}, i = 1, 2, \cdots N^{\text{anchor}}; j = 1, 2, \cdots N^{\text{label}} \right\} \tag{7}$$

So, the reference anchor points in an image belong to a positive sample point set, a negative sample point set, and a general anchor point set. Figure 5 shows the crack key points (green), positive sample point set (red), and negative sample point set (blue) when $\varepsilon_\text{N} = \varepsilon_\text{P}$.

(**a**) Original crack image

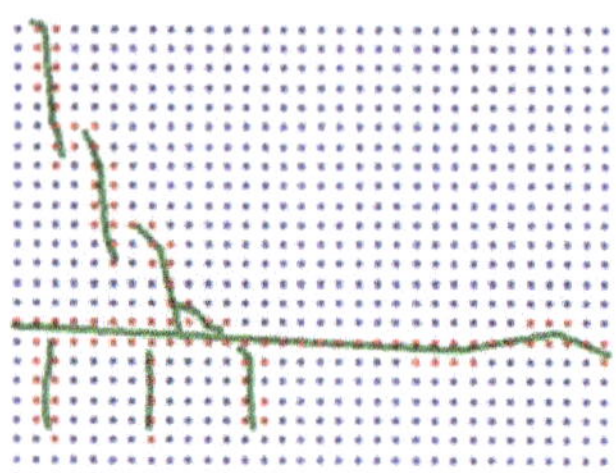

(**b**) Crack key points and anchor points

Figure 5. Positive and negative sample points.

3. Crack Detection Model KP-CraNet

To consider the influence on crack detection exerted by the linear and global features of cracks, a crack detection model based on crack key point, KP-CraNet, has been built. The model contains three hierarchical submodels. The bottom submodel is a feature extraction model based on crack key points of deep convolutional neural networks, which is also called a key point feature extraction network or basic network. The second submodel uses feature pyramid fusion and a reinforce network [38] (FPN feature fusion and reinforce network) to fuse and reinforce the features of extracted key points. The third submodel is a feature filtration network, which filters the features of crack key points, the results of which will lead to an area of a crack key point as an approximate location of a crack.

3.1. The Network Frame of KP-CraNet

Figure 6 shows the network frame of KP-CraNet. The bottom submodel for crack key point feature extraction adopted the ResNet [39] network framework, choosing the last three feature layers with different sizes, R1, R2, and R3, as the feature fusion input and reinforce networks. In this network, the FPN reinforcement network is used for global and local feature fusion. The output results are dominated by the current layer features, and the upper layer feature layer is incorporated into the current layer using inverse convolution, outputting a total of five layers. The first three layers have the same size as R1, R2, and R3, while the last two layers are based on the R3 layer, again convolved and downsampled to obtain a smaller layer to better express the global features. Finally, the five feature layers are subjected to feature filtering. The features of feature layers 1, 3, and 5, which are in decreasing size order, are inputted into the feature screening submodel, and the probability of predicting the anchor point as a crack key point is used to screen the positive sample points sequentially to detect cracked and noncracked areas for the final detection of cracks.

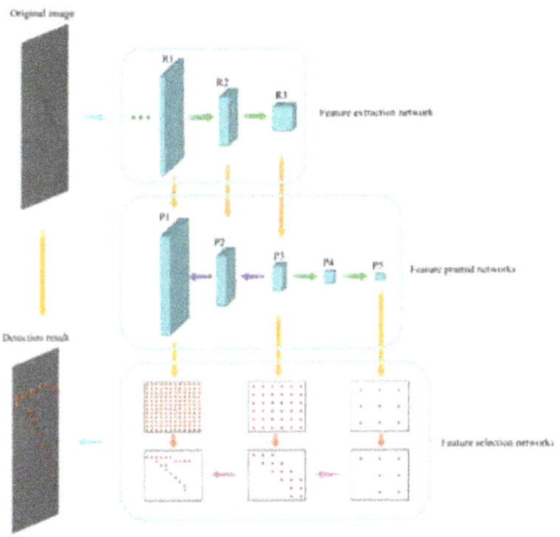

Figure 6. Model network structure.

3.2. Crack Key Point Feature Extraction Network

The key point feature extraction network uses deep convolutional neural networks, usually adopting ResNet or VGG or other models commonly used in the computer vision field. Crack detection experiments show few differences between the ResNet and VGG models, even when increasing the network depth. Therefore, ResNet18, which requires lower depth, was chosen to decrease the training difficulty and improve the prediction efficiency.

The input to the lowest layer is the image, and the information contained in each pixel of it is the smallest unit of the feature, called the smallest local feature. The topmost layer feature is the crack identification result, and all intermediate result layers from the bottommost to the topmost are called feature layers. For any current layer, whether it is convolution or pooling, the feature points within the range of 3 × 3 convolution kernels or 2 × 2 pooling kernels of the current layer are weighed and summed, and the activation function is used to obtain new information about the feature points. Thus, each feature point of the next layer contains multiple feature point information of the corresponding position of the current layer. The feature information is continuously downscaled and integrated, gradually transitioning from local feature information to global feature information.

Figure 7a,b shows the convolution and pooling layers. The rear, lighter part is the current layer feature map. The front, darker part is the next layer. The images demonstrate how framed feature points in the current layer descend to the next layer, which means the next layer contains information on points in the current layer, so the local feature information is gradually integrated with the layers, adding up. As the feature layer increases, the area of information contained in feature points becomes larger. The red dots in Figure 8 are feature points, and the yellow-green background is the feature area. In a higher-level feature map, the feature points reflect global features.

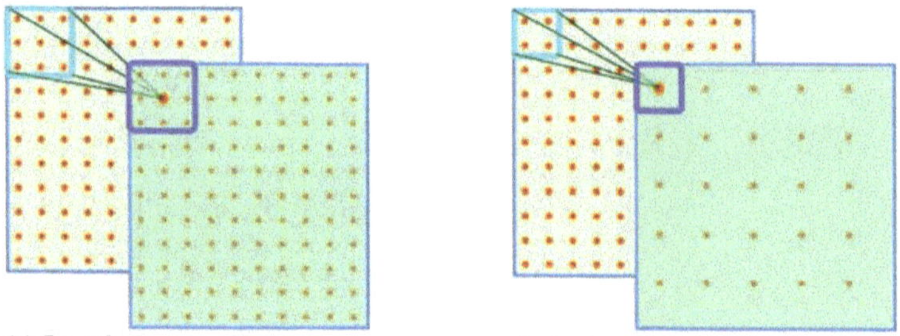

(**a**) Convolution operation (**b**) Pooling operation

Figure 7. The extraction process of convolution and pooling layers.

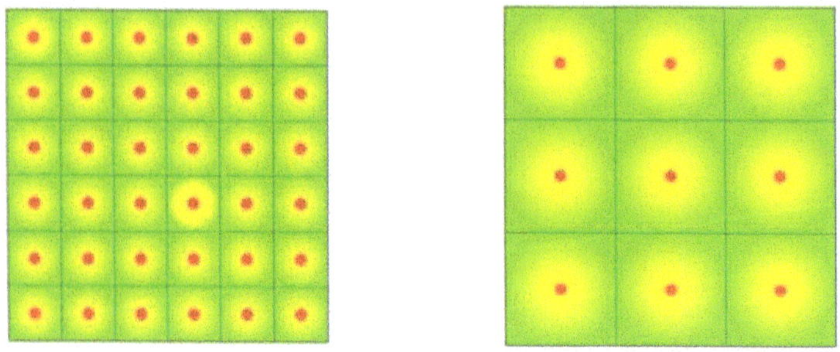

(**a**) Feature points and area of current layer (**b**) Feature points and area of next layer

Figure 8. Feature points and feature area.

3.3. *FPN Feature Fusion and Reinforce Network*

Three maps of features from the feature extraction network contain both global and local feature information. Although the specific location of cracks should be precisely positioned by a local feature map, we also need global features as a reference to avoid the errors caused by tree branch shadows and stains that look similar to cracks in pictures. Although, in the feature extraction of single-track networks, higher feature layers contain some of the features of lower layers, the same layers may contain the feature information of different layers in different images because of the varying crack rate. Therefore, it is important to consider both global and local features. Feature information in different layers should be further fused and reinforced to obtain integrated feature information that contains features through all layers. FPN feature fusion and reinforce network is an effective method of feature fusion and reinforces various sizes that are extensively applied in the object detection field. After the convolution and fusion of the three feature maps extracted by the network, three new feature layers are generated, with convolution and pooling operations repeated twice to form a higher global feature layer, so the FPN feature fusion and reinforce network would generate five feature layers in total. Apart from containing more information, it can also process images of different sizes, learn the feature rules automatically, and finally, generate three reinforced feature maps and two with global feature information. We selected the first, third, and fifth layers in sequence as the final extracted feature layer with detailed local features, transition features, and global features.

3.4. Feature Filtration Network

The outputs of FPN feature fusion and reinforce network need another convolution operation for feature processing and finally output an anchor point prediction result the same size as the corresponding original feature layers. Different layers contain different numbers of local or global features, so the prediction results of each layer can be regarded as the prediction results for different features' extent, i.e., whether the corresponding area contains a crack.

Predictions that consider global features will have higher prediction accuracy, while predictions considering local features will have higher localization accuracy. Thus, starting with global features, the feature points that do not have cracks are gradually filtered out based on the results of the current layer prediction. Finally, the identification results are obtained, and the prediction accuracy and crack location accuracy are guaranteed. The process of crack detection shown in Figure 9 exemplifies how a feature filtration network works.

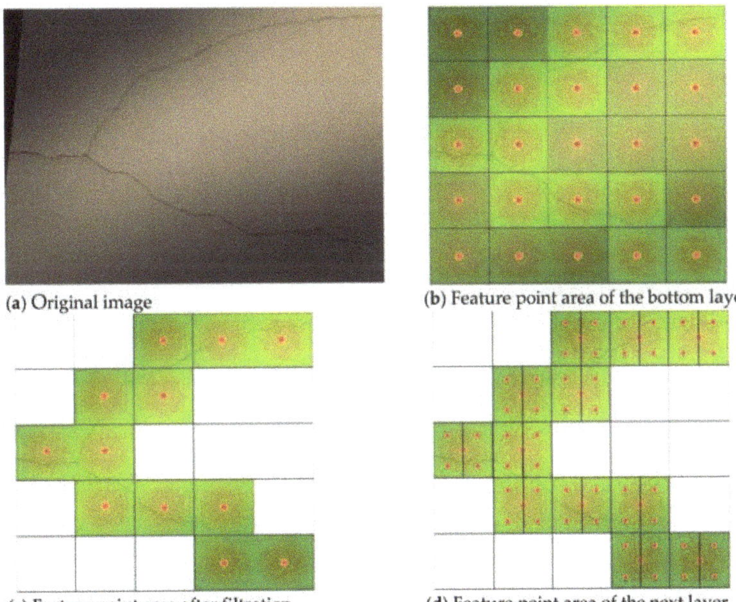

(a) Original image **(b)** Feature point area of the bottom layer **(c)** Feature point area after filtration **(d)** Feature point area of the next layer

Figure 9. Feature filtration network.

Firstly, we input the prediction results of the bottom feature layer that contains the most comprehensive global information, as shown in Figure 9b, into the feature filtration network. Due to the relatively large distance between feature points, each feature point contains comprehensive global feature information. The prediction result of the current layer helps to distinguish between the crack area and noncrack area, as shown in Figure 9b. We removed the feature points in noncrack areas and kept those in the crack area, as shown in Figure 9c.

Based on this filtering result, the feature points corresponding to the next feature layer were retained. As shown in Figure 9d, the features and anchor points of this layer were screened, and after the screening, only the anchor points of the current layer in the crack-containing area were retained again; this process was repeated continuously. According to the network structure, three times in sequence were filtered, and the points finally retained were the anchor points of the positive sample of the lowest local features at the maximum resolution set and the minimum step size. These were the key points of the identified cracks.

It is important to note that the above method is completely different from directly deciding whether an area contains cracks. The method depends totally on convolution or pooling operations, whose weight is obtained via the model's automatic learning.

3.5. Feature Filtration Network

In model training, the input for each layer level is the anchor point prediction result of the previous layer of features and the eigenvalue of the current layer of features. The output is the anchor point prediction result of the current layer, which is used to determine whether each anchor point is a positive sample point of the current layer. Due to the different sizes of the input images at different layers, or the different sizes of the feature layers, the tolerance of the critical point determination for identifying cracks is also different. So, the thresholds ε_P, ε_N, and loss function are determined by the size of feature maps.

For reticulated reference anchor points at equal intervals, if their anchor point stride is S pixels, then the maximum distance between a crack key point and its nearest anchor point is $\sqrt{2}S/2$ pixels. As shown in Figure 10, the yellow and green points are fracture key points, and the furthest possible location from the nearest anchor point to that fracture key point is the four red points around it. $\varepsilon_P = 0.8S$ and $\varepsilon_N = 1.5S$ can be extracted to define positive and negative sample anchor points in every layer. When the possibility that the reference anchor point is a positive sample point is higher than the set threshold, the reference anchor points are regarded as positive sample points of the current layer.

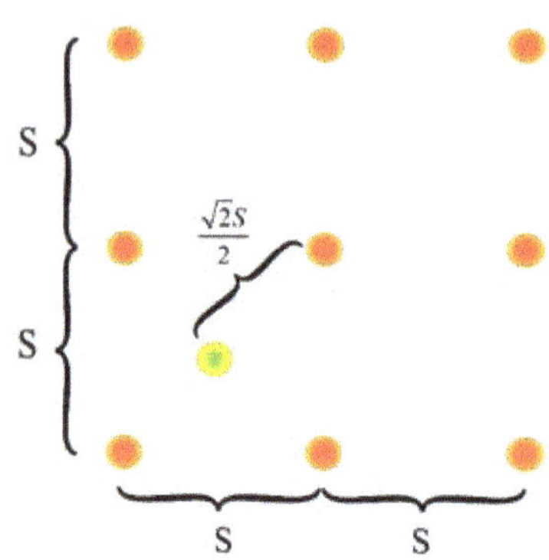

Figure 10. Set the distant threshold of positive and negative sample points.

A loss function is introduced in each layer to speed up the algorithm convergence and improve the judgment accuracy during training. For this classification problem, a binary cross-entropy loss function can generally be used. However, for image crack detection problems, because a crack occupies only a small part of the whole image, there are many more negative sample points than positive ones, which will lead to error. Therefore, a loss function in RetinaNet [40] is applied to balance the positive and negative sample points, using the Focal Loss function as in Equation (8):

$$f_{\text{loss}} = \begin{cases} -\alpha(1-y)^{\gamma}\log y, & \text{when } \tilde{y} = 1 \\ -(1-\alpha)y^{\gamma}\log(1-y), & \text{when } \tilde{y} = 0 \end{cases} \quad (8)$$

In order to balance positive and negative samples, regulatory factors α and β were introduced into the formula. y is the predicted value, i.e., the possibility that reference anchor points are crack key points of the feature layer. $\tilde{y}$ is the actual value of the same reference anchor points: the positive sample point is 1, while the negative one is 0.

4. Model Training and Evaluation

4.1. Data Collection

The dataset used in this research was derived from the surface crack images of various wall surfaces on a university campus in Shanghai. First, we used a camera to take multiple 6720 × 4480 pixel images, divided them into 6 × 6 blocks, and used bilinear interpolation to unify each block image to a size of 1024 × 768. From these images, 1349 images containing cracks were selected; 90% of them were used as the training set and 10% as the test set (https://github.com/csga11/craData, accessed on 24 November 2021).

4.2. Training Parameters

The crack detection model in this paper was based on the Pytorch 1.0 deep learning framework. The GPU used for training and testing was NVIDIA GTX1080. The initial parameter weights of the feature extraction network all used the weights of the Pytorch official ResNet network pretraining model. During training, the batch size was 16, the number of iterations was 100, and the learning rate was 0.0001. In order to improve the training effect and robustness of the model, conventional online dataset expansion methods, such as random inversion, filtering, and brightness enhancement, were used during the model training.

4.3. Assessment Criteria

The model assessment criteria need to be determined to evaluate whether a crack detection model is effective. For each anchor point, the distance threshold method was applied to determine whether a sample point was positive or negative. As the possibility of positive sample points was calculated as the output, a possibility threshold ε_b had to be set for determining the sample points, i.e., the positive and negative sample point set were as in Equation (9):

$$E^{\text{P}} = \left\{p_i^{\text{anchor}} | y_i > \varepsilon_b\right\}, E^{\text{N}} = \left\{p_i^{\text{anchor}} | y_i < \varepsilon_b\right\}, \quad (9)$$

where y_i is the predicted possibility value of the anchor point i.

The commonly used accuracy rate in the object detection area was also adopted to represent the correct prediction rate among all the image detection results as in Equation (10):

$$P = \frac{num\{p_i | p_i \in E^{\text{T}}, p_i \in E^{\text{P}}\}}{num\{E^{\text{P}}\}} \quad (10)$$

In Equation (11), $num\{\}$ represents the number of elements in the set.

Meanwhile, the introduction of another conventional indicator-recall rate, the proportion of all actual crack points that are correctly identified, describes whether all the key points of the crack are correctly identified. For any key point of a crack, as long as there is an anchor point identified as a positive sample point in the circle whose radius is step S, the key point of the crack is considered to be correctly detected. Therefore, the recall rate was defined as in Equation (11):

$$R = \frac{num\{p_j | p_i \in \mathrm{P^{anchor}}, p_j \in \mathrm{P^{label}}, \mathrm{dis}(p_i, p_j) < S\}}{num\{\mathrm{P^{label}}\}} \tag{11}$$

It can be seen that, in order to ensure accuracy, forecasts should be as few and precise as possible. However, in order to ensure the recall rate, the prediction should be as complete as possible, so the two indicators are contradictory to a certain extent. According to actual engineering needs, these two parameters can usually be adjusted. Here, the weighted balance F1 score of the two is adopted as the final identification evaluation index. The recall rate is defined as in Equation (12):

$$\mathrm{F1} - \mathrm{score} = \frac{2PR}{P + R} \tag{12}$$

This research proposes the concept of a distance distribution map to reveal the distance distribution of the identification point from the nearest crack key point. The abscissa is the distance between the anchor point and the key point of the nearest crack, and the ordinate is the number of anchor points at the above distance. The results of all the images in the test set were plotted in one graph. The quality of crack detection was evaluated by analyzing the number of anchor points within a certain distance.

4.4. Analysis of Detection

The correctness of the model design needs to be tested first to demonstrate the effectiveness and applicability of the method. We tested three models without a feature filtration subnetwork, then output three layers, P5, P4, and P3. P5 represents the highest layer for global features. After testing the two characteristic screening subnetwork models, P5 is the screening output of P4, and P5 and P4 are the screening output of P3. We set the threshold rate using the experimental results of the above five models, as shown in Table 1.

Table 1. The assessment result of each output layer when the threshold rate set as 0.5.

Output Layer	Distance Threshold	Accuracy Rate	Recall Rate	F1 Score
P5	128	0.869	0.992	0.895
P4	64	0.814	0.894	0.853
P3	32	0.786	0.882	0.831
P5 + P4	64	0.819	0.909	0.862
P5 + P4 + P3	32	0.881	0.833	0.856

When the output layer is a single feature layer, and the sizes of feature layers increase (from P5 to P3), the correctness of prediction gradually drops. The F1 score decreased from 0.895 to 0.831, which showed that the global feature exerted a notable influence on crack detection. So, the global feature should be adequately considered.

However, after adding the feature screening subnetwork, comparing the results of P4 with P5 + P4 and P3 with P5 + P4 + P3 shows that the model has better detection results after adding the feature screening subnetwork. The F1 score increased from 0.853 to 0.862 and from 0.831 to 0.856.

Figure 11 presents a distance distribution map of the detection results of the single-track output layer model. The intensive blue strips are a histogram of the distance between anchor points and the nearest crack key point. This shows that, as the distance between anchor points and the nearest crack key point increases, the number of anchor points reduces. For the orange anchor point segmenting vertical lines in the image, its vertical coordinates are the threshold of positive/negative sample determination for the current feature layer: positive to the left, negative to the right. The curves in different colors are anchor point segmenting lines under different possibility thresholds, and any line divides all the anchor points (blue strips) within the distance into two parts: the upper parts represent the positive prediction points, and the lower represent the negative prediction points. Any point in the curve represents the number of positive points in the corresponding distance, which can be called a positive detection curve.

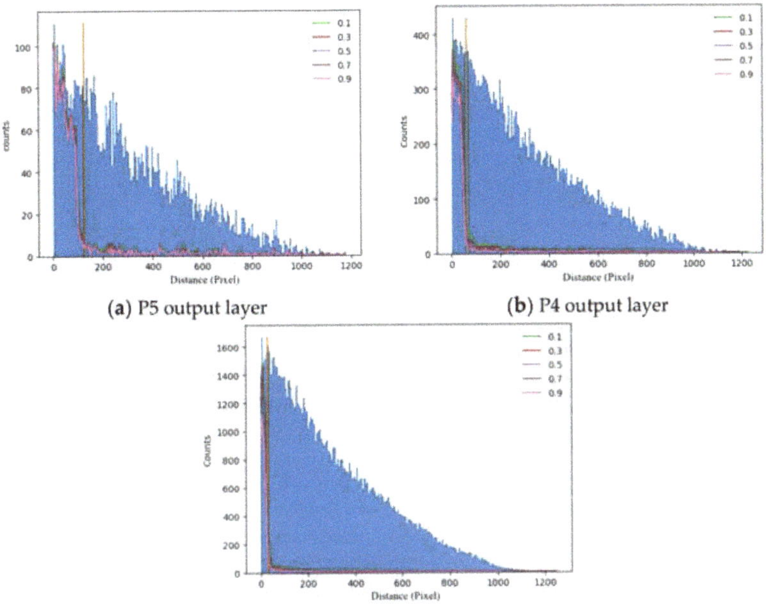

(**a**) P5 output layer (**b**) P4 output layer

(**c**) P3 output layer

Figure 11. The histogram of the shortest distance between anchor points and marked points of single-track output layer.

Any one of the identifying positive case curves and vertical dividing lines divides the blue area into four regions. In the upper left region, the points that are actually positive case anchors are identified as negative case anchors; in the lower left region, the points that are actually positive case anchors are identified as positive case anchors; in the upper right region, the points that are actually negative case anchors are identified as negative case anchors; and in the lower right region, the points that are actually negative case anchors are identified as positive case anchors. The percentage of points in the lower left and upper right areas should be as high as possible.

Therefore, for the P5 output layer shown in Figure 11a, the positive prediction curves under each possibility threshold almost coincide; the threshold selection has little influence on the result, which means that the global feature exerts a relatively significant influence on model detection and the detection of the crack key point is accurate. The prediction rate of anchor points marked as positive is close to 1, while the prediction rate of anchor points marked as negative is close to 0, which is an ideal prediction result. Figure 11b,c demonstrate that, with the decrease in stride, the detection result of the positive/negative sample point becomes more sensitive to the possibility threshold, and more detection errors appear, because images with high resolution emphasize the local features, leading to the misjudgment of anchor points in areas such as cracks. This illustrates that global features are more helpful for effective crack detection, yet the relatively long stride complicated the accurate location of cracks. The introduction of a filtration subnetwork requires a correction of the distance distribution map, because the number of anchor points is influenced by the feature filtration subnetwork and the size of its output. Changing the vertical ordinate from the number of anchor points to the anchor point frequency would reduce the influence of the model anchor point to an extent.

Figure 12 presents the distance distribution map of a single-track output layer and introduces the feature filtration layer under a different possibility threshold. It shows that the curves become steep near the possibility threshold after introducing the feature filtration network, especially the output of P3; more points are detected near the crack key points, and fewer are detected away from the crack key points. The figure illustrates that the feature filtration network has a significant effect, verifying the suitability of the networks we presented.

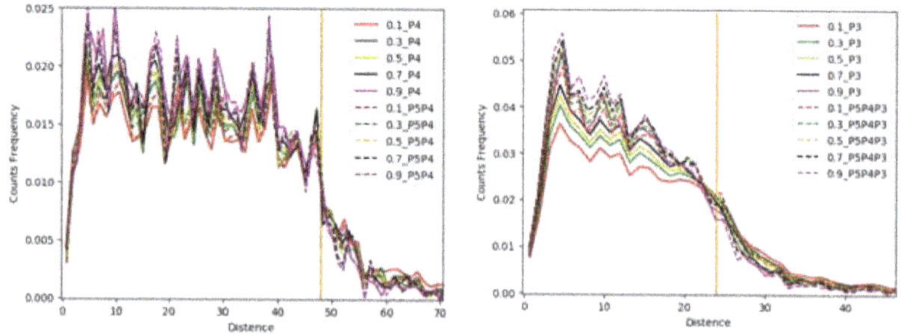

(**a**) P4 output size (**b**) P3 output size

Figure 12. Frequency histogram of distance between single output layer and filtration layer anchor point.

Finally, we chose P5 and P3 as feature filtration layers, while P1 was an output layer. The final results of different crack detection were as shown in Figure 13. This figure proves that the application of our detection model can lead to satisfactory crack detection results. Moreover, the detection experiments discovered that the model had good generalizability, enabling the model to detect manual marking errors. Figure 14a was used to identify cracks. Figure 14a contains much interference, such as cracks, shadows, and numerical values. The model detection result is shown in Figure 14b. The comparison of the two figures shows that the model can detect cracks in a complex environment and meets the basic requirements of crack identification in a practical engineering environment. The suitability of the model design proposed in this research has been verified again. It is worth mentioning that the method proposed in this paper utilizes points but not crack characteristics in the marking stage. Therefore, width recognition of cracks is not accurate enough. We should consider modifying and optimizing this method in future research.

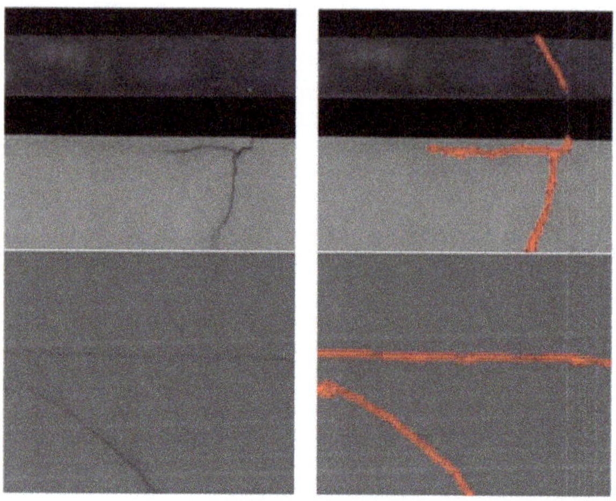

Figure 13. *Cont.*

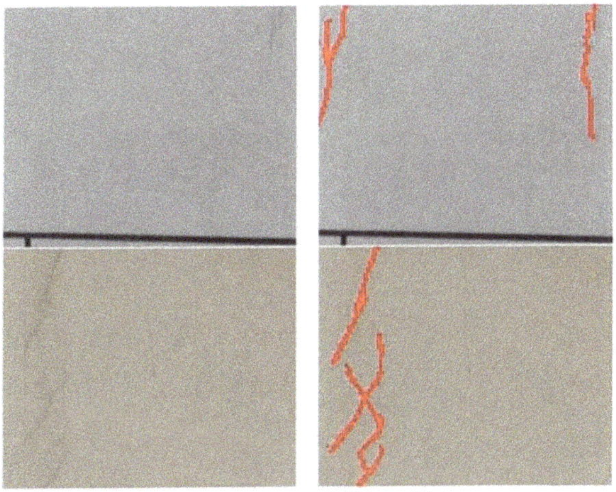

(a) Original image (b) Detected image

Figure 13. Detection results of different cracks.

(a) Original wall (b) Detection result

Figure 14. Detection of cracks on wall with interference factors.

5. Conclusions

Based on the characteristics of cracks, this paper defines the concept of crack key points, combined with the anchor mechanism in computer vision technology, and proposes a new crack identification method—the reference anchor point method. This research established a new model of image crack detection based on deep learning. Through the analysis of the detection network model and the crack detection experimental results, the following conclusions were obtained:

- This research proposed a crack characterization method, combined with the features of image cracks based on key points of cracks. Its detection accuracy is controllable, which can lead to pixel-level recognition effects and can greatly improve detection efficiency based on meeting the accuracy requirements of engineering. When the computer is configured with NVIDIA GeForce GTX 1080, the recognition time of a single photo is 30 ms.
- The advantages of characterizing image cracks based on the key points of cracks are explained. By designing algorithms such as fixed-distance decentralization and a reference anchor point method, the judgment conditions of positive and negative examples are clarified so that the crack image mark data based on the key points of the crack are suitable for model training.

- The image crack detection model KP-CraNet is established. From the perspective of global and local features, the principle of model detection is discussed, and the network structure of the model is introduced. The results show that crack key points greatly improve the crack detection effect.

A new model evaluation method is proposed. The distance distribution map is used to evaluate the model detection effect based on the key points of the fracture. This research evaluated the model's detection effect through a distance distribution map and the accuracy rate, recall rate, and F1 score. It is shown that the identification model has strong crack identification and robustness.

Author Contributions: Conceptualization, D.W. and H.C.; methodology, H.C.; software, H.C.; validation, D.W., H.C. and J.C.; formal analysis, D.W.; investigation, H.C.; resources, J.C.; data curation, J.C.; writing—original draft preparation, H.C.; writing—review and editing, D.W.; visualization, H.C.; supervision, J.C.; project administration, J.C.; funding acquisition, D.W. All authors have read and agreed to the published version of the manuscript.

Funding: This research was funded by Natural Science Foundation of Shanghai, grant number 18ZR1414500.

Data Availability Statement: Data is available at https://github.com/csga11/craData (accessed on 24 November 2021).

Conflicts of Interest: The authors declare no conflict of interest.

References

1. Yao, Y.; Tung, S.-T.E.; Glisic, B. Crack detection and characterization techniques—An overview. *Struct. Control Health Monit.* **2014**, *21*, 1387–1413. [CrossRef]
2. Qian, G.; Gu, S.; Jiang, J. The dynamic behaviour and crack detection of a beam with a crack. *J. Sound Vib.* **1990**, *138*, 233–243. [CrossRef]
3. Rangaraj, R.; Pokale, B.; Banerjee, A.; Gupta, S. Investigations on a particle filter algorithm for crack identification in beams from vibration measurements. *Struct. Control Health Monit.* **2015**, *22*, 1049–1067. [CrossRef]
4. Kirschke, K.; Velinsky, S.A. Histogram-Based Approach for Automated Pavement-Crack Sensing. *J. Transp. Eng.* **1992**, *118*, 700–710. [CrossRef]
5. Kim, B.; Cho, S. Image-based concrete crack assessment using mask and region-based convolutional neural network. *Struct. Control Health Monit.* **2019**, *26*, e2381. [CrossRef]
6. Otsu, N. A Threshold Selection Method from Gray-Level Histograms. *IEEE Trans. Syst. Man Cybern.* **2007**, *9*, 62–66. [CrossRef]
7. Canny, J. A Computational Approach to Edge Detection. *IEEE Trans. Pattern Anal. Mach. Intell.* **1986**, *PAMI-8*, 679–698. [CrossRef]
8. Abdel-Qader, I.; Abudayyeh, O.; Kelly, M.E. Analysis of edge-detection techniques for crack identification in bridges. *J. Comput. Civ. Eng.* **2003**, *17*, 255–263. [CrossRef]
9. Yan, M.; Bo, S.; Xu, K.; Yuyao, H. ICEMI '07. 8th International Conference on Y B T-E M AND I. Pavement Crack Detection and Analysis for High-Grade Highway. In Proceedings of the 8th International Conference on Electronic Measurement and Instruments, Xi'an, China, 16–18 August 2007.
10. Oullette, R.; Browne, M.; Hirasawa, K. Genetic Algorithm Optimization of a Convolutional Neural Network for Autonomous Crack Detection. In Proceedings of the 2004 Congress on Evolutionary Computation, Portland, OR, USA, 19–23 June 2004; pp. 516–521.
11. Oliveira, H.; Correia, P. Automatic Road Crack Segmentation Using Entropy and Image Dynamic Thresholding. In Proceedings of the 2009 17th European Signal Processing Conference, Glasgow, UK, 24–28 August 2009.
12. Moon, H.; Kim, J.H. Inteligent Crack Detecting Algorithm on the Concrete Crack Image Using Neural Network. In Proceedings of the 28th ISARC, Seoul, Korea, 29 June–2 July 2011; pp. 1461–1467.
13. Koch, C.; Georgieva, K.; Kasireddy, V.; Akinci, B.; Fieguth, P. A review on computer vision based defect detection and condition assessment of concrete and asphalt civil infrastructure. *Adv. Eng. Inf.* **2015**, *29*, 196–210. [CrossRef]
14. Choudhary, G.; Sayan, D. Crack Detection in Concrete Surfaces Using Image Processing, Fuzzy Logic, and Neural Networks. In Proceedings of the 2012 IEEE Fifth International Conference on Advanced Computational Intelligence (ICACI), Nanjing, China, 18–20 October 2012; pp. 404–411.
15. Jahanshahi, M.R. An innovative methodology for detection and quantification of cracks through incorporation of depth perception. *Mach. Vis. Appl.* **2013**, *24*, 227–241. [CrossRef]
16. O'Byrne, M.; Schoefs, F.; Gosh, B.; Pakrashi, V. Texture Analysis Based Damage Detection of Ageing Infrastructural Elements. Computer-Aided Civil and Infrastructure Engineering. *Wiley* **2013**, *28*, 162–177.

17. Nguyen, H.-N.; Kam, T.-Y.; Cheng, P.-Y. An Automatic Approach for Accurate Edge Detection of Concrete Crack Utilizing 2D Geometric Features of Crack. *J. Signal Process. Syst.* **2014**, *77*, 221–240. [CrossRef]
18. Civera, M.; Fragonara, L.; Surace, C. Video Processing Techniques for the Contactless Investigation of Large Oscillations. In Proceedings of the Journal of Physics: Conference Series, Milan, Italy, 29–30 November 2018; IOP Publishing: Bristol, UK, 2019; Volume 1249, p. 012004.
19. Civera, M.; Zanotti Fragonara, L.; Surace, C. Using video processing for the full-field identification of backbone curves in case of large vibrations. *Sensors* **2019**, *19*, 2345. [CrossRef]
20. Hinton, G.; Osindero, S.; The, Y. A Fast Learning Algorithm for Deep Belief Nets. *Neural Comput.* **2006**, *18*, 1527–1554. [CrossRef]
21. Zhang, L.; Yang, F.; Zhang, Y.; Zhu, Y.J. Road crack detection using deep convolutional neural network. In Proceedings of the 2016 IEEE International Conference on Image Processing (ICIP), Phoenix, AZ, USA, 25–28 September 2016; pp. 3708–3712.
22. Cha, Y.; Choi, W.; Büyüköztürk, O.O. Deep Learning-Based Crack Damage Detection Using Convolutional Neural Networks. *Comput.-Aided Civ. Infrastruct. Eng.* **2017**, *32*, 361–378. [CrossRef]
23. Zhang, A.; Wang, K.C.P.; Li, B.; Yang, E.; Dai, X.; Peng, Y.; Fei, Y.; Liu, Y.; Li, J.; Chen, C. Automated Pixel-Level Pavement Crack Detection on 3D Asphalt Surfaces Using a Deep-Learning Network. *Comput.-Aided Civ. Infrastruct. Eng.* **2017**, *32*, 805–819. [CrossRef]
24. Dorafshan, S.; Thomas, R.; Maguire, M. Comparison of deep convolutional neural networks and edge detectors for image-based crack detection in concrete. *Constr. Build. Mater.* **2018**, *186*, 1031–1045. [CrossRef]
25. Liu, Y.; Yao, J.; Lu, X.; Xie, R.; Li, L. Deep Crack: A deep hierarchical feature learning architecture for crack segmentation. *Neurocomputing* **2019**, *338*, 139–153. [CrossRef]
26. Fei, Y.; Wang, K.C.P.; Zhang, A.; Chen, C.; Li, J.; Liu, Y.; Yang, G.; Li, B. Pixel-Level Cracking Detection on 3D Asphalt Pavement Images Through Deep-Learning- Based CrackNet-V. *IEEE Trans. Intell. Transp. Syst.* **2020**, *21*, 273–284. [CrossRef]
27. Zou, Q.; Zhang, Z.; Li, Q.; Qi, X.; Wang, Q.; Wang, S. Deep Crack: Learning Hierarchical Convolutional Features for Crack Detection. *IEEE Trans. Image Process.* **2019**, *28*, 1498–1512. [CrossRef]
28. Lee, J.S.; Hwang, S.H.; Choi, I.Y.; Choi, Y. Estimation of crack width based on shape sensitive kernels and semantic segmentation. *Struct Control Health Monit.* **2020**, *27*, e2504. [CrossRef]
29. Cao, M.-T.; Tran, Q.-V.; Nguyen, N.-M.; Chang, K.-T. Survey on performance of deep learning models for detecting road damages using multiple dashcam image resources. *Adv. Eng. Inform.* **2020**, *46*, 101182. [CrossRef]
30. Zhou, X.; Wang, D.; Krähenbühl, P. Objects as Points. *arXiv* **2019**, arXiv:1904.07850.
31. Hei, L.; Yun, T.; Russakovsky, O.; Deng, J. Corner Net-Lite: Efficient Keypoint Based Object Detection. *arXiv* **2019**, arXiv:1904.08900.
32. Duan, K.; Bai, S.; Xie, L.; Qi, H.; Huang, Q.; Tian, Q. Center Net: Key point Triplets for Object Detection. *arXiv* **2019**, arXiv:1904.08189.
33. Lee, D.; Kim, J.; Lee, D. Robust Concrete Crack Detection Using Deep Learning-Based Semantic Segmentation. International. *J. Aeronaut. Space Sci.* **2019**, *20*, 287–299. [CrossRef]
34. Fan, Z.; Wu, Y.; Lu, J.; Li, W. Automatic Pavement Crack Detection Based on Structured Prediction with the Convolutional Neural Network. *arXiv* **2018**, arXiv:1802.02208.
35. Civera, M.; Fragonara, L.Z.; Surace, C. An experimental study of the feasibility of phase-based video magnification for damage detection and localisation in operational deflection shapes. *Strain* **2020**, *56*, e12336. [CrossRef]
36. Amhaz, R.; Chambon, S.; Idier, J.; Baltazart, V. Automatic crack detection on 2D pavement images: An algorithm based on minimal path selection. *IEEE Trans. Intell. Transp. Syst.* **2016**, *17*, 2718–2729. [CrossRef]
37. Ren, S.; He, K.; Girshick, R.; Sun, J. Faster R-CNN: Towards Real-Time Object Detection with Region Proposal Networks. *arXiv* **2015**, arXiv:1506.01497. [CrossRef]
38. Lin, T.-Y.; Dollár, P.; Girshick, R.; He, K.; Hariharan, B.; Belongie, S. Feature Pyramid Networks for Object Detection. *arXiv* **2016**, arXiv:1612.03144.
39. He, K.; Zhang, X.; Ren, S.; Sun, J. Deep Residual Learning for Image Recognition. In Proceedings of the IEEE Computer Society Conference on Computer Vision and Pattern Recognition, Seattle, WA, USA, 27–30 June 2016; pp. 770–778.
40. Lin, T.Y.; Goyal, P.; Girshick, R.; He, K.; Dollár, P. Focal loss for dense object detection. In Proceedings of the 2017 IEEE international conference on computer vision, Venice, Italy, 22–29 October 2017; pp. 2980–2988.

applied sciences

MDPI

Article

Ceramic Cracks Segmentation with Deep Learning

Gerivan Santos Junior *, Janderson Ferreira, Cristian Millán-Arias, Ramiro Daniel, Alberto Casado Junior and Bruno J. T. Fernandes

Polytechnic School of Pernambuco, University of Pernambuco, Recife 50720-001, Brazil; jrb@ecomp.poli.br (J.F.); ccma@ecomp.poli.br (C.M.-A.); rdbr@poli.br (R.D.); acasado@poli.br (A.C.J.); bjtf@ecomp.poli.br (B.J.T.F.)
* Correspondence: gcsj@ecomp.poli.br

Abstract: Cracks are pathologies whose appearance in ceramic tiles can cause various damages due to the coating system losing water tightness and impermeability functions. Besides, the detachment of a ceramic plate, exposing the building structure, can still reach people who move around the building. Manual inspection is the most common method for addressing this problem. However, it depends on the knowledge and experience of those who perform the analysis and demands a long time and a high cost to map the entire area. This work focuses on automated optical inspection to find faults in ceramic tiles performing the segmentation of cracks in ceramic images using deep learning to segment these defects. We propose an architecture for segmenting cracks in facades with Deep Learning that includes an image pre-processing step. We also propose the Ceramic Crack Database, a set of images to segment defects in ceramic tiles. The proposed model can adequately identify the crack even when it is close to or within the grout.

Keywords: deep learning; segmentation; ceramics; cracks; image

Citation: Junior, G.S.; Ferreira, J.; Millán-Arias, C.; Ruiz, D.; Junior, A.C.; Fernandes, B.J.T. Ceramic Cracks Segmentation with Deep Learning. *Appl. Sci.* **2021**, *11*, 6017. https://doi.org/10.3390/app11136017

Academic Editor: Antonio Fernández-Caballero

Received: 7 May 2021
Accepted: 26 May 2021
Published: 28 June 2021

1. Introduction

In civil construction, buildings must be able to withstand the action of degradation agents for a predetermined or predicted time [1]. The building's facades include the cladding system that serves to protect the building from external degradation agents, in addition to providing functional and aesthetic comfort to its users [2]. Pathological manifestations are common at these points, and they occur more frequently in ceramic materials, which are used on a large scale in buildings. Besides, these manifestations arise in other types of materials, such as mortar and stone. They can be related to several factors such as excessive load, humidity variation, thermal variation, biological agents, material incompatibility, and atmospheric agents [3]. These manifestations compromise the essential function of protection, which aims to protect the coated surfaces against the agents that cause deterioration that can present themselves in different ways. Thus, the consequences can range from aesthetic problems or performance of coating to risks of accidents with people, substantially aggravated by the height of the buildings [4].

The main types of pathological manifestations associated with ceramic facade coverings are cracks, efflorescence, detachment, and those resulting from biological processes. Among these, the fissure is the most found in the literature since it compromises the building safety, puts at risk the people that travel around it, and presents a more critical aesthetic aspect [3,5–7].

A fissure's main characteristic is the rupture appearance on the ceramic plate surface or body, causing the loss of the facade's integrity and uncovers some of its components, the plates, or joints. When the fissure happens, a detachment of the substrate plate is generated [4].

Image processing techniques (IPTs) are currently applied in civil engineering for images collected from inspections. These techniques emerged to detect cracks in the civil infrastructure, partially reducing the work done by human beings, and used several image processing techniques to extract characteristics of cracks in the surfaces of the images [8]. However, many structure analyses and inspections are carried out manually, and this requires a lot of knowledge, experience, and time from those who will perform this activity, thus making the activity long and time-consuming.

Automatic crack detection is essential in places that are difficult to access due to height or scale, to avoid exposing people to dangerous situations, and to speed up the inspection process [9]. On the other hand, applying procedures takes time due to the complexity of the work performed, including the installation of scaffolding, observation of a large area, and even the use of elevators or Bosun's chair.

Therefore, to create an automatic crack detection solution in ceramics, we focus on an image segmentation methodology. Such methodology includes a pre-processing step because many ceramics have textures in opposition to concrete that has linearity in its texture. The proposed methodology also includes a deep learning model to solve the problem of crack detection in ceramics, which in the future may be coupled to drones to carry out these inspections in a less manual, faster, and less dependent on human action with specific knowledge for the area. This solution also allows the location of the crack to be identified by means of the segmented image, since the crack is segmented, showing its exact location, because in facade inspections, in addition to identifying that there is a crack, which other works do through classification, it is important to know where it is located. We can generate an overlay of the images, highlighting where the crack is for future analysis through segmentation. We created a database to implement the segmentation models that contain images of defective ceramic plates and the basic crack truth in each image.

In summary, this work has as a novelty the detection of cracks through segmentation. The proposed methodology identifies the exact location of the crack in the image. A pre-processing step in the input enables such identification to increase the prominence of the cracks in the raw images. Thus, it facilitates the model's learning process. As a novelty, the work also brings a database of cracks in ceramics that can be used to improve future research.

This paper is organized as follows. We present related works in Section 2 and the proposed approach to address the segmentation problem in Section 3. In Section 4, the metrics, the loss function, and the experimental configuration are described, and we also present our set of images for segmentation of ceramic cracks. In Section 5, the results are described, and the experiments are discussed. Finally, in Section 6, we present our main conclusions and describe future research.

2. Theoretical Foundation

Image segmentation is a process that aims to divide images into regions or objects of interest that are homogeneous. This activity is the initial step in image processing applications, such as pattern recognition and image analysis. Image analysis includes characterization and representation of objects and measurement of resources. This process is mainly used to find objects and shapes [10]. There are currently several types of segmentation, usually based on formats, pixel characteristics, histograms, and movement. Each type supports common features in pixels or a group of pixels. [10,11]. The evolution of deep learning to various kinds of computer vision problems in the literature encourages this work to build a crack segmentation model based on deep learning.

IPTs had a significant advance in the last years. However, other problems have no solution found by IPTs yet, such as the real world's perception (lack of context of images, images with shadow, textures, variation in lighting), shading, and lighting variation. In parallel to this, there has been an exponential growth in the use of convolutional neural networks (CNN), and they have obtained better results for these problems. CNNs, too, have been used to classify cracks and fissures [8,9,12–15]. However, none of the proposed CNNs deals specifically with coating ceramics.

There are several studies in the literature on crack detection [8,9,11,12,16–18], but regarding cracking in ceramics there are few works and each surface has its specificity.

Young-Jin Cha in [8] applied a vision-based method using a deep architecture of convolutional neural networks (CNNs) to detect concrete cracks without calculations as defect characteristics. He aimed to create a model that could solve the problem without the use of processing image techniques. Moreover, Young-Jin Cha compared the obtained results with traditional methods based on edge detection. CNNs obtained the best performance, but the work is not related to identifying the location of the crack in the image, but in an evaluation of an intact or cracked part, which for a facade inspection is not ideal, and it is necessary to identify where the crack is located. In this work, it was resolved using segmentation, managing to extract the exact location of the fissure

In another study, Silva W.R.L [12] aimed to increase the level of automation in the inspection of concrete infrastructure when combined with unmanned aerial vehicles. The crack detection model developed is based on an image classification algorithm of the deep learning convolutional neural network (CNN). A relatively heterogeneous dataset has been provided. The authors claimed that deep learning allows the development of a concrete crack detection system responsible for several conditions, such as different light, surface finish, and humidity that a concrete surface can display. In this work, the model VGG16 [19] was used as a backend to the transference-learning technique. Silva's best experiment produced a model with an accuracy of 92.27%. However, Silva's work deals with image classification, stating whether or not there are cracks in concrete structures. Moreover, it does not make clear where the crack is located, which is essential for automated inspection of structures.

In Ahmed Mahgoub Ahmed Talab [18], the authors presented a new approach in image processing to detect cracks in the images of concrete structures. The method involves three steps. Firstly, changing the image to grayscale to use the Sobel method to detected edges and find the cracks. Second, determining an appropriate threshold in a binary image and classifying all pixels into two categories: background and foreground, and obtaining the region's area. Finally, using the filter area and changing the area if it is smaller than the specified number. Third, after applying the Sobel filter to eliminate residual noise, performing the Otsu method to detect large cracks. The article describes a method for detecting crack patterns in cement using image processing techniques. According to the author, this method's advantage is the precise and accurate detection of cracks in the images. The experimental work shows that the method is better than other widely used techniques. However, it does not use deep learning, and it is limited only to the use of image processing, which has the advantage of the low computational cost. Moreover, the work from Talab does not present the same generalization capacity as CNNs. In opposition, the methodology proposed in this work includes the combination of deep learning with image processing, improving the generalization in crack detection.

3. Crack Segmentation of Ceramic Surface

This work presents an architecture for segmenting cracks in facades with Deep Learning that was named the CCS model (Segmentation Model for cracks in ceramics) that includes a pre-processing step and a deep neural network for segmentation proposal followed by a threshold operation, as shown in Figure 1. The output is a binary image that brings white lines where the cracks were located, and, through overlapping images, it is possible to highlight the cracks in the original image. At CCS, pre-processing is done in the database before using the segmentation model. The pre-processed image with its label is adopted as input to perform the model's training. After training, only the original image is needed to run this network.

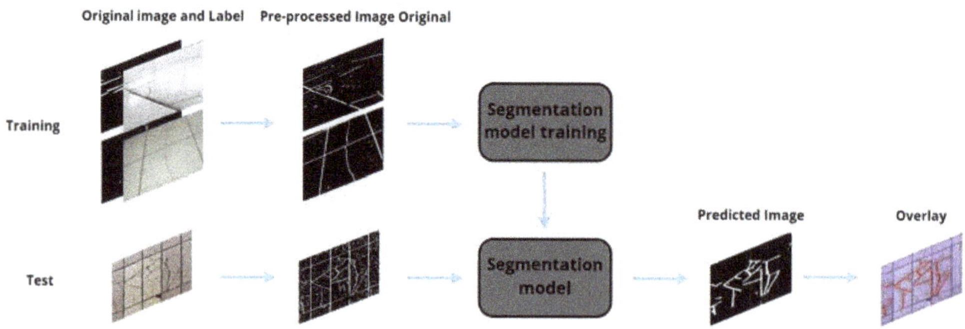

Figure 1. Architecture for crack segmentation in CCS model.

3.1. Data Pre-Processing

Pre-processing became necessary due to differences in context in the images, as just a grayscale image binarization is not enough, as much of the area of interest in the image is lost, and in some cases, the cracks present in the image disappearedm as shown in Figure 2. With that, it is necessary to apply some techniques in addition to binarization. Several experiments were carried out regarding the detection of lines, edges, and objects through computer vision to find a generic pre-processing for this problem.

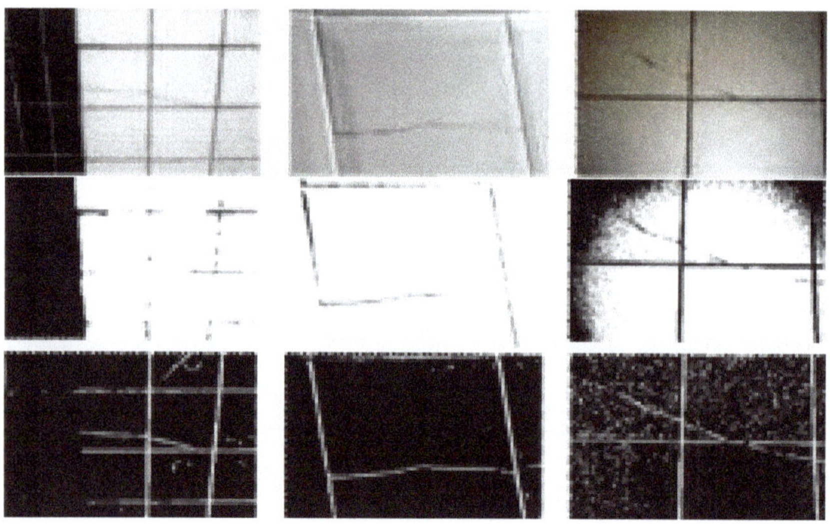

Figure 2. Example of a pre-procesing: (**up**) Original image, (**middle**) simple binarization, and (**down**) performed pre-processing on the original image.

The techniques used are listed below:

1. Histogram equalization;
2. Gaussian filter (with kernel 3×3);
3. Light and contrast adjust;
4. Inversion;
5. Erode and dilate functions (with kernel 5×5);
6. Finally, Otsu thresholding.

As seen in Figure 2, the pre-processing result highlights the area of interest, becoming much more in evidence, thus facilitating the neural network's performance in the extraction of characteristics.

This configuration was obtained through several attempts to highlight the images' cracks, using digital image processing techniques until reaching an acceptable result, where it obtained a better learning result by the model. The histogram equalization was used to change the image values' distribution, allowing the sharp differences to be reduced and accentuating details not previously visible. The Gaussian filter was used to soften the image, blurring it to remove noise, using a 3×3 kernel to make a smaller effect since the kernel's size influences the blurring power. Light and contrast adjustment was used to correct images with excessive lighting problems, not impacting those with standard lighting. The pixel inversion used in the images was necessary to comply with the standard established in the labeling of the images, where it was decided that everything white would be cracked surfaces and what was black would be parts of the ceramic, so the inversion made what was black turn white, and whatever was white turned black, since the grayscale highlighted the cracks in black. The erosion and dilation process was applied to solve the discontinuity of some cracks that broke during the blurring process, using a 5×5 kernel to continue the cracks, and lastly, a threshold was used. Some other techniques were tested but, in analysis, no significant change was seen for the objective that wanted to be achieved (highlighted in the fissures) and did not significantly influence the learning of the model.

3.2. Segmentation Model

The segmentation model uses the U-Net, proposed by Ronneberger et al. [20], which stands out in the segmentation problems due to the better performance, even with few images for training. The peculiar name of U-Net is due to the "U" shape of its architecture. The network input is the image that needs to be segmented. The output is the image label, a label that represents the model's expected output.

The network has a typical convolutional network architecture; however, it has two complementary paths, the contracting path (left side) and the expansive path (right side). The contracting path handles executing controls to extract characteristics from the image. This process reduces the dimensionality and increases the filters applied to extract features, generating a map for each level. On the other hand, the expansive path handles reducing the filters and increasing the dimensionality. A concatenation process is performed with the correspondingly cropped feature map from the contracting path to reach the segmented image's formation.

The contraction path is a typical convolutional network architecture. It contains nine learning convolutional layers and four max pooling operations after every three convolutions [20]. We applied two 3×3 convolutions in our implementations, each followed by a rectified linear unit (ReLU) and a 2×2 max pooling operation with stride 2 for downsampling. The number of feature channels is doubled for each downsampling step. A cropping process is made during the expansive path to avoid the loss of border pixels in every convolutional operation [20]. In our implementations, the feature map is halved by a 2×2 up-convolution [20] in an up-sampling process, followed by concatenation with the correspondingly cropped feature map from the contracting path, and two 3×3 convolutions. A ReLU operation follows each convolution. Cropping is necessary due to the loss of edge pixels in every convolution. A 1×1 convolution is used in the final layer to map each feature vector of the component to the desired number of classes.

3.3. Threshold

As a final step, it is necessary to apply a simple threshold to perform a binarization of the image and ensure that the end of the image values is of 0 (zero) or 1 (one), with 0 points for no cracks and 1 point for cracks in the semantic segmentation of pixel by pixel. For each pixel, the limit value used was 0.5, where, if the pixel value is less than the limit, it will be set to 0. Otherwise, it will be defined with the maximum value defined, in this case, 1.

4. Methodology

This section describes the methodology followed in this article. First, we describe the loss function used in the segmentation models. Next, we describe the database, metrics, and the experimental setup to obtain the results.

4.1. Loss Function

In this work, we used the Jaccard distance as our loss function. The Jaccard distance measures dissimilarity between sample sets. This function is complementary to the Jaccard coefficient or intersection over union. The loss function is calculated as:

$$L(A, B) = 1 - \frac{|A \cap B|}{|A \cup B|},$$

where A and B, in the CCS model, are binary images of the same size.

4.2. Ceramic Cracks Database

We propose a ceramic crack database with 167 ceramic crack images. The images were collected by students of the University of Pernambuco from the civil engineering department. The database consists of images of a fixed resolution of 256 × 256 in RGB format without any pre-processing. Each element is labeled with a binary image of segmented cracks. The data has various characteristics, like different sizes, angles, illumination, distances, or even materials and textures. The database has images of building facades with ceramics with cracks of different shapes, both superficial and more profound. Besides, the database has images of ceramics with different colors and textures, which enrich its diversity and give more information to the model used. Figure 3 shows examples of the database images, the first line (up) shows the original images resize by 256 × 256, and the second line (down) shows the respective segmented label.

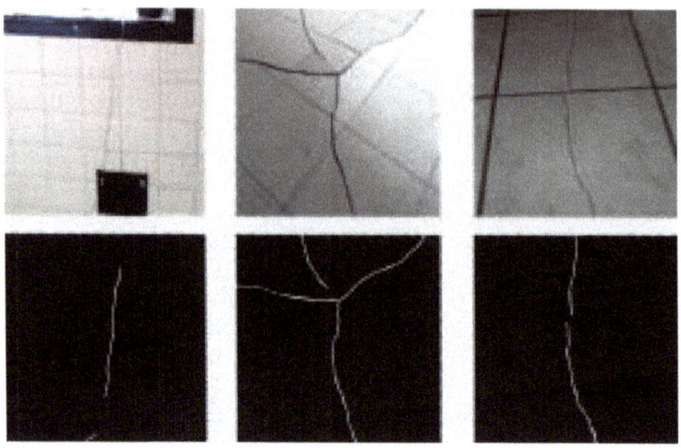

Figure 3. Examples of images collected to assemble the database and their respective ground truth (black and white).

A label corresponding to the images is required to perform the training of the segmentation networks. Those labels are the expected results of the network. They were manually generated for all the collected data and featured white for the original image regions characterized by a crack and black for all other regions. The database will be available for public use in future works related to the ceramic crack segmentation problem at the link https://github.com/gerivansantos/ceramic-cracks-dataset (accessed on 2 June 2021).

4.3. Metrics

The described metrics compare the previously mentioned approaches and evaluate which provided the best solution. For this purpose, the metrics selected are the Intersection over Union (IoU), Precision, Recall, the Kappa Coefficient, and Specificity.

4.3.1. Intersection over Union

IoU or Jaccard coefficient is a measurement commonly used to validate semantic segmentation, and it is direct and effective. It is the intersection between predicted segmentation and ground truth divided by the union between the two, as demonstrated in Figure 4. This metric oscillates between 0 and 1, or 0 and 100%, wherein 0 indicates no intersection and 1 indicates an intersection equal to the union.

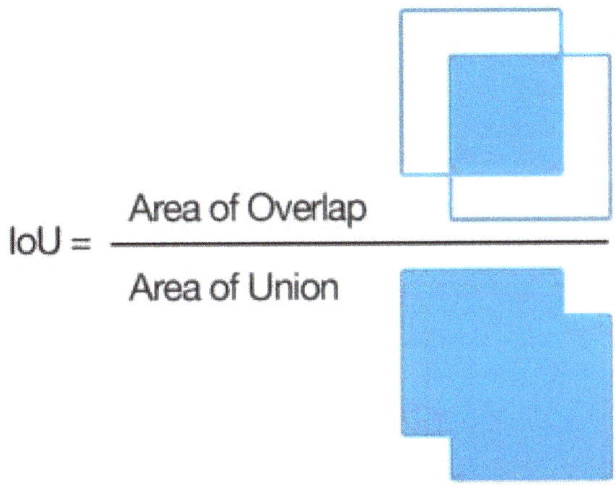

Figure 4. Representation of the Intersection over Union.

4.3.2. Precision and Recall

Precision demonstrates the percentage of the relevance of the results. Conversely, recall illustrates the percentage of relevant results that are correctly classified by the algorithm. It relates the number of correct positive predictions to all positive predictions. The following equations calculate precision and recall:

$$Precision = \frac{TruePositives}{TruePositives + FalsePositives}$$

$$Recall = \frac{TruePositives}{TruePositives + FalseNegatives}$$

4.3.3. Kappa Coefficient

The Kappa Coefficient is a statistic that evaluates the relation between two sets of data, calculated as follows:

$$k = \frac{P_0 - P_e}{1 - P_e},$$

where P_0 is the relative acceptance rate and P_e is the hypothetical acceptance rate. Thus, the closer k is to 1.0, the more the two data sets are related.

4.3.4. Specificity

Specificity was also measured, and it is defined as the proportion of real negatives predicted to be negatives (True Negatives), as illustrated in the following equation. It implies another portion of real negatives that were predicted to be positives (False Positives), which must equal 1 when summed with specificity. Another existing metric is sensitivity, which measures the proportion of correctly classified real negatives.

$$Specificity = \frac{TrueNegatives}{TrueNegatives + FalsePositives}$$

4.3.5. Confusion Matrix

The Confusion Matrix contains information on the real data and a classifying system's predictions, and it is commonly used to evaluate such a system. It is a table with four different relations between real and predicted values: True Positives (TP), correctly predicted positives; True Negative (TN), correctly predicted negatives; False Positives (FP), type 1 error, incorrectly predicted positives; False Negatives (FN), type 2 error, incorrectly predicted negatives. Thus, it is useful in measuring Precision, Recall, and Specificity.

4.4. Experimental Setup

This paper sets a benchmark over the proposed database, using state-of-the-art models. The 70% of the data is randomly allocated for training and 30% towards testing in the several experiments performed, which were used as comparative parameters for the database. The models selected to make a comparison with the proposed model were variations of implementations of U-Net [20] and LinkNet [21]. The criteria for its selection are the relevance in the image segmentation literature and the good accuracy in solving the proposed problem. Data augmentation was also applied to improve the generalization of the model.

The used architecture, backbone, and weight initialization method are illustrated in Figure 5. The backbone is the network architecture implemented in each model. In this work, we use the different architectures like resnet34 [22], resnet50 [22] and vgg16 [19], which are 34 layers deep, 50 layers deep and 16 layers of deep, respectively. The initialization of the weights is carried out randomly and using the weights of the pre-trained neural network with the ImageNet database [23].

Input images are set with size 224 × 224 to U-Net and LinkNet models. We apply the Adam algorithm, a stochastic gradient descent method based on the adaptive estimation of first-order and second-order moments [24]. In our experiments, we set the hyperparameters to the Adam algorithm with $\beta_1 = 0.9$, $\beta_2 = 0.999$, and $\epsilon = 10^{-07}$, with a learning rate of $\alpha = 0.001$.

A total of 20 experiments are performed 30 times to ensure that the results were statistically significant, and from them were extracted the metrics used to compare the results. Additionally, since fine-tuning presents good results in deep learning applications, we analyzed its efficiency in this work approach.

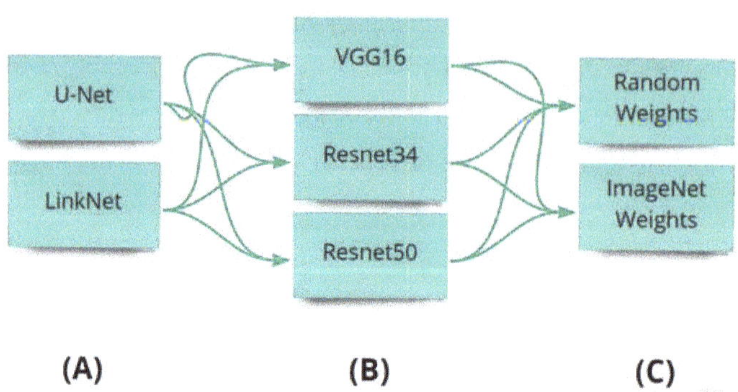

Figure 5. Conducted experiments, where (**A**) are models, (**B**) are backbones, and (**C**) are weight initialization, where "None" is the initialization of the weights carried out randomly.

5. Results and Discussions

The models were selected according to their success in the segmentation of images, as previously presented. After that, 20 fine-tuning experiments were performed, altering the models' backbone and the weight initialization. Table 1 outlines each model's three best results, utilizing the IoU metric.

Table 1. Result of the metrics for the U-Net and LinkNet models, and the different network architectures. The weights of a pre-trained network with ImageNet are used to initialize each of the models.

Metrics	Models						
	U-Net				LinkNet		
	CCS †	**resnet50 †**	**resnet50**	**resnet34**	**resnet34**	**resnet34 †**	**vgg16**
IoU	0.865	0.685	0.681	0.675	0.697	0.684	0.672
Precision	0.999	0.713	0.727	0.720	0.727	0.711	0.704
Recall	0787	0.946	0.933	0.929	0.946	0.922	0.897
Kappa	0.724	0.805	0.808	0.803	0.814	0.794	0.775
Specificity	0.999	0.988	0.988	0.988	0.988	0.987	0.988

† For these models, weights were randomly initialized.

In terms of average IoU, our model CCS overcame the IoU value from the other models. We obtain an index of 86.5% in CCS model, and in the U-Net and LinkNet models, the value is around 68%. The obtained average precision in our approach in the U-Net model reaches a value of 99.9%. The U-Net and LinkNet models demonstrate that the resnet50 and resnet34 models, when initialized with the ImageNet weights, present an average precision of 72.7%. Regarding the average recall, to U-Net with resnet50 backbone, without weight initialization, and LinkNet with restnet34, initialized with the ImageNet weights, obtained percentages were equal to 94.6%. This value overcomes the average recall of our approach, with a value of 78.7%. Our approach shows a high value of accuracy and a low recall. However, most of our predicted labels are correct.

The best average kappa coefficient is obtained by a LinkNet model with a resnet34 backbone, using the ImageNet weights, with 81.4%. All the models to U-Net and LinkNet achieve a value of kappa coefficient around 80%. The lower value is from our approach, with 72.4%, followed by the LinkNet model with vgg16 and ImageNet initialization. We observe that the models with vgg16 architecture reach the low values of the kappa coefficient. The average value of specificity for the U-Net and LinkNet models reached a value of around 98%. Our approach obtains a value of 99.9%, overcoming the other values; nonetheless, it does not seem to show a significant difference.

Regarding the confusion matrix, our approach correctly classified 99% of the pixels that belong to the crack (Figure 6a). However, 27% of the pixels that are not from the crack are classified as part of it. Using the U-Net model, resnet50 with the random initialization, 95% of the pixels are correctly classified as cracks (Figure 6b). For all other implementations (Figure 6c,d), 93% of the crack is correctly classified. The Linknet model, with resnet34 and initialization with ImageNet, correctly classify 95% of the crack pixels (Figure 6e), surpassing the other implementations (Figure 6f,g) with values of 90% and 84%.

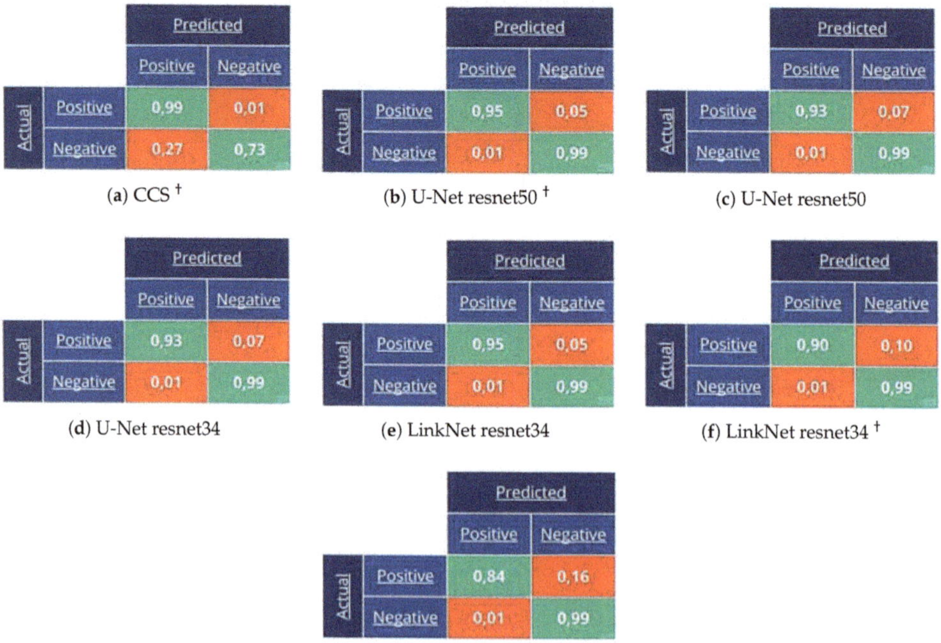

(**a**) CCS †   (**b**) U-Net resnet50 †   (**c**) U-Net resnet50

(**d**) U-Net resnet34   (**e**) LinkNet resnet34   (**f**) LinkNet resnet34 †

(**g**) Linknet vgg16

Figure 6. Confusion matrix from the prediction to (**a–d**) U-Net model, where (**a**) is CCS model, (**e–g**) LinkNet model. The weights of a pre-trained network with ImageNet are used to initialize each of the models (For the models †, weights were randomly initialized).

Qualitative analysis can be illustrated with the results presented in Figure 7. The Figure shows an example image (Figure 7a) and its ground truth (Figure 7b). The original image is inputted to the network, having the expected output (Figure 7c–i). It is possible to observe mistakes in segmentation labeling in some regions. Figure 7c,h shows a thick segmentation in comparison with the other predictions, overestimating the region where the crack is. A fine segmentation can lose crack representation. In Figure 7e, some parts of the crack are not identified. It should be noted that the models above can segment the cracks that are above or near the grouts. In some IPTs, such as fissures, they are easily confused with grouts. During the training, some analyses were made, and it was observed that by increasing the number of epochs in the training, the model was able to learn more, including understanding when the cracks were overlapping in the grout reducing the problem of not identifying cracks near or overlapping the grout.

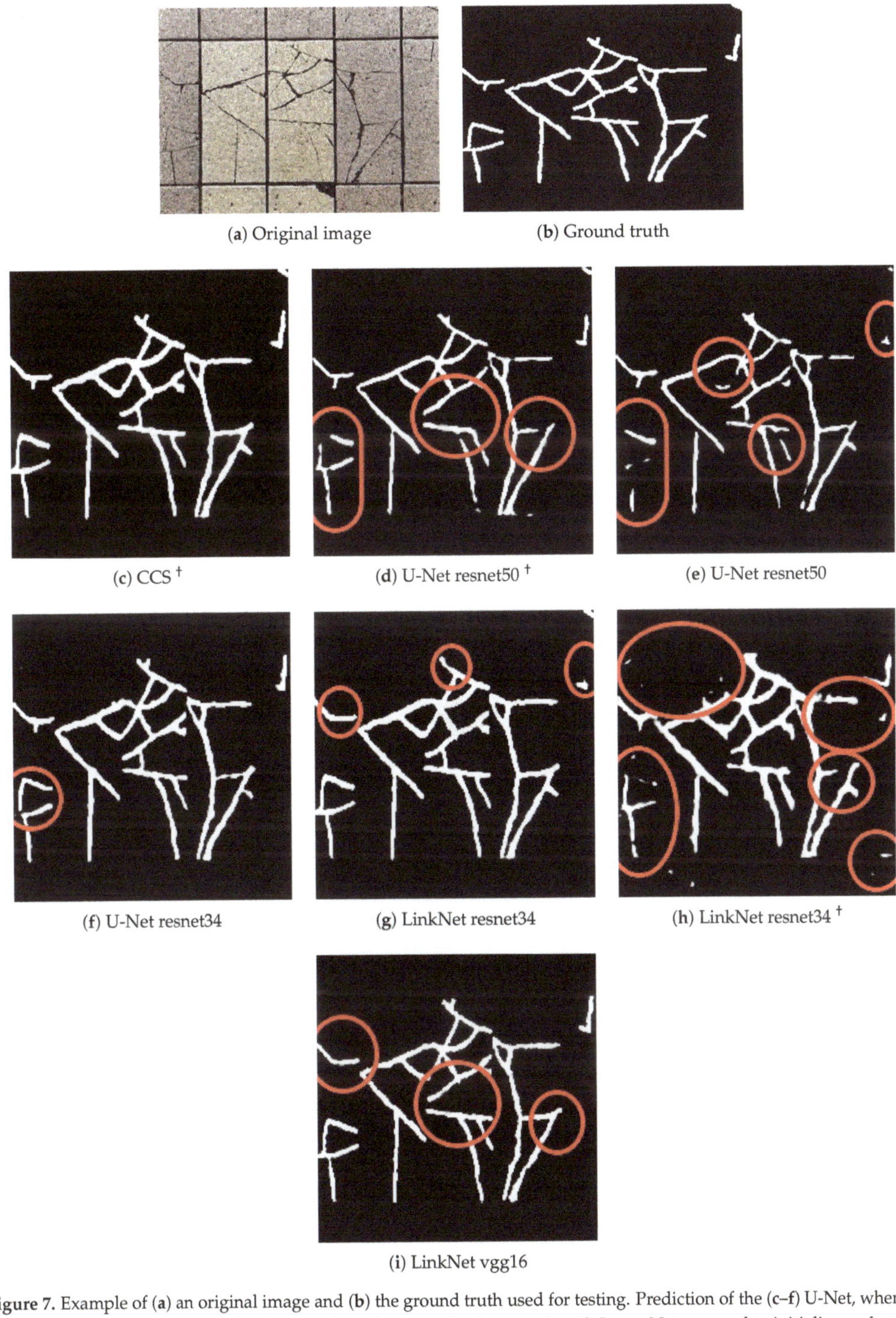

(a) Original image (b) Ground truth
(c) CCS † (d) U-Net resnet50 † (e) U-Net resnet50
(f) U-Net resnet34 (g) LinkNet resnet34 (h) LinkNet resnet34 †
(i) LinkNet vgg16

Figure 7. Example of (**a**) an original image and (**b**) the ground truth used for testing. Prediction of the (**c–f**) U-Net, where (**c**) is CCS, and (**g**,**h**) LinkNet models. The weights of a pre-trained network with ImageNet are used to initialize each of the models (For the models †, and weights were randomly initialized).

6. Conclusions

In this work, we analyze deep learning models' capabilities in the segmentation of cracks in ceramics tiles. We propose a pre-processing methodology to improve the performance of models to ceramic crack segmentation. Besides, we present the Ceramic Cracks database, a set of images with ceramic tiles with cracks destined for crack segmentation. Our results show that it is possible to identify cracks in ceramic images, although there are a few minor errors. The crack is identified with a high precision value in the model using a pre-processing methodology. The U-Net and LinkNet models achieve good results, using the resnet50 and resnet34 as backbones, respectively, and the weights of a pre-trained network with ImageNet to initialize. By increasing the number of epochs during training, the models manage to segment cracks even when they are in the tiles' grout.

Thus, other researchers can use the proposed study and database to improve their segmentation issues using computer vision. This paper then contributes to a new segmentation problem and a new database for crack segmentation in ceramic tiles. Future works are expected to compare the proposed solution's efficiency to other deep learning segmentation models and update the database, increasing the number of instances. We also intend to study the computational cost of the proposed solution and other solutions in the literature. On the other hand, we expect to implement this work in drones for optical facade inspection, which allows a more efficient inspection at a low cost.

Author Contributions: Conceptualization, G.S.J., R.D., A.C.J., and B.J.T.F.; data curation, G.S.J., R.D., A.C.J. and B.J.T.F.; formal analysis, G.S.J. and B.J.T.F.; funding acquisition, B.J.T.F.; investigation, G.S.J.; methodology, G.S.J., J.F. and C.M.-A. ; project administration, G.S.J., B.J.T.F. and A.C.J.; resources, G.S.J., B.J.T.F., A.C.J.; software, G.S.J. and J.F.; supervision, B.J.T.F. and A.C.J.; validation, G.S.J., J.F. and B.J.T.F.; visualization, G.S.J.; writing—original draft, G.S.J.; writing—review and editing, J.F., C.M.-A., R.D., A.C.J. and B.J.T.F. All authors have read and agreed to the published version of the manuscript.

Funding: This study was financed in part by the Coordenação de Aperfeiçoamento de Pessoal de Nível Superior-Brasil (CAPES)-Finance Code 001, and the Brazilian agencies FACEPE and CNPq.

Conflicts of Interest: The authors declare no conflict of interest.

References

1. Possan, E.; Demoliner, C.A. Desempenho, durabilidade e vida útil das edificações: Abordagem geral. *Revista Técnico-Científica* **2013**, 5–7. Available online: http://creaprw16.crea-pr.org.br/revista/Sistema/index.php/revista/article/view/14 (accessed on 2 June 2021).
2. Costa e Silva, A.J.d.; Franco, L.S. Descolamentos dos revestimentos cerâmicos de fachada na cidade do Recife. In *Boletim Técnico No 285 do Departamento de Engenharia de Construção Civil de Escola Politécnica*; University de São Paulo: São Paulo, Brazil, 2001.
3. Galletto, A.; Andrello, J.M. Patologia em Fachadas com Revestimentos Cerâmicos. In Proceedings of the CINPAR—IX Congresso Internacional sobre Patología y Recuperación de Estruturas, João Pessoa, Brazil, 2–5 June 2013.
4. Luz, M.d.A. Manifestações patológicas em Revestimentos Cerâmicos de Fachada em Tres Estudos de caso na Cidade de Balneário Camboriú. Master's Thesis, Universidade Federal de Santa Catarina, Florianópolis, Brazil, 2004.
5. Horsth, A.A.; Barrelete, F.N.; Junior, J.A.M.; Miranda, L.R. Patologias nos Revestimentos CerÂMicos em Fachada. *REMAS-Rev. Educ. Meio Ambiente e Saúde* **2018**, *8*, 27–38.
6. De Freitas, A.H.C.; França, P.M.; França, T.M. Patologia de Fachadas. *Rev. Pensar Eng.* **2013**, *1*.
7. Toledo, L.B.F.d. Patologia em revestimentos cerâmicos de fachadas de edifícios: Estudo regionalizado para a cidade de Divinópolis-MG. Master's Thesis, Curso de Engenharia Civil, Universidade Federal de Minas Gerais, Belo Horizonte, Brazil, 2007.
8. Cha, Y.J.; Choi, W.; Büyüköztürk, O. Deep learning-based crack damage detection using convolutional neural networks. *Comput. Aided Civ. Infrastruct. Eng.* **2017**, *32*, 361–378. [CrossRef]
9. Yang, L.; Li, B.; Li, W.; Jiang, B.; Xiao, J. Semantic metric 3d reconstruction for concrete inspection. In Proceedings of the IEEE Conference on Computer Vision and Pattern Recognition Workshops, Salt Lake City, UT, USA, 18–22 June 2018; pp. 1543–1551.
10. Acton, S.T.; Ray, N. Biomedical image analysis: Segmentation. *Synth. Lect. Image Video Multimed. Process.* **2009**, *4*, 1–108. [CrossRef]
11. Akagic, A.; Buza, E.; Omanovic, S.; Karabegovic, A. Pavement crack detection using Otsu thresholding for image segmentation. In Proceedings of the 2018 41st International Convention on Information and Communication Technology, Electronics and Microelectronics (MIPRO), Opatija, Croatia, 21–25 May 2018; pp. 1092–1097.

12. Silva, W.R.L.d.; Lucena, D.S.d. Concrete Cracks Detection Based on Deep Learning Image Classification. *Proceedings* **2018**, *2*, 489 [CrossRef]
13. Choi, W.; Cha, Y.J. SDDNet: Real-time crack segmentation. *IEEE Trans. Ind. Electron.* **2019**, *67*, 8016–8025. [CrossRef]
14. Liu, Y.; Yao, J.; Lu, X.; Xie, R.; Li, L. DeepCrack: A deep hierarchical feature learning architecture for crack segmentation. *Neurocomputing* **2019**, *338*, 139–153. [CrossRef]
15. Lau, S.L.; Chong, E.K.; Yang, X.; Wang, X. Automated pavement crack segmentation using u-net-based convolutional neural network. *IEEE Access* **2020**, *8*, 114892–114899. [CrossRef]
16. Coudray, N.; Karathanou, A.; Chambon, S. Multi-resolution approach for fine structure extraction: Application and validation on road images. In Proceedings of the International Joint Conference, VISIGRAPP 2010, Angers, France, 17–21 May 2010.
17. Qu, Z.; Ju, F.R.; Guo, Y.; Bai, L.; Chen, K. Concrete surface crack detection with the improved pre-extraction and the second percolation processing methods. *PLoS ONE* **2018**, *13*, e0201109. [CrossRef] [PubMed]
18. Talab, A.M.A.; Huang, Z.; Xi, F.; HaiMing, L. Detection crack in image using Otsu method and multiple filtering in image processing techniques. *Optik* **2016**, *127*, 1030–1033. [CrossRef]
19. Simonyan, K.; Zisserman, A. Very deep convolutional networks for large-scale image recognition. *arXiv* **2014**, arXiv:1409.1556.
20. Ronneberger, O.; Fischer, P.; Brox, T. U-net: Convolutional networks for biomedical image segmentation. In *International Conference on Medical Image Computing and Computer-Assisted Intervention*; Springer: Cham, Switzerland, 2015; pp. 234–241.
21. Chaurasia, A.; Culurciello, E. Linknet: Exploiting encoder representations for efficient semantic segmentation. In Proceedings of the 2017 IEEE Visual Communications and Image Processing (VCIP), St. Petersburg, FL, USA, 10–13 December 2017; pp. 1–4.
22. He, K.; Zhang, X.; Ren, S.; Sun, J. Deep residual learning for image recognition. In Proceedings of the IEEE Conference on Computer Vision and Pattern Recognition, Las Vegas, NV, USA, 26 June–1 July 2016; pp. 770–778.
23. Deng, J.; Dong, W.; Socher, R.; Li, L.J.; Li, K.; Fei-Fei, L. Imagenet: A large-scale hierarchical image database. In Proceedings of the 2009 IEEE Conference on Computer Vision and Pattern Recognition, Miami, FL, USA, 20–25 June 2009; pp. 248–255.
24. Kingma, D.P.; Ba, J. Adam: A method for stochastic optimization. *arXiv* **2014**, arXiv:1412.6980.

applied sciences

MDPI

Article

Automating Visual Blockage Classification of Culverts with Deep Learning

Umair Iqbal [1,*], Johan Barthelemy [1], Wanqing Li [2] and Pascal Perez [1]

1 SMART Infrastructure Facility, University of Wollongong, Wollongong 2500, Australia; johan@uow.edu.au (J.B.); pascal@uow.edu.au (P.P.)
2 School of Computing and Information Technology, University of Wollongong, Wollongong 2500, Australia; wanqing@uow.edu.au
* Correspondence: ui010@uowmail.edu.au

Abstract: Blockage of culverts by transported debris materials is reported as the salient contributor in originating urban flash floods. Conventional hydraulic modeling approaches had no success in addressing the problem primarily because of the unavailability of peak floods hydraulic data and the highly non-linear behavior of debris at the culvert. This article explores a new dimension to investigate the issue by proposing the use of intelligent video analytics (IVA) algorithms for extracting blockage related information. The presented research aims to automate the process of manual visual blockage classification of culverts from a maintenance perspective by remotely applying deep learning models. The potential of using existing convolutional neural network (CNN) algorithms (i.e., DarkNet53, DenseNet121, InceptionResNetV2, InceptionV3, MobileNet, ResNet50, VGG16, EfficientNetB3, NASNet) is investigated over a dataset from three different sources (i.e., images of culvert openings and blockage (ICOB), visual hydrology-lab dataset (VHD), synthetic images of culverts (SIC)) to predict the blockage in a given image. Models were evaluated based on their performance on the test dataset (i.e., accuracy, loss, precision, recall, F1 score, Jaccard Index, region of convergence (ROC) curve), floating point operations per second (FLOPs) and response times to process a single test instance. Furthermore, the performance of deep learning models was benchmarked against conventional machine learning algorithms (i.e., SVM, RF, xgboost). In addition, the idea of classifying deep visual features extracted by CNN models (i.e., ResNet50, MobileNet) using conventional machine learning approaches was also implemented in this article. From the results, NASNet was reported most efficient in classifying the blockage images with the 5-fold accuracy of 85%; however, MobileNet was recommended for the hardware implementation because of its improved response time with 5-fold accuracy comparable to NASNet (i.e., 78%). Comparable performance to standard CNN models was achieved for the case where deep visual features were classified using conventional machine learning approaches. False negative (FN) instances, false positive (FP) instances and CNN layers activation suggested that background noise and oversimplified labelling criteria were two contributing factors in the degraded performance of existing CNN algorithms. A framework for partial automation of the visual blockage classification process was proposed, given that none of the existing models was able to achieve high enough accuracy to completely automate the manual process. In addition, a detection-classification pipeline with higher blockage classification accuracy (i.e., 94%) has been proposed as a potential future direction for practical implementation.

Keywords: convolutional neural networks; visual blockage of culverts; intelligent video analytic; image classification

Citation: Iqbal, U.; Barthelemy, J.; Li, W.; Perez, P. Automating Visual Blockage Classification of Culverts with Deep Learning. *Appl. Sci.* **2021**, *11*, 7561. https://doi.org/10.3390/app11167561

Academic Editors: Nikos D. Lagaros and Oscar Reinoso García

Received: 2 July 2021
Accepted: 16 August 2021
Published: 18 August 2021

1. Introduction

Cross-drainage structures (e.g., culverts, bridges) are prone to blockage by debris and are reported as one of the leading causes of flash floods in urban areas [1–7]. The

1998 and 2011 floods in Wollongong, Australia [1,8–11] and the 2007 floods in Newcastle, Australia [1,12] are typical examples where blockage of cross drainage hydraulic structures contributed to triggering the flash flooding. Project 11: blockage of hydraulic structures [4] was initiated under the Australian rainfall and run-off (ARR) [13] framework to study the blockage behaviour and design considerations of hydraulic structures. Under this project, Wollongong City Council (WCC) proposed the guidelines to consider the hydraulic blockage in the hydraulic structures design process [2,4,14–17]. However, because of the unavailability of relevant supporting data from peak flooding events, proposed guidelines were not adaptive and were based on the post flood visual assessments, which many researchers believe is not the correct representation of blockage during the peak flooding events [1,2,14]. The guidelines suggested that any culvert with an opening diagonal of 6 m or more is not prone to blockage. However, this claim was only supported by post flood visual assessments and was not considered economically efficient to implement.

Initially, the blockage was defined as the percentage occlusion of hydraulic structure opening; however, many argued that hydraulic blockage and visual blockage are two separate concepts. The hydraulic blockage is more complex and has no established relationship with visual blockage. The hydraulic blockage is associated with the interaction of debris with the culvert and its corresponding effect on fluid dynamics around the culvert; however, due to the highly non-linear and uncertain behavior of debris, it is difficult to model and predict the hydraulic blockage using conventional means. Culvert blockage management is categorized under the broader "The Smart Stormwater Management" project [18] which aims to make use of data analytic and Internet of Things (IoT) for efficient stormwater management. Detection of blockage (i.e., StopBlock) to avoid flash floods is one of the critical components of this project. From a management and maintenance perspective, making use of multi-dimensional information (i.e., visual blockage status, type of debris material, percentage of blocked openings) extracted using computer vision algorithms may prove helpful in making timely decisions, as suggested in literature [7,19]. As of now, to assess the visual blockage at culverts, manual visual inspections by flood management teams are performed to decide if a culvert needs maintenance towards avoiding the overtopping of flow and flash flooding. However, this process is inefficient in terms of required human resources and unsafe during peak flood events. This paper attempts to address the problem from a different perspective and proposes the use of visual information extracted using automated analysis in better management of blockage at cross drainage hydraulic structures and automating the process of manual culvert visual blockage status classification.

This paper investigated the potential of convolutional neural network (CNN) algorithms towards classifying culvert images as "clear" or "blocked" as an automated solution for visual blockage inspections of culverts. Existing CNN models (i.e., DarkNet53 [20], DenseNet121 [21], InceptionResNetV2 [22], InceptionV3 [23], MobileNet [24], ResNet50 [25], VGG16 [26], EfficientNetB3 [27], NASNet [28]) pre-trained over ImageNet, and conventional machine learning approaches (i.e., SVM, RF, xgboost) were implemented for the culvert blockage classification task using data from three different sources (i.e., Images of culvert openings and blockage (ICOB), visual hydrology-lab dataset (VHD), synthetic images of culverts (SIC)), and performance was compared based on the standard evaluation measures. As a summary, the followings are the main contributions of this research:

1. Developed a culvert blockage visual dataset using multiple sources, including real culvert images from WCC records, simulated lab-scale hydrology experiments and computer-generated synthetic images;
2. Explored the potential of existing deep learning CNN and conventional machine learning models for classifying blocked culvert images as a potential solution towards automating the manual visual classification process of culverts for making blockage maintenance-related decisions;
3. Highlighted the challenges of culvert blockage visual dataset and inferred important insights to help improving the classification performance in future;

4. Proposed a detection-classification pipeline to achieve higher blockage classification accuracy for practical implementation. Furthermore, a partial automation framework based on the class prediction probability is introduced using a single deep learning model to assist the visual inspection process.

The rest of the paper is organized as follows: Section 2 presents the theoretical background of the implemented CNN models in this investigation. Section 3 provides information about the dataset used in this study for culvert blockage classification. Section 4 outlines the experimental protocols adopted to perform the experiments. Furthermore, this section provides information about the evaluation measures used to assess the classification performance of the implemented models. Section 5 presents the results of the experiments and also reports the critical insights from the investigation. Section 6 presents a brief introduction to the detection-classification pipeline towards improving the blockage classification performance. Section 7 concludes the study and reports the highlighted outcomes from the experiments. Furthermore, the section lists potential future applications of the presented research.

2. Deep Learning Models

This section presents the theoretical background of the implemented deep learning models for culvert visual blockage classification. For the presented investigation, one model from each common and state of the art category of deep learning models was selected to demonstrate the diversity of applied approaches. A brief introduction, model concept, model architecture, and the fundamental mathematics is outlined for each model.

2.1. DarkNet53

Redmon and Farhadi [20] proposed you only look once (YOLOV3) in the year 2018, where they used DarkNet53 CNN architecture as the feature extractor. DarkNet53 is the variant of DarkNet19 (i.e., feature extractor CNN in YOLOV2) but with an increased number of convolutional layers and residual connections in between. The structure of the DarkNet53 model consists of successive (3×3) and (1×1) convolutional layers. DarkNet53 is much deeper than DarkNet19 and achieved better performance than DarkNet19, ResNet50, and ResNet102 for the ImageNet challenge. Model structure best utilizes the graphical processing unit (GPU), which makes it faster.

2.2. ResNet

He et al. [25] proposed a novel residual learning framework to facilitate the training of extremely deep networks. Rather than learning unreferenced functions, authors proposed the reformulation of layers as residual learning functions with reference to inputs of the layer. The residual learning concept helped in optimizing the deep networks and made it possible to achieve higher accuracy from deep models. Mathematically, let us say $H(x)$ denotes the desired mapping function, in residual learning, stacked non-linear layers fit another mapping function $F(x) := H(x) - x$. x denotes the inputs to the layer.

2.3. MobileNet

Howard et al. [24] proposed a category of CNN called MobileNets for cutting edge hardware applications with the idea of using depthwise separable convolutions towards building the deep networks. Two global hyperparameters were introduced to develop problem-specific models with accuracy and latency adjustments. Depthwise separable convolution is the type of factorized convolution that splits the standard convolution process of convolving and combining into two layers. At the first layer, depthwise convolution is performed, while at the second layer, a 1×1 pointwise convolution is performed to combine the outputs from the depthwise convolution layer. All layers in the network are followed by a BatchNormalization and ReLu non-linearity. A depthwise convolution for a single filter per input channel can be expressed mathematically as in Equation (1).

$$\hat{\mathbf{G}}_{k,l,m} = \sum_{i,j} \hat{\mathbf{K}}_{i,j,m} \cdot \mathbf{F}_{k+i-1,l+j-1,m} \quad (1)$$

where $\hat{\mathbf{K}}$ denotes the depthwise convolutional kernel, $\mathbf{F}$ denotes the feature map and $\hat{\mathbf{G}}$ denotes the filtered output feature map.

2.4. InceptionV3 and InceptionResNet

Szegedy et al. [22,23] introduced the idea of inception module towards reducing the computational cost of the network without significantly affecting the generalized performance. InceptionV3 [23] and InceptionResNet [22] are improved versions of the proposed inception module. In InceptionV3, the idea of replacing large filters with small asymmetric filters was introduced and a 1×1 convolution was used as a bottleneck before the large filters. Concurrent placement of 1×1 filter resulted in cross-channel correlation. On the other hand, in the InceptionResNet model, Szegedy et al. [22] integrated both inception and residual concepts where concatenated filters were replaced by the residual connections. InceptionResNet was able to more quickly converge and achieved accelerated training performance.

2.5. VGG16

Simonyan and Zisserman [26] investigated the performance of deep convolutional networks by making architectural changes. The main idea was to replace the higher dimension filters with 3×3 filters and increase the depth of the network. This resulted in improving the computational cost with a significantly smaller trade-off in accuracy. From experimental investigations, authors reported that smaller filters were able to induce similar features as larger dimension filters. Padding was used to maintain the spatial resolution. The idea of increasing the depth of the network with smaller resolution filters demonstrated significant success for large scale classification and localization tasks. However, an increase in depth to a large scale resulted in an increased number of trainable parameters.

2.6. DensNet121

Huang et al. [21] proposed densely connected convolutional networks called DenseNet by extending the concept of residual connections in the traditional networks. The authors proposed the idea of connecting each layer in the network to every other layer in the feedforward direction. This way, each layer will have the feature maps of all preceding layers at its input. In terms of the number of layer connections, a traditional network with L layers have L connections, while a densely connected convolutional network will have $L(L+1)/2$ connections. Densely connected networks have advantages including better feature propagation, feature reuse, a significant reduction in the number of network parameters, and improving the vanishing-gradient problem. A key difference between residual networks and densely connected networks is that in a densely connected network, feature maps from preceding layers are combined by concatenation rather than summation before feeding it to the next layer.

Mathematically, if a network consists of L number of layers each with a non-linear transformation through a composite function F_l, the output x_l for the densely connected layer can be represented as in Equation (2).

$$x_l = F_l([x_0, x_1, \ldots, x_{l-1}]) \quad (2)$$

where $[x_0, x_1, \ldots, x_{l-1}]$ denotes the concatenation of the feature maps from the previous layers.

2.7. NASNet

Zoph et al. [28] proposed a new category of convolutional networks called NASNet based on the idea of directly training the architecture over the desired dataset. In order to overcome the issue of computational cost for relatively larger datasets, authors proposed to search for an architectural building for a smaller dataset, often called proxy dataset and then transferring it to a larger dataset. The search space which enables the transfer from a smaller dataset to a larger dataset is referred to as NASNet search space inspired from neural architecture search (NAS) [29]. Furthermore, to improve the generalization of the NASNet model, the authors proposed a novel normalization approach called ScheduledDropPath. In NAS, a recurrent neural network (RNN) controller samples the child architectures, which are trained over proxy datasets and based on the training accuracy, the controller improves the architecture. The main contribution of this approach is the decoupling of architecture complexity from depth.

2.8. EfficientNet

Tan and Le [27] proposed a novel compound coefficient based scaling of deep neural networks. Based on this idea, a new category of networks called EfficientNet is introduced, which is built on NAS. The idea of uniformly scaling the model in all dimensions, including width, depth, and resolution, is implemented. Balanced scaling up of models resulted in higher accuracy. Mathematically, if the intention is to extend the computation power to 2^n times, a model can be scaled up in depth, width, and resolution as α^n, β^n, and γ^n, respectively.

3. Dataset

The dataset used in this research consisted of images of culverts (i.e., blocked, clear) from three different sources (i.e., ICOB, VHD, SIC). Overall, the dataset consisted of 3848 images. Details about each subset of the dataset are presented as follows.

3.1. Images of Culvert Openings and Blockage (ICOB)

This dataset consisted of real images of culverts and referred to as "Images of Culvert Openings and Blockage (ICOB)". Primary sources of images included WCC historical records, online records, and custom captured local culvert images. WCC records were scrutinized using a Microsoft ACCESS based application for filtering the culvert images with visible openings. The dataset contained images with a high level of variation from each other (intra-class variation) in terms of culvert types, blockage accumulation, presence of debris materials, illumination conditions, culvert viewpoint variations, scale variations, resolution, and backgrounds. This high level of diversity within a relatively small dataset makes it a challenging dataset for visual analysis, even for a binary classification problem. In total, there were 929 images in ICOB with 487 images in the "clear" class and 442 images in the "blocked" class. Figure 1 shows the sample instances from each class of ICOB.

Figure 1. Sample instances of clear (First row) and blocked (Second row) culverts from ICOB.

3.2. Visual Hydrology-Lab Dataset (VHD)

This dataset consisted of simulated images of culverts captured from a controlled hydrology lab experiment. A comprehensive in-lab investigation of blockage of carried out by performing series of experiments using scaled physical models of culverts under multiple flooding conditions. Experiments were recorded using two high definition (HD) cameras with a different view of the culvert and images of culverts in blockage and clear condition were extracted, referred to as "Visual Hydrology-Lab Dataset (VHD)". VHD consisted of a diversity of images including four different culvert configurations (i.e., single circular, double circular, single box, double box), different blockage types (i.e., urban, vegetative, mixed), different simulated lighting conditions, different camera viewpoints, and different flood levels controlled by inlet water discharge. Limitations of the dataset included reflections from the water surface and flume walls, identical background and scaling, and clear water. For this investigation, in total, 1630 images were used with 1526 images from the "blocked" class while 104 images from the "clear" class. Figure 2 shows the sample images of each class from VHD.

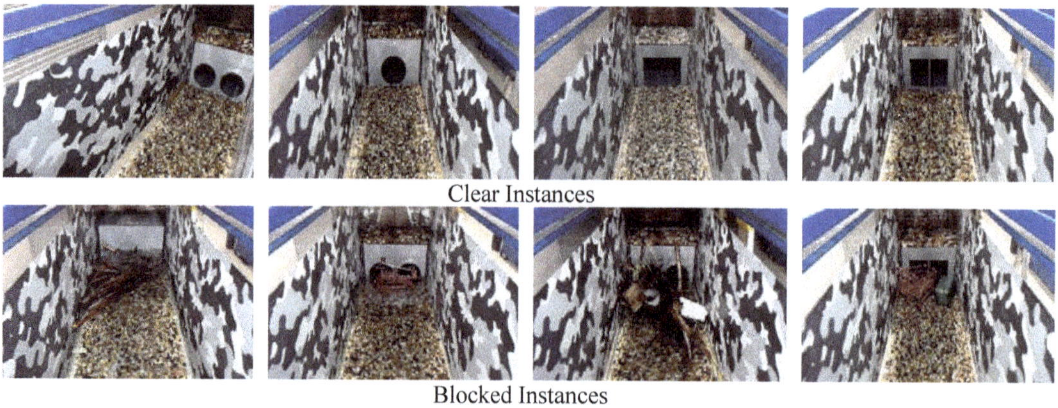

Figure 2. Sample instances of clear (First row) and blocked (Second row) culverts from VHD.

3.3. Synthetic Images of Culverts (SIC)

This dataset consisted of synthetic images of culverts generated using a three-dimensional (3D) computer application based on a gaming engine (i.e., Unity3D) specifically designed to simulate multiple culvert blockage scenarios. Application has the capability to generate virtually countless blockage scenarios by dragging different debris materials into the scene and placing them in desired orientation/location. Images of different simulated blockage scenarios were captured using batch capture functionality and are referred to as "Synthetic Images of Culvert (SIC)". Dataset offered the diversity in terms of debris type (i.e., urban, vegetative, mixed), culvert types (i.e., pipe, single circular, double circular, single box, double box, triple box), camera viewpoints, time of day, and water levels. Limitations of the dataset include single natural background and non-realistic effects/animations. For this investigation, 1289 images were used with 1140 images from the "blocked" class while 149 images from the "clear" class. Figure 3 shows the sample images of each class from SIC.

3.4. Labeling Criteria

Dataset was manually labeled for binary classification of a given image with culvert as "clear" or "blocked". A culvert being visually blocked or clear is not as simple and may require defining detailed criteria in collaboration with flood management officers; however, for this article, simple occlusion based criteria was used. Following subjective annotation criteria was used for labeling.

- If all of the culvert openings are visible, classify it as "clear";
- If any of the culvert openings is visually occluded by debris material or foreground object (e.g., debris control structure, vegetation, tree), classify it as "blocked".

Figure 3. Sample instances of clear (First row) and blocked (Second row) culverts from SIC.

4. Experimental Setup and Evaluation Measures

Experiments were planned to investigate the performance of existing CNN models for binary classification of culvert images as blocked or clear. Pre-trained CNN models with ImageNet weights were used for this investigation and implemented using Keras with Tensorflow at the backend. Images of dimension $224 \times 224 \times 3$ were used as input to model except for NASNet where $331 \times 331 \times 3$ and InceptionResNetV2 where $229 \times 229 \times 3$ was used. Data augmentations techniques including samplewise standard deviation normalization, horizontal flip, vertical flip, rotation, width shift, and height shift were used in the simulations for improved performance. All the models were tuned with a dropout of 0.2, ReLU activation, and batch normalization. Stochastic gradient descent (SGD) optimizer with a constant learning rate of 0.01 and categorical entropy loss was used. Each model was trained for 30 epochs. For this investigation, the test dataset consisted of selected real images from ICOB (i.e., 91 from blocked, 98 from clear). The rest of the dataset was divided using conventional train:val split with an 80:20 ratio. In addition to the conventional train:val:test dataset split, the 5-fold cross-validation approach has also been implemented and compared towards providing better insight into the performance of deep learning models. The idea of classifying deep visual features extracted from CNN models using conventional machine learning approaches was also implemented. The simulations were performed using Nvidia GeForce RTX 2060 GPU with 6 GB memory and 14 Gbps memory speed. Models were trained at full precision using floating point (FP-32) optimization.

The performance of the models was measured in terms of their test accuracy, test loss, precision score, recall score, F1 score, Jaccard Index, ROC curves, and processing times. Each of the evaluation metrics is defined briefly as follows.

- **Loss:** Loss is the simplest of the measure to evaluate model training and testing performance. It is the measure of how much instances are classified incorrectly and is the ratio of number of incorrect predictions over total predictions. Minimum value of loss indicated better performance;
- **Accuracy:** In contrast to loss, accuracy is the measure that how much percentage of data instances are classified correctly. It is the ratio of number of correct predictions over total predictions. High value of accuracy represents better performance;
- **Precision Score:** Precision measures the ability of a model to not to classify a negative instance as positive. It answers the question that from all the positive predicted

instances by model, how many were actually positive. The equation below presents the expression for the precision score:

$$\text{Precision Score} = \frac{\text{True Positive (TP)}}{\text{True Positive (TP)} + \text{False Positive (FP)}}$$

- **Recall Score:** Recall answers the question that from all the positive instances, how many were correctly classified by the model. Expression for recall score is given as follows:

$$\text{Recall Score} = \frac{\text{True Positive (TP)}}{\text{True Positive (TP)} + \text{False Negative (FN)}}$$

- **F1 Score:** F1 score is the single measure which combines both precision and recall by harmonic mean and range between 0 and 1. Higher F1 score indicates the better performance of model. Expression for F1 score is given as follows:

$$\text{F1 Score} = 2 \times \frac{\text{Precision} \times \text{Recall}}{\text{Precision} + \text{Recall}}$$

- **Jaccard Index:** In context of classification, Jaccard similarity index score measures the similarity between predicted labels and actual labels. Mathematically, let $\hat{y}$ denotes the predicted label and y denotes the actual label, then J index can be expressed as follows. Higher J index indicates better performance of model.

$$\text{Jaccard Index} = \frac{|\hat{y} \cap y|}{|\hat{y} \cup y|} = \frac{|\hat{y} \cap y|}{|\hat{y}| + |y| - |\hat{y} \cap y|}$$

In addition, confusion matrices were plotted to assess the Type I and Type II errors. Type I (False positive (FP)) and Type II (False negative (FN)) errors [30] are commonly used terms in machine learning and the main goal of the model is to minimize one of these two errors, depending on the context that which error is more critical in the given task. By definition, a Type I error is concluding the existence of a relationship, while in fact it does not exist (e.g., classifying an image as "blocked" while there is no blockage). Similarly, a Type II error is the rejection of the existence of relationship while, in fact, it exists (e.g., classifying an image as "clear" while there is a blockage). For the given culvert blockage context, Type II error is more critical to be minimized in comparison to Type I error because having notified as blocked while there is no blockage is tolerable in comparison to having notified as clear while there is a blockage. Type II error will result in damages because it may be very late for the response team to clear the blockage before the diversion of flow. Finally, the performance of implemented deep learning models was benchmarked against the conventional machine learning models (i.e., SVM, RF, xgboost) to demonstrate the effectiveness of CNN's for images (i.e., matrix) type dataset.

5. Results and Discussion

Implemented CNN models were evaluated as per defined measures in Section 4 and results were compared. Figure 4 shows the training performance of implemented models in terms of training loss for conventional dataset split and the training times. From the figure, it can be observed that other than the DarkNet53 and VGG16, all models training behaviour was similar with loss following the negative exponential curve and converging to a minimum value. However, unusual training behavior was observed for DarkNet53 and VGG16 where models failed to learn the training examples and loss did not decrease significantly over the training epochs. In terms of training times, as expected NASNet was the slowest to train (i.e., 414 s per epoch), while MobileNet was the fastest to train (i.e., 22 s per epoch) based on their respective complexity.

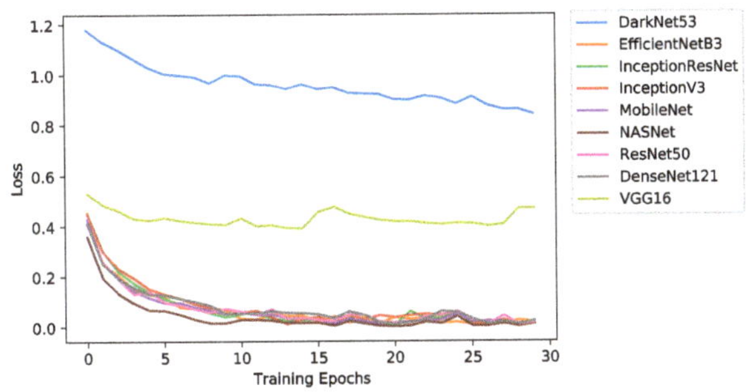

(**a**) Training loss

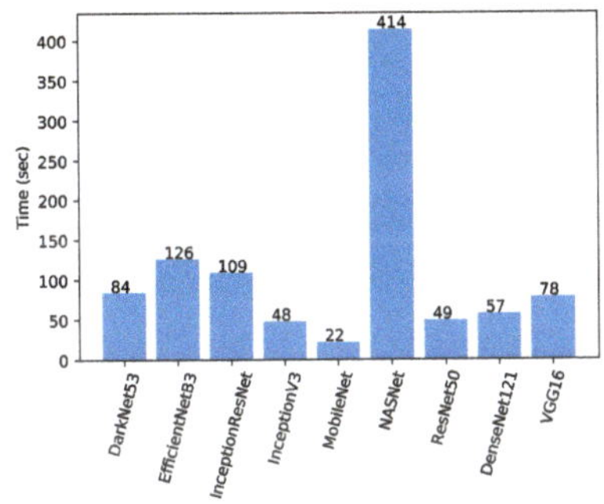

(**b**) Training time per epoch

Figure 4. Training performance of implemented CNN models for culvert visual blockage classification.

Table 1 presents the empirical results of all implemented models using conventional dataset split and 5-fold cross-validation when evaluated for test dataset in terms of accuracy, loss, precision, recall, F1 score, and Jaccard Index. Furthermore, the table benchmarks the results of deep learning models against conventional machine learning algorithms. From the results, NASNet was reported as the best among all others with an F1 score of 0.84 and 0.85 for conventional and 5-fold, respectively. MobileNet was reported as the second-best for conventional dataset split with an F1 score of 0.81, while InceptionV3 was reported second best for 5-fold with an F1 score of 0.80. DarkNet and VGG performed worst with the F1 scores of 0.61 and 0.48 for 5-fold cross-validation. When benchmarked against the conventional machine learning algorithms, it can be clearly observed that deep learning models performed significantly better. However, for the case where the idea of classifying deep visual features using conventional machine learning models, comparable performance to standard CNN models was achieved. 5-fold accuracy of 77% was achieved as best for the case where MobileNet extracted visual features were classified using an SVM conventional machine learning classifier.

Performance of deep learning models was also assessed using ROC curves as given in Figure 5. ROC plot confirmed that NASNet outperformed other models with an area under the curve (auc) of 0.92. Figures 6 and 7 show the confusion matrices for both conventional and 5-fold cross-validation experiments, respectively, to observe the Type I and Type II errors. For the conventional case presented in Figure 6, it can be observed that NASNet performed best in terms of the lowest Type II error of the only 10%; however, Type I error was reported 21%. On the other hand, MobileNet was reported with balanced Type I and Type II errors (19% and 18%). A similar trend was observed for the 5-fold cross-validation experiment where NASNet was reported with the lowest Type II error (i.e., 12%) and EfficientNetB3 was reported balanced Type I and Type II errors (19% and 22%).

Overall, comparatively similar performance was reported for both conventional and 5-fold experiments except for the case of VGG where 5-fold cross-validation performance was degraded significantly (see Figure 8a).

Table 1. Classification performance of implemented artificial intelligence models for visual blockage detection.

	Test Accuracy		Test Loss/Log Loss		Precision Score		Recall Score		F1 Score		Jaccard Index		FLOPs
	Conventional	5-Fold	Conventional	5-Fold	Conventional	5-Fold	Conventional	5-Fold	Conventional	5-Fold	Conventional	5-Fold	
Deep Learning Models													
DarkNet53	0.61	0.63	1.20	1.21	0.63	0.65	0.62	0.61	0.61	0.61	0.44	0.46	14.2 G
DenseNet121	0.77	0.79	0.47	0.57	0.77	0.80	0.77	0.79	0.77	0.79	0.62	0.66	5.7 G
InceptionResNetV2	0.79	0.77	0.58	0.65	0.79	0.78	0.80	0.78	0.80	0.77	0.66	0.64	13.3 G
InceptionV3	0.76	0.80	0.74	0.64	0.76	0.80	0.76	0.80	0.76	0.80	0.62	0.66	5.69 G
MobileNet	0.81	0.78	0.51	0.59	0.81	0.79	0.81	0.79	0.81	0.78	0.69	0.65	**1.15 G**
ResNet50	0.78	0.79	0.70	0.62	0.78	0.76	0.78	0.76	0.78	0.79	0.64	0.65	7.75 G
VGG16	0.71	0.57	0.58	0.79	0.72	0.43	0.70	0.58	0.70	0.48	0.55	0.41	30.7 G
EfficientNetB3	0.78	0.79	**0.46**	0.57	0.78	0.80	0.78	0.79	0.78	0.79	0.64	0.66	1.97 G
NASNet	**0.84**	**0.85**	0.58	**0.55**	**0.85**	**0.85**	**0.84**	**0.85**	**0.84**	**0.85**	**0.73**	**0.73**	47.8 G
Conventional Machine Learning Algorithms													
SVM	0.55	0.63	15.53	12.61	0.70	0.64	0.57	0.63	0.46	0.63	0.38	0.47	NA
RF	0.47	0.57	18.27	14.80	0.47	0.57	0.48	0.57	0.40	0.56	0.31	0.40	NA
xgboost	0.50	0.58	17.18	14.62	0.58	0.58	0.52	0.57	0.40	0.57	0.34	0.41	NA
Deep CNN Visual Features Classification using Conventional Machine Learning Approaches													
ResNet50 Features + SVM	0.82	0.74	6.31	9.11	0.82	0.74	0.82	0.74	0.82	0.74	0.69	0.58	NA
ResNet50 Features + RF	0.76	0.73	8.24	9.29	0.78	0.73	0.76	0.73	0.76	0.73	0.61	0.58	NA
ResNet50 Features + xgboost	0.78	0.72	7.36	9.64	0.79	0.72	0.79	0.72	0.79	0.72	0.65	0.56	NA
MobileNet Feature + SVM	0.84	0.77	5.25	7.88	0.85	0.77	0.85	0.77	0.85	0.77	0.74	0.63	NA
MobileNet Feature + RF	0.76	0.75	8.06	8.41	0.77	0.78	0.77	0.76	0.77	0.75	0.62	0.61	NA
MobileNet Feature + xgboost	0.72	0.66	9.64	11.74	0.73	0.66	0.72	0.66	0.72	0.66	0.56	0.49	NA

From the FP instances in Figure 9, it was observed that for the cases where there are more than two openings and only one opening was blocked, the algorithm classified it as clear. This insight led to a suggestion in the change of labeling criteria. A better approach could be to label the image as blocked if half or more than half of the openings are blocked; otherwise, label it as clear. Furthermore, if there is no debris material present in the image and occlusion is due to some foreground object not similar to debris in visual appearance, the image should be labeled as clear. From the FN instances in Figure 9, it was observed that for the cases where the image contained contents with a visual appearance similar to blockage material, the image was classified as blocked. This indicated the existence of background clutter/noise problems for this investigation. Background clutter hypothesis was also verified by the intermediate CNN layers activation and heatmaps as given in Figures 10 and 11, respectively. From the layers intermediate activation, it can be observed that at the initial layer, the model retained almost all the visual information as in the input image. However, as the layers go deeper, the model tends to encode higher-level features, such as borders, lines, and edges. Going further deeper results in activation which are not visually interpretable and possess more information related to the class of the input image. Heatmaps for selected FN cases presented in Figure 11 confirmed the hypothesis of background clutter. It can be observed that in most cases, the focus was more on the background contents rather than the culvert opening. Interestingly, in the case of the box culvert, the reflection of light through the culvert was considered by the model as the background which resulted in false classification.

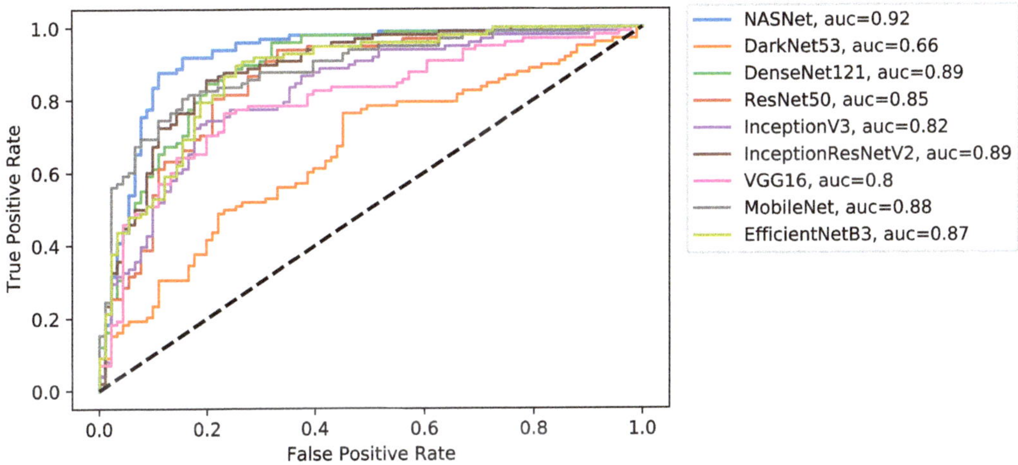

Figure 5. ROC curves for implemented deep learning models.

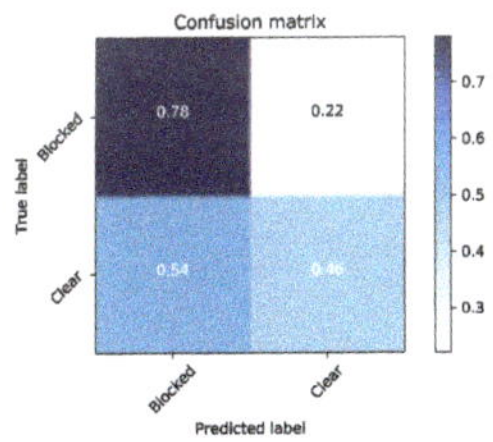

(**a**) DarkNet53

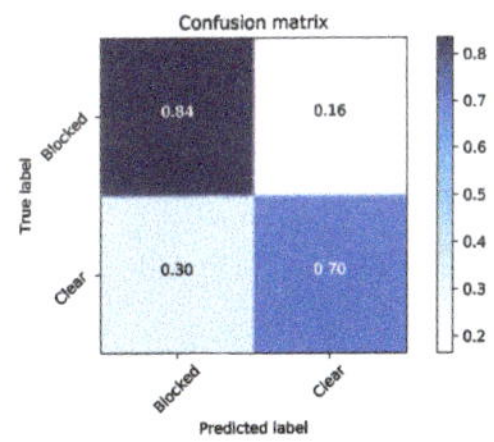

(**b**) DenseNet121

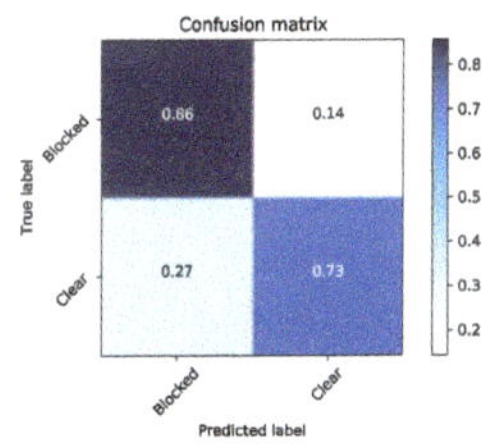

(**c**) InceptionResNetV2

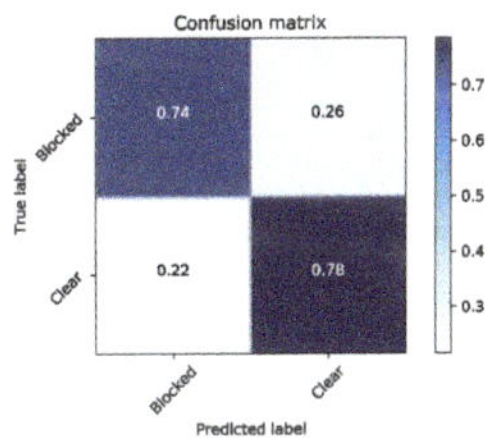

(**d**) InceptionV3

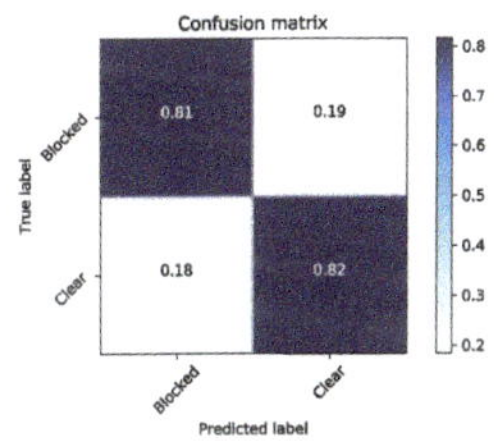

(**e**) MobileNet

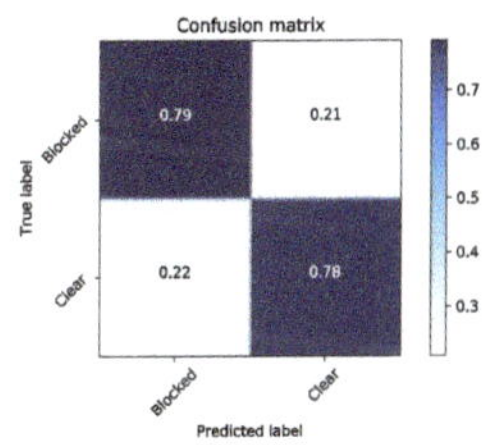

(**f**) ResNet50

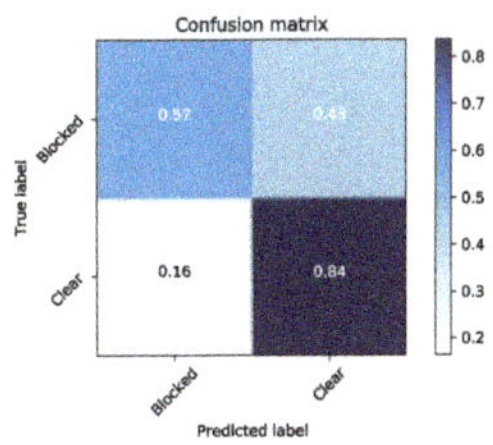

(**g**) VGG16

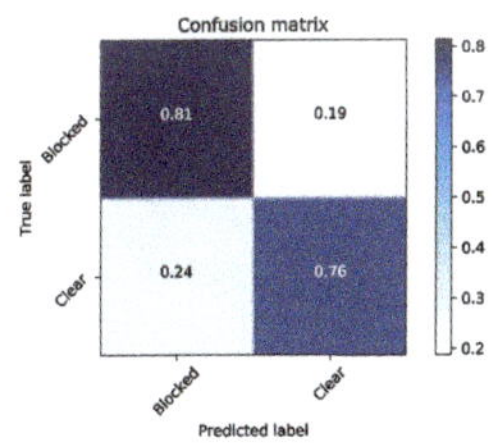

(**h**) EfficientNetB3

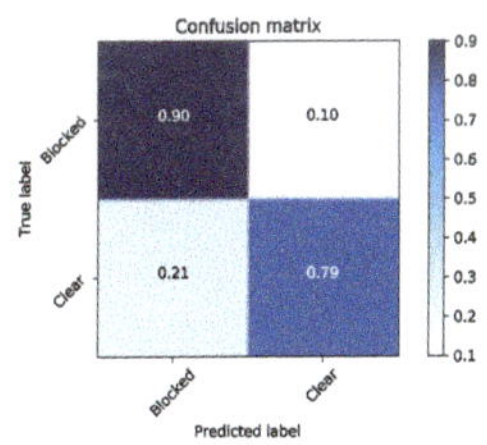

(**i**) NASNet

Figure 6. Confusion matrices of implemented CNN models for blockage detection (Conventional Train:Val:Test split).

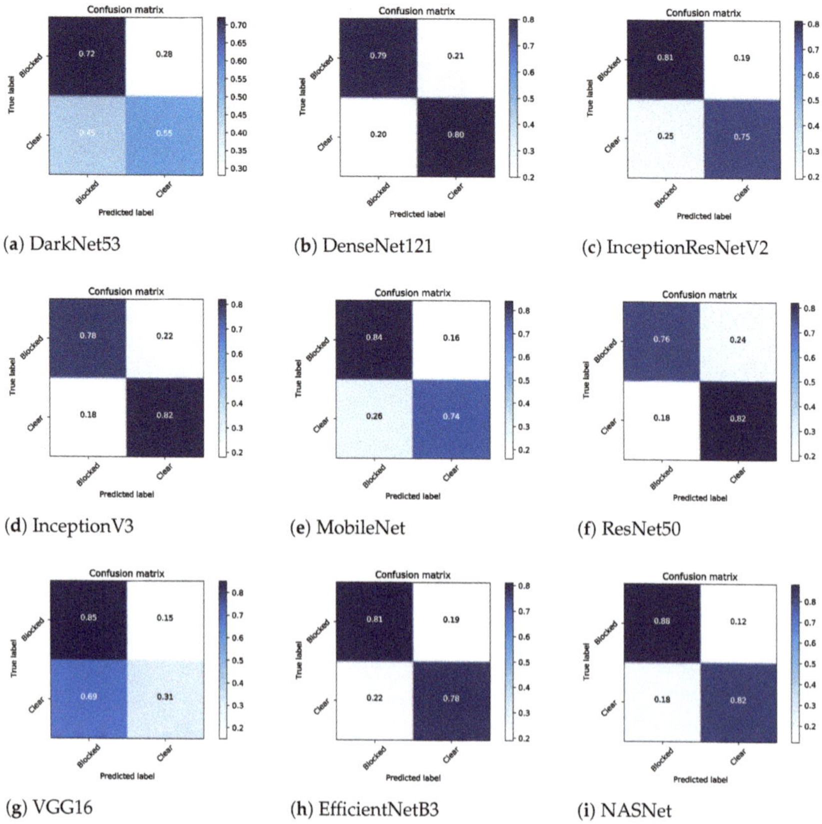

(**a**) DarkNet53 (**b**) DenseNet121 (**c**) InceptionResNetV2

(**d**) InceptionV3 (**e**) MobileNet (**f**) ResNet50

(**g**) VGG16 (**h**) EfficientNetB3 (**i**) NASNet

Figure 7. Confusion matrices of implemented CNN models for blockage detection (5-fold cross validation).

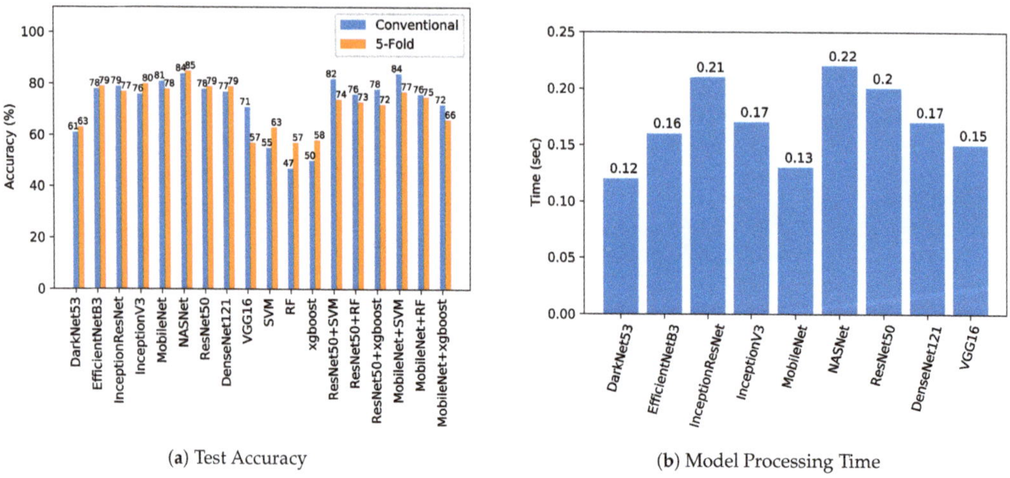

(**a**) Test Accuracy (**b**) Model Processing Time

Figure 8. Graphical comparison of implemented CNN models for test performance.

Figure 9. Selected false positive (First row) and false negative (Second row) instances.

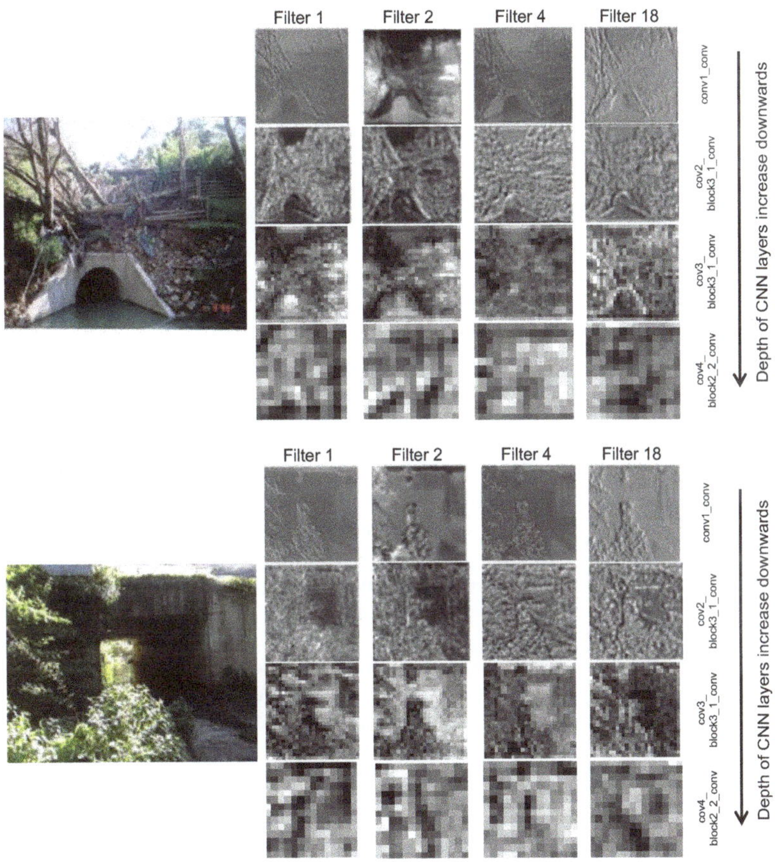

Figure 10. Selected intermediate ResNet50 CNN layer activation.

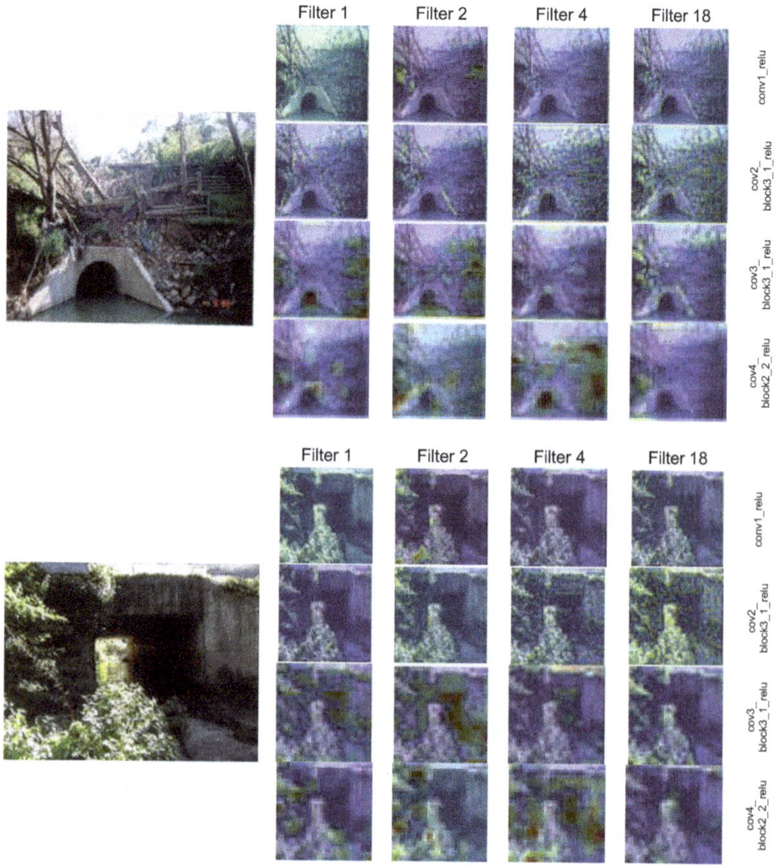

Figure 11. Selected ResNet50 CNN layers heatmaps.

Given that, existing CNN models, conventional machine learning approaches, and classification of deep CNN visual features using conventional machine learning approaches did not achieve high enough accuracy (i.e., maximum 5-fold accuracy of 85% achieved for NASNet model) that they can be deployed to replace the manual visual inspection of culverts. However, they can potentially be used to partially automate the process of manual visual inspection of culverts. Along with the predicted blockage class of a given culvert (i.e., blocked, clear), the value of class prediction probability can help in estimating the model confidence of the prediction. Partial automation can be achieved by setting a threshold on prediction probability (i.e., 80%) to filter only those images for manual inspection for which prediction probability is less than a set threshold. Figure 12 shows the conceptual block diagram of the proposed framework for partial automation of the culvert visual blockage inspection process.

Implemented CNN models were also compared for their processing times to investigate the relative response times. The purpose of these analyses was to investigate the hardware implementability of proposed models for real-world applications. Model inference time and image processing time were calculated as two measures to compare the models. Three different size images were used; image 1 of 2048×1536, image 2 of 3264×2448, and image 3 of 4032×3024. From Table 2 and Figure 8, it can be observed that MobileNet and DarkNet53 were fastest among others while the NASNet model was the slowest. In terms of accuracy, NASNet was the most accurate; however, MobileNet also exhibited comparatively good accuracy (i.e., 78% in comparison to 85% for NASNet) and was recommended as a suitable choice to implement for on-board processing. Figure 8 shows the graphical comparison of implemented models in terms of test accuracy and processing times. It is important to mention that reported processing times are for relative comparison between models and not the actual measure of cutting edge hardware performance. However, given the availability of efficient computing hardware, such as Nvidia Jetson TX2 [31] and Nvidia Jetson Nano [32], it is highly probable to implement any of the implemented models for real-world applications (e.g., pedestrian detection [33], wildlife tracking [34]).

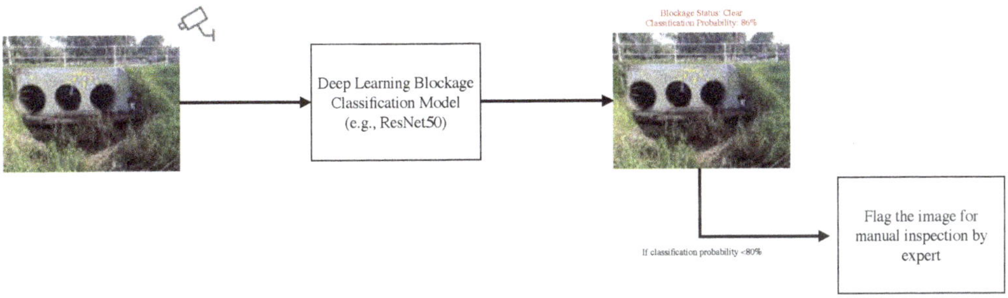

Figure 12. Conceptual block diagram of framework for partial automation of visual blockage classification.

Table 2. CNN model processing times for three different size images.

	Model Processing Time (s)	Total Execution Time (s)		
		Image 1	Image 2	Image 3
DarkNet53	0.05	0.12	0.2	0.35
DenseNet121	0.09	0.17	0.24	0.39
InceptionResNetV2	0.14	0.21	0.29	0.44
InceptionV3	0.09	0.17	0.24	0.39
MobileNet	0.06	0.13	0.21	0.36
ResNet152	0.13	0.20	0.28	0.43
ResNet50	0.08	0.15	0.23	0.38
VGG16	0.08	0.15	0.23	0.38
EfficientNetB3	0.09	0.16	0.24	0.39
NASNet	0.15	0.22	0.30	0.45

6. Detection-Classification Pipeline for Visual Blockage Detection

In light of the reported insights from the presented experiments in Section 5, a detection-classification pipeline is a potential future work in the development process to address the background clutter issue. The idea is to detect the culvert openings from the image using the object detection model (i.e., Faster R-CNN [35]) at the first stage and classify the detected culvert openings as "blocked" or "clear" using a deep learning classification model (i.e., ResNet50). Figure 13 shows the conceptual block diagram of the detection-classification pipeline. A preliminary system with such a pipeline has already been deployed on a cutting edge hardware for testing purposes [18]. For culvert opening detection, m@AP50 of 0.95 has been achieved using the Faster R-CNN model while an improved 94% blockage classification accuracy has been achieved using the ResNet50 model.

Figure 13. Conceptual block diagram of detection-classification pipeline for visual blockage detection.

7. Conclusions and Future Directions

The idea of using visual analytic for the culvert blockage analysis has been successfully pitched by implementing existing CNN models for culvert blockage classification. Dataset from three different sources (i.e., ICOB, VHD, SIC) has been developed with a diversity of clear and blocked culvert instances for training the convolutional neural network (CNN) models. From the analysis, it has been observed that the NASNet model performed best among all in terms of classification performance; however, it was the slowest in relative comparison of processing times. Based on the classification performance and processing times, MobileNet was recommended model to be deployed for real-world applications. Deep learning models were benchmarked against conventional machine learning algorithms and reported significantly improved performance. From the false positive (FP) and false negative (FN) instances, background noise and oversimplified labeling criteria were found potential factors for degraded performance. A detection-classification pipeline was proposed with higher blockage classification accuracy (i.e., 94%) as a potential solution for real-world implementation. Furthermore, a partial automation framework based on the model class prediction probability was introduced to facilitate the manual visual inspections of culverts. A visual attention based approach and problem-specific CNN design are potential future directions of this research. Furthermore, study the impact of high-resolution images on the accuracy and developing a hybrid model taking into account information from multiple sensors are the potential concepts that can be investigated in future.

Author Contributions: Conceptualization and methodology, U.I., J.B., W.L. and P.P.; investigation, U.I.; writing—original draft preparation, U.I.; writing—review and editing, U.I. and J.B.; supervision, J.B., W.L. and P.P. All authors have read and agreed to the published version of the manuscript.

Funding: This work was supported by Wollongong City Council (WCC) in partnership with Shellharbour, Kiama and Shoalhaven Councils, Lendlease and the University of Wollongong's SMART Infrastructure Facility. This program also received funding from the Australian Government under the Smart Cities and Suburbs Program (SCS69244). We also gratefully acknowledge the support of NVIDIA Corporation with the donation of the Titan RTX GPU used for this research.

Acknowledgments: I would like to thank WCC for providing resources and support to carry out this study. Furthermore, I would like to thank the University of Wollongong (UOW) and the Higher Education Commission (HEC) of Pakistan for funding my Ph.D. studies.

Conflicts of Interest: The authors declare no conflicts of interest.

References

1. French, R.; Jones, M. Culvert blockages in two Australian flood events and implications for design. *Australas. J. Water Resour.* **2015**, *19*, 134–142. [CrossRef]
2. French, R.; Rigby, E.; Barthelmess, A. The non-impact of debris blockages on the August 1998 Wollongong flooding. *Australas. J. Water Resour.* **2012**, *15*, 161–169. [CrossRef]
3. Blanc, J. An Analysis of the Impact of Trash Screen Design on Debris Related Blockage at Culvert Inlets. Ph.D. Thesis, School of the Built Environment, Heriot-Watt University, Edinburgh, UK, 2013.

4. Weeks, W.; Witheridge, G.; Rigby, E.; Barthelmess, A.; O'Loughlin, G. *Project 11: Blockage of Hydraulic Structures*; Technical Report P11/S2/021; Engineers Australia, Water Engineering: Barton, ACT, Australia, 2013.
5. Roso, S.; Boyd, M.; Rigby, E.; VanDrie, R. Prediction of increased flooding in urban catchments due to debris blockage and flow diversions. In Proceedings of the 5th International Conference on Sustainable Techniques and Strategies in Urban Water Management (NOVATECH), Lyon, France, 6–10 June 2004; Water Science and Technology: Lyon, France, 2004; pp. 8–13.
6. Wallerstein, N.; Thorne, C.R.; Abt, S. *Debris Control at Hydraulic Structures, Contract Modification: Management of Woody Debris in Natural Channels and at Hydraulic Structures*; Technical Report; Department of Geography, Nottingham University (United Kingdom): Nottingham, UK, 1996.
7. Iqbal, U.; Perez, P.; Li, W.; Barthelemy, J. How Computer Vision can Facilitate Flood Management: A Systematic Review. *Int. J. Disaster Risk Reduct.* **2021**, *53*, 102030. [CrossRef]
8. Barthelmess, A.; Rigby, E. Culvert Blockage Mechanisms and their Impact on Flood Behaviour. In Proceedings of the 34th World Congress of the International Association for Hydro-Environment Research and Engineering, Brisbane, Australia, 26 June–1 July 2011; Engineers Australia: Barton, ACT, Australia, 2011; pp. 380–387.
9. Rigby, E.; Silveri, P. Causes and effects of culvert blockage during large storms. In Proceedings of the Ninth International Conference on Urban Drainage (9ICUD), Portland, OR, USA, 8–13 September 2002; Engineers Australia: Portland, OR, USA, 2002; pp. 1–16.
10. Van Drie, R.; Boyd, M.; Rigby, E. Modelling of hydraulic flood flows using WBNM2001. In Proceedings of the 6th Conference on Hydraulics in Civil Engineering, Hobart, Australia, 28–30 November 2001; Institution of Engineers Australia: Hobart, Australia, 2001; pp. 523–531.
11. Davis, A. An Analysis of the Effects of Debris Caught at Various Points of Major Catchments during Wollongong's August 1998 Storm Event. Bachelor's Thesis, University of Wollongong, Wollongong, Australia, 2001
12. WBM, B. *Newcastle Flash Flood 8 June 2007 (the Pasha Bulker Storm) Flood Data Compendium*; Prepared for Newcastle City Council; BMT WBM: Broadmeadow, Australia, 2008.
13. Ball, J.; Babister, M.; Nathan, R.; Weinmann, P.; Weeks, W.; Retallick, M.; Testoni, I. *Australian Rainfall and Runoff—A Guide to Flood Estimation*; Commonwealth of Australia: Barton, ACT, Australia, 2016
14. French, R.; Jones, M. Design for culvert blockage: The ARR 2016 guidelines. *Australas. J. Water Resour.* **2018**, *22*, 84–87. [CrossRef]
15. Rigby, E.; Silveri, P. The impact of blockages on flood behaviour in the Wollongong storm of August 1998. In Proceedings of the 6th Conference on Hydraulics in Civil Engineering: The State of Hydraulics, Hobart, Australia, 28–30 November 2001; Engineers Australia: Barton, ACT, Australia, 2001; pp. 107–115
16. Ollett, P.; Syme, B.; Ryan, P. Australian Rainfall and Runoff guidance on blockage of hydraulic structures: Numerical implementation and three case studies. *J. Hydrol.* **2017**, *56*, 109–122.
17. Jones, R.H.; Weeks, W.; Babister, M. *Review of Conduit Blockage Policy Summary Report*; WMA Water: Sydney, NSW, Australia, 2016.
18. Barthelemy, J.; Amirghasemi, M.; Arshad, B.; Fay, C.; Forehead, H.; Hutchison, N.; Iqbal, U.; Li, Y.; Qian, Y.; Perez, P. Problem-Driven and Technology-Enabled Solutions for Safer Communities: The case of stormwater management in the Illawarra-Shoalhaven region (NSW, Australia). In *Handbook of Smart Cities*; Augusto, J.C., Ed.; Springer: Berlin, Germany, 2020; pp. 1–28.
19. Arshad, B.; Ogie, R.; Barthelemy, J.; Pradhan, B.; Verstaevel, N.; Perez, P. Computer Vision and IoT-Based Sensors in Flood Monitoring and Mapping: A Systematic Review. *Sensors* **2019**, *19*, 5012. [CrossRef] [PubMed]
20. Redmon, J.; Farhadi, A. Yolov3: An incremental improvement. *arXiv* **2018**, arXiv:1804.02767.
21. Huang, G.; Liu, Z.; Van Der Maaten, L.; Weinberger, K.Q. Densely connected convolutional networks. In Proceedings of the IEEE Conference on Computer Vision and Pattern Recognition, Honolulu, HI, USA, 21–26 July 2017; pp. 4700–4708.
22. Szegedy, C.; Ioffe, S.; Vanhoucke, V.; Alemi, A. Inception-v4, inception-resnet and the impact of residual connections on learning. *arXiv* **2016**, arXiv:1602.07261.
23. Szegedy, C.; Vanhoucke, V.; Ioffe, S.; Shlens, J.; Wojna, Z. Rethinking the inception architecture for computer vision. In Proceedings of the IEEE Conference on Computer Vision and Pattern Recognition, Las Vegas, NV, USA, 27–30 June 2016; pp. 2818–2826.
24. Howard, A.G.; Zhu, M.; Chen, B.; Kalenichenko, D.; Wang, W.; Weyand, T.; Andreetto, M.; Adam, H. Mobilenets: Efficient convolutional neural networks for mobile vision applications. *arXiv* **2017**, arXiv:1704.04861.
25. He, K.; Zhang, X.; Ren, S.; Sun, J. Deep residual learning for image recognition. In Proceedings of the IEEE Conference on Computer Vision and Pattern Recognition, Las Vegas, NV, USA, 27–30 June 2016; pp. 770–778.
26. Simonyan, K.; Zisserman, A. Very deep convolutional networks for large-scale image recognition. *arXiv* **2014**, arXiv:1409.1556.
27. Tan, M.; Le, Q.V. Efficientnet: Rethinking model scaling for convolutional neural networks. *arXiv* **2019**, arXiv:1905.11946.
28. Zoph, B.; Vasudevan, V.; Shlens, J.; Le, Q.V. Learning transferable architectures for scalable image recognition. In Proceedings of the IEEE Conference on Computer Vision and Pattern Recognition, Salt Lake City, UT, USA, 18–23 June 2018; pp. 8697–8710.
29. Zoph, B.; Le, Q.V. Neural architecture search with reinforcement learning. *arXiv* **2016**, arXiv:1611.01578.
30. Banerjee, A.; Chitnis, U.; Jadhav, S.; Bhawalkar, J.; Chaudhury, S. Hypothesis testing, type I and type II errors. *Ind. Psychiatry J.* **2009**, *18*, 127. [CrossRef] [PubMed]
31. Cui, H.; Dahnoun, N. Real-Time Stereo Vision Implementation on Nvidia Jetson TX2. In Proceedings of the 2019 8th Mediterranean Conference on Embedded Computing (MECO), Budva, Montenegro, 10–14 June 2019; pp. 1–5.

32. Basulto-Lantsova, A.; Padilla-Medina, J.A.; Perez-Pinal, F.J.; Barranco-Gutierrez, A.I. Performance comparative of OpenCV Template Matching method on Jetson TX2 and Jetson Nano developer kits. In Proceedings of the 2020 10th Annual Computing and Communication Workshop and Conference (CCWC), Las Vegas, NV, USA, 6–8 January 2020; pp. 0812–0816.
33. Barthélemy, J.; Verstaevel, N.; Forehead, H.; Perez, P. Edge-computing video analytics for real-time traffic monitoring in a smart city. *Sensors* **2019**, *19*, 2048. [CrossRef] [PubMed]
34. Arshad, B.; Barthelemy, J.; Pilton, E.; Perez, P. Where is my Deer?—Wildlife Tracking And Counting via Edge Computing And Deep Learning. In Proceedings of the 2020 IEEE Sensors, Rotterdam, The Netherlands, 25–28 October 2020; pp. 1–4.
35. Ren, S.; He, K.; Girshick, R.; Sun, J. Faster R-CNN: Towards real-time object detection with region proposal networks. *arXiv* **2015**, arXiv:1506.01497.

applied sciences

MDPI

Article

Using Artificial Neural Network Models to Assess Hurricane Damage through Transfer Learning

Landon Calton † and Zhangping Wei *

Department of Physics & Physical Oceanography, University of North Carolina Wilmington, Wilmington, NC 28403, USA; lcalton@ncsu.edu
* Correspondence: weiz@uncw.edu
† Current address: Department of Electrical and Computer Engineering, North Carolina State University, Raleigh, NC 27606, USA.

Abstract: Coastal hazard events such as hurricanes pose a significant threat to coastal communities. Disaster relief is essential to mitigating damage from these catastrophes; therefore, accurate and efficient damage assessment is key to evaluating the extent of damage inflicted on coastal cities and structures. Historically, this process has been carried out by human task forces that manually take post-disaster images and identify the damaged areas. While this method has been well established, current digital tools used for computer vision tasks such as artificial intelligence and machine learning put forth a more efficient and reliable method for assessing post-disaster damage. Using transfer learning on three advanced neural networks, ResNet, MobileNet, and EfficientNet, we applied techniques for damage classification and damaged object detection to our post-hurricane image dataset comprised of damaged buildings from the coastal region of the southeastern United States. Our dataset included 1000 images for the classification model with a binary classification structure containing classes of *floods* and *non-floods* and 800 images for the object detection model with four damaged object classes *damaged roof*, *damaged wall*, *flood damage*, and *structural damage*. Our damage classification model achieved 76% overall accuracy for ResNet and 87% overall accuracy for MobileNet. The F1 score for MobileNet was also 9% higher than the F1 score of ResNet at 0.88. Our damaged object detection model achieved predominant predictions of the four damaged object classes, with MobileNet attaining the highest overall confidence score of 97.58% in its predictions. The object detection results highlight the model's ability to successfully identify damaged areas of buildings and structures from images in a time span of seconds, which is necessary for more efficient damage assessment. Thus, we show that this level of accuracy for our damage assessment using artificial intelligence is akin to the accuracy of manual damage assessments while also completing the assessment in a drastically shorter time span.

Keywords: hurricane; building damage; damage classification; damage detection; artificial intelligence; transfer learning

Citation: Calton, L.; Wei, Z. Using Artificial Neural Network Models to Assess Hurricane Damage through Transfer Learning. *Appl. Sci.* **2022**, *12*, 1466. https://doi.org/10.3390/app12031466

Academic Editors: Amerigo Capria, Nikos D. Lagaros and Vagelis Plevris

Received: 23 December 2021
Accepted: 20 January 2022
Published: 29 January 2022

1. Introduction

Coastal storms and hazard events are often analyzed to address dangers faced by coastal communities around the world. Many potential threats to communities residing in coastal areas are captured with a comprehensive plan for risk analysis. In 2018, a preliminary risk analysis estimated almost $17 billion in damages across the state of North Carolina in the wake of Hurricane Florence [1]. As a result, accurate and efficient evaluations of damage from coastal hazards such as hurricanes are necessary to provide data for addressing post-disaster relief efforts. Damage assessment is a primary tool for understanding the levels of damage to coastal populations in the aftermath of a hazard event. Knowledge of damage is further applied to models for risk assessment to mitigate damage from future hazards [2].

Efficient relief plans and proper allocation of relief funding to the affected areas are impractical without accurate data. Traditionally, post-disaster data have been collected through methods involving individuals or teams making initial observations and assessments of damage. These people capture photographs of the damage in door-to-door assessments or windshield surveys (e.g., [3,4]). Remote validations are a supplemental tool used during the damage assessment process which increases the swiftness of the manual evaluations. These desktop assessments replace onsite validations when the risk for preliminary damage assessment staff is high and images of the damaged area are readily available [4]. However, these validations still rely on humans to identify damaged structures and verify damage assessments making them prone to a significant level of inaccuracy.

In a myriad of classification tasks, artificial neural network technology has proven to be significantly more efficient in performing the same work as a human to a higher level of accuracy. Machine learning techniques possess particular advantages over humans in tasks that incorporate a large data set from multiple events of highly similar situations [2]. Hurricanes provide a multitude of events for data collection that can be used by artificial neural network models to perform damage assessment. There are usually two types of data capturing hurricane damage to buildings. The first type is satellite imagery (e.g., [5,6]), and the other type is ground-level images/photos (e.g., [7]) taken by drones or other similar ways. Both data types have been used for damage assessment. For example, Weber and Kané [8] used the Mask R-CNN [9] to predict both building locations and damage level based on pre-disaster and post-disaster images of xBD database [6]. Furthermore, Hao et al. [10] developed a multi-class deep learning model with an attention mechanism to assess damage levels of buildings given a pair of satellite images depicting a scene before and after a disaster using the xView2 dataset [11]. Cheng et al. [12] developed a stacked convolutional neural network architecture to train on an in-house visual dataset from Hurricane Dorian that was collected using an unmanned aerial vehicle. An effective hurricane damage assessment model should train on both aerial and ground-level image data to increase adaptability for emergency damage assessment of a future coastal hazard.

Social media has been explored as a primary source of data for hurricane damage assessment because of the swift integrability these platforms provide to automated damage assessments (e.g., [13–15]). Hao and Wang [16] used five machine learning classifiers that take social networking platform images and output the damage types and severity levels presented in images. Leveraging social media platforms to train damage assessment models has shown success with rapid operation capabilities.

The transfer learning approach to developing artificial neural network models for hurricane damage assessment has also been recently explored. Most of these studies focus on using transfer learning on pre-trained convolutional neural network (CNN) models with aerial images of hurricane damage to buildings (e.g., [17–20]). Liao et al. [21] uses transfer learning on two well-established CNNs, AlexNet and VGGNet, to create classification models for the two-dimensional orthomosaic images gathered from unpiloted aerial systems. These and other similar studies limit the source of the training dataset, making the classification models useful only for functional datasets comprised of aerial images taken by satellite or drone. Our work incorporates both aerial and ground-level images for hurricane damage classification and detection of damaged buildings to create a more operational damage assessment framework to apply to future coastal hazards.

Incorporating transfer learning for building damage assessment is affected by the transferability of the learned features and information from the source domain to the target domain used for testing the model. Domain adaption when using transfer learning arises when there are discrepancies among images in the source domain and between the source and target domains (e.g., [22,23]). These discrepancies are a result of how remote sensing captures images with varying sensors, locations, times, and perspectives. This issue with domain invariance extends to the transferability of information derived from different coastal hazards. A CNN-based model was shown to reach high classification performance when training on the same damage type for different disasters [24]. The source and target domains in our study do not present any major discrepancies. Rather, our damage classification and damage detection models focus on a single coastal hazard that causes multiple types of damage to enhance the efficacy of damage assessment.

There are several challenges in the area of building damage assessment using artificial neural network models. First, machine learning training requires a considerable amount of input data in order to sufficiently assess the damage or classify the damage levels from images (e.g., [5,6]). Second, in-house machine learning model development requires a significant amount of effort to achieve high accuracy. This study focuses on the building damages due to hurricanes in the U.S. southeast area, and we improve the efficiency of assessing hurricane damage to buildings by applying neural network models for damage classification and object detection. We address the first challenge by developing our in-house building damage dataset using internet search engines, and we address the second challenge by utilizing the advanced artificial intelligence models for computer vision, MobileNet [25], ResNet [26], and EfficientNet [27], through transfer learning.

This paper is organized as follows: Section 2 presents the development of our in-house building damage images including data collection, data statistics, and data pre-processing. Section 3 reviews the background of three artificial intelligence models that were used as the base of transfer learning for building damage assessment and explains the transfer learning workflow for both damage classification and damage detection. Section 4 presents the training metrics, the damage classification results, and the damage detection results, further discussing the transfer learning results among three models. Finally, the conclusion and significance of this study are stated in Section 5.

2. Building Damage Dataset

This section first presents the development of our in-house building damage image dataset. Then, we explain the data statistics for damage classification and damage detection.

2.1. Data Collection and Preparation

This study primarily focuses on the hurricane damage to buildings in the U.S. southeast region. We sourced the data from an internet search specifying criteria for photos related to hurricane damage, and a few thousand images taken from hurricane damage in Florida, Georgia, North Carolina, and South Carolina were prepared for a preliminary data cleanup. Each image in our in-house dataset was further examined for types of buildings and structures contained in the images to ensure they were characteristic of the U.S. east coast region.

The raw dataset was further processed for two tasks: damage classification and damaged object detection. For the first task, we examined the data to be used in the classification model and identified potential classes for image categorization. This step involved the additional cleanup and removal of remaining images that were duplicates or would not be a candidate for one of the image classes. For the second task, we also examined individual images to be used in the object detection model and removed those that did not capture a damaged structure. After final examination of both versions of the data, the images were ready for pre-processing before inputting them into the neural network models using transfer learning.

2.2. Data Statistics

The next step required dividing the dataset into a set for the classification model and a second set for the object detection model. The main difference among the two datasets was that the set applied to object detection required images only containing buildings, and the set applied to classification was independent from only using images containing buildings. Images contained in both sets of data are of varying pixel resolution and unaltered from the original source.

Historical hurricanes usually brought about significant flood damage due to storm surges and heavy precipitation. Thus, the damage classification research in this work aims to determine if there are floods in the image. To this purpose, we selected 1000 images from our dataset and divided them into two categories, *floods* and *non-floods*, as indicated in Table 1. The motivation is to examine if neural network models can perform binary classification on our dataset. Flood damage is characterized by flood waters in the images, and it can occur in various ways. Typical *floods* damages in our dataset include (1) flooded buildings, houses, and communities, (2) flooded streets, (3) flooded vehicles, and (4) flooded coastal areas. The *non-floods* images are related to hurricane damage, but they do not include floods in them; these images needed to be characteristic of areas and buildings damaged from hurricanes because the purpose of our classification models is to exemplify their success learning from data that would be used for traditional hurricane damage assessment. Finally, the binary classification task does not require additional data processing other than sorting the images into two categories.

Unlike the data preparation for damage classification, machine learning object detection requires the preparation of labeled data, which guides the model to learn common features in a specific type of object. The pre-processing image labeling in this work was accomplished by using the open-source annotation tool, LabelImg [28]. This tool allowed us to take an input image in our dataset and create bounding boxes around the areas of interest in the image corresponding to an annotation label. The position of the bounding box and the label were then exported for neural network model training. The object detection dataset consisted of 800 images that were annotated, and annotation labels are the damaged objects as listed in Table 2. Four categories of objects were identified from our hurricane damage dataset, and the features associated with each of them are briefly explained below.

- *Damaged roof*. The bounding box label highlights a roof that has the whole roof, some shingles, or parts of the roof damaged. The bounding box label typically encompasses the entire roof in the image; however, if the entire roof is not visible then the damaged area and any additional parts of the roof that are visible were included.
- *Damaged wall*. The labeling bounding box highlights a damaged building wall or windows within a wall. Damage to walls/windows could range from areas with minor disintegration of brick or glass structure to entire loss of the wall or window structure.
- *Flood damage*. The bounding box label highlights flood waters in an image. The flood water can occur in various places as explained in the binary classification dataset. Due to this sporadic nature, in some images, multiple bounding box labels were used to encapsulate the entirety of the flood water.
- *Structural damage*. The bounding box label highlights a building suffering from structural damage, e.g., the disintegration of the roof and/or any floor(s) within the building, complete loss of multiple walls/structures, or the collapse of the whole building.

It should be pointed out that the total number of samples in Table 2 is 958, which is greater than the total number of annotated images, i.e., 800. The difference is due to the fact that multiple objects were annotated/observed in a single image, resulting in a larger number of objects than the number of images.

Table 1. Summary of images for the flood damage classification task. The selected 1000 images were divided into two categories for binary classification.

Damage Classification Types	# of Samples	Percentage
floods	463	46.3%
non-floods	537	53.7%

Table 2. Summary of images for the building damage detection task. The selected 800 images were divided into four categories to characterize damage inflicted upon different kinds of objects or structures by historical hurricanes. The total number of object samples of 958 is larger than the number of images (i.e., 800) due to the fact that an individual image can include multiple damaged objects.

Damage Detection Types	# of Samples	Percentage
damaged roof	365	45.625%
damaged wall	281	35.125%
flood damage	167	20.875%
structural damage	145	18.125%

3. Transfer Learning

The previous section showed that our in-house hurricane damage dataset only has about 800–1000 images. To be able to develop effective hurricane damage assessment machine learning models using such a small dataset, we utilize a machine learning technique, transfer learning, in this work. This section first presents the background information about transfer learning. Then, we review the existing neural network models that were used in this study, and we focus on the typical model architecture. Next, we present the transfer learning workflows used in this study.

3.1. The Fundamentals of Transfer Learning

Transfer learning is a machine learning technique that leverages feature representations from a pre-trained artificial neural network model to train a new target model on a different, usually smaller size dataset. The crucial step for implementing transfer learning is to use learned weights/parameters from a pre-trained neural network model, which is a saved model that was previously trained on a large dataset, e.g., the ImageNet [29] and Coco [30] datasets. This choice is justified by the fact that if the original dataset is large enough and general enough, then the spatial hierarchy of features learned by the pre-trained models can effectively act as a generic model of the visual world; thus, its features are useful for many different computer vision problems, even though these new topics involve completely different classes than those of the original task [31].

3.2. Artificial Neural Network Models

This study utilizes three well-established neural network models in computer vision, namely, ResNet [26], MobileNet [25], and EfficientNet [32]. These networks were selected for several reasons. First of all, we aim to explore the efficiency of varying neural network architectures for hurricane damage assessment. To that end, collating results from multiple models would provide deeper insight than results obtained from one model trained on a single neural framework. Second, these models have been pre-trained using large image datasets, and their pre-trained weights are freely available. The following sub-sections provide a brief review of the selected three neural network models with a focus on their typical model architecture.

3.2.1. ResNet

The ResNet architecture was developed with a deep residual learning framework to directly address the issue of degradation of training accuracy in deeper networks that begin to converge [26]. Identity shortcut connections within the ResNet architecture do not rely on parameters and allow a continuous flow of information between layers as well as additional learning of residual functions. Thus, the residual net framework allows for easier optimization of the residual mapping and increased accuracy from enhanced depth of the residual nets [26]. The 50-layer ResNet contains a 3-layer bottleneck design that results in a more efficient model when paired with the identity shortcuts. We incorporated the 50-layer ResNet architecture into our model to match the target input resolution and simplistic model structure.

3.2.2. MobileNet

The MobileNet architecture focuses on streamlining the convolution layers through depthwise separable convolutions to build an attenuated deep neural network [25]. The separation of the convolution into two layers, one for filtering the inputs and one for combining the outputs with the depthwise convolution, significantly decreases the magnitude of computation and model scale. This in turn generally leads to low latency for incorporating the MobileNet model into classification and object detection.

Additionally, the MobileNet architecture makes use of two global hyper-parameters: a width multiplier and a resolution multiplier. The width multiplier aims to shrink each layer of the network in a uniform fashion, while the resolution multiplier is applied to the input image which results in reducing each subsequent layer by the same parameter [25]. We incorporated MobileNet V1 to match the target input resolution and maintain consistency with the choice of primitive model architectures.

3.2.3. EfficientNet

The EfficientNet architecture was created through a focus on prioritizing efficiency while maintaining state-of-the-art accuracy. Traditionally, convolutional neural networks are scaled up from a baseline model to improve the accuracy of detections/classifications; more training data and model layers generally produce more accurate predictions. EfficientNet uses compound scaling of the network's dimensions (width, depth, and resolution) to achieve high accuracy while striving to be the most efficient CNN [32].

The EfficientDet network is an extension of the EfficientNet architecture that was created specifically for object detection applications; it uses the EfficientNet model architecture as a base network. This variant of EfficientNet focuses on compound scaling paired with a weighted bi-directional feature pyramid network (BiFPN) to connect subsequent layers of the model together for the most successful optimizations of model efficiency and accuracy [27].

3.3. *Transfer Learning Workflow for Flood Damage Classification*

Figure 1 shows the workflow to train ResNet and MobileNet for flood damage classification. Both ResNet and MobileNet were trained using the ImageNet dataset, which includes more than 1 million images and 1000+ target classes or labels. Our transfer learning work utilizes the pre-trained knowledge in these two models, i.e., model weights that characterize typical features in images in the real world. Specifically, our hurricane damage classification transfer learning work consists of the following steps.

- Obtain the pre-trained neural network model and its weights;
- Remove the top layer which is used to predict the original 1000 classes;
- Freeze other layers in the pre-trained model to avoid destroying any of the extracted feature information;
- Add new and trainable layers on top of the frozen layers. These layers learn to turn the old features into the target predictions (i.e., *floods* and *non-floods* images) using a new dataset;
- Train the new layers on our in-house hurricane damage dataset related to flood damage.

It should be pointed out that there are no *floods* nor *non-floods* classes in the ImageNet dataset. As a result, the primary goal to use these pre-trained models is feature extraction. Our flood damage dataset was configured with a 60/20/20 split for training, validation, and testing purposes. Furthermore, data augmentation was used to increase our training samples to reduce model overfitting. The data augmentation technique randomly transforms training samples to yield believable-looking images, and it helps expose the model to more aspects of the data for better generalization [31].

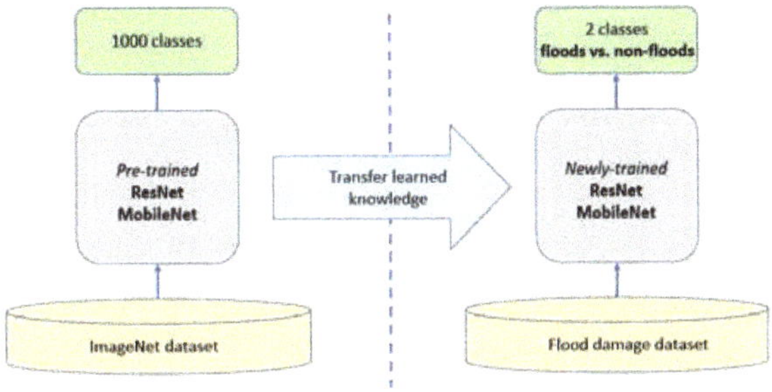

Figure 1. The transfer learning workflow for binary classification into floods and non-floods by using pre-trained ResNet and MobileNet.

3.4. *Transfer Learning Workflow for Hurricane Damage Detection*

Figure 2 shows the object detection workflow for training ResNet, MobileNet, and EfficientNet. All three networks were pre-trained on the same image resolutions (640 × 640 pixels) and the COCO 2017 dataset [30] which contains 330,000 images and 80 object categories. Each of the three models were also configurable to begin with the same training parameters. Therefore, the batch size was set to four images, and each model training instance was terminated after the twenty-thousandth epoch. Our damage detection dataset was configured with a 50/50 split for training and testing purposes when pre-processed into each of our models. This left 400 images and their corresponding annotations for model training and the other 400 images and their corresponding annotations for testing each of the models' predictions. Our transfer learning work leverages the pre-trained model weights for features extracted from the typical objects contained within the COCO 2017 dataset. More specifically, our hurricane damage detection transfer learning work consists of the following steps.

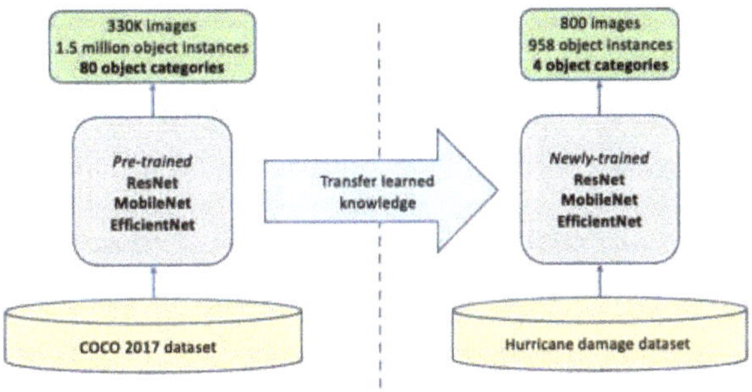

Figure 2. The transfer learning workflow for object detection of four object categories by using pre-trained ResNet, MobileNet, and EfficientNet.

- Initialize training with pre-trained neural network model and extracted feature weights;
- Configure a new pipeline with specified training parameters for our model;
- Use the pre-trained model checkpoint as the starting point for adding new, trainable layers that contain predictions of the four distinct object categories in our dataset;
- Train the new layers on our in-house hurricane damage dataset related to building damage.

Because the four object categories from Table 2 do not appear in the COCO 2017 dataset, the primary objective for using the pre-trained models is feature extraction.

3.5. *Computing Environment*

This machine learning research was conducted using Google Colab, a free Jupyter notebook environment that runs entirely in the cloud. The computing environment was configured as follows.

- The CPU model name is Intel(R) Xeon(R) CPU @ 2.00 GHz;
- The clock speed of the CPU is 2 K MHz, and the CPU cache size is 39,424 KB;
- The Graphics Processing Units (GPU) card is NVIDIA Tesla P100. It is based on the NVIDIA Pascal GPU architecture, and it has 3584 NVIDIA CUDA cores. The GPU memory is 16 GB. A single GPU card was used in this study.

4. Results and Discussion

4.1. Metrics and Prediction Skills

The metrics utilized in tracking the training behavior of the classification models and object detection models differ and are presented in the following sections. Additionally, there are specific prediction skills primarily used in determining the success in evaluating the classification models.

4.1.1. Classification Metrics

The metrics used to track the progress of classification models during training are loss and accuracy. Cross-entropy loss is the particular formulation in Equation (1) where the index i is the i-th training example in a dataset, y_i is the ground-truth label for the i-th training example, and $\hat{y}_i$ is the prediction for the i-th training example [33]. Cross-entropy loss is much larger for false predictions with a high level of confidence, resulting in those predictions being more denounced. Cross-entropy loss is used in many classifier models such as MobileNet and ResNet.

$$\text{Cross-Entropy Loss} = -\frac{1}{N}\sum_{i=1}^{N}(y_i \log(\hat{y}_i) + (1-y_i)\log(1-\hat{y}_i)). \quad (1)$$

Accuracy is defined by Equation (2) for binary classification models in terms of the four possible predictions: true positive (TP), true negative (TN), false positive (FP), and false negative (FN). This metric simply measures the percentage of correct predictions made during the validation step of training when considering the total number of predictions.

$$\text{Accuracy} = \frac{TP+TN}{TP+TN+FP+FN} \quad (2)$$

In addition to the training metrics, the metrics used for evaluating the classification models include precision, recall, and the F1 score. Precision is defined by Equation (3) and measures the percentage of correct positive predictions when considering the total number of positive predictions made.

$$\text{Precision} = \frac{TP}{TP+FP} \quad (3)$$

Recall is defined by Equation (4) and measures the percentage of positive predictions made when considering the total amount of positive samples.

$$\text{Recall} = \frac{TP}{TP+FN} \quad (4)$$

The F1 score is an equally weighted combination of both precision and recall. Equation (5) describes the formulation of the F1 score which implies that both FP and FN predictions are considered in determining the value. This characteristic of the F1 score makes it a well-balanced measure of model performance.

$$\text{F1 score} = \frac{2}{\frac{1}{\text{Precision}}+\frac{1}{\text{Recall}}} = \frac{2*(\text{Precision*Recall})}{\text{Precision+Recall}}. \quad (5)$$

4.1.2. Object Detection Metrics

The training metrics used to track the progress of the object detection models deal with the associated loss parameters for distinct training operations—the three major operations being classification, localization, and regularization. Classification loss is associated with the determination of the target object class [34]. Classification loss is represented as a combination of the cross-entropy loss from Equation (1) and the SoftMax activation function in Equation (6)

$$\sigma(z)_j = \frac{e^{z_j}}{\sum_{k=1}^{K} e^{z_k}} \tag{6}$$

where z is a vector input containing K elements corresponding to the possible object classes, j is the index variable for the input vector z, and z_j is the j-th element of z. The denominator of Equation (6) contains the normalization term that ensures $\sigma(z)_j$ is a valid probability distribution where all j elements sum to 1, allowing the predicted object classes to be converted to probabilities before computing the cross-entropy loss [33]. The localization loss is associated with bounding box regression to pinpoint the target object through training another head with an independent loss function [34]. This loss function must account for given samples/instances of bounding box coordinates represented as y_i and the target coordinates of the ground-truth bounding box represented as $\hat{y}_i$ in Equation (7). This localization loss is characterized as a Mean Square Error (MSE).

$$\text{MSE} = \frac{\sum_{i=1}^{n}(y_i - \hat{y}_i)^2}{n}. \tag{7}$$

The third type of loss, regularization loss, aims to reduce overfitting in the neural network by penalizing certain values of the weights in each layer. The result is a constrained range of values for these weights that purportedly reduces the memory capacity of the model without sacrificing model performance. Regularization is formulated in two distinct fashions (and usually implemented as a combination of both) with L1 and L2 regularization. L1 and L2 are shown in Equation (8) with the weight value w, total number of weights in a given layer n, and the regularization hyperparameter λ

$$\begin{aligned} \text{L1} &= \text{MSE} + \lambda \sum_{i=1}^{n} |w_i|, \\ \text{L2} &= \text{MSE} + \lambda \sum_{i=1}^{n} w_i^2. \end{aligned} \tag{8}$$

It is clear from Equation (8) that L1 is a function of a scaled sum of the magnitude of each weight value, and L2 is a function of a scaled sum of each weight value squared.

Finally, the total loss function is used as a generalized metric for evaluating the training performance. It is a weighted sum of the classification loss, the localization loss, and the regularization loss parameters that are calculated by the model. The model weights for classification loss and localization loss were kept equal at a value of 1.0 while the regularization loss weight was set to a much smaller fraction of the previous weights. This was standard for configuring the training of all three damage detection models. It should be added that there are two opposed structures for the heads being trained to evaluate the loss parameters mentioned above: the convolution head and fully connected head. The former is more appropriate and has better results for the classification task, while the latter is more advantageous at conducting bounding box regression [35].

4.2. Model Training

We present the transfer learning model training and validation metrics in this section. Validation occurs in the training process to evaluate the model's predictions on the validation dataset which contains images the model has not encountered during training. This gives an objective estimate of the model's accuracy and loss to compare to the training accuracy and training loss. Figure 3 shows the training metrics for flood damage classification using the cross-entropy loss and the accuracy defined in Equations (1) and (2), respectively. Figure 3a shows that the training loss and the validation loss using the ResNet model both converge to a value of approximately 0.5, indicating that the model does not experience overfitting or underfitting issues. On the other hand, training loss for the MobileNet model is around 0.05, while its validation loss is similar to that of ResNet, 0.5. Figure 3b shows training accuracy and validation accuracy for the two base models. Accuracy measures the ratio of correct prediction (including true *floods* damage prediction and true *non-floods* damage prediction) to the total number of predictions. The accuracy for the ResNet model converges to a value between 0.7 and 0.75. The accuracy using MobileNet is slightly different between training and validation. The training accuracy is close to 0.975, while the validation accuracy is about 0.9. Overall, the training metrics comparison shows that flood damage classification using the MobileNet model has a similar validation loss compared to that of ResNet, but it has a better accuracy.

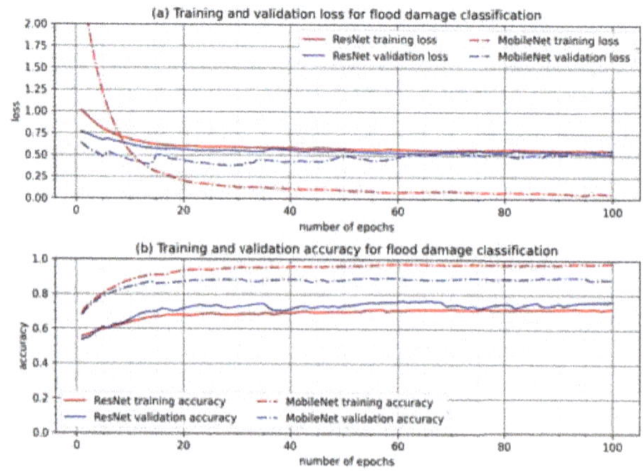

Figure 3. The training and validation cross-entropy loss and accuracy for the flood damage classification using ResNet and MobileNet.

Figure 4 shows the training metrics for hurricane damage detection which utilize the cross-entropy, MSE, L1, and L2 loss functions in Equations (1), (7) and (8). Figure 4a shows that EfficientNet converges to an approximate value of 0.3, ResNet converges to an approximate value of 0.18, and MobileNet converge to the lowest value of training classification loss at approximately 0.05. All three models achieve values for classification loss ≤ 0.3, which is generally accepted for concluding model training. However, ResNet experiences a sharp spike in classification loss between 0 and 500 epochs. This can be attributed to the early point in the training process where there has not been a significant relation of extracted features within the model's network to image features for the four classes in our model.

Figure 4b shows that ResNet converges to a training localization loss of approximately 0.8, while MobileNet and EfficientNet both converge to an approximate value of 0.01 for training localization loss. EfficientNet also has a unique path of convergence for this parameter, as it begins the training process with a localization loss value of approximately 0.05 and finishes with a value of 0.01 which characterizes its loss curve to be constant relative to the loss curves for ResNet and MobileNet. This result can be attributed to the BiFPN structure of the EfficientDet model [27] which optimizes the accuracy of predictions, specifically bounding box regression in this case.

Figure 4c shows that EfficientNet converges to an approximate value of 0.08, ResNet converges to an approximate value of 0.2, and MobileNet converges to the largest value of approximately 0.56 for regularization loss. The relatively high value for MobileNet can be attributed to less effective modeling of the regularization loss with the second term in both L1 and L2 of Equation (8). This term likely contributes a larger value for the MobileNet layer weights than the layer weights for EfficientNet and ResNet. Additionally, the MobileNet layer architecture lacks multiple two dimensional convolutions as implemented in the ResNet and EfficientNet layer architectures. The addition of supplemental convolution layers would likely further reduce overfitting in each model. Thus, the regularization loss would naturally decrease in accordance.

Figure 4d shows that ResNet and EfficientNet both converge to an approximate value of 0.4 while MobileNet converges to an approximate value of 0.6 for the total loss. The increased value of total loss for MobileNet can be attributed to the regularization loss which converges to a relatively large value in comparison to the localization loss and the classification loss for MobileNet. Thus, the significantly higher value of regularization loss for MobileNet skews the value of total loss, despite the model's significantly lower values for classification loss and localization loss.

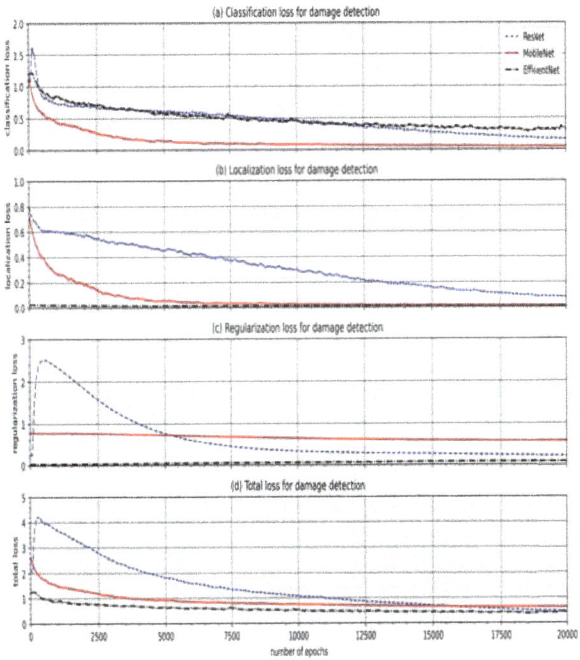

Figure 4. The training metrics for damage detection using classification loss, localization loss, regularization loss, and total loss up to the twenty-thousandth (20 k) epoch.

4.3. Damage Classification

After the completion of the transfer learning model training using the base models of ResNet and MobileNet on our in-house flood damage dataset, the newly trained models are tested with images they did not see during the training stage. Among the test dataset, the number of *floods* images is 92, and the number of *non-floods* images is 107. It is noted that the *floods* and *non-floods* images are slightly imbalance. The same set of *floods* images and *non-floods* images was tested by the newly trained models using ResNet and MobileNet. In this study, the *floods* class was the positive class, and the *non-floods* class was the negative class, resulting in the following four possible predictions:

- TN / True negative: an image was *non-floods* and predicted as *non-floods*;
- TP / True positive: an image was *floods* and predicted as *floods*;
- FN / False negative: an image was *floods* and predicted as *non-floods*;
- FP / False positive: an image was *non-floods* and predicted as *floods*.

Figure 5 shows the confusion matrix for both ResNet and MobileNet. Each row in a confusion matrix represents a true label (i.e., an actual class), while each column in a confusion matrix represents a predicted label. In this study, the first row and the first column have the label of *non-floods*, while the second row and the second column have the label of *floods*. The diagonal elements of the confusion matrix represent that the predicted label is equal to the true label, while off-diagonal elements are those that are mislabeled by the classifier. The values in each of the elements are normalized by the total number of images for each class. It is expected that a perfect classifier would have only true positives (lower right) and true negatives (top left). Figure 5 shows both ResNet and MobileNet are able to classify *non-floods* and *floods* images with the larger percentages along the diagonal elements. An accurate prediction means that a *non-floods* image was predicted as *non-floods*, and a *floods* image was predicted as *floods*. The prediction accuracy of ResNet is about 76%, and the accuracy of MobileNet is about 87%. Both classifiers show that true positive predictions (85% by ResNet and 89% by MobileNet) are higher than true negative predictions (66% by ResNet and 85% by MobileNet). A further comparison shows that the transfer learning model using MobileNet outperforms the one trained using ResNet for both true positive and true negative predictions. Specifically, the true positive prediction percentage of MobileNet is about 89%, while the true positive prediction percentage of ResNet is about 85%. Similarly, the true negative prediction percentage of MobileNet is about 85%, while the true negative prediction percentage of ResNet is about 66%.

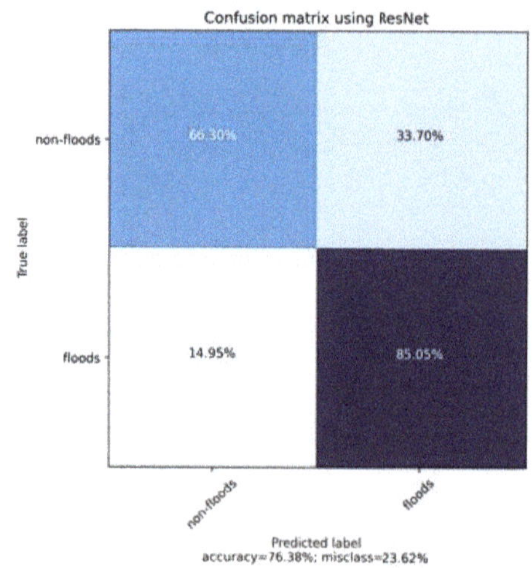

(a) ResNet

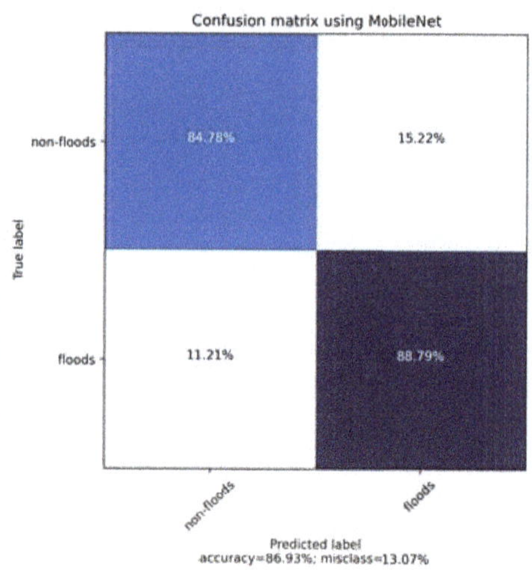

(b) MobileNet

Figure 5. The confusion matrix for flood damage classification using (**a**) ResNet and (**b**) MobileNet. The locations of four possible predictions are: TN (top left), FP (top right), FN (lower left), and TP (lower right).

Model classification performance is further evaluated by using precision, recall, and F1 score, as defined in Equations (3)–(5), respectively. The result is summarized in Table 3. The precision metric measures the accuracy of the true positive predictions (i.e., *floods* label) by dividing the true positive predictions over the sum of true positive and false positive predictions. The precision values for ResNet and MobileNet are 0.75 and 0.87, respectively. This indicates that MobileNet shows a higher accuracy (about 12%) than ResNet in terms of *floods* image classification. The second metric recall measures the true positive rate (i.e., the ratio of *floods* images that are correctly detected by the classifiers. The result shows that the recall of MobileNet is about 4% higher than that of ResNet. The last metric examined in this study is the F1 score, which is the harmonic mean of precision and recall, as defined in Equation (5). As a result, the classifier only obtains a high F1 score if both precision and recall are high. The F1 score result shows that MobileNet obtains a F1 score that is about 9% higher than that obtained by the ResNet-based classifier. In short, both the confusion matrix comparison in Figure 5 and metric comparison in Table 3 show that the flood damage classification models developed through transfer learning are accurate. Furthermore, the classifier using MobileNet as the base model performs better than a transfer learning model developed on the basis of ResNet.

Table 3. Summary of flood damage classification using transfer learning on the basics of ResNet and MobileNet.

ResNet			MobileNet		
Precision	Recall	F1-score	Precision	Recall	F1-score
0.75	0.85	0.79	0.87	0.89	0.88

4.4. Damage Detection

In the following sections, we compare the predictions from each model on a set of four test images that the model has not seen previously with each image containing one of the four specific classes of damage. The scores in Tables 4–7 are confidence/probability scores associated with each predicted type of damage. Each confidence score is assigned by the model to a different bounding box prediction as a measure of how likely the detected object in the image belongs to the predicted class. In other words, the confidence score is a measure of the model's ability to isolate a damaged area within an image and correctly identify that damage using the model's trained classifiers. Higher confidence scores are associated with the most accurate predictions of damaged areas in the image. The top three confidence scores were taken from each inference run on the corresponding image, and the top confidence score is associated with the bounding box prediction displayed on the images in Figures 6–9.

4.4.1. Damaged Roof Comparison

In Table 4, the top predictions for MobileNet and EfficientNet correctly classify the object as a *damaged roof*, and MobileNet achieves the higher confidence score of 90.93% for that predicted class. The bounding box location in Figure 6c predicted by EfficientNet more accurately encompasses the entire *damaged roof* structure in comparison to the bounding box location in Figure 6b predicted by MobileNet. On the other hand, the top two predictions made by ResNet are inaccurately classified as *structural damage*, and the bounding box location in Figure 6a is also inaccurate due to the different image features associated with the *structural damage* class. The same applies to the third prediction made by ResNet of the *flood damage* class and the third prediction made by EfficientNet of the *structural damage* class.

Table 4. Summary of confidence scores and associated object categories for the top three predictions made on the image of a *damaged roof* for each of the three models.

	ResNet		MobileNet		EfficientNet	
	Score	**Type**	**Score**	**Type**	**Score**	**Type**
#1	28.12%	structural damage	90.93%	damaged roof	62.85%	damaged roof
#2	21.99%	structural damage	32.73%	damaged roof	47.72%	damaged roof
#3	12.47%	flood damage	21.75%	damaged roof	15.97%	structural damage

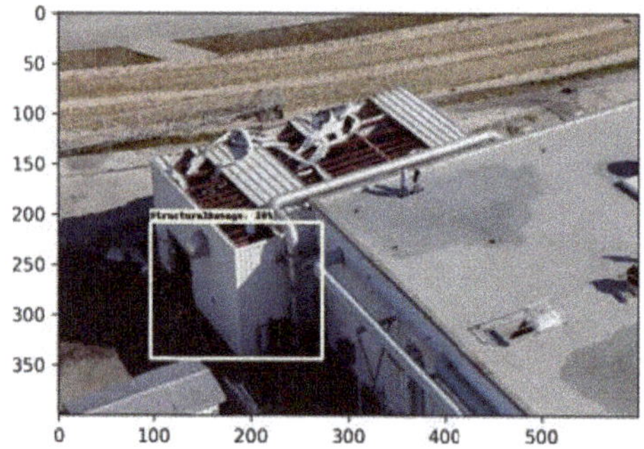

(**a**) ResNet

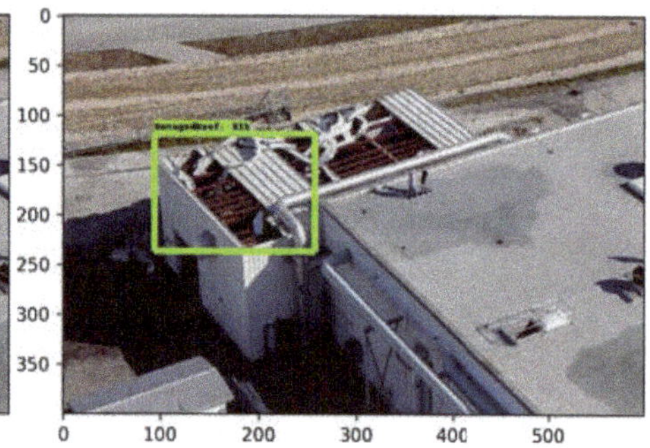

(**b**) MobileNet

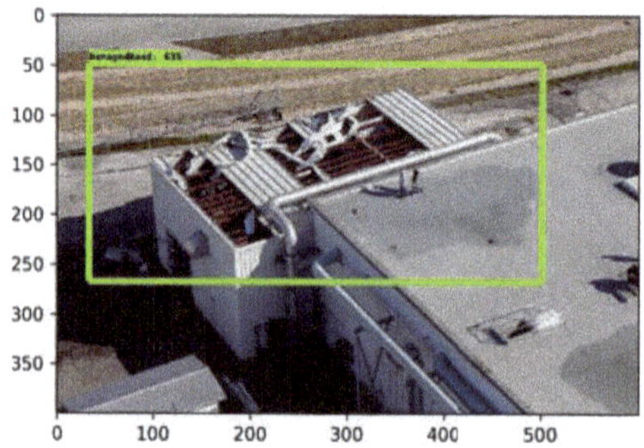

(**c**) EfficientNet

Figure 6. Inference carried out on the image of a *damaged roof* for each of the three models. The green bounding boxes correspond to the correct predictions of a *damaged roof*, and the grey bounding box corresponds to the incorrect prediction of *structural damage* from the first row of Table 4.

4.4.2. Damage Wall Comparison

In Table 5, the top predictions for ResNet, MobileNet, and EfficientNet correctly classify the object as a *damaged wall*, and MobileNet again achieves the highest confidence score of 97.58% for that predicted class. The bounding box location in Figure 7b predicted by MobileNet more accurately encompasses the entire *damaged wall* structure in comparison to the bounding box location in Figure 7a predicted by MobileNet and Figure 7c predicted by EfficientNet. However, the second and third predictions made by ResNet and MobileNet are inaccurately classified as either *structural damage* or a *damaged roof*, likely due to similar features of a *damaged wall* in this image that the model has learned to extract in predicting the *structural damage* and *damaged roof* classes as well.

Table 5. Summary of confidence scores and associated object categories for the top three predictions made on the image of a *damaged wall* for each of the three models.

	ResNet		MobileNet		EfficientNet	
	Score	**Type**	**Score**	**Type**	**Score**	**Type**
#1	75.00%	damaged wall	97.58%	damaged wall	55.22%	damaged wall
#2	23.44%	structural damage	15.07%	structural damage	18.41%	damaged wall
#3	20.54%	damaged roof	11.56%	structural damage	13.46%	damaged wall

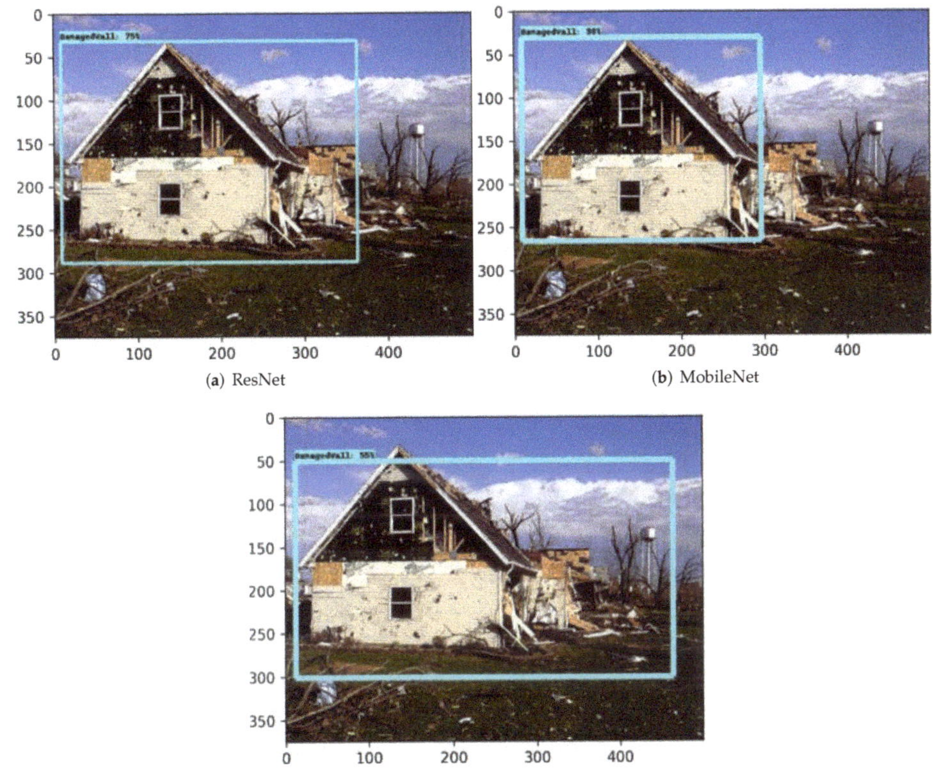

(**a**) ResNet (**b**) MobileNet (c) EfficientNet

Figure 7. Inference carried out on the image of a *damaged wall* for each of the three models. The blue bounding boxes correspond to the correct prediction of a *damaged wall* from the first row of Table 5.

4.4.3. Flood Damage Comparison

In Table 6, the top predictions for ResNet, MobileNet, and EfficientNet correctly classify the object as *flood damage*, and MobileNet again achieves the highest confidence score of 97.45% for that predicted class. The bounding box location in Figure 8b predicted by MobileNet more accurately encompasses the entire *flood damage* area in comparison to the bounding box location in Figure 8a predicted by ResNet and Figure 8c predicted by EfficientNet. However, the second and third predictions made by ResNet and MobileNet are inaccurately classified as either a *damaged wall* or a *damaged roof*, likely due to some features of the building in the background of the image that would be extracted to predict those classes. The same applies to the second prediction made by EfficientNet of the *structural damage* class.

Table 6. Summary of confidence scores and associated object categories for the top three predictions made on the image of *flood damage* for each of the three models.

	ResNet		MobileNet		EfficientNet	
	Score	**Type**	**Score**	**Type**	**Score**	**Type**
#1	52.46%	flood damage	97.45%	flood damage	48.79%	flood damage
#2	18.73%	damaged wall	24.39%	damaged wall	10.45%	structural damage
#3	12.63%	damaged roof	6.14%	damaged roof	10.30%	flood damage

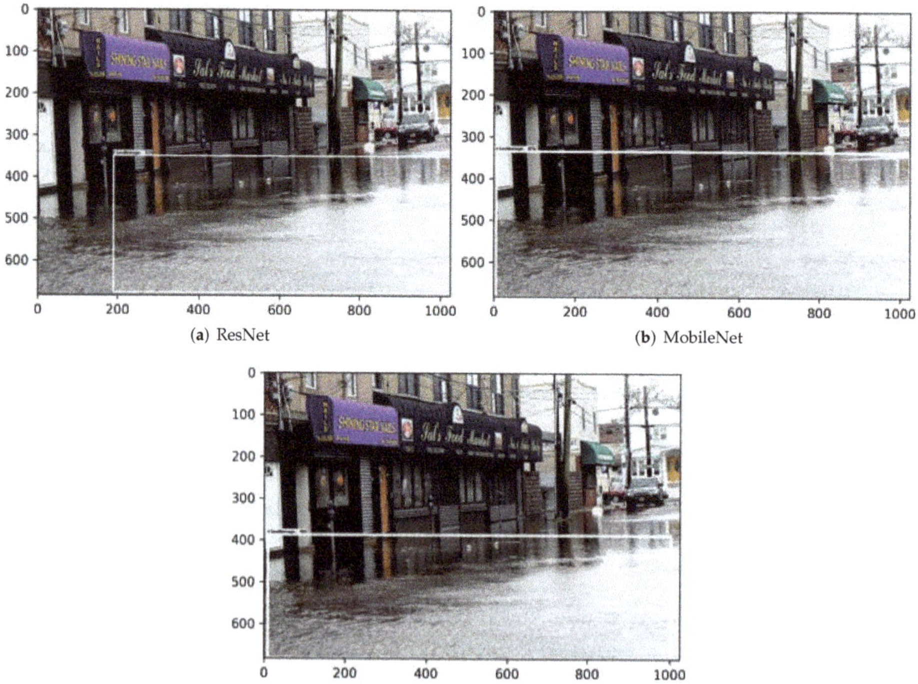

(**a**) ResNet (**b**) MobileNet (**c**) EfficientNet

Figure 8. Inference carried out on the image of *flood damage* for each of the three models. The white bounding boxes correspond to the correct predictions of *flood damage* from the first row of Table 6.

4.4.4. Structural Damage Comparison

In Table 7, the top predictions for ResNet, MobileNet, and EfficientNet correctly classify the object as *structural damage*, and MobileNet actually achieves the lowest confidence score of 42.85% for that predicted class. ResNet and EfficientNet achieve similar confidence scores of 68.11% and 67.96%, respectively, for their top prediction. Although the bounding box location in Figure 9b predicted by MobileNet more accurately encompasses the entire *structural damage* area in comparison to the bounding box location in Figure 9a predicted by ResNet and Figure 9c predicted by EfficientNet, the second and third predictions made by ResNet and MobileNet are inaccurately classified as either a *damaged wall* or a *damaged roof*, very likely due to the similar features of *structural damage* in this image that the model has learned to extract in predicting the *damaged roof* and *damaged wall* classes as well. The same applies to the third prediction made by EfficientNet of the *flood damage* class.

Table 7. Summary of confidence scores and associated object categories for the top three predictions made on the image of *structural damage* for each of the three models.

	ResNet		MobileNet		EfficientNet	
	Score	**Type**	**Score**	**Type**	**Score**	**Type**
#1	68.11%	structural damage	42.85%	structural damage	67.96%	structural damage
#2	13.43%	damaged wall	11.74%	damaged roof	25.56%	structural damage
#3	12.26%	structural damage	6.30%	damaged roof	18.86%	flood damage

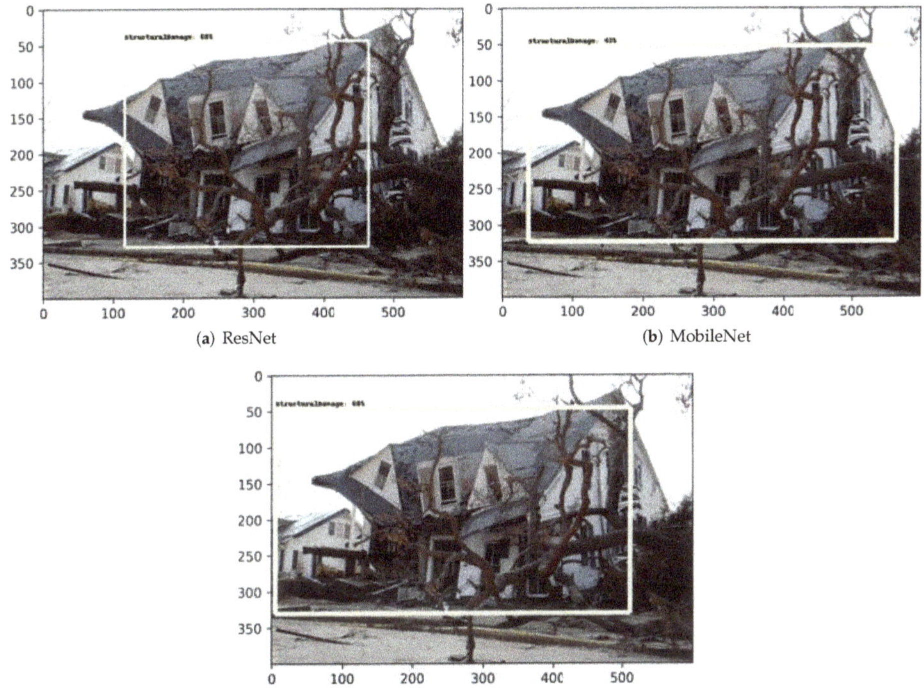

(**a**) ResNet (**b**) MobileNet (**c**) EfficientNet

Figure 9. Inference carried out on the image of *structural damage* for each of the three models. The tan colored bounding boxes correspond to the correct predictions of *structural damage* from the first row of Table 7.

4.4.5. Overall Performance of the Damage Detection Models

It can be seen from the results of the inference for each type of damage that the most accurate damage detector is the model trained on the MobileNet architecture. The MobileNet model most consistently achieved the highest confidence scores when predicting each type of damage with the highest overall confidence score of 97.58% when predicting the damaged roof in Figure 6b. This is likely a result of the unique structure of the MobileNet architecture; the depthwise separable convolutions allow for these detections to have an improved computation and overall model scale in comparison to the EfficientNet and ResNet models. Additionally, the width and resolution multipliers that are incorporated into the model architecture likely give the MobileNet model a significant advantage in scaling the layers for a more-tailored fit to each object class. Thus, each classifier corresponding to the four object classes predicts the type of damage after it is located in the image with high accuracy.

However, the damage detection model with the most consistent classifier was actually EfficientNet. As can be seen in Tables 4–7, EfficientNet was the most consistent model with classifying the predicted damage as the correct object/label. Out of the top three predictions among each of the four types of damage, EfficientNet incorrectly classified the damage a total of three times, while MobileNet and ResNet incorrectly classified the damage six times and eight times, respectively. This result is likely due to the primary advantage of the EfficientNet architecture being its efficiency of predictions due to the use of compound scaling. In turn, this efficiency is optimized by the BiFPN mentioned previously; increased efficiency of accurate predictions leads to the EfficientNet model producing more consistently accurate classifications of the damage—but at the cost of losing a certain degree of accuracy in predicting the precise location of the damage, thus leading to a lower confidence score for EfficientNet in general.

Since our in-house dataset for damage detection was not included in the original dataset used to develop all three AI models (ResNet, MobileNet, and EfficientNet), then our models are subject to developing a certain level of negative transfer [36,37]. The effect of negative transfer on model performance leads to less accurate predictions made by each model. Lower confidence scores can be attributed to the negative impact each model endured from transferring learned features of each pre-trained model to the target domain of our dataset.

5. Conclusions

This study has developed transfer-learning-based artificial intelligence models to assess building damages due to hurricanes in the U.S. southeast region. We developed our in-house building damage image dataset and subset it into (i) damage classification (i.e., *floods* vs. *non-floods*) and (ii) damaged object detection including *damaged roof, damaged wall, flood damage*, and *structural damage*. We developed transfer learning workflows that take advantage of feature extraction from three advanced neural network models in computer vision (i.e., EfficientNet, ResNet, and MobileNet) and successfully retrained these models for building damage assessment. Finally, we evaluated the classification and object detection performance among the different models. Our major findings and contributions include

- The transfer learning based flood damage classification models were developed using ResNet and MobileNet. A binary classification was carried out to detect *floods* and *non-floods* images. Several methods were used to evaluate the performance of the transfer learning models. The confusion matrix comparison showed both ResNet and MobileNet are able to correctly classify *floods* and *non-floods* with a relatively high accuracy. Specifically, the overall accuracy is about 76% using ResNet and 87% using MobileNet. Three metrics (precision, recall, and F1 score) were further calculated and compared between two models. The result obtained using MobileNet as the base model is consistently better than that using ResNet. For example, the F1 score, a harmonic mean of precision and recall, is about 0.88 using MobileNet. It is about 9% higher than the F1 score using ResNet (0.79). Overall, this study showed that hurricane flood damage to buildings can be correctly classified using artificial intelligence models developed using transfer learning techniques on the basis of advancing machine learning models in computer vision.
- The transfer-learning-based damage detection models were developed using ResNet, MobileNet, and EfficientNet. Four damage types were captured in four object classes: *damaged roof, damaged wall, flood damage*, and *structural damage*. Two methods were primarily used to evaluate the performance of the transfer learning models for damage detection. The top three confidence scores and associated object class were tabulated for each model, showing that each model was capable of predicting the correct object class in the image; the MobileNet model consistently achieved the highest confidence score and proved to be the more accurate model in detecting hurricane damage. Then, the images of each type of damage were displayed with the top bounding box prediction for each model. Likewise, MobileNet consistently achieved the most accurate localizations of the detected damage in each image. Therefore, this study showed that various types of damage from hurricanes can be accurately detected using artificial intelligence models developed through transfer learning to further advance machine learning applications in computer vision.

From creating our in-house damage assessment framework, we were able to show that a significant level of accuracy for damage classification can be achieved using transfer learning techniques on a pre-trained neural network. Given the relatively small and broad range of images used for the input data set, our classification model displayed a high degree of versatility that could be used during a spectrum of hurricane and other coastal hazard events. The object detection results highlight the model's ability to successfully identify damaged areas of buildings and structures from test data in a time span of seconds, which is necessary for more efficient damage assessment.

Our work can be improved with further research into applying transfer learning techniques to create classification and object detection models trained on post-disaster imagery. Using these machine learning models would significantly reduce the time required for damage assessment. Therefore, relief plans created in the wake of a future coastal hazard would save hours to days of time required to determine the total damage incurred. As a result, impacted coastal communities would be able to receive more reliable and prompt relief from direct implementation of artificial intelligence technology such as our classification and object detection models.

Author Contributions: L.C.: methodology, software, validation, formal analysis, investigation, resources, data curation, writing—original draft, writing—review and editing, and visualization. Z.W.: conceptualization, methodology, software, validation, formal analysis, investigation, resources, data curation, writing—original draft, writing—review and editing, and visualization. All authors have read and agreed to the published version of the manuscript.

Funding: This research was partly supported by the research momentum fund and the faculty start-up fund provided by the University of North Carolina Wilmington.

Data Availability Statement: The data presented in this study are available for download at https://doi.org/10.15139/S3/DPNPBM.

Acknowledgments: All test images used in this article were openly licensed and public domain works obtained from the Creative Commons Search tool (https://search.creativecommons.org/ (accessed on 28 July 2021)).

Conflicts of Interest: The authors declare no conflict of interest.

References

1. Cooper, R. Hurricane Florence Recovery Recommendations. 2018. Available online: https://www.osbm.nc.gov/media/824/open (accessed on 15 March 2021).
2. Guikema, S. Artificial Intelligence for Natural Hazards Risk Analysis: Potential, Challenges, and Research Needs. *Risk Anal.* **2020**, *40*, 1117–1123. [CrossRef] [PubMed]
3. Massarra, C.C. Hurricane Damage Assessment Process for Residential Buildings. Master's Thesis, Louisiana State University, Baton Rouge, LA, USA, 2012.
4. FEMA. *FEMA Preliminary Damage Assessment Guide*; FEMA: Washington, DC, USA, 2020.
5. Lam, D.; Kuzma, R.; McGee, K.; Dooley, S.; Laielli, M.; Klaric, M.; Bulatov, Y.; McCord, B. xview: Objects in context in overhead imagery. *arXiv* **2018**, arXiv:1802.07856.
6. Gupta, R.; Hosfelt, R.; Sajeev, S.; Patel, N.; Goodman, B.; Doshi, J.; Heim, E.; Choset, H.; Gaston, M. xbd: A dataset for assessing building damage from satellite imagery. *arXiv* **2019**, arXiv:1911.09296.
7. Roueche, D.B.; Lombardo, F.T.; Krupar, R.; Smith, D.J. *Collection of Perishable Data on Wind-and Surge-Induced Residential Building Damage During Hurricane Harvey (TX)*; DesignSafe-CI: Austin, TX, USA, 2018.
8. Weber, E.; Kané, H. Building disaster damage assessment in satellite imagery with multi-temporal fusion. *arXiv* **2020**, arXiv:2004.05525.
9. He, K.; Gkioxari, G.; Dollár, P.; Girshick, R. Mask r-cnn. In Proceedings of the IEEE International Conference on Computer Vision, Venice, Italy, 27–29 October 2017; pp. 2961–2969.
10. Hao, H.; Baireddy, S.; Bartusiak, E.R.; Konz, L.; LaTourette, K.; Gribbons, M.; Chan, M.; Comer, M.L.; Delp, E.J. An attention-based system for damage assessment using satellite imagery. *arXiv* **2020**, arXiv:2004.06643.
11. Gupta, R.; Goodman, B.; Patel, N.; Hosfelt, R.; Sajeev, S.; Heim, E.; Doshi, J.; Lucas, K.; Choset, H.; Gaston, M. Creating xBD: A dataset for assessing building damage from satellite imagery. In Proceedings of the IEEE/CVF Conference on Computer Vision and Pattern Recognition Workshops, Long Beach, CA, USA, 16–17 June 2019; pp. 10–17.
12. Cheng, C.S.; Behzadan, A.H.; Noshadravan, A. Deep learning for post-hurricane aerial damage assessment of buildings. *Comput.-Aided Civ. Infrastruct. Eng.* **2021**, *36*, 695–710. [CrossRef]
13. Hao, H.; Wang, Y. Leveraging multimodal social media data for rapid disaster damage assessment. *Int. J. Disaster Risk Reduct.* **2020**, *51*, 101760. [CrossRef]
14. Imran, M.; Alam, F.; Qazi, U.; Peterson, S.; Ofli, F. Rapid Damage Assessment Using Social Media Images by Combining Human and Machine Intelligence. *arXiv* **2020**, arXiv:2004.06675.

15. Zhang, Y.; Zong, R.; Wang, D. A Hybrid Transfer Learning Approach to Migratable Disaster Assessment in Social Media Sensing. In Proceedings of the 2020 IEEE/ACM International Conference on Advances in Social Networks Analysis and Mining (ASONAM), The Hague, The Netherlands, 7–10 December 2020; pp. 131–138. [CrossRef]
16. Hao, H.; Wang, Y. Hurricane damage assessment with multi-, crowd-sourced image data: A case study of Hurricane Irma in the city of Miami. In Proceedings of the 17th International Conference on Information System for Crisis Response and Management (ISCRAM), Valencia, Spain, 19–22 May 2019.
17. Li, Y.; Hu, W.; Dong, H.; Zhang, X. Building Damage Detection from Post-Event Aerial Imagery Using Single Shot Multibox Detector. *Appl. Sci.* **2019**, *9*, 1128. [CrossRef]
18. Presa-Reyes, M.; Chen, S.C. Assessing Building Damage by Learning the Deep Feature Correspondence of before and after Aerial Images. In Proceedings of the 2020 IEEE Conference on Multimedia Information Processing and Retrieval (MIPR), Shenzhen, China, 6–8 August 2020; pp. 43–48. [CrossRef]
19. Pi, Y.; Nath, N.D.; Behzadan, A.H. Convolutional neural networks for object detection in aerial imagery for disaster response and recovery. *Adv. Eng. Inform.* **2020**, *43*, 101009. [CrossRef]
20. Pi, Y.; Nath, N.D.; Behzadan, A.H. Disaster impact information retrieval using deep learning object detection in crowdsourced drone footage. In Proceedings of the International Workshop on Intelligent Computing in Engineering, Berlin, Germany, 1–4 July 2020; pp. 134–143.
21. Liao, Y.; Mohammadi, M.E.; Wood, R.L. Deep Learning Classification of 2D Orthomosaic Images and 3D Point Clouds for Post-Event Structural Damage Assessment. *Drones* **2020**, *4*, 24. [CrossRef]
22. Wang, X.; Li, Y.; Lin, C.; Liu, Y.; Geng, S. Building damage detection based on multi-source adversarial domain adaptation. *J. Appl. Remote Sens.* **2021**, *15*, 036503. [CrossRef]
23. Li, Y.; Hu, W.; Li, H.; Dong, H.; Zhang, B.; Tian, Q. Aligning Discriminative and Representative Features: An Unsupervised Domain Adaptation Method for Building Damage Assessment. *IEEE Trans. Image Process.* **2020**, *29*, 6110–6122. [CrossRef] [PubMed]
24. Valentijn, T.; Margutti, J.; van den Homberg, M.; Laaksonen, J. Multi-Hazard and Spatial Transferability of a CNN for Automated Building Damage Assessment. *Remote Sens.* **2020**, *12*, 2839. [CrossRef]
25. Howard, A.G.; Zhu, M.; Chen, B.; Kalenichenko, D.; Wang, W.; Weyand, T.; Andreetto, M.; Adam, H. MobileNets: Efficient Convolutional Neural Networks for Mobile Vision Applications. *arXiv* **2017**, arXiv:1704.04861.
26. He, K.; Zhang, X.; Ren, S.; Sun, J. Deep residual learning for image recognition. In Proceedings of the IEEE Conference on Computer Vision and Pattern Recognition, Las Vegas, NV, USA, 27–30 June 2016; pp. 770–778.
27. Tan, M.; Pang, R.; Le, Q.V. EfficientDet: Scalable and Efficient Object Detection. In Proceedings of the IEEE/CVF Conference on Computer Vision and Pattern Recognition, Seattle, WA, USA, 13–19 June 2020.
28. Tzutalin. LabelImg. 2015. Available online: https://github.com/tzutalin/labelImg (accessed on 15 March 2021).
29. Deng, J.; Dong, W.; Socher, R.; Li, L.J.; Li, K.; Li, F.-F. Imagenet: A large-scale hierarchical image database. In Proceedings of the 2009 IEEE Conference on Computer Vision and Pattern Recognition, Miami, FL, USA, 20–25 June 2009; pp. 248–255.
30. Lin, T.; Maire, M.; Belongie, S.J.; Bourdev, L.D.; Girshick, R.B.; Hays, J.; Perona, P.; Ramanan, D.; Dollár, P.; Zitnick, C.L. Microsoft COCO: Common Objects in Context. *arXiv* **2014**, arXiv:1405.0312.
31. Chollet, F. *Deep Learning with Python*; Simon and Schuster: New York, NY, USA, 2017.
32. Tan, M.; Le, Q.V. EfficientNet: Rethinking Model Scaling for Convolutional Neural Networks. *arXiv* **2019**, arXiv:1905.11946.
33. Parmar, R. Common loss functions in machine learning. Available online: https://towardsdatascience.com/common-loss-functions-in-machine-learning-46af0ffc4d23 (accessed on 22 January 2022).
34. Jiang, S.; Qin, H.; Zhang, B.; Zheng, J. Optimized Loss Functions for Object detection: A Case Study on Nighttime Vehicle Detection. *arXiv* **2020**, arXiv:2011.05523.
35. Wu, Y.; Chen, Y.; Yuan, L.; Liu, Z.; Wang, L.; Li, H.; Fu, Y. Rethinking Classification and Localization in R-CNN. *arXiv* **2019**, arXiv:1904.06493.
36. Zhang, W.; Deng, L.; Wu, D. Overcoming Negative Transfer: A Survey. *arXiv* **2020**, arXiv:2009.00909.
37. Wang, Z.; Dai, Z.; Póczos, B.; Carbonell, J.G. Characterizing and Avoiding Negative Transfer. *arXiv* **2018**, arXiv:1811.09751.

applied sciences

MDPI

Article

Wind Power Forecasting with Deep Learning Networks: Time-Series Forecasting †

Wen-Hui Lin [1], Ping Wang [1,*], Kuo-Ming Chao [2], Hsiao-Chung Lin [1], Zong-Yu Yang [1] and Yu-Huang Lai [1]

1 Green Energy Technology Research Center, Faculty of Department of Information Management, Kun Shan University, Tainan 710303, Taiwan; linwh@mail.ksu.edu.tw (W.-H.L.); fordlin@mail.ksu.edu.tw (H.-C.L.); s109000200@g.ksu.edu.tw (Z.-Y.Y.); s106001738@g.ksu.edu.tw (Y.-H.L.)
2 Engineering and Computing, School of MIS, Coventry University, Coventry CV1 5FB, UK; csx240@coventry.ac.uk
* Correspondence: pingwang@mail.ksu.edu.tw; Tel.: +886-6-205-0545

† This paper is an extended version of our paper published in 7th IEEE International Conference on Applied System Innovation 2021 (IEEE ICASI2021), Chiayi, Taiwan, 24–25 September 2021.

Abstract: Studies have demonstrated that changes in the climate affect wind power forecasting under different weather conditions. Theoretically, accurate prediction of both wind power output and weather changes using statistics-based prediction models is difficult. In practice, traditional machine learning models can perform long-term wind power forecasting with a mean absolute percentage error (MAPE) of 10% to 17%, which does not meet the engineering requirements for our renewable energy project. Deep learning networks (DLNs) have been employed to obtain the correlations between meteorological features and power generation using a multilayer neural convolutional architecture with gradient descent algorithms to minimize estimation errors. This has wide applicability to the field of wind power forecasting. Therefore, this study aimed at the long-term (24–72-h ahead) prediction of wind power with an MAPE of less than 10% by using the Temporal Convolutional Network (TCN) algorithm of DLNs. In our experiment, we performed TCN model pretraining using historical weather data and the power generation outputs of a wind turbine from a Scada wind power plant in Turkey. The experimental results indicated an MAPE of 5.13% for 72-h wind power prediction, which is adequate within the constraints of our project. Finally, we compared the performance of four DLN-based prediction models for power forecasting, namely, the TCN, long short-term memory (LSTM), recurrent neural network (RNN), and gated recurrence unit (GRU) models. We validated that the TCN outperforms the other three models for wind power prediction in terms of data input volume, stability of error reduction, and forecast accuracy.

Keywords: renewable energy; wind power forecasting; deep learning network; temporal convolutional network; long short-term memory

Citation: Lin, W.-H.; Wang, P.; Chao, K.-M.; Lin, H.-C.; Yang, Z.-Y.; Lai, Y.-H. Wind Power Forecasting with Deep Learning Networks: Time-Series Forecasting. *Appl. Sci.* **2021**, *11*, 10335. https://doi.org/10.3390/app112110335

Academic Editor: Nikos D. Lagaros

Received: 20 October 2021
Accepted: 31 October 2021
Published: 3 November 2021

1. Introduction

With the increasingly serious global warming crisis and the burning of fossil fuels inducing air pollution and climate change, concerned parties have begun to invest in the development and application of renewable energy. European countries such as Denmark, Germany, and Sweden have invested in renewable energy through smart power grids, in which power suppliers and regional suppliers provide two-way complementary power supply and demand. The key technology of a smart power grid is power forecasting in relation to renewable energy, which is a clean power supply.

Many techniques have been applied to wind power forecasting to solve various problems, such as the fluctuations in power from wind farms for very short-term, short-term (from 30 min to day-ahead), medium-term (from day-ahead to month-ahead), and long-term (more than month-ahead) [1].

Wind power forecasting prediction models can be classified using the following three approaches: (1) the physical approach, in which weather changes are considered as deterministic events [1], (2) the statistical approach, in which weather changes are considered as a random process [2,3], and (3) the hybrid approach, which constitutes a weighted aggregation of the other two prediction models [4–9]. Compared with these three methods for wind power prediction problems, deep learning network (DLN) approaches, such as Boltzmann machines (RBM), long short-term memory (LSTM), temporal convolutional networks (TCN), and convolutional neural networks (CNN) have exhibited superior results and are generally considered as an alternative solution for wind power prediction [10,11]. These wind power forecasting schemes are summarised as Table 1.

Table 1. Four Major Approaches for Wind Power Forecasting.

	Features	Limitations
Physical methods [1]	• Physical methods for wind forecasting use numerical weather prediction (NWP) to predict weather, considering the effects of atmosphere, local terrain, and wind farm layout factors.	• Needs a lot of weather experts to handle numerical weather data prediction. • In case that accuracy of NWP is poor, the wind power generation forecasting becomes inaccurate.
Statistical methods [2,3]	• Applies statistical methods to find the relationships between weather features and the predicted power. • Statistical methods include Bayesian, regression, and auto regression integrated moving average (ARIMA) models.	• A specific statistical method cannot handle complex weather conditions affected by atmosphere and environment factors. • Thus, enhanced learning schemes such as random trees and GDBT are proposed in order to increase the accuracy for wind power prediction.
Hybrid methods [4–9]	• Aggregate different weights of models to improve model performance by preserving advantages of each approach, such as combination of fuzzy logic approach, artificial neural network (ANN) and support vector machine (SVM), where SVM and fuzzy logic approach can complement each other and ensure superior results.	• These hybrid models have problems with stable prediction as their complex learning architecture may cause low efficiency, long training times and even under-fitting.
Deep learning methods [10,11]	• Use convolution operation to extract the features of time series data and predict the output using classification results. • Multilayer neural networks for multiclass classification exhibited superior results in wind power forecasting applications. • Compared with the traditional ANNs, deep learning neural networks do not need extra unsupervised networks or data preprocessing (e.g., decomposition).	• The performance of the DLN model is constrained by the quality of data input and neural architecture design. • To avoid constraints from data inputs, researchers have begun to study and propose new (NWP + DLN) models recently.

These developed models can perform long-term day-ahead wind power forecasting; however, forecasting schemes with a mean absolute percentage error (MAPE) between 12% and 17% [12,13] do not meet the engineering requirements. Thus, the development of an accurate and robust approach for wind power forecasting under varying climate conditions is still a challenge. Considering the increasing role of wind power in the renewable energy system, the research gaps and opportunities for wind power predicting are summarised as:

1. Practically, most existing approaches to forecasting do not model the uncertainty of wind well. Thus, a high-accuracy wind power model needs high resolution weather data inputs generated by an NWP model, which is not a trivial task.
2. Typically, deep learning-based neural networks for day ahead wind power forecasting outperform traditional neural networks such as ANN in renewable power

forecasting problems, as these deep learning networks (DLNs) do not need extra data pre-processing, i.e., decomposition, in order to retrieve features from datasets.

Typically, wind power forecasting is subject to a power level output classification problem related to different spatial temporal weather data. Four major types of DLNs for time series data have been applied to wind power forecasting from the time-series sequence data input, namely the recurrent neural network (RNN), long short-term memory (LSTM), gated recurrence unit (GRU), and temporal convolutional network (TCN).

RNNs were the first neural networks to assist in analyzing and learning sequences of data. However, some problems with RNNs were raised during model training, including slow computation times on account of their recurrent nature. Particularly when using Relu or Tanh as the activation function, long sequence inputs (i.e., gradient exploding and vanishing problems) become difficult to process [14]. LSTM was later proposed to solve the gradient exploding and vanishing problems. Typically, LSTM is capable of learning lengthy time dependencies by using the forget, input, and output gates in the module. Similarly, LSTMs have some weaknesses, for example difficulty in applying the dropout algorithm. Dropout is a regularization method in which input and recurrent connections to the LSTM units are probabilistically excluded from activation and weight updates when training a network [15,16]. GRU is a type of RNN that, in certain cases, has advantages over LSTM. GRU uses less memory and is faster than LSTM, although LSTM is more accurate when using datasets with longer sequences [17–19].

In this study, we intend to answer the question of which deep machine learning methods for time series data input can predict day-ahead wind power generation with the smallest error. To investigate the forecast accuracy of day-ahead for wind turbines measured with a performance evaluation index (i.e., MAPE), we developed a feature-based learning model for wind power forecasting and trained TCNs [20–23] to learn meteorological features and identify the output class of power generation. We applied a multilayer neural convolutional architecture with gradient descent algorithms to minimize the estimation model error.

Four major types of sequence-to-sequence DLN models for wind power forecasting were compared to assess model performance. The experimental results demonstrated that the TCN outperforms canonical recurrent networks, LSTMs, RNNs, and GRUs across a diverse range of experiments and datasets. Thus, the TCN provides an effective means of accurately predicting power generation under varying climate conditions.

In summary, the primary contributions of this study are as follows:

- The optimal parameters of the models were investigated using evolutionary algorithms (EAs) in order to minimize convergence loss in the learning process.
- Four crucial architecture parameters for developed wind power prediction models were analysed, incorporating the differential evolution (DE) algorithm [16–18] in the learning process of the TCN model, namely, (i) number of filters, (ii) activation function, (iii) optimizer, and (iv) dilatation coefficient, in order to determine the initial model architecture for model training, according to the natural feature of TCN.
- In our experiment, the prediction error of the TCN model for wind power prediction decreased most steadily among the four models, followed by LSTM, GRU, and RNN.
- With an increasing amount of historical data, the prediction error (MAPE) of the TCN-based model decreased significantly; the 72 h forecast error of the 1-week, 1-month, and 1-year training datasets was 66.43%, 10.93%, and 5.13%, respectively.
- Compared with LSTM, GRU, and RNN models, the TCN model created long effective memory in the deep learning framework and exhibited a lower forecast error to predict 24-, 48-, and 72-h ahead of wind power generation, which is more suitable for sequence modeling based on sequence-to-sequence applications.

The remainder of this article is organized as follows. Section 2 provides a review of other relevant studies in this field, and Section 3 introduces the proposed TCN-based model for wind power forecasting. The results and performance analysis are presented in Section 4. Finally, Section 5 provides the concluding remarks.

2. Literature Review

This section provides an overview of deep neural networks (DNNs) in relation to their processing of time-series data and the application of the differential evolution (DE) algorithm to improve the forecast accuracy of wind power generation.

2.1. DNNs for Processing Time-Series Data

To address the issues involved in wind power forecasting, researchers have developed DNNs, which include the RNN, LSTM, GRU, and TCN and can be applied to address complex nonlinear relations between wind power output and climate data.

RNNs can manage several types of sequence problems, including speech and text recognition, language-to-language translation, handwriting recognition, and sequence data analysis (i.e., forecasting). Generally, RNNs are the best candidate for sequence-to-sequence learning because their internal memory gates obtain outstanding results in natural language processing and other applications. However, RNNs have limited testing with wind time-series data, as well as long memory requirements. LSTMs were later designed to avoid the vanishing gradient that occurs with long sequences. A simplified version of the LSTM, the GRU was applied to resolve simple problems using shorter sequences. In 2016, Lea et al. [20] first proposed temporal convolutional networks (TCNs) for video-based action segmentation. In practice, TCNs have all the advantages of LSTMs as well as extended memory processing input based on dilated convolution architecture and residual connections, with higher classification accuracy than LSTMs.

TCN architecture is based on dilated casual convolutions that enable an exponentially large receptive field. This is more suitable for sequence modeling based on sequence-to-sequence applications that require long effective memory, such as long- or medium-term wind power forecasting [14]. Dilated convolution is a means of increasing the receptive view of the network exponentially, as well as linear parameter accretion [21]. Thus, TCNs are considered a better-adapted architectures thanks to their simplicity, autoregressive prediction, and flexibility for sequence modeling, with a large long memory.

Many researchers have demonstrated that TCNs effectively perform sequence-to-sequence tasks, such as machine translation or speech synthesis in text-to-speech systems. Bai [21] conducted a systematic evaluation of generic convolutional and recurrent networks for sequence modeling and reported that the TCN outperformed canonical recurrent networks across a broad range of standard tasks. Four deep learning network schemes for wind power forecasting are summarised in Table 2.

Table 2. Deep Learning Approaches for Wind Power Forecasting.

	Features	Limitations
RNN [14]	• Used for mapping inputs to outputs of varying types and lengths, and are fairly generalized in their applications such as text translation and voice recognition.	• RNNs have a major setback called the exploding/vanishing gradient, which causes difficulties in learning long-range dependencies. • RNNs become severely difficult to train as the number of parameters becomes extremely large.
LSTM [15,16]	• LSTMs are a novel, efficient, gradient-based method for handling complex, artificial long-time-lag tasks. • Essentially, LSTMs are a special kind of RNN capable of learning long-term dependencies.	• LSTMs require a lot of memories and time in order to be trained for real-world applications. • LSTMs can solve the problem of vanishing gradients; however, they fail to remove it completely.
GRU [17–19]	• GRUs reduce the number of gating units on the LSTM model and optimize the network structure, which is now widely used in industrial practice.	• GRU models have problems with slow convergence rate and low learning efficiency, resulting in too long a training time, and even under-fitting.
TCN [20–23]	• TCNs consist of dilated, causal 1D convolutional layers with the same input and output lengths to create a powerful forecasting model in distinct domains.	• Many studies show that TCNs exhibit better performance than RNNs in domain applications, while avoiding the drawback of the exploding/vanishing gradient problem in RNN models.

TCNs exhibit outstanding behavior with sequences of undetermined length, and the TCN architecture can assist engineers in managing information flows in incredibly long sequences. Consequently, TCNs have stronger learning capabilities and exhibit equal or better results compared to those of the RNN, LSTM, and GRU.

To meet the large memory requirements for DLNs, TCNs use one-dimensional (1D) separable convolutions to factorize a standard convolution into a depth-wise and pointwise convolution. Typically, the TCN consists of three parts: dilated causal convolutions, nonlinear activation, and residual connections. A causal convolutional network is used with 1-dimensional fully convolutional network architecture. A key characteristic is that the output at time t is only convolved with the elements that occurred before t. In 2020, Yan et al. [23] used a TCN for weather state predictions in a comparative experiment conducted with an LSTM. Notably, the results demonstrated that the TCN outperformed other models, including the RNN, LSTN, and GRU, in prediction tasks with time-series data. As shown in Figure 1, the TCN used the 1-dimensional convolutional neural network for short-term wind power prediction, showing that it not only retained the powerful ability of feature learning from both the weather data and electric power output, but was also suitable for processing large volumes of time series data.

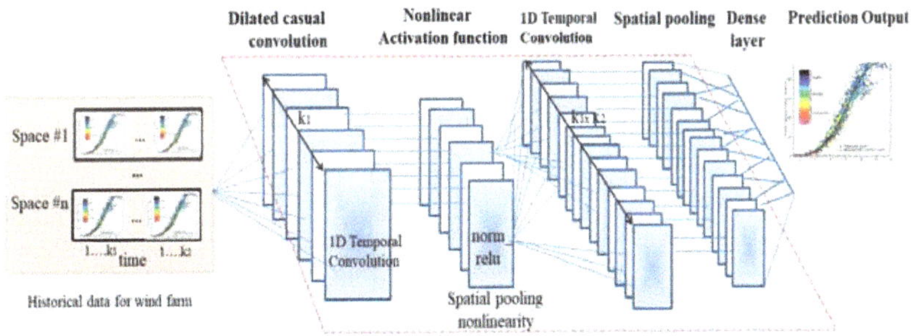

Figure 1. Basic architecture of a temporal convolutional network for wind power prediction.

2.2. Differential Evolution Algorithm

In the design of DNNs for processing time-series data, the optimal parameters of the developed model are identified from training data in order to achieve high predictive precision in the model output.

In supervised machine learning algorithms, in order to minimize the convergence loss of the model in the learning process the optimal parameters of the model can be investigated using evolutionary algorithms (EAs). Practically, the EA algorithm is an effective and efficient approach for solving global numerical optimization problems, avoiding overfitting, and preventing the gradient descent algorithms from converging prematurely on a local suboptimal solution. EAs constitute a smart approach to solving constrained multiobjective optimization problems. In practice, EAs are a family of nature-inspired algorithms widely used for solving complex optimization problems which can be used for assisting developers in determining the optimal parameters of the training model. The differential evolution (DE) algorithm [24–28] is a branch of EA that follows the general procedures of EAs.

In detail, DE is a metaheuristic method that optimizes a problem by iteratively attempting to provide an improved candidate solution with regard to a set measure of quality. DE was introduced by Storn and Price in the 1990s [16], and is applied to solve multiobjective optimization with constraints. Typically, metaheuristic methods can search large spaces for candidate solutions. DE is particularly used for multidimensional real-valued functions; however, it does not use the gradient of the problem being optimized and therefore does not require the optimization problem to be differentiable. Thus, DE can be used on optimization problems that are not continuous, noisy, or changeable over time.

The three basic operators of the DE algorithm are the mutation, crossover, and selection operators. The fundamental idea behind DE is a scheme for producing trial vectors according to the manipulation of target vector and difference vector. If the trail vector yields a lower objective function than a predetermined population member, the newly generated trail vector will replace the vector and be compared in the following generation [28].

After the initialization process, DE forms a loop of the mutation, crossover, and selection processes until the termination condition is satisfied [24]. The processes of these operators are described as follows:

(i) Initialization

Suppose that each individual of the population is denoted as $X_i = [x_{ij}] = (x_{i,1}, \ldots, x_{i,j}, \ldots, x_{i,D})$, where $i = 1, \ldots, N$, N is the number of the solution as well as $j = 1, \ldots, D$, and D represents the number of the dimension. X_i is limited by $X_{min} = (x^G_{min}, \ldots, x^G_{min}, \ldots, x^G_{min})$ and $X_{max} = (x^G_{max}, \ldots, x^G_{max}, \ldots, x^G_{max})$, which is specified by the user. G is the generation number.

An individual of the population can be defined as follows:

$$X_i = (x_{i,1}, \ldots, x_{i,j}, \ldots, x_{i,D}), i = 1, \ldots, N. j = 1, \ldots, D \tag{1}$$

First, the initialization population randomly selects the initial parameter values uniformly based on the intervals $[X_{min}, X_{max}]$. The commonly used initialization method for individuals is

$$X_i^G = rand(0,1) \cdot (X_{max} - X_{\text{min}}) + X_{min} \tag{2}$$

where $rand(0, 1)$ represents the generation of a uniformly distribution random number located in [0, 1].

(ii) Mutation

The *DE* algorithm adopts the mutation strategy, in which a mutant vector is created for each individual V_i^G (also called the trial vector) in each generation G. For a given parameter vector V_i^G, three vectors are selected randomly: X^G_{r1}, X^G_{r2}, and X^G_{r3}, such that the indices i, r_1, r_2, and r_3 are distinct. First, the weighted difference of two of the vectors is added to formV_i^G

$$V_i^G = X^G_{r1} + F \cdot \left(X^G_{r2} - X^G_{r3}\right), \tag{3}$$

where F is the scaling factor that controls the amplification of the differential evolution, i.e., mutation scale; its value is located in [0, 2]. Small values of F will lead to smaller mutation step sizes. Consequently, it will take longer for the algorithm to converge. Conversely, large values of F enable exploration, but can lead to the algorithm overshooting good optima. Thus, the value has to be small enough to enhance local exploration but also large enough to maintain diversity [25]. A well-known DE mutation operation is described as follows [26,27]:

$$V_i^G = V^G_{best} + F \cdot \left(V^G_{r1} - V^G_{r2}\right) \tag{4}$$

$$V_i^G = V^G_{r1} + F \cdot \left(V^G_{r2} - V^G_{r3}\right) + F \cdot \left(V^G_{r4} - V^G_{r5}\right) \tag{5}$$

$$V_i^G = V_i^G + F \cdot \left(V^G_{r1} - V^G_{ri}\right) + F \cdot \left(V^G_{r2} - V^G_{r3}\right), \tag{6}$$

where r_1, r_2, r_3, r_4, and r_5 are the distinct integers randomly generated from the range of [1, N] and are not equal to i, i.e., ($r_1 \neq r_2 \neq r_3 \neq r_4 \neq r_5 \neq i$). V^G_{best} is the best individual with the highest fitness value (objective value) at generation G.

(iii) Crossover

After mutation, a trial vector $X_i^G = (X_{i,1}^G, X_{i,2}^G, \ldots, X_{i,D}^G)$ is generated for each individual according to a binomial crossover operator on X_i^G and V_i^G, as follows:

$$U_{i,j}^G = \begin{cases} V_{i,j}^G & if\ (rand_{i,j}(0,1) \leq CR\ or\ j == j_{rand}) \\ X_{i,j}^G & otherwise \end{cases} \tag{7}$$

In this equation, *rand* is a uniformly distributed random integer in the range of [1, *D*], which is generated for each individual. *CR* is the crossover rate, which is restricted in the range of [0, 1]. *CR* controls the number of elements that will adjust. Larger values of *CR* will lead to deriving more variation in the new population, therefore increasing it also increases exploration [25].

If the *j*-th variable $U_{i,j}^G$ of the trial vector U_i^G violates the boundary constraints, it is reset as follows:

$$U_{i,j}^G = X_j^{min} + rand(0,1) \cdot \left(X_j^{max} - X_j^{min}\right). \tag{8}$$

(iv) Selection

The selection operator determines whether the target or trial vector survives and enters the next generation based on their fitness values. For a minimization problem, the decision vector with the lower fitness value (objective value) can enter the next generation, which can be expressed as follows:

$$X_i^{G+1} = \begin{cases} V_i^G & if\ (fit(U_i^G) \leq fit(X_i^G) \\ X_i^G & otherwise \end{cases} \tag{9}$$

The process is repeated with the expectation, though it is not guaranteed, that a satisfactory solution will eventually be discovered.

3. Wind Power Forecasting Model with Temporal Convolutional Networks

In this section, a TCN-based approach for a long-term wind power forecasting model is presented. A detailed workflow of the TCN model design for 24–72 h wind power forecasting is described herein. The following three subphases comprise the TCN model development process with DE for determining the optimal parameters of the proposed model: (i) architecture design, (ii) determination of the architecture parameters of the model, and (iii) the overall process for model development.

3.1. Architectural Design for TCN

Inspired by [20], we incorporated the convolutional network architecture involved in casual convolution with residual connections to construct a stable TCN-based prediction model for 24–72 h wind power forecasting. Dilated convolution is used to select which values of the neurons from the previous layer contribute to those in the next layer. Thus, the dilated convolution operation captures both local and temporal information.

The dilated convolution function, *F*(*s*), is provided by [21]

$$F(s) = (x *_d f)(s) = \sum_{i=0}^{k-1} f(i) \cdot x_{s-d.i}, \tag{10}$$

where x_s is the current input sequence data at time *t*, *d* is the dilation factor parameter, and *f* is a filter of size *k*.

The TCN model can be defined as follows [23]:

$$x_t^l = \sigma(W_x^l \cdot F_d\left(x_t^{l-1}\right) + b_x^l + x_t^{l-1}), \tag{11}$$

where $F_d(.)$ is the dilated convolution function of *d* factor, x_t^l is the value of the neuron of the *l*-th layer at time *t*, W_x^l and b_x^l are the weights and bias corresponding to the *l*-th layer, and σ is the activation function. The dilated residual block in our project is detailed in Figure 2.

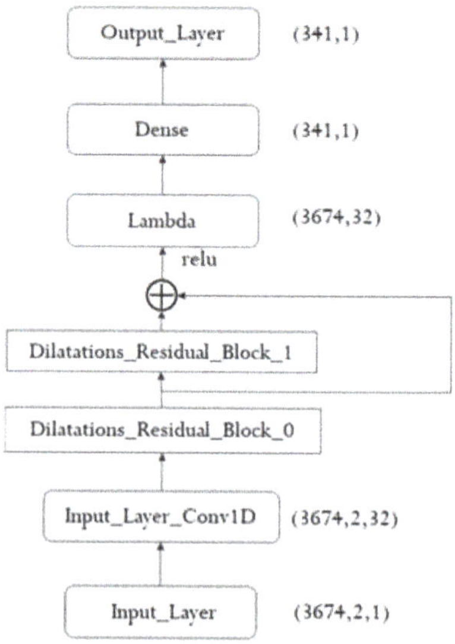

Figure 2. TCN-based architecture for wind power forecasting.

The system must use the residual block to the convolutional layers when deep and large TCNs are employed in order to achieve further stabilization. As presented in Figure 3, the residual connections constituted the addition of the data input to the output before applying the activation function; the residual block (d = 16) is used between each layer in the TCN to accelerate convergence and enable the training of deeper models.

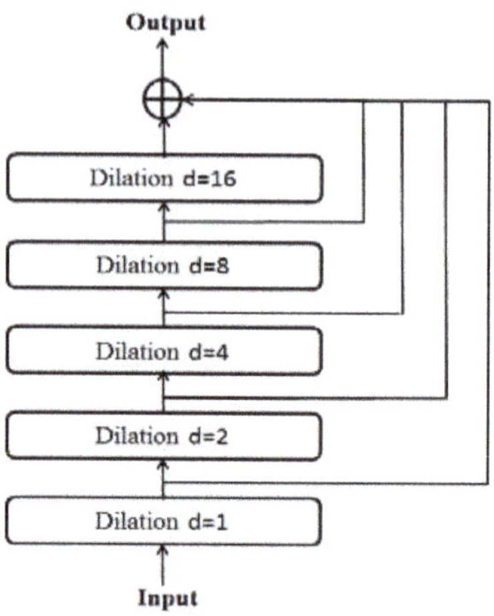

Figure 3. Detailed diagram of the dilated residual block.

3.2. Parameter Selection for TCN Model Using Evolutionary Algorithm

The architecture design for TCN and optimal parameters for the developed wind power prediction model with the DE search mechanism are analyzed in this section.

3.2.1. Preliminary Architecture Design

In this step, four crucial architecture parameters were selected by transferring learning cases in the TCN predictor [20–23]; the original parameters of TCN models developed for wind power prediction models were obtained from P. Rémy at GitHub [29], namely, (i) number of filters, (ii) activation function, (iii) optimizer, and iv) dilatation coefficient, in order to decide the initial model architecture for model training.

(i) Filter size

In practice, the cost function is a measure of the inaccuracy of the model in terms of the difference between predicted values and real measured values. Following the analysis of three filter sizes (8, 16, and 32), as described in Figure 4, the filter size of 32 exhibited the smallest convergence error of the cost function after 100 iterations of simulation and was the optimal choice for the designed TCN-based prediction model.

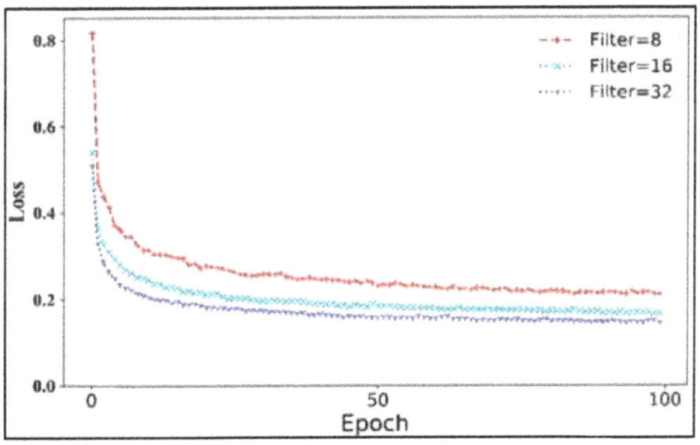

Figure 4. Convergence error of cost function with three different filter sizes.

(ii) Activation function

In artificial neural networks, the activation function of a node defines the output of that node based on the input or set of inputs. Generally, nonlinear activation functions allow such networks to compute complex problems using a few nodes. In the experiment, we analyzed two types of nonlinear activation function, the norm_relu and Tanh*Sigmoid activation function used in WaveNet. The corresponding convergence error of the cost function for these activation functions is presented in Figure 5; norm_relu was selected as the activation function of the model.

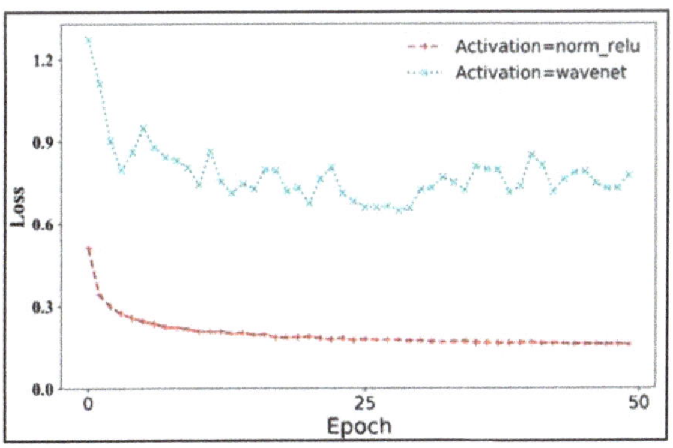

Figure 5. Convergence error of cost function for two nonlinear activation functions.

(iii) Optimizer

To minimize loss during model training, an optimizer was adopted to improve the accuracy of the model through adjustment of the filter weights. We assessed three popular optimizers in the TCN model, Adam, SGD, and RMSprop; the corresponding convergence error of cost function is detailed in Figure 6. The Adam optimizer was chosen as the ideal optimizer for model training.

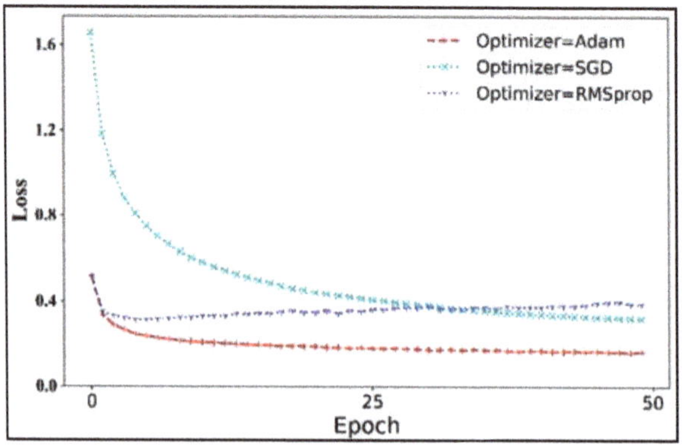

Figure 6. Convergence error of cost function for three types of optimizer.

(iv) Dilatation coefficient

Following the analysis of three combinations of dilatation coefficients (1, 2, 4; 1, 2, 4, 8; 1, 2, 4, 8, 16) in the model input architecture, the third set (1, 2, 4, 8,16) achieved the most optimal results in relation to the error convergence of the cost function, as depicted in Figure 7.

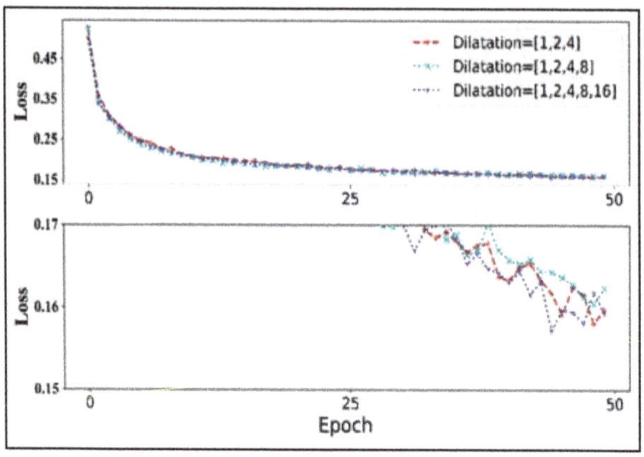

Figure 7. Convergence error of cost function with three sets of dilatation coefficients.

The architectural components of the proposed TCN model for transferring learning from model training are listed in Table 3.

Table 3. Architecture of the proposed TCN model.

	Number of Filter	Kernel_Size	Dilatations	Number of Stack	Optimizer
Parameter Value	32	10	[1, 2, 4, 8, 16]	2	Adam

To confirm the appropriate architecture of the model design, different model architecture parameters were experimentally investigated. In this experiment, 83.3% (500,000 records) of the data from the open dataset of wind farm in Turkey (2018) [30] was randomly selected to serve as the training dataset, and the remaining 17.7% (100,000 records) was used for testing. The average accuracy of the results of experimental training and testing is summarized in Table 4. Model accuracy was above 96.4% based on different parameter combinations, with a low convergence loss result for the cost function. Thus, the architectural parameters of Table 3 for the TCN model were validated.

Table 4. Accuracy analysis of the TCN model architecture.

	Maximum	Minimum	Average
Training accuracy	98.2%	97.1%	97.8%
Test accuracy	96.7%	92.9%	95.1%
Average	97.45%	95.0%	96.4%

3.2.2. Analysis of Architecture Design

Following the architecture analysis step of detailed model design, experiments were conducted with model training in order to validate the optimal parameters of the developed TCN model based on training samples. In the experiment, differential evolution (DE) methods [24–28] explored these solutions to handle the hyper-parameter tuning of the TCN model for predicting wind power output in order to reach the satisfactory prediction accuracy for different weather conditions. Essentially, the DE method is a population-based stochastic search process using the distance and direction information from the current population to conduct its search. Inspired by [25], we selected the DE method to solve our problem because the historical data for wind power generation are generally not continuous, noisy, or changeable over time; thus, the gradient of the problem being optimized is not used.

For all experiments, 50 independent runs were conducted for each test function. The parameter settings for the DE algorithm are listed in Table 5. As shown in Table 5, the following parameters were chosen for the application of DE: population size $NP = 10$; scaling_rate $F = 0.5$, crossover_rate $CR = 0.3$, generation number $G = 30$, and maximum iteration = 500. Optimization was terminated at the pre-specified number of generations.

Table 5. The parameter settings for the architectural analysis.

Algorithm	Parameter Setting
DE	$NP = 10, F = 0.5, CR = 0.3, G = 30$

Then, nine trial vectors V_i^G and difference vectors (Table 6) were generated for each generation G according to the mutation, crossover and selection operations using Equations (1)–(9). If the trail vector yielded a lower error value of objective function than a predetermined population member, the newly generated trail vector replaced the target vector and was compared in the following generation.

Table 6. The parameter settings for the nine trial vectors.

Target Vector	Filter_Size	Dilatation	Number of Stacks	Activation Function	Optimizer
#1	8	[1, 2, 4, 8]	2	norm_relu	RMSporp
#2	32	[1, 2, 4, 8, 16]	2	wavenet	RMSporp
#3	16	[1, 2, 4, 8]	4	norm_relu	Adam
#4	8	[1, 2, 4, 8]	3	norm_relu	SGD
#5	32	[1, 2, 4, 8]	3	wavenet	RMSporp
#6	8	[1, 2, 4, 8]	2	wavenet	SGD
#7	16	[1, 2, 4, 8, 16]	3	norm_relu	SGD
#8	16	[1, 2, 4, 8]	2	wavenet	SGD
#9	32	[1, 2, 4]	4	norm_relu	Adam

For analysis of the optimal parameters of the TCN model, DE allows for the process of mutation, crossover, and selection until the termination condition is satisfied, using Equations (1)–(9); the analysis process is shown in Figure 8.

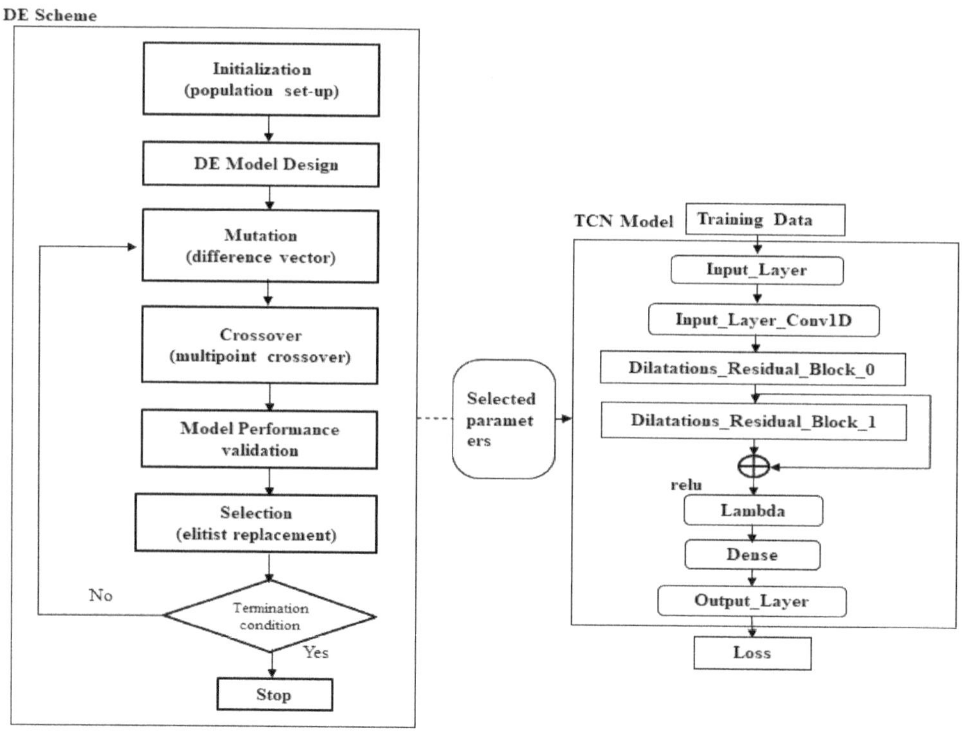

Figure 8. Differential evolution used to decide the optimal parameters of TCN model training.

In our experiment, the error value and error reduction speed of the loss function were adopted to evaluate whether the appropriate model parameters had been selected. If the convergence error decrease was not smooth (i.e., a bouncing phenomenon), the cost function or model parameters required adjustment. The convergence error of cost function in the training process for TCN model is illustrated in Figure 9.

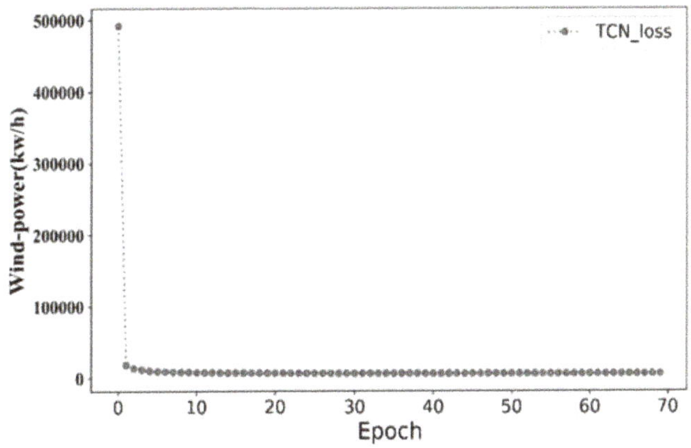

Figure 9. Convergence error training of the TCN model.

As shown in Figure 9, a stable convergence was reached after a stable error descent was exhibited in the experiments. Overall, the experimental results for the selected optimal parameters of the TCN model are listed in Table 7.

Table 7. Selected optimal parameters of the TCN model.

Algorithm	Layers	Total Params/Kernels	AF/LF	Optimizer	Dilations	Number of Stack
TCN model for wind power prediction	TCN	45,761/16	norm_Relu	Adam	[1, 2, 4, 8]	4
	Input Layer	0	-			
	Initial_Conv	(11,16)				
	Dilated ConvLayer	(16,161,16)	norm_Relu/MSE			
	Dropout layer	0				
	Conv Layer	(16,17,16)	norm_Relu/MSE			
	OutputDense Layer	(17,1)	linear			

AF = activation function, LF = loss function, MSE = mean squared error.

3.3. Overall Process of the Model Operations

Four experimental DNN models for wind power forecasting incorporating RNN, LSTM, GRU, and TCN were employed to verify the performance of model training. The execution process of model development is illustrated in Figure 10. The proposed TCN model comprised the following three subphases in the model operation process: (i) data preprocessing, (ii) model training, and (iii) model validation.

Step 1. Data preprocessing

Before performing model pretraining, engineers must perform data processing for wind farm datasets that contain real weather observations and wind turbine power outputs with anomaly data. Each record in the wind power dataset includes items such as wind speed, wind direction, temperature, humidity, height of wind turbine for model pretraining, and some null fields where linear proportions of the neighbouring observation data are noted in advance.

Following the study of patterns in high-dimension data using principal component analysis, the prediction experiment employed two key model parameters, wind speed and wind direction, which were applied to model training.

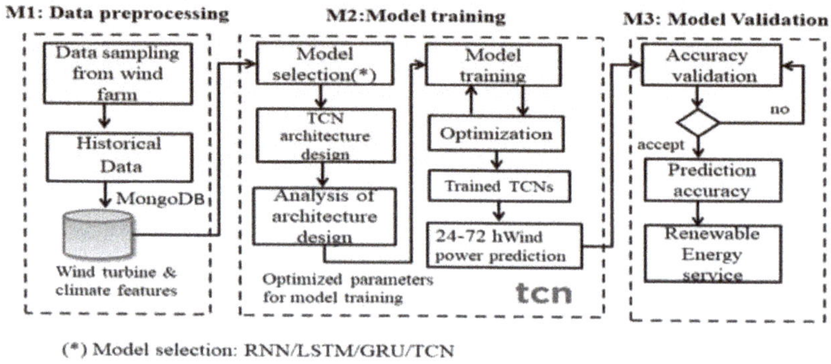

Figure 10. Experiment execution process.

Step 2. Model training

Step 2.1 Model pretraining

In the pretraining phase, the proposed model incorporated the gradient descent optimization algorithm in order to fine-tune the model parameters for transfer learning using error derivatives of back-propagation with the optimizer for all layers. Then, a series of experiments were pretrained to investigate the performance of the TCN-based classifier using the Scada dataset, where the learning results were regarded as a basis of the optimal model parameters, including number of filters, dilatation coefficient, activation function, epochs, and prediction accuracy (Table 2).

Step 2.2 Model fine-tuning

During the fine-tuning of the model, we adopted a cross-validation scheme to evaluate the predicted accuracy of the model and overcome the problem of over-training using various n-folds of the cross-validation scheme. For example, k = 5 indicates that 80% of the dataset collected was used in the training experiment, with the remaining 20% used for alternative testing that was repeated five times. In the model validation phase, the system provided a quick response for wind power forecasting using the weights of the neural nets employed in the trained TCN model learning.

Step 3. Model validation

To test the robustness of the proposed model, the trained TCN model associated with the test dataset was adopted to examine the model performance. Finally, the MAPE was selected to evaluate the power prediction performance of the proposed model as follows:

$$MAPE = \frac{1}{n}\sum_{i=1}^{n}\left|\frac{O_i - \widehat{O_i}}{W_i}\right|, \tag{12}$$

where W_i represents the installed capacity of the wind turbine, O_i is the real value, and $\widehat{O_i}$ is estimated output.

In the experiment, an open historical dataset from the Scada wind farm in Turkey, including real weather observations and wind turbine power outputs, was used in model pretraining. Each record of the Scada dataset can be found in [30].

The detailed algorithm for wind power prediction with the TCN-based model is described by PDL as follows (Algorithm 1).

Algorithm 1 Pseudocode of the TCN-based model for wind power prediction.

Input: 1. Historical weather data and wind turbine power outputs from Scada wind farm in Turkey, containing five parameters sampled every 10 min, with a total of 52,560 samples listed. 2. Model parameters of the proposed TCN model including the batch set, kernel, and epochs.
Output: predicted accuracy of wind power generation
1: Initialize the model parameters of the model
2: Set the value of the epochs to 50
3: Assign the stop condition value (ε) as 0.0001
4: Training loop
5: While (the number of epochs) do
6: Determine the optimal parameters of TCN model, as given in Equations (1)–(9)
7: Perform the wind power prediction, as given in Equations (10) and (11)
8: Return (model_file)
9: The training results of the model_file include: (1) filter size, (2) activation function, (3) optimizer, (4) dilatation coefficient, (5) final loss of the cost function, and (6) output result of the training process ($\widetilde{O}_i$)
10: Return train (output_file)
11: End loop
12: Test phase
13: Accuracy prediction with loss of cost function by using specific parameters from the model
14: return predict (accuracy)
15: End

4. Results

In this section, the performance of the proposed TCN-based model for wind power prediction is demonstrated by means of an example. The experiments were conducted using the Python programming language and TensorFlow, which is an open source software library for numerical computation. Moreover, TensorFlow incorporates numerical libraries such as Pandas, NumPy, and Matplotlib for computation. The parallelisation of the multicore architecture increased the computation speed of the TCN model. The multicore architecture included an AMD Ryzen Threadripper processor (3.4 GHz) with 32 GB RAM, a 64-bit Ubuntu 14.04 operating system, an Nvidia GeForce GTX 1080 graphics card (GPU), graphics core computing, and the MongoDB 2.2.6 database. The experimental environment is described in Table 8.

Table 8. Experimental environment for TCN-based prediction model.

IP	Programming Language	Operating System	Numerical Library
192.168.1.10 (AMD Ryzen Threadripper, 1920X, 3.4G)	Python 3.5	Ubuntu 14.04 LTS 64	Tensorflow-gpu 1.1.3
			Pandas 0.23.4
			Numpy 1.1.8
			Matplotlib 3.3.2

4.1. *Case Study: Performance Analysis for TCN-Based Model (Scada Wind Farm, Turkey)*

Step 1. Data preprocessing phase

In the experiment, an open historical dataset from Scada wind farm in Turkey, including real weather observations and wind turbine power outputs, was used in model pretraining. Each record in the Scada dataset contained five parameters sampled every 10 min, with a total of 52,560 samples listed [31]. Notably, some null fields contained linear proportions of the neighbouring observation data in advance. Our training dataset comprised samples from 1 January 2018, to 26 December 2018, and the test dataset used samples from three days, namely December 27 to 29, 2018.

Scada Systems measured and saved data including wind speed, wind direction, generated power, etc. This file was taken from a Scada Systems wind turbine working and generating power in Turkey. The data in the file are listed as follows: [30]

1. Date/Time: 10 min intervals
2. LV ActivePower (kW): The power generated by the turbine for that moment.
3. Wind Speed (m/s): The wind speed at the hub height of the turbine.
4. Theoretical Power Curve (KWh): The theoretical power values that the turbine generates with that wind speed, which is given by the turbine manufacturer.
5. Wind Direction (°): The wind direction at the hub height of the turbine

Step 2. Model training phase

To examine model efficiency, four deep neural models were incorporated for series data processing, namely, RNN, LSTM, GRU, and TCN, in order to conduct wind power forecasting 72 h ahead of time. We set the initial values of TCN model parameters as in Algorithm 1, and the experiment parameters for the RNN, LSTM and GRU models as in Table 9.

Table 9. Experimental parameters for RNN, LSTM and GRU prediction models.

Parameter / Model	Output Unit	Optimizers	Learning Rate (lr)	Layers
RNN	64	RMSporp	0.002	3
LSTM	64	RMSporp	0.002	2
GRU	32	RMSporp	0.01	2

Step 3. Model validation phase

Two experiments with training datasets of different sizes (i.e., one month and one year) were conducted to verify the effectiveness of the four DNN-based models for wind power forecasting.

In the first experiment, the one-month dataset was used to train four DNNs. The LSTM model had the lowest prediction error (MAPE = 3.8%), followed by the GRU (9.09%) and the TCN (10.93%) models, with the RNN model exhibiting the poorest performance (11.21%), as detailed in Figure 11.

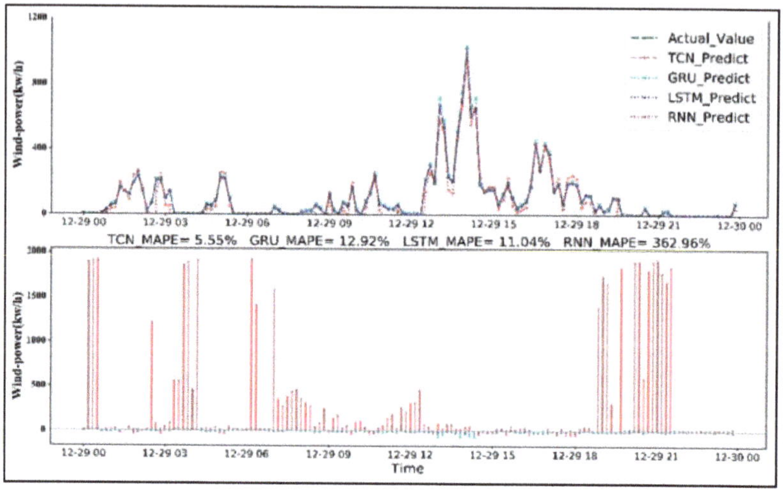

Figure 11. 72-h wind power prediction (Experiment I).

In the second experiment, the one-year historical data were used to pretrain the four DNNs; the prediction results for 24, 48, and 72 h are presented in Figures 12–14. For 72 h ahead of time, the prediction error of the TCN model indicated the highest accuracy (MAPE = 5.13%), followed by the GRU (6.25%), LSTM (9.12%), and RNN (173.87%) models.

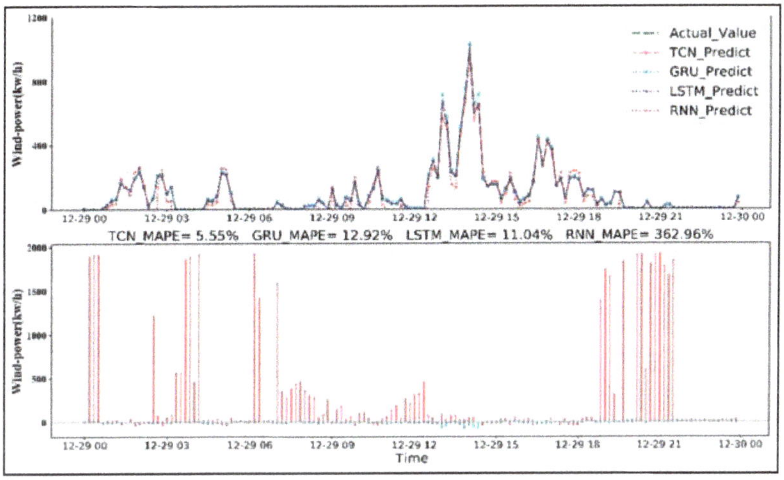

Figure 12. 24-h wind power forecasting (Experiment II).

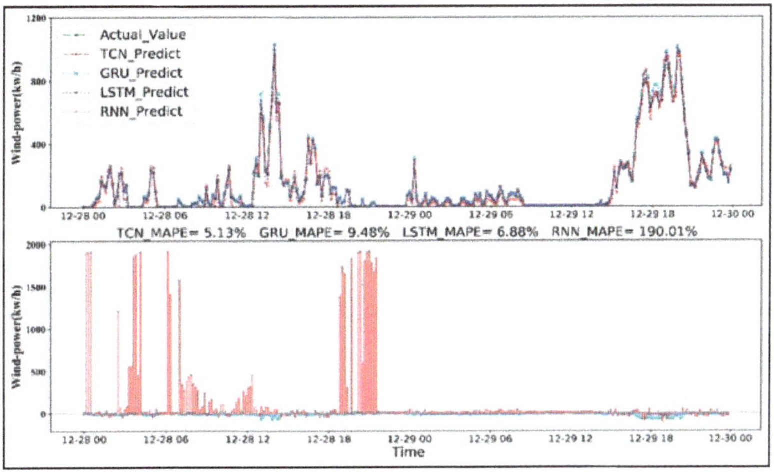

Figure 13. 48-h wind power forecasting (Experiment II).

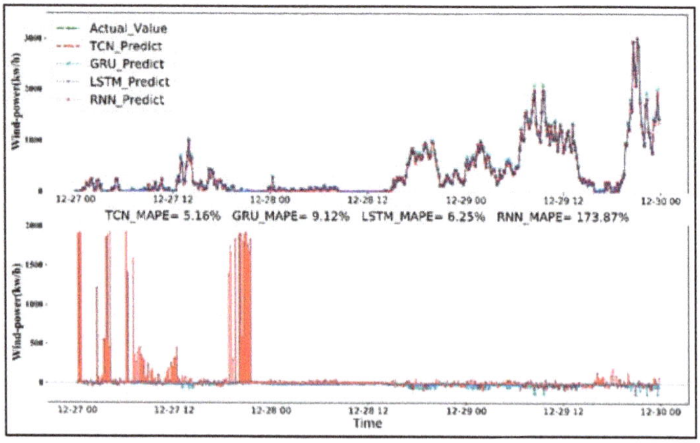

Figure 14. 72-h wind power forecasting (Experiment II).

4.2. Method Comparisons

Theoretically, the convergence error of the cost function decreased in each iteration of model training, as detailed in Figure 15. As presented in Figure 15, the convergence error gradually converged and decreased with the increasing number of iterations (epochs) when three of the prediction models (TCN, LSTM, and GRU) were applied; the RNN model convergence error did not converge, and the prediction error did not decrease. RNNs are thus not suitable for wind power forecasting from large amounts of temporal–spatial data series inputs. The convergence error of the TCN model decreased more than that of the LSTM close to the twentieth epoch, and continued to decrease steadily with the increasing number of iterations. The prediction error of the TCN model decreased most steadily among the four models, followed by LSTM and then GRU.

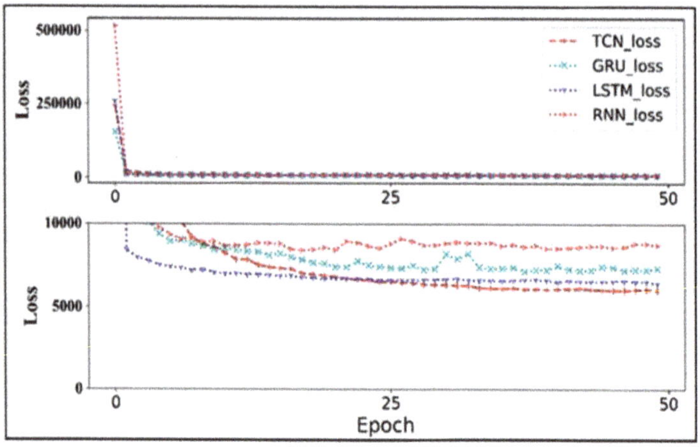

Figure 15. Convergence error of the cost function.

The performance of the four prediction models was affected by the varying amounts of input training data. Therefore, performance analysis must be assessed with different amounts of historical data for long-term prediction. In our experiment, different amounts of historical data were used for model pretraining and the output results of the modules were sorted; the stability comparison of the convergence errors is detailed in Table 10.

With increasing amounts of historical data, the prediction error (MAPE) of the TCN-based model decreased significantly, and the 72-h forecast error of the one-week, one-month, and one-year training datasets was 66.43%, 10.93%, and 5.13%, respectively. In the experiment, the TCN model exhibited consistent, stable and good prediction results by using selected parameters from the DE algorithm.

Table 10. Performance comparison of four DNN models for wind power forecasting.

Performance / Period	Excellent	Good	Medium	Bad
Week	RNN	GRU	LSTM	TCN
Month	LSTM	GRU	TCN	RNN
Year	TCN	LSTM	GRU	RNN

In summary, the wind power forecast error (MRE) of the proposed TCN-based model was near 5.13% based on one-year historical data in different climatic scenarios. Compared to the accuracy of other projects in wind power forecasting, the European team's SafeWind project in 2011 achieved a forecasting error of 17%. In 2017, the predicted error improved to 11%; a project of the BSI electric power company reached within 10% in 2019. As shown in Table 11, the proposed TCN-based approach provides a lower prediction error with higher prediction accuracy than those of real projects in studies of wind power forecasting [12,13,31].

Table 11. Performance comparison of other real projects for wind power forecasting.

Research Team	Prediction Cycle	Prediction Intervals	Prediction Error (MAPE)
ANEMOS, ANEMOS.plus & SafeWind, 2011 [31]	Short-term	within 36 h	17–35%
ANEMOS.plus & SafeWind, 2017 [12]	Short-term	within 48 h	11–14%
UK Power Networks 2016 [13]	Medium-term	within 120 h	10%
Proposed model	Medium-term	within 72 h	Near 5%

5. Conclusions

This study presented a TCN-based model for day-ahead wind power prediction based on a casual convolution architecture with residual connections, in order to learn correlations between meteorological features and wind power generation. The proposed scheme effectively solves the long-distance dependency problem, as demonstrated by the input of large amounts of temporal–spatial series data such as one-year wind power data. The experimental results indicate that TCN models have the capability for feature extraction of long-term sequence data, and exhibit the same or higher prediction accuracy compared to LSTM and GRU models. Overall, the proposed TCN-based approach provides a lower convergence error with higher prediction accuracy than those of other models employed in other studies of wind power forecasting [12,13,31].

Author Contributions: Conceptualization, P.W. and K.-M.C.; methodology, W.-H.L.; resources, P.W.; formal analysis, W.-H.L.; data curation, W.-H.L.; writing—original draft, Z.-Y.Y., Y.-H.L. and W.-H.L.; writing-review and editing, P.W.; software, H.-C.L. and Y.-H.L.; validation, H.-C.L. and W.-H.L.; visualization, Y.-H.L. and Z.-Y.Y.; project administration, P.W.; funding acquisition, P.W. All authors have read and agreed to the published version of the manuscript.

Funding: This research was funded by the Ministry of Science and Technology of Taiwan under Grant Nos. MOST 110-2410–H-168-003, and Taiwan's Ministry of Education (MOE) under Grant No. MOE 2000-109CC5-001.

Institutional Review Board Statement: Not applicable.

Informed Consent Statement: Informed consent was obtained from all subjects involved in the study.

Data Availability Statement: The data presented in this study are available on request from the corresponding author.

Acknowledgments: This work was jointly supported by Taiwan's Ministry of Science and Technology under Grant No. MOST 110-2410–H-168-003 and Taiwan's Ministry of Education (MOE) under Grant No. MOE 2000-109CC5-001.

Conflicts of Interest: The authors declare no conflict of interest

References

1. Alexiadis, M.; Dokopoulos, P.; Sahsamanoglou, H. Wind speed and power forecasting based on spatial correlation models. *IEEE Trans. Energy Convers.* **1999**, *14*, 836–842. [CrossRef]
2. Cadenas, E.; Rivera, W.; Campos-Amezcua, R.; Heard, C. Wind speed prediction using a univariate ARIMA model and a multivariate NARX model. *Energies* **2016**, *9*, 109. [CrossRef]
3. Mohandes, M.; Halawani, T.; Rehman, S.; Hussain, A.A. Support vector machines for wind speed prediction. *Renew. Energy* **2004**, *29*, 939–947. [CrossRef]
4. Moreno, S.; da Silva, R.G.; Mariani, V.C.; Coelho, L. Multi-step-ahead wind speed forecasting based on a hybrid decomposition method and temporal convolutional networks. *Energy* **2021**, *238*, 121981. [CrossRef]
5. Donadio, L.; Fang, J.; Porté-Agel, F. Numerical weather prediction and artificial neural network coupling for wind energy forecast. *Energies* **2021**, *14*, 338. [CrossRef]
6. Hong, Y.Y.; Rioflorido, C.L.P.P. A hybrid deep learning-based neural network for 24-h ahead wind power forecasting. *Appl. Energy* **2019**, *250*, 530–539. [CrossRef]
7. Mana, M.; Astolfi, D.; Castellani, F.; Meißner, C. Day-ahead wind power forecast through high-resolution mesoscale model: Local computational fluid dynamics versus artificial neural network downscaling. *J. Sol. Energy Eng.* **2020**, *142*, 034502. [CrossRef]
8. Emeksiz, C.; Tan, M. Multi-step wind speed forecasting and Hurst analysis using novel hybrid secondary decomposition approach. *Energy* **2022**, *238*, 121764. [CrossRef]
9. Jiang, P.; Liu, Z.; Niu, X.; Zhang, L. A combined forecasting system based on statistical method, artificial neural networks, and deep learning methods for short-term wind speed forecasting. *Energy* **2021**, *217*, 119361. [CrossRef]
10. Neshat, M.; Nezhad, M.M.; Abbasnejad, E.; Mirjalili, S. A deep learning-based evolutionary model for short-term wind speed forecasting: A case study of the Lillgrund offshore wind farm. *Energy Convers. Manag.* **2021**, *236*, 114002. [CrossRef]
11. Sareta, K. Short-term wind speed forecasting system using deep learning for wind turbine applications. *Int. J. Electr. Comput. Eng.* **2020**, *10*, 5779–5784.
12. Kariniotakis, G. *Renewable Energy Forecasting, from Models to Applications*; Cambridge Elsevier Science & Technology: Cambridge, UK, 2017.
13. UK Power Networks. KASM SDRC 9.3: Installation of Forecasting Modules, November 2016. Available online: https://innovation.ukpowernetworks.co.uk/wp-content/uploads/2019/05/Installation-of-forecasting-modules.pdf (accessed on 8 February 2021).
14. Gupta, D. Fundamentals of Deep Learning—Introduction to Recurrent Neural Networks. 2019. Available online: https://www.analyticsvidhya.com/blog/2017/12/introduction-to-recurrent-neural-networks/ (accessed on 20 March 2021).
15. Hochreiter, S. Long-short term memory. *Neural Comput.* **1997**, *9*, 1735–1780. [CrossRef] [PubMed]
16. Kandpal, A. Generating Text Using an LSTM Network, Codeburst Web Site. 2018. Available online: https://machinelearningmastery.com/text-generation-lstm-recurrent-neural-networks-python-keras/ (accessed on 8 February 2021).
17. Firat, O.; Oztekin, I. Learning Deep Temporal Representations for Brain Decoding. In Proceedings of the First International Workshop, MLMMI 2015, Lille, France, 11 July 2015; pp. 25–34.
18. Ding, L.; Xu, C. TricorNet: A. Hybrid Temporal Convolutional and Recurrent Network for Video Action Segmentation. *arXiv* **2017**, arXiv:1705.07818.
19. Wang, X.; Xu, J.; Shi, W.; Liu, J. OGRU: An Optimized Gated Recurrent Unit Neural Network. *J. Phys. Conf. Ser.* **2019**, *1325*, 012089. [CrossRef]
20. Lea, C.; Flynn, M.D.; Vidal, R.; Reiter, A.; Hager, G.D. Temporal Convolutional Networks for Action Segmentation and Detection. In Proceedings of the 2017 IEEE Conference on Computer Vision and Pattern Recognition (CVPR), Honolulu, HI, USA, 21–26 July 2017; pp. 1003–1012.
21. Bai, S.; Kolter, J.Z.; Koltun, V. An Empirical Evaluation of Generic Convolutional and Recurrent Networks for Sequence Modeling. *arXiv* **2018**, arXiv:1803.01271.

22. Torres, J.F.; Hadjout, D.; Sebaa, A.; Martínez-Álvare, F.Z.; Troncoso, A. Deep Learning for Time Series Forecasting: A Survey. *Big Data* **2021**, *9*, 3–21. [CrossRef] [PubMed]
23. Yan, J.; Mu, L.; Wang, L.; Ranjan, R. Temporal Convolutional Networks for the Advance Prediction of ENSO. *Sci. Rep.* **2020**, *10*, 8055. [CrossRef]
24. Storn, R.; Price, K.V. Differential Evolution: A Simple and Efficient Heuristic for Global Optimization over Continuous Space. *J. Glob. Optim.* **1997**, *11*, 341–359. [CrossRef]
25. Georgioudakis, M.; Plevris, V. A Comparative Study of Differential Evolution Variants in Constrained Structural Optimization. *Front. Built Environ. Comput. Methods Struct. Eng.* **2020**, *6*, 1–14. [CrossRef]
26. Zhang, J.; Sanderson, A.C. JADE: Adaptive Differential Evolution with Optional External Archive. *IEEE Trans. Evol. Comput.* **2009**, *13*, 945–958. [CrossRef]
27. Yu, W.J.; Shen, M.; Chen, W.N.; Zhan, Z.H.; Gong, Y.J.; Lin, Y. Differential Evolution with Two-level Parameter Adaption. *IEEE Trans. Cybern.* **2014**, *44*, 1080–1099. [CrossRef] [PubMed]
28. Li, X.; Yin, M. Application of Differential Evolution Algorithm on Self-Potential Data. *PLoS ONE* **2012**, *7*, e51199. [CrossRef] [PubMed]
29. Rémy, P. Keras-tcn, GitHub. Available online: https://github.com/philipperemy/keras-tcn (accessed on 15 February 2021).
30. Erisen, B. Wind Turbine Scada Dataset: 2018 Scada Data of a Wind Turbine in Turkey. Available online: https://www.kaggle.com/berkerisen/wind-turbine-scada-dataset (accessed on 7 January 2021).
31. Giebel, G.; Brownsword, R.; Kariniotakis, G.; Denhard, M.; Draxl, C. *The State-of-the-Art in Short-Term Prediction of Wind Power: A Literature Overview*, 2nd ed. 2011. Available online: http://ecolo.org/documents/documents_in_english/wind-predict-ANEMOS.pdf (accessed on 10 February 2021).

applied sciences

MDPI

Article

Relational Graph Convolutional Network for Text-Mining-Based Accident Causal Classification

Zaili Chen [1,2,†], Kai Huang [2,3,†], Li Wu [1,*], Zhenyu Zhong [1] and Zeyu Jiao [2,*]

1 Faculty of Engineering, China University of Geosciences, Wuhan 430074, China; zl.chen@cug.edu.cn (Z.C.); zy.zhong@giim.ac.cn (Z.Z.)
2 Guangdong Key Laboratory of Modern Control Technology, Institute of Intelligent Manufacturing, Guangdong Academy of Sciences, Guangzhou 510070, China; kirehuang@gmail.com
3 School of Economics and Management, Beihang University, Beijing 100191, China
* Correspondence: lwu@cug.edu.cn (L.W.); zy.jiao@giim.ac.cn (Z.J.)
† These authors contributed equally to this work.

Abstract: Accident investigation reports are text documents that systematically review and analyze the cause and process of accidents after accidents have occurred and have been widely used in the fields such as transportation, construction and aerospace. With the aid of accident investigation reports, the cause of the accident can be clearly identified, which provides an important basis for accident prevention and reliability assessment. However, since accident record reports are mostly composed of unstructured data such as text, the analysis of accident causes inevitably relies on a lot of expert experience and statistical analyses also require a lot of manual classification. Although, in recent years, with the development of natural language processing technology, there have been many efforts to automatically analyze and classify text. However, the existing methods either rely on large corpus and data preprocessing methods, which are cumbersome, or extract text information based on bidirectional encoder representation from transformers (BERT), but the computational cost is extremely high. These shortcomings make it still a great challenge to automatically analyze accident investigation reports and extract the information therein. To address the aforementioned problems, this study proposes a text-mining-based accident causal classification method based on a relational graph convolutional network (R-GCN) and pre-trained BERT. On the one hand, the proposed method avoids preprocessing such as stop word removal and word segmentation, which not only preserves the information of accident investigation reports to the greatest extent, but also avoids tedious operations. On the other hand, with the help of R-GCN to process the semantic features obtained by BERT representation, the dependence of BERT retraining on computing resources can be avoided.

Keywords: accident causal classification; accident investigation reports; text mining; R-GCN; BERT

Citation: Chen, Z.; Huang, K.; Wu, L.; Zhong, Z.; Jiao, Z. Relational Graph Convolutional Network for Text-Mining-Based Accident Causal Classification. *Appl. Sci.* **2022**, *12*, 2482. https://doi.org/10.3390/app12052482

Academic Editors: Nikos D. Lagaros and Vagelis Plevris

Received: 30 January 2022
Accepted: 24 February 2022
Published: 27 February 2022

Publisher's Note: MDPI stays neutral with regard to jurisdictional claims in published maps and institutional affiliations.

1. Introduction

Accident investigation reports are usually text documents formed by professional investigators or teams through visits, conversations, viewing video surveillance and analyzing recorded data after accidents occur [1] and have been widely used in aviation, construction, transportation and other fields [2]. The process and consequences of the accident recorded in the reports can be leveraged by experts to analyze the cause of the accident, which is of great significance for preventing the recurrence of the accident or forming the accident response plan [3]. However, the current analysis of accident investigation reports mainly relies on expert experience to manually determine the cause of the accident, which requires a lot of work, and the accuracy is affected by the subjective experience of experts [4]. On 29 October 2018, an Indonesian Lion Air Boeing 737 MAX8 plane carrying 189 passengers and crew was flying from Jakarta's Soekarno Hatta International Airport to Penang Port, Bangka Belitung Province. The plane lost contact 13 min after takeoff and was later confirmed to have crashed in the waters off Karawang, West Java province [5].

Although experts have been investigating the cause of the accident as soon as possible after the accident, unfortunately, on 10 March 2019, another Ethiopian Boeing 737 MAX8 with 157 passengers and crew on board suffered the same accident [6]. If the causes of some accidents can be identified as early as possible, for example, the cause of the accident can be preliminary determined based on the records of the accident and it is possible to take appropriate measures in advance to avoid the occurrence of the accident [7].

Although the possible causes of the accident are hidden in the accident investigation report, analyzing the possible accident causes from textual records is extremely challenging and usually requires an analysis performed by an expert team composed of scholars, engineers, designers, etc., which, to a certain extent, leads to the long process of accident cause analysis [8]. Therefore, a naive idea is to build an expert system to automatically analyze textual records in accident investigation reports, which is essentially a text classification problem, that is, by constructing suitable models to mine the information in the text and classify the text into different categories [9]. Text mining is the process of extracting effective, novel, useful, understandable, valuable knowledge scattered in text documents and using this knowledge to better organize information [10].

The rapid development of artificial intelligence technology [11–13], especially natural language processing (NLP) and text mining technology, makes it possible to analyze accident investigation reports on a large scale and automatically [14]. With the help of these emerging technologies, time consumption and human error in determining the accident causes would be minimized [15] and the efficiency of analyzing would be significantly improved. A great deal of work has been conducted in existing studies to apply different models to the accident causal classification. According to the different ways of constructing models, existing research can be divided into the methods based on statistics and machine learning [16–19] and the methods based on deep neural networks [20–23]. The methods based on statistics and machine learning are mainly utilized to manually determine a series of text features, such as the term frequency, keyword search, N-grams [24], etc. These methods transform the original unstructured text data into structured feature vectors by artificially determining some features that can represent the key information of the document and, at the same time, create new features based on the existing data. The authors of [25] adopted a variety of machine learning and text mining methods, such as support vector machine (SVM) and Naive Bayes (NB). By combining them into a more powerful learning algorithm through ensemble learning methods, results showing an accuracy of 1.0, a recall rate of 0.96 and a F1-score of 0.96 were obtained. Zhang et al. [26] utilized five baseline models to classify the cause of the accident, including SVM, linear regression (LR), K-nearest neighbor (KNN), decision tree (DT) and NB, and the weight of each classifier in the integrated model was optimized by the sequential quadratic programming (SQP) algorithm. In general, the classification results of simple statistical and machine learning methods, such as keyword search or SVM, largely depend on the quality of feature selection and have a high misidentification rate in the analysis of accident causes [27].

The methods based on deep neural networks usually map the terms in the text to the word vector space, process the word vector and classify it with the help of the structure of the neural network, which has gradually become the mainstream of sequential data processing in recent years. Zhang et al. [20] exploited Word2Vec to skip the gram model to learn the word embedding from the corpus of a specific domain and embedded the learned words into the mixed structured deep neural network for accident report classification. Zhong et al. [22] proposed a latent Dirichlet assignment (LDA) algorithm model to identify risk topics and utilized convolutional neural networks (CNNs) to automatically classify hazards. Meanwhile, a word co-occurrence network (WCN) was generated to determine the relationship between hazards and word cloud (WC) technology was used for the quantitative analysis of keywords to provide a visual overview of hazard accident records. Heidarysafa et al. [28] employed deep learning methods and powerful word embedding (such as Word2Vec and GloVe) to classify accident cause values in the main cause field using text in the narrative. The results show that these methods not only can accurately classify the causes of accidents according to the report description, but can also find the important inconsistencies in the accident report. A deep neural network is essentially a polynomial regression model, which is better characterized by the stacking of multi-layer neural units than a shallow classifier such as SVM [29]. This superb characteristic also enables the model to have the ability of processing text data and implement accident causal classification.

However, the existing methods either rely on large corpus and data preprocessing methods, which are cumbersome, or extract text information based on bidirectional encoder representation from transformers (BERT) [30], but the computational cost is extremely high. These shortcomings make it still a great challenge to automatically analyze accident investigation reports and extract the information therein. To address the aforementioned problems, this study proposes a text-mining-based accident causal classification method based on a relational graph convolutional network (R-GCN) and pre-trained BERT. On the one hand, the proposed method avoids preprocessing such as stop word removal and tokenization, which not only preserves the information of accident investigation reports to the greatest extent, but also avoids tedious operations. On the other hand, with the help of a R-GCN to process the semantic features obtained by BERT representation, the dependence of BERT retraining on computing resources can be avoided. The main contributions can be summarized as follows:

- A text-mining-based accident causal classification method based on a R-GCN and pre-trained BERT is proposed.
- The pre-trained BERT was adopted to avoid preprocessing in traditional text mining and ensure efficient text feature extraction.
- The R-GCN was utilized to avoid the expensive retraining of BERT and enable classification of accident investigation reports.
- To eliminate prediction errors that may be caused by domain GAP when embedding text features based on BERT, a gate mechanism was introduced into the R-GCN architecture.
- The proposed method gets rid of preprocessing such as tokenization and stop word removal and can quickly classify accident causes without relying on expert experience.

2. Methodology

2.1. Overall Scheme of the Proposed Method

The overall scheme of the proposed text-mining-based accident causal classification method is shown in Figure 1, which mainly includes two stages, the text feature extraction stage and the text classification stage. The text feature extraction stage is mainly based on the pre-trained BERT to map the text into a high-dimensional space to obtain a series of embedded text features. In the text classification stage, on the basis of the extracted text features, the R-GCN is utilized to obtain the corresponding category of accident causes.

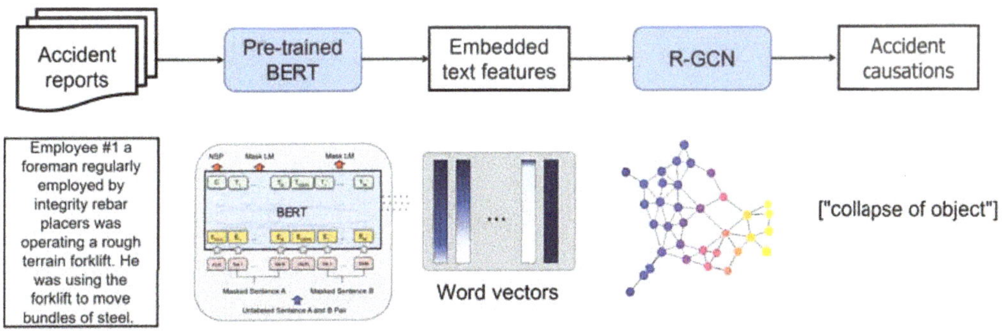

Figure 1. Overall scheme of the proposed method.

2.2. Pre-Trained BERT

Text data usually contain a lot of symbols and numbers to make it easier for readers to understand the meaning of the text, but it is difficult for computers to process and understand them [31]. In traditional methods [32–34], data cleaning is usually performed through a series of preprocessing methods, such as text cleaning, stop word removal, tokenization, data division and word embedding, in order to extract key information in the text. These preprocessing methods not only rely on pre-built corpora, but also lead to the loss of contextual semantic information in the original sentence during the preprocessing. With the excellent performance of the transformer model [35] in NLP, text information is mapped into a high-dimensional space to achieve the quantitative representation of text features. On this basis, BERT is proposed as a pre-trained language representation model. It emphasizes that the traditional one-way language model or the method of shallow splicing of two one-way language models for pre-training is no longer used as before, but a new masked language model (MLM) is exploited to generate deep bidirectional linguistic representation, as shown in Figure 2. BERT aims to pre-train deep bidirectional representations by jointly conditioning the context in all layers. Therefore, the pre-trained BERT representation can be fine-tuned with an additional output layer, suitable for the construction of state-of-the-art models for a wide range of tasks, such as question answering and language inference, without requiring significant architectural modifications for specific tasks.

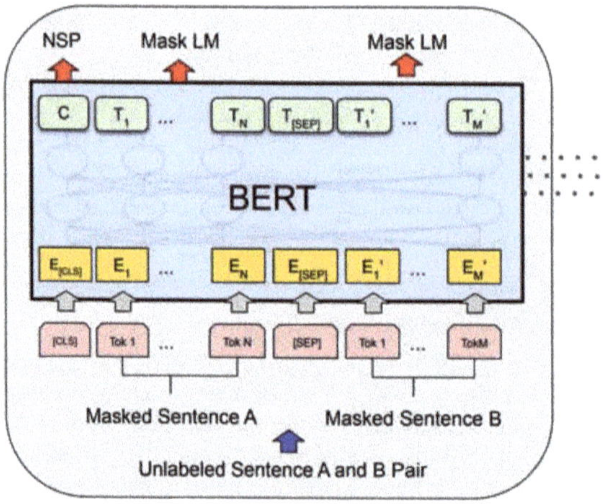

Figure 2. Architecture of BERT in [30].

Although BERT can adaptively learn word-to-word association information in texts in an unsupervised manner, retraining BERT on new datasets is expensive and computationally intensive [36]. While considering the number of accident investigation reports, it is unrealistic to repeatedly retrain BERT, but a pre-trained BERT on large datasets can cover common accident investigation report texts. All that remains to be conducted is to use an appropriate method to mine the text features output by BERT and obtain the accident cause category from the accident investigation reports.

2.3. R-GCN

A graph convolutional network (GCN) [37,38] is a topological network model based on graph theory, which was originally proposed to deal with non-Euclidean data. On the basis of a graph neural network (GNN), the convolution operation in GCNs is performed to realize the differentiable information transfer process of adjacent graph nodes. The transmitted information is usually the hidden state of the node itself, which is essentially high-dimensional feature vectors. GCNs naturally have the advantage of processing text data [39]. Every word, symbol and datum in the text can be regarded as a node of the network. Based on the word co-occurrence relationship and the relationship between document words, a text graph can be built for a specific corpus and then a text graph convolutional network (text-GCN) model can be built. Let us suppose that a directed graph $\mathcal{G} = (\mathcal{V}, \mathcal{E})$ has nodes $v_i \in \mathcal{V}$ and edges $(v_i, v_j) \in \mathcal{E}$. According to the definition by Kipf et al. [37], each node v_i contains a self-loop edge, namely, $(v_i, v_i) \in \mathcal{E}$. Let $X \in R^{n\times m}$ be a matrix containing the eigenvectors of n nodes, where m is the dimension of the eigenvectors and each row of $x_v \in R^m$ is the eigenvector of node v. Let A be the adjacency matrix of graph $\mathcal{G}$ and D be the degree matrix of $\mathcal{G}$, where $D_{ii} = \sum_j A_{ij}$. The diagonal element of A is 1 due to the presence of self-loops. One convolutional layer of the GCN can only capture near-domain information. When multiple GCN layers are stacked, larger domain information is aggregated. For a single-layer GCN, the k-dimensional node feature matrix $L^{(1)} \in R^{n\times k}$ is calculated as follows:

$$L^{(1)} = \rho(\tilde{A}XW_0), \tag{1}$$

where $\tilde{A} = D^{-\frac{1}{2}}AD^{-\frac{1}{2}}$ is the normalized symmetric adjacency matrix and $W_0 \in R^{m\times k}$ is the weight matrix. As mentioned earlier, higher-order neighborhood information can be incorporated by stacking multiple GCN layers.

$$L^{j+1} = \rho\left(\tilde{A}L^jW_j\right), \tag{2}$$

where j represents the number of layers and $L^0 = X$.

Therefore, the forward propagation process in the R-GCN can be defined as

$$h_i^{(l+1)} = \mathrm{ReLU}\left(\sum_{u\in\mathcal{N}(v_i)} \frac{1}{c_i} W^{(l)} h_u^{(l)}\right), \tag{3}$$

where $\mathcal{N}_r(v_i)$ represents the set of neighbor nodes whose relationship is r for node i, l denotes the layer number and c_i is a normalization constant. It should also be noted that the bias term is ignored in the formula and the bias is added to the calculation to promote the convergence of the model during training.

By constructing a large heterogeneous text graph containing word nodes and document nodes, global word co-occurrences can be explicitly modeled and graph convolutions can be easily applied. The number of text graph nodes $|v|$ is equal to the number of documents (corpus size) plus the number of distinct words in the corpus (vocabulary size). The text-GCN simply lets the feature matrix $X = I$ be the identity matrix, meaning that each word or document is represented as a one-hot vector as input to the text-GCN. Edges are established between nodes based on word occurrences in the document (document node–word node edges) and word co-occurrences in the entire corpus (word node–word node edges). The weight of an edge between a document node and a word node is the term frequency–inverse document frequency (TF–IDF) of that word in the document. However, due to the prior information of the syntactic structure between sentences, the traditional GCN can only represent text as an isomorphic graph and the relationship between different words may be different, which also means that the topological structure of the text is essentially heterogeneous graph [40]. Schlichtkrull et al. [41] proposed an R-GCN structure to address this heterogeneous graph problem, where different edges have different definitions of relations. Based on the above method, the update method of node vi in the graph is as follows:

$$h_i^{(l+1)} = \mathrm{ReLU}\left(\sum_{r\in R}\sum_{u\in\mathcal{N}_r(v_i)} \frac{1}{c_{i,r}} W_r^{(l)} h_u^{(l)}\right), \tag{4}$$

where $c_{i,r}$ is a regularization constant, where the value of $c_{i,r}$ is $|N_i^r|$; $W_r^{(l)}$ is a linear transformation function, which transforms the neighbor nodes of the same type of edge using a parameter matrix $W_r^{(l)}$. Following the definition of text syntactic structure relationship in Reference [42], the relationship between texts in accident investigation reports can be divided into three types, including related, irrelevant and self-loop. Figure 3 gives an example of an analytic syntactic structure. When constructing a syntactic graph, information is also allowed to flow in the opposite direction of the syntactic dependency arc, i.e., from the dependency arc to the head.

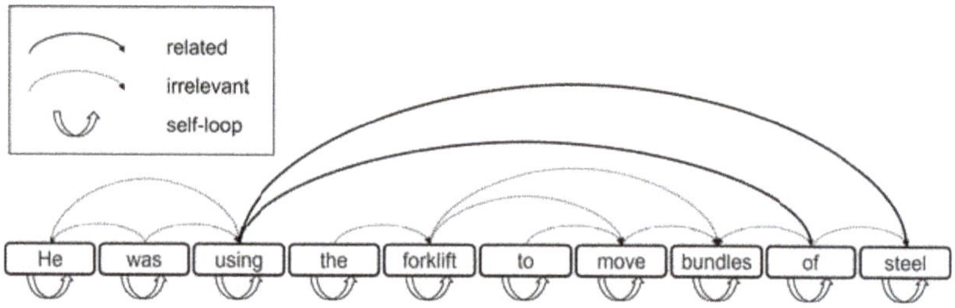

Figure 3. Example of an analytic syntactic structure. It should be noted that the syntactic structure in the figure is only an illustration and the relationships between words are not all listed. Moreover, in practical use, the input of the R-GCN is not the original text itself, but the text features embedded in the text after pre-training BERT.

Considering that the predicted grammatical information may be wrong due to the domain gap when embedding text features based on the pre-trained BERT, some mechanisms are needed to reduce the influence of false dependent edges. To this end, the gate mechanism [43,44] was introduced into the R-GCN architecture. The gate mechanism dynamically assigns a weight between 0 and 1 to the dissemination of information from different nodes. By multiplying this weight into the forward pass, the impact of incorrectly embedded features on the final result is reduced. The weight of the gate mechanism can be calculated as follows:

$$g_{u,v}^{(l)} = \text{Sigmoid}\left(h_u^{(l)} \cdot W_{r,g}\right). \tag{5}$$

Updating these weights by backpropagation, the R-GCN with a gate mechanism can be computed by

$$h_i^{(l+1)} = \text{ReLU}\left(\sum_{r \in R} \sum_{u \in \mathcal{N}_r(v_i)} g_{u,v_i}^{(l)} \frac{1}{c_{i,r}} W_r^{(l)} h_u^{(l)}\right). \tag{6}$$

2.4. Pre-Trained BERT Combined with R-GCN

The pre-trained BERT and the R-GCN with the gate mechanism were introduced in the previous article. How to combine the two has become the only unsolved problem. Marcheggiani and Titov [45] utilized a GCN to integrate syntactic information into sequential neural networks and transformed the syntactic prior into a syntactic dependency graph, which was digested using the GCN. This architecture combines syntactic structure with BERT embeddings for text classification tasks. Following this idea, by concatenating the text features and syntactic structure information of the pre-trained BERT embedding, it can be regarded as containing all the information of the text. As shown in Figure 4, on the basis of the original BERT structure, by placing the R-GCN with gate mechanism in parallel, the embedded text features and syntactic structures can be extracted simultaneously, concatenating these two together to form features that can be used for text classification.

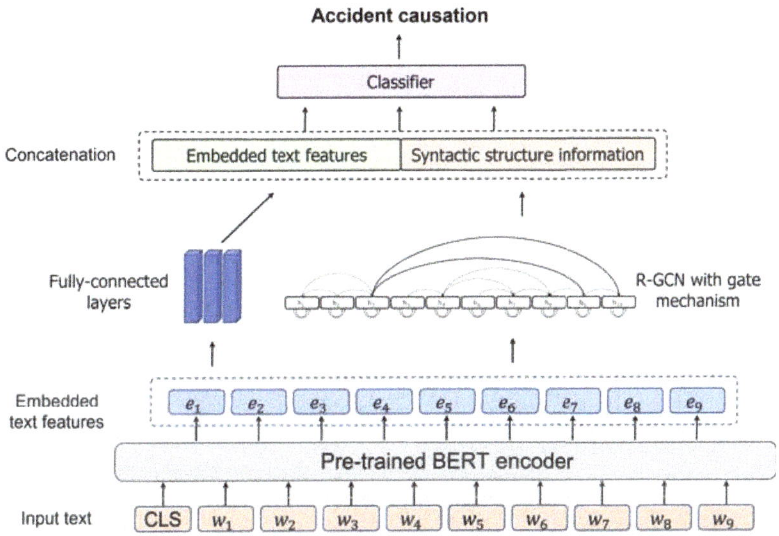

Figure 4. Structure of pre-trained BERT combined with R-GCN.

The text input to BERT adds a classification mark, such as 'CLS' in Figure 4, before the first sentence, so that the corresponding vector of this bit in the last layer of BERT can be used as the semantic representation of the whole sentence, which can be used for downstream classification tasks. Compared with other words in the text, this symbol without obvious semantic information more "fairly" fuses the semantic information of each word in the text, so as to better represent the semantics of the whole sentence. In addition, the text feature vectors embedded by the pre-trained BERT are coupled through a fully connected network, so that the vectors of all th words are weighted and fused to obtain features that can represent text word information. The reason for concatenating the output of the R-GCN with the embedded text features is that the graph convolution of the GCN model is actually a special form of Laplacian smoothing [46], which may mix features of vertices and make them indistinguishable.

3. Experimental Details

This section gives details of the experiment in this study, including the datasets and pre-trained models used, training settings, evaluation metrics and experimental platforms.

3.1. Dataset and Pre-Trained Model

The primary accident investigation report data used in this study were construction site accident data collected from the Occupational Safety and Health Administration (OSHA) open source database [47]. It contains the textual records of 16,323 construction site accidents that occurred from 1983 to 2016. However, the document only provides a detailed description of the event, including the causal factors and events that led to the incident. Therefore, this study adopts the labeled dataset provided by Goh and Ubeynarayana [9]. Goh and Ubeynarayana manually annotated parts of the original OSHA dataset. A new construction site accident was created with 1000 accident causal categories annotated and a total of 11 construction accident causes were derived. This dataset has been widely used in accident cause analysis [48]. The 11 accident causes were assigned different indexes and the number of various accident causes in the 1000 data were also counted; they are shown in Table 1.

Table 1. Labeled cause distribution of dataset provided by Goh and Ubeynarayana [9].

Index	Cause	Labeled Number
1	Traffic	63
2	Collapse of object	212
3	Falls	236
4	Caught in/between objects	68
5	Struck by moving objects	134
6	Others	43
7	Exposure to chemical substance	29
8	Fires and explosion	47
9	Electrocution	108
10	Struck by falling object	43
11	Exposure to extreme temperatures	17

When building the pre-trained BERT model, a large-scale general language understanding evaluation (GLUE) benchmark [49] is adopted, which is an ensemble of multiple natural language understanding tasks. Based on the work by Devlin et al. [30], a pre-trained BERT model was directly used for the text representation of accident investigation reports in this study.

3.2. Training Settings

During the training, Adam [50] was exploited as the optimizer with a $\beta_1 = 0.9$ and $\beta_2 = 0.999$ and we applied a high weight decay of 0.1. The initial learning rate was set to 10^{-4} and the batch size for the training was set to 512. It should be noted that the BERT used in the model was not fine-tuned and retrained, but directly adopted with its network parameters fixed—the 'bert-large-uncased' version [30] of BERT to generate raw embedded text features. Besides, batch normalization and drop out were also leveraged in all fully connected layers. Following the setting of [42], a layer of the R-GCN with a gate mechanism was utilized to capture immediate syntactic neighbor information. In addition, given that the data were still very limited, five-fold cross-validation was utilized to achieve better generalization performance and more accurate model performance estimates.

3.3. Evaluation Metrics

In the training stage of the model, the performance evaluation criteria used were the precision of accident report classification, recall rate, F1-score and average weighted F1-score. Precision is the ultimate criterium of the predicted result. It can be calculated by using Equation (7) and is obtained by dividing the true result by the sum of the true and false positive values.

$$\text{Precision} = \frac{TP}{TP + FP} \tag{7}$$

Recall is a measure of how well each unique label fits into the predicted results. It can be seen, from Equation (8), that the recall rate is the sum of the real result divided by the real value and the false negative value.

$$\text{Recall} = \frac{TP}{TP + FN} \tag{8}$$

The F1-score is the harmonic mean of precision and recall rate, in which the F1-score reaches the best value when 1 and the worst value when 0. Formula for obtaining F1-score is shown in Equation (9).

$$\text{F1-score} = \frac{2(\text{Precision} * \text{Recall})}{(\text{Precision} + \text{Recall})} \tag{9}$$

When unbalanced classes appear in the dataset, an average weighted F1-score is required. Count the number of cases and the total number of case classes that involve support for a particular tag. The average weighted F1-score can be computed by

$$\text{avg F1}_{\text{weighted}} = \sum_{i=1}^{N}\left(\frac{S_i}{T} * \text{F1}_i\right), \tag{10}$$

where S_i is the number of cases supported by label i and T is the total number of the dataset.

When constructing the model, in order to balance the precision and recall of the model, the training objective of the model is selected to maximize the average weighted F1 score.

3.4. Experimental Platforms

All the experiments were conducted on an Intel i7-6700 CPU at 4.0 GHz with a 16 GB RAM and a Nvidia P100 GPU with a 16 GB memory. The programming language was Python 3.6 and the integrated development environment was Anaconda 3. Several open source libraries, including SpaCy, Jieba and Deep Graph Library (DGL), were also used. Among them, DGL was used to convert each dependency graph into a DGL graph object. The R-GCN model was also implemented based on the DGL.

4. Experimental Results

Through five-fold cross-validation, the original labeled dataset was equally divided into the same quintiles, i.e., each part contained 200 accident records. For each training, four of them were used as the training set and the remaining one was regarded as the test set, ensuring that each part was treated as the test set throughout the validation process. The results of each cross-validation were measured through the average F1-score to evaluate the performance of the whole model and the model with the highest average F1-score was adopted as the final built model combining the pre-trained BERT and a R-GCN. The results of the five-fold cross-validation are illustrated in Table 2.

Table 2. Results of the five-fold cross-validation.

	Fold 1	Fold 2	Fold 3	Fold 4	Fold 5
AvgF1-score	0.69	0.64	0.72	0.77	0.75

As shown in Table 2, the accident causal classification model combining the pre-trained BERT and the R-GCN could achieve an average F1-score of up to 0.77. In order to more intuitively show the performance of the constructed model on each type of accident, we show the confusion matrix of the adopted model in Table 3. The corresponding procedure for calculating the average F1-score of Fold 4 is shown in Table 4.

It can be seen, from the confusion matrix in Table 3, that the model used can achieve accurate classification of most texts, but there is still a certain error for types that also contain fall or object. This shows that the proposed text-mining-based accident cause analysis could be roughly classified, but the specific accident cause analysis results still need to be further improved. To further demonstrate the advantages of the proposed model, comparative experiments were performed to numerically evaluate the improvement of the proposed method over previous methods. Traditional text mining methods, including decision tree, k-nearest neighbors (KNN), Naive Bayes and logistic regression, were also adopted to classify text in the OSHA dataset. Deep-learning-based networks, including long short-term memory (LSTM), gate recurrent unit (GRU) and symbiotic organisms search–gate recurrent unit (SGRU) [48], were also compared. Furthermore, to explore the role of the gate mechanism in the proposed method, the results of an ablation experiment were also analyzed. It should be noted that all experiments were performed on Fold 4. Due to space constraints, only the final average results, not class-by-class results, are shown in Table 5.

Table 3. Confusion matrix of the adopted model.

		Prediction														
		1	2	3	4	5	6	7	8	9	10	11	Total	TP	FN	Recall
Ground truth	1	9	1	0	1	0	0	0	0	0	1	0	12	9	3	0.75
	2	0	34	0	1	4	0	1	0	0	0	0	40	34	6	0.85
	3	1	0	39	0	0	0	1	0	0	8	1	50	39	11	0.78
	4	0	0	1	12	1	0	0	1	0	0	0	15	12	3	0.80
	5	0	2	0	0	19	0	0	0	0	2	0	23	19	4	0.83
	6	0	1	0	2	0	3	0	1	1	1	0	9	3	6	0.33
	7	0	1	0	0	0	0	5	0	0	0	0	6	5	1	0.83
	8	0	0	1	0	0	0	0	7	0	0	2	10	7	3	0.70
	9	0	0	0	0	2	2	2	0	15	0	0	21	15	6	0.71
	10	0	0	4	0	0	0	0	0	0	6	0	10	6	4	0.60
	11	0	0	0	0	1	0	0	0	0	0	3	4	3	1	0.75
	Total	10	39	45	16	27	5	9	9	16	18	6	200			
	TP	9	34	39	12	19	3	5	7	15	6	3				
	FP	1	5	6	4	8	2	4	2	1	12	3				
	Precision	0.90	0.87	0.87	0.75	0.70	0.60	0.56	0.78	0.94	0.33	0.50				

Table 4. Corresponding procedure for calculating the average F1-score of Fold 4.

	Precision	Recall	F1-Score	Number of Cases	AvgF1-Score
1	0.90	0.75	0.82	12	
2	0.87	0.85	0.86	40	
3	0.87	0.78	0.82	50	
4	0.75	0.80	0.77	15	
5	0.70	0.83	0.76	23	
6	0.60	0.33	0.43	9	0.77
7	0.56	0.83	0.67	6	
8	0.78	0.70	0.74	10	
9	0.94	0.71	0.81	21	
10	0.33	0.60	0.43	10	
11	0.50	0.75	0.60	4	

Table 5. Results of the comparison experiment and the ablation experiment.

	Average Precision	Average Recall	AvgF1-Score
Decision trees	0.48	0.55	0.51
KNN	0.49	0.52	0.50
Naive Bayes	0.57	0.54	0.55
Logistic regression	0.47	0.87	0.61
LSTM	0.58	0.64	0.61
GRU	0.70	0.61	0.65
SGRU	0.73	0.69	0.71
Ours w/o gate mechanism	0.74	0.72	0.73
Ours	0.79	0.76	0.77

From the results in Table 5, it can be seen that, although the proposed method still has some limitations, it achieved a 6% improvement of the average F1-score compared to existing research. At the same time, it can also be found from the results of the ablation experiments that the gate mechanism played a key role in the entire model. By eliminating the possible errors of the pre-trained BERT, the model achieved a 4% improvement.

5. Conclusions and Future Works

This study proposes a text mining method combining the pre-trained BERT and a R-GCN to automatically explore accident causal information in accident investigation reports. The proposed method avoids the tedious preprocessing steps of previous text mining methods and extracts text features by employing a pre-trained BERT to embed words from text reports into a high-dimensional vector space. Then, by using a R-GCN with a gate mechanism, the syntactic structure in the text is also processed into high-dimensional vectors. By concatenating these two features and with the help of the classifier, it is possible to understand both the word and syntax of the accident investigation reports. Compared with methods such as text frequency alone, it is more accurate and concise. Compared with retraining BERT to extract text and syntax features at the same time, it is very cheap and fast. The experimental results show that the proposed method could achieve an average F1-score as high as 0.77, which exceeds existing methods and has important practical significance for accident causal classification.

However, it is undeniable that, although the existing methods have made certain breakthroughs compared with previous studies, the classification accuracy still needs to be improved. The methods proposed at present can only assist in the analysis of accident causes from accident investigation reports to a certain extent and cannot completely replace experienced experts. Especially in accident cause analyses, once the accident causal classification is wrong, it may bring unnecessary investment or mislead accident prevention. This is a key breakthrough in future work. In the future, the accuracy of accident causal classification can be improved by enriching the accident causal dataset and adding relevant labels. At the same time, on the basis of this study, we can further explore the BERT-based text information encoding method to build a more efficient expert system.

Author Contributions: Conceptualization, Z.J.; methodology, K.H. and Z.C.; validation, L.W. and Z.Z.; investigation, L.W., Z.J. and Z.Z.; writing—original draft preparation, K.H. and Z.C.; writing—review and editing, L.W. and Z.Z.; funding acquisition, Z.J. and Z.Z. All authors have read and agreed to the published version of the manuscript.

Funding: Figure 2 was modified from 'BERT: Pre-training of Deep Bidirectional Transformers for Language Understanding' (https://aclanthology.org/N19-1423/, accessed on 15 January 2022), licensed under a Creative Commons Attribution 4.0 International License. This work was financially supported by GDAS' Project of Science and Technology Development (grant no. 2021GDASYL-20210103090) and GDAS' Project of Science and Technology Development (grant nos. 2019GDASYL-0502007 and 2020GDASYL-20200302015).

Institutional Review Board Statement: Not applicable.

Informed Consent Statement: Not applicable.

Data Availability Statement: Not applicable.

Acknowledgments: All authors extend their sincerest thanks to the reviewers. Thanks to Yingjie Cai, Department of Electrical Engineering, Chinese University of Hong Kong, for her guidance.

Conflicts of Interest: The authors declare no conflict of interest.

References

1. Williams, H.; Edwards, A.; Hibbert, P.; Rees, P.; Evans, H.P.; Panesar, S.; Carter, B.; Parry, G.; Makeham, M.; Jones, A.; et al. Harms from discharge to primary care: Mixed methods analysis of incident reports. *Br. J. Gen. Pract.* **2015**, *65*, e829–e837. [CrossRef] [PubMed]
2. Reason, J. *Managing the Risks of Organizational Accidents*; Routledge: London, UK, 2016.
3. Nixon, J.; Braithwaite, G.R. What do aircraft accident investigators do and what makes them good at it? Developing a competency framework for investigators using grounded theory. *Saf. Sci.* **2018**, *103*, 153–161. [CrossRef]
4. Jiao, Z.; Lei, H.; Zong, H.; Cai, Y.; Zhong, Z. Potential Escalator-related Injury Identification and Prevention Based on Multi-module Integrated System for Public Health. *arXiv* **2021**, arXiv:2103.07620.
5. Kahfie, I.; Ramadan, M.; Rafi, S.; Perawati, D. The Crash Of Boeing 737 Max 8 And It's Effect On Costumer Trust: Case On Lion Air Passenger. *Adv. Transp. Logist. Res.* **2019**, *2*, 764–769.
6. Johnston, P.; Harris, R. The Boeing 737 MAX saga: Lessons for software organizations. *Softw. Qual. Prof.* **2019**, *21*, 4–12.
7. Zhang, J.; Wan, C.; He, A.; Zhang, D.; Soares, C.G. A two-stage black-spot identification model for inland waterway transportation. *Reliab. Eng. Syst. Saf.* **2021**, *213*, 107677. [CrossRef]
8. Topuz, K.; Delen, D. A probabilistic Bayesian inference model to investigate injury severity in automobile crashes. *Decis. Support Syst.* **2021**, *150*, 113557. [CrossRef]
9. Goh, Y.M.; Ubeynarayana, C. Construction accident narrative classification: An evaluation of text mining techniques. *Accid. Anal. Prev.* **2017**, *108*, 122–130. [CrossRef] [PubMed]

10. Hotho, A.; Nürnberger, A.; Paaß, G. A brief survey of text mining. In *Ldv Forum*; Citeseer: Princeton, NJ, USA, 2005; Volume 20, pp. 19–62.
11. Jiao, Z.; Jia, G.; Cai, Y. A new approach to oil spill detection that combines deep learning with unmanned aerial vehicles. *Comput. Ind. Eng.* **2019**, *135*, 1300–1311. [CrossRef]
12. Cai, Y.; Li, B.; Jiao, Z.; Li, H.; Zeng, X.; Wang, X. Monocular 3D object detection with decoupled structured polygon estimation and height-guided depth estimation. In Proceedings of the AAAI Conference on Artificial Intelligence, New York, NY, USA, 7–12 February 2020; Volume 34, pp. 10478–10485.
13. Cai, Y.; Chen, X.; Zhang, C.; Lin, K.Y.; Wang, X.; Li, H. Semantic Scene Completion via Integrating Instances and Scene in-the-Loop. In Proceedings of the IEEE/CVF Conference on Computer Vision and Pattern Recognition, Nashville, TN, USA, 20–25 June 2021; pp. 324–333.
14. Baclic, O.; Tunis, M.; Young, K.; Doan, C.; Swerdfeger, H.; Schonfeld, J. Artificial intelligence in public health: Challenges and opportunities for public health made possible by advances in natural language processing. *Can. Commun. Dis. Rep.* **2020**, *46*, 161. [CrossRef]
15. Kotsiantis, S.B.; Zaharakis, I.; Pintelas, P. Supervised machine learning: A review of classification techniques. *Emerg. Artif. Intell. Appl. Comput. Eng.* **2007**, *160*, 3–24.
16. ZHANG, Y.k.; LI, H.j. Text classification of accident news based on category keyword. *J. Comput. Appl.* **2008**, *28*, 139–140.
17. Sebastiani, F. Machine learning in automated text categorization. *ACM Comput. Surv. (CSUR)* **2002**, *34*, 1–47. [CrossRef]
18. Kwok, J.T.Y. Automated text categorization using support vector machine. In Proceedings of the International Conference on Neural Information Processing (ICONIP), Kitakyushu, Japan, 21–23 October 1998; Citeseer: Princeton, NJ, USA, 1998.
19. Caropreso, M.F.; Matwin, S.; Sebastiani, F. Statistical phrases in automated text categorization. *Cent. Natl. Rech. Sci.* **2000**, *47*, 1–18.
20. Zhang, F. A hybrid structured deep neural network with Word2Vec for construction accident causes classification. *Int. J. Constr. Manag.* **2019**, 1–21. [CrossRef]
21. Brown, D.E. Text mining the contributors to rail accidents. *IEEE Trans. Intell. Transp. Syst.* **2015**, *17*, 346–355. [CrossRef]
22. Zhong, B.; Pan, X.; Love, P.E.; Sun, J.; Tao, C. Hazard analysis: A deep learning and text mining framework for accident prevention. *Adv. Eng. Inform.* **2020**, *46*, 101152. [CrossRef]
23. Soltanzadeh, A.; Mohammadfam, I.; Mahmoudi, S.; Savareh, B.A.; Arani, A.M. Analysis and forecasting the severity of construction accidents using artificial neural network. *Saf. Promot. Inj. Prev.* **2016**, *4*, 185–192.
24. Paul, D.B. Experience with a stack decoder-based hmm csr and back-off n-gram language models. In Proceedings of the Workshop Speech and Natural Language, Pacific Grove, CA, USA, 19–22 February 1991.
25. Ubeynarayana, C.; Goh, Y. An Ensemble Approach for Classification of Accident Narratives. In *Computing in Civil Engineering 2017*; The American Society of Civil Engineers: Reston, VA, USA, 2017; pp. 409–416.
26. Zhang, F.; Fleyeh, H.; Wang, X.; Lu, M. Construction site accident analysis using text mining and natural language processing techniques. *Autom. Constr.* **2019**, *99*, 238–248. [CrossRef]
27. Chen, L.; Vallmuur, K.; Nayak, R. Injury narrative text classification using factorization model. *BMC Med. Inform. Decis. Mak.* **2015**, *15*, S5. [CrossRef]
28. Heidarysafa, M.; Kowsari, K.; Barnes, L.; Brown, D. Analysis of Railway Accidents' Narratives Using Deep Learning. In Proceedings of the 2018 17th IEEE International Conference on Machine Learning and Applications (ICMLA), Orlando, FL, USA, 17–20 December 2018; IEEE: Piscataway, NJ, USA, 2018; pp. 1446–1453.
29. Cheng, X.; Khomtchouk, B.; Matloff, N.; Mohanty, P. Polynomial regression as an alternative to neural nets. *arXiv* **2018**, arXiv:1806.06850.
30. Devlin, J.; Chang, M.-W.; Lee, K.; Toutanova, L.K. Bert: Pre-training of deep bidirectional transformers for language understanding. In Proceedings of the NAACL-HLT, Minneapolis, MN, USA, 2–7 June 2019; pp. 4171–4186.
31. Rajput, A. Natural language processing, sentiment analysis, and clinical analytics. In *Innovation in Health Informatics*; Elsevier: Amsterdam, The Netherlands, 2020; pp. 79–97.
32. Xiong, G.; Zhang, J.; Yuan, X.; Shi, D.; He, Y. Application of symbiotic organisms search algorithm for parameter extraction of solar cell models. *Appl. Sci.* **2018**, *8*, 2155. [CrossRef]
33. Karatzoglou, A.; Jablonski, A.; Beigl, M. A Seq2Seq learning approach for modeling semantic trajectories and predicting the next location. In Proceedings of the 26th ACM SIGSPATIAL International Conference on Advances in Geographic Information Systems, Seattle, WA, USA, 6–9 November 2018; pp. 528–531.
34. Zulqarnain, M.; Ghazali, R.; Ghouse, M.G.; Mushtaq, M.F. Efficient processing of GRU based on word embedding for text classification. *JOIV Int. J. Inform. Vis.* **2019**, *3*, 377–383. [CrossRef]
35. Vaswani, A.; Shazeer, N.; Parmar, N.; Uszkoreit, J.; Jones, L.; Gomez, A.N.; Kaiser, Ł.; Polosukhin, I. Attention is all you need. In *Advances in Neural Information Processing Systems*; MIT Press: Cambridge, MA, USA, 2017; pp. 5998–6008.
36. Vasantharajan, C.; Thayasivam, U. Towards Offensive Language Identification for Tamil Code-Mixed YouTube Comments and Posts. *SN Comput. Sci.* **2022**, *3*, 1–13. [CrossRef]
37. Kipf, T.N.; Welling, M. Semi-supervised classification with graph convolutional networks. *arXiv* **2016**, arXiv:1609.02907.
38. Jiao, Z.; Jia, G.; Cai, Y. Ensuring Computers Understand Manual Operations in Production: Deep-Learning-Based Action Recognition in Industrial Workflows. *Appl. Sci.* **2020**, *10*, 966. [CrossRef]

39. Lin, Y.; Meng, Y.; Sun, X.; Han, Q.; Kuang, K.; Li, J.; Wu, F. BertGCN: Transductive Text Classification by Combining GCN and BERT. *arXiv* **2021**, arXiv:2105.05727.
40. Cao, R.; Chen, L.; Chen, Z.; Zhao, Y.; Zhu, S.; Yu, K. LGESQL: Line Graph Enhanced Text-to-SQL Model with Mixed Local and Non-Local Relations. *arXiv* **2021**, arXiv:2106.01093.
41. Schlichtkrull, M.; Kipf, T.N.; Bloem, P.; Berg, R.V.D.; Titov, I.; Welling, M. Modeling relational data with graph convolutional networks. In Proceedings of the European Semantic Web Conference, Heraklion, Greece, 3–7 June 2018; Springer: Berlin/Heidelberg, Germany, 2018; pp. 593–607.
42. Xu, Y.; Yang, J. Look again at the syntax: Relational graph convolutional network for gendered ambiguous pronoun resolution. *arXiv* **2019**, arXiv:1905.08868.
43. Ryu, S.; Lim, J.; Hong, S.H.; Kim, W.Y. Deeply learning molecular structure-property relationships using attention-and gate-augmented graph convolutional network. *arXiv* **2018**, arXiv:1805.10988.
44. Du, C.; Wang, J.; Sun, H.; Qi, Q.; Liao, J. Syntax-type-aware graph convolutional networks for natural language understanding. *Appl. Soft Comput.* **2021**, *102*, 107080. [CrossRef]
45. Marcheggiani, D.; Titov, I. Encoding sentences with graph convolutional networks for semantic role labeling. *arXiv* **2017**, arXiv:1703.04826.
46. Li, Q.; Han, Z.; Wu, X.M. Deeper insights into graph convolutional networks for semi-supervised learning. In Proceedings of the Thirty-Second AAAI Conference on Artificial Intelligence, New Orleans, LA, USA, 2–7 February 2018.
47. Occupational Safety and Health Administration. *Fatality and Catastrophe Investigation Summaries*; Occupational Safety and Health Administration: Washington, DC, USA, 2016.
48. Cheng, M.Y.; Kusoemo, D.; Gosno, R.A. Text mining-based construction site accident classification using hybrid supervised machine learning. *Autom. Constr.* **2020**, *118*, 103265. [CrossRef]
49. Wang, A.; Singh, A.; Michael, J.; Hill, F.; Levy, O.; Bowman, S.R. GLUE: A multi-task benchmark and analysis platform for natural language understanding. *arXiv* **2018**, arXiv:1804.07461.
50. Kingma, D.P.; Ba, J. Adam: A method for stochastic optimization. *arXiv* **2014**, arXiv:1412.6980.

applied sciences

Article

Forecasting the Bearing Capacity of the Driven Piles Using Advanced Machine-Learning Techniques

Mohammed Amin Benbouras [1,2,*], Alexandru-Ionuţ Petrişor [3,4], Hamma Zedira [5], Laala Ghelani [5] and Lina Lefilef [6]

1 Technology Department, École Normale Supérieure d'Enseignement Technologique de Skikda (ENSET), Skikda 21001, Algeria
2 Central Public Works Laboratory (LCTP), Algiers 16006, Algeria
3 Doctoral School of Urban Planning, "Ion Mincu" University of Architecture and Urbanism, 010014 Bucharest, Romania; alexandru.petrisor@uauim.ro
4 National Institute for Research and Development in Tourism, 50741 Bucharest, Romania
5 Civil Engineering Department, University of Abbes Laghrour, Khenchela 40051, Algeria; zedirahamma2003@yahoo.fr (H.Z.); ghilanilaala@yahoo.fr (L.G.)
6 Department of English Language and Literature, Mohamed Seddik Ben Yahia University, Jijel 18000, Algeria; lefileflina@gmail.com
* Correspondence: mouhamed_amine.benbouras@g.enp.edu.dz

Abstract: Estimating the bearing capacity of piles is an essential point when seeking for safe and economic geotechnical structures. However, the traditional methods employed in this estimation are time-consuming and costly. The current study aims at elaborating a new alternative model for predicting the pile-bearing capacity based on eleven new advanced machine-learning methods in order to overcome these limitations. The modeling phase used a database of 100 samples collected from different countries. Additionally, eight relevant factors were selected in the input layer based on the literature recommendations. The optimal inputs were modeled using the machine-learning methods and their performance was assessed through six performance measures using a *K*-fold cross-validation approach. The comparative study proved the effectiveness of the DNN model, which displayed a higher performance in predicting the pile-bearing capacity. This elaborated model provided the optimal prediction, i.e., the closest to the experimental values, compared to the other models and formulae proposed by previous studies. Finally, a reliable and easy-to-use graphical interface was generated, namely "BeaCa2021". This will be very helpful for researchers and civil engineers when estimating the pile-bearing capacity, with the advantage of saving time and money.

Keywords: pile-bearing capacity; machine learning; deep neural network; *K*-fold cross-validation approach; sensitivity analysis

Citation: Benbouras, M.A.; Petrişor, A.-I.; Zedira, H.; Ghelani, L.; Lefilef, L. Forecasting the Bearing Capacity of the Driven Piles Using Advanced Machine-Learning Techniques. *Appl. Sci.* **2021**, *11*, 10908. https://doi.org/10.3390/app112210908

Academic Editors: Nikos D. Lagaros and Vagelis Plevris

Received: 10 October 2021
Accepted: 16 November 2021
Published: 18 November 2021

1. Introduction

Pile foundations are used to transmit construction loads deep into the ground in order to ensure structure stability [1,2]. Furthermore, computing the bearing capacity of piles is essential when designing economic and safe geotechnical structures [3]. To date, numerous approaches have been conceived for the sake of creating alternative methods and techniques that contain numerical, experimental, and analytical approaches aiming at predicting the bearing capacity of piles [4–6]. Among the most frequently used methods is the Cone Penetration Test (*CPT*), known for producing accurate results in a variety of situations [7,8]. This is probably due to the fact that *CPT*-based methods have been modeled in harmony with the *CPT* results, which were proven to estimate more effective different geotechnical properties, and make more precise pile capacity predictions [6]. Other semi-empirical methods have been widely utilized, such as Meyerhof's formula, which could yield an acceptable pile-bearing capacity [4]. On the other hand, the High-Strain Dynamic Load Test (*HSDLT*) and the Static Load Test (*SLT*) have been employed considerably for predicting the pile-bearing capacity [9]. The *HSDT* is preferable to the

SLT, because it operates with a faster, more advanced, and economic technology [2]. This quality supports its paramount importance addressed by the American Standards Test Methods to standardize the *HSDT* method [1]. The literature on bearing capacity values revealed a relatively close accuracy in both the *HSDT* and the *SLT* [1]. Momeni et al. [10] added that *HSDT* is faster and more economic compared to SLT, but it generally requires several *HSDT* tests for each project to obtain a reliable result [11]. Hence, increasing the number of *HSDT* tests is extremely undesirable since it may increase the total project budget. Moreover, other empirical researchers have proposed traditional methods for estimating the bearing capacity [12–15]. The quality of easiness and common usage has made these methods very important. However, determining the bearing capacity of bored and driven piles by means of the aforementioned methods is found to be time-consuming and costly [16]. This is probably due to the complex behavior of piles, heterogeneity of the soil around piles, material and shape of piles, and their installation. Accordingly, all the proposed methods/models in the literature yielded ineffective predictions [17]. On the other hand, currently, due to emerging new easy-to-use performance software such as PLAXIS, utilizing finite element analysis for which the system is discretized into a number of meshes to obtain axial capacity is of interest [18]. For this reason, numerical methods based on the finite element approach have recently become well-known for the evaluation of bearing capacity, yielding effective results [19,20]. Recently, the application of some new advanced techniques, namely "artificial intelligence (*AI*)" or "machine learning (*ML*)", has witnessed a spectrum of interest, and they provided exceptional results in solving several issues by learning from the available data [21,22].

Subsequently, the use of machine-learning methods to predict pile-bearing capacity has witnessed considerable development since the early 1990s [21–24]. Several studies are now able to estimate the pile-bearing capacity with a higher degree of precision in comparison to traditional methods. Among the fundamental research dealing with the pile-bearing capacity, Nawari et al. have used one hidden layer of the *ANN* model by investigating a database consisting of 25 test data. The chosen input parameters included the *SPT-N* values and geometrical properties. The *ANN* model efficiently predicted the pile-bearing capacity compared to traditional methods [25]. Furthermore, Mahnesh has predicted the pile-bearing capacity by using Support Vector Machines and Generalized Regression Neural Network with an input layer containing dynamic stress-wave data [26]. He concluded that the Generalized Regression Neural Network was the best model with a high correlation coefficient (0.977). In addition, Milad et al. have developed an effective model based on Artificial Neural Network, genetic programming, and linear regression methods to predict the bearing capacity of piles by learning from 100 samples. They utilized the Flap number, basic properties of the surrounding soil, pile geometry, and pile-soil friction angle as an input layer. The suggested *ANN* model has better stability compared to the other methods [27]. Jahed et al. used hybrid *PSO–ANN* to predict the bearing capacity of rock-socketed piles, by taking into consideration soil length to socket length ratio, total length to diameter ratio, uniaxial compressive strength, and standard penetration test. The proposed *PSO–ANN* model has demonstrated its efficiency since it produced a high correlation coefficient (R = 0.9685) [1]. Moayedi et al. have used *ANFIS, GP*, and *SA–GP* for modeling a database consisting of 50 tests. The chosen input parameters included the pile length, pile cross-sectional area, hammer weight, pile set, and drop height. The *SA–GP* model efficiently predicted the pile-bearing capacity compared to other methods [28]. Shaik et al. have predicted the pile-bearing capacity by using *ANFIS* and *ANFIS–GMDH–PSO* with an input layer containing *CPT* and pile loading test results [29]. They have proven that the metaheuristic hybrid *ANFIS–GMDH–PSO* model is the best one, with a high correlation coefficient (0.998) [29]. Harandizadeh et al. have used hybrid *MLP–GWO* and *ANFIS–GWO* to predict the bearing capacity of piles from the input layer, including pile area, pile length, flap number, average cohesion, and friction angle, average soil-specific weight, and average pile-soil friction angle. The proposed *MLP–GWO* model has demonstrated that its efficiency yielded a high correlation coefficient (R = 0.991) [30].

Table 1 summarizes more than ten studies that have used machine-learning models to predict the pile-bearing capacity.

Table 1. Proposed machine-learning models in the literature to estimate the pile-bearing capacity.

Authors	Inputs	Methods	Database	References
Nawari et al. (1999)	*SPT-N* values and geometrical properties	Neural Network	25	[25]
Mahnesh (2011)	Dynamic stress-wave data	Support Vector Machines and Generalized Regression Neural Network	105	[26]
Milad et al. (2015)	Flap number, basic properties of the surrounding soil, pile geometry, and pile-soil friction angle	Artificial Neural Network, Genetic Programming and Linear Regression	100	[27]
Jahed et al. (2017)	Soil length to socket length ratio, total length to diameter ratio, uniaxial compressive strength, and standard penetration test	hybrid *PSO–ANN*	132	[1]
Moayedi and Jahed (2018)	Internal friction angle of soil located in shaft and tip, pile length, effective vertical stress at pile toe and pile area	*ICA-ANN*	59	[31]
Yong et al. (2021)	Pile length, pile cross-sectional area, hammer weight, pile set, and drop height	*ANFIS*, *GP*, and *SA–GP*	50	[2]
Shaik et al. (2019)	Internal friction angle of soil located in shaft and tip, effective vertical stress at pile toe, pile area, and pile length	*ICA-ANN* and *ANFIS*	59	[29]
Kardani et al. (2020)	Shear resistance angle at the shaft of the pile, soil shear resistance angle at the tip of the pile, length of pile, cross-sectional area of the pile, and effective stress at the tip of the pile	*Decision tree, k-nearest neighbor, Multilayer Perceptron Artificial Neural Network, Random Forest, Support Vector Regressor*, and *Extreme Gradient Boosting*	59	[32]
Harandizadeh et al. (2021)	CPT and pile loading test results	*ANFIS* and *ANFIS–GMDH–PSO*	72	[30]
Moayedi et al. (2020)	Pile diameter, pile length, relative density, embedment ratio, and both the pile end resistance and base resistance	*GA-ANFIS* and *PSO-ANFIS*	20	[28]
Liu et al. (2020)	Laboratory and in situ testing results	*ANFIS*, *ANN*, and *GA-ANN*	43	[33]
Dehghanbanadaki et al. (2021)	Pile area, pile length, flap number, average cohesion and friction angle, average soil-specific weight, and average pile-soil friction angle	*MLP–GWO* and *ANFIS–GWO*	100	[34]

According to the authors' knowledge, previous studies have been limited mostly to the use of *ANN*, *ANFIS*, and *SVM* methods for predicting the pile-bearing capacity, although recent studies have shown that other techniques could have yielded more effective and accurate results [35–37]. Furthermore, they assessed the predictive capability of suggested models depending on only one split to check the data learning validity. Consequently, the ability of their proposed model to overcome over-fitting and under-fitting problems cannot be assured. Moreover, the majority of published papers have proposed machine-learning models in the form of mathematical equations, which are hard to duplicate in future studies. Admittedly, this practice has very little value for other researchers and civil engineers in the field. Conveniently, to overcome these limitations, investigators have presented their optimal models in the form of a programmed interface or a simple script by a well-known programming language such as Python or Matlab for generating the proposed model. This will make it available to anyone interested in the problem of modeling regardless of their proficiency level.

The current study contributes to providing a new alternative model for predicting the pile-bearing capacity based on 12 advanced machine-learning methods, which are applied for the first time for this aim. Furthermore, a high-performance method to estimate the generalization capability of the learning model, and to check the validity of the model for other cases, has been used, namely "*K*-fold cross-validation analysis". Finally, in order to treat the hard usage problem of machine-learning models in future studies, the proposed optimal model was used afterwards to develop a *GUI* public interface. Consequently, the suggested "BeaCa2021" interface is very handy and easy-to-use by civil engineers and researchers, by offering plenty of benefits such as reliability, easiness, and lowering the budget used to predict the pile-bearing capacity from relevant and easily obtained parameters without the need to operate expensive in situ tests.

2. Materials and Methods

2.1. Overview of the Methodology

Several advanced machine learning methods, such as Extreme Deep Neural Network (*DNN*), Extreme Learning Machine (*ELM*), Support Vector Regression (*SVR*), *LASSO* regression (*LASSO*), Random Forest (*RF*), Ridge Regression (*Ridge*), Partial Least Square Regression (*PLSR*), Stepwise Regression (*Stepwise*), Kernel Ridge (*KRidge*), Genetic Programming (*GP*), and Least Square Regression (*LSR*), have been used to learn from 100 samples collected from previous studies [27]. Multiple input parameters, including the pile material, average cohesion (kN/m^2), average friction angle (°), average soil-specific weight (kN/m^3), average pile-soil friction angle (°), flap number, pile area (m^2), and pile length (m), have been used. Firstly, the aforementioned advanced machine-learning methods have been utilized for modeling the input parameters, and their effectiveness was assessed through various statistical indicators. To evaluate the predictive ability of the optimal model, the *k*-fold cross-validation approach, which is based on five splits, has been employed. Afterward, in order to know which input variables have the biggest effect on the pile-bearing capacity through the proposed model, a sensitivity analysis has been performed via the step-by-step method. Finally, a reliable, easy-to-use, and the graphical interface was designed based on our optimal model in order to help civil engineers and researchers to easily predict the pile-bearing capacity in future studies.

2.2. Database

Choosing the Neural Network inputs is deemed to be the most significant phase for achieving accurate predictions. The selected relevant inputs should cover various aspects of the understudied problem. Besides, several factors have been selected, such as soil characteristics, pile-soil contact characteristics, and geometry and pile characteristics, which can affect the pile-bearing capacity. This study used data from 100 static load-bearing tests on the ultimate bearing capacity (*UBC*) of both the steel- and the concrete-driven piles from various countries, such as Iran, Mexico, and India [38–42]. The input parameters, including pile material, average cohesion (kN/m^2), average friction angle (°), average soil unit weight (kN/m^3), average pile-soil friction angle (°), flap number, pile area (m^2), and pile length (m), were selected as optimal input parameters. We have supposed that the cohesion, angle of shearing resistance, and soil unit weight were the parameters characterizing the soil condition, whereas the pile area and pile length are the parameters characterizing the pile geometric size. In addition, the pile-soil friction angle is the parameter describing the pile material. Finally, the flap number was assumed to symbolize all other hidden effective factors in measuring the pile-bearing capacity [27]. The considered output was obtained from the static bearing capacity, which used static testing in the fully drained condition (long term). Two types of materials (concrete and steel piles, see Table S1) were used in this study. The data samples in both the training and validation phase have been randomly selected and completely detached. Table 2 shows the input and output parameters used in our study.

Table 2. Input and output parameters of the proposed model.

Code	Parameter Type	Type of Variable	Subdivision	Variable
X1	Input	Qualitative	*X1* = 1 (Steel) *X1* = 2 (Concrete)	Pile material
X2	Input	Quantitative		Average cohesion (kN/m^2)
X3	Input	Quantitative		Average friction angle (°)
X4	Input	Quantitative		Average soil-specific weight (kN/m^3)
X5	Input	Quantitative		Average pile-soil friction angle (°)
X6	Input	Quantitative		Flap number
X7	Input	Quantitative		Pile area (m^2)
X8	Input	Quantitative		Pile length (m)
Y	Output	Quantitative		Pile capacity (kN)

2.3. Machine-Learning Methods

In the present paper, numerous machine-learning approaches have been utilized in order to perform a consistent study and to suggest an effective model. Many studies have revealed the effectiveness of the machine-learning methods, which have shown impressive results in the abroad fields. Hence, only the utilized methods are mentioned below, followed by some relevant references, which could be observed by the concerned readers to perfectly understand each method. The methods used were Deep Neural Network (*DNN*) [43,44], Extreme Learning Machine (*ELM*) [45], Random Forest (*RF*) [46], Support Vector Regression (*SVR*) [47], Partial Least Square Regression (*PLSR*) [48], LASSO regression (*LASSO*) [49], Kernel Ridge Regression (*KRidge*) [50], Ridge Regression (*Ridge*) [51], Genetic Programming (*GP*) [43], and Stepwise Regression (*Stepwise*) [52]. Matlab has been applied for modeling the algorithms corresponding to each method, except for GP, where the HeuristicLab Interface has been utilized [53]. The controlling parameters of the *ELM*, *DNN*, *SVR*, *RF*, *LASSO*, *PLS*, *Ridge*, *KRidge*, *Stepwise*, and *GP* algorithms used in this study are listed in Table 3. It is worth mentioning that the trial-and-error method has been applied in most *ML* approaches used in our study. This method is based on changing the controlling parameters of each technique and computing the mean square error in order to find the best parameters. Nevertheless, the controlling parameters of other methods, such as *ELM*, *PLS*, *Ridge*, and *KRidge*, are based on the aforementioned literature recommendations.

Table 3. Initial parameter settings for the algorithms.

Algorithms	Algorithm Parameters	Value
ELM	Hidden layers Hidden neurons Activation function Regulation parameter	$H = 1$ $N = 12$ 'linear' C = 0.02
DNN	Hidden layers Hidden neurons in the first layer Hidden neurons in the second layer Activation function in the first layer Activation function in the second layer	$H = 2$ $N1$ = [1–20] $N2$ = [1–20] 'Tansig' 'Tansig'
SVR	Regulation parameter C Regulation parameter lambda Kernel function	Series of C Series of lambda 'rbf'
RF	nTrees mTrees	nTrees = 100 mTrees = 26
LASSO	Lambda	series of lambda
PLS	PLS components	NumComp = 3 for PSO NumComp = 4 for GT and FS
Ridge	Regularization parameter lambda	lambda = 1
KRidge	Regularization parameter lambda Kernel function Parameter for kernel	lambda = 1 'linear' sigma = 2×10^{-7}
GP	Function set Population size Number of generations Genetic operators	$+, -, \times, \div$, power, ln, sqrt, sin, cos, tan 100 up to 500 1000 Reproduction, crossover, mutation

2.4. Statistical Performance Indicators

The estimation precision of the suggested models was assessed through several statistical performance indicators and by utilizing graphical presentation. The statistical performance indicators are mean absolute error (*MAE*), root mean square error (*RMSE*), index of scattering (*IOS*), coefficient of determination (R^2), Pearson correlation coefficient (*R*), and index of agreement (*IOA*). They are expressed as follows [54,55]:

1. Mean absolute error (*MAE*):

$$MEA = \frac{1}{N}\sum_{i=1}^{N}|Y_{tar,i} - Y_{out,i}| \quad (0 < MAE < \infty) \tag{1}$$

2. Root mean square error (*RMSE*):

$$RMSE = \sqrt{\frac{1}{N}\sum_{i=1}^{N}(Y_{tar,i} - Y_{out,i})^2} \quad (0 < RMSE < \infty) \tag{2}$$

3. Index of scattering (*IOS*):

$$IOS = \frac{\sqrt{\frac{1}{N}\sum_{i=1}^{N}(Y_{tar,i} - Y_{out,i})^2}}{\overline{Y_{tar}}} \quad (0 < RMSE < \infty) \tag{3}$$

4. Coefficient of determination (R^2):

$$R^2 = 1 - \frac{\sum_{i=1}^{N}(Y_{tar,i} - Y_{out,i})^2}{\sum_{i=1}^{N}\left(Y_{tar,i} - \overline{Y_{tar}}\right)^2} \quad (-\infty < NSE < 1) \tag{4}$$

5. Pearson correlation coefficient (R):

$$R = \frac{\sum_{i=1}^{N}\left(\left(Y_{tar,i} - \overline{Y_{tar}}\right)\left(Y_{out,i} - \overline{Y_{out}}\right)\right)}{\sqrt{\sum_{i=1}^{N}\left(\left(Y_{tar,i} - \overline{Y_{tar}}\right)^2\left(Y_{out,i} - \overline{Y_{out}}\right)^2\right)}} \quad (-1 < R < 1) \tag{5}$$

6. Index of agreement (IOA):

$$IOA = 1 - \frac{\sum_{i=1}^{N}(Y_{tar,i} - Y_{out,i})^2}{\sum_{i=1}^{N}\left(\sum_{i=1}^{N}\left|Y_{out,i} - \overline{Y_{tar}}\right| + \sum_{i=1}^{N}\left|Y_{tar,i} - \overline{Y_{tar}}\right|\right)^2} \quad (0 < IOA < 1) \tag{6}$$

where $Y_{tar,i}$, $Y_{out,i}$, $\overline{Y_{tar}}$, and $\overline{Y_{out}}$ characterize the target, output, mean of the target, and mean of output pile-bearing capacity values for N data samples, respectively. Moreover, the suggested machine-learning model possessed the minimum value of *RMSE*, *IOS*, and *MAE*, and the peak value of *IOA*, R^2, and R presents the optimal one and the closest to the experimental values.

Therefore, after choosing the optimal model based on statistical performance indicators, its predictive capability was evaluated by utilizing the K-fold cross-validation approach. The latter is an advanced approach, which revealed more accuracy and robustness when assessing the ability of the optimal model to overcome over-fitting and under-fitting problems in data learning [56,57]. The approach relies on dividing the database into k equal splits. Hence, for each split, $K-1$-folds are utilized for the training phase and the last one for validation. This procedure is reiterated successively until the use of all splits for the validation step [58,59]. The key benefit of this approach is that all the data are modeled in both the training and the validation steps [57]. Breiman and Spector have confirmed that K = 10- or K = 5-fold cross-validation is the best choice for assessing the model [56]. In our study, we selected K-fold cross-validation with K = 5 for assessing the predictive ability of the best model.

2.5. Methodology

In order to select the optimal model to predict the pile-bearing capacity using the aforementioned parameters as an input, the methodology followed the following phases:

1. Creating a geotechnical database, collected from different countries such as Iran, Mexico, and India. In this step, 100 static load-bearing tests on the UBC of steel- and concrete-driven piles were collected as datasets.
2. Modeling the chosen inputs by means of numerous machine-learning methods. The *ELM, DNN, SVR, RF, LASSO, PLS, Ridge, K Ridge, Stepwise,* and *GP* methods have been employed in this step for suggesting 11 models.
3. Defining the optimal model for estimating the pile-bearing capacity value using important statistical performance indicators such as *MAE*, *RMSE*, *IOS*, R^2, R, and *IOA*.
4. Evaluating the predictive capability of the optimal model to overcome under-fitting and over-fitting problems by utilizing the K-fold cross-validation approach with K = 5.
5. Performing a sensitivity analysis by using the step-by-step method to define the most or least influential input on the bearing capacity via the proposed model.
6. Designing a reliable, easy-to-use, and graphical interface based on our optimal model.

The research methodology for defining the optimal model to predict the pile-bearing capacity is systematically illustrated in Figure 1.

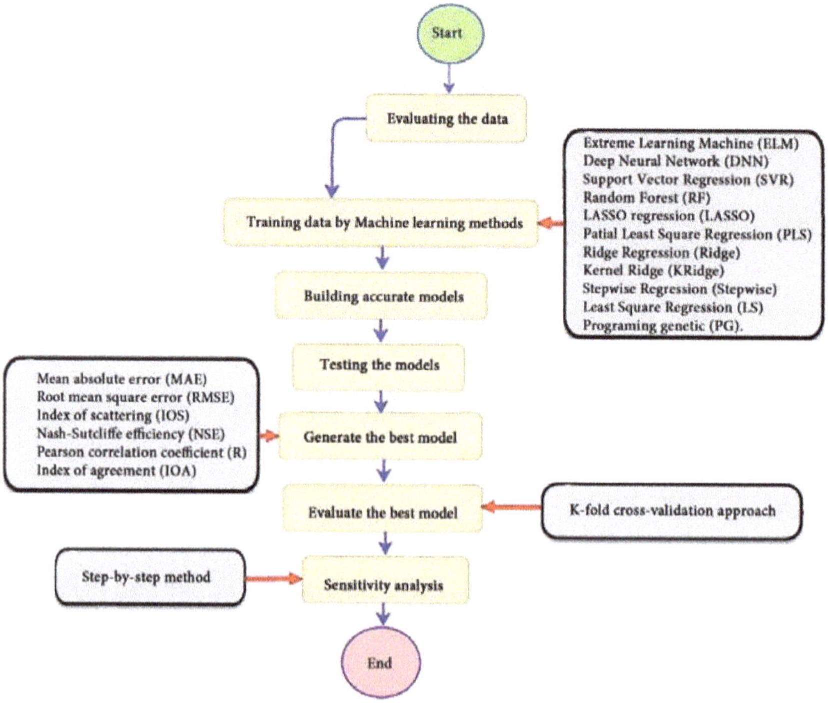

Figure 1. Flowchart describing the key steps for the methodology of research to estimate the pile-bearing capacity.

3. Results

3.1. Database Compilation

In the present paper, a database of 100 samples has been collected from previous studies, resulting in a dataset containing diverse data, considered as satisfactory for an efficient study. For the purpose of a precise modeling step, we have tried to make the dataset balanced for both concrete and steel material samples in both the training and validation phase. Furthermore, the data samples in both phases have been randomly chosen and completely detached. Table 4 shows the descriptive statistics of the user database, computed by using *SPSS*, including the range, minimum, maximum, mean, standard deviation (SD), variance, skewness, and kurtosis. The skewness values prove that all the parameters were equally distributed. Furthermore, the findings indicated that the dataset comprises a wide range of data. Consequently, the gathered database could be very handy when seeking to develop new empirical equations and models, as well as in evaluating the predictive capability of published formulae.

3.2. Correlation between Bearing Capacity and Input Parameters

To statistically estimate the relationship between the pile-bearing capacity and input parameters, SPSS software has been utilized. The correlation matrix between them is displayed in Figure 2, which shows a descriptive summary of the data distribution. The findings show a positive correlation between the pile-bearing capacity and other inputs, except for *X2*, *X4*, and *X5*, which appear to have a negative correlation (see Figure 2). This highlights that the decrease in these parameters tends to proportionally decrease the pile-bearing capacity. Moreover, Pearson correlation coefficient (*R*) and its significance between the pile-bearing capacity and other inputs is presented in Table 5. The findings prove that the significance is less than 0.05, except for *X3*, *X4*, and *X5*, showing that most correlations are statistically significant. Hence, according to Smith's classification (1986) [43], the pile-bearing capacity is significantly correlated with the input parameters, excluding X3, X4, and X5, which are poorly correlated. The results point out that these factors can have a complex nonlinear relationship with the pile-bearing capacity. Besides, in order to precisely model this complex phenomenon, new sophisticated machine-learning approaches should be developed.

Table 4. Descriptive statistics of the collected samples (Std. Error = standard error, SD = standard deviation).

	Range	Minimum	Maximum	Mean		SD	Variance	Skewness		Kurtosis	
	Statistic	Statistic	Statistic	Statistic	Std. Error	Statistic	Statistic	Statistic	Std. Error	Statistic	Std. Error
X2	148.00	0.00	148.00	32.3741	3.28447	32.84	1078.77	2.011	0.241	4.570	0.478
X3	36.62	0.00	36.62	25.5803	0.96535	9.653	93.191	−1.310	0.241	0.855	0.478
X4	8.11	5.38	13.49	10.2029	0.18409	1.840	3.389	−0.406	0.241	0.262	0.478
X5	6.86	10.14	17.00	13.6823	0.16987	1.698	2.885	0.073	0.241	−0.076	0.478
X6	2277.00	14.00	2291.00	494.99	60.23	602.32	362,794.16	1.502	0.241	1.286	0.478
X7	1.52	0.07	1.59	0.4327	0.04656	0.46562	0.217	1.128	0.241	−0.233	0.478
X8	83.80	14.20	98.00	27.1120	1.86024	18.60	346.048	2.761	0.241	6.962	0.478
Y	51,560.00	540.00	52,100.00	5133.12	929.01	9290.14	86,306,843.19	4.043	0.241	16.258	0.478

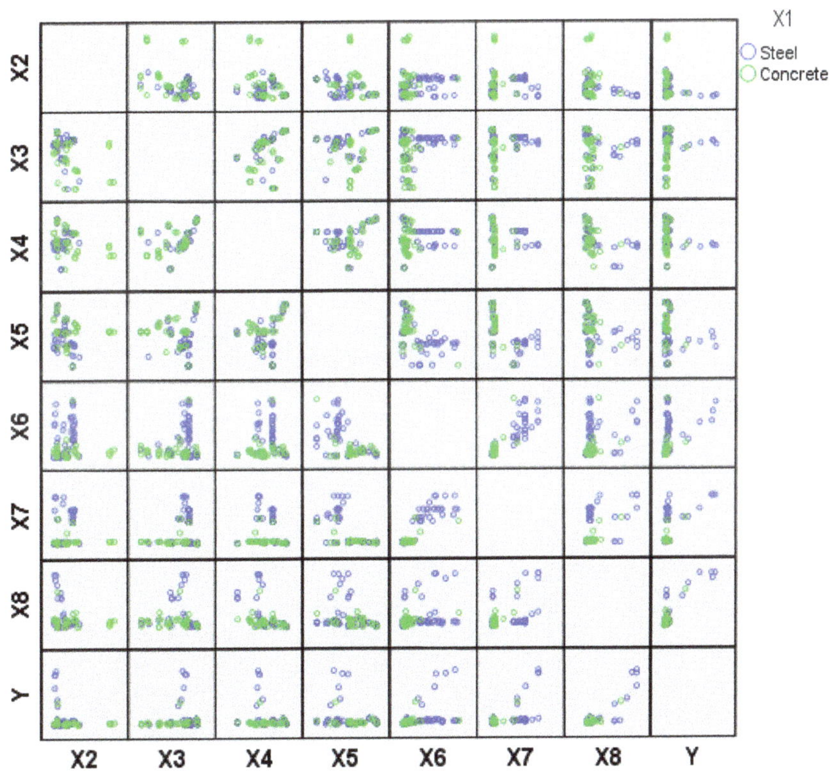

Figure 2. The correlation matrix between the pile-bearing capacity and soil parameters (green points: concrete material; blue points: steel material).

Table 5. Matrix of the correlation between the geotechnical parameters (**: Correlation is significant at the 0.01 level; *: Correlation is significant at the 0.05 level).

		X2	X3	X4	X5	X6	X7	X8	Y
X2	Pearson Correlation	1	−0.370 **	−0.234 *	−0.221 *	0.086	0.038	−0.229 *	−0.229 *
	Significance		0.000	0.019	0.027	0.396	0.707	0.022	0.022
	N	100	100	100	100	100	100	100	100
X3	Pearson Correlation	−0.370 **	1	0.463 **	0.011	0.206 *	0.259 **	−0.063	0.099
	Significance	0.000		0.000	0.916	0.040	0.009	0.531	0.326
	N	100	100	100	100	100	100	100	100
X4	Pearson Correlation	−0.234 *	0.463 **	1	0.270 **	0.124	0.051	−0.433 **	−0.138
	Significance	0.019	0.000		0.007	0.218	0.612	0.000	0.172
	N	100	100	100	100	100	100	100	100
X5	Pearson Correlation	−0.221 *	0.011	0.270 **	1	−0.489 **	−0.555 **	−0.189	−0.142
	Significance	0.027	0.916	0.007		0.000	0.000	0.059	0.159
	N	100	100	100	100	100	100	100	100
X6	Pearson Correlation	0.086	0.206 *	0.124	−0.489 **	1	0.876 **	0.335 **	0.449 **
	Significance	0.396	0.040	0.218	0.000		0.000	0.001	0.000
	N	100	100	100	100	100	100	100	100
X7	Pearson Correlation	0.038	0.259 **	0.051	−0.555 **	0.876 **	1	0.446 **	0.563 **
	Significance	0.707	0.009	0.612	0.000	0.000		0.000	0.000
	N	100	100	100	100	100	100	100	100
X8	Pearson Correlation	−0.229 *	−0.063	−0.433 **	−0.189	0.335 **	0.446 **	1	0.866 **
	Significance	0.022	0.531	0.000	0.059	0.001	0.000		0.000
	N	100	100	100	100	100	100	100	100
Y	Pearson Correlation	−0.229 *	0.099	−0.138	−0.142	0.449 **	0.563 **	0.866 **	1
	Significance	0.022	0.326	0.172	0.159	0.000	0.000	0.000	
	N	100	100	100	100	100	100	100	100

On the other hand, we were generally interested in the correlation between inputs if the multicollinearity phenomenon existed. This could appear between certain independent variables with a high *R*, causing problems when fitting the model and interpreting the results, and reducing the statistical power of the regression model. However, the correlation coefficient presented in Table 5 indicates a moderate *R* between inputs, indicating a moderate multicollinearity, but it is not severe enough to require corrective measures. For this reason, there was no interest in the correlation coefficient between input variables.

3.3. Bearing Capacity Prediction through AI Models

To define the optimal machine-learning model, the first step consists of selecting the optimal input parameters that have a high influence on the target value, and the second step is to determine the best machine-learning methods. To begin with, in order to define the suitable input parameters, eight factors have been used following the literature recommendations. Afterward, we attempted to determine the optimal *ANN* model for predicting the pile-bearing capacity depending on six statistical measures. The performance of each model for the selected optimal input in both concrete and steel piles is presented in Table 6. Six performance measures have been used to compare the proposed models in order to select the best one, in terms of the mean absolute error (*MAE*), root mean square error (*RMSE*), index of scattering (*IOS*), coefficient of determination (R^2), Pearson correlation coefficient (*R*), and index of agreement (*IOA*). The data were divided into two parts, i.e., 80% for the training and 20% for the validation. As Table 6 demonstrates, the target values were modeled via the machine-learning methods, where the parameters of the methods have been fixed (as presented in Table 4) and compared using the six performance measures in order to find the best model. The different models produced the values: *MAE* (0.1650×10^3 to 3.0424×10^3), *RMSE* (0.2140×10^3 to 4.2390×10^3), *IOS* (0.0755 to 0.7737), *R* (0.9315 to 0.9977), R^2 (0.8676 to 0.9954), and *IOA* (0.9360 to 0.9988) in concrete piles. Similarly, in the steel piles, we obtained *MAE* (0.1870×10^3 to 3.1064×10^3), *RMSE* (0.3100×10^3 to 4.3966×10^3), *IOS* (0.0448 to 0.9081), *R* (0.8478 to 0.9997), R^2 (0.7187 to 0.9994), and *IOA* (0.9118 to 0.9998). The results indicate that the best performance was obtained from the *DNN* model trained by the Tan-Sigmoid function. This model is said to be the most appropriate one because it displays the highest accuracy in terms of *MAE* ($0.1650 \times 10^3/0.1870 \times 10^3$), *RMSE* ($0.214 \times 10^3/0.31 \times 10^3$), *IOS* (0.0755/0.0448), *R* (0.9977/0.9997), R^2 (0.9954/0.9994), and *IOA* (0.9988/0.9998) in both concrete/steel piles. Finally, the most appropriate *DNN* model displayed the higher values of performance measures criteria in both the training and validation phase. Furthermore, this model is closely followed by the *GP* model, which shows an acceptable accuracy as it ranked second. Moreover, the results showed the poor performance of the *ELM* model in predicting the pile-bearing capacity. With respect to the performance of machine-learning models during the training phase, the performance hierarchy follows the following order: *DNN, GP, RF, Kridge, SVR, LS, Ridge, Step, PLS, Lasso,* and *ELM*. Finally, the scatter plots between the target and the output bearing capacity value of each model are presented in Appendix A (Figures A1–A11).

3.4. Evaluating the Best Fitted Model Using the K-Fold Cross-Validation Approach

The 5-fold cross-validation approach was effectively utilized to evaluate the predictive capability of the optimal model. It is worthy to note that the aforementioned studies interested in predicting the pile-bearing capacity have assessed the predictive capability of their optimal models based on one single split. Consequently, the ability of the models to overcome the over-fitting and under-fitting problems could not be verified. Figure 3 displays the performance measures of the optimal *DNN* models utilizing 5-fold cross-validation based on the validation data for each split. The results clearly indicate the fulfillment of the *DNN* model. Additionally, the fact that the correlation coefficient ranged between 0.9777 and 0.9998 for data validation in the 5 splits proved the predictive capability of the optimal *DNN* model to learn existing data, generate novel validation data, and overcome over-fitting and under-fitting problems.

Table 6. Performance indicators values of the AI models for predicting the pile-bearing capacity in both concrete and steel piles (bold: the optimal model).

	$MAE \times 10^3$	$RMSE \times 10^3$	*IOS*	*R*	R^2	*IOA*
Concrete piles						
DNN	0.1650	0.2140	0.0755	0.9977	0.9954	0.9988
ELM	3.0424	4.2390	0.7737	0.9320	0.8686	0.9610
Lasso	2.4324	3.5390	0.6637	0.9620	0.9254	0.9700
PLS	2.5524	3.6390	0.6837	0.9688	0.9386	0.9700
RF	1.1024	2.1690	0.3837	0.9880	0.9761	0.9912
Kridge	2.2930	3.5917	0.6816	0.9433	0.8899	0.9641
Ridge	2.4268	3.6145	0.6876	0.9409	0.8853	0.9636
LS	2.3093	3.5867	0.6824	0.9414	0.8863	0.9656
Step	2.4738	3.6421	0.6970	0.9352	0.8746	0.9626
SVR	1.9787	4.0984	0.7734	0.9315	0.8676	0.9360
GP	0.5966	0.9612	0.1731	0.9975	0.9951	0.9961
Steel piles						
DNN	0.1870	0.3100	0.0448	0.9997	0.9994	0.9998
ELM	3.1064	4.3966	0.9081	0.8478	0.7187	0.9118
Lasso	2.7149	3.6962	0.7527	0.8990	0.8082	0.9437
PLS	2.6329	3.6973	0.7763	0.8966	0.8038	0.9398
RF	1.1213	2.3475	0.4893	0.9875	0.9751	0.9712
Kridge	2.2482	3.6937	0.7342	0.8993	0.8088	0.9441
Ridge	2.3820	3.7165	0.7402	0.8969	0.8044	0.9436
LS	2.2646	3.6887	0.7350	0.8974	0.8054	0.9456
Step	2.4291	3.7441	0.7496	0.8912	0.7943	0.9426
SVR	1.9340	4.2004	0.8260	0.8875	0.7876	0.9160
GP	0.5518	1.0632	0.2257	0.9975	0.9951	0.9965

3.5. Comparison between the Proposed Models and Empirical Formulae

To test the effectiveness of the suggested *DNN* model, a comparative study was performed using 12 empirical models proposed in the literature of predicting the bearing capacity, as presented in Table 7. It should be noted that no author has shared the mathematical equations of the proposed *ML* model to compare results with the same database. Published research was limited in presenting modeling results. Therefore, we cannot validate the proposed models using the current collected dataset. Consequently, the current study was limited to compare the proposed models based on the correlation coefficient. It is needless to say that the correlation coefficient is an important indicator when assessing the prediction precision, as the best model is represented by a prediction value close to 1. The results of the comparative study indicated that the proposed *DNN* model in our study is the best-performing model, with maximum accuracy (0.9996 for all data). Furthermore, our model is closely followed by the *ANN* model which was proposed by Milad et al. [27], and it showed an acceptable accuracy as it ranked secondly. Moreover, the results revealed the poor performance of the ANN model proposed by Nawari et al. [25] in the bearing capacity. With respect to the performance of machine-learning models, the hierarchy follows the following order: Milad et al. [27], Liu et al. [32], Yong et al. [28], Moayedi et al. [34], Dehghanbanadaki et al. [30], Mahnesh [26], Kardani et al. [2], Jahed et al. [1], Shaik et al. [33], Moayedi and Jahed [31], Harandizadeh et al. [29], and Nawari et al. [25]. We believe that the reasonable ground standing behind the high accuracy found in our suggested model is due to deep learning (more than one hidden layer). The latter could offer the necessary flexibility for modeling complex functions in many cases.

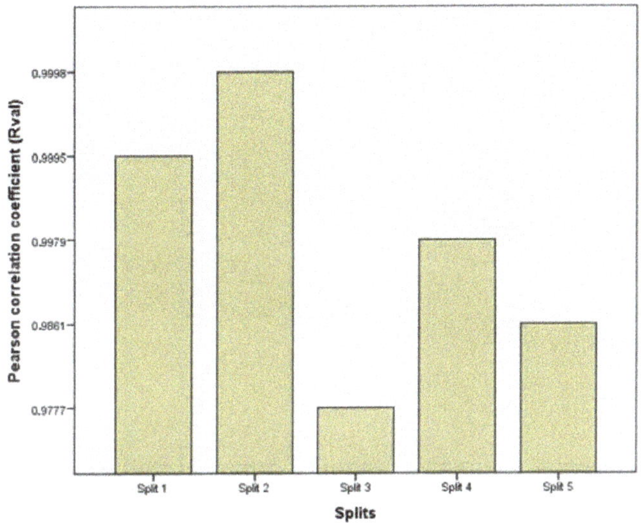

Figure 3. Performance measures of the DNN model using the *K*-fold cross-validation, with $K = 5$.

Table 7. Comparison between the proposed DNN model and some of the empirical models found in the literature.

Authors	Sample Size	Best Methods	Correlation Coefficient	References
Nawari et al. (1999)	25	*ANN*	0.91	[25]
Mahnesh (2011)	105	Generalized Regression Neural Network	0.977	[26]
Milad et al. (2015)	100	Neural Network	0.9995	[27]
Jahed et al. (2017)	132	*PSO–ANN*	0.9685	[1]
Moayedi and Jahed (2018)	59	*ICA-ANN*	0.96369	[31]
Yong et al. (2021)	50	*GP*	0.997	[28]
Shaik et al. (2019)	59	*ANFIS*	0.967	[33]
Kardani et al. (2020)	59	Extreme Gradient Boosting	0.975	[2]
Harandizadeh et al. (2021)	72	*ANFIS–GMDH–PSO*	0.94	[29]
Moayedi et al. (2020)	20	*GA–ANFIS*	0.9935	[34]
Liu et al. (2020)	43	*GA-ANN*	0.998	[32]
Dehghanbanadaki et al. (2021)	100	*MLP–GWO*	0.991	[30]
Our study	100	Deep Neural Network	0.9996	

3.6. Sensitivity Analysis

In order to know what input variables have a significant effect on the pile-bearing capacity, with the assistance of the DNN model, a sensitivity analysis was performed by utilizing the step-by-step technique [60]. In this method, each normalized input parameter varies at a constant rate, one at a time, while the other variables are held constant. Diverse constant rates (0.3, 0.6, and 0.9) were chosen in this study. For every input, the percentage of variation in the output, as a result of the variation in the input, was computed. The sensitivity of each input was computed based on Equation (7):

$$\text{Sensitivity level of } X_j(\%) = \frac{1}{K}\sum_{i=1}^{K}\left(\frac{\%\ change\ in\ output}{\%\ change\ in\ input}\right)_i \quad (7)$$

where K refers to the number of the datasets used in the study (K = 100). The outcomes of the sensitivity analysis of the proposed *DNN* model are illustrated in Figure 4. It can be noticed that the pile-bearing capacity was significantly influenced by the pile area, and its sensibility ratio ranged between 26.3% and 38.06%. This parameter is closely followed by the pile length, which showed a moderate sensitivity level that ranged between 15% and 19%. In addition, the cohesion and friction angle had a moderate effect on the pile-bearing capacity, with a sensibility ratio ranging between 9% and 15%. Finally, other parameters had little effect on the target values.

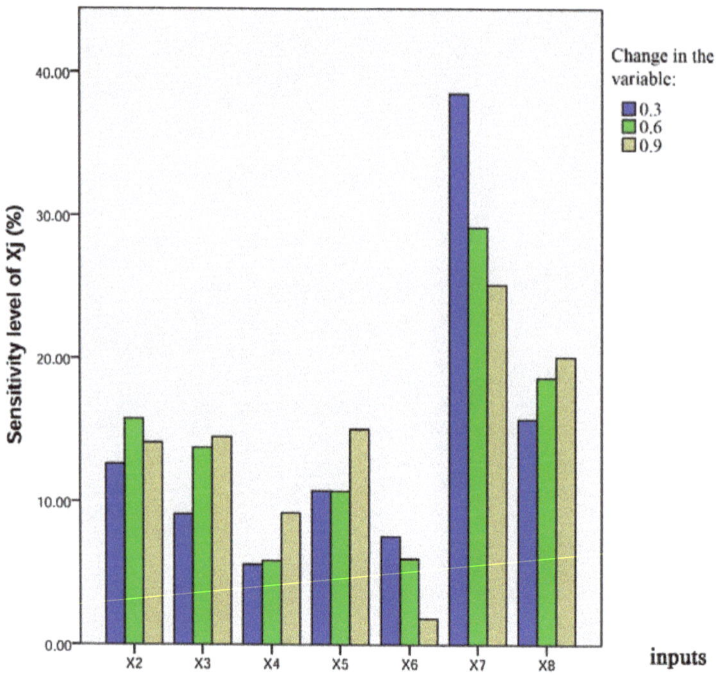

Figure 4. Results of the sensitivity analysis of the proposed model.

3.7. Graphical User Interface (GUI) Design "BeaCa2021"

It is a common practice in the majority of published papers using machine-learning modeling to present models in the form of mathematical equations, which suffer from their hard fitting in future studies. Seemingly, this practice has very little value for other researchers and civil engineers in the field. In order to make it useful, the proposed machine-learning architecture should be presented either in the form of a programmed interface such as Matlab or in a simple script employing a known programming language such as Python for generating the proposed model [55]. In such a case, the machine-learning model can be readily used and is thus available to anyone interested in the problem of modeling. In this study, a reliable, graphical, and easy-to-use interface was designed based on our optimal *DNN* model, as presented in Figure 5. The proposed optimal model was afterward used to develop a *GUI* public interface. The designed interface, called "BeaCa2021", was programmed by Matlab software. The reason for choosing this name is due to "Bea" relative to "Bearing", "Ca" relative to Capacity, and 2021, the year this interface was designed. In addition, BeaCa2021 includes the most relevant input parameters on the bearing capacity. Initially, the user must define the pile material type (either steel or concrete). Secondly, the user is required to introduce the other input parameters: average cohesion, average friction angle, average soil-specific weight, average pile-soil friction angle, flap number, pile area, and pile length. Finally, by clicking Run, the prediction result appears in the outputs. The suggested BeaCa2021 interface will be very useful to civil engineers and researchers, by helping them to predict the bearing capacity, which is deemed as one of the most complex parameters to determine.

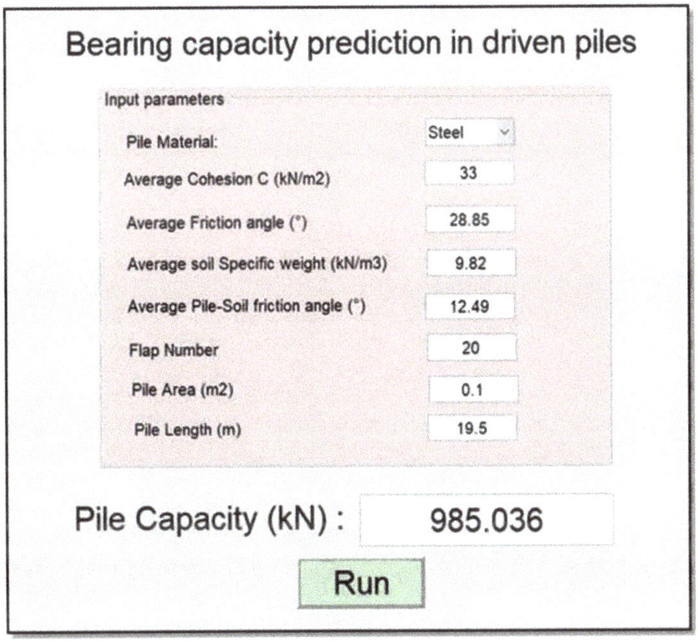

Figure 5. BeaCa2021 interface.

4. Discussion

In the current study, a very important contribution in the geotechnical community has been introduced for the sake of enhancing the performance of the pile-bearing capacity model. It is worth mentioning here that the model quality is influenced by the method utilized. Hence, other unused advanced machine-learning methods demonstrated efficient results in other areas. Consequently, in the current study, we examined the usage of twelve advanced machine-learning methods, such as Deep Neural Network (*DNN*), Extreme Learning Machine (*ELM*), Support Vector Regression (*SVR*), LASSO regression (*LASSO*), Random Forest (*RF*), Ridge Regression (*Ridge*), Partial Least Square Regression (*PLS*), Stepwise Regression (Stepwise), Kernel Ridge (*KRidge*), Genetic Programming (*GP*), and Least Square Regression (*LSR*), to predict the pile-bearing capacity. According to the authors' knowledge, the use of the aforementioned machine-learning methods in predicting the pile-bearing capacity is very rare. Therefore, this study began with collecting a wide range of data consisting of 100 static load-bearing tests on the *UBC* of both steel- and concrete-driven piles from different countries, such as Iran, Mexico, and India. Afterward, we selected eight relevant factors based on the literature recommendations, such as average cohesion (kN/m^2), average friction angle (°), average soil-specific weight (kN/m^3), average pile-soil friction angle (°), flap number, pile area (m^2), and pile length (m). Based on that, eleven advanced machine-learning methods (*DNN, ELM, SVR, LASSO, RF, Ridge, PLS, Stepwise, KRidge, GP,* and *LS*) were applied for modeling the selected optimal input set for the first time. The findings clearly indicate that the Deep Neural Network (*DNN*) presents the most appropriate model, which yielded the minimum values of error metrics (*MEA, RMSE,* and *IOS*) and the higher values of R^2, *R*, and *IOA* compared to other models. Furthermore, the newly developed model was assessed by the *K*-fold cross-validation method and compared to other proposed models from the literature based on the correlation coefficient. The conclusion drawn is that the optimal *DNN* model could produce new data without causing over-fitting or under-fitting, plus being much more precise than the other proposed empirical models. Moreover, the last part in the current study consisted of the sensitivity analysis, which provided an overview of the most influential parameters on the pile-bearing capacity according to the proposed model. The findings indicate that the pile area was the most influential factor on the pile-bearing capacity. Pile length also had a considerable effect. In addition, the cohesion and friction angle demonstrated a moderate effect on the pile-bearing capacity, with a sensibility ratio ranging between 9% and 15%. Finally, the proposed optimal model was then used to develop a *GUI* public interface in order to facilitate its usage in the future. A reliable, easy-to-use, and graphical interface, named "BeaCa2021, presented in the current study, was programmed via Matlab software. The essential advantage of "BeaCa2021" is to help researchers and civil engineers interested in the problem of modeling regardless of their proficiency, by offering them plenty of benefits, such as reliability, easiness, and lowering the budget used for predicting the pile-bearing capacity from relevant and easily obtained parameters without the need to operate expensive in situ tests.

The results obtained in the current study also proved that the performance of the pile-bearing capacity model was considerably enhanced by using new machine-learning methods. The model prediction by the *DNN* was improved by 8.91% with the *ANN* method proposed by Nawari et al. [25], 3.58% with the *PSO–ANN* method proposed by Jahed et al. [1], and 0.86% with the *MLP–GWO* method proposed by Dehghanbanadaki et al. [30]. The obtained results are logical because deep learning is generally employed either in the prediction or in the problematic classification, which can reduce the bias and variance plus avoiding over-fitting and under-fitting problems, as opposed to the traditional *ANN* methods, to improve their predictive capability. According to these data, we can infer that the *DNN* method, which was employed in this study for the first time for the purpose of modeling the pile-bearing capacity, could yield more effective and accurate results than the other machine-learning methods.

Despite the multiple extraordinary findings of this study, a number of important limitations need to be addressed. The fundamental limitation would be the fact that the sample size was relatively small, which may affect the precision of the pile-bearing capacity. This may lead to the proposed model's inability to generalize the new conditions or circumstances that were not used in the training data stage. Besides, researchers generally utilize large and diverse data collected by transferring knowledge between them. This is an important issue to build on in future research, i.e., to rely on the data gathered from multiple countries to enhance its learning and, therefore, produce a better model. Additionally, further studies using meta-heuristic algorithms for the prediction of pile-bearing capacity are strongly recommended. We mention, for example, the Particle Swarm Optimization (*PSO*) and Gravitational Search Algorithm (*GSA*), Bee Colony Algorithm (*BCA*), Bio-geography-Based Optimization (*BBO*), Whale Optimization Algorithm (*WOA*), Ant Colony Optimization (*ACO*), and Grey Wolf Optimizer (*GWO*). These algorithms have shown high-performance results when combined with machine-learning techniques, leading to improving their learning, and therefore rapidly converging to the best solution. The application of these meta-heuristic algorithms combined with machine-learning methods has shown impressive results in the abroad fields [55,61].

5. Conclusions

This study relied on a considerable number of steel- and concrete-driven pile data collected from different countries, such as Iran, Mexico, and India. The comparison of the results' assessment between the different proposed models revealed the superiority of the *DNN* model proposed in our study, which yielded the highest accuracy in terms of *MAE, RMSE, IOS, R*, R^2, and *IOA* in both the training/validation phases. The findings indicate that this model has a high correlation coefficient, ranging between 0.9777 and 0.9998 for the validation data in the 5 splits of the *k*-fold cross-validation approach, meaning that there was no over-fitting or under-fitting. Furthermore, the results indicated that the aforementioned *DNN* model is more effective compared to other empirical models proposed in the literature. The sensitivity analysis results proved that pile area had the most significant effect on the prediction of the pile-bearing capacity. Pile lengths had a moderate influence and were ranked second. In addition, cohesion and friction angle had little effect on the pile-bearing capacity. Finally, the proposed optimal model was then used to develop a *GUI* public interface with Matlab software, named "BeaCa2021". The fundamental benefit of "BeaCa2021" is to help researchers and practicing civil engineers, regardless of their proficiency, interested in the problem of modeling, to estimate the pile-bearing capacity with the benefits of gaining time and money.

This work has opened up several questions that need further investigations to overcome certain limitations. Firstly, there is a need to use more data from other countries to enhance the learning phase, which is needed to develop the BeaCa2021 in the future. Secondly, we propose the usage of meta-heuristic algorithms combined with machine-learning methods for predicting the pile-bearing capacity in future studies. These algorithms have demonstrated high-performance results when used with machine-learning techniques, leading to improved learning.

Supplementary Materials: The following are available online at https://www.mdpi.com/article/10.3390/app112210908/s1, Table S1: The Database about Two Types of Materials.

Author Contributions: Conceptualization, M.A.B.; methodology, M.A.B. and H.Z.; software, M.A.B. and L.G.; validation, M.A.B. and A.-I.P.; formal analysis, M.A.B. and L.L.; investigation, M.A.B.; resources, M.A.B.; data curation, M.A.B.; writing—original draft preparation, M.A.B., L.L., and A.-I.P.; writing—review and editing, M.A.B., L.L., and A.-I.P.; visualization, M.A.B. and A.-I.P.; supervision, M.A.B., H.Z., and A.-I.P.; project administration, M.A.B. and L.G.; funding acquisition, M.A.B. and A.-I.P. All authors have read and agreed to the published version of the manuscript.

Funding: This research received no external funding.

Institutional Review Board Statement: Not applicable.

Informed Consent Statement: Not applicable.

Conflicts of Interest: The authors declare no conflict of interest.

Appendix A

The scatter plots between target and output pile-bearing capacity values by the advanced machine-learning models.

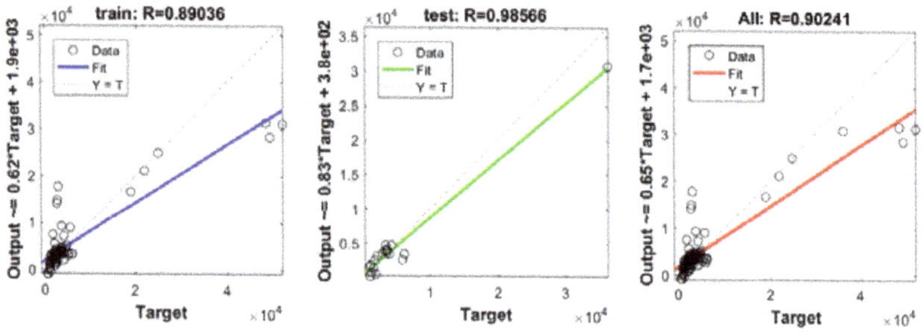

Figure A1. Scatter plots between target and output values by the *SVR* model.

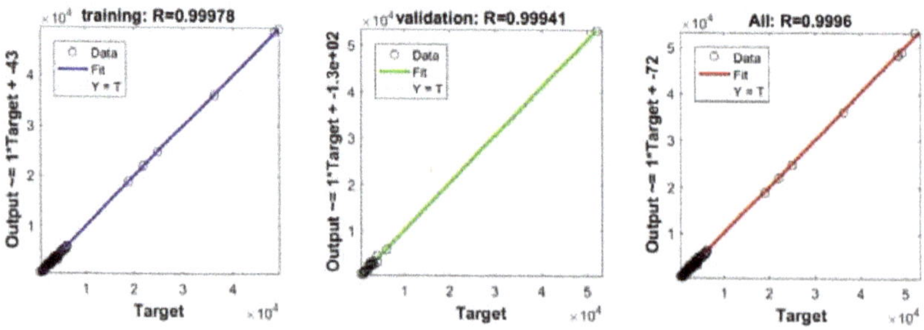

Figure A2. Scatter plots between target and output values by the *DNN* model.

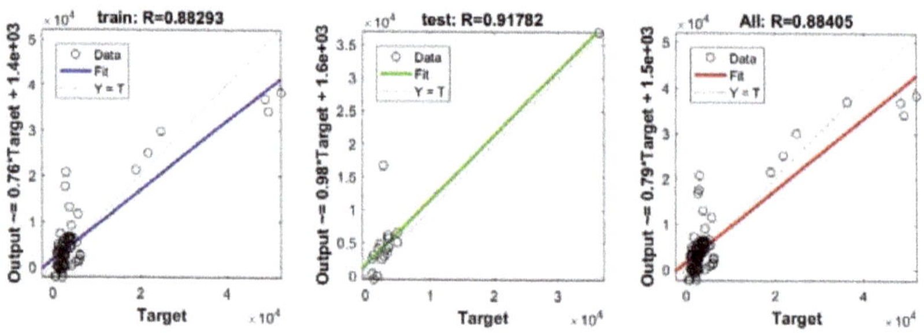

Figure A3. Scatter plots between target and output values by the *ELM* model.

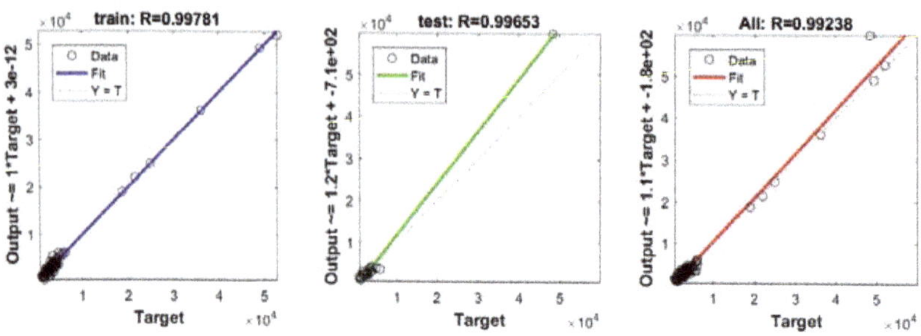

Figure A4. Scatter plots between target and output values by the *GP* model.

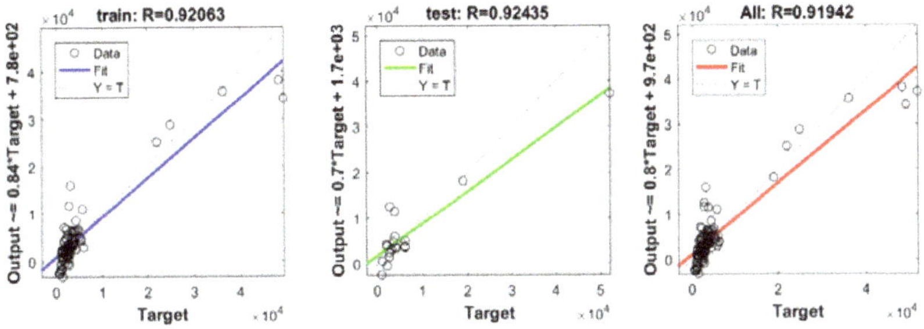

Figure A5. Scatter plots between target and output values by the *KRidge* model.

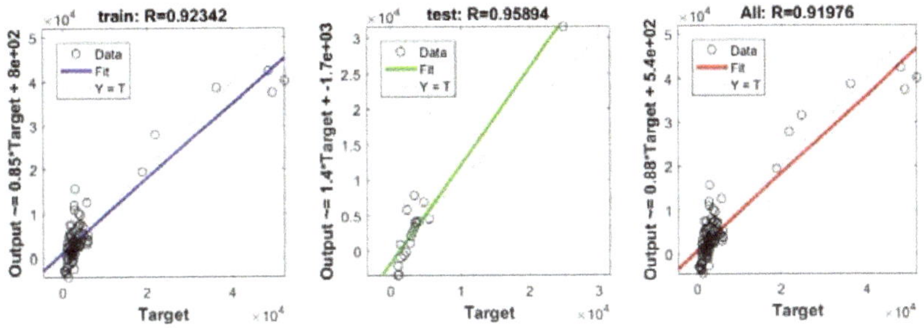

Figure A6. Scatter plots between target and output values by the *LASSO* model.

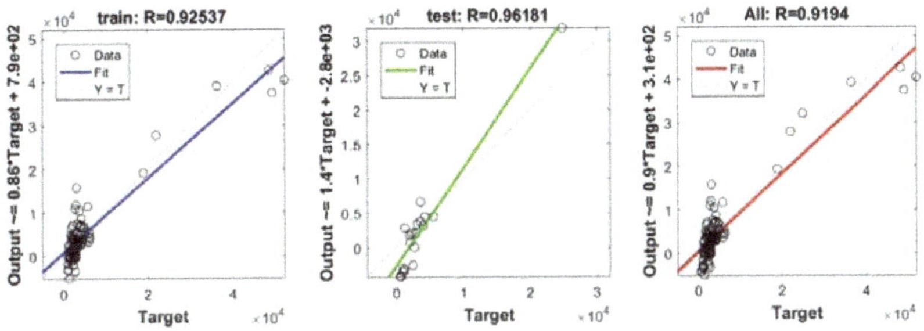

Figure A7. Scatter plots between target and output values by the *PLS* model.

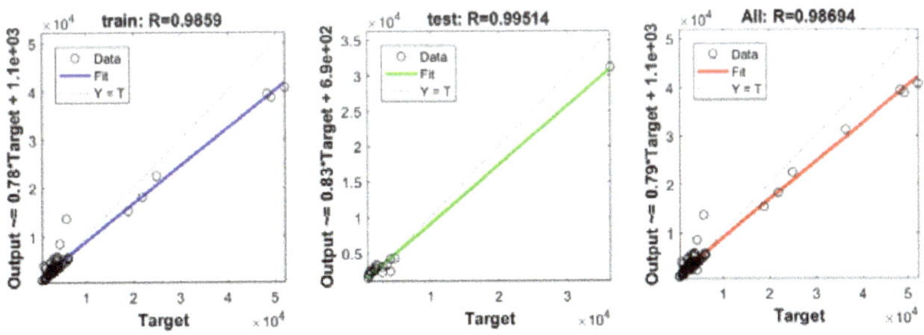

Figure A8. Scatter plots between target and output values by the *RF* model.

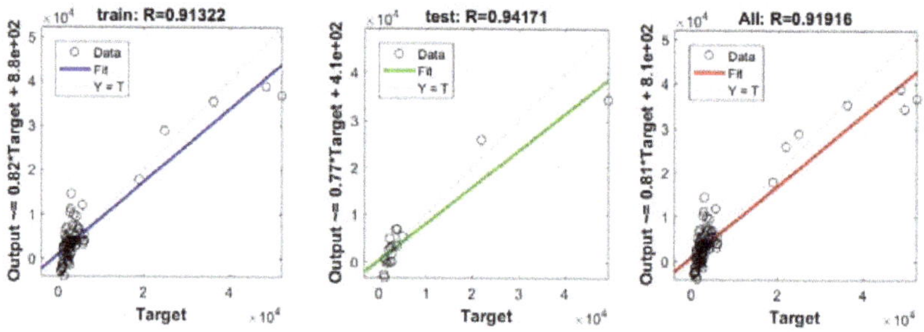

Figure A9. Scatter plots between target and output values by the *Ridge* model.

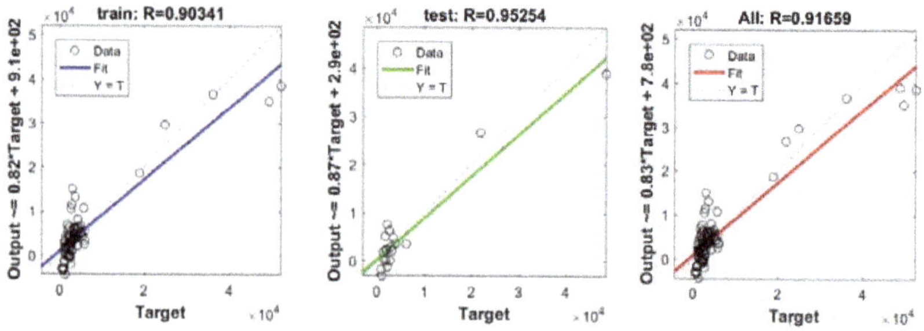

Figure A10. Scatter plots between target and output values by the *STEP* model.

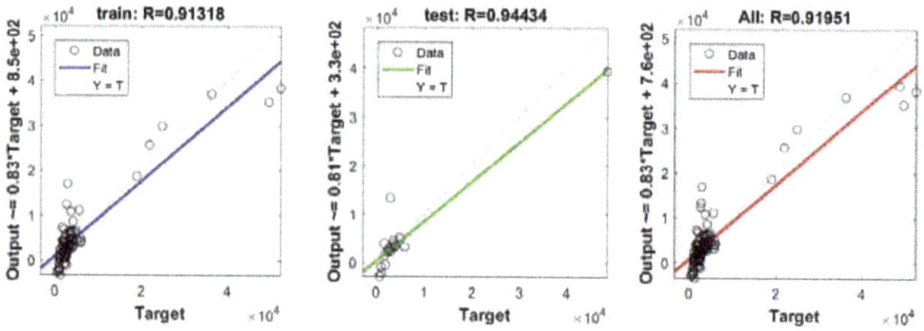

Figure A11. Scatter plots between target and output values by the *LS* model.

References

1. Jahed Armaghani, D.; Shoib, R.S.N.S.B.R.; Faizi, K.; Rashid, A.S.A. Developing a hybrid PSO–ANN model for estimating the ultimate bearing capacity of rock-socketed piles. *Neural Comput. Appl.* **2017**, *28*, 391–405. [CrossRef]
2. Yong, W.; Zhou, J.; Jahed Armaghani, D.; Tahir, M.M.; Tarinejad, R.; Pham, B.T.; Van Huynh, V. A new hybrid simulated annealing-based genetic programming technique to predict the ultimate bearing capacity of piles. *Eng. Comput.* **2021**, *37*, 2111–2127. [CrossRef]
3. Niazi, F.S.; Mayne, P.W. CPTu-based enhanced UniCone method for pile capacity. *Eng. Geol.* **2016**, *212*, 21–34. [CrossRef]
4. Meyerhof, G.G. Bearing Capacity and Settlement of Pile Foundations. *J. Geotech. Eng. Div.* **1976**, *102*, 197–228. [CrossRef]
5. Coyle, H.M.; Castello, R.R. New Design Correlations for Piles in Sand. *J. Geotech. Eng. Div.* **1981**, *107*, 965–986. [CrossRef]
6. Shahin, M.A. Intelligent computing for modeling axial capacity of pile foundations. *Can. Geotech. J.* **2010**, *47*, 230–243. [CrossRef]
7. Cai, G.; Liu, S.; Puppala, A.J. Reliability assessment of CPTU-based pile capacity predictions in soft clay deposits. *Eng. Geol.* **2012**, *141–142*, 84–91. [CrossRef]
8. Cai, G.; Liu, S.; Tong, L.; Du, G. Assessment of direct CPT and CPTU methods for predicting the ultimate bearing capacity of single piles. *Eng. Geol.* **2009**, *104*, 211–222. [CrossRef]
9. Eslami, A.; Heidarie Golafzani, S. Relevant data-based approach upon reliable safety factor for pile axial capacity. *Mar. Georesour. Geotechnol.* **2020**, *39*, 1373–1386. [CrossRef]

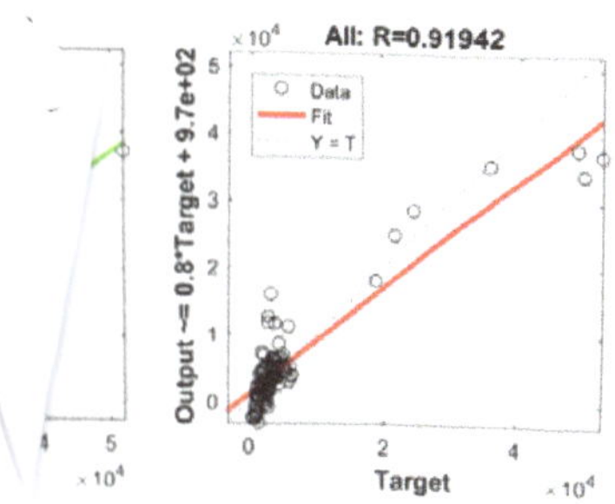

…ues by the *KRidge* model.

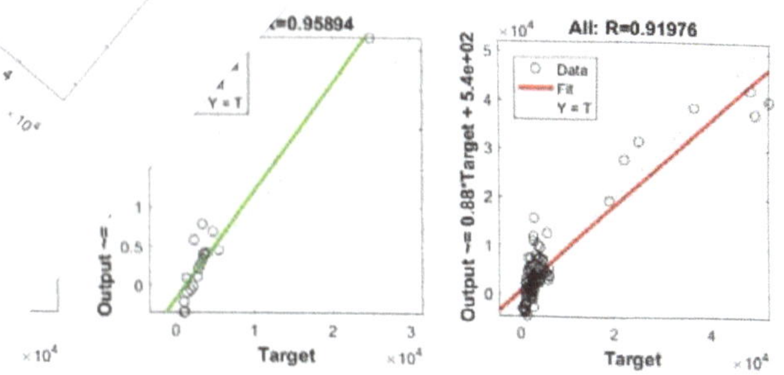

…plots between target and output values by the *LASSO* model.

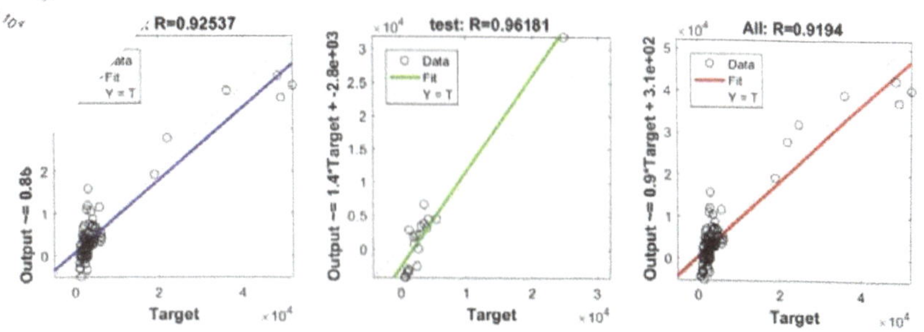

Figure A7. Scatter plots between target and output values by the *PLS* model.

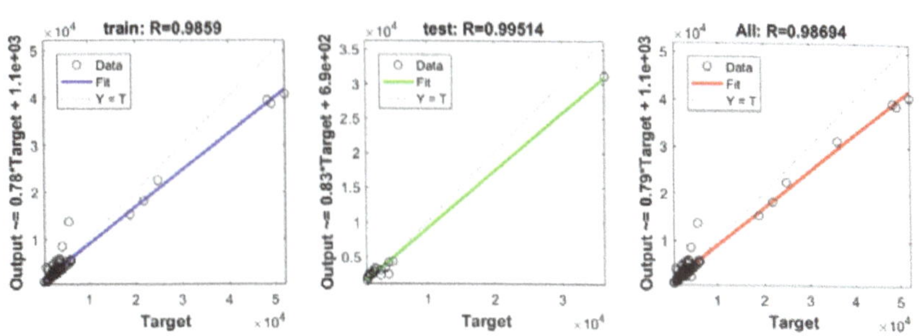

Figure A8. Scatter plots between target and output values by the *RF* model.

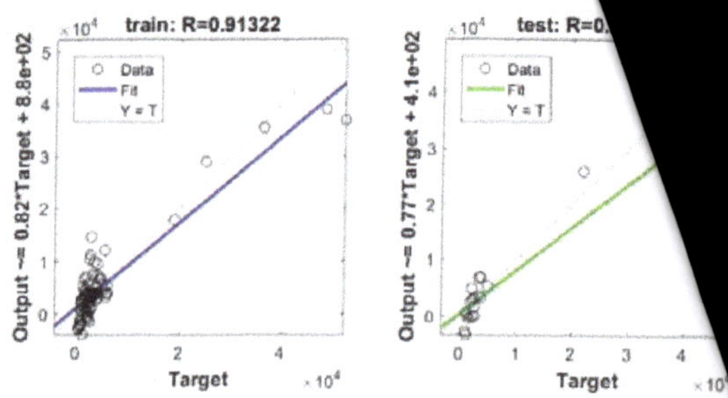

Figure A9. Scatter plots between target and output values by th

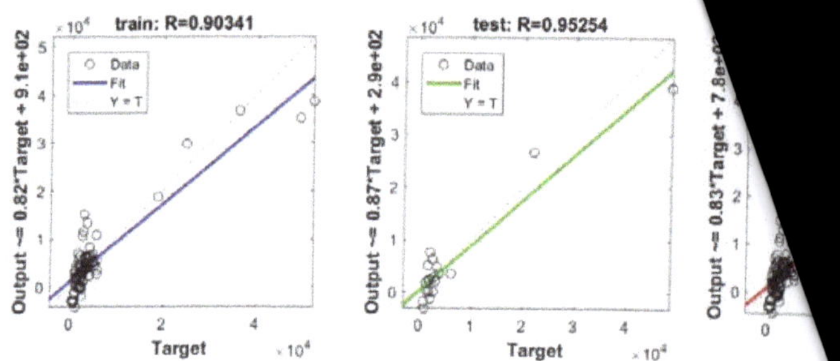

Figure A10. Scatter plots between target and output values by the *STEP* model.

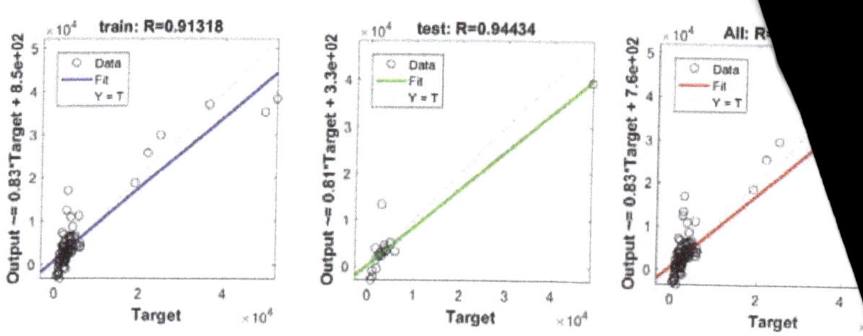

Figure A11. Scatter plots between target and output values by the *LS* model.

References

1. Jahed Armaghani, D.; Shoib, R.S.N.S.B.R.; Faizi, K.; Rashid, A.S.A. Developing a hybrid PSO–ANN model for estimating th ultimate bearing capacity of rock-socketed piles. *Neural Comput. Appl.* **2017**, *28*, 391–405. [CrossRef]
2. Yong, W.; Zhou, J.; Jahed Armaghani, D.; Tahir, M.M.; Tarinejad, R.; Pham, B.T.; Van Huynh, V. A new hybrid simulated annealing-based genetic programming technique to predict the ultimate bearing capacity of piles. *Eng. Comput.* **2021**, *37*, 2111–2127. [CrossRef]
3. Niazi, F.S.; Mayne, P.W. CPTu-based enhanced UniCone method for pile capacity. *Eng. Geol.* **2016**, *212*, 21–34. [CrossRef]
4. Meyerhof, G.G. Bearing Capacity and Settlement of Pile Foundations. *J. Geotech. Eng. Div.* **1976**, *102*, 197–228. [CrossRef]
5. Coyle, H.M.; Castello, R.R. New Design Correlations for Piles in Sand. *J. Geotech. Eng. Div.* **1981**, *107*, 965–986. [CrossRef]
6. Shahin, M.A. Intelligent computing for modeling axial capacity of pile foundations. *Can. Geotech. J.* **2010**, *47*, 230–243. [CrossRef]
7. Cai, G.; Liu, S.; Puppala, A.J. Reliability assessment of CPTU-based pile capacity predictions in soft clay deposits. *Eng. Geol.* **2012**, *141–142*, 84–91. [CrossRef]
8. Cai, G.; Liu, S.; Tong, L.; Du, G. Assessment of direct CPT and CPTU methods for predicting the ultimate bearing capacity of single piles. *Eng. Geol.* **2009**, *104*, 211–222. [CrossRef]
9. Eslami, A.; Heidarie Golafzani, S. Relevant data-based approach upon reliable safety factor for pile axial capacity. *Mar. Georesour. Geotechnol.* **2020**, *39*, 1373–1386. [CrossRef]

10. Momeni, E.; Nazir, R.; Jahed Armaghani, D.; Maizir, H. Prediction of pile bearing capacity using a hybrid genetic algorithm-based ANN. *Measurement* **2014**, *57*, 122–131. [CrossRef]
11. D18 Committee. *Test Method for High-Strain Dynamic Testing of Deep Foundations*; ASTM International: West Conshohocken, PA, USA, 2008.
12. Berezantzev, V.G. Design of deep foundations. In Proceedings of the Proc. 5th ICSMFE, Montreal, QC, Canada, 11–13 September 1965.
13. Hansen, J.B. Simple statical computation of permissible pileloads. *Christ. Nielsen Post* **1951**, *12*, 14–17.
14. De Beer, E.E. Etude des fondations sur pilotis et des fondations directes. *Ann. Trav. Publics Belqique* **1945**, *46*, 1–78.
15. Vesic, A.S. Design of pile foundations. In *NCHRP Synthesis of Highway Practice*; Transportation Research Board: Washington, DC, USA, 1977.
16. Abu-Farsakh, M.Y.; Titi, H.H. Assessment of Direct Cone Penetration Test Methods for Predicting the Ultimate Capacity of Friction Driven Piles. *J. Geotech. Geoenvironmental Eng.* **2004**, *130*, 935–944. [CrossRef]
17. Kordjazi, A.; Pooya Nejad, F.; Jaksa, M.B. Prediction of ultimate axial load-carrying capacity of piles using a support vector machine based on CPT data. *Comput. Geotech.* **2014**, *55*, 91–102. [CrossRef]
18. Maizir, H.; Suryanita, R.; Jingga, H. Estimation of pile bearing capacity of single driven pile in sandy soil using finite element and artificial neural network methods. In Proceedings of the International Conference on Engineering and Technology, Computer, Basic and Applied Sciences ECBA, Osaka, Japan, 28–29 November 2016.
19. Graine, N.; Hjiaj, M.; Krabbenhoft, K. 3D failure envelope of a rigid pile embedded in a cohesive soil using finite element limit analysis. *Int. J. Numer. Anal. Methods Geomech.* **2021**, *45*, 265–290. [CrossRef]
20. Conte, E.; Pugliese, L.; Troncone, A.; Vena, M. A Simple Approach for Evaluating the Bearing Capacity of Piles Subjected to Inclined Loads. *Int. J. Geomech.* **2021**, *21*, 04021224. [CrossRef]
21. Debiche, F.; Kettab, R.M.; Benbouras, M.A.; Benbellil, B.; Djerbal, L.; Petrisor, A.-I. Use of GIS systems to analyze soil compressibility, swelling and bearing capacity under superficial foundations inalgiers region, ALGERIA. *Urban. Arhit. Constr.* **2018**, *9*, 357–370.
22. Ikeagwuani, C.C. Estimation of modified expansive soil CBR with multivariate adaptive regression splines, random forest and gradient boosting machine. *Innov. Infrastruct. Solut.* **2021**, *6*, 199. [CrossRef]
23. Shahin, M.A.; Jaksa, M.B.; Maier, H.R. State of the art of artificial neural networks in geotechnical engineering. *Electron. J. Geotech. Eng.* **2008**, *8*, 1–26.
24. Benbouras, M.A.; Kettab, R.M.; Zedira, H.; Petrisor, A.-I.; Debiche, F. Dry density in relation to other geotechnical proprieties of Algiers clay. *Rev. Şcolii Dr. Urban.* **2017**, *2*, 5–14.
25. Nawari, N.O.; Liang, R.; Nusairat, J. Artificial intelligence techniques for the design and analysis of deep foundations. *Electron. J. Geotech. Eng.* **1999**, *4*, 1–21.
26. Mahesh, P. Modeling pile capacity using generalized regression neural network. In Proceedings of the Indian Geotechnical Conference, Kochi, India, 15–17 December 2011; No. N-027. pp. 811–814.
27. Milad, F.; Kamal, T.; Nader, H.; Erman, O.E. New method for predicting the ultimate bearing capacity of driven piles by using Flap number. *KSCE J. Civ. Eng.* **2015**, *19*, 611–620. [CrossRef]
28. Moayedi, H.; Raftari, M.; Sharifi, A.; Jusoh, W.A.W.; Rashid, A.S.A. Optimization of ANFIS with GA and PSO estimating α ratio in driven piles. *Eng. Comput.* **2020**, *36*, 227–238. [CrossRef]
29. Shaik, S.; Krishna, K.S.R.; Abbas, M.; Ahmed, M.; Mavaluru, D. Applying several soft computing techniques for prediction of bearing capacity of driven piles. *Eng. Comput.* **2019**, *35*, 1463–1474. [CrossRef]
30. Harandizadeh, H.; Jahed Armaghani, D.; Khari, M. A new development of ANFIS–GMDH optimized by PSO to predict pile bearing capacity based on experimental datasets. *Eng. Comput.* **2021**, *37*, 685–700. [CrossRef]
31. Moayedi, H.; Jahed Armaghani, D. Optimizing an ANN model with ICA for estimating bearing capacity of driven pile in cohesionless soil. *Eng. Comput.* **2018**, *34*, 347–356. [CrossRef]
32. Kardani, N.; Zhou, A.; Nazem, M.; Shen, S.-L. Estimation of Bearing Capacity of Piles in Cohesionless Soil Using Optimised Machine Learning Approaches. *Geotech. Geol. Eng.* **2020**, *38*, 2271–2291. [CrossRef]
33. Liu, L.; Moayedi, H.; Rashid, A.S.A.; Rahman, S.S.A.; Nguyen, H. Optimizing an ANN model with genetic algorithm (GA) predicting load-settlement behaviours of eco-friendly raft-pile foundation (ERP) system. *Eng. Comput.* **2020**, *36*, 421–433. [CrossRef]
34. Dehghanbanadaki, A.; Khari, M.; Amiri, S.T.; Armaghani, D.J. Estimation of ultimate bearing capacity of driven piles in c-φ soil using MLP-GWO and ANFIS-GWO models: A comparative study. *Soft Comput.* **2021**, *25*, 4103–4119. [CrossRef]
35. Alavi, A.H.; Gandomi, A.H.; Mollahasani, A.; Bazaz, J.B. Linear and tree-based genetic programming for solving geotechnical engineering problems. In *Metaheuristics in Water, Geotechnical and Transport Engineering*; Elsevier: Amsterdam, The Netherlands, 2013; pp. 289–310. ISBN 978-0-12-398296-4.
36. Narendra, B.S.; Sivapullaiah, P.V.; Suresh, S.; Omkar, S.N. Prediction of unconfined compressive strength of soft grounds using computational intelligence techniques: A comparative study. *Comput. Geotech.* **2006**, *33*, 196–208. [CrossRef]
37. Rezania, M.; Javadi, A.A. A new genetic programming model for predicting settlement of shallow foundations. *Can. Geotech. J.* **2007**, *44*, 1462–1473. [CrossRef]
38. Stockard, D.M. *Case Histories-Pile Driving in the Gulf of Mexico*; OnePetro: Houston, TX, USA, 1979.

39. Stockard, D.M. *Case Histories: Pile Driving Offshore India*; OnePetro: Houston, TX, USA, 1986.
40. Tucker, L.M.; Briaud, J.-L. *Analysis of the Pile Load Test Program at the Lock and Dam 26 Replacement Project*; Texas A&M University College Station Dept of Civil Engineering: College Station, TX, USA, 1988.
41. Gupta, R.C. Estimating Bearing Capacity Factors and Cone Tip Resistance. *J. Jpn. Geotech. Soc.* **2002**, *42*, 117–127. [CrossRef]
42. Fellenius, B.H.; Altaee, A. Pile dynamics in geotechnical practice—six case histories. In *Deep Foundations 2002: International Perspective Theory Design, Construction Perform*; American Society of Civil Engineers: Reston, VA, USA, 2012; pp. 619–631. [CrossRef]
43. Benbouras, M.A.; Kettab Mitiche, R.; Zedira, H.; Petrisor, A.I.; Mezouar, N.; Debiche, F. A new approach to predict the Compression Index using Artificial Intelligence Methods. *Mar. Georesources Geotechnol.* **2019**, *37*, 704–720. [CrossRef]
44. Benbouras, M.A.; Kettab, R.M.; Zedira, H.; Debiche, F.; Zaidi, N. Comparing Nonlinear Regression Analysis And Artificial Neural Networks To Predict Geotechnical Parameters From Standard Penetration Test. *Urban. Archit. Constr. Arhit. Constr.* **2018**, *9*, 275–288.
45. Huang, G.-B.; Zhu, Q.-Y.; Siew, C.-K. Extreme learning machine: Theory and applications. *Neurocomputing* **2006**, *70*, 489–501. [CrossRef]
46. Biau, G.; Scornet, E. A random forest guided tour. *Test* **2016**, *25*, 197–227. [CrossRef]
47. Awad, M.; Khanna, R. Support vector regression. In *Efficient Learning Machines: Theories, Concepts, and Applications for Engineers and System Designers*; Awad, M., Khanna, R., Eds.; Apress: Berkeley, CA, USA, 2015; pp. 67–80. ISBN 978-1-4302-5990-9.
48. Vinzi, V.E.; Chin, W.W.; Henseler, J.; Wang, H. Editorial: Perspectives on partial least squares. In *Handbook of Partial Least Squares: Concepts, Methods and Applications*; Springer Handbooks of Computational Statistics; Esposito Vinzi, V., Chin, W.W., Henseler, J., Wang, H., Eds.; Springer: Berlin/Heidelberg, Germany, 2010; pp. 1–20. ISBN 978-3-540-32827-8.
49. Hebiri, M.; Lederer, J. How Correlations Influence Lasso Prediction. *IEEE Trans. Inf. Theory* **2013**, *59*, 1846–1854. [CrossRef]
50. Douak, F.; Melgani, F.; Benoudjit, N. Kernel ridge regression with active learning for wind speed prediction. *Appl. Energy* **2013**, *103*, 328–340. [CrossRef]
51. Hoerl, A.E.; Kennard, R.W. Ridge Regression—1980: Advances, Algorithms, and Applications. *Am. J. Math. Manag. Sci.* **1981**, *1*, 5–83. [CrossRef]
52. Jennrich, R.I.; Sampson, P.F. Application of Stepwise Regression to Non-Linear Estimation. *Technometrics* **1968**, *10*, 63–72. [CrossRef]
53. Wagner, S.; Kronberger, G.; Beham, A.; Kommenda, M.; Scheibenpflug, A.; Pitzer, E.; Vonolfen, S.; Kofler, M.; Winkler, S.; Dorfer, V.; et al. Architecture and design of the HeuristicLab optimization environment. In *Advanced Methods and Applications in Computational Intelligence*; Topics in Intelligent Engineering and Informatics; Klempous, R., Nikodem, J., Jacak, W., Chaczko, Z., Eds.; Springer International Publishing: Berlin/Heidelberg, Germany, 2014; pp. 197–261. ISBN 978-3-319-01436-4.
54. Tikhamarine, Y.; Malik, A.; Pandey, K.; Sammen, S.S.; Souag-Gamane, D.; Heddam, S.; Kisi, O. Monthly evapotranspiration estimation using optimal climatic parameters: Efficacy of hybrid support vector regression integrated with whale optimization algorithm. *Environ. Monit. Assess.* **2020**, *192*, 696. [CrossRef] [PubMed]
55. Amin Benbouras, M.; Petrisor, A.-I. Prediction of Swelling Index Using Advanced Machine Learning Techniques for Cohesive Soils. *Appl. Sci.* **2021**, *11*, 536. [CrossRef]
56. Breiman, L.; Spector, P. Submodel Selection and Evaluation in Regression. The X-Random Case. *Int. Stat. Rev. Rev. Int. Stat.* **1992**, *60*, 291–319. [CrossRef]
57. Oommen, T.; Baise, L.G. Model development and validation for intelligent data collection for lateral spread displacements. *J. Comput. Civ. Eng.* **2010**, *24*, 467–477. [CrossRef]
58. Goetz, J.N.; Brenning, A.; Petschko, H.; Leopold, P. Evaluating machine learning and statistical prediction techniques for landslide susceptibility modeling. *Comput. Geosci.* **2015**, *81*, 1–11. [CrossRef]
59. Amin, B. Predicting Shear Stress Parameters in Consolidated Drained Conditions Using Artificial Intelligence Methods. *Basic Appl. Sci.-Sci. J. King Faisal Univ.* **2021**, *22*, 1–7. [CrossRef]
60. Liong, S.-Y.; Lim, W.-H.; Paudyal, G.N. River Stage Forecasting in Bangladesh: Neural Network Approach. *J. Comput. Civ. Eng.* **2000**, *14*, 1–8. [CrossRef]
61. El Amin Bourouis, M.; Zadjaoui, A.; Djedid, A. Contribution of two artificial intelligence techniques in predicting the secondary compression index of fine-grained soils. *Innov. Infrastruct. Solut.* **2020**, *5*, 96. [CrossRef]

applied sciences

MDPI

Article

Dam Deformation Interpretation and Prediction Based on a Long Short-Term Memory Model Coupled with an Attention Mechanism

Yan Su [1], Kailiang Weng [1], Chuan Lin [1,*] and Zeqin Chen [2]

[1] College of Civil Engineering, Fuzhou University, Fuzhou 350108, China; suyan@fzu.edu.cn (Y.S.); wkl961029@126.com (K.W.)
[2] Electric Power Research Institute of State Grid Fujian Electric Power Co. Ltd., Fuzhou 350007, China; chzq1991@163.com
* Correspondence: linchuan@fzu.edu.cn

Abstract: An accurate dam deformation prediction model is vital to a dam safety monitoring system, as it helps assess and manage dam risks. Most traditional dam deformation prediction algorithms ignore the interpretation and evaluation of variables and lack qualitative measures. This paper proposes a data processing framework that uses a long short-term memory (LSTM) model coupled with an attention mechanism to predict the deformation response of a dam structure. First, the random forest (RF) model is introduced to assess the relative importance of impact factors and screen input variables. Secondly, the density-based spatial clustering of applications with noise (DBSCAN) method is used to identify and filter the equipment based abnormal values to reduce the random error in the measurements. Finally, the coupled model is used to focus on important factors in the time dimension in order to obtain more accurate nonlinear prediction results. The results of the case study show that, of all tested methods, the proposed coupled method performed best. In addition, it was found that temperature and water level both have significant impacts on dam deformation and can serve as reliable metrics for dam management.

Keywords: dam deformation; attention mechanism; long short-term memory; dam safety monitoring; prediction

Citation: Su, Y.; Weng, K.; Lin, C.; Chen, Z. Dam Deformation Interpretation and Prediction Based on a Long Short-Term Memory Model Coupled with an Attention Mechanism. *Appl. Sci.* **2021**, *11*, 6625. https://doi.org/10.3390/app11146625

Academic Editor: Nikos D. Lagaros

Received: 9 June 2021
Accepted: 15 July 2021
Published: 19 July 2021

1. Introduction

As a crucial social engineering infrastructure, dams must be operated safely to guarantee the needs of a steadily growing national economy are met. Unfortunately, due to the inherent physical limitations of dam materials, dams often have unhealthy structural responses such as dam body cracking and abnormal deformation [1]. In order to reduce the probability of engineering failures, most dams are equipped with precise health monitoring systems to evaluate their operational behavior and health through real-time measurements of multiple structural and environmental indicators. Among the many monitoring indicators, dam deformation is easy to measure and intuitively reflects the overall structural response state [2]. In order to improve the effectiveness of management strategies, research focused on accurately predicting dam deformation has increased in recent years. This area of research commonly uses simulations, and the most commonly used forecasting models can be categorized as mathematical statistical models or artificial intelligence models.

Hydrostatic-seasonal-time (HST) can be considered a representative flagship statistical regression model, it quantitatively interprets the influencing factors behind dam deformation based on the assumptions of mechanical theory and, then, performs a linear approximation fitting using the observed data. It was originally proposed by Willm et al. [3] to forecast deformation of concrete dams and has since been widely implemented. However, there is a strong correlation between dam water level and ambient temperature, which directly influences environmental loads and dam integrity, but the HST model does not consider local air temperature, which will be detrimental to the prediction accuracy under extreme weather conditions. To make up for these shortcomings, Penot et al. [4] proposed the hydrostatic-seasonal-time-temperature (HSTT) model by correcting the thermal component based on the actual air temperature. Another common approach has been to replace the thermal component with the actual temperature inside dams, which prompted the advent of the hydrostatic-thermal-time (HTT) model. Mata et al. [5] used a combined principal component analysis (PCA) method to include dam temperature in the model construction and applied the HTT model to explain the observed displacement of a concrete dam more accurately and with a lower residual standard deviation. In addition, there is a certain delay before a dam responds to changes in load, this can be observed in the influence of water level on pore water pressure and temperature on the thermal field of dams. The most popular solution is to add moving averages or gradients of original variables to the model to supplement the delayed information. For example, Popovici et al. [6] added the moving averages of air temperature for the previous 3, 10, and 30 days and the water level for the previous 3 days as variables to improve the performance of the dam deformation prediction model.

Mathematical statistical models generally output linear relationships between impact factors and target variables. The coefficients are determined by the least squares method using a building process that is simple and easy to understand. However, in practice, the relationship between dam deformation and impact factors is rarely linear and the capacity of the above models to capture nonlinear features and generalize is insufficient. To address this deficit, artificial intelligence algorithms based on machine learning have gradually attracted more attention in dam deformation prediction. The application of artificial intelligence models in dam safety monitoring systems has since become another important research subject, using approaches such as support vector machine (SVM), random forest (RF), and gaussian process (GP). The machine learning algorithm captures the characteristics of observed data through specific algorithm steps and uses the extracted characteristic information to continuously update the model to achieve the best fit. Through various complex processing operations, machine learning algorithms can obtain highly accurate predictive models that meet the management needs of safety monitoring. Mata [7] introduced a prediction algorithm based on artificial neural networks (ANN) to map the relationship between the load and concrete dam deformation and compared it with the multiple linear regression (MLR) model. The results showed that the ANN model provided a better fit than the traditional statistical model under extreme temperatures. In addition, Kao et al. [8] showed that the information provided by small static deformations can be enhanced by ANN-based methods. Furthermore, they developed a threshold level method for diagnosing the health of dams, and the impact of different factors on the health of dams was analyzed in detail. Recently, Ranković et al. [9] constructed an SVM nonlinear autoregressive model with exogenous inputs to predict the nonlinear behavior of a dam's structure. The safety measures protecting dams can be improved by being able to accurately predict the displacements of dams. Kang et al. [10] demonstrated the accuracy of a dam deformation prediction model based on the GP method, which added the average air temperature and temperature lag information as input variables to predict the radial displacement of a concrete dam. In subsequent prediction comparisons, their GP model had the smallest error value. More recently, combinations of multiple machine learning algorithms have received increasing attention. Ren et al. [11] used a fruit fly optimization algorithm to upgrade the SVM and applied it to the hysteresis correction of dam deformation impact factors. Subsequently, Su [12] proposed an SVM model with a wavelet based kernel function that made full use of the discrete transformation of the wavelet function. Li et al. [13] used the PCA method to extract the effective information from the dam temperature data as the input for the SVM model, effectively filtering redundant information from the input variables. However, the high-dimensional nonlinear tasks and the characteristic representation of time-varying dam deformation undoubtedly present a huge challenge for traditional shallow learning, meaning that the prediction accuracy of traditional machine learning algorithms is becoming increasingly unable to meet the needs of many engineering management tasks. In recent years, another branch of artificial intelligence technology, deep learning, has been vigorously developed in various industries, these approaches include convolutional neural networks (CNN), which are applied in image processing [14–16] and speech recognition [17–19], and long short-term memory (LSTM) models, which are applied in time series processing [20–22]. Liu et al. [23] proposed an approach coupling PCA with LSTM to make short-term and long-term predictions of dam observation data. Qu et al. [24] compared LSTM and SVM prediction algorithms for dam deformation monitoring. Xu et al. [25] decomposed dam deformation time series into linear and nonlinear parts, then used traditional statistical models to fit the linear part, while LSTM was used to capture the sequence features of the nonlinear part. Deep learning uses a layered structure to embody abstract non-linear relationships and superimposes this structure to improve the expressive ability to map complex relationships. Each layer transfers information to another, with the output of the current layer being used as the input of the next layer, until the final output is obtained. After multiple layers of feature extraction and complex information representation, a sequence feature representation model can be obtained. This layered architecture makes deep learning highly customizable, allowing it to achieve a better prediction accuracy than traditional shallow learning. Furthermore, most studies focus on improving the accuracy of predictive models, but ignore the interpretation and evaluation of input variables, which can be considered using deep learning models.

To better consider the influence of time dimension, this paper coupled an attention mechanism with an LSTM network to develop a dam deformation prediction algorithm. The attention mechanism in the time dimension can preferentially allocate the limited information processing resources in the short term to key data, while the LSTM network can extract long-term change trends from dam deformation time series. This coupled model is able to obtain more accurate dam deformation prediction results while also enriching variable interpretation in the time dimension of the prediction model. During actual monitoring, the physical deformation sensor can be affected by environmental (external) factors or internal factors, which can produce abnormal data due to equipment error. Therefore, the density-based spatial clustering of applications with noise (DBSCAN) density-based clustering algorithm is introduced to eliminate equipment-based abnormal values in real time to ensure that the observed data meet the subsequent modeling accuracy requirements. Then, the relative importance of each input variable is obtained through the variable importance measure data processing method, which not only enriches the information interpretation of the model, but also screens redundant information to reduce the difficulty of modeling. The performance of the proposed model was verified by the real-world concrete gravity dam deformation data. The main contributions of this paper can be summarized as follows:

1. This paper proposed and tested a DBSCAN method to filter the dam deformation time series data. The method effectively removed the equipment based abnormal values caused by environmental factors or equipment failures, thereby smoothing the random measurement errors in the observed data, which improved prediction accuracy.
2. The importance of input variables to the dam deformation prediction model was analyzed to interpret and evaluate the model. This resulted in a useful and efficient qualitative measure of dam deformation, which improved prevention and control of abnormal structural responses.
3. A coupled model was developed to better address the needs of dam deformation prediction. An attention mechanism focuses on the important variables in the short-term time dimension, while the LSTM model captures long-term change characteristics. This algorithm is very suitable for the prediction of dam deformation by accounting for time lag.

The remainder of this paper is organized as follows: Section 2 describes the detailed process of the established mathematical model of dam deformation and the preprocessing method. Section 3 introduces the selection of input variables, and the design and operation of the comparative experimental model. Section 4 elaborates on a case study from which actual monitoring data was collected. Section 5 explains and analyzes the input variables and evaluates the performance of the proposed coupled method. Finally, conclusions and future research directions are provided in Section 6.

2. Methods

2.1. Modeling Dam Deformation

Dam deformation is a key indicator reflecting the structural health of the dam. During project operation, polynomial functions are often used to approximate the dam deformation as:

$$\delta = \delta_H + \delta_T + \delta_\theta \tag{1}$$

where δ denotes the dam deformation, and the subscripts H, T, and θ in the formula respectively represent the components of the dam deformation caused by hydrostatic pressure, temperature, and aging effect over time.

2.1.1. Hydrostatic Pressure Component

A simplified two-dimensional model of a homogeneous gravity dam is taken as an analysis example, as shown in Figure 1, we obtain the following explanation through the mechanical relationship between water level and deformation. Under hydraulic load, the horizontal displacement δ_H generated at any measuring point of the dam is composed of four parts, as shown in Figure 1: the displacement δ_{1H} caused by the deformation of foundation due to internal forces on foundation surface and the displacement δ_{2H} caused by rotation of foundation surface due to the gravity of the reservoir water, the displacement δ_{3H} caused by the rotation of reservoir water pressure acting on dam body, and the shear horizontal displacement δ_{4H} caused by the internal force of reservoir water pressure acting on dam body. Therefore, the dam deformation is expressed by:

$$\delta = \delta_{1H} + \delta_{2H} + \delta_{3H} + \delta_{4H} \tag{2}$$

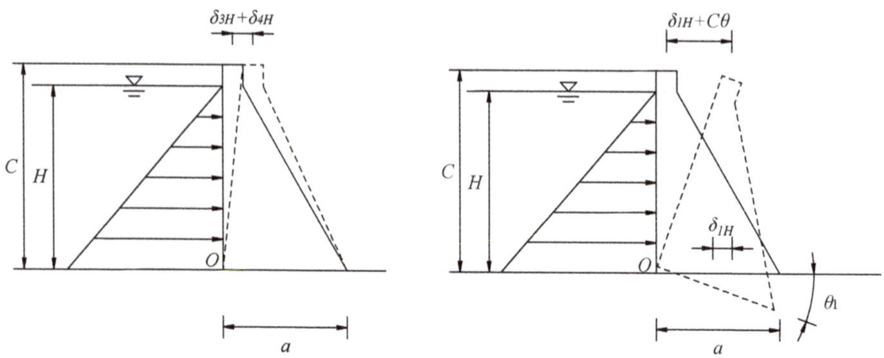

(a) (b)

Figure 1. *Cont.*

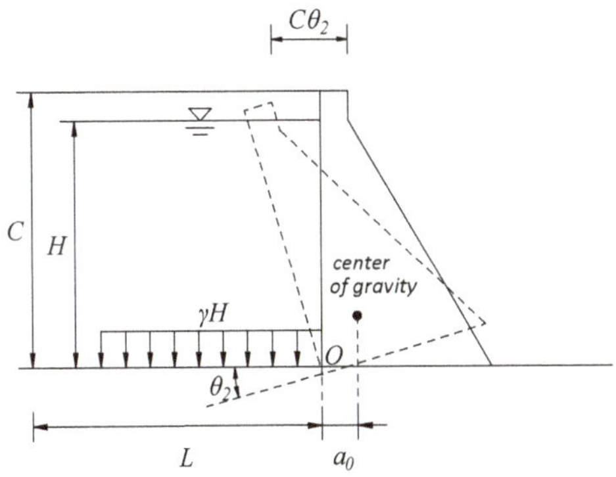

(c)

Figure 1. Schematic of the influence of hydraulic load on horizontal dam displacement: (**a**) horizontal displacement of dam foundation; (**b**) horizontal displacement caused by rotation of the dam foundation; and (**c**) horizontal displacement caused by dam rotation and shear horizontal movement. (unit: m).

According to F.Vogt theory [26], the water pressure to the bottom of a dam causes the horizontal displacement δ_{1H} and rotation angel θ_1 of dam foundation as:

$$\delta_{1H}= (\gamma K_1/6E_f a)H^3 + (\gamma K_2/2E_f)H^2. \quad (3)$$

$$\theta_1= (\gamma K_3/6E_f a^2)H^3 + (\gamma K_1/2E_f a)H^2. \quad (4)$$

where E_f denotes the elastic modulus of bedrock; γ is unit weight of water; the width of dam bottom is a; H presents the upstream water depth; and K_1, K_2, and K_3 are the coefficients that depend on the Poisson's ratio of the bedrock and the length-to-width ratio of the equivalent rectangle at the bottom of the dam [27].

If the length of the reservoir L is very large, the weight of water acting on reservoir in front of dam will deform reservoir bank and cause rotation θ_2 at the bottom of dam as:

$$\theta_2= -H(1+\mu_f)\ln(n + \sqrt{n^2+1})/\pi E_f \quad (5)$$

where μ_f is the Poisson ratio of bedrock and n is half the ratio of the width of water in front of dam to the distance a_0 from the center of gravity of the dam to the heel of the dam.

Therefore, the horizontal displacement δ_{2H} of crest caused by rotation of dam foundation θ as:

$$\delta_{2H} = c\theta= (\theta_1 + \theta_2)c \quad (6)$$

The horizontal displacement δ_{3H} of dam crest caused by the rotation θ_3 of dam body as:

$$\delta_{3H} = \int_0^c \int_0^c Mdydy/E_d I \quad (7)$$

where $M = \gamma H^3/6$ is the bending moment caused by water pressure, which is proportional to the third power of water depth; E_d is the elastic modulus of dam concrete material; I is the moment of inertia of the horizontal section of dam and c is dam height.

The dam shear horizontal displacement δ_{4H} as:

$$\delta_{4H} = \int_0^c 2k(1+\mu_d)Qdy/AE_d \quad (8)$$

where $Q = \gamma h^2/2$ is shear force caused by water pressure, which is proportional to the third power of the water depth; A is the horizontal area section of dam; k is the shear force distribution coefficient on the section, about 1.2; and μ_d is the Poisson's ratio of the concrete material of dam.

From the above analysis, it can be found that the horizontal displacement of dam crest caused by the water depth H in front of the dam is a function of H, H^2, and H^3. Among them, the deflection displacement caused by the bending moment M is mainly related to H^3, the tangential displacement caused by the shearing force Q is mainly related to H^2, and the slope-deflection at the tilt of reservoir bottom caused by the weight of water is related to H. Therefore, the hydrostatic pressure component in the horizontal displacement of a concrete gravity dam can be established as a mathematical model as follows:

$$\delta_H = a_0 + \sum_{i=1}^{3} a_i H^i \tag{9}$$

where a_0 and a_i are regression coefficients.

In addition, if the downstream water level changes greatly and the upstream and downstream water level difference is not obvious, the impact of downstream water level on monitoring should be considered, as:

$$\delta_H = a_0 + \sum_{i=1}^{3} a_{1i} H_1^i + \sum_{i=1}^{3} a_{2i} H_2^i \tag{10}$$

where H_1 is the water level of upstream and H_2 is the water level of downstream.

2.1.2. Temperature Component

Dam deformation is also affected by temperature, which is the displacement caused by temperature changes in the concrete of dam and rock foundation. When the dam has been in service for many years, the hydration heat of the concrete material has been dissipated, and the internal temperature of the dam body reaches a quasi-stable temperature field. At this time, it only depends on the boundary temperature variation, and the temperature component presents a simple harmonic periodic change. In order to qualitatively analyze the temperature component, multi-cycle harmonics can be selected as a factor to simplify and simulate it as:

$$\delta_T = \sum_{i=1}^{2} [b_{1i} \sin(2\pi i t/365) + b_{2i} \cos(2\pi i t/365)] \tag{11}$$

2.1.3. Time Component

The time component is an irreversible component that developed in a certain direction with the passage of time. Under the influence of a variety of factors, the dam body and rock foundation undergo plastic deformation, which leads to the rapid change of the time-effect displacement at the initial stage and gradually stabilizes in the later stage. According to existing research, linear functions and logarithmic functions can be used to model with time effects as:

$$\delta_\theta = c_1 t + c_2 \ln t \tag{12}$$

where c_1 and c_2 are regression coefficients.

2.2. Density-Based Spatial Clustering of Applications with Noise

In practical engineering, sensor monitoring is usually subject to harmful factors such as harsh environments or network transmission errors, resulting in abnormal values of equipment in data information. The density-based spatial clustering of applications with the noise (DBSCAN) method is introduced to reduce these random errors and eliminate device outliers, which has been proven to be capable of handling large database [28–30]. The DBSCAN is a density-based clustering algorithm. It assigns any point to a specified radius area and calculates the total number of points in the area to obtain the density of the specified point. When the density is higher than the preset maximum point set threshold, these points will be constructed into clusters. Based on this concept, a huge feature space is grouped into multiple regions with high density, so as to achieve the purpose of screening out random outliers of the monitoring equipment. We choose the DBSCAN method because it has the ability to divide clusters of arbitrary shapes in the noise spatial database, such as linear, elliptical shapes, etc. The detailed process of DBSCAN is described in Algorithm 1, where n and m are the number of samples and clusters, respectively. k represents any observed data in the original data set. k_i represents the i-th observed data in the original dataset. x represents the number of observed data existing in the radius of k_i. N represents the set of all observed points in each cluster. y represents the number of observed data existing in the radius of k_i'.

Algorithm 1 Equipment outlier filtering

Inputs: Dataset $D^{(n\times1)}$, radius ε, domain density threshold $MinPts$
Outputs: Density-based cluster set $D_C{}^{(m\times1)}$
  Marking k in $D^{(n\times1)}$ as UNVISITED, $k \in [1, n]$
  For $k_i$in $D^{(n\times1)}$:
    If k_i is UNVISITED
      Marking k_i as VISITED
      If x>= $MinPts$
        Create a new cluster C and add k_i to C to form a list set N
        For each point k_i' in N
          If k_i' is UNVISITED
            Marking k_i' is VISITED
            If y>= $MinPts$ and each one in y $\notin$ any cluster
              Adding them into C
            End if
          End if
        End for
      Else
        Marking k_i as outlier
      End if
    End if
  End for

2.3. *Variable Importance Measures*

The artificial intelligence learning model is a training model similar to a black box, and there is usually a problem that it is difficult to understand the training meta-model. Therefore, we introduce the random forest (RF) algorithm to calculate the relative importance of any combination of variables. The RF can reduce the average impurity of input variables, which is a very important part to improve the accuracy and interpretability of model. The main idea of RF is to ensure that the weight of the corresponding input variable is output under the condition of all input variables after receiving a given input. In theory, this method can be applied to any kind of dam deformation monitoring tasks. The detailed steps of RF model to calculate the relative importance of variables are described in Algorithm 2, where x and y represent the number of impact factors and observed data, respectively. ξ is the number of combinations containing different numbers of variables, and each combination must contain at least one variable. X_α is the α-th input variable set. $\hat{y}^{(\alpha)}$ represents the dam deformation prediction result obtained by X_α. N is the number of decision trees in the set random forest. $\hat{y}_i^{(\alpha)}$ is the dam deformation prediction result obtained by the i-th decision tree using X_α. $\overline{y}$ is the average value of the observed data for dam deformation.

Algorithm 2 Variable importance measure for predictive model interpretation

Inputs: Observed dataset $X^{(x \times y)}, Y^{(y \times 1)}$
Outputs: Variable relative importance VI
Divide X into α sets of tensors, $\alpha \in [1, \xi]$
For $\alpha = 1, \ldots, \xi$:
$\hat{y}^{(\alpha)} = f(X_\alpha)$
$\sigma_\alpha = \sum_{i=1}^{n} \left| \hat{y}_i^{(\alpha)} - \overline{y} \right| / n$
End for
For $\alpha = 1, \ldots, \xi$:
$VI_\alpha = \sigma_\alpha / \sum_{\alpha=1}^{\xi} \sigma_\alpha \times 100\%$
End for

2.4. Long Short-Term Memory Networks Couple with Attention

Long short-term memory (LSTM) networks can effectively capture the nonlinear characteristics of time series through storing historical data streams. However, due to the diversified complexity of dam deformation monitoring, the time lag of influencing factors is particularly prominent. For example, an increase in temperature or water level does not immediately lead to dam deformation, but a gradual process, which manifests as a delayed deformation response. This factor increases the complexity of dam deformation prediction in the time dimension, so this paper develops a dam deformation prediction method based on the attention mechanism coupled with LSTM model.

LSTM model is essentially a device that superimposes and stores information. It adjusts the destination of information in the model unit through the special structure called gate. LSTM model has three gate structures: the forget gate determines the information that needs to be removed in the current cell unit; the input gate determines which information in the newly input information fragment is updated to the current cell unit and the output gate collects the information memorized by the cell unit to obtain the final output result. The calculation process expression of each LSTM cell unit is as follows:

$$f_t = \sigma(W_f \cdot [h_{t-1}, x_t] + b_f) \tag{13}$$

$$i_t = \sigma(W_i \cdot [h_{t-1}, x_t] + b_i) \tag{14}$$

$$\widetilde{c}_t = \tanh(W_c \cdot [h_{t-1}, x_t] + b_c) \tag{15}$$

$$c_t = f_t * c_{t-1} + i_t * \widetilde{c}_t \tag{16}$$

$$o_t = \sigma(W_o \cdot [h_{t-1}, x_t] + b_o) \tag{17}$$

$$h_t = o_t * \tanh(c_t) \tag{18}$$

where f_t is the operating threshold of the forget gate at time t; h_t is the output value of the network at time t; h_{t-1} is the output value of the network at time $t-1$; i_t is the operating threshold of the input gate at time t; $\widetilde{c}_t$ is the candidate value that needs to be updated at time t; c_t is the new cell unit state at time t; o_t is the operating threshold of the output gate at time t; W_f, W_i, W_c, and W_o are the weight matrix of the forget gate, input gate, cell state, and output gate, respectively; b_f, b_i, b_c, and b_o are the bias of the forget gate, input gate, cell state, and output gate, respectively; and σ and $tanh$ are the activation functions.

However, the conventional LSTM networks are unable to perform quantitative impact analysis on the input data. Thanks to the concept of attention mechanism, it can achieve the focus on key information. It imitates the way humans process information and enable the algorithm to focus on core variables by means of activation functions. Therefore, a novel LSTM algorithm structure is proposed, coupled with an attention mechanism to predict dam deformation. The attention mechanism adds a matrix with the same dimension as the input tensor, weights the hidden state in the time dimension, and then outputs the attention vector value of each time step. The attention mechanism can be added in two positions, in front of the LSTM layer and behind the LSTM layer, as shown in Figure 2. In actual application, the target series has a complex amount of information, so the position added by the attention mechanism should be selected according to the actual effect. The attention mechanism calculates the weight e_t^i in each time dimension, and outputs the attention weight a_t^i between zero and one through the activation function, as shown below:

$$e_t^i = V^T \tanh(W_s S_{t-1} + W_h h_t) \tag{19}$$

$$a_t^i = \text{softmax}(e_t^i) = \exp(e_t^i) / \sum_{k=1}^{T} \exp(e_t^k) (t = 1, 2, \ldots, T) \tag{20}$$

where S_{t-1} is the attention layer input at the time $t-1$, and V^T and W_s are the weight matrices that can be trained.

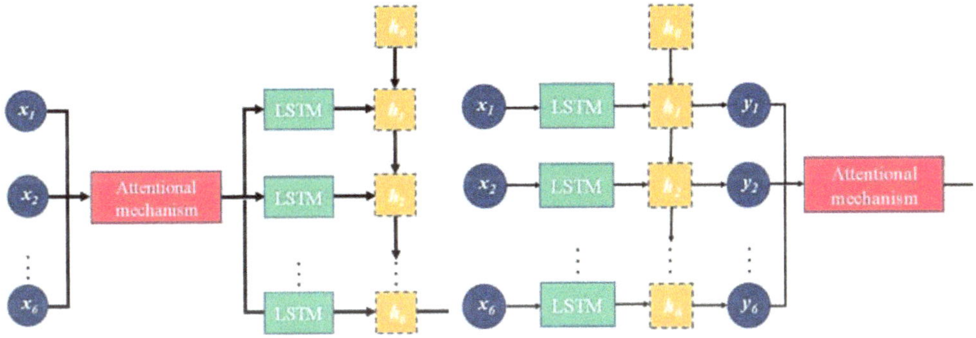

(a) (b)

Figure 2. Schematic of the LSTM coupled with the attention mechanism: (**a**) attention mechanism preceding the LSTM layer and (**b**) attention mechanism following the LSTM layer.

Then, we use this matrix vector to multiply the value in the time dimension to get the input of hidden layer c_t^i and, then, we update the hidden layer state of the LSTM model together with the previous unit cell state. The calculation formula is as follows:

$$c_t = \sum_{k=1}^{T} a_t^k h_k \tag{21}$$

$$S_t = LSTM(Y_{t-1}, c_t, S_{t-1}) \tag{22}$$

where Y_{t-1} is the attention layer output at the time $t-1$.

3. Model Implementation

This section mainly introduces the specific realization of the coupled algorithm for dam deformation prediction requirements, as shown in Figure 3. Since it uses a large number of historical data samples to test its validity, it can provide guidance and suggestions for the use of actual projects. Meanwhile, the evaluation metrics of the model are introduced.

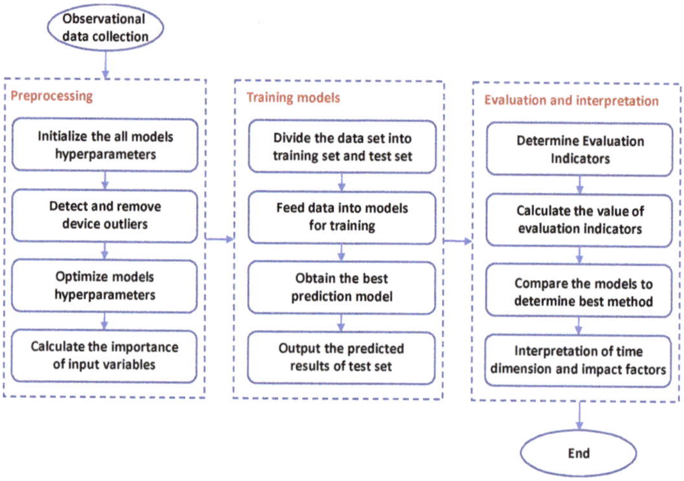

Figure 3. Flowchart of dam deformation prediction based on the coupled algorithm.

3.1. Selection of Input Variables

As discussed in Section 2.1. dam deformation is composed of water level component, temperature component and time component, so we consider these three components as input variables of the coupled model. Since the heat of hydration of the concrete material has been dissipated and the dam has a stable temperature field, the annual harmonic and half-year harmonic are used to express the temperature component. Finally, we generate a set of variables like Equation (23).

$$\begin{aligned} x &= \{H, H^2, H^3, \sin(2\pi/365), \cos(2\pi/365), \sin(4\pi/365), \cos(4\pi/365), \theta, \ln\theta\} \\ &= \{H_1, H_2, H_3, T_1, T_2, T_3, T_4, t_1, t_2\} \end{aligned} \tag{23}$$

In order to ensure the convergence of the coupled model or speed up its convergence, all input variables of Equation (23) need to be normalized as in Equation (24). Where μ is mean of sample data and σ is standard deviation of sample data.

$$x_i = \frac{x_i - \mu}{\sigma} \tag{24}$$

3.2. Design of Comparison Schemes and Tuning Parameters

Design of a comparison scheme using eight algorithm models, these eight algorithm models are HST model, support vector machine using poly kernel (SVM-poly), support vector machine using radial basis function kernel (SVM-rbf), RF model, multilayer perceptron (MLP) model, standard LSTM, the attention mechanism is coupled to the model before LSTM layer, and the attention mechanism is coupled to the model after the LSTM layer. All of the algorithms are implemented in the Python 3.6 environment. The data processing is conducted with the Numpy and Pandas packages of Python. The modelling process using MLR, SVM, and RF is carried out using the package of Scikit-Learn in Python. Our MLP and LSTM networks are developed with Keras on top of Google TensorFlow.

For the HST model, it is essentially linear regression, and its regression coefficients are calculated by the least square method. For artificial intelligence models, each model has its own unique hyperparameters that need to be set. We introduce the method of grid search (GS) [31] to tune the hyperparameters on the data of training set. It arranges and combines the possible values of each hyperparameter from the pre-declared parameter interval, and lists all possible combinations to generate a grid. Then, use each combination for training and use ten-fold cross-validation to evaluate performance. After the fitting function has tried all the parameter combinations, a best hyperparameter combination is returned. SVM-poly model has two important parameters: the penalty coefficient C and the specified degree d. SVM-rbf model has two important parameters: penalty coefficient C and influence radius γ. RF model has three important parameters: the total number of trees *n_estimators*, the maximum number of features a single tree can have *max_features* and the minimum number of sample data owned by leaves *min_sample_leaf*. In order to suppress the overfitting phenomenon of LSTM and MLP models, it is necessary to add a dropout layer. Therefore, both the MLP and LSTM models have two important parameters: the number of hidden layer units u and the dropout rate. However, different from MLP, LSTM also needs to declare the size of the sliding time window w, which is how long the previous data needs to be considered when predicting the dam deformation in the next time period. Therefore, the partial autocorrelation function (PACF) of original dam deformation is calculated, as shown in Figure 4. It can be found that when the deformation lag is greater than eight, the values of PACF stabilize within the 95% confidence interval. Meanwhile, tuning the sliding time window size of the LSTM. When the optimal range of w is determined, testing the performance of the LSTM model with different values of w on the training set, and the results are shown in Figure 5. At first, due to insufficient effective information, the performance of model changes greatly. As the value of w increases, the model receives more information, which makes the performance of model gradually improve and eventually stabilize. Through Figures 4 and 5, when the LSTM model uses a larger w, the performance of LSTM will not be affected by redundant information. In order to simplify the process of determining specific hysteresis of the impact factor, we can choose a larger w. The final experimental results are listed in Table 1.

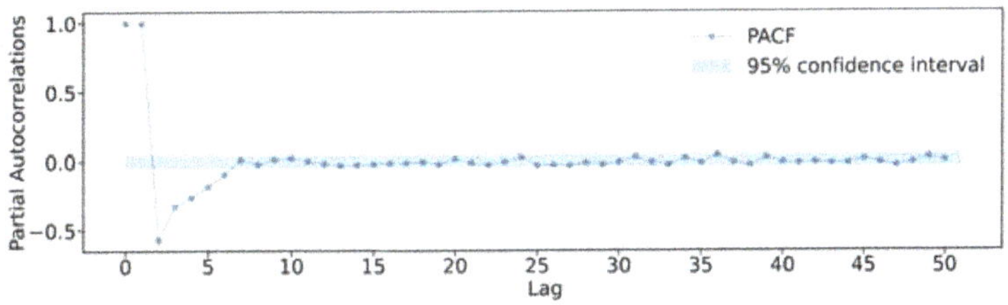

Figure 4. The PACF results of the original observed data.

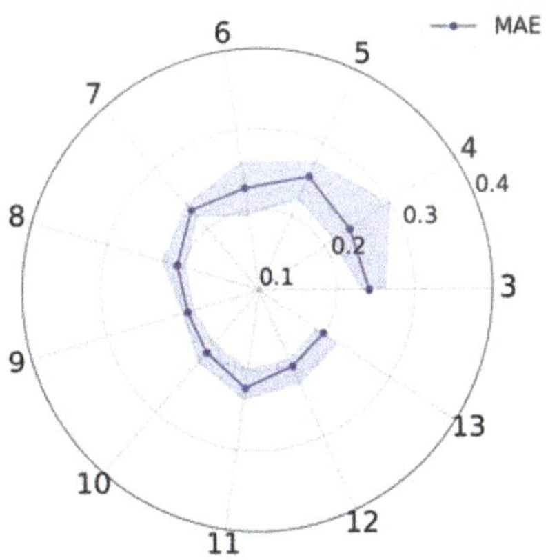

Figure 5. MAE values of the LSTM model with different sliding time window size. The solid lines represent the average value from ten repeated simulations and the upper and lower boundaries of the band represent the maximum and minimum from the ten simulations.

Table 1. Basic information of model parameter tuning.

Models	Hyperparameters	Search Range	Optimal Value	Training: MAE	Test: MAE
SVM-poly	C d	(15, 30) [1,2,3,4,5]	21 1	1.5504	1.3880
SVM-rbf	C γ	(15, 25) (0.1, 0.5)	20 0.2	0.5598	1.2799
RF	*n_estimators* *max_features* *min_sample_leaf*	(25, 40) (0,1) [1,2,3,4,5]	29 0.5 3	0.3987	1.2282
MLP	u *Dropout rate*	[64, 512] [0.4–0.8]	128 0.5	0.6561	1.3364
LSTM	u *Dropout rate* w	(64, 521) [0.4–0.8] [3, 15]	256 0.5 9	0.1752	0.2037

3.3. Evaluation Criteria

To evaluate the performance of each model, the mean absolute error (MAE), root mean square error (RMSE), and maximum absolute error (AE_{max}) are selected as the evaluation indicators of models. The relevant calculation formulas are shown in Equations (25)–(27). For these three indicators, the smaller the value, the better the performance of the prediction model.

$$MAE = \frac{1}{n}\sum_{i=1}^{n}|y_i - \hat{y}_i| \tag{25}$$

$$RMSE = \sqrt{\frac{1}{n}\sum_{i=1}^{n}(y_i - \hat{y}_i)} \tag{26}$$

$$AE_{\max} = \max|y_i - \hat{y}_i|, i = 1, 2, \ldots, n \tag{27}$$

where n is the number of sample data, y_i is the original dam deformation data, and $\hat{y}_i$ is the predicted dam deformation data.

4. Case Study

4.1. Case Description

This study uses an integral roller compacted concrete gravity dam located in China as a case to verify the performance of the proposed coupled model. The dam axis is arranged in a broken line, the dam crest elevation is 145 m, the maximum dam height is 63 m, and the dam crest length is 196.62 m. The length of dam sections on the left and right banks are respectively 58.92 and 62.70 m, and there is a longitudinal drainage and grouting gallery in the dam body. The dam began to be closed for water storage in 1993 and passed the completion acceptance in 1995. In order to ensure the safe operation of the project, automatic monitoring equipment is used to continuously monitor the response of the dam structure. Meanwhile, for the realization of the horizontal displacement monitoring, nine measuring points are set on the top of the dam, numbered E01 to E09, as shown in Figure 6. The main distribution of measuring points is one on the left and right sections of the dam, two on the left and right abutments of the dam, five on the overflow section.

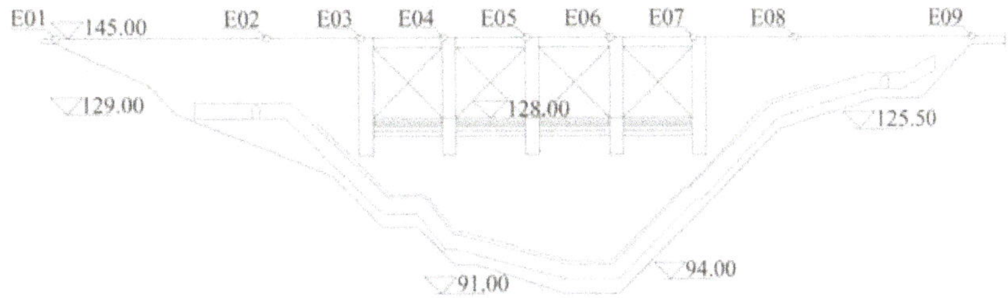

Figure 6. The layout of horizontal displacement measuring points on the dam.

The environmental monitoring of water level is divided into upstream water level and downstream water level. The upstream water level measuring station is located in front of the upstream dam, and the downstream water level measuring station is located near the tail water outlet of the powerhouse. The upstream and downstream water levels are monitored in real time by self-registering water level gauges, supplemented by manual observation of water gauges.

Monitoring equipment collects data on dam deformation and related environmental variables once a day. In this study, since the E06 measuring point has the most obvious variation for the impact factor, it was chosen as an example for analysis. The DBSCAN method is used to preprocess the monitoring data from 1998 to 2021, and the dam deformation prediction performance of the proposed coupled model is studied.

4.2. Data Analysis

For the problem of dam deformation, it was first necessary to assess whether the influence of downstream water level should be considered. The data of the upstream and downstream water levels are presented in Figure 7. The downstream water level was generally stable compared to the upstream water level and the two water levels were quite different, so the influence of changes in the downstream water level were not included in the dam deformation model.

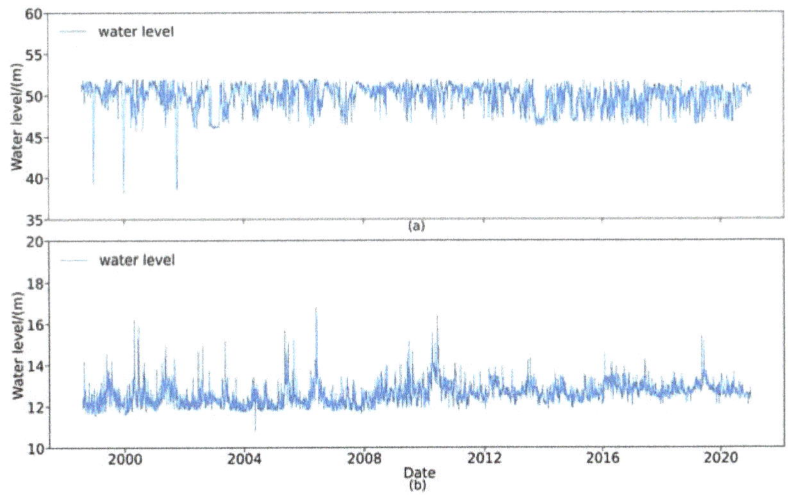

Figure 7. Water level data collected from automated monitoring equipment: (**a**) upstream water level and (**b**) downstream water level.

In addition to the standard dam related variables, there are many influencing factors that are difficult to remove, such as lightning strikes and interference by magnetic fields during monitoring. However, using the DBSCAN method to preprocess the observed data can effectively prevent equipment based abnormal values caused by such factors from contaminating the data set. The original observation data collected from the automated monitoring equipment are shown in Figure 8a, from which we can clearly find several equipment based abnormal values. The DBSCAN method uses clustering to identify equipment based outliers of the same level into a single category and then eliminate them. The results after removing the outliers can be seen in Figure 8b. The data processed through the DBSCAN method were smoother than the original observation data, and most of the abnormal values due to equipment error were eliminated.

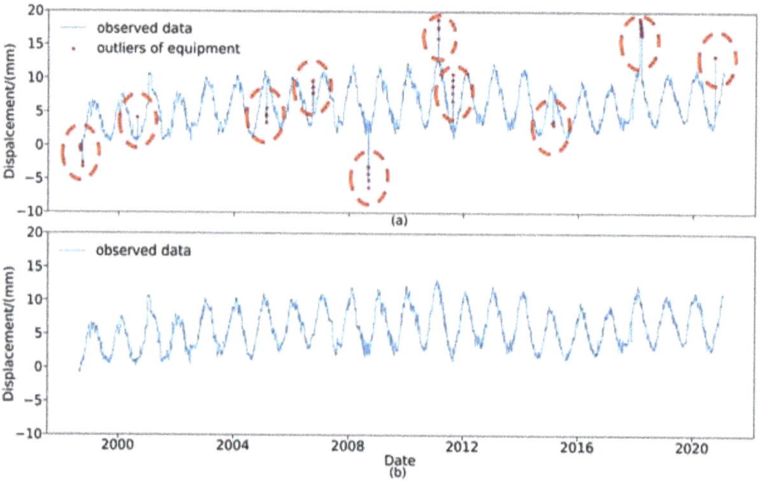

Figure 8. Dam horizontal displacement data collected from automated monitoring equipment: (**a**) the original data and (**b**) the data processed through DBSCAN.

The basic stats of the original data and the data processed using the DBSCAN method are listed in Table 2. The average values of the processed data were similar to the average values of the original data, but the standard deviation was reduced. This result showed that the DBSCAN method can effectively filter device caused outliers and help suppress any interference by such data on the prediction.

Table 2. Summary statistics of the original data and the data processed through DBSCAN.

Attribute	The Original Data	The Data Processed through DBSCAN
Max	19.0791	13.2493
Min	−6.2839	−0.8423
Mean	5.9806	6.0159
Median	5.7828	5.7860
Standard	3.0866	2.8342

5. Results and Discussions

5.1. Importance of Input Variables and Model Interpretation

For a prediction model, the choice of input variables will directly affect the accuracy of the prediction results, so it needs to be determined if the input variables can fully represent the nonlinear relationship of dam deformation and whether the input variables are redundant. To do this we adopted the RF model to conduct a deep analysis of the input variables and determine the relative importance of each.

Figure 9 shows the relative importance of the nine input variables as calculated by the RF model using the GS method to optimize parameters. It can be seen that the temperature and water level factors were most relevant to dam deformation, with a relative importance of 60.97% and 21.29%, respectively. While the time component has the lowest bearing on the prediction results. From the perspective of hydraulic structural engineering, these calculated importance values demonstrated that temperature plays a central role in the deformation responses of dam structures, followed by water level, which is in line with the actual situation of the project and is reasonable. Although the time component had little influence on the model, the aging effect of the dam material and its inherent rheological characteristics may need to be analyzed as input variables to improve the nonlinear model.

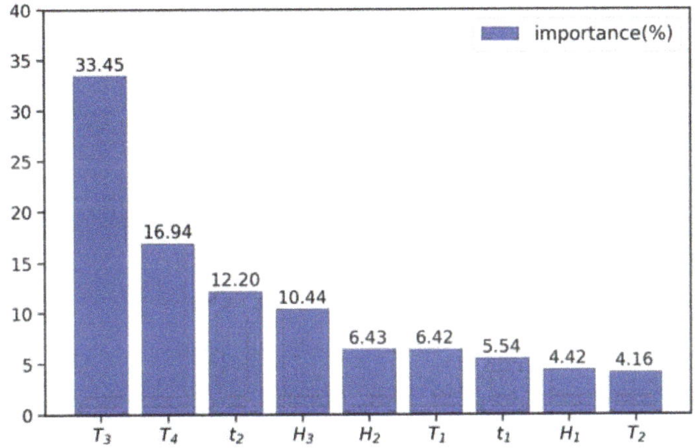

Figure 9. The relative importance of input variables determined by the RF model.

Using the sequential backward selection method, the input variables are sorted from top to bottom according to importance. As can be seen from Figure 9, the order of importance of the input variables was T_3, T_4, t_2, H_3, H_2, T_1, t_1, H_1, and T_2. In this order, the number of input variables fed to the RF model for training and testing is gradually increased. In order to meet the test requirements, 70% of the original data set is used as the training set and the remaining 30% data is used as the test set. Since the input sequence of the data set will also affect the test results, a ten-fold cross-validation is used to ensure the stability and reliability of the test results. The MAE values of the test results are provided in Table 3, and shown in Figure 10 to better visualize the different combinations. There were obvious fluctuations in the MAE values of the test and training sets at the beginning, and has not yet converged. This was because the model weighs the input variables and MAE values, and the effective information provided by the few input variables may not have been enough to represent the original time series. When the process continued and more variables were input, the MAE value significantly declined and the test set and training set began to stabilize. When using four, five, and six input variables, the MAE values of the test sets were basically stable at around 0.95, while those of the training set started to stabilize around 0.35. However, as the input variables continued to increase, the MAE values of the test and training sets increased again, and the increasing trend continued with additional variables. This was because the additional input variables contained redundant data that were not conducive to the prediction results, resulting in new features being captured by the model that contradicted those of the earlier data. Therefore, to ensure that the amount of information contained in the input variables is sufficient, while ensuring that MAE is as small as possible, a sixth plan should be selected to characterize dam deformation.

Table 3. Evaluation of simulation results for different input factors.

Serial Number	Search Range									MAE	
	T_3	T_4	t_2	H_3	H_2	T_1	t_1	H_1	T_2	Training	Test
1	√									1.2022	1.0520
2	√	√								1.1023	1.1563
3	√	√	√							0.8505	1.0482
4	√	√	√	√						0.3405	0.9541
5	√	√	√	√	√					0.3802	0.9581
6	√	√	√	√	√	√				0.3453	0.9550
7	√	√	√	√	√	√	√			0.3990	1.0633
8	√	√	√	√	√	√	√	√		0.4280	1.0584
9	√	√	√	√	√	√	√	√	√	0.4392	1.0821

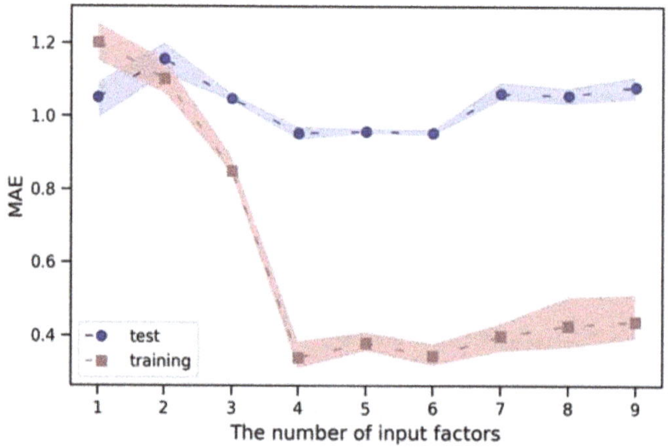

Figure 10. The MAE values of each group under ten-fold cross-validation.

5.2. Performance of Prediction Accuracy and Interpretation in Time Dimension

The LSTM model coupled with the attention mechanism was compared to a variety of advanced algorithms. As was done for the proposed model, all models were trained with 70% of the original data set and tested with the remaining 30%. The evaluation index value of each model is listed in Table 4, where the best performing model is marked in bold. The evaluation indexes showed that, except for SVM-poly model which had too low a fitting accuracy for the high-dimensional data of the training set, all of the models fit the observation data well. The LSTM models exhibited the best fitting performances, and the model in which the LSTM layer was preceded by an attention mechanism performed better than all other models. The RMSE, MAE, and AEmax values of this model were all significantly lower than the other models, indicating its prediction results were the best.

Table 4. Performance comparison of different models.

Model	Training			Test		
	RMSE	MAE	AE_{max}	RMSE	MAE	AE_{max}
Attention before LSTM	0.1337	**0.1248**	3.0304	**0.0498**	**0.1535**	**2.6235**
Attention after LSTM	0.0758	0.1801	**3.0295**	0.0646	0.1846	2.7444
LSTM	**0.0498**	0.1752	3.1269	0.0798	0.2037	3.1708
RF	0.2662	0.3987	2.3540	2.2030	1.2282	4.1766
SVM-rbf	0.5777	0.5598	2.7555	2.3969	1.2799	4.4931
SVM-poly	3.3283	1.5504	4.0004	2.6943	1.3880	3.7492
MLP	0.6858	0.6561	3.5570	2.5970	1.3364	4.5004
HST	0.8714	0.7424	3.5900	3.6562	1.7142	4.7921

To further compare the stabilities of the prediction results, the prediction residuals of all models were plotted, shown in Figure 11. The plot showed that the residual distribution of the HST model was right-skewed, and the prediction result gradually tended towards the average level of the training model over time. This was because linear fitting is more inclined towards the training data and does not consider non-linear effects of influencing factors, making linear fitting unable to accurately predict displacement caused by extreme weather. Although the distribution of the SVM-poly model was not skewed, its accuracy was not sufficient to meet the needs of dam deformation prediction through the median and quartiles. The SVR-rbf and MLP models showed improved prediction accuracy, but there were still problems with overfitting. The RF model did not exhibit overfitting, due to its own architecture, but it did exhibit generalization errors [32]. Therefore, the RF model had obvious right skewness, and would eventually tend to predict the average level like the HST model. Due to the introduction of time-dependent storage, the resulting LSTM has greatly improved the prediction accuracy and has a strong ability to represent nonlinear features, but it is still unable to predict peaks, as shown in Figure 12.

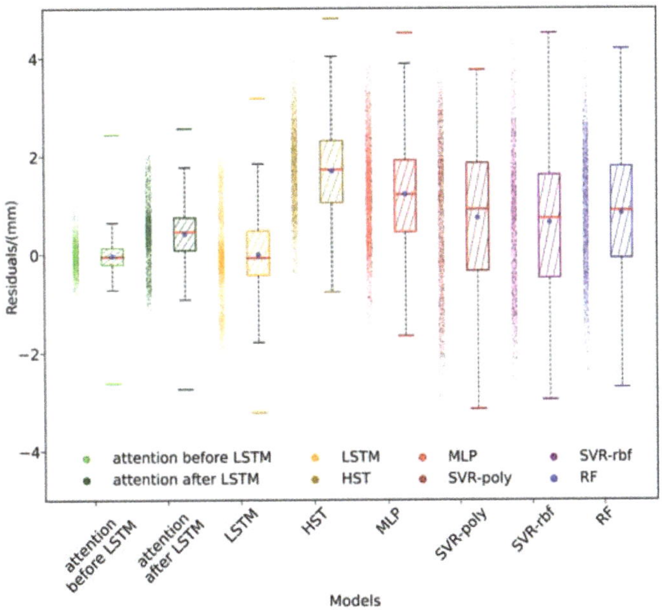

Figure 11. Comparison of the inherent stability and predictive performance of different models.

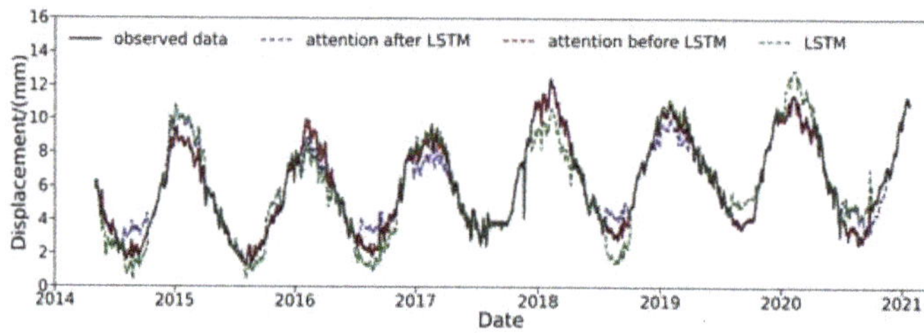

Figure 12. Predictive fitting curve of the LSTM and coupled models.

Meanwhile, in order to verify the reliability of the proposed model, drawing residuals plot of the better model, as shown in Figure 13. The results show that the proposed model has weaker residual autocorrelation than LSTM model, and its residual value distribution is more uniform. Although there are a few residual points of the proposed model on the residual graph with irregular distribution, most of residual points are evenly spread on both sides of 0. The residual distribution demonstrates unpredictability and randomness. Therefore, the LSTM model coupled with attention mechanism can fully capture the available information in the dam deformation influence factor.

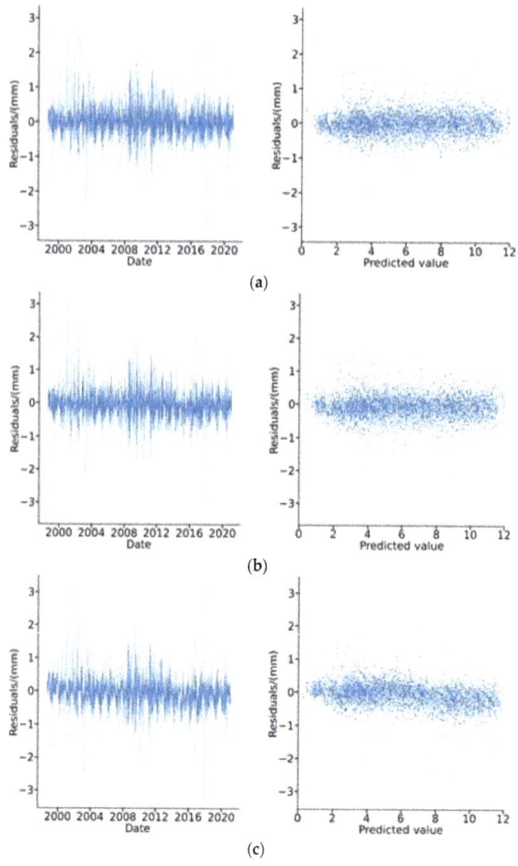

Figure 13. Residuals distribution plot of different models: (**a**) attention mechanism is coupled before LSTM; (**b**) attention mechanism is coupled after LSTM; and (**c**) LSTM.

The attention mechanism helps to automatically focus on important factors in the time dimension. This addition significantly improved the ability of the LSTM model to predict the peaks in the observed data. Furthermore, with the attention mechanism, the outliers in the forecasting data were relatively minor and the distribution of the residuals was also relatively concentrated. However, the attention mechanism performance depended on its coupling position, preceding or following the LSTM, and the results showed that when it is coupled before the LSTM layer it had better prediction accuracy. For dam deformation predictions, recent data has a stronger reference value and, as shown in Figure 14, attention disorder occurs when the attention mechanism is coupled after the LSTM layer. This is because after the data is processed by the LSTM layer it will be mapped into a complex high-dimensional tensor which makes it difficult for the attention mechanism to accurately focus on important factors. These observations demonstrated that the attention mechanism should be placed before the LSTM layer to better predict the structural response of dams using complex nonlinear factors and time delay information.

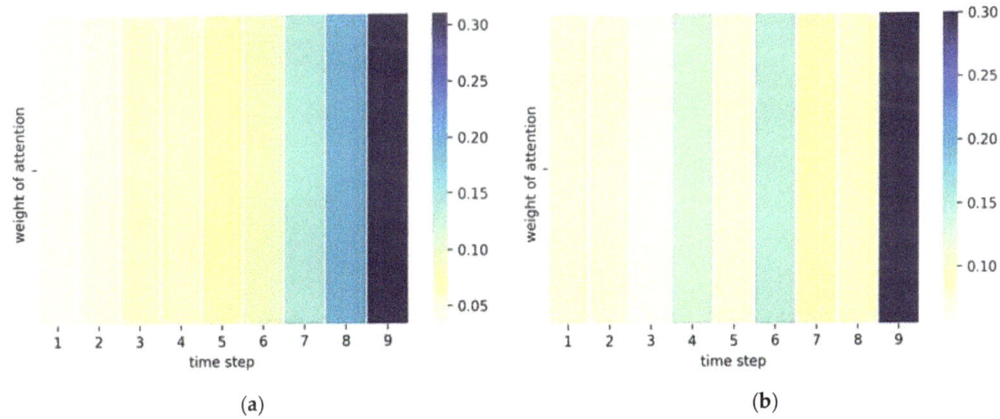

Figure 14. Heat map of attention weight: (**a**) the attention mechanism is coupled before LSTM and (**b**) the attention mechanism is coupled after LSTM.

6. Conclusions

The accuracy of the dam deformation prediction algorithm is vital to dam safety monitoring. This paper proposed a coupled LSTM/attention mechanism model that used GS and cross-validation methods to adaptively identify the most important parameters. Real world monitoring data of deformations of a concrete gravity dam were used as the research object to test the model and its performance was compared with several other advanced methods. The RF model was used to calculate the importance of input variables and any extraneous influencing factors were screened out while maintaining a sufficiently reasonable description of the dam deformation response. In addition, abnormal data points caused by various factors interfering with the monitoring equipment were eliminated using the DBSCAN method. According to this comprehensive study, the following conclusions were drawn:

1. The results showed that the DBSCAN method is suitable for the detection of equipment based abnormal values. The processed data had an average value that was similar to that of the original data, but the variance and random errors were greatly reduced.
2. The RF model identified the most important variables needed to provide a reasonable explanation for dam deformation to be input into the model. The results revealed that the temperature was a particularly important factor in dam deformation, the

importance of which was more than 50%, followed by water level, while the time component had the weakest influence.
3. The time-lag effect in dam monitoring plays an important role in predicting dam de-formation. When the model contained a time sliding window, the accuracy of the results was significantly improved, the residual distribution was relatively concentrated, and the outliers were greatly reduced.
4. The addition of the attention mechanism enabled the model to focus on important factors in the time dimension and improved the prediction of extreme values. The attention mechanism should be added before the LSTM layer to prevent the high-dimensional mapping of data processed by the LSTM layer from causing attention disorder.

In the future, the effects of different normalization methods for input variables on the robustness of attention mechanism need to be discussed. Meanwhile, we will explore the dynamic effects of time lag and spatial characteristics of dam deformation to achieve coordinated dynamic predictions of dam deformation using multiple measurement points, which will analyze the dam safety as a whole and realize data reduction.

Author Contributions: K.W. conceived and designed the experiments and performed the modeling. Y.S. provided relevant technical support and produced the first draft. C.L. and Z.C. analyzed the data and reviewed and edited the paper. All authors discussed the basic structure of the manuscript and read and approved the manuscript. All authors have read and agreed to the published version of the manuscript.

Funding: This work was funded by the Young Scientist Program of Fujian Province Natural Science Foundation (Grant No. 2020J05108), and Talent Introduction Scientific Start-up Foundation of Fuzhou University (Grant No. 510890).

Institutional Review Board Statement: Not applicable.

Informed Consent Statement: Not applicable.

Data Availability Statement: Not applicable.

Acknowledgments: The authors would like to acknowledge the Electric Power Research Institute of State Grid Fujian Electric Power Co. Ltd. for dam deformation data.

Conflicts of Interest: The authors declare no conflict of interest.

References

1. Cao, E.; Bao, T.; Gu, C.; Li, H.; Liu, Y.; Hu, S. A Novel Hybrid Decomposition—Ensemble Prediction Model for Dam Deformation. *Appl. Sci.* **2020**, *10*, 5700. [CrossRef]
2. Lin, C.; Li, T.; Chen, S.; Liu, X.; Lin, C.; Liang, S. Gaussian process regression-based forecasting model of dam deformation. *Neural Comput. Appl.* **2019**, *31*, 8503–8518. [CrossRef]
3. Willm, G.; Beaujoint, N. Les méthodes de surveillance des barrages au service de la production hydraulique d'Electricité de France, problèmes anciens et solutions nouvelles (The methods of surveillance of dams to serve hydraulic Production at Élec-tricité de France: Old problems and new solutions), Q34/R30. In Proceedings of the 9th International Congress on Large Dams (ICOLD), Istanbul, Turkey, 4–8 September 1967.
4. Penot, I.; Daumas, B.; Fabre, J.P. Monitoring behaviour. *Water Power Dam. Const.* **2005**, *57*, 24–27.
5. Mata, J.; Tavares de Castro, A.; Sá da Costa, J. Constructing statistical models for arch dam deformation. *Struct. Control Health Monit.* **2014**, *21*, 423–437. [CrossRef]
6. Popovici, A.; Ilinca, C.; Ayvaz, T. The performance of the neural networks to model some response parameters of a buttress dam to environment actions. In Proceedings of the 9th ICOLD European Club Symposium, Venice, Italy, 10–12 April 2013.
7. Mata, J. Interpretation of concrete dam behaviour with artificial neural network and multiple linear regression models. *Eng. Struct.* **2011**, *33*, 903–910. [CrossRef]
8. Kao, C.Y.; Loh, C.H. Monitoring of long-term static deformation data of Fei-Tsui arch dam using artificial neural network-based approaches. *Struct. Control Health Monit.* **2013**, *20*, 282–303. [CrossRef]
9. Ranković, V.; Grujović, N.; Divac, D.; Milivojević, N. Development of support vector regression identification model for pre-diction of dam structural behaviour. *Struct. Saf.* **2014**, *48*, 33–39. [CrossRef]
10. Kang, F.; Li, J. Displacement Model for Concrete Dam Safety Monitoring via Gaussian Process Regression Considering Extreme Air Temperature. *J. Struct. Eng.* **2020**, *146*, 5019001. [CrossRef]

11. Ren, Q.; Li, M.; Song, L.; Liu, H. An optimized combination prediction model for concrete dam deformation considering quantitative evaluation and hysteresis correction. *Adv. Eng. Inform.* **2020**, *46*, 101154. [CrossRef]
12. Su, H.; Li, X.; Yang, B.; Wen, Z. Wavelet support vector machine-based prediction model of dam deformation. *Mech. Syst. Signal Process.* **2018**, *110*, 412–427. [CrossRef]
13. Li, Y.; Bao, T.; Shu, X.; Chen, Z.; Gao, Z.; Zhang, K. A Hybrid Model Integrating Principal Component Analysis, Fuzzy C-Means, and Gaussian Process Regression for Dam Deformation Prediction. *Arab. J. Sci. Eng.* **2021**, *46*, 4293–4306. [CrossRef]
14. Ren, Y.; Zhu, C.; Xiao, S. Small Object Detection in Optical Remote Sensing Images via Modified Faster R-CNN. *Appl. Sci.* **2018**, *8*, 813. [CrossRef]
15. van der Burgh, H.K.; Schmidt, R.; Westeneng, H.-J.; de Reus, M.A.; Berg, L.H.V.D.; Heuvel, M.P.V.D. Deep learning predictions of survival based on MRI in amyotrophic lateral sclerosis. *NeuroImage Clin.* **2017**, *13*, 361–369. [CrossRef]
16. Bello-Cerezo, R.; Bianconi, F.; Di Maria, F.; Napoletano, P.; Smeraldi, F. Comparative Evaluation of Hand-Crafted Image Descriptors vs. Off-the-Shelf CNN-Based Features for Colour Texture Classification under Ideal and Realistic Conditions. *Appl. Sci.* **2019**, *9*, 738. [CrossRef]
17. Gelly, G.; Gauvain, J.-L. Optimization of RNN-Based Speech Activity Detection. *IEEE/ACM Trans. Audio, Speech, Lang. Process.* **2017**, *26*, 646–656. [CrossRef]
18. Kim, M.J.; Cao, B.; Mau, T.; Wang, J. Speaker-Independent Silent Speech Recognition from Flesh-Point Articulatory Movements Using an LSTM Neural Network. *IEEE/ACM Trans. Audio Speech Lang. Process.* **2017**, *25*, 2323–2336. [CrossRef]
19. Song, E.; Soong, F.K.; Kang, H.-G. Effective Spectral and Excitation Modeling Techniques for LSTM-RNN-Based Speech Synthesis Systems. *IEEE/ACM Trans. Audio Speech Lang. Process.* **2017**, *25*, 2152–2161. [CrossRef]
20. Yuan, X.; Chen, C.; Lei, X.; Yuan, Y.; Adnan, R.M. Monthly runoff forecasting based on LSTM–ALO model. *Stoch. Environ. Res. Risk Assess.* **2018**, *32*, 2199–2212. [CrossRef]
21. Su, Y.; Weng, K.; Lin, C.; Zheng, Z. An Improved Random Forest Model for the Prediction of Dam Displacement. *IEEE Access* **2021**, *9*, 9142–9153. [CrossRef]
22. Liu, Y.; Guan, L.; Hou, C.; Han, H.; Liu, Z.; Sun, Y.; Zheng, M. Wind Power Short-Term Prediction Based on LSTM and Discrete Wavelet Transform. *Appl. Sci.* **2019**, *9*, 1108. [CrossRef]
23. Liu, W.; Pan, J.; Ren, Y.; Wu, Z.; Wang, J. Coupling prediction model for long-term displacements of arch dams based on long short-term memory network. *Struct. Control Health Monit.* **2020**, *27*, e2548. [CrossRef]
24. Qu, X.; Yang, J.; Chang, M. A Deep Learning Model for Concrete Dam Deformation Prediction Based on RS-LSTM. *J. Sens.* **2019**, *2019*, 1–14. [CrossRef]
25. Xu, G.; Jing, Z.; Mao, Y.; Su, X. A Dam Deformation Prediction Model Based on ARIMA-LSTM. In Proceedings of the 2020 IEEE Sixth International Conference on Big Data Computing Service and Applications (BigDataService), Oxford, UK, 3–6 August 2020; Institute of Electrical and Electronics Engineers (IEEE): New York, NY, USA, 2020; pp. 205–211.
26. Hartley, G.A.; McNeice, G.M.; Stensch, W. Vogt Boundary for Finite Element Arch Dam Analysis. *J. Struct. Div.* **1974**, *100*, 51–62. [CrossRef]
27. Li, Z.Z. *Dam Safety Monitoring*; China Electric Power Press: Beijing, China, 1997; pp. 216–220.
28. Zhou, A.; Zhou, S.; Cao, J.; Fan, Y.; Hu, Y. Approaches for scaling DBSCAN algorithm to large spatial databases. *J. Comput. Sci. Technol.* **2000**, *15*, 509–526. [CrossRef]
29. Ester, M.; Kriegel, H.P.; Sander, J.; Xu, X. A Density-Based Algorithm for Discovering Clusters in Large Spatial Databases with Noise. In Proceedings of the 2nd International Conference on Knowledge Discovery and Data Mining KDD-96, Portland, OR, USA, 2–4 August 1996; pp. 226–231.
30. Ester, M.; Kriegel, H.P.; Sander, J.; Xu, X. Clustering for mining in large spatial databases. *KI J.* **1998**, *12*, 18–24.
31. Bergstra, J.; Bengio, Y. Random search for hyper-parameter optimization. *J. Mach. Learn. Res.* **2012**, *13*, 281–305.
32. Breiman, L. Random forests. *Mach. Learn.* **2001**, *45*, 5–32. [CrossRef]

mathematics

MDPI

Article

A Comparative Study of SSA-BPNN, SSA-ENN, and SSA-SVR Models for Predicting the Thickness of an Excavation Damaged Zone around the Roadway in Rock

Guoyan Zhao, Meng Wang and Weizhang Liang *

School of Resources and Safety Engineering, Central South University, Changsha 410083, China; gyzhao@csu.edu.cn (G.Z.); mwanglh@csu.edu.cn (M.W.)
* Correspondence: wzlian@csu.edu.cn

Abstract: Due to the disturbance effect of excavation, the original stress is redistributed, resulting in an excavation damaged zone around the roadway. It is significant to predict the thickness of an excavation damaged zone because it directly affects the stability of roadways. This study used a sparrow search algorithm to improve a backpropagation neural network, and an Elman neural network and support vector regression models to predict the thickness of an excavation damaged zone. Firstly, 209 cases with four indicators were collected from 34 mines. Then, the sparrow search algorithm was used to optimize the parameters of the backpropagation neural network, Elman neural network, and support vector regression models. According to the optimal parameters, these three predictive models were established based on the training set (80% of the data). Finally, the test set (20% of the data) was used to verify the reliability of each model. The mean absolute error, coefficient of determination, Nash–Sutcliffe efficiency coefficient, mean absolute percentage error, Theil's U value, root-mean-square error, and the sum of squares error were used to evaluate the predictive performance. The results showed that the sparrow search algorithm improved the predictive performance of the traditional backpropagation neural network, Elman neural network, and support vector regression models, and the sparrow search algorithm–backpropagation neural network model had the best comprehensive prediction performance. The mean absolute error, coefficient of determination, Nash–Sutcliffe efficiency coefficient, mean absolute percentage error, Theil's U value, root-mean-square error, and sum of squares error of the sparrow search algorithm–backpropagation neural network model were 0.1246, 0.9277, −1.2331, 8.4127%, 0.0084, 0.1636, and 1.1241, respectively. The proposed model could provide a reliable reference for the thickness prediction of an excavation damaged zone, and was helpful in the risk management of roadway stability.

Keywords: excavation damaged zone; prediction; sparrow search algorithm; BP neural network; Elman neural network; support vector regression

MSC: 86-10

Citation: Zhao, G.; Wang, M.; Liang, W. A Comparative Study of SSA-BPNN, SSA-ENN, and SSA-SVR Models for Predicting the Thickness of an Excavation Damaged Zone around the Roadway in Rock. *Mathematics* **2022**, *10*, 1351. https://doi.org/10.3390/math10081351

Academic Editors: Nikos D. Lagaros and Vagelis Plevris

Received: 7 March 2022
Accepted: 16 April 2022
Published: 18 April 2022

1. Introduction

After the excavation of roadway, the initial stress in the surrounding rock mass is redistributed. When the stress is greater than the strength of the surrounding rock, the rock mass will be damaged. Then, a ringlike broken zone can be formed around the excavated space; this is called the excavation damaged zone (EDZ) [1,2]. The thickness of the EDZ can not only be used to judge the stability of the roadway, but can also be adopted in the support design [3–5]. In addition, due to the weakening in the rock strength, an EDZ can also be utilized for nonexplosive continuous mining in deep hard-rock mines [6]. Therefore, predicting the thickness of the EDZ around a roadway is significant.

Since the concept of the EDZ was proposed, many scholars have conducted plenty of research to determine its size or thickness. These methods can be mainly summarized as the onsite measurement technique, the numerical simulation method, the empirical formula, and the machine-learning (ML) algorithm. Among them, the onsite measurement technique is the most direct method to determine the thickness of an EDZ, and includes digital panoramic drilling camera technology [7], ground-penetrating radar [8], ultrasonic detection technology [9], the borehole imaging method [10], the complex resistivity method [11], and microseismic monitoring [12,13]. Although the results of these measurements are accurate, the operation is complicated and vulnerable to the site conditions. With the rapid development of rock mechanics and computers, numerical-simulation methods became popular to determine the thickness of an EDZ. Liu et al. [14] used the ANSYS software to analyze the influencing factors and distribution law of an EDZ around a rectangular roadway. Sun et al. [15] studied the formation mechanism of a butterfly-shaped EDZ by combining the force and elastic wave theory, and then used Midas/GTS-FLAC3D simulation technology to determine the range of the EDZ. Perras et al. [16] used the finite element method in the Phase2 software to determine the thickness of an EDZ. Wan et al. [17] used 3DEC to determine the thickness of an EDZ, and the simulation results were consistent with the measured values. Although a numerical simulation is low-cost and convenient to operate, many assumptions exist in the simulation process that lead to idealized results and affect the accuracy. According to field-engineering experience and theoretical analysis, some empirical formulas were proposed to calculate the thickness of an EDZ. Yan [18] proposed an empirical formula for the thickness prediction of an EDZ based on the wave velocity of the rock and rock mass. Wang [19] combined the elastoplastic theory and measured data to determine the range of the EDZ in the Chazhen Tunnel. Chen et al. [20] deduced the radius of an EDZ based on the Hoek–Brown criterion and elastoplastic solution of a circular hole. Based on similar simulation tests and field experience, Dong [4] proposed a relationship between the stress and the rock strength to calculate the thickness of an EDZ. Zhao [21] derived the quantitative relationship between the thickness of an EDZ and its influencing factors based on a dimensional-analysis method. In addition, the zonal disintegration phenomenon, which indicates the alternation of fractured and intact zones, appears in deep roadways. Shemyakin [22] proposed the concept of zonal disintegration and deduced the empirical formula for the thickness of the discontinuous zone. After Myasnikov [23] proposed a non-Euclidean continuum model to describe the stress-field distribution, some scholars [24–26] used that non-Euclidean model to investigate the zonal disintegration phenomenon in a surrounding rock mass, and obtained the corresponding formulas. Although an empirical formula is easy to understand, it ignores the effects of joints and mining. Currently, there is no universally accepted empirical formula for predicting the thickness of an EDZ.

Considering ML can well deal with nonlinear and complex problems [27,28], it shows great potential to predict the thickness of an EDZ. Asadi et al. [29] used artificial neural networks in the thickness prediction of an EDZ. Zhou [30] verified that the support vector machine (SVM) could reliably estimate the range of an EDZ. In addition, some scholars adopted intelligent optimization techniques to improve traditional ML algorithms for the determination of an EDZ's thickness. For example, Hu [31] used a layered fish school to improve the SVM; Ma [32] combined the particle-swarm algorithm (PSO) and the least-squares support vector machine; Yu [33] integrated PSO and a Gaussian process model; and Liu [34] employed the wavelet-relevance vector machine. In addition, ML is being increasingly used in other civil engineering fields, and achieves an excellent prediction performance [35–37]. For example, Mangalathu et al. [38] used ML to classify the building damage caused by earthquakes; Ruggieri et al. [39] adopted ML to analyze the vulnerability of existing buildings. ML can not only obtain reliable prediction results, but can also save time and economic costs. However, numerous EDZ cases are needed to improve its credibility.

After comparing it with other types of approaches, the ML method was preferentially chosen to predict the thickness of an EDZ. An important reason is that it has strong self-learning and adaptive capabilities based on big data and can find implicit relationships between indicators. Nevertheless, it is essential to determine the favorable parameters of ML, because they directly affect its predictive performance [40]. The sparrow search algorithm (SSA) proposed by Xue [41] in 2020 is an efficient swarm-intelligence optimization algorithm. Compared with other optimization algorithms, the SSA has a higher search efficiency and a simpler operation. SSA considers all possible situations of a sparrow population, so that the sparrows in the population are close to the global optimal value, and converge [42]. At the same time, SSA has a high convergence speed, a good stability, a strong global search ability, and few parameters. In addition, backpropagation neural network (BPNN), Elman neural network (ENN), and support vector regression (SVR) models have shown extraordinary capabilities in solving prediction problems, and have been widely used in various engineering fields [43–45]. Therefore, using SSA to optimize the parameters of BPNN, ENN, and SVR models is more competitive.

The goal of this study was to use SSA-BPNN, SSA-ENN, and SSA-SVR models to predict the thickness of an EDZ. Firstly, an EDZ database including 209 cases was established. Secondly, SSA-BPNN, SSA-ENN, and SSA-SVR models were proposed for the thickness prediction of an EDZ. Thirdly, seven indexes were used to evaluate the performance of each model. Finally, all models were compared and analyzed, and the best model was determined.

2. Data Collection

To establish a reliable predictive model, a total of 209 cases from 34 mines were collected [46–50]. The locations of these mines are shown in Figure 1. It can be seen that the types of these mines were different, and included coal mines, gold mines, phosphate mines, and lead–zinc mines. In addition, these mines were in different regions, which indicated that the collected dataset was complex to some extent.

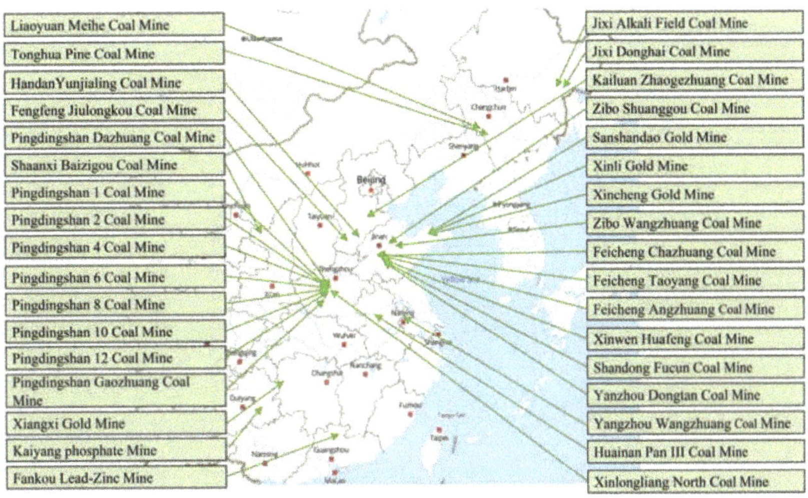

Figure 1. Locations of the selected mines.

The dataset statistics of each indicator are shown in Table 1, where A_1 indicates the embedding depth, A_2 indicates the drift span, A_3 indicates the surrounding rock strength, A_4 indicates the joint index, and A_5 indicates the EDZ thickness. The complete data can be found in Appendix A Table A1. It should be noted that EDZ indicates the ruptured zone around the roadway, but not the zonal disintegration. For a better understanding, the structure of an EDZ and some indicators are illustrated in Figure 2. The meaning of these indicators is indicated in Table 2.

Table 1. Statistics for each indicator.

Indicator	Max	Min	Mean	Standard Deviation
A_1 (m)	1159.00	97.00	499.20	242.94
A_2 (m)	10.00	2.40	3.71	1.13
A_3 (MPa)	158.83	7.50	30.80	31.12
A_4	5.00	1.00	2.89	1.15
A_5 (m)	3.45	0.30	1.56	0.61

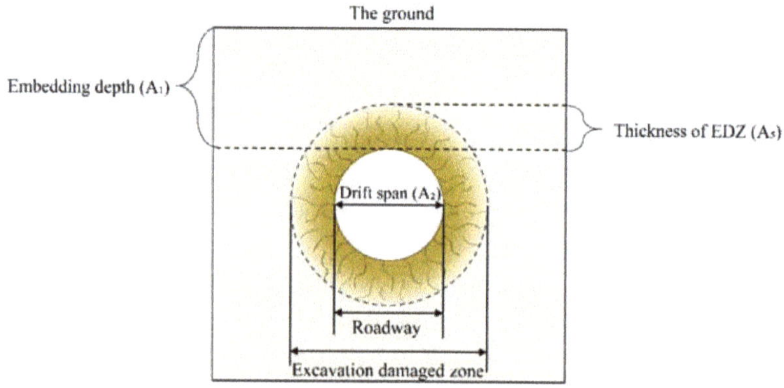

Figure 2. Excavation damaged zone around the roadway.

Table 2. Meanings of these indicators.

Indicator	Meaning
A_1 (m)	Indicates the depth of the roadway from the ground.
A_2 (m)	Indicates the width of the roadway.
A_3 (MPa)	Indicates the uniaxial compressive strength of the surrounding rock.
A_4	Indicates the development degree of joints in the surrounding rock.

Each sample contained four indicators and the thickness of the EDZ. The thickness of an EDZ is affected by many factors, such as the strength of surrounding rock mass, in situ stress, size and shape of the roadway, excavation method, time effect, and other environmental factors. First of all, the strength of the surrounding rock mass reflects the ability of the rock mass to resist damage, and is inversely proportional to the thickness of EDZ. Therefore, the indicators A_3 and A_4 were selected. Second, considering that the thickness of an EDZ is proportional to the in situ stress around the roadway, the indicator A_1 was chosen. Third, because different roadway sizes have diverse influences on an EDZ, A_2 was used as an indicator. Fourth, since the thickness of the EDZ used in this study was a stable value, the time effect could be ignored. When considering the influence of other factors, such as temperature, groundwater, and excavation method, they were deemed too complicated to quantify, and were not considered in this study.

To quantitatively describe the correlation between indicators and the thickness of an EDZ, the Pearson correlation coefficient was calculated, as shown in Figure 3. In this figure, red represents a positive correlation, blue represents a negative correlation, and the depth of color indicates the strength of correlation. It can be seen that the thickness of an EDZ and these four indicators had different correlation degrees, which showed that these indicators were relatively independent. Therefore, these four indicators were used as the input variables.

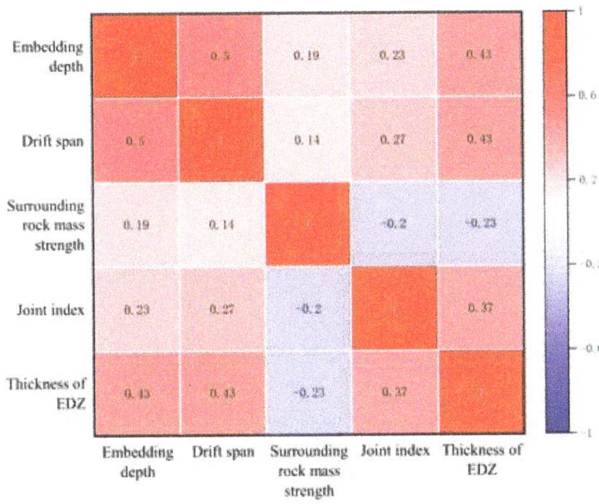

Figure 3. The correlation heat map of each indicator.

The advantages of these selected indicators can be summarized as: (1) they could reflect the main factors affecting the formation of an EDZ; (2) their values were easy to obtain; and (3) the information described by these indicators was independent.

3. Methodology

The structure of the proposed methodology is shown in Figure 4. Firstly, the original data were randomly divided into a training set (80%) and test set (20%). Secondly, the SSA was used to optimize the parameters of the BPNN, ENN and SVR models. Thirdly, the training set was adopted to train the optimized model. Fourthly, the test set was employed to analyze the accuracy of each model, and seven indexes, including the mean absolute error (MAE), coefficient of determination (R^2), Nash–Sutcliffe efficiency coefficient (NSC), mean absolute percentage error (MAPE), Theil's U value, root-mean-square error (RMSE), and the sum of squares error (SSE), were used to evaluate each model's performance. Finally, the optimal model was determined based on their comprehensive performance. The whole process was implemented in the MATLAB software. This section introduces the principles of the different models and the performance-evaluation indexes in detail.

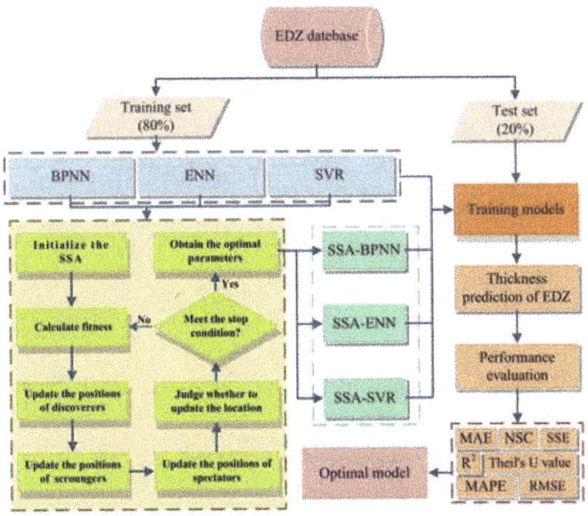

Figure 4. Structure of the proposed methodology.

3.1. Sparrow Search Algorithm (SSA)

Xue [41] proposed the SSA, which was inspired by the behavior strategy of a sparrow population. It solves the global optimization problems by simulating the behavior characteristics of sparrows, and provides a new approach to solving practical problems with a large number of local optimal values. SSA has a faster solution speed, a better stability, and convergence accuracy. In addition, because randomness is introduced in the search process, it can avoid falling into local solutions, and solve global optimization problems more effectively [51–53].

According to the original foraging principle of sparrow populations, a discoverer–scrounger model was established [54]. The interrelationships between individuals in the sparrow population are shown in Figure 5. Generally, the discoverer S1 is responsible for finding food and safe areas, while the scrounger S2 tracks the location of S1 to obtain food, and their roles are constantly changing [55,56]. S3 represents the sparrow at the edge of feeding area. It may leave the location and find another place because it is in the most dangerous position. S4 is responsible for detecting the safety of surrounding environment, and other sparrows also pay attention to S4 while eating.

Figure 5. Interrelationships between individuals in the sparrow population.

During the foraging and eating process of sparrow groups, individuals monitor each other while constantly observing changes in the surrounding environment [57,58]. If S4 sends a hazard signal to the population, the entire group will scatter away immediately. In addition, sparrows at the edge of community are more likely to be attacked by natural enemies than those at the center, so they will spontaneously and constantly adjust their positions to ensure safety.

According to the above idea, the SSA model can be established. Assuming that the sparrow population is in the space of $N \times D$, it can be defined by:

$$X = \begin{bmatrix} X_1 \\ X_2 \\ \vdots \\ X_n \\ \vdots \\ X_N \end{bmatrix} = \begin{bmatrix} x_{1,1} & x_{1,2} & \cdots & x_{1,d} & \cdots & x_{1,D} \\ x_{2,1} & x_{2,2} & \cdots & x_{2,d} & \cdots & x_{2,D} \\ \vdots & \vdots & \vdots & \vdots & \vdots & \vdots \\ x_{n,1} & x_{n,2} & \cdots & x_{n,d} & \cdots & x_{n,D} \\ \vdots & \vdots & \vdots & \vdots & \ddots & \vdots \\ x_{N,1} & x_{N,2} & \cdots & x_{N,d} & \cdots & x_{N,D} \end{bmatrix}, n = 1, 2, 3, \ldots.., N \quad (1)$$

where x is the position of sparrows, D indicates the spatial dimension, and N represents the number of total sparrows.

The fitness value indicates the energy reserve, which is defined as:

$$F_X = \begin{bmatrix} f_1 \\ f_2 \\ \vdots \\ f_n \\ \vdots \\ f_N \end{bmatrix} = \begin{bmatrix} f[x_{1,1} & x_{1,2} & \cdots & x_{1,d} & \cdots & x_{1,D}] \\ f[x_{2,1} & x_{2,2} & \cdots & x_{2,d} & \cdots & x_{2,D}] \\ \vdots & \vdots & \vdots & \vdots & \vdots & \vdots \\ f[x_{n,1} & x_{n,2} & \cdots & x_{n,d} & \cdots & x_{n,D}] \\ \vdots & \vdots & \vdots & \vdots & \ddots & \vdots \\ f[x_{N,1} & x_{N,2} & \cdots & x_{N,d} & \cdots & x_{N,D}] \end{bmatrix}, n = 1,2,3,\ldots\ldots, N \quad (2)$$

Generally, the discoverer S1 has a larger foraging range than the scrounger S2, and updates its position constantly. The update process can be calculated as:

$$X_{nd}^{t+1} = \begin{cases} X_{nd}^t \cdot \exp\left(\frac{-n}{\alpha \cdot P}\right), & R_2 < ST \\ X_{nd}^t + Q \cdot L, & R_2 \geq ST \end{cases} \quad (3)$$

where t denotes the current number of iterations; P refers to the maximum number of iterations; α is a random number of $[0,1]$; R_2 is the warning threshold, and $R_2 \in [0,1]$; ST indicates the safety value, and $ST \in [0.5,1]$; Q represents a random number that follows the normal distribution; L shows a $1 \times d$ matrix in which each element is 1; and X_{nd} signifies the position of a sparrow.

When $R_2 < ST$, it indicates the current foraging area is safe, and the sparrows can continue to eat, so the foraging range can be expanded. When $R_2 \geq ST$, the spectators find the predator and immediately issue an alarm signal, then all sparrows will scatter away immediately.

The location of scroungers S2 is also updated accordingly because the central location is more secure. The update equation is indicated as:

$$X_{n,d}^{t+1} = \begin{cases} Q \cdot \exp\left(\frac{X_{worst}^t - X_{n,d}^t}{n^2}\right), & n > \frac{N}{2} \\ X_{best}^{t+1} + \left|X_{n,d}^t - X_{best}^{t+1}\right| \cdot A^+ \cdot L, & n \leq \frac{N}{2} \end{cases} \quad (4)$$

where X_{worst}^t represents the global worst position; $X_{n,d}^t$ is the best position occupied by the discoverer; and A represents a $1 \times$ d matrix in which the elements are randomly assigned to be 1 or -1, and $A^+ = A^T\left(AA^T\right)^{-1}$.

When $n > \frac{N}{2}$, it means that the nth scrounger with a poor fitness value is hungry, and should fly in other directions to find food.

The spectators S4 generally account for 10% to 20% of the population. When danger approaches, they will scatter away and move to a new location. The position-update equation is:

$$X_{n,d}^{t+1} = \begin{cases} X_{best}^t + \beta\left(X_{n,d}^t - X_{best}^t\right), & f_n \neq f_g \\ X_{n,d}^t + K \cdot \left(\frac{X_{n,d}^t - X_{worst}^t}{|f_n - f_w| + e}\right), & f_n = f_g \end{cases} \quad (5)$$

where X_{best}^t represents the global safest position; β and $K \in [-1,1]$ are both control parameters of step length, while β is a random number that follows the normal distribution with a mean value of 0 and variance of 1, and $K \in [-1,1]$ represents the direction of sparrow movement; f_n, f_g, and f_w respectively represent the current fitness value of a sparrow, the global optimal value, and the global worst value; and e is a constant to prevent the denominator from being 0.

When $f_n > f_g$, it means the sparrow is at the edge of the population and is vulnerable to predators. When $f_n = f_g$, it indicates the sparrow in the middle of the population is aware of the danger and needs to be close to other sparrows to reduce the probability of being preyed upon.

3.2. Sparrow Search Algorithm–Back Propagation Neural Network (SSA-BPNN) Model

The calculation process of a BPNN is similar to nonlinear mapping, which uses multiple neurons to form a multilayer feedback model. The characteristics of the data are obtained through the continuous iteration of the algorithm. BPNN has a high learning adaptability, and can predict unknown data based on a previously learned pattern [59–63].

However, a BPNN has the defect of being easy to converge to a local minimum when fitting nonlinear functions. The SSA provides a new method to solve the parameter-optimization problem of a BPNN. In the SSA-BPNN model, the SSA reduces the error by continuously adjusting the weight and threshold of each layer, and improves the convergence speed.

The steps for an SSA to optimize a BPNN are as follows:

Step 1: The relevant parameters of the BPNN are initialized;

Step 2: The relevant parameters of the sparrow population are initialized, and the maximum number of iterations P is defined;

Step 3: Based on the fitness values, the sparrows are sorted to generate initial population positions. The mean-square error (MSE) is selected as the fitness function;

Step 4: According to Equations (3)–(5), the positions of discoverer S1, scrounger S2, and spectator S4 are updated;

Step 5: The current updated position is obtained. If the new position is better than the old position from a previous iteration, the update operation is performed; otherwise, the iterative process continues until the condition is met. Finally, the best individual and fitness values are obtained;

Step 6: The global optimal individual is used as the weight of the BPNN, and the global optimal solution is adopted as the threshold of the BPNN;

Step 7: When the number of iterations is reached or the error is met, the calculation process stops; otherwise, the program re-executes beginning at step 3.

3.3. Sparrow Search Algorithm–Elman Neural Network (SSA-ENN) Model

An ENN is a dynamic recurrent neural network that adds local memory units on the basis of a traditional feedforward network [64,65]. In addition to the hidden layer, it inserts an undertaking layer to the original grid that is used as a one-step delay operator to record dynamic information. Therefore, it obtains the ability to adapt to time-varying characteristics. Compared with traditional neural networks, it has better learning capabilities and can be used to solve problems including optimization, fitting, and regression.

However, an ENN has the randomness problem with initial weights and thresholds, which affects the accuracy of its predictions. In this study, an SSA was used to optimize the initial weights and thresholds of an ENN to improve the overall predictive performance.

The stages of SSA optimization of an ENN are as follows:

Stage 1: Initialize the relevant parameters of ENN and SSA;

Stage 2: Calculate the fitness of initial population and sort the results. The best and worst individuals can be determined. MSE is selected as the fitness function;

Stage 3: According to Equations (3)–(5), the positions of sparrows S1, S2, and S4 are updated based on fitness ranking;

Stage 4: Calculate the fitness value. The position of each sparrow is updated constantly. If the stop condition is met, the iterative process stops. Otherwise, the above process should be repeated;

Stage 5: Obtain the optimal weights and thresholds of the ENN.

3.4. Sparrow Search Algorithm–Support Vector Regression (SSA-SVR) Model

An SVR model mainly includes two steps. First, the nonlinear data is mapped into a high-dimensional space through the kernel function to make the data linearly separable. Then, the data is processed based on the principle of structural risk minimization.

Two important parameters in the SVR model include the penalty parameter c and the kernel function parameter g. Among them, c represents the error tolerance and g indicates the learning ability. The higher the value of c, the smaller the tolerance to error, and the more likely to overfit. In addition, g affects the prediction accuracy directly. Therefore, it is necessary to determine the optimal c and g during the training process.

The stages of SSA optimization of an SVR are as follows:

Stage 1: Build and initialize the SVR model;

Stage 2: Initialize the parameters of SSA, and determine the range of c and g;

Stage 3: Calculate the fitness of the initial population and determine the best and worst individuals. MSE is selected as the fitness function;

Stage 4: Update the positions of sparrows S1, S2, and S4 based on Equations (3)–(5);

Stage 5: Calculate the fitness value and update the position of the sparrows. The iterative process will break when the stop condition is met. Otherwise, the above steps will be repeated;

Stage 6: Obtain the optimal c and g, which are then used for model training.

3.5. Model Evaluation Indexes

In order to evaluate model performance, seven indexes, including MAE, R^2, NSC, MAPE, Theil's U value, RMSE, and SSE, were adopted.

MAE represents the average error between predicted value and actual value. The calculation equation is [66]:

$$\mathrm{MAE} = \frac{1}{n}\sum_{i=1}^{n}|f_i - y_i| \tag{6}$$

where n is the number of samples; f_i and y_i are the predicted value and actual value of the ith sample, respectively; and $\overline{y}$ denotes the average of the actual values.

R^2 is used to indicate the correlation between two variables. The calculation equation is [33]:

$$\mathrm{R}^2 = \frac{(n\sum_{i=1}^{n} f_i y_i - \sum_{i=1}^{n} f_i \sum_{i=1}^{n} y_i)^2}{(n\sum_{i=1}^{n}(f_i)^2 - (\sum_{i=1}^{n} f_i)^2)(n\sum_{i=1}^{n}(y_i)^2 - (\sum_{i=1}^{n} y_i)^2)} \tag{7}$$

NSC is used to describe the predictive efficiency. The calculation equation is [67]:

$$\mathrm{NSC} = 1 - \frac{\sum_{i=1}^{n}(y_i - f_i)^2}{\sum_{i=1}^{n}(y_i - \overline{y})^2} \tag{8}$$

Theil's U value is used to indicate the prediction accuracy. The calculation equation is [68]:

$$\mathrm{Theil's\ U} = \frac{\sqrt{\frac{1}{n}\sum_{i=1}^{n}(y_i - f_i)^2}}{\sqrt{\frac{1}{n}\sum_{i=1}^{n} y_i^2} + \sqrt{\frac{1}{n}\sum_{i=1}^{n} f_i^2}} \tag{9}$$

MAPE denotes the average value of the relative error. The calculation equation is [69]:

$$\mathrm{MAPE} = \frac{1}{n}\sum_{i=1}^{n}\left|\frac{f_i - y_i}{y_i}\right| \times 100\% \tag{10}$$

RMSE is used to describe the deviation between the predicted value and actual value. The calculation equation is [70]:

$$\mathrm{RMSE} = \sqrt{\frac{1}{n}\sum_{i=1}^{n}(f_i - y_i)^2} \tag{11}$$

SSE is used to calculate the sum of the squared error. The calculation equation is [71]:

$$\text{SSE} = \sum_{i=1}^{n} (y_i - f_i)^2 \quad (12)$$

4. Results and Analysis

4.1. Results of SSA-BPNN Model

A three-layer grid was used to establish the SSA-BPNN model. Based on the input and output of the model, the number of nodes in the input and output layers were determined as 5 and 1. In addition, the number of nodes in the hidden layer was obtained based on the MSE of the training set. By setting the number of nodes from 3 to 15, the MSE of the training set was obtained, as shown in Figure 6. It can be seen that the MSE was the smallest when the number of nodes was 5. Therefore, the grid structure of the SSA-BPNN model was chosen as $4-5-1$, as shown in Figure 7.

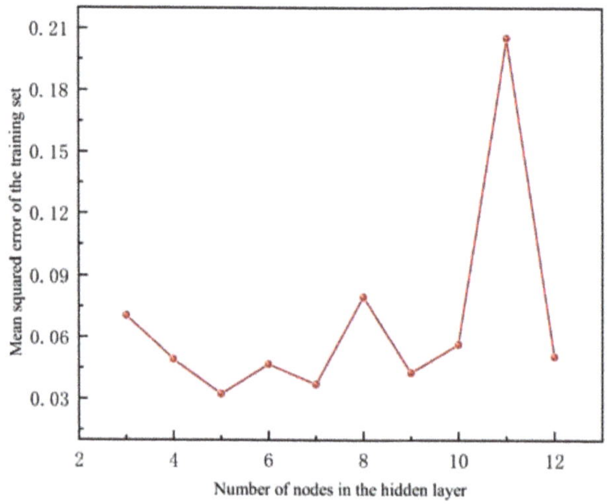

Figure 6. MSE of different number of hidden layer nodes for SSA-BPNN model.

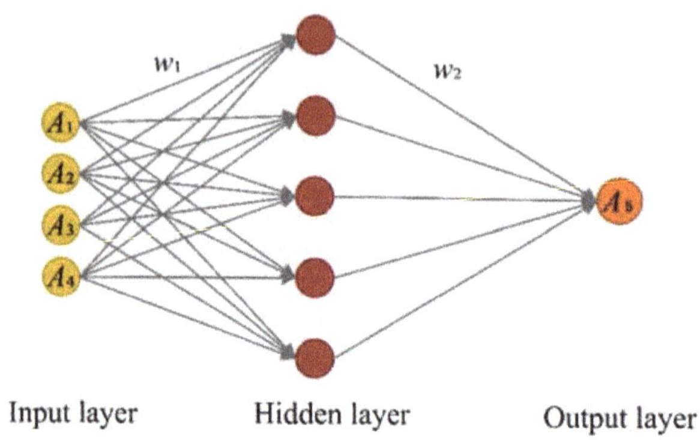

Figure 7. Grid structure of SSA-BPNN model.

In order to obtain a better predictive performance, other grid parameters, such as the number of trainings, minimum error of training target, initial population size of the SSA, and maximum evolutionary generation, were optimized by trial and error and set to

1000, 0.0001, 30, and 100, respectively. Then, the SSA was used to optimize the weights and thresholds of the BPNN model, and the optimized weight matrices w_1 and w_2 were obtained as follows:

$$w_1 = \begin{bmatrix} 0.1089 & -0.2795 & 0.2414 & -0.7100 \\ 0.0890 & -11.3880 & 7.6001 & 1.5449 \\ 0.3197 & 0.4036 & -0.0070 & -0.4563 \\ -6.0878 & 0.7283 & 0.1796 & 1.3281 \\ -0.6581 & -2.0707 & 11.9725 & 10.5624 \end{bmatrix} \tag{13}$$

$$w_2 = \begin{bmatrix} -2.6242 & -0.2653 & -4.0481 & -0.2858 & -0.4481 \end{bmatrix} \tag{14}$$

After the model was trained based on the training set, the test set was used to evaluate the predictive performance of the SSA-BPNN model. The relationship between the actual and predicted values is shown in Figure 8. It can be seen that their values had a good correlation.

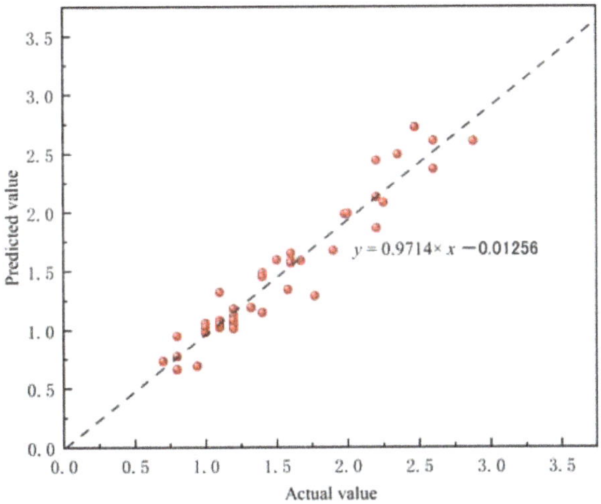

Figure 8. Predictive performance of SSA-BPNN model.

In addition, the predictive performance of the BPNN model before and after the SSA optimization was compared. Seven performance evaluation indexes, including MAE, R^2, NSC, MAPE, Theil's U value, RMSE, and SSE, were calculated, as shown in Table 3. According to these index values, the SSA-BPNN model performed better than the BPNN model. The absolute error of the different samples was determined, as demonstrated in Figure 9. Overall, the absolute error of the SSA-BPNN model was smaller and more stable than that of the BPNN model. At the same time, the maximum error of the BPNN model was larger. Therefore, SSA improved the predictive performance of the BPNN model to some extent.

Table 3. Evaluation index values of BPNN and SSA-BPNN models.

Model	MAE	R^2	NSC	MAPE	Theil's U Value	RMSE	SSE
BPNN	0.2169	0.7425	−3.8026	15.4075%	0.0144	0.2873	3.4665
SSA-BPNN	0.1246	0.9277	−1.2331	8.4127%	0.0084	0.1636	1.1241

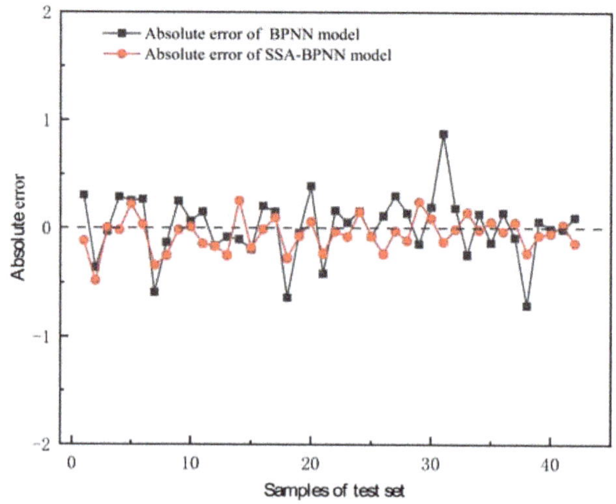

Figure 9. Absolute errors of BPNN and SSA-BPNN models.

4.2. Results of SSA-ENN Model

To establish the optimal grid structure, SSA-ENN models with different numbers of hidden layer nodes were built. The training error was determined when the number of nodes was selected from 3 to 12, as shown in Figure 10. When the number of hidden layer nodes was 10, the error was the smallest. Therefore, the grid structure of the SSA-ENN model was selected as $4-10-10-1$, as shown in Figure 11.

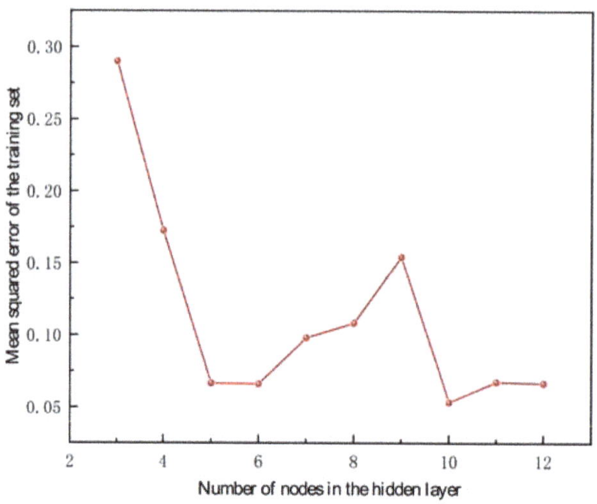

Figure 10. MSE of different number of hidden layer nodes for SSA-ENN model.

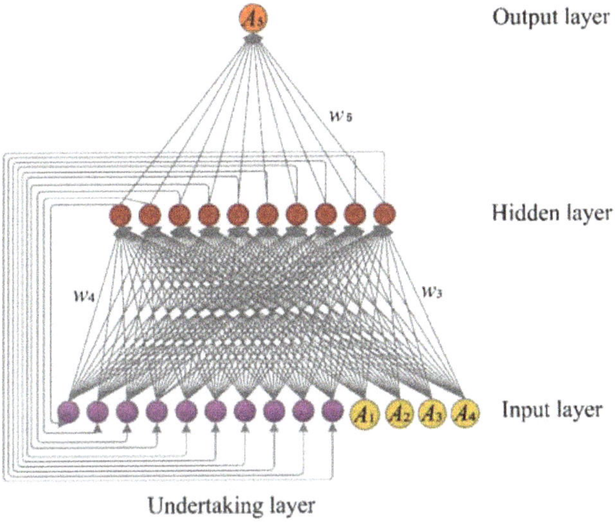

Figure 11. Grid structure of SSA-ENN model.

For training the model, other grid parameters, such as the learning rate, minimum error of training target, and initial population size of SSA, were optimized by trial and error and set to 0.01, 0.0001, and 30, respectively. Then, the weights and thresholds of the ENN were optimized by the SSA based on the training set, and the optimized weight matrices w_3, w_4, and w_5 were obtained as follows:

$$w_3 = \begin{bmatrix} -2.0388 & -1.5616 & -0.1513 & 1.3243 \\ 1.0527 & -0.0656 & -0.7647 & -2.0608 \\ -2.6026 & -0.0921 & -1.1416 & -3.2409 \\ 0.2770 & 0.4636 & -1.1443 & -1.8673 \\ -0.4656 & -1.4285 & -0.0750 & -0.7841 \\ -1.9833 & 0.1517 & -0.6582 & -0.5568 \\ -0.5375 & -1.6703 & 0.1149 & 0.6181 \\ -0.3918 & 0.7763 & -2.5244 & -1.8498 \\ 0.0175 & -2.2129 & -1.1439 & -1.3245 \\ -2.1403 & -0.4759 & -0.2951 & -2.0509 \end{bmatrix} \quad (15)$$

$$w_4 = \begin{bmatrix} -0.6309 & -0.5862 & -0.636 & -0.6344 & -0.71 & -0.6323 & -0.665 & -0.7285 & -0.6339 & -0.6346 \\ -0.6356 & -0.6129 & -0.632 & -0.7323 & -0.6546 & -0.6458 & -0.627 & -0.7207 & -0.759 & -0.6144 \\ -0.582 & -0.7272 & -0.683 & -0.7091 & -0.6001 & -0.6132 & -0.654 & -0.6353 & -0.6318 & -0.5988 \\ -0.708 & -0.613 & -0.635 & -0.6603 & -0.6202 & -0.6294 & -0.636 & -0.6121 & -0.6181 & -0.597 \\ -0.5904 & -0.633 & -0.644 & -0.668 & -0.6803 & -0.6779 & -0.635 & -0.6116 & -0.6568 & -0.6448 \\ -0.6247 & -0.6524 & -0.631 & -0.6138 & -0.6359 & -0.7771 & -0.667 & -0.6259 & -0.634 & -0.6419 \\ -0.6149 & -0.6013 & -0.691 & -0.6621 & -0.6069 & -0.7022 & -0.678 & -0.633 & -0.5831 & -0.691 \\ -0.6466 & -0.6066 & -0.643 & -0.6517 & -0.7294 & -0.7349 & -0.668 & -0.6627 & -0.6243 & -0.6588 \\ -0.6707 & -0.5838 & -0.645 & -0.6155 & -0.6464 & -0.6233 & -0.64 & -0.5822 & -0.6532 & -0.6794 \\ -0.6237 & -0.631 & -0.602 & -0.644 & -0.658 & -0.6256 & -0.63 & -0.638 & -0.6506 & -0.702 \end{bmatrix} \quad (16)$$

$$w_5 = [0.9634\ 1.5589\ 1.4220\ -1.5235\ -1.1393\ 0.7611\ -1.3683\ 0.3639\ 1.3427\ -1.8511] \quad (17)$$

The test set was used to evaluate the predictive performance of the SSA-ENN model. The relationship between the actual and predicted values is shown in Figure 12. It can be observed that the predicted value was very close to the actual value.

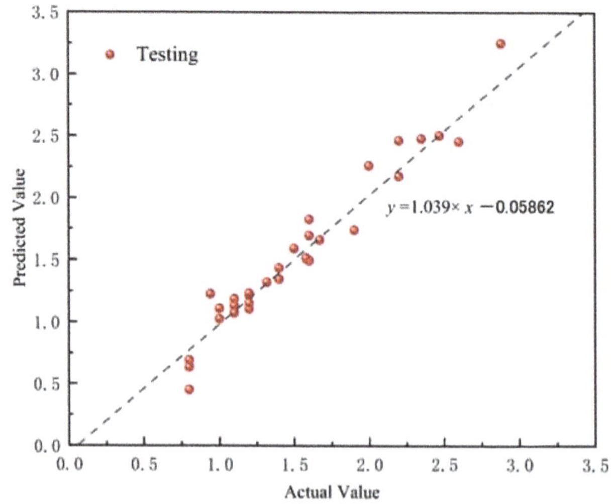

Figure 12. Predictive performance of SSA-ENN model.

In order to compare the predictive performance of the ENN and SSA-ENN models, their performance evaluation indexes were calculated, as shown in Table 4. All index values indicated that the SSA-ENN performed better than the ENN. In particular, the R^2 increased from 0.6824 to 0.9204, and the MAPE decreased from 15.5604% to 8.6297%. The absolute error of each sample in the test set was obtained, as shown in Figure 13. It can be seen that the error of the SSA-ENN was around zero, but the error of the ENN was larger and relatively unstable. Therefore, the ENN model optimized by the SSA had a better accuracy.

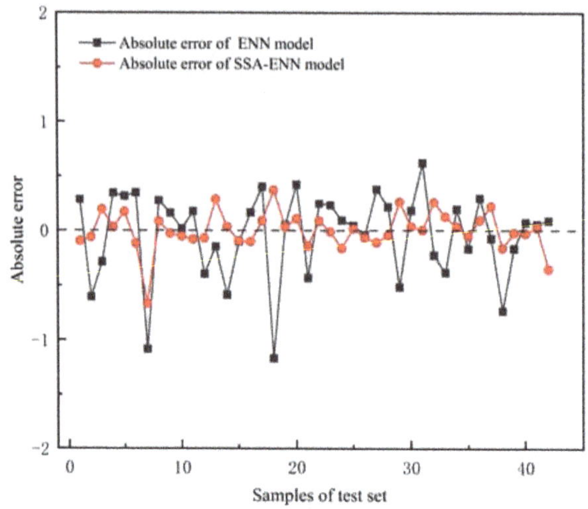

Figure 13. Absolute errors of ENN and SSA-ENN models.

Table 4. Evaluation index values of ENN and SSA-ENN models.

Model	MAE	R^2	NSC	MAPE	Theil's U Value	RMSE	SSE
ENN	0.2210	0.6824	−5.4164	15.5604%	0.0175	0.3429	4.9377
SSA-ENN	0.1215	0.9204	−1.3872	8.6297%	0.0088	0.1735	1.2646

4.3. Results of SSA-SVR Model

The SSA was used to optimize the parameters c and g of the SVR model. The population size of the SSA, the maximum number of iterations, and the cross-validation fold were set to 20, 100, and 5, respectively. Then, the best values of c and g were determined as 58.0379 and 2.2764, and the corresponding MAE of the cross-validation was 0.0569. The structure of the SSA-SVR model is shown in Figure 14 Then, the test set was adopted to assess the predictive performance of the SSA-SVR model. The fitting relation of the actual and predicted values is shown in Figure 15.

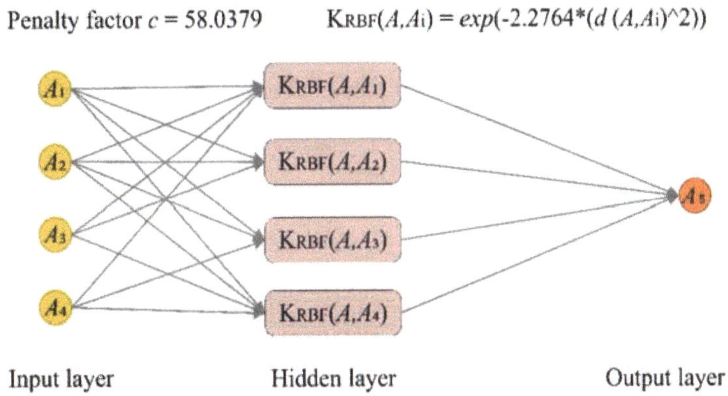

Figure 14. Structure of SSA-SVR model.

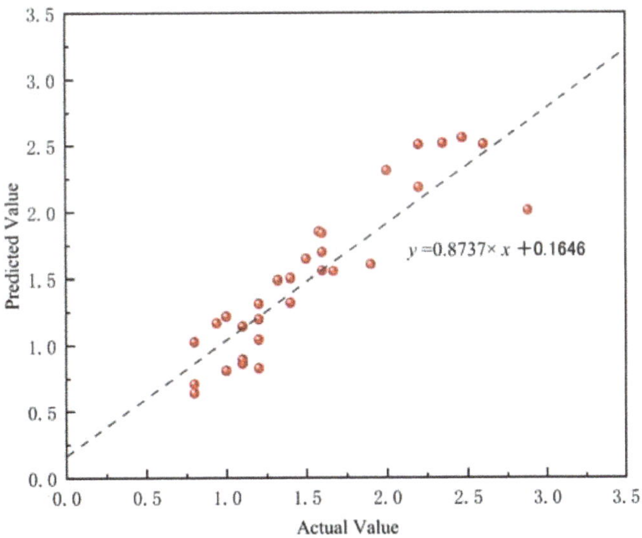

Figure 15. Predictive performance of SSA-SVR model.

The variation of fitness (MSE) with the number of iterations is shown in Figure 16. It can be seen that as the number of iterations increased, the fitness decreased. Especially in the first five generations, the fitness decreased rapidly. At the 12th iteration, the optimal fitness value was obtained as 0.0569.

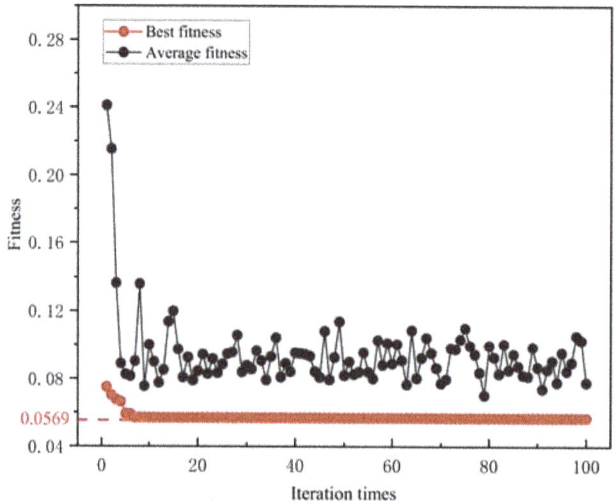

Figure 16. Fitness value of the SSA-SVR model.

The traditional SVR model was also used for a comparison. Based on the cross-validation results, the values of c and g were determined to be 1024 and 0.0078, respectively, and the MAE was 0.2647. Seven performance evaluation indexes for the SVR and SSA-SVR models were calculated, as shown in Table 5. Based on these index values, the SSA-SVR model had a better predictive performance. The absolute error corresponding to each sample is displayed in Figure 17. It can be seen that the prediction results of the SSA-SVR model were closer to the actual value. Therefore, the predictive performance of the SVR model was improved by the SSA.

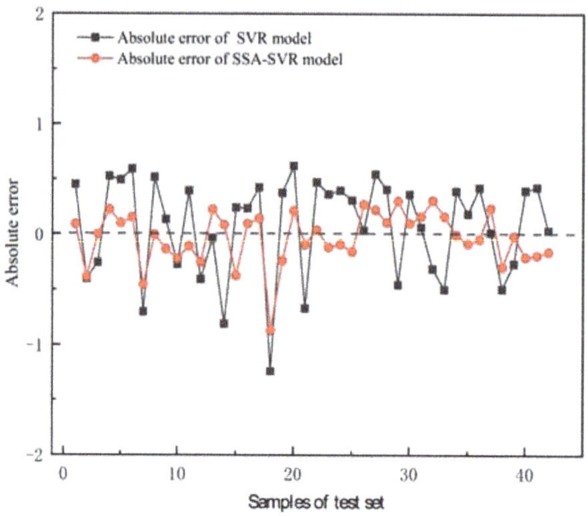

Figure 17. Absolute errors of SVR and SSA-SVR models.

Table 5. Evaluation index values of SVR and SSA-SVR models.

Model	MAE	R^2	NSC	MAPE	Theil's U Value	RMSE	SSE
SVR	0.3978	0.3699	−2.7614	28.3285%	0.0225	0.4568	8.7640
SSA-SVR	0.1856	0.8261	−4.6656	12.8829%	0.0122	0.2386	2.3907

5. Discussion

The main purpose of this study was to select an appropriate model to predict the thickness of an EDZ. Although these models were comprehensively evaluated based on seven evaluation indexes, as shown in Tables 3–5, it was necessary to determine their ranking results. In order to obtain the predictive performance of each model more intuitively, a score method was proposed. The specific scoring principle was that the best model in each index was given 5 points, the second-ranked model was given 4.5 points, and the lower-ranked models' scores were sequentially reduced by 0.5 points. The radar chart of the model score corresponding to each index is shown in Figure 18. It can be seen that most of the scores for each model index were basically at the same level. According to the area of the radar chart, it can clearly be seen that the SSA-BPNN and SSA-ENN models performed better. Based on the scores of various indexes, a stacked chart was obtained, as shown in Figure 19. It can be seen that the scores of the models optimized by SSA increased significantly, which illustrated the importance of parameter optimization with the SSA. According to the total scores, the ranking results were determined as SSA-BPNN > SSA-ENN > SSA-SVR > BPNN > ENN >SVR. The SSA-BPNN model was more suitable for predicting the thickness of an EDZ.

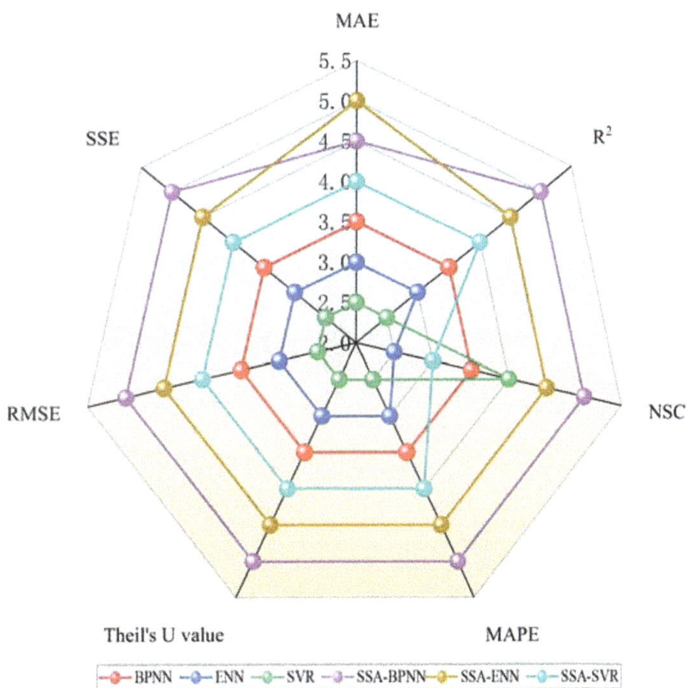

Figure 18. Radar chart of model scores corresponding to each index.

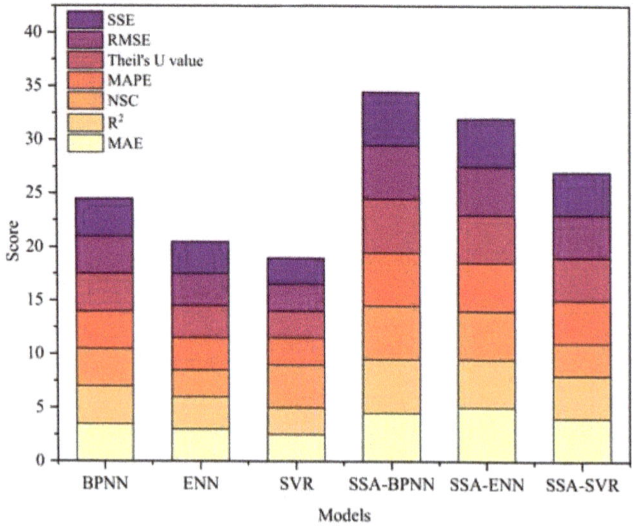

Figure 19. Stacked chart of index scores for all models.

In addition, the predicted and actual values of each model were compared, as shown in Figure 20. In the figure, it can be seen that the predicted results of the SSA-BPNN model were closer to the actual value. Combined with the scoring principle, we determined that the SSA-BPNN model had the highest accuracy.

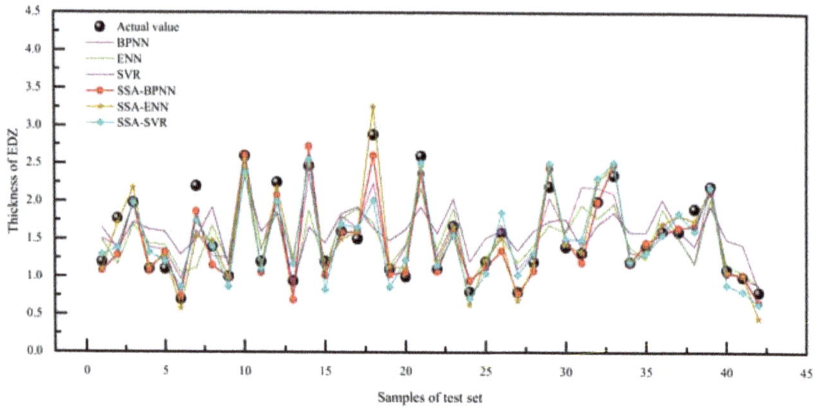

Figure 20. Predicted and actual values of each model.

In order to further verify the reliability of the proposed model, it was necessary to compare it with the empirical-formula method. Zhao et al. [21] proposed an empirical formula to determine the thickness of an EDZ as follows:

$$A_5 = 0.0145 A_1 {A_4}^{0.9324} \left(\frac{\gamma A_1}{A_3}\right)^{0.4459} \left(\frac{\sigma_{H\max}}{A_3}\right)^{-0.4308} \left(\frac{A_2}{A_1}\right)^{0.5334} \quad (18)$$

where γ is the unit weight of rock and $\sigma_{H\max}$ is the maximum horizontal principal stress.

This empirical formula adopts six indicators, such as A_1, A_2, A_3, A_4, γ, and $\sigma_{H\max}$. Because some of its indicators are the same as those in this study, this empirical formula could be adopted to compare with the proposed model. The data in reference [21] were used to evaluate the predictive performance, and the predicted results are shown in Table 6. It can be seen that the overall prediction performance of the SSA-BPNN model was better than that of the empirical formula.

Table 6. Predictive performance of SSA-BPNN model and empirical formula.

Mine	A_1/m	A_2/m	A_3/MPa	A_4	γ/(KN/m^3)	$\sigma_{H\max}$/MPa	A_5/m	MAPE of Empirical Formula	MAPE of SSA-BPNN Model
Maluping Phosphate Mine	660	4.5	34.37	4	27.2	34.49	1.65	9.34%	9.10%
Maluping Phosphate Mine	660	4.0	147.89	5	32.2	34.49	2.34	5.13%	6.27%
Sanshandao Gold Mine	600	3.8	71.26	3	27.1	32.45	1.10	7.27%	2.35%
Jinchuan Nickel Mine	1000	4.60	39.19	3	28.6	50.80	1.93	7.77%	4.68%

Although the proposed models could obtain satisfactory results, there were still some limitations:

(1) The dataset of EDZ cases was relatively small. The accuracy of a regression model heavily relies on the quantity and quality of the dataset. If the dataset is small, the model may overfit, which will affect its generalization and reliability. Although this study integrated most of the cases in the existing literature, the dataset was still relatively small. Therefore, establishing a more comprehensive EDZ database would be helpful to predict the thickness of an EDZ more efficiently using the proposed models.

(2) Only four indicators were selected for the thickness prediction of an EDZ. Due to the complexity of EDZ formation, the thickness of an EDZ is affected by various factors. Other indicators, such as the roadway shape, the presence of underground water, and the excavation method, may also have influences on the prediction results. Therefore, it is necessary to investigate the influences of more indicators in the future.

6. Conclusions

Determining the thickness of an EDZ is a crucial issue in the design of roadway support. This study proposed SSA-BPNN, SSA-ENN, and SSA-SVR models for the thickness prediction of EDZ. A dataset including 209 cases from 34 mines was collected to establish the predictive models. An SSA was used to optimize the parameters of the BPNN, ENN, and SVR models. MAE, R^2, NSC, MAPE, Theil's U value, RMSE, and SSE were used to evaluate model performance. According to these index values, the ranking result of each model was determined to be: SSA-BPNN > SSA-ENN > SSA-SVR > BPNN > ENN > SVR. Overall, the SSA improved the predictive performance of the traditional BPNN, ENN, and SVR models. The proposed models obtained satisfactory results and were more suitable for the thickness prediction of an EDZ. The SSA-BPNN model had the best comprehensive performance. The MAE, R^2, NSC, MAPE, Theil's U value, RMSE, and SSE were 0.1246, 0.9277, −1.2331, 8.4127%, 0.0084, 0.1636, and 1.1241, respectively. The prediction results provided an important reference for the determination of EDZ thickness.

In the future, a more comprehensive and higher-quality EDZ database should be developed. In addition, it is necessary to analyze the influences of other indicators, especially the excavation method, on the prediction results. Considering the complexity of an EDZ, other swarm-intelligence or ML algorithms can be used for comparison.

Author Contributions: Conceptualization, G.Z. and W.L.; methodology, M.W.; software, M.W.; validation, G.Z., W.L. and M.W.; formal analysis, M.W.; investigation, W.L.; resources, G.Z.; data curation, M.W.; writing—original draft preparation, M.W.; writing—review and editing, G.Z., W.L. and M.W.; visualization, W.L.; supervision, G.Z.; project administration, G.Z.; funding acquisition, G.Z. All authors have read and agreed to the published version of the manuscript.

Funding: This research was funded by the National Key Research and Development Program of China (2018YFC0604606).

Institutional Review Board Statement: Not applicable.

Informed Consent Statement: Not applicable.

Data Availability Statement: The data presented in this study are available upon request from the corresponding author.

Conflicts of Interest: The authors declare no conflict of interest.

Abbreviations

Full Name	**Abbreviation**
Excavation damaged zone	EDZ
Sparrow search algorithm	SSA
Backpropagation neural network	BPNN
Elman neural network	ENN
Support vector regression	SVR
Mean absolute error	MAE
Coefficient of determination	R2
Nash–Sutcliffe efficiency coefficient	NSC
Mean absolute percentage error	MAPE
Root-mean-square error	RMSE
Sum of squares error	SSE
Particle swarm algorithm	PSO
Support vector machine	SVM
Machine learning	ML

Appendix A

The complete database of the EDZ cases is shown in Table A1.

Table A1. Database of EDZ cases.

Samples	A1/m	A2/m	A3/MPa	A4	A5/m
1	800.00	4.00	48.00	3.00	1.62
2	800.00	3.60	67.00	4.00	2.14
3	650.00	6.00	16.80	5.00	3.45
4	220.00	3.40	21.30	4.00	1.31
5	876.40	3.40	13.80	3.00	1.41
6	321.00	3.00	13.30	1.00	2.41
7	420.00	3.20	9.10	4.00	1.40
8	800.00	3.60	70.60	3.00	1.84
9	340.00	3.40	18.80	1.00	1.52
10	315.00	2.80	11.20	1.00	1.77
11	740.00	3.30	32.42	3.00	2.00
12	1154.00	4.20	15.94	2.00	1.53
13	720.00	4.80	102.40	2.00	1.10
14	125.00	2.80	13.30	2.00	0.70
15	680.00	3.30	30.49	3.00	2.10
16	199.00	3.60	10.73	3.00	1.30
17	252.00	5.20	15.08	4.00	2.65
18	660.00	3.60	21.96	5.00	2.20
19	140.00	3.60	13.40	2.00	0.50
20	342.50	3.20	15.94	2.00	0.39
21	280.00	2.80	12.70	2.00	0.80
22	665.00	3.60	10.90	2.00	2.94

Table A1. *Cont.*

Samples	A1/m	A2/m	A3/MPa	A4	A5/m
23	510.00	8.00	54.87	3.00	1.40
24	350.00	3.20	10.50	1.00	1.93
25	689.00	3.00	15.10	4.00	1.80
26	321.00	2.60	13.30	3.00	1.10
27	615.00	3.60	25.64	3.00	1.50
28	510.00	4.00	64.54	2.00	1.10
29	740.00	4.00	22.70	5.00	2.25
30	249.00	3.20	16.80	1.00	2.46
31	961.00	4.00	12.76	3.00	1.57
32	362.00	2.60	62.40	2.00	0.60
33	680.00	3.80	25.64	5.00	2.35
34	180.00	2.80	110.20	1.00	0.89
35	342.50	3.20	13.80	3.00	0.55
36	403.00	2.90	12.60	1.00	1.90
37	150.00	3.60	14.60	2.00	0.60
38	510.00	3.70	12.60	4.00	1.40
39	470.00	3.60	9.10	2.00	3.26
40	876.40	3.60	12.76	3.00	1.48
41	660.00	4.40	12.50	5.00	2.20
42	610.00	3.00	34.00	3.00	1.35
43	869.00	4.00	67.00	4.00	1.78
44	340.00	3.40	18.80	3.00	1.30
45	450.00	3.00	11.20	1.00	2.11
46	350.00	3.20	10.50	3.00	1.20
47	680.00	4.20	47.00	3.00	1.40
48	400.00	5.32	21.71	3.00	1.62
49	740.00	2.60	32.42	2.00	1.50
50	520.00	3.80	11.90	2.00	2.59
51	480.00	3.00	10.35	3.00	1.20
52	470.00	4.00	10.10	2.00	2.97
53	300.00	4.50	20.00	3.00	1.50
54	300.00	4.50	20.00	4.00	1.60
55	340.00	3.00	73.60	1.00	0.99
56	961.00	3.60	15.94	2.00	1.35
57	850.00	3.80	48.00	1.00	1.04
58	690.00	4.60	47.00	3.00	1.50
59	1154.00	4.20	13.80	3.00	1.69
60	178.00	2.60	23.80	3.00	1.20
61	710.00	6.00	16.80	4.00	3.20
62	125.00	2.80	13.30	1.00	1.91
63	630.00	4.80	52.45	2.00	1.20
64	310.00	3.20	28.00	1.00	1.33
65	800.00	5.10	28.22	4.00	2.43
66	218.00	3.89	38.11	3.00	0.70
67	837.00	3.80	70.00	1.00	0.67
68	525.00	3.20	15.80	4.00	1.60
69	220.00	3.40	7.80	4.00	1.50
70	97.00	2.60	11.20	3.00	1.20
71	97.00	2.60	11.20	1.00	1.70
72	125.00	2.80	13.30	3.00	1.30
73	700.00	3.60	16.80	5.00	2.55
74	450.00	3.40	9.10	2.00	2.36
75	685.00	3.60	25.64	3.00	1.78
76	1154.00	4.20	12.76	3.00	1.77
77	961.00	4.00	10.00	5.00	2.46
78	700.00	3.80	25.64	3.00	1.78
79	460.00	3.20	101.60	1.00	0.99
80	410.00	3.00	26.78	3.00	1.00
81	665.00	3.60	10.90	4.00	1.70

Table A1. *Cont.*

Samples	A1/m	A2/m	A3/MPa	A4	A5/m
82	400.00	5.32	15.30	3.00	1.90
83	876.40	3.60	10.00	5.00	2.34
84	700.00	3.60	48.00	3.00	1.40
85	125.00	3.40	13.30	3.00	1.00
86	705.00	3.80	16.80	5.00	2.85
87	316.00	8.00	78.43	4.00	1.90
88	392.00	2.80	14.50	2.00	0.80
89	340.00	3.20	32.20	2.00	0.70
90	357.00	3.20	10.50	3.00	1.10
91	665.00	4.40	10.90	4.00	1.70
92	293.00	3.50	11.90	1.00	1.93
93	329.67	2.40	44.72	3.00	0.70
94	362.00	2.60	58.00	2.00	0.80
95	869.00	3.80	70.00	3.00	1.21
96	480.00	3.00	15.00	3.00	1.00
97	620.00	3.60	21.96	5.00	2.12
98	410.00	3.20	13.30	1.00	2.32
99	244.00	3.40	11.20	3.00	1.00
100	690.00	4.80	9.20	4.00	2.10
101	420.00	3.20	9.10	1.00	2.85
102	293.00	3.50	11.90	3.00	1.10
103	660.00	3.60	21.96	4.00	2.20
104	675.00	3.80	25.64	4.00	2.10
105	296.00	3.40	22.40	3.00	1.20
106	296.00	3.40	22.40	1.00	1.67
107	236.00	3.00	14.30	1.00	2.00
108	410.00	3.60	13.30	4.00	1.40
109	313.00	10.00	38.00	3.00	1.76
110	340.00	3.20	32.20	1.00	1.31
111	252.00	5.20	17.23	4.00	2.41
112	450.00	3.60	13.30	2.00	2.58
113	520.00	3.80	11.90	4.00	1.70
114	296.00	3.40	22.40	4.00	1.40
115	720.00	4.80	86.30	2.00	1.20
116	268.00	3.40	11.96	4.00	1.40
117	490.00	3.70	12.50	4.00	1.50
118	321.00	3.00	13.30	3.00	1.10
119	450.00	3.60	13.30	4.00	1.60
120	410.00	3.20	13.30	4.00	1.40
121	720.00	10.00	102.40	2.00	1.20
122	192.70	2.40	35.83	4.00	0.85
123	400.00	5.32	10.61	3.00	2.10
124	868.00	3.80	60.00	1.00	0.98
125	872.00	4.00	48.00	2.00	1.49
126	610.00	3.60	22.40	4.00	1.75
127	420.00	3.20	9.10	4.00	1.70
128	510.00	8.00	54.87	3.00	1.50
129	345.00	3.00	65.00	2.00	0.70
130	200.00	3.00	15.12	4.00	0.90
131	510.00	4.00	64.54	2.00	1.20
132	342.50	3.40	12.76	3.00	0.62
133	470.00	3.60	9.10	5.00	2.10
134	1056.00	4.00	15.94	2.00	1.44
135	403.00	2.90	12.60	3.00	1.30
136	876.40	3.40	15.94	2.00	1.26
137	467.00	3.40	10.10	4.00	1.80
138	322.00	4.40	14.30	4.00	1.50
139	208.00	3.40	42.41	2.00	0.38
140	268.00	3.40	11.96	1.00	2.05

Table A1. Cont.

Samples	A1/m	A2/m	A3/MPa	A4	A5/m
141	249.00	3.40	16.80	3.00	1.00
142	178.00	2.60	23.80	1.00	1.12
143	520.00	4.20	25.43	3.00	1.30
144	690.00	4.80	32.00	2.00	1.70
145	1159.00	4.50	13.80	4.00	2.21
146	460.00	3.20	101.60	1.00	0.40
147	310.00	3.20	28.00	3.00	0.80
148	780.00	3.00	70.60	1.00	0.65
149	450.00	3.40	9.10	5.00	2.00
150	180.00	2.80	110.20	1.00	0.30
151	550.00	3.40	12.50	5.00	2.10
152	467.00	3.40	10.10	2.00	2.31
153	231.00	3.00	18.30	2.00	0.70
154	340.00	3.20	19.80	3.00	1.30
155	685.00	3.20	52.00	2.00	1.20
156	305.00	3.20	10.10	4.00	1.30
157	321.00	2.60	9.20	3.00	1.20
158	420.00	3.60	14.30	1.00	1.98
159	670.00	3.60	16.80	5.00	2.35
160	370.00	3.50	10.50	1.00	2.00
161	236.00	3.00	14.30	3.00	1.20
162	510.00	3.20	12.60	4.00	1.60
163	1154.00	4.50	10.00	5.00	2.73
164	630.00	4.80	48.97	3.00	1.42
165	520.00	4.20	63.54	3.00	1.10
166	373.00	2.50	14.60	2.00	0.90
167	305.00	3.20	10.10	1.00	1.98
168	384.00	3.50	8.50	3.00	1.20
169	500.00	3.00	38.50	4.00	1.77
170	640.00	3.60	25.64	4.00	1.98
171	420.00	3.60	14.30	3.00	1.10
172	480.00	2.80	10.00	3.00	1.10
173	343.00	3.20	32.20	2.00	0.70
174	325.07	2.40	17.15	5.00	2.20
175	213.00	10.00	48.00	2.00	1.40
176	480.00	3.80	64.50	3.00	1.00
177	1056.00	4.00	10.00	5.00	2.60
178	348.00	3.20	7.50	3.00	1.20
179	600.00	3.60	16.80	5.00	2.25
180	362.00	2.60	62.40	1.00	0.94
181	510.00	3.70	12.60	1.00	2.47
182	310.00	2.80	13.80	3.00	1.20
183	465.00	4.00	9.50	4.00	1.60
184	961.00	3.60	13.80	3.00	1.50
185	322.00	4.40	14.30	2.00	2.88
186	315.00	2.80	11.20	3.00	1.10
187	370.00	3.50	10.50	3.00	1.00
188	630.00	4.00	21.96	5.00	2.60
189	410.00	3.20	13.30	3.00	1.10
190	1056.00	4.00	12.76	3.00	1.67
191	246.00	3.20	38.20	3.00	0.80
192	435.00	2.80	15.20	3.00	1.20
193	690.00	4.20	32.00	2.00	1.58
194	276.00	2.60	15.90	2.00	0.80
195	428.00	3.60	16.50	3.00	1.20
196	680.00	2.60	22.70	5.00	2.20
197	420.00	3.70	9.10	4.00	1.40
198	400.00	5.32	66.31	4.00	1.32
199	740.00	3.80	38.00	4.00	2.00

Table A1. *Cont.*

Samples	A1/m	A2/m	A3/MPa	A4	A5/m
200	680.00	3.30	22.70	5.00	2.35
201	450.00	3.00	11.20	3.00	1.20
202	292.00	3.40	12.50	4.00	1.40
203	1056.00	4.00	13.80	3.00	1.60
204	676.00	4.00	35.00	3.00	1.60
205	680.00	2.60	36.14	3.00	1.90
206	470.00	4.00	10.10	5.00	2.20
207	264.00	3.20	11.20	3.00	1.10
208	249.00	3.20	16.80	3.00	1.00
209	340.00	3.00	73.60	2.00	0.80

References

1. Jing, H.; Meng, Q.; Zhu, J.; Meng, B.; Wei, L. Advances in theories and technologies for stability control of the loose zone of surrounding rock in deep roadways. *J. Min. Safety Eng.* **2020**, *37*, 429–442.
2. Dong, F.; Guo, Z.; Lan, B. The theory of supporting broken zone in surrounding rock. *J. Univ. Sci. Technol.* **1991**, *2*, 64–71.
3. Chen, J.; Hu, H.; Zhang, Y. Dynamic and static analysis of mechanism of loosen zone in surrounding rock of tunnels. *J. Geotech. Eng.* **2011**, *33*, 1964–1968.
4. Dong, F.; Song, H.; Guo, Z. Tunnel wall rock loose circle support theories. *J. China Coal Soc.* **1994**, *19*, 21–23.
5. Zhao, G.; Wu, H. Support vector machine model of loose ring thickness prediction. *J. Guangxi Univ.* **2013**, *38*, 444–450.
6. Wang, S.F.; Sun, L.C.; Huang, L.Q.; Li, X.B.; Shi, Y.; Yao, J.R.; Du, S.L. Non-explosive mining and waste utilization for achieving green mining in underground hard rock mine in china. *Trans. Nonferrous Met. Soc. China* **2019**, *29*, 1914–1928. [CrossRef]
7. Jing, H.; Li, Y.; Liang, J. The mechanism and practice of borehole camera test surrounding rock loose zone. *J. Univ. Sci. Technol.* **2009**, *38*, 645–649.
8. Song, H.; Wang, C.; Jia, Y. The principle and practice of using ground penetrating radar to test the loose zone of surrounding rock. *J. Univ. Sci. Technol.* **2002**, *4*, 43–46.
9. Shen, F.; Zhong, W.; Liu, G.; Guo, W. Test and analysis on loose circle of surrounding rock to large distortion region of tong sheng tunnel. *J. Eng. Geophys.* **2011**, *8*, 366–369.
10. Shin, S.W.; Martin, C.D.; Park, E.S.; Christianson, R. Methodology for estimation of excavation damaged zone around tunnels in hard rock. In Proceedings of the 1st Canada/United States Rock Mechanics Symposium, Vancouver, BC, Canada, 27–31 May 2007.
11. Kruschwitz, S.; Yaramanci, U. Detection and characterization of the disturbed rock zone in claystone with the complex resistivity method. *J. Appl. Geophys.* **2004**, *57*, 63–79. [CrossRef]
12. Zou, H.; Xiao, M. Research on evaluation method of underground cavern excavation loose zone. *J. Rock Mech. Geotech.* **2010**, *29*, 513–519.
13. Ma, J.; Zhao, G.; Dong, L.; Chen, G.; Zhang, C. A comparison of mine seismic discriminators based on features of source parameters to waveform characteristics. *Shock Vib.* **2015**, *2015*, 919143. [CrossRef]
14. Liu, G.; Song, H. Numerical simulation of influencing factors of surrounding rock loose zone. *Min. Metall. Eng.* **2003**, *23*, 1–3.
15. Sun, X.; Chang, Q.; Shi, X.; Li, X. Thickness measurement and distribution law of surrounding rock loose zone in large section semicircular arch coal road. *Coal Sci. Technol.* **2016**, *44*, 1–6.
16. Perras, M.A.; Diederichs, M.S. Predicting excavation damage zone depths in brittle rocks. *J. Rock Mech. Geotech.* **2016**, *8*, 60–74. [CrossRef]
17. Wan, C.; Li, X.; Ma, C. Optimization of support technology for deep soft rock roadway based on field measurement of excavation damage zone. *Min. Metall. Eng.* **2012**, *32*, 12–16.
18. Yan, I. Derivation of excavation damaged zone's thickness and plastic zone's scope of tunnel. *Jpn. Railw. Tech. Res. Rep.* **1974**, *900*, 1–38.
19. Wang, R.; Liang, Y.; Qin, W. Measurement and research on broken zone of surrounding rock in Cha Zhen tunnel. In Proceedings of the 4th International Conference on Civil Engineering, Architecture and Building Materials (CEABM), Haikou, China, 24–25 May 2014; pp. 1321–1326.
20. Chen, Q.; Huang, X.; Xie, X. Derivation and improvement of the radius of the loose circle of surrounding rock based on Hoek-Brown criterion. *J. Appl. Mech.* **2015**, *32*, 304–310.
21. Zhao, G.Y.; Liang, W.Z.; Wang, S.F.; Hong, C.S. Prediction model for extent of excavation damaged zone around roadway based on dimensional analysis. *Rock Soil Mech.* **2016**, *37*, 273–278.
22. Shemyakin, I.; Fisenko, G.L.; Kurlenya, M.V. Zonal disintegration of rocks around underground workings. III. Theoretical notions. *Sov. Min. Sci.* **1987**, *23*, 1–6. [CrossRef]
23. Myasnikov, V.P.; Guzev, M.A. Thermomechanical model of elastic-plastic materials with defect structures. *Theor. Appl. Fract. Mec.* **2000**, *33*, 165–171. [CrossRef]

24. Zhou, X.P.; Song, H.F.; Qian, Q.H. Zonal disintegration of deep crack-weakened rock masses: A non-Euclidean model. *Theor. Appl. Fract. Mec.* **2011**, *55*, 227–236. [CrossRef]
25. Bao, Y.; Jiang, B.; Ni, T. Theory and numerical simulation of deep rock mass based on a non-euclidean model. *Sci. Program.-Neth.* **2022**, *2022*, 4492406. [CrossRef]
26. Guzev, M.A.; Odintsev, V.N.; Makarov, V.V. Principals of geomechanics of highly stressed rock and rock massifs. *Tunn. Undergr. Sp. Tech.* **2018**, *81*, 506–511. [CrossRef]
27. Liang, W.; Sari, A.; Zhao, G.; McKinnon, S.D.; Wu, H. Short-term rockburst risk prediction using ensemble learning methods. *Nat. Hazards* **2020**, *104*, 1923–1946. [CrossRef]
28. Liang, W.; Sari, A.; Zhao, G.; McKinnon, S.D.; Wu, H. Probability estimates of short-term rock burst risk with ensemble classifiers. *Rock Mech. Rock Eng.* **2021**, *54*, 1799–1814. [CrossRef]
29. Asadi, A.; Abbasi, A.; Asadi, E. Prediction of excavation damaged zone in underground blasts using artificial neural networks. In Proceedings of the International European Rock Mechanics Symposium (EUROCK), Saint Petersburg, Russia, 22–26 May 2018; pp. 461–464.
30. Zhou, J.; Li, X.B. Evaluating the thickness of broken rock zone for deep roadways using nonlinear SVMs and multiple linear regression model. In Proceedings of the ISMSSE 2011, 1st International Symposium on Mine Safety Science and Engineering (ISMSSE), Beijing, China, 26–29 October 2011; pp. 972–981.
31. Hu, J.; Wang, K.; Xia, Z. Layered fish school optimization support vector machine to predict the thickness of the loose zone of roadway surrounding rock. *Metal. Mine* **2014**, *11*, 31–34.
32. Ma, W. Research on prediction of loose zone of roadway surrounding rock based on parameter optimization LSSVM. In Proceedings of the 9th National Symposium on Numerical Analysis and Analytical Methods of Geotechnical Mechanics, Wuhan, China, 28 October 2007; pp. 470–474.
33. Yu, Z.; Shi, X.; Zhou, J.; Huang, R.; Gou, Y. Advanced prediction of roadway broken rock zone based on a novel hybrid soft computing model using gaussian process and particle swarm optimization. *Appl. Sci.* **2020**, *10*, 6031. [CrossRef]
34. Liu, Y.; Ye, Y.; Wang, Q.; Liu, X.; Wang, W. Predicting the loose zone of roadway surrounding rock using wavelet relevance vector machine. *Appl. Sci.* **2019**, *9*, 6024. [CrossRef]
35. Liang, W.; Luo, S.; Zhao, G.; Wu, H. Predicting hard rock pillar stability using GBDT, XGBoost, and LightGBM algorithms. *Mathematics* **2020**, *8*, 765. [CrossRef]
36. Cardellicchio, A.; Ruggieri, S.; Leggieri, V.; Uva, G. View VULMA: Data set for training a machine-learning tool for a fast vulnerability analysis of existing buildings. *Data* **2022**, *7*, 4. [CrossRef]
37. Harirchian, E.; Jadhav, K.; Kumari, V.; Lahmer, T. ML-EHSAPP: A prototype for machine learning-based earthquake hazard safety assessment of structures by using a smartphone app. *Eur. J. Environ. Civ. Eng.* **2021**, 1–21. [CrossRef]
38. Mangalathu, S.; Sun, H.; Nweke, C.C.; Yi, Z.X.; Burton, H.V. Classifying earthquake damage to buildings using machine learning. *Earthq. Spectra* **2020**, *36*, 183–208. [CrossRef]
39. Ruggieri, S.; Cardellicchio, A.; Leggieri, V.; Uva, G. Machine-learning based vulnerability analysis of existing buildings. *Automat. Constr.* **2021**, *132*, 103936. [CrossRef]
40. Dong, L.; Shu, W.; Sun, D.; Li, X.; Zhang, L. Pre-alarm system based on real-time monitoring and numerical simulation using internet of things and cloud computing for tailings dam in mines. *IEEE Access* **2017**, *5*, 21080–21089. [CrossRef]
41. Xue, J.; Shen, B. A novel swarm intelligence optimization approach: Sparrow search algorithm. *Syst. Sci. Control Eng.* **2020**, *8*, 22–34. [CrossRef]
42. Tang, C. Research on Forecast of Tourist Volume Based on Web Search Index and SSA-BP Model. Master's Thesis, Jiangxi University of Finance and Economics, Nanchang, China, 2021.
43. Hoang, N.; Choi, Y.; Xuan-Nam, B.; Trung, N.-T. Predicting blast-induced ground vibration in open-pit mines using vibration sensors and support vector regression-based optimization algorithms. *Sensors* **2020**, *20*, 132.
44. Cleophas, T.J.; Cleophas, T.F. Artificial intelligence for diagnostic purposes: Principles, procedures and limitations. *Clin. Chem. Lab. Med.* **2010**, *48*, 159–165. [CrossRef]
45. Salah, L.B.; Fourati, F. Systems modeling using deep ELMAN neural network. *Eng. Technol. Appl. Sci.* **2019**, *9*, 3881–3886. [CrossRef]
46. Jing, Y.; Wang, S.; Lu, J. Thickness prediction of loose zone in rock excavation and criterion of non-explosive mechanized mining. *Gold Sci. Technol.* **2021**, *29*, 525–534.
47. Xu, G.; Jing, H. Research on intelligent prediction of loose zone of surrounding rock in coal mine roadway. *J. Univ. Sci. Technol.* **2005**, *2*, 23–26.
48. Pang, J.; Guo, L. Comprehensive classification of coal roadway surrounding rock stability and reasonable selection of bolt support forms. *J. Univ. Sci. Technol.* **1998**, *18*, 9–15.
49. Zhu, Z.; Zhang, H.; Chen, Y. Prediction model of loosening zones around roadway based on MPSO-SVM. *Comput. Eng. Appl.* **2014**, *50*, 1–5.
50. Jing, H. Study on the Intelligence Prediction of Thickness of Broken Rock Zone for Coal Mine Roadways and Its Application. Master's Thesis, Tongji University, Shanghai, China, 2004.
51. Zhu, Y.; Yousefi, N. Optimal parameter identification of PEMFC stacks using adaptive sparrow search algorithm. *Int. J. Hydrogen Energ.* **2021**, *46*, 9541–9552. [CrossRef]

52. Morgan, N.B.; Baco, A.R. Recent fishing footprint of the high-seas bottom trawl fisheries on the northwestern hawaiian ridge and emperor seamount chain: A finer-scale approach to a large-scale issue. *Ecol. Indic.* **2021**, *121*, 107051. [CrossRef]
53. Yuan, J.; Zhao, Z.; Liu, Y.; He, B.; Wang, L.; Xie, B.; Gao, Y. DMPPT Control of photovoltaic microgrid based on improved sparrow search algorithm. *IEEE Access* **2021**, *9*, 16623–16629. [CrossRef]
54. Tuerxun, W.; Chang, X.; Hongyu, G.; Zhijie, J.; Huajian, Z. Fault diagnosis of wind turbines based on a support vector machine optimized by the sparrow search algorithm. *IEEE Access* **2021**, *9*, 69307–69315. [CrossRef]
55. Johnson, C.A.; Giraldeau, L.A.; Grant, J.W.A. The effect of handling time on interference among house sparrows foraging at different seed densities. *Behaviour* **2001**, *138*, 597–614. [CrossRef]
56. Liker, A.; Barta, Z. The effects of dominance on social foraging tactic use in house sparrows. *Behaviour* **2002**, *139*, 1061–1076. [CrossRef]
57. Bautista Sopelana, L.; Alonso, J.C.; Alonso, J. Foraging site displacement in common crane flocks. *Anim. Behav.* **1998**, *56*, 1237–1243. [CrossRef]
58. Lendvai, A.Z.; Barta, Z.; Liker, A.; Bokony, V. The effect of energy reserves on social foraging: Hungry sparrows scrounge more. *Proc. Biol. Sci.* **2004**, *271*, 2467–2472. [CrossRef] [PubMed]
59. Mohamad, E.T.; Jahed Armaghani, D.; Momeni, E.; Alavi Nezhad Khalil Abad, S.V. Prediction of the unconfined compressive strength of soft rocks: A PSO-based ANN approach. *Bull. Eng. Geol. Environ.* **2014**, *74*, 745–757. [CrossRef]
60. Hasanipanah, M.; Noorian-Bidgoli, M.; Jahed Armaghani, D.; Khamesi, H. Feasibility of PSO-ANN model for predicting surface settlement caused by tunneling. *Eng. Comput.-Ger.* **2016**, *32*, 705–715. [CrossRef]
61. Ersoy, O.K.; Deng, S.W. Parallel, self-organizing, hierarchical neural networks with continuous inputs and outputs. *IEEE Trans. Neural. Netwo.* **1995**, *6*, 1037–1044. [CrossRef]
62. Gudise, V.G.; Venayagamoorthy, G.K. Comparison of particle swarm optimization and backpropagation as training algorithms for neural networks. In Proceedings of the 2003 IEEE Swarm Intelligence Symposium (SIS 03), Indianapolis, IN, USA, 24–26 April 2003; pp. 110–117.
63. Burks, T.F.; Shearer, S.A.; Gates, R.S.; Donohue, K.D. Backpropagation neural network design and evaluation for classifying weed species using color image texture. *Trans. Asabe* **2000**, *43*, 1029–1037. [CrossRef]
64. Wang, J.; Zhang, W.; Li, Y.; Wang, J.; Dang, Z. Forecasting wind speed using empirical mode decomposition and Elman neural network. *Appl. Soft Comput.* **2014**, *23*, 452–459. [CrossRef]
65. Elman, J. Finding structure in time. *Cogn. Sci.* **1990**, *14*, 179–211. [CrossRef]
66. Kisi, O.; Cimen, M. A wavelet-support vector machine conjunction model for monthly streamflow forecasting. *J. Hydrol.* **2011**, *399*, 132–140. [CrossRef]
67. Barzegar, R.; Adamowski, J.; Moghaddam, A.A. Application of wavelet-artificial intelligence hybrid models for water quality prediction: A case study in Aji-Chay River, Iran. *Stoch. Env. Res. Risk Access.* **2016**, *30*, 1797–1819. [CrossRef]
68. Nash, J.E.; Sutcliffe, J.V. River flow forecasting through conceptual models' part I—A discussion of principles. *J. Hydrol.* **1970**, *10*, 282–290. [CrossRef]
69. De Myttenaere, A.; Golden, B.; Le Grand, B.; Rossi, F. Mean absolute percentage error for regression models. *Neurocomputing* **2016**, *192*, 38–48. [CrossRef]
70. Chai, T.; Draxler, R.R. Root mean square error (RMSE) or mean absolute error (MAE)?—Arguments against avoiding RMSE in the literature. *Geosci. Model Dev.* **2014**, *7*, 1247–1250. [CrossRef]
71. Roslan, N.; Yahya, Z.R. Reconstruction of font with cubic bezier using differential evolution. *Sains Malays.* **2015**, *44*, 1203–1208. [CrossRef]

applied sciences

MDPI

Article

Multiclassification Prediction of Clay Sensitivity Using Extreme Gradient Boosting Based on Imbalanced Dataset

Tao Ma [1], Lizhou Wu [2,*], Shuairun Zhu [1] and Hongzhou Zhu [2]

[1] College of Environment and Civil Engineering, Chengdu University of Technology, Chengdu 610059, China; opmatao@163.com (T.M.); zhushuairun@163.com (S.Z.)

[2] State Key Laboratory of Mountain Bridge and Tunnel Engineering, Chongqing Jiaotong University, Chongqing 400074, China; zhuhongzhouchina@cqjtu.edu.cn

* Correspondence: lzwu@cqjtu.edu.cn

Abstract: Predicting clay sensitivity is important to geotechnical engineering design related to clay. Classification charts and field tests have been used to predict clay sensitivity. However, the imbalanced distribution of clay sensitivity is often neglected, and the predictive performance could be more accurate. The purpose of this study was to investigate the performance that extreme gradient boosting (XGboost) method had in predicting multiclass of clay sensitivity, and the ability that synthetic minority over-sampling technique (SMOTE) had in addressing imbalanced categories of clay sensitivity. Six clay parameters were used as the input parameters of XGBoost, and SMOTE was used to deal with imbalanced classes. Then, the dataset was divided using the cross-validation (CV) method. Finally, XGBoost, artificial neural network (ANN), and Naive Bayes (NB) were used to classify clay sensitivity. The F1 score, receiver operating characteristic (ROC), and area under the ROC curve (AUC) were considered as the performance indicators. The results revealed that XGBoost showed the best performance in the multiclassification prediction of clay sensitivity. The F1 score and mean AUC of XGBoost were 0.72 and 0.89, respectively. SMOTE was useful in addressing imbalanced issues, and XGBoost was an effective and reliable method of classifying clay sensitivity.

Keywords: clay sensitivity; imbalanced categories; SMOTE; XGBoost

Citation: Ma, T.; Wu, L.; Zhu, S.; Zhu, H. Multiclassification Prediction of Clay Sensitivity Using Extreme Gradient Boosting Based on Imbalanced Dataset. *Appl. Sci.* **2022**, *12*, 1143. https://doi.org/10.3390/app12031143

Academic Editor: Daniel Dias

Received: 4 December 2021
Accepted: 19 January 2022
Published: 21 January 2022

1. Introduction

Soft clays are widely distributed near lakes, rivers, and coastal areas in countries such as Sweden, Norway, Canada, Thailand, and China [1–3]. For grain size, clay is a fine-grained mineral (<2 μm in size), which is the main component of soil [4]. Clay minerals belong to the family of phyllosilicates and provide information on formation conditions and diagenesis [4]. Additionally, clay can be used as an additive for green processing technology and sustainable development, such as medical materials and treatment, agriculture, building materials, adsorbents of organic pollutants in soil, water, and air, etc. [5–10]. For engineers, clays are characterized by high compressibility, low shear strength, and high sensitivity. The sensitivity is defined as the ratio of the unconfined compressive strength of the undisturbed samples to the strength of the remolded samples [11–13].

Nowadays, in situ and laboratory testing and classification charts are often used to predict clay sensitivity. Cone Penetration Tests (CPTu) and Field Vane Tests (FVT) are commonly carried out to obtain the shear strength and classify clay sensitivity [14–18]. Yafrate et al. [19] employed full-flow penetrometers to evaluate the remolded soil strength and clay sensitivity. Abbaszadeh Shahri et al. [20] proposed a Unified Soil Classification System (USCS) to assess soils classification and used high-resolution files to detect potential sensitive clays. Different soil classification charts are widely used to determine clay sensitivity or types [13,21]. For example, Robertson [22] proposed a few updated charts to predict soil type based on CPTu data. Gylland et al. [23] used pore pressure ratio and modified cone resistance to build a set of diagrams identifying sensitive and quick clays.

However, in situ tests are costly and time consuming [14], and the construction conditions are complicated. It is difficult to accurately measure the sensitivity of highly sensitive and quick clays. It is a great challenge if the site clays are not textbook soils in classification charts, which makes clays sensitivity difficult to determine [13]. Therefore, advanced methods are required to resolve this issue.

Artificial Intelligence (AI) approaches, such as machine learning (ML) and deep learning, are being rapidly developed. Machine learning methods are data-driven tools that learn from the relationships of existing data [24]. Hence, ML does not assume a statistical model [25–27]. Additionally, these techniques have been widely applied in engineering [28–33]. For example, Zhang et al. [34] used XGBoost and Bayesian optimization to predict the shear strength of soft clays. Machine learning methods outperform traditional methods [27,28]. Moreover, XGBoost is an excellent ensemble method, better than conventional ML, and shows great potential in geotechnical engineering [34].

The high sensitivity of clays is one of the main properties in soft clay engineering, and considerably influences the safety of such structures [35]. For example, the shear strength decreases due to the excavation disturbance of soft clays, which is related to the sensitivity [3]. Landslides are also related to the sensitivity parameter of geomaterials [36,37]. Therefore, clay sensitivity must be predicted to ensure the safety of geotechnical engineering. There are a few different methods for clay sensitivity classification [11,38,39], as shown in Figure 1. The Canadian Foundation Engineering Manual [38] is used to simplify the classification issue.

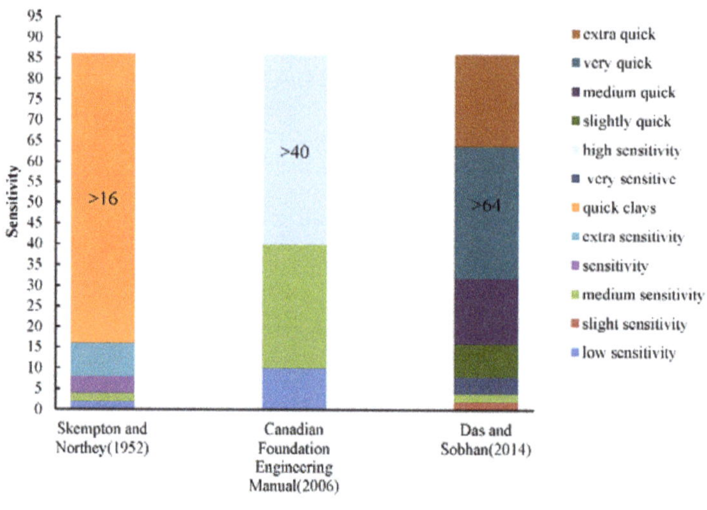

Figure 1. Different methods for clay sensitivity classification.

Few studies have focused on sensitivity classification for soft clay using machine learning methods. Shahri et al. [37] used an optimized ANN to predict clay sensitivity. It is the first time that machine learning methods have been used to predict clay sensitivity. However, it would be more convenient and direct if the result were a category value because clay sensitivity is a category value in Figure 1. Godoy et al. [13] used different machine learning methods to identify quick and highly sensitive clays, and they found that classification accuracy was improved significantly despite limited training data. However, these approaches have a few shortcomings. First, there are more than two sensitivity categories in different classification charts (Figure 1), but few studies have investigated the multiclassification prediction of sensitivity. One previous study [13] only considered two different sensitivity categories, including quick and highly sensitive clays. Therefore, the multiclassification of clay sensitivity should be further investigated. Second, the distribution of different sensitivity categories influencing machine learning methods is imbalanced, which reduces the prediction accuracy [40]. Therefore, SMOTE has been used to address the imbalanced classes of the dataset and improve the accuracy of the results [40]. Furthermore, SMOTE has been successfully applied in geotechnical engineering [41].

The objective of this study was to investigate the potential of XGBoost and SMOTE with regard to predicting the categories of clay sensitivity based on imbalanced datasets. First, the considered dataset, XGBoost, and SMOTE are introduced. Next, SMOTE is used to address the imbalanced categories, and the input parameters include the vertical preconsolidation pressure (VPP), liquid limit (LL), plastic limit (PL), effective vertical pressure (EVP), depth (Dep), and moisture content (W). Then, the F1 score and the area under the receiver operating characteristic (ROC) curve (AUC) are considered as performance indicators to evaluate the proposed model. Finally, the predictive accuracy of XGBoost is compared with that of other methods.

2. Materials and Methods

2.1. XGBoost

The XGBoost method provides an advanced boosted tree model [42] and is a common machine learning method with high accuracy. This method implements a new regularized learning objective, which is simpler than the regularized greedy forest model. The predicted function of XGBoost is defined as follows:

$$\hat{y} = \sum_{k=1}^{K} f_k(x_i), \qquad f_k \in \Gamma \tag{1}$$

where $\hat{y}_i$ is the predicted output value of the ith instance; K is the number of regression trees; f_k is the tree structure; x_i is the feature vector of the ith sample; and Γ is the space of regression trees.

To overcome overfitting problems, a penalty function is used to smooth the learning weights as follows:

$$Obj = \sum_{i=1}^{n} l(y_i, \hat{y}_i) + \sum_{k=1}^{K} \Omega(f_k) \tag{2}$$

$$\Omega(f_k) = \gamma T + \frac{1}{2}\lambda \sum_{j=1}^{T} w_j^2 \tag{3}$$

where Obj is the regularized objective function; $\sum_{i=1}^{n} l(y_i, \hat{y}_i)$ is the loss function, which measures the model accuracy; $\Omega(f_k)$ is a penalty function handling overfitting; y_i is the real value of the ith sample; γ is the complexity cost of introducing additional leaves; T is the number of leaves; λ is a regular item parameter; and w_j^2 is the weight of the jth leaf node.

The additive method is used to train the model as follows [42]:

$$Obj^{(t)} = \sum_{i=1}^{n} l\left(y_i, \hat{y}^{(t-1)} + f_t(x_i)\right) + \Omega(f_t) \tag{4}$$

where $\hat{y}_i{}^{(t)}$ is the prediction of the ith sample at the ith iteration, and f_t is applied to minimize the objective. To rapidly optimize the loss function of the first term in Equation (4), a second-order Taylor expansion can be written as

$$Obj^{(t)} = \left[l\left(y_i - \hat{y}_i{}^{(t-1)}\right) + g_i f_t(x_i) + \frac{1}{2} h_i f_t^2(x_i)\right] + \Omega(f_t) \tag{5}$$

where $g_i = \partial_{\hat{y}}(t-1) l\left(y_i, \hat{y}^{(t-1)}\right)$ and $h_i = \partial^2_{\hat{y}^{(t-1)}} l\left(y_i - \hat{y}^{(k-1)}\right)$ are the first and second order gradient statistics of the loss function, respectively. The constant term is removed to obtain the objective function of the ith step as follows:

$$Obj^{(t)} = \sum_{i=1}^{n}\left[g_i f_t(x_i) + \frac{1}{2}h_i f_t^2(x_i)\right] + \Omega(f_t) \quad (6)$$

The parameters in Equation (6) can be continuously updated until the conditions are satisfied. More details on XGBoost can be found in ref. [42]. XGBoost has been used to predict shear strength of clay, and the results demonstrate that XGBoost is a promising tool for predicting geotechnical parameters [34]. The potential of XGBoost for predicting multicategory of clay sensitivity will be investigated.

2.2. SMOTE

The Synthetic Minority Over-sampling Technique (SMOTE) was first proposed by Chawla et al. [40] and is used to solve the classification problems of imbalanced datasets. A dataset is imbalanced if the distribution of categories is unequal, which results in low classification accuracy. A few methods have been proposed to address the imbalance issue, such as re-sampling the original dataset, over-sampling the minority categories, and under-sampling the majority class [43]. However, these methods do not considerably enhance the accuracy of minority class recognition [20]. The SMOTE creates synthetic examples instead of performing typical over-sampling. First, for each sample x_l in the minority class, the Euclidean Distance between x_l and other samples of the minority class is calculated, and the k neighbors of x_l are obtained. Then, a random sample x_m from the k neighbors is selected. Finally, the new sample x_n can be expressed as follows:

$$x_n = x_l + \lambda(x_l - x_m) \quad (7)$$

where λ is a random number in the range of 0–1.

SMOTE has been utilized to solve imbalanced rock mass classification in tunneling engineering [41]. The categories of clay sensitivity comprise more than two classes (Figure 1), which may cause imbalanced problems. Therefore, SMOTE could be used to address imbalanced classes.

3. Preprocessing Data

3.1. Description of Data

The clay dataset was obtained from F-CLAY/7/216 and S-CLAY/7/168 [14,44]. F-CLAY/7/216 consists of 216 samples and was compiled through field vane tests (FVT) in Finland; S-CLAY/7/168 was compiled through 168 FVT tests in Sweden and Norway. Therefore, there are 384 samples in total. Each sample contains six attributes, namely, LL, PL, W, EVS, Dep, and VPP. The sensitivity (St) is the predicted value. The distribution of the six attributes is shown in Figure 2, and the statistical information of the input attributes is listed in Table 1. Python 3.6 and the scikit-learn 0.23.2 library [45] were used to prepare the data and train the model.

Table 1. Statistical information of clay parameters.

	Depth (m)	LL (%)	PL (%)	W (%)	EVS (kPa)	VPP (kPa)	S_t
mean	6.97	68.37	28.49	76.47	48.72	79.82	16.27
std	3.95	23.86	7.97	23.32	27.33	48.54	13.26
min	0.50	22.00	2.73	17.27	6.86	15.2	2.00
50%	6.00	68.72	27.02	75.00	43.08	64.88	11.00
max	24.00	201.81	73.92	180.11	212.87	315.64	64.00

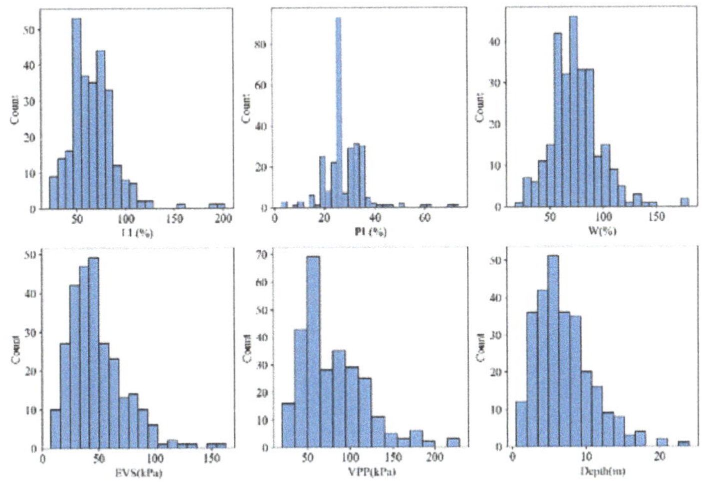

Figure 2. Distribution of input parameters.

3.2. Data Preparation and Performance

3.2.1. Analysis of Clay Dataset

This study referred to the Canadian Foundation Engineering Manual [38] for the classification of clay sensitivity. The category distribution is shown in Figure 3, where it can be seen that the proportion of low sensitivity is close to the proportion of medium sensitivity. However, the proportion of high sensitivity is particularly smaller compared with other categories and only accounts for 9.54%. Hence, it is difficult to classify the minority class [41]. To deal with the imbalanced classes, SMOTE is used to over-sample the high sensitivity category.

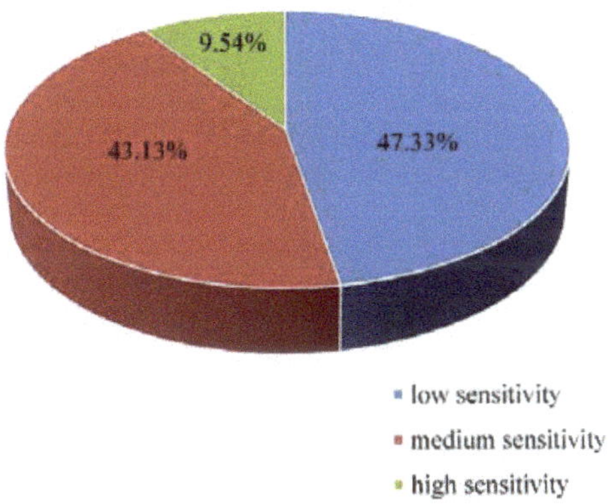

Figure 3. Distribution of sensitivity categories.

A heat map is used to show the correlation coefficient among the attributes [34]. The correlation coefficient is calculated using the Pearson coefficient [24]. The heat map of clay attributes is shown in Figure 4, where there is no strong linear relationship between the clay attributes and the sensitivity. However, machine learning methods can effectively solve the above-mentioned nonlinear issues [26].

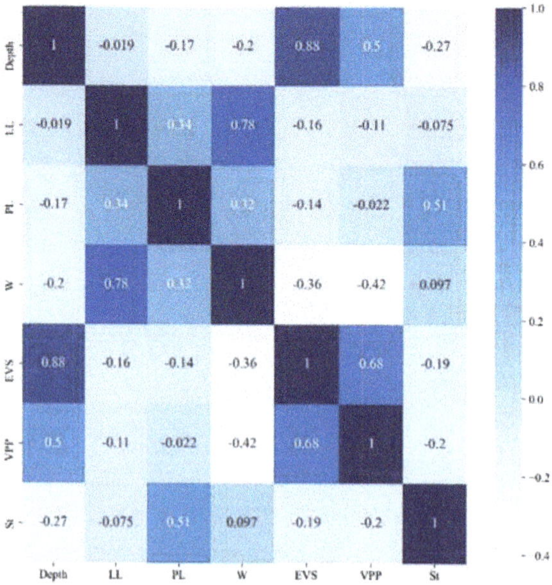

Figure 4. Heat map of clay parameters.

3.2.2. Cross-Validation

The processed data used in machine learning are commonly separated into training and validation datasets [46]. The machine learning method is trained on the training dataset, and then the accuracy is tested on the test dataset. However, small datasets may potentially cause bias. Therefore, *k*-fold cross-validation (CV) is used to divide the datasets [47,48]. CV does not only increase the training times but also improves the robustness of the model. The CV method divides the datasets into *k* mutually exclusive subsets, that is, $D = D_1 \cup D_2 \cup \cdots D_k$, $D_i \cap D_j = \varnothing (i \neq j)$. Next, the set of $k-1$ subsets is used as the training set, the remaining subset is acted as the validation dataset, and training and validation can be conducted for *k* times. Then, *k* results are obtained, and the validation results are the mean value of the *k* results. Figure 5 shows the 5-fold CV.

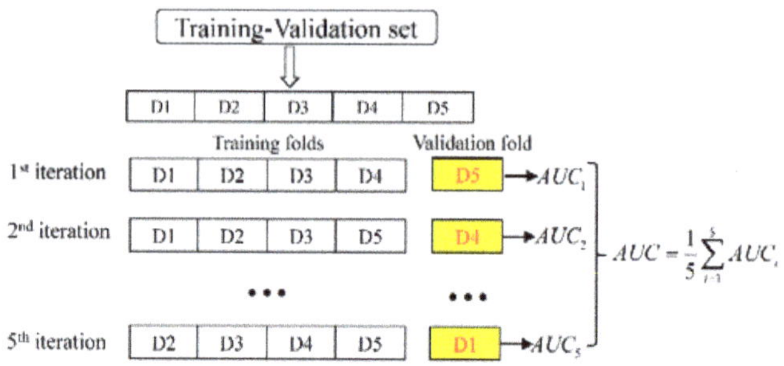

Figure 5. Schematic diagram of 5-fold CV.

3.3. *Performances*

The confusion matrix, F1 score, ROC, and AUC are considered as the evaluation indicators.

3.3.1. Confusion Matrix

The confusion matrix consists of the True Positive (*TP*), False Positive (*FP*), True Negative (*TN*), and False Negative (*FN*) [49], as shown in Figure 6. In the confusion matrix,

TP and *TN* are a correct classification, while *FP* and *FN* are an erroneous one. A value closer to 1 indicates higher accuracy.

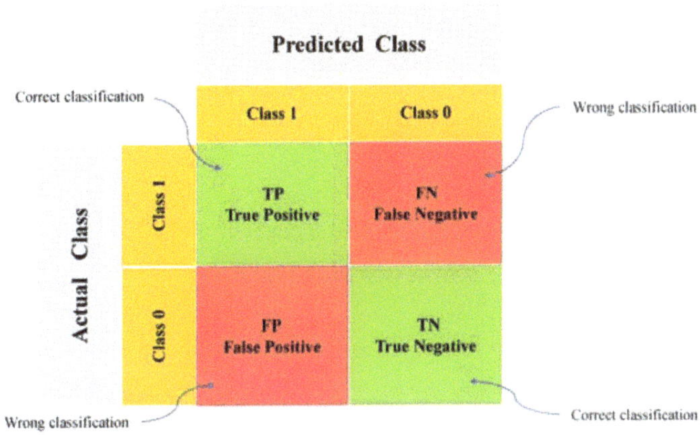

Figure 6. Confusion matrix.

3.3.2. F1 Score

The F1 score is the harmonic mean of Precision (*P*) and Recall (*R*). Precision and Recall are shown in Figure 7. Precision, Recall, and F1 score are defined as follows [50]:

$$F_{score} = \frac{2 \times P \times R}{P + R} \tag{8}$$

$$P = \frac{TP}{TP + FP} \tag{9}$$

$$R = \frac{TP}{TP + FN} \tag{10}$$

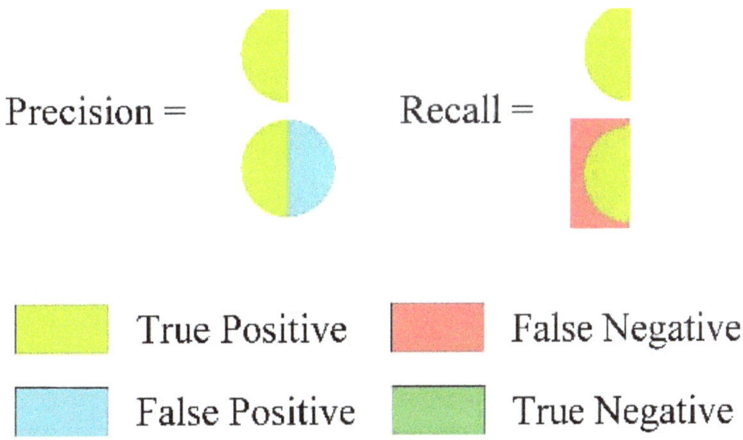

Figure 7. Precision and recall (adapted from [50]).

3.3.3. AUC and ROC

The ROC graph helps visualize the classification algorithms [51]. The y-axis of the ROC graphs is the true positive rate (TPR), while the x-axis is the false positive rate (FPR) (Figure 8). The TPR and FPR are defined as follows:

$$TPR = \frac{TP}{TP + FN} \tag{11}$$

$$FPR = \frac{FP}{TN + FP} \tag{12}$$

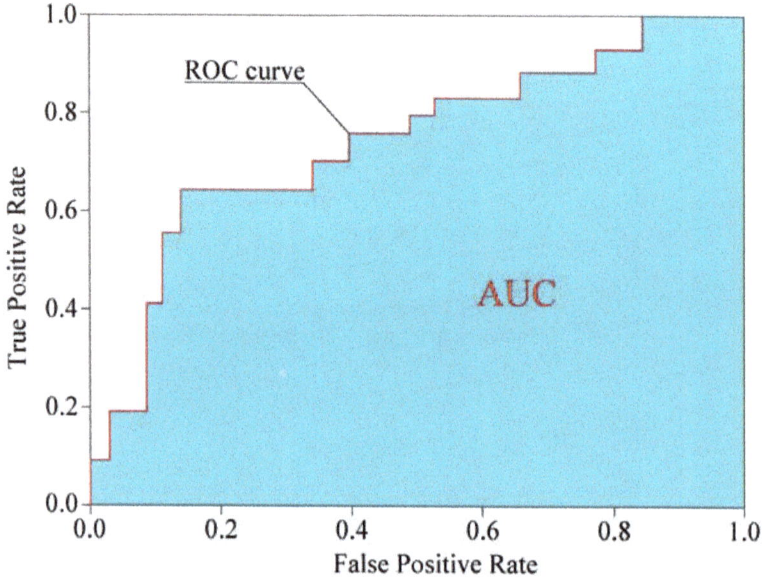

Figure 8. AUC and ROC diagram.

Because the ROC curve is two-dimensional, a single scalar value, such as AUC, can evaluate the algorithms [51,52]. The AUC value is in the range of 0–1. An AUC value closer to 1 indicates a better fit for the model.

3.3.4. Evaluation Methods

In this study, the XGBoost method was compared with ANN and NB. To further investigate SMOTE, the data without SMOTE were also used as the training data of XGBoost, which is referred to as XGBoost_NoSmote. The dataset is divided into 260 training–validation sets (70%) and 112 test sets (30%). Gridsearch is a custom method used to optimize parameters [45]. To ensure that models could achieve their own best performance, CV and gridsearch are used to optimize the hyperparameters on a training–validation set, and the final performance of a given model is evaluated on a test set. These methods are different from previous studies [13,37] that did not incorporate parameter optimization. The flow chart of the method is shown in Figure 9. Table 2 lists the optimal parameters of models.

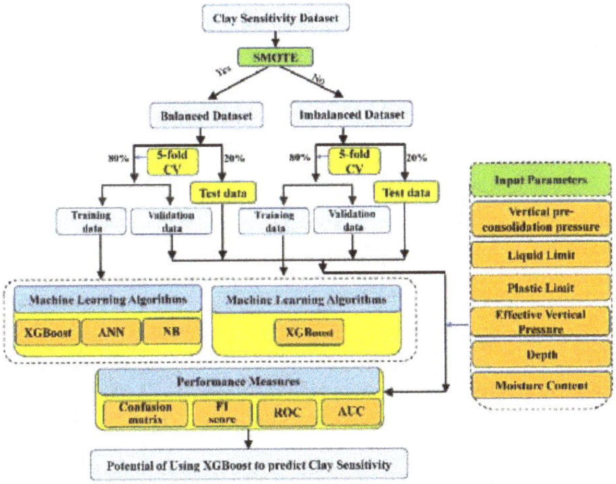

Figure 9. Flow chart of proposed method.

Table 2. Optimal parameters of models.

Model	Parameters	Value
XGBoost	n_estimators	360
	learning_rate	0.002
	max_depth	6
	min_child_weight	1
	gamma	0.2
	colsample_bytree	0.5
	subsample	0.8
ANN	learning_rate_init	0.0001
	activation	tanh
	hidden_layer_sizes	(100, 100, 100)
	max_iter	260
NB	priors	3
	var_smoothing	10^{-9}
XGBoost_NoSmote	n_estimators	360
	learning_rate	0.005
	max_depth	5
	min_child_weight	1
	gamma	0.3
	colsample_bytree	0.7
	subsample	0.8

4. Results

4.1. Confusion Matrix and F1 Score Results

Figure 10 shows the confusion matrix of different machine learning methods, and labels 0–2 represent high sensitivity, low sensitivity, and medium sensitivity, respectively. When the number in the matrix is closer to 1, the model fits better. The XGBoost matrix is larger than that of the other models, which indicates that XGBoost outperforms ANN and NB. Furthermore, the matrix value of XGBoost without SMOTE is smaller than that of XGBoost with SMOTE.

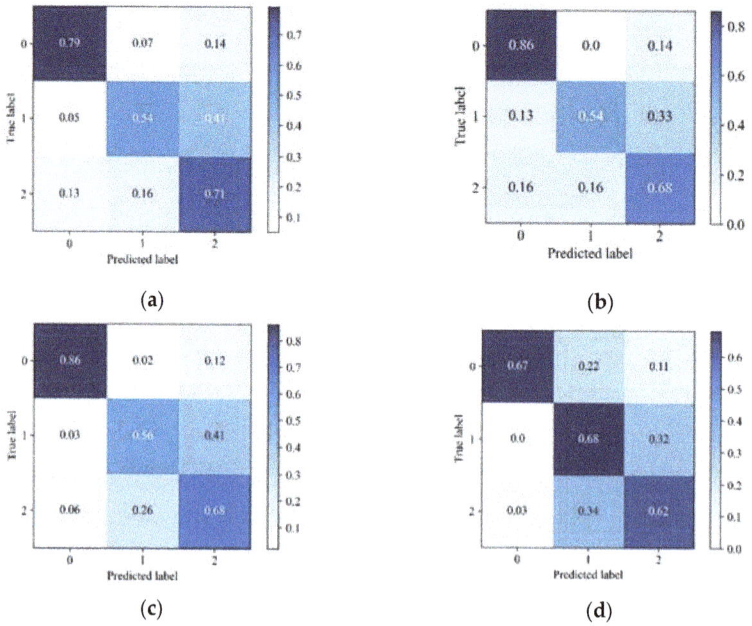

Figure 10. Confusion matrix of different machine learning methods: (**a**) ANN; (**b**) NB; (**c**) XGBoost; (**d**) XGBoost_NoSmote.

Figure 11 presents the F1 score of different machine learning methods. An F1 score closer to 1 indicates that the model has better performance. Figure 11 shows that XGBoost achieved the best F1 score, Recall, and Precision (0.72, 0.72, and 0.73, respectively), followed by NB (0.68, 0.69, and 0.70, respectively), and XGBoost_NoSmote (0.68, 0.66, and 0.71, respectively). ANN had the worst performance in terms of the F1 score, Recall, and Precision (0.61, 0.62, and 0.63, respectively). Using SMOTE, the performance of XGBoost on the F1 score, Recall, and Precision improved by 6.9%, 9.1%, and 2.8%, respectively.

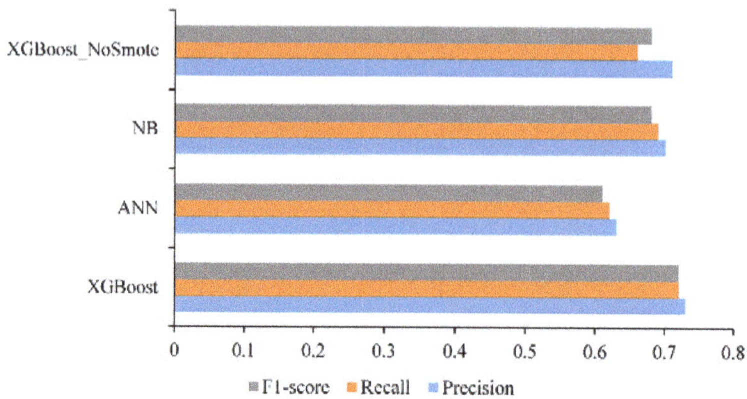

Figure 11. F1 score of different machine learning methods.

4.2. ROC and AUC Results

Figure 12 shows ROC and AUC of different machine learning methods. The XGBoost method achieved the best AUC score for all classes, but it achieved the worst AUC score on the data without SMOTE. All indices of the models are listed in Table 3. The XGBoost method achieved the highest F1 score, followed by NB, ANN, and XGBoost_NoSmote. For all AUC evaluations, XGBoost achieved the best effect. Compared with XGBoost_NoSmote, the AUC score of XGBoost improved by 5.43%, 14.47%, and 10.81%, respectively. For medium and low sensitivity classification, XGBoost performed the best, but the performance of ANN and NB was slightly inferior compared with XGBoost. Finally, the performance of XGBoost_NoSmote was poor.

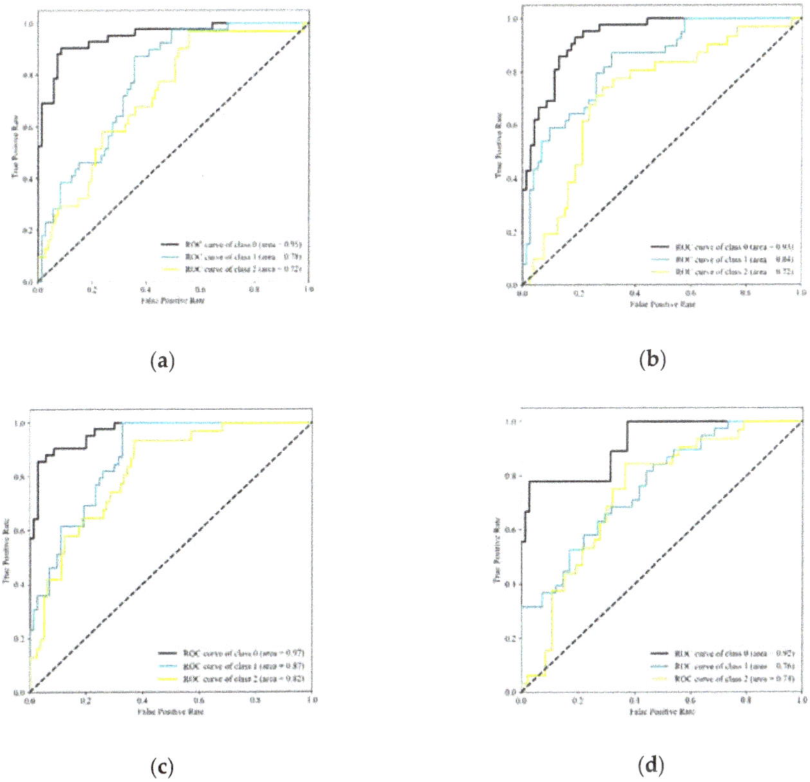

Figure 12. ROC and AUC of different machine learning methods: (**a**) ANN; (**b**) NB; (**c**) XGBoost; (**d**) XGBoost_NoSmote.

Table 3. Evaluation measures of different models.

Evaluation Measures	Models			
	XGBoost	**ANN**	**NB**	**XGBoost_NoSmote**
Precision	0.73	0.63	0.70	0.71
Recall	0.72	0.62	0.69	0.66
F1 score	0.72	0.61	0.68	0.68
AUC of high sensitivity	0.97	0.95	0.93	0.92
AUC of medium sensitivity	0.82	0.72	0.72	0.74
AUC of low sensitivity	0.87	0.78	0.84	0.76
Mean AUC of classification	0.89	0.82	0.83	0.81

4.3. Compared with Previous Studies

The results of Sections 4.1 and 4.2 indicate that XGBoost performs best on all evaluation indices, and the performance is identical to refs. [33,53], which proves that XGBoost is a powerful tool to classify the properties of engineering materials. This study also produces some valid and new results that are different from other studies. The results in this study are category values and are consistent with the clay sensitivity standards [37], as shown in Figure 1. Second, the results of XGBoost are better than those of XGBoost_NoSmote, which indicates that SMOTE is good at solving imbalanced problems, and it is the first time that SMTOE has been adopted to address imbalanced clay sensitivity problems. Moreover, the results of Godoy et al. [13] are a binary classification, i.e., the clays are divided into quick and highly sensitive groups. In this study, multiclass of clay sensitivity is predicted using XGBoost, and the results are more satisfactory.

5. Discussion

Clay sensitivity is not only important to the safety of geotechnical engineering, but it also influences the feasibility of such projects. However, it is difficult to classify clay sensitivity, which is greatly influenced by disturbance. There are more than two sensitivity categories under different classification charts. Moreover, the distribution of clay sensitivity categories is often imbalanced. Therefore, new methods are needed. However, XGBoost is rarely used for multiclassification problems, such as predicting the sensitivity categories of soft clays. Additionally, it is necessary to deal with imbalanced problems.

In this study, NB, ANN, and XGBoost were used to predict the multiple classes of clay sensitivity. SMOTE was applied to address the imbalanced classes of data. Additionally, a set of performance indices were developed to evaluate accuracy.

The evaluation indices of XGBoost incorporating SMOTE were better compared with those of other machine learning methods, based on the classification results. Compared with XGBoost_NoSmote, the mean AUC of classification for XGBoost improved by 9.9%, which indicates that SMOTE improves the model performance of imbalanced datasets. The best classification performance was achieved for high sensitivity, followed by low sensitivity and medium sensitivity. This study did not only investigate multiclassification with regard to clay sensitivity, but it also employed SMOTE to handle imbalanced issues. The results prove that the combination of XGBoost and SMOTE is a simple and quick way to classify imbalanced clay datasets. Furthermore, more accurate indices for evaluating the model performance, such as the AUC and F1 score, were applied to assess the models.

However, this study has a few limitations. For the AUC of medium sensitivity (Table 3), all models performed slightly worse compared with other categories. The possible reason is that the medium sensitivity is between the low sensitivity and high sensitivity categories, which results in a particularly unclear boundary and affects the model performance. Other studies have also proven that the overlap between different classes can influence classification performance [54,55]. Therefore, new methods should be developed to solve these issues.

6. Conclusions

In this study, NB, ANN, and XGBoost were employed to predict the multiple classes of clay sensitivity and evaluate their performance, respectively. SMOTE was first applied to address the imbalanced classes of clay sensitivity. The conclusions demonstrated that the XGBoost achieved the best performance, according to all performance indices. The results obtained by XGBoost were better than those of XGBoost_NoSmote, which means that SMOTE can improve model performance. Therefore, the proposed XGBoost is an effective and low-cost method for the multiclassification prediction of clay sensitivity, and the proposed SMOTE is useful for addressing the imbalanced classes of clay datasets. However, models may perform better on the AUC of medium sensitivity. It is recommended that SMOTE could be improved according to the distribution of clay sensitivity. Additionally, XGBoost predicts more than three clay sensitivity categories, which makes the classification results more delicate.

Author Contributions: Conceptualization, T.M. and L.W.; methodology, T.M.; software, T.M. and S.Z.; validation, T.M.; formal analysis, T.M.; writing—original draft preparation, T.M. and L.W.; writing—review and editing, L.W. and H.Z.; visualization, T.M. and S.Z.; supervision, L.W. and H.Z.; funding acquisition, L.W. All authors have read and agreed to the published version of the manuscript.

Funding: This research was funded by the National Natural Science Foundation of China (No. 41790432 and 51908093), and the National Key Research and Development Program of China (Grant No. 2021YFB2600026).

Institutional Review Board Statement: Not applicable.

Informed Consent Statement: Not applicable.

Data Availability Statement: The data presented in this study are available on request from the corresponding author.

Conflicts of Interest: The authors declare no conflict of interest. The funders had no role in the design of the study; in the collection, analyses, or interpretation of data; in the writing of the manuscript, or in the decision to publish the results.

References

1. Likitlersuang, S.; Surarak, C.; Wanatowski, D.; Oh, E.; Balasubramaniam, A. Finite element analysis of a deep excavation: A case study from the Bangkok MRT. *Soils Found.* **2013**, *53*, 756–773. [CrossRef]
2. Arasan, S.; Akbulut, R.K.; Isik, F.; Bagherinia, M.; Zaimoglu, A.S. Behavior of polymer columns in soft clayey soil: A preliminary study. *Geomech. Eng.* **2016**, *10*, 95–107. [CrossRef]
3. Hu, J.; Ma, F. Failure Investigation at a Collapsed Deep Open Cut Slope Excavation in Soft Clay. *Geotech. Geol. Eng.* **2018**, *35*, 665–683. [CrossRef]
4. Jlassi, K.; Krupa, I.; Chehimi, M.M. Overview: Clay Preparation, Properties, Modification. *Clay-Polym. Nanocomposites* **2017**, 1–28. [CrossRef]
5. Paiva, L.B.; Morales, A.R.; Valenzuela Díaz, F.R. Organoclays: Properties, preparation and applications. *Appl. Clay Sci.* **2008**, *42*, 8–24. [CrossRef]
6. Zhou, C.H.; Zhao, L.Z.; Wang, A.Q.; Chen, T.H.; He, H.P. Current fundamental and applied research into clay minerals in China. *Appl. Clay Sci.* **2016**, *119*, 3–7. [CrossRef]
7. Zahid, I.; Ayoub, M.; Abdullah, B.B.; Nazir, M.H.; Zulqarnain; Kaimkhani, M.A.; Sher, F. Activation of nano kaolin clay for bio-glycerol conversion to a valuable fuel additive. *Sustainability* **2021**, *13*, 2631. [CrossRef]
8. Doğan-Sağlamtimur, N.; Bilgil, A.; Szechyńska-Hebda, M.; Parzych, S.; Hebda, M. Eco-friendly fired brick produced from industrial ash and natural clay: A study of waste reuse. *Materials* **2021**, *14*, 877. [CrossRef] [PubMed]
9. Otunola, B.O.; Ololade, O.O. A review on the application of clay minerals as heavy metal adsorbents for remediation purposes. *Environ. Technol. Innov.* **2020**, *18*, 100692. [CrossRef]
10. Abdallah, Y.K.; Estévez, A.T. 3d-printed biodigital clay bricks. *Biomimetics* **2021**, *6*, 59. [CrossRef]
11. Skempton, A.W.; Northey, R.D. The sensitivity of clays. *Geotechnique* **1952**, *3*, 30–53. [CrossRef]
12. Terzaghi, K.; Peck, R.B.; Mesri, G. *Soil Mechanics in Engineering Practice*, 3rd ed.; John Wiley & Sons: New York, NY, USA, 2016; pp. 20–25.
13. Godoy, C.; Depina, I.; Thakur, V. Application of machine learning to the identification of quick and highly sensitive clays from cone penetration tests. *J. Zhejiang Univ. Sci. A* **2020**, *21*, 445–461. [CrossRef]
14. D'Ignazio, M.; Phoon, K.K.; Tan, S.A.; Länsivaara, T.T. Correlations for undrained shear strength of finish soft clays. *Can. Geotech. J.* **2016**, *53*, 1628–1645. [CrossRef]
15. Eslami, A.; Fellenius, B.H. Pile capacity by direct CPT and CPTu methods applied to 102 case histories. *Can. Geotech. J.* **1997**, *34*, 886–904. [CrossRef]
16. Gao, Y.B.; Ge, X.N. On the sensitivity of soft clay obtained by the field vane test. *Geotech. Test. J.* **2016**, *39*, 282–290. [CrossRef]
17. Meijer, G.; Dijkstra, J. A novel methodology to regain sensitivity of quick clay in a geotechnical centrifuge. *Can. Geotech. J.* **2013**, *50*, 995–1000. [CrossRef]
18. Schneider, J.A.; Randolph, M.F.; Mayne, P.W.; Ramsey, N.R. Analysis of Factors Influencing Soil Classification Using Normalized Piezocone Tip Resistance and Pore Pressure Parameters. *J. Geotech. Geoenviron. Eng.* **2008**, *134*, 1569–1586. [CrossRef]
19. Yafrate, N.; DeJong, J.; DeGroot, D.; Randolph, M. Evaluation of Remolded Shear Strength and Sensitivity of Soft Clay Using Full-Flow Penetrometers. *J. Geotech. Geoenviron. Eng.* **2009**, *135*, 1179–1189. [CrossRef]
20. Abbaszadeh Shahri, A.; Malehmir, A.; Juhlin, C. Soil classification analysis based on piezocone penetration test data-A case study from a quick-clay landslide site in southwestern Sweden. *Eng. Geol.* **2015**, *189*, 32–47. [CrossRef]
21. Moreno-Maroto, J.M.; Alonso-Azcárate, J.; O'Kelly, B.C. Review and critical examination of fine-grained soil classification systems based on plasticity. *Appl. Clay Sci.* **2021**, *200*, 105955. [CrossRef]

22. Robertson, P.K. Cone penetration test (CPT)-based soil behaviour type (SBT) classification system—An update. *Can. Geotech. J.* **2016**, *53*, 1910–1927. [CrossRef]
23. Gylland, A.S.; Sandven, R.; Montafia, A.; Pfaffhuber, A.A.; Kåsin, K.; Long, M. Cptu classification diagrams for identification of sensitive clays. In *Advances in Natural and Technological Hazards Research*; Springer: Cham, Switzerland, 2017.
24. Zhang, W.; Wu, C.; Li, Y.; Wang, L.; Samui, P. Assessment of pile drivability using random forest regression and multivariate adaptive regression splines. *Georisk* **2021**, *15*, 27–40. [CrossRef]
25. Dickson, M.E.; Perry, G.L.W. Identifying the controls on coastal cliff landslides using machine-learning approaches. *Environ. Model. Softw.* **2016**, *76*, 117–127. [CrossRef]
26. Pourghasemi, H.R.; Rahmati, O. Prediction of the landslide susceptibility: Which algorithm, which precision? *Catena* **2018**, *162*, 177–192. [CrossRef]
27. Li, S.H.; Wu, L.Z.; Luo, X.H. A novel method for locating the critical slip surface of a soil slope. *Eng. Appl. Artif. Intell.* **2020**, *94*, 103733. [CrossRef]
28. Zhu, S.; Wu, L.; Huang, J. Application of an improved P(m)-SOR iteration method for flow in partially saturated soils. *Comput. Geosci.* **2021**, 1–15. [CrossRef]
29. Pham, B.T.; Son, L.H.; Hoang, T.A.; Nguyen, D.M.; Tien Bui, D. Prediction of shear strength of soft soil using machine learning methods. *Catena* **2018**, *166*, 181–191. [CrossRef]
30. Mishra, P.; Samui, P.; Mahmoudi, E. Probabilistic design of retaining wall using machine learning methods. *Appl. Sci.* **2021**, *11*, 5411. [CrossRef]
31. Huang, Z.; Zhang, D.; Zhang, D. Application of ANN in Predicting the Cantilever Wall Deflection in Undrained Clay. *Appl. Sci.* **2021**, *11*, 9760. [CrossRef]
32. Li, S.H.; Luo, X.H.; Wu, L.Z. A new method for calculating failure probability of landslide based on ANN and a convex set model. *Landslides* **2021**, *18*, 2855–2867. [CrossRef]
33. Wu, L.Z.; Li, S.H.; Huang, R.Q.; Xu, Q. A new grey prediction model and its application to predicting landslide displacement. *Appl. Soft Comput. J.* **2020**, *95*, 106543. [CrossRef]
34. Zhang, W.; Wu, C.; Zhong, H.; Li, Y.; Wang, L. Prediction of undrained shear strength using extreme gradient boosting and random forest based on Bayesian optimization. *Geosci. Front.* **2021**, *12*, 469–477. [CrossRef]
35. Chen, R.P.; Li, Z.C.; Chen, Y.M.; Ou, C.Y.; Hu, Q.; Rao, M. Failure Investigation at a Collapsed Deep Excavation in Very Sen-sitive Organic Soft Clay. *J. Perform. Constr. Facil.* **2015**, *29*, 04014078. [CrossRef]
36. Gylland, A.; Long, M.; Emdal, A.; Sandven, R. Characterisation and engineering properties of Tiller clay. *Eng. Geol.* **2013**, *164*, 86–100. [CrossRef]
37. Abbaszadeh Shahri, A. An Optimized Artificial Neural Network Structure to Predict Clay Sensitivity in a High Landslide Prone Area Using Piezocone Penetration Test (CPTu) Data: A Case Study in Southwest of Sweden. *Geotech. Geol. Eng.* **2016**, *34*, 86–100. [CrossRef]
38. *Canadian Foundation Engineering Manual*, 4th ed.; Canadian Geotechnical Society: Prince George, BC, Canada, 2006; pp. 17–20.
39. Das, B.; Sobhan, K. *Principles of Geotechnical Engineering*, 8th ed.; CENGAGE Learning: Stanford, CA, USA, 2014; pp. 450–466.
40. Chawla, N.V.; Bowyer, K.W.; Hall, L.O.; Kegelmeyer, W.P. SMOTE: Synthetic minority over-sampling technique. *J. Artif. Intell. Res.* **2002**, *16*, 321–357. [CrossRef]
41. Liu, Q.; Wang, X.; Huang, X.; Yin, X. Prediction model of rock mass class using classification and regression tree integrated AdaBoost algorithm based on TBM driving data. *Tunn. Undergr. Sp. Technol.* **2020**, *106*, 103595. [CrossRef]
42. Chen, T.; Guestrin, C. XGBoost: A scalable tree boosting system. In Proceedings of the ACM SIGKDD International Conference on Knowledge Discovery and Data Mining, San Francisco, CA, USA, 13–17 August 2016. [CrossRef]
43. Japkowicz, N. The Class Imbalance Problem: Significance and Strategies. In Proceedings of the 2000 International Conference on Artificial Intelligence, Acapulco, Mexico, 11–14 April 2000.
44. Ching, J.; Phoon, K.K. Transformations and correlations among some clay parameters-The global database. *Can. Geotech. J.* **2014**, *51*, 663–685. [CrossRef]
45. Pedregosa, F.; Varoquaux, G.; Gramfort, A.; Michel, V.; Thirion, B.; Grisel, O.; Blondel, M.; Prettenhofer, P.; Weiss, R.; Du-bourg, V.; et al. Scikit-learn: Machine learning in Python. *J. Mach. Learn. Res.* **2011**, *12*, 2825–2830.
46. Alkroosh, I.; Nikraz, H. Predicting axial capacity of driven piles in cohesive soils using intelligent computing. *Eng. Appl. Artif. Intell.* **2012**, *25*, 618–627. [CrossRef]
47. Kohavi, R. A Study of Cross-Validation and Bootstrap for Accuracy Estimation and Model Selection. *Int. Jt. Conf. Artif. Intell.* **1995**, *14*, 1137–1145.
48. Wong, T.T. Performance evaluation of classification algorithms by k-fold and leave-one-out cross validation. *Pattern Recognit.* **2015**, *48*, 2839–2846. [CrossRef]
49. Trajdos, P.; Kurzynski, M. Weighting scheme for a pairwise multi-label classifier based on the fuzzy confusion matrix. *Pattern Recognit. Lett.* **2018**, *103*, 60–67. [CrossRef]
50. Xue, Y.; Bai, C.; Qiu, D.; Kong, F.; Li, Z. Predicting rockburst with database using particle swarm optimization and extreme learning machine. *Tunn. Undergr. Sp. Technol.* **2020**, *98*, 103287. [CrossRef]
51. Fawcett, T. An introduction to ROC analysis. *Pattern Recognit. Lett.* **2006**, *27*, 861–874. [CrossRef]

52. Huang, J.; Ling, C.X. Using AUC and accuracy in evaluating learning algorithms. *IEEE Trans. Knowl. Data Eng.* **2005**, *17*, 299–310. [CrossRef]
53. Pei, L.; Sun, Z.; Yu, T.; Li, W.; Hao, X.; Hu, Y.; Yang, C. Pavement aggregate shape classification based on extreme gradient boosting. *Constr. Build. Mater.* **2020**, *256*, 119356. [CrossRef]
54. García, V.; Mollineda, R.A.; Sánchez, J.S. On the k-NN performance in a challenging scenario of imbalance and overlapping. *Pattern Anal. Appl.* **2008**, *11*, 269–280. [CrossRef]
55. López, V.; Fernández, A.; García, S.; Palade, V.; Herrera, F. An insight into classification with imbalanced data: Empirical re-sults and current trends on using data intrinsic characteristics. *Inf. Sci.* **2013**, *250*, 113–141. [CrossRef]

applied sciences

MDPI

Article

A Two-Phase Approach for Predicting Highway Passenger Volume

Yun Xiang [1], Jingxu Chen [2,*], Weijie Yu [2,*], Rui Wu [3], Bing Liu [4], Baojie Wang [5] and Zhibin Li [2]

1 College of City Construction, Jiangxi Normal University, No. 99 Ziyang Avenue, Nanchang 330022, China; yunxiang@jxnu.edu.cn
2 School of Transportation, Southeast University, No. 2 Southeast University Road, Nanjing 211189, China; lizhibin@seu.edu.cn
3 School of Computer Science and Engineering, Southeast University, No. 2 Southeast University Road, Nanjing 211189, China; rhys@seu.edu.cn
4 School of Information Technology and Electrical Engineering, The University of Queensland, St Lucia, Brisbane, QLD 4072, Australia; bing.liu@uq.edu.au
5 College of Transportation Engineering, Chang'an University, Xi'an 710054, China; wangbj2@163.com
* Correspondence: chenjingxu@seu.edu.cn (J.C.); 213153324@seu.edu.cn (W.Y.)

Abstract: With the continuous process of urbanization, regional integration has become an inevitable trend of future social development. Accurate prediction of passenger volume is an essential prerequisite for understanding the extent of regional integration, which is one of the most fundamental elements for the enhancement of intercity transportation systems. This study proposes a two-phase approach in an effort to predict highway passenger volume. The datasets subsume highway passenger volume and impact factors of urban attributes. In Phase I, correlation analysis is conducted to remove highly correlated impact factors, and a random forest algorithm is employed to extract significant impact factors based on the degree of impact on highway passenger volume. In Phase II, a deep feedforward neural network is developed to predict highway passenger volume, which proved to be more accurate than both the support vector machine and multiple regression methods. The findings can provide useful information for guiding highway planning and optimizing the allocation of transportation resources.

Keywords: intercity transportation; highway passenger volume; urban attributes; two-phase approach

Citation: Xiang, Y.; Chen, J.; Yu, W.; Wu, R.; Liu, B.; Wang, B.; Li, Z. A Two-Phase Approach for Predicting Highway Passenger Volume. *Appl. Sci.* **2021**, *11*, 6248. https://doi.org/10.3390/app11146248

Academic Editor: Paola Pellegrini

Received: 29 May 2021
Accepted: 1 July 2021
Published: 6 July 2021

1. Introduction

Recently, with the continuous process of urbanization, regional integration has become an inevitable trend of future social development in many developing countries [1,2]. In this situation, establishing a convenient and efficient intercity transportation system is a prerequisite for supporting regional integration, in which accurate prediction of passenger volume is one of the most fundamental elements required for the enhancement of intercity transportation systems [3–6].

The primary concern of passenger volume prediction is to extract relevant impact factors and build appropriate models. Firstly, multiple impact factors related to urban attributes, such as gross domestic product (GDP) and population, determine the absolute value and spatial distribution of passenger volume [7,8]. Consequently, extracting significant impact factors and further analyzing their relationship with passenger volume is recognized as a prerequisite for accurately predicting the passenger volume. Secondly, the prediction models attracted wide attention and the performance of different models was evaluated in past research. Some typical models, including multiple logit models, machine learning models, and deep learning models have been developed based on the historical passenger volume [9,10]. Nevertheless, the predicted accuracy of the existing models was largely affected by the dataset size of historical passenger volume [11]. Hence, the models with historical data cannot perform an accurate prediction if lacking sufficient data, which is quite common for intercity transportation.

There are two key steps in the prediction of intercity passenger volume: (1) extracting the significant impact factors, (2) developing a deep learning model to achieve the prediction. Thus, it is practical to develop a two-phase approach to predicting intercity passenger volume based on impact factors reflecting urban attributes and deep learning models. As the highway is always an important intercity mode of transport with a high mode share, this study took the highway as the research object. Phase I made a correlation analysis to remove the highly correlated impact factors and developed a random forest (RF) algorithm to extract the significant impact factors of highway passenger volume; Then, Phase II developed a deep feedforward neural network (DFNN) to predict highway passenger volume. To overcome the existing limitations on predicting intercity passenger volume, the primary contributions of this study are as follows:

(1) A total of 69 impact factors of urban attributes were collected from 280 administrative districts in China, which provides a macroscopic dataset for the prediction of highway passenger volume and overcomes the limitations of traditional travel surveys and questionnaires that only focus on a single city or single transportation corridor;
(2) Multiple urban attributes, including urban economy, population, industry, income and consumption, and resource and environment, were modeled together. Furthermore, A total of 30 significant impact factors of highway passenger volume were extracted by the RF algorithm, which improves the traditional process based on subjective experience and avoids the omission of significant factors;
(3) A deep learning method, DFNN, was developed to predict highway passenger volume, which proved to be more accurate than the SVM and multiple regression methods and can provide more reliable information for optimizing traffic structure and reducing waste of traffic resources.

The remainder of this study is organized as follows. Section 2 gives as overview of the related literature. In Section 3, the data source is introduced, and the impact factors of urban attributes are collected and presented. Section 4 presents the underlying principle of the RF and DFNN algorithm. Section 5 presents the process of extracting the significant impact factors. In Section 6, the DFNN is developed to predict highway passenger volume, which is further compared with two benchmark methods. Finally, Section 7 draws conclusions and gives an outlook on future research.

2. Literature Review

This section concludes the existing research on the above two phases: (1) extracting the significant impact factors of intercity passenger volume, (2) developing models to achieve an accurate prediction. Furthermore, the limitations of existing research are itemized at the end.

The first phase is to extract the significant impact factors. Multiple impact factors related to urban attributes, including urban economic level, urban industrial structure, population, etc., were widely studied to understand their relationship with intercity passenger volume. Firstly, the urban economic level proved to be one of the necessary impact factors of intercity passenger volume [12–14]. Traffic demand for business and tourism in intercity transportation increases with the development of the urban economy. The impact factors reflecting the urban economic level were found to be per-capita gross domestic product (GDP), per-capita income, industrial structure, etc., and it was verified that they had a strong correlation with intercity passenger volume [15,16]. Moreover, both population structure and population size affect the intercity passenger volume significantly. Limtanakool et al. [17] took population density and land use as variables and found that a higher population density and mixed degree of land use have a positive impact on passenger volume of public modes in medium- and long-distance trips. A similar conclusion was also reached by related research [18]. Although the impact factors related to economic level and population have been widely studied in the existing research, those related to the quality of residents' lives, resources, and the environment were rarely studied because they are hard to be quantified with one or several indicators and the corresponding dataset is difficult to obtain [19–21]. This problem indicates that the relative research on extracting significant impact factors of intercity passenger volume is incomplete and causes the inaccurate prediction of intercity passenger volume, especially for some tourism-driven cities and resource-driven cities.

The second phase is to develop a model to achieve an accurate prediction of intercity passenger volume. In the existing studies, multiple logit models, such as the multinomial logit model [22,23], Box–Cox logit model [24], and nested logit model [25], were developed to study the mode choice of intercity trips and deduce the intercity passenger volume of various modes by calculating the intercity travel rate of surveyed samples [26,27]. Moreover, intercity passenger volume was predicted by introducing the impact factors. Harker et al. [28] proposed a network equilibrium model with considerations of market price and economic mechanism to predict the intercity freight volume. Li et al. [29] predicted the passenger volume of intercity railway with multiple indicators of passenger demand, regional economy, and regional traffic infrastructure, with an average predicted error of 3.37%. Another practical approach to predicting intercity passenger volume is based on the historical passenger volume. Xie et al. [30] analyzed the spatiotemporal characteristics of intercity passenger volume and predicted intercity passenger volume on holiday, with a predicted error of 6.43%. Recently, deep learning and machine learning algorithms, represented by various neural networks, have become remarkable at predicting intercity passenger volume by using cellular signaling data and location-based data [4,22–32]. Numerous studies have shown that predicted accuracy can be significantly improved by deep learning algorithms [33].

It is noted that the difficulties in obtaining the dataset of intercity passenger volume have been widely emphasized in past studies, especially for some intercity passenger modes of transportation that have additional requirements for an urban population, geographical location, or urban scale, such as airways, railways, and waterways. This means that the prediction of intercity passenger volume can be only conducted in a few cities [34]. In contrast, the highway has better accessibility and connects to all kinds of cities, expanding the study scope of predicting intercity passenger volume [35]. As previously stated, intercity passenger volume is largely determined by impact factors. Thus, the process of extracting significant impact factors at first, and then analyzing the interaction between intercity passenger volume and impact factors with deep learning algorithms, is practical for predicting intercity passenger volume but has rarely been studied in the existing research.

From the above analysis, the relationship between intercity passenger volume and urban attributes has been widely studied, and some typical models have been developed to predict passenger volume. Nevertheless, some limitations still exist in previous research and need further improvement, which are listed as follows:

(1) Due to the restrictions of the research data, most existing research predicted intercity passenger volume from a single city or transportation corridor. As a result, the current achievements are difficult to apply to intercity transportation between all kinds of cities.
(2) Existing research only focuses on common urban attributes such as the population or the economy. However, more urban attributes related to the quality of residents' lives, resources, and environment were neglected for lacking the available data and quantitative indicators, causing the inaccurate prediction of intercity passenger volume, especially in some tourism-driven cities and resource-driven cities. Moreover, the selection process of significant attributes also received less attention.
(3) Microcosmic datasets collected from traffic surveys have been widely used for studying the choice of transportation mode in intercity trips but is not practical to predict intercity passenger volume. In contrast, the macroscopic datasets of urban attributes provided a novel approach to predict the intercity passenger volume, but have rarely been used in the existing literature.

3. Data Source

In this study, the dataset, including highway passenger volume and impact factors of urban attributes, was obtained from China's urban statistical yearbook. In China, the urban statistical yearbook is regularly published online to evaluate the social and economic levels. The statistical yearbook covers multiple aspects of urban attributes, including society, economy, etc. People can download the statistical yearbook for academic research, providing a novel macroscopic dataset with the prediction of highway passenger volume.

Considering the possible complex-relevance between impact factors of urban attributes, it is necessary to select appropriate impact factors for the convenience of data processing. The selection principles in this study are summarized as follows: (1) The selected impact factors can well reflect the urban attributes and have a significant impact on intercity passenger volume. (2) The selected impact factors can be quantifiable and comparable. (3) The selected impact factors can be provided by the urban statistical yearbook and easily accessible. It is noteworthy that some non-quantifiable factors can be comparable by converting into different levels. Yet in this study, most non-quantifiable factors have a high correlation with the existing quantifiable factors. Furthermore, subjective judgment and personal preference are often included in the non-quantifiable level division, which inevitably brings errors into the process. Accordingly, this study only focuses on the prediction of highway passenger volume with the quantifiable impact factors.

Based on the above principles, a total of 69 impact factors of urban attributes were selected from China's urban statistical yearbook. To facilitate data processing, the selected impact factors of urban attributes were divided into five categories, namely, urban economic level, urban population size and structure, per-capita income and consumption, resource and environment, and urban industrial structure. The selected impact factors of urban attributes and their information are summarized in Table A1 in Appendix A.

As the data in the statistical yearbook is aggregated from the whole district or city, the authors took the administrative district as the basic unit of data collection. As a result, 3444 samples, including the selected 69 impact factors and highway passenger volume, from 280 administrative districts, were collected. The recorded date is from 2003 to 2014, covering 12 years, because there is a unified statistical standard during this period and the statistical data changed smoothly without a sharp increase or decrease. In which. The highway passenger volume was set as the unique dependent variable, and impact factors were set as the alternative independent variables for predicting highway passenger volume.

4. Methodology

The flow diagram of the proposed two-phase approach and associated designed framework is shown in Figure 1. Firstly, the raw dataset, including highway passenger volume and impact factors, was collected. Then, the two-phase approach was proposed. Phase I extracted the significant impact factors with the RF algorithm and Phase II predicted highway passenger volume with the DFNN. Finally, the typical machine learning algorithm, support vector machine (SVM), was also developed for predicting highway passenger volume and compared with the DFNN, because it has a better ability to solve machine learning problems with a small sample size. Moreover, the traditional multiple regression, which is widely used for discerning the relationship between dependent variables and multiple independent variables, served as the benchmark for the prediction of highway passenger volume. All predicted models were evaluated by calculating errors, including mean absolute error (MAE) and root mean squared error (RMSE).

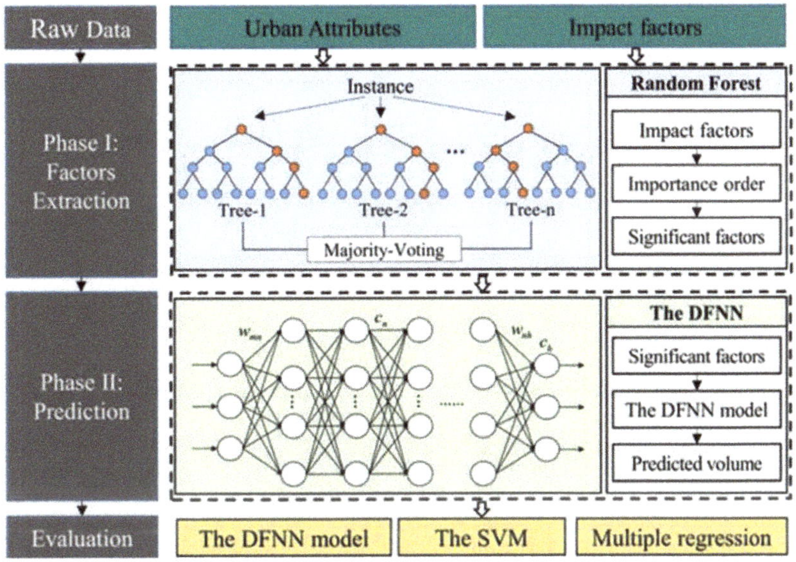

Figure 1. The flow diagram of the designed framework.

The fundamentals of the two primary methods used in this study are briefly discussed as follows, including the RF algorithm and the DFNN. Moreover, the evaluating indicators, MAE and RMSE, are introduced as well.

4.1. *Random Forest Algorithm*

In this study, the RF algorithm was used in Phase I to extract significant impact factors. The RF algorithm is a classifier established with multiple decision trees randomly, which has better robustness to noise and an excellent ability to maintain accuracy even if partial features are missing compared to other tree-based models [36,37]. Moreover, existing research has proved that the RF algorithm can efficiently analyze the complex interaction among features and pick out the significant features. As a result, it is widely used for removing the variables with a high correlation or low importance degree [38].

For any impact factor in Table 1, its importance degree can be calculated with the RF algorithm. After that, the selection of significant impact factors follows two processes: (1) Remove the impact factors that are highly correlated with others. (2) Determine the removed proportion and remove impact factors with a low importance degree.

The above processes of the RF algorithm, including calculating importance degree and selecting significant impact factors, were repeatedly conducted until the number of selected significant factors is less than the set value. Finally, the selected impact factors were set as the independent variables for predicting highway passenger volume.

4.2. *Deep Feedforward Neural Network*

Recently, the neural network is widely used in the prediction of traffic volume and proposes the development of deep learning [39–41]. The DFNN is a deep learning model comprised of an input layer, several hidden layers, and an output layer [42–44]. The quantity of hidden layers defines the depth of the architecture [45]. The topological structure of the DFNN is shown in Figure 2.

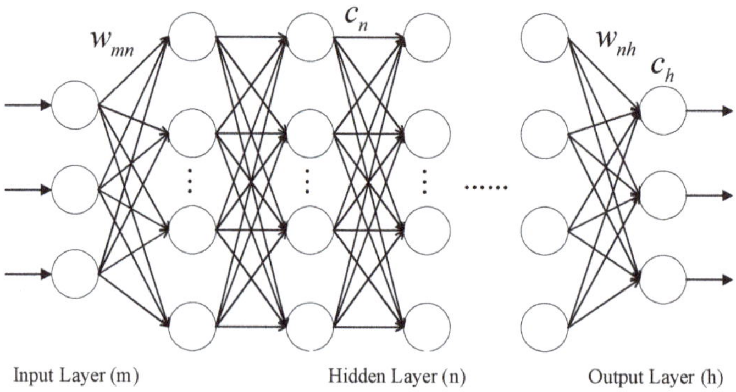

Figure 2. The topological structure of the DFNN algorithm.

The theory of the DFNN is available in past research [44–46]. In this section, we introduce the activation function and objective function used in the DFNN algorithm.

Firstly, the rectified linear unit (ReLU) function was selected as the activation function of hidden layers and the output layer, considering that the ReLU function has a higher computing efficiency because it only activates a fraction of the neurons in each epoch. The ReLU function has been proven to be effective at avoiding gradient vanishing and overfitting, and serves as the preferred choice when developing a neural network to solve multiple problems except for the binary classification [46,47]. The ReLU function is shown in Equation (1).

$$f(x) = \begin{cases} 0 & x < 0 \\ x & x \geq 0 \end{cases} \tag{1}$$

Then, the objective function was built by minimizing the loss function of mean square error, as in Equation (2).

$$\min \frac{1}{N} \sum_{i=1}^{N} (y_i - \hat{y}_i)^2 + \lambda \cdot R(\theta) \tag{2}$$

where y_i represents the actual highway passenger volume and $\hat{y}_i$ represents the predicted highway volume. N is the number of predicted samples. $R(\cdot)$ is a regularized constraint, represented by the $L2$ norm of the parameter θ, which is solved by the gradient descent method. λ is the coefficient of regularized constraint $R(\cdot)$.

4.3. Evaluating Indicators

To better evaluate the deviation of predicted results and assess the predicted method's performance, two indicators, MAE and RMSE, were calculated in this study. They are defined by Equations (3) and (4), respectively.

$$MAE = \frac{1}{N} \sum_{i=1}^{N} |y_i - \hat{y}_i| \tag{3}$$

$$RMSE = \sqrt{\frac{1}{N} \sum_{i=1}^{N} (y_i - \hat{y}_i)^2} \tag{4}$$

where y_i and $\hat{y}_i$ represent the actual highway passenger volume and the predicted highway passenger volume, respectively. N is the number of predicted samples. Both MAE and RMSE represent the degree of deviation between the actual and predicted highway passenger volume. The smaller the value of MAPE and RMSE, the more accurate the predicted result.

5. Phase I: Extraction of Significant Factors

In Phase I, the RF algorithm was used for removing the highly correlated impact factors and extracting the significant impact factors. Specifically, impact factors with a high importance degree were retained and those with a low importance degree were removed. The RF algorithm has the advantage of showing the extraction of significant factors step by step and the extracted significant impact factors are interpretable, compared with some auto-encoder methods like neural networks. Finally, a dataset of significant impact factors was built for predicting highway passenger volume.

Firstly, the correlation coefficients between impact factors were calculated by correlation analysis, and fifteen groups of highly correlated impact factors were found based on the calculated correlation coefficients, which are shown in Table 1.

Table 1. Groups of highly correlated impact factors.

Group	Highly Correlated Impact Factors	Group	Highly Correlated Impact Factors
1	NSS, NSP, NSSP, TP	8	DLB, HD
2	RT, SC, DRSC, TSP	9	GIO, DGIO
		10	IFA, DIFA, IRE, DIRE
3	DLA, DCAB	11	WS, WCS
4	FC, PFI, PFE, DPFI, DPFE	12	AEC, ECI, HEC
5	DB, DDB	13	NOB, PB, NT
6	HD, DHD	14	AGL, APGL, GCA
7	LB, DLB	15	NH, NBH, DNBH

Then, the importance degree of highly correlated impact factors in each group was calculated with the RF algorithm, as shown in Figure 3. The horizontal axis represents impact factors in each group, and the vertical axis represents the corresponding importance degree. Only the impact factor with the largest importance degree in each group was retained, and other impact factors were removed. Consequently, 28 impact factors, including NSS, NSSP, NSP, SC, DRSC, TSP, DCAB, DPFI, DPFE, PFI, FC, DDB, DHD, DLB, LB, DGIO, IFA, DIFA, DIRE, WS, ECI, AEC, PB, NT, GCA, AGL, DNBH, and NH, were removed and the other 41 impact factors were retained. Then, the importance degree of the remaining impact factors was calculated again and sorted in order, as shown in Figure 4.

In this study, the removed proportion was set at 10%. Therefore, impact factors with importance degree rankings in the bottom 10% were removed. According to Figure 4a, the removed impact factors included RP, CPR, VISR, and DNH, and the remaining 37 impact factors were retained for the subsequent data processing.

Similarly, the importance degree of impact factors was calculated repeatedly and sorted in order, and impact factors whose importance degree ranked in the bottom 10% were removed until the importance degree of the remaining impact factors reached 0.01. The above process was repeated twice. PCGRP, IRE, LA, and PFE, and DPD, PTPT, and CLPGR were removed during these two processes, respectively, as seen in Figure 4b,c. Finally, a total of 30 impact factors were retained, and are shown in Table 2. The category of resource and environment had more retained factors than any other, indicating that this category has a significant impact on highway passenger volume. Moreover, the importance degrees of HD, GDP, WCS, NOB, RT, HEC, TP, and TI rank in the top 25%, meaning that these eight factors significantly impact highway passenger volume.

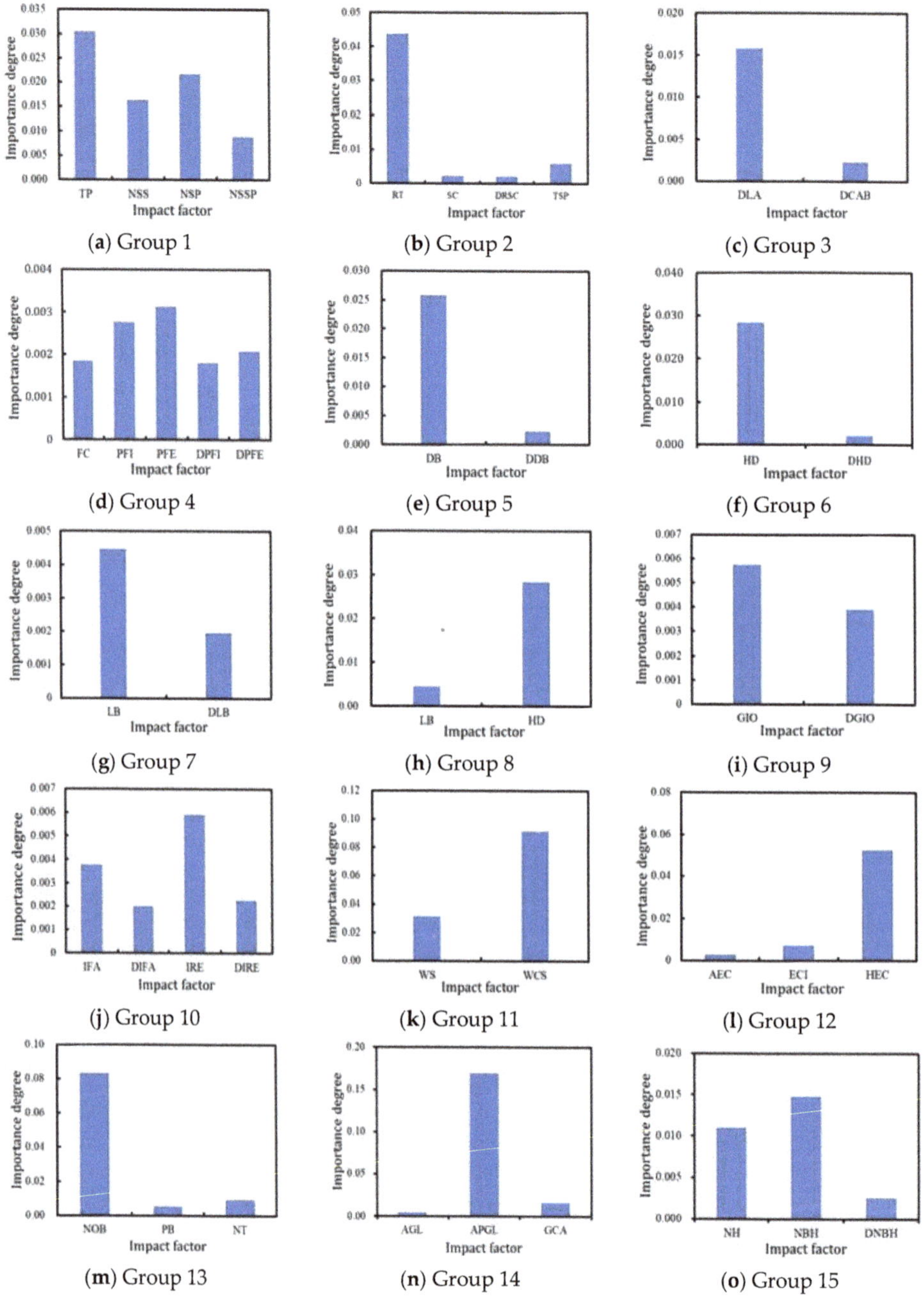

Figure 3. The importance degrees of significantly correlated variables in each group.

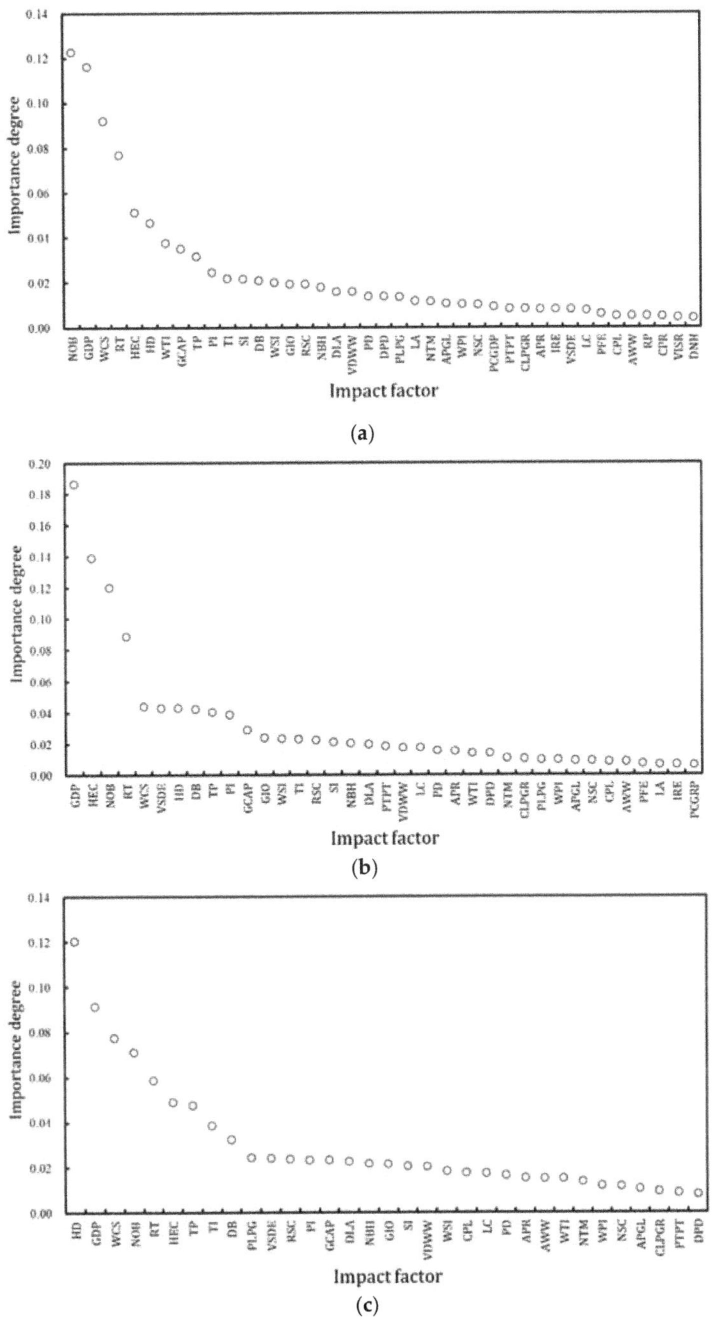

Figure 4. Importance degrees of impact factors. (**a**) The first iteration. (**b**) The second iteration. (**c**) The third iteration.

Table 2. The extraction result of significant impact factors.

Category	Included Impact Factors
Urban economic level	GDP, RSC, RT, GIO
Urban population size and structure	TP, NSC, WPI, WSI, WTI, PD, PLPG
Per-capita income and consumption	AWW, DB, HD, WCS, HEC
Urban industrial structure	PI, SI, TI
Resource and environment	DLA, LC, NOB, APR, APGL, GCAP, NBH, NTM, CPL, VDWW, VSDE

6. Phase II: Model Prediction and Evaluation

6.1. Model Prediction

With the significant impact factors selected by Phase I as input variables, Phase II developed the DFNN to predict highway passenger volume. The primary concern of developing DFNN is to determine the appropriate quantity of hidden layers and neurons in each hidden layer. In this study, the grid search method was adopted, whose initial range for the number of hidden layers was set from 1 to 10 and that for the number of neurons was set from 1 to 140. Taking MAE as an evaluating index, the result of the grid search method is shown in Figure 5.

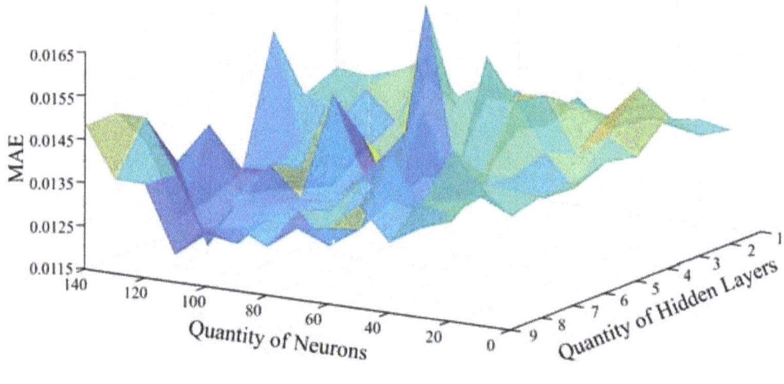

Figure 5. The result of grid search method for determining hidden layers and neurons.

The quantity of hidden layers and neurons with the minimum MAE is selected. Finally, the quantity of hidden layers is set to 9, and the quantity of neurons in each hidden layer is set to 120 in the DFNN of this study. Moreover, the quantity of neurons in the input layer and the output layer is set to 30 and 1, respectively, because there are 30 independent variables and 1 dependent variable.

Additionally, multiple epochs are needed for improving the predicted accuracy of the DFNN. Consequently, we continuously increased the epoch and calculated the loss of training set and verification set. When the loss of four consecutive epochs is less than 0.0001, it is considered that the training process has reached convergence and can be stopped. The loss of the training process is shown in Figure 6. Finally, the epoch of the DFNN in this study was set to 12.

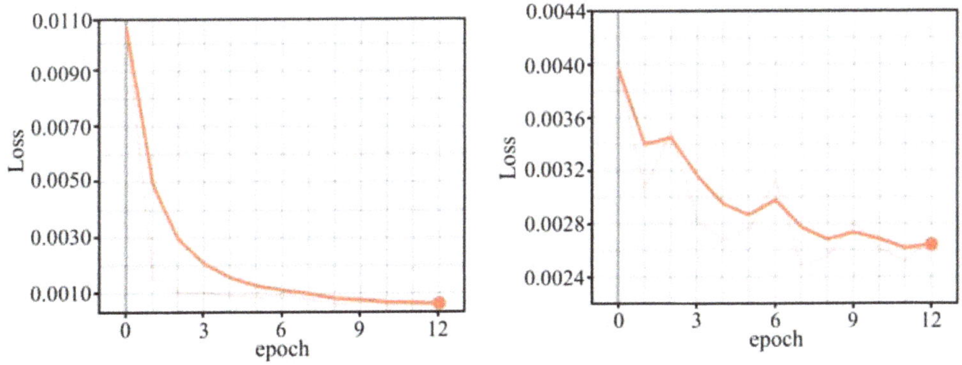

(**a**) the training set (**b**) the verification set

Figure 6. The result of loss for determining epoch.

Afterward, the significant impact factors were input in the developed DFNN, and the highway passenger volume was predicted. Then, evaluating indicators were calculated, showing that the MAE and RMSE of predicted highway volume from the DFNN are 2066.31 persons per day and 4176.37 persons per day, respectively.

6.2. Model Evaluation

To further evaluate the performance of the DFNN, the traditional SVM and multiple regression were used for comparison. For the SVM, the RBF kernel function whose penalty coefficient is set as 1000, and the Gamma coefficient is set as 0.001, was selected by adopting the grid search method based on the alternative sets of the kernel function, penalty coefficient, and gamma coefficient, as shown in Table 3.

Table 3. The alternative sets of parameters in the SVM.

Kernel Function	Set of Penalty Coefficients
RBF	[0.001, 0.01, 0.1, 1, 10, 100, 1000]
Linear Function	[0.001, 0.01, 0.1, 1, 10, 100, 1000]
Kernel Function	**Set of Gamma Coefficients**
RBF	[0.0001, 0.001, 0.1, 1, 10, 100, 1000]
Linear Function	-

The final predicted result is shown in Table 4, both MAE and RMSE of the DFNN are less than those of the SVM and multiple regression. The DNFF reduces the MAE and RMSE by 8.49% and 2.20%, respectively, compared with the multiple regression. The DFNN reduces MAE and RMSE by 2.90% and 1.15%, respectively, compared with the SVM. The result indicates that the DFNN is more accurate in predicting highway volume than the SVM and multiple regression.

Table 4. Model comparison between MAE and RMSE.

Model	MAE	RMSE
Multiple regression	2258.05	4270.29
SVM algorithm	2128.03	4225.06
DFNN	2066.31	4176.37

7. Conclusions

This study overcomes the limitations of existing research on predicting highway passenger volume. The main work and results of this study are as follows:

(1) A two-phase approach, in which Phase I extracts the significant impact factors and Phase II develops a deep learning model to achieve the prediction, was proposed to predict the highway passenger volume with the dataset of multiple urban attributes;
(2) Phase I extracted a dataset with 30 significant factors reflecting urban economic level, urban population size and structure, per-capita income and consumption, urban industrial structure, and resource and environments with the RF algorithm and proved that they have a significant impact on highway passenger volume.
(3) Phase II developed the deep learning method, DFNN, to predict the highway passenger volume with a mean absolute error of 2066.31 persons per day, improving the predicted accuracy by 8.49% compared to the multiple regression and 2.20% compared to the SVM algorithm.

This study contributes to proposing a novel approach for predicting highway passenger volume, but limitations still exist and are worth further study. Recently, deep learning algorithms have been proposed and are expected to be utilized for further improving the predicted accuracy of highway passenger volume as well as increasing the interpretability. As the statistical yearbook only publishes the annual statistics, it is difficult to make a detailed analysis of highway passenger volume in quarters or months. Moreover, it is possible to find data mutation caused by the change of statistical caliber in the statistical yearbook, which affects the predicted accuracy. Therefore, other new datasets can be considered to introduce into future research for more accurate analysis.

Author Contributions: Conceptualization, Y.X. and W.Y.; methodology, Y.X. and J.C.; software, R.W. and J.C.; data acquisition, W.Y. and B.L.; data analysis, Y.X.; writing—original draft preparation, Y.X. and W.Y.; writing—review and editing, J.C., B.W. and Z.L. All authors have read and agreed to the published version of the manuscript.

Funding: This research was supported by the National Natural Science Foundation of China (No. 71901059), the Natural Science Foundation of Jiangsu Province in China (BK20180402), the General Project of Humanities and Social Sciences Research of the Ministry of Education (19YJCZH152), and the Fundamental Research Funds for the Central Universities (2242021R10126, 2242021R10068).

Institutional Review Board Statement: Not applicable.

Informed Consent Statement: Not applicable.

Data Availability Statement: The data presented in this study are available on request from the first author.

Acknowledgments: The authors would like to thank the students from the school of computer science and engineering of Southeast University for their assistance with the data collection.

Conflicts of Interest: The authors declare no conflict of interest.

Appendix A

Table A1. The selected impact factors of urban attributes.

Category	Impact Factors	Symbol	Units
Urban Economic Level	Regional Gross Domestic Product	GDP	yuan
	Per-capita Regional Gross Domestic Product	PCGDP	yuan
	Total Sales of Retail Commodities	SC	yuan
	Total Retail Sales of Consumer Goods of the City	RSC	yuan
	Total Retail Sales of Consumer Goods of the Districts	DRSC	yuan

Table A1. *Cont.*

Category	Impact Factors	Symbol	Units
	Public Financial Income of the City	PFI	yuan
	Public Financial Expenditure of the City	PFE	yuan
	Public Financial Income of the Districts	DPFI	yuan
	Public Financial Expenditure of the Districts	DPFE	yuan
	Foreign Capital Used in the Year	FC	dollar
	Investment in Fixed Assets of the City	IFA	yuan
	Investment in Fixed Assets of the Districts	DIFA	yuan
	Investment in Real Estate of the City	IRE	yuan
	Investment in Real Estate of the Districts	DIRE	yuan
	Revenue of Postal Business	RP	yuan
	Revenue of Telecommunication Business	RT	yuan
	Gross Industrial Output Value of the City	GIO	yuan
	Gross Industrial Output Value of the Districts	DGIO	yuan
	Electricity Consumption of Industry	ECI	KW·h
Urban Population Size and Structure	Total Population of the City	TP	–
	Number of Students in the Colleges or Universities	NSC	–
	Number of Students in the Secondary School	NSS	–
	Number of Students in the Primary School	NSP	–
	Number of Students in the Primary–Secondary School	NSSP	–
	Number of Workers in the Primary Industry	WPI	–
	Number of Workers in the Secondary Industry	WSI	–
	Number of Workers in the Third Industry	WTI	–
	Number of Workers in the Transportation, Storage and Postal Services	TSP	–
	Population Density of the City	PD	/Km^2
	Population Density of the Districts	DPD	/Km^2
	Population Using Liquefied Petroleum Gas	PLPG	–
Per-capita income and Consumption	Average Wage of Workers	AWW	yuan
	Deposit Balance of Financial Institutions of the City	DB	yuan
	Deposit Balance of Financial Institutions of the Districts	DDB	yuan
	Deposit Balance of Household of the City	HD	yuan
	Deposit Balance of Household of the Districts	DHD	yuan
	Loan Balance of Financial Institutions of the City	LB	yuan
	Loan Balance of Financial Institutions of the Districts	DLB	yuan
	Water Consumption of Society	WCS	ton
	Electricity Consumption of Household	HEC	KWh
	Consumption of Liquefied Petroleum Gas for Resident	CLPGR	ton
	Total Water Supply	WS	ton
	All the Electricity Consumption of the Society	AEC	KWh
Urban Industrial Structure	The proportion of Primary Industry	PI	%
	The proportion of Secondary Industry	SI	%
	The proportion of Third Industry	TI	%
Resource and Environment	Administrative Land Area of the City	LA	Km^2
	Administrative Land Area of the Districts	DLA	Km^2
	Construction Area of Buildings of the Districts	DCAB	Km^2
	Land Area for Construction	LC	Km^2
	Actual Urban Road Area	CPR	m^2
	Number of Operating Public Buses	NOB	veh
	Total Passenger Volume of Public Buses in the Year	PB	–
	Number of Operating Taxis	NT	veh
	Number of Buses for Ten Thousand People	PTPT	veh
	Average Per-capita Road	APR	m^2

Table A1. *Cont.*

Category	Impact Factors	Symbol	Units
	All the Green Land Area	AGL	Km^2
	All the Green Land Area of Parks	APGL	Km^2
	Green Land Area of Construction Area	GCA	Km^2
	The Proportion of Green Land of Construction Area	GCAP	%
	Number of Hospitals of the City	NH	–
	Number of Hospitals of the Districts	DNH	–
	Number of Hospital Beds of the City	NBH	–
	Number of Hospital Beds of the Districts	DNBH	–
	Number of Theatres and Movie Theatres	NTM	–
	Total Collection of Books in Public Libraries	CPL	–
	Industrial Discharge of Waste Water	VDWW	ton
	Industrial Sulfur Dioxide Emission	VSDE	ton
	Removal Amount of Industrial Smoke and Dust	VISR	ton

References

1. Lin, L.; Hao, Z.; Post, C.J.; Mikhailova, E.A.; Yu, K.; Yang, L.; Liu, J. Monitoring Land Cover Change on a Rapidly Urbanizing Island Using Google Earth Engine. *Appl. Sci.* **2020**, *10*, 7336. [CrossRef]
2. Bong, A.; Premaratne, G. Regional Integration and Economic Growth in Southeast Asia. *Glob. Bus. Rev.* **2018**, *19*, 1403–1415. [CrossRef]
3. Liu, J.; Wu, N.; Qiao, Y.; Li, Z. A scientometric review of research on traffic forecasting in transportation. *IET Intell. Transp. Syst.* **2021**, *15*, 1–16. [CrossRef]
4. Chen, J.; Li, D.; Zhang, G.; Zhang, X. Localized Space-Time Autoregressive Parameters Estimation for Traffic Flow Prediction in Urban Road Networks. *Appl. Sci.* **2018**, *8*, 277. [CrossRef]
5. Xiang, Y.; Xu, C.; Yu, W.; Wang, S.; Hua, X.; Wang, W. Investigating Dominant Trip Distance for Intercity Passenger Transport Mode Using Large-Scale Location-Based Service Data. *Sustainability* **2019**, *11*, 5325. [CrossRef]
6. Li, X.; Tang, J.; Hu, X.; Wang, W. Assessing intercity multimodal choice behavior in a Touristy City: A factor analysis. *J. Transp. Geogr.* **2020**, *86*, 102776. [CrossRef]
7. Soltani, A.; Allan, A. Analyzing the Impacts of Microscale Urban Attributes on Travel: Evidence from Suburban Adelaide, Australia. *J. Urban Plan. Dev.* **2006**, *132*, 132–137. [CrossRef]
8. Miao, D.; Wang, W.; Xiang, Y.; Hua, X.; Yu, W. Analysis on the Influencing Factors of Traffic Mode Choice Behavior for Regional Travel in China. In *CICTP 2020*; American Society of Civil Engineers (ASCE): Virginia, VA, USA, 2020; pp. 3969–3980.
9. Nikravesh, A.Y.; Ajila, S.A.; Lung, C.-H.; Ding, W. Mobile Network Traffic Prediction Using MLP, MLPWD, and SVM. In Proceedings of the 2016 IEEE International Congress on Big Data (BigData Congress), Washington, DC, USA, 5–8 December 2016; Institute of Electrical and Electronics Engineers (IEEE): San Francisco, CA, USA, 2016; pp. 402–409.
10. Gu, Y.; Lu, W.; Xu, X.; Qin, L.; Shao, Z.; Zhang, H. An improved Bayesian combination model for short-term traffic prediction with deep learning. *IEEE Trans. Intell. Transp. Syst.* **2019**, *21*, 1332–1342. [CrossRef]
11. Lin, L.; Handley, J.; Gu, Y.; Zhu, L.; Wen, X.; Sadek, A.W. Quantifying uncertainty in short-term traffic prediction and its application to optimal staffing plan development. *Transp. Res. Part C Emerg. Technol.* **2018**, *92*, 323–348. [CrossRef]
12. Brueckner, J.K. Airline Traffic and Urban Economic Development. *Urban Stud.* **2003**, *40*, 1455–1469. [CrossRef]
13. Caceres, N.; Romero, L.M.; Morales, F.J.; Reyes, A.; Benitez, F.G. Estimating traffic volumes on intercity road locations using roadway attributes, socioeconomic features and other work-related activity characteristics. *Transportation* **2018**, *45*, 1449–1473. [CrossRef]
14. Chen, W.; Liu, W.; Ke, W.; Wang, N. Understanding spatial structures and organizational patterns of city networks in China: A highway passenger flow perspective. *J. Geogr. Sci.* **2018**, *28*, 477–494. [CrossRef]
15. Antipova, A.; Wang, F.; Wilmot, C. Urban land uses, socio-demographic attributes and commuting: A multilevel modeling approach. *Appl. Geogr.* **2011**, *31*, 1010–1018. [CrossRef]
16. Low, J.M.; Lee, B.K. A Data-Driven Analysis on the Impact of High-Speed Rails on Land Prices in Taiwan. *Appl. Sci.* **2020**, *10*, 3357. [CrossRef]
17. Limtanakool, N.; Dijst, M.; Schwanen, T. The influence of socioeconomic characteristics, land use and travel time considera-tions on mode choice for medium- and longer-distance trips. *J. Transp. Geogr.* **2006**, *14*, 327–341. [CrossRef]
18. De Witte, A.; Hollevoet, J.; Dobruszkes, F.; Hubert, M.; Macharis, C. Linking modal choice to motility: A comprehensive review. *Transp. Res. Part A Policy Pract.* **2013**, *49*, 329–341. [CrossRef]
19. Tian, Y.; Yao, X. Urban form, traffic volume, and air quality: A spatiotemporal stratified approach. *Environ. Plan. B Urban Anal. City Sci.* **2021**, 2399808321995822. [CrossRef]
20. Li, Z.; Wang, Y.; Zhao, S. Study of Intercity Travel Characteristics in Chinese Urban Agglomeration. *Int. Rev. Spat. Plan. Sustain. Dev.* **2015**, *3*, 75–85. [CrossRef]

21. Lee, D.; Derrible, S.; Pereira, F.C. Comparison of Four Types of Artificial Neural Network and a Multinomial Logit Model for Travel Mode Choice Modeling. *Transp. Res. Rec. J. Transp. Res. Board* **2018**, *2672*, 101–112. [CrossRef]
22. Bhatta, B.P.; Larsen, O.I. Errors in variables in multinomial choice modeling: A simulation study applied to a multinomial logit model of travel mode choice. *Transp. Policy* **2011**, *18*, 326–335. [CrossRef]
23. Huang, B.; Fioreze, T.; Thomas, T.; Van Berkum, E. Multinomial logit analysis of the effects of five different app-based incentives to encourage cycling to work. *IET Intell. Transp. Syst.* **2018**, *12*, 1421–1432. [CrossRef]
24. Jourquin, B. Mode choice in strategic freight transportation models: A constrained Box–Cox meta-heuristic for multivariate utility functions. *Transp. A Transp. Sci.* **2021**, 1–21. [CrossRef]
25. Elmorssy, M.; Onur, T.H. Modelling Departure Time, Destination and Travel Mode Choices by Using Generalized Nested Logit Model: Discretionary Trips. *Int. J. Eng.* **2020**, *33*, 186–197. [CrossRef]
26. Rahmat, O.K. Modeling of intercity transport mode choice behavior in Libya: A binary logit model for business trips by private car and intercity bus. *Aust. J. Basic Appl. Sci.* **2013**, *7*, 302–311.
27. Wang, R.; Zhang, T.; Liu, S.; Zhang, Z. Prediction of Passenger Traffic Volume Sharing Rate Based on Logit Model. In Proceedings of the 3rd International Conference on Information Technology and Intelligent Transportation Systems (ITITS 2018), Xi'an, China, 15–16 September 2018; p. 296.
28. Harker, P.T.; Friesz, T.L. Prediction of intercity freight flows, I: Theory. *Transp. Res. Part B Methodol.* **1986**, *20*, 139–153. [CrossRef]
29. Li, H.-L.; Lin, M.-K.; Wang, Q.-C. Passenger Flow Prediction Model of Intercity Railway Based on G-BP Network. In *Lecture Notes in Electrical Engineering*; Springer Science and Business Media LLC: Berlin/Heidelberg, Germany, 2020; pp. 859–870.
30. Xie, B.; Sun, Y.; Huang, X.; Yu, L.; Xu, G. Travel Characteristics Analysis and Passenger Flow Prediction of Intercity Shuttles in the Pearl River Delta on Holidays. *Sustainability* **2020**, *12*, 7249. [CrossRef]
31. Lv, Y.; Duan, Y.; Kang, W.; Li, Z.; Wang, F.-Y. Traffic Flow Prediction With Big Data: A Deep Learning Approach. *IEEE Trans. Intell. Transp. Syst.* **2014**, *16*, 1–9. [CrossRef]
32. Moreira-Matias, L.; Gama, J.; Ferreira, M.; Mendes-Moreira, J.; Damas, L. Predicting Taxi–Passenger Demand Using Streaming Data. *IEEE Trans. Intell. Transp. Syst.* **2013**, *14*, 1393–1402. [CrossRef]
33. Yin, X.; Wu, G.; Wei, J.; Shen, Y.; Qi, H.; Yin, B. *Deep Learning on Traffic Prediction: Methods, Analysis and Future Directions*; IEEE: New York City, NY, USA, 2021.
34. Tortum, A.; Yayla, N.; Gökdağ, M. The modeling of mode choices of intercity freight transportation with the artificial neural networks and adaptive neuro-fuzzy inference system. *Expert Syst. Appl.* **2009**, *36*, 6199–6217. [CrossRef]
35. Allard, R.F.; Moura, F. The Incorporation of Passenger Connectivity and Intermodal Considerations in Intercity Transport Planning. *Transp. Rev.* **2016**, *36*, 251–277. [CrossRef]
36. Le, H.T.; West, A.; Quinn, F.; Hankey, S. Advancing cycling among women: An exploratory study of North American cyclists. *J. Transp. Land Use* **2019**, *12*, 355–374. [CrossRef]
37. Pal, M. Random forest classifier for remote sensing classification. *Int. J. Remote. Sens.* **2005**, *26*, 217–222. [CrossRef]
38. Sun, J.; Sun, J. Real-time crash prediction on urban expressways: Identification of key variables and a hybrid support vector machine model. *IET Intell. Transp. Syst.* **2016**, *10*, 331–337. [CrossRef]
39. Xu, C.; Ji, J.; Liu, P. The station-free sharing bike demand forecasting with a deep learning approach and large-scale datasets. *Transp. Res. Part C Emerg. Technol.* **2018**, *95*, 47–60. [CrossRef]
40. Xie, Z.; Zhu, J.; Wang, F.; Li, W.; Wang, T. Long short-term memory based anomaly detection: A case study of China railway passen-ger ticketing system. *IET Intell. Transp. Syst.* **2020**. [CrossRef]
41. Liu, P.; Zhang, Y.; Kong, D.; Yin, B. Improved Spatio-Temporal Residual Networks for Bus Traffic Flow Prediction. *Appl. Sci.* **2019**, *9*, 615. [CrossRef]
42. Oliveira, T.P.; Barbar, J.S.; Soares, A.S. Computer network traffic prediction: A comparison between traditional and deep learning neural networks. *Int. J. Big Data Intell.* **2016**, *3*, 28. [CrossRef]
43. Glorot, X.; Bengio, Y. Understanding the difficulty of training deep feedforward neural networks. In Proceedings of the Thirteenth International Conference on Artificial Intelligence and Statistics, Chia Laguna Resort, Sardinia, Italy, 13–15 May 2010; pp. 249–256.
44. Gupta, T.K.; Raza, K. Optimizing Deep Feedforward Neural Network Architecture: A Tabu Search Based Approach. *Neural Process. Lett.* **2020**, *51*, 2855–2870. [CrossRef]
45. Loiseau, P.; Boultifat, C.N.E.; Chevrel, P.; Claveau, F.; Espié, S.; Mars, F. Rider model identification: Neural networks and quasi-LPV models. *IET Intell. Transp. Syst.* **2020**, *14*, 1259–1264. [CrossRef]
46. Agarap, A.F. Deep learning using rectified linear units (relu). *arXiv* **2018**, arXiv:1803.08375.
47. Eckle, K.; Schmidt-Hieber, J. A comparison of deep networks with ReLU activation function and linear spline-type methods. *Neural Netw.* **2019**, *110*, 232–242. [CrossRef] [PubMed]

applied sciences

MDPI

Article

An Intelligent Time-Series Model for Forecasting Bus Passengers Based on Smartcard Data

Ching-Hsue Cheng [1], Ming-Chi Tsai [2,*] and Yi-Chen Cheng [1]

1 Department of Information Management, National Yunlin University of Science & Technology, Touliou, Yunlin 640, Taiwan; chcheng@yuntech.edu.tw (C.-H.C.); hsue6771@gmail.com (Y.-C.C.)
2 Department of Business Administration, I-Shou University, Kaohsiung City 84001, Taiwan
* Correspondence: mct@isu.edu.tw; Tel.: +886-7-6577711 (ext. 5923)

Abstract: Public transportation systems are an effective way to reduce traffic congestion, air pollution, and energy consumption. Today, smartcard technology is used to shorten the time spent boarding/exiting buses and other types of public transportation; however, this does not alleviate all traffic congestion problems. Accurate forecasting of passenger flow can prevent serious bus congestion and improve the service quality of the transportation system. To the best of the current authors' knowledge, fewer studies have used smartcard data to forecast bus passenger flow than on other types of public transportation, and few studies have used time-series lag periods as forecast variables. Therefore, this study used smartcard data from the bus system to identify important variables that affect passenger flow. These data were combined with other influential variables to establish an integrated-weight time-series forecast model. For different time data, we applied four intelligent forecast methods and different lag periods to analyze the forecasting ability of different daily data series. To enhance the forecast ability, we used the forecast data from the top three of the 80 combined forecast models and adapted their weights to improve the forecast results. After experiments and comparisons, the results show that the proposed model can improve passenger flow forecasting based on three bus routes with three different series of time data in terms of root-mean-square error (RMSE) and mean absolute percentage error (MAPE). In addition, the lag period was found to significantly affect the forecast results, and our results show that the proposed model is more effective than other individual intelligent forecast models.

Keywords: passenger flow; integrated-weight time-series model; public transportation systems; long short-term memory network

Citation: Cheng, C.-H.; Tsai, M.-C.; Cheng, Y.-C. An Intelligent Time-Series Model for Forecasting Bus Passengers Based on Smartcard Data. *Appl. Sci.* **2022**, *12*, 4763. https://doi.org/10.3390/app12094763

Academic Editor: Feng Guo

Received: 12 March 2022
Accepted: 4 May 2022
Published: 9 May 2022

1. Introduction

Public transportation is considered to be an effective solution to traffic congestion and environmental pollution. The Federal Transit Administration (FTA) also believes that public transportation is an effective way to reduce traffic congestion, air pollution, energy consumption, and private vehicle use [1]. The use rate of buses accounted for 46% of all public transportation use in 2016 by people aged over 15 years according to the Taiwan Ministry of Transportation survey [2].

Taiwan's EasyCard Company promoted the smartcard system in 2002 based on the idea of "one card in hand, unimpeded travel". It was the first card to be issued for Taipei mass rapid transit and was then expanded to the Taiwan railway, Taiwan high-speed railway, and various other types of public transportation. Smartcards can collect information about vehicle routes, schedules, and real-time driving conditions through the automatic fare collection (AFC) system for vehicle monitoring, which can greatly improve public transportation efficiency and safety. The AFC system, when referring to the transportation system [3], is also called the smartcard system. The smartcard system is regarded as a dynamic and real-time data source for the public transportation system. It has attracted a significant amount of attention from researchers, and many studies have used smartcard data [3,4].

Although the smartcard system helps passengers greatly reduce their transaction time and shortens the time taken to board/exit buses, it also helps the bus industry collect large amounts of data to improve their service quality. Despite the utilization of the bus system, serious traffic congestion still occurs. Traffic flow describes the number of vehicles passing through a specific road section within a predetermined time interval [5]. It is different from traffic congestion, which is caused by excessive travel demand by people, resulting in abnormal traffic flow. There have been many studies on passenger flow predictions. In addition, smartcard systems can increase the convenience of users and help bus operators formulate practical route planning and reform timetables and related policies; however, this all depends on the use of smartcard system data to accurately forecast bus passenger flow. Accurately forecasting passenger flow can help cities implement transportation policies, strengthen local construction, reduce excessive energy consumption and carbon emissions, and improve urban ecosystems to achieve sustainable development.

The accessibility of the urban bus system is greater than for other modes of public transport, as this system utilizes the road network; however, passenger demands are affected by a number of factors such as crowding and different weather conditions. Tang et al. [6] confirmed prediction models would be better if the weather conditions were considered. The number of bus rides varies depending on the time of day, but there are still expected peak periods. For example, there will be many passengers during peak hours on weekdays and working days and at times when leisure activities are taking place during the holidays. We must consider passengers' needs, but external factors are also important.

In the past ten years, many successful traffic flow forecast methods have been proposed, especially deep learning methods. Li et al. [7] proposed a dynamic radial basis function neural network to predict short-term passenger flow through the Beijing subway. Ke et al. [8] proposed a fusion convolution long-term short-term memory network to forecast short-term passenger demand for ride services. Xu et al. [9] used a combined seasonal autoregressive integrated moving average with a support vector regression model to forecast the demand for the aviation industry. Deep learning methods have led to great progress in transportation research, but there have been few studies on forecasting buses' passenger flow compared with other types of public transportation. In our study, we collected data from the smartcard system for the bus industry and considered other external factors that affected the ride. In this article, we propose an integrated-weight time-series model to forecast passenger flow and detail our comparison with the listing methods. In summary, the goals of this study are as follows:

(1) To identify the important attributes that affect passenger flow from a total of 42 attributes in the smartcard system;
(2) To add other variables that affect passenger flow, such as climate, time, space, and lag period, to establish a prediction model;
(3) To apply multilayer perceptron (MLP), support vector regression (SVR), radial basis function (RBF) neural network, and long short-term memory network (LSTM) methods to forecast passenger flow with different types of time data series (weeks, weekdays, and holidays);
(4) To propose an integrated-weight time-series forecast model that uses forecast data from the top three of the 80 intelligent forecast models as the adaptive factors;
(5) To provide results that can be used as a reference by the government, industry, and related personnel.

The remaining sections are organized as follows: Section 2 is a literature review. In Section 3, we describe the research model and discuss the research design and methodology. Section 4 shows the results and findings. Finally, Section 5 presents the implications, limitations, and future work.

2. Literature Review

This section introduces related work on forecasting passenger flow using smartcard data, time series forecasting, and intelligent forecast methods.

2.1. Forecasting Passenger Flow by Smartcard Data

The smartcard is popular and convenient and can store a large amount of transaction data. Therefore, in the past decade, researchers have paid more and more attention to smartcard data. Ma et al. [10] used one-month data from smart bus cards to analyze the patterns of commuters in the area and the spatial distribution of movement. Eom et al. [11] applied the smartcard data from a five-day working week to learn about various social roles, such as the distribution of students and office workers in Seoul. Tao et al. [12] used smartcard data to visually compare the spatial-temporal trajectories of bus rapid transit trips and other bus trips.

To investigate factors relevant to forecasting passenger flow, Briand et al. [13] applied a Gaussian mixture model based on weather, time, and space to regroup passengers according to their public transportation habits in terms of time. Arana et al. [14] analyzed the impact of weather conditions on the number of public bus trips taken for shopping and personal business. Tang and Thakuriah [15] used the unemployment rate, gasoline prices, weather conditions, transportation services, and socioeconomic factors to implement a quasi-experimental design to examine changes in the monthly average number of bus passengers on weekdays.

The literature on passenger flow forecasting in bus services can be divided into long-term and short-term forecasts. Traditional long-term passenger flow forecasting usually involves the use of regression techniques to estimate future travel demand [16]. The regression model is used to establish the relationship between the number of passengers and influencing factors, which includes demographic, economic, and land use information [17,18]. For short-term passenger flow forecasting, models based on statistics and computational intelligence have been studied extensively [19,20].

There has been much research on passenger flow forecasting, but most has not included bus passenger flow forecasting. We present some of the research techniques and methods that have been used in previous studies. Sun et al. [21] proposed a hybrid model based on wavelet analysis and the support vector machine to evaluate the historical passenger flow through the Beijing subway. Xie et al. [22] applied seasonal decomposition and a least squares support vector to find the best hybrid method for the short-term prediction of airline passengers. Liu and Chen [23] proposed a passenger flow prediction model using deep learning where an autoencoder deeply and abstractly extracts the nonlinear features in many hidden layers and a back-propagation algorithm is applied to train the model.

2.2. Time Series Forecasting

Forecasting passenger flow is a time-series research field because bus data points (including smartcard and meteorology data) are indexed in time order and are therefore time-series data. A time series is a sequence of discrete time data, and the use of a time series model can help organizations understand the underlying causes of trends and systemic patterns over time. As such, the following section briefly introduces relevant knowledge about time series. Time series are data arranged in time order, and the first time series model to be developed was the linear autoregressive moving average model (ARIMA), which was proposed by Box and Jenkins in 1970. The ARIMA model [24] consists of three components, and each component helps model different types of patterns. The autoregressive (AR) component attempts to explain the patterns between any time period and previous lag periods; the moving average (MA) component can adapt the new forecasts to previous forecast errors (error feedback term); and the integrated (I) component indicates trends or other integrative processes in the data.

Traffic flow data are time series of periodic and irregular fluctuations, and many studies have used time-series methods to predict traffic flow. Hou et al. [25] combined

ARIMA with a wavelet neural network to overcome the limitations of using ARIMA for short-term forecasting of traffic flow. Xu et al. [9] used seasonal differences to eliminate nonstationary seasonal ARIMA and combined these data with support vector regression to predict the demand of the aviation industry. Wang et al. [26] used wavelet analysis to detect abnormal passenger flow to estimate the sudden traffic peak and then used a multiple regression model to estimate the peak time. Finally, they used seasonal ARIMA to estimate passenger flow.

In terms of an intelligent time-series model, AR neural network (ARNN) is a classic intelligent time-series model that uses a neural network to learn AR coefficients [27]. Further, we can collect many influential variables and lag periods of the dependent variable, and then we use MLP, SVR, RBF neural network, and LSTM network to train their parameters for building intelligent time series models. From the practical viewpoint, it is critical to properly handle weights in time series, and weighted time-series models include weighting on recent observations, important variables, and the better forecasting methods. For example, Hajirahimi and Khashei [28] proposed a weighted sequential hybrid model to calculate each model weight to construct a final hybrid output for time series forecasting; Tsai et al. [29] proposed a multifactor fuzzy time-series fitting model to weight the three significant variables; Jiang et al. [30] presented a weighted time-series forecasting model to weight recent observations.

2.3. Intelligent Forecast Methods

This section introduces four intelligent forecast methods, and they are applied to forecast the collected data in this study: multilayer perceptron, support vector regression, radial basis function neural network, and long short-term memory network.

(1) Support vector regression (SVR)

The support vector machine (SVM) is a supervised learning algorithm for data classification and regression analysis that was developed by Vapnik and colleagues [31]. The SVM is used for classification problems, known as support vector classification (SVC) problems, and regression problems, known as support vector regression (SVR) problems. The main purpose of SVR is to find the best separation hyperplane to separate clustered data to solve nonlinear problems. SVR is quite good when dealing with small samples and can handle high-dimensional attributes without relying on all available data for classification, but its disadvantage is that its efficiency is very low for a large number of forecast samples. In addition, the SVM needs to find a suitable kernel function, such as linear, polynomial, sigmoid, or radial basis functions, and it is sensitive to missing data. The SVM can be used to solve problems in many fields, such as text classification, image classification, and time-series prediction.

In forecasting traffic passenger flow, SVR is suitable for nonlinear and complex models. Castro-Neto et al. [32] considered that SVR cannot be fully trained with real-time data. To address this, they developed online SVR models. To reduce the computational complexity of the SVM, Xie et al. [22] proposed the combined seasonal decomposition and least squares support vector to get the best hybrid method for short-term forecasting of airline passengers.

(2) Multilayer perceptron neural network (MLP)

Perceptron is a type of artificial neural network invented by Rosenblatt [33]. It can be regarded as the simplest form of feedforward neural network, and it is a binary linear classifier. An MLP consists of at least three layers of nodes (input layer, hidden layer, and output layer). Except for the input nodes, each node is a neuron that uses a nonlinear activation function. The MLP uses supervised backpropagation learning, and its multilayer structure and nonlinear activation function distinguish it from linear perception. The MLP is a nonlinear learning model that can be processed in parallel and has good fault tolerance. It can be used as a real-time online learning model with associative memory, adaptivity, and self-learning ability. To make the output of the MLP as close to the actual target value as possible, a set of optimal weight values must be found in the training model, and one would need to determine the number of neurons used in each hidden layer.

Past studies have used MLP models to forecast multifactor problems. Ma et al. [34] used the MLP to forecast the network-wide co-movement patterns of all traffic flows, and they used ARIMA to postprocess the residual of the MLP. Tsai et al. [35] proposed a multiple temporal unit MLP to forecast short-term passenger demand.

(3) Radial basis function (RBF) network

The radial basis function (RBF) network proposed by Broomhead and Lowe has an input layer, a hidden layer, and an output layer [36]. In an RBF network, the nonlinear transformation is from the input layer to the hidden layer, and then the linear transformation is from the hidden layer to the output layer. This can achieve mapping from the input layer space to the output layer space, approximate any nonlinear function, and deal with difficult problems. The RBF network is conceptually similar to the k-nearest neighbor (k-NN) algorithm. In the self-organizing learning stage, basis function centers can be obtained; in the supervised learning stage, the weight between the hidden layer and the output layer is obtained, and each parameter can be learned quickly, thus overcoming the local minima problem.

To solve the central problem of the RBF function and the number of neurons in the hidden layer, Li et al. [7] proposed a new dynamic radial basis function (RBF) network to predict outbound passenger traffic. Li et al. [37] proposed a multiscale radial basis function (MSRBF) network to address the issue of when the number of input vectors is large, there may be a large number of candidates in the initial model. The MSRBF network can be applied to forecast irregular fluctuations in subway passenger flow.

(4) Long Short-Term Memory (LSTM)

LSTM is a time recurrent neural network (RNN) that was first proposed by Hochreiter and Schmidhuber [38]. Due to its unique design structure, an LSTM is suitable for processing and predicting important events with exceedingly long intervals and delays in time series. An LSTM network is a special type of regression neural network that uses a forget gate, an input gate, and an output gate to control the storage units. LSTM overcomes the problems of a RNN through gradient disappearance and gradient explosion. LSTM applications include time-series forecasting, language modeling, machine translation, image captions, and handwriting recognition.

In forecasting passenger flow, Ke et al. [8] proposed a fusion convolutional long short-term memory network (FCL-Net) for forecasting short-term passenger demand. Xu et al. [39] developed a long short-term memory network to forecast bike-sharing trip production and attractions at different time intervals.

3. Proposed Model

Passenger flow forecasting is a nonlinear, nonstationary time series problem, and a good forecast result mainly depends on having a large amount of high-quality data and a large number of methods. Nowadays, there are many passenger flow forecast models; however, some issues can be improved to enhance performance, such as:

1. Passenger flow forecasting is a periodic pattern, and many forecast models have been proposed to address this pattern. Previous studies have shown that datasets of no more than one month can be used to predict passenger flow at intervals of 5 or 15 min, and some studies use longer time datasets to predict passenger flow, such as daily or weekly intervals. To avoid the impact of extreme passenger flow, some studies do not consider the data collected on national holidays or weekends, and some studies treat special data as another forecast model for separate training. There is some room for improvement to obtain satisfactory results.
2. To solve the shortcomings of the model, more and more studies are taking advantage of different methods that complement each other and proposing hybrid models to forecast passenger flow. These hybrid models mainly combine traditional algorithms

and neural networks, but their nature still has limitations. Hence, hybrid models can be further strengthened to obtain the dynamics and forecasts of passenger flow.

3. Most research in this area has focused on passenger flow forecasting for railways, high-speed railways, and subways. Compared with other public transportation systems, there have been fewer studies forecasting bus passenger flow based on smartcard data, and few studies have considered time-series lag periods as forecast variables.
4. Previous research on the spatio–temporal nature of smartcard data has been widely conducted, and different attributes have been used in these studies; however, these studies have rarely discussed why these attributes have been selected and which attributes should be used in these methods. The selection and combination of input attributes is an important bridge between methods and forecast results.

Based on the discussion above, current forecast models of bus passenger flow still have limitations in terms of attribute selection, methods, and public transportation models. To process these limitations, we propose an integrated-weight time-series model for forecasting bus passengers using smartcards. First, the proposed model considers the attributes of time, space, and the lag period and uses four intelligent forecast models (multilayer perceptron, support vector regression, RBF network, and LSTM network) to forecast passenger flow for different time series (weeks, weekdays, and holidays). Second, the forecast data from the top three of the 80 combined forecast models (8 lag periods $\times$ 10 algorithms) were used as adaptive factors in the proposed model to enhance the forecast results.

The proposed time series forecasting model was revised from adaptive expectations theory [40,41]. Adaptive expectations theory is an economic theory that gives importance to past events when predicting future outcomes—a hypothetical process through which people can form expectations of what will happen in the future based on what has happened in the past. In a more complex and adaptive expectation model, different weights can be assigned to past values, and we can look at how different the fluctuations are from the predicted fluctuations.

To quickly understand the proposed model, Figure 1 shows a detailed explanation of the procedure to clarify the research process and computational steps involved. The proposed procedure, from top to bottom, includes data collection, data preprocessing, lag period testing, building a time-series forecast model, and evaluation and comparison.

Computational steps

The proposed procedure has five steps (see Figure 1). A detailed breakdown of the five steps is provided in the following sections.

Step 1: Data collection

In this step, two types of data were collected:

(1) One type was smartcard data from a bus industry in Kaohsiung City, Taiwan; the data were collected over a total of 669 days, including 2,865,763 records from January 2018 to October 2019, 17 bus lines (routes), and 137 bus stations. There were 42 attributes in the collected data (see Table 1), covering 15 administrative districts of Kaohsiung City in Taiwan. Regarding data location, the longitude range is 22.58706 to 22.792377, and the latitude range is 120.32016 to 120.29944.

(2) The other data type was meteorological data because the number of passengers boarding is often affected by many external factors, especially weather, which has always affected the travel behavior of passengers. Many researchers have presented the impact of weather conditions on passenger flow [13–15]. We collected weather data from the Kaohsiung Meteorological Bureau.

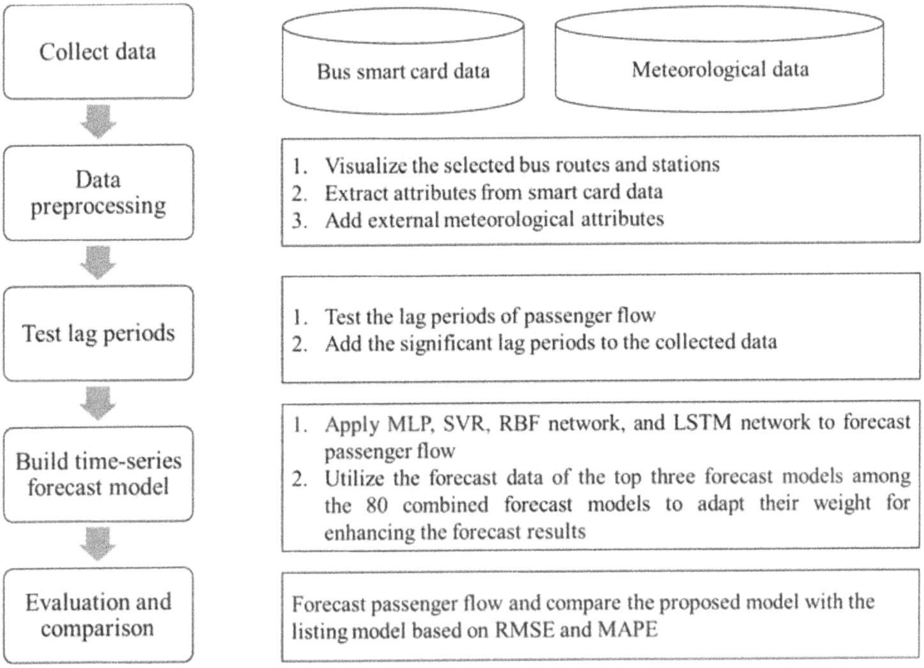

Figure 1. Proposed procedure.

Table 1. Original 42 data attributes.

Bus schedule number	Trading time for boarding	Card payment amount for exiting
Station number	Types of trading	Benefit points discount for exiting
Station name	Voice code for boarding	Free
Driver number	Boarding station code	Cash
Driver name	Boarding station name	Penalty fine
Bus number	Transferring discount amount	Making up the fare difference
Route number	Onboard card payment amount	Company subsidy amount
Route name	Boarding by benefit points discount	Transaction file name for boarding
Card number	Trading date for exiting	Transaction file name for exiting
Service type	Trading time for exiting	Outbound/return
Trade tickets	Types of trading for exiting	Counting status
Fare	Voice code for exiting	Counting date
Smartcard payment amount	Station code for exiting	Transferring group code
Trading date for boarding	Station name for exiting	Smartcard company

Step 2: Data preprocessing

We calculated the total number of bus passengers (22 months) for each route based on smartcard data. Figure 2 shows the total number of passengers for each route. Among the 17 routes, Route 1, Route 7, and Route 52 had the top three numbers of passengers: 508,997, 545,915, and 272,968, respectively.

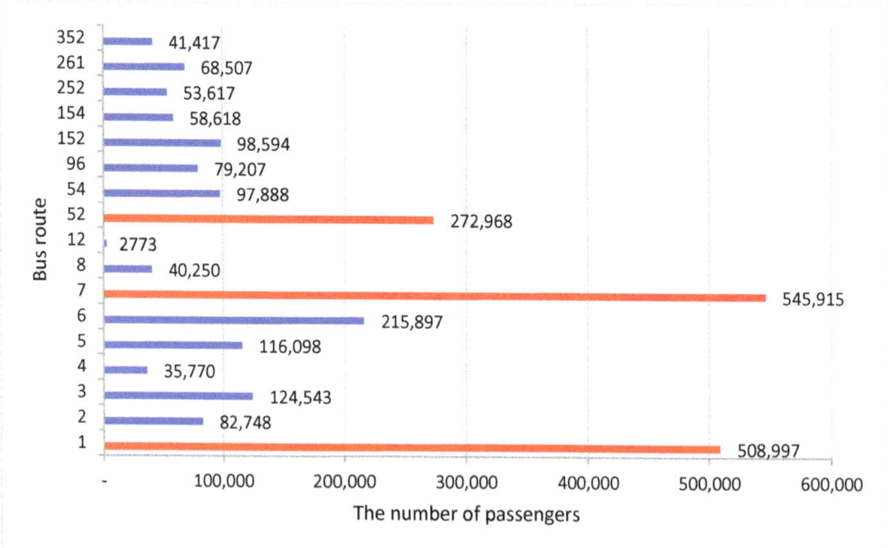

Figure 2. The number of passengers for each route.

Step 2.1: Extraction of attributes from smartcard data

This step involved the extraction of different time attributes from smartcard data as follows:

From the selected top three routes, the data from the three popular routes were divided into seven days (weeks), weekdays, and holidays. Weekdays were Monday to Friday (455 days), and holidays were Saturdays and Sundays (214 days). The total number of days was 669 (455 weekdays + 214 holidays), and a different day was used as an additional attribute. We extracted seven attributes from the smartcard data, including months, days, weeks, bus lines, bus stations, station passengers, and the number of passengers. Table 2 lists all the attributes used in this study in detail.

To check whether the number of rides was periodic, we plotted three figures to show the changes in the number of passengers for the top three routes based on the number of passengers per day on different days, as shown in Figures 3–5.

Figure 3 shows the number of passengers on three routes per day by week, which shows that the number of rides was periodic. Only Route 1 showed peak ride times, January 1 (New Year Day) and December 31 (New Year Eve), which are both national holidays. These long holidays are suitable times for going home or traveling, and as such, there is large-scale passenger flow. Route 7 has many bus stations, and the first half of the route is the same as Route 1; therefore, the number of passengers was found to have a similar periodicity to Route 1. For Route 52, because the bus station is different from the other two routes, the number of rides was found to be less similar to the other two routes, but it still showed periodicity.

Figure 4 shows the changes in the number of passengers per day for the three routes on weekdays. For a few days in the weekday period, peak passenger numbers occur, such as after 1 January (New Year's Day) and before the national holiday on 28 February, as people go home early before the holidays. In addition, 25 December (Christmas) is a religious holiday and a time when various industries launch marketing activities. As a result, people celebrating the festive season go out to purchase discounted goods.

Figure 5 shows the changes in the number of passengers on the three routes during the holidays. Compared with Figure 3, the weekly data show that 1 January (New Year's Day) and 31 December (New Year's Eve) are days with many passengers, and those two days are also holidays; therefore, the trend for the numbers of passengers on the three routes showed a similar periodicity for holiday data.

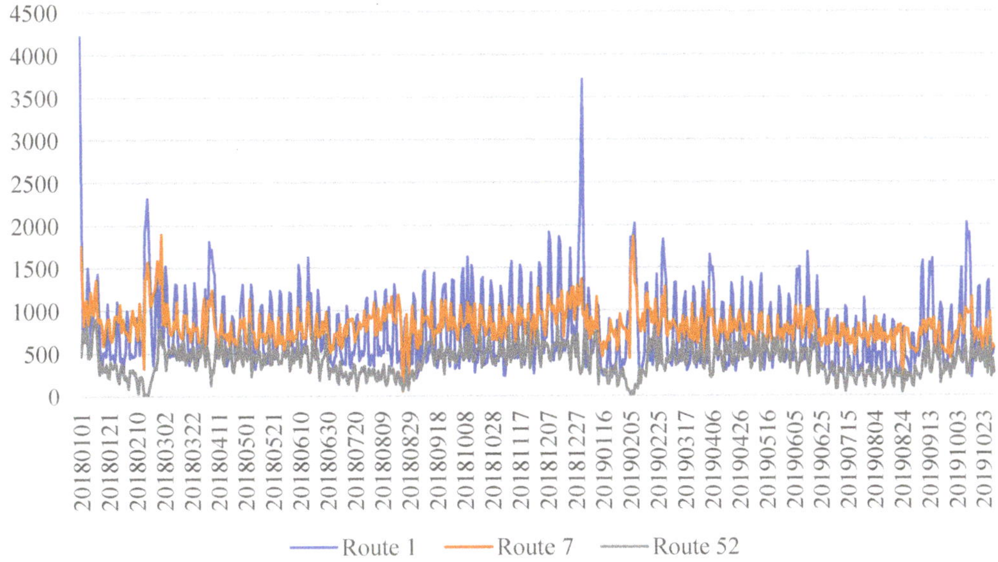

Figure 3. The number of passengers per day for the top three routes by week.

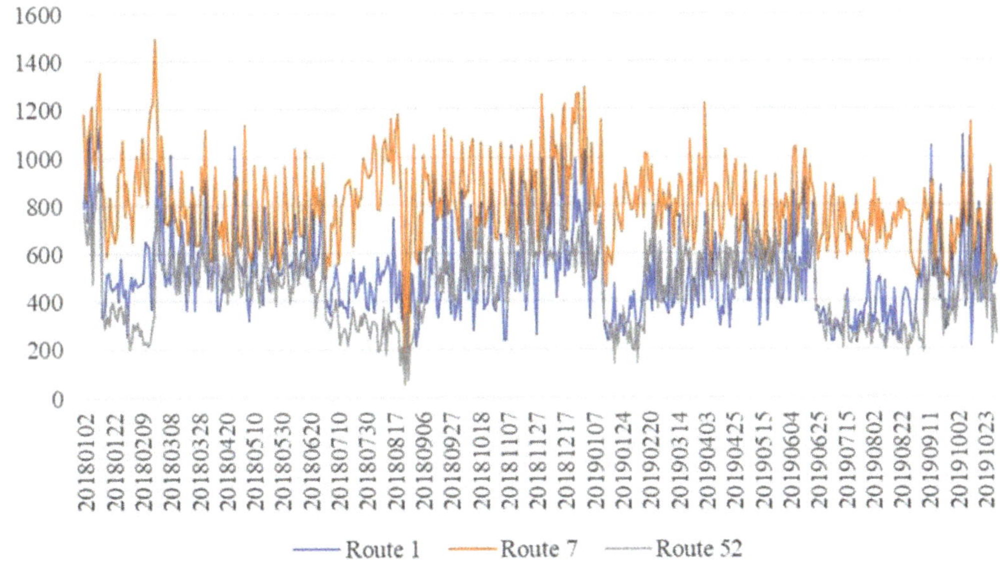

Figure 4. The number of passengers per day for the top three routes on weekdays.

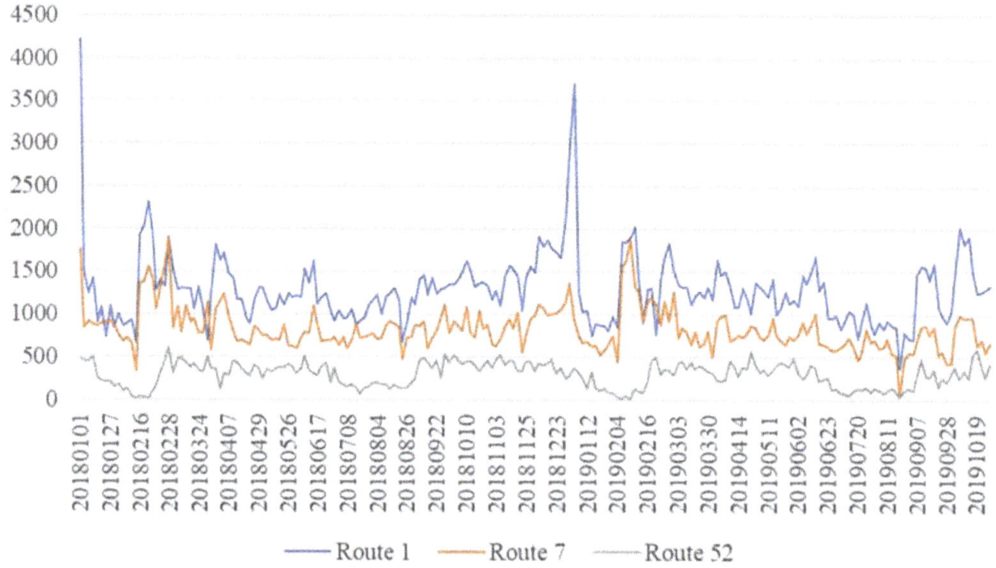

Figure 5. The number of passengers per day for the top three routes on holidays.

Step 2.2: Addition of external meteorological attributes

Previous studies [13–15] presented the impacts of weather factors on passenger flow. This study collected six types of meteorological data, including temperature, humidity, wind speed, rainfall, sunshine, and ultraviolet radiation. From the practical application of bus passenger flow forecasting, we collected data on 15 meteorological attributes, described in the meteorological type of Table 2.

Step 3: Test lag periods

In a time series, the autoregressive model is the output variable that is linearly dependent on its own previous value and random term. In order to test whether passenger flow has a time lag, this study used a partial autocorrelation function (PACF) to test how many passenger flows are significant at the 0.05 significance level, as the PACF is most useful for identifying the amount of lag in an autoregressive model. Further, lag-n is defined as follows: from the original data, the series values are moved forward n periods [42]. For example, lag 1 is moved forward 1 period; lag 10 is moved forward 10 periods. The test results are listed in Table 3 to show the lag periods for the number of passengers travelling on the top three routes based on the number of passengers per day for different daily data series (seven days, weekdays, and holidays). Table 3 shows that the largest lag period was seven for seven-day data; the largest lag period for the weekend data was six, and the holiday data had two lag periods.

Table 2. Attributes used in this study.

Type	Attribute	Description
Smartcard	Month	"1" denotes Jan, "2" means Feb, ... , "12" represents Dec
	Day	"1" denotes the first day for each month, "2" means the second day for each month, ... , "31" denotes the last day for each month.
	Week	"1" denotes Monday, "2" means Tuesday, ... , "7" represents Sunday.
	Bus line	The attribute was applied to visualize the heat map.
	Bus station	The attribute was applied to visualize the heat map.
	Station passengers	The attribute represents the passengers boarding at each bus station, which was applied to calculate the number of passengers at all stations for each day
	Passengers	Number of passengers on the bus line for each day
Meteorology	Temp	Average temperature, degrees Celsius, °C
	Tmax	Maximum temperature, degrees Celsius, °C
	Tmin	Minimum temperature, degrees Celsius, °C
	RH	Relative humidity, percent %
	RH_min	Minimum relative humidity, percent %
	WS	The wind speed was taken as the average value 10 min before the observation point, meters per second (m/s).
	WS_max	The maximum wind speed was taken as the maximum instantaneous wind speed within 1 h before the observation point, meters per second (m/s).
	Precp	The precipitation was taken as the total rainfall in a day, milliliters per day.
	Precp_hr	Total number of rainy hours in a day, number of hours
	Precp_10max	Maximum precipitation within ten minutes of the day, milliliters per ten minutes.
	Precp_hrmax	Maximum precipitation within an hour of the day, milliliters per hour.
	SunS	Sunshine hours, number of hours
	SunS_rate	The sunshine rate is a percentage ratio of the recorded bright sunshine duration and daylight duration in a day, percent %.
	GloblRad	Global radiation refers to a value used to measure the solar radiation energy for a given time and area, megajoules per square meter and per day, MJ/m^2.
	UVImax	The maximum ultraviolet index refers to the international measurement standard for the solar ultraviolet (UV) radiation intensity at a certain place on a certain day; the index value from 0 to 11+ is divided into five levels.
Lag period	Lag 1	A Lag 1 autocorrelation is the correlation between values that are one time period apart.
	Lag 2 Lag 3 Lag 4 Lag 5 Lag 6 Lag 7	a Lag k autocorrelation is a correlation between values that are *k* time periods apart, where *k* = 2, 3, 4, 5, 6, 7.

Table 3. Number of lag periods for each route in different time data.

	Week (Seven Days)	Weekday	Holiday
Route 1	Lag: 1, 2, 4, 6, 7	Lag:1, 2, 4, 5	Lag:1
Route 7	Lag: 1, 2, 3, 4, 7	Lag:1, 3, 4, 5, 6	Lag:1, 2
Route 52	Lag: 1, 2, 3, 4, 6, 7	Lag:1, 2, 3, 4, 5	Lag:1, 2

Step 4: Establishment of an integrated-weight time-series forecasting model

After extracting/adding the attributes of Steps 2 and 3, all attributes are listed in Table 2. This study proposed an integrated-weight time-series forecasting model to improve forecasting performance. That is, this step applied the four intelligent forecast methods to train these parameters of multi-variables and lag periods based on different data series and forecast passenger numbers. Further, we input the collected data with time order to train their parameters by SVR, MLP, RBF network, and LSTM; hence, these models are called intelligent time-series models. In addition, the four intelligent forecast methods (MLP, SVR, RBF network, and LSTM) were separated into 10 models according to the hidden layer, activation function, and kernel functions, as shown in Table 4.

Table 4. Abbreviation of ten intelligent forecast models.

Model Abbreviation	Full Name
MLP_1_ lin	MLP with 1 hidden layer and linear activation function
MLP_1_ log	MLP with 1 hidden layer and logistic activation function
MLP_2_ lin	MLP with 2 hidden layer and linear activation function
MLP_2_log	MLP with 2 hidden layer and logistic activation function
SVM_lin	SVR with linear kernel function
SVM_pol	SVR with polynomial kernel function
SVM_rbf	SVR with RBF kernel function
SVM_sig	SVR with sigmoid kernel function
RBF net	Radial basis function network
LSTM	Long short-term memory

The proposed model is based on the concept of adaptive expectation theory [40,41] to adapt the forecast data of the top three of the 80 combined forecast models (10 intelligent forecast models with 8 different lag periods, as shown in Table 5). The equation used in the proposed model is as follows.

$$F(t) = \alpha \times first(t) + \beta \times second(t) + \gamma \times third(t) + T(t-1) \quad (1)$$

where

$F(t)$ is the forecast of the number of passengers at time t,
$T(t-1)$ denotes the actual number of passengers at time $(t-1)$,
$first(t)$ represents the forecast of the best model for the number of passengers at time t,
$second(t)$ is the forecast of the second-best model for the number of passengers at time t,
$third(t)$ denotes the forecast of the third-best model for the number of passengers at time t,
α is the parameter of $first(t)$,
β represents the parameter of $second(t)$,
γ denotes the parameter of $third(t)$, and the range of α, β, and γ is from -1 to 1 (-1 means a negative correlation, and 1 represents a positive correlation).

Table 5. Results of 80 combined models for Route 1 on week data.

Algorithm		No lag	Lag 1	Lag 2	Lag 3	Lag 4	Lag 5	Lag 6	Lag 7
LSTM	RMSE	346.983	323.136	314.640	313.599	311.582	**270.137**	**246.327**	**235.838**
	MAPE	55.297	51.271	54.108	53.805	46.663	42.201	**30.713**	**29.091**
MLP_1_lin	RMSE	419.087	457.096	440.914	464.523	487.629	465.230	444.713	441.920
	MAPE	50.414	40.028	42.689	39.118	**38.048**	39.447	42.019	42.427
MLP_1_log	RMSE	413.770	422.451	437.535	442.538	454.443	413.121	425.137	473.599
	MAPE	53.621	48.826	43.573	42.355	40.548	54.963	46.756	38.588
MLP_2_lin	RMSE	417.011	436.597	606.395	496.521	520.526	451.678	455.457	481.922
	MAPE	51.544	43.844	58.925	38.235	39.951	41.111	40.371	38.435
MLP_2_log	RMSE	423.232	456.767	463.386	427.124	433.571	448.466	435.913	466.868
	MAPE	48.477	40.148	39.307	46.883	44.817	41.475	43.736	38.926
RBF net	RMSE	407.477	406.968	436.204	416.700	413.988	406.394	405.928	400.344
	MAPE	67.738	65.773	44.045	51.765	54.012	65.769	64.782	65.439
SVR_lin	RMSE	415.328	438.604	444.973	460.935	440.719	434.932	435.827	435.218
	MAPE	52.564	43.338	41.875	39.496	43.074	45.093	43.985	43.888
SVR_pol	RMSE	428.524	413.265	424.142	434.607	419.408	417.863	422.436	423.037
	MAPE	46.397	54.061	41.223	44.322	50.989	52.713	49.107	48.805
SVR_rbf	RMSE	437.040	414.256	422.609	415.376	419.644	421.224	420.811	423.770
	MAPE	43.773	53.330	48.703	52.607	50.366	50.886	50.222	48.335
SVR_sig	RMSE	416.775	445.232	448.546	446.987	500.864	441.484	406.373	410.386
	MAPE	51.681	41.895	41.223	41.478	38.510	43.311	62.067	54.371

Model abbreviations are shown in Table 4, and the bold print denotes the top three of the 80 combined models in terms of the RMSE and MAPE.

To optimize the parameters for α, β, and γ, the top three forecasted data points were used to adapt these parameters based on the minimum root mean square error (RMSE) and average absolute percentage error (MAPE) by using Equation (1). First, we set a feasible step iteration (step = 0.001) to produce the best parameters for α, β, and γ. Because the right-hand side of equation (1) has the actual number of passengers at the previous time $t-1$, the parameters of the first three forecast data points would fall between plus and minus one.

Step 5: Evaluation and comparison

In order to evaluate the proposed model and compare it with the listed models, in this step, the minimum RMSE and MAPE criteria were used for the evaluation and comparison. In terms of data, the data from three routes with three different types of days were compared experimentally based on an 8:2 ratio of training and testing data with time order. The data collected for each route were as follows: weekly data 669 = 535 training data + 134 testing data; weekday data 455 = 364 training data + 91 testing data; weekend data 214 = 171 training data + 23 testing data. The performance indicators used were the root mean square error (RMSE) and average absolute percentage error (MAPE). The equations are shown as Equations (2) and (3).

$$\text{RMSE} = \sqrt{\frac{\sum_{t=1}^{n} |F(t) - T(t)|^2}{n}} \quad (2)$$

$$\text{MAPE} = \frac{100\%}{n} \sum_{t=1}^{n} \left| \frac{F(t) - T(t)}{T(t)} \right| \quad (3)$$

where $T(t)$ is the actual number of passengers at time t, $F(t)$ is the forecasted number of passengers at time t, and n is the number of data points.

4. Experimental Comparison

From the procedure proposed in Section 3, the initial data analysis and acquisition of necessary attributes were selected. This section describes the experiments implemented to compare the proposed model with the listed models based on the minimum RMSE and MAPE criteria and then gives some findings.

4.1. Experimental Results

The experiments involved nine data series based on three routes with three different day types; each data series was partitioned into 80% training data and 20% testing data by time sequence. Based on Steps 4 and 5, which were presented in Section 3, this section presents the experimental results.

(1) Route 1

The results of 80 combined models for Route 1 for weekly data are shown in Table 5. In terms of RMSE, the top three best results were LSTM with lag 7, LSTM with lag 6, and LSTM with lag 5. For MAPE, the top three best results were LSTM with lag 7, LSTM with lag 6, and MLP_1_lin with lag 4 for the MAPE. We used Equation (1) to adapt the forecast data of the top three best forecast models to the minimal RMSE and MAPE. Similarly, the other two data series for Route 1 were experimented with based on Steps 4 and 5. Finally, the forecast data for the top three forecast models under the minimal RMSE were collected, and the results for Route 1 are shown in Table 6. The results show that the proposed model performed better than the listed models for Route 1, and the top three models were LSTM intelligent forecast methods.

(2) Route 7

As with the Route 1 experiment, we applied the proposed model to adapt the forecast data from the top three best forecast models for three data series of Route 7 using the minimal RMSE. The results are shown in Table 7. Table 7 shows that the proposed model was better than the listed models for Route 7 in terms of RMSE and MAPE.

Table 6. Results of proposed model for Route 1 based on minimal RMSE.

	Week			Weekday			Holiday	
LSTM lag 7	RMSE	235.838	LSTM lag 4	RMSE	175.697	LSTM lag 4	RMSE	255.503
	MAPE	**29.091**		MAPE	34.973		MAPE	21.913
LSTM lag 6	RMSE	246.327	LSTM lag 1	RMSE	177.303	LSTM lag 5	RMSE	255.907
	MAPE	30.713		MAPE	34.003		MAPE	21.469
LSTM lag 5	RMSE	270.137	LSTM lag 2	RMSE	177.751	LSTM lag 7	RMSE	259.302
	MAPE	42.201		MAPE	36.563		MAPE	21.085
Proposed	RMSE	**199.882**	Proposed method	RMSE	**115.963**	Proposed method	RMSE	**171.627**
	MAPE	54.534		MAPE	**33.068**		MAPE	**20.426**

Model abbreviations are shown in Table 4, and the bold digits denote the optimal results among the four models for RMSE and MAPE.

Table 7. Results of proposed model for Route 7 based on minimal RMSE.

	Week			Weekday			Holiday	
LSTM (no lag)	RMSE	142.964	RBF net lag 4	RMSE	131.499	LSTM (no lag)	RMSE	148.165
	MAPE	27.182		MAPE	15.400		MAPE	36.102
LSTM lag 5	RMSE	143.360	MLP_2_log lag 1	RMSE	131.794	LSTM lag 2	RMSE	157.255
	MAPE	27.397		MAPE	15.607		MAPE	37.800
LSTM lag 4	RMSE	144.51	MLP_2_log lag 5	RMSE	133.274	LSTM lag 1	RMSE	164.564
	MAPE	27.964		MAPE	15.657		MAPE	40.155
Proposed	RMSE	**93.682**	Proposed method	RMSE	**82.124**	Proposed method	RMSE	**110.650**
	MAPE	**26.728**		MAPE	**15.295**		MAPE	**35.097**

Model abbreviation is shown in Table 4, and the bold digits denote the optimal results among the four models for RMSE and MAPE.

(3) Route 52

Similarly, we only list the results for three data series of Route 52 in terms of the minimal RMSE, as shown in Table 8. The results show that the proposed model performs better than the listed models for Route 7 in terms of RMSE and MAPE.

Table 8. Results of proposed model for Route 52 based on minimal RMSE.

	Week			Weekday			Holiday	
LSTM	RMSE	117.833	LSTM	RMSE	107.196	LSTM	RMSE	121.741
lag 7	MAPE	**37.901**	lag 1	MAPE	29.631	lag 4	MAPE	52.935
LSTM	RMSE	119.873	LSTM	RMSE	109.904	LSTM	RMSE	121.767
lag 6	MAPE	38.999	lag 7	MAPE	28.371	lag 6	MAPE	53.773
LSTM	RMSE	131.381	LSTM	RMSE	110.222	LSTM	RMSE	121.943
lag 5	MAPE	48.909	lag 2	MAPE	29.768	lag 5	MAPE	52.555
Proposed	RMSE	**79.963**	Proposed	RMSE	**78.179**	Proposed	RMSE	**60.968**
	MAPE	38.126		MAPE	**26.025**		MAPE	**41.958**

Model abbreviation is shown in Table 4, and the bold digits denote the optimal results among the four models for RMSE and MAPE.

4.2. Findings and Discussion

The experimental results show that the proposed model is better than the listed models based on the minimum RMSE and MAPE criteria. However, there are some other findings to be discussed, as follows.

(1) Key attributes

In the forecast experiments, this study used the forecast data from the top three forecast models to adapt the optimal forecast. Simultaneously, we obtained the attributes of the top three forecast models. Additionally, we used the top three forecast models to rank the smartcard, lag periods, and meteorological attributes based on their impacts on passenger numbers. Then, we took the common attributes (at least two of the same attributes of the top three models) as the key attributes. The ordering of the key attributes of bus routes for different time series are shown in Table 9. Based on the ordering of key attributes, we identified the following features:

Routes 1 and 7: The three different time series for Route 1 have the same top three key attributes: Precp_10max, Precp, and Precp_hrmax. The top three key attributes in the weekly data for Route 7 are the same as for Route 1. This means that rainfall is an important factor for passenger flow on Route 1 and in the weekly data for Route 7. The top two key attributes in the weekday and holiday data for Route 7 are lag 1 and week, which shows that the passenger flow through Route 7 on weekdays and holidays is dependent on the week (Monday, Tuesday, ... , Sunday) and the passenger flow in the previous period.

Because the bus stations on Routes 1 and 7 are close to the university, high-speed rail, theme park, and tourist attractions, the collected data reveal that most passengers on these routes are students and tourists. Passengers want to go to schools and theme parks, and most of them will take the two routes. In addition, Route 7 has a highway transit station and Buddha memorial station. These two stations are important transportation and tourist attractions; hence, climate attributes (such as SunS, SunS_rate, and Tmin) influence the activity of tourists on Route 7 for the holiday data series.

Route 52: The bus stations on Route 52 are close to the university, high-speed rail, hospital, and detention center. These bus stations are used by students, patients, government employees, and their families, and the students, patients, and government employees are off duty on holiday; hence, the passenger flow is influenced by climate attributes (such as Precp_10max, Precp_hrmax, and Precp_hr) on the weekdays.

Three Routes: From Table 9, it can be seen that the three attributes (smartcard, meteorology, and lag period) affect the passenger flow, but, in different data series, the different attributes have different degrees of influence. Overall, the passenger flow through Route 1 and the weekly data for Route 7 have more attributes in common than other data.

Table 9. Key attributes of bus routes for different time data.

Route	Dataset	Ordering of Attribute Importance
Route 1	week	Precp_10max > Precp > Precp_hrmax > lag 7 > lag 1 > Precp_hr > week > lag 5 > lag 2 > RH > month > lag 6 > Tmax > UVImax
	weekday	Precp_10max > Precp > Precp_hrmax > Precp_hr > lag 1 > week > Temp > WS > Tmax
	holiday	Precp_hrmax > Precp > Precp_10max > lag 1 > Precp_hr > week > month > SunS_rate > Tmax
Route 7	week	Precp > Precp_hrmax > Precp_10max > Precp_hr > lag 1 > GloblRad > WS > RH_min > Temp > Tmin > RH > SunS_rate > UVImax
	weekday	lag 1 > week > month > Temp > Tmin > GloblRad > Precp_hr > RH_min > Precp > Precp_10max
	holiday	lag 1 > week > SunS > SunS_rate > Tmin > month > GloblRad > Precp_hr > RH_min > WS > lag 2
Route 52	week	lag 1 > week > lag 2 > SunS_rate > Tmax > GloblRad > WS_max > UVImax > Month > Precp > SunS
	weekday	Precp > lag 1 > Precp_10max > Precp_hrmax > Precp_hr > RH > Tmin > WS_max > Day > lag 5
	holiday	lag 1 > SunS_rate > lag 2 > SunS > Precp > Precp_10max > GloblRad > Tmax > Precp_hr

(2) Lag period

From the lag period test results shown in Table 3, it can be seen that the number of lag periods is consistent with the different time-series data. We can see that the lag period is seven for the weekly data (a week has seven days); the holiday data are organized by Saturdays and Sundays and have a lag period of two; the lag period for the weekday data is five. We checked whether the lag period is consistent with the number of lags for the top three models; if the same lag periods exist, then the data have seasonal variation (seasonality). Seasonality means that the time series data have periodic, repetitive, and predictable patterns [43]. We summarized the data from Tables 6–8, and the lag period results for the top three models for each route for different time series are shown in Table 10.

Based on Table 10, it can be seen that the weekly data for Routes 1 and 52 are consistent with the number of days in a week, as the number of lags is seven. This means that the weekly time-series data for Routes 1 and 52 have a weekly seasonal variation (seasonality), as shown in 6. Thus, it is necessary to add the passenger flow lag periods to forecast the number of passengers.

Table 10. Lag period of top three models for each route in different time data.

	Week (Seven Days)	**Weekday**	**Holiday**
Criteria	RMSE	RMSE	RMSE
	LSTM lag 7	LSTM lag 4	LSTM lag 4
Route 1	LSTM lag 6	LSTM lag 1	LSTM lag 5
	LSTM lag 5	LSTM lag 2	LSTM lag 7
	LSTM (no lag)	RBF net lag 4	LSTM (no lag)
Route 7	LSTM lag 5	MLP_2_log lag 1	**LSTM lag 2**
	LSTM lag 4	MLP_2_log lag 5	LSTM lag 1
	LSTM lag 7	LSTM lag 1	LSTM lag 4
Route 52	LSTM lag 6	LSTM lag 7	LSTM lag 6
	LSTM lag 5	LSTM lag 2	LSTM lag 5

Bold text denotes that the lag period of the top three models is consistent with the number of days in a week.

(3) Model performance

From Table 10, we can see that most of the top three models used LSTM, because LSTM is suitable for processing and predicting important events with very long intervals and delays in the time series [44]. Furthermore, the proposed model was found to be better than the listed models for each route for different time series, as shown in Tables 6–8. Therefore, the proposed model has some advantages: (1) The proposed model incorporates the smartcard, meteorology, and lag period attributes; (2) To enhance the forecast performance, this study proposed an integrated-weight time-series model to adapt the data from the top three of the 80 combined forecast models; (3) Bus data (including smartcard and meteorology data) are time series data, and the time series model helps organizations to understand the underlying causes of trends and systemic patterns over time. Therefore, we propose the use of an intelligent time series model to forecast passenger numbers.

(4) Sensitivity analysis

A sensitivity analysis can determine the associations between attributes. It facilitates more accurate forecasting and is the process of adjusting only one input and studying how it affects the overall model [45]. Based on sensitivity analysis involving the removal of attributes, we removed the meteorology attributes, the first key attribute, and the second key attribute to show the forecasting ability and robustness of the proposed model. This study was based on the ordering of key attributes for the three routes, as shown in Table 9, and we used the weekly datasets of the three routes to conduct the sensitivity analysis. The results show that the proposed model without meteorological data has a larger RMSE than the model with data from all attributes for the three routes, as shown in Table 11. Further, removing the first key attribute generates a larger error, and removing the second key attribute also increases the error. Based on the sensitivity analysis, we can confirm that the meteorological data are important when building the proposed model, and the first and second key attributes affect the proposed model.

Table 11. Results of sensitivity analysis based on RMSE.

	Route 1	Route 7	Route 52
Full attributes	199.882	93.682	79.963
Removal of meteorology attributes	231.714	144.949	113.959
Removal of first key attribute	228.594	136.121	121.006
Removal of second key attribute	228.066	140.402	116.717

5. Conclusions

Since 2000, Taiwan has been implementing the AFC system, also called the smartcard system, in the transportation system. The widespread use of smartcards helps passengers greatly reduce their transaction time and helps companies collect a large amount of information. Although there is a smartcard system, serious traffic jams still occur. Therefore, a good passenger flow forecast could be used to reduce traffic congestion, increase passenger convenience, and assist enterprises with formulating route planning, resetting timetables, and constructing other policies. In addition, a good passenger flow forecast can help cities reduce excessive energy consumption and carbon emissions and improve urban ecosystems to achieve sustainable development.

In order to achieve better prediction performance, we carried out the following steps:

(1) We proposed an integrated-weight time-series forecast model to forecast passenger flow. We used real smartcard data to verify that the proposed model has good predictive capabilities, rather than using simulated data to show the research results. The experiments showed that the proposed model performed better than the listed models for each route for different time series, as shown in Tables 6–8.

(2) In terms of the verification data, we focused on the top three routes with the most passengers out of the 17 routes—Route 1, which showed the largest fluctuations; Route 7, which has the largest number of passengers; and Route 52, which has the least number of passengers of the top three routes—as shown in Figure 2.

(3) In terms of attribute screening, this study used smartcard data and time attributes as well as 15 external weather attributes. In addition, as the number of passengers varies with time, this is a time-series forecasting problem; hence, seasonal trends had to be considered. Therefore, we added the number of lags to the forecast of passenger flow.

(4) As shown in Tables 6–8, we found that the data for each route could be partitioned by time (weeks, weekdays, and holidays) to improve the forecast result. Based on the key attributes shown in Table 9 and the lag periods of the top three models shown in

Table 10, the number of lags affected forecast results. Furthermore, most of the top three models are in the LSTM family, which presents a better forecast.

In terms of future work, we have two suggestions to further enhance this topic by making the results less conservative and improving the forecasting performance: (1) other attributes could be used in these forecast models, and (2) other methods (such as deep learning algorithms) could be applied to this topic.

Author Contributions: Conceptualization, C.-H.C. and M.-C.T.; methodology, C.-H.C.; software, Y.-C.C.; validation, C.-H.C. and M.-C.T.; formal analysis, M.-C.T.; investigation, Y.-C.C.; resources, C.-H.C.; data curation, M.-C.T.; writing—original draft preparation, Y.-C.C.; writing—review and editing, C.-H.C.; visualization, M.-C.T.; supervision, C.-H.C.; project administration, M.-C.T. All authors have read and agreed to the published version of the manuscript.

Funding: This research received no external funding.

Institutional Review Board Statement: Not applicable.

Informed Consent Statement: Not applicable.

Data Availability Statement: Not applicable.

Acknowledgments: The authors acknowledge partial support from Ministry of Science and Technology, Taiwan (Grant No. MOST-2221-E-224-037-).

Conflicts of Interest: The authors declare no conflict of interest.

References

1. FTA. Transit's Role in Environmental Sustainability. 2015. Available online: https://www.transit.dot.gov/regulations-and-guidance/environmental-programs/transit-environmental-sustainability/transit-role (accessed on 10 August 2020).
2. TMTC. Public Transport Market Share from Taiwan's Ministry of Transportation and Communications. 2016. Available online: https://www.motc.gov.tw/uploaddowndoc?file=public/201707031545021.pdf&filedisplay=201707031545021.pdf&flag=doc (accessed on 10 August 2020).
3. Trépanier, M.; Tranchant, N.; Chapleau, R. Individual trip destination estimation in a transit smart card automated fare collection system. *J. Intell. Transport. Syst.* **2007**, *11*, 1–14. [CrossRef]
4. Cheon, S.H.; Lee, C.; Shin, S. Data-driven stochastic transit assignment modeling using an automatic fare collection system. *Transp. Res. Part C Emerg. Technol.* **2019**, *98*, 239–254. [CrossRef]
5. Transportation Research Board. *HCM 2010. Highway Capacity Manual*; Transportation Research Board: Washington, DC, USA, 2010.
6. Tang, T.; Liu, R.; Choudhury, C. Incorporating weather conditions and travel history in estimating the alighting bus stops from smart card data. *Sustain. Cities Soc.* **2020**, *53*, 101927. [CrossRef]
7. Li, H.; Wang, Y.; Xu, X.; Qin, L.; Zhang, H. Short-term passenger flow prediction under passenger flow control using a dynamic radial basis function network. *Appl. Soft Comput.* **2019**, *83*, 105620. [CrossRef]
8. Ke, J.; Zheng, H.; Yang, H.; Chen, X. Short-term forecasting of passenger demand under on-demand ride services: A spatio-temporal deep learning approach. *Transp. Res. Part C Emerg. Technol.* **2017**, *85*, 591–608. [CrossRef]
9. Xu, S.; Chan, H.; Zhang, T. Forecasting the demand of the aviation industry using hybrid time series SARIMA-SVR approach. *Transp. Res. Part E Logist. Transp. Rev.* **2019**, *122*, 169–180. [CrossRef]
10. Ma, X.; Liu, C.; Wen, H.; Wang, Y.; Wu, Y. Understanding commuting patterns using transit smart card data. *J. Transp. Geogr.* **2017**, *58*, 135–145. [CrossRef]
11. Eom, J.K.; Choi, J.; Park, M.S.; Heo, T.Y. Exploring the catchment area of an urban railway station by using transit card data: Case study in Seoul. *Cities* **2019**, *95*, 102364. [CrossRef]
12. Tao, S.; Corcoran, J.; Mateo-Babiano, I.; Rohde, D. Exploring Bus Rapid Transit passenger travel behaviour using big data. *Appl. Geogr.* **2014**, *53*, 90–104. [CrossRef]
13. Briand, A.; Côme, E.; Trépanier, M.; Oukhellou, L. Analyzing year-to-year changes in public transport passenger behaviour using smart card data. *Transp. Res. Part C Emerg. Technol.* **2017**, *79*, 274–289. [CrossRef]
14. Arana, P.; Cabezudo, S.; Peñalba, M. Influence of weather conditions on transit ridership: A statistical study using data from Smartcards. *Transp. Res. Part A Policy Pract.* **2014**, *59*, 1–12. [CrossRef]
15. Tang, L.; Thakuriah, P. Ridership effects of real-time bus information system: A case study in the City of Chicago. *Transp. Res. Part C Emerg. Technol.* **2012**, *22*, 146–161. [CrossRef]
16. Horowitz, R. Legal notes. *J. Futures Mark.* **1984**, *4*, 229–230. [CrossRef]
17. Taylor, B.; Miller, D.; Iseki, H.; Fink, C. Nature and/or nurture? Analyzing the determinants of transit ridership across US urbanized areas. *Transp. Res. Part A Policy Pract.* **2009**, *43*, 60–77. [CrossRef]

18. Chan, S.; Miranda-Moreno, L. A station-level ridership model for the metro network in Montreal, Quebec. *Can. J. Civ. Eng.* **2013**, *40*, 254–262. [CrossRef]
19. Karlaftis, M.; Vlahogianni, E. Statistical methods versus neural networks in transportation research: Differences, similarities and some insights. *Transp. Res. Part C Emerg. Technol.* **2011**, *19*, 387–399. [CrossRef]
20. Ma, Z.; Xing, J.; Mesbah, M.; Ferreira, L. Predicting short-term bus passenger demand using a pattern hybrid approach. *Transp. Res. Part C Emerg. Technol.* **2014**, *39*, 148–163. [CrossRef]
21. Sun, Y.; Leng, B.; Guan, W. A novel wavelet-SVM short-time passenger flow prediction in Beijing subway system. *Neurocomputing* **2015**, *166*, 109–121. [CrossRef]
22. Xie, G.; Wang, S.; Lai, K. Short-term forecasting of air passenger by using hybrid seasonal decomposition and least squares support vector regression approaches. *J. Air Transp. Manag.* **2014**, *37*, 20–26. [CrossRef]
23. Liu, L.; Chen, R.C. A novel passenger flow prediction model using deep learning methods. *Transp. Res. Part C Emerg. Technol.* **2017**, *84*, 74–91. [CrossRef]
24. Box, G.E.P.; Jenkins, G.M. *Time Series Analysis: Forecasting and Control*, revised ed.; Holden Day: San Francisco, CA, USA, 1976.
25. Hou, Q.; Leng, J.; Ma, G.; Liu, W.; Cheng, Y. An adaptive hybrid model for short-term urban traffic flow prediction. *Phys. A Stat. Mech. Its Appl.* **2019**, *527*, 121065. [CrossRef]
26. Wang, H.; Li, L.; Pan, P.; Wang, Y.; Jin, Y. Early warning of burst passenger flow in public transportation system. *Transp. Res. Part C Emerg. Technol.* **2019**, *105*, 580–598. [CrossRef]
27. Triebe, O.; Laptev, N.P.; Rajagopal, R. AR-Net: A simple Auto-Regressive Neural Network for time-series. *arXiv* **2019**, arXiv:1911.12436.
28. Hajirahimi, Z.; Khashei, M. Weighted sequential hybrid approaches for time series forecasting. *Phys. A Stat. Mech. Its Appl.* **2019**, *531*, 121717. [CrossRef]
29. Tsai, M.-C.; Cheng, C.-H.; Tsai, M.-I. A Multifactor Fuzzy Time-Series Fitting Model for Forecasting the Stock Index. *Symmetry* **2019**, *11*, 1474. [CrossRef]
30. Jiang, Y.; Ye, Y.; Wang, Q. Study on Weighting Function of Weighted Time Series Forecasting Model in the Safety System. In Proceedings of the 2011 Asia-Pacific Power and Energy Engineering Conference, Wuhan, China, 25–28 March 2011; pp. 1–4. [CrossRef]
31. Cortes, C.; Vapnik, V. Support-vector networks. *Mach Learn* **1995**, *20*, 273–297. [CrossRef]
32. Castro-Neto, M.; Jeong, Y.; Jeong, M.; Han, L.D. Online-SVR for short-term traffic flow prediction under typical and atypical traffic conditions. *Expert Syst. Appl.* **2009**, *36*, 6164–6173. [CrossRef]
33. Rosenblatt, F. *The Perceptron—A Perceiving and Recognizing Automaton*; Report 85-460-1; Cornell Aeronautical Laboratory: Buffalo, NY, USA, 1957.
34. Ma, T.; Antoniou, C.; Toledo, T. Hybrid machine learning algorithm and statistical time series model for network-wide traffic forecast. *Transp. Res. Part C Emerg. Technol.* **2020**, *111*, 352–372. [CrossRef]
35. Tsai, T.; Lee, C.; Wei, C. Neural network based temporal feature models for short-term railway passenger demand forecasting. *Expert Syst. Appl.* **2009**, *36*, 3728–3736. [CrossRef]
36. Broomhead, D.S.; Lowe, D. Multivariable functional interpolation and adaptive networks. *Complex Syst.* **1988**, *2*, 321–355.
37. Li, Y.; Wang, X.; Sun, S.; Ma, X.; Lu, G. Forecasting short-term subway passenger flow under special events scenarios using multiscale radial basis function networks. *Transp. Res. Part C Emerg. Technol.* **2017**, *77*, 306–328. [CrossRef]
38. Hochreiter, S.; Schmidhuber, J. Long Short-Term Memory. *Neural Comput.* **1997**, *9*, 1735–1780. [CrossRef] [PubMed]
39. Xu, C.; Ji, J.; Liu, P. The station-free sharing bike demand forecasting with a deep learning approach and large-scale datasets. *Transp. Res. Part C Emerg. Technol.* **2018**, *95*, 47–60. [CrossRef]
40. Cagan, P. The monetary dynamics of hyper-inflation. In *Studies in the Quantity Theory of Money*; Friedman, M., Ed.; University of Chicago Press: Chicago, IL, USA, 1956.
41. Kmenta, J. *Elements of Econometrics*, 2nd ed.; Macmillan: New York, NY, USA, 1986.
42. Brockwell, P.J.; Davies, R.A. *Time Series: Theory and Methods*, 2nd ed.; Springer: New York, NY, USA, 1991.
43. Hyndman, R.J.; Athanasopoulos, G. *Forecasting: Principles and Practice*, 2nd ed.; OTexts: Melbourne, Australia, 2018; Available online: https://otexts.org/fpp2/ (accessed on 15 January 2021).
44. Zhang, Y.; Xiong, R.; He, H.; Pecht, M.G. Long short-term memory recurrent neural network for remaining useful life prediction of lithium-ion batteries. *IEEE Trans. Veh. Technol.* **2018**, *67*, 5695–5705. [CrossRef]
45. Yang, J.; Kim, J.; Jiménez, P.A.; Sengupta, M.; Dudhia, J.; Xie, Y.; Golnas, A.; Giering, R. An efficient method to identify uncertainties of WRF-Solar variables in forecasting solar irradiance using a tangent linear sensitivity analysis. *Sol. Energy* **2021**, *220*, 509–522. [CrossRef]

applied sciences

MDPI

Article

Development of a User-Centric Bridge Visual Defect Quality Control Assisted Mobile Application: A Case of Thailand's Department of Highways

Pravee Kruachottikul [1], Nagul Cooharojananone [2,*], Gridsada Phanomchoeng [3,4] and Kittikul Kovitanggoon [2]

1 University Technology Center (UTC), Chulalongkorn University, Bangkok 10330, Thailand; pravee.k@gmail.com
2 Department of Mathematics and Computer Science Faculty of Science, Chulalongkorn University, Bangkok 10330, Thailand; kovitang.cern@gmail.com
3 Department of Mechanical Engineering Faculty of Engineering, Chulalongkorn University, Bangkok 10330, Thailand; gridsada.p@chula.ac.th
4 Wireless Communication Ecosystem Research Unit, Chulalongkorn University, Bangkok 10330, Thailand
* Correspondence: nagul.c@chula.ac.th; Tel.: +662-218-5156

Abstract: Digital innovations have changed the way many industries operate, but the construction industry has been slow to adopt these technologies. However, challenges such as low productivity, project overruns, labor shortages, and inefficient performance management have motivated Thailand's Department of Highways to adopt digital innovations to build a competitive advantage. Because this industry requires a large work force, obstacles to collaboration can result in ineffective project management. We aimed to improve collaboration on bridge inspections that typically requires the involvement of many people, personal judgement, and extensive travel to survey bridges across the country. One major challenge is to standardize human judgement. To address these challenges, we developed a user-centric bridge visual defect quality control mobile application to improve collaboration and assist field technicians to conduct visual defect inspection. Our results can be used as a case study for other construction firms to embrace digital transformation technologies. This research also demonstrates the new-product development process using the new technology in known markets innovation development and technology acceptance model. We offer several recommendations for future research, including other infrastructure applications.

Keywords: construction; collaborative platform; bridge defect inspection; project management; structural health monitoring; mobile application; new-product development

Citation: Kruachottikul, P.; Cooharojananone, N.; Phanomchoeng, G.; Kovitanggoon, K. Development of a User-Centric Bridge Visual Defect Quality Control Assisted Mobile Application: A Case of Thailand's Department of Highways. *Appl. Sci.* **2021**, *11*, 9555. https://doi.org/10.3390/app11209555

Academic Editors: Nikos D. Lagaros and Vagelis Plevris

Received: 22 August 2021
Accepted: 24 September 2021
Published: 14 October 2021

1. Introduction

Recently, many industries in Thailand have begun to adopt digital transformation to stay competitive. For example, traditional car rental is transitioning into a car sharing business using Internet of Things (IoT) and mobile phone application technology [1]. The construction industry is also undergoing digital transformation to improve its ability to address challenges such as low productivity, project overruns, labor shortages, and inefficient performance management [2] Suboptimal collaboration among project managers, administrative staff, and on-site laborers is a common problem in large construction firms. Ineffective collaboration leads to inefficient project management, delays, and increased costs. Miscommunication, human error, poorly standardized work processes, and data loss, also contribute to inefficient project management. Bridge inspection requires many staff operating in different locations to perform visual inspection, and to update bridge information. This routine process involves inspection planning, problem identification, and considerable human judgment that is difficult to standardize [3]. The analysis of large digital data sets may improve project management, safety, energy management, decision-making design, resource management, and control costs [4].

Bridge substructure defect inspection is ideally suited for automated or semi-automated digital applications due to frequent human errors and a chronic shortage of skilled inspectors as shown by the study of V. Gattulli and L. Chiaramote [5]. A 2021 study by P. Kruachottikul et al. described the use of mobile phone applications to improve quality control, reduce miscommunication, increase transparency, and improve collaboration in Thailand's major construction organizations. Such applications enable users to remotely monitor their projects in real-time, support multimedia attachments, and reduce communication time [6]. In addition, our previous study described how artificial intelligence (AI) can assist users with bridge substructure defect inspection and severity prediction [7]. Recently, more study on the various bridge defect detection with AI have been performed in the laboratory, but not in real environment [8]. Here, we continue this work with the development of a mobile phone application with a built-in AI tool to improve quality control (QC) for bridge defect inspection quality control assessment. This tool can also be used to assist QC field technicians to build bridge information databases, detect defects, perform severity ranking, communicate with team members, and assign new tasks. In addition to technical development, we also describe the product development process from concept to implementation and propose factors that influence intention to use under the framework of the technology acceptance model.

The overarching objective is to create value to Thailand's construction industry through digital transformation technology. In doing so, in collaboration with Thailand's Department of Highways, our study can improve project management, reduce workload for its limited experts, and improve collaboration among team members. Importantly, by helping to improve road and bridge infrastructure, these technologies may also reduce accidents, a serious problem in Thailand which ranked second in deaths from traffic accidents in a 2018 World Health Organization report [9]. Therefore, we investigated the use of digital technology to conduct nondestructive bridge defect inspection to evaluate the condition of reinforced concrete [10].

Improving project management through better collaboration and quality control can increase productivity and efficiency, leading to cost savings and competitive advantages. Collaboration is about teamwork and communication. Teamwork is individuals working together to achieve a common goal and is developed through continuous communication among team members. Each team member seeks to clarify their roles and responsibilities and reduce ambiguity. Team members then concentrate on their individual tasks and coordinate in sub-groups where mutual expectations are developed. Therefore, the published literature on teamwork, project management, and visual inspection were reviewed and summarized.

First, teamwork is achieved through cooperation and coordination facilitated by effective communication. When team members understand each other and are able to synchronize their activities on a given schedule, the team can avoid delays, budget overruns, and reduce worksite accidents. Together, communication, coordination, cooperation, and synchronicity are key elements of team performance [11]. Communication is also an essential part of project management. The balance of frequent informal and formal communication affects the performance on the project, by influencing the degree of collaboration and the level of trust within the team [12]. Consequently, communication is a major determinant for project success or failure [13]. Instant messaging is a form of computer-mediated communication (CMC) that involves sending digital messages through computer networks [14].

Second, project management has become an important topic for an improvement in construction industry, where communication as information exchange in the broader organization is key [15]. As project teams face challenges from increasing competition and higher productivity targets, it is essential that team members are capable of utilizing CMC effectively. General forms of CMC are, for example, email, instant messaging, collaboration tools, and social networking. Instant messaging, or chat, is the application for sending and receiving short text messages, images, or files between people through a smart phone or computer. Instant messaging, through time, has evolved from teen sensation to an important modern communication tool in daily business. Although email can technically do the same function, instant messaging has some advantages that are team members can send and receive information instantaneously. Moreover, cloud services can be applied as a central repository for construction data as an end-to-end solution for timely decision-making, [16] because construction projects usually involve several project teams, handling different project locations, and dealing with various business report models.

Finally, automated optical inspection offers significant advantages over human eyes in terms of fatigue, completeness, speed, and accuracy. Technological advances in sensors, hardware, software, and data transfer speeds have improved automated visual inspection and enabled high standard quality control product inspection and certification. These digital technologies also generate data that can be used for statistical analysis. These tools can be adopted by the construction industry to automate visual inspection processes to improve competitiveness in road pavement, bridge construction, and other infrastructure projects. Pavement maintenance requires the updated information on road conditions. Because humans are subjective, have different levels of expertise and experience, and can suffer from eye fatigue, problems with inter-rater variability are common. Semi-automated pavement inspection systems have been used since 2000, but still rely heavily on human inspection [17,18]. Additionally, pavement inspections are not only concerned with acquiring images and designing image processing algorithms, [19] but also require integrating data related to the surface stages of structures from sensors to analyze the visual quality of the pavement. For example, a mobile phone to provide information of the location and way to communicate between involved parties. Large concrete structures such as bridges differ in age, performance, and condition. Various infrastructure monitoring programs and systems have been integrated to monitor the structural safety and serviceability to provide information for decision making and the allocation of resources [20].

This study explores the use of a mobile application for bridge substructure inspection, a task that requires visual inspection as an initial step that is usually performed manually according to the inspector's judgment and an evaluation manual. However, this process is subjective, time consuming, and involves safety risks; limitations that can be addressed by the application of digital technology [21].

2. Materials and Methods

2.1. New Product Development Process

The next-generation stage-gate development system involves the conception, development, and launch of new products [22]. It comes with the "Triple A" system concept that promotes an innovation-driven product development process to be adaptive, agile, and accelerated according to each development phase. This is also similar to that of the lean startup concept that encourages development of a minimal viable product (MVP) that is sufficient to generate user feedback to confirm the concept or prompt a pivot [23]. This study combines the stage-gate development system with a human-centered design innovation [24] for a new technology-known market (NT-KM) to discover unmet needs useful for restructuring product processes and to help inform the direction of the late design process. NT-KM starts with identifying the design scope and direction of an innovative product and/or its functionality in existing markets. Moreover, the research method tends to be semi-structured with many open-ended questions and sometimes through direct observation or physical models. The design direction is developed through a detailed understanding of the existing processes to clearly define the desired outcome or capability.

2.2. Technology Acceptance Model

The unified theory of acceptance and use of technology (UTAUT) [25] is applicable for identifying the key factors that influence behavioral intention to use. The objective of the UTAUT is to support management by providing a useful tool to assess the success of new technology introductions and understand the drivers of user acceptance, thereby enabling them to design interventions targeted at populations of users that may be less inclined to adopt and use new systems. The UTAUT framework was modified, as shown in Figure 1, by adding a bridge visual defect quality inspection (VIS) factor that was designed though group interviews with ten lead users from Thailand's Department of Highways. VIS influences the performance expectancy (PER), effort expectancy (EFF), and social influence (SOC) of UTAUT, and leads to a behavioral intention to use (BEH). Additionally, VIS consists of an image acquisition to capture the bridge and identify defects (VIS1), environment setup (VIS2), bridge defect information assistance (VIS3), and bridge history information (VIS4). After the user acceptance framework was developed, a correlation analysis based on a quantitative analysis involving 400 questionnaires was performed to confirm the validity of previously identified factors, and to gather information to build a user-driven mobile application. This analysis identified only VIS-PER-BEH and VIS-EFF-BEH to be significant (Table 1). VIS-SOC-BEH was not significant, suggesting that a user's behavioral intention relies on personal factors and not from social influence. The analysis also determined that VIS affects PER the most and that PER has the biggest impact on BEH. In other words, user intention occurs once users try the product themselves, and not when other people in an organization convince them to use it. This suggests that allowing the user to test the product in a free trial could be an effective strategy. Additionally, user intention to trial the product is more likely to occur if they recognize that the product helps them improve their work performance. This also suggests that product features that include image acquisition to capture the bridge and identify defects (VIS1), environment setup assistance (VIS2), bridge defect information assistance (VIS3), and bridge history information (VIS4), can assist user collaboration and inspection issues explicitly. After all factors are confirmed, they are taken into account to design the user experience and user interface (UXUI) of the mobile application.

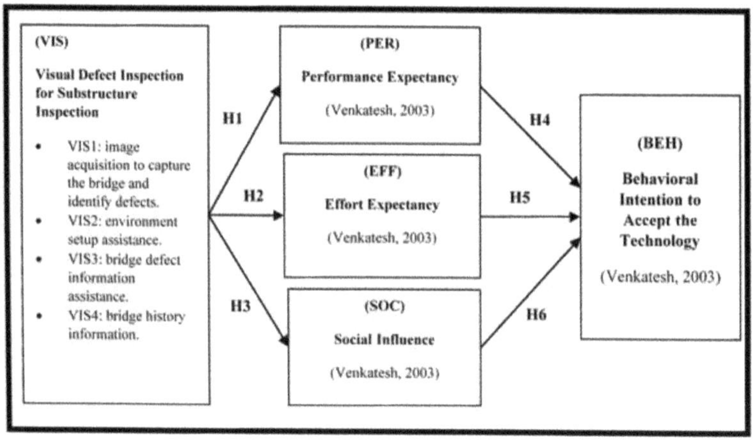

Figure 1. Technology acceptance model using modified UTAUT with VIS factor.

Table 1. Technology acceptance model framework results.

Factor	Hypothesis	df	Mean Square	SD	F	Sig.
VIS > PER	H1	39	0.473	0.464	2.529	0.000*
VIS > EFF	H2	39	0.337	0.462	1.688	0.008*
VIS > SOC	H3	39	0.416	0.525	1.598	0.016*
PER > BEH	H4	8	0.593	0.520	5.036	0.000*
EFF > BEH	H5	10	0.466	0.520	3.960	0.000*
SOC > BEH	H6	11	0.113	0.520	0.962	0.482

2.3. *Functional and Non-Functional Requirements*

This section explains user requirements and key findings for functional and non-functional requirements. The initial requirements were derived from our previous study [3] as described in items 1 to 6 in Table 2, and then combined with the key findings (items 7 and 8) from our present study of the technology acceptance model.

Table 2. User requirements and key findings for bridge inspection mobile application.

User Requirements and Key Findings
1. The application can communicate in the same way as instant messaging.
2. The application can report to project managers in real-time.
3. The application can upload multimedia files, such as images or videos.
4. The application contains a progress status for each subtask.
5. The application is able to identify the responsible person for each task.
6. The application can send the report via email.
7. The application can be trialed for free before committing to use.
8. The application trial convinces the user that it has the potential to improve the user's work performance. This suggests that product features such as image acquisition to capture the bridge (VIS1), identify defects (VIS2), environment setup assistance (VIS3), and bridge defect information and bridge history information (VIS4), will support user collaboration and inspection issues.

In addition, Thailand's Department of Highways' inspection procedure flow chart for bridge substructure was modified by adding a mobile application with assisted technology for the visual inspection process. (Figure 2) [25].

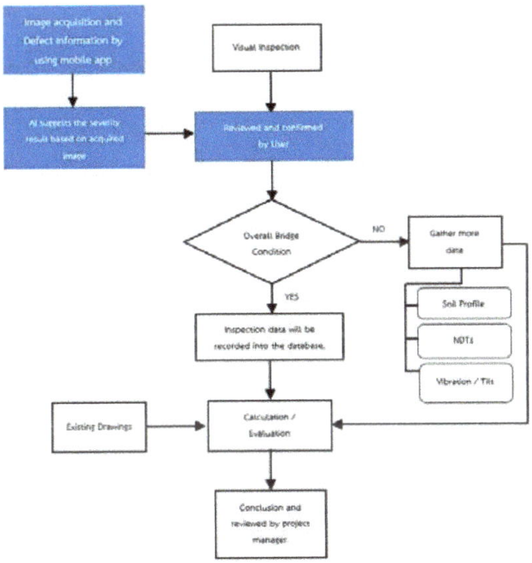

Figure 2. Modified inspection process of reinforced concrete bridge substructure by using mobile application with visual defect inspection assisted technology.

Next, the system is designed to automatically synchronize data with the cloud server system via a mobile phone network (Figure 3).

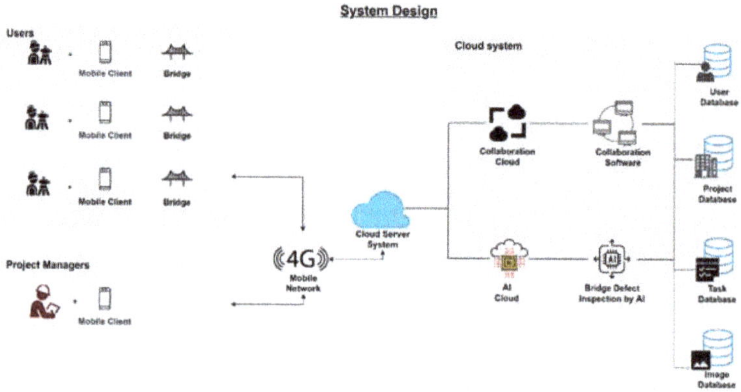

Figure 3. A system design for bridge inspection mobile application.

2.4. Use Case Diagram

After in-depth interviews, document analysis, and on-site monitoring, a use case diagram was created (Figure 4). Users were divided into project managers (expert engineers) and users (bridge inspectors). Each group can access different functions according to their roles. Project managers can manage all functions in the application, including adding projects and bridges that need to be inspected, and assign users specific tasks during the inspection processes. Project managers have authority to confirm or reject the AI-assisted inspection result. All data from each inspection is stored on a cloud server and can be used to make maintenance plans and to improve AI accuracy. Users are limited to task functions. They can add and edit the bridge and its damage information and upload an image for AI analysis. A comment system for each task facilitates communication between users and project managers.

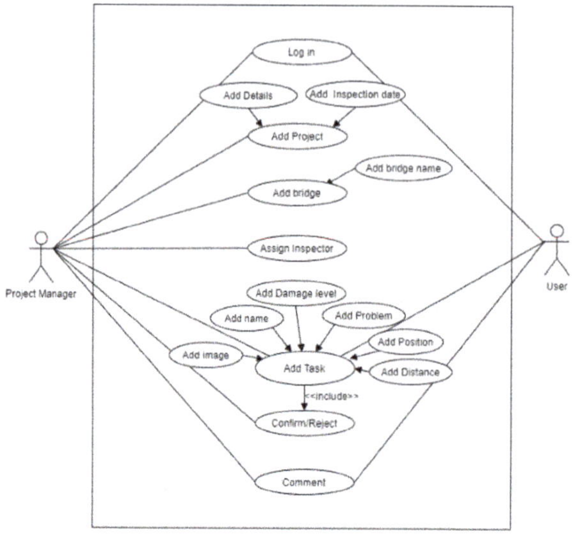

Figure 4. A user case diagram.

2.5. Hierarchy of Application Screen

The application starts with the default log-in page, where a user can either log in or create a new account. After logging in, three screens can be selected as All Project, Notification, and Setting screen. Then, under the All Project screen, the user can add or edit a project, view all bridge details, and view assigned task lists from the Task Screen. Finally, under Task Screen, the user can add task, assign, or edit accordingly. Hierarchy of application screen is shown in Figure 5.

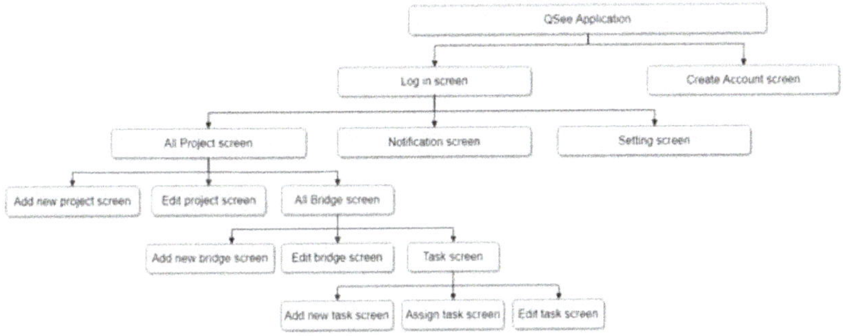

Figure 5. Hierarchy of application screen.

2.6. Backend Design

This mobile application backend design is based on microservice architecture because it is a collection of small services, making the service management straightforward and easier to add services in the future. (Figure 6) A variety of Amazon web services (AWS) are employed to develop the application (API Gateway, Cognito, Cloudwatch, AWS Lambda, DynamoDB, S3, and AWS IAM) [25]. The user interface (UI) of the application was developed using the react native framework [26]. The UI communicates with backend software using AWS Amplify via Amazon API Gateway, involving API method data management such as GET, POST, PUT, and DELETE that read, write, update, and delete data via APIs. There are four functions (user, project, task, and notification) with their corresponding APIs, created on AWS Lambda for managing project collaboration information. The Expo [27] backend server was chosen to handle the notification part on account of multi-platform message transmission via Firebase Cloud Messaging for Android and Apple push notification service for iOS [28]. The developed image processing software, CNN and ANN, are also executed on AWS Lambdas. For data storage, Amazon DynamoDB, which is the NoSQL database, creates five tables to store the text data and Amazon S3 Public Bucket stores images. Any procedures related to the authentication system, such as log-in, register, credential request, and status checking, are required to connect with Amazon Cognito. Lambda Trigger function is used as post-confirmation for creating the user DynamoDB database.

2.7. Microservice Development

The development of a collection of small services in the microservice architecture can best be explained with a sequence diagram. The sequence diagram of the AI assisted bridge visual defect inspection microservice collection was derived from our prior study [7], which starts from the user uploading an image and defect information through the system. This original image is stored in Amazon S3 bucket whereas the defect information is recorded on the image table. Image processing function processes the original image by enhancing and filtering to increase contrast and remove noise. The processed image is stored in Amazon S3 bucket. The defect inspection function analyzes a processed image to identify the defect section and send these to defect classification function to categorize the types of defects. The severity prediction function evaluates the seriousness of the defect to help the project manager make decisions. This study proposes an alternative sequence diagram of the user collaboration as shown in Figure 7b. It indicates that when users upload or modify the task information, the application credentials, paths, and parameters are sent to the API gateway. The system will call a task function with the task path to store the task info in the task table and return these parameters back to the application UI.

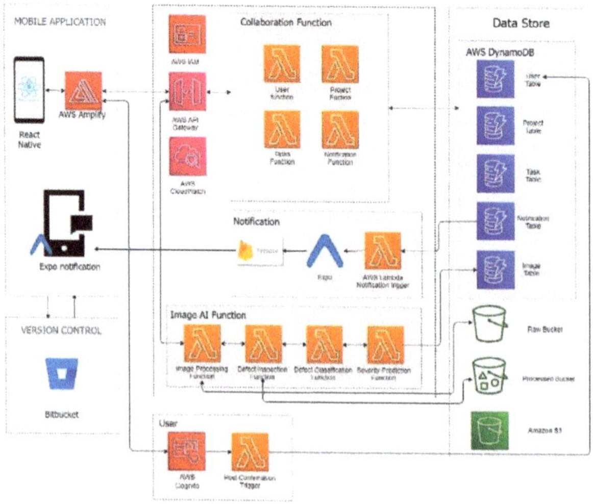

Figure 6. Backend design.

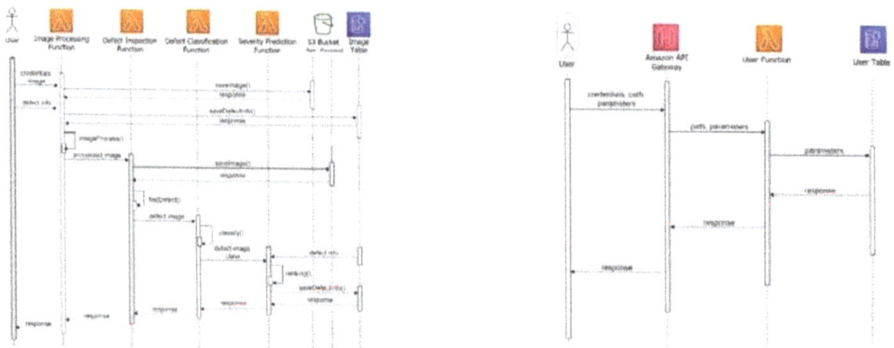

(a) (b)

Figure 7. (**a**) Sequence diagram of AI assisted bridge visual defect inspection. (**b**) Sequence diagram of user collaboration.

3. Results

3.1. Mobile Application

First, a user communication requirement was analyzed, the web board feature was chosen, and then implemented at the task detail page. This feature allows the user and project manager to communicate based on each task topic, so it is easy to track the progress and see the historical conversation. After the new project is created, the user can add a new task in a web board function. When creating a new task, the user must add an image and task detail, and can insert a new comment or reply to the existing topic. When a real-time report is required, a notification feature can be added to inform the project manager via multiple channels such as in-app notification or email. At the task summary and detail page, the user can open, read, or edit issues including progress status, detail, and web board. A person in-charge of a specific task can be assigned by the user. Furthermore, a task status shown in the task detail page is used to indicate whether the task is on-going or finished. Visual inspection assistance was implemented to support users to make more accurate decisions, reduce inspection time, and provide an improved inspection experience. Thus, the image acquisition tool supports users to acquire the bridge images easily in different environments and the vision AI assisted feature automatically identifies defects and predicts the severity result (severe or non-severe). The result is sent to users to verify and confirm the results, saving time and enabling prescreening of results. Users trialed the product and suggested the addition of self-learning tutorial tools, a more intuitive UXUI, and using the mobile application as a core platform. Next, the mobile application wireframe (Figure 8) and the application page schematic as a visual blueprint were developed using the earlier UXUI to improve visual understanding of the application during the research development phase. Its advantages are to optimize the layout and content placement easily, and solve navigation and functionality problems in an adjustable format.

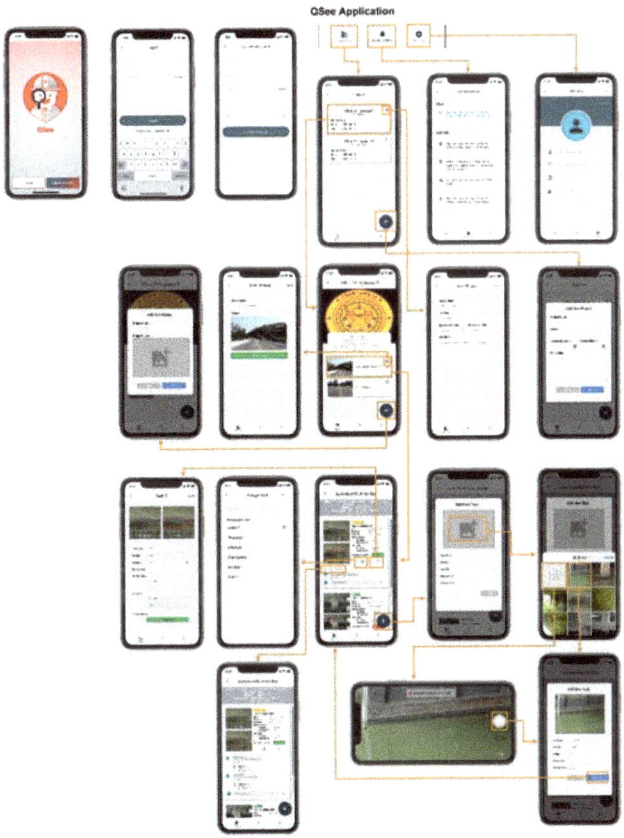

Figure 8. Wireframe of the mobile application.

3.2. Application Employment

The system is designed as a communication tool with AI embedded to assist the project managers and bridge inspectors. Following systematic instruction of the application as shown in Figure 9 was demonstrated to inspection crews. Project managers creates new projects and assign bridges to be inspected according to the inspection plan from the headquarter responding to their local divisions. Bridge inspectors are assigned. People involved in the process are notified by the application. Bridge defects are examined by taking defect images and recording the defect details. These data are sent through Cloud via 4G mobile network to be stored and analyzed by the assist AI. Project managers review the defect information and can communicate with bridge inspectors to discuss in-depth details on each defect with application web board system. With these organized and AI assisted data, maintenance plan can be created efficiently and optimally.

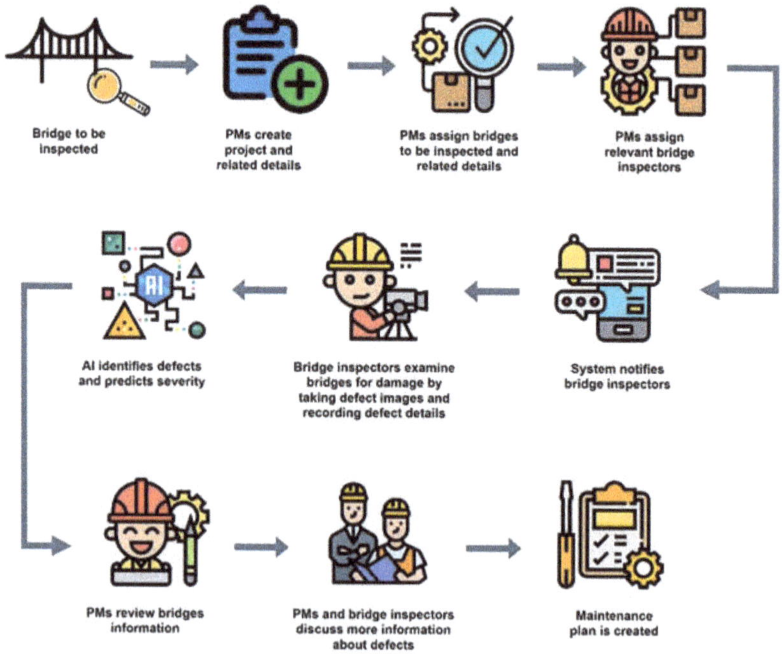

Figure 9. System overview of the application.

3.3. User Satisfaction Results

After development was completed, a prototype was delivered for a hands-on trial in a controlled environment by 14 project managers and bridge inspectors from the Thailand Department of Highways. After the trial, satisfaction surveys and in-depth interviews were conducted to elicit and measure feedback. Four domains were assessed: functionality, user interface design, user experience, and overall performance. User satisfaction in each domain was scored from 0 (unsatisfactory) to 5, indicating highly satisfied.

The average satisfaction result of each domain was 4.00, 3.714, 4.00, and 4.142, respectively, and the total score was 4.024 (80.48%). (Table 3) This means that the implemented features in this application that are required for behavioral intention to use, were built and tested by another group of users who were highly satisfied. However, the lowest score of user interface design (Table 4) can be improved in the future. According to the concept of stage-gate process, the improvement will be performed after collecting and analyzing more user validation feedback.

Table 3. User satisfaction for bridge inspection mobile application with 14 trial users.

	Score (Max. 5.00)	Percentage
Functionality	4.00	80%
User interface design	3.714	74.28%
User Experience	4.00	80%
Overall Performance	4.142	82.84%

Table 4. User Interface.

Log in	Project Overview	List of Bridge under Project	Notification
	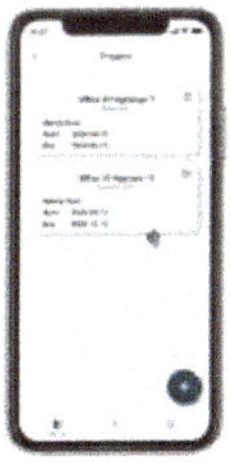		
Add/Edit Project	**Add/Edit Bridge**	**Add New Task with Image**	**Add Task Detail**
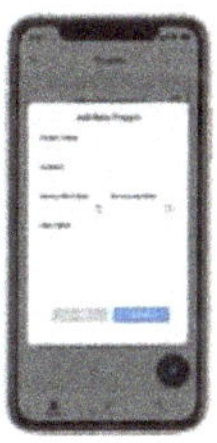			
Task Detail for Inspector with comment	**Assigned Person**	**Edit Task**	**User Setting**
	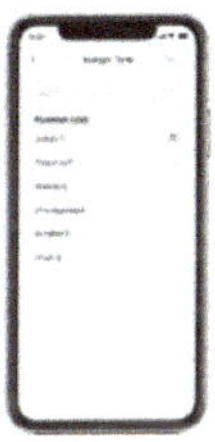	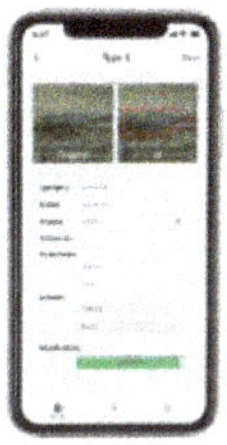	

4. Discussion

This study was conducted in collaboration with the Thailand Department of Highways and is an example of high-impact digital transformation that can improve public service and safety. We developed a prototype mobile application to assist field technicians to conduct visual bridge defect inspection and improved collaboration among team members. To develop a successful user driven application, we applied the unified theory of acceptance and use of technology (UTAUT) to identify the key factors that influence behavioral intention to use of the mobile application. The fundamental specifications were derived from our previous research on mobile applications for construction process quality control [6], AI for visual bridge defect-inspection system [7], and related literature reviews in the area of teamwork, project management, and visual inspection. Building on this body of work, the current study proposes a modified technology acceptance model UTAUT research framework, which identifies parameters that influence user intention to use the product by adding a new VIS factor that leads to either performance expectancy or effort efficacy.

As a result, two valuable insights were obtained. First, the user expects the technology to improve their routine task performance. Second, the intention to use will occur when prospective users trial the product and see the potential benefit of the technology. This suggests that allowing the user to test the product in a free trial could be an effective strategy. Next, the product concept was developed including the system architecture, use case diagram, user interface, and backend design. This process was repeated several times according to the stage-gate process "Triple A" concept. After the prototype was constructed, it was used for a field trial in a controlled environment and a user satisfaction survey was conducted. The overall satisfaction score of 14 users was 4.024 of 5.00, reflecting the high satisfaction for user behavioral intention to use the product. However, satisfaction scores were lowest (3.714 of 5.00) with the user interface design section and so the UXUI will be improved with fine-tuning after the analyses of additional user feedback.

Although, few aspects of the new system need to be improved before replacing traditional methods, there is an agreement that this new system can significantly enhance the performance of the overall inspection processes. It might be slowly replaced or parallelly implemented with the traditional methods but eventually will be a main tool for bridge inspection as a government digital transformation policy.

5. Conclusions

In summary, trial users were highly satisfied with the first version of the Bridge Visual Defect Quality Control Assisted Mobile Application. Although there are details that need to be improved, engineers at Thailand's Department of Highways expressed a strong interest in using this application to reduce workload and improve quality. There are more than 10,000 bridges in Thailand and their maintenance is a priority. This mobile application could be the first in a series of steps transforming bridge maintenance systems to full digitalization, uniting multiple operations into a single system for better project management. There are many other potential opportunities for research using a collaborative platform and visual inspection assisted technology in the construction industry. With the continued cooperation of the Thailand Department of Highways, future applications may be applied in areas such as road pavement, highway infrastructure inspection, or even overall transportation maintenance system.

Author Contributions: Due to this research article consists of several authors, each individual contribution is stated in this part. P.K. contributed to this research in terms of overall conceptualization, research methodology and experiment, and writing. N.C. contributed in terms of research methodology, supervision, review, and editing. G.P. contributed in terms of supervision, review, and editing. Lastly, K.K. contributed in terms of research methodology and experiment. All authors have read and agreed to the published version of the manuscript.

Funding: This research received no external funding.

Institutional Review Board Statement: Not applicable.

Informed Consent Statement: Informed consent was obtained from all subjects involved in the study.

Data Availability Statement: Not applicable.

Acknowledgments: The authors would like to thank the Bureau of Road Research and Development, Department of Highways, Thailand Ministry of Transportation, for its collaboration. They also express gratitude to Solitech Innovations and Technologies Co., Ltd., Bangkok, Thailand, for the software and hardware used in this research.

Conflicts of Interest: The authors declare no conflict of interest.

References

1. Sakarin, S.; Phanomchoeng, G. The Development of Car Tracking System for Car Sharing Business by Using OBD-II Port. In Proceedings of the 2021 International Symposium on Electrical, Electronics and Information Engineering (ISEEIE 2021), Seoul, South Korea, 19–21 February 2021.
2. *Global Digital Transformation on the Building and Construction Sector, Forecast to 2025*; Frost & Sullivan: San Antonio, TX, USA, 2018; 75p.

3. McRobbie, S.G.; Wright, M.A.; Chan, A. Can technology improve routine visual bridge inspections? In *Proceedings of the Institution of Civil Engineers—Bridge Engineering*; ICE Publishing: London, UK, 2015; Volume 168, pp. 197–207. [CrossRef]
4. Ismail, S.A.; Bandi, S.; Maaz, Z.N. An Appraisal into the Potential Application of Big Data in the Construction Industry. *Int. J. Built Environ. Sustain.* **2018**, *5*, 145–154. [CrossRef]
5. Gattulli, V.; Chiaramonte, L. Condition Assessment by Visual Inspection for a Bridge Management System. *Comput. Aided Civ. Infrastruct. Eng.* **2005**, *20*, 95–107. [CrossRef]
6. Kruachottikul, P.; Cooharojananone, N.; Phanomchoeng, G.; Wongsanao, K.; Thabsuwan, S.; Suwannatrai, K.; Pongsuttisatta, W.; Ganyaporngul, S.; Kovitanggoon, K.; Vajarodaya, N. Construction Process Quality Control Assisted Mobile Application. In Proceedings of the 2021 IEEE 8th International Conference on Industrial Engineering and Applications, Chengdu, China, 23–26 April 2021; p. 6.
7. Kruachottikul, P.; Cooharojananone, N.; Phanomchoeng, G.; Chavarnakul, T.; Kovitanggoon, K.; Trakulwaranont, D. Deep learning-based visual defect-inspection system for reinforced concrete bridge substructure: A case of Thailand's department of highways. *J. Civ. Struct. Health Monit.* **2021**. [CrossRef]
8. Dong, C.; Li, L.; Yan, J.; Zhang, Z.; Pan, H.; Catbas, F.N. Pixel-Level Fatigue Crack Segmentation in Large-Scale Images of Steel Structures Using an Encoder–Decoder Network. *Sensors* **2021**, *21*, 4135. [CrossRef] [PubMed]
9. *Global Status Report on Road Safety 2018: Summary*; World Health Organization: Geneva, Switzerland, 2018.
10. Brysse, D. *Non-Destructive Evaluation of Reinforced Concrete Structures*; Woodhead Publishing: Sawston, UK, 2010.
11. Chiocchio, F.; Grenier, S.; O'Neill, T.; Savaria, K.; Willms, J. The effects of collaboration on performance: A multilevel validation in project teams. *Int. J. Proj. Organ. Manag.* **2012**, *4*, 1–37. [CrossRef]
12. Bond-Barnard, T.; Fletcher, L.; Steyn, H. Exploring the influence of instant messaging and video conferencing on the quality of project communication. *Acta Structilia* **2016**, *23*. [CrossRef]
13. Müller, R. Determinants for external communications of IT project managers. *Int. J. Proj. Manag.* **2003**, *21*, 345–354. [CrossRef]
14. Liang, Y.; Walther, J.B. Computer Mediated Communication. In *International Encyclopedia of the Social & Behavioral Sciences*, 2nd ed.; Wright, J.D., Ed.; Elsevier: Oxford, UK, 2015; pp. 504–509.
15. Pollack, J.; Adler, D. Emergent trends and passing fads in project management research: A scientometric analysis of changes in the field. *Int. J. Proj. Manag.* **2015**, *33*, 236–248. [CrossRef]
16. Bello, S.A.; Oyedele, L.O.; Akinade, O.O.; Bilal, M.; Davila Delgado, J.M.; Akanbi, L.A.; Ajayi, A.O.; Owolabi, H.A. Cloud computing in construction industry: Use cases, benefits and challenges. *Autom. Constr.* **2021**, *122*, 103441. [CrossRef]
17. Wang, K.C.P. Designs and Implementations of Automated Systems for Pavement Surface Distress Survey. *J. Infrastruct. Syst.* **2000**, *6*, 24–32. [CrossRef]
18. German, S.; Brilakis, I.; DesRoches, R. Rapid entropy-based detection and properties measurement of concrete spalling with machine vision for post-earthquake safety assessments. *Adv. Eng. Inform.* **2012**, *26*, 846–858. [CrossRef]
19. Medina, R.; Gómez-García-Bermejo, J.; Zalama, E. Automated Visual Inspection of Road Surface Cracks. In Proceedings of the 27th ISARC, Bratislava, Slovakia, 25–27 June 2010; International Association for Automation and Robotics in Construction (IAARC): Bratislava, Slovakia, 2010; pp. 154–164. [CrossRef]
20. Koch, C.; Paal, S.; Rashidi, A.; Zhu, Z.; König, M.; Brilakis, I. Achievements and Challenges in Machine Vision-Based Inspection of Large Concrete Structures. *Adv. Struct. Eng.* **2014**, *17*, 303–318. [CrossRef]
21. Zhu, Z.; German, S.; Brilakis, I. Detection of large-scale concrete columns for automated bridge inspection. *Autom. Constr.* **2010**, *19*, 1047–1055. [CrossRef]
22. Robert, G.C.; Kleinschmidt, E.J. Stage-Gate Process for New Product Success. *Innov. Manag. U* **2001**, *3*, 8.
23. Ries, E. *The Lean Startup: How Today's Entrepreneurs Use Continuous Innovation to Create Radically Successful Businesses*; Crown Business: New York, NY, USA, 2011.
24. Praima, C. *Human Centered Design Innovation*; Institute of Design, IIT Technology, Chulalongkorn University: Bangkok, Thailand, 2000.
25. Venkatesh, V.; Morris, M.; Davis, G.; Davis, F. User Acceptance of Information Technology: Toward a Unified View. *MIS Q.* **2003**, *27*, 425–478. [CrossRef]
26. Reactnative.dev. React Native. Available online: www.reactnative.dev (accessed on 12 April 2021).
27. Expo.io. Expo Notification API. Available online: www.expo.io/versions/latest/sdk/notifications (accessed on 12 April 2021).
28. Apple. Apple Push Notification. Available online: www.developer.apple.com/notifications (accessed on 12 April 2021).

symmetry

MDPI

Article

The Cooperative Control of Subgrade Stiffness on Symmetrical Bridge–Subgrade Transition Section

Yang Zhang [1], Rui Li [2,*] and Jun Chen [3]

1 Jinhua Highway and Transportation Management Center, Jinhua 321000, China; zhangyangzj@outlook.com
2 Highway School, Chang'an University, Xi'an 710064, China
3 Jinhua Wucheng District Highway and Transportation Management Center, Jinhua 321000, China; chenjunzj@outlook.com
* Correspondence: lirui@chd.edu.cn

Abstract: In the field of civil engineering and architecture, the concept of symmetry has been widely accepted. The bridge can be treated as a typical symmetrical structure of civil engineering buildings. Among them, the Subgrade can be identified as an important part to bear the vehicle loads. Severe pavement problems and bridge service capabilities will be caused by problems of the bridge–subgrade transition section. Therefore, setting the rigid–flexible transition is an important method to solve this problem. The bridge–subgrade transition section has been set at both ends of the bridge, which can be regarded as a typical symmetrical structure. Based on nonlinear finite element numerical simulation and synergistic theory, the cooperative control problems of the bridge–subgrade transition section were studied in this work. The change rule of the stiffness of the transition section was discussed and the influence of stiffness variation of the bridge–subgrade transition section on the stress state of the structure was also analyzed. Furthermore, the influence of subgrade stiffness change on the stress and strain field was analyzed. A permanent strain prediction model was established and stiffness or subsidence difference coordination control was also discussed.

Keywords: bridge–subgrade transition section; rigidity–flexibility transition; permanent deformation; stiffness difference; settlement difference

Citation: Zhang, Y.; Li, R.; Chen, J. The Cooperative Control of Subgrade Stiffness on Symmetrical Bridge–Subgrade Transition Section. *Symmetry* **2022**, *14*, 950. https://doi.org/10.3390/sym14050950

Academic Editors: Victor A. Eremeyev and Sergei D. Odintsov

Received: 17 March 2022
Accepted: 26 April 2022
Published: 6 May 2022

1. Introduction

The bridge–subgrade transition section is the connection between bridge and pavement. The transition section is set at the bridge head and tail, so it can be regarded as a typical symmetrical structure. This section can balance the stiffness difference and settlement difference of the two structures and increase the pavement's integral continuity. Due to the lack of processing bridge–subgrade transition sections during the engineering design process, construction control is not effective, and the problems of operation or maintenance technology are unreasonable. It will lead to the following two kinds of problems in transition section, widely. First, the pavement structure of the transition section produces a settlement or fracture in retailing backwall. Second, the vehicles have obvious bumps when they pass the transition section, which is the so-called "bump at bridge-head". Because the bridge–subgrade transition section is set symmetrically, the "bump at bridge-head" occurs when the driver drives into and out of the bridge.

Wahls thought the step height of 1.2 cm would produce the bump effect [1]. Strak et al. found the step height of 2.5 cm would produce the bump effect [2,3]. This problem would bring many undesirable results such as affecting driving comfort, forcing vehicle deceleration, influencing pavement service performance and transportation efficiency, affecting service life of the vehicle, and leading to traffic accidents [4]. The problems of the bridge–subgrade transition section have become one of the important factors that affect pavement service ability seriously. Therefore, it is necessary to put forward a more reasonable design method to solve the problem of the bridge–subgrade transition section.

There are many factors that can cause the vehicle dumping effect. Researchers have paid a lot of effort to solve the "bump at bridge-head", and foundation treatment, subgrade treatment, and surface treatment are common preventive measures [5]. However, many practical projects showed that a simple "rigid–flexible transition" design of bridge–subgrade transition section cannot fundamentally solve the transitional problem. In this work, the change of stress and strain caused by loads in subgrade and base was explored to solve the vehicle dumping problem. The difference between the stiffness and the settlement was analyzed [6,7], and the effect of the stiffness variation of the subgrade on the stress state of the transition section was discussed by using finite element analysis [8,9].

2. Numerical Model and Parameters

2.1. Geometric Parameter

The numerical model of bridge–subgrade transition section was established. The length of the model was 26 m and the width was 4 m. Also, the pavement type of this model was semi-rigid asphalt, of which the thick of upper layer, middle layer, lower layer, and base were 4 cm, 6 cm, 8 cm, and 40 cm, respectively, as shown in Figure 1. In order to achieve a continuous change of subgrade stiffness longitudinally along the pavements in the transition section, the subgrade structure was divided into 22 parts in this paper. The length of both ends was 3 m, and the rest of the parts were all 1 m, as shown in Figure 2. When conducting parametric analysis, each part was endowed with the material parameters of continuous change to describe the nature of the transition section stiffness change.

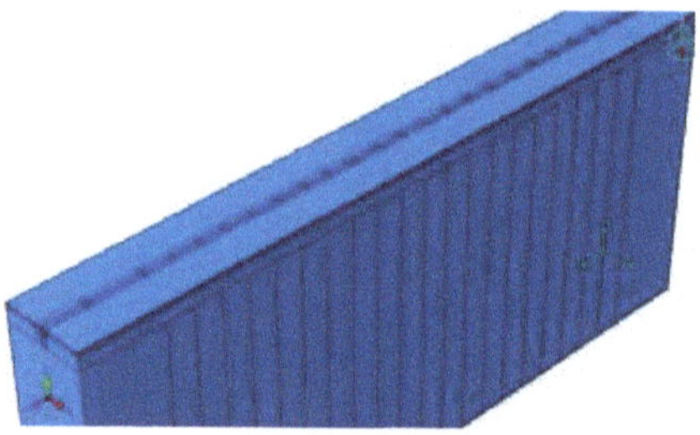

Figure 1. Numerical model of bridge–subgrade transition section.

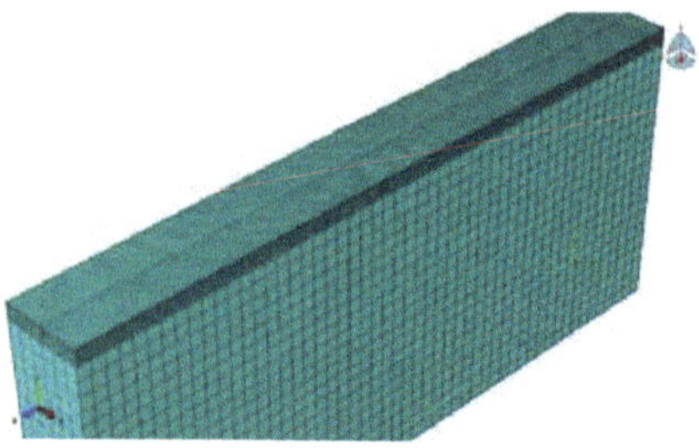

Figure 2. Numerical model grid of subgrade structure.

The large difference in stiffness between abutment and backfill at the junction of pavement and bridge, and the stress response generated by vehicle load is transmitted to the junction. However, most of the energy is reflected back on the surface of the abutment, and only a small part is transmitted to the inside abutment. The stress response of the abutment was not considered in this work of establishing numerical model. It is assumed that all the loads transmitted to the junction can be reflected back to the subgrade. Therefore, horizontal constraint conditions are adopted at the junction of pavement and bridge in the model.

2.2. Simplified Model of Load

Using the existing theory to analyze pavement stress state, the vehicle load was simplified to a circular uniform distributed load. A large number of investigations and experimental research results showed that the shape of the wheel contacting with the pavement was closer to a rectangle and indicated a non-uniform characteristic phenomenon [10]. Based on the comprehensive research results, Figure 3 can be selected as the simplified model of wheel load, the double-type load was simplified into distributed pressure of two rectangular loads. The dimensions were 0.22 m × 0.16 m. The distance between two rectangular centers was 0.32 m, and load intensity was 0.7 MPa.

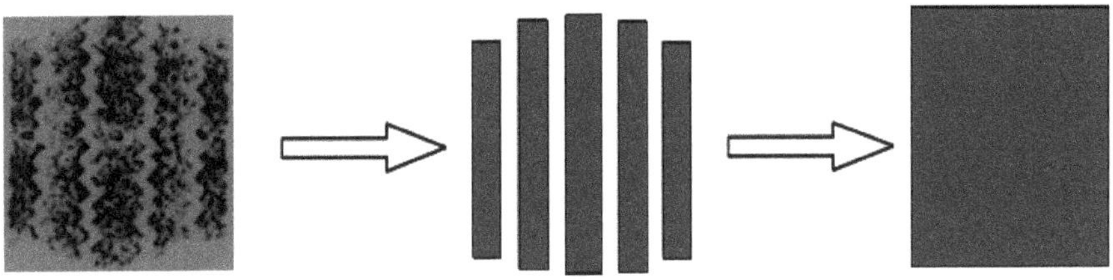

Figure 3. Simplified model of wheel load.

2.3. Material Parameters

Assuming that each of the layer structures was line elastomer, the elastic modulus and Poisson's ratio were control parameters. This paper focused on the influence of subgrade stiffness on the pavement structure, and a combination structure of semi-rigid pavement, structural grouping, and material parameters were shown in Table 1.

The subgrade material was used as the analytical parameter of the material in this paper. It was assumed that the variation of subgrade stiffness is from 30,000 MPa to 40 MPa and corresponding Poisson's ratio ranges from 0.2 to 0.35, as shown in Table 1.

Table 1. Material parameters in the calculation model.

Structural Layer	Material and Thickness	Elastic Modulus E	Poisson's Ratio μ
Upper layer	Asphalt concrete ac-10/4 cm	E = 1500 MPa	μ = 0.3
Middle surface	Asphalt concrete ac-16/6 cm	E = 1200 MPa	μ = 0.3
Bottom layer	Asphalt concrete ac-20/8 cm	E = 1000 MPa	μ = 0.3
Semi-rigid base	Cement stabilized crushed stone/40 cm	E = 1500 MPa	μ = 0.3
Subgrade	/	E = 30,000~10,000 MPa	μ = 0.2
		E = 10,000~5000 MPa	μ = 0.25
		E = 5000~1000 MPa	μ = 0.3
		E = 1000~40 MPa	μ = 0.35
Abutment	Cement concrete	E = 30,000 MPa	μ = 0.2

3. The Influence of Changes in the Way the Stiffness of Bridge–Subgrade Transition Section on Structural Stress State

3.1. Selecting the Calculation Index

When ignoring the stiffness difference between the abutments and bridge–subgrade transition section, the stiffness of the internal bridge–subgrade transition section varies continuously. Although there is no stress concentration phenomenon, stiffness changes within a certain range and it can also cause stress redistribution in the transition section of the pavement structure. Considering that the pavement structure has different stiffness sensitivities, the stress state of pavement may exceed the strength of pavement structure when the stiffness has decreased to a certain value. Based on the results, according to the stress characteristics, the stress magnitude, and variation range of each structural pavement layers, four indicators can be selected as the pavement surface deflection value. Tensile strain at the bottom of the middle layer, tensile stress at the bottom of the base layer, and subgrade top surface compressive strain were selected to describe the stress state change in the pavement structure of the transition section when the subgrade stiffness decreases along the pavement by the longitudinal.

3.2. The Influence of Stiffness Linear Change on Structural Stress State

At present, the backfilling scheme was mainly in the form of an inverted trapezoid and a positive trapezoid, which was based on stiffness with the backfill material thickness in order to achieve uniform continuous reduction. However, the ratio of stiffness difference and the bridge–subgrade transition section will affect the stress state of pavement and the filling cost directly. It is necessary to explore the transition form and proportion of the transition section. In this paper, the minimum value of the subgrade stiffness near abutment and the bridge–subgrade transition section were 500 MPa and 40 MPa, respectively. In addition, using a linear proportional transition, the filling length of the bridge–subgrade transition sections were analyzed by 20, 15, 10 and 5-m, as shown in Figure 4a–e.

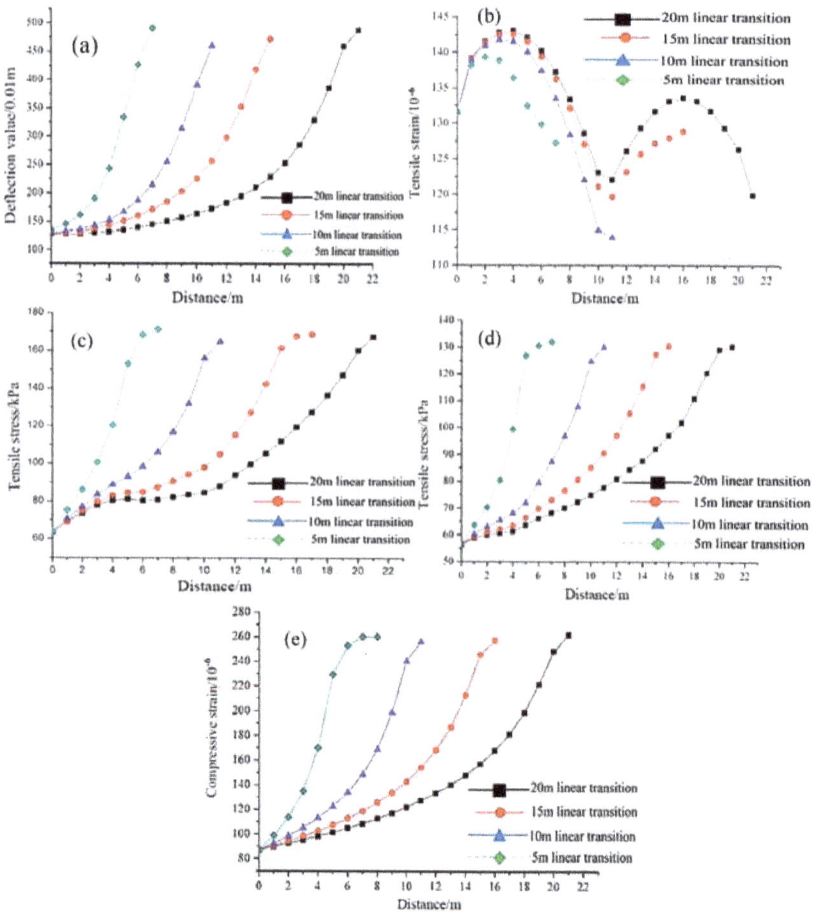

Figure 4. (**a**) Surfacing deflection under the linear change of stiffness, (**b**) Tension strain of the middle plane under the linear change of stiffness, (**c**) Transverse tensile stress of the bottom of semi-rigid base under the linear change of stiffness, (**d**) Longitudinal tensile stress of subgrade under linear change of stiffness, (**e**) The compressive strain of subgrade under the linear change of stiffness.

In addition to the different variations of tensile strain of the middle layer, the stress state of each layer structure in the transition section has a similar variation rule. First, pavement surface deflection, the tensile stress of the base layer, and the compressive strain of the top of subgrade were all increased with the distance from the abutment. Second, with the length of the transition section decrease, the stiffness transition ratio and the variation rate of the pavement structure stress state increase. However, the stress state of the pavement structure under different stiffness transition ratios has little difference when compared to the pavement stiffness. Therefore, the stress state of the pavement structure is mainly affected by the stiffness of the subgrade and less by the variation rate subgrade stiffness which directly affected the variation rate of stress state of the pavement structure. Third, the variation rate of the stress and strain were increased with the distance from the abutment. As the stiffness was large, the sensitivity of the stress state was weak and as the stiffness of the pavement gradually decreases, the sensitivity was strong.

In a word, tension stress on the bottom of the middle surface course was affected largely by the boundary condition. The oversize length–width ratio of the numerical model was the main reason for this result.

3.3. The Influence of the Nonlinear Stiffness Ratio to the Structural Stress State

Simulation results showed that the variation of pavement stress state and stiffness change have a good correlation with the form of the denary logarithm. Although the homogeneous continuous change of stiffness can occur, the stress state of the pavement doesn't achieve a homogeneous continuous change. In actual pavement construction, the ideal pavement stress state should change continuously and homogeneously along the pavement longitudinal line. Therefore, this study attempts to realize a pavement structure stress state that varies linearly when the stiffness of the pavement bridge transition section changes nonlinearly in the same change interval along the longitudinal direction. When the subgrade stiffness along the longitudinal pavements are in the range of 500~40 MPa, continuous transition was realized in the 20 m and 10 m length range and in the form of the denary logarithm, as shown in Figure 5a–d.

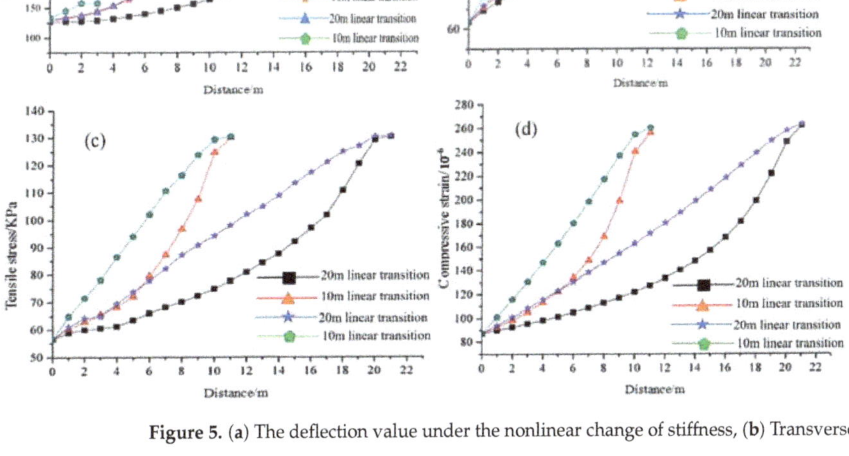

Figure 5. (**a**) The deflection value under the nonlinear change of stiffness, (**b**) Transverse tensile stress

at the bottom of base under the nonlinear change of stiffness, (**c**) Longitudinal tensile stress of subgrade bottom under the nonlinear change of stiffness, (**d**) Compressive strain on the top of subgrade under the nonlinear change of stiffness.

4. The Influence of the Stiffness Variation Mode of the Transition Section on the Stress and Strain Field

Permanent deformation of aggregates and soil was affected by internal factors and environmental factors, it was also affected by load factors and major load size and load times were considered as primary load factors. For foundation settlement, its additional load contains vehicle loads and superstructure gravity loads, and because of the limited influence depth of vehicle load, the vehicle load can be simplified as the equivalent height of the soil. The additional load on the subgrade was a uniformly distributed strip load and vehicle load imposed on the pavement, vehicle load exists in a limited subgrade depth, the load stress values at different layers varied greatly. In addition, the vehicle load in subgrade often exists in the form of dynamic loads and impact loads, which cause the complicated subgrade settlement problems. The boundary conditions of the bridge transition section were complex. The subgrade modulus changes are orderly along the longitudinal line of pavement, which make the stress field and strain field of the subgrade in the bridge transition section different from other general sections.

4.1. *The Influence of Variation of Subgrade Stiffness on Stress and Strain Field*

When the resilient modulus of the subgrade is in the range of 40~2000 MPa and the length of transition section is within 20-m, the distribution of stress field is in the base and subgrade. Note that the abscissa "depth" was not the true depth value, it was defined that the top surface of the base was the starting point-zero. Then the depth of boundary of the bottom of subgrade and surface of basement layer was 0.4 m. For the need of drawing and reading, the abscissa was only an approximate coordinate in Figure 6a. The stress field distribution of base and subgrade in a real depth coordinate is shown in Figure 6b,c.

It can be seen from Figure 7a–c that resilience modulus of subgrade in the bridge transition section decreases linearly as the distance increases. However, the variation in the stiffness of subgrade has little influence on vertical stress in the top surface of the basement layer. With the distance from the top surface to the base layer increasing (i.e., the depth increasing), the impact of changes in subgrade stiffness on the stress field in the base layer and subgrade increases. As the depth increases, the base layer and subgrade stress field decline rapidly. At the same time, the concept of "subgrade work area" can verify that the impact of stress on the subgrade vehicle load occurs within a limited depth, typically 0.8 m [11]. When this model subgrade depth is 2 m, vehicle load and subgrade stiffness' effect on the stress field was very small and can be ignored.

Variation of strain and stress field were significantly different from each other, and the performance was more complex. There was a big difference in the distribution of the base layer and the subgrade strain field. With the increase of the distance that the abutment or subgrade stiffness decreases, the stress field in the substratum and subgrade decreases. The strain field in the substratum also showed the same trend, but the strain field in the subgrade increased. Second, with the increase of depth, the stress field was gradually reduced. When the subgrade stiffness was small, the strain field meet this rule. When the subgrade stiffness decreases to a certain value, the strain field on the top of the subgrade will break the rule. At the same time of the emergence of a maximum strain, the strain field in the subgrade will be stronger than the stress field in the substratum. Moreover, the maximum value of medium pressure stress exists on the surface of the base and the maximum value of the compressive strain appears in a certain depth under the substratum. This model appeared at 10 cm. Finally, with the reduction of stiffness, the difference of compressive stress in different depths decreases and the compressive strain increases.

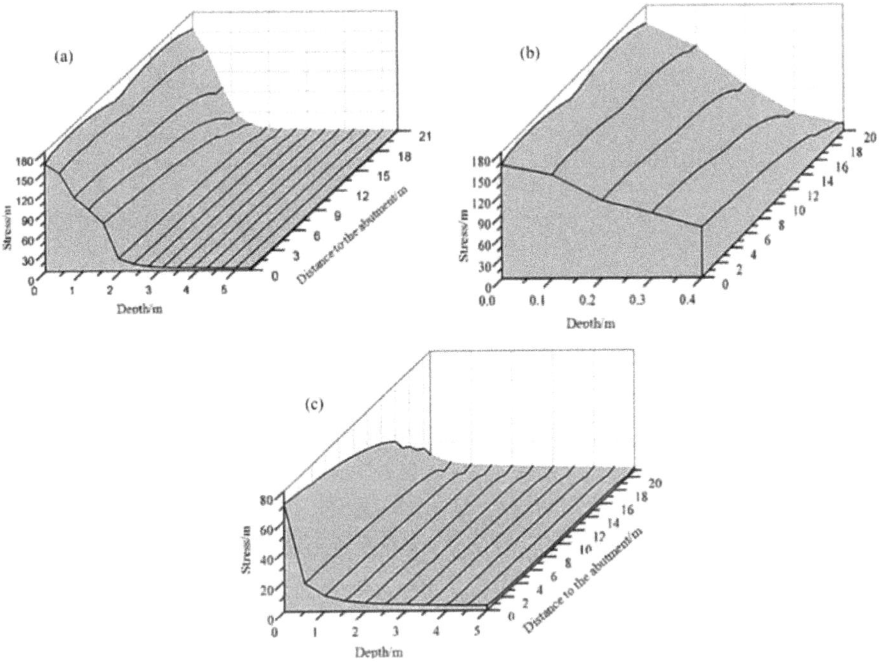

Figure 6. (**a**) Stress field of base and subgrade when the resilient modulus of subgrade is changed, (**b**) Distribution of stress field in base, (**c**) Distribution of stress field in subgrade.

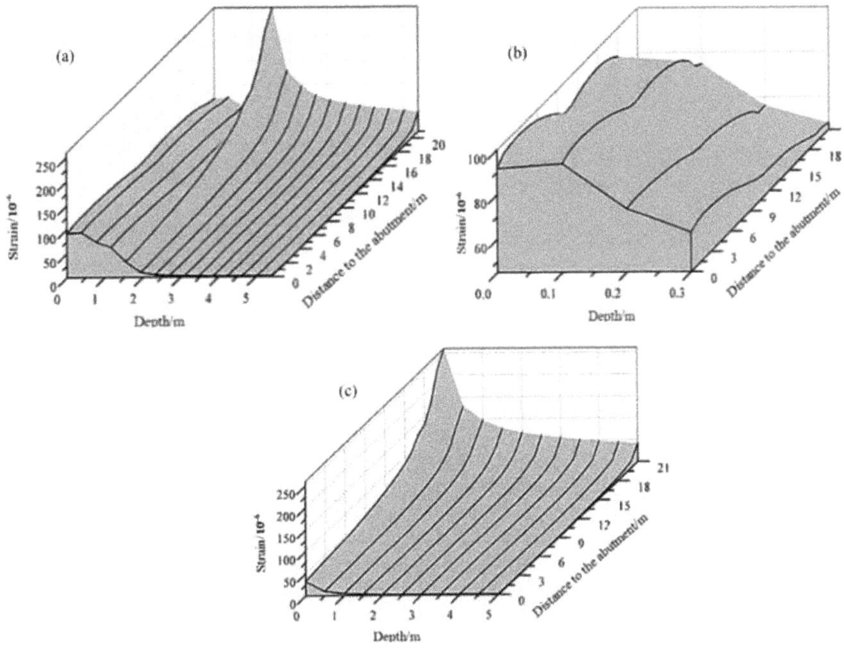

Figure 7. (**a**) Strain field of base and subgrade when the subgrade modulus is changed, (**b**) Strain field in base, (**c**) Strain field in subgrade.

4.2. Effect of Subgrade Stiffness Variation on Stress and Strain Field

In chapter 4.2, the rule of stress and strain field in the base and subgrade was introduced. During the structural designs of the bridge–subgrade transition section, the subgrade stiffness or the resilient modulus of subgrade may be changed. For this reason, compared with the maximum subgrade resilient modulus, respectively, as 2000 MPa, 1000 Mpa, and 500 MPa, the corresponding stress and strain field were changed with the linear variation in 20 m, 15 m, and 10 m.

Figure 8a is a "stiffness equivalent" approach (i.e., the corresponding stress value was drawn by the subgrade stiffness value, the equivalent stress field was obtained). Ignoring the boundary conditions of the numerical model itself, different horizons of the transition section and the vertical stress field at different locations were only related to the subgrade stiffness (modulus of resilience) and had little relation with the rate of change of stiffness.

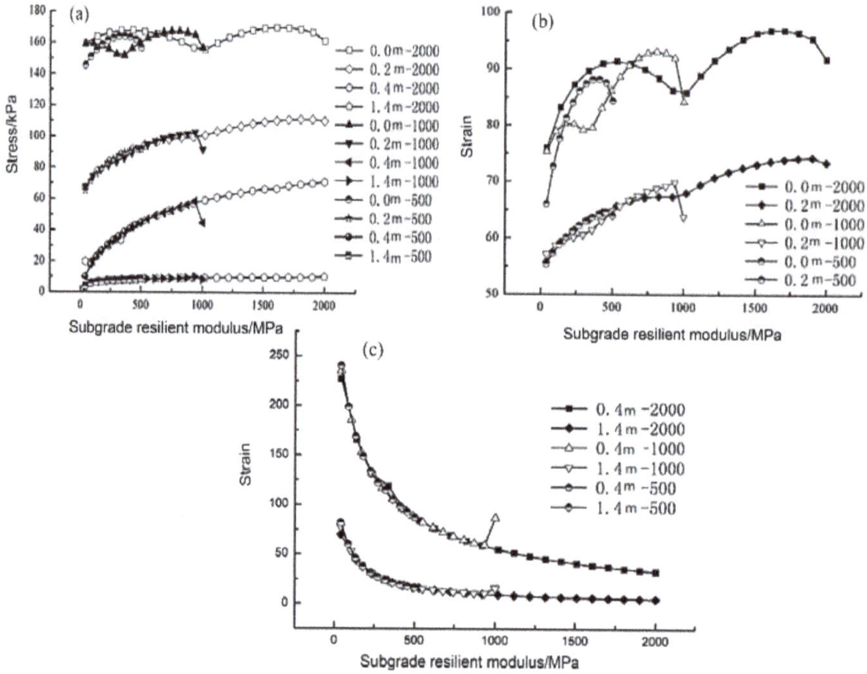

Figure 8. (**a**) Distribution of stress field when the subgrade modulus is changed, (**b**) Distribution of strain field in base when the subgrade modulus is changed, (**c**) Distribution of strain field in subgrade when the subgrade modulus is changed.

Figure 8b,c corresponds to the base and the distribution rule of the strain field in the subgrade. It can be concluded that the strain field also satisfies the same rule and the compressive strain values at different positions and locations on the transition section were related to the stiffness (modulus) of subgrade at its location and had little relation with the change of stiffness (modulus of resilience).

The rule of the stress and strain field along with the change of the subgrade stiffness in the bridge–subgrade transitional section was shown once again. Stiffness difference and settlement difference can be solved by collaborative design of structures and materials. However, stiffness needs continuous change. Too much difference can result in stress redistribution, especially a stress concentration phenomenon. The functional relationship between the compressive stress value and the subgrade stiffness value, the compressive strain value, and the subgrade stiffness value need to be further determined and modified.

When the subgrade stiffness (modulus of resilience) of bridge–subgrade transition section was a linear or nonlinear variation, the stress and strain field were also related to the size of stiffness. However, stiffness variation was relative (i.e., stiffness variation within a certain distance or between two adjacent points). The variation of the stress and strain field can only be judged by the change of the adjacent two points or a certain distance. Analysis of the rule of the reaction from Figure 9a, the relationship between strain and stiffness was an exponential function, which also fit the reasoning above. The subgrade stiffness was nonlinear variation, and the change of stress state on pavement structure was more balanced. Figure 9b,c once again verifies this inference.

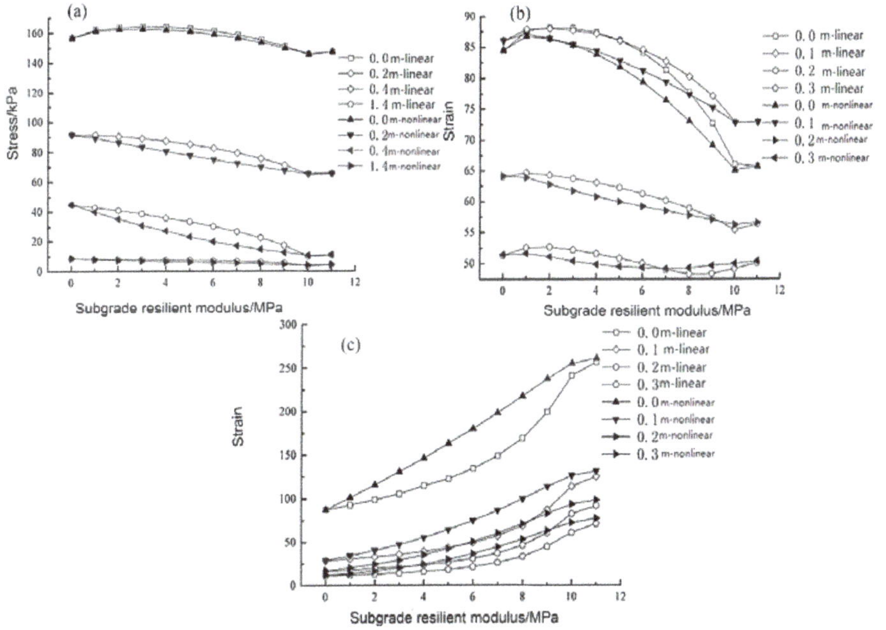

Figure 9. (**a**) The state of stress field when subgrade stiffness varies in different ways, (**b**) The state of strain field in base when subgrade stiffness varies in different ways, (**c**) The state of strain field in subgrade when subgrade stiffness varies in different ways.

From the above three figures, we can conclude that when the subgrade stiffness (modulus of resilience) forms a nonlinear function change, the stress and strain field in base and subgrade approximate linear variation along with the increase of the longitudinal distance of the pavement; and the change was balanced and continuous.

In a further comparison, the results of non-linear changes in the stress field are slightly smaller than the linear variation in the stress field, and the strain field in the base also fits this rule. In the subgrade strain field, the results of nonlinear changes are slightly larger than the result of a linear change. This phenomenon was determined by the subgrade stiffness value and the sensitivity of stress and strain to the change of stiffness.

4.3. Effect of Load Size on the Stress and Strain Field

The distribution of stress and strain in a bridge–subgrade transition section was influenced by its structure and boundary conditions and the size of the load. The result of the previous paper was based on the vehicle load size of 0.7 MPa. Considering the real traffic load, the load was selected as 0.7 MPa, 0.8 MPa, and 0.9 MPa. The condition to analyze the effect of the load size on the stress and strain field was also provided in this paper.

As can be seen from Figure 10a–c, the change of the load size does not change the rule of the stress field and the strain field in base and subgrade. When the load increases, the stress and strain in base and subgrade were obviously increased. As the depth increases, the influence of the load size on the stress and strain field was weakened. The influence of load size on the stress and strain field in the subgrade was very limited. In addition, the data shown in this paper is within 2 m depth.

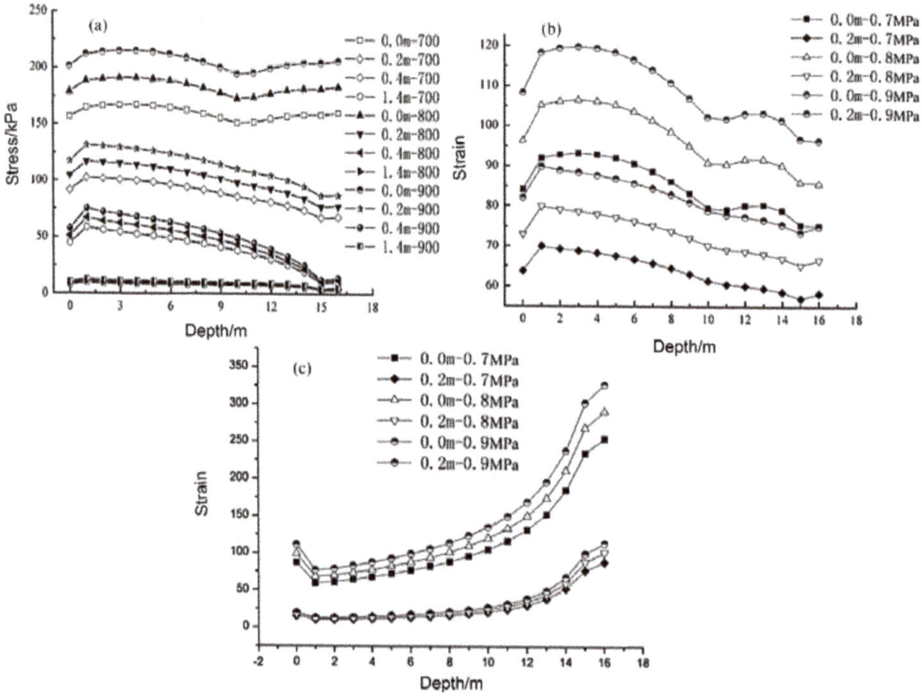

Figure 10. (**a**) The influence of load on stress field, (**b**) The influence of load on base strain field, (**c**) The influence of load on subgrade strain field.

4.4. The Relation between Stress and Strain Field

In the previous paper, it has been revealed that the change of the stress and strain field in the base and subgrade was affected by the change of the stiffness and the load in the subgrade. Further comprehensive explanation of the rule was shown as follows.

As can be seen from Figure 11a–c, the stress and strain field of the base have similar variation, which is compressive stress and strain increasing with the distance increased from the abutment and reducing with the subgrade stiffness decreases. The stress and strain field of subgrade have the opposite variation, which is the stress field decreasing with the subgrade stiffness decreases and the opposite situation with the strain field. The strain field in base and the strain field in the subgrade have the opposite variation, but the rule of the stress field variation is the same. When the subgrade stiffness changed by a nonlinear rule along the longitudinal pavement, the stress and strain field were changed more evenly and continuously.

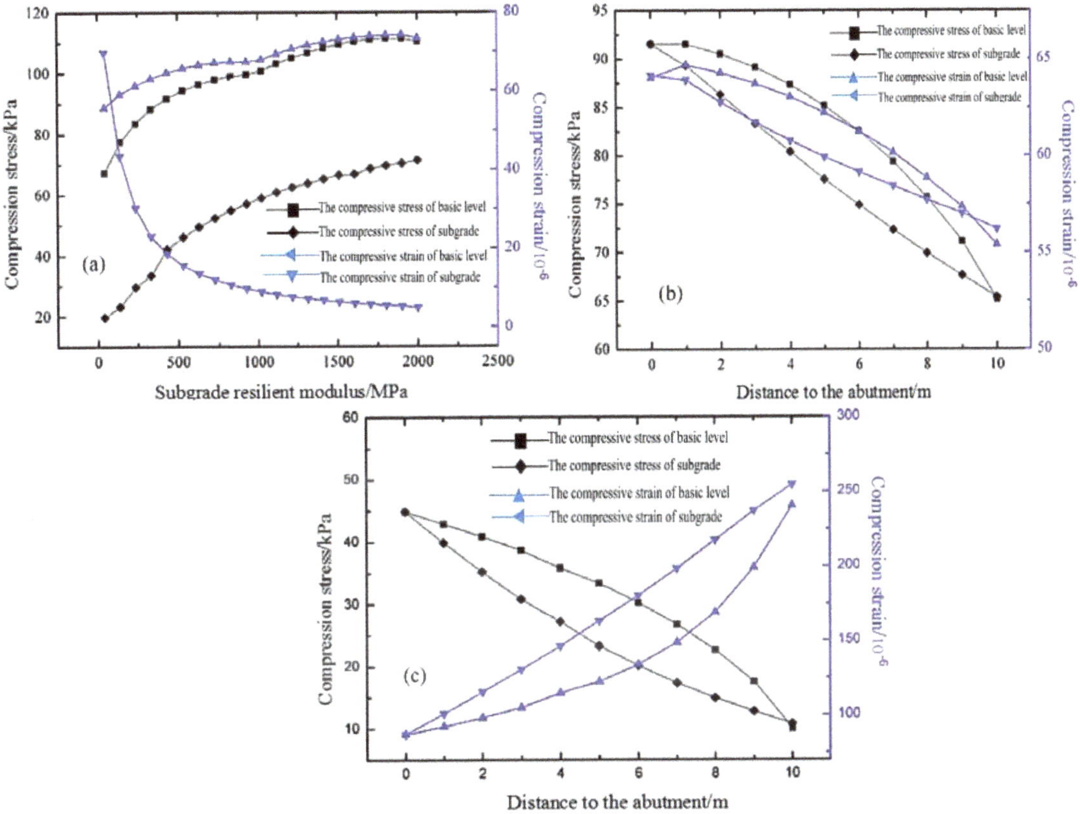

Figure 11. (**a**) Stress and strain in subgrade and base, (**b**) Stress and strain field in base, (**c**) Stress and strain field in subgrade.

In general, the stress and strain field of base were relatively stable and the sensitivity to the subgrade stiffness was weak. If the change of stress field caused by the stiffness of subgrade was less than 50 kPa at a certain depth, and the strain field changes less than 50 micro-strain, the stress and strain field at a certain depth in the subgrade were much higher than this standard. When the resilient modulus of subgrade was lower than 500 MPa, the resilient modulus of subgrade near the abutment should be higher than 500 MPa. In addition, when the resilient modulus of subgrade of transition section was lower than 500 MPa, the modulus change rate should be reduced appropriately to ensure that the stress and strain field were changed continuously and evenly.

5. The Coordinated Control of the Stiffness and Permanent Deformation in Bridge–Subgrade Transition Section

5.1. The Main Factors Influencing Permanent Deformation of Aggregate and Soil

The filling materials of the subgrade and the base in the bridge–subgrade transition section are granular materials, and nonlinear elastic plastic materials. The strain under the load was composed of two parts (i.e., resilience strain and permanent strain) [12,13]. Although the permanent stress of aggregate and soil variables by each load was very small, it gradually accumulated in many repeated traffic loads, resulting in over large post-construction settlement in the bridge–subgrade transition section.

The repeated stress level and loading times are the most important factors affecting the permanent deformation and accumulation of aggregates and soil. Many repeated loading tests of three axes have proved that the axial permanent stress variable and sum of accumulation increased with the increase of the partial stress. In addition, with the repeated action of the load, the permanent strain of aggregate and soil can produce tiny increments. The permanent strain tends to be stable or destroyed when it reaches a certain amount of accumulation. Under the expected service life or the traffic loads, a point in the spatial structure of transition section of the stress level and loading times is constant. Therefore, we can make the permanent stress change in the controllable range by choosing reasonable materials or predicting the permanent deformation of the bridge–subgrade transition section.

5.2. The Prediction Model of Permanent Strain

By studying the permanent strain long-term character of aggregate and soil, many scholars have established the constitutive model which can predict permanent strain accumulation. These models mainly consider the permanent strain in different natures and states of aggregate and soil, the rule of gradual accumulation along with the increase of the number of loads, and the important role of different stress conditions and levels in the accumulation of permanent strain [14]. The following contents were mainly introduced in the model in the same load action times and stress conditions.

By relating the accumulation of permanent strain with the number of loads and the corresponding stress levels (partial stress q and confining stress σ_3), Kim used 13 kinds of stress combinations on an aggregate of base to conduct a constant confining pressure stress three axis test and established the following model [15]:

$$\varepsilon_{1p} = aq^b \sigma_3^c N^d \left(R^2 = 0.843\right) \tag{1}$$

$$\varepsilon_{1p} = a\left(\frac{q}{\sigma_3}\right)^b N^d \left(R^2 = 0.167\right) \tag{2}$$

where: a, b, c, d are the model parameters obtained by regression analysis.

By using the experimental methods of staged repeated loading, Gidel concluded that the permanent strain increased with the increase of the average principal stress, ratio of the partial stress, and the average principal stress, correlating with the stress path length and height of stress ratio. A permanent strain model was established which was composed of the number of role functions and the role stress functions [16].

$$\varepsilon_{1p} = A\left[1 - \left(\frac{N}{100}\right)^{-b}\right] \varepsilon_0 \left(\frac{L_{\max}}{p_a}\right)^n \left(m + \frac{s}{p_{\max}} - \frac{q_{\max}}{p_{\max}}\right)^{-1} \tag{3}$$

where: ε_0, m, n, s—test parameters;

p_a—reference stress (1000 kPa);

$L_{\max}$—stress path length. $L_{\max}^2 = p_{\max}^2 + q_{\max}^2$.

The COST337 project of the European Commission considers that establishing a relationship between the permanent strain and the resilient modulus can easily calculate the cumulative amount of permanent deformation by incremental damage mode [17].

$$\varepsilon_p = a\varepsilon_r^b N^c \tag{4}$$

$$\text{or } \varepsilon_p = a\varepsilon_r^b \left(\frac{\sigma_r}{p_a}\right) N^d \tag{5}$$

where: σ_r—rebound stress;

p_a—reference stress, taken as the atmospheric pressure (100 kPa); a, b, c, d—coefficient of restitution.

According to Uzan, a function can be used between the number of times and the ratio of permanent strain and rebound strain in cohesive soil subgrade [18].

$$\lg\frac{\varepsilon_p}{\varepsilon_r} = a_0 + b_0 \lg N \tag{6}$$

where: ε_p, ε_r—permanent strain and rebound strain;

N—action times;

a_0, b_0—constant.

The permanent deformation model used by Tseng is shown as follows [19].

$$\frac{\varepsilon_p}{\varepsilon_r} = (\frac{\varepsilon_0}{\varepsilon_r})e^{-(\frac{\rho}{N})^{\beta}} \tag{7}$$

where: ε_0, ρ, β—the parameters relating to properties of materials.

The Mechanics-Empirical method of pavement design experience guide uses the summation of stratified strain method to estimate permanent deformation of granular layer and subgrade based on the permanent strain model established by Tseng. The correction model was put forward to predict the permanent deformation of granular layer and subgrade layer [20].

$$\delta_p(N) = \beta_C(\frac{\varepsilon_0}{\varepsilon_r})e^{-(\frac{\rho}{N})}\varepsilon_V h \tag{8}$$

$$\lg\beta = -0.61119 - 0.017638\omega_c \tag{9}$$

$$\lg(\frac{\varepsilon_0}{\varepsilon_r}) = 0.5(e^{\rho^{\beta}} \times a_1 E_r^{b_1}) + 0.5\left[e^{(\frac{\rho}{10^7})^{\beta}} \times a_2 E_r^{b_2}\right] \tag{10}$$

$$\rho = 10^7\left[\frac{c_0}{1-(10^7)^{\beta}}\right]^{-\beta} \tag{11}$$

$$\omega_c = 51.712\left[(\frac{E_r}{2555})^{\frac{1}{0.64}}\right]^{-0.3586\times d_W^{0.1192}} \tag{12}$$

$$c_0 = \ln(\frac{a_1 E_r^{b_1}}{a_2 E_r^{b_2}}) \tag{13}$$

where: $\delta_p(N)$—permanent deformation of the layer after loading N times;

ε_0, ρ, β—property parameters in the layer material;

ε_r—the resilience strain in the laboratory tests for obtain the material properties;

ε_V—the average vertical rebound strain in the layer was obtained from the basic response model;

h—thickness of the layer;

β_C—calibration coefficients in the model;

ω_c—water content;

E_r—resilient modulus of the layer;

d_W—deep underground water level;

a_1, b_1, a_2 and b_2—coefficient, corresponding to 1.0942×10^{-18}, 3.520049, 0.03162278 and 0.5.

5.3. *Cooperative Control the Difference of Stiffness and Settlement*

The post-construction settlement of bridge–subgrade transition period is mainly composed of three parts: foundation settlement, consolidation settlement of subgrade and compression deformation of subgrade. However, there was a great difference in actual engineering that the contribution ratio of the three parts hadn't been given a reasonable range. Therefore, the common method is to enable the three parts to achieve "zero settlement".

The permanent strain accumulation rule of aggregate and soil was connected with repeated load levels and action times. The present experimental studies have shown that a critical stress level can make a granular layer and a subgrade tend to be stable under the action of repeated loads, called the "stability limit". When the maximum stress in the granular layer and the subgrade soil do not exceed this stability limit, the granular layer or substrate show resilience after repeated loading. Meanwhile, the accumulation of permanent deformation can be controlled within a limited range.

At present, in the main method of pavement design, the Shell and AI method [21], the permanent deformation of the subgrade was effectively controlled by controlling the compressive strain and the compressive stress on the top of the subgrade. Compressive strain on the top of the subgrade method was mainly based on the direct ratio of material plastic and elastic strain. The elastic strain was controlled within the effective range, and the plastic strain can be effectively controlled [22]. If the elastic strain level of the subgrade was limited, the plastic strain and the permanent deformation of subgrade and pavement can be limited.

The analysis in the previous work shown that there was a good correlation between the variation of subgrade stiffness and the stress [23]. The strain field of subgrade and the size of stress and strain field at a certain depth in subgrade was only related to the stiffness of the subgrade. The strain field in subgrade increases with the decrease of the subgrade stiffness. The function characteristic of bridge–subgrade transition section requires that the transition section to achieve the settlement transition by longitudinal along the pavement, therefore, the permanent deformation of the subgrade can be realized by controlling the change rule of subgrade stiffness.

6. Conclusions

Pavement surface deflection value, tensile strain at the bottom of the middle plane, tensile stress at the bottom of the base layer, and subgrade top surface compressive strain were chosen as the indices of stress state analysis on the transition section. The influence of the subgrade stiffness variation on stress state was discussed in this paper. Based on the analysis, the following conclusions can be made: first, when the subgrade stiffness of bridge–subgrade transition section linear decreased longitudinally along the pavement, the four indicators had a better correlation. In addition, when the stiffness value was low, the rate of change of each index increased, which is caused by the sensitivity of stress state increasing with the decrease of the stiffness value. Second, the stress and strain level in the stress state of the pavement structure were mainly related to the subgrade stiffness and the rate of stiffness change was mainly related to the rate of subgrade stiffness change. Third, when the subgrade stiffness has nonlinear variation in the form of an exponential function, the stress state change within the pavement structure is more uniform and continuous than a stiffness linear variation. It can better ensure the continuity of the overall structure on the pavement.

On the other hand, the variation rule of the vertical compressive stress and vertical compressive strain in the base and subgrade under wheel load was described when the subgrade stiffness changed in the bridge–subgrade transition section. The stress field in subgrade and base was only related to the subgrade stiffness. The rate of stress and strain field in subgrade and base was related to the rate of stiffness variation. When the stiffness was small, the rate of change was large. The variation rule of the stress field was similar to the strain field in base layer, both decrease with the decrease of the subgrade stiffness. The rule of change of the stress and strain field in the subgrade was the opposite, the stress field in the subgrade decreases with the decrease of the stiffness. The strain field increases with the decrease of the stiffness. The stress field and strain field were limited in the propagation depth of the subgrade. The depth of the model was 2 m.

By mastering the variation rule of the strain field in the subgrade, the accumulation of permanent deformation can be predicted scientifically. Theoretical analysis shows that coordination design of the stiffness difference and settlement difference can be realized in the bridge–subgrade transition section by regulating the change rule of subgrade stiffness.

Author Contributions: Formal analysis, Y.Z.; data curation, R.L.; funding acquisition, J.C. All authors have read and agreed to the published version of the manuscript.

Funding: This research was funded by the Science and technology planning project of Zhejiang Provincial Highway and Transportation Management Center (No. 211821200009) and the Science and technology planning project of Zhejiang Provincial Department of Transportation (No. 211821220089).

Institutional Review Board Statement: Not applicable.

Informed Consent Statement: Not applicable.

Data Availability Statement: Not applicable.

Conflicts of Interest: The authors declare no conflict of interest.

References

1. Wahls, H.E. *Design and Construction of Bridge Approaches*; Transportation Research Board: Washionton, DC, USA, 1990.
2. Stark, T.D.; Olson, S.M.; Long, J.H. *Differential Movement at the Embankment/Structure Interface-Mitigation and Rehabilitation*; Final Report; National Technical Information Service: Urbana, IL, USA, 1995.
3. Zhang, H.; Changshun, H.U.; Gao, J. Method for Analyzing Dynamic Vehicle-roadway Interaction in Roadway-bridge Transition Sections. *Cent. South Highw. Eng.* **2005**.
4. Wang, H.; Rath, P.; Buttlar, W.G. Recycled asphalt shingle modified asphalt mixture design and performance evaluation. *J. Traffic Transp. Eng.* **2020**, *7*, 205–214. [CrossRef]
5. Ba, R. *The Subgrade Disease Characteristics of Bridge-Subgrade Transition Section and Treatment Method Research*; Chang'an University: Xi'an, China, 2005.
6. Liu, D. *Study on the Theory and Application of Rigid—Flexible Transition in Bridge Jump*; Hunan University: Hunan, China, 2001.
7. Yu, H.J. *Study on Reasonable Cross-Sectional Connection of Expressway Bridge (Road) Tunnel Transition Section*; Chang'an University: Xi'an, China, 2010.
8. Wu, S. *Study on Shearing Fatigue Property and Evaluation of Interlayer Contact Condition between Base Course and Surface Course for Semi-Rigid Base Asphalt Pavement*; Chang'an University: Xi'an, China, 2013.
9. Zhang, H. *Study on Top-Down Crack Propagation Law near Load Formation Mechanism and Area of Asphalt Pavement*; Chang'an University: Xi'an, China, 2011.
10. Hu, X.D. *Stress Response Analysis of Asphalt Pavement under Measured Tire Ground Pressure of Heavy Vehicle*; Tongji University: Shanghai, China, 2006.
11. Deng, X. *Road Subgrade and Pavement Engineering*, 3rd ed.; China Communications Press: Beijing, China, 2008.
12. Sun, M. *Research on the Prediction Structrual Performance of Asphalt Pavement*; Chang'an University: Xi'an, China, 2013.
13. Barksdale, R.D. Laboratory evaluation of rutting in basecourse materials. In Proceedings of the Third International Conference on the Structural Design of Asphalt Pavements Grosvenor House, Park Lane, London, UK, 11–15 September 1972.
14. Yao, Z. *Structral Design of Asphalt Pavement*; China Communications Press: Beijing, China, 2011.
15. Kim, I.T.; Kwon, J.; Tutumluer, E. Rutting of Airport Pavement Granular Layers. In *Airfield Pavements. Challenges and New Technologies*; ASCE Press: Reston, VA, USA, 2004; pp. 334–347.
16. Gidel, G. A new approach for investigating the permanent deformation behaviour of unbound granular material using the repeated load triaxial apparatus. *Bull. Lab. Ponts Chaussees* **2001**, *233*, 5–21.
17. Zhang, Y.J.; Tan, S.L.; Lang, J.C.; Su, S.M. A novel water-based acrylic luminescent coating with long lasting photoluminescence. *J. Funct. Mater. Devices* **2002**, *8*, 415–417.
18. Uzan, J. Permanent Deformation in Flexible Pavements. *J. Transp. Eng.* **2004**, *130*, 6–13. [CrossRef]
19. Tseng, K.-H.; Lytton, R.L. Prediction of Permanent Deformation in Flexible Pavement Materials. In *Implication of Aggregates in the Design, Construction, and Performance of Flexible Pavements*; ASTM International: West Conshohocken, PA, USA, 1989; Volume 1016, pp. 154–172.
20. NCHRP. *Guide for Mechanistic-Empirical Design of New and Rehabilitated Pavement Structures*; NCHRP: Washington, DC, USA, 2004.
21. Huang, Y.H. *Pavement Analysis and Design*; Pearson: London, UK, 1993.
22. Nassiraei, H.; Rezadoost, P. Stress concentration factors in tubular T/Y-joints strengthened with FRP subjected to compressive load in offshore structures. *Int. J. Fatigue* **2020**, *140*, 105719. [CrossRef]
23. Peng, J.; Zhang, J.; Li, J.; Yao, Y.; Zhang, A. Modeling humidity and stress-dependent subgrade soils in flexible pavements. *Comput. Geotech.* **2020**, *120*, 103413. [CrossRef]

applied sciences

MDPI

Article

Data-Driven Reinforcement-Learning-Based Automatic Bucket-Filling for Wheel Loaders

Jianfei Huang, Dewen Kong, Guangzong Gao, Xinchun Cheng and Jinshi Chen *

Key Laboratory of CNC Equipment Reliability, Ministry of Education, School of Mechanical and Aerospace Engineering, Jilin University, Changchun 130022, China; jfhuang19@mails.jlu.edu.cn (J.H.); dwkong@jlu.edu.cn (D.K.); gaoguangzong_jlu@163.com (G.G.); chengxc19@mails.jlu.edu.cn (X.C.)
* Correspondence: chenjinshi8304@163.com

Abstract: Automation of bucket-filling is of crucial significance to the fully automated systems for wheel loaders. Most previous works are based on a physical model, which cannot adapt to the changeable and complicated working environment. Thus, in this paper, a data-driven reinforcement-learning (RL)-based approach is proposed to achieve automatic bucket-filling. An automatic bucket-filling algorithm based on Q-learning is developed to enhance the adaptability of the autonomous scooping system. A nonlinear, non-parametric statistical model is also built to approximate the real working environment using the actual data obtained from tests. The statistical model is used for predicting the state of wheel loaders in the bucket-filling process. Then, the proposed algorithm is trained on the prediction model. Finally, the results of the training confirm that the proposed algorithm has good performance in adaptability, convergence, and fuel consumption in the absence of a physical model. The results also demonstrate the transfer learning capability of the proposed approach. The proposed method can be applied to different machine-pile environments.

Keywords: data-driven model; reinforcement learning; wheel loaders; automatic bucket-filling

Citation: Huang, J.; Kong, D.; Gao, G.; Cheng, X.; Chen, J. Data-Driven Reinforcement-Learning-Based Automatic Bucket-Filling for Wheel Loaders. *Appl. Sci.* **2021**, *11*, 9191. https://doi.org/10.3390/app11199191

Academic Editors: Nikos D. Lagaros and Vagelis Plevris

Received: 12 July 2021
Accepted: 24 September 2021
Published: 2 October 2021

1. Introduction

Construction machinery has a pivotal role in the building and mining industry, which makes a great contribution to the world economy [1]. The wheel loader is one of the most common mobile construction machinery and is often used to transport different materials at production sites [2].

The automation of wheel loaders, which has received great attention over the past three decades, can improve safety and reduce costs. Dadhich et al. [3] proposed five steps to full automation of wheel loaders: manual operation, in-sight tele-operation, tele-remote operation, assisted tele-remote operation, and fully autonomous operation. Despite extensive research in this field, fully automated systems for wheel loaders have never been demonstrated. Remote operation is considered a step towards fully automated equipment, but it has led to a reduction in productivity and fuel efficiency [4].

In the working process of wheel loaders, bucket-filling is a crucial part, as it determines the weight of the loaded materials. Bucket-filling is a relatively repetitive task for the operators of wheel-loaders and is suitable for automation. Automatic bucket-filling is also required for efficient remote operation and the development of fully autonomous solutions [5]. The interaction condition between the bucket and the pile strongly affects the bucket-filling. However, due to the complexity of the working environment, the interaction condition is unknown and constantly changing. The difference in working materials also influences the bucket-filling. A general automatic bucket-filling solution is still a challenge for different piles.

In this paper, a data-driven RL-based approach is proposed for automatic bucket-filling of wheel loaders to achieve low costs and adapt to changing conditions. The Q-learning algorithm can learn from different conditions and is used to learn the optimal action in different states by maximizing the expected sum of rewards. Aiming to achieve low costs, an indirect RL is employed. Indirect RL requires a virtual environment constructed from data or the known knowledge, and the agent learns from interacting with the virtual environment instead of the real environment. Direct RL needs to interact with the real environment, and the agent of direct RL learns from interacting with the real environment. Compared to direct RL, indirect RL can more efficiently take advantage of samples by planning [6]. In addition, the parameters of Q-learning in source tasks are partially transferred to the Q-learning of target tasks to demonstrate the transfer learning capability of the proposed approach. Considering the nonlinearity and complexity of interactions between the bucket and pile [7], the data obtained from field tests are utilized to build a nonlinear, non-parametric statistical model for predicting the state of the loader bucket in the bucket-filling process. The prediction model is used to train the Q-learning algorithm to validate the proposed algorithm.

The main contributions of this paper are summarized as follows:

(1) A data-based prediction model for the wheel loader is developed.
(2) A general automatic bucket-filling algorithm based on Q-learning is presented and the transfer ability of the algorithm is demonstrated. The proposed automatic bucket-filling algorithm does not require a dynamic model and can adapt for different changing conditions with low costs.
(3) The performance of the automatic bucket-filling algorithm and expert operators on loading two different materials is compared.

The rest of this paper is summarized below. Section 2 presents the related existing research. Section 3 states the problem and develops the prediction model. Section 4 details the experimental setup and data processing. Section 5 explains the automatic bucket-filling algorithm based on Q-learning and presents the state and reward. Section 6 discusses the experimental results and evaluates the performance of our model by comparing it with real operators. Lastly, the conclusions are drawn in Section 7.

2. Related Works

Numerous researchers have attempted to use different methods to achieve automatic bucket-filling. These studies can be summarized into the following three categories, which are: (1) physical model-based, (2) neural networks-based, and (3) reinforcement learning (RL)-based. This section will review related works in these three aspects, respectively.

Most relevant research attempted to realize automatic bucket-filling via physical-model-based control [8]. Meng et al. [9] applied Coulomb's passive earth pressure theory to establish a model of bucket force during the scooping process for load-haul-dump machines. The purpose of developing the model was to calculate energy consumption, and the trajectory was determined through optimizing the minimum energy consumption in theory. Shen and Frank [10,11] used the dynamic programming algorithm to solve the optimal control of variable trajectories based on the model of construction machinery. The control results are compared to an extensive empirical measurement done on a wheel loader. The results show that the fuel efficiency is higher compared to the fuel efficiency measured among real operators. These works require accurate machine models, so they are prone to collapse under conditions of modeling errors, wear, and change. An accurate model of the bucket-pile interaction is difficult to build because the working condition is unpredictable, and the interaction forces between the bucket and material are uncertain and changing. When the machine and materials change, the model needs to be rebuilt. Therefore, the model-based approach is not a generic automatic bucket-filling solution for various the bucket-pile environments.

In recent years, non-physical-model-based approaches [12] have been employed in the autonomous excavation of loaders and excavators. With the development of artificial intelligence, neural networks have been used in non-model-based approaches. A time-delayed neural network trained on expert operator data has been applied to execute the bucket-filling task automatically [13]. The results show that time-delayed neural network

(TDNN) architecture with input data obtained from the wheel loader successfully performs the bucket-filling operation after an initial period (100 examples) of imitation learning from an expert operator. The TDNN algorithm is used to compare with the expert operator and performs slightly worse than the expert operator with 26% longer bucket-filling time. Park et al. [14] utilized an Echo-State Networks-based online learning technique to control the position of hydraulic excavators and compensate for the dynamics changes of the excavators over time. Neural network-based approaches do not require any machine and material models. However, these approaches require a large amount of labeled data obtained from expert operators for training, which is too costly.

Reinforcement learning (RL) is capable of learning effectively through interaction with complex environments without labeled data. The learning procedure of RL includes perceiving the environmental state, taking related actions to influence the environment, and evaluating an action by the reward from the environment [15]. Reinforcement learning not only achieved surprising performance in GO [16] and Atari games [17], but has also been widely used for autonomous driving [18] and energy management [19]. The application of RL in construction machinery automation is mainly based on real-time interaction with the real or simulation environment. Hodel et al. [20] applied RL-based simulation methods to control the excavator to perform the bucket-leveling task. Kurinov et al. [21] investigated the application of an RL algorithm for excavator automation. In the proposed system, the agent of the excavator can learn a policy by interacting with the simulated model. Because simulation models are not derived from the real world, RL-based simulation cannot learn features of the real world well. Dadhich et al. [5] used RL to achieve the automatic bucket-filling of wheel loaders through real-time interaction with the real environment. However, interacting with the real environment to train the RL algorithm is costly and time-consuming.

3. Background and Modeling

3.1. Working Cycle

Wheel loaders are used to remove material (sand, gravel, etc.) from one site to another or an adjacent load receiver (dump truck, conveyor belt, etc.). Although there are many repetitive operation modes in the working process of wheel loaders, the different working cycles increase the complexity of data analysis. For wheel loaders, the representative short loading cycle, sometimes also dubbed the V-cycle, is adopted in this experiment, as illustrated in Figure 1. The single V-cycle is divided into six phases, namely, V1 forward with no load (start and approach the pile), V2 bucket-filling (penetrates the pile and load), V3 backward with full load (Retract from the pile), V4 forward and hoisting (approach to the dumper), V5 dumping, and V6 backward with no load (Retract from the dumper), as shown in Table 1. This article only focuses on the automation of the bucket-filling process (V2), which highly affected the overall energy efficiency and productivity of a complete V-cycle. The bucket-filling process (V2) accounts for 35–40% of the total fuel consumption per cycle [22]. In the bucket-filling process, the operator needs to modulate three actions simultaneously: a forward action (throttle), an upward action (lift), and a rotating action of the bucket (tilt) to obtain a large bucket weight.

Table 1. Basic parameters of the experimental wheel loader.

Parameters	Value	Units
Length	6600	mm
width	2750	mm
Height	3470	mm
Maximum traction	175	kN
rated power	162	kW
Rated load	5	t

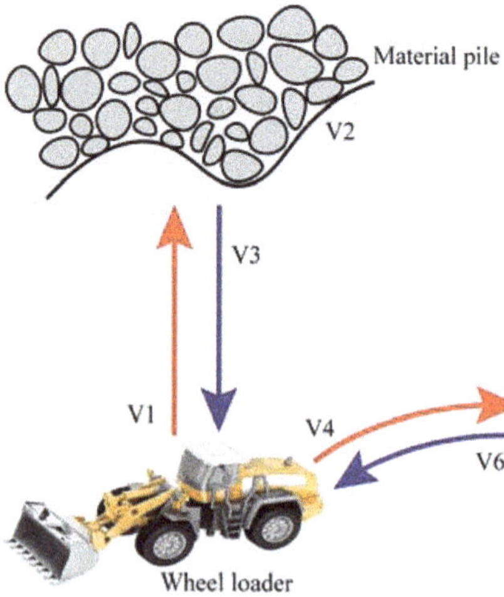

Figure 1. V-cycle of wheel loaders.

3.2. Problem Statement

The working process of scooping can be split into three stages: approach, fill, and exit the pile, as shown in Figure 2. In the first stage, wheel loaders move towards the pile of earth and the bucket penetrates the soil. In the second stage, the operator simultaneously adjusts the lift, tilt, and throttle to navigate the bucket tip through the earth pile and load as much material as possible within a short period. The throttle controls the engine speed, while the lift and tilt levers command valves in the hydraulics system that ultimately control the motion of the linkage's lift and tilt cylinder, respectively. In the third phase, the bucket is tilted until the breakout is involved and the bucket exits the pile. The scoop phase is treated as a stochastic process where the input is the wheel loader state, and the output is the action. The goal is to find a policy using RL that maps the wheel loader state to action.

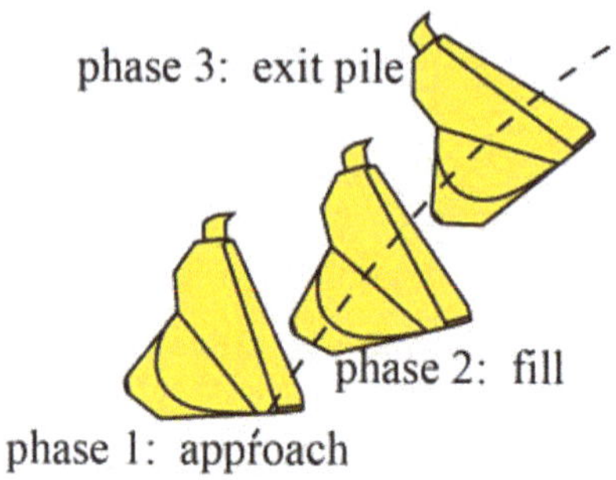

Figure 2. The three phases in the scooping process.

3.3. Prediction Model

The Markov property is a prerequisite for reinforcement learning. In the actual operation process, the operator mainly executes the next actions according to the current state of the loader. Thus, the wheel loader state of the scooping process at the next moment is considered not to be related to the past, but to the current state, which satisfies the Markov property. Therefore, the interaction between the wheel loader bucket and the continuously changing pile can be modeled as a Finite Markov decision process (FMDP) which is expressed by a quadruple $F(S, A, P, R)$, consisting of the set of possible states S, the set of available actions A, the transition probability P, and the reward R. The state

$s \in S$ includes the velocity, the tilt cylinder pressure, and the lift cylinder pressure. The actions consist of lift, tilt, and throttle commands which are all discrete. The ranges of lift, tilt, and throttle commands are from 0 to 160, 0 to 230, and 0 to 100, respectively. Besides, as the pile's shape and loaded material vary randomly, the change of the pile is considered as a stochastic process, which also satisfies the Markov property. Therefore, the problem of automatic bucket-filling for wheel loaders is considered as a finite Markov decision process (FMDP).

To achieve indirect RL, a prediction model needs to be constructed to predict the wheel loader state at the next moment according to the current state and action during the scooping phase. In this paper, changes in the wheel loader state are regarded as a series of discrete dynamic stochastic events and described with a Markov chain. The transition probability can be expressed as:

$$P(S_j|S_i) = \frac{N_{i,j}}{N_i}, \tag{1}$$

where N_{ij} is the number of times the wheel loader state transits from S_i to S_j, and N_i is the total number of times the wheel loader state transits from S_i to all possible states.

The prediction model of the wheel loader state can be expressed as:

$$P(S_j, r|S_i, a) = \frac{N_{i,j}^{a,r}}{N_i^a}, \tag{2}$$

where $P(S_j, r|S_i, a)$ denotes the probability of state transits from S_i to S_j and to get a reward r when action a is taken in state S_i, N_i^a is the total number of times the wheel loader state transits from S_i to all possible states when action a is taken, and $N_{i,j}^{a,r}$ is the total number of times the wheel loader state transits from S_i to S_j when action a is taken and gets the reward r.

Python is used to construct the prediction model. We read the experimental data in sequence. The current state S_t and action a are stored as a key of the Python dictionary, and the value corresponding to the key is another dictionary whose keys are the next state S_{t+1} and reward r, and values are $P(S_{t+1}, r|S_t, a)$. According to the current state S_t and action a, the next state S_{t+1} and reward r are selected randomly with probability.

The prediction model can approximate the real working environment, as it is built using the real data obtained from tests. Besides, the prediction model not only covers the working information of wheel loaders, but also reflects the environmental effect. The sampling frequency is important because the complexity of the model can be controlled by adjusting the sampling frequency. The high sampling frequency will increase the complexity of the model and the computation load, while the low sampling frequency might cause model distortion.

4. Experiment and Sampling

4.1. Experimental Setup

The experimental wheel loader is shown in Figure 3. it is equipped with pressure sensors, displacement sensors, and GPS. The Liugong ZL50CN wheel loader is taken as the experiment machine. The basic parameters of the wheel loader are listed in Table 1. In order to verify whether the proposed automatic bucket-filling algorithm can converge to the optimal strategy on the data model based on different piles, we collected data from two types of piles, which are shown in Figure 4. It has been proven that the Q-learning algorithm with lookup tables are guaranteed to converge to the optimal solution. Small coarse gravel (SCG) mainly contains particles up to 25 mm, while medium coarse gravel (MCG) mainly contains particles up to 100 mm.

Figure 3. Experimental wheel loader.

Figure 4. Two types of piles. (**a**) Small coarse gravel. (**b**) Medium coarse gravel.

4.2. Data Acquisition and Processing

According to the working characteristics of wheel loaders in a working cycle, the V-cycle is divided by extracting the working condition features of the actuator and walking device, including suspension, axle housing, tires, and rims. The mapping between the collected data and the working state is realized by dividing the V-cycle, as shown in Figure 5. The data in the scooping phase were selected to develop the prediction model. There was no benchmark dataset. For different piles, we collected 51 sets of data to build prediction models. The sampling frequency was 200 HZ. Due to the high dimension of the state vector, it is difficult to present the complete prediction model in a figure. Therefore, transition probability maps for the individual elements of the state vector are drawn and transition probability maps of different materials are similar, as shown in Figures 6 and 7. Figures 6 and 7 based on Equation (1) reflect the virtual environment built from real data and the changing trend of states corresponding to different piles at the next moment.

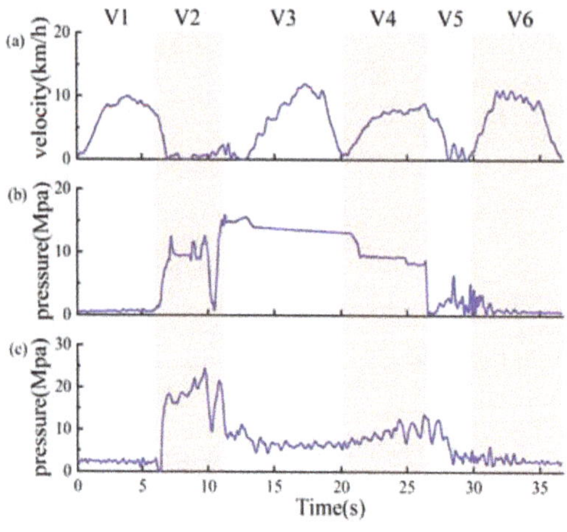

Figure 5. Schematic diagram of working condition division. (**a**) Velocity of wheel loader. (**b**) Lift cylinder pressure. (**c**) Tilt cylinder pressure.

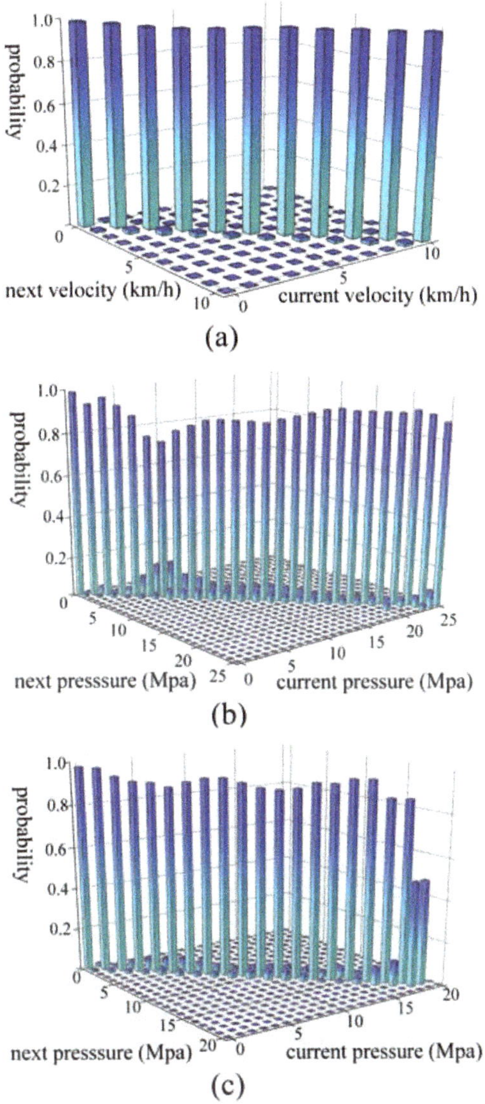

Figure 6. Transition probability for small coarse gravel. (**a**) Velocity of wheel loader. (**b**) Lift cylinder pressure. (**c**) Tilt cylinder pressure.

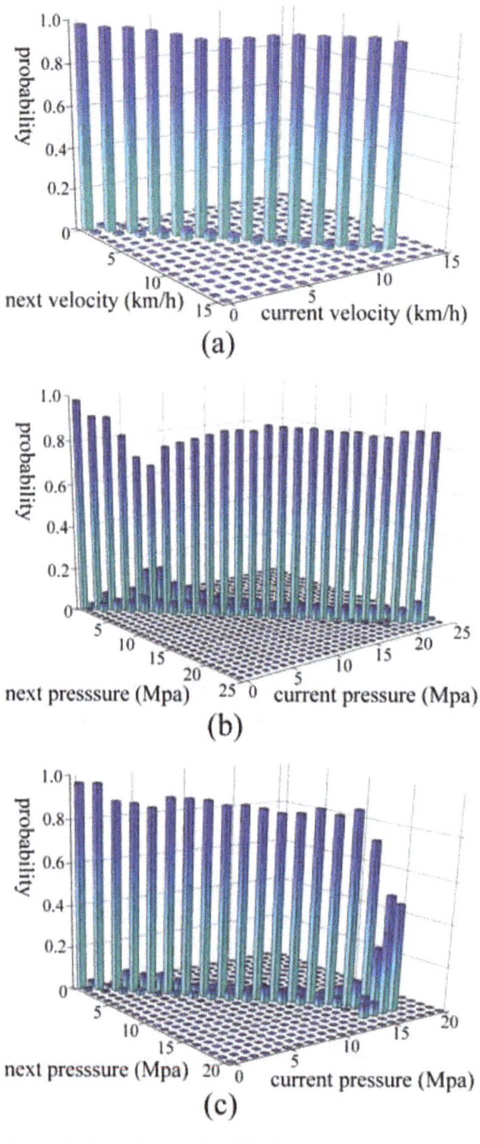

Figure 7. Transition probability for medium coarse gravel. (**a**) Velocity of wheel loader. (**b**) Lift cylinder pressure. (**c**) Tilt cylinder pressure.

5. Automatic Bucket-Filling Algorithm

5.1. Automatic Bucket-Filling Algorithm Based on Q-Learning

Reinforcement learning is learning what to do—how to map situations to actions—so as to maximize the expected discounted long-term reward. The two most important distinguishing features of reinforcement learning are trial-and-error search and delayed reward. The learner and decision-maker are called the agent. At each time-step, the agent takes action a according to the current environmental state S_t and the policy π which is a mapping from perceived states to actions. Therefore, as a consequence of action, the environmental state transits from S_t to S_{t+1} and the agent gets a reward r. The agent and environment generate the trajectories $(S_1; A_1; R_1)$, $(S_2; A_2; R_2)$, ..., $(S_T; A_T; R_T)$ [23], until an episode is over. The basic architecture of RL is shown in Figure 8.

Q-learning is a widely used RL algorithm. Similar to other classical RL methods, the goal of Q-learning is to obtain an optimal policy that maximizes the long-term reward. In the Q-learning algorithm, the agent receives the reward and updates the Q-function corresponding to the action-state. The Q-function represents the expected estimated accumulated reward for the action-state pair under a policy. For example, $Q(S_t, A_t)$ is denoted as the expected long-term reward starting from state S_t, taking action A_t. By continuous exploitation and exploration, the agent will eventually obtain the optimal Q-function (Q_*) which determines the action selection policy. The optimal policy $\pi^*(S_t)$ can be calculated by the following equation:

$$\pi^*(S_t) = \arg\max_{a \in A} Q_*(S_t, A), \tag{3}$$

where $Q_*(S_t, A)$ is the maximum Q-function over all policies. The optimal policy $\pi^*(S_t)$ is to select the action that maximizes the $Q_*(S_t, A)$.

The Bellman equation of the optimal Q-function (Q_*) is:

$$Q_*(S_t, A_t) = \sum_{S_{t+1}, r} P(S_{t+1}, r|S_t, A_t)\left[r + \gamma \max_{A_{t+1}} Q_*(S_{t+1}, A_{t+1})\right], \tag{4}$$

where $\gamma \in [0, 1]$ is the discount factor that determines the present value of future rewards.

The Q-learning algorithm is designed by the Bellman equation and contraction mapping theorem. Q-learning is defined by

$$Q(S_t, A_t) \leftarrow Q(S_t, A_t) + \alpha\left[R_{t+1} + \gamma \max_a Q(S_{t+1}, a) - Q(S_{t+1}, A_t)\right], \tag{5}$$

where $\alpha \in [0, 1]$ is the learning rate which reflects the influence of the new experience on the current estimation $Q(S_t, A)$.

Q-learning starts with an initial $Q(S_1, A_1)$ for each state-action pair. At each time-step, the agent performs an action based on a commonly used exploration method ε greedy strategy that selects the greedy action with probability $1 - \varepsilon$, but every once in a while, it selects randomly from all the actions with equal probability ε independently of the action-value estimates. Each time an action a is taken in state s, then the reward r is fed back from the environment and the next state s' is observed, thus the Q-value is updated with a combination of its current value and the Temporal-Difference Error (TDE)(date-drive). The pseudo-code of the Q-learning algorithm is shown in Figure 9. The code can be found in supplementary materials.

In this study, Q-learning based on a prediction model was used to optimize the choice of actions. The Q-learning architecture in automatic bucket-filling is illustrated in Figure 10. By using the real data, the environmental characteristics can be abstracted into the prediction model. Q-learning is trained on the prediction model until the algorithm converges. By interacting with the prediction model built from the collected real data, the agent is able to learn the working characteristics of wheel loaders and the optimal strategy. Based on the prediction model, the Q-function can be updated via the learning process of the agent. To investigate the transfer ability of the proposed algorithm, the automatic bucket-filling algorithm is first trained on a bucket-pile interaction model and then the Q-function learned from the previous model is transferred to the target task to enhance the learning efficiency and the learning rate on the bucket-pile interaction model of the target task.

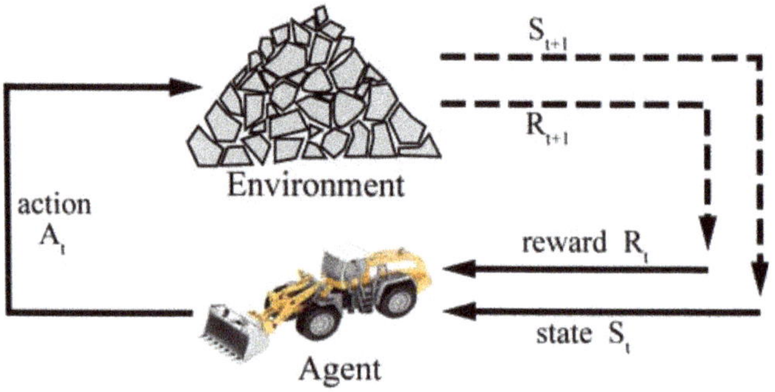

Figure 8. Basic architecture of RL.

```
Initialize Q (s, a) arbitrarily
Repeat for each episode:
    Initialize s
    Repeat for each step of episode
        take action a at the state s according to the policy π (ε-greedy )
        Perform action a
        Observe reward r and s'
        Q(s,a) ← Q(s,a) + α[ r + γ max_a Q(s',a) − Q(s,a)]
        s ← s'
    end
end
```

Figure 9. Pseudo-code of the Q-learning algorithm.

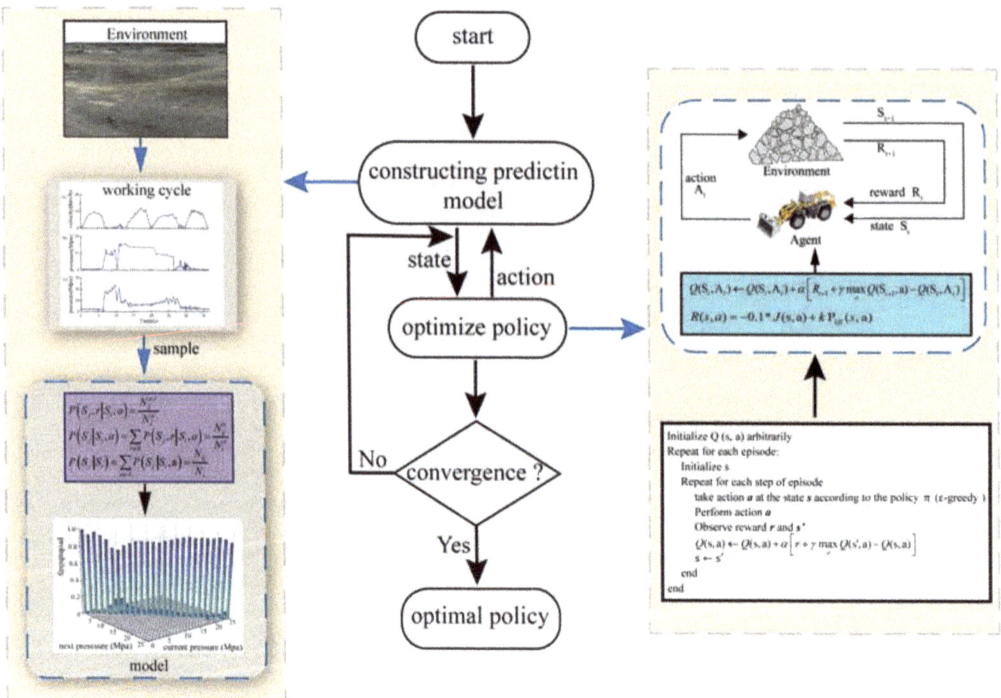

Figure 10. Prediction model-based Q-learning architecture in automatic bucket-filling.

5.2. State and Reward Representation

The appropriate state and reward function should be set to optimize action selection. The state needs to be able to reflect the characteristics of the environment when the agent interacts with the environment and the dimensionality should not be too high to avoid the curse of dimensionality. Lift force is the most important feature affecting the lift and tilt commands [13] and lift force is related to the lift cylinder pressure. Besides, the velocity of wheel loaders is of significance to the choice of throttle command. The tilt cylinder pressure can be used as a redundant feature. Thus, in this study, we defined the state using a three-dimensional vector consisting of velocity, tilt cylinder pressure, and lift cylinder pressure, which are expressed as:

$$S = \left\{ s = \left[V_{loader}, P_{lift}, P_{tilt} \right] \right\}, \tag{6}$$

where V_{loader} is the velocity of wheel loaders, P_{lift} is the lift cylinder pressure, and P_{tilt} is the tilt cylinder pressure.

It can be seen from Figure 11 that the bucket-soil interaction force mainly depends on the amount of loaded soil. Therefore, the bucket-soil interaction force can directly translate into the amount of loaded soil and is important for the bucket-filling of wheel loaders. In order to encourage the loader to improve the bucket digging force and fuel economy during the training process, the reward function should be composed of the negative value of fuel consumption and the bucket-pile interaction force. Because the bucket-soil interaction force is difficult to measure directly and has a positive correlation with lift cylinder pressure, we use the lift cylinder pressure as a part of the reward function to represent digging force. The bigger bucket digging force demands increased fuel consumption. Therefore, a trade-off is necessary between the fuel consumption and bucket digging force. The reward function is expressed as follows:

$$R(s,a) = -0.1 * J(s,a) + kP_{lift}(s,a), \tag{7}$$

where $J(s,a)$ is the fuel consumption from the current state s to the next state s' when the agent takes action a, $P_{lift}(s,a)$ is the lift cylinder pressure of next state s', and k is a constant to control the priority of the fuel economy and bucket weight and $k = 0.2$.

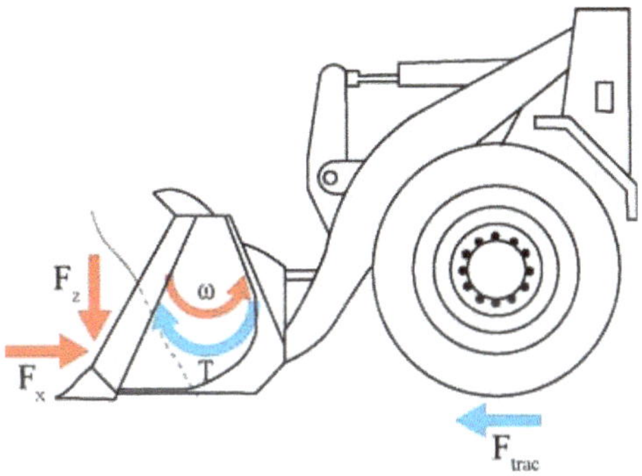

Figure 11. Schematic picture of the forces of the bucket.

6. Results and Discussions

In this section, the proposed automatic bucket-filling algorithm is utilized to learn the policy on prediction models and present the results. We choose 0.15 as the learning rate of Q-learning, ε in the ε greedy policy is 0.1, and the discount factor γ = 0.15.

Wheel loaders have a complicated working environment and are used to transport different materials. In order to verify the convergence of the algorithm on the diverse environment, the reward curves based on different prediction models are depicted in Figure 12. It can be observed that the proposed automatic bucket-filling algorithm can converge to the optimal policy that maximizes reward, indicating that the agent was learning the policy correctly under different prediction models. This shows that the proposed algorithm can be adapted to different bucket-pile models, thus dealing with the complex and changing working environment of wheel loaders in the absence of a complex dynamic model.

The digging force reward per episode is obtained by accumulating the lift cylinder pressure of each step and is used to approximate the change of digging force. As can be seen from Figure 13, compared with the algorithm interacting with the small coarse gravel model, the algorithm interacting with the medium coarse gravel model can converge to a smaller value of digging force reward. A larger digging force usually leads to higher fuel consumption. Thus, loading small coarse gravel has higher fuel consumption compared to medium coarse gravel on this data-based prediction model, as shown in Figure 14. This finding suggests that the prediction model can truly reflect the interaction between the bucket and the material to a certain extent.

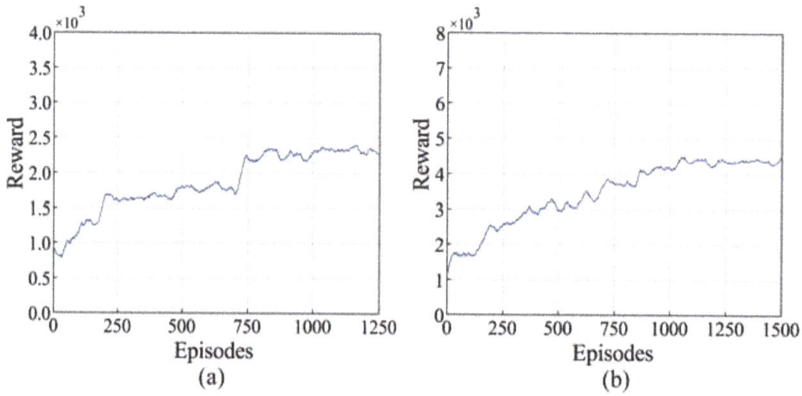

Figure 12. Reward per episode on different prediction models. (**a**) Medium coarse gravel. (**b**) Small coarse gravel.

The results of fuel consumption of the agent in different models are shown in Figure 14. The data used to build the model come from the real environment. The wheel loader operated by a human operator is the same as the wheel loader used to obtain the data. In addition, the working environment of the wheel loader is also the same. Therefore, the agent we trained has the same operating object and operating environment as the human operator. A comparison with humans is used as a generally accepted method of machine learning algorithm testing [11,13]. Physical-model-based methods require a physical model. However, the diversity between the physical model and the wheel loader used to obtain the data is great. In addition, the environment constructed for the physical model is also very different from the environment constructed in the article. Therefore, physical-model-based methods and the method proposed in this article have different operating objects and environments. In addition, deep learning-based methods mainly predict actions based on previous actions and states. As deep learning-based methods mainly solve the prediction problem, root mean square error (RMSE) is used as the evaluation indicator, which is different from our paper. Therefore, the fuel consumption measured by the human operator is used to compare with the fuel consumption of the agent. Table 2 shows the average fuel consumption of loading different piles and the variance of fuel consumption in the recorded bucket-filling phase. In Figure 14b, there is a relatively stable convergence, while in Figure 14a, the curve fluctuates violently. A possible explanation for this is that the prediction model built by data with higher variance is more complex and variable. Therefore, the agent will encounter more situations in each episode, resulting in the oscillation of the convergence curve. In addition to this, the convergence values of fuel consumption of agents on medium coarse gravel model and small coarse gravel model are around 33.3 mL and 45.6 mL, respectively, and improve by 8.0% and 10.6% compared to the average fuel consumption measured by real operators because Q-learning can learn the optimal action in different states.

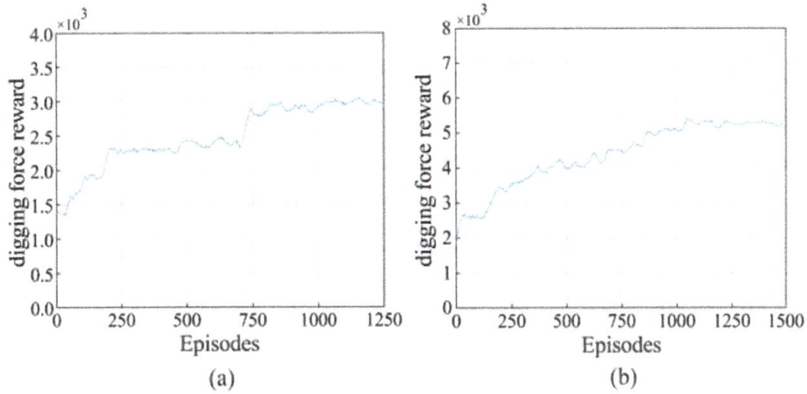

Figure 13. Digging force reward of agent on different prediction models. (**a**) Medium coarse gravel. (**b**) Small coarse gravel.

Table 2. The comparison of fuel consumption.

Pile	Average Fuel Consumption of Recorded Bucket-Filling Phase (mL)	Variance	Convergence Value (mL)	Improvement (%)
medium coarse gravel	36.2	61.6	33.3	8.0
small coarse grave	51.0	42.8	45.6	10.6

The transfer learning ability can help the algorithm to improve the learning performance on new bucket-filling tasks, thereby saving training costs. In this paper, transfer Q-learning refers to the Q-learning that has been trained in other tasks and learned relevant knowledge.

Figure 15 shows the convergence curve of rewards for Q-learning and transfer Q-learning in the different bucket-pile interaction models. In Figure 15a, Q-learning is only trained on the MCG-pile model, and transfer Q-learning is first trained on the SCG-pile model and then trained on the MCG-pile model. In Figure 15b, Q-learning is only trained on the SCG-pile model, and transfer Q-learning is first trained on the MCG-pile model and then trained on the SCG-pile model. The convergence rate (learning rate efficient) of Q-learning and transfer Q-learning in two bucket-pile interaction models are compared, as presented in Table 3. Using Q-learning as the benchmark, the convergence speed of transfer Q-learning in the medium coarse gravel and small coarse gravel model is improved by 30.3% and 34.1%, respectively. This means that the proposed algorithm has a good transfer learning capability. This improvement can be ascribed to the fact that Q-learning stores the learned knowledge in the Q-function and the transfer Q-learning transfers the Q-function learned from the source task to the Q-function of the target task. Therefore, the agent no longer needs to learn the basic action characteristics in the bucket-filling phase.

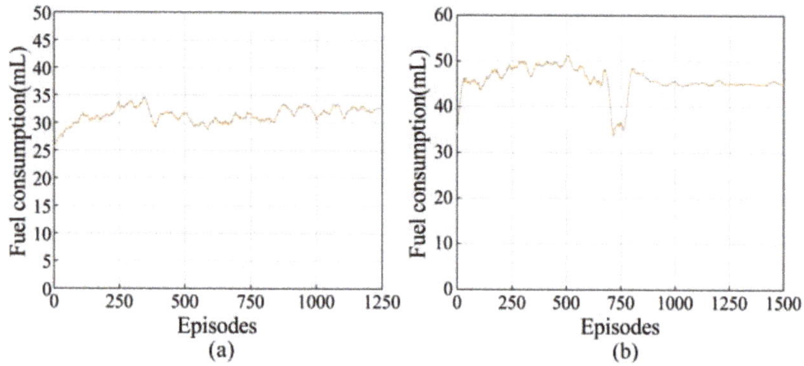

Figure 14. Fuel consumption of agent on different interaction prediction models. (**a**) Medium coarse gravel. (**b**) Small coarse grave.

When the two piles have similar characteristics, such as in category and shape, transfer Q-learning might have better performance in the new bucket-filling task due to the similarity of the optimal Q-function in two tasks [24]. Finally, the amount of data used to build the interaction model also has an impact on the performance of transfer Q-learning on the prediction model. The more data there is, the more states and actions the developed prediction model contains. Therefore, different prediction models have more identical states and actions, and the Q-function of the target task can learn more knowledge from the Q-function of the source task. However, the transfer learning method potentially does not work or even harm the new tasks [25] when the piles or environment are greatly different.

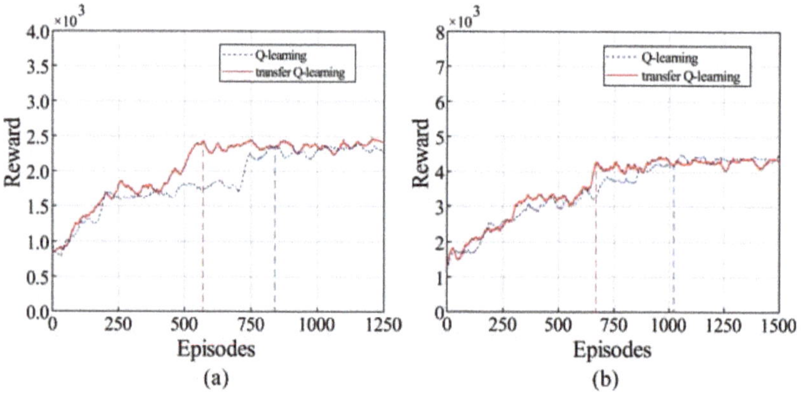

Figure 15. Comparison between Q-learning and transfer Q-learning. (**a**) Q-learning is only trained on the MCG-pile interaction prediction model, and transfer Q-learning is first trained on the SCG-pile interaction prediction model and then trained on the MCG-pile interaction prediction model. (**b**) Q-learning is only trained on the SCG-pile model, and transfer Q-learning is first trained on the MCG-pile model and then trained on the SCG-pile interaction prediction model.

Table 3. The comparison of Q-learning and transfer Q-learning on convergence speed.

Pile	Convergence Episode of Q-Learning	Convergence Episode of Transfer Q-Learning	Improvement (%)
medium coarse gravel	825	575	30.3
small coarse grave	1025	675	34.1

7. Conclusions

This paper investigated the automatic bucket-filling algorithm based on RL for wheel loaders, and the algorithm was tested. The data-driven prediction model was established using previously obtained excavation data of two piles. The transfer Q-learning-based automatic bucket-filling algorithm was proposed, and the algorithm was trained on the prediction model. The results of training show that the proposed algorithm has good performance of adaptability and convergence even without parameters of wheel loaders. Moreover, the proposed algorithm has good performance in fuel consumption, with 8.0% and 10.6% reduction compared to the average fuel consumption measured by real operators on two piles. Transfer learning is used to transfer the parameter of Q-learning in the source task to the target task. The results show the promising performance of the proposed method on an automatic bucket-filling task. The proposed data-driven RL-based approach in this paper has generality, which means that this approach can be applied to different machine-pile environments. Furthermore, compared to most previous solutions for the automation of bucket-filling, the approach proposed in this paper does not require a dynamic model and has the advantages of no direct interaction with the real environment and transfer ability. In future research, the method proposed in this paper will be applied to the real wheel loaders and compared with other methods to further enhance the performance of the reinforcement-learning-based automatic bucket-filling algorithm.

Supplementary Materials: The following are available online at https://www.mdpi.com/article/10.3390/app11199191/s1.

Author Contributions: Conceptualization, J.H. and D.K.; methodology, J.H. and G.G.; software, J.H.; validation, J.H. and X.C.; formal analysis, D.K.; investigation, X.C.; resources, J.C.; data curation, G.G. and X.C.; writing—original draft preparation, J.H.; writing—review and editing, J.C. and D.K.; visualization, X.C. and G.G.; supervision, G.G. and J.C.; project administration, J.C.; funding acquisition, J.C. and D.K. All authors have read and agreed to the published version of the manuscript.

Funding: This research was funded by the National Natural Science Foundation of China grant number 51875239.

Institutional Review Board Statement: Not applicable.

Informed Consent Statement: Not applicable.

Conflicts of Interest: The authors declare no conflict of interest. The funders had no role in the design of the study; in the collection, analyses, or interpretation of data; in the writing of the manuscript, or in the decision to publish the results.

References

1. Bogue, R. What are the prospects for robots in the construction industry? *Ind. Robot.* **2018**, *45*, 1–6. [CrossRef]
2. Bobbie, F.; Lennart, S.; Reno, F.; Anders, F. On Increasing Fuel Efficiency by Operator Assistant Systems in a Wheel Loader. In Proceedings of the International Conference on Advanced Vehicle Technologies and Integration (VTI 2012), ChangChun, China, 16–19 July 2012; pp. 155–161.
3. Dadhich, S.; Bodin, U.; Andersson, U. Key challenges in automation of earth-moving machines. *Autom. Construct.* **2016**, *68*, 212–222. [CrossRef]
4. Dadhich, S.; Bodin, U.; Sandin, F.; Andersson, U. From Tele-Remote Operation to Semi-Automated Wheel-Loader. *Int. J. Electr. Electron. Eng. Telecommun.* **2018**, 178–182. [CrossRef]
5. Dadhich, S.; Sandin, F.; Bodin, U.; Andersson, U.; Martinsson, T. Adaptation of a wheel loader automatic bucket-filling neural network using reinforcement learning. In Proceedings of the International Joint Conference on Neural Networks, Glasgow, UK, 19–24 July 2020; pp. 1–9.
6. Shitole, V.; Louis, J.; Tadepalli, P. Optimizing Earth Moving Operations Via Reinforcement Learning. In Proceedings of the 2019 Winter Simulation Conference (WSC), National Harbor, MD, USA, 8–11 December 2019; pp. 2954–2965.
7. Sandzimier, R.J.; Asada, H.H. A Data-Driven Approach to Prediction and Optimal Bucket-Filling Control for Autonomous Excavators. *IEEE Robot. Autom. Lett.* **2020**, *5*, 2681–2688. [CrossRef]
8. Feng, H.; Yin, C.; Ma, W.; Yu, H.; Cao, D. Parameters identification and trajectory control for a hydraulic system. *ISA Trans.* **2019**, *92*, 228–240. [CrossRef] [PubMed]

9. Meng, Y.; Fang, H.; Liang, G.; Gu, Q.; Liu, L. Bucket Trajectory Optimization under the Automatic Scooping of LHD. *Energies* **2019**, *12*, 3919. [CrossRef]
10. Shen, W.; Jiang, J.; Su, X.; Reza Karimi, H. Control strategy analysis of the hydraulic hybrid excavator. *J. Frankl. Inst.-Eng. Appl. Math.* **2015**, *352*, 541–561. [CrossRef]
11. Frank, B.; Kleinert, J.; Filla, R. Optimal control of wheel loader actuators in gravel applications. *Autom. Construct.* **2018**, *91*, 1–14. [CrossRef]
12. Fernando, H.; Marshall, J.A.; Larsson, J. Iterative Learning-Based Admittance Control for Autonomous Excavation. *J. Intell. Robot. Syst.* **2019**, *96*, 493–500. [CrossRef]
13. Dadhich, S.; Sandin, F.; Bodin, U.; Andersson, U.; Martinsson, T. Field test of neural-network based automatic bucket-filling algorithm for wheel-loaders. *Autom. Construct.* **2019**, *97*, 1–12. [CrossRef]
14. Park, J.; Lee, B.; Kang, S.; Kim, P.Y.; Kim, H.J. Online Learning Control of Hydraulic Excavators Based on Echo-State Networks. *IEEE Trans. Autom. Sci. Eng.* **2017**, *14*, 249–259. [CrossRef]
15. Sutton, R.S.; Barto, A.G. *Reinforcement Learning: An Introduction*; MIT Press: Cambridge, MA, USA, 1998.
16. Schrittwieser, J.; Antonoglou, I.; Hubert, T.; Simonyan, K.; Sifre, L.; Schmitt, S.; Guez, A.; Lockhart, E.; Hassabis, D.; Graepel, T.; et al. Mastering Atari, Go, chess and shogi by planning with a learned model. *Nature* **2020**, *588*, 604–609. [CrossRef] [PubMed]
17. Mnih, V.; Kavukcuoglu, K.; Silver, D.; Rusu, A.A.; Veness, J.; Bellemare, M.G.; Graves, A.; Riedmiller, M.; Fidjeland, A.K.; Ostrovski, G.; et al. Human-level control through deep reinforcement learning. *Nature* **2015**, *518*, 529–533. [CrossRef] [PubMed]
18. Zhu, M.; Wang, X.; Wang, Y. Human-like autonomous car-following model with deep reinforcement learning. *Transp. Res. Pt. C-Emerg. Technol.* **2018**, *97*, 348–368. [CrossRef]
19. Zhang, W.; Wang, J.; Liu, Y.; Gao, G.; Liang, S.; Ma, H. Reinforcement learning-based intelligent energy management architecture for hybrid construction machinery. *Appl. Energy* **2020**, *275*, 115401. [CrossRef]
20. Hodel, B.J. Learning to Operate an Excavator via Policy Optimization. *Procedia Comput. Sci.* **2018**, *140*, 376–382. [CrossRef]
21. Kurinov, I.; Orzechowski, G.; Hamalainen, P.; Mikkola, A. Automated Excavator Based on Reinforcement Learning and Multibody System Dynamics. *IEEE Access* **2020**, *8*, 213998–214006. [CrossRef]
22. Filla, R. Representative testing of emissions and fuel consumption of working machines in reality and simulation. In Proceedings of the SAE 2012 Commercial Vehicle Engineering Congress, Rosemont, IL, USA, 2–3 October 2012; SAE Technical Paper Series; Volume 8.
23. Azulay, O.; Shapiro, A. Wheel Loader Scooping Controller Using Deep Reinforcement Learning. *IEEE Access* **2021**, *9*, 24145–24154. [CrossRef]
24. Wang, Y.; Liu, Y.; Chen, W.; Ma, Z.M.; Liu, T.Y. Target transfer Q-learning and its convergence analysis. *Neurocomputing* **2020**, *392*, 11–22. [CrossRef]
25. Spector, B.; Belongie, S. Sample-Efficient Reinforcement Learning through Transfer and Architectural Priors. *arXiv* **2018**, arXiv:1801.02268.

applied sciences

MDPI

Article

Hoek-Brown Failure Criterion-Based Creep Constitutive Model and BP Neural Network Parameter Inversion for Soft Surrounding Rock Mass of Tunnels

Chao Chen, Tianbin Li *, Chunchi Ma *, Hang Zhang, Jieling Tang and Yin Zhang

State Key Laboratory of Geohazard Prevention and Geoenvironment Protection, College of Environment and Civil Engineering, Chengdu University of Technology, Chengdu 610059, China; xchenchaox@126.com (C.C.); zhanghang_nn720@163.com (H.Z.); tangjieling@stu.cdut.edu.cn (J.T.); zhangyin0503@gmail.com (Y.Z.)
* Correspondence: ltb@cdut.edu.cn (T.L.); machunchi17@cdut.edu.cn (C.M.)

Abstract: This paper summarizes the main factors affecting the large deformation of soft rock tunnels, including the lithology combination, weathering effect, and underground water status, by reviewing the typical cases of largely-deformed soft rock tunnels. The engineering geological properties of the rock mass were quantified using the rock mass block index (*RBI*) and the absolute weathering index (*AWI*) to calculate the geological strength index (*GSI*). Then, the long-term strength σ_r and the elastic modulus E_0 of the rock mass were calculated according to the Hoek–Brown failure criterion and substituted into the creep constitutive model based on the Nashihara model. Finally, the creep parameters of the surrounding rock mass of the Ganbao tunnel were inverted and validated by integrating the on-site monitoring and BP neural network. The inversion results were consistent with the measured convergence during monitoring and satisfied the engineering requirements of accuracy. The method proposed in this paper can be used to invert the geological parameters of the surrounding rock mass for a certain point, which can provide important mechanical parameters for the design and construction of tunnels, and ensure the stability of the surrounding rock mass during the period of construction and the safety of the lining structure during operation.

Keywords: tunnel engineering; soft rock; creep parameter; parameter inversion; BP neural network

Citation: Chen, C.; Li, T.; Ma, C.; Zhang, H.; Tang, J.; Zhang, Y. Hoek-Brown Failure Criterion-Based Creep Constitutive Model and BP Neural Network Parameter Inversion for Soft Surrounding Rock Mass of Tunnels. *Appl. Sci.* **2021**, *11*, 10033. https://doi.org/10.3390/app112110033

Academic Editor: Nikos D. Lagaros

Received: 16 August 2021
Accepted: 19 October 2021
Published: 26 October 2021

1. Introduction

Tunnel projects in western China often encounter soft rocks with well-developed bedding, such as carbonaceous phyllite, sericite phyllite, schist, carbonaceous slate, sandy slate, and carbonaceous shale. Under high in situ stress, the laminated soft surrounding rock masses are vulnerable to large and rapid deformation and local destruction [1–3]. In these conditions, tunnels frequently suffer from large deformation hazards.

Laminated rock mass, frequently seen in engineering practice and presenting oriented grouped bedding, has more heterogeneous mechanical properties than normal rock mass. Many researchers have developed constitutive models for laminated surrounding rock masses. For instance, Jia et al. [4] applied the microscopic element method with the constitutive model based on damage mechanics and statistical theory to the simulation of rock tunnel stability using the finite element method (FEM). Li et al. [5] developed a three-dimensional creep constitutive model for transversely isotropic rock mass, based on the Burgers viscoelastic model. Li et al. [6] proposed three basic creep patterns of shale and a general methodology for developing the anisotropic creep model. These constitutive models have been used to probe the deformation and failure characteristics of laminated rock masses with varied dip angles. However, research on the constitutive theory of laminated surrounding rock masses in tunnels is fairly limited and cannot provide the theoretical guidance for the engineering design of actual tunnels. In terms of the mechanisms of laminated rock mass deformation, previous studies have mainly investigated patterns of stability [7], deformation [8–10], mechanical behavior [11], and mechanical properties [12]. The main triggers for large tunnel deformation through highly-dipped laminated soft rock under high in situ stress are high structural stress, unfavorable rock occurrences, and low rock strength [13]. On-site tunnel monitoring has shown that the deformation–failure zone of the surrounding rock mass of laminated soft rock tunnels is concentrated along the direction perpendicular to the rock bedding, instead of the direction of the maximum principal stress; the local deformation of laminated soft rock tunnels is affected by the topography, rock mass structure, and in situ stress [14]. Four mechanisms have been proposed to explain large tunnel deformation, namely, along-bedding sliding, flexural deformation, toppling deformation, and plastic extrusion. For each mechanism, a specific anchoring support plan should be developed [15]. Numerical simulation has been performed to comprehensively analyze the stress environment, the deviator stress distribution [16–19], and the plastic zone distribution [20,21] of the surrounding rock mass during tunnel excavation. Such simulations, incorporating the rock characteristics of deterioration, stabilization, and accelerated creep [22], reveal the characteristics of the surrounding rock mass failure with varied surrounding rock mass strengths, buried depths, and roof strengths [23], which provide guiding values to support the design of the analogous tunnels.

Due to the complexity of geology, geotechnical materials are typically characterized by discontinuity and heterogeneity, and the rheological parameters of the actual surrounding rock mass of the tunnel are often hard to measure. There are two main methods to obtain creep parameters: one is to calculate creep parameters through laboratory testing of rock combined with a creep constitutive model, but there is a large deviation between the laboratory results for rock and on-site rock mass parameters. Second, to obtain the mechanical parameters of the surrounding rock mass through displacement back analysis based on monitoring data [24]. Wenzheng Cao [25] proposed a novel back analysis program based on a BP neural network, which can realize automatic correction and the adjustment of parameters and adapt to most tunnel projects. Qingdong Wu [26] introduced a support vector machine (SVM) and an artificial neural network (ANN) to predict tunnel surrounding rock mass displacement, and compared and analyzed the results of the two methods. Xianghui Deng [27] established a tunnel risk assessment model by combining a fuzzy method with a BP neural network based on historical data from 50 tunnels. Numerical simulation has been widely combined with back-propagation (BP) neural network analysis to calculate the physical and mechanical parameters of surrounding rock mass, which has achieved good application performance [28–30].

This paper proposed a solution for obtaining the creep parameters of the surrounding rock mass in highway tunnels after excavation. First, the long-term strength of the engineering rock mass was obtained using the Hoek–Brown failure criterion, and a creep constitutive model for the rock mass was developed. Then, back-propagation analysis was performed using the on-site monitoring data to obtain the initial values of the creep parameters of the tunnel's surrounding rock mass. Subsequently, training samples were generated via orthogonal experiments. Then, a BP neural network and numerical simulation were integrated to inversely calculate the tunnel's surrounding rock mass creep parameters that matched the reality.

2. Factors Affecting the Deformation of Soft Rock Tunnels under High In Situ Stresses

The engineering data from some typical largely-deformed soft rock tunnels in China are summarized in Table 1, including the buried depth, formation lithology, in-situ stress, deformation magnitude, and characteristics of deformation and failure.

Table 1. Deformation-failure characteristics of tunnels' surrounding rock masses and support structures under complex geological conditions in western China.

Tunnel Name	Tunnel Length /m	Buried Depth /m	Lithology	Max Principal Stress σ_1^{max} /MPa	Strength of Rock /MPa	Maximum Deformation /mm	Deformation Characteristics	Strain ε = Tunnel Closure/Tunnel Diameter × 100
Maoxian Tunnel, Chenglan Railway	25,000	675	Sericite phyllite, carbonaceous phyllite	27.52	1.95	510	Compressive large deformation occurs, with prolonged deformation growth and notable time-dependency; horizontal convergence exceeds crown settlement, due to steep inclination of the surrounding rock mass	8.05
Wuqiaoling Tunne, Lanxin Railway	20,050	1100	Phyllite, slate	32.8	0.7~2.5	1209	It penetrates the compressive fault; the overall stability of the surrounding rock mass is low; large deformation, early rapid deformation growth, and prolonged duration of deformation are observed due to intensive compression.	9.69
Zhegushan Tunnel, 317 National Rd.	4423	1000	Thin layers of carbonaceous phyllite	17~20	12	300	The relatively large magnitude and prolonged duration of surrounding rock mass deformation are manifested as support breakage, steel arch twisting, and their intrusion into tunnel clearance; the tunnel is prone to collapse.	9.31
Maoyushan Tunnel, Lanyu Railway	8503	700	Thin bedded slate	21.28	5.63	540	Large rapid deformation, with notable rheological effects; severe twisting and fracturing of the steel arch; horizontal convergence far larger than crown settlement	8.94
Gonghe Tunnel, Yusha Expressway	4779	1000	Sandy shale	29.86	11.4	200	Longitudinal cracking and steel frame bending occurs at the initial support of the right spandrel and left arch foot, indicating severe biased compression of the surrounding rock mass	7.52

Surrounding rock masses that are found with large deformations are mostly soft rock with a highly developed bedding texture, such as shale, slate, and phyllite (Table 1). The bedding plane profoundly affects the strength and deformation characteristics of tunnel-surrounding rock mass and is considered the key factor determining the mechanical behavior and deformation-failure characteristics of anisotropic rock mass. In general, the compressional deformation of the surrounding rock mass of tunnels in laminated soft rock occurs mainly at the two sides along the normal direction of the weakest bedding plane. Moreover, the smaller the bedding spacing is, the worse the integrity and self-stabilizing ability of the surrounding rock mass are, which leads to increased proneness to local large deformation and thus a relatively large compressional load on the support structure and highly-deformed lining and surrounding rock mass in local areas. The factors affecting the large deformation of the surrounding rock mass can be summarized as follows:

Based on lithology, the surrounding rock mass can be grouped into two types, namely, plastic and brittle rock. The former consists of soft rock and is typically characterized by low mechanical strength, softening and swelling on wetting, and low slaking resistance, which are all unfavorable for the load-bearing and stability of the tunnel's surrounding rock mass. On the contrary, the latter is of hard rock and generally has a high load-bearing capacity, low softening and swelling tendencies on wetting, and high slaking resistance, which are favorable for the load-bearing and stability of the tunnel's surrounding rock mass. The lithology combination can be generally summarized into four cases, namely, the consistent lithology, hard interbeds (in much softer surrounding rock mass), soft interbeds (in much harder surrounding rock mass), and soft–hard alternating lithology. Different lithology combinations are associated with varied deformation strengths. A soft interbed in the surrounding rock mass is equivalent to a weak part in the surrounding rock mass. In most cases, deformation-failure occurs first in the soft interbed during tunnel excavation. If a hard interbed exists in the surrounding rock mass, it will often suppress the rock deformation, due to its more rigid mechanical properties than those of the surrounding rock mass. Therefore, the deformation magnitude is often relatively small at the hard interbed.

The weathering effect is a common geological phenomenon in nature, which can weaken the cementation between rock particles, form damage cracks, reduce surface roughness, and worsen the physical and mechanical properties of rock. Rapid weathering of soft rock in tunnels will reduce the stability of the surrounding rock mass and affect the safety of the tunnel construction. After the tunnel excavation, the weak layered surrounding rock mass will form internal cracks due to the unloading effect; then, an effective seepage channels are formed, groundwater seeps through these channels, reducing the effective stresses and thus the shear strength along discontinuities and therefore the strength of the rock mass. Therefore, the presence of underground water makes the rock mass more prone to large deformation. The varied underground water seepage pathways or inconsistent thickness of the broken rock zone in the surrounding rock mass after excavation may result in local large deformation of the surrounding rock mass.

Due to the complex geological conditions outlined above, obvious asymmetry is observed for the deformation occurring at the two sides of the central axis, crown, and inverted arch, and local deformation is a common phenomenon for the practice of tunnel engineering. Therefore, it is necessary to quantitatively analyze the engineering geology of each part of the tunnel's surrounding rock mass and accurately invert the creep parameters at each local point from the on-site deformation monitoring data, so as to guide the design and construction of tunnels.

3. Rock Strength and Creep Constitutive Model Based on the Hoek–Brown Failure Criterion

The classic rock mass failure criterion considers rock as a continuous homogeneous medium, which greatly simplifies the reality. A constitutive model for the rock mass with multiple structural planes was developed via superimposition of the constitutive equations of the rock mass with a single joint, which cannot accurately capture the complex mechanical characteristics of the tiny fissures and multiple groups of structural planes in the rock mass [31,32]. Hence, the semi-empirical semi-theoretical approach was used to characterize the laminated rock mass in engineering practice.

In 1980, E. Hoek and E. T. Brown derived the correlation expression among the limiting principal stresses for the failure of rock (mass) (namely, the Hoek–Brown failure criterion) via a trial-and-error process, based on the statistical analysis of massive data from rock triaxial tests and field testing of rock mass, and the Griffith theory. Then, E. Hoek further proposed the generalized Hoek–Brown empirical failure criterion. The Hoek–Brown empirical failure criterion can reflect the effects of various factors of the rock mass—such as the rock strength, the number of structural planes, and the in situ stress—on the rock mass strength, and overcome the disadvantages of the conventional theoretical equation. The generalized Hoek–Brown failure criterion can be expressed as below:

$$\sigma_r = \sigma_3 + \sigma_c \left(m_b \frac{\sigma_3}{\sigma_c} + s \right)^a \tag{1}$$

where σ_r is the maximum principal stress when the rock fails, namely, the rock strength; σ_3 is the minimum principal stress when the rock fails; σ_c is the uniaxial compressive strength of the intact rock; m_b, s, and a are all empirical parameters; m_b represents the hardness of the rock, with a value of 0.0000001–25 (0.0000001 for severely-disturbed rock and 25 for intact hard rock); s represents the broken degree of the rock mass, with a value of 0–1 (zero for completely broken rock and one for intact rock); a is related to the rock mass quality.

For rock mass with good quality, its strength characteristics are mainly controlled by the rock particle strength, due to the tight packing of rock particles. The restricted Hoek–Brown empirical failure criterion is more applicable to this case and $a = 0.5$. On the contrary, for rock mass with poor quality, the packing of fragments in the rock mass is loosened by shearing or weathering, which results in loss of the tensile strength of the rock mass (zero cohesion). Under such circumstances, if there is no confinement, the rock mass will collapse and a should be assigned to other values. The expressions of the parameters in Equation (1) are listed below:

$$m_b = m_i \exp\left(\frac{GSI - 100}{28 - 14D} \right) \tag{2}$$

$$s = \exp\left(\frac{GSI - 100}{9 - 3D} \right) \tag{3}$$

$$a = \frac{1}{2} + \frac{1}{6}\left(e^{-GSI/15} - e^{-20/3} \right) \tag{4}$$

$$0 < \sigma_3 < \frac{\sigma_c}{4} \tag{5}$$

where D is the disturbance factor (zero for the undisturbed rock mass, and one for completely disturbed rock mass). m_i indicates the values of constants for intact rock, which can be determined by laboratory testing (uniaxial compression test and conventional triaxial compression test). Based on laboratory data and engineering experience, E. Hoek et al. [33,34] developed a comprehensive and detailed m_i value table that was able to cover a variety of rocks; the m_i index of intact phyllite in this paper refers to this m_i value table.

Afterwards, E. Hoek and E. J. Brown further derived the estimations of the relevant mechanical parameters of the rock mass [35], based on Equation (1):

$$\sigma_{cr} = \sqrt{s}\sigma_c \tag{6}$$

$$\sigma_{tr} = \frac{1}{2}\sigma_c \left(m_b - \sqrt{m_b^2 + 4s} \right) \tag{7}$$

$$E_{\rm rm} = 10^5 \frac{1 - D/2}{1 + \exp[(75 + 25D - GSI)/11]} \quad (8)$$

Equation (1) shows that the *GSI* value is the key to determining the mechanical parameters of the Hoek–Brown failure criterion for the rock mass. However, E. Hoek proposed only a general range for the texture type and weathering conditions of the rock mass and offered no quantitative calculation method.

Rock quality designation (*RQD*) is the most commonly used index to describe the structural characteristics of rock mass, which is defined as the percentage of the sum of the lengths of intact core pieces longer than 0.1 m relative to the chosen length of an evaluated drill core ($RQD = (\frac{\text{the lengths of intact core pieces} \geq 10 \text{ cm per footage}}{\text{the length of an evaluated drill core}}(\%))$; however, the definition of *RQD* requires that the quality of core pieces must be "hard and sound", meaning it can only calculate the percentage of the core with a length beyond 10 cm in the total cored length, which is not suitable for integrity evaluations when characterizing the multi-joint rock mass.

This paper introduces the rock mass block index (*RBI*) and the absolute weathering index (*AWI*) to quantify the *GSI*. The *RBI*, proposed by Hu et al. [36], can thoroughly characterize the block dimension, texture type, and structural packing of blocks for the rock mass. The *RBI* defines the percentages of drilling cores with the measured lengths of 3–10 cm, 10–30 cm, 30–50 cm, 50–100 cm, and >100 cm as the weights, and the value of the *RBI* is the sum of the weights multiplied by the corresponding coefficients.

Although the rock mass has the same *RQD* value, the *RBI* values can be different. For example, the rock mass of *RQD* = 90% can be 10~30 cm with a mosaic structure, or 30~50 cm with a block structure, or 50~100 cm with a block structure, or even more than 100 cm with intact structure. Under the same *RQD*, the larger the *RBI* is (from a mosaic structure to an intact structure), the more intact the rock mass is, and the *RBI* can be regarded as an extension of *RQD*. The calculation formula of *RBI* is shown below:

$$RBI = 3C_{r3} + 10C_{r10} + 30C_{r30} + 50C_{r50} + 100C_{r100} \quad (9)$$

where C_{r3}, C_{r10}, C_{r30}, C_{r50}, and C_{r100} are the acquisition rates (weights in percentages) of the cores with the lengths of 3–10 cm, 10–30 cm, 30–50 cm, 50–100 cm, and >100 cm, respectively. The details of the rock mass texture representation by *RBI*, associated with the field description, are shown in Table 2.

Table 2. Representation of rock mass textures by *RBI*.

Texture Type	*RBI*	Rock Mass Characteristics
Laminated mosaic texture	30–10	Relatively intact, barely, or partially disturbed, often developing 3 groups of structural planes with the spacing of 30–50 cm
Mosaic texture	10–3	Less intact, mostly disturbed, broken yet with tightly packed fragments, generally developing 3–4 groups of structural planes with the spacing of 10–30 cm
Broken texture	3–1	Broken rock mass, sufficiently disturbed, composed of fragments or thin layers, with extensive structural planes presenting spacing generally smaller than 10 cm
Loose texture	1–0	Extremely crushed rock mass, extremely disturbed, composed of loose rock blocks, and angular fragments with crushed debris

The mechanical properties of rock are also related to the weathering level. Chemical weathering can alter the mineral composition of rock and thus alter its physical and mechanical performance.

Parker [37] proposed an index of weathering for silicate rocks, based on the proportions of the alkali and alkaline earth metals present. The bond strengths of the most mobile of the major elements with oxygen are used as weighting factors in the index; the proposed index is defined by the following expression:

$$WI = \left[\frac{a(\text{Na})}{0.35} + \frac{a(\text{Mg})}{0.9} + \frac{a(\text{K})}{0.25} + \frac{a(\text{Ca})}{0.7}\right] \times 100 \quad (10)$$

where a(X) (X = Na, Mg, K, Ca) indicates the atomic proportion of element X, defined as the atomic percentage divided by the atomic weight, and the denominator is the bond strength of element X with oxygen, which represents the stability of the element in the weathering process. From fresh rock to weathered rock, the index *WI* is gradually reduced.

The weathering index of rock mass AWI_0 [38] can be expressed as below:

$$AWI_0 = WI/WI' \quad (11)$$

where *WI* stands for the weathering index of the weathered rock, whereas *WI′* represents that of the fresh rock. Thus, a higher value of *AWI* indicates fresher rock with a lower level of chemical weathering and correspondingly better mechanical properties. On the contrary, a higher *AWI* value denotes a high level of chemical weathering. Obviously, *AWI* quantitatively characterizes the weathering conditions of the engineering rock mass. Based on the research on the relationship between the weathering index and the weathering level of the rock [39], the weathering level of rock based on *AWI* is shown in Table 3/Figure 1. In order to conveniently determine the absolute weathering index AWI_0 of the rock mass, can use the ratio of uniaxial compressive strength of fresh rock to weathered rock.

Table 3. Representation of the rock mass weathering characteristics by *AWI*.

Weathering Condition	AWI_0	Weathering Characteristics
Non-weathered	>0.90	Very good fracture surface: very coarse, fresh, indicating well-sealed fresh rock matrix and no weathering
Slightly weathered	0.90–0.75	Good fracture surface: coarse, relatively fresh, with the presence of rush and the slight alteration of minerals with low weathering resistance, indicating slight weathering
Weakly weathered	0.75–0.55	Ordinary fracture surface: smooth with no filling, partial alteration of minerals with low weathering resistance, indicating weak weathering
Intensively weathered	0.55–0.35	Poor fracture surface: the presence of slickensides, covering of tight films or filling of angular debris on the surface, high alteration of minerals with low weathering resistance, indicating intensive weathering
Extremely weathered	≤0.35	Very poor fracture surface: the presence of slickensides, and soft clay films or clay filling; the vast majority of minerals with low weathering resistance are altered; indicating extreme weathering

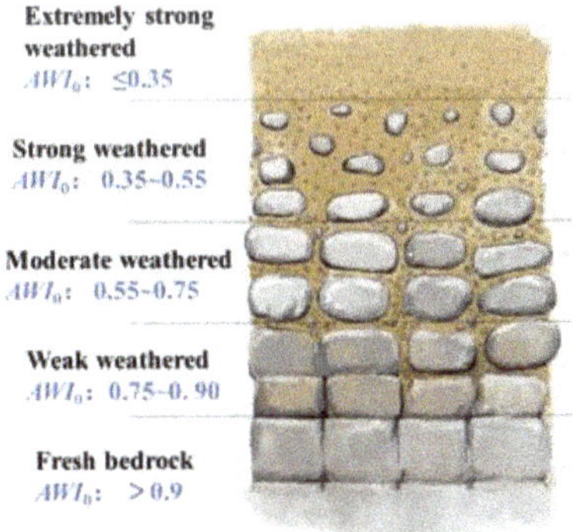

Figure 1. Schematic diagram of rock weathering.

A correction factor λ is introduced to describe the effects of underground water on the weathering conditions of the jointed rock mass (Table 4). Accordingly, the rock mass weathering index, considering the effects of underground water, can be expressed as $AWI = \lambda AWI_0$.

Table 4. Recommended values for the correction factor λ to capture the effects of underground water on the weathering conditions of rock mass.

Production Status of Underground Water	*RBI* Values of Jointed Rock Mass			
	30–10	10–3	3–1	1–0
Humid or dripping	0.95	0.95–0.89	0.89–0.83	0.83–0.76
Rain-like or spring-like production with water pressure < 0.1 MPa; or unit water production rate < 10 L/min·m	0.95–0.89	0.89–0.83	0.83–0.76	0.76–0.71
Rain-like or spring-like production with water pressure > 0.1 MPa; or unit water production rate > 10 L/min·m	0.89–0.83	0.83–0.76	0.76–0.71	0.71–0.67

By integrating the quantitative indexes in Tables 2–4 for the texture and weathering conditions of the rock mass, the geological strength index (*GSI*) quantification method (Figure 2) can be obtained.

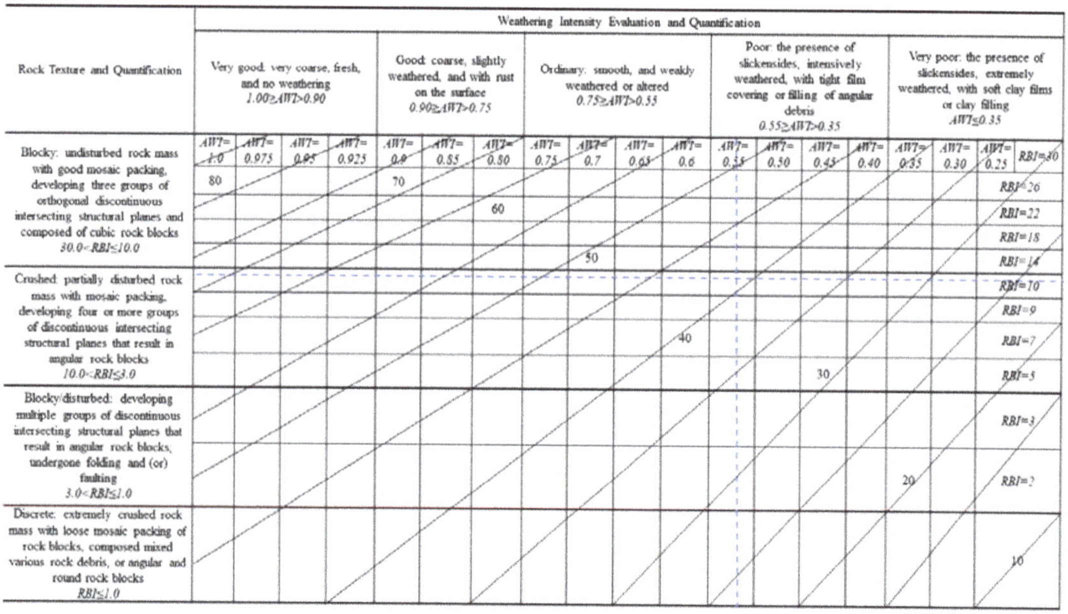

Figure 2. *GSI* quantification of rock mass [31]. Note: The following example is used to demonstrate how to use the above table. In the case of *AWI* = 0.52 and *RBI* = 9.7, two vertical and horizontal lines are drawn according to the index values (the dashed lines in the table); the intersection point is found between 40 and 45; the *GSI* value at the point is calculated via linear interpolation (the result is 41).

The widely applied Nishihara model can characterize the decelerating, steady-state, and accelerating creep phases of rock. Extensive experiments have shown that notable damage of the rock occurs only in the accelerating creep phase. The rock strength σ_r, presented upon the failure of the rock, is the long-term strength for rock creep. The creep elements of the rock mass model are shown in Figure 3.

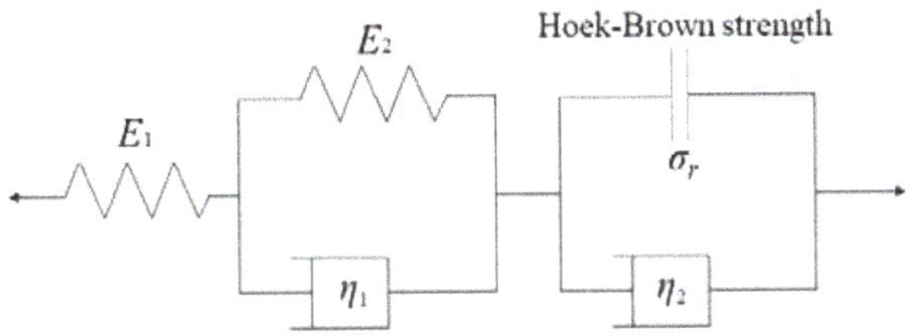

Figure 3. Combination of the creep elements for rock mass.

The creep equation can be expressed as below:

$$\varepsilon = \begin{cases} \frac{\sigma}{E_0} + \frac{\sigma}{E_1}\left[1 - \exp\left(-\frac{E_1}{\eta_1}t\right)\right] & (\sigma \le \sigma_r) \\ \frac{\sigma}{E_0} + \frac{\sigma}{E_1}\left[1 - \exp\left(-\frac{E_1}{\eta_1}t\right)\right] + \frac{\sigma - \sigma_r}{\eta_2}t & (\sigma > \sigma_r) \end{cases} \quad (12)$$

where E_0 is the elastic modulus; E_1 is the viscoelastic modulus; η_1 and η_2 are the viscosity coefficients; σ is the load stress; ε is the total strain; σ_r is the long-term strength calculated using Equation (1) (the Hoek–Brown failure criterion); and t is the creep time.

The strength parameters of the laminated jointed rock mass at each position (Figure 4) can be obtained according to the Hoek–Brown failure criterion. Then, they are input into the numerical simulator via the creep constitutive model (Equation (12)) to calculate the deformation of the surrounding rock mass (Figure 5).

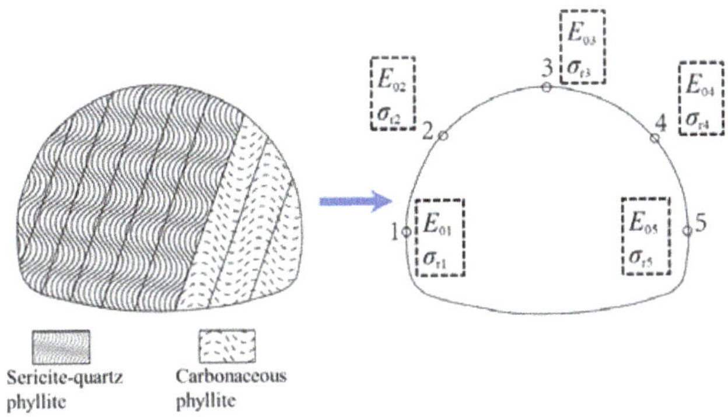

Figure 4. Surrounding rock mass strength parameters based on the Hoek–Brown failure criterion.

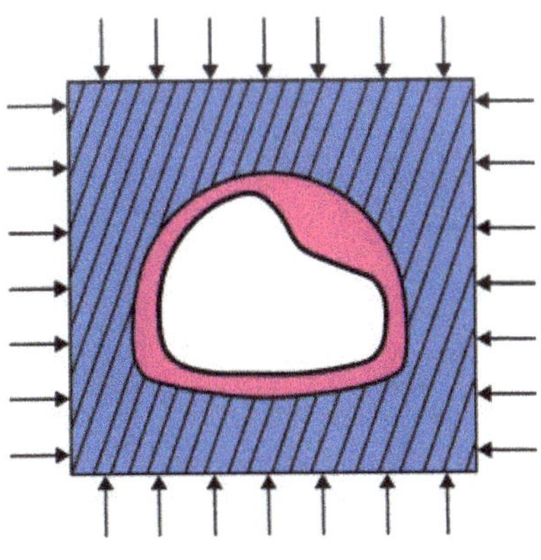

Figure 5. Schematic diagram of the surrounding rock mass deformation.

4. Field Application

The stress state of the surrounding rock mass changes after starting the excavation of the tunnel. Hence, the support design based on the surrounding rock mass parameters obtained in the previous investigation is inappropriate, in particular for a tunnel engineering design with a high buried depth, large span, and highly-developed structural planes. To provide a basis for theoretical reference and design, this paper inverts and validates the surrounding rock mass parameters by integrating the on-site monitoring with a BP neural network.

The BP neural network is a typical nonlinear algorithm, composed of the input, output, and several (one or more) hidden layers (Figure 6), and each layer has several nodes. The connection of nodes between layers is represented by the weight. The BP neural network with one hidden layer is the traditional shallow neural network, whereas that containing multiple hidden layers is the deep learning neural network (Figure 7).

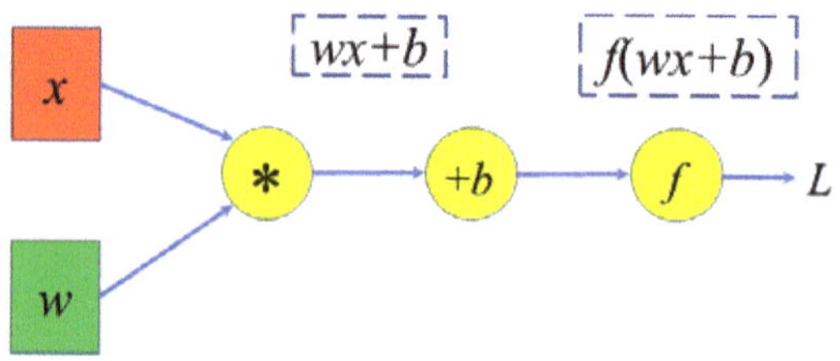

Figure 6. Schematic diagram of the BP neural network structure. Note: The weight w represents the connection strength; the bias b indicates whether or not the node is easy to activate (threshold).

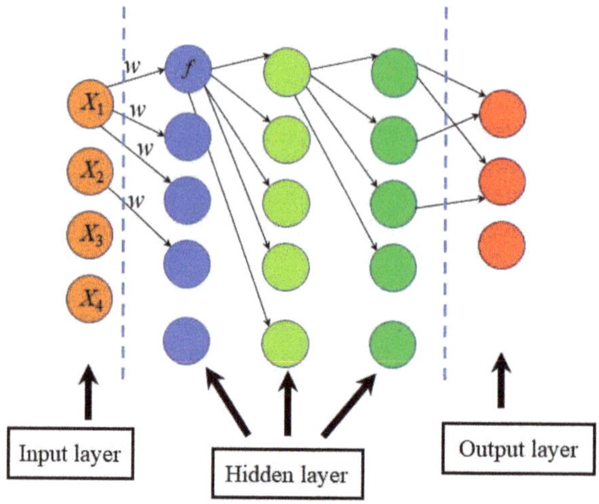

Figure 7. Deep BP neural network structure.

The core steps for training the BP neural network are illustrated below (Figure 8), in which the solid line represents forward propagation and the dashed line represents back propagation. The forward propagation means that the data (information or signal) are first imported into the input end, delivered along the network direction, and multiplied by the corresponding weight. The products of the input data and the corresponding weights are then summed, and the results are input into the activation function for calculation. Then calculation results are delivered to the next node as the input. The calculation is performed successively until the final output is obtained.

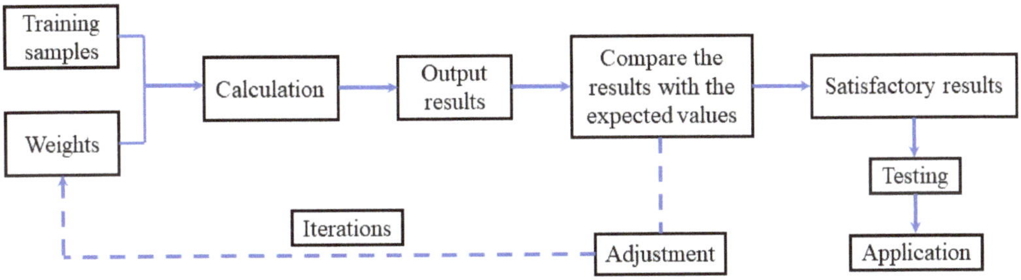

Figure 8. Steps for training the BP neural network.

Here we summarize the inversion workflow of the creep parameters of the surrounding rock mass, as shown below (Figure 9).

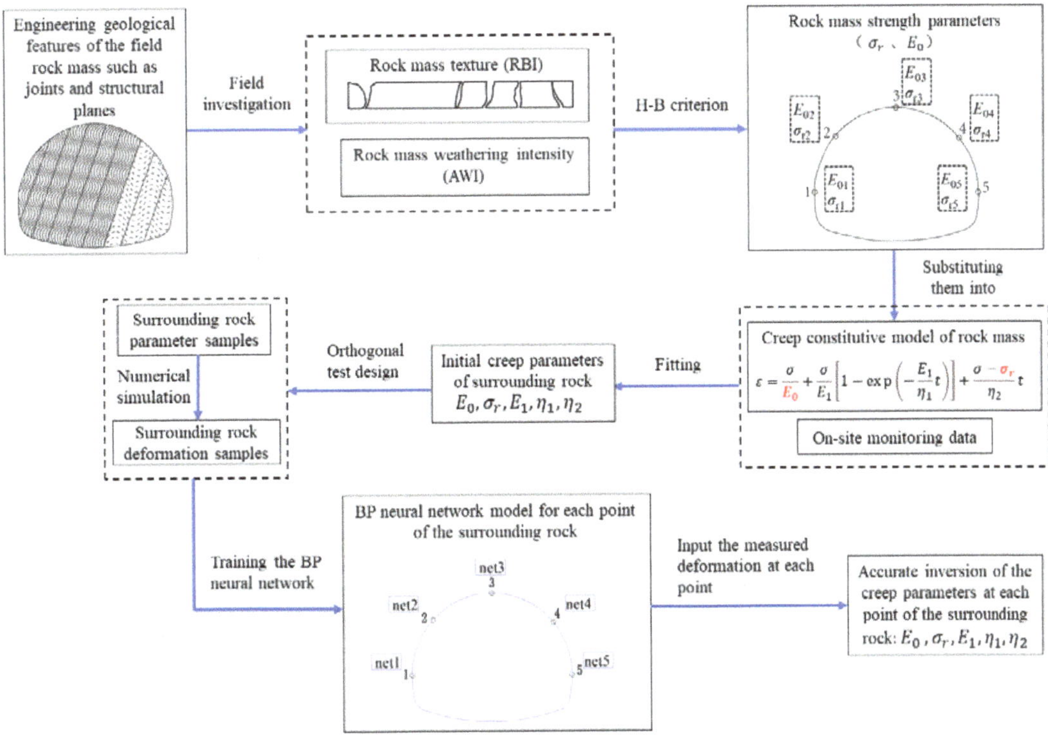

Figure 9. Inversion workflow for the creep parameters of the surrounding rock mass.

The Ganbao tunnel project of the Wenchuan–Maerkang expressway was taken as an engineering case study. It is an ultra-long left-right separated tunnel project (the left tunnel is 4777 m long and the right one is 4796 m long). The longitudinal profile of the tunnel is illustrated in Figure 10, in which the red surrounding rock mass mainly represents the sericite phyllite; the yellow part represents the sericite phyllite interbedded by carbonaceous phyllite and metasandstone; the blue part shows the carbonaceous phyllite and sericite phyllite interbedded by metasandstone; the green part represents the sericite phyllite and carbonaceous phyllite; and the black part indicates the carbonaceous phyllite. The tunnel was excavated using the drilling and blasting method. A large deformation occurs at the mile sections K141 + 540~K141 + 860 and K143 + 485~K143 + 650, which results in the circumferential cracking of multiple points along the initially supported arch and local cracking and spalling of the sidewall and tunnel crown (Figure 11). The monitoring section is located at K143 + 621 and starts monitoring after 12 h of excavation. The overburden near the monitoring section consists of collapsed debris deposits, sericite phyllite and carbonaceous phyllite (from top to bottom, Figure 12),the lithology presented on the face of the surrounding rock mass of the largely-deformed tunnel section (Figure 13) is dominated by the carbonaceous phyllite and sericite phyllite, mainly the Grade-V surrounding rock mass. The rock mass is found with a laminated texture and has a buried depth of 646 m, with a vertical stress of 21.7 MPa and a horizontal stress of 8.3 MPa. Field observations indicate minimal effects of underground water, and thus the correction factor λ for the effects of underground water on the weathering conditions of the rock mass is set as one.

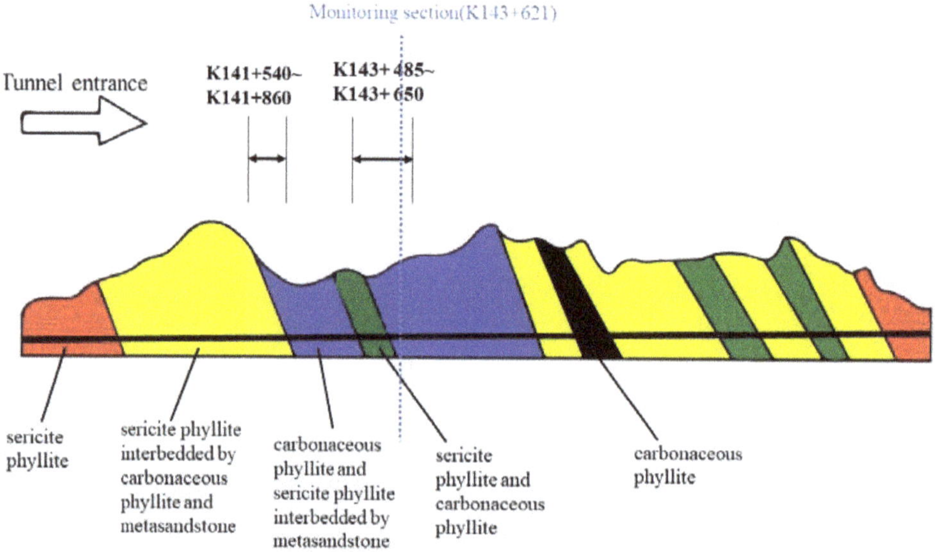

Figure 10. Longitudinal profile of the tunnel.

Figure 11. The twisted steel arch and back haunch fillet of the arch, and rockfall of the initial support.

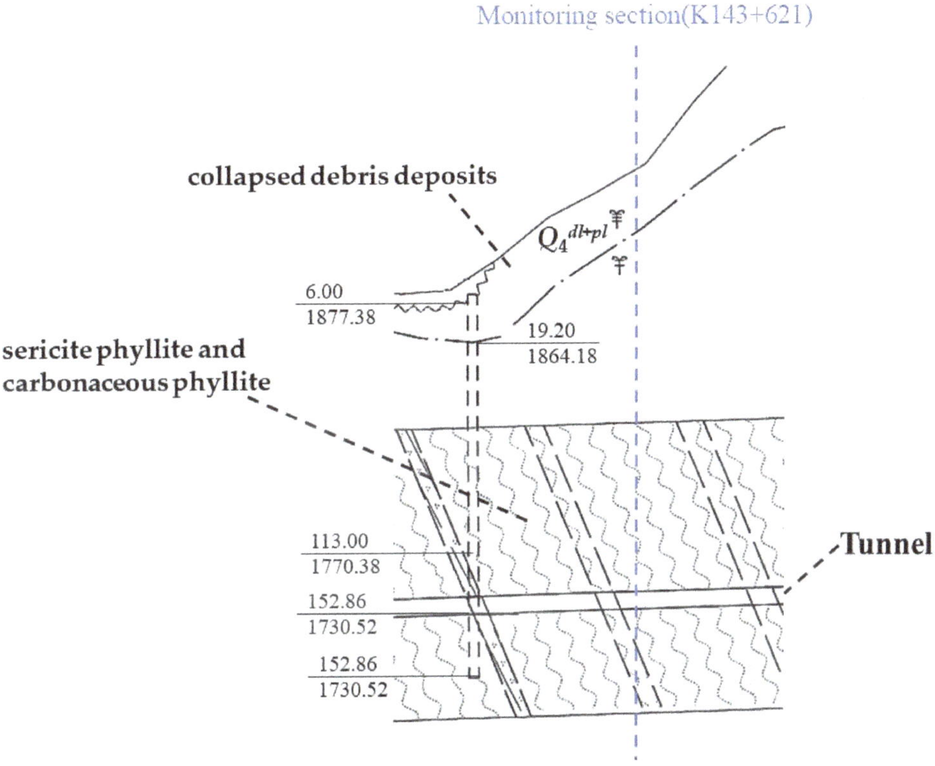

Figure 12. The overburden near monitoring section K143 + 621.

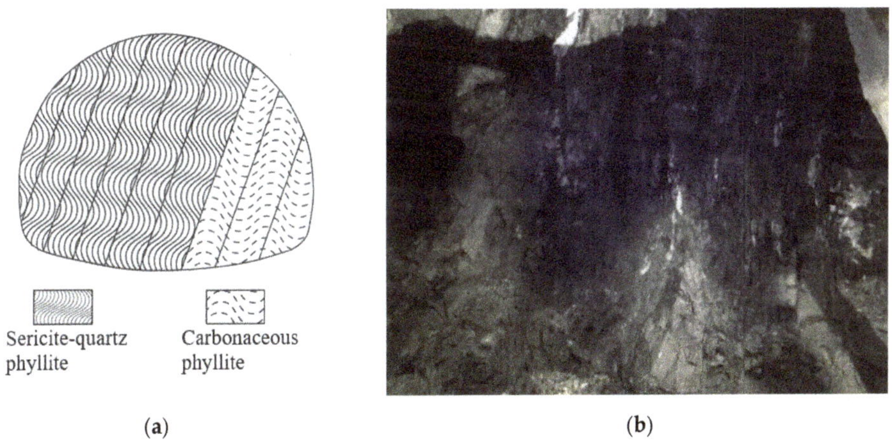

Figure 13. Tunnel face (unexcavated) at K143 + 630. (**a**) Geological sketch picture. (**b**) Photo of the tunnel face.

The pre-excavation boreholes are arranged on the tunnel face (Figure 14), the rock cores are collected via pre-excavation boreholes for investigation, and the *RBI* value of rock mass (Table 5) for each borehole is calculated using Equation (9) and the core length measured on site. For the monitoring of surrounding rock mass deformation after excavation, the on-site monitoring point placement is illustrated in Figure 14.

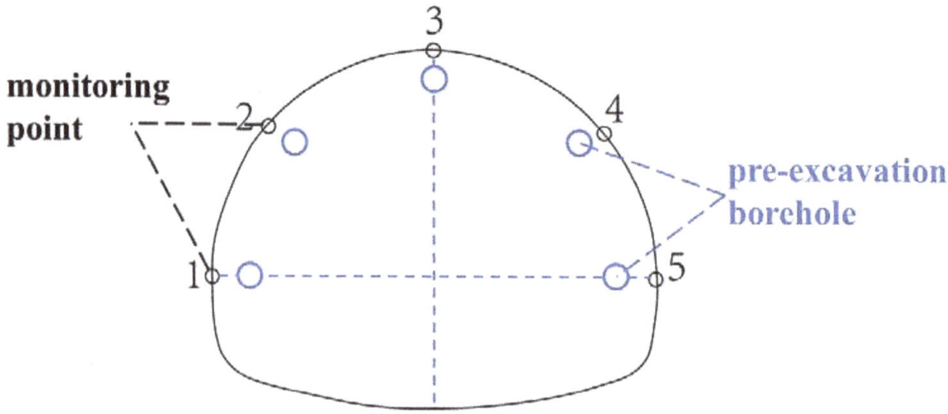

Figure 14. Monitoring point and pre-excavation borehole placement in the tunnel (K143 + 621).

Table 5. Cores collected via pre-excavation boring at each monitoring point and the RBI values.

Monitoring Point	Cores	RBI Values
1		6.763
2		5.626
3		3.748
4		2.8
5		4.331

Taking Point 4 as an example, the field investigation shows that the RBI of the surrounding rock mass cores at Point 4 is about 2.8; the AWI is about 0.65. According to Table 4, the *GSI* value is 33. For the intact phyllite rock, the Hoek–Brown constant $m_i = 10$ and the uniaxial compressive strength $\sigma_c = 35$ MPa. From Equation (5), we have $\sigma_{3max} = \frac{\sigma_c}{4} = 8.75$ MPa and from Equations (2)–(4), we have $m_b = 0.916$, s = 0.000587, and $a = 0.518$.

Finally, according to Equation (1), the uniaxial comprehensive strength of the rock mass $\sigma_r = 16.33$ MPa and the elastic modulus $E_0 = 1.11$ GPa. The basic parameters at the other points are summarized in Table 6.

Table 6. Basic parameters of the surrounding rock mass at each monitoring position.

Monitoring Point	*RBI*	*AWI*	*GSI*	m_b	s	a	σ_r (MPa)	$\sigma_3=\frac{\sigma_c}{4}$ (MPa)	E_0 (GPa)
1	6.763	1.570	79.708	2.212	0.001418	0.27	36.8	45.29	2.69
2	5.626	1.306	66.305	1.840	0.001179	0.22	32.28	37.67	2.24
3	3.748	0.870	44.169	1.226	0.000786	0.15	21.57	25.10	1.49
4	2.8	0.65	33	0.916	0.000587	0.11	16.33	18.75	1.11
5	4.331	1.006	51.050	1.417	0.000908	0.17	22.8	29.01	1.73

The fitting of the creep constitutive model (Equation (11)) to the surrounding rock mass monitoring data (Figure 15) was performed using the least square method to calculate the creep parameters of the rock mass at each monitoring point. The calculation results are shown in Table 7.

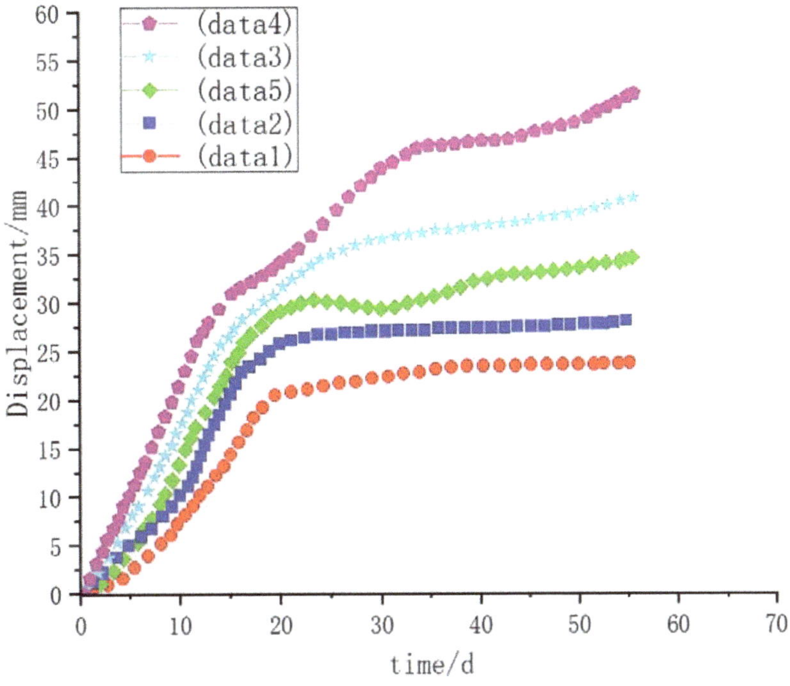

Figure 15. Measured displacement vs. time at each monitoring point.

Table 7. Creep parameters of the rock mass at each monitoring point.

Point No.	E_0/GPa	σ_r (MPa)	E_1/GPa	η_1/(GPa·d)	η_2/(GPa·d)
1	2.69	36.8	0.8	16.38	93.71
2	2.24	32.28	0.71	11.22	88.63
3	1.49	21.57	0.5	8.78	6.55
4	1.11	16.33	0.4	7.87	5.62
5	1.73	22.8	0.6	10.11	78.52

Since the factors affecting the deformation of the surrounding rock mass are varied in the different positions of the tunnel, a neural network was constructed and trained for each point to improve the inversion accuracy. The single-variable method was used for training, which means when the neural network of one point is being trained, the parameters of the other points are constant. Here, Point 4 was taken as an example to demonstrate the training of the neural network. As stated above, the parameters of the other points were fixed, and the learning sample dataset of the surrounding rock mass mechanical properties was built based on the orthogonal test design (Figure 16). The sample parameters in the training dataset were substituted into the numerical simulation model developed using Midas (Figure 17), and the forward calculation was performed to obtain the surrounding rock mass deformation magnitude at each point (Table 8). We used 80% of the data in Table 8 as the training dataset, whereas we used the other 20% as the testing dataset.

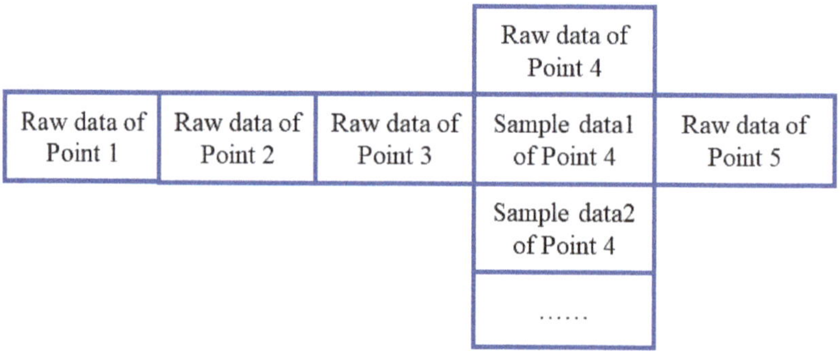

Figure 16. Training the dataset at a certain point in a single-variable approach.

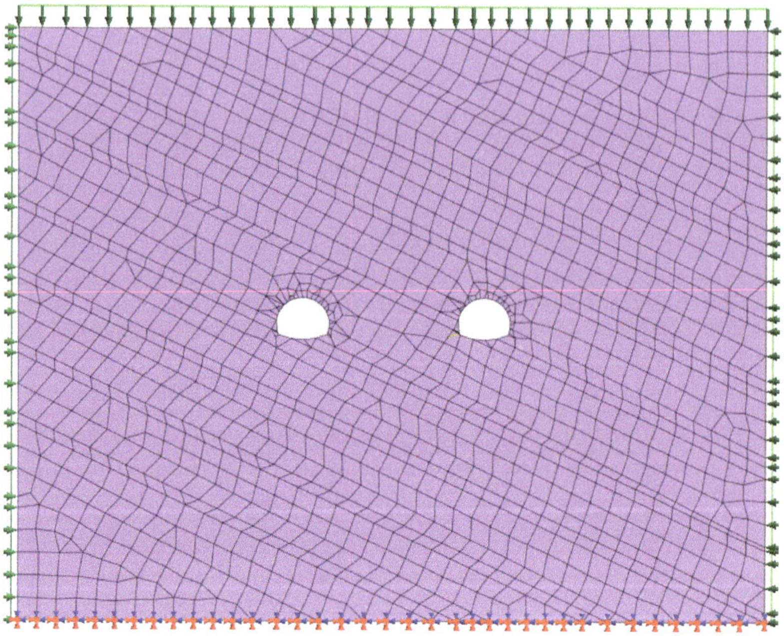

Figure 17. Numerical simulation model of the tunnel.

Table 8. The orthogonal test design plan and calculation results for Point 4.

No.	E_0 /GPa	σ_r /MPa	E_1 /GPa	η_1 /(GPa·d)	η_2 /(GPa·d)	**Displacement at Point 1 /mm**	**Displacement at Point 2 /mm**	**Displacement at Point 3 /mm**	**Displacement at Point 4 /mm**	**Displacement at Point 5 /mm**
1	2.69	36.80	0.80	16.38	93.71	7.31	9.11	13.11	27.94	11.50
2	2.69	41.80	1.30	21.38	98.71	12.51	15.59	22.43	47.80	19.67
3	2.69	46.80	1.80	26.38	103.71	11.02	13.74	19.76	42.11	17.33
4	2.69	51.80	2.30	31.38	108.71	9.84	12.27	17.64	37.60	15.47
5	2.69	56.80	2.80	37.38	113.71	8.90	11.09	15.95	34.00	13.99
6	2.24	36.80	1.30	26.38	108.71	15.83	19.73	25.85	33.66	22.41
7	2.24	41.80	1.80	31.38	113.71	14.31	17.84	23.37	30.43	20.26
8	2.24	46.80	2.30	37.38	93.71	11.76	14.66	19.21	25.01	16.65
9	2.24	51.80	2.80	16.38	98.71	17.73	22.10	28.95	37.70	25.10
10	2.24	56.80	0.80	21.38	103.71	10.91	13.60	17.82	23.20	15.45
11	1.49	36.80	1.80	37.38	98.71	10.60	13.21	18.74	30.89	16.44
12	1.49	41.80	2.30	16.38	103.71	8.71	10.85	15.40	25.38	13.51
13	1.49	46.80	2.80	21.38	108.71	14.90	18.57	26.35	43.43	23.11
14	1.49	51.80	0.80	26.38	113.71	13.13	16.36	23.22	38.26	20.36
15	1.49	56.80	1.30	31.38	93.71	11.72	14.61	20.73	34.16	18.18
16	1.11	36.80	2.30	21.38	113.71	23.73	28.05	40.76	51.49	34.56
17	1.11	41.80	2.80	26.38	93.71	20.91	24.71	35.91	45.37	30.45
18	1.11	46.80	0.80	31.38	98.71	18.67	22.07	32.06	40.51	27.19
19	1.11	51.80	1.30	37.38	103.71	16.88	19.95	28.99	36.62	24.58
20	1.11	56.80	1.80	16.38	108.71	13.87	16.39	23.82	30.09	20.20
21	1.73	36.80	2.80	31.38	103.71	18.26	21.85	25.08	31.69	21.27
22	1.73	41.80	0.80	37.38	108.71	16.31	19.51	18.12	22.89	15.37
23	1.73	46.80	1.30	16.38	113.71	14.74	17.64	16.39	20.70	13.89
24	1.73	51.80	1.80	21.38	93.71	12.12	14.49	13.46	17.01	11.42
25	1.73	56.80	2.30	26.38	98.71	20.73	24.80	33.23	41.46	28.48

Section K143 + 621 was chosen for numerical simulation. Its buried depth was 646 m, we set the size of the model as long × wide = 360 m × 150 m, the fixed end restraint was set at the bottom, the overlying pressure (γH = 33.6 KN/m^3 × (646 − 150) m = 16.67 MPa) was added above the model, the horizontal stress (8.3 MPa) was added on both sides of the model, and the surrounding rock masses all adopted the M-C failure criterion, as shown in Figure 17. The physical parameters of the surrounding rock mass, dominated by sericite and carbonaceous phyllite, were: average density ρ = 3.42 g/cm^3, average elastic modulus E = 5.52 GPa, Poisson ratio ν = 0.38, internal cohesion c = 175 kPa, internal friction angle φ = 27°.

Using the BP neural network toolkit in MATLAB, the forward training of the rock mass parameters and deformation at Point 4 (Table 7) was performed, which produced the neural network model of Point 4, referred to as net4. Similarly, the training samples for the other point were generated and so were their neural network models, namely net1, net2, net3, and net5, respectively. Finally, the deformation measured at each point was input into the corresponding neural network model and the surrounding rock mass parameters were inverted. For the purpose of validation, the inverted mechanical parameters of the surrounding rock mass were substituted into the Midas numerical model to determine the deformation at each point of the surrounding rock mass (Table 9).

Table 9. Parameter inversion of surrounding rock mass and error analysis.

Monitoring Point	Measured Deformation /mm	Output: Inverted Mechanical Parameters of the Surrounding Rock Mass					Simulated Deformation /mm	Relative Error /%
		E_0 /GPa	σ_r /MPa	E_1 /GPa	η_1 /(GPa·d)	η_2 /(GPa·d)		
1	23.73	3.44	35.35	0.92	16.85	104.49	26.67	12.37
2	28.05	2.91	31.89	0.81	12.39	97.20	30.69	9.41
3	40.76	2.12	21.67	0.59	10.01	9.13	45.57	11.81
4	51.50	1.71	16.91	0.48	9.19	8.11	58.54	13.67
5	34.56	2.39	22.63	0.71	11.15	88.00	37.51	8.53

The relative error between the measured displacement and the numerically simulated displacement obtained through inversion was within 15%, indicating the good consistency and high applicability of the proposed method. Therefore, one can build a monitoring system for the largely-deformed surrounding rock mass section, according to the on-site information (such as the lithology combination, underground water status, bedding thickness, rock mass texture, surrounding rock mass integrity, maximum principal stress, and on-site measured rock deformation) for the purposes of gaining real-time information of the tunnel's surrounding rock mass parameter, adjusting the support plan in a timely manner, and providing references for design and construction.

5. Conclusions

Due to the concealment of tunnel engineering, it is difficult to obtain the parameters of the surrounding rock mass after excavation. In this study, we combined the traditional Nashihara model with the Hoek–Brown failure criterion and developed the creep constitutive model for the jointed rock mass. Then, the creep parameters of the surrounding rock mass of the Ganbao tunnel were inverted and validated by integrating the on-site monitoring and the BP neural network. The following conclusions were drawn:

(1) By reviewing the typical cases of large deformation in soft rock tunnels, the main influential factors can be summarized as the lithology combination, weathering effect, and underground water status. With the classical rock mass failure criterion, it is hard to thoroughly incorporate the geological characteristics of the actual rock mass and therefore the semi-empirical semi-theoretical Hoek–Brown approach is more fit-for-purpose.

(2) The geological characteristics of the engineering rock mass were quantitatively characterized using two indexes, namely, the rock mass block index (*RBI*) and the absolute weathering index (AWI). Following the Hoek–Brown criterion, the long-term strength σ_r and elastic modulus E_0 of the rock were obtained and then substituted into the rock mass creep constitutive model based on the Nashihara model and the Hoek–Brown failure criterion. By doing so, the original five creep parameters that needed to be determined in the creep equation were reduced to three, which simplifies the calculation of the constitutive model, while well reflecting the engineering, geological, and creep characteristics of the rock mass on site.

(3) Considering the fact that the actual engineering geology varies at different positions of the tunnel's surrounding rock mass, a specific BP neural network model was built for each monitoring point. Then, the rock mass parameters at each point were inverted from the on-site measured deformation. At last, the inverted parameters were input into the numerical model to calculate the deformation at each point, which was compared with the corresponding measured deformation. The resultant errors

were all within 15%, satisfying the engineering requirements and demonstrating the reliability of the proposed method.

(4) The values of the rock mass *GSI* (Figure 2) were all determined according to the engineering geology handbook, relevant standards, and existing literature. Therefore, these values can be adjusted as per the field condition. The inversion of the surrounding rock mass parameters is highly affected by the basic parameters of the engineering geological characteristics and the in-situ stress field of the tunnel's surrounding rock mass. Hence, during applications of the proposed method, the basic parameters need to be accurately measured to improve the accuracy of the inversion.

(5) This method is suitable for tunneling in unsupported rock mass or plainly supported tunnels after excavation.

Author Contributions: Writing—original draft preparation, C.C.; writing—review and editing, T.L.; methodology, C.M.; investigation, H.Z.; software, J.T.; data curation, Y.Z.; funding acquisition, T.L. All authors have read and agreed to the published version of the manuscript.

Funding: This research was funded by National Natural Science Foundation of China, grant number U19A20111.

Institutional Review Board Statement: Not involved.

Informed Consent Statement: Not involved.

Data Availability Statement: The study did not report any data.

Acknowledgments: This work was supported by the National Natural Science Foundation of China (NO. U19A20111). The financial support is gratefully acknowledged. Special thanks are due to Tianbin Li and Chunchi Ma for their help in improving this article.

Conflicts of Interest: The authors declare no conflict of interest.

References

1. He, M.; Jing, H.; Sun, X. *Mechanics of Soft Rock Engineering*; Science Press: Beijing, China, 2002.
2. Kong, P.; Jiang, L.S.; Shu, J.M.; Sainoki, A.; Wang, Q.B. Effect of fracture heterogeneity on rock mass stability in a highly heterogeneous underground roadway. *Rock Mech. Rock Eng.* **2019**, *52*, 4547–4564. [CrossRef]
3. Chen, Z.; He, C.; Xu, G.; Ma, G.; Wu, D. A case study on the asymmetric deformation characteristics and mechanical behavior of deep-buried tunnel in phyllite. *Rock Mech. Rock Eng.* **2019**, *52*, 4527–4545. [CrossRef]
4. Jia, P.; Tang, C.; Yang, T.; Wang, S. Numerical Stability Analysis of Surrounding rock mass Mass Layered by Structural Planes with Different Obliquities. *J. Northeast Univ.* **2006**, *27*, 1275–1278.
5. Li, L.; Guan, J.; Xiao, M.; Liu, H.; Tang, K. A creep constitutive model for transversely isotropic rocks. *Rock Soil Mech.* **2020**, *41*, 2922–2930, 2942.
6. Li, C.; Wang, J.; Xie, H. Anisotropic creep characteristics and mechanism of shale under elevated deviatoric stress. *J. Pet. Sci. Eng.* **2020**, *185*, 106670. [CrossRef]
7. Li, A.; Xu, N.; Dai, F.; Gu, G.; Hu, Z.; Liu, Y. Stability analysis and failure mechanism of the steeply inclined bedded rock masses surrounding a large underground opening. *Tunn. Undergr. Space Technol.* **2018**, *77*, 45–58. [CrossRef]
8. Song, D.; Chen, J.; Cai, J. Deformation monitoring of rock slope with weak bedding structural plane subject to tunnel excavation. *Arab. J. Geosci.* **2018**, *11*, 251. [CrossRef]
9. Gao, F.; Guo, J. Stress mechanism and deformation monitoring of bias tunnel. *Int. J. Simul. Syst. Sci. Technol.* **2016**, *17*, 14.1–14.4. [CrossRef]
10. Yin, C.; Li, H.; Che, F.; Li, Y.; Hu, Z.; Liu, D. Susceptibility mapping and zoning of highway landslide disasters in China. *PLoS ONE* **2020**, *15*, e0235780. [CrossRef]
11. Liu, C.H.; Li, Y.Z. Analytical study of the mechanical behavior of fully grouted bolts in bedding rock slopes. *Rock Mech. Rock Eng.* **2017**, *50*, 2413–2423. [CrossRef]
12. Vergara, M.R.; Kudella, P.; Triantafyllidis, T. Large scale tests on jointed and bedded rocks under multi-stage triaxial compression and direct shear. *Rock Mech. Rock Eng.* **2015**, *48*, 75–92. [CrossRef]
13. Li, L.; Tan, Z.S.; Guo, X.L.; Wu, Y.; Luo, N. Large deformation of tunnels in steep dip strata of interbedding phyllite under high geostresses. *Chin. J. Rock Mech. Eng.* **2017**, *36*, 1611–1622.
14. Li, X.H.; Xia, B.W.; Li, D.; Han, C.R. Deformation characteristics analysis of layered rockmass in deep buried tunnel. *Rock Soil Mech.* **2010**, *31*, 1163–1167.
15. Fu, X. *Numerical Simulation of Asymmetric Large-Deformation Energy-Releasing Bolt Support for Layered Soft Rock Tunnel*; Chengdu Univerisity of Technology: Chengdu, China, 2020.

16. Li, Z.; Shan, R.; Wang, C.; Yuan, H.; Wei, Y. Study on the distribution law of stress deviator below the floor of a goaf. *Geomech. Eng.* **2020**, *21*, 301–313. [CrossRef]
17. Tian, M.; Han, L.; Meng, Q.; Ma, C.; Zong, Y.; Mao, P. Physical model experiment of surrounding rock mass failure mechanism for the roadway under deviatoric pressure form mining disturbance. *KSCE J. Civ. Eng.* **2020**, *24*, 1103–1115. [CrossRef]
18. Wang, Z.-J.; Luo, Y.-S.; Guo, H.; Tian, H. Effects of initial deviatoric stress ratios on dynamic shear modulus and damping ratio of undisturbed loess in China. *Eng. Geol.* **2012**, *143–144*, 43–50. [CrossRef]
19. Kroon, M.; Faleskog, J. Numerical implementation of a J2 and J3 dependent plasticity model based on a spectral decomposition of the stress deviator. *Comput. Mech.* **2013**, *52*, 1059–1070. [CrossRef]
20. Wang, Q.; Pan, R.; Jiang, B.; Li, S.; He, M.; Sun, H.; Wang, L.; Qin, Q.; Yu, H.; Luan, Y. Study on failure mechanism of roadway with soft rock in deep coal mine and confined concrete support system. *Eng. Fail. Anal.* **2017**, *81*, 155–177. [CrossRef]
21. Wang, W.; Zhang, C.; Wei, S.; Zhang, X.; Guo, S. Whole section anchor–grouting reinforcement technology and its application in underground roadways with loose and fractured surrounding rock mass. *Tunn. Undergr. Space Technol.* **2016**, *51*, 133–143. [CrossRef]
22. Xu, G.; He, C.; Chen, Z.; Yang, Q. Transversely isotropic creep behavior of phyllite and its influence on the long-term safety of the secondary lining of tunnels. *Eng. Geol.* **2020**, *278*, 105834. [CrossRef]
23. Li, X.; Ju, M.; Yao, Q.; Zhou, J.; Chong, Z. Numerical investigation of the effect of the location of critical rock block fracture on crack evolution in a gob-side filling wall. *Rock Mech. Rock Eng.* **2016**, *49*, 1041–1058. [CrossRef]
24. Zhou, J.; Wei, Q.; Liu, G. Back analysis on rock mechanics parameters for highway tunnel by BP neural network method. *Chin. J. Rock Mech. Eng.* **2004**, *23*, 941–945.
25. Cao, W.; Jiang, Y.; Sakaguchi, O.; Li, N.; Han, W. Predication of Displacement of Tunnel Rock Mass Based on the Back-Analysis Method-BP Neural Network. *Geotech. Geol. Eng.* **2021**, *2021*, 1–14. [CrossRef]
26. Wu, Q.; Yan, B.; Zhang, C.; Wang, L.; Ning, G.; Yu, B. Displacement prediction of tunnel surrounding rock mass: A comparison of support vector machine and artificial neural network. *Math. Probl. Eng.* **2014**, *2014*, 351496. [CrossRef]
27. Deng, X.; Xu, T.; Wang, R. Risk evaluation model of highway tunnel portal construction based on BP fuzzy neural network. *Comput. Intell. Neurosci.* **2018**, *2018*, 8547313. [CrossRef]
28. Wen, H.; Yin, J.; Qin, Z.; Xie, R. Application of BP Neural Network to the Back Analysis of Mechanical Parameters of Tunnel Surrounding rock mass. *J. Yangtze River Sci. Res. Inst.* **2013**, *30*, 47–51, 56.
29. He, Y.; Sun, X.; Zhang, Y.; Guo, H.; Li, Q. Intelligent fusion model and analysis method for rock parameter inversion of water diversion tunnel. *J. Hydroelectr. Eng.* **2021**, *40*, 114–126.
30. Ma, F.; Jia, S. Back analysis of elastoplastic parameters of surrounding rock mass for roadway in mudstone and its long-term stability prediction. *Rock Soil Mech.* **2014**, *7*, 1987–1994.
31. Amadei, B. Strength of a regularly jointed rock mass under biaxial and axisymmetric loading conditions. *Int. J. Rock Mech. Min. Sci. Geomech. Abstr.* **1988**, *25*, 3–13. [CrossRef]
32. Xiao, S.; Yang, S. *Rock Mass Mechanics*; Geological Publishing House: Beijing, China, 1987.
33. Hoek, E.; Brown, E.T. Practical estimates of rock mass strength. *Int. J. Rock Mech. Min. Sci.* **1997**, *34*, 1165–1186. [CrossRef]
34. Marinos, P.; Hoek, E. Estimating the geotechnical properties of heterogeneous rock masses such as flysch. *Bull. Eng. Geol. Environ.* **2001**, *60*, 85–92. [CrossRef]
35. Hoek, E.; Brown, E.T. The Hoek–Brown failure criterion and GSI–2018 edition. *J. Rock Mech. Geotech. Eng.* **2019**, *11*, 445–463. [CrossRef]
36. Hu, X.; Zhong, P.; Ren, Z. Rock-mass block index and its engineering practice significance. *J. Hydraul. Eng.* **2002**, *33*, 80–83.
37. Parker, A. An index of weathering for silicate rocks. *Geol. Mag.* **1970**, *107*, 501–504. [CrossRef]
38. Su, Y.; Feng, L.; Li, Z.; Zhao, M. Quantification of elements for geological strength index in Hoek-Brown criterion. *Chin. J. Rock Mech. Eng.* **2009**, *28*, 36–43.
39. Li, R.; Wu, L. Research on characteristic indexes of weathering intensity of rocks. *Chin. J. Rock Mech. Eng.* **2004**, *23*, 3830.

applied sciences

MDPI

Article

Integrating InSAR Observables and Multiple Geological Factors for Landslide Susceptibility Assessment

Yan-Ting Lin [1], Yi-Keng Chen [2], Kuo-Hsin Yang [2], Chuin-Shan Chen [2] and Jen-Yu Han [2,*]

1 National Center for Research on Earthquake Engineering, National Applied Research Laboratories, Taipei 106, Taiwan; yanting@ncree.narl.org.tw
2 Department of Civil Engineering, National Taiwan University, Taipei 106, Taiwan; ykchen0814@gmail.com (Y.-K.C.); khyang@ntu.edu.tw (K.-H.Y.); dchen@ntu.edu.tw (C.-S.C.)
* Correspondence: jyhan@ntu.edu.tw; Tel.: +886-2-33664347

Abstract: Due to extreme weather, researchers are constantly putting their focus on prevention and mitigation for the impact of disasters in order to reduce the loss of life and property. The disaster associated with slope failures is among the most challenging ones due to the multiple driving factors and complicated mechanisms between them. In this study, a modern space remote sensing technology, InSAR, was introduced as a direct observable for the slope dynamics. The InSAR-derived displacement fields and other in situ geological and topographical factors were integrated, and their correlations with the landslide susceptibility were analyzed. Moreover, multiple machine learning approaches were applied with a goal to construct an optimal model between these complicated factors and landslide susceptibility. Two case studies were performed in the mountainous areas of Taiwan Island and the model performance was evaluated by a confusion matrix. The numerical results revealed that among different machine learning approaches, the Random Forest model outperformed others, with an average accuracy higher than 80%. More importantly, the inclusion of the InSAR data resulted in an improved model accuracy in all training approaches, which is the first to be reported in all of the scientific literature. In other words, the proposed approach provides a novel integrated technique that enables a highly reliable analysis of the landslide susceptibility so that subsequent management or reinforcement can be better planned.

Keywords: landslide potential; InSAR; spatial factors; machine learning; slope unite

Citation: Lin, Y.-T.; Chen, Y.-K.; Yang, K.-H.; Chen, C.-S.; Han, J.-Y. Integrating InSAR Observables and Multiple Geological Factors for Landslide Susceptibility Assessment. *Appl. Sci.* **2021**, *11*, 7289. https://doi.org/10.3390/app11167289

Academic Editor: Francesco Fiorillo

Received: 24 June 2021
Accepted: 5 August 2021
Published: 8 August 2021

1. Introduction

In Asian subtropical monsoon regions, July to September is a season of strong typhoons. High rainfall intensity usually causes serious landslide events in mountainous areas [1]. It is necessary to predict landslide occurrence and behavior and adopt appropriate prevention policies and methods to improve disaster relief effectiveness and reduce casualties and property loss during and after disasters. Landslide prediction aims to predict the possibility of the occurrence of landslides in a specific area; available data are commonly used, including conditional factors and historical landslides. These data are collected from landslide inventories and static instruments, and their values are shown in spatial analysis [2]. However, traditional landslide prediction, such as mathematical evaluation models, lacks information about the temporal probability of landslides, i.e., time-series landslide behavior. Landslide displacement time-series data can directly reflect ground surface deformation and stability characteristics. Therefore, they have been recently used to develop landslide prediction models. Generally, these time-series data are collected from one-point survey equipment, such as surface extensometers and GPS devices [3]. However, field GPS surveying projects, which depend on only one or two temporarily installed reference stations, have many disadvantages [4]. In practice, steadily obtaining survey data using these single reference stations is often difficult because of poor performance or failure. Therefore, the use of only the single-point method in landslide surveys would limit the cost-effectiveness.

In recent years, remote sensing technology has effectively detected large-scale landslide-sensitive areas and generated landslide inventories, which are crucial for predicting landslides before they occur or recur, especially in far or barely accessible areas [5]. In daytime satellite images without shadows and clouds, landslide positions can be identified through noticeable radiometric contrasts between land cover types [6]. Optical sensors cover the electromagnetic spectrum from 390 nm to 1 mm, including the visible and infrared bands. Such devices can measure the visual properties in the spectral characteristics of the land surface, which can then be used to detect and map landslides. Researchers can also combine time-series satellite images with digital elevation models (DEMs) to acquire 3D terrain, which can be used to visually detect and predict potential landslides.

However, affected by monsoons, typhoons, and thunderstorms, mountainous areas are usually shrouded in clouds at times; thus, the use of satellite images to monitor landslide disasters could be limited by weather conditions. Compared with optical sensors, synthetic aperture radar (SAR) sensors use a longer wavelength—microwaves; having all-weather and all-day operational capability, SAR sensors can penetrate cloud cover and reduce the limitation imposed by the atmosphere to remotely evaluate the accurate range and severity of landslide disasters in almost real-time [7]. Although some particular meteorological situations, such as thick rain cells, may disturb the backscattering coefficient, SAR remains more powerful than optical sensors for long-term landslide observation [8]. Spaceborne SAR, such as Envisat, ALOS PALSAR, RADARSAT, TerraSAR-X, and Sentinel-1, provide high spatial resolutions and can clearly observe target objects in full-time and in almost all-weather conditions.

Numerous applications of SAR data to ground displacement detection have demonstrated their usefulness for landslide characterization and mapping [9]. Differential SAR interferometry (DInSAR) is a commonly used method of ground deformation measurement, and it can efficiently generate or update landslide inventory [10], which is critical information about landslide behavior for landslide susceptibility assessment. DInSAR calculates the phase variation of two SAR images acquired in the same region at different times. Long-term InSAR observations are calculated as the deformation-induced phase shift through the backscattered microwave signal between several coherent acquisitions. The landslide behavior of time-series information, which depends on the millimetric measurement accuracy and the metric spatial resolution, is obtained under most atmospheric conditions [11].

Landslide prediction methods can be classified into three types: image analysis, mathematical evaluation models, and machine learning methods [12]. Image analysis uses geographic information systems, which can collect, store, manage, and analyze geographical data. The risk of landslides can be predicted by analyzing disaster data, such as history of landslides and land. The probability of landslides varies because it is based on the number of data layers used for analysis. Mathematical evaluation models use a single evaluation equation that is combined with the physical concepts of mechanics and hydrographic data, such as rainfall, runoff, and infiltration data, for landslide susceptibility assessment [13]. The use of such models is easy for simulation and fits a wide range of environments. However, mathematical evaluation models require detailed data of the geotechnical engineering and geological aspects of slope failure at sites [14], which makes these models costly and impractical for large-scale areas.

In recent years, machine learning and data mining techniques, such as support vector machine, artificial neural network, and decision tree (DT) models, have been applied for landslide susceptibility modeling [15]. These methods incorporate different factors that might cause landslides to evaluate the probability of landslide occurrence. Machine learning algorithms enrich the quality and accuracy of generated susceptibility maps. Researchers use and compare various machine learning models on the basis of different data [16–19], integrate different machine learning models to improve accuracy [20–23], or develop new algorithms that are based on traditional machine learning models to strengthen landslide prediction results [24–26]. These techniques perform better than do classical methods. Most machine learning techniques achieve overall success rates of 75% to 95% [27]. Although many applications have demonstrated the feasibility of data-driven models for capturing nonlinear relationships and modeling the dynamic processes of landslides on the basis of historical model data, certain limitations remain [28]. As shown, landslide behavior involves temporal dependencies. However, common machine learning models ignore this intrinsic temporal dependency, which involves the effect of preceding actions on present actions in the model [29,30]. The solution proposed by this study is to combine spatial-temporal data, including InSAR observables, as a landslide susceptibility factor with other traditional geological and land cover factors into a model that can improve the prediction accuracy of potential landslides. To our knowledge, integrating InSAR observables and multiple geological factors for landslide susceptibility analysis is an effective and pioneered contribution for landslide potential prediction research.

2. Methods

This research method effectively estimates the landslide potential of slopes through four steps: (1) segmentation of slope units, (2) numerical indexing of related spatial factors, (3) correlation between spatial factors and slope landslides, and (4) use of machine learning methods. A displacement prediction analysis model was constructed following the above process. Finally, a confusion matrix was used to verify the results of the displacement prediction analysis. The overall research method and procedure are shown in Figure 1.

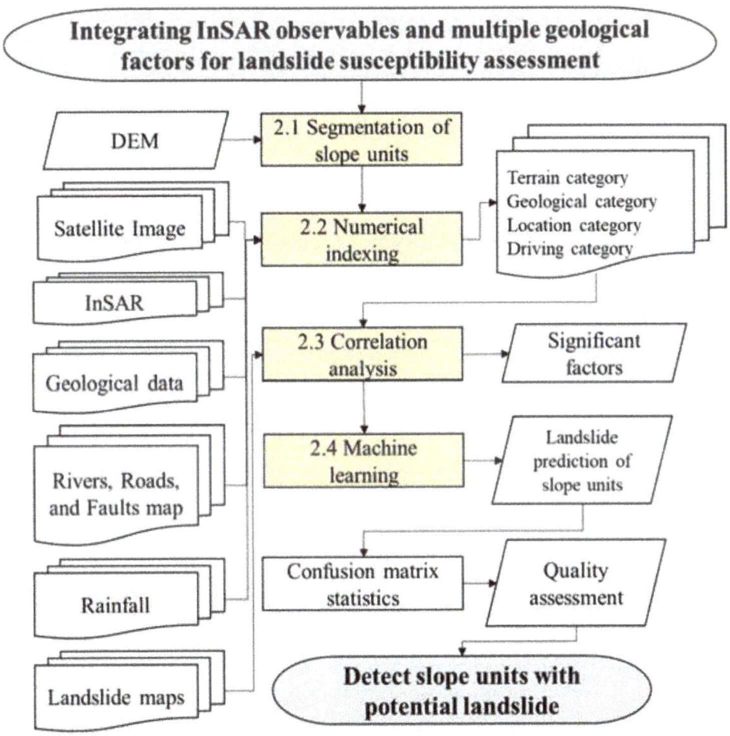

Figure 1. Flowchart of landslide susceptibility analysis based on the spatial factors with machine learning approach.

2.1. *Segmentation of Slope Units*

This study used the slope unit as the basis of analysis to show the topographic characteristics of each slope. These slope units serve as a framework for the subsequent geographical interpretation of environmental spatial factors. The method of slope unit segmentation refers to the catchment overlap concept proposed by Xie et al. [31], as shown in Figure 2. First, the water catchment area in a DEM is identified through the hydrology module in the software ArcGIS, and the water line is turned into a ridge line by flipping the DEM, which is divided into two slope units (left and right). When the hydrology module identifies small catchment areas, the default flow accumulation value is set to 500 as the threshold value for dividing the river area. Then, the slope units are cut out, and each area becomes less than 30 ha. With the aid of a shadow map, aspect map, slope map, river map, and satellite orthophoto overlay, the overlap between each slope unit is confirmed.

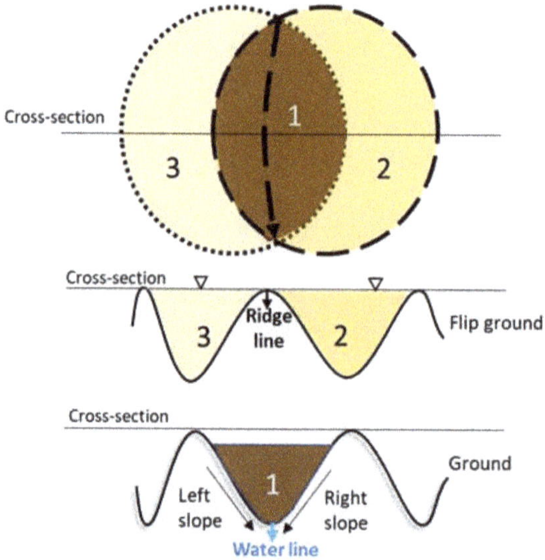

Figure 2. Schematic of dividing the slope units with the overlap method of catchment areas (modified from [31]).

2.2. *Numerical Indexing of Related Spatial Factors*

In this study, the spatial factors were divided into four categories: terrain, location, geological, and driving. The terrain category represents the geometric changes in surface elevation and coverage distribution, including elevation, slope, aspect, terrain roughness, profile curvature, vegetation index, and the displacement velocity gradient of InSAR. The location category shows the distance of influencing factors, including roads, rivers, and geo-faults. The geological category reflects the strength, folds, and dip slopes of rock formations. The driving category is the rainfall factor. The index calculations of these factors are described below. It should be mentioned that these spatial factors were first selected based on suggestions reported in the relevant studies in the literature [16–20]. A significance test was then performed to identify the most influential factors that have the high correlation with the landslides in the study areas. The results and discussion on the significance test of spatial factors are presented in Section 3.2.

2.2.1. Terrain Category

- Elevation, slope, and aspect

On the basis of the framework of the slope unit, the highest elevation in each unit was extracted and represented as the elevation factor, as shown in Equation (1). According to the height change caused by the horizontal movement distance, the slope factor is expressed by a tangent function on average, as indicated in Equation (2). The aspect factor refers to the direction of the maximum elevation change in the slope unit. It is calculated by the angle with the true north direction, as shown in Equation (3), where the true north direction is 0°, and the angle increases to 360° in the clockwise direction.

$$I_{elevation} = \max(Z_i) \tag{1}$$

$$I_{slope} = \tan\overline{\theta_s} = \frac{\Delta\overline{Z}}{\Delta\overline{L}} \tag{2}$$

$$I_{aspect} = \frac{180^\circ}{\pi}\tan^{-1}[\max(\theta_s)] \tag{3}$$

where Z_i is elevation, $\Delta\overline{Z}$ is the mean elevation difference, $\Delta\overline{L}$ is the mean horizontal distance, and θ_s is the main slope angle.

- Terrain roughness

Terrain roughness represents the degree of height change. When the undulating terrain faces the effect of large gravity, the smaller resistance force makes the slope have a higher possibility of landslide. The elevation standard deviation σ is used to describe the degree of elevation change in the slope unit (Equation (4)).

$$\sigma = \sqrt{\frac{\sum_i (Z_i - \overline{Z})^2}{n_s - 1}} \tag{4}$$

where $\overline{Z}$ is the average elevation in a slope unit, and n_s is the number of grids in the slope unit.

- Profile curvature

The profile curvature is expressed as the slope steepness. This study used the spatial analysis module of the software ArcMap to calculate the profile curvature of each slope unit on the basis of a 3 × 3 moving grid, which is the default grid size in ArcMap. A negative (positive) value of the curvature represents a convex (concave) slope.

- Vegetation index

Plants can effectively stabilize the rock and soil on slopes, but the exposed soil area may suffer from repeated landslide and displacement problems. Hence, the vegetation index is defined as the proportion of vegetation area in the slope unit, as shown in (Equation (5)).

$$I_{veg.} = \frac{A_{veg.}}{A_s} \tag{5}$$

where $A_{veg.}$ is the area of the vegetation and A_s is the area of the slope unit.

- Annual displacement velocity gradient of InSAR

InSAR technology calculates the phase difference to estimate the displacement of the ground through more than two periods of SAR observations. The InSAR-derived ground displacement can be regarded as a direct observation of ground stability and was thus proposed as an essential index for landslide susceptibility analysis in this study. However, the original displacements from InSAR observations suffer from various influencing factors, such as vegetation changes and orbital variations of SAR satellites. In order to reduce the periodical or systematic noises due to those uncontrollable factors and to extract a meaningful index for evaluating the ground stability, the annual velocity gradients derived from InSAR displacement fields were used in this study. First, the annual displacement information of InSAR is placed in the range from −1 to 1 by mean normalization, which is shown in Equation (6), to unify the scale and reduce the systematic error of InSAR data.

$$Z_{\overline{S_i}} = \frac{Z_{S_i} - \mu}{\max(Z_{S_i}) - \min(Z_{S_i})} \tag{6}$$

where $Z_{\overline{S_i}}$ is the normalized InSAR displacement value, Z_{S_i} is the annual displacement of InSAR, and μ is the average annual displacement.

The annual displacement velocity of InSAR is obtained as the slope value in first-order linear fitting (Equation (7)). These discrete observation points are interpolated with a regular grid size of 20 m to present the field of annual displacement velocity. For highlighting the displacement positions, the field gradient is calculated with a 3 × 3 moving window, the same as for computing the profile curvatures. The index calculation is expressed as Equation (8).

$$Z_{\overline{S_i}} = V\Delta t + \Delta Z \tag{7}$$

$$I_{InSAR} = \nabla_V \mathbf{f}(V) \tag{8}$$

where V is the annual displacement speed of InSAR, Δt is annual observation time, and ΔZ is the difference in annual displacement.

2.2.2. Location Category

Potential displacements are affected by the distances between slope units and location factors. In this study, three location factors were selected for analysis, namely, the river distance, road distance, and fault distance. Through each shortest distance from the centroid of the slope units to the three location factors, the formula of the location factors $I_{location}$ is expressed by Equation (9).

$$I_{location(rivers,\ roads,\ faults)} = \min\left(\sqrt{(X_c - X_l)^2 + (Y_c - Y_l)^2}\right) \tag{9}$$

where (X_c, Y_c) is the centroid coordinates of slope units, and (X_l, Y_l) is the coordinates of location factors (including rivers, roads, and faults).

2.2.3. Geological Category

- Rock Mass Strength

Rock masses with weaker strength are prone to landslides due to their difficulty in resisting the disturbance of external forces. Franklin used the degree of rock structure fracture and single compressive strength to classify the rock mass strength into seven levels [32]. In this study, the slope unit was superimposed on the environmental geological map produced by the Central Geological Survey of Taiwan, and the corresponding rock mass strength information was used as the rock mass strength index.

- Folds

When a rock is squeezed into curved folds, the fold layer becomes prone to landslides. In this study, the fold factor is defined as the number of folds in the slope unit, as shown in Equation (10).

$$I_{fold} = \sum n_f \tag{10}$$

where n_f is the number of folds in a slope unit.

- Dip Slopes

Dip slopes mean that a stratum has the same inclination as that of the slope; a slope landslide may be formed by sliding along the layer. In this study, the dip slope index is defined as the ratio of the dip slope area to the slope unit area, as shown in Equation (11).

$$I_{dip\ slope} = \frac{A_d}{A_s} \tag{11}$$

where A_d is the area of the dip slope and A_s is the area of the slope unit.

2.2.4. Driving Category (Rainfall)

The density of rainfall data collected by rainfall stations is much lower in mountainous areas than that in urban areas. Relevant studies have mostly used distance as an interpolation reference to obtain the rainfall in a whole area through grid interpolation. This study considered the distance and elevation factors of rainfall stations and added the aspect factor to construct a rainfall interpolation model, as shown in Equation (12). In this model, the elevation parameter α, distance parameter β, and aspect parameter γ are obtained through the least squares adjustment, and the parameter weight is shown in Equation (13).

$$\mathrm{I_{rain}} = \alpha\left(\sum_{\mathrm{iH}} \mathrm{W_{iH}} \cdot \mathrm{R_i}\right) + \beta\left(\sum_{\mathrm{iL}} \mathrm{W_{iL}} \cdot \mathrm{R_i}\right) + \gamma\left(\sum_{\mathrm{i\theta}} \mathrm{W_{i\theta}} \cdot \mathrm{R_i}\right) \tag{12}$$

$$\mathrm{W_{iH}} \propto \frac{1}{\Delta \mathrm{H}^2};\ \mathrm{W_{iL}} \propto \frac{1}{\Delta \mathrm{L}^2};\ \mathrm{W_{i\theta}} \propto \frac{1}{\Delta\theta^2} \tag{13}$$

where $\mathrm{I_{rain}}$ is the rainfall index, $\mathrm{W_{iH}}$ is the elevation weight, $\mathrm{W_{iL}}$ is the distance weight, $\mathrm{W_{i\theta}}$ is the aspect weight, $\mathrm{R_i}$ is the rainfall observation at Station i, α is the elevation parameter, β is the distance parameter, γ is the aspect parameter, ΔH is the elevation difference, ΔL is the distance difference, and $\Delta\theta$ is the aspect difference.

2.3. *Correlation between Spatial Factors and Slope Landslides*

Significant factors were detected through the spatial factors and the displacement correlation score. The Spearman method was adopted to arrange the data in order of numerical value, thereby improving the limitation of the normal distribution assumption in the correlation analysis. The correlation coefficient γ_s is distributed between 1 and -1; a positive (negative) value indicates a positive (negative) correlation. The closer the coefficient value to 0, the more unlikely it is to affect the displacement. Its sequential linear relationship is described in Equation (14).

$$\gamma_s = 1 - \frac{6\sum \Delta_i^2}{n(n^2-1)} \tag{14}$$

where γ_s is the correlation coefficient, Δ is the difference between the spatial factor and displacement, and n is the number of samples.

Finally, a significance test was conducted through the correlation coefficient to check the significance of each factor. This test is shown in Equation (15).

$$t = 1 - \frac{\gamma_s - \rho_0}{\sqrt{\frac{1-\gamma_s^2}{n-2}}} \tag{15}$$

where ρ_0 is 0, and it is the null hypothesis (indicating no correlation). If the significance level t is greater than 0.99, the null hypothesis will be rejected; that is, the factor is correlated with the displacement.

2.4. Use of Machine Learning Methods

Machine learning is applied to establish prediction models, which are used in landslide potential and displacement prediction, by inputting the spatial factors and displacement observations. Widely used machine learning algorithms for classification prediction include naive Bayes, DT, random forest, adaptive boosting (AdaBoost), and extreme gradient boosting (XGBoost).

- Naive Bayes

As the probability model of naive Bayes assumes that the factors are independent of each other and conform to a Gaussian distribution, naive Bayes classification helps clarify a large number of complex classification problems. The early-stage spatial factors correspond to the landslide and nonlandslide slope units, and they are regarded as training samples to establish a prediction model. The later-observed spatial factors are inputted into the model to determine the landslide probability of each slope unit. The naive Bayes prediction model is based on the probability density function of the Bayesian classification method [33], as shown in Equation (16).

$$P(w_i|x) = \frac{P(x|w_i)P(w_i)}{P(x)}, \begin{cases} j \neq i \\ j = 1,2 \end{cases} \quad (16)$$

where $P(w_i|x)$ is the probability of the classifying w_i occurring in the slope unit x, $P(x|w_i)$ is the probability of the slope unit x occurring in the classifying w_i, $P(w_i)$ is the probability of classifying w_i, and $P(x)$ is the probability of the slope unit x.

- DT

A DT assumes that the factors are independent of each other, and the category probability of the DT path is defined by the factor characteristics [34]. This algorithm adopts a dichotomy method, which is similar to a double-forked tree branch, to calculate the Gini coefficient value at the node. Finally, the gain value in each path is summed, and the largest accumulator will be predicted to belong to a category, as shown in Equation (17).

$$\text{gain} = \sum p_i \left(1 - \sum p_i^2\right) \quad (17)$$

where p_i is the probability. If the node has only one category, p_i will be 0. If the numbers of two categories are the same, p_i is 0.5.

- Random forest

Random forest is a collection of multiple DTs and adds the use of bagging. The observation data are taken out of the number of samples and trained as n types of classifiers. According to the sample difference in each DT, the random uncertainty of the data is considered. Under the same weight, the classifier uses the summed majority as the best classification tree to predict the classification [35]. Equation (18) represents the probability of the c-th factor in the t-th DT, and the average probability value gc of the category is obtained according to the sum of multiple DTs. Finally, the category of the slope unit x is determined according to the maximum g_c value (Equation (19)).

$$P(c|v_i(x)) = \frac{P(c|v_i(x))}{\sum_l^{n=1} P(c_l|v_i(x))} \quad (18)$$

$$g_c(x) = \frac{1}{t}\sum_{i=1}^{t} \hat{P}(c|v_i(x)) \quad (19)$$

where P is the probability, c-th is the category, v is the node, l is the number of categories, t is the number of DTs, and g_c is the average probability of the c-th category.

- AdaBoost

Boosting increases the weight of wrong data in a classification model, and the wrong information is trained to strengthen the identification. The derived new classifier will reduce the chance of early error [36]. The iterative process of the AdaBoost calculation is extremely sensitive to noise and abnormal data; therefore, these should be reduced so that the process can focus on difficult-to-classify feature factors. AdaBoost analysis initially assumes that the sample weights are equal. After the k-th iteration, samples are selected on the basis of the weight W_k to train the classifier Ck, as expressed by Equation (20).

$$\begin{cases} D = \{x^1, y_1, \cdots, x^n, y_n\} \\ W_k(i) = \frac{1}{n}, i = 1 \cdots, n \end{cases} \tag{20}$$

where D is the sample category, (x^i, y_i) is the sample information, n is the number of samples, and W_k is the weight distribution of all samples in the k-th iteration.

The classification error E_k confirms the correctness of the classification and updates the weight W_{k+1}, as shown in Equation (21). The iterative calculation of classification is completed when the error E_k is less than the preset threshold.

$$W_{k+1}(i) \leftarrow \frac{W_k(i)}{Z_k} \times \begin{cases} e^{-\frac{1}{2}\ln\frac{1-E_k}{E_k}}, \text{ if } y_k(x^i) = y_i \\ e^{\frac{1}{2}\ln\frac{1-E_k}{E_k}}, \text{ if } y_k(x^i) \neq y_i \end{cases} \tag{21}$$

where W_{k+1} is the updated weight, Z_k is the normalization coefficient, E_k is the error, and y_k is the prediction category.

- XGBoost

The XGBoost function is composed of two components: the prediction error of boosting and the complexity of DT. The feature factors are combined and branched into a DT, and a new boost function is learned from the previous calculation residuals [37]. In Equation (22), the first component calculates the error between the prediction and actual observation, and the other component indicates the complexity of the regularized DT, which covers the number of nodes and the node probability value.

$$f = \sum_{i=1}^{n} E(y_i, y_{ki}) + \sum_{k=1}^{K} \Omega(f_k) \tag{22}$$

where E is the error between the prediction and actual observation and $\Omega(f_k)$ is the complexity of the DT.

3. Results

The experiment based on the slope unit was conducted for the following two parts of test analysis. In the first part, the correlation analysis of the spatial factor and the landslide unit was adopted to detect the significant spatial factor. In the second part, the spatial factor indicators and landslide units observed from 2007 to 2009 were applied to run the machine learning models. Then, the 2010 spatial factors were inputted into those models, and the landslide slope units were estimated. The prediction was compared with the landslide location announced by the Central Geological Survey of Taiwan's Ministry of Economic Affairs (MOEA) through a confusion matrix to verify the feasibility of this study.

3.1. Study Areas

Experimental cases in Siaolin Village and the Putunpunas River area (Kaohsiung, Taiwan) were selected to verify this study method. Both areas continued to experience a large number of landslides after the typhoon Morakot in 2009. In the Siaolin Village area, there were 128 slope units (covering 15.81 km^2), and Provincial Highway 29 is the main external traffic road. In the Putunpunas River area, there were 349 slope units (covering 61.21 km^2), and the Southern Cross-Island Highway presents a north–south vertical, as shown in Figure 3.

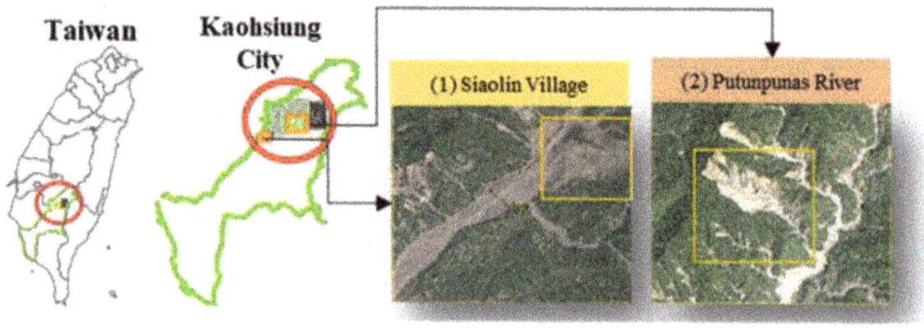

Figure 3. Geographical locations of experimental areas—(1) Siaolin village; (2) Putunpunas River.

The observation time of the spatial factors ranged from hours to years. For establishing a common timescale, a year was deemed the basis of unit time, and the observed data time was a total of four years (from 2007 to 2010). The 14 spatial factors used were the elevation, slope, aspect, terrain roughness, profile curvature, vegetation index, annual displacement velocity gradient of InSAR, water distance, road distance, fault distance, rock mass strength, folds, dip slopes, and an annual rainfall, as shown in Figure 4.

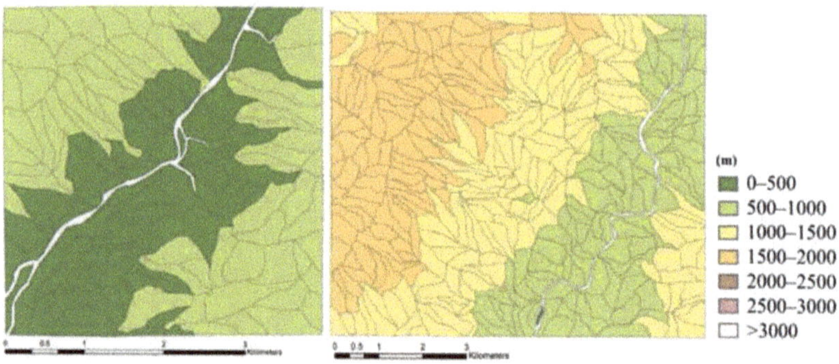

(**a**) elevation

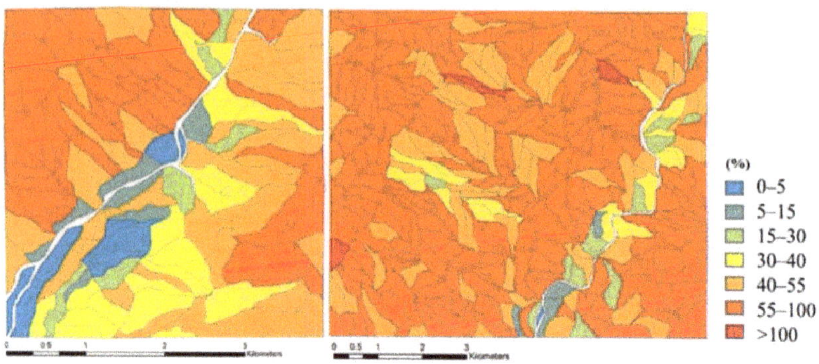

(**b**) slope

Figure 4. *Cont.*

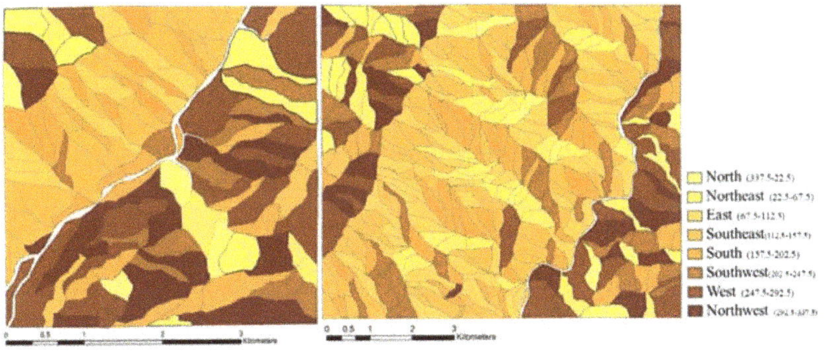

(**c**) aspect

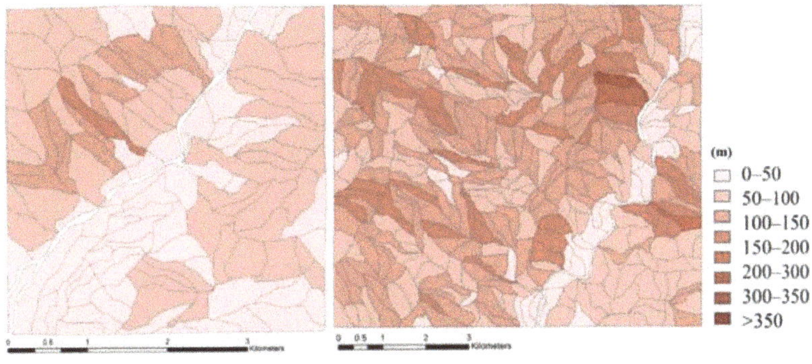

(**d**) terrain roughness

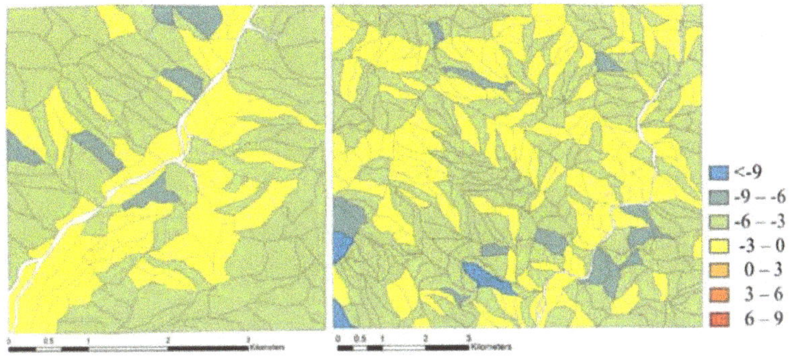

(**e**) profile curvature

Figure 4. *Cont.*

(**f**) vegetation index

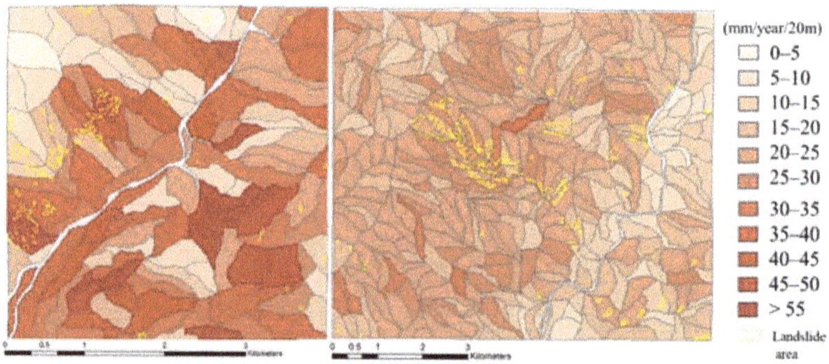

(**g**) annual displacement velocity gradient of InSAR

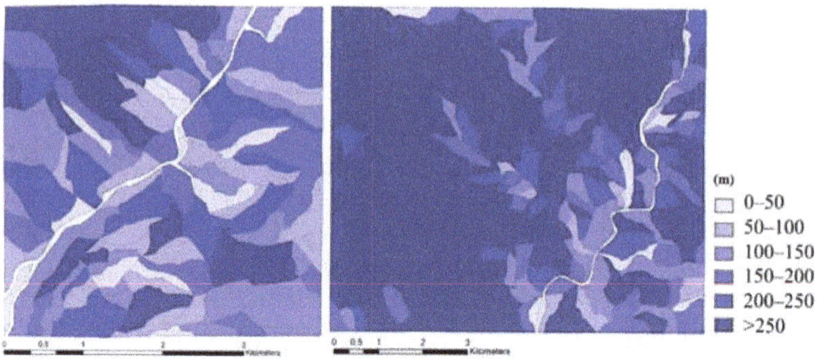

(**h**) water distance

Figure 4. *Cont.*

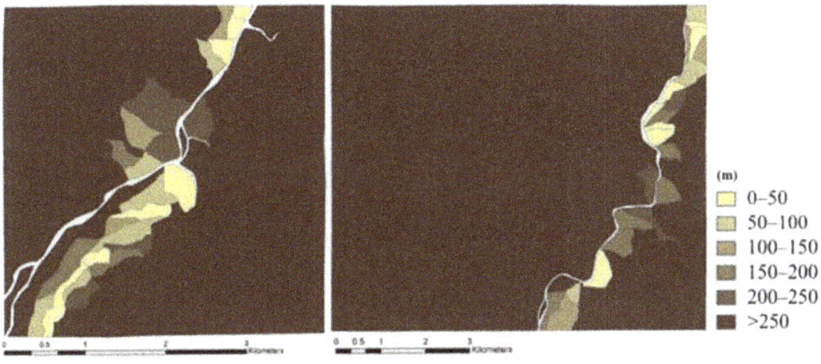

(**i**) road distance

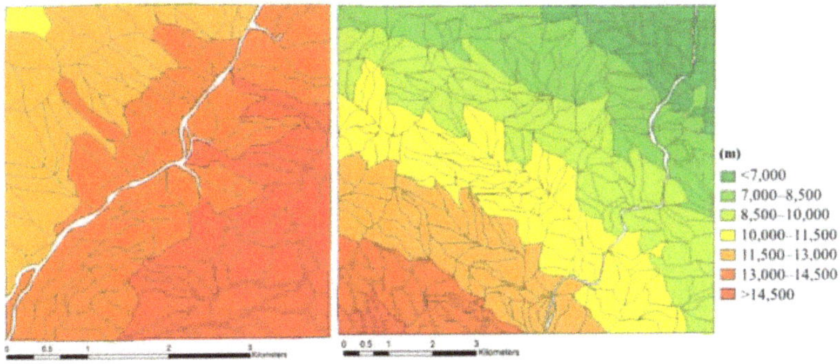

(**j**) fault distance

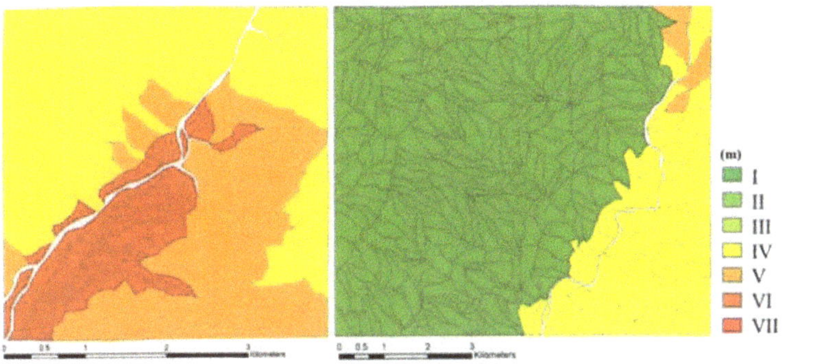

(**k**) rock mass strength

Figure 4. *Cont.*

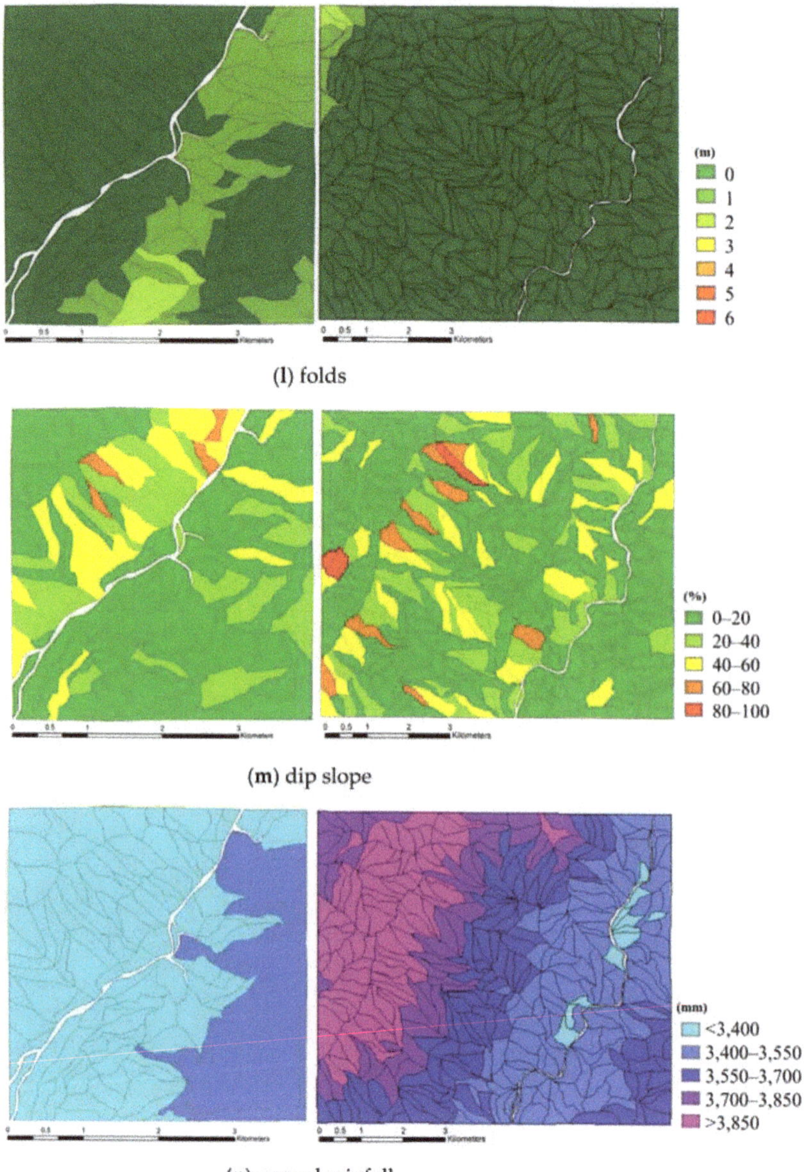

Figure 4. Fourteen spatial factors used in this study. These observations in 2017 are for Siaolin Village (**left**) and Putunpunas River (**right**).

3.2. Significance Test of Spatial Factors

The factor scales were unified from 1 to −1 through numerical standardization to solve the inconsistency of the factor value distribution. Then, the correlation between the spatial factors and landslides based on the slope units was examined. The correlation coefficient values were expressed as positive or negative. As seen in Figure 5, the correlation coefficients of Siaolin Village (yellow bar) were between −0.47 and 0.43, and those of Putunpunas River (dark-blue bar) were between −0.42 and 0.36. Hypothesis significance testing was performed, and the probability of obtaining the test resulted in the *p*-value, as shown in Table 1. Then, the significant spatial factors were screened on the basis of a 99% reliability as the test threshold. There were five significant spatial factors in the Siaolin Village area (rock mass strength, aspect, terrain roughness, slope, and dip slopes) and six significant spatial factors in the Putunpunas River area (rock mass strength, aspect, vegetation index, water distance, terrain roughness, and dip slopes).

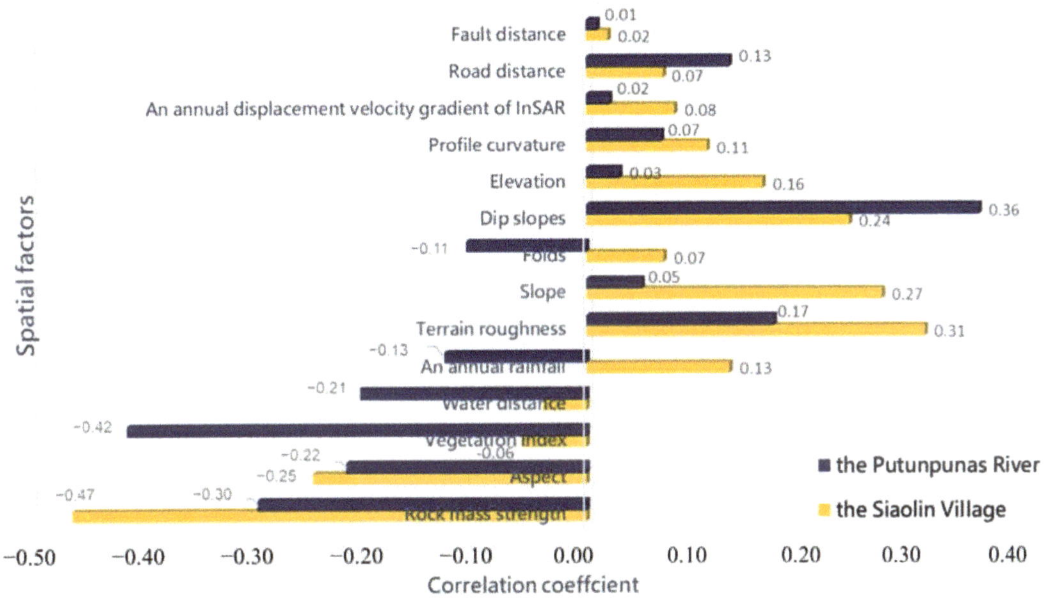

Figure 5. Histogram of correlation coefficient between landslides and the 14 spatial factors in the slope units.

Table 1. Correlation coefficients and *p*-values, quantified according to the relationship between landslides and the 14 spatial factors in the slope units.

Spatial Factor	Siaolin Village		Putunpunas River	
	Correlation Coefficient	*p*-Value	Correlation Coefficient	*p*-Value
Rock mass strength	−0.47	1.00	−0.30	1.00
Aspect	−0.25	1.00	−0.22	1.00
Vegetation index	−0.06	0.49	−0.42	1.00
Water distance	−0.04	0.37	−0.21	1.00
Annual rainfall	0.13	0.85	−0.13	0.98
Terrain roughness	0.31	1.00	0.17	1.00
Slope	0.27	1.00	0.05	0.66
Folds	0.07	0.57	−0.11	0.95
Dip slopes	0.24	0.99	0.36	1.00
Elevation	0.16	0.93	0.03	0.37
Profile curvature	0.11	0.81	0.07	0.78
Annual displacement velocity gradient of InSAR	0.08	0.62	0.02	0.29
Road distance	0.07	0.59	0.13	0.98
Fault distance	0.02	0.28	0.01	0.23

3.3. ML Prediction and Verification

According to the five machine learning methods used in this research, the relevant parameters were set as shown in Table 2. In these machine learning calculations, three years of spatial factor data (from 2007 to 2009) were used as input for learning, and the landslide prediction of the slope units was based on the 2010 spatial factors. Finally, the landslide location announced by the Central Geological Survey (MOEA, Taiwan) in 2010 was used to verify the accuracy of slope unit prediction.

Table 2. Parameters and settings required for the machine learning methods.

ML	Parameters	Values
Naive Bayes	Smoothing	10^{-9}
DT	Criterion	Gini
	The maximum of depth	20
	The minimum of samples split	10
	The minimum of samples leaf	5
Random Forest	Criterion	Gini
	The maximum of depth	20
	The minimum of samples split	2
	The minimum of samples leaf	5
	The number of estimators	100
AdaBoost	Criterion	Gini
	The maximum of depth	20
	The minimum of samples split	2
	The minimum of samples leaf	5
	The number of estimators	10
	Algorithm	SAMME
	Learning rate	0.1
XGBoost	The maximum of depth	5
	The number of estimators	1000
	Learning rate	0.1
	The minimum of child weight	1
	Gamma number	0
	Subsample number	0.8
	Colsample bytree	0.8
	Objective binary	Logistic
	nthread	4

The prediction accuracy of machine learning prediction is shown in Table 3. From the correct rate, the addition of the InSAR factor increased the accuracy of prediction by 0% to 6%. For Siaolin Village, the random forest method had the highest prediction accuracy rate (82.95%), followed by XGBoost (79.31%), AdaBoost (78.49%), naive Bayes (70.93%), and DT (68.02%). Putunpunas River showed a similar trend; the best prediction was observed from the random forest method (80.51%), followed by XGBoost (78.80%), AdaBoost (75.64%), DT (68.19%), and naive Bayes (68.19%).

The prediction results of the best learning method (random forest) were used to compare and evaluate the predicted classification through confusion matrixes. In Figure 6, the correctly predicted landslide slope units are colored red, and the correctly predicted noncollapsed slope units are colored cyan. In addition, the erroneously predicted landslide slope units are marked with green diagonal stripes, and the erroneously predicted nonlandslide slope units are marked with red diagonal stripes.

Table 3. Average prediction accuracies before and after including InSAR data in different ML methods.

ML	Siaolin Village		Putunpunas River	
	With InSAR (%)	Without InSAR (%)	With InSAR (%)	Without InSAR (%)
Naive Bayes	70.93	70.85	68.19	68.19
DT	68.02	62.02	75.45	75.07
Random Forest	82.95	79.84	80.52	78.79
AdaBoost	78.49	77.52	75.64	75.64
XGBoost	79.31	75.97	78.80	75.80

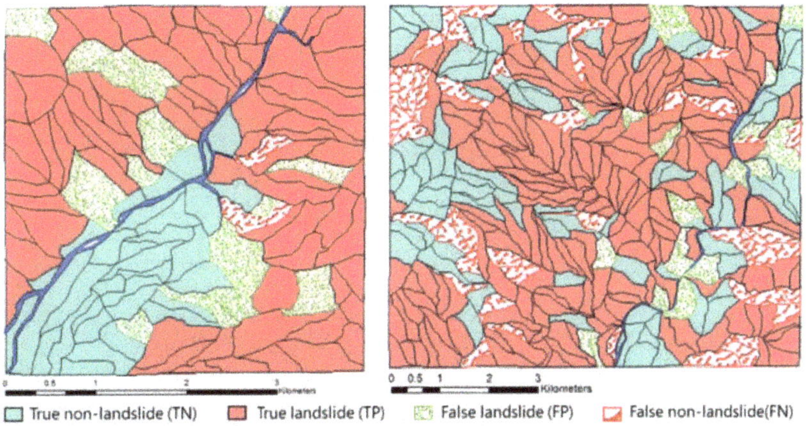

Figure 6. Visual illustration of the landslide prediction results—(**left**) Siaolin Village; (**right**) Putunpunas River area.

The confusion matrixes of Siaolin Village and Putunpunas River are shown in Tables 4 and 5. In Siaolin Village, the correct prediction rates of landslide and noncollapsed slope units were 78.72% and 94.30%, respectively; the average accuracy rate of the overall prediction was 82.95%. In Putunpunas River, the correct prediction rates of landslide and noncollapsed slope units were 89.67% and 66.18%, respectively; the average accuracy rate of the overall prediction was 80.52%.

Table 4. Confusion matrix for the case of the Siaolin Village analysis.

Actual \ Predictied	Lanslide	Nonlanslide	Average
Lanslide	74 (TP)	2 (FN)	-
Nonlanslide	20 (FP)	33 (TN)	-
Correct rate (%)	78.72	94.30	82.95

Table 5. Confusion matrix for the case of the Putunpunas River analysis.

Actual \ Predictied	Lanslide	Nonlanslide	Average
Lanslide	191 (TP)	46 (FN)	-
Nonlanslide	22 (FP)	90 (TN)	-
Correct rate (%)	89.67	66.18	80.52

4. Discussion

Throughout the time series, the relevant spatial observation data showed changes in slopes. This study used these environmental observation data to construct the spatial factor indicators on the basis of the slope unit conditions. Significant spatial factors were then determined from the correlation analysis. According to the spatial characteristics of the slope units, the machine learning methods were applied to construct the calculation models, and the landslide potential of the slope units was evaluated.

This study was implemented with two experimental cases: Siaolin Village and Putunpunas River (Kaohsiung, Taiwan). The experiment collected four-year spatial data (topography, locations, geology, driving categories, and landslide locations) from 2007 to 2010. Then, these data were used to construct the 14 spatial factors through indexed analysis. A common timescale (year) was established for the analysis to resolve the differences in timescales of the various spatial factors. The spatial factor datasets from 2007 to 2009 served as the input for the correlation analysis and machine learning, and the 2010 spatial factor data were used to calculate the output of potential evaluation. In the Siaolin Village area, the significant spatial factors were the rock mass strength, aspect, terrain roughness, slope, and dip slopes; the significant spatial factors in the Putunpunas River area were the rock mass strength, aspect, vegetation index, water distance, terrain roughness, and dip slopes. These significant factors in both study areas were all in the geological category, including rock mass strength, terrain roughness, and dip slopes. Obviously, the geological conditions in these areas highly influence the landslide trend.

The machine learning algorithms used in this research achieved accuracies of 60–80% in landslide classification. Among them, the random forest method exhibited the best calculation in Siaolin Village, where it yielded a prediction accuracy rate of 82.95%; its prediction accuracy rate in Putunpunas River was 80.50%. The random forest method effectively performed independent training for high-dimensional, multi-feature factors. In addition, the random forest algorithm exhibited strong anti-interference capabilities, such as an imbalance in the number of classifications and missing parts of the feature data, so it could avoid excessive parameter setting and reduce overfitting problems. Moreover, the addition of the InSAR factor increased the accuracy of prediction up to 6%.

To further verify the proposed approach, the model established based on the training data from the two study areas was applied to another area in northern Taiwan. In December 2020, a landslide covering a slope area of around 4000 m^2 and 10,000 m^3 in earth volume occurred in this region. By feeding the local spatial factors into the model, the landslide susceptibility of each slope unit was obtained. Figure 7a,b illustrate the validation results from using 13 spatial factors (excluding InSAR data) and 14 spatial factors (including InSAR data), respectively. It shows that a medium (50–75%) landslide potential was obtained for the landslide area if only the geological factors were considered. However, when the InSAR data were included, the model gave a high (>75%) landslide potential for that slope unit. In other words, the InSAR data provided an essential contribution for improving the prediction accuracy, as also revealed in the two study areas previously mentioned. Furthermore, it should be stressed that the model used here was established based on the training data in the two study areas in southern Taiwan, but it can still perform well in this validation case in northern Taiwan. This gives an encouraging indication that the model established based on the proposed methodology is valid not only in the study areas but could be also applicable elsewhere.

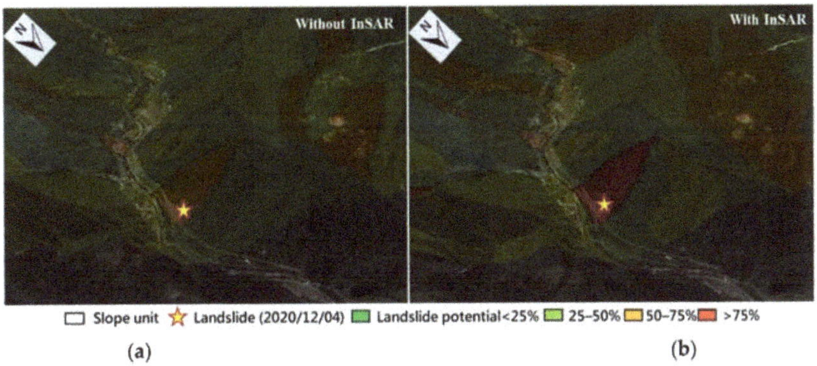

Figure 7. Landslide susceptibility analysis for the 2020 landslide case in northern Taiwan: (**a**) with InSAR data; (**b**) without InSAR data.

Overall, this research reveals that InSAR observables and multiple geological factors should be integrated for landslide susceptibility analysis with machine learning technology. Future studies can refine the current timescale of annual observations into months or days to enhance the calculation accuracy. Furthermore, mechanical factors, such as fluid shearing forces and soil slippage, can be considered to improve the prediction model.

5. Conclusions

Slope instability is affected by the topography and geological conditions, and artificial construction, such as tree cutting for planting cash crops and building roads, increases the vulnerability of the landform. The prevailing extreme climate now promotes the possibility of landslide disasters in the event of short-term heavy rainfall. This study introduced the modern InSAR technology, terrain, geological, and rainfall observation data to construct spatial factors based on slope units. Through Spearman correlation analysis and verification, significant impact factors in the experimental areas were detected. More importantly, machine learning was applied for the first time to construct prediction models combining spatial factors and landslide issues. Finally, two field experiments confirmed the feasibility of the landslide susceptibility prediction analysis proposed in this study. The results prove that a better-than-80% model accuracy can be achieved by the Random Forest algorithm, and the InSAR observable is able to increase the accuracy of prediction for all training models. Relevant management will be able to follow the potential landslide slope unit to provide vegetation restoration and slope reinforcement. Eventually, this novel strategy will provide the benefits of prevention and rescue for slope landslide disasters in a forward-looking manner. Finally, it should be noted that this study only used the landslide cases in Taiwan as examples. Further studies can be conducted using the proposed methodology for the cases with various geological and climatic conditions around the world using the training data in that region.

Author Contributions: Conceptualization, Y.-T.L. and J.-Y.H.; methodology, Y.-T.L., K.-H.Y., C.-S.C. and J.-Y.H.; software, Y.-T.L. and Y.-K.C.; validation, Y.-T.L., K.-H.Y., C.-S.C. and J.-Y.H.; formal analysis, Y.-T.L. and J.-Y.H.; writing, Y.-T.L. and Y.-K.C.; visualization, Y.-T.L. and Y.-K.C.; supervision, Y.-K.C., C.-S.C. and J.-Y.H. project administration, J.-Y.H.; funding acquisition, J.-Y.H. All authors have read and agreed to the published version of the manuscript.

Funding: This research is supported by CECI Engineering Consultants Inc. under contact number 06109C9007.

Institutional Review Board Statement: Not applicable.

Informed Consent Statement: Not applicable.

Data Availability Statement: Data is available from the corresponding author upon reasonable request.

Acknowledgments: The authors would like to thank the four anonymous reviewers for their constructive comments that helped to improve the original manuscript. The authors would also like to express their gratitude to the CECI Engineering Consultants Inc. in Taiwan for providing the InSAR data used in this study.

Conflicts of Interest: The authors declare no conflict of interest.

References

1. Lin, S.C.; Ke, M.C.; Lo, C.M. Evolution of landslide hotspots in Taiwan. *Landslides* **2017**, *14*, 1491–1501. [CrossRef]
2. Ma, Z.; Mei, G.; Piccialli, F. Machine learning for landslides prevention: A survey. *Neural Comput. Appl.* **2020**, 1–27. [CrossRef]
3. Benoit, L.; Briole, P.; Martin, O.; Thom, C.; Malet, J.P.; Ulrich, P. Monitoring landslide displacements with the Geocube wireless network of low-cost GPS. *Eng. Geol.* **2015**, *195*, 111–121. [CrossRef]
4. Wang, G. GPS Landslide Monitoring: Single Base vs. Network Solutions—A case study based on the Puerto Rico and Virgin Islands Permanent GPS Network. *J. Geod. Sci.* **2011**, *1*, 191–203. [CrossRef]
5. Ciuffi, P.; Bayer, B.; Berti, M.; Franceschini, S.; Simoni, A. Deformation Detection in Cyclic Landslides Prior to Their Reactivation Using Two-Pass Satellite Interferometry. *Appl. Sci.* **2021**, *11*, 3156. [CrossRef]
6. Guzzetti, F.; Mondini, A.C.; Cardinali, M.; Fiorucci, F.; Santangelo, M.; Chang, K.T. Landslide inventory maps: New tools for an old problem. *Earth-Sci. Rev.* **2012**, *112*, 42–66. [CrossRef]
7. Dabiri, Z.; Hölbling, D.; Abad, L.; Helgason, J.K.; Sæmundsson, Þ.; Tiede, D. Assessment of Landslide-Induced Geomorphological Changes in Hítardalur Valley, Iceland, Using Sentinel-1 and Sentinel-2 Data. *Appl. Sci.* **2020**, *10*, 5848. [CrossRef]
8. Ramirez, R.; Lee, S.R.; Kwon, T.H. Long-Term Remote Monitoring of Ground Deformation Using Sentinel-1 Interferometric Synthetic Aperture Radar (InSAR): Applications and Insights into Geotechnical Engineering Practices. *Appl. Sci.* **2020**, *10*, 7447. [CrossRef]
9. Ventisette, C.D.; Righini, G.; Moretti, S.; Casagli, N. Multitemporal landslides inventory map updating using spaceborne SAR analysis. *Int. J. Appl. Earth Obs. Geoinf.* **2014**, *30*, 238–246. [CrossRef]
10. Shirani, K.; Pasandi, M. Detecting and monitoring of landslides using persistent scattering synthetic aperture radar interferometry. *Environ. Earth Sci.* **2019**, *78*, 42. [CrossRef]
11. Carlà, T.; Intrieri, E.; Raspini, F.; Bardi, F.; Farina, P.; Ferretti, A.; Colombo, D.; Novali, F.; Casagli, N. Perspectives on the prediction of catastrophic slope failures from satellite InSAR. *Sci. Rep.* **2019**, *9*, 14137. [CrossRef]
12. Utomo, D.; Chen, S.F.; Hsiung, P.A. Landslide Prediction with Model Switching. *Appl. Sci.* **2019**, *9*, 1839. [CrossRef]
13. Chen, H.E.; Chiu, Y.Y.; Tsai, T.L.; Yang, J.C. Effect of Rainfall, Runoff and Infiltration Processes on the Stability of Footslopes. *Water* **2020**, *12*, 1229. [CrossRef]
14. Bui, D.T.; Tuan, T.A.; Klempe, H.; Pradhan, B.; Revhaug, I. Spatial prediction models for shallow landslide hazards: A comparative assessment of the efficacy of support vector machines, artificial neural networks, kernel logistic regression, and logistic model tree. *Landslides* **2016**, *13*, 361–378.
15. Al-Najjar, H.A.H.; Pradhan, B. Spatial landslide susceptibility assessment using machine learning techniques assisted by additional data created with generative adversarial networks. *Geosci. Front.* **2021**, *12*, 625–637. [CrossRef]
16. Nsengiyumva, J.B.; Valentino, R. Predicting landslide susceptibility and risks using GIS-based machine learning simulations, case of upper Nyabarongo catchment. *Geomat. Nat. Hazards Risk* **2020**, *11*, 1250–1277. [CrossRef]
17. Nhu, V.H.; Zandi, D.; Shahabi, H.; Chapi, K.; Shirzadi, A.; Al-Ansari, N.; Singh, S.K.; Dou, J.; Nguyen, H. Comparison of Support Vector Machine, Bayesian Logistic Regression, and Alternating Decision Tree Algorithms for Shallow Landslide Susceptibility Mapping along a Mountainous Road in the West of Iran. *Appl. Sci.* **2020**, *10*, 5047. [CrossRef]
18. Shha, S.; Saha, A.; Hembram, T.K.; Pradhan, B.; Alamri, A.M. Evaluating the Performance of Individual and Novel Ensemble of Machine Learning and Statistical Models for Landslide Susceptibility Assessment at Rudraprayag District of Garhwal Himalaya. *Appl. Sci.* **2020**, *10*, 3772. [CrossRef]
19. Lin, Y.T.; Yen, H.Y.; Chang, N.H.; Lin, H.M.; Han, J.Y.; Yang, K.H.; Chen, C.S.; Zheng, H.K.; Hsu, J.Y. Prediction of Landslides Using Machine Learning Techniques Based on Spatio-temporal Factors and InSAR Data. *J. Chin. Inst. Civ. Hydraul. Eng.* **2021**, *33*, 93–104.
20. Yu, L.; Cao, Y.; Zhou, C.; Wang, Y.; Huo, Z. Landslide Susceptibility Mapping Combining Information Gain Ratio and Support Vector Machines: A Case Study fromWushan Segment in the Three Gorges Reservoir Area, China. *Appl. Sci.* **2019**, *9*, 4756. [CrossRef]
21. Hu, X.; Zhang, H.; Mei, H.; Xiao, D.; Li, Y.; Li, M. Landslide Susceptibility Mapping Using the Stacking Ensemble Machine Learning Method in Lushui, Southwest China. *Appl. Sci.* **2020**, *10*, 4016. [CrossRef]
22. Chen, W.; Shahabi, H.; Zhang, S.; Khosravi, K.; Shirzadi, A.; Chapi, K.; Pham, B.T.; Zhang, T.; Zhang, L.; Chai, H.; et al. Landslide Susceptibility Modeling Based on GIS and Novel Bagging-Based Kernel Logistic Regression. *Appl. Sci.* **2018**, *8*, 2540. [CrossRef]

23. Chen, W.; Sun, Z.; Han, J. Landslide Susceptibility Modeling Using Integrated Ensemble Weights of Evidence with Logistic Regression and Random Forest Models. *Appl. Sci.* **2019**, *9*, 171. [CrossRef]
24. Wang, H.; Zhang, L.; Yin, K.; Luo, H.; Li, J. Landslide identification using machine learning. *Geosci. Front.* **2021**, *12*, 351–364. [CrossRef]
25. Truong, X.L.; Mitamura, M.; Kono, Y.; Raghavan, V.; Yonezawa, G.; Truong, X.Q.; Do, T.H.; Bui, D.T.; Lee, S. Enhancing Prediction Performance of Landslide Susceptibility Model Using Hybrid Machine Learning Approach of Bagging Ensemble and Logistic Model Tree. *Appl. Sci.* **2018**, *8*, 1046. [CrossRef]
26. Pourghasemi, H.R.; Rahmati, O. Prediction of the landslide susceptibility: Which algorithm, which precision? *Catena* **2018**, *162*, 177–192. [CrossRef]
27. Korup, O.; Stolle, A. Landslide prediction from machine learning. *Geol. Today* **2014**, *30*, 26–33. [CrossRef]
28. Niu, X.; Ma, J.; Wang, Y.; Zhang, J.; Chen, H.; Tang, H. A Novel Decomposition-Ensemble Learning Model Based on Ensemble Empirical Mode Decomposition and Recurrent Neural Network for Landslide Displacement Prediction. *Appl. Sci.* **2021**, *11*, 4684. [CrossRef]
29. Samia, J.; Temme, A.; Bregt, A.; Wallinga, J.; Guzzetti, F.; Ardizzone, F.; Rossi, M. Do landslides follow landslides? Insights in path dependency from a multi-temporal landslide inventory. *Landslides* **2017**, *14*, 547–558. [CrossRef]
30. Wu, J.; Zeng, W.; Yan, F. Hierarchical Temporal Memory method for time-series-based anomaly detection. *Neurocomputing* **2018**, *273*, 535–546. [CrossRef]
31. Xie, M.; Esaki, T.; Zhou, G. GIS-Based Probabilistic Mapping of Landslide Hazard Using a Three-Dimensional Deterministic Model. *Nat. Hazards* **2004**, *33*, 265–282. [CrossRef]
32. Franklin, J.A. Safety and economy in tunneling. In Proceedings of the 10th Canadian Rock Mechanics Symposium, Kingston, ON, Canada, 2–4 September 1975; Volume 1, pp. 27–53.
33. Rish, I. An empirical study of the naive Bayes classifier. In Proceedings of the IJCAI 2001 Workshop on Empirical Methods in Artificial Intelligence, Seattle, WA, USA, 4–6 August 2001; Volume 3, pp. 41–46.
34. Safavian, S.R.; Landgrebe, D. A survey of decision tree classifier methodology. *IEEE Trans. Syst. Man Cybern.* **1991**, *21*, 660–674. [CrossRef]
35. Ho, T.K. Random decision forests. In Proceedings of the 3rd International Conference on Document Analysis and Recognition, Montreal, QC, Canada, 14–16 August 1995; Volume 1, pp. 278–282.
36. Freund, Y.; Schapire, R.E. A decision-theoretic generalization of on-line learning and an application to boosting. *J. Comput. Syst. Sci.* **1997**, *55*, 119–139. [CrossRef]
37. Friedman, J.H. Stochastic gradient boosting. *Comput. Stat. Data Anal.* **2002**, *38*, 367–378. [CrossRef]

mathematics

MDPI

Article

Thermoelastic Coupling Response of an Unbounded Solid with a Cylindrical Cavity Due to a Moving Heat Source

Ashraf M. Zenkour [1,2,*], Daoud S. Mashat [1] and Ashraf M. Allehaibi [1,3]

1 Department of Mathematics, Faculty of Science, King Abdulaziz University, P.O. Box 80203, Jeddah 21589, Saudi Arabia; dmashat@kau.edu.sa (D.S.M.); amlehaibi@uqu.edu.sa (A.M.A.)
2 Department of Mathematics, Faculty of Science, Kafrelsheikh University, Kafrelsheikh 33516, Egypt
3 Department of Mathematics, Jamoum University College, Umm Al-Qura University, Jamoum, Makkah 21955, Saudi Arabia
* Correspondence: zenkour@kau.edu.sa or zenkour@sci.kfs.edu.eg

Abstract: The current article introduces the thermoelastic coupled response of an unbounded solid with a cylindrical hole under a traveling heat source and harmonically altering heat. A refined dual-phase-lag thermoelasticity theory is used for this purpose. A generalized thermoelastic coupled solution is developed by using Laplace's transforms technique. Field quantities are graphically displayed and discussed to illustrate the effects of heat source, phase-lag parameters, and the angular frequency of thermal vibration on the field quantities. Some comparisons are made with and without the inclusion of a moving heat source. The outcomes described here using the refined dual-phase-lag thermoelasticity theory are the most accurate and are provided as benchmarks for other researchers.

Keywords: G–N; L–S and CTE theories; cylindrical hole; dual-phase-lag; moving velocity

Citation: Zenkour, A.M.; Mashat, D.S.; Allehaibi, A.M. Thermoelastic Coupling Response of an Unbounded Solid with a Cylindrical Cavity Due to a Moving Heat Source. *Mathematics* **2022**, *10*, 9. https://doi.org/10.3390/math10010009

Academic Editors: Nikos D. Lagaros and Vagelis Plevris

Received: 27 November 2021
Accepted: 17 December 2021
Published: 21 December 2021

1. Introduction

The thermoelasticity theory is adopted in various applications to obtain interesting formulations due to a variety of microphysical processes. The starting point of the classical coupled thermoelasticity (CTE) model was founded by Duhamel [1]. While Biot [2] formulated the CTE theory by considering the second law of thermodynamics. One of the first generalized theories is established by Lord and Shulman (L–S) [3] by including a thermal relaxation parameter. While Green and Lindsay [4] developed another generalized model by including two thermal relaxation parameters. Such generalized theories with one or more thermal relaxation parameters are also stated as hyperbolic thermoelasticity theories [5]. Green and Nagdhi (G–N) [6–8] formulated three various theories of thermoelasticity in an unusual way. In addition, Tzou [9,10] presented a modern generalized one which is called a dual-phase-lag (DPL) theory. A lot of research is presented to include and modify Tzou's model (see, e.g., [11–15]).

Many problems found in the literature are concerned with the thermoelastic response of unbounded bodies with cylindrical cavities. Chandrasekharaiah and Srinath [16] applied the G–N II model to analyze axisymmetric thermoelastic communications in an unbounded solid including a cylindrical hole. Allam et al. [17] discussed thermal distribution field quantities of a half-space containing a circular cylindrical cavity in the framework of a G–N model. Ezzat and El-Bary [18,19] used a fractional-order of both thermo-viscoelasticity and magneto-thermoelasticity theories to deal with an unbounded perfect conducting media having a cylindrical hole in the existence of an axial uniform magnetic field. Sharma et al. [20] tried to solve the dynamic formulation of an elasto-thermo-diffusion infinite cylindrical hole under various boundary conditions. Kumar and Mukhopadhyay [21] presented the impacts of three-phase-lags (TPLs) on thermoelastic communications under step input in temperature on a cylindrical hole in an infinite body. Mukhopadhyay and Kumar [22] dealt with the thermoelastic communications in an infinite solid with a cylindrical hole based on a two-temperature L–S model. Kumar et al. [23] described the thermoelastic communications based on a hyperbolic two-temperature L–S model in an unbounded body including a cylindrical hole. Sarkar and Mondal [24] examined transient behavior in a two-temperature model in an infinite body with a cylindrical hole under a time-dependent moving heat source.

A lot of infinite bodies having cylindrical cavities may be exposed to either continuous heat source, ramp-type heating effect, or thermal shock. Sharma et al. [25] considered one-dimensional elasto-thermo-diffusive communications in an infinite solid containing a cylindrical hole under the action of a continuous heat source utilizing the L–S theory. Mukhopadhyay and Kumar [26] analyzed the thermoelastic communications in a medium having a cylindrical hole under a ramp-type heating effect using various models. Xia et al. [27] used the L–S model to develop a generalized thermoelastic diffusion theory for the dynamic response of an unbounded body having a cylindrical hole and its surface undergoing a thermal shock. In addition, Xiong and Tian [28] discussed the thermoelastic analysis of an unbounded medium with a cylindrical hole whose surface undergoes time-dependent thermal shock due to G–N II and III theories.

Some interesting problems are concerned with thermoelastic communications in elastic infinite media with cylindrical holes and subjected to moving heat sources. Abouelregal [29] obtained the induced fields in such an unbounded body having a cylindrical hole under a traveling heat source and harmonically varying heat based on the dual-phase-lag theory. Youssef [30] presented the analysis of thermoelastic communications in an elastic infinite body with a cylindrical cavity a moving heat source with a uniform velocity that thermally shocked at the bounding surface. Youssef [31] used the G–N II theory to develop a two-temperature model for an infinite medium having constant elastic parameters. Shaw and Mukhopadhyay [32] presented thermoelastic communications in a micro-stretched body in the existence of a traveling heat source. Sarkar and Lahiri [33] solved a 1D problem for a thermoelastic infinite medium under a moving plane of heat source. Youssef [34] presented a two-temperature fractional-order theory for an infinite medium. Xia et al. [35] studied a semi-infinite medium under a traveling heat source by utilizing the finite element method in the time domain in the context of the L–S model. Abbas [36] solved the problem of thermoelastic communication in a clamped microscale beam under a moving heat source based on G–N III theory. Youssef [37] discussed the thermoelastic communications in an unbounded solid having a cylindrical hole in the existence of moving heat sources utilizing the L–S model.

In this paper, the problem of an unbounded solid containing a cylindrical cavity is studied. The governing equations are carried out based on the refined dual-phase-lag (RDPL) thermoelasticity theory. The general solution gained is utilized to a certain problem once the bounding plane of the cavity is exposed to a traveling heat source. The inverse Laplace transforms is calculated numerically to obtain the field quantities. Some comparisons will be tabulated and shown graphically to study the benefit of different theories and estimate the effect of different parameters.

2. Fundamental Equations

Let us discuss a thermoelastic coupling response of an unbounded solid containing a cylindrical hole due to a traveling heat source using a unified dual-phase-lag theory. The cylindrical coordinates system (r, φ, z) is selected to deal with such a problem in which z-axis is sitting alongside the axis of the cylindrical cavity.

The displacement vector $\vec{u}$ of the present, an axially symmetric cylindrical cavity is summarized as

$$u_r = u(r,t), u_\varphi = u_z = 0. \tag{1}$$

The strains can be expressed as

$$e_{rr} = \frac{\partial u}{\partial r}, \quad e_{\varphi\varphi} = \frac{u}{r}, \quad e = e_{rr} + e_{\varphi\varphi} = \frac{1}{r}\frac{\partial}{\partial r}(ru). \tag{2}$$

The dynamic equation in the non-existence of body force is stated as

$$(2\mu+\lambda)\left(\nabla^2 u-\frac{u}{r^2}\right)-\gamma\frac{\partial\theta}{\partial r}=\rho\frac{\partial^2 u}{\partial t^2}, \tag{3}$$

where ∇^2 is the Laplacian operator. It satisfies the formula

$$\nabla^2(*)=\frac{1}{r}\frac{\partial}{\partial r}\left(r\frac{\partial(*)}{\partial r}\right). \tag{4}$$

The constitutive equations for the coupled thermoelastic solid with omitting the volume forces can be stated as

$$\begin{aligned}\sigma_{rr}&=2\mu\tfrac{\partial u}{\partial r}+\sigma_{zz},\\ \sigma_{\varphi\varphi}&=2\mu\tfrac{u}{r}+\sigma_{zz},\\ \sigma_{zz}&=\lambda e-\gamma\theta,\\ \sigma_{zr}&=\sigma_{r\varphi}=\sigma_{\varphi z}=0.\end{aligned} \tag{5}$$

The heat conduction equation in the refined form is represented by

$$kL_\theta\nabla^2\theta=L_q\left(\rho C_e\frac{\partial\theta}{\partial t}+\gamma T_0\frac{\partial e}{\partial t}-Q\right), \tag{6}$$

where L_θ and L_q are higher-order time derivative operators given by

$$L_\theta=1+\sum_{n=1}^{N}\frac{\tau_\theta^n}{n!}\frac{\partial^n}{\partial t^n},\ L_q=\varrho+\sum_{n=1}^{N}\frac{\tau_q^n}{n!}\frac{\partial^n}{\partial t^n}, \tag{7}$$

which represents one of the modified coupled forms of the heat transport equation presented in [38–41]. The above equation represents the more general when N has some +ve integers larger than zero. Different particular cases can be considered along with Equation (6) as follows:

(i) Coupled dynamical thermoelasticity (CTE theory) [2]: $\tau_\theta=\tau_q=0$ and $\varrho=1$

$$k\nabla^2\theta=\rho C_e\frac{\partial\theta}{\partial t}+\gamma T_0\frac{\partial e}{\partial t}-Q. \tag{8}$$

(ii) Lord and Shulman generalized thermoelasticity theory (L–S theory) [3]: $\tau_\theta=0$, $\tau_q=\tau_0$ and $\varrho=1$

$$k\nabla^2\theta=\left(1+\tau_q\frac{\partial}{\partial t}\right)\left(\rho C_e\frac{\partial\theta}{\partial t}+\gamma T_0\frac{\partial e}{\partial t}-Q\right). \tag{9}$$

(iii) Green and Naghdi generalized thermoelasticity theory (G–N theory) without energy dissipation [6–8]: $\tau_\theta=0$, $\tau_q=1$, $N=1$, $\varrho=0$, and $k\to k^*$

$$k^*\nabla^2\theta=\frac{\partial}{\partial t}\left(\rho C_e\frac{\partial\theta}{\partial t}+\gamma T_0\frac{\partial e}{\partial t}-Q\right). \tag{10}$$

(iv) Simple generalized thermoelasticity theory with dual-phase-lag (SDPL theory) [12–15]: $\tau_q\geq\tau_\theta>0$, $N=1$ and $\varrho=1$

$$k\left(1+\tau_\theta\frac{\partial}{\partial t}\right)\nabla^2\theta=\left(1+\tau_q\frac{\partial}{\partial t}\right)\left(\rho C_e\frac{\partial\theta}{\partial t}+\gamma T_0\frac{\partial e}{\partial t}-Q\right). \tag{11}$$

(v) Refined generalized thermoelasticity theory with dual-phase-lag (RDPL theory) [12–15]: $\tau_q\geq\tau_\theta>0$, $N>1$ and $\varrho=1$,

$$k\left(1+\sum_{n=1}^{N}\frac{\tau_\theta^n}{n!}\frac{\partial^n}{\partial t^n}\right)\nabla^2\theta=\left(1+\sum_{n=1}^{N}\frac{\tau_q^n}{n!}\frac{\partial^n}{\partial t^n}\right)\left(\rho C_e\frac{\partial\theta}{\partial t}+\gamma T_0\frac{\partial e}{\partial t}-Q\right). \tag{12}$$

3. Problem Construction

It is appropriate to introduce the following non-dimensional variables in the next sections:

$$\begin{gathered}\{r',u'\}=c_0\eta\{r,u\},\left\{t',\tau_\theta',\tau_q'\right\}=\eta c_0^2\{t,\tau_\theta,\tau_q\},\\ \sigma_{ii}'=\frac{\sigma_{ii}}{\gamma T_0},\theta'=\frac{\theta}{T_0},c_0^2=\frac{\lambda+2\mu}{\rho},\quad \eta=\frac{\rho C_e}{k},\end{gathered} \tag{13}$$

and setting $Q'=Q/\left(kT_0c_0^2\eta^2\right)$. All governing equations, with the directions above dimensionless variables, are lowered to (dropping the dashed for suitability)

$$\sigma_{rr}=c_1\frac{\partial u}{\partial r}+\sigma_{zz}, \tag{14}$$

$$\sigma_{\varphi\varphi}=c_1\frac{u}{r}+\sigma_{zz}, \tag{15}$$

$$\sigma_{zz}=c_2e-\theta, \tag{16}$$

$$\nabla^2u-\frac{u}{r^2}-\frac{1}{c_3}\frac{\partial\theta}{\partial r}=\frac{\partial^2u}{\partial t^2}, \tag{17}$$

$$\left(\nabla^2L_\theta-L_q\frac{\partial}{\partial t}\right)\theta-\varepsilon L_q\left(\frac{\partial e}{\partial t}\right)=-L_qQ, \tag{18}$$

where

$$c_1=\frac{2\mu}{\gamma T_0},\quad c_2=\frac{\lambda}{\gamma T_0},\quad c_3=\frac{\lambda+2\mu}{\gamma T_0}=c_1+c_2,\quad \varepsilon=\frac{\gamma}{\rho C_e}. \tag{19}$$

The heat source is shifting along the radial direction with a uniform velocity ϑ, which be able to be defined by the formula

$$Q=Q_0\delta(r-\vartheta t). \tag{20}$$

4. Closed-Form Solution

The complete solutions will be given by solving Equations (17) and (18) to get firstly the temperature θ and the dilatation e. Then, the corresponding displacement and stresses can be given in terms of temperature and dilatation. For this purpose, we utilize the following initial conditions:

$$u(r,0)=\left.\frac{\partial u}{\partial t}\right|_{t=0}=0,\quad \theta(r,0)=\left.\frac{\partial\theta}{\partial t}\right|_{t=0}=0,\quad R\le r<\infty. \tag{21}$$

We supply the above homogenous initial conditions by additional boundary conditions. The present medium will be considered as inactive and the surface of the cylindrical cavity under harmonically varying heat and traction free. These conditions are defined as

- The surface of the cylindrical cavity is exposed to a harmonically varying heat

$$(R,t)=\theta_0H(t)\cos(\omega t),\quad t>0. \tag{22}$$

- The mechanical boundary condition is considered as the surface of the cylindrical cavity is traction free

$$\sigma_{rr}(R,t)=0. \tag{23}$$

It is to be noted that when the angular frequency of thermal vibration $\omega \to 0$, the problem tends to be a thermal shock one. In addition to the above initial and boundary conditions, we take into account the following regularity conditions

$$u(r,t) = 0, \quad \theta(r,t) = 0, \quad r \to \infty. \tag{24}$$

Laplace transform is taken for Equations (20)–(23), under the homogeneous initial conditions that appeared in Equation (28), one gets:

$$\overline{\sigma}_{rr} = c_1 \frac{d\overline{u}}{dr} + \overline{\sigma}_{zz}, \tag{25}$$

$$\overline{\sigma}_{\varphi\varphi} = c_1 \frac{\overline{u}}{r} + \overline{\sigma}_{zz}, \tag{26}$$

$$\overline{\sigma}_{zz} = c_2 \overline{e} - \overline{\theta}. \tag{27}$$

Taking the discrepancy of both sides of Equation (24) then one gets

$$c_3 \left(\nabla^2 - s^2\right)\overline{e} - \nabla^2 \overline{\theta} = 0, \tag{28}$$

while the other governing equation of heat conduction became

$$\left(\nabla^2 - s\varpi_0\right)\overline{\theta} - \varepsilon s \varpi_0 \overline{e} = -\varpi_1 e^{-\left(\frac{s}{\vartheta}\right)r}, \tag{29}$$

where

$$\varpi_0 = \frac{\overline{L}_q}{\overline{L}_\theta}, \quad \varpi_1 = \frac{Q_0}{|\vartheta|}\varpi_0, \quad \overline{L}_\theta = 1 + \sum_{n=1}^{N} \frac{\tau_\theta^n}{n!} s^n, \quad \overline{L}_q = \varrho + \sum_{n=1}^{N} \frac{\tau_q^n}{n!} s^n. \tag{30}$$

The equations occurred in Equations (36)–(38) be able to be identified in a fourth-order ordinary non-homogenous differential equation in the dilatation $\overline{e}$ as in the form

$$\left(\nabla^4 - \beta_1 \nabla^2 + \beta_0\right)\overline{e}(r) = g(r), \tag{31}$$

where

$$g(r) = -\varpi_2 \left(s - \frac{\vartheta}{r}\right) e^{-\left(\frac{s}{\vartheta}\right)r}, \tag{32}$$

and the coefficients β_i are given by

$$\beta_0 = s^3 \varpi_0, \beta_1 = s(s + \varpi_3), \varpi_2 = \frac{\varpi_1 s}{C_3 \vartheta^2}, \varpi_3 = \varpi_0 \left(1 + \frac{\varepsilon}{c_3}\right). \tag{33}$$

Equation (31) is extremely complicated since it is done in a polar coordinate system. It can be re-considered as

$$\left(\nabla^2 - \zeta_1^2\right)\left(\nabla^2 - \zeta_2^2\right)\overline{e}(r) = g(r), \tag{34}$$

where ζ_j^2 are the roots of the equation

$$\zeta^4 - \beta_1 \zeta^2 + \beta_0 = 0. \tag{35}$$

These roots ζ_j are given respectively by

$$\zeta_{1,2}^2 = \frac{1}{2}\left(\beta_1 \pm \sqrt{\beta_1^2 - 4\beta_0}\right). \tag{36}$$

Equation (34) represents the next modified Bessel's equation of zero-order

$$\left[\frac{1}{r}\frac{\partial}{\partial r}\left(r\frac{\partial}{\partial r}\right)-\zeta_1^2\right]\left[\frac{1}{r}\frac{\partial}{\partial r}\left(r\frac{\partial}{\partial r}\right)-\zeta_2^2\right]\bar{e}(r)=g(r), \quad (37)$$

which gets a solution underneath the regularity conditions: $\bar{u}$, $\bar{\theta}\to 0$ as $r\to\infty$. So, the general solution of $\bar{e}$ is represented by

$$\bar{e}(r)=\sum_{j=1}^{2}B_jK_0(\zeta_j r)+\bar{e}_p, \quad (38)$$

where B_j are integration parameters, $K_0(\zeta_j r)$ is modified Bessel's function of the first kind, and $\bar{e}_p$ is a particular solution. It is provided by

$$\bar{e}_p=K_0(\zeta_2 r)\int\frac{g(r)K_0(\zeta_1 r)}{W(r)}\mathrm{d}r-K_0(\zeta_1 r)\int\frac{g(r)K_0(\zeta_2 r)}{W(r)}\mathrm{d}r, \quad (39)$$

in which $W(r)$ is the Wronskian

$$W(r)=\zeta_1K_0(\zeta_2 r)K_1(\zeta_1 r)-\zeta_2K_0(\zeta_1 r)K_1(\zeta_2 r). \quad (40)$$

Using the solution of $\bar{e}$ in Equation (39), one gets the solution of $\bar{\theta}$ as

$$\bar{\theta}=c_3\left[\sum_{j=1}^{2}\check{\zeta}_jB_jK_0(\zeta_j r)+\bar{\theta}_p\right], \quad (41)$$

where

$$\check{\zeta}_j=1-\frac{s^2}{\zeta_j^2},\quad \bar{\theta}_p=\bar{e}_p-s^2\int\bar{u}_p\mathrm{d}r,\quad \bar{u}_p=\frac{1}{r}\int r\bar{e}_p\mathrm{d}r. \quad (42)$$

The radial displacement be able to be stated for the regularity condition $\bar{u}\to 0$ as $r\to\infty$ from the formula

$$\bar{e}(r)=\mathcal{D}\bar{u}(r),\quad \mathcal{D}=\frac{\mathrm{d}}{\mathrm{d}r}+\frac{1}{r}, \quad (43)$$

in the form

$$\bar{u}(r)=-\sum_{j=1}^{2}\frac{1}{\zeta_j}B_jK_1(\zeta_j r)+\bar{u}_p, \quad (44)$$

where $\bar{u}_p$ is already given in Equation (42).

At this point, the solution is finished. It is sufficient to decide the two parameters B_j from the boundary conditions offered in Equations (22) and (23). Then, it is simple to carry out the stresses in terms of radial displacement and temperature. According to Equations (25)–(27), the stresses may be given in Laplace state as

$$\bar{\sigma}_1=\sum_{j=1}^{2}\left[c_3(1-\check{\zeta}_j)K_0(\zeta_j r)+\frac{c_1}{\zeta_j r}K_1(\zeta_j r)\right]B_j+c_1\frac{\mathrm{d}\bar{u}_p}{\mathrm{d}r}+c_2\bar{e}_p-c_3\bar{\theta}_p, \quad (45)$$

$$\bar{\sigma}_2=\sum_{j=1}^{2}\left[(c_2-c_3\check{\zeta}_j)K_0(\zeta_j r)-\frac{c_1}{\zeta_j r}K_1(\zeta_j r)\right]B_j+c_1\frac{\bar{u}_p}{r}+c_2\bar{e}_p-c_3\bar{\theta}_p, \quad (46)$$

$$\bar{\sigma}_3=\sum_{j=1}^{2}(c_2-c_3\check{\zeta}_j)K_0(\zeta_j r)B_j+c_2\bar{e}_p-c_3\bar{\theta}_p. \quad (47)$$

So, the current analytical solution is previously given for the RDPL theory in the Laplace domain. To complete the solution in the physical domain we should consider the function $f(t)$ as an inversion of the Laplace transform $\overline{f}(s)$ as

$$f(t) = \frac{e^{qt}}{t}\left[\frac{1}{2}\overline{f}(q) + \mathrm{Re}\left\{\sum_{l=1}^{L}(-1)^l\overline{f}\left(q + \frac{il\pi}{t}\right)\right\}\right], \tag{48}$$

where q is a random constant, Re denotes the real part, i implies the imaginary quantity unit and L implies a suitably huge integer. For quicker combination, numerous numerical analyses have shown that the evaluation of q satisfies the connection $qt \approx 4.7$ [35]. Utilizing the numerical technique mentioned, to reverse the statements of temperature θ, radial displacement u, dilatation e, radial stress σ_1, hoop stress σ_2, and axial stress σ_3.

5. Validation of Results

Several presentations will be offered to put into recommendation the impact of numerous models on the variable quantities. The material properties of the unbounded body having a cylindrical hole due to a traveling heat source are identified according to the following values of parameters:

$$\lambda = 7.76 \times 10^{10}\ \mathrm{N\,m^{-2}},\ \mu = 3.86 \times 10^{10}\ \mathrm{N\,m^{-2}},\ k = 386\ \mathrm{W\,m^{-1}\,K^{-1}},$$

$$\rho = 8954\ \mathrm{kg\,m^{-3}},\ \alpha_t = 1.78 \times 10^{-5}\ \mathrm{K^{-1}}$$

$$C_e = 383.1\ \mathrm{J\,kg^{-1}\,K^{-1}},\ T_0 = 293\ \mathrm{K},\ k^* = 1.2\ \mathrm{W\,m^{-1}\,K^{-1}}.$$

Numerical results are obtained (except otherwise stated) for $\theta_0 = 10$, $Q_0 = 1$, $\vartheta = 17$, $\omega = 20$, $\tau_q = 0.02$, $\tau_\theta = 0.018$, $t = 0.03$, and the inner radius $R = 1$.

5.1. First Validation Example

Results of the field quantities due to different thermoelasticity theories with dual-phase-lag are reported in Tables 1 and 2 at the position $r = 1.2$. The impact of the velocity of heat source ϑ on all field variables of different thermoelasticity theories with are presented at dimensionless time $t = 0.03$, and for two values of the angular frequency of thermal vibration $\omega = 0$ (Table 1) and $\omega = 20$ (Table 2). Additional outcomes of variable quantities are outlined in Figures 1–24 across the radial direction of an unbounded body with a cylindrical hole.

Table 1. Effect of the velocity of heat source ϑ on the field variables of different thermoelasticity theories with $t = 0.03$, $r = 1.2$, $\omega = 0$.

ϑ	Variable	CTE	G–N	L–S	SDPL		RDPL	
					N = 1	N = 3	N = 4	N = 5
17	$\overline{\theta}$	3.5206798	0.0074277	4.5558651	3.4642684	3.6460648	3.9083885	4.2446350
	e^*	−6.4643864	−0.7652114	−10.4814114	−6.5686154	−6.8240864	−7.0147188	−7.2539067
	u^*	−50.6571091	0.0540479	0.0234225	−37.4998105	−4.5146261	5.7141745	11.3103719
	$\overline{\sigma}_1$	16.5744843	−0.0362606	−4.6655407	11.3944541	−1.9149801	−6.2492888	−8.8146180
	$\overline{\sigma}_2$	−23.7087643	0.0104058	−4.5968524	−18.4174099	−5.4752409	−1.6683198	0.2210782
	$\overline{\sigma}_3$	−3.5517066	−0.0111005	−4.6061725	−3.4957955	−3.6788182	−3.9420569	−4.2794514
20	$\overline{\theta}$	4.1072506	−0.0029614	4.4306514	4.0800238	4.3206529	4.6090319	4.9665501
	e^*	−6.5105278	0.3051538	−10.1923716	−6.6189071	−6.8807134	−7.0727326	−7.3110451
	u^*	−0.2306539	0.0382336	0.0231236	2.4286536	2.5890897	1.9194700	1.3751839
	$\overline{\sigma}_1$	−4.0778064	−0.0093305	−4.5374406	−5.1097870	−5.4167623	−5.4405303	−5.5837524
	$\overline{\sigma}_2$	−4.2302789	0.0196396	−4.4703704	−3.1454027	−3.3234490	−3.8791992	−4.4544390
	$\overline{\sigma}_3$	−4.1384990	0.0044260	−4.4795715	−4.1117924	−4.3536781	−4.6429788	−5.0016408

$e^* = 10^4\overline{e}$, $u^* = 10^2\overline{u}$.

Table 2. Effect of the velocity of heat source ϑ on the field variables of different thermoelasticity theories with $t = 0.03, r = 1.2, \omega = 20$.

ϑ	Variable	CTE	G–N	L–S	SDPL		RDPL	
					$N = 1$	$N = 3$	$N = 4$	$N = 5$
17	$\overline{\theta}$	3.2835398	0.0074277	4.5528182	3.2447447	3.4508927	3.7188299	4.0555735
	e^*	−6.2860288	−0.7652114	−10.4740672	−6.3932523	−6.6542973	−6.8464693	−7.0855337
	u^*	−50.6572509	0.0540479	0.0234210	−37.4999432	−4.5147434	5.7140628	11.3102642
	$\overline{\sigma}_1$	16.8133885	−0.0362606	−4.6624229	11.6157097	−1.7181357	−6.0580749	−8.6239015
	$\overline{\sigma}_2$	−23.4708247	0.0104058	−4.5937709	−18.1970973	−5.2793005	−1.4779982	0.4109050
	$\overline{\sigma}_3$	−3.3137105	−0.0111005	−4.6030904	−3.2754301	−3.4828312	−3.7516908	−4.0895818
20	$\overline{\theta}$	3.8701106	−0.0029614	4.4276045	3.8605001	4.1254808	4.4194734	4.7774886
	e^*	−6.3321702	0.3051538	−10.1850275	−6.4435440	−6.7109243	−6.9044831	−7.1426721
	u^*	−0.2307957	0.0382336	0.0231221	2.4285209	2.5889724	1.9193583	1.3750762
	$\overline{\sigma}_1$	−3.8389023	−0.0093305	−4.5343228	−4.8885315	−5.2199179	−5.2493164	−5.3930359
	$\overline{\sigma}_2$	−3.9923393	0.0196396	−4.4672888	−2.9250901	−3.1275086	−3.6888775	−4.2646122
	$\overline{\sigma}_3$	−3.9005030	0.0044260	−4.4764894	−3.8914270	−4.1576911	−4.4526127	−4.8117711

$e^* = 10^4\overline{e}$, $u^* = 10^2\overline{u}$.

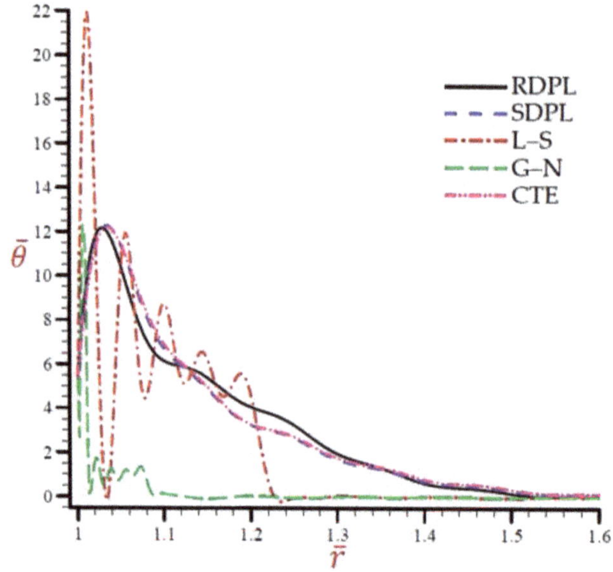

Figure 1. The temperature $\overline{\theta}$ across the radial direction of the cylindrical cavity conferring to all theories.

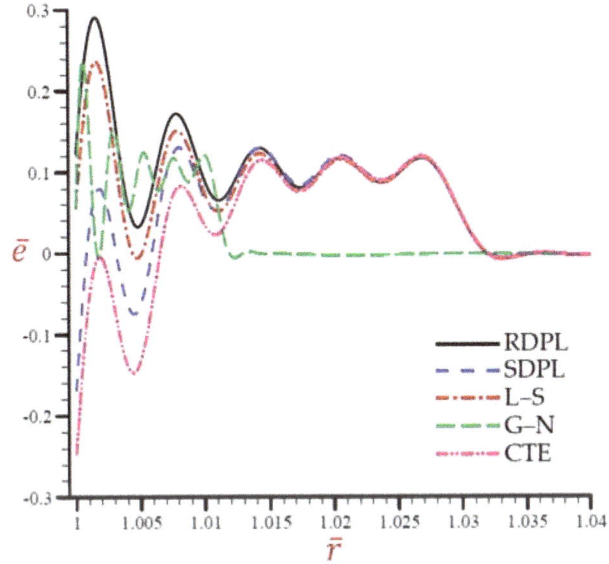

Figure 2. The dilatation $\bar{e}$ across the radial direction of the cylindrical cavity conferring to all theories.

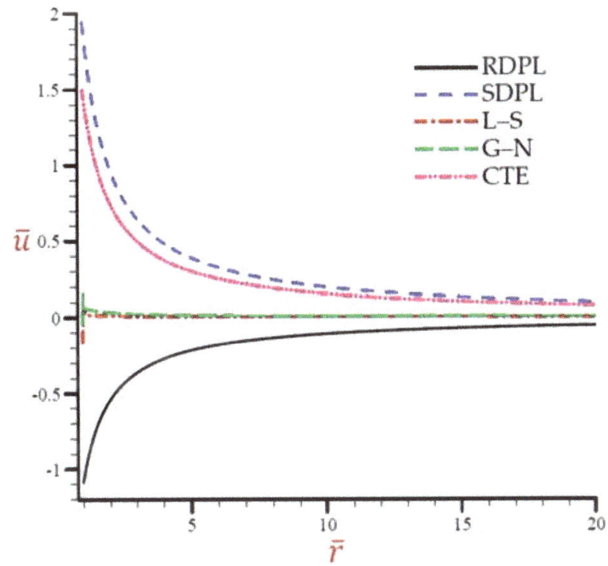

Figure 3. The radial displacement $\bar{u}$ across the radial direction of the cylindrical cavity conferring to all theories.

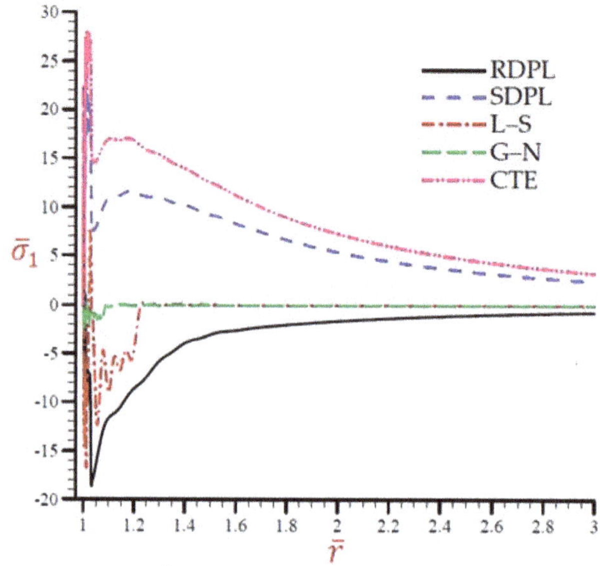

Figure 4. The radial stress $\overline{\sigma}_1$ across the radial direction of the cylindrical cavity conferring to all theories.

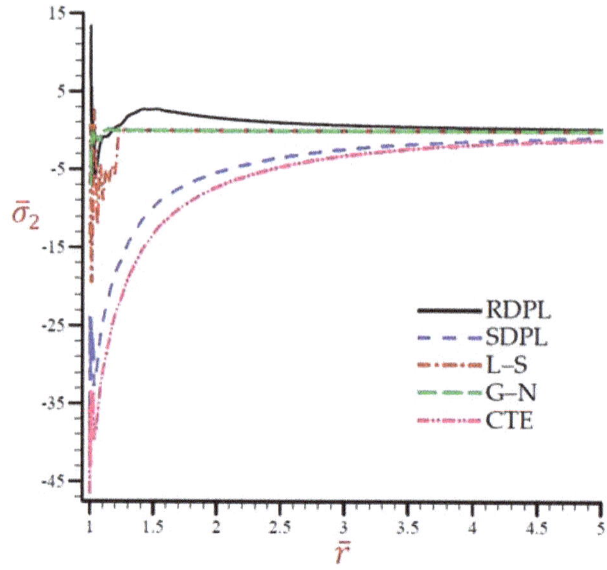

Figure 5. The hoop stress $\overline{\sigma}_2$ across the radial direction of the cylindrical cavity conferring to all theories.

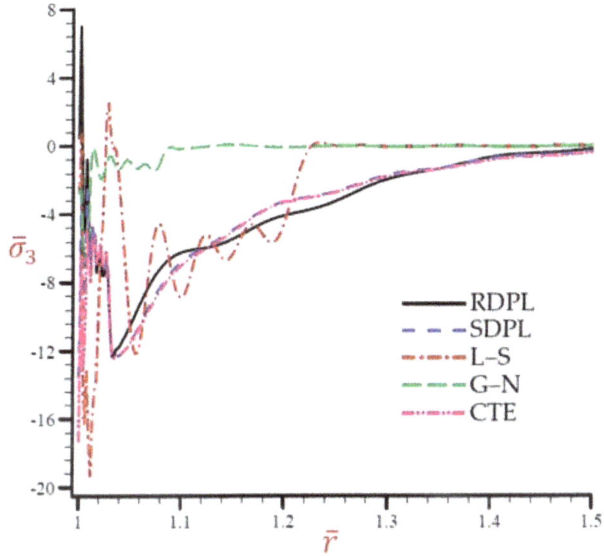

Figure 6. The axial stress $\bar{\sigma}_3$ across the radial direction of the cylindrical cavity conferring to all theories.

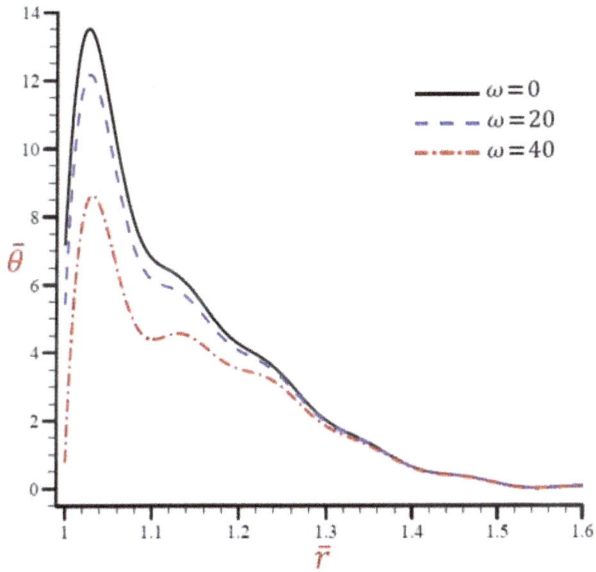

Figure 7. Effect of ω on temperature $\bar{\theta}$ across the radial direction of the cylindrical cavity utilizing the RDPL model.

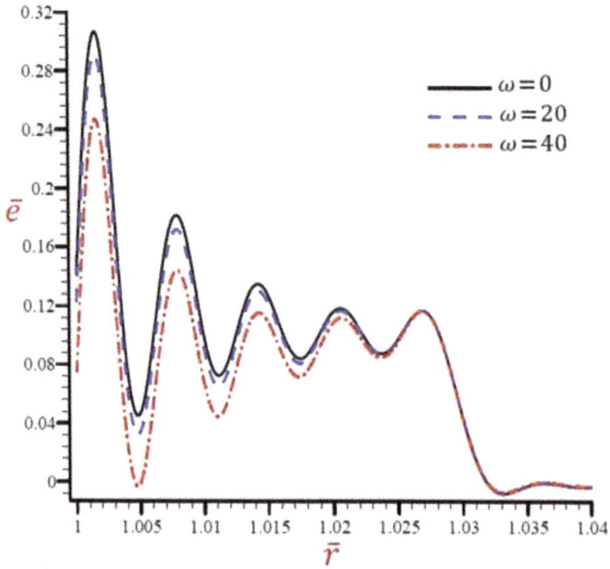

Figure 8. Effect of ω on dilatation $\bar{e}$ across the radial direction of the cylindrical cavity utilizing the RDPL model.

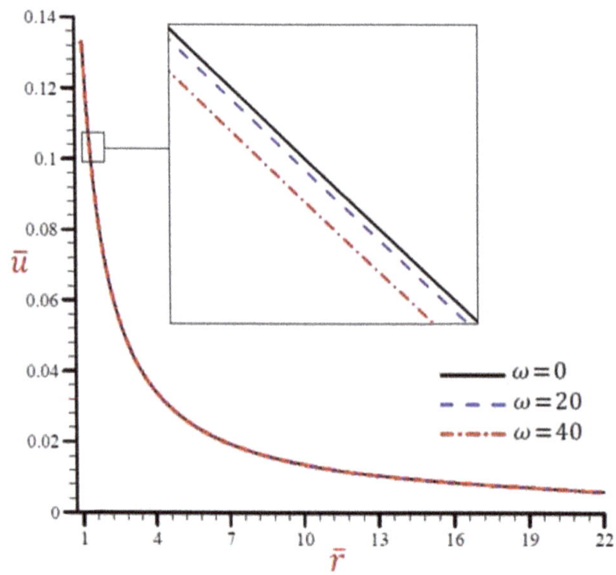

Figure 9. Effect of ω on radial displacement $\bar{u}$ across the radial direction of the cylindrical cavity utilizing the RDPL model.

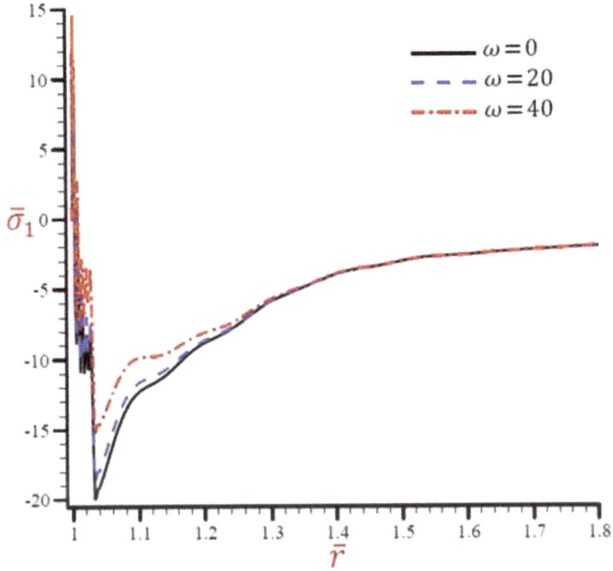

Figure 10. Effect of ω on radial stress $\overline{\sigma}_1$ across the radial direction of the cylindrical cavity utilizing the RDPL model.

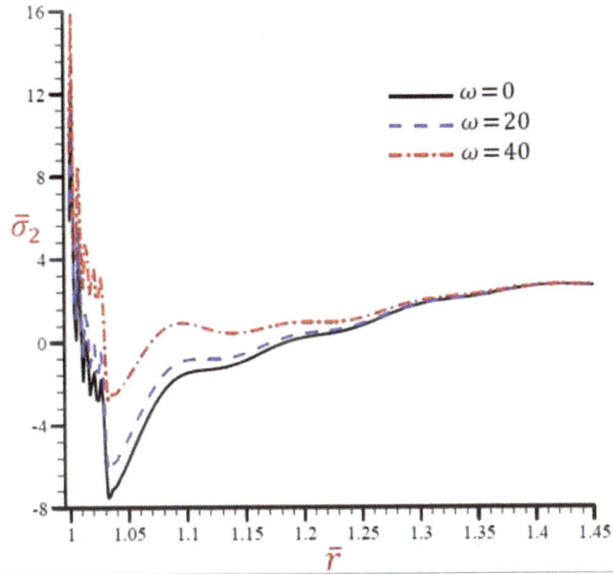

Figure 11. Effect of ω on radial stress $\overline{\sigma}_2$ across the radial direction of the cylindrical cavity utilizing the RDPL model.

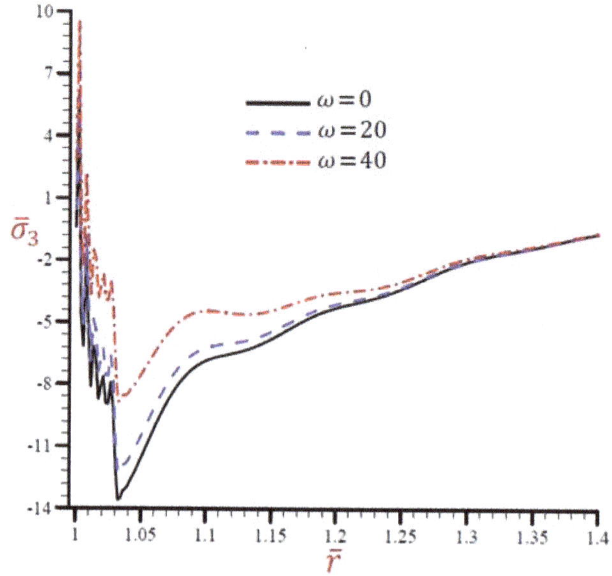

Figure 12. Effect of ω on radial stress $\overline{\sigma}_3$ across the radial direction of the cylindrical cavity utilizing the RDPL model.

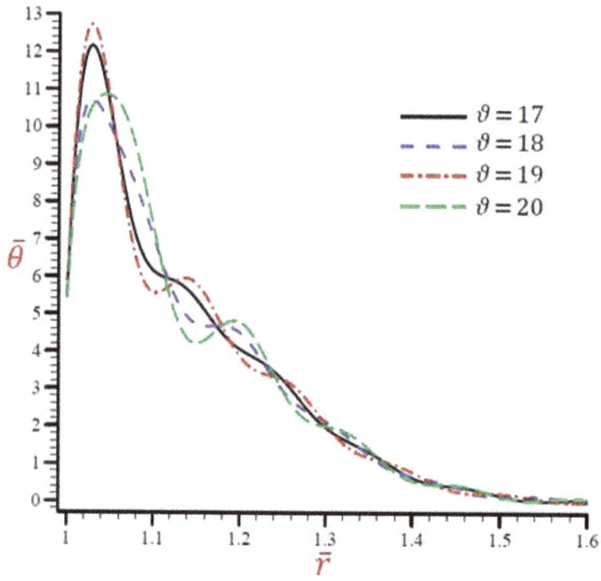

Figure 13. Effect of ϑ on temperature $\overline{\theta}$ across the radial direction of the cylindrical cavity utilizing the RDPL model.

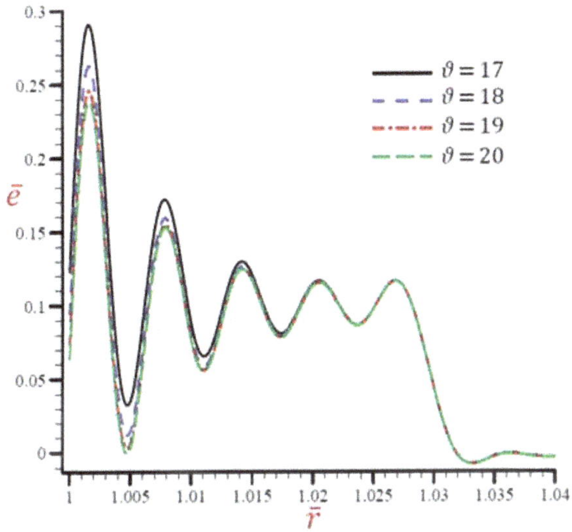

Figure 14. Effect of ϑ on dilatation $\bar{e}$ across the radial direction of the cylindrical cavity utilizing the RDPL model.

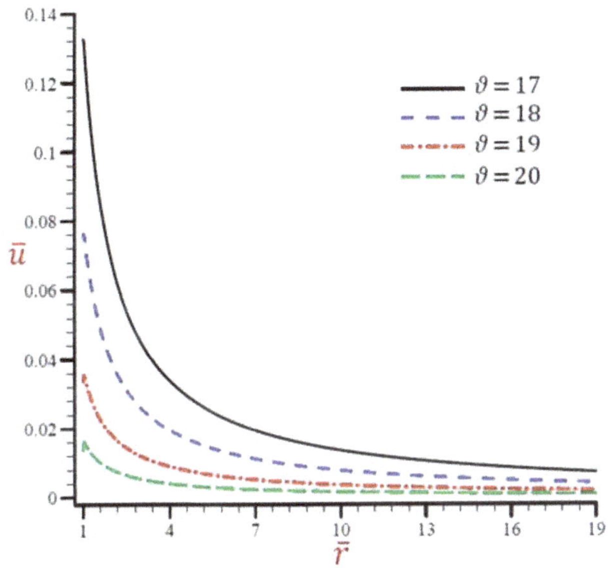

Figure 15. Effect of ϑ on radial displacement $\bar{u}$ across the radial direction of the cylindrical cavity utilizing the RDPL model.

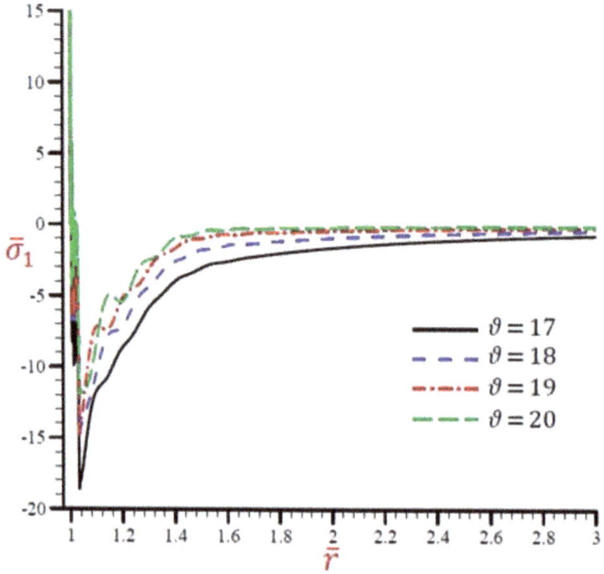

Figure 16. Effect of ϑ on radial stress $\overline{\sigma}_1$ across the radial direction of the cylindrical cavity utilizing the RDPL model.

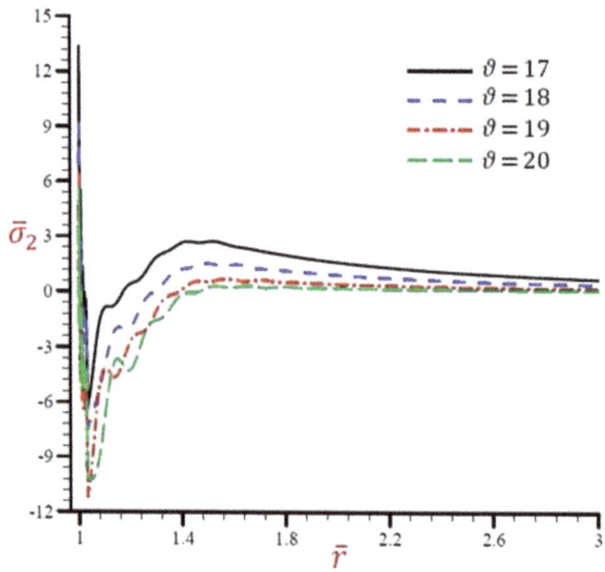

Figure 17. Effect of ϑ on radial stress $\overline{\sigma}_2$ across the radial direction of the cylindrical cavity utilizing the RDPL model.

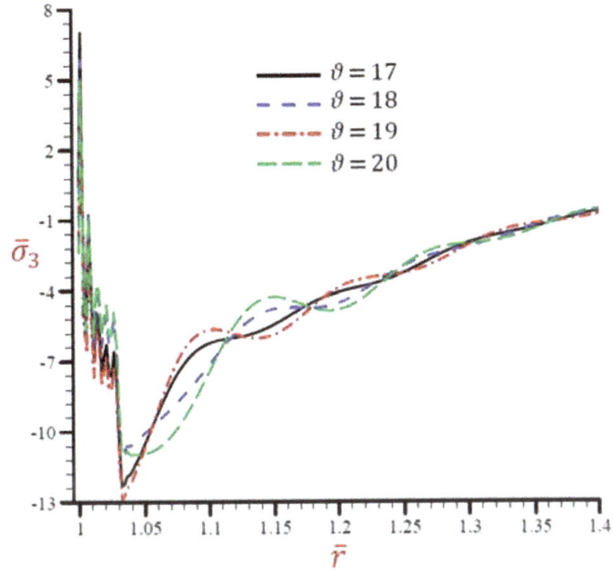

Figure 18. Effect of ϑ on radial stress $\bar{\sigma}_3$ across the radial direction of the cylindrical cavity utilizing the RDPL model.

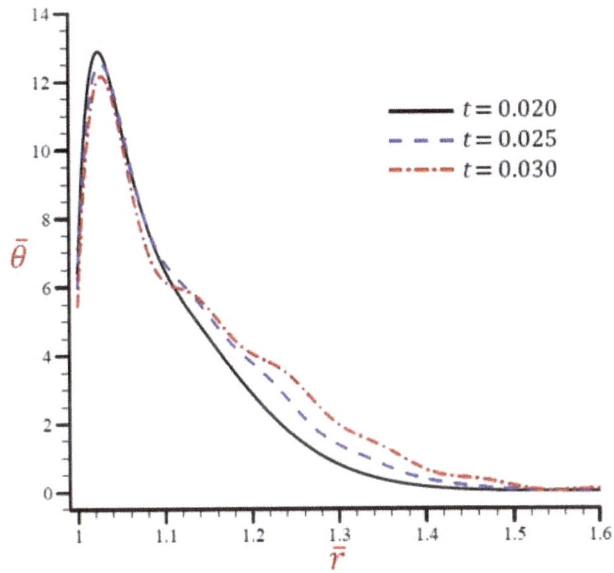

Figure 19. Effect of t on temperature $\bar{\theta}$ across the radial direction of the cylindrical cavity utilizing the RDPL model.

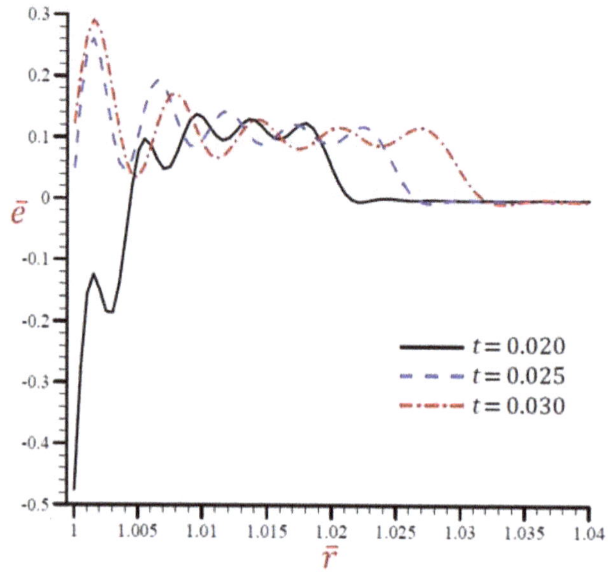

Figure 20. Effect of t on dilatation $\bar{e}$ across the radial direction of the cylindrical cavity utilizing the RDPL model.

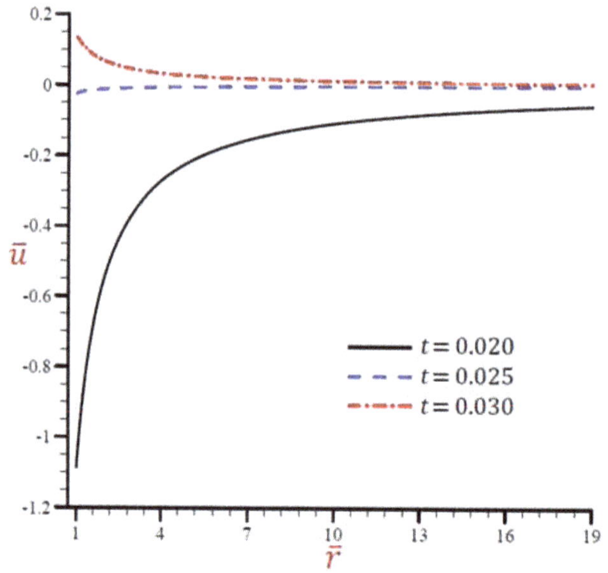

Figure 21. Effect of t on radial displacement $\bar{u}$ across the radial direction of the cylindrical cavity utilizing the RDPL model.

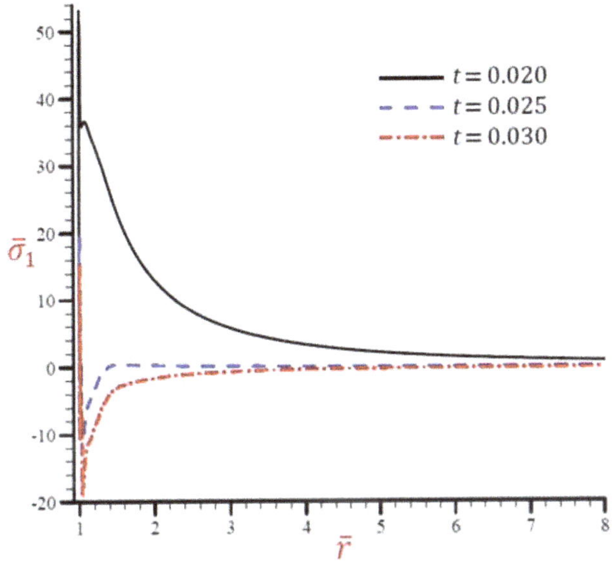

Figure 22. Effect of t on radial stress $\bar{\sigma}_1$ across the radial direction of the cylindrical cavity utilizing the RDPL model.

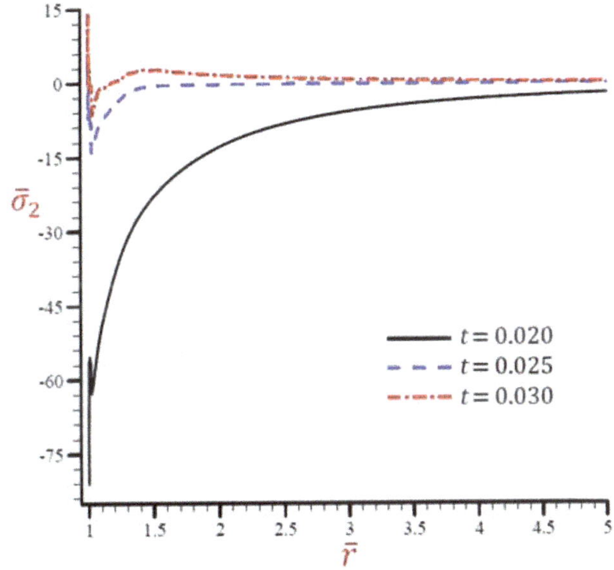

Figure 23. Effect of t on radial stress $\bar{\sigma}_2$ across the radial direction of the cylindrical cavity utilizing the RDPL model.

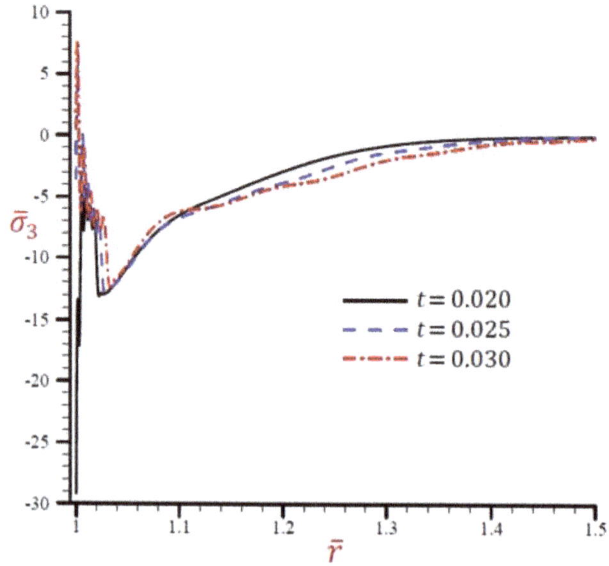

Figure 24. This is a figure. Schemes follow the same formatting Effect of t on radial stress $\overline{\sigma}_3$ across the radial direction of the cylindrical cavity utilizing the RDPL model.

The outcomes described in Tables 1 and 2 will be provided as benchmarks for other researchers. It is established from these tables that:

- The G–N theory gives the smallest absolute field of all field quantities.
- The other theories CTE and L–S give suitable results for the field quantities.
- Three values $N = 3, 4,$ and 5 have been used for the RDPL theory while the simple dual-phase-lag (SDPL) theory is described with $N = 1$.
- The most accurate results are given by using the RDPL theory.
- For the RDPL theory the temperature, displacement, and hoop stress are slightly increasing with the increase in many terms N, while the dilatation, radial stress, and axial stress are slightly decreasing. The increasing and decreasing amounts may be un-sensitive when $N \geq 5$.

5.2. Second Validation Example

Now, Figures 1–6 show the impact of all theories on the field quantities with fixed time $t = 0.03$, velocity of heat source $\vartheta = 17$, and angular frequency of thermal vibration $\omega = 20$. The rest of the figures are presented based on the refined dual-phase-lag (RDPL) theory with $N = 5$ to investigate the influence of different parameters on the field quantities.

The variation of the temperature $\overline{\theta}$ across the radial direction of the cylindrical cavity according to all theories is presented in Figure 1. Similar graphs of the rest of the field quantities are produced in Figures 2–6. Figure 1 shows that the temperature CTE, L–S, and the SDPL theories vibrate around the path of the RDPL theory. In addition, the G–N theory vibrates around the path of the RDPL theory, but in a small range. The temperature of the G–N theory may early vanish than the other theories. Figure 2 shows that the dilatation $\overline{e}$ of CTE, L–S, G–N, and SDPL theories vibrate around the path of the RDPL theory. Figure 3 shows that the radial displacements $\overline{u}$ of the L–S and G–N theories may vanish during the radial direction. The radial displacements of the CTE theory are the lowest ones while those of the RDPL theory are the biggest ones. The displacement of all theories may vanish as r increases. Figure 4 shows that the radial stress $\overline{\sigma}_1$ of the L–S and G–N theories may vanish during the radial direction when $r > 1.2$. The radial stresses of the RDPL theory are the lowest ones while those of the CTE theory are the greatest ones. The radial stresses of all theories may vanish as r increases. Figure 5 shows that the hoop stress $\overline{\sigma}_2$ of the L–S and G–N theories may vanish during the radial direction when $r > 1.3$. The hoop stresses of the CTE theory are the smallest ones while those of the RDPL theory are the greatest ones. The hoop stresses of all theories may vanish as r increases. Finally, Figure 6 shows that the axial stress $\overline{\sigma}_3$ of CTE, L–S, and the SDPL theories vibrate around the path of the RDPL theory. In addition, the G–N theory vibrates around the path of the RDPL theory, but in a small range. The axial stress of the G–N theory may early vanish than the other theories. It is concluded from Figures 1–6 that the outcomes of the RDPL theory are the most truthful.

5.3. *Additional Applications*

5.3.1. Effect of Angular Frequency of Thermal Vibration

Now, we discuss the impact of the angular frequency of thermal vibration ω on the field quantities using the RDPL theory. Figure 7 shows the impact of ω on the temperature $\overline{\theta}$ along the radial direction of the cylindrical cavity. Similar graphs of the rest of the field quantities are presented in Figures 8–12. It is clear in Figure 7 that the temperature increases as ω decreases. The temperature vanishes as r increases irrespective of the value of ω. Figure 8 shows that the dilatation $\overline{e}$ vibrates across the radial direction of the cylindrical cavity. The wave magnitude increases as ω decreases. The radial displacement $\overline{u}$ directly decreases along the radial directional of the cylindrical cavity in Figure 9. It is clear that at a fixed position $\overline{u}$ increases as ω decreases.

The radial stress $\overline{\sigma}_1$ across the radial direction of the cylindrical cavity due to the RDPL theory is drawn in Figure 10. The radial stress $\overline{\sigma}_1$ is rabidly vibrating across the radial direction in a small range $1 \leq r < 1.035$. The radial stress $\overline{\sigma}_1$ is increasing with the increase in ω when $1.035 \leq r < 1.3$. After that the values of $\overline{\sigma}_1$ are coincident to vanish as r increases. Once again, the hoop $\overline{\sigma}_2$ and axial $\overline{\sigma}_3$ stresses along the radial direction of the cylindrical using the RDPL theory are plotted in similar graphs of the radial stress $\overline{\sigma}_1$ in Figures 11 and 12. Both hoop and axial stresses are rabidly vibrating along the radial direction in a small range $1 \leq r < 1.03$. The hoop and axial stresses increase as ω increases when $1.03 \leq r < 1.45$. After that the values of $\overline{\sigma}_2$ and $\overline{\sigma}_3$ are coincident to vanish as r increases.

5.3.2. Effect of Velocity of Heat Source

The effects of the velocity of heat source ϑ on all field variables based upon the RDPL theory are presented in Figures 13–18. Figure 13 shows the effect of ϑ on the temperature $\overline{\theta}$ across the radial direction of the cylindrical cavity. Similar graphs of the rest of the field quantities are presented in Figures 14–18. It is clear in Figure 13 that the temperature $\overline{\theta}$ vibrates across the radial direction for different values of the velocity of heat source ϑ. The temperature vanishes as r increases and this irrespective of the value of ϑ. Figure 14 shows that the dilatation $\overline{e}$ vibrates across the radial direction of the cylindrical cavity. The wave magnitude increases as ϑ decreases. The radial displacement $\overline{u}$ directly decreases along the radial directional of the cylindrical cavity in Figure 15. It is clear that at a fixed position $\overline{u}$ increases as ϑ decreases.

The radial stress $\overline{\sigma}_1$ across the radial direction of the cylindrical cavity using the RDPL theory is drawn in Figure 16 for distinct values of ϑ. The radial stress $\overline{\sigma}_1$ is rabidly vibrating increases the radial direction in a small range $1 \leq r < 1.05$. The radial stress $\overline{\sigma}_1$ increases as ϑ increases when $r \geq 1.05$. The values of $\overline{\sigma}_1$ are coincident to vanish as r increases. Once again, the hoop $\overline{\sigma}_2$ and axial $\overline{\sigma}_3$ stresses along the radial direction of the cylindrical using the RDPL theory are plotted in similar graphs of the radial stress $\overline{\sigma}_1$ in Figures 11 and 12. Both hoop and axial stresses are rabidly vibrating along the radial direction in a small range $1 \leq r < 1.03$. The hoop and radial stresses increase as ω increases when $1.03 \leq r < 1.45$. After that the values of $\overline{\sigma}_2$ and $\overline{\sigma}_3$ are coincident to vanish as r increases. In addition, the hoop stress $\overline{\sigma}_3$ across the radial direction of the cylindrical using the RDPL theory is plotted for different ϑ in similar graphs of the radial stress $\overline{\sigma}_1$ in Figure 17. The hoop stress is rabidly vibrating along the radial direction in a small range $1 \leq r < 1.05$. It increases as

ϑ decreases when $r \geq 1.05$. For large values of r the values of $\overline{\sigma}_2$ will be coincident and maybe vanished. However, the axial stresses $\overline{\sigma}_3$ is drawn across the radial direction of the cylindrical using the RDPL theory for distinct values of ϑ in Figure 18. The axial stress is rabidly vibrating along the radial direction in a small range $1 \leq r < 1.032$ while it slowly vibrating after that in a large range $1.032 \leq r < 1.4$. For large values of r the values of $\overline{\sigma}_3$ will be coincident and maybe vanished.

5.3.3. Effect of Dimensionless Time

The effects of dimensionless time t on all field variables based upon the RDPL theory are presented in Figures 19–24. Figure 19 shows the effect of t on the temperature $\overline{\theta}$ across the radial direction of the cylindrical cavity. Similar graphs of the rest of field quantities are presented in Figures 20–24. It is clear in Figure 19 that the temperature $\overline{\theta}$ vibrates across the radial direction for different values of t. The temperature $\overline{\theta}$ is no longer increasing and has its maximum values at $r = 1.04$. The temperature vanishes as r increases and this is irrespective of the value of the dimensionless time. Figure 20 shows that the dilatation $\overline{e}$ vibrates across the radial direction of the cylindrical cavity nicely. The wave magnitude increases as t increases. In Figure 21, the radial displacement $\overline{u}$ is rapidly increasing along the radial directional of the cylindrical cavity when $t = 0.02$ while $\overline{u}$ is slowly increasing when $t = 0.025$. In addition, $\overline{u}$ is slowly decreasing when $t = 0.03$ It is clear that at a fixed position $\overline{u}$ increases as t increases.

The radial stress $\overline{\sigma}_1$ across the radial direction of the cylindrical utilizing the RDPL theory is plotted in Figure 22 for different values of t. The radial stress $\overline{\sigma}_1$ vibrates in a very small range, then it increases for $t = 0.025$ and 0.03 while it decreases when $t = 0.02$. At any fixed position, the radial stress $\overline{\sigma}_1$ increases as t decreases. The hoop stress $\overline{\sigma}_2$ is plotted across the radial direction of the cylindrical using the RDPL theory in Figure 23 for distinct values of t. It vibrates in a very small range, then it increases for $t = 0.02$ while it decreases when $t = 0.025$ and 0.03. At any fixed position, the hoop stress $\overline{\sigma}_2$ increases as t increases.

Finally, the axial stress $\overline{\sigma}_3$ is plotted across the radial direction of the cylindrical using the RDPL theory in Figure 24 for distinct values of t. It rabidly vibrates in a very small range $1 \leq r < 1.035$, then it slowly vibrates and increases to vanish at large values of r.

6. Conclusions

The refined dual-phase-lag theory is presented to get novel and accurate outcomes of the variable quantities such as temperature, dilatation, displacement, and stresses. The multi-time derivatives heat equation is illustrated in the present formulation. The constitutive equations for the stresses in cylindrical coordinates are added to discuss the thermoelastic coupling response of an unbounded body with a cylindrical hole due to a traveling heat source. From the unified model, one can construct other theories concerning coupled dynamical thermoelasticity (CTE theory), Lord and Shulman generalized thermoelasticity theory (L–S theory), Green and Naghdi generalized thermoelasticity theory (G–N theory) without energy dissipation as well as a simple generalized thermoelasticity theory with dual-phase-lag (SDPL theory). The system of two highly-time-derivatives differential coupled equations is solved, and all field variables are gained for the thermoelastic coupling response of an unbounded medium with a cylindrical cavity. Different validation examples and applications are presented to compare all theories with the refined one. A sample set of plots are illustrated along the radial direction of the cylindrical cavity. Two tables are reported for a validation example to serve as benchmark results for future comparisons with other investigators. The reported and illustrated results show various behaviors of all field quantities and the effects of the velocity of heat source, angular frequency of thermal vibration, and dimensionless time parameters. The G–N theory gives suitable results in a small range. However, the RDPL theory yields modified and accurate results.

Author Contributions: Conceptualization, A.M.Z., D.S.M., and A.M.A.; data collection, A.M.A.; methodology, A.M.Z. and A.M.A.; software, A.M.Z. and A.M.A.; validation, A.M.Z. and D.S.M.;

writing—original draft preparation, A.M.A. and D.S.M.; writing—review and editing, A.M.Z. All authors have read and agreed to the published version of the manuscript.

Funding: This research was funded by the Deanship of Scientific Research (DSR) at King Abdulaziz University, Jeddah, under grant no. (DG: 18-130-1441).

Institutional Review Board Statement: Not applicable.

Informed Consent Statement: Not applicable.

Data Availability Statement: Not applicable.

Acknowledgments: This project was funded by the Deanship of Scientific Research (DSR) at King Abdulaziz University, Jeddah, under grant no. (DG: 18-130-1441). The authors, therefore, acknowledge with thanks DSR for technical and financial support.

Conflicts of Interest: The authors declare no conflict of interest.

Nomenclature

α_t	thermal expansion coefficient $\left(\mathrm{K}^{-1}\right)$
C_e	specific heat at uniform strain $\left(\mathrm{J\,kg^{-1}\,K^{-1}}\right)$
δ_{ij}	Kronecker delta function
$e_{\varphi\varphi}$	hoop strain
e_{rr}	radial strain
e	dilatation
e_{ij}	linear strain tensor
ϑ	the velocity of heat source $(\mathrm{m\,s^{-1}})$
$\gamma \equiv (3\lambda + 2\mu)\alpha_t$	thermal modulus $\left(\mathrm{N\,m^{-2}\,K^{-1}}\right)$
$H(t)$	Heaviside unit step function
k	coefficient of heat conductivity $\left(\mathrm{W\,m^{-1}\,K^{-1}}\right)$
k^*	rate of thermal conductivity of an isotropic material $\left(\mathrm{W\,m^{-1}\,K^{-1}}\right)$
λ, μ	Lame's constants $\left(\mathrm{N\,m^{-2}}\right)$
ρ	material density $\left(\mathrm{kg\,m^{-3}}\right)$
R	The radius of the cylindrical cavity (m)
(r, φ, z)	cylindrical coordinates system
σ_{ij}	stress tensor components $\left(\mathrm{N\,m^{-2}}\right)$
$\sigma_{\varphi z}, \sigma_{zr}, \sigma_{r\varphi}$	shear stresses $\left(\mathrm{N\,m^{-2}}\right)$
$\sigma_{\varphi\varphi}$	hoop stress $\left(\mathrm{N\,m^{-2}}\right)$
σ_{rr}	radial stress $\left(\mathrm{N\,m^{-2}}\right)$
σ_{zz}	axial stress $\left(\mathrm{N\,m^{-2}}\right)$
s	Laplace parameter
$\theta = T - T_0$	temperature change (K)
θ_0	thermal constant (K)
T_0	environment temperature (K)
τ_q	phase-lag of heat flux (s)
τ_θ	phase-lag of temperature gradient (s)
τ_0	first relaxation time (s)
ω	angular frequency of thermal vibration $\left(\mathrm{rad\,s^{-1}}\right)$
Q_0	strength of heat source $\left(\mathrm{W\,m^{-3}}\right)$
δ	delta function
$\vec{q}$	heat flux vector $\left(\mathrm{W\,m^{-2}}\right)$
u_r	radial displacement (m)
u_ϕ	hoop displacement (m)
u_z	axial displacement (m)

References

1. Duhamel, J.M.C. Second memoire sur les phenomes thermomechaniques. *J. École Polytech.* **1837**, *15*, 1–57.
2. Biot, M.A. Thermoelasticity and irreversible thermodynamics. *J. Appl. Phys.* **1956**, *27*, 240–253. [CrossRef]
3. Lord, H.W.; Shulman, Y. A generalized dynamical theory of thermoelasticity. *J. Mech. Phys. Solids* **1967**, *15*, 299–309. [CrossRef]
4. Green, A.E.; Lindsay, K.A. Thermoelasticity. *J. Elast.* **1972**, *2*, 1–7. [CrossRef]
5. Chandrasekharaiah, D.S. Hyperbolic thermoelasticity: A review of recent literature. *Appl. Mech. Rev.* **1998**, *51*, 705–729. [CrossRef]
6. Green, A.E.; Naghdi, P.M. A re-examination of the basic postulates of thermomechanics. *Proc. R. Soc. A* **1991**, *432*, 171–194.
7. Green, A.E.; Naghdi, P.M. On undamped heat waves in an elastic solid. *J. Therm. Stresses* **1992**, *15*, 253–264. [CrossRef]
8. Green, A.E.; Naghdi, P.M. Thermoelasticity without energy dissipation. *J. Elast.* **1993**, *31*, 189–208. [CrossRef]
9. Tzou, D.Y. A unified approach for heat conduction from macro- to micro-scales. *J. Heat Transf.* **1995**, *117*, 8–16. [CrossRef]
10. Tzou, D.Y. *Macro- to Microscale Heat Transfer: The Lagging Behavior*; Taylor & Francis: Washington, DC, USA, 1997.
11. Zenkour, A.M. Thermo-diffusion of solid cylinders based upon refined dual-phase-lag models. *Multid. Model. Mater. Struct.* **2020**, *16*, 1417–1434. [CrossRef]
12. Zenkour, A.M. Wave propagation of a gravitated piezo-thermoelastic half-space via a refined multi-phase-lags theory. *Mech. Adv. Mater. Struct.* **2020**, *27*, 1923–1934. [CrossRef]
13. Zenkour, A.M. Thermoelastic diffusion problem for a half-space due to a refined dual-phase-lag Green-Naghdi model. *J. Ocean Eng. Sci.* **2020**, *5*, 214–222. [CrossRef]
14. Zenkour, A.M. Thermal-shock problem for a hollow cylinder via a multi-dual-phase-lag theory. *J. Therm. Stresses* **2020**, *43*, 687–706. [CrossRef]
15. Zenkour, A.M. Exact coupled solution for photothermal semiconducting beams using a refined multi-phase-lag theory. *Opt. Laser Tech.* **2020**, *128*, 106233. [CrossRef]
16. Chandrasekharaiah, D.S.; Srinath, K.S. Axisymmetric thermoelastic interactions without energy dissipation in an unbounded body with cylindrical cavity. *J Elast.* **1997**, *46*, 19–31. [CrossRef]
17. Allam, M.N.; Elsibai, K.A.; Abouelregal, A.E. Thermal stresses in a harmonic field for an infinite body with a circular cylindrical hole without energy dissipation. *J. Therm. Stresses* **2002**, *25*, 57–67. [CrossRef]
18. Ezzat, M.A.; El-Bary, A.A. Fractional order theory to an infinite thermo-viscoelastic body with a cylindrical cavity in the presence of an axial uniform magnetic field. *J. Electromag. Waves Appl.* **2017**, *31*, 495–513. [CrossRef]
19. Ezzat, M.A.; El-Bary, A.A. Application of fractional order theory of magneto-thermoelasticity to an infinite perfect conducting body with a cylindrical cavity. *Microsyst. Technol.* **2017**, *23*, 2447–2458. [CrossRef]
20. Sharma, J.N.; Sharma, N.K.; Sharma, K.K. Diffusion in generalized thermoelastic solid in an infinite body with cylindrical cavity. *Proc. Natl. Acad. Sci. India Sect. A Phys. Sci.* **2013**, *83*, 353–364. [CrossRef]
21. Kumar, R.; Mukhopadhyay, S. Effects of three phase lags on generalized thermoelasticity for an infinite medium with a cylindrical cavity. *J. Therm. Stresses* **2009**, *32*, 1149–1165. [CrossRef]
22. Mukhopadhyay, S.; Kumar, R. Thermoelastic interactions on two-temperature generalized thermoelasticity in an infinite medium with a cylindrical cavity. *J. Therm. Stresses* **2009**, *32*, 341–360. [CrossRef]
23. Kumar, R.; Prasad, R.; Kumar, R. Thermoelastic interactions on hyperbolic two-temperature generalized thermoelasticity in an infinite medium with a cylindrical cavity. *Eur. J. Mech. A Solids* **2020**, *82*, 233–245. [CrossRef]
24. Sarkar, N.; Mondal, S. Transient responses in a two-temperature thermoelastic infinite medium having cylindrical cavity due to moving heat source with memory-dependent derivative. *ZAMM J.* **2019**, *99*, 1–19. [CrossRef]
25. Sharma, J.N.; Kumari, N.; Sharma, K.K. Diffusion in a generalized thermoelastic solid in an infinite body with a cylindrical cavity. *J. Appl. Mech. Tech. Phys.* **2013**, *54*, 819–831. [CrossRef]
26. Mukhopadhyay, S.; Kumar, R. A problem on thermoelastic interactions in an infinite medium with a cylindrical hole in generalized thermoelasticity III. *J. Comput. Appl. Mech.* **2008**, *31*, 455–475. [CrossRef]
27. Xia, R.-H.; Tian, X.-G.; Shen, Y.-P. The influence of diffusion on generalized thermoelastic problems of infinite body with a cylindrical cavity. *Int. J. Eng. Sci.* **2009**, *47*, 669–679. [CrossRef]
28. Xiong, Q.-L.; Tian, X.-G. Thermoelastic study of an infinite functionally graded body with a cylindrical cavity using the Green-Naghdi model. *J. Therm. Stresses* **2012**, *35*, 718–732. [CrossRef]
29. Abouelregal, A.E. Generalized thermoelastic infinite transversely isotropic body with a cylindrical cavity due to moving heat source and harmonically varying heat. *Meccanica* **2013**, *48*, 1731–1745. [CrossRef]
30. Youssef, H.M. Two-temperature generalized thermoelastic infinite medium with cylindrical cavity subjected to moving heat source. *Arch. Appl. Mech.* **2010**, *80*, 1213–1224. [CrossRef]
31. Youssef, H.M. State-space approach to two-temperature generalized thermoelasticity without energy dissipation of medium subjected to moving heat source. *Appl. Math. Mech.* **2013**, *34*, 63–74. [CrossRef]
32. Shaw, S.; Mukhopadhyay, B. Moving heat source response in a thermoelastic microelongated solid. *J. Eng. Phys. Thermophys.* **2013**, *86*, 716–722. [CrossRef]
33. Sarkar, N.; Lahiri, A. Interactions due to moving heat sources in generalized thermoelastic half-space using L-S model. *Int. J. Appl. Mech. Eng.* **2013**, *18*, 815–831. [CrossRef]
34. Youssef, H.M. State-space approach to fractional order two-temperature generalized thermoelastic medium subjected to moving heat source. *Mech. Adv. Mater. Struct.* **2013**, *20*, 47–60. [CrossRef]

35. Xia, R.-H.; Tian, X.-G.; Shen, Y.-P. Dynamic response of two-dimensional generalized thermoelastic coupling problem subjected to a moving heat source. *Acta Mech. Solida Sin.* **2014**, *27*, 300–305. [CrossRef]
36. Abbas, I.A. A GN model for thermoelastic interaction in a microscale beam subjected to a moving heat source. *J. Comput. Acta Mech.* **2015**, *226*, 2527–2536. [CrossRef]
37. Youssef, H.M. Generalized thermoelastic infinite medium with cylindrical cavity subjected to moving heat source. *Mech. Res. Commun.* **2009**, *36*, 487–496. [CrossRef]
38. Zenkour, A.M. Thermal diffusion of an unbounded solid with a spherical cavity via refined three-phase-lag Green–Naghdi models. *Indian J. Phys.* 2021, *in press*. [CrossRef]
39. Kutbi, M.A.; Zenkour, A.M. Refined dual-phase-lag Green–Naghdi models for thermoelastic diffusion in an infinite medium. *Waves Rand. Complex Media* 2021, *in press*. [CrossRef]
40. Zenkour, A.M.; El-Mekawy, H.F. On a multi-phase-lag model of coupled thermoelasticity. *Int. Commun. Heat Mass Transf.* **2020**, *116*, 104722. [CrossRef]
41. Sobhy, M.; Zenkour, A.M. Modified three-phase-lag Green–Naghdi models for thermomechanical waves in an axisymmetric annular disk. *J. Therm. Stresses* **2020**, *43*, 1017–1029. [CrossRef]

applied sciences

MDPI

Article

Challenges of Data Refining Process during the Artificial Intelligence Development Projects in the Architecture, Engineering and Construction Industry

Seokjae Heo, Sehee Han, Yoonsoo Shin and Seunguk Na *

Department of Architectural Engineering, College of Engineering, Dankook University, 152 Jukjeon-ro, Yongins-si 16890, Gyeonggi-do, Korea; mill@dankook.ac.kr (S.H.); edu.hansh@gmail.com (S.H.); shinys@dankook.ac.kr (Y.S.)
* Correspondence: drseunguk@dankook.ac.kr; Tel.: +82-31-8005-3727

Abstract: The paper examines that many human resources are needed on the research and development (R&D) process of artificial intelligence (AI) and discusses factors to consider on the current method of development. Labor division of a few managers and numerous ordinary workers as a form of light industry appears to be a plausible method of enhancing the efficiency of AI R&D projects. Thus, the research team regards the development process of AI, which maximizes production efficiency by handling digital resources named 'data' with mechanical equipment called 'computers', as the digital light industry of the fourth industrial era. As experienced during the previous Industrial Revolution, if human resources are efficiently distributed and utilized, no less progress than that observed in the second Industrial Revolution can be expected in the digital light industry, and human resource development for this is considered urgent. Based on current AI R&D projects, this study conducted a detailed analysis of necessary tasks for each AI learning step and investigated the urgency of R&D human resource training. If human resources are educated and trained, this could lead to specialized development, and new value creation in the AI era can be expected.

Keywords: digital light industry; fourth Industrial Revolution; artificial intelligence; human resource development; work index; architecture; engineering and construction industry

Citation: Heo, S.; Han, S.; Shin, Y.; Na, S. Challenges of Data Refining Process during the Artificial Intelligence Development Projects in the Architecture, Engineering and Construction Industry. *Appl. Sci.* **2021**, *11*, 10919. https://doi.org/10.3390/app112210919

Academic Editors: Nikos D. Lagaros and Vagelis Plevris

Received: 30 October 2021
Accepted: 15 November 2021
Published: 18 November 2021

1. Introduction

For half a century, entrepreneurs of South Korea have transformed the originally agriculture-focused country into one with a focus on light and heavy industries, and following the third Industrial Revolution era, South Korea gained a state-of-the-art electronic industry. Furthermore, some South Korean corporate companies that emerged after the 2000s are now top ranking on the global scale. Not only have they established artificial intelligence (AI) research and development (R&D) centers, but they have also aggressively developed professional human resources in order to become global leading companies in the fourth Industrial Revolution era. Notably, amongst construction and transportation industries, the AI-based indoor mapping and positioning technologies developed by Naver Labs are acknowledged as top-tier technologies [1], and KAKAO BRAIN has developed a technology that allows world-class performance of learning of images without the image labeling process [2,3]. According to King et al. [4], AI will soon be applied to all the industrial sectors around the globe at a fast pace.

The South Korean government-initiated Data Dam project focused on digital infrastructure investment since the mid-2020s [5]. The South Korean government announced that the Data Dam project would have a total budget of KRW 292.5 billion (equivalent to approximately USD 252 million) for the first half of 2021 and is planning to collect the training data from 84 areas, including vision, geographic information, healthcare, and construction. The aim of this project is to enable research institutes and private companies to focus on AI-related tasks, and the government will develop and establish AI learning datasets, which will take approximately 80% of the total development process time and lead to a delay in AI R&D projects. Consequently, it is expected that this project will have a positive effect on the creation of mass employment as well as on innovations in AI research.

Figure 1 depicts a schematic diagram of data acquisition, refinement, and verification during the execution of AI R&D projects. The set of data in the AI R&D project could be divided into one using image-based data and another using time series data. The overall process of constructing a dataset goes through the process of raw data acquisition, segmentation, and labeling. In general, after collecting the raw data, the refinement process to adjust the size and shape and the labeling process, including annotation and segmentation, are carried out to create the learning dataset based on the two types of data. Additionally, the refined data are re-classified into the weighted model referred to as the brain of AI. The final process of AI dataset construction is quality control to determine what to use in order to ensure the production of optimum results [6–19]. While most previous studies divided the construction of the dataset process into the acquisition of raw data, pre-conditioning, data refinement, data labeling, and the composition of the dataset, the process can be simplified into data acquisition, refinement, and labeling steps, excluding the pre-conditioning and composition of the dataset. This is because AI R&D is conducted not by academia but by industrial processes once the data are acquired, and labeling work is carried out to refine the procedure for cost and time reduction. Moreover, subdivided processes performed at research institutes are not followed, as pre-developed models are chosen rather than the development of models for learning on their own, and they swiftly undergo refinement and verification after acquiring data, such as suitable images or videos [20,21]. Verification stated here refers to a process that confirms whether refinement was appropriately carried out by utilizing manpower or automation tools, not forming a validation set generally cited in AI research.

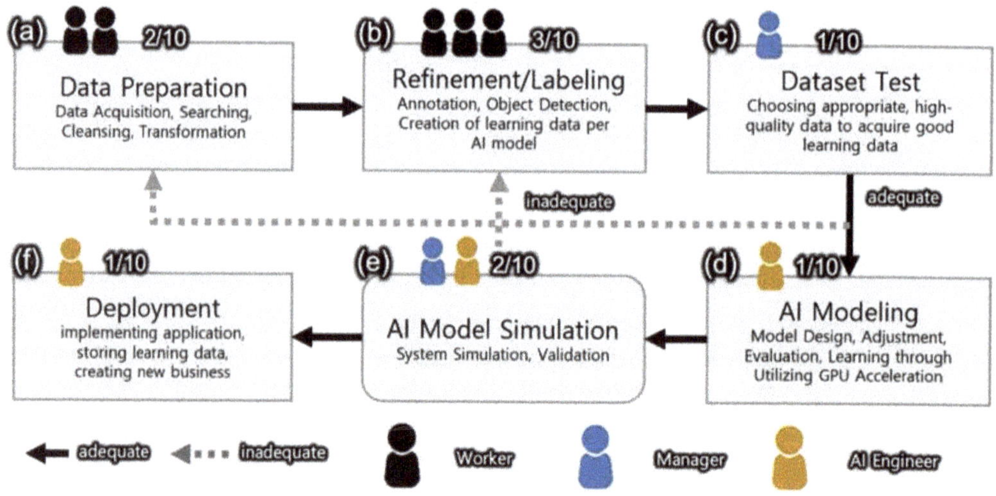

Figure 1. The current AI R&D process in academia and in the industrial field, and manpower status required for development (if 10 personnel are needed for total input, 50% of them are for data acquisition, processing, refinement, and labeling).

It seems that AI R&D could be carried out easily and fast if the above procedures are followed, but there remain two factors to consider in AI research. Firstly, a great deal of high-quality research manpower and equipment are needed at the learning level. In order to verify the performance of training models, numerous variable studies need to be carried out, and researchers who can understand the model structure and change equipment to reduce the training time, as well as manpower to revise labeling, are required. Moreover, a number of studies compared the performance of various AI algorithms in order to verify the optimal model for the construction projects [22–24] For example, Chakraborty et al. [23] tested six different types of machine learning algorithm for predicting the construction costs. Currently, South Korea lacks high-quality manpower that would facilitate a proper understanding and establishment of AI structures, and available personnel are concentrated only in some specific fields. Notably, in minor fields, such as construction and civil engineering, the demand for R&D continues to emerge, but their progress is slow due to a lack of human resources. Problems with development equipment also exist. Hardware, particularly the number of GPUs, is important in the development of the latest AI models [25]. AI calculation techniques using many GPUs in parallel exist, and, hence, the higher the quantity, the easier the variable research. However, arithmetic disciplines researching minor fields such as construction and civil engineering seem to conduct research with equipment with a budget of USD 4000~10,000 as of 2020 (as shown in Table 1), and they also use services provided by Google or Amazon without equipment [26,27]. As discussed above, AI research may be delayed due to the low performance of computing equipment in non-mainstream sectors.

Table 1. Status and cost of AI research equipment for architectural and civil engineering studies at universities in South Korea (2020).

University	Major	Amount of Manpower	Hardware	Cost
D	Civil Engineering	3	RTX3080 × 4	USD 12,000
	Environmental Engineering	3	RTX2090 × 3	USD 6000
S	Environmental Engineering	5	RTX2060 × 4	USD 4000
H	Construction	2	Google API	USD 10/month

Secondly, more manpower would be required upon data acquisition and refinement than in the training level. There are limits to searching on the web or reusing existing resources for the acquisition of data, and for special fields, procuring training data itself may be difficult. For instance, in order to acquire data on structure damage due to disasters such as earthquakes and typhoons, research teams must use their time and financial resources at appropriate times and acquire data directly on site. If they are lucky enough to acquire training data after this, refinement is necessary to study the latest AI models, and this requires revision of each data item with human interference. Notably, data refinement that requires expertise may be delayed due to limited manpower.

Although companies carry out AI research by focusing on learning rather than data acquisition and refinement for cost reduction, the outcome is imperfect, as they are unable to procure satisfactory training data. Luckily, South Korea is investing a considerable amount not only in AI model development research but also data acquisition and refinement. In February 2021, the National Information Society Agency (NIA) attempted to standardize the development method by publishing a manual on building data and refinement [28,29]. However, the types of data to be acquired consisted of images and video, and simply increasing the quality without considering the development purpose was prioritized; as a result, it is difficult to utilize this model in methods other than the original acquisition purpose. If these data are meaningfully used later, the problem of re-refining the source data by using manpower and financial resources again occurs. Therefore, the aim of this study is to discuss the manpower-dependent problem of the current image-based AI research, its improvement, and its future.

2. Current Research on Data Acquisition, Refinement, and Quality Evaluation

2.1. AI Research Imitating Human Visual Information

In general, an ordinary image or a cut-frame image from a video clip is a good material for AI to learn. As shown in Figure 2, until 2008, during which the early stage of data acquisition development occurred, AI R&D was conducted in terms of simple image classification. During this period, data acquisition did not require many human resources to create training data since labeling consisted of simply adding text labels to images. Later, object detection was developed, which made it possible to enhance the accuracy of AI. In object detection, it is an essential task to annotate images, indicating the parts needed to learn from the images [7]. During the early stage of annotation, the data used round boxes for image classification, but efficiency was enhanced following the development of the ability to identify several objects on one image, as shown in Figure 3. Nonetheless, if the image classification data are used without considering annotation, as shown in Figure 3, the advantages of annotation cannot be exploited, which leads to no significant changes in the accuracy of AI. Consequently, researchers developing object detection had to acquire new image data corresponding to the method shown in Figure 3b, and they had difficulties in reconducting atypical refinement accordingly with the AI model to be applied.

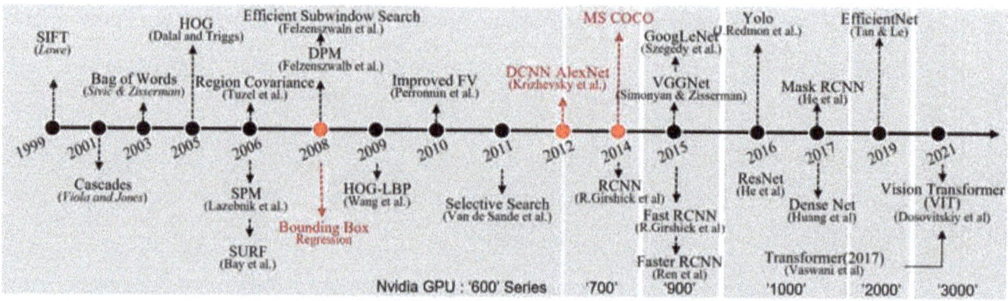

Figure 2. Milestones of object detection techniques and the evolution of hardware (GPU).

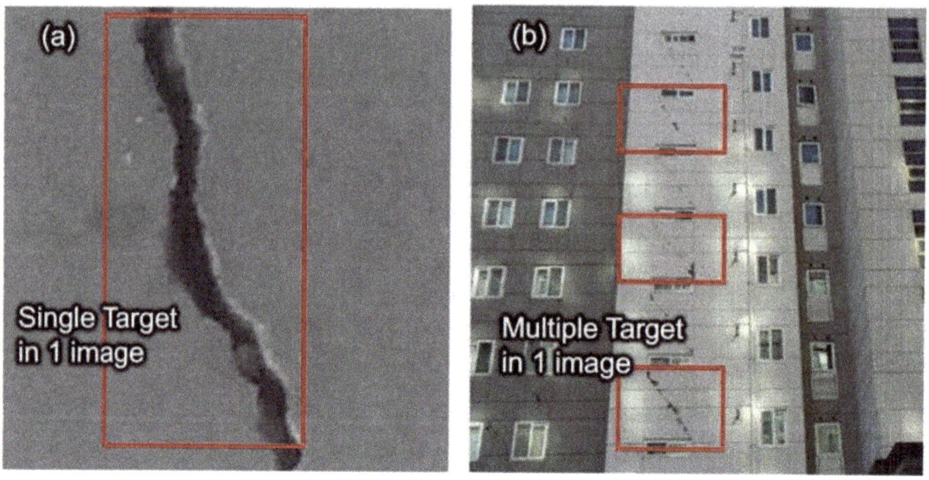

Figure 3. Difference between (**a**) an example of learning data collected in the era when image classification techniques were dominant and (**b**) learning data collected after object detection was developed. (**a**) Data can also be used for object detection, but they are less efficient and may lead to inaccurate results.

In 2012, researchers who attempted to enhance the accuracy of an AI model developed instance segmentation, a type of object detection that categorizes classes per pixel. Most researchers nowadays apply this method, and it shows excellent performance [6–19,30–34]. Since instance segmentation specifies parts on the image needed for learning by pixel, labeling is mostly performed in polygonal form, as shown in Figure 4d. This case requires more workload than required by the existing round boxing. To make up for this, the so-called auto-labeling methods started to be developed, ultimately reaching the level where labeling is carried out on parts without the need to click on certain features, such as trees and cars [35]. Nevertheless, many errors occurred, so revision by personnel was needed, and this required additional acquisition and learning on images in order to perfectly recognize the target. An additional amount of acquisition differs in terms of target accuracy, and figures may vary depending on the engineer. Acquisition methods are ramified as online collection through web surfing and offline collection by shooting [29]. Online collection is carried out by conducting searches on the image server of the web established by people or by conducting automatic web searches through surfing. Due to problems regarding copyrights, the industry prefers offline acquisition, and R&D institutes combine on- and off-line methods to reduce the development schedule and costs.

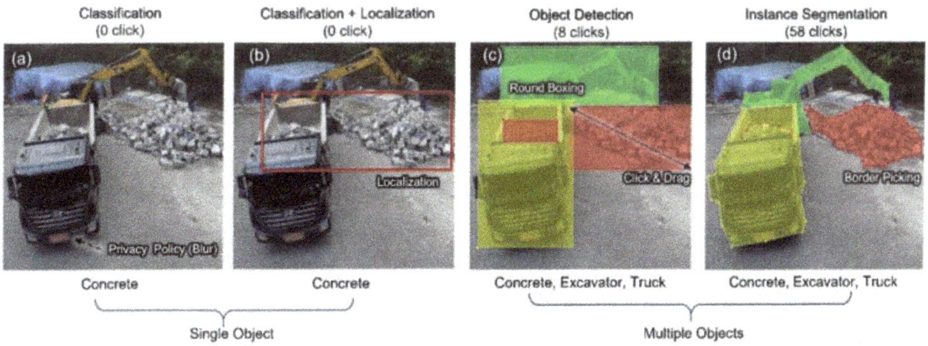

Figure 4. (**a**) Classification (0 click); (**b**) Classification + Localization (0 click); (**c**) Object Detection (8 clicks); (**d**) Instance Segmentation (58 clicks). A typical artificial intelligence method to perceive images. The number in the parentheses represents the number of mouse clicks. License plates and human faces were mosaicked or blurred for privacy. Following the development of the ability to perceive various objects on one image, the number of mouse clicks as well as the refinement time increased, whereas the volume of the training dataset decreased, leading to quickened learning speed and, thus, the need for comparative studies on this topic.

The acquired data are called source data and are in an incompatible state for AI training at the current status quo. After conducting data cleansing on the source data needed for training, the data must be revised to correspond with the model to be developed, and the task incurred here is called data refinement. In brief, refinement includes various steps, from creating a category folder and classifying images to adjusting the image size, changing the resolution, blurring for privacy, and binarizing to simplify color, and the images go through labeling processes such as image boxing to indicate the target object on the image, and segmentation. All tasks have a proper program tool that is not yet automatized, because the tasks cannot be immediately completed. The most frequently used tasks during refinement are changing the image size, adjusting the resolution, and changing the name of the file; labeling can be carried out again after these tasks are completed. During these procedures, manpower is constantly deployed. Furthermore, some cases may require expert knowledge on refinement and labeling depending on the research field, such as construction, civil engineering, and medical care. However, research results of AI in specific fields show low performance or are delayed because highly educated personnel are costly and limited.

2.2. Research on Quality Assurance of Acquired Data and AI Models

Quality assurance can be categorized into the tasks of extracting appropriate data for training to enhance the accuracy of AI models and conducting a performance evaluation of the developed AI models. [28] First, performance evaluation indicators on the classified model, such as accuracy, confusion matrix, precision, recall, F1 Score, ROC AUC, and mAP, are developed based on a method of judging the similarity between the actual and estimated data. In 2012, during which the so-called embryonic stage of AI occurred, quality assurance of training data was considered unimportant since the focus was only on the mechanical training model, but the necessity of quality evaluation rose as the amount of data continued to increase, and the importance of refinement and labeling was ascertained. If quality evaluation is included as a general business process, quality evaluation could be carried out by at least those who have knowledge on AI development, which infers that AI engineers would additionally have to bear the quality evaluation task of data before the training period. According to recent research, there are tendencies that prove the accuracy enhancement of the model only with a quality training data set, which would solve if research on training data evaluation indicators are done and the related manpower are educated from now on, but lack of AI development personnel would rise as a problem in the short term.

3. What Is the Digital Light Industry?

3.1. Manpower-Dependent Source Data Acquisition Method

AI studies are currently being actively carried out in many fields, but the interest in acquiring training data is relatively small. Acquisition is mostly divided in collection via direct photography or video-shooting and collection via web surfing. Many research engineers are mistaken on the notion that the data acquisition process is not difficult and unimportant due to the incorrect assumptions that infinite data on the Internet can be used and that all objects in the surrounding environment are usable references. In fact, in the case of specialized fields, in order to acquire a datum, bachelor's degree knowledge and verification in accordance with it are needed. Even if one is fortunate enough to acquire the data, the specialist limit must be considered. In the medical field, for example, where there are numerous clinical cases, effort must be made to prevent leakage of personal information when acquiring image data of wounds, conditions, etc. In the case of the construction field, attention should be paid to not intrude on private property, and all other fields should give attention to copyrights, personal information leakage, and security on the image and acquired data. Thus, although acquisition can be smoothly carried out, blurring parts containing personal information or deleting data violating copyrights from the collected source data in order to abide by R&D ethics regulations should be conducted, which can only be done manually by professionals. If the data are to be used for commercial purposes, much more detailed work is needed.

The digital surfing technique using search engines may appear easy and simple compared to the direct acquisition method, but its usage is being reduced by professional engineers due to its many downsides. In order to transform data into source data after acquisition, inspection of each datum is needed due to overlapping or irrelevant data, format variation, resolution quality, and copyright issues, and research on this subject is required just as much as it is on direct acquisition. Many current beginners who are acquiring AI data believe that numerous training data exist on the web and prefer web-based data acquisition, but when a certain quantity of data is acquired, they become aware of this method's limit and eventually turn their research toward direct acquisition. However, both methods are means of data acquisition that cannot be overlooked, and a considerable amount of manpower and financial resources are needed in this process due to its trial-and-error nature.

3.2. *Necessity for Professional Manpower in Data Processing/Refinement/Labeling*

As stated in the previous section, processing and refinement of image data are generally conducted manually using manpower as shown in Table 2. If categorizing the acquired data is considered processing and refining, those in charge of the database should classify images one by one at this step. The next steps, namely, cleansing, size transformation, binarizing, and naming, can be solved with a simple program code of the programmer, but the database manager can individually implement the procedures if program personnel are not present. Round boxing and segmentation, which are conducted to utilize the latest AI method, may require professional knowledge depending on the data type, and they may be inefficient in terms of cost if all manpower is gentrified. This issue can be solved by using a method in which a manager with professional knowledge has many ordinary workers, followed by processing and refinement.

Table 2. Parts to be considered for verification in data acquisition and cleansing.

Parts to Verify	Processing Method
Copyright infringement	Data deletion
Personal information infringement	Blur image, delete personal information
Security facilities and public facilities	Acquire data upon prior permission
Discrimination against a particular region, society, or race	Set up procedure to avoid inequality
Possibility for safety-related accident outbreak	Prior check on accident risks and provide safety education

During this step, a special program for labeling may be needed. Non-profit research institutes mostly use refinement programs based on open-source software, such as 'Labelme' [36] or 'VGG annotator' [37], whereas business entities utilize self-developed programs and require manpower to manage these. Lastly, in order to manage refined data and utilize them for learning purposes, manpower for professionally developing AI is needed, and specialists with knowledge of at least five knowledge standards in data processing, refinement, and labeling are required, as in Table 3. It can be clearly observed that programs that automatically assist labeling are released as technology development progresses [35], and their usages are expected to continue to increase in diversity. However, at this point, human interference must not occur in any of the steps from data acquisition and processing to refinement, and it could be inferred that the AI R&D industry is highly human dependent.

Table 3. Research process, human resources, and level of academic knowledge required for AI research and development prior to learning.

Step	Occupational Group	Level of Academic Knowledge
Data acquisition	DB manager	Low level of computer programming major
	Worker	Major not required
Data refinement	Programmer	General level of programming major
Data labeling (annotation)	Labeling Manager	High level of special field major
	Worker	Major not required
Refinement/labeling program development	Programmer	High level of programming major
Training data transformation	AI Engineer	High level of AI major

3.3. Acquisition/Process Methods for Low Versatility Data

In general, AI development is not carried out with one-time data refinement. To reach the level of commercialization, training data inevitably have to go through several passes of re-processing, refinement, and labeling, and uneven data management may occur due to the participation of numerous workers. There are three main reasons for re-processing and refinement: the first reason is labeling being carried out in a manner that is not in accordance with the engineer and data managers' instructions, even if it is performed by professionals. Here, a series of refinements should be conducted again in accordance with the development purpose. The second reason is low accuracy during the training period or when the classified model categorizes unwanted objects. Unlike the problem occurring during model training, the cause cannot be determined in this case, and re-processing and refinement of overall data may be required, which greatly affects the project schedule. This could be due to low precision in the refinement and labeling process (labeling unnecessary data and inserting different data from the development intention) or lack of training data figures. The third reason is the development of a new AI network. Since the training data are composed of a bundle of categories, it is not difficult to add or delete additional categories, but it is difficult to use the bundled data, as they are for the development of AI with a new purpose. For instance, construction material could be a product on-site or rubbish in a dumpsite depending on its usage. Although AI perceiving materials has been successfully developed, it should be able to differentiate between new and used products for resource management, and, in this case, the existing category should be revised. That is, as human beings see and judge, data refinement should be carried out accordingly so as to discern other categories depending on the situation, even if the product is the same. The results of the research team's trial suggest that the workload lessened considerably when newly labeling from the source data rather than editing the existing labeled data. Research and projects on AI that are currently being developed do not consider subcategories such as the issues above, and as they collect and refine training data, such data are not expected to be used in further developments and would ultimately be considered waste.

3.4. Necessity of Acquisition/Refinement/Labeling Manager

The acquisition/refinement/labeling manager could be defined as the individual who possesses professional knowledge on the specialized field and is capable of making decisions regarding the use of data in training. The advantages gained from hiring such personnel are a shortened development schedule and increased accuracy. If the performance of the model trained with refined data in the absence of a manager is low, it is difficult to find where the cause lies among the training material and AI network. In this case, the AI engineers have to perform the task to identify errors within training data and to improve the network simultaneously, thus leading to the research development schedule being prolonged by twice the amount of time. On the other hand, if the data refinement manager has knowledge of data quality and provides refined data, the AI engineer can concentrate solely on developing the AI network, which may allow for a reduction in R&D time.

4. Solutions

As discussed in Section 3, many human resources are required for AI R&D. For AIs in specialized fields, the accuracy of the model can only be ensured by highly educated personnel, but such manpower is usually lacking. Moreover, the acquisition and refinement of data, which correspond to the immediate development process but with no consideration in expansion, are used only once in a manner similar to disposable products. In order to avoid such losses, it is necessary to train human resources to consider and manage the expansion from the beginning of the development.

4.1. Solution to the Manpower-Dependent R&D Method

The majority of people in South Korea attend university, and their computer skills are the world's best. Hence, general human resources required for AI R&D can be considered abundant. However, AI R&D is currently being carried out in an inefficient manner due to notable manpower shortage, and this issue will only be resolved when AI is popularized, which represents the beginning step of the fourth Industrial Revolution. Moreover, Korea is expected to be able to quickly become a powerhouse in this field, as the most efficient R&D is currently being conducted in this country.

Manpower-related issues can be solved by an AI expert selecting a detailed learning method, data acquisition, refinement, and labeling. Although it is ideal to plan for the consideration of further extendibility, an expert cannot do this alone and, thus, requires AI personnel and an additional general manager. In order to solve this problem, a new type of manpower distribution for the AI development process is suggested, as shown in Figure 5. The structure and figures of manpower distribution are empirical values obtained by the research team. Aside from their own expertise, general managers play a supporting role, ensuring that R&D is successfully processed by performing tasks in relation to their understanding of the scheme of AI professionals and managing manpower. Should there be insufficient supporting human resources, acquisition, refinement, and labeling managers need to be chosen through professional training and administering manpower, and AI engineers should provide assistance in focusing on enhancement in the capacity of the model. Ultimately, no differences in terms of division could be observed during the Industrial Revolution, and because productivity has dramatically risen as a consequence, AI R&D is expected to produce the same results. For this, large scale learning programs and support are needed for ordinary researchers to become advanced ones through proper education.

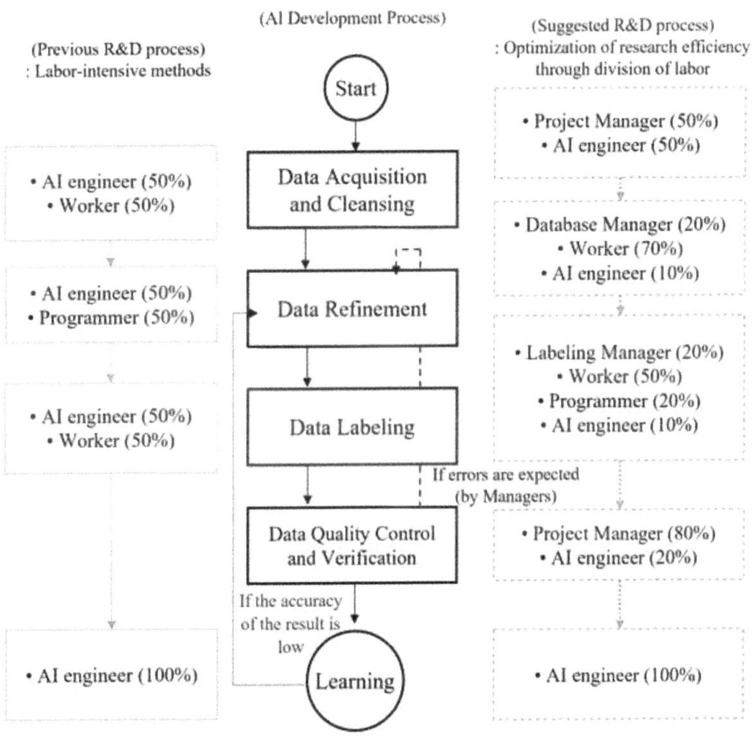

Figure 5. The current R&D method for the human resource structure used in the AI development process and the improved manpower distribution method proposed by the researchers.

4.2. Solution to Concerns Regarding Data Waste

The number of images per project acquired by the Korean government's Data Dam establishment project is about 200 k~500 k. The cost of this is around KRW 1.8 billion. The likelihood of using these data for other development purposes should be up for reconsideration.

According to a case study of the research team, the same amount of effort required for the existing refined quantity was also required even when adding one simple classification category. Regardless of the selected AI development methods, new category classification and labeling methods should be applied to the model to be developed, and even datasets developed after the consideration of extendibility require half the amount of the pre-existing manpower. Therefore, data should be acquired and refined so as to possess generality by establishing detailed categories, organizing data with extendibility reflected, adopting labeling techniques, starting with a broad development range from the beginning, and reducing cost and time losses.

5. Case Study and Suggestion Regarding Work Index

Three case studies were carried out to support the research team's argument that AI R&D is similar to a type of light industry. The case studies were carried out in a similar manner to that of the development method performed by general researchers, and the manpower and hours consumed per development step can be observed in Table 4. Equation (1) presented below is suggested with image quantity, input manpower, and work hours.

$$\text{Work index} = \frac{\text{Total amount of data}}{\text{Degree of input manpower} \times \text{Work hours}} \tag{1}$$

Table 4. 'Work index' on the basis of required manpower and hours depending on the AI R&D method.

Case	Method	Amount of Data	Step	Manpower /Hour	Working Time	Metrics per Data	Avg.
A. Crack	Classification	20,000	Acquisition/labeling	1	200	1.4	1.4
B. Construction waste	Instant segmentation	866	Acquisition	3	48	6.0	3.3
			Labeling	4	180	1.2	
			Relabeling	4	60	3.6	
C. Rebar		726	Acquisition	3	48	5.0	4.1
			Labeling	4	110	1.65	

Through this equation, the degree of resource consumption in comparison with the data can be quantitatively calculated as a value from 1 to 10 depending on the development method of AI. If a high amount of manpower and hours are input, the index shows a low value; if data generated from other further R&D processes are analyzed, the adequacy of the resource input for data acquisition, refinement, and labeling development can be verified.

5.1. Development on Concrete Surface Crack Recognition

In order to use the object detection method, a total of 20,000 photos showing concrete cracks in 224 × 224 pixels were collected. Concrete cracks were acquired by photographing actual architecture, and the cracks on the image were cropped and saved. It took 200 h under one worker/hour to complete data acquisition and cropping. Special labeling was not required; thus, the work index was 1.4 when compared to a considerable amount of learning data. As a result of learning, the AI models could perceive cracks. While acquiring learning data, the images were amassed around horizontal or vertical cracks, so diagonal cracks were identified, but those intersecting in X were unrecognizable.

5.2. Development in Perceiving Constructional Waste

In order to develop an AI model that differentiates 5 out of 18 constructional wastes being generated from construction sites, namely, concrete, brick, lumber, board, and mixed waste, data were acquired and refined through instant segmentation. A total of 866 images were acquired, and those taken on-site were cropped, collected online at a resolution of 512 × 512 pixels, and refined to 100 kb per photo. Moreover, pixels around the object boundaries were extracted using the polygonal method so as to differentiate the classes for wastes on the images. It took 48 h under three workers/hour to conduct data acquisition, and this time includes all aspects from planning the data acquisition process to taking pictures on-site as well as moving and saving the pictures on a data server. Refinement and labeling were simultaneously carried out, which took 180 h under four workers/hour. The work index for each development step was different, as shown in Table 4, and it could be observed that many resources were used for data acquisition in this research. Since inaccuracy was observed in some parts after learning, around 60 h under four workers/hour were additionally required for relabeling (segmentation) based on the acquired source data. Relabeling required only one-third of the resources needed to complete the previous work, and it was possible to differentiate five types of wastes after learning with the relabeled data, but it was ultimately decided to increase the amount of learning data due to cases in which board and lumber were being confused for each other.

5.3. Development of Model Identifying the Quantity of Rebar

A model to identify individual bundles of rebar was developed to manage automatized materials on a construction site. Within images of 512 × 512 pixels, 50–250 bundles of rebar were photographed, and the cross-section of rebar was extracted using the polygonal method so as to perceive each bundle of rebar separately. It took 48 h under three manpower/hour to acquire 726 images, and it took 100 h under four manpower/hour to separately segment around 120 k rebars. As a result of learning, the extraction of rebar quantity from the photo, which was achieved by identifying each rebar cross-section individually, was confirmed. However, if more than 300 bundles of rebar were included in one image, some were difficult to differentiate on 512 × 512 pixels, even with the naked eye; thus, a notable decrease in accuracy was observed in some parts with low resolution. This case study had the tendency to a show high work index compared to that of others, and this is considered to be due to no additional labeling being conducted and the development being efficiently carried out by a skilled research team.

6. Conclusions

The fourth Industrial Revolution refers to an intellectualized society with various technologies, such as AI, big data, block chains, and robotics. This study focused on AI among technologies in the fourth Industrial Revolution and examined the development process's similarities to that of light industry. Compared with the heavy industry, the production efficiency of the light industry is dependent upon manpower and production stages. All processes for AI R&D, from data acquisition to refinement, learning, and quality evaluation, need to be carried out by personnel, an observation that is also highly applicable to the light industry. However, since previous business entities/institutes processed development by only focusing on the AI learning step, manpower supply was not prioritized, which led to low assessment of the importance of human resources in national R&D projects, thus leading to a reality with improper AI R&D. Therefore, this study conducted a detailed analysis of necessary tasks for each AI learning step and examined the urgency of R&D personnel training.

6.1. Data Acquisition Step

Previously, images or time series data were simply copied into the digital form for data acquisition. However, for R&D with commercialization purposes or being conducted in specific professional fields, it is difficult to indiscriminately use existing data due to copyrights and personal information protection, and a process to check the data individually after acquisition is mandatory. A tool to automatize this process exists but cannot be considers as completed; thus, human interference is necessary throughout all processes.

6.2. Data Refinement/Labeling Step

It was verified through a case study that the accuracy of the AI model and reduction in the research period were expectable only with the enhancement of refinement technology. It was observed that a refinement manager and developing personnel who possess academic knowledge of development field were necessary, and it was important that the data were refined by the personnel who received professional education on development rather than by the public recruited through the existing cloud sourcing method. In addition, a refinement manager should be hired to manage R&D projects, and AI engineers should be able to focus on their own tasks, thereby leading to maximized efficiency of labor division.

6.3. Quality Evaluation Step

According to the research results, it appears that managers could carry out quality evaluation for each step. This could lead to quality evaluation focused only on the research purpose, and reutilization of learning data would be impossible. Although it is inessential at this point, reutilization of learning data could serve as a shortcut to reduce the development period for researching and developing similar AI when the golden age of AI research is reached. Hereupon, the necessity of professional manpower to predict development in other fields and manage data would become important.

If a sufficient number of personnel are educated and trained to lead specialized development projects, as per the results above, new value creation in the AI era can be expected. Such an AI industry would increase job opportunities in the short term, and job opportunities in general are not considered to decrease due to AI until a particular point is reached. Labor division and development of mechanisms that emerged during second Industrial Revolution period brought explosive productivity enhancement. The fourth and second Industrial Revolutions show similarities in terms of natural resources (data) and mechanisms (computer). A considerable number of examples verified the increased productivity achieved by using AI, and if human resources are efficiently distributed and utilized, advancement in mankind similar to that observed during the second Industrial Revolution is expected.

Author Contributions: Conceptualization S.H. (Seokjae Heo) and S.N.; methodology S.H. (Sehee Han), Y.S. and S.N.; formal analysis S.N., S.H. (Seokjae Heo), and S.H. (Sehee Han); resources S.H. (Sehee Han); project administration S.N.; resource, S.H. (Seokjae Heo) and Y.S.; funding acquisition S.H. (Seokjae Heo); writing—original draft S.N. and Y.S.; writing—review and editing, S.N. and Y.S.; visualization: S.H. (Seokjae Heo) and Y.S. All authors have read and agreed to the published version of the manuscript.

Funding: This work was supported by the National Research Foundation of Korea (NRF) grant funded by the Korea government Ministry of Education (No. NRF-2019R1A6A301091459 and NRF-2020R1C1C1005406).

Institutional Review Board Statement: Not applicable.

Informed Consent Statement: Not applicable.

Data Availability Statement: The data used to support the results in this study are included within the article. In addition, some of the data in this research are supported by the references mentioned in the article. If you have any queries regarding the data, the data will be available from the correspondence of the research upon request.

Conflicts of Interest: We declare that there are no conflict of interests among all the authors of this research.

References

1. Revaud, J.; Heo, M.; Rezende, R.S.; You, C.; Jeong, S.-G. Did it change? Learning to detect point-of-interest changes for proactive map updates. In Proceedings of the IEEE/CVF Conference on Computer Vision and Pattern Recognition, Long Beach, CA, USA, 15–20 June 2019.
2. Kim, B.; Lee, J.; Kang, J.; Kim, E.-S.; Kim, H.J. HOTR: End-to-End Human-Object Interaction Detection with Transformers. In Proceedings of the IEEE/CVF Conference on Computer Vision and Pattern Recognition, Nashville, TN, USA, 19–25 June 2021.
3. Roh, B.; Shin, W.; Kim, I.; Kim, S. Spatially consistent representation learning. In Proceedings of the IEEE/CVF Conference on Computer Vision and Pattern Recognition, Nashville, TN, USA, 19–25 June 2021.
4. King, T.M.; Arbon, J.; Santiago, D.; Adamo, D.; Chin, W.; Shanmugam, R. AI for testing today and tomorrow: Industry perspectives. In Proceedings of the 2019 IEEE International Conference On Artificial Intelligence Testing (AITest), Newark, CA, USA, 4–9 April 2019.
5. NIA National Information Society Agency. NIA2020-032. 2020. Available online: https://www.nia.or.kr/ (accessed on 27 September 2021).
6. Dosovitskiy, A.; Beyer, L.; Kolesnikov, A.; Weissenborn, D.; Zhai, X.; Unterthiner, T.; Dehghani, M.; Minderer, M.; Heigold, G.; Gelly, S.; et al. An image is worth 16x16 words: Transformers for image recognition at scale. *arXiv* **2020**, arXiv:2010.11929.
7. Felzenszwalb, P.; McAllester, D.; Ramanan, D. A discriminatively trained, multiscale, deformable part model. In Proceedings of the 2008 IEEE Conference on Computer Vision and Pattern Recognition, Anchorage, AK, USA, 23–28 June 2008.
8. Girshick, R. Fast r-cnn. In Proceedings of the IEEE International Conference on Computer Vision, Santiago, Chile, 7–13 December 2015.
9. He, K.; Gkioxari, G.; Dollar, P.; Girshick, R. Mask r-cnn. In Proceedings of the IEEE International Conference on Computer Vision, Venice, Italy, 22–29 October 2017.
10. Huang, G.; Liu, Z.; van der Maaten, L.; Weinberger, K.Q. Densely connected convolutional networks. In Proceedings of the IEEE Conference on Computer Vision and Pattern Recognition, Honolulu, HI, USA, 21–26 July 2017.
11. He, K.; Zhang, X.; Ren, S.; Sun, J. Deep residual learning for image recognition. In Proceedings of the IEEE Conference on Computer Vision and Pattern Recognition, Las Vegas, NV, USA, 27–30 June 2016.
12. Krizhevsky, A.; Sutskever, I.; Hinton, G.E. Imagenet classification with deep convolutional neural networks. *Adv. Neural Inf. Process. Syst.* **2012**, *25*, 1097–1105. [CrossRef]
13. Lin, T.-Y.; Maire, M.; Belongie, S.; Hays, J.; Perona, P.; Ramanan, D.; Dollár, P.; Zitnick, C.L. Microsoft coco: Common objects in context. In *European Conference on Computer Vision*; Springer: Dordrecht, The Netherlands, 2014.
14. Lowe, D.G. Object recognition from local scale-invariant features. In Proceedings of the Seventh IEEE International Conference on Computer Vision, Corfu, Greece, 20–27 September 1999.
15. Redmon, J.; Divvala, S.; Girshick, R.; Farhadi, A. You only look once: Unified, real-time object detection. In Proceedings of the IEEE Conference on Computer Vision and Pattern Recognition, Las Vegas, NV, USA, 27–30 June 2016.
16. Ren, S.; He, K.; Girshick, R.; Sun, J. Faster r-cnn: Towards real-time object detection with region proposal networks. *Adv. Neural Inf. Process. Syst.* **2015**, *28*, 91–99. [CrossRef] [PubMed]
17. Simonyan, K.; Zisserman, A. Very deep convolutional networks for large-scale image recognition. In Proceedings of the 3rd International Conference on Learning Representations, San Diego, CA, USA, 7–9 May 2015.
18. Tan, M.; Le, Q. EfficientNet: Rethinking model scaling for convolutional neural networks. In Proceedings of the 36th International Conference on Machine Learning, Long Beach, CA, USA, 10–15 June 2019.
19. Vaswani, A.; Shazeer, N.; Parmar, N.; Uszkoreit, J.; Jones, L.; Gomez, A.N.; Kaiser, L.; Polosukhin, I. Attention is all you need. In *Advances in Neural Information Processing Systems*; Curran Associates, Inc.: Red Hook, NY, USA, 2017.
20. Kim, J.M.; Hyeon, S.G.; Chae, J.H.; Do, M.S. Road Crack Detection based on Object Detection Algorithm using Unmanned Aerial Vehicle Image. *J. Korea Inst. Intell. Transp. Syst.* **2019**, *18*, 155–163. [CrossRef]
21. Lee, B.-Y.; Yi, S.-T.; Kim, J.-K. Surface crack evaluation method in concrete structures. *J. Korean Soc. Nondestruct. Test.* **2007**, *27*, 173–182.
22. Abioye, S.O.; Oyedele, L.O.; Akanbi, L.; Ajayi, A.; Delgado, J.M.D.; Bilal, M.; Akinade, O.O.; Ahmed, A. Artificial intelligence in the construction industry: A review of present status, opportunities and future challenges. *J. Build. Eng.* **2021**, *44*, 103299. [CrossRef]
23. Chakraborty, D.; Elhegazy, H.; Elzarka, H.; Gutierrez, L. A novel construction cost prediction model using hybrid natural and light gradient boosting. *Adv. Eng. Inform.* **2020**, *46*, 101201. [CrossRef]
24. Elhegazy, H.; Chakraborty, D.; Elzarka, H.; Ebid, A.M.; Mahdi, I.M.; Haggag, S.Y.A.; Rashid, I.A. Artificial Intelligence for Developing Accurate Preliminary Cost Estimates for Composite Flooring Systems of Multi-Storey Buildings. *J. Asian Archit. Build. Eng.* **2021**, 1–13. [CrossRef]
25. Baji, T. Evolution of the GPU Device widely used in AI and Massive Parallel Processing. In Proceedings of the 2018 IEEE 2nd Electron Devices Technology and Manufacturing Conference, Kobe, Japan, 13–16 March 2018.
26. Na, S.; Heo, S.-J.; Han, S. Construction Waste Reduction through Application of Different Structural Systems for the Slab in a Commercial Building: A South Korean Case. *Appl. Sci.* **2021**, *11*, 5870. [CrossRef]
27. Seok-jae, H.; Dae-ho, M.; Bang-yeon, L.; Sang-hyun, L. Detection of Plane Member Deformation using Moire Fringe and Recognition using Machine Learning. *J. Wind. Eng. Inst. Korea* **2018**, *22*, 25–32.

28. Telecommunications Technology Association. *AI Learning Data Quality Management Guidelines v1.0*; Ministry of Science and ICT: Gwacheon, Korea, 2021.
29. Telecommunications Technology Association. *AI Dataset Construction Guidebook*; Ministry of Science and ICT: Gwacheon, Korea, 2021.
30. Girshick, R.; Donahue, J.; Darrell, T.; Malik, J. Rich feature hierarchies for accurate object detection and semantic segmentation. In Proceedings of the IEEE Conference on Computer Vision and Pattern Recognition, Columbus, OH, USA, 23–28 June 2014.
31. Tuzel, O.; Porikli, F.; Meer, P. Region covariance: A fast descriptor for detection and classification. In Proceedings of the 9th European Conference on Computer Vision, Graz, Austria, 7–13 May 2006.
32. Lazebnik, S.; Schmid, C.; Ponce, J. Beyond bags of features: Spatial pyramid matching for recognizing natural scene categories. In Proceedings of the 2006 IEEE Computer Society Conference on Computer Vision and Pattern Recognition, New York, NY, USA, 17–22 June 2006.
33. Dalal, N.; Triggs, B. Histograms of oriented gradients for human detection. In Proceedings of the 2005 IEEE Computer Society Conference on Computer Vision and Pattern Recognition, San Diego, CA, USA, 20–25 June 2005.
34. Viola, P.; Jones, M. Rapid object detection using a boosted cascade of simple features. In Proceedings of the 2001 IEEE Computer Society Conference on Computer Vision and Pattern Recognition, Kauai, HI, USA, 8–14 December 2001.
35. Acuna, D.; Ling, H.; Kar, A.; Fidler, S. Efficient interactive annotation of segmentation datasets with polygon-rnn++. In Proceedings of the 2018 IEEE Conference on Computer Vision and Pattern Recognition, Salt Lake City, UT, USA, 18–22 June 2018.
36. Russell, B.C.; Torralba, A.; Murphy, K.P.; Freeman, W.T. LabelMe: A database and web-based tool for image annotation. *Int. J. Comput. Vis.* **2008**, *77*, 157–173. [CrossRef]
37. Dutta, A.; Gupta, A.; Zissermann, A. *VGG Image Annotator (VIA)*; University of Oxford: Oxford, UK, 2016.

applied sciences

Article

A Hierarchical Generative Embedding Model for Influence Maximization in Attributed Social Networks

Luodi Xie [1], Huimin Huang [2,*] and Qing Du [3]

[1] School of Computer Science, Sun Yat-sen University, Guangzhou 510275, China; xield@mail2.sysu.edu.cn
[2] School of Data Science and Artificial Intelligence, Wenzhou University of Technology, Wenzhou 325035, China
[3] School of Software, South China University of Technology, Guangzhou 510641, China; duqing@scut.edu.cn
* Correspondence: huanghm45@gmail.com

Abstract: Nowadays, we use social networks such as Twitter, Facebook, WeChat and Weibo as means to communicate with each other. Social networks have become so indispensable in our everyday life that we cannot imagine what daily life would be like without social networks. Through social networks, we can access friends' opinions and behaviors easily and are influenced by them in turn. Thus, an effective algorithm to find the top-K influential nodes (the problem of influence maximization) in the social network is critical for various downstream tasks such as viral marketing, anticipating natural hazards, reducing gang violence, public opinion supervision, etc. Solving the problem of influence maximization in real-world propagation scenarios often involves estimating influence strength (influence probability between two nodes), which cannot be observed directly. To estimate influence strength, conventional approaches propose various humanly devised rules to extract features of user interactions, the effectiveness of which heavily depends on domain expert knowledge. Besides, they are often applicable for special scenarios or specific diffusion models. Consequently, they are difficult to generalize into different scenarios and diffusion models. Inspired by the powerful ability of neural networks in the field of representation learning, we designed a hierarchical generative embedding model (HGE) to map nodes into latent space automatically. Then, with the learned latent representation of each node, we proposed a HGE-GA algorithm to predict influence strength and compute the top-K influential nodes. Extensive experiments on real-world attributed networks demonstrate the outstanding superiority of the proposed HGE model and HGE-GA algorithm compared with the state-of-the-art methods, verifying the effectiveness of the proposed model and algorithm.

Keywords: influence maximization; influence strength; network embedding; social networks

Citation: Xie, L.; Huang, H.; Du, Q. A Hierarchical Generative Embedding Model for Influence Maximization in Attributed Social Networks. *Appl. Sci.* **2022**, *12*, 1321. https://doi.org/10.3390/app12031321

Academic Editors: Nikos D. Lagaros and Vagelis Plevris

Received: 24 November 2021
Accepted: 20 January 2022
Published: 26 January 2022

1. Introduction

Fueled by the spectacular growth of the internet and Internet of Things, plenty of social networks such as Facebook, Twitter and WeChat have sprung up, changed the mode of interaction between people, and accelerated the development of viral marketing. Originally from the idea of word-of-mouth advertising, viral marketing takes advantage of trust among close social circles of friends, colleagues or families to promote a new product, i.e., when a friend relationship affects a user making decisions on item selection [1,2]. Motivated by applications to early viral marketing, a new study area of influence diffusion has thrived. Therein, the problem of influence maximization is to select a fixed size set of seed nodes in a network to maximize the influence spread according to a specially designed influence diffusion model. Figure 1 gives a toy example of social influence. The nodes v_1,v_2,v_3 in black are the seed nodes which are initially active, and the nodes in gray color are newly activated by the seed nodes. In terms of viral marketing, for example, if user v_1,v_2,v_3 bought a product, their friends in a given social network will likely buy this product because of the friend-to-friend influence.

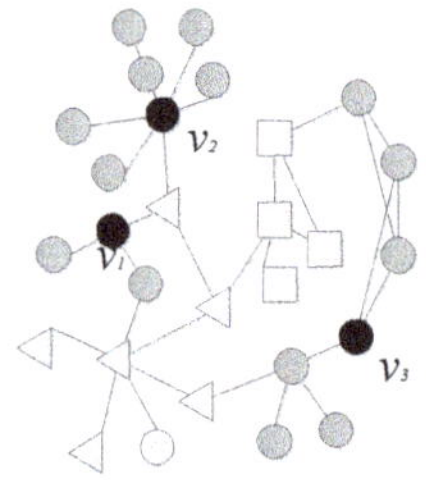

Figure 1. A toy example of social influence. The seed nodes are v_1,v_2,v_3, represented in black, and the nodes influenced by the seed nodes are in gray.

The applications of influence maximization are prevalent in the real world and include such purposes as viral marketing, anticipating natural hazards, reducing gang violence, public opinion supervision etc., whereas the time complexity of solving the problem of influence maximization is NP hard [3]. This inspires a lot of studies on influence diffusion and influence maximization algorithms. In early time, most studies have focused on the influence maximization algorithms themselves under a general independent cascade (IC) diffusion model or a general linear threshold (LT) diffusion model [4,5], including greedy-based algorithms [3,6–10] and heuristic-based algorithms [11–13]. In terms of running time, heuristic-based algorithms are generally more efficient than various greedy-based algorithms, but they do not have any theoretical guarantee.

As in the real world, influence propagation often involves latent variables that are directly unobservable, including influence strength. Estimating influence strength is a fundamental problem for influence diffusion. Some models to estimate influence strength have been proposed, including the topic model [14,15], probabilistic methods [16,17] and models based on deep learning [18,19]. Although existing models and methods have achieved a lot, they demonstrate a number of major drawbacks: (1) conventional approaches propose various humanly devised rules to extract features of user interactions, the effectiveness of which heavily depends on domain expert knowledge. (2) They only consider general network structure and node attributes as factors underlying social network influence, but in the real world, the structure of a large-scale network is more complex and often involves communities, community tensility and node depth in a clustering tree. (3) They are often applicable for special scenarios or specific diffusion models, not for generalized conditions.

Like most scenarios in real-world propagation [16], influence strength, in this paper, is assumed to be directly unobservable. To predict influence strength, we need to compute the similarity of node pairs, with the motivation originating from previous studies that showed that the more equivalent the network structure and node attributes of two nodes are, the more likely they make similar judgments, even if they have no direct connection reciprocally [20]. Inspired by the powerful ability of neural networks in the field of representation learning, we design a deep hierarchical network embedding model, HGE, to map nodes into latent space automatically, preserving network structures and node attributes as much as possible. The correspondence between the general attributed network and the hierarchical latent space is illustrated in Figure 2. Then, with the learned latent representation of each node, we measure the similarity of node pairs and regard it as the influence probability between two nodes, since the underlying factors of influence probability are network structures and node attributes [20]. Next, the top-K influential nodes can be computed through a greedy-based maximization algorithm. The overview of the proposed method is illustrated in Figure 3.

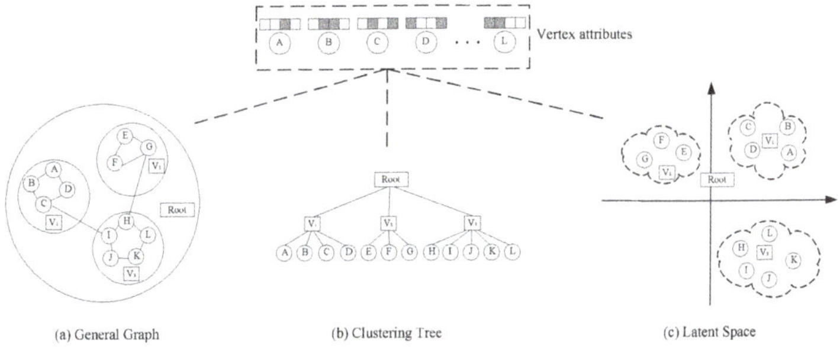

Figure 2. The correspondence between the general attributed network and the hierarchical latent space.

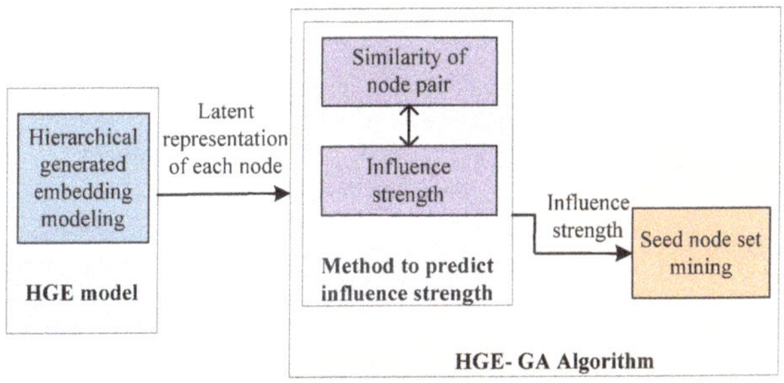

Figure 3. Overview of the system architecture.

Social networks usually have millions of users, with a large amount of user-related information, community structures, and complex hierarchical network structures. When embedding nodes into a low-dimensional vector space, in order to preserve network structures as much as possible, hierarchical structures should be taken into account. Furthermore, a community is formed of nodes which have similar attributes while repelling each other due to attribute differences measured by the tensility of the community. In other words, a community is a node set with tensility, and in order to capture this tensility, we embed a community as a Gaussian distribution. Specifically, we propose a hierarchical generative embedding model which integrates hierarchical community structure, node attributes and general network structure into a unified generative framework.

To summarize, we make the following contributions:

(1) We study the problem of incorporating hierarchical generative embedding into influence strength prediction for the first time;

(2) In order to predict influence strength, we propose a novel model, the HGE model, which automatically maps nodes into latent space, avoiding humanly devising rules to extract features of user interactions. The proposed HGE model integrates hierarchical community structure, node attributes and general network structure into a unified generative framework that ensures the preservation of node characteristics and captures the granularity of hierarchical characteristics as well as community tensility in an attributed network effectively. Furthermore, we propose a new algorithm, HGE-GA, to predict influence strength and compute the top-K influential nodes effectively. The proposed HGE-GA algorithm can be generalized to different scenarios or diffusion models;

(3) We evaluate our method on various datasets and tasks, and experimental results show that the proposed model and algorithm significantly outperform the state-of-the-art approaches.

The rest of this paper is organized as follows. Section 2 reviews related work. Section 3 presents notations and problem formulation. Section 4 details the proposed method.

Section 5 lists the experiment setup. Section 6 presents the experimental results & analysis. Section 7 concludes the paper.

2. Related Work

Influence Diffusion. The studies of influence diffusion focus on the key factors for influence diffusion, such as diffusion models of a single entity or multiple entities, as well as the influence probability between two nodes, which is more relevant to our work. Estimating influence probability between two nodes (or influence strength) is a fundamental problem for influence diffusion, since in real-world propagation scenarios, the variables such as influence strength and infection time may be not directly observed. Some models to estimate influence strength have been proposed. One line of research is based on topic extraction and lays emphasis on incorporating the topic model into influence diffusion. Ref. [15] proposes a latent variable model, which captures community-level topic interest, item-topic relevance and community membership distribution of each user and further infers user-to-user influence strength. Ref. [14] proposes a novel model, COLD, to model community-level influence diffusion. Another line of studies utilize various probabilistic methods such as Bayesian inference [16] and EM (Expectation maximization) [17] to stimulate influence diffusion and learn the parameters in the generative model. Recently, researchers have endeavored to build models of influence prediction using deep learning, which can predict social influence automatically and are no longer limited to expert knowledge when extracting user features or network features. Ref. [18] proposes a recurrent neural variational model (RNV) to dynamically track entity correlations and cascade correlations. Ref. [19] proposes the DeepInf model to predict social influence, which detects the dynamics of social influence and integrates network embedding and graph attention mechanisms into the model. Ref. [21] proposes a reinforcement learning framework to discover effective network sampling heuristics by leveraging automatically learned node and graph representations.

To the best of our knowledge, the existing models to predict the influence probability between two nodes usually extracts features of user interactions with various humanly devised rules, which heavily depends on expert knowledge of this domain. Besides, the underlying factors that affect social network influence, such as communities, community tensility and node depth in a clustering tree, have not been taken into account in the existing models when computing the influence probability of node pairs. Moreover, the existing models are often applicable for special scenarios or specific diffusion models, not for generalized conditions.

Network Embedding. Network embedding techniques aim at inferring representations, also called embeddings, of entities in the networks. The basic idea of network embedding is to embed the nodes into a low-dimensional vector space in which the similarity between nodes can be measured and network structures can be preserved as much as possible. Many downstream learning tasks can benefit from this form of representation, such as link prediction, node classification, node clustering and network visualization.

Existing models of network embedding are concerned about the technology of mapping or dimension reduction. In order to obtain effective low-dimensional embeddings, numerous algorithms have been proposed. Inspired by Word2Vec [22], deep learning has been migrated to network embedding. DeepWalk [23], Node2Vec [24] and Line [25] extract a number of sequential nodes in the network by random walk. Nodes in a sequence are metaphorically equivalent to words in language models. Skip-Gram [22] is then employed to obtain the embeddings of nodes. After obtaining the embeddings of nodes, the similarity of nodes can be calculated by the inner-product of their embeddings. LINE [25] uses first-order and second-order proximities and trains the model via negative sampling. SDNE [26] adopts a deep auto-encoder to preserve both the first-order and second-order proximities.

As one of the most important features of social networks, community structure is usually integrated into network embedding as auxiliary information [27–30]. M-NMF [27] incorporates the community structure into network embedding and exploits the consensus relationship between the representations of nodes and community structure. M-NME then jointly optimize NMF (Network Matrix Factorization)-based representation learning model and modularity-based community detection model in a unified framework. GraphAGM [28] combines AGM [31] and GAN [30] in a unified framework which can generate the most likely motifs with graph structure awareness in a computationally efficient way. NECS [29] preserves the high-order vertex proximity and incorporates the community structure of networks in vertex representation learning.

Hierarchical Network Embedding. The community structures in complex networks are often hierarchical [32–36]. Inspired by the natural hierarchical structure of a galaxy with its stars, planets, and satellites, ref. [33] proposes the galaxy network embedding (GEM) model. GEM captures hierarchical structures by forming an optimization problem, including pairwise proximity, horizontal relationships and vertical relationships. In order to solve various defects caused by the exponential decay of the radius in the GEM model, Long et al. [36] first apply subspace to hierarchical network embedding and propose the SpaceNE model, which preserves proximity between pairwise nodes and between communities. Ref. [35] proposes NetHiex, a network embedding algorithm which can capture the different levels of granularity and alleviate data scarcity. Ma et al. [37] propose the MINES framework, which can embed multi-dimensional networks with a hierarchical structure to low-dimensional vector spaces, with the learned representations for each dimension containing the hierarchical information. Nickel et al. [38] use hyperbolic space instead of Euclidean space to capture hierarchical structures in networks. The learned embeddings are difficult to convert into Euclidean space, which is inappropriate for some downstream tasks of machine learning.

However, these hierarchical embedding algorithms suffer from some defects: (1) They embed each node or community as a vector in the latent space, which ignores the tensility of the community. Essentially, a community is a set of nodes or smaller communities with similar attributes and tensility, and a simple vector can not represent this relationship of affiliation. The tensility of a community has been defined in Definition 2. (2) They do not consider node depth in a clustering tree when dealing with vertical relationships in a hierarchical network. Thus, it is inappropriate to use the traditional method of measuring vertical relationships in a hierarchical network to compute vector distance in latent space. The granularity of hierarchical structures should be taken into account, and a personalized method to measure vector distance in latent space should be designed. The granularity of hierarchical structures has been defined in Definition 3.

3. Notations and Problem Formulation

In this section, we first introduce the definitions and notations, and then formulate the problem studied in this paper.

Definition 1 (attributed network). *An attributed network is an undirected graph $\mathcal{G} = (\mathcal{V}, \mathcal{E}, \mathcal{A})$, where $\mathcal{V}$ is the set of network nodes, $\mathcal{E}$ is the set of edges and $\mathcal{A} \in \{0,1\}^{|\mathcal{V}| \times F}$ is a binary-valued attribute matrix recording node-attribute associations, with F being the number of attributes. The element $A_{i,a} \in \mathcal{A}$ is a binary-valued attribute weight, indicating that attribute a associates with node i if $A_{i,a} = 1$.*

Definition 2 (the tensility of community). *A community is a set composed of nodes with similar attributes. These nodes form a community due to similar attributes and repel each other in the community due to differences in attributes. The tensility of the community is used to measure the difference of node attributes in the community.*

Definition 3 (the granularity of hierarchical characteristics). *In hierarchical networks, vertexes at different levels have different granularities of characteristics. As shown in Figure 4, the university in the top level has the characteristics of name, address etc., and the colleges in the second level have the characteristics of majors, number of students etc. The characteristics of nodes* *at different levels have different granularities, while the granularity of the characteristics is the same for the nodes at the same level. It is the granularity of hierarchical characteristics that causes the vertical constraint of hierarchical structure properties (see Section 4.1.1).*

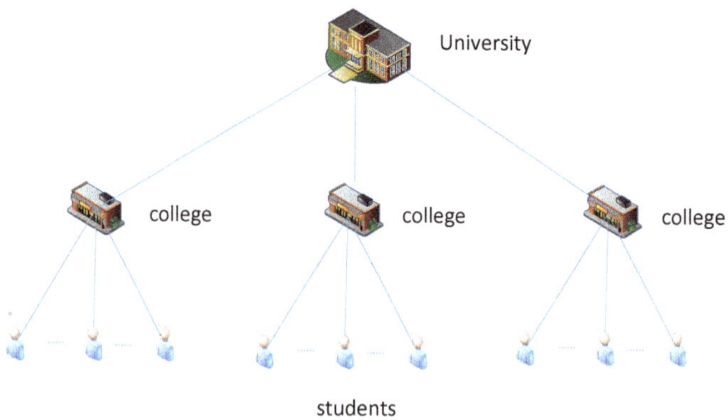

Figure 4. Illustration of vertexes at different levels having different granularity features.

The hierarchical clustering tree of G is denoted as T with a depth of L. $\mathcal{V}^l$ is the vertex set at the l-th level of T, and $v_i^l \in \mathcal{V}^l$ is the i-th vertex at the l-th level. $Ch(v)$ denotes the child vertex set of the vertex v, $fa(v)$ denotes the father vertex of v and $\mathbf{\Phi}(v)$ denotes the embedding of vertex v. In our paper, we use 'vertex' to represent a user or community in the hierarchical network, while using 'node' to represent a user in the general network, especially when $l = L$, v_i^l represents a leaf vertex (user) at the bottom of the clustering tree. For ease of reference, we summarize the basic notations in Table 1.

Table 1. Main notation used across the whole paper.

Notation	Description
$\mathcal{G}$	An undirected graph
$\mathcal{V}$	Set of nodes
$\mathcal{E}$	Set of edges
T	A hierarchical clustering tree
L	The depth of the clustering tree
$Ch(v)$	The child vertex set of vertex v
$fa(v)$	The father vertex of v
v_i^l	The i-th vertex at the l-th level
$\mathbf{\Phi}_v$	The embedding of vertex v

With the definitions and notations described above, we formulate the problem studied.

Problem 1 (Node Embedding). *Given an undirected graph $\mathcal{G} = (\mathcal{V}, \mathcal{E}, \mathcal{A})$, the goal is to learn the latent representation of nodes with the mapping function Ξ:*

$$\mathcal{G} \xrightarrow{\Xi} \mathbf{\Phi}_{\mathcal{V}} \tag{1}$$

such that the information of hierarchical network structure, general network and node attributes can be preserved as much as possible by $\mathbf{\Phi}_{\mathcal{V}} \in \mathcal{R}^{N \times D}$, where $\mathbf{\Phi}_{\mathcal{V}}$ is the latent embedding matrix for all nodes and D is the dimension size of the embeddings.

Problem 2 (Influence Maximization in Attributed Social Networks). *Given an IC diffusion model, an undirected graph $\mathcal{G} = (\mathcal{V}, \mathcal{E}, \mathcal{A})$ and the latent embedding matrix for all nodes $\mathbf{\Phi}_{\mathcal{V}}$, the goal is to find the node set $\mathcal{S}$ with an algorithm $\mathcal{F}$:*

$$\mathcal{G}, \mathbf{\Phi}_{\mathcal{V}} \xrightarrow{\mathcal{F}} \mathcal{S} \tag{2}$$

such that the expected spread of node set $\mathcal{S}$, i.e., $\sigma(\mathcal{S})$, is maximized and $|\mathcal{S}| = K$. The expected spread of node set $\mathcal{S}$ is defined as the total number of active nodes after the influence spread ends, including the newly activated nodes and the initially active nodes.

4. The Proposed Method

Like most scenarios in real-world propagation [16], the influence probability of node pairs, in this paper, is assumed to be directly unobservable. To predict the influence probability of node pairs, we first embed the nodes into a low-dimensional vector space by jointly modeling the horizontal constraint, vertical constraint, affiliation constraint and node attributes. After obtaining the embedding of each node, we use the two-normal form to measure the similarity of the node pair. Then we predict the influence probability of the node pair with the measured similarity of the node pair. We now describe the proposed method in the following subsections: the hierarchical generative embedding model is described in Section 4.1, the learning procedure is described in Section 4.2, and the HGE-GA algorithm is described in Section 4.3.

4.1. Hierarchical Generative Embedding Model

The proposed HGE model focuses on preserving the following characteristics of the network: (1) Horizontal structure characteristics. Nodes belonging to the same community are more similar than nodes belonging to different communities (in terms of blood relations, blood brothers are closer than cousins). (2) Vertical structure characteristics. The more children a node has, the more genes it inherits from its father; that is, the more similar it is to its father. (3) Affiliation relationship characteristics. Users in the same community often have both similarities and differences between them. Differences between nodes are defined as the tensility of node and should be preserved. (4) General network structure characteristics. In a general network, the more similar the nodes are, the closer they are in the embedding space. (5) Node attribute characteristics. Node attributes provide rich information regarding node characteristics, which should be taken advantage of for preserving node characteristics in latent space.

We will detail the HGE model from three aspects. The first one is how to preserve the hierarchical network structure, the second one is how to capture general network structure properties, and the third one is how to acquire node attribute features.

4.1.1. Hierarchical Structure Properties

Similar to most embedding methods [23–25], we preserve the network structure through the distance of the vertex in the hidden space. A community is a node set with tensility, and in order to capture this tensility, we embed a community as a Gaussian distribution. The greater the variance, the greater tensility of the community, that is, the difference between nodes contained in the community will be greater. Therefore, we use the KL distance to measure the similarity between communities. More specifically, suppose $\mathbf{\Phi}_{v_1} = \mathcal{N}(\mu_1, \Sigma_1)$, $\mathbf{\Phi}_{v_2} = \mathcal{N}(\mu_2, \Sigma_2)$. Then, the distance between v_1 and v_2 can be calculated in two ways, as follows:

$$\delta(v_1, v_2) = D_{KL}(\mathcal{N}_1 \| \mathcal{N}_2) = \frac{1}{2}[tr(\Sigma_2^{-1}\Sigma_1) + (\mu_2 - \mu_1)^T \Sigma_2^{-1}(\mu_2 - \mu_1) - L - log\frac{\Sigma_2}{\Sigma_1}]. \tag{3}$$

where $tr(\Sigma_2^{-1}\Sigma_1)$ is the trace of matrix $\Sigma_2^{-1}\Sigma_1$, v_1 and v_2 are communities, and L is a constant.

$$\delta(v_1, v_2) = \|\mu_1 - \mu_2\|_2 \tag{4}$$

where v_1 and v_2 are nodes.

Unlike normal networks, hierarchical networks contain a large number of hierarchical relationships. To capture this complex hierarchical relationship, we force the embedding of a vertex to submit to two constraints, the horizontal and the vertical:

Horizontal constraint: Generally, community-based embedding methods [27,28] consider that the nodes belonging to the same community are more similar than those belonging to different communities. Thus, we intend to extend this constraint to hierarchical networks and name such a constraint as the horizontal constraint. The horizontal constraint can be defined as follows: For each layer l in clustering tree T, for all vertex-pairs v_i^l, v_j^l at layer l with $fa(v_i^l) = fa(v_j^l)$, for all v_k^l at layer l with $fa(v_i^l) \neq fa(v_k^l)$ and $fa(v_j^l) \neq fa(v_k^l)$, we have

$$\delta(v_i^l, v_j^l) < \delta(v_i^l, v_k^l) \tag{5}$$

$$\delta(v_i^l, v_j^l) < \delta(v_j^l, v_k^l) \tag{6}$$

Vertical constraint: From the biological point of view, a clustering tree can be likened to a gene tree, in which the lower vertexes propagate from the upper vertexes. The more children a vertex has, the more genes it inherits from its father, that is, the more similar it is to its father. We use a personalized distance rank method to describe the vertical constraint: for all vertices m,

$$\delta(v_1, v_m) < \delta(v_2, v_m) < \cdots < \delta(v_n, v_m) \tag{7}$$

if $n(v_1) > n(v_2) \cdots > n(v_n)$, where $v_1, v_2, \cdots, v_n$ are children of v_m, i.e., $v_1, v_2, \cdots, v_n \in Ch(v_m)$, and $n(v_1), n(v_2), \cdots, n(v_n)$, respectively, denote the number of the children of vertices $v_1, v_2 \cdots v_n$.

Affiliation constraint: In hierarchical networks, bottom-node clustering forms small communities, while small-community clustering forms large communities. Therefore, the parent node can be regarded as the feature set of all child nodes. Our HGE model describes the formation process of hierarchical networks from the perspective of generation: as the attribute set of the entire hierarchical network, the root node generates the child node of the next layer according to the attribute classification and repeats this top-down generation process to obtain the entire hierarchical network. In order to capture this hierarchical affiliation (generation) relationship, in the HGE model, nodes are embedded as Gaussian distributions. The mean value represents the position in the embedding space and is generated by the distribution of the parent node, and the variance represents the tensility of the node, which can also represent the differences among all child nodes. In particular, the leaf node has no tensility; that is, the variance of the embedding is 0.

A community is a small group of users who have the same hobbies, occupations, social relations and so on. Users in the same community often have similar attributes. Traditional community-based embedding methods always focus on how to capture the similarity between nodes in the same community [27], which ignores the differences between nodes. The affiliation constraint can be defined as follows: for vertex v_i^l and his father vertex v_j^{l-1}, $\mathbf{\Phi}_i^l = \mathcal{N}(\mu_i, \Sigma_i)$, $\mathbf{\Phi}_j^{l-1} = \mathcal{N}(\mu_j, \Sigma_j)$, where the μ_i is sampled from the hyper-distribution $\mathbf{\Phi}_j^{l-1}$, i.e., $\mu_i \sim \mathcal{N}(\mu_j, \Sigma_j)$. In particular, when $i = T$ (vertex i is at the bottom layer), $\mathbf{\Phi}_i^l = \mathcal{N}(\mu_i, 0)$.

4.1.2. General Network Structure Properties

In order to preserve the structure of the general network, we keep the embedding of nodes to satisfy the first-order proximity and the second-order proximity. The first-order proximity posits that, for each pair of nodes v_1 and v_2 linked by an edge(v_1, v_2), the similarity between v_1 and v_2 should be greater than that of two nodes without an edge connection, and the second-order proximity keeps each node pair(v_1, v_2) that share the same neighborhood similar.

4.1.3. Node Attribute Properties

Given a node i with an attribute a, the logistic model is used with a mean value of the multidimensional Gaussian distribution of node i, i.e., μ_i, as input features to predict the probability of node i associating with attribute a:

$$R_{i,a} = \sigma(W_a^T \mu_i + b), \tag{8}$$

where W_a is the logistic weight factor, b is the bias and $\sigma(.)$ is a logistic function defined as $\sigma(x) = (1 + e^{-x})^{-1}$.

4.2. *Learning Procedure*

In this subsection, we will introduce the learning procedure of the HGE model. Like most unsupervised embedding models, we formulate network structure preservation as an optimization problem. By optimizing the lower bound of the corresponding loss function, the HGE model converges. Our loss function can be divided into three independent parts, corresponding to the capture of the structure of the hierarchical network, general network and node attributes.

4.2.1. Hierarchical Network Optimization

First, the similarity of the same community node is captured by optimizing the lower bound of the loss function. Nodes in the same community can be closer in the hidden space, while nodes in different communities will be pushed farther apart. This can be defined as follows:

$$\mathcal{L}_{hor} = \sum_{l=3}^{T} \sum_{\substack{i,j,k \in \mathcal{S}_l, \\ fa(v_i^l) = fa(v_j^l) \neq fa(v_k^l)}} \delta(v_i^l, v_j^l)^2 + exp^{-\delta(v_i^l, v_k^l)}, \tag{9}$$

where $\mathcal{S}_l$ is the vertex set of layer l, and T is the depth of the clustering tree.

In order to capture the hierarchical vertical relationship, we use the following loss function to achieve the personalized distance ranking in Equation (7), as shown below:

$$\mathcal{L}_{ver} = \sum_{l=1}^{T-1} \sum_{\substack{i,j \in \mathcal{S}_l, \\ fa(v_i^l) = fa(v_j^l)}} \delta(v_i^l, fa(v_i^l))^2 - exp^{\delta(v_j^l, fa(v_i^l))}, \tag{10}$$

when $n(v_i^l) < n(v_j^l)$.

To capture hierarchical affiliation, the HGE model assumes that the mean of the node embedding is subordinated to the Gaussian distribution corresponding to its father vertex. We use maximum likelihood to approximate the embedding of the father node. For an arbitrary vertex i, suppose that vertices $v_1, v_2, \ldots, v_m$ are children of vertex i with $\mathbf{\Phi}_{v_1} = \mathcal{N}(\mu_1, \Sigma_1), \mathbf{\Phi}_{v_i} = \mathcal{N}(\mu_i, \Sigma_i), \ldots, \mathbf{\Phi}_{v_m} = \mathcal{N}(\mu_m, \Sigma_m)$. Then, the maximum likelihood function can be defined as follows:

$$\begin{aligned} L(\mu, \Sigma) &= \prod_{i=1}^{m} f(\mu_i; \mu, \Sigma) \\ &= \prod_{i=1}^{m} (2\pi)^{-\frac{n}{2}} |\Sigma|^{-\frac{1}{2}} \exp(-\frac{1}{2}(\mu_i - \mu)^T \Sigma^{-1} (\mu_i - \mu)) \\ &= (2\pi)^{-\frac{mn}{2}} |\Sigma|^{-\frac{m}{2}} \exp(-\frac{1}{2} \sum_{i=1}^{m} (\mu_i - \mu)^T \Sigma^{-1} (\mu_i - \mu)) \end{aligned} \tag{11}$$

where n is the dimension of μ.

The logarithmic likelihood function is:

$$\begin{aligned}\ln L(\mu,\Sigma) = \ln(2\pi)^{-\frac{mn}{2}} + \ln|\Sigma|^{-\frac{m}{2}} \\ + \ln\exp(-\frac{1}{2}\sum_{i=1}^{m}(\mu_i-\mu)^T\Sigma^{-1}(\mu_i-\mu)) \\ = C - \frac{m}{2}\ln|\Sigma| - \frac{1}{2}\sum_{i=1}^{m}(\mu_i-\mu)^T\Sigma^{-1}(\mu_i-\mu)),\end{aligned} \quad (12)$$

where C is a constant.

We use the negative logarithmic maximum likelihood as the loss function to capture hierarchical affiliation:

$$\mathcal{L}_{aff} = -\ln L(\mu,\Sigma) \quad (13)$$

To sum up, the objective function to capture the hierarchical network structure can be defined as follows:

$$\mathcal{L}_{hie} = \mathcal{L}_{hor} + \mathcal{L}_{ver} + \lambda_1\mathcal{L}_{aff}, \quad (14)$$

where λ_1 is a trade-off parameter to balance the three parts of the loss function $\mathcal{L}_{hierarchical}$. We can optimize the parameters such that the loss $\mathcal{L}_{hierarchical}$ is minimized; thus, the three constraints we proposed can be satisfied.

4.2.2. General Network Optimization

Following the LINE model [25], we construct our unsupervised training loss based on the first-order and second-order proximities, which is an efficient unsupervised learning objective for graph data and can capture both the direct and indirect similarities between nodes in a general network.

$$\mathcal{L}_1 = -\sum_{(v_i,v_j)\in\mathcal{E}} w_{i,j}\log p_1(v_i,v_j) \quad (15)$$

$$\mathcal{L}_2 = -\sum_{(v_i,v_j)\in\mathcal{E}} w_{i,j}\log p_2(v_j|v_i) \quad (16)$$

$$\mathcal{L}_{ns} = \mathcal{L}_1 + \mathcal{L}_2 \quad (17)$$

where the $\mathcal{L}_1$ and $\mathcal{L}_2$ are the first-order and second-order objectives, and Ons is the node similarity objectives of our model. The probability p_1 and p_2 are computed as:

$$p_1(v_i,v_j) = \frac{1}{1+\exp(-(\mu_i)^T\cdot\mu_j)}, \quad (18)$$

$$p_2(v_j|v_i) = \frac{\exp((\mu_j)^T\cdot\mu_i)}{\sum_{k=1}^{V}\exp((\mu_k)^T\cdot\mu_i)}, \quad (19)$$

The $\mathcal{L}_2$ can further be optimized by the negative sampling.

4.2.3. Node Attribute Optimization

Incorporating Equation (8), the loss function of this part is defined as follows:

$$\mathcal{L}_A = D_{KL}(\hat{R}||R) = -\sum_{i\in\mathcal{V}}\sum_{A_{i,a}\in\mathcal{A}} A_{i,a}\log R_{i,a}, \quad (20)$$

where $A_{i,a} \in \mathcal{A}$ is a binary-valued attribute weight indicating that attribute a associates with node i if $A_{i,a} = 1$, $\hat{R}$ is the empirical distribution of the attribute weight with the empirical distribution of node i associating with attribute a, i.e., $\hat{R}_{i,a}$, being simply set as $A_{i,a}$. Note that all the attributes in this paper are binary-valued. Although $\hat{R}_{i,a}$ is applicable for not only binary-valued attributes but also real-valued attributes, we focus primarily on binary-valued attributes in this paper.

To summarize, the final objective $\mathcal{L}_{final}$ of our model can be written by the sum of the $\mathcal{L}_{hie}$, $\mathcal{L}_{ns}$ and $\mathcal{L}_{A}$:

$$\mathcal{L}_{final} = \mathcal{L}_{hie} + \mathcal{L}_{ns} + \mathcal{L}_{A}, \tag{21}$$

in which the $\mathcal{L}_{final}$ can be optimized by the Adam method.

To ensure that embedding can preserve both vertical and horizontal features after training, the whole embedding process is executed from the bottom up. For each node v_i^l at layer l, the learned representation can be obtained after optimizing the horizontal loss of layer l and the vertical loss at layer $i+1$ with Adam.

4.3. HGE-GA Algorithm

It has been proposed by previous work on social network influence [20] that the more equivalent the network structure of two nodes is, the more likely they make similar judgments, even without an edge between them, because these two nodes connect to other nodes more identically. It has also been revealed by previous studies on social networks that the opinions and behaviors of nodes are affected by node attributes [20], i.e., similar attributes of nodes cause similar behaviors. Therefore, the similarity of network structures and node attributes are the two factors underlying the influence probability between two nodes. We measure the similarity of node pairs and predict the influence probability of node pairs with the embedding of nodes inferred from the proposed HGE model. With the predicted influence probability of node pairs, we utilize a general greedy strategy to compute the top-K influential nodes in the network. The proposed HGE-GA algorithm is outlined in Algorithm 1, and will be detailed in this subsection.

Algorithm 1 HGE-GA Algorithm

Input: social network $\mathcal{G} = (\mathcal{V}, \mathcal{E}, \mathcal{A})$, the embeddings of all nodes $\mathbf{\Phi}$, seed set size K.
Output: Seed node set $\mathcal{S}$.

```
1: predicts influence probability of arbitrary pairwise nodes, p_{u,v};
2: for j = 1...K do
3:    u* ← argmax_{u*∈U} σ(S ∪ u*) − σ(S);
4:    S = S ∪ u*;
5: end for
6: return S;
```

Traditional network representation methods embed nodes into a low-dimensional vector space, and they always use the inner product to measure the similarities of node pairs. Then the relationships between node pairs can be captured. However, this method cannot capture the relationships among neighborhoods of nodes in some cases. As shown in Figure 5, for a node v_1 and its two neighborhoods u_1 and u_2, two relationships are represented, as follows:

$$\begin{aligned} v_1 \cdot u_1 &= v_1 \cdot u_2, \\ u_1 \cdot u_2 &= 0 \end{aligned} \tag{22}$$

As neighbors of node v_1, nodes u_1 and u_2 are similar in attributes, but the method based on the inner product cannot capture the relationship between u_1 and u_2 in the latent space. Similarly, the relationship between node v_1 and v_2 can not be captured by using the inner product.

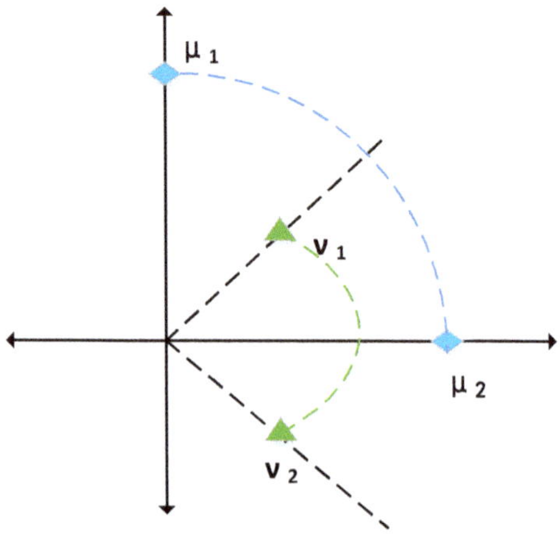

Figure 5. A case for which methods based on the inner product cannot capture the relationship between nodes. As we can see, the relationship between u_1 and u_2 and the relationship between node v_1 and v_2 can not be captured.

To avoid the case shown in Equation (26), we use the two-normal form, which satisfies the critical triangle inequality. The norm of a vector can simply be understood as its length or the corresponding distance between two points. The two-normal form is a common way to measure distance between vectors. Using the two-normal form, we define the relationship between node i and node j as follows:

$$g(\mathbf{\Phi}_i, \mathbf{\Phi}_j) = \left\|\mathbf{\Phi}_i - \mathbf{\Phi}_j\right\|_2 \tag{23}$$

The above relationship with the two-normal form adequately captures the similarity of two nodes. We regard the similarity of two nodes as the influence probability between two nodes, since the similarity of the network structure and node attributes are the two factors underlying the influence probability between two nodes. Thus, the influence probability between node i and j, i.e., $p_{i,j}$ can be predicted as follows.

$$p_{i,j} = g(\mathbf{\Phi}_i, \mathbf{\Phi}_j) = \left\|\mathbf{\Phi}_i - \mathbf{\Phi}_j\right\|_2 \tag{24}$$

After obtaining the influence probability of arbitrary pairwise nodes, under a general independent cascade (IC) diffusion model, we utilize a general greedy strategy to compute the top-K influential nodes, as Algorithm 1 shows.

5. Experiment Setup

We quantitatively evaluate the performance of the proposed HGE model in downstream learning tasks (vertex classification, link prediction, network visualization) and compare the proposed algorithm, HGE-GA, with the state-of-art influence maximization algorithms using the metric of expected spread on several large-scale real-world datasets. In this section, we will detail the experimental setup, including research questions (Section 5.1), datasets (Section 5.2), baselines (Section 5.3), evaluation metrics and experimental settings (Section 5.4).

5.1. Research Questions

We aim at answering the following research questions to evaluate the performance of the proposed HGE model and HGE-GA algorithm.
(RQ1) How does the proposed HGE model perform in terms of traditional graph-mining tasks, e.g., vertex classification, link prediction and network visualization, compared with baselines?

(RQ2) Can the proposed influence maximization algorithm, HGE-GA, outperform the state-of-art influence maximization algorithms with respect to the expected spread?
(RQ3) Does the two-normal form to measure distance contribute to the improvement of the performance of the proposed algorithm?

5.2. Datasets

We use four datasets, the detailed properties of which are shown in Table 2.

- Cora [39]: The Cora dataset is a citation networks which is composed of a large number of academic articles. Nodes are publications and edges are citation links;
- DBLP: The DBLP dataset consists of bibliography data in computer science. Each paper may cite or be cited by other papers, which naturally forms a citation network;
- BlogCatalog: This dataset is a social relationship network which is crawled from the BlogCatalog website. BlogCatalog is composed of bloggers and their social relationships. The labels of nodes indicate the interests of the bloggers;
- Flickr: Flickr is a social network where users can share pictures and videos. In this dataset, each node is a user, and each side is a friend relationship between users. In addition, each node has a label that identifies the user's interest group;
- Pubmed: Pubmed is a public search database that provides biomedical paper and abstract search services. In this dataset, nodes are publications and edges are citation links;

Table 2. Statistics of the datasets used in the experiments.

	#Nodes	#Edges	#Attributes	#Labels
Cora	2708	5429	1433	7
DBLP	17,716	105,734	1639	4
BlogCatalog	5196	171,743	8189	6
Flickr	7575	239,738	12,047	9
Pubmed	19,717	44,338	500	3

5.3. Baselines

We evaluated the performance of the proposed HGE model compared with the following baselines:

- GraphSAGE [40]: an attributed network embedding model which leverages node attributes and generates embeddings by sampling and aggregating features from a node's local neighborhood;
- AANE [41]: a model which learns attributed network embedding efficiently by decomposing complex modeling and optimization into sub-problems;
- M-NMF [27]: a single-layer community structure preserving baseline, which integrates the community information through a matrix factorization;
- GNE [33]: a multi-layer community structure preserving baseline, which embeds communities onto surfaces of spheres;
- SpaceNE [36]: a method which applies subspace to the hierarchical network embedding model and preserves the proximity between pairwise nodes and between communities;

Moreover, we evaluate the performance of the proposed HGE-GA algorithm compared with the following baselines:

- PMIA [12]: This is a heuristic algorithm that defines the maximum influence in-arborescence (MIIA) and leverages sequence submodularity to compute the influence spread;
- IMM [9]: This utilizes a classical statistical tool, martingale, as well as RR sets (reverse reachable sets), and can provide higher efficiency in practice;
- TSH-GA : Zhou et al. [20] propose a method to compute the influence probability of node pairs, which considers social ties, general network structure, and node attributes

as factors underlying influence but executes feature extraction of these three factors by typically hand-crafted rules. TSH-GA is an influence maximization algorithm jointing the method to compute the influence probability of node pairs proposed in [20] and the general greedy algorithm;

- HGE-1N-GA: This is a variant of our HGE-GA which uses the one-normal form to compute the similarity of two nodes.

5.4. Evaluation Metrics and Experimental Settings

Evaluation metrics for the proposed HGE model. We evaluate the performance of link prediction of HGE in terms of the area under the curve (AUC) and average precision (AP). We evaluate the performance of vertex classification in terms of the F1-measure (F1), which is defined as $F1 = \frac{2 \cdot Precision \cdot Recall}{Precision + Recall}$.

Evaluation metric for the proposed HGE-GA algorithm. We evaluate the overall performance of our HGE-GA algorithm by adopting the expected spread as the evaluation metric. This is an extensively used metric for the influence maximization problem. The expected spread is defined as the expected number of active nodes in the network, including the newly activated nodes and the set of seed nodes that were initially active (seed node set) after the influence spread is over (there are no more nodes being activated), given the seed set size K, which is an integer.

Experimental settings. All embedding sizes are 64, and the number of training epochs is 1000. The initial learning rate of Adam is set as 0.001. All results are obtained by averaging 10 experiments. We implement all the baselines by the codes released by the authors. The parameters of the baselines are tuned to be optimal or set according to the corresponding literature. All algorithms are run under the IC diffusion model. Furthermore, for the two algorithms PMIA and IMM, which do not have the step of predicting influence probability of node pairs, we set the propagation probability from node u to node v, i.e., $p(u, v)$, as $1/i$, where i denotes the number of incoming edges of node v. This method of setting influence probability is extensively adopted in previous literature [12].

6. Results and Analysis

6.1. Performance of Vertex Classification of HGE Model

Vertex classification is one of the most important tasks to detect the performance of embedding models. In this section, we perform vertex classification task to evaluate the performance of the learned embeddings and compare with the baseline methods. We choose three datasets (DBLP, BlogCatalog and Flickr) which have ground-truth classes. To be specific, we first sample a small number of nodes as training data, and the rest is used as test data. Similar to [23], we use one-vs-rest logistic regression for node classification, and the training data size is 10%; the results are reported in Table 3. As can be seen in Table 3, the proposed HGE model performs the best in the DBLP and BlogCatalog datasets, and shows competitive performance on the Flickr dataset compared with other network embedding baselines. This proves that the proposed model can better capture the network structures.

Table 3. Node classification performance.

Method	DBLP	BlogCatalog	Flickr
	F1	F1	F1
AANE	0.702	0.515	0.517
GraphSAGE	0.731	0.625	0.649
M-NMF	0.524	0.604	0.587
GNE	0.741	0.623	0.631
SpaceNE	0.767	0.654	**0.651**
HGE	**0.807**	**0.669**	0.635

6.2. Performance of Link Prediction of the HGE Model

Link prediction aims to predict the future interactions of nodes in a network. In this subsection, we compare the proposed HGE model with the baselines on the link prediction task. As in [26], we create a validation/test set that contains 5%/10% randomly selected edges, respectively, and an equal number of randomly selected non-edges. To measure the performance of the link prediction task, we report the area under the ROC curve (AUC) and the average precision (AP) scores for each method.

Table 4 shows the link prediction performance of the proposed HGE model and the baselines on the five datasets mentioned in Section 5.2. Our HGE model significantly outperforms the baselines across all datasets, which demonstrates that modelling the hierarchical network structure, general network structure as well as node attributes in latent space is effective to learn better node embeddings.

Table 4. Link prediction performance with embedding size L = 64.

Method	Cora		DBLP		BlogCatalog		Flickr		Pubmed	
	AUC	AP	AUC	AP	AUC	AP	AUC	AP	AUC	AP
AANE	0.834	0.812	0.761	0.789	0.771	0.776	0.678	0.639	0.842	0.837
GraphSAGE	0.869	0.892	0.803	0.811	0.807	0.819	0.802	0.817	0.812	0.825
M-NMF	0.867	0.861	0.841	0.837	0.716	0.710	0.728	0.736	0.816	0.825
GNE	0.944	0.941	0.935	0.944	0.805	0.802	0.828	0.837	0.945	0.942
SpaceNE	0.927	0.914	0.927	0.933	0.815	0.827	0.908	0.917	0.936	0.958
HGE	**0.979**	**0.974**	**0.986**	**0.989**	**0.836**	**0.845**	**0.929**	**0.913**	**0.977**	**0.972**

6.3. Performance of Network Visualization of the HGE Model

Network visualization is also one of the most important means to detect the quality of the learned embedding. Network visualization maps a network into two-dimensional space. In this section, for better visualization, we visualize a sub-network which is selected from the BlogCatalog dataset with 175 nodes and 150 edges.

From Figure 6, we find that our HGE model preserves the default hierarchical structure, and distributes the nodes of each layer more uniformly in a fan-shaped area.

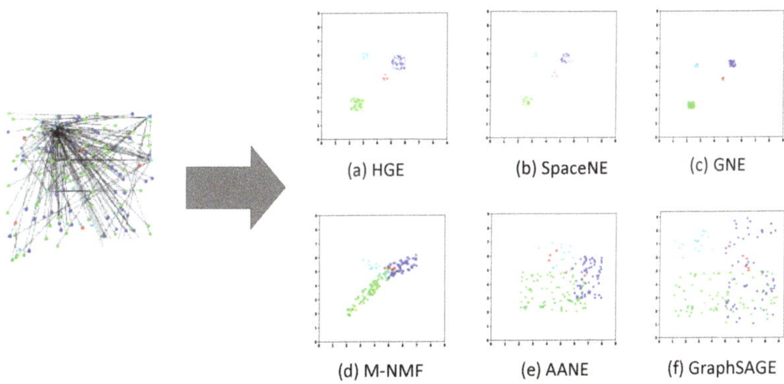

Figure 6. The experimental results of network visualization in two-dimensional space.

6.4. Overall Performance of HGE-GA Algorithm

We adopt the metric of expected spread to evaluate the overall performance of all influence maximization algorithms. To compare the proposed HGE-GA algorithm with the baselines, we change the size of the seed node set K to be 1, 20, 30, 40, and 50. The experimental results are reported in Figure 7a–e. It can be observed from Figure 7a–e that:

(1) The expected spread increases for all algorithms when enlarging the size of the seed node set. The proposed HGE-GA algorithm outperforms all the baselines significantly for

all seed sizes on the five datasets. For instance, on the Pubmed dataset, when the seed set size $K = 50$, the value of the expected spread of the proposed HGE-GA algorithm is 1445, compared with 721 for PMIA, 819 for IMM, 1052 for TSH-GA, and 1384 for HGE-1N-GA;

(2) On all five datasets, all of the attribute-related algorithms, TSH-GA, HGE-1N-GA and HGE-GA, outperform the non-attribute-related algorithms, PMIA and IMM, in terms of the performance of the expected spread. This is because the PMIA and IMM algorithms do not have the step of predicting the influence probability of node pairs, with influence probability being set uniformly. Moreover, this demonstrates that homophily is a cause of similar behaviors and is useful for predicting future behaviors of users. Consequently, the expected spread can be increased obviously when node attributes are taken into account in modeling;

(3) HGE-1N-GA and HGE-GA outperform TSH-GA on the five datasets for the metric of expected spread. The reason is that HGE-1N-GA and HGE-GA consider the general network structure, hierarchical network structure as well as node attributes to learn users' latent feature representations for predicting the influence probability of node pairs, but TSH-GA uses hand-crafted rules to extract the features of users' interactions, network structure and node homophily, heavily depending on the domain expert's knowledge and without considering hierarchical network structure;

(4) The expected spread of HGE-GA is superior than HGE-1N-GA. The reason is obvious: HGE-1N-GA uses the one-normal form to compute the similarity of two nodes, while HGE-GA leverages the two-normal form to compute similarity, which can capture the relationships between nodes better.

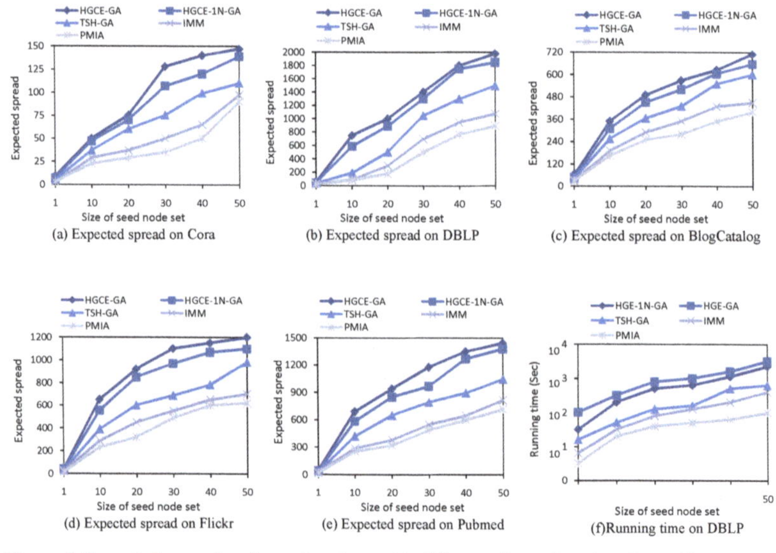

Figure 7. Expected spread and running time with different sizes of seed node set K.

The running times of all algorithms are reported in Figure 7f. This figure only presents the experimental results on the DBLP dataset due to space constraints. Please note that the DBLP dataset is the biggest dataset in the five datasets considering both the number of nodes and the number of edges. Similar trends of performance of the algorithms are found for the other four datasets. From Figure 7b,f, it can be observed that, for the DBLP dataset, compared with the baseline algorithms, the proposed algorithm, HGE-GA, obtains a better performance of the expected spread by spending more time. However, in general, the proposed algorithm HGE-GA spends an acceptable running time to obtain 54.5%, 45.5%, 24.2% more expected spread, respectively, than the PMIA, IMM and TSH-GA algorithms when the seed set size K is set to 50.

7. Conclusions

In this paper, we study the problem of influence maximization in attributed social networks, assuming the influence strength is directly unobservable. From the deep learning perspective, we formulate the problem and propose a hierarchical generative embedding model, HGE, to map nodes into latent space automatically, incorporating hierarchical community structure, node attributes and general network structure into a unified deep generative framework. Then, with the learned latent representation of each node, we propose a HGE-GA algorithm to predict influence strength and find the seed node set. The experimental results show that the proposed HGE model is able to learn representations of nodes in attributed networks far more effectively than state-of-the-art models in terms of several downstream applications, such as vertex classification, link prediction and network visualization. It is also verified by the experimental results that the proposed HGE-GA algorithm significantly outperforms the state-of-the-art algorithms for influence maximization in attributed social networks.

As to future work, we intend to extend our HGE model to temporal attributed networks, which is more challenging because this kind of attributed network evolves over time and vertexes need to be embedded dynamically.

Author Contributions: Conceptualization, H.H.; data curation, L.X. and Q.D.; formal analysis, L.X.; methodology, L.X. and H.H.; project administration, H.H.; software, Q.D.; supervision, H.H.; validation, H.H.; visualization, L.X. and Q.D.; writing—original draft, L.X.; writing—review & editing, H.H.; funding acquisition, H.H. All authors have read and agreed to the published version of the manuscript.

Funding: This research was funded by Wenzhou Science and Technology Planning Project #2021R0082.

Institutional Review Board Statement: Not applicable.

Informed Consent Statement: Not applicable.

Data Availability Statement: Not applicable.

Conflicts of Interest: The authors declare no conflict of interest. The funders had no role in the design of the study; in the collection, analyses, or interpretation of data; in the writing of the manuscript, or in the decision to publish the results.

References

1. Huang, H.; Meng, Z.; Shen, H. Competitive and complementary influence maximization in social network: A follower's perspective. *Knowl.-Based Syst.* **2021**, *213*, 106600. [CrossRef]
2. Fedushko, S.; Peráček, T.; Syerov, Y.; Trach, O. Development of Methods for the Strategic Management of Web Projects. *Sustainability* **2021**, *13*, 742. [CrossRef]
3. Kempe, D.; Kleinberg, J.; Tardos, É. Maximizing the spread of influence through a social network. In Proceedings of the Ninth ACM SIGKDD International Conference on Knowledge Discovery and Data Mining, Washington, DC, USA, 24–27 August 2003; pp. 137–146.
4. Pei, S.; Morone, F.; Makse, H.A. Theories for influencer identification in complex networks. In *Complex Spreading Phenomena in Social Systems*; Springer: Berlin/Heidelberg, Germany, 2018; pp. 125–148.
5. Sheshar, S.; Srivastva, S.D.; Verma, M.; Singh, J. Influence Maximization Frameworks, Performance, Challenges and Directions on Social Network: A Theoretical Study. *J. King Saud-Univ.-Comput. Inf. Sci.* 2021, *in press*. [CrossRef]
6. Leskovec, J.; Krause, A.; Guestrin, C.; Faloutsos, C.; VanBriesen, J.; Glance, N. Cost-effective outbreak detection in networks. In Proceedings of the 13th ACM SIGKDD International Conference on Knowledge Discovery and Data Mining, San Jose, CA, USA, 12–15 August 2007; pp. 420–429.
7. Chen, W.; Wang, Y.; Yang, S. Efficient influence maximization in social networks. In Proceedings of the 15th ACM SIGKDD International Conference on Knowledge Discovery and Data Mining, Paris, France, 28 June–1 July 2009; pp. 199–208.
8. Goyal, A.; Lu, W.; Lakshmanan, L.V. Celf++ optimizing the greedy algorithm for influence maximization in social networks. In Proceedings of the 20th International Conference Companion on World Wide Web, Hyderabad, India, 28 March–1 April 2011; pp. 47–48.
9. Tang, Y.; Shi, Y.; Xiao, X. Influence maximization in near-linear time: A martingale approach. In Proceedings of the 2015 ACM SIGMOD International Conference on Management of Data, Melbourne, VIC, Australia, 31 May–4 June 2015; pp. 1539–1554.

10. Cheng, S.; Shen, H.; Huang, J.; Zhang, G.; Cheng, X. Staticgreedy: Solving the scalability-accuracy dilemma in influence maximization. In Proceedings of the 22nd ACM International Conference on Information & Knowledge Management, Francisco, CA, USA, 27 October–1 November 2013; pp. 509–518.
11. Galhotra, S.; Arora, A.; Roy, S. Holistic influence maximization: Combining scalability and efficiency with opinion-aware models. In Proceedings of the 2016 International Conference on Management of Data, San Francisco, CA, USA, 26 June–1 July 2016; pp. 743–758.
12. Chen, W.; Wang, C.; Wang, Y. Scalable influence maximization for prevalent viral marketing in large-scale social networks. In Proceedings of the 16th ACM SIGKDD International Conference on Knowledge Discovery and Data Mining, Washington, DC, USA, 24–28 July 2010; pp. 1029–1038.
13. Jiang, Q.; Song, G.; Gao, C.; Wang, Y.; Si, W.; Xie, K. Simulated annealing based influence maximization in social networks. In Proceedings of the AAAI Conference on Artificial Intelligence, San Francisco, CA, USA, 7–11 August 2011; Volume 25.
14. Hu, Z.; Yao, J.; Cui, B.; Xing, E. Community level diffusion extraction. In Proceedings of the 2015 ACM SIGMOD International Conference on Management of Data, Melbourne, VIC, Australia, 31 May–4 June 2015; pp. 1555–1569.
15. Huang, H.; Shen, H.; Meng, Z.; Chang, H.; He, H. Community-based influence maximization for viral marketing. *Appl. Intell.* **2019**, *49*, 2137–2150. [CrossRef]
16. Shaghaghian, S.; Coates, M. Bayesian inference of diffusion networks with unknown infection times. In Proceedings of the 2016 IEEE Statistical Signal Processing Workshop (SSP), Palma de Mallorca, Spain, 26–29 June 2016; pp. 1–5.
17. Barbieri, N.; Bonchi, F.; Manco, G. Topic-aware social influence propagation models. *Knowl. Inf. Syst.* **2013**, *37*, 555–584. [CrossRef]
18. Huang, H.; Meng, Z.; Liang, S. Recurrent neural variational model for follower-based influence maximization. *Inf. Sci.* **2020**, *528*, 280–293. [CrossRef]
19. Qiu, J.; Tang, J.; Ma, H.; Dong, Y.; Wang, K.; Tang, J. Deepinf: Social influence prediction with deep learning. In Proceedings of the 24th ACM SIGKDD International Conference on Knowledge Discovery & Data Mining, London, UK, 19–23 August 2018; pp. 2110–2119.
20. Zhou, F.; Jiao, R.J.; Lei, B.Y. A linear threshold-hurdle model for product adoption prediction incorporating social network effects. *Inf. Sci.* **2015**, *307*, 95–109. [CrossRef]
21. Kamarthi, H.; Vijayan, P.; Wilder, B.; Ravindran, B.; Tambe, M. Influence Maximization in Unknown Social Networks: Learning Policies for Effective Graph Sampling. In Proceedings of the 19th International Conference on Autonomous Agents and MultiAgent Systems, Auckland, New Zealand, 9–13 May 2020; pp. 575–583.
22. Mikolov, T.; Chen, K.; Corrado, G.; Dean, J. Efficient estimation of word representations in vector space. *arXiv* **2013**, arXiv:1301.3781.
23. Perozzi, B.; Al-Rfou, R.; Skiena, S. Deepwalk: On learning of social representations. In Proceedings of the 20th ACM SIGKDD International Conference on Knowledge Discovery and Data Mining, New York, NY, USA, 24–27 August 2014; pp. 701–710.
24. Grover, A.; Leskovec, J. node2vec: Scalable feature learning for networks. In Proceedings of the 22nd ACM SIGKDD International Conference on Knowledge Discovery and Data Mining, San Francisco, CA, USA, 13–17 August 2016; pp. 855–864.
25. Tang, J.; Qu, M.; Wang, M.; Zhang, M.; Yan, J.; Mei, Q. Line: Large-scale information network embedding. In Proceedings of the 24th International Conference on World Wide Web, Florence, Italy, 18–22 May 2015; pp. 1067–1077.
26. Wang, D.; Cui, P.; Zhu, W. Structural deep network embedding. In Proceedings of the 22nd ACM SIGKDD International Conference on Knowledge Discovery and DATA Mining, San Francisco, CA, USA, 13–17 August 2016; pp. 1225–1234.
27. Wang, X.; Cui, P.; Wang, J.; Pei, J.; Zhu, W.; Yang, S. Community preserving network embedding. In Proceedings of the Thirty-First AAAI Conference on Artificial Intelligence, San Francisco, CA, USA, 4–9 February 2017.
28. Jia, Y.; Zhang, Q.; Zhang, W.; Wang, X. Communitygan: Community detection with generative adversarial nets. In Proceedings of the World Wide Web Conference, San Francisco, CA, USA, 13–17 May 2019; pp. 784–794.
29. Li, Y.; Wang, Y.; Zhang, T.; Zhang, J.; Chang, Y. Learning network embedding with community structural information. In Proceedings of the 28th International Joint Conference on Artificial Intelligence, Macao, China, 10–16 August 2019; pp. 2937–2943.
30. Goodfellow, I.; Pouget-Abadie, J.; Mirza, M.; Xu, B.; Warde-Farley, D.; Ozair, S.; Courville, A.; Bengio, Y. Generative adversarial nets. In Proceedings of the Advances in Neural Information Processing Systems, Montreal, QC, USA, 8–13 December 2014; pp. 2672–2680.
31. Yang, J.; Leskovec, J. Community-affiliation graph model for overlapping network community detection. In Proceedings of the 2012 IEEE 12th International Conference on Data Mining, Brussels, Belgium, 10–13 December 2012; pp. 1170–1175.
32. Clauset, A.; Moore, C.; Newman, M.E. Structural inference of hierarchies in networks. In Proceedings of the ICML Workshop on Statistical Network Analysis, Pittsburgh, PA, USA, 29 June 2006; pp. 1–13.
33. Du, L.; Lu, Z.; Wang, Y.; Song, G.; Wang, Y.; Chen, W. Galaxy Network Embedding: A Hierarchical Community Structure Preserving Approach. In Proceedings of the IJCAI, Stockholm, Sweden, 13–19 July 2018; pp. 2079–2085.
34. Sales-Pardo, M.; Guimera, R.; Moreira, A.A.; Amaral, L.A.N. Extracting the hierarchical organization of complex systems. *Proc. Natl. Acad. Sci. USA* **2007**, *104*, 15224–15229. [CrossRef] [PubMed]
35. Ma, J.; Cui, P.; Wang, X.; Zhu, W. Hierarchical taxonomy aware network embedding. In Proceedings of the 24th ACM SIGKDD International Conference on Knowledge Discovery & Data Mining, London, UK, 19–23 August 2018; pp. 1920–1929.

36. Long, Q.; Wang, Y.; Du, L.; Song, G.; Jin, Y.; Lin, W. Hierarchical Community Structure Preserving Network Embedding: A Subspace Approach. In Proceedings of the 28th ACM International Conference on Information and Knowledge Management, Beijing, China, 3–7 November 2019; pp. 409–418.
37. Ma, Y.; Ren, Z.; Jiang, Z.; Tang, J.; Yin, D. Multi-Dimensional Network Embedding with Hierarchical Structure. In Proceedings of the Eleventh ACM International Conference on Web Search and Data Mining 2018, Los Angeles, CA, USA, 5–9 February 2018; pp. 387–395.
38. Nickel, M.; Kiela, D. Poincaré embeddings for learning hierarchical representations. In Proceedings of the Advances in Neural Information Processing Systems, Long Beach, CA, USA, 4–9 December 2017; pp. 6338–6347.
39. Bojchevski, A.; Günnemann, S. Deep Gaussian Embedding of Graphs: Unsupervised Inductive Learning via Ranking. In Proceedings of the International Conference on Learning Representations, Vancouver, BC, Canada, 30 April–3 May 2018.
40. Hamilton, W.L.; Ying, Z.; Leskovec, J. Inductive Representation Learning on Large Graphs. In Proceedings of the Advances in Neural Information Processing Systems 30: Annual Conference on Neural Information Processing Systems, Long Beach, CA, USA, 4–9 December 2017; pp. 1024–1034.
41. Huang, X.; Li, J.; Hu, X. Accelerated Attributed Network Embedding. In Proceedings of the 2017 SIAM International Conference on Data Mining, Houston, TX, USA, 27–29 April 2017; pp. 633–641.

applied sciences

MDPI

Article

A Co-Embedding Model with Variational Auto-Encoder for Knowledge Graphs

Luodi Xie [1], Huimin Huang [2,*] and Qing Du [3]

[1] School of Computer Science, Sun Yat-Sen University, Guangzhou 510000, China; xield@mail2.sysu.edu.cn
[2] School of Data Science and Artificial Intelligence, Wenzhou University of Technology, Wenzhou 325000, China
[3] School of Software, South China University of Technology, Guangzhou 510000, China; duqing@scut.edu.cn
* Correspondence: huanghm45@gmail.com

Abstract: Knowledge graph (KG) embedding has been widely studied to obtain low-dimensional representations for entities and relations. It serves as the basis for downstream tasks, such as KG completion and relation extraction. Traditional KG embedding techniques usually represent entities/relations as vectors or tensors, mapping them in different semantic spaces and ignoring the uncertainties. The affinities between entities and relations are ambiguous when they are not embedded in the same latent spaces. In this paper, we incorporate a co-embedding model for KG embedding, which learns low-dimensional representations of both entities and relations in the same semantic space. To address the issue of neglecting uncertainty for KG components, we propose a variational auto-encoder that represents KG components as Gaussian distributions. In addition, compared with previous methods, our method has the advantages of high quality and interpretability. Our experimental results on several benchmark datasets demonstrate our model's superiority over the state-of-the-art baselines.

Keywords: knowledge graph; embedding; variational auto-encoder

Citation: Xie, L.; Huang, H.; Du, Q. A Co-Embedding Model with Variational Auto-Encoder for Knowledge Graphs. *Appl. Sci.* **2022**, *12*, 715. https://doi.org/10.3390/app12020715

Academic Editors: Nikos D. Lagaros and Vagelis Plevris

Received: 29 November 2021
Accepted: 6 January 2022
Published: 12 January 2022

1. Introduction

Knowledge graph (KG) embeddings are low-dimensional representations for entites and relations. This approach can benefit a range of downstream tasks, such as semantic parsing [1,2], knowledge reasoning [3], and question answering [4,5]. Embeddings are supposed to contain semantic information and should be able to deal with multiple linguistic relations.

At present, research on knowledge graph embedding occurs mainly along three main lines. One of these lines of research includes studies based on translation. TransE [6] was the first model to introduce translation-based embedding, which represents entities and relationships in the same space, and regards the relationship vector r as the translation between the head entity vector h and the tail entity vector t, that is, $h + r \approx t$. Since transE cannot handle one-to-many, many-to-one, and many-to-many relationships (1-to-N, N-to-1, N-to-N), TransH [7] is proposed to enable an entity to have different representations when involved in various relations. In the TransR model [8], an entity is a complex of multiple attributes, and different relationships focus on different attributes of the entity. Another line of research includes studies based on semantic matching. RESCAL [9] obtains its latent semantics by using a vector to represent each entity. Each relationship is represented as a matrix that is used to model the interaction of potential relationships. It defines the scoring function of the triple (h, r, t) as a bilinear function. DistMult [10] simplifies RESCAL by restricting the relationship matrix to a diagonal matrix, which greatly improves training efficiency. ComplEx [11] extends DistMult by introducing complex number domain embedding to better model asymmetric relationships. In ComplEx, the embedding of entities and relationships no longer exists in real space, but in complex space. The third line of research includes studies based on graph convolutional neural networks. ConvE [12] employs a multi-layer convolutional network, which enables expressive feature learning, while remaining highly parameter-efficient. Unlike previous works, which focused on shallow, fast models that can scale to large knowledge graphs, ConvE uses 2D convolution over embeddings and multiple layers of nonlinear features to model KGs. Subsequently, the ConvKB [13] model has been used to explore the global relationships among same-dimensional entries of the entity and relation embeddings. However, neither of them models the interactions between various positions of entities and relations. R-GCN [14] is another convolutional network designed for KBs, generalized from GCN [15] for uni-relational data.

A typical KG embedding technique has two necessary elements: (i) an encoder to generate KG embeddings and (ii) a scoring function to measure plausibility for each fact. Entities are usually represented as vectors in low-dimensional space, whereas relations are represented as an operation between entities, resulting in vectors for translational operations [6] or matrices for linear transformation [16]. By doing so, the embedding of KG components can be used to enhance the performance in many downstream tasks. Despite the success those previous algorithms have achieved, we note that those methods have the following defects(1) the n-dimensional representation of the KG component can be regarded as a single point, neglecting the uncertainties for entities and relations; (2) they represent entities as vectors located in low-dimensional space and relations as an operation between entities [6,16], thus ignoring the affinities between entities and relations as they are embedded in different semantic spaces.

To address the issues mentioned above, we propose a co-embedding model for KG, learning low-dimensional representations for entities and relations in the same semantic space so that the affinities between them can be effectively captured. Moreover, we introduce a variational auto-encoder to infer the representations of KG components as Gaussian distributions. The mean of the distributions indicatesthe position in semantic space, and the variance of the distributions indicates the uncertainty for each KG components.

Compared with previous works that regard relations as an operation between entities, co-embedding of entities and relations in the same semantic space can improve the performance of KG representation. For example, in Freebase, the relation 'Perfession' is used in (El Lissitzky, Perfession, Architect) and (Vlad. Gardin, Perfession, Screen Writer) uses two distinct semantic information categories, corresponding to a scientist and a writer, resulting in the finding that the resulting representations calculated using the two triples are not the same. The co-embedding model embeds entities and relations at the same semantic space, thus providing high-quality embeddings for both of them.

In summary, our contributions in this work are as follows:

1. We propose a co-embedding model for knowledge graphs, which learns low-dimensional representations for KG components, including entities and relations in the same semantic space, as a result of measuring their affinities effectively.
2. To address the issue of neglecting uncertainty, we introduce a variational auto-encoder into our model, which represents KG components as Gaussian distributions. The variational auto-encoder consists of two parts: (1) an inference model to encode KG components into latent vector spaces, (2) a generative model to reconstruct random variables from latent embeddings.
3. We conduct experiments on real-world datasets to evaluate the performance of our model in link prediction. The experimental result demonstrates that our model outperforms the state-of-the-art baselines.

2. Related Work

2.1. Knowledge Graph Representation

Knowledge representation is a technique that aims at learning low-dimensional representations for KG entities and relations, consisting of two critical steps: (1) constructing a scoring function measuring plausibility for triples, and (2) embedding KG components in continuous vector spaces.

TransE [6], the most representative method in KG embedding, represents entities as vectors and relations as an operations between entities, i.e., $\mathbf{h} + \mathbf{r} \approx \mathbf{t}$. The scoring function is defined as the distance between entities and relations in latent space, written as $f_{\mathbf{r}}(\mathbf{h}, \mathbf{t}) = - \parallel \mathbf{h} + \mathbf{r} - \mathbf{t} \parallel_{1/2}$. However, TransE fails to deal with one-to-many, many-to-one, and many-to-many relations [7,8]. For example, given a relation holding two facts $(\mathbf{h}, \mathbf{r}, \mathbf{t1})$ and $(\mathbf{h}, \mathbf{r}, \mathbf{t2})$, we can infer $\mathbf{t1} = \mathbf{t2}$ even though they are totally different entities. To overcome the above defects, Z. Wang, J. Zhang, J. Feng, and Z. Chen proposed TransH [7] to obtain distinct representations for entities when dealing with different relations, by projecting entity representations onto a hyperplane, resulting in a normal vector. e.g., $\mathbf{h}_{\perp} = \mathbf{h} - \mathbf{w}_{\mathbf{r}}^{\top} \cdot \mathbf{h} \cdot \mathbf{w}_{\mathbf{r}}$, with $\mathbf{w}^{\mathbf{r}}$ as the normal vector.

TransE and its extensions represent both entities and relations as deterministic points in vector spaces, ignoring the uncertainty for KG components. To resolve this problem, some recent works have introduced uncertainty into KG embedding by representing KG components as distributions, e.g., KG2E [17], proposed by Shizhu He, Kang Liu, and Guoliang Ji and Jun Zhao, represents both entities and relations as distributions via Gaussian embedding. Inspired by the previous works, we tackle the embedding problem for KG by modeling both entities and relations as Gaussian distributions and representing them in the semantic space to effectively measure the affinities between them.

2.2. Gaussian Embedding

Gaussian embedding [18] is a method to embed word types into the space of Gaussian distributions, and learn the embeddings directly in that space, which represents words not as low-dimensional vectors, but as densities over a latent space, directly representing notions of uncertainty and enabling a richer geometry in the embedded space.

In word representation, embedding an object as a single point can not naturally express uncertainty about the target concepts with which the input may be associated, and the relationships between points are normally measured by distances required to obey the triangle inequality. Point vectors are typically compared via their dot products, cosine-distance, or Euclean distance, none of which provide asymmetric comparisons between objects. To overcome the limitations in representing objects as points, Gaussian embedding is proposed to learn representations in the space of Gaussian distributions, advocating for density-based distributed embeddings.

In Gaussian embedding, we learn both means and variances from data, representing them as densities over a latent space instead of low-dimensional vectors. As Gaussians innately represent uncertainty and have a geometric interpretation as an inclusion between families of ellipses, our method adopts KL divergence between Gaussian distributions to measure the relationship between objects, which is straightforward to calculate.

Mapping to a density provides many advantages, including better capturing uncertainty about a representation and its relationships, providing asymmetric comparisons between objects, which is more effective than dot product or cosine similarity, and which enables more expressive parameterization of decision boundaries.

2.3. Variational Auto-Encoder

Variational Auto-encoders [19], abbreviated as VAEs, are proposed to learn probability distributions of data. A typical VAE model is made up of two computational neural networks, an inference model to encode observations into latent variables and a generative model to decode from latent deterministic representations to random variables. Given a dataset $X = \{x^i\}_{i=1}^N$, the VAE regards data as random numbers generated via two steps: (1) the latent variable z_i is sampled from prior distribution $p_\theta(z_i)$, and (2) the random variable x_i is generated by the conditional distribution $p_\theta(x|z)$, where θ is the prior distribution parameter. Using the stochastic gradient variational Bayes (SGVB) estimator and reparameterization, we can learn approximate distributions for each data point via the VAE.

In the VAE, we treat the encoder and decoder as a whole and train them at the same time. The goal of training is to maximize the evidence lower bound of the likelihood function. Specifically, we first input random variables (randomly initialized node embedding) to the encoder, obtain the output, and calculate the encoder error, then we use the output of the encoder as the input of the decoder and calculate the reconstruction error of the decoder. The two parts of the errors are added together as the overall error of the network and propagated backward, thus realizing the simultaneous training of the encoder and the decoder.

In recent years, the VAE algorithm and its variations have been studied and applied in many downstream tasks such as semi-supervised classification [20], clustering [21,22], and image generation [23].

3. Notations and Problems

In this section, we introduce the notation used and define our studied problem.

3.1. Notations

In this paper, we define scalars as normal alphabets (e.g., the output dimension of latent variables: D), sets as typeface alphabets (e.g., the set of entities: $\mathcal{E}$), and vectors as lowercase alphabets (e.g., the representation of head entities: **h**). A triple in KG is denoted by τ, whereas it can be written as $\tau = (\mathbf{h}, \mathbf{r}, \mathbf{t})$. Our main notations are shown in Table 1.

Table 1. Main notations in our paper.

Symbol	Description
$\mathcal{G}$	a knowledge graph
$\mathcal{E}$	set of entities
$\mathcal{R}$	set of relations
$\mathcal{O}$	set of triples
$M = \lvert\mathcal{E}\rvert$	size of entities
$N = \lvert\mathcal{R}\rvert$	size of relations
$W = \lvert\mathcal{O}\rvert$	size of triples
D	dimension of latent variables
$\mathbf{O} \in \mathbb{R}^{W \times 3}$	observed data for triples
$\mathbf{Z}^{\mathcal{E}} \in \mathbb{R}^{M \times D}$	latent representation matrix for entities
$\mathbf{Z}^{\mathcal{R}} \in \mathbb{R}^{N \times D}$	latent representation matrix for relations

Given a knowledge graph $\mathcal{G}$, We represent the set of entities as $\mathcal{E}$ and the set of relations as $\mathcal{R}$, whereas $\mathcal{G}$ can be defined as $\mathcal{G} = (\mathcal{E}, \mathcal{R}, \mathcal{O})$, where $\mathcal{O}$ is the set of triples denoted as $\tau = (\mathbf{h}, \mathbf{r}, \mathbf{t}), \mathbf{h}, \mathbf{t} \in \mathcal{E}$ and $\mathbf{r} \in \mathcal{R}$.

3.2. Problem Definition

Using the notation mentioned above, we define the problem of co-embedding in KG as follows.

Problem 1. *The Co-embedding Model for KG Embedding. Given a knowledge graph $\mathcal{G} = (\mathcal{E}, \mathcal{R}, \mathcal{O})$, our goal is to learn the representations of KG components, including entities and relations, in the same semantic space as that of a transformation Ξ.*

$$\mathcal{G} \xrightarrow{\Xi} \mathbf{Z}^{\mathcal{E}}, \mathbf{Z}^{\mathcal{R}}, \tag{1}$$

where $\mathbf{Z}^{\mathcal{E}} \in \mathbb{R}^{M \times D}$ and $\mathbf{Z}^{\mathcal{R}} \in \mathbb{R}^{N \times D}$, respectively. The i-th row vector in $\mathbf{Z}^{\mathcal{E}}$, written as $\mathbf{z}_i^{\mathcal{E}}$, is denoted as the resulting embedding of the i-th entity, and the j-th row vector in $\mathbf{Z}^{\mathcal{R}}$ written as $\mathbf{z}_j^{\mathcal{R}}$ is denoted as the resulting embedding of the j-th relation.

4. Model

To address the issues we mentioned above, we propose the co-embedding model, learning representations for both entities and relations as Gaussian distributions in the same semantic space, as Gaussians innately represent uncertainty. To obtain high-quality Gaussian embeddings for both entities and relations, we introduce VAE into our model, learning the distributions from training triples in KG via a stochastic gradient variational Bayes [19] estimator. We introduce the details in the following subsections.

4.1. Variational Lower Bound

For a KG represented as $\mathcal{G} = (\mathcal{E}, \mathcal{R}, \mathcal{O})$, the embeddings of KG components can be represented as $\mathbf{Z}^{\mathcal{E}}, \mathbf{Z}^{\mathcal{R}}$ in latent spaces. To embed both entities and relations in the same semantic space, we first define the log-likelihood of $\mathcal{O}$, notated as the set of triples in KG, as:

$$\begin{aligned} \log p(\mathbf{O}) &= \log p(\mathbf{H}, \mathbf{R}, \mathbf{T}) \\ &= \log p(\mathbf{H}) + \log p(\mathbf{R}) + \log p(\mathbf{T}) \end{aligned} \tag{2}$$

where $\mathbf{O} \in \mathbb{R}^{W \times 3}$ and $\mathbf{H}$, $\mathbf{R}$, and $\mathbf{T}$ are components in $\mathbf{O}$. The log-likelihood of KG components, represented as $\log p(\mathbf{H})$, $\log p(\mathbf{R})$, and $\log p(\mathbf{T})$, can be derived using the Bayesian algorithm:

$$\begin{aligned} \log p(\mathbf{H}) &= \log \sum_{i=0}^{D} \left\{ p_\theta(\mathbf{H} \mid \mathbf{Z}_i^{\mathcal{E}}) \cdot p_\theta(\mathbf{Z}_i^{\mathcal{E}}) \right\} \\ &= \log \sum_{i=0}^{D} \left\{ p_\theta(\mathbf{Z}_i^{\mathcal{E}} \mid \mathbf{H}) \cdot p_\theta(\mathbf{H}) \right\} \\ &= \log \sum_{i=0}^{D} \left\{ p_\theta(\mathbf{Z}_i^{\mathcal{E}} \mid \mathbf{H}) \cdot \frac{p_\theta(\mathbf{Z}_i^{\mathcal{E}}, \mathbf{H}) \cdot q_\phi(\mathbf{Z}_i^{\mathcal{E}} \mid \mathbf{H})}{q_\phi(\mathbf{Z}_i^{\mathcal{E}} \mid \mathbf{H}) \cdot p_\theta(\mathbf{Z}_i^{\mathcal{E}} \mid \mathbf{H})} \right\} \\ &= q_\phi(\mathbf{Z}^{\mathcal{E}} \mid \mathbf{H}) \cdot \log \frac{q_\phi(\mathbf{Z}^{\mathcal{E}} \mid \mathbf{H})}{p_\theta(\mathbf{Z}^{\mathcal{E}} \mid \mathbf{H})} \\ &\quad + q_\phi(\mathbf{Z}^{\mathcal{E}} \mid \mathbf{H}) \cdot \log \frac{p_\theta(\mathbf{Z}^{E}, \mathbf{H})}{q_\phi(\mathbf{Z}^{\mathcal{E}} \mid \mathbf{H})} \\ &= D_{KL}(q_\phi(\mathbf{Z}^{\mathcal{E}} \mid \mathbf{H}) \parallel p_\theta(\mathbf{Z}^{\mathcal{E}} \mid \mathbf{H})) + \mathcal{L}(\theta, \phi; \mathcal{E}) \\ &\geq \mathcal{L}(\theta, \phi; \mathbf{H}) \end{aligned} \tag{3}$$

The conditional probability $q_\phi(\mathbf{Z}^{\mathcal{E}} \mid \mathbf{H})$ is the variational posterior to approximate the true posterior $p(\mathbf{Z}^{\mathcal{E}} \mid \mathbf{H})$, where the parameter ϕ is estimated in the inference model. In Equation (3), the second RHS term $\mathcal{L}(\theta, \phi; \mathcal{E})$ is called the evidence lower bound (ELBO) on the marginal likelihood of the variables $\mathcal{E}$:

$$\begin{aligned} \mathcal{L}(\theta, \phi; \mathbf{H}) &= \mathbb{E}_{q_\phi(\mathbf{Z}^{\mathcal{E}} \mid \mathbf{H})} \left[-\log q_\phi(\mathbf{Z}^{\mathcal{E}} \mid \mathbf{H}) + \log p_\theta(\mathbf{H}, \mathbf{Z}^{\mathcal{E}}) \right] \\ &= -D_{KL}(q_\phi(\mathbf{Z}^{\mathcal{E}} \mid \mathbf{H}) \parallel p_\theta(\mathbf{Z}^{\mathcal{E}})) \\ &\quad + \mathbb{E}_{q_\phi(\mathbf{Z}^{\mathcal{E}} \mid \mathbf{H})} \left[\log p_\theta(\mathbf{H} \mid \mathbf{Z}^{\mathcal{E}}) \right] \end{aligned} \tag{4}$$

where the D_{KL} term denotes the Kullback–Leibler divergence, a measure of how one probability distribution is different from a second. Respectively, we have:

$$\begin{aligned}\mathcal{L}(\theta,\phi;\mathbf{R}) &= -D_{KL}(q_\phi(\mathbf{Z}^{\mathbf{R}} \mid \mathbf{R}) \parallel p_\theta(\mathbf{Z}^{\mathbf{R}})) \\ &\quad + \mathbb{E}_{q_\phi(\mathbf{Z}^{\mathbf{R}}|\mathbf{R})}\left[\log p_\theta(\mathbf{R} \mid \mathbf{Z}^{\mathbf{R}})\right] \\ \mathcal{L}(\theta,\phi;\mathbf{T}) &= -D_{KL}(q_\phi(\mathbf{Z}^{\mathbf{E}} \mid \mathbf{T}) \parallel p_\theta(\mathbf{Z}^{\mathbf{E}})) \\ &\quad + \mathbb{E}_{q_\phi(\mathbf{Z}^{\mathbf{E}}|\mathbf{T})}\left[\log p_\theta(\mathbf{T} \mid \mathbf{Z}^{\mathbf{E}})\right]\end{aligned} \tag{5}$$

Substituting Equations (3)–(5) into Equation (2), the variational lower bound of $\log \mathcal{O}$ can be represented with the parameters θ and ϕ:

$$\begin{aligned}\log p(\mathbf{O}) &= \log p_\theta(\mathbf{H}) + \log p_\theta(\mathbf{R}) + \log p_\theta(\mathbf{T}) \\ &\geq \mathcal{L}(\theta,\phi;\mathbf{H}) + \mathbf{L}(\theta,\phi;\mathcal{R}) + \mathcal{L}(\theta,\phi;\mathbf{T}) \\ &= \mathcal{L}(\theta,\phi;\mathbf{O})\end{aligned} \tag{6}$$

where

$$\begin{aligned}\mathcal{L}(\theta,\phi;\mathbf{O}) = &- D_{KL}(q_\phi(\mathbf{Z}^{\mathcal{E}} \mid \mathbf{H}) \parallel p_\theta(\mathbf{Z}^{\mathcal{E}})) \\ &+ \mathbb{E}_{q_\phi(\mathbf{Z}^{\mathcal{E}}|\mathbf{H})}\left[\log p_\theta(\mathbf{H} \mid \mathbf{Z}^{\mathcal{E}})\right] \\ &- D_{KL}(q_\phi(\mathbf{Z}^{\mathcal{R}} \mid \mathbf{R}) \parallel p_\theta(\mathbf{Z}^{\mathcal{R}})) \\ &+ \mathbb{E}_{q_\phi(\mathbf{Z}^{\mathcal{R}}|\mathbf{R})}\left[\log p_\theta(\mathcal{R} \mid \mathbf{Z}^{\mathcal{R}})\right] \\ &- D_{KL}(q_\phi(\mathbf{Z}^{\mathcal{E}} \mid \mathbf{T}) \parallel p_\theta(\mathbf{Z}^{\mathcal{E}})) \\ &+ \mathbb{E}_{q_\phi(\mathbf{Z}^{\mathcal{E}}|\mathbf{T})}\left[\log p_\theta(\mathbf{T} \mid \mathbf{Z}^{\mathcal{E}})\right]\end{aligned} \tag{7}$$

In Equation (7), the conditional probabilities $q(\mathbf{Z}^{\mathcal{E}} \mid \mathbf{H})$, $q(\mathbf{Z}^{\mathcal{R}} \mid \mathbf{R})$ and $q(\mathbf{Z}^{\mathcal{E}} \mid \mathbf{T})$ can be regarded as probabilistic encoders to embed real data into latent space. Similarly, the conditional probabilities $p(\mathbf{H} \mid \mathbf{Z}^{\mathcal{E}})$, $p(\mathbf{R} \mid \mathbf{Z}^{\mathcal{R}})$ and $p(\mathbf{T} \mid \mathbf{Z}^{\mathcal{E}})$ can be regarded as probabilistic decoders, producing corresponding data from latent vector representations. To approximate the real distributions of KG components, we assume that the prior distributions and the variational posterior distributions are Gaussian distributions.

$$\begin{aligned}&p(\mathbf{Z_i}^{\mathcal{E}}) = \mathcal{N}(\mathbf{0},\mathbf{I}) \\ &p(\mathbf{Z_j}^{\mathcal{R}}) = \mathcal{N}(\mathbf{0},\mathbf{I}) \\ &q_\phi(\mathbf{Z_h}^{\mathcal{E}} \mid \mathbf{H}) = \mathcal{N}(\overline{\mathbf{E}},\sigma_{\mathcal{E}}^2 \cdot \mathbf{I}) \\ &q_\phi(\mathbf{Z_r}^{\mathcal{R}} \mid \mathbf{R}) = \mathcal{N}(\overline{\mathbf{R}},\sigma_{\mathcal{R}}^2 \cdot \mathbf{I}) \\ &q_\phi(\mathbf{Z_t}^{\mathcal{E}} \mid \mathbf{T}) = \mathcal{N}(\overline{\mathbf{E}},\sigma_{\mathcal{E}}^2 \cdot \mathbf{I})\end{aligned} \tag{8}$$

Assuming priors and variational posteriors to be Gaussian distributions, the D_{KL} terms in Equation (7) can be formed computationally. In addition, we adopt the Monte Carlo gradient estimator to deal with the $\mathbb{E}_{q_\phi}$ terms:

$$\begin{aligned}\mathcal{L}(\theta,\phi;\mathbf{O}) = &\frac{1}{L}\sum_{i=0}^{L}\sum_{(h_i,r_i,t_i)\in\mathcal{O}}^{W}(\log p_\theta(t_i \mid \mathbf{Z}_{t_i}^{\mathcal{E}}) \\ &+ \log p_\theta(h_i \mid \mathbf{Z}_{h_i}^{\mathcal{E}}) + \log p_\theta(r_i \mid \mathbf{Z}_{r_i}^{\mathcal{R}})) \\ &+ \frac{1}{M}\sum_{e_i\in\mathcal{E}}^{M}\sum_{d=0}^{D}(\mu_{e_i,d}^2 + \sigma_{e_i,d}^2 - \log\sigma_{e_i,d}^2 - 1) \\ &+ \frac{1}{N}\sum_{r_i\in\mathcal{R}}^{N}\sum_{d=0}^{D}(\mu_{r_i,d}^2 + \sigma_{r_i,d}^2 - \log\sigma_{r_i,d}^2 - 1),\end{aligned} \tag{9}$$

where D is the output dimension of latent variables, L is the sampling number in the Monte Carlo estimator, and M, N, and W are the number of entities, relations, and triples. We also adopt the reparameterization trick mentioned in the VAE section to generate samples.

$$\begin{aligned} \mathbf{z}^{\mathcal{E}}_{h_i} &= \overline{\mathbf{h_i}} + \sigma^2_{h_i} \odot \epsilon, \text{ with } h_i \in \mathbf{H}, \epsilon \sim \mathcal{N}(\mathbf{0}, \mathbf{I}) \\ \mathbf{z}^{\mathcal{R}}_{r_i} &= \overline{\mathbf{r_i}} + \sigma^2_{r_i} \odot \epsilon, \text{ with } r_i \in \mathcal{R}, \epsilon \sim \mathcal{N}(\mathbf{0}, \mathbf{I}) \\ \mathbf{z}^{\mathcal{E}}_{t_i} &= \overline{\mathbf{t_i}} + \sigma^2_{t_i} \odot \epsilon, \text{ with } t_i \in \mathbf{T}, \epsilon \sim \mathcal{N}(\mathbf{0}, \mathbf{I}) \end{aligned} \tag{10}$$

4.2. Learning

To optimize the parameters in Equation (9), we apply two neural networks in VAE: (1) An inference model f_ϕ with parameter ϕ to map observation data into latent vector spaces. (2) A generative model g_θ with parameter θ to produce random variables from latent embeddings.

Inference model f_ϕ. To encode KG components to Gaussian embeddings, we apply two fully-connected layers to map the entities and relations to the means and log-variances in their resulting Gaussian embeddings. One of the benefits of encoding log-variance instead of variance is that it enables us to avoiding using activation functions, since the variance σ^2 must be a positive number.

$$\begin{aligned} (\overline{\mathbf{h_i}}, \log \sigma^2_{\mathbf{h_i}}) &= f_{\phi_1}(\mathbf{h_i}) \\ (\overline{\mathbf{r_i}}, \log \sigma^2_{\mathbf{r_i}}) &= f_{\phi_2}(\mathbf{r_i}) \\ (\overline{\mathbf{t_i}}, \log \sigma^2_{\mathbf{t_i}}) &= f_{\phi_3}(\mathbf{t_i}) \end{aligned} \tag{11}$$

where $\phi = [\phi_1, \phi_2, \phi_3]$ and μ and $\log \sigma^2$ are the means and log-variances of learned Gaussian embeddings of KG components:

$$\begin{aligned} q_\phi(\mathbf{z}^{\mathcal{E}}_{\mathbf{h_i}} \mid \mathbf{h_i}) &= \mathcal{N}(\overline{\mathbf{h_i}}, \sigma^2_{\mathbf{h_i}} \cdot \mathbf{I}) \\ q_\phi(\mathbf{z}^{\mathcal{R}}_{\mathbf{r_i}} \mid \mathbf{r_i}) &= \mathcal{N}(\overline{\mathbf{r_i}}, \sigma^2_{\mathbf{r_i}} \cdot \mathbf{I}) \\ q_\phi(\mathbf{z}^{\mathcal{E}}_{\mathbf{t_i}} \mid \mathbf{t_i}) &= \mathcal{N}(\overline{\mathbf{t_i}}, \sigma^2_{\mathbf{t_i}} \cdot \mathbf{I}) \end{aligned} \tag{12}$$

We apply the reparameterization trick mentioned in Equation (10) to obtain the deterministic variables $\mathbf{Z}^{\mathcal{E}}_h$, $\mathbf{Z}^{\mathcal{R}}_r$, and $\mathbf{Z}^{\mathcal{E}}_t$, transformed from latent random variables, with a noise term ϵ from $\mathcal{N}(\mathbf{0}, \mathbf{I})$, which benefit from gradient propagation between the inference model and the generative model. We compute the loss of the inference model by measuring the KL divergence between those conditional probabilities and $\mathcal{N}(\mathbf{0}, \mathbf{I})$.

Generative model g_θ. The generative model decodes from deterministic values to random variables. For example, given resulting embeddings $\mathbf{Z}^{\mathcal{E}}$ and $\mathbf{Z}^{\mathcal{R}}$ from a KG represented as $\mathcal{G} = (\mathcal{E}, \mathcal{R}, \mathcal{O})$, our goal is to reconstruct random variables for each triple $(\mathbf{h_i}, \mathbf{r_i}, \mathbf{t_i}) \in \mathcal{O}$, where:

$$p_\theta(\mathbf{h_i}, \mathbf{r_i}, \mathbf{t_i} \mid \mathbf{z}^{\mathcal{E}}_{\mathbf{h_i}}, \mathbf{z}^{\mathcal{R}}_{\mathbf{r_i}}, \mathbf{z}^{\mathcal{E}}_{\mathbf{t_i}}) = g_\theta(\mathbf{z}^{\mathcal{E}}_{\mathbf{h_i}}, \mathbf{z}^{\mathcal{R}}_{\mathbf{r_i}}, \mathbf{z}^{\mathcal{E}}_{\mathbf{t_i}}) \tag{13}$$

The random distributions of those components can be defined as:

$$\begin{aligned} p_{\theta_1}(\mathbf{h_i} \mid \mathbf{z}^{\mathcal{E}}_{\mathbf{h_i}}) &= \mathcal{N}(\overline{\mathbf{z_{h_i}}}, \sigma^2_{\mathbf{z_{h_i}}} \cdot \mathbf{I}) \\ p_{\theta_2}(\mathbf{r_i} \mid \mathbf{z}^{\mathcal{R}}_{\mathbf{r_i}}) &= \mathcal{N}(\overline{\mathbf{z_{r_i}}}, \sigma^2_{\mathbf{z_{h_i}}} \cdot \mathbf{I}) \\ p_{\theta_3}(\mathbf{t_i} \mid \mathbf{z}^{\mathcal{E}}_{\mathbf{t_i}}) &= \mathcal{N}(\overline{\mathbf{z_{t_i}}}, \sigma^2_{\mathbf{z_{h_i}}} \cdot \mathbf{I}) \end{aligned} \tag{14}$$

where $\theta = [\theta_1, \theta_2, \theta_3]$, and the reconstruction loss of the generative model can be measured based on the binary cross entropy (BCE) between the generative variables and the real data.

5. Experiment

5.1. Data Sets

In this work, we conducted experiments and evaluated the related methods using real-world databases of KG, commonly used in previous works: WordNet [24] and Freebase [25]. WordNet is an extensive lexical database of English. Nouns, verbs, adjectives, and adverbs are grouped into sets of cognitive synonyms (synsets), each expressing a distinct concept, and with interlinked sysnets employing conceptual-semantic and lexical relations. Freebase is a large collaborative knowledge base consisting of data compiled mainly by its community members. It is an online collection of structured data harvested from many sources, including individual wiki contributions. The most representative dataset in WorldNet is WN18, and FB15k in Freebase.

In those datasets, WN18 contains 18 relations and 40,943 entities, whereas FB15k contains 1345 relations and 14,951 entities. However, both of them suffer from test leakage through inverse relations: a large number of test triples can be obtained simply by inverting triples in the training set. Therefore, we introduced FB15k-237, a subset of FB15k, in which reversible relations were removed. Similarly, WN18 was corrected by WN18RR. Therefore, we selected WN18RR and FB15k-237 as datasets in our experiments.

5.2. Experimental Setup

We compared our models with serveral KG embedding algorithms in our experiments:

1. TransE [6]. TransE was the first model to introduce translation-based embedding, which interprets relations as the translations operating on entities.
2. DistMult [10]. DistMult is based on the bilinear model, where each relation is represented by a diagonal rather than a full matrix. DistMult enjoys the same scalable properties as TransE and it achieves superior performance over TransE.
3. ComplEx [11]. ComplEx extends DistMult by introducing complex-valued embeddings so as to better model asymmetric relations. It has been proven that HolE is subsumed by ComplEx as a special case.
4. ConvE [12]. ConvE is a multi-layer convolutional network model for link prediction [24] of KGs, and it reports state-of-the-art results for several established datasets. Unlike previous work which has focused on shallow, fast models that can scale to large knowledge graphs, ConvE uses 2D convolution over embeddings and multiple layers of nonlinear features to model KGs.
5. ConvKB [13]. ConvKB applies the global relationships among same-dimensional entries of the entity and relation embeddings, so that ConvKB generalizes the transitional characteristics in the transition-based embedding models.
6. R-GCN [14]. R-GCN applies graph convolutional networks to relational knowledge bases, creating a new encoder for link prediction and entity classification tasks.

The experimental results from those baselines were obtained from the codes provided by the authors. In our method, we made configurations by selecting a learning rate α among [0.01, 0.05, 0.10] and an output dimension D among [100, 200, 400]. For WN18RR, the configuration was as follows. The learning rate α was 0.01 and the output dimension D was 400, with 3000 training iterations using the Adam [27] optimizer. For FB15k-237, the configuration was as fallows. The learning rate α was 0.10, and the output dimension D was 200, with 1000 training iterations using the SGD optimizer. We trained the model until it converged.

5.3. Link Prediction

Link prediction, aiming at predicting the missing KG components for incomplete triples, is a typical task in KG embedding. e.g., predicting the head entity for a given triple $(*, \mathbf{r}, \mathbf{t})$ or predicting the tail entity for a given triple $(\mathbf{h}, \mathbf{r}, *)$. Following the protocols in [6], we evaluated the performance of our model. Given a test triple, we replaced the head or tail with all available entities and ranked them by measuring the scoring function defined in the methods section. Based on the ranking lists, we report the proportion of correct entities in the top N ranked entities, where N = 1, 3, and 10, denoted as Hits@1, Hits@3, and Hits@10.

$$MRR = \frac{1}{|M|}\sum_{i=0}^{M}\frac{1}{rank(e_i)}$$
$$MR = \frac{1}{|M|}\sum_{i=0}^{M} rank(e_i) \tag{15}$$

We also record the average reciprocal rank of correct entities (denoted as MRR) and the average rank of correct entities (denoted as MR) for link prediction, where the function $rank(e_i)$ transforms to the rank of e_i. A good embedding algorithm should obtain a relatively low mean rank and a relatively high mean reciprocal rank.

5.4. Results and Analysis

In this subsection, we report the ability of our model to represent uncertainty, and the experimental results regarding link prediction.

Qualitative Analysis Before evaluating the performance in specific task compared with other methods, we need to discuss the ability of our model to represent uncertainty in KG embedding.

In our method, we measure the uncertainty of KG components by the variances of their embeddings, where an entity/relation with a higher level of uncertainty has a large covariance. We discuss the relations in FB15k-237 with '/education' as the domain, providing a (log) determinant and trace of their covariances as shown in Table 2, from which we have made the following observations:

1. Our method has the ability to measure the uncertainty in KG embedding. The covariance of Gaussian embedding can effectively describe the uncertainties by calculating the determinants and traces of the covariances.
2. The relations with more semantic information (the number of associated heads and tails, type of relation) have larger uncertainty. For example, the 'major_field_of_study' relation has the largest uncertainty, and the 'educational_insitution' relation has the smallest uncertainty in those relations.

Table 2. The relations with /education/ as the domain and their determinants and traces of the corresponding covariances, sorted by descending order of traces.

Relation	#Head	#Tail	Type	log (det)	Trace
major_field_of_study	225	77	m-n	−338.8	38.1
student	183	292	1-n	−340.6	34.8
institution	22	222	m-n	−376.2	32.8
colors	85	19	m-n	−400.9	26.9
fraternities_sororities	20	3	m-1	−406.9	24.9
campuses	13	13	1-1	−411.9	21.3
currency	5	3	m-1	−423.4	19.8
educational_institution	13	13	1-1	−430.6	18.7

6. Results

We compared our method with the state-of-the-art baselines mentioned above, including TransE, DisMult, ComplEx, ConvKB, and R-GCN. First of all, the codes in the baseline we used are provided by other authors. All models were fully trained, and the data sets used were public. Our models, in both the Hits@3 and Hits@10 metrics for this dataset, achieved superior results, which proves that the embedding obtained using our proposed method is of high quality. The experimental results regarding link prediction are shown in Table 3. We observe that:

1. The experimental results on FB15k-237 and WN18RR indicate that our method can learn high-quality representations in KG.
2. Our method outperformed other baselines in terms of the Hits@3 and Hits@10 metrics, but its performance was poor in terms of mean reciprocal rank and the Hits@1 metric on WN18RR. This may be because WN18RR contains a large number of entities and several relations, so most methods can only judge the correctness of a triple but cannot rank it in the top position.
3. On FB15k-237, our method outperformed other baselines in terms of the Hits@3, Hits@10, and mean reciprocal rank metrics, and came second in terms of the Hits@1 and mean rank metrics. The improvements observed in FB15k-237 were greater than those in WN18RR, showing that FB15k-237 contains more relations and thus the uncertainties in its components are larger than those in WN18RR, which indicates that our method can learn valid representations with uncertainties in KG.

Table 3. Experimental results for WN18RR and FB15k-237 test sets. Hits@N values are presented as percentages. The best score is in bold and the second best score is underlined.

	WN18					**FB15k-237**				
	MR	**MRR**	**HITS@N**			**MR**	**MRR**	**HITS@N**		
			1	**3**	**10**			**1**	**3**	**10**
TransE (Bordes et al., 2013) [6]	2300	0.243	4.27	44.1	53.2	323	0.279	19.8	37.6	44.1
DistMult (Yang et al., 2015) [10]	7000	0.444	41.2	47	50.4	512	0.281	19.9	30.1	44.6
ComplEx (Trouillon et al., 2016) [11]	7882	0.449	40.9	46.9	53	546	0.278	19.4	29.7	45
ConvE (Dettmers et al., 2018) [12]	4464	**0.456**	**41.9**	47	53.1	245	0.312	**22.5**	34.1	49.7
ConvKB (Nguyen et al., 2018) [13]	**1295**	0.265	5.82	44.5	55.8	**216**	0.289	19.8	32.4	47.1
R-GCN (Schlichtkrull et al., 2018) [14]	6700	0.123	8	13.7	20.7	600	0.164	10	18.1	30
Our work	1963	0.236	11.4	**48.0**	**57.6**	240	**0.518**	21.8	**42.0**	**52.1**

7. Conclusions

In this paper, we propose the co-embedding model to learn the latent representations of both entities and relations in the same semantic space, embedding them as Gaussian distributions. To obtain high-quality embeddings, we introduced the variational auto-encoder, an auto-encoder model consisting of a probabilistic encoder and a probabilistic decoder, into our model. One of the assets of the technique is that the affinities between entities and relations can be measured effectively since they are embedded in the same semantic space, and we also explain the transformation from observation values to latent representations via the two models using the variational auto-encoder. In our experiments, we evaluated the performance of the co-embedding model and other baselines on several benchmark datasets. From these experimental results, we can conclude that our method can learn high-quality representations of KG components.

In the future, we plan to extend our method by assuming the priors with other distributions and optimizing the variational lower bounds in an effective way.

Author Contributions: Conceptualization, H.H.; data curation, L.X. and Q.D.; formal analysis, L.X.; methodology, L.X. and H.H.; project administration, H.H.; software, Q.D.; supervision, H.H.; validation, H.H.; visualization, L.X. and Q.D.; writing—original draft, L.X.; writing—review and editing, H.H.; funding acquisition, H.H. All authors have read and agreed to the published version of the manuscript.

Funding: This research was funded by Wenzhou Science and Technology Planning Project #2021R0082.

Institutional Review Board Statement: Not applicable.

Informed Consent Statement: Not applicable.

Data Availability Statement: Not applicable.

Conflicts of Interest: The authors declare no conflict of interest. The funders had no role in the design of the study; in the collection, analyses, or interpretation of data; in the writing of the manuscript, or in the decision to publish the results.

References

1. Berant, J.; Chou, A.; Frostig, R.; Liang, P. Semantic Parsing on Freebase from Question-Answer Pairs. In Proceedings of the 2013 Conference on Empirical Methods in Natural Language Processing, Seattle, WA, USA, 18–21 October 2013; pp. 1533–1544.
2. Heck, L.; Hakkani-Tür, D.; Tur, G. Leveraging Knowledge Graphs for Web-Scale Unsupervised Semantic Parsing. In Proceedings of the International Speech Communication Association, Lyon, France, 25–29 August 2013.
3. Wang, W.Y.; Mazaitis, K.; Lao, N.; Mitchell, T.; Cohen, W.W. Efficient Inference and Learning in a Large Knowledge Base: Reasoning with Extracted Information using a Locally Groundable First-Order Probabilistic Logic. *arXiv* **2014**, arXiv:cs.AI/1404.3301.
4. Bordes, A.; Weston, J.; Usunier, N. Open Question Answering with Weakly Supervised Embedding Models. *arXiv* **2014**, arXiv:cs.CL/1404.4326.
5. Bordes, A.; Chopra, S.; Weston, J. Question Answering with Subgraph Embeddings. In Proceedings of the 2014 Conference on Empirical Methods in Natural Language Processing (EMNLP), Doha, Qatar, 25–29 October 2014; pp. 615–620. [CrossRef]
6. Bordes, A.; Usunier, N.; Garcia-Duran, A.; Weston, J.; Yakhnenko, O. Translating Embeddings for Modeling Multi-relational Data. In *Advances in Neural Information Processing Systems 26*; Burges, C.J.C., Bottou, L., Welling, M., Ghahramani, Z., Weinberger, K.Q., Eds.; Curran Associates, Inc.: Red Hook, NY, USA, 2013; pp. 2787–2795.
7. Wang, Z.; Zhang, J.; Feng, J.; Chen, Z. Knowledge graph embedding by translating on hyperplanes. In Proceedings of the AAAI Conference on Artificial Intelligence, Quebec, QC, Canada, 27–31 July 2014; Volume 28.
8. Lin, Y.; Liu, Z.; Sun, M.; Liu, Y.; Zhu, X. Learning Entity and Relation Embeddings for Knowledge Graph Completion. In Proceedings of the Twenty-Ninth AAAI Conference on Artificial Intelligence, Austin, TX, USA, 25–30 January 2015; pp. 2181–2187.
9. Nickel, M.; Tresp, V.; Kriegel, H.P. A three-way model for collective learning on multi-relational data. In Proceedings of the ICML, Bellevue, WA, USA, 28 June–2 July 2011.
10. Yang, B.; tau Yih, W.; He, X.; Gao, J.; Deng, L. Embedding Entities and Relations for Learning and Inference in Knowledge Bases. *arXiv* **2014**, arXiv:cs.CL/1412.6575.
11. Trouillon, T.; Welbl, J.; Riedel, S.; Gaussier, É.; Bouchard, G. Complex Embeddings for Simple Link Prediction. *arXiv* **2016**, arXiv:cs.AI/1606.06357.
12. Dettmers, T.; Minervini, P.; Stenetorp, P.; Riedel, S. Convolutional 2D Knowledge Graph Embeddings. *arXiv* **2017**, arXiv:cs.LG/1707.01476.
13. Nguyen, D.Q.; Nguyen, T.D.; Nguyen, D.Q.; Phung, D. A Novel Embedding Model for Knowledge Base Completion Based on Convolutional Neural Network. In Proceedings of the 2018 Conference of the North American Chapter of the Association for Computational Linguistics: Human Language Technologies, New Orleans, LA, USA, 1–6 June 2018; Volume 2, pp. 327–333. [CrossRef]
14. Schlichtkrull, M.; Kipf, T.N.; Bloem, P.; van den Berg, R.; Titov, I.; Welling, M. Modeling Relational Data with Graph Convolutional Networks. *arXiv* **2017**, arXiv:stat.ML/1703.06103.
15. Kipf, T.N.; Welling, M. Semi-supervised classification with graph convolutional networks. *arXiv* **2016**, arXiv:1609.02907.
16. Paccanaro, A.; Hinton, G.E. Learning distributed representations of concepts using linear relational embedding. *IEEE Trans. Knowl. Data Eng.* **2001**, *13*, 232–244. [CrossRef]
17. He, S.; Liu, K.; Ji, G.; Zhao, J. Learning to Represent Knowledge Graphs with Gaussian Embedding. In Proceedings of the 24th ACM International on Conference on Information and Knowledge Management, CIKM'15, New York, NY, USA, 19–30 October 2015; pp. 623–632. [CrossRef]
18. Vilnis, L.; McCallum, A. Word Representations via Gaussian Embedding. *arXiv* **2014**, arXiv:cs.CL/1412.6623.
19. Kingma, D.P.; Welling, M. Auto-Encoding Variational Bayes. *arXiv* **2013**, arXiv:stat.ML/1312.6114.
20. Kingma, D.P.; Rezende, D.J.; Mohamed, S.; Welling, M. Semi-Supervised Learning with Deep Generative Models. *arXiv* **2014**, arXiv:cs.LG/1406.5298.
21. Jiang, Z.; Zheng, Y.; Tan, H.; Tang, B.; Zhou, H. Variational Deep Embedding: An Unsupervised and Generative Approach to Clustering. *arXiv* **2016**, arXiv:cs.CV/1611.05148.
22. Makhzani, A.; Shlens, J.; Jaitly, N.; Goodfellow, I.; Frey, B. Adversarial Autoencoders. *arXiv* **2015**, arXiv:cs.LG/1511.05644.
23. Dosovitskiy, A.; Brox, T. Generating Images with Perceptual Similarity Metrics based on Deep Networks. *arXiv* **2016**, arXiv:cs.LG/1602.02644.
24. Miller, G.A. WordNet: A Lexical Database for English. *Commun. ACM* **1995**, *38*, 39–41. [CrossRef]
25. Bollacker, K.; Evans, C.; Paritosh, P.; Sturge, T.; Taylor, J. Freebase: A Collaboratively Created Graph Database for Structuring Human Knowledge. In Proceedings of the 2008 ACM SIGMOD International Conference on Management of Data, SIGMOD'08, Vancouver, BC, Canada, 9–12 June 2008; pp. 1247–1250. [CrossRef]

MDPI
St. Alban-Anlage 66
4052 Basel
Switzerland
Tel. +41 61 683 77 34
Fax +41 61 302 89 18
www.mdpi.com

MDPI Books Editorial Office
E-mail: books@mdpi.com
www.mdpi.com/books

www.ingramcontent.com/pod-product-compliance
Lightning Source LLC
Chambersburg PA
CBHW061446220926
51850CB00004B/27
9783036550831